# Ocean Dynamics

Dirk Olbers · Jürgen Willebrand · Carsten Eden

# Ocean Dynamics

Springer

Prof. Dr. Dirk Olbers
Alfred-Wegener-Institut für Polar- und Meeresforschung
Postfach 12 01 61
27515 Bremerhaven
Germany
*dirk.olbers@awi.de*

Prof. Dr. Carsten Eden
University Hamburg
Institut für Meereskunde
Bundesstrasse 53
20146 Hamburg
Germany
*carsten.eden@zmaw.de*

Prof. Dr. Jürgen Willebrand
Leibniz-Institut für Meereswissenschaften
IFM-GEOMAR
Düsternbrookerweg 20
24105 Kiel
Germany
*jwillebrand@ifm-geomar.de*

ISBN 978-3-642-23449-1 ISBN 978-3-642-23450-7 (eBook)
DOI 10.1007/978-3-642-23450-7
Springer Heidelberg Dordrecht London New York

Library of Congress Control Number: 2012935922



Cover illustration: Dirk Olbers

Printed on acid-free paper

Springer is part of Springer Science+Business Media (www.springer.com)

# Foreword

The oceans cover more than two thirds of the surface of the Earth. They have a profound influence on our climate, our weather, and the Earth's ecosystems. They store and transport heat, $CO_2$, nutrients, and, of course, water, as well as other important components determining the conditions of life on our planet. Through their large storage capacity for heat, the oceans act as a flywheel, dampening short-term natural variations in weather and climate. But, by the same process, they also convert short-term weather fluctuations into longer term climate variations, in the same way as a long swing exposed to short gusts of wind slowly develops longer-term random oscillations (or large molecules in a sea of small molecules exhibit Brownian motion). Thus, a clear insight into the dynamics of the ocean is a prerequisite for understanding our present climate, including both the mean climate state and the superimposed natural climate variability. This applies still more, of course, to our attempts to assess the impacts of human activities on climate and to predict the evolution of climate in the future. To resolve these questions, it is essential that we are able to distinguish conceptually and in observations between natural and human-induced climate change. This is again critically dependent on understanding of the dynamics of the oceans.

The widespread recognition of the important role of the oceans for climate has led to a strong increase in ocean research in recent years. Much of the enhanced effort has been directed towards the development of more detailed numerical models of the ocean circulation, a focus which has appeared particularly rewarding in view of the parallel rapid increase in supercomputer power. However, the oceans represent a highly complex system, characterized by many different space and time scales. Although the physics is well understood at the basic level of the underlying fluid dynamic equations, the range of space and time scales involved in the coupling of the oceans to the rest of the climate system span far too many orders of magnitude to be captured by even the most advanced super-computer. Thus, approximations are necessary. But in order to introduce meaningful approximations, and to assess their implications and ranges of validity, a thorough understanding of the underlying dynamics of the oceans is essential.

This book is the most thorough and carefully researched representation of the dynamics of the ocean that I am aware of. It begins at the most fundamental level of the fluid dynamic equations and, from there, gradually develops a series of pictures of the ocean system obtained by focusing on different aspects of the dynamics. Each approximation is carefully and rigorously introduced. The treatise leads, finally, in the last chapters and Appendix B, to a number of state-of-the-art ocean models designed for particular applications. But the emphasis is always on the fundamentals, rather than the numerical models.

One cannot fail to be impressed. However, when first asked to write a foreword, I demurred. The level of fundamental rigor lies so far beyond the level at which I have been working as a climate scientist that I felt (and still feel) unable to provide due justice to the detailed, careful mathematics on which the book is based. But on second thoughts I thought that precisely because I have always been a rather broad-brush scientist with a somewhat cavalier attitude towards mathematical rigor, it is fitting that I pay homage to this truly impressive work of fundamental scientific analysis.

This is all the more appropriate, since the many of the topics in the book have evolved from the joint beginnings of the first two authors and myself in the early years of the Max Planck Institute of Meteorology in Hamburg. In the second half of the 1970's we were struggling to develop a realistic numerical global ocean circulation model that we could couple to our available global atmospheric model to create a useful global climate model. Our thinking was very much along the lines of this treatise: first identify the modes of motion of the ocean circulation relevant for the space and time scales of climate variability that we were interested in; then develop numerical models for the different ocean modes based on this space and time scale decomposition; finally, couple the different subsystems together to produce a numerically efficient global ocean model. We profited much from the lectures of Pierre Welander and the lively discussions on ocean dynamics during his many sabbaticals at MPI. However, the project turned out to be far more complex than we had naively envisaged, and our goals were never achieved. Fortunately, the day was saved through the numerical creativity of Ernst Maier-Reimer, another member of the MPI ocean modeling team who produced a global large scale geostrophic (LSG) numerical ocean model using a novel implicit filtering scheme, which we were then able to couple to our atmospheric model, producing one of the first realistic global climate models. But this channeled the future MPI ocean modeling efforts into large numerical models, rather than in the direction of careful dynamical analysis pursued in this book.

It is thus a source of considerable gratification to see that Dirk Olbers and Jürgen Willebrand, after leaving the MPI to head theoretical oceanography departments at their respective institutes in Bremerhaven and Kiel, have, in collaboration with their colleague Carsten Eden, successfully pursued and realized our original dream. The present book does indeed lead from first principles to the basic modes of motion of realistic oceans to an example of a highly efficient but simplified ocean model (BARBI, modestly hidden in an appendix) based on intelligent mode decomposition and filtering.

Unfortunately, the sheer volume of detailed analysis that needed to be carefully presented to describe the dynamics of the ocean itself precluded the discussion of further interesting problems related to the role of the oceans in coupled systems. Examples include the storage and transport of $CO_2$ and other substances in the ocean, interactions with sea-ice, natural modes of oscillation of the coupled ocean-atmosphere system and the related question of decadal-scale climate predictability, the dynamics of the air-sea interface and ocean waves, and methods of satellite remote sensing of the sea surface. For climate scientists and other researchers interested in these broader questions, the present book, through its rigorous treatment focusing on the complex dynamics of the ocean itself – the key component in all of these examples – will provide an invaluable reference text. But still more important, it should lay the foundation for a deeper appreciation of the complexities of the dynamics of the ocean for the next generation of physical oceanographers striving to understand the role of the oceans in the Earth system.

Hamburg, November 2011 *Klaus Hasselmann*

# Preface

This book is directed to graduate students of physical oceanography and neighboring fields like meteorology, geophysics or general physics, and to anybody interested in a thorough discussion of ocean dynamics. Based on the well-known fundaments of fluid mechanics, thermodynamics, and wave theory, the first three parts of the book provide a detailed derivation of the basic physical laws describing the motions in the ocean, common approximations which are made to simplify the discussion of e. g. the large-scale fluid dynamics of the ocean, and a comprehensive treatment of linear wave theory. The following part on the theory of turbulence in the ocean attempts to reach for newer results, in particular regarding the role of eddies for the large-scale dynamics. In the next part, classical concepts and models of ocean circulation are combined with newer material. Finally, an appendix reviews some of the needed mathematical tools and the models which are used in the book. While far from being complete, we have included as much as possible of what we think is important to understand the physics of the ocean, aiming for a high accuracy both in physical argumentation and mathematical derivation.

In the last decades, increasing interest in climate change has fostered research with respect to the role of the ocean in the climate system and has changed the field of physical oceanography from a small group of largely ignored academical experts, into a highly recognized arena of scientific discussion, which sometimes even takes place in the media. At the same time, the increasing performance of computers allowed more and better resolved integrations of numerical ocean models. In this book we have not addressed the field of numerical ocean modeling. However, we believe that for both, the scientific discussion and a thorough interpretation of numerical models, knowledge of the material presented in this book should be of value.

The book is based on material from a series of lectures to graduate oceanography students at the University Kiel and to graduate physics students at Bremen University which has evolved over the years. While the reader of this book does not need any prior knowledge about physical oceanography, we assume a sound physical and mathematical basis comparable to that of a Bachelor in physics. In the notation we mostly follow the conventions in the oceanographic literature. Relevant variables are generally introduced when they arise in the context of the discussion; a list of symbols and their meaning is given in Appendix .

We want to express our specific gratitude to our colleagues, Drs. Bach Lien Hua, Hans Burchard, Sergey Danilov, Sybren Drijfhout, Theo Gerkema, Stephen Griffies, Leo Maas, Trevor McDougall, Geoffrey Vallis, Christoph Völker, Jörg-Olaf Wolff, and Carl Wunsch, who commented on an earlier version of the manuscript and, most importantly, expressed highly welcome words of encouragement. Our thanks go to Andrea Bleyer for proofreading and checking the text. We are indebted to Frauke Thiele-Wolf for many of the figures, also to Frauke Nevoigt for drawing some of the figures. Michael Brüdgam, Nils Brüggemann und Jan Viebahn read preliminary versions and provided some figures.

Over many years our work has been supported by three major German institutes doing ocean research: the Alfred Wegener Institute for Polar and Marine Research in Bremerhaven, the Leibniz Institute of Marine Sciences in Kiel, and the Institute of Oceanography in Hamburg. Financial support by the Alfred Wegener Institute, and the Clusters of Excellence "Future Ocean" and "CliSAP" is also greatly acknowledged.

July 2011

*Dirk Olbers*
*Jürgen Willebrand*
*Carsten Eden*

# Contents

Chapters and sections with an asterisk indicate material that is not written for a beginner and may be omitted in a first reading.

# List of Symbols

In a text covering various aspects of ocean dynamics, it is unfortunately not possible to use a notation which is at the same time both simple and unambiguous. To facilitate reading, as far as possible conventional notation is used in this book. However, it occurs that a symbol has different meanings in different parts of the book. The following list contains symbols and the corresponding variables which appear, with the same meaning, in more than one chapter. Also listed are the physical unit and the section of first appearance or explanation.

| Symbol | Variable, Unit | Section |
|---|---|---|
| $\nabla$ | $= (\partial/\partial x, \partial/\partial y, \partial/\partial z)$ gradient (or *nabla*) operator, in $m^{-1}$ | 1.1 (see also Appendix A.1.1) |
| $a$ | Earth radius, in m | 4.2 |
| $\boldsymbol{B}$ | baroclinic vector, in $m\,s^{-2}$ | 2.11.2 |
| Bu | Burger-number | 5.2.1 |
| $b$ | buoyancy, in $m\,s^{-2}$ | 4.1.6 |
| $C$ | Cox-number | 11.3.3 |
| $\boldsymbol{c}_g$ | wave group velocity, in $m\,s^{-1}$ | 6.2.1 |
| $c_p$ | specific heat, in $J\,kg^{-1}\,K^{-1}$ | 1.2.7 |
| $c_s$ | sound velocity, in $m\,s^{-1}$ | 2.6.4 |
| $E$ | internal energy per mass, in $J\,kg^{-1} = m^2\,s^{-2}$ | 1.2.3 |
| $\mathcal{E}$ | evaporation, in $kg\,m^{-2}\,s^{-1}$ | 2.2 |
| $E_k$ | turbulent kinetic energy, in $m^2\,s^{-2}$ | 11.3.1 |
| Ek | Ekman-number | 2.8.1 |
| $\boldsymbol{\mathcal{F}}$ | frictional force per mass, in $kg\,m^{-2}\,s^{-2}$ | 2.8 |
| $f$ | Coriolis parameter, in $s^{-1}$ | 4.2.5 |
| $\mathcal{G}_S$ | salt source, in $g\,m^{-3}\,s^{-1}$ | 2.8 |
| $\mathcal{G}_\theta$ | temperature source, in $K\,kg\,m^{-3}\,s^{-1}$ | 2.8 |
| $\mathcal{G}_\rho$ | density source, in $kg\,m^{-3}\,s^{-1}$ | 2.8 |
| $g$ | gravity acceleration, in $m\,s^{-2}$ | 2.3.4 |
| $H$ | enthalpy per mass, in $J\,kg^{-1}$ | 1.2.3 |
| $H, h$ | water depth, in m | 4.2.3 |
| $\boldsymbol{J}_S$ | salt flux, in $g\,m^{-2}\,s^{-1}$ | 2.2 |
| $\boldsymbol{J}_T$ | (sensible) heat flux, in $W\,m^{-2}$ | 2.4.1 |

| | | |
|---|---|---|
| $\boldsymbol{J}_{\mathrm{rad}}$ | radiative heat flux, in $\mathrm{W\,m^{-2}}$ | 2.4.1 |
| $\boldsymbol{J}_{\mathrm{chem}}$ | chemical energy flux, in $\mathrm{W\,m^{-2}}$ | 2.4.1 |
| $E_{\mathrm{k}}$ | kinetic energy per mass, in $\mathrm{m^2\,s^{-2}}$ | 2.4.2 |
| $E_{\mathrm{p}}$ | potential energy, in $\mathrm{m^2\,s^{-2}}$ | 4.1 |
| $K$ | turbulent diffusivity, in $\mathrm{m^2\,s^{-2}}$ | 11.2 |
| $\boldsymbol{k}$ | $= (k_1, k_2, k_3)$ wave number, in $\mathrm{m^{-1}}$ | 6.2 |
| $N$ | Brunt–Väisälä frequency , in $\mathrm{s^{-1}}$ | 2.9.2 |
| $\mathcal{P}$ | precipitation, in $\mathrm{kg\,m^{-2}\,s^{-1}}$ | 2.2 |
| Pe | Peclet-number | 2.8.1 |
| Pr | Prandtl-number | the box on p. 356 |
| $p$ | pressure, in $\mathrm{Pa = N\,m^{-2}}$, also in dbar ($1\,\mathrm{dbar} = 10^4\,\mathrm{Pa}$) | 1.2.2 |
| $Q$ | potential vorticity, in $\mathrm{m^{-1}\,s^{-1}}$ (for $\psi = \rho_{\mathrm{p}}$) | 2.11.3 |
| $R$ | Rossby radius, in m | 5.1 (see also 8.1.1) |
| Re | Reynolds-number | 2.8.1 |
| Ri | Richardson-number | 7.7.3 |
| Ro | Rossby-number | 2.11.1 |
| $R_\beta$ | $\beta$-Rossby-number | 5.1 |
| $r$ | distance from Earth center, in m | 2.3.4 |
| $S$ | salinity, in g salt/kg seawater | 1.2.1 |
| $T$ | temperature, in K (also in °C) | 1.2.2 |
| $t$ | time, in s | 1.2.2 |
| $\boldsymbol{u}$ | $= (u, v, w) = (u_1, u_2, u_3)$ velocity vector, in $\mathrm{m\,s^{-1}}$ | 1.1 |
| $\boldsymbol{u}_{\mathrm{g}}$ | $= (u_{\mathrm{g}}, v_{\mathrm{g}})$ geostrophic velocity (horizontal), in $\mathrm{m\,s^{-1}}$ | 14.1.1 |
| $\upsilon$ | specific volume, in $\mathrm{m\,kg^{-1}}$ | 1.2.2 |
| $w_{\mathrm{e}}$ | Ekman-velocity, in $\mathrm{m\,s^{-1}}$ | 14.1.4 |
| $\boldsymbol{x}$ | $= (x, y, z) = (x_1, x_2, x_3)$ coordinate vector, in m | 1.1 |
| $\alpha$ | thermal expansion, in $\mathrm{K^{-1}}$ | 1.2.6 |
| $\beta$ | $= \partial f / \partial y$, change of Coriolis parameter in northward direction, in $\mathrm{m^{-1}\,s^{-1}}$ | 5.1 |
| $\gamma$ | haline expansion, in $(\mathrm{g\,kg^{-1}})^{-1}$ | 1.2.6 |
| $\epsilon$ | energy dissipation, in $\mathrm{m^2\,s^{-3}}$ | 2.4.2 |
| $\zeta$ | surface elevation, in m | 2.2 |
| $\eta$ | entropy per mass, in $\mathrm{J\,kg^{-1}\,K^{-1}}$ | 1.2.4 |
| $\eta$ | relative vorticity, in $\mathrm{s^{-1}}$ | 2.11.1 |
| $\theta$ | potential/conservative temperature, in K (also in °C) | 2.6.3/2.6.2 |
| $\kappa$ | isothermal compressibility, in $\mathrm{m\,s^2\,kg^{-1}}$ | 1.2.6 |
| $\hat{\kappa}$ | adiabatic compressibility, in $\mathrm{m\,s^2\,kg^{-1}}$ | 2.6.4 |
| $\kappa_{\mathrm{m}}$ | molecular kinematic viscosity, in $\mathrm{m^2\,s^{-1}}$ | 2.5.1 |
| $\kappa_{\mathrm{S}}$ | molecular salt diffusivity, in $\mathrm{m^2\,s^{-1}}$ | 2.5.1 |
| $\kappa_{\mathrm{T}}$ | mol. temperature conductivity, in $\mathrm{m^2\,s^{-1}}$ | 2.5.1 |
| $\lambda$ | longitude, in ° | 2.3.4 |
| $\mu$ | chemical potential per mass, in $\mathrm{J\,g^{-1}}$ | 1.2.5 |
| $\xi$ | elevation (of density surface), in m | 2.9 |
| $\rho$ | *in-situ* density, in $\mathrm{kg\,m^{-3}}$ ($1\,\mathrm{g/cm^3} = 10^3\,\mathrm{kg/m^3}$) | 1.2.1 |
| $\rho_{\mathrm{b}}$ | density, state of rest w. mean background thermohaline stratification | 4.1 |
| $\rho_{\mathrm{c}}$ | density, state of rest with constant temperature and salinity | 4.1 |

| | | |
|---|---|---|
| $\rho_{\mathrm{p}}$ | potential density, in kg m$^{-3}$ | 2.6 |
| $\rho_{\mathrm{r}}$ | density, state of rest w. arbitrary thermohaline stratification | 2.9 |
| $\rho_0$ | constant density, typical for seawater | 4.1 |
| $\hat{\rho}$ | $= \rho - \rho_{\mathrm{b}}$, deviation from mean background thermohaline stratification | 5.2 |
| $\tilde{\rho}$ | $= \rho - \rho_{\mathrm{c}}$, deviation from state of rest | 4.1 |
| $\Phi$ | geopotential, in m$^2$ s$^{-2}$ | 2.3.4 |
| $\varphi$ | latitude, in ° | 2.3.4 |
| $\psi$ | stream function, in m$^2$ s$^{-1}$ | 3.2 |
| $\boldsymbol{\Omega}$ | Earth rotation vector, in s$^{-1}$ | 2.3.3 |
| $\boldsymbol{\omega}$ | vorticity vector, in s$^{-1}$ | 1.1 |
| $\omega$ | wave frequency, in s$^{-1}$ | 3.2 |

# Part I
# Fundamental Laws

The fundaments of ocean physics are based on the concepts of hydrodynamics and thermodynamics. Molecular structures of the media are not considered explicitly (except in radiation processes which, however, are not treated in this book). Physical properties of small but finite volume elements are defined according to the *continuum hypothesis*, e. g. a mean velocity or a mean concentration of constituents are defined locally at each point of the fluid domain. Such mean values are well defined because a volume of only 1 $\mu m^3$ of sea water or air contains more than $10^{10}$ molecules. The average separation $\lambda$ between molecules is much smaller than any relevant scale $L$ over which variations of variables are considered. In water $\lambda$ is only about $10^{-8}$ cm, whereas the smallest scale in microstructure measurements in the ocean is typically a few millimeters. With very good accuracy. the corresponding state variables can thus be regarded as continuous fields in space and time.

The equations, which describe the evolution of the state of the ocean, are the macroscopic conservation theorems for partial masses, momentum, and internal energy, as used in conventional hydrodynamics and thermodynamics. These macroscopic theorems can in principle be derived from a statistical theory in which the basic physical laws, governing the microscopic state of single molecules, are embedded (see textbooks on statistical mechanics as e. g. Huang, 1987). The molecular structure of the fluid also determines the values of important parameters appearing in these equations, such as molecular diffusivities for momentum and partial masses, in terms of the forces acting between molecules.

In this part we follow instead a phenomenological derivation of hydrodynamics, based on the empirical knowledge that mass, momentum and energy of small (material) volume elements moving in the fluid are conserved. The thermodynamics is also formulated for such volume elements, assuming that the properties are locally in a thermodynamical equilibrium. Thermodynamic and molecular properties then have to be taken from empirical findings. The macroscopic theory is built in this *Lagrangian* point of view, but the evolution equations are transformed to the *Eulerian* form of a field theory which is more useful in applications.

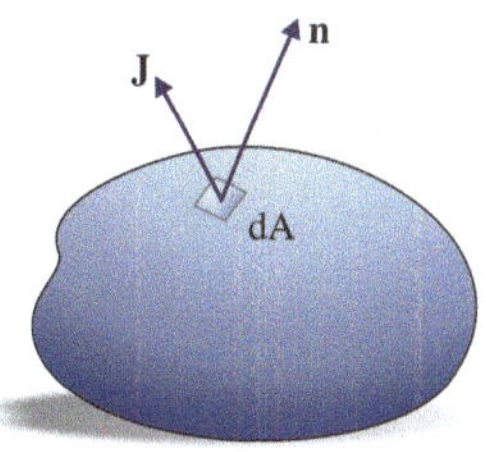


A fluid parcel (also called fluid particle) is a small volume element moving with the local velocity, in case of seawater this is the barycentric velocity of the pure water and the salt constituents. The fluid parcel conserves its mass but may exchange material by diffusive processes with the surrounding fluid (pure water in/out and salt out/in). The sketch shows the parcel with a vectorial flux $\boldsymbol{J}$ of a substance through a surface element $\mathrm{d}A$ with normal $\boldsymbol{n}$. Further explanations are given in Chapter 2.

# 1 Preliminaries

In this chapter some aspects of the kinematical description of oceanic flows are discussed. Furthermore, the equilibrium thermodynamics of seawater is reviewed. The results are prerequisites for the derivation of conservation laws in Chapter 2.

The concept of a *fluid element* is essential for the formulation of hydrodynamics. A fluid element is a small portion of the fluid, marked at some initial time. Its mass is small, and the linear dimension related to the volume of the fluid element is small compared to all relevant scales of motion. We only consider fluid elements with a mass which is constant in time, so that the fluid element is to some extent analogous to a mass point in mechanics. Note, however, that individual molecules of the fluid element may be exchanged with the environment (cf. the treatment of salt diffusion in Section 2.2). Although mass and volume of the fluid element are finite, mathematically it is often practical to treat both as infinitesimally small. We also refer to these fluid elements as fluid 'particles' or 'parcels'. Section 1.1 describes the kinematics of a moving fluid particle, whereas Section 1.2 reviews the thermodynamic properties of a seawater parcel at rest.

## 1.1 Flow Kinematics

In the present section we outline some properties of fluid kinematics. Kinematics deals with the motion of the fluid without account of its dynamical balance. Forces affecting the fluid motion are considered in Section 2.3. The description of hydrodynamic fields most conveniently uses vector calculus. The mathematical tools necessary to describe vector and tensor fields arising in this context are summarized in Appendix A.1. For the notation see the box on p. 4.

### 1.1.1 *Lagrangian and Eulerian Representation*

In the LAGRANGIAN[1] framework of hydrodynamics the motion of the fluid is described by following the trajectory, i. e. the location $\boldsymbol{x} = (x_1, x_2, x_3)$ of each single

[1] JOSEPH LOUIS LAGRANGE, *1736 in Turin, †1813 in Paris, mathematician and physicist.

D. Olbers, J. Willebrand, C. Eden, *Ocean Dynamics*,
DOI 10.1007/978-3-642-23450-7_1, © Springer-Verlag Berlin Heidelberg 2012

**1. Notation for Fields and Derivatives**

The notation for scalar functions $\chi(x_1, x_2, x_3, t) = \chi(\boldsymbol{x}, t)$ and vectors functions $\boldsymbol{u}(\boldsymbol{x}, t)$ is standard, and denotes their functional dependence on position and time (or on other arguments when needed). Tensor functions are denoted as $\mathsf{A}(\boldsymbol{x}, t) \cdots$. Partial derivatives of such functions with respect to the coordinates $x_i$ and time $t$ are denoted by $\partial/\partial x_i$ and $\partial/\partial t$, respectively, and likewise for other arguments. Moreover, we use the vector notation $\boldsymbol{\nabla}$ for the spatial gradient, $\boldsymbol{\nabla}\cdot$ for the divergence, and $\boldsymbol{\nabla}\times$ for the curl. The total or substantial derivative $\mathrm{D}/\mathrm{D}t$ is defined by (1.3) and used throughout this book. It acts on variables which are functions of time and space. Derivatives of variables which are functions of a single independent variable, e. g. time $t$, are usually expressed as $\mathrm{d}/\mathrm{d}t$. Occasionally, time derivatives are denoted by a dot over the symbol.

fluid element as function of time $t$. Fluid elements (or parcels) are identified by an index $\boldsymbol{a}$, such that the trajectory of parcel $\boldsymbol{a}$ through the fluid is given by

$$\boldsymbol{x} = \boldsymbol{X}(\boldsymbol{a}, t)$$

A convenient choice for the parcel's index $\boldsymbol{a}$ is the position vector of the parcel at some initial time $t_0$,

$$\boldsymbol{a} = \boldsymbol{X}(\boldsymbol{a}, t_0)$$

The Lagrangian framework corresponds to the kinematics of a mass point in mechanics. The velocity of the parcel with index $\boldsymbol{a}$ is given by

$$\boldsymbol{u}^{\mathrm{L}}(\boldsymbol{a}, t) = \frac{\partial}{\partial t}\boldsymbol{X}(\boldsymbol{a}, t) \tag{1.1}$$

and is referred to as Lagrangian velocity. The acceleration is the second time derivative of $\boldsymbol{X}(\boldsymbol{a}, t)$. Similarly, any scalar property $\chi$ of the fluid can be represented as a Lagrangian variable by attaching it to the parcels in the form $\chi = \chi(\boldsymbol{a}, t)$. A sketch of the parcel motion is given in Figure 1.1.

The kinematic relation between the position $\boldsymbol{X}(\boldsymbol{a}, t)$ and the velocity $\boldsymbol{u}^{\mathrm{L}}(\boldsymbol{a}, t)$

$$\boldsymbol{X}(\boldsymbol{a}, t) = \boldsymbol{a} + \int_{t_0}^{t} \boldsymbol{u}^{\mathrm{L}}(\boldsymbol{a}, t')\mathrm{d}t'$$

follows by integration of (1.1). Inversion yields the parcel's index $\boldsymbol{a}$ in terms of the position vector $\boldsymbol{X} = \boldsymbol{x}$ at time $t$,

$$\boldsymbol{a} = \boldsymbol{A}(\boldsymbol{x}, t) \tag{1.2}$$

This relation identifies the parcel that is situated at time $t$ at the position $\boldsymbol{x}$ in the fluid. It may be used to transform to EULERIAN[2] fields which describe the fluid properties

[2] LEONHARD EULER, *1707 in Basel, †1783 St. Petersburg, mathematician and physicist. He worked in Berlin and St. Petersburg. The attribution of the development of hydrodynamics is a bit intricate. The Eulerian representation was actually introduced by d'Alembert who gave the complete equations of motion for an incompressible fluid – the so-called Euler equations – in 1747, while Euler generalized it in 1752. Euler was the first who considered the equations of motion in rotating coordinates. The Lagrangian representation was not developed by Lagrange. In his memoir of 1781 he contributed to methods of solving the equations, written in Lagrangian coordinates. They were used by Euler already in 1751. Also Laplace used the Lagrangian representation in his theory of tides in 1776.

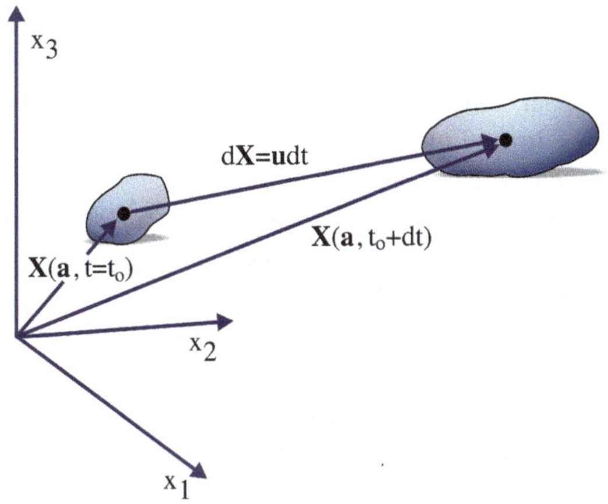


Fig. 1.1 Sketch of Lagrangian trajectory of a fluid element starting at the initial position $\boldsymbol{X} = \boldsymbol{a}$. The initial position $\boldsymbol{a}$ is also used to label the element and to differentiate it from others

at a fixed position and their changes in the course of time. Thus

$$\begin{aligned} \boldsymbol{u}^{\mathrm{E}}(\boldsymbol{x},t) &= \boldsymbol{u}^{\mathrm{L}}(\boldsymbol{a},t) = \boldsymbol{u}^{\mathrm{L}}(\boldsymbol{A}(\boldsymbol{x},t),t) \\ \chi^{\mathrm{E}}(\boldsymbol{x},t) &= \chi^{\mathrm{L}}(\boldsymbol{a},t) = \chi^{\mathrm{L}}(\boldsymbol{A}(\boldsymbol{x},t),t) \end{aligned}$$

are the Eulerian velocity and property fields, respectively. Knowledge of the Eulerian velocity $\boldsymbol{u}^{\mathrm{E}}(\boldsymbol{x},t)$ enables as well to determine the trajectory of each parcel. It follows from

$$\frac{\partial \boldsymbol{X}}{\partial t} = \boldsymbol{u}^{\mathrm{E}}(\boldsymbol{X},t)$$

with the initial condition $\boldsymbol{X}(t_0) = \boldsymbol{a}$. Likewise, the rate of change of a parcel's property can be expressed in either form,

$$\begin{aligned} \frac{\partial}{\partial t}\chi^{\mathrm{L}}(\boldsymbol{a},t) &= \left.\frac{\partial \chi^{\mathrm{E}}}{\partial t}\right|_{\boldsymbol{X}} + \left.\frac{\partial \chi^{\mathrm{E}}}{\partial x_i}\right|_{\mathrm{t}} \frac{\partial X_i}{\partial t} = \frac{\partial \chi^{\mathrm{E}}(\boldsymbol{x},t)}{\partial t} + \boldsymbol{u}^{\mathrm{L}}(\boldsymbol{a},t)\cdot\nabla\chi^{\mathrm{E}}(\boldsymbol{x},t) \\ &= \frac{\partial \chi^{\mathrm{E}}(\boldsymbol{x},t)}{\partial t} + \boldsymbol{u}^{\mathrm{E}}(\boldsymbol{x},t)\cdot\nabla\chi^{\mathrm{E}}(\boldsymbol{x},t) \end{aligned}$$

Note that we use here and in the following the Einstein summation convention: if the same index appears twice in one term of an equation, the summation over this index is implied. We now define the operator

$$\frac{\mathrm{D}}{\mathrm{D}t} \equiv \frac{\partial}{\partial t} + \boldsymbol{u}^{\mathrm{E}}\cdot\nabla \tag{1.3}$$

and arrive at

$$\frac{\partial}{\partial t}\chi^{\mathrm{L}}(\boldsymbol{a},t) \equiv \frac{\mathrm{D}}{\mathrm{D}t}\chi^{\mathrm{E}}(\boldsymbol{x},t) = \frac{\partial}{\partial t}\chi^{\mathrm{E}}(\boldsymbol{x},t) + \boldsymbol{u}^{\mathrm{E}}\cdot\nabla\chi^{\mathrm{E}}(\boldsymbol{x},t) \tag{1.4}$$

often referred to as *Euler's relation*. Both expressions in (1.4) describe a temporal change of the parcel's property, however, in the Eulerian form (right hand sides of (1.4)) the parcel context of the Lagrangian form (left hand side of (1.4)) is lost. We shall refer to the above operator $\mathrm{D}/\mathrm{D}t$ as the 'material' or 'substantial' derivative (see also the box on p. 4). While the term 'material' is custom, as mentioned before, the matter in a 'material' fluid parcel is not necessarily conserved because of diffusion of partial masses. In the Eulerian form, the material derivative consists of a local rate of change and the change implied by the advection of fluid through the point of observation. If the material rate of change vanishes for a specific property, this property is conservative – a parcel carries a constant value of that property for all times. Nevertheless, the property observed at any fixed (Eulerian) position $\boldsymbol{x}$ may change in the course of time, because of advection of parcels with different properties.

The Lagrangian and Eulerian formulations of fluid mechanics are completely equivalent. While the first is more intuitive, the latter is generally more convenient. This becomes evident when the dynamical evolution equations are written completely in Lagrangian coordinates; see Section 2.12. We shall be mostly concerned with the Eulerian framework and abandon the index 'E' in the rest of this book, thus e. g. $\boldsymbol{u}(\boldsymbol{x},t) \equiv \boldsymbol{u}^{\mathrm{E}}(\boldsymbol{x},t)$ is referred to as the velocity field which in Section 2.2 will be identified with the velocity of the center of partial masses constituting the fluid. The derivation of conservation equations of fluid properties outlined in the following sections will use the Eulerian form, some arguments will however be borrowed from the Lagrangian framework.

### 1.1.2 Deformation and Rotation

The velocity field $\boldsymbol{u}(\boldsymbol{x},t)$ completely quantifies the macroscopic flow of the fluid. At any fixed time, the velocity vectors define a family of *streamlines* which are everywhere tangential to $\boldsymbol{u}(\boldsymbol{x},t)$, i. e. we have $\boldsymbol{u}(\boldsymbol{x},t) \times \mathrm{d}\boldsymbol{x} = 0$ where $\mathrm{d}\boldsymbol{x}$ is an infinitesimal line element along the streamline at the position $\boldsymbol{x}$, or

$$\frac{\mathrm{d}x}{u(\boldsymbol{x},t)} = \frac{\mathrm{d}y}{v(\boldsymbol{x},t)} = \frac{\mathrm{d}z}{w(\boldsymbol{x},t)}$$

in a Cartesian coordinate system with $\boldsymbol{u} = (u, v, w)$ and $\boldsymbol{x} = (x, y, z)$. Streamlines, taken at a particular time, and parcel trajectories, filling out the space with time progressing, are congruent if the flow is steady (meaning that it does not change in the course of time). In an unsteady flow, the streamline pattern taken at time $t$ and the trajectories with initial condition taken at the same time are initially tangent but generally diverge at later times.

These families of lines are used to present a 'global' view of the fluid motion. Zooming in locally, a fluid parcel at $\boldsymbol{x}$ is translated by the motion an infinitesimal distance $\boldsymbol{u}(\boldsymbol{x},t)\delta t$ during the time interval $\delta t$ along the corresponding streamline (and trajectory). But taking parcels on neighboring streamlines, we notice that these could be displaced in a slightly different way. The distortion of a pattern of infinitesimal neighboring parcels is contained in the gradient $\nabla\boldsymbol{u}$ of the velocity field which is tensor with (3×3) Cartesian components $\partial u_i/\partial x_j$. Even simple structures of this tensor may lead to complex distortions of the fluid. Note that some important elements of vector and tensor algebra are summarized in Appendix A.1.

The velocity vector gradient may be decomposed into a symmetric and an anti-symmetric part,

$$\frac{\partial u_i}{\partial x_j} = \frac{1}{2}\left(\frac{\partial u_i}{\partial x_j} + \frac{\partial u_j}{\partial x_i}\right) + \frac{1}{2}\left(\frac{\partial u_i}{\partial x_j} - \frac{\partial u_j}{\partial x_i}\right) = D_{ij} + R_{ij} \tag{1.5}$$

where

$$\begin{aligned} D_{ij} &= \frac{1}{2}\left(\frac{\partial u_i}{\partial x_j} + \frac{\partial u_j}{\partial x_i}\right) = D_{ji} \quad \text{is the deformation tensor} \\ R_{ij} &= \frac{1}{2}\left(\frac{\partial u_i}{\partial x_j} - \frac{\partial u_j}{\partial x_i}\right) = -R_{ji} \quad \text{is the rotation tensor} \end{aligned} \tag{1.6}$$

The above names already indicate the role of the corresponding tensors, which will be discussed in this section. Other names are rate of strain tensor[3] for $D_{ij}$ and vorticity tensor for $R_{ij}$.

Two parcels at the positions $\boldsymbol{x}_0$ and $\boldsymbol{x}_0 + \boldsymbol{\ell}(t)$, separated by an infinitesimal vector $\boldsymbol{\ell}(t)$ at time $t$ (see Figure 1.2), move relative to each other with the velocity $\delta \boldsymbol{u}(\boldsymbol{x}_0, t) = \boldsymbol{u}(\boldsymbol{x}_0 + \boldsymbol{\ell}(t), t) - \boldsymbol{u}(\boldsymbol{x}_0, t) = \boldsymbol{\ell}(t) \cdot \nabla \boldsymbol{u}(\boldsymbol{x}_0, t) + O(\boldsymbol{\ell}^2)$. At a later time $t_0 + \delta t$ the parcels are separated by $\boldsymbol{\ell}(t + \delta t) = \boldsymbol{\ell}(t) + \delta \boldsymbol{u} \delta t$; so the distance vector $\boldsymbol{\ell}(t)$ between them is governed by

$$\frac{\mathrm{d}\boldsymbol{\ell}}{\mathrm{d}t} = \frac{\mathrm{D}\boldsymbol{\ell}}{\mathrm{D}t} = \boldsymbol{\ell} \cdot \nabla \boldsymbol{u} = (\mathsf{D} + \mathsf{R}) \cdot \boldsymbol{\ell} \tag{1.7}$$

using (1.5). The above evolution equation needs more explanation. In fact, $\boldsymbol{\ell}$ is a Lagrangian property: it is the distance vector of two parcels. We attach the Lagrangian index $\boldsymbol{x}_0$ to $\boldsymbol{\ell}$ writing $\boldsymbol{\ell} = \boldsymbol{\ell}(\boldsymbol{x}_0, t)$ and replace the time derivative by the Lagrangian operator $\mathrm{D}/\mathrm{D}t$ (see also the box on p. 4).

Apparently, the distance vector is affected by both tensor contributions to the velocity gradient. There are other important quantities which are affected by only the one or the other. We elucidate the two contributions to the gradient tensor separately.

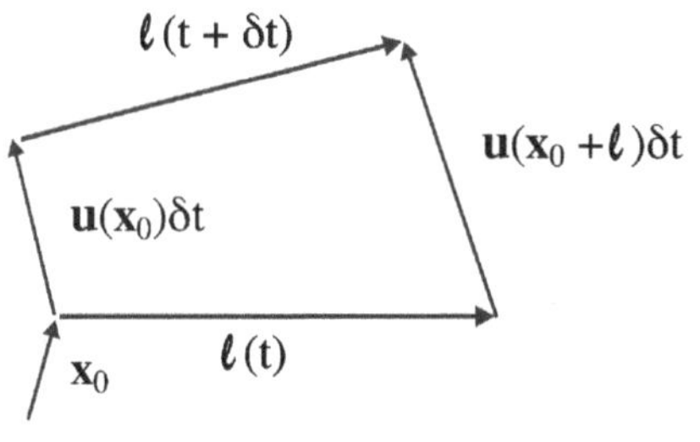


Fig. 1.2 Sketch of the motion of two Lagrangian points in the fluid separated by the distance vector $\boldsymbol{\ell}(t)$

[3] The strain tensor is often defined as the tensor of displacement gradients corresponding to D.

### Deformation

The evolution of the distance $\sqrt{\boldsymbol{\ell}^2}$ between the parcels is governed by the deformation tensor D only,

$$\frac{\mathrm{D}\boldsymbol{\ell}^2}{\mathrm{D}t} = 2\boldsymbol{\ell} \cdot (\mathsf{D} \cdot \boldsymbol{\ell} + \mathsf{R} \cdot \boldsymbol{\ell}) = 2\boldsymbol{\ell} \cdot \mathsf{D} \cdot \boldsymbol{\ell}$$

since the quadratic form $\boldsymbol{\ell} \cdot \mathsf{R} \cdot \boldsymbol{\ell}$ of the antisymmetric rotation tensor vanishes identically. The rate of change of the volume $V = (\boldsymbol{a} \times \boldsymbol{b}) \cdot \boldsymbol{c}$ of an infinitesimal material volume element spanned by three vectors $\boldsymbol{a}$, $\boldsymbol{b}$ and $\boldsymbol{c}$ is given by

$$\begin{aligned}\frac{\mathrm{D}V}{\mathrm{D}t} &= \left(\frac{\mathrm{D}\boldsymbol{a}}{\mathrm{D}t} \times \boldsymbol{b}\right) \cdot \boldsymbol{c} + \left(\boldsymbol{a} \times \frac{\mathrm{D}\boldsymbol{b}}{\mathrm{D}t}\right) \cdot \boldsymbol{c} + (\boldsymbol{a} \times \boldsymbol{b}) \cdot \frac{\mathrm{D}\boldsymbol{c}}{\mathrm{D}t} \\ &= (\mathsf{D} \cdot \boldsymbol{a} \times \boldsymbol{b}) \cdot \boldsymbol{c} + (\boldsymbol{a} \times \mathsf{D} \cdot \boldsymbol{b}) \cdot \boldsymbol{c} + (\boldsymbol{a} \times \boldsymbol{b}) \cdot \mathsf{D} \cdot \boldsymbol{c} \end{aligned} \tag{1.8}$$

Again, the contributions from the asymmetric part of $\nabla \boldsymbol{u}$ are canceled. From the symmetric part only the diagonal tensor elements $D_{ii}$ remain, as shown by an elementary evaluation of (1.8). Hence

$$\frac{\mathrm{D}V}{\mathrm{D}t} = V D_{ii} = V \nabla \cdot \boldsymbol{u} \tag{1.9}$$

Here, $\nabla \cdot \boldsymbol{u}$ is the divergence of the velocity field, in Cartesian coordinates given by $\nabla \cdot \boldsymbol{u} = \partial u_i / \partial x_i$. If the flow is divergent at the respective position, i. e. $\nabla \cdot \boldsymbol{u} > 0$, then the fluid parcel expands in volume. If the flow is convergent, i. e. $\nabla \cdot \boldsymbol{u} < 0$, then the parcel's volume shrinks. Equation (1.9) is most easily proven if the tensor D is diagonal: expanding the scalar triple products in (1.8) into the determinant form, it is found that each single sum to the determinants contains only one of the diagonal entries of D, and the three triple products then combine to the form (1.9).

Since D is symmetric, a coordinate system where D is diagonal can always be found. It is the system spanned by the three eigenvectors of D (for eigenvalue problems see the box on p. 9); hence the new coordinates are along the principal axes of D. In these rotated coordinates the deformation tensor is given by

$$\mathsf{D} = (D_{ij}) = \begin{pmatrix} \lambda^{(1)} & 0 & 0 \\ 0 & \lambda^{(2)} & 0 \\ 0 & 0 & \lambda^{(3)} \end{pmatrix}$$

where $\lambda^{(\alpha)}, \alpha = 1, 2, 3$ are the eigenvalues of D. Note that this coordinate transformation applies to the reference point $\boldsymbol{x}_0$, i. e. all quantities, including the $\lambda^{(\alpha)}$ thus depend on the position. On the other hand, the trace $D_{ii} = \sum_i \lambda^{(i)}$ of a tensor is invariant under orthogonal transformations. The evolution equation (1.9) for the volume of a fluid parcel is thus invariant as well. The interpretation of this relation is now straightforward in the coordinate system of the principal axes: each side $a$, $b$, $c$ of the volume element is stretched by a factor $1 + \lambda^{(\alpha)}\delta t$, respectively, in the time $\delta t$. The expanded volume is thus $a(1 + \lambda^{(1)}\delta t)b(1 + \lambda^{(2)}\delta t)c(1 + \lambda^{(3)}\delta t) = abc\left[1 + \sum_i \lambda^{(i)}\delta t + O(\delta t^2)\right]$.

The deformation is thus completely determined by the three *principal strain rates* $\lambda^{(\alpha)}$ describing stretching along the principle axes. Distance vectors pointing in other directions experience stretching *and* rotation, however. The deviator $\boldsymbol{D}^\star$ of the deformation tensor, defined by

$$D^\star_{ij} = D_{ij} - S\delta_{ij} \quad \text{with} \quad S = \frac{1}{3}\sum_i D_{ii} = \frac{1}{3}\sum_i \lambda_i = \frac{1}{3}\nabla\cdot\boldsymbol{u} \tag{1.10}$$

has zero trace and thus has no effect on the volume change. Only the isotropic part $(S\delta_{ij})$ is responsible for the change of volume. The deviator will be met again in Section 2.3 where it will be related to molecular friction in the momentum budget.

### Rotation

The rotation tensor R, given by

$$(R_{ij}) = \frac{1}{2}\begin{pmatrix} 0 & -\omega_3 & \omega_2 \\ \omega_3 & 0 & -\omega_1 \\ -\omega_2 & \omega_1 & 0 \end{pmatrix}$$

is antisymmetric and thus has only three independent components $\omega_i$, $(i = 1, 2, 3)$, defining a vector $\boldsymbol{\omega} = (\omega_1, \omega_2, \omega_3)$. More accurately, $\boldsymbol{\omega}$ is a pseudovector (see Appendix A.1.6). The contribution to the rate of change of the distance vector can thus

**2. Eigenvalue Problems**

The real $n \times n$ matrix $\mathbf{A}$ has an eigenvalue $\lambda$ with corresponding (right) eigenvector $\boldsymbol{a}$ if $(\mathbf{A} - \lambda\mathbf{I})\boldsymbol{a} = 0$. In order that this problem has nontrivial solutions, the determinant of the matrix must vanish,

$$\det(\mathbf{A} - \lambda\mathbf{I}) = 0$$

which is the *characteristic polynomial* of $\mathbf{A}$. It is of $n$th order in $\lambda$; so there are in general $n$ solutions $\lambda^{(i)}, \boldsymbol{a}^{(i)}, i = 1, \ldots, n$ to the eigenvalue problem. Eigenvectors with different eigenvalues are linearly independent but not necessarily orthogonal. Eigenvectors may be normalized to unity because they are defined only up to a factor.

**If all eigenvalues are different**, the set of normalized eigenvectors forms a basis in which $\mathbf{A}$ becomes diagonal, i. e. grouping the eigenvectors to a matrix $\mathbf{T} = [\boldsymbol{a}^{(1)}, \ldots, \boldsymbol{a}^{(n)}]$ the transformation generates the eigenvalue diagonal matrix $\mathbf{T}^{-1}\mathbf{A}\mathbf{T} = \mathbf{\Lambda} = \text{diag}(\lambda^{(1)}, \ldots, \lambda^{(n)})$. The right-eigenvalue problem may thus be written in matrix form $\mathbf{A}\mathbf{T} = \mathbf{\Lambda}T$. It is associated with a left-eigenvalue problem $\mathbf{T}^{-1}\mathbf{A} = \mathbf{\Lambda}\mathbf{T}^{-1}$ which generates a set of left eigenvectors $\boldsymbol{b}^{(i)}, i = 1, \ldots, n$ which are mutually orthogonal to the right eigenvectors, $\boldsymbol{a}^{(i)}\boldsymbol{b}^{(j)} = \delta_{ij}$. The sets of eigenvalues for the left and right eigenvalue problems are identical. Further properties:

- $\det(\mathbf{A}) = \prod_{i=1}^{n}\lambda^{(i)}$ and $\text{trace}(\mathbf{A}) = \sum_{i=1}^{n}\lambda^{(i)}$
- If $\mathbf{A}$ is symmetric then:
  1. All eigenvalues are real.
  2. Eigenvectors with different eigenvalues are orthogonal, i. e. $\mathbf{T}^{-1} = \mathbf{T}^{\mathrm{T}}$.
- If $\mathbf{A}$ is not symmetric, complex eigenvalues may occur but always in conjugate pairs. Then also the eigenvectors and $\mathbf{T}$ and $\mathbf{T}^{-1}$ are complex.
- If $\mathbf{A} = \mathbf{B}\mathbf{C}$ with symmetric $\mathbf{B}$ and $\mathbf{C}$ and invertible $\mathbf{B}$, hence $(\mathbf{C} - \lambda\mathbf{B}^{-1})\boldsymbol{a} = 0$, then $\mathbf{A}$ has real eigenvalues. Most eigenvalue problems arising in physics are of this form.

**If there are multiple eigenvalues** and if $\mathbf{A}$ is of diagonal similarity (or reducible; i. e. it can be transformed to a diagonal form $\mathbf{T}^{-1}\mathbf{A}\mathbf{T} = \mathbf{\Lambda}$), then a complete set of eigenvectors is contained in the transformation matrix $\mathbf{T}$. There are, however, irreducible cases where $\mathbf{A}$ cannot be put into a diagonal form and less than $n$ eigenvectors exist. It is always possible to find a $\mathbf{T}$ which transforms $\mathbf{A}$ to a Jordan form $\mathbf{T}^{-1}\mathbf{A}\mathbf{T}$ which has zeros below the diagonal, the eigenvalues (with their multiplicity) in the diagonal and generally nonzero elements above the diagonal.

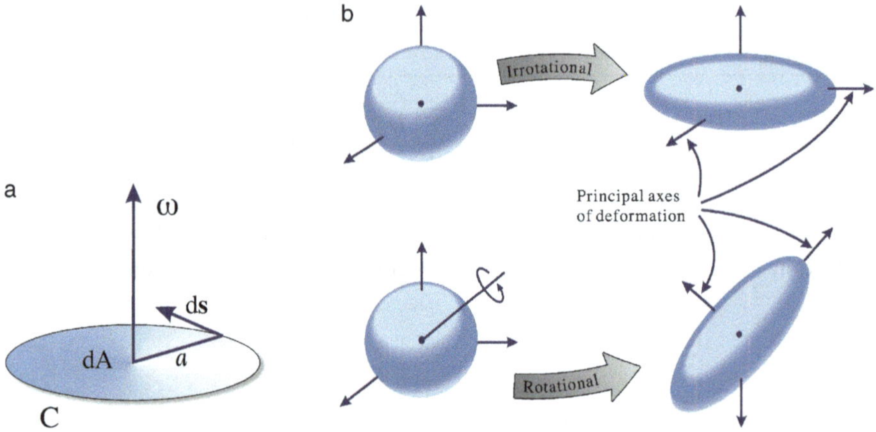


Fig. 1.3 **a** A sketch of the local rotation. **b** A sketch visualizing the distortions due to dilation and rotation of a fluid parcel. Redrawn after Aris (1989)

be expressed as the vectorial product of $\boldsymbol{\omega}$ and $\boldsymbol{\ell}$,

$$\mathsf{R} \cdot \boldsymbol{\ell} = \frac{1}{2}\boldsymbol{\omega} \times \boldsymbol{\ell} \tag{1.11}$$

describing pure (rigid body) rotation of $\boldsymbol{\ell}$ with angular velocity $\boldsymbol{\omega}/2$. Written in terms of the velocity gradient, the vector $\boldsymbol{\omega}$ becomes

$$\boldsymbol{\omega} = \left(\frac{\partial u_3}{\partial x_2} - \frac{\partial u_2}{\partial x_3}, \quad \frac{\partial u_1}{\partial x_3} - \frac{\partial u_3}{\partial x_1}, \quad \frac{\partial u_2}{\partial x_1} - \frac{\partial u_1}{\partial x_2}\right) = \nabla \times \boldsymbol{u} \tag{1.12}$$

which is the curl of the velocity vector and termed *vorticity vector*. The effect of the rotation tensor on the local relative motion is elucidated by considering the fluid in an infinitesimal disk with radius $r$ and normal vector $\boldsymbol{n}$, centered around the reference point (see Figure 1.3a)). The fluid motion generates an average angular velocity $u_{\text{tangential}}/r$ which we estimate as $u_{\text{tangential}} \approx (r/2)\boldsymbol{\omega} \cdot \boldsymbol{n} = (r/2)\omega_\perp$ following STOKES'[4] integral theorem (see Appendix A.1.3). The disk thus rotates with $\omega_\perp/2$, half the normal component of the local vorticity vector.

We summarize this section by noting that any arbitrary state of motion may be decomposed in each point into a uniform translation, a dilation along three mutually orthogonal axes, and a rigid rotation, according to

$$\boldsymbol{u}(\boldsymbol{x}, t) = \boldsymbol{u}(\boldsymbol{x}_0, t) + \boldsymbol{\ell} \cdot \mathsf{D} + \boldsymbol{\ell} \cdot \mathsf{R}$$

with $\boldsymbol{\ell} = \boldsymbol{x} - \boldsymbol{x}_0$ and D and R referring to the point $\boldsymbol{x}_0$. Figure 1.3b) shows a sketch of the distortions due to dilation and rotation, associated with the deformation and rotation tensors.

## 1.2 Thermodynamics of Sea Water

In this section, we discuss the fundamental laws of thermodynamics which form the basis for the consideration of energy conservation in fluids in Chapter 2, by applying the thermodynamic laws to a moving infinitesimal material fluid element. The

[4] GEORGE GABRIEL STOKES, *1819 in Skreen, †1903 in Cambridge, mathematician and physicist.

**3. Variations in the Composition of Seawater**

A constant mass ratio of the salt components in seawater is a very good assumption for dynamical purposes. This assumption will be used in this book. However, it turns out that there are small variations in the salt components, which are measurable (with some effort) but difficult to deal with. It is even not obvious how to differentiate freshwater from dissolved salt: Consider for instance a certain amount of mass of the gas $CO_2$ dissolving in sea water: it is decomposed into different fractions, i. e. $CO_2$, $H_2CO_3$, $HCO_3^-$, $CO_3^{2-}$, $H^+$, $OH^-$ and $H_2O$. This means that some of the freshwater turns into salt during the dissolving process; the amount, furthermore, depends on ambient temperature and pressure.
The largest deviation in the worlds ocean of the composition of sea water from a reference composition (referring to the North Atlantic) can be seen in the North Pacific Ocean with a value in terms of salt concentration of about $0.025\,\mathrm{g\,kg^{-1}}$. The reason for this deviation appears to be mostly related to the microbial demineralization of biological material in the interior ocean. A variety of variables describing the salt concentration and corrections for the variable composition of seawater have been put forward so far; a summary and some recommendations can be found in IOC, SCOR and IAPSO (2010).

discussion will also allow to infer the direction of molecular transports of heat and salt. Furthermore, we will introduce thermodynamic properties such as thermal expansion, specific heat, etc. which will also turn out to be important for the dynamics of fluids.

The concept of *thermodynamic equilibrium* plays a central role in thermodynamics: a system is in thermodynamic equilibrium if there is mechanical equilibrium (i. e. if all forces are balanced), if there is thermal equilibrium (i. e. if the temperature is uniform), and if there is chemical equilibrium (i. e. if there is no change of structure and chemical composition). While almost any interesting hydrodynamic process is usually not in thermodynamic equilibrium, one can normally assume that *locally*, i. e. for a sufficiently small volume (i. e. fluid element), thermodynamic equilibrium prevails to a sufficiently good approximation, due to the small space-time-scales of molecular processes.

Different types of thermodynamic systems can be distinguished. *Open* systems exchange both matter and energy with the environment. *Closed* systems exchange energy but not matter, *adiabatic* systems exchange matter but not energy, and *adiabatically closed* (or *isolated*) systems exchange neither energy nor matter. While a fluid element in the ocean always constitutes an open system, it is nevertheless often very useful to consider the idealized case of an adiabatically closed system.[5]

### 1.2.1 Salt Concentration and Salinity

Seawater consists of pure water and a number of dissolved salts. The main components are chloride (about 55% by mass), sodium (30%), sulfate (8%), magnesium (4%), potassium and calcium (each 1%) and many further components (in fact, any element worth talking about can be found in seawater at a certain concentration). The mass ratio of the salt components is very nearly constant throughout most of the world's oceans (compare the box on p. 11), and for almost all dynamical purposes it is sufficient to consider their total mass, respectively their concentration. Hence we have the variables

[5] Adiabatic systems with mass exchange are rarely of interest, and in the following the term 'adiabatic' is used to describe 'adiabatically closed' systems.

| | | |
|---|---|---|
| $s$ | salt concentration | in kg salt/kg seawater |
| $\rho_s$ | density of salt | in kg salt/$m^3$ |
| $w$ | content of freshwater | in kg freshwater/kg seawater |
| $\rho_w$ | density of freshwater | in kg freshwater/$m^3$ |

Notice that $s$ and $w$ are concentrations in the conventional sense. They are dimensionless, e. g. $s = 0.035$ is a typical value for the open ocean. The definitions imply

$$\rho_s = s\rho\,, \quad \rho_w = w\rho = (1-s)\rho \quad \text{and} \quad \rho = \rho_s + \rho_w \tag{1.13}$$

for the oceanic partial masses. The density variable $\rho$ refers to the *total* mass of the system, and the relations (1.13) have to be modified if more than two partial masses contribute to the total mass. Usually the additional components have negligible concentrations compared to those considered above. In principle, one also should account for sea ice concentration as an additional variable. Ice formation may occur locally in the interior of the ocean in supercooled seawater in form of small ice plates (the ocean's equivalent of rain drops) which, however, experience rapid updraught to the surface. We, therefore, assume that sea ice normally occurs only very close the ocean–atmosphere interface, so that the associated phase transitions can be considered separately.

Instead of using the salt concentration $s$, it is common to use the *salinity* $S$ with the numerical value $S = 1000\,s$ such that $s = 0.035$ kg salt/kg seawater corresponds to a salinity of $S = 35$ g/kg. This choice follows the oceanographic convention; it has, however, the disadvantage that many thermodynamic relations are more conveniently formulated in terms of the salt concentration $s$, as can be seen already from the relations (1.13). In Section 1.2 we will use, therefore, both variables simultaneously, but in the following sections only the salinity $S$ will be used.

The definition of salt concentration, respectively salinity, as a mass fraction is natural from a theoretical viewpoint. For practical reasons, other definitions such as practical salinity or absolute salinity have been introduced, which can be calculated from conductivity and temperature. See IOC, SCOR and IAPSO (2010) for an extensive discussion.

### 1.2.2 Additive State Variables

Consider a system with $n$ chemically different constituents $i = 1, \ldots, n$ with partial masses $M_i$ and a total mass $M = \sum M_i$ in a volume $V$. Let $m_i = M_i/M$ be the respective concentration and $\rho_i = M_i/V = m_i\rho$ the density of constituent $i$, so that

$$\sum_i m_i = 1\,, \quad \sum_i \mathrm{d}m_i = 0 \quad \text{and} \quad \sum_i \rho_i = \rho \tag{1.14}$$

where $\rho = M/V$ is the density of the sum of partial masses, and $\mathrm{d}m_i$ is an infinitesimal increment of the concentration $m_i$.

The thermodynamic state of a system with $n$ constituents can be characterized by its temperature $T$, pressure $p$, and $n-1$ concentrations $m_i$, since one of the $m_i$ is redundant due to the first relation in (1.14). Hence we have $n+1$ independent thermodynamic *state variables*.

In the case of seawater, we have two constituents, salt (concentration $m_s$) and pure water (concentration $m_w$), and we choose the salt concentration $s$ – respectively

**4. The Additivity Relation for Extensive Variables**

The volume $V$ will depend on the partial masses $M_i$ as well as on temperature $T$ and pressure $p$, $V = V(M_1, \dots, M_n, T, p)$, and is an *extensive* (or *additive*) quantity. An extensive variable depends on the amount of material in the region considered (e. g. total volume, total mass or total energy) whereas an intensive variable is independent of the amount of matter but has a value at each point of the space (e. g. temperature, pressure or density).
Consider the general extensive variable $W$. It must double if we double all partial masses $M_i$ for given (constant) $T$ and $p$. In general, the relation

$$W(\alpha M_1, \dots, \alpha M_n, T, p) \equiv \alpha W(M_1, \dots, M_n, T, p)$$

must hold for an arbitrary value of $\alpha$. This relation states that $W$ is a homogeneous function (of grade one) of the partial masses $M_i$. Differentiation with respect to $\alpha$, and then putting $\alpha = 1$ yields the additivity relation

$$W(M_1, \dots, M_n, T, p) = \sum_i \frac{\partial W}{\partial M_i} M_i \tag{B4.1}$$

The corresponding variable $W$ per unit mass, $w = W/M$, accordingly satisfies the additivity relation

$$w(m_1, \dots, m_n, T, p) = \sum_i w_i m_i \quad \text{with} \quad w_i(m_1, \dots, m_n, T, p) = \frac{\partial w}{\partial m_i} \tag{B4.2}$$

The quantities $w_i = \partial W/\partial M_i = \partial w/\partial m_i$ can be identified as the specific quantity of the constituent $i$, referred to the respective mass $M_i$. The $w_i$ may still depend on the concentrations $m_1, \dots, m_n$, but no longer on the partial masses $M_1, \dots, M_n$. Note that $w$ becomes now an *intensive* variable. Taking $W$ to be the total volume $V$, $\upsilon = V/M = 1/\rho$ is the *specific volume*. Note that $\rho_i \upsilon_i < 1$ and $\sum_i \rho_i \upsilon_i = 1$.

salinity $S$ – as state variable (in addition to temperature $T$ and pressure $p$). For the specific volume $\upsilon = 1/\rho$, it follows

$$\upsilon(m_s, m_w, T, p) = m_s \upsilon_s + m_w \upsilon_w \quad \text{with} \quad \upsilon_s = \left.\frac{\partial \upsilon}{\partial m_s}\right|_{m_w} \quad \text{and} \quad \upsilon_w = \left.\frac{\partial \upsilon}{\partial m_w}\right|_{m_s}$$

from the general additivity relation (B4.2) for the case of seawater with the specific volumes of salt $\upsilon_s$ and pure water $\upsilon_w$ and where the derivative with respect to each constituent is such that the other is kept constant. Hence the change of specific volume $\upsilon$ with respect to changes in salt concentration $s$ is given by

$$\frac{\partial \upsilon}{\partial s} = \frac{\partial}{\partial s} \upsilon(m_s, m_w, T, p) = \frac{\partial \upsilon}{\partial m_s}\frac{\partial m_s}{\partial s} + \frac{\partial \upsilon}{\partial m_w}\frac{\partial m_w}{\partial s}$$

We now specify $s = m_s$ as state variable so that $m_w = 1 - s$. Note that $m_s = s$ but $\partial/\partial m_s \neq \partial/\partial s$ because $m_w$ is constant when taking the derivative with respect to $m_s$, whereas both $m_s$ and $m_w$ change in correspondence when taking the derivative with respect to $s$. We find for $\upsilon = \upsilon(s, T, p)$ the rule

$$\left(\frac{\partial \upsilon}{\partial s}\right)_{T,p} = \upsilon_s - \upsilon_w \tag{1.15}$$

where the subscripts $T, p$ indicate the variables that are kept constant during differentiation[6]. The relation (1.15) states that the changes in $\upsilon$ by changes in salt

[6] From now on we shall drop this notation, which is customary in thermodynamics, and assume that differentiation with respect to one of the independent state variables $s, T, p$ (respectively $S, T, p$) is always such that the other two are kept constant, unless stated otherwise.

concentration $s$ are given by changes in the specific volume of salt or freshwater, i. e. $\upsilon$ increases by adding salt or removing freshwater.

Since only the additivity relation (B4.2) for the specific volume has been used in deriving (1.15), corresponding relations must hold for *all* additive state variables. All energies and the entropy (to be introduced below) are additive variables which must satisfy relations analogously to (B4.1) or, in their intensive form, to (B4.2), or, for the case of seawater, to (1.15). From now on, we will use the intensive form for all variables unless otherwise stated. In other words, these variables are defined for a unit mass. This definition enables as to proceed in a straightforward way from the differentials in thermodynamics to the (material) Lagrangian rate of change and further to the Eulerian form of a field equation in which all state variables are locally defined, so that e. g. $\upsilon(\boldsymbol{x}, t)$ is the specific volume at time $t$ and the spatial point represented by the three-dimensional position vector $\boldsymbol{x}$.

### 1.2.3 First Law of Thermodynamics

The first law of thermodynamics introduces the concept of an *internal energy* $E$ (energy per mass, in $\mathrm{J\,kg^{-1}} = \mathrm{m^2\,s^{-2}}$). In seawater, the internal energy (per mass) $E$ contains the kinetic energy of water molecules and the binding/solution energy of freshwater and of salt ions, which we will refer to as chemical energy. In a closed system, internal energy may be changed by work on the volume[7], i. e. $-p\mathrm{d}\upsilon$ if surface effects such as e. g. surface tension are ignored, and through energy exchange. Thus we have

$$\mathrm{d}E = -p\mathrm{d}\upsilon + \delta Q \quad \text{(closed system)} \tag{1.16}$$

Here $\mathrm{d}E$ is the change in internal energy, and $\delta Q$ denotes the *total* heating per mass which is added, resulting both from external fluxes (e. g. radiation and heat flux) as well as from internal conversions (as e. g. heat production by friction). Whereas $\mathrm{d}E$ refers to the change of the state variable $E$, the heating rate $\delta Q$ is not connected to a state variable ($\delta Q$ is an infinitesimal increment but not a differential). The forms of energy that have to be included in $E$ depend on which constituents are included in the systems and which processes are relevant. The thermal energy of molecular motion is always relevant, likewise the binding energy of molecules. Other forms of energy may be relevant in phase transitions (which are not considered here).

The internal energy per mass $E = E(m_1, \ldots, m_n, T, p)$ is an additive variable. As discussed in the box on p. 13, it satisfies the additivity relation

$$E = \sum_i m_i E_i \quad \text{with} \quad E_i = \frac{\partial E}{\partial m_i} \tag{1.17}$$

where $E_i(m_1, \ldots, m_n, T, p)$ is the internal energy of constituent $i$, referred to its respective mass. The first law may also be formulated in terms of the *enthalpy* (per mass)

$$H = E + p\upsilon \tag{1.18}$$

[7] This derives from the simple argument *work* = *force* × *change of distance* = *force*/*area* × *change of distance* × *area* = *force*/*area* × *change of volume*.

Since the enthalpy is also an additive variable, we have again

$$H = \sum_i m_i H_i \quad \text{with} \quad H_i = \frac{\partial H}{\partial m_i} = E_i + p\upsilon_i \tag{1.19}$$

where $H_i(m_1, \ldots, m_n, T, p)$ is the specific enthalpy of constituent $i$. With the total differential of the enthalpy $\mathrm{d}H = \mathrm{d}E + p\mathrm{d}\upsilon + \upsilon \mathrm{d}p$ one finds

$$\mathrm{d}H = \upsilon \mathrm{d}p + \delta Q \quad \text{(closed system)} \tag{1.20}$$

The change in enthalpy $H$ thus corresponds to the energy $\delta Q$ added at constant pressure ($\mathrm{d}p = 0$). In contrast, the change of internal energy corresponds to the energy added at constant volume ($\mathrm{d}\upsilon = 0$). Note that enthalpy is, with respect to the following discussion, more convenient than internal energy since we might assume that a fluid element exchanges energy with its environment at constant pressure, while the volume (density) of the fluid element might change during or because of this exchange (cf. also the box on p. 16).

Fluid elements do not just exchange energy but also matter with their environment. If we now allow for mass exchange, i. e. if the system is open, we must account for the different enthalpies of the constituents. For a mass exchange at constant pressure, a change $\mathrm{d}m_i$ of constituent $i$ changes the enthalpy of the system by an amount $H_i \mathrm{d}m_i$. Hence for an *open* system we reformulate the first law in the form

$$\mathrm{d}H = \delta Q + \upsilon \mathrm{d}p + \sum_i H_i \mathrm{d}m_i \quad \text{(open system)} \tag{1.21}$$

or equivalently

$$\mathrm{d}E = \delta Q - p\mathrm{d}\upsilon + \sum_i H_i \mathrm{d}m_i \quad \text{(open system)} \tag{1.22}$$

From (1.22) it follows immediately that $\mathrm{d}E = \delta Q - p\mathrm{d}\upsilon + H_\mathrm{s}\mathrm{d}m_\mathrm{s} + H_\mathrm{w}\mathrm{d}m_\mathrm{w}$ for seawater. Using relation (1.15) for additive variables and changing the independent variable from salt concentration $s$ to salinity $S$, one obtains[8]

$$\mathrm{d}E = \delta Q - p\mathrm{d}\upsilon + \frac{\partial H}{\partial S}\mathrm{d}S \quad \text{or} \quad \mathrm{d}H = \delta Q + \upsilon \mathrm{d}p + \frac{\partial H}{\partial S}\mathrm{d}S \tag{1.23}$$

as a convenient form of the first law for seawater. Note that $\delta Q$ contains all internal irreversible and all external energy exchanges, except those connected to the diffusive exchange $\mathrm{d}S$ of salinity, which are explicitly included in the last term of (1.23). On the other hand, it will turn out in Section 2.4 that the effect of salt exchanges on the internal energy changes of a fluid element takes a slightly different form as suggested by the last term in (1.23) (compare with Section 2.4.3).

In principle, melting/freezing leads also to a substantial energy exchange (heat of melting) and could be included in $\delta Q$. However, since this process occurs almost always very near to the ocean's surface, it is commonly included in the formulation of energy and mass exchange at the boundary rather than as a source/sink in the interior ocean. Note that this is different in the atmosphere[9], where phase changes in the interior are important (i. e. in clouds).

[8] Note that when one changes the independent variable from salt concentration $s$ to salinity $S$, derivatives such as $\partial H/\partial s$ change by a factor of 1,000 but $(\partial H/\partial s)\mathrm{d}s = (\partial H/\partial S)\mathrm{d}S$.

[9] Differences and congruences of oceanic and atmospheric thermodynamics are discussed in Section 2.10.

**5. Alternative Derivation of the First Law for Open Systems**

Applying the first law for closed systems (1.16) to open systems is not immediately possible, since the exchange of matter with the environment will change the internal energy, and, in particular, the total and specific volume of the system. It is, however, possible to account for that exchange by adjusting the differentials of internal energy and volume in (1.16) (see e. g. Warren, 2006). According to the additivity rule of the box on p. 13 these adjustments due to the matter exchange $dm_i$ are given by

$$\mathrm{d}E - \sum_i E_i \mathrm{d}m_i \quad \text{and} \quad \mathrm{d}v - \sum_i v_i \mathrm{d}m_i$$

Using the adjusted differential in (1.16) yields

$$\mathrm{d}E = \delta Q - p\mathrm{d}v + \sum_i (E_i + pv_i)\mathrm{d}m_i = \delta Q - p\mathrm{d}v + \sum_i H_i \mathrm{d}m_i$$

implementing the definition of the specific enthalpy (1.19) of constituent $i$. Hence we derive the first law for open systems (1.22) again.

### *1.2.4 Second Law of Thermodynamics*

If energy and mass are exchanged in such a way that the system can be brought back into its initial state without leaving changes in the environment, the process is called *reversible*, otherwise *irreversible*. Natural processes are generally irreversible because dissipative changes occur during the process. Reversibility is thus an idealization.

The second law of thermodynamics states that when a small amount of energy $\delta Q$ is added to a closed system at absolute temperature $T$, the quantity $\delta Q/T$ is equivalent to the change of a state variable, which is the *entropy* $\eta$ (entropy per mass, thus in $\mathrm{m}^2\,\mathrm{s}^{-2}\,\mathrm{K}^{-1}$) given by

$$\mathrm{d}\eta = \frac{\delta Q}{T} \quad \text{(closed system)} \tag{1.24}$$

The entropy is also an intensive additive state variable, hence the additive rule of the box on p. 13 applies

$$\eta(m_1, \ldots, m_n, T, p) = \sum_i m_i \eta_i \quad \text{with} \quad \eta_i = \frac{\partial \eta}{\partial m_i} \tag{1.25}$$

where $\eta_i(m_1, \ldots, m_n, T, p)$ denotes the specific entropy of the component $i$. For an open system with mass exchange like a fluid element, we must also account for the different entropies of the mass entering or leaving system. In analogy to (1.21) we then obtain the second law in the form

$$\mathrm{d}\eta = \frac{\delta Q}{T} + \sum_i \eta_i \mathrm{d}m_i \quad \text{(open system)} \tag{1.26}$$

From the first law (1.22) and the second law (1.26) we see that external masses bring or take their respective enthalpy (not internal energy) and entropy when entering or leaving an open system, because this process occurs at constant pressure.

As stated above, exchanges of energy and mass in a system are reversible when they take place in a way, such that the system can be brought back into its initial state without any changes in the environment. The second law (1.26) holds for both reversible and irreversible processes, as discussed by De Groot and Mazur (1984). It is normally assumed to be valid even for small deviations from thermodynamic equilibrium and can hence be applied to fluid motions. It is conceptually useful to decompose energy and mass fluxes into reversible and irreversible components, $\delta Q = \delta Q^{\text{rev}} + \delta Q^{\text{irr}}$ and $\mathrm{d}m_i = \mathrm{d}m_i^{\text{rev}} + \mathrm{d}m_i^{\text{irr}}$. The second law, in addition to the requirement that the entropy is a state variable, states that irreversible processes always lead to an increase in entropy; hence one must have $\delta Q^{\text{irr}} > 0$ and $\sum_i \eta_i \mathrm{d}m_i^{\text{irr}} > 0$. This property will be used in Section 2.5 below to determine the direction of certain molecular fluxes.

Processes that conserve entropy, i. e. satisfy $\mathrm{d}\eta = 0$, are called *isentropic*. Processes in an adiabatically closed system with $\delta Q = 0$ and $\mathrm{d}m_i = 0$ are always isentropic. However, reversible processes (which have $\delta Q^{\text{irr}} = 0$ and $\mathrm{d}m_i^{\text{irr}} = 0$) are not necessarily isentropic, and likewise isentropic processes need not be reversible. The two concepts are only equivalent if there are no external (irreversible) sources for energy and mass. This is largely the case in the interior of the ocean but not in the atmosphere, due to radiation and condensation/evaporation processes.

Applying the general relation (1.26) to seawater, we have $\mathrm{d}\eta = \delta Q/T + \eta_{\text{s}}\mathrm{d}m_{\text{s}} + \eta_{\text{w}}\mathrm{d}m_{\text{w}}$. Since the entropy satisfies the additivity relation (1.15), the second law takes for seawater the form

$$\mathrm{d}\eta = \frac{\delta Q}{T} + \frac{\partial \eta}{\partial S}\mathrm{d}S \tag{1.27}$$

Note that salinity exchanges in the interior ocean are normally diffusive and hence irreversible. We will use the second law in the form (1.27) in Section 2.5 to infer the direction of molecular heat and salt fluxes.

### 1.2.5 Thermodynamic Potentials

Combining the thermodynamic laws (1.22) and (1.26) with elimination of $\delta Q$ yields the GIBBS[10] relation

$$T\mathrm{d}\eta = \mathrm{d}E + p\mathrm{d}\upsilon - \sum_i \mu_i \mathrm{d}m_i \quad \text{or} \quad T\mathrm{d}\eta = \mathrm{d}H - \upsilon\mathrm{d}p - \sum_i \mu_i \mathrm{d}m_i \tag{1.28}$$

is also sometimes called the fundamental thermodynamic relation. Here the variable $\mu_i$ has been defined as

$$\mu_i = E_i + p\upsilon_i - T\eta_i \tag{1.29}$$

It is called the *chemical potential* of the constituent $i$. Note that (1.28) also holds for systems with external mass fluxes. Multiplication of (1.29) by $m_i$ and summation yields

$$E = T\eta - p\upsilon + \sum_i m_i \mu_i \tag{1.30}$$

[10] JOSIAH WILLARD GIBBS, *1839 in New Haven/Connecticut, †1903 in New Haven/Connecticut, physicist.

combining (1.30) with (1.28) one then derives the Gibbs–Duhem relation

$$\eta \mathrm{d}T - \upsilon \mathrm{d}p + \sum_i m_i \mathrm{d}\mu_i = 0 \tag{1.31}$$

which relates changes in the chemical potentials $\mu_i$ to changes in $T$ and $p$.

The internal energy $E(m_1, \ldots, m_n, T, p)$ and the enthalpy $H(m_1, \ldots, m_n, T, p)$ are called *thermodynamic potentials*. This is because they allow to determine certain state variables by differentiation, e. g. the specific volume is given by $\partial H/\partial p$ at constant $\eta$ and $m_i$ which can be seen from (1.28) for $\mathrm{d}\eta = 0$ and $\mathrm{d}m_i = 0$. Additional potentials which can be equally useful are the HELMHOLTZ[11] *free energy* (or work function) $F(m_1, \ldots, m_n, T, p)$ and the *free enthalpy* (or Gibbs function) $G(m_1, \ldots, m_n, T, p)$ which are defined by

$$F = E - T\eta = \sum m_i F_i \tag{1.32}$$

$$G = E + p\upsilon - T\eta = H - T\eta = \sum m_i \mu_i \tag{1.33}$$

All thermodynamic potentials are additive, which is easily shown from the additivity rules (B4.2), (1.17), and (1.25). The four potentials $E$, $H$, $F$ and $G$ all have the dimension energy per mass but a rather different physical interpretation. The internal energy $E$ describes the sum of all energies participating in energy transformations, in particular energy of molecular motions and intermolecular attraction potential, as well as the 'chemical' energy of the dissolution of salt in seawater. The enthalpy $H$ describes the heat content at constant pressure. The Gibbs function is the chemical potential of the mixture (in other texts $G$ is frequently denoted by $\mu$; we will not follow this convention). Note that $\mathrm{d}F = -\eta \mathrm{d}T - p\mathrm{d}\upsilon$, hence the free energy $F$ is the energy that is available for conversion into work under isothermal conditions. From $\mathrm{d}G = -\eta \mathrm{d}T + \upsilon \mathrm{d}p$ we learn that $G$ is constant for isothermal-isobaric processes, it is actually the only thermodynamic potential with this property.

All thermodynamic properties of a system can be derived from any of the four potentials as a function of the state variables. Using the Gibbs relation (1.28) and the definitions (1.32) and (1.33), the total change of any potential can be expressed in the following form

$$\mathrm{d}E(m_1, \ldots, m_n, \eta, \upsilon) = \sum \mu_i \mathrm{d}m_i + T\mathrm{d}\eta - p\mathrm{d}\upsilon \tag{1.34}$$

$$\mathrm{d}H(m_1, \ldots, m_n, \eta, p) = \sum \mu_i \mathrm{d}m_i + T\mathrm{d}\eta + \upsilon \mathrm{d}p \tag{1.35}$$

$$\mathrm{d}F(m_1, \ldots, m_n, T, \upsilon) = \sum \mu_i \mathrm{d}m_i - \eta \mathrm{d}T - p\mathrm{d}\upsilon \tag{1.36}$$

$$\mathrm{d}G(m_1, \ldots, m_n, T, p) = \sum \mu_i \mathrm{d}m_i - \eta \mathrm{d}T + \upsilon \mathrm{d}p \tag{1.37}$$

These *fundamental relations* (1.34)–(1.37) describe the change in each thermodynamic potential in terms of changes in the mass ratios and in two state variables which may be considered the 'canonical' independent state variables for the respective potential because the formulation of derivatives is particularly simple (see the diagram in Figure 1.4). In oceanography and meteorology, temperature $T$ and pressure $p$ are the most convenient independent state variables as they can be easily measured – in contrast to entropy and specific volume.

[11] HERMANN LUDWIG FERDINAND VON HELMHOLTZ, *1821 in Potsdam, †1894 in Charlottenburg, physicist and physiologist.

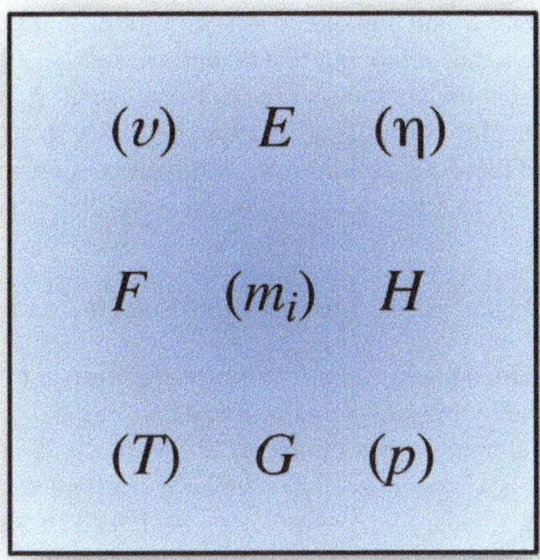


Fig. 1.4 Thermodynamic potentials with associated independent (canonical) variables indicated in the neighboring parentheses, as e. g. $G = G(m_i, T, p)$

It follows that the free enthalpy $G$, which has $T$ and $p$ as its canonical variables, contains the most convenient description of thermodynamic properties. It is, therefore, also used to express the thermodynamic properties of seawater (Feistel, 2008). However, occasionally, other potentials may also be convenient and will be used in this book instead of $G$. Using (1.29), the chemical potentials of the individual components are related according to

$$E_i = \mu_i - p\upsilon_i + T\eta_i \tag{1.38}$$

$$H_i = \mu_i + T\eta_i \tag{1.39}$$

$$F_i = \mu_i - p\upsilon_i \tag{1.40}$$

$$G_i = \mu_i \tag{1.41}$$

where the last relation follows immediately from (1.33). From the Gibbs relation (1.28) and using the definition of the chemical potential (1.29), the additivity rule for internal energy (1.17) and entropy (1.25), one obtains

$$\sum_i m_i\,(T\mathrm{d}\eta_i - \mathrm{d}E_i - p\mathrm{d}\upsilon_i) = 0$$

The terms in the sum depend on $m_i, T, p$. The equality must hold for arbitrary concentrations $m_i$ which can only be satisfied if the equality holds for each component

Table 1.1 State variables and the thermodynamic coefficients defined in the Sections 1.2.6 and 1.2.7 expressed in terms of the Gibbs function $G(S, T, p)$ and its derivatives

| | | | |
|---|---|---|---|
| Specific volume | $\upsilon = \partial G/\partial p$ | Thermal expansion | $\alpha = (\partial^2 G/\partial T\partial p)/(\partial G/\partial p)$ |
| Chemical potential | $\mu = \partial G/\partial S$ | Haline contraction | $\gamma = -(\partial^2 G/\partial S\partial p)/(\partial G/\partial p)$ |
| Entropy | $\eta = -\partial G/\partial T$ | Isothermal compressibility | $\kappa = -(\partial^2 G/\partial p^2)/(\partial G/\partial p)$ |
| Enthalpy | $H = G - T\partial G/\partial T$ | Specific heat | $c_p = -T\partial^2 G/\partial T^2$ |

**6. Units of State Variables**

We use temperature $T$, salinity $S$ and pressure $p$ as state variables. It should be noted that in this chapter $T$ refers to thermodynamic temperature with units in kelvin (K) with $273.15\,\mathrm{K} = 0°\mathrm{C}$, where °C refers to the Celsius temperature scale. However, in all figures of this section, $T$ is shown for the relevant oceanographic range of temperatures using the Celsius temperature scale, but it should be kept in mind that all equations refer to the thermodynamic temperature. We have discussed the units of salinity in Section 1.2.1. For pressure, there is a large variety of units, examples are

$$1\,\mathrm{Pa} = 1\,\mathrm{N\,m^{-2}} = 10^{-4}\,\mathrm{dbar} = 10^{-5}\,\mathrm{bar} = 10.197 \times 10^{-6}\,\mathrm{at} = 9.8692 \times 10^{-6}\,\mathrm{atm}$$

where at denotes technical atmosphere and atm atmosphere. The unit dbar is common amongst oceanographers, since by the hydrostatic relation $\partial p/\partial z = -g\rho$, which determines the overwhelming part of the vertical pressure variations in the ocean, (hydrostatic) pressure relates the depth by $1\,\mathrm{dbar} \approx 1\,\mathrm{m}$. It is also common to subtract a constant from the pressure variable, given by the pressure of the 'standard atmosphere', $p_0 = 1.01325 \times 10^5\,\mathrm{Pa}$.

individually so that the differentials of specific potentials are related by

$$\mathrm{d}E_i = T\mathrm{d}\eta_i - p\mathrm{d}\upsilon_i \tag{1.42}$$

and likewise by

$$\mathrm{d}H_i = T\mathrm{d}\eta_i + \upsilon_i\mathrm{d}p \tag{1.43}$$

$$\mathrm{d}F_i = -\eta_i\mathrm{d}T - p\,\mathrm{d}\upsilon_i \tag{1.44}$$

$$\mathrm{d}\mu_i = -\eta_i\mathrm{d}T + \upsilon_i\,\mathrm{d}p \tag{1.45}$$

Equations (1.42)–(1.45) can be considered as the component-wise versions of the fundamental relations for the thermodynamics potentials (1.34)–(1.37).

According to the definition (1.33) for $G$, the Gibbs function for seawater is given by

$$G(S,T,p) = \sum_{i=1}^{2} m_i\mu_i = \mu_s s + \mu_w(1-s)$$

The physical units of $T$ and $p$ are discussed in the box on p. 20. As the other thermodynamic potentials, it satisfies the additivity relation

$$\left(\frac{\partial G}{\partial s}\right)_{Tp} = \mu_s - \mu_w \equiv \hat{\mu} \tag{1.46}$$

The difference between the chemical potentials of salt and pure water, $\hat{\mu} = \mu_s - \mu_w$ is often (albeit somewhat imprecisely) termed chemical potential of seawater. For the corresponding term in the Gibbs relation (1.28) we have

$$\sum_{i=1}^{2} \mu_i\mathrm{d}m_i = \mu_s\mathrm{d}S + \mu_w\mathrm{d}(1-S) = \hat{\mu}\mathrm{d}s = \mu\mathrm{d}S$$

where we have introduced two chemical potentials, one referring ($\hat{\mu}$) to salt concentration and one ($\mu$) referring to salinity, which take different numerical values. With (1.34) we obtain the Gibbs relation e. g. in the form

$$\mathrm{d}E = T\mathrm{d}\eta - p\mathrm{d}\upsilon + \mu\mathrm{d}S \tag{1.47}$$

while for the other thermodynamic potentials corresponding relations hold. The Gibbs relation (1.47) states that any change in salinity, e. g. by diffusion, leads to a change in internal energy (and likewise enthalpy). As will be seen in Section 2.6 below, the contribution of salinity on the energy budgets, however, is generally small. Differential changes in the Gibbs function follow from (1.36) and are given by

$$dG = -\eta dT + \upsilon dp + \mu dS \tag{1.48}$$

where the derivatives are seen to be

$$\mu = \frac{\partial G}{\partial S}\,, \quad \eta = -\frac{\partial G}{\partial T}\,, \quad \upsilon = \frac{\partial G}{\partial p} \tag{1.49}$$

Note that similar relations hold for the other thermodynamic potentials, where, however, other state variables (than $T$, $S$ or $p$ for the case of the Gibbs function) are held constant. Numerous thermodynamic relations can be derived from (1.49), in particular the MAXWELL[12] relations

$$\frac{\partial \eta}{\partial S} = -\frac{\partial \mu}{\partial T}\,, \quad \frac{\partial \mu}{\partial p} = \frac{\partial \upsilon}{\partial S}\,, \quad \frac{\partial \upsilon}{\partial T} = -\frac{\partial \eta}{\partial p} \tag{1.50}$$

All oceanographically relevant thermodynamic variables of seawater follow directly from $G(S, T, p)$, as shown in Table 1.1. A very accurate determination of $G(S, T, p)$ which is based on high-precision measurements of sound velocity, freezing temperature, expansion coefficients, specific heats, etc. has been given by Feistel (2003, 2008). An updated version based on absolute rather than practical salinity can be found in IOC, SCOR and IAPSO (2010).

We finally note in this section that the Gibbs function $G(S, T, p)$ has an arbitrariness to some degree: replacing $G$ by

$$G^* = G + A + BT + (C + DT)S \tag{1.51}$$

with arbitrary constants $A, B, C, D$ defines an equally valid Gibbs function $G^*$. The fundamental Gibbs relation (1.48) holds for any choice of the constants, however with corresponding changes in entropy $\eta$ and chemical potential $\mu$. It is, therefore, not very useful to display $\eta$ and $\mu$. However, the first derivative of $G$, i. e. the specific volume $\upsilon = \partial G/\partial p$ is not affected by (1.51) and shown in Figure 1.5.

### 1.2.6 Equation of State

For each substance with $n$ constituents there is an equation relating the specific volume $\upsilon$ or density $\rho = 1/\upsilon$ with $n + 1$ other independent state variables. This is the *equation of state* for the particular substance. For seawater it takes the form[13]

$$\rho = \rho(S, T, p) = F(S, T, p) \tag{1.52}$$

[12] JAMES CLERK MAXWELL, *1831 in Edinburgh, †1879 in Cambridge, physicist and mathematician.

[13] In the following $F$ denotes the equation of state; the previous notation for the Helmholtz free energy will not be used anymore.

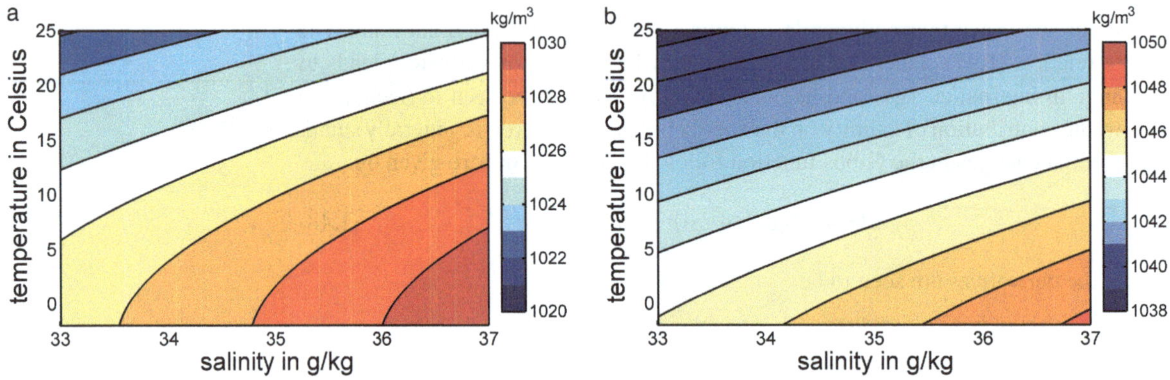


Fig. 1.5 Density of seawater as function of $S$, $T$ at 0 dbar (**a**) and 1,000 dbar (**b**)

The equation of state is determined by the molecular structure of the substance. While analytical (exact) expressions for the function $F$ do not exist, empirical expressions with a relative accuracy of $(3-5)\times 10^{-6}$ are available in various polynomial approximations (e. g. Fofonoff and Millard Jr., 1983; Jackett et al., 2006). All empirical expressions are based on measured properties of seawater. These properties have been embedded into the above mentioned construction of the Gibbs function, and formally the equation of state of seawater then follows by differentiation,

$$\upsilon = \frac{1}{\rho} = \frac{\partial G}{\partial p} \tag{1.53}$$

For more than 90% of the ocean the salinity is in the range $34\,\mathrm{g\,kg^{-1}} < S < 35\,\mathrm{g\,kg^{-1}}$ and the temperature is in the range $-2°\mathrm{C} < T < 10\,°\mathrm{C}$. Density changes in the ocean by at most a few percent, $\Delta\rho/\rho \cong 10^{-3}$–$10^{-2}$. It is, therefore, customary to define the density variable $\rho - 1{,}000\,\mathrm{kg\,m^{-3}}$. The density is mainly dependent on pressure. Only a small part of the density variations, mainly the one arising from temperature and salinity variations (see Figure 1.5), turns out to be dynamically relevant, as will be discussed in Chapter 4 below.

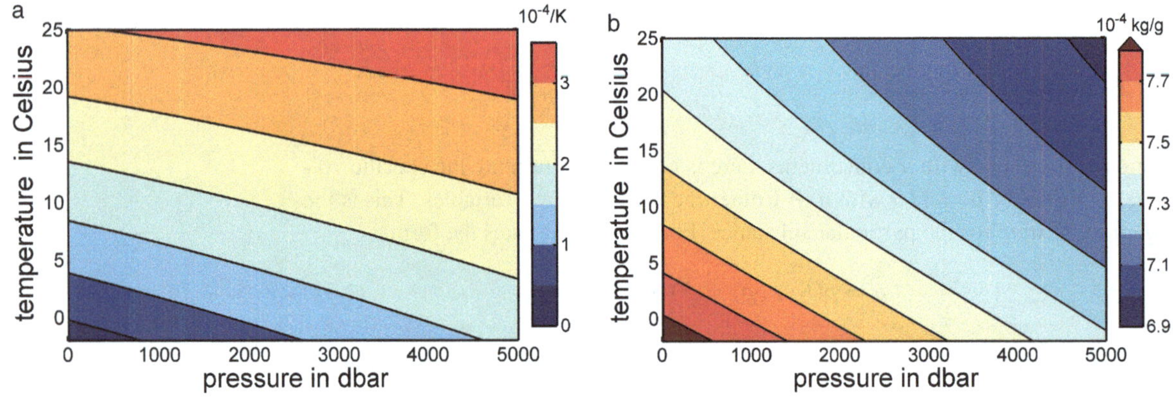


Fig. 1.6 **a** Thermal expansion coefficient $\alpha(S,T,p)$ and **b** haline contraction coefficient $\gamma(S,T,p)$ at $S = 35\,\mathrm{g\,kg^{-1}}$

The derivatives of the state equation with respect to the independent variables

$$\alpha = -\frac{1}{\rho}\left(\frac{\partial \rho}{\partial T}\right)_{S,p} \qquad \text{coefficient of thermal expansion} \tag{1.54}$$

$$\gamma = \frac{1}{\rho}\left(\frac{\partial \rho}{\partial S}\right)_{T,p} \qquad \text{coefficient of haline contraction} \tag{1.55}$$

$$\kappa = \frac{1}{\rho}\left(\frac{\partial \rho}{\partial p}\right)_{S,T} \qquad \text{isothermal compressibility} \tag{1.56}$$

appear in many of the thermodynamic equations considered in the following chapters. All these derivatives can be expressed by the Gibbs function as outlined in Table 1.1. The coefficient of thermal expansion $\alpha$ is mainly dependent on pressure and temperature, whereas the one describing haline contraction, $\gamma$, is nearly constant (see Figure 1.6). The isothermal compressibility is mainly dependent on pressure and temperature. A further discussion of these coefficients (as well as of the heat capacities introduced below) can be found in Section 2.6.4.

### 1.2.7 Specific Heat

The heat capacity $c$ (per unit mass, also named *specific heat*) describes for constant mass ratio the energy or heat change, respectively, per unit temperature change

$$\delta Q = c\mathrm{d}T$$

The value of $c$ depends on how the heat is exchanged. For constant pressure and constant composition ($\mathrm{d}p = 0$, $\mathrm{d}m_i = 0$) one has from the first law (1.21) $\mathrm{d}H = c_p\mathrm{d}T$ or

$$c_p = \left(\frac{\partial H}{\partial T}\right)_{p,m_i} = T\frac{\partial \eta}{\partial T} = -T\frac{\partial^2 G}{\partial T^2} \tag{1.57}$$

This is the specific heat at constant pressure. If the volume rather than the pressure is kept constant during the heat exchange ($\mathrm{d}\upsilon = 0$, $\mathrm{d}m_i = 0$), it follows from the first law formulated for $\mathrm{d}E$ as in (1.22) that $\mathrm{d}E = c_\upsilon\mathrm{d}T$ and hence

$$c_\upsilon = \left(\frac{\partial E}{\partial T}\right)_{\upsilon,m_i} \tag{1.58}$$

which is the specific heat at constant volume. In this case, changes of pressure and of temperature are related according to

$$\mathrm{d}\upsilon = \frac{\partial \upsilon}{\partial T}\mathrm{d}T + \frac{\partial \upsilon}{\partial p}\mathrm{d}p = 0$$

and it follows that

$$c_\upsilon = \left(\frac{\partial E}{\partial T}\right)_{\upsilon,m_i} = \left(\frac{\partial E}{\partial T}\right)_{p,m_i} - \frac{\partial E}{\partial p}\frac{\partial \upsilon/\partial T}{\partial \upsilon/\partial p} = c_p - \frac{\alpha^2 \upsilon T}{\kappa}$$

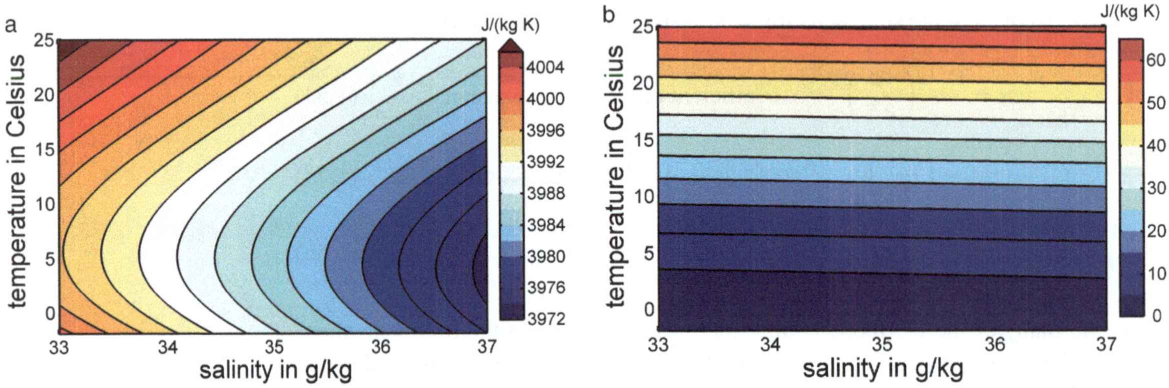


**Fig. 1.7** Specific heat of seawater $c_p$ (**a**) and difference $c_p - c_v$ (**b**) as a function of $S$ and $T$ at surface pressure, in $\mathrm{J\,kg^{-1}\,K^{-1}}$

The second relation follows from $E = G - pv + T\eta$, the derivatives of the Gibbs function (1.49) and the definitions of the derivatives of the equation of state given by (1.54) and (1.56). Thus we always have $c_v < c_p$.

For seawater, the Gibbs function is conveniently used to evaluate the specific heat at constant pressure $c_p = c_p(S, T, p)$ by (1.57) as function of the state variables. Its values at surface pressure are approximately $4{,}000\,\mathrm{J\,kg^{-1}\,K^{-1}}$ within a range of $\pm 1\%$ (the pressure dependence of $c_p$ is not relevant for the energy budgets considered later). The specific heat increases with temperature and decreases with salinity (see Figure 1.7). The difference between $c_p$ and $c_v$ is below $1\%$; it is relevant mainly for thermodynamic considerations.

# 2 Conservation Laws for Moving Fluids

In this chapter the basic conservation laws for mass, momentum, and energy are presented, including a discussion of appropriate boundary conditions. From the basic laws, budgets of other variables such as temperature and vorticity are derived. The result of the discussion will be a set of equations suitable to describe oceanic motions at all scales.

The equations which govern the evolution of fluid properties and its motion are differential equations, derived from elementary conservation laws of physics. They are usually formulated as *conservation equations* and arise if a property $\chi(\boldsymbol{x}, t)$ (units property per mass) respectively $\rho(\boldsymbol{x}, t)\chi(\boldsymbol{x}, t)$ (units property per volume) of a substance is changed due to sources and sinks, represented by $C(\boldsymbol{x}, t)$ (units property per volume and time), and due to transport by a flux $\boldsymbol{J}(\boldsymbol{x}, t)$ (this is a vector in the direction of transport, with the magnitude giving the units property per area and time). If $C$ is nonzero, there, is actually no conservation of the property, not locally and generally not in the integral sense. Even in case of vanishing sources/sinks there is no local conservation unless the divergence of the flux vector vanishes in addition to $C$. An important aspect is the condition of fluxes through the boundary of the domain, governed by the normal component $\boldsymbol{J} \cdot \boldsymbol{n}$ where $\boldsymbol{n}$ is the outward normal vector (see the figure on page 1). Hence $\chi$ is generally not constant, but still we will call the equation governing $\chi(\boldsymbol{x}, t)$ a conservation equation. Fluid dynamics and thermodynamics largely deal with conservation equations.

## 2.1 General Form of Conservation Equations

Consider a volume $V$, which is *fixed* in space and bounded by a surface $A$, and an arbitrary scalar fluid property $\chi$ (in units per mass) or $\rho\chi$ (in units per volume). The total amount of the property $\chi$ in the volume, given by $\int_V \rho\chi \mathrm{d}V$, may change in time in one of two ways:

- **Transport through the surface:**
  The volume transport (volume per unit time) through a surface element $\mathrm{d}\boldsymbol{A} = \boldsymbol{n}\mathrm{d}A$ ($\boldsymbol{n}$ outward normal unit vector) is $\boldsymbol{u} \cdot \mathrm{d}\boldsymbol{A}$, the outward transport of $\chi$ is thus $\rho\chi\boldsymbol{u} \cdot \mathrm{d}\boldsymbol{A}$. Here $\rho\chi\boldsymbol{u}$ is the *advective flux* of $\chi$ (unit $[\chi]\,\mathrm{m}^{-2}\,\mathrm{s}^{-1}$). In addition,


D. Olbers, J. Willebrand, C. Eden, *Ocean Dynamics*,
DOI 10.1007/978-3-642-23450-7_2, 

a nonadvective flux $\boldsymbol{J}_\chi$ may occur, e. g. by diffusion, heat conduction, radiation etc. The total outward transport across $A$ is hence given by

$$\oint_A (\rho\chi\boldsymbol{u} + \boldsymbol{J}_\chi) \cdot \mathrm{d}\boldsymbol{A} \tag{2.1}$$

- **Interior sources and sinks:**
  Let $C_\chi$ denote the net sources/sinks of $\chi$, measured in $\chi$-units per time and volume, through internal processes within the volume. Examples are heat sources, radioactive decay, chemical reaction, consumption of $\chi$ material etc. The change within the volume is then given by

$$\int_V C_\chi \mathrm{d}V$$

Both quantities, the flux $\boldsymbol{J}_\chi$ and the sources/sinks $C_\chi$, have to be specified for the corresponding fluid property. They contribute to the total rate of change of the $\chi$-content in the volume according to

$$\frac{\partial}{\partial t}\int_V \rho\chi \mathrm{d}V = -\oint_A (\rho\chi\boldsymbol{u} + \boldsymbol{J}_\chi) \cdot \mathrm{d}\boldsymbol{A} + \int_V C_\chi \mathrm{d}V \tag{2.2}$$

For a fixed volume the time rate of change may be placed inside the integral over the volume, and with GAUSS'[1] theorem (see Appendix A.1) the surface integral in (2.2) may be rewritten as a volume integral according to

$$\oint_A (\rho\chi\boldsymbol{u} + \boldsymbol{J}_\chi) \cdot \mathrm{d}\boldsymbol{A} = \int_V \nabla \cdot (\rho\chi\boldsymbol{u} + \boldsymbol{J}_\chi) \mathrm{d}V$$

so that (2.2) can be written as

$$\int_V \left[\frac{\partial}{\partial t}(\rho\chi) + \nabla \cdot (\rho\chi\boldsymbol{u} + \boldsymbol{J}_\chi) - C_\chi\right] \mathrm{d}V = 0 \tag{2.3}$$

As this holds for an *arbitrary* volume, the integrand has to vanish identically, i. e.

$$\frac{\partial}{\partial t}(\rho\chi) = -\nabla \cdot (\rho\chi\boldsymbol{u} + \boldsymbol{J}_\chi) + C_\chi \tag{2.4}$$

which is the general conservation equation for the $\chi$ property in the so-called *flux form*. The right-hand side contains the divergence of the total $\chi$-flux $\rho\chi\boldsymbol{u} + \boldsymbol{J}_\chi$, as well as interior sources/sinks contained in $C_\chi$. It should be mentioned that the advective part of the flux – despite its nonlinear character – describes a conceptually simple process because the fluid velocity $\boldsymbol{u}$ and the property $\chi$ belong to the state vector, and thus usually to the resolved part of the system. The challenge is to obtain the nonadvective fluxes and the sources and sinks as a correct physical, mathematical and resolved form. For many properties of the ocean $\boldsymbol{J}_\chi(\boldsymbol{x}, t)$ and $C_\chi(\boldsymbol{x}, t)$ are not well known. In Part IV we describe the most elementary attempts to construct parameterizations of turbulent fluxes in the fluid dynamical and thermodynamics equations.

[1] CARL FRIEDRICH GAUSS, *1777 in Braunschweig, †1855 in Göttingen, mathematician.

### General Boundary Condition

The air-sea interface is of fundamental importance for ocean dynamics. With the exception of tides, nearly all oceanic motions are forced by the exchange of momentum, energy, and water between atmosphere and ocean across the sea surface. Momentum exchange with the solid earth across the bottom boundary is also important, whereas the water, heat and salt exchanges at the bottom play only a limited role for oceanic motions.

The considerations leading to conservation equations for energy, mass, momentum, etc. within the ocean must also be valid at the boundary between ocean and atmosphere. The general conservation equation for a *fixed* volume in its integral form was given in (2.2). As the air-sea boundary in general will move in time, we now consider an infinitesimal volume instead, which is moving with the velocity $\boldsymbol{u}_I$ of the air-sea interface and which includes the interface at all times. More specific, the volume is bounded by a surface $A$ which consists of a surface $I$ just above the interface between ocean and atmosphere on the one side, and corresponding surface $I'$ inside the fluid and the associated side boundaries (see Figure 2.1). The following analysis applies as well at any interface on the fluid boundaries or inside the fluid. At the bottom where $I$ is the bottom surface, however, $\boldsymbol{u}_I$ is to be taken as zero. At the surface of the test volume, only the difference velocity $\boldsymbol{u} - \boldsymbol{u}_I$ between the fluid and the moving volume is relevant for the budget inside the volume. Notice that $\boldsymbol{u}_I \neq \boldsymbol{u}$ in general since the surface $I$ is not necessarily a material surface. With the same notation as before it follows that we have now

$$\frac{\partial}{\partial t}\int_V (\rho\chi)\mathrm{d}V = -\oint_A \left[\rho\chi(\boldsymbol{u}-\boldsymbol{u}_I) + \boldsymbol{J}_\chi\right]\cdot \mathrm{d}\boldsymbol{A} + \int_V C_\chi \mathrm{d}V \tag{2.5}$$

where $\mathrm{d}\boldsymbol{A}$ is the surface element pointing in the outward normal direction. The contributions of the volume integrals in (2.5) can be made small compared to those arising from the surface integrals by taking a pillbox of sufficiently small height. Neglecting thus the volume integral for an infinitesimal pillbox height, one has

$$\oint_A \left[\rho\chi(\boldsymbol{u}-\boldsymbol{u}_I) + \boldsymbol{J}_\chi\right]\cdot \mathrm{d}\boldsymbol{A} = 0 \tag{2.6}$$

The integrand in (2.6) describes the sum of the advective flux of $\chi$ (relative to the moving boundary) and the diffusive flux. As the outward normal vector points into opposite directions on both sides of the interface, the condition (2.6) requires in the limit of very small height that the normal component of the flux vector is continuous

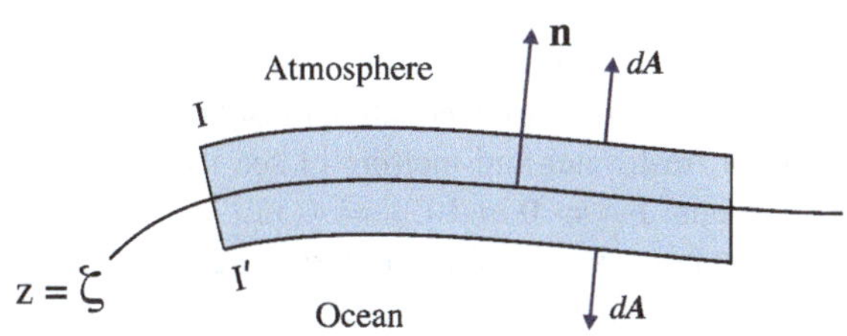


Fig. 2.1 Sketch of the ocean-atmosphere boundary and the test volume described in the text

across the interface, and one obtains the general boundary condition

$$\left[\rho\chi\left(\boldsymbol{u}-\boldsymbol{u}_I\right)+\boldsymbol{J}_\chi\right]\cdot\boldsymbol{n}=\mathcal{J}_\chi^{\mathrm{tot}} \tag{2.7}$$

Here $\boldsymbol{n}$ is the unit vector normal to the boundary, and $\mathcal{J}_\chi^{\mathrm{tot}}$ denotes the normal component of the *total* flux of the substance $\chi$ in the other system, e. g. the atmosphere or the solid earth. Note that $\mathcal{J}_\chi^{\mathrm{tot}}$ may also include flux components associated with phase transitions at the boundary, such as evaporation at the sea surface.

The velocity $\boldsymbol{u}-\boldsymbol{u}_I$ in (2.7) can be expressed in terms of the state variables. For the air-sea interface, the location of the boundary can be implicitly represented by $I(\boldsymbol{x}_\mathrm{h},z,t)=z-\zeta(\boldsymbol{x}_\mathrm{h},t)=0$. The normal vector $\boldsymbol{n}$, taken positive upwards from the sea surface, is

$$\boldsymbol{n}=\frac{\nabla I}{|\nabla I|}=\frac{(-\nabla_\mathrm{h}\zeta,1)}{\sqrt{1+(\nabla_\mathrm{h}\zeta)^2}} \tag{2.8}$$

On the surface $I$, one has $\mathrm{d}I=\partial I/\partial t\,\mathrm{d}t+\nabla I\cdot\mathrm{d}\boldsymbol{x}=0$ and $\mathrm{d}\boldsymbol{x}/\mathrm{d}t=\boldsymbol{u}_I$, and hence $\partial I/\partial t+\boldsymbol{u}_I\cdot\nabla I=0$. Then, the normal velocity of the moving boundary is found as

$$\boldsymbol{u}_I\cdot\boldsymbol{n}=-\frac{\partial I/\partial t}{|\nabla I|}=\frac{\partial\zeta/\partial t}{\sqrt{1+(\nabla_\mathrm{h}\zeta)^2}} \tag{2.9}$$

and the difference to the normal component of the fluid velocity is thus given by

$$(\boldsymbol{u}_I-\boldsymbol{u})\cdot\boldsymbol{n}=\frac{\partial\zeta/\partial t+\boldsymbol{u}_\mathrm{h}\cdot\nabla_\mathrm{h}\zeta-w}{\sqrt{1+(\nabla_\mathrm{h}\zeta)^2}} \tag{2.10}$$

For the ocean bottom, with $I(\boldsymbol{x}_\mathrm{h},z)=z+h(\boldsymbol{x}_\mathrm{h})=0$ where $h(\boldsymbol{x}_\mathrm{h})$ is the ocean depth referred to the mean sea level, the normal vector $\boldsymbol{n}$ defined in (2.8) is pointing from the bottom into the fluid. Equation (2.7) then applies with

$$(\boldsymbol{u}_I-\boldsymbol{u})\cdot\boldsymbol{n}=-\boldsymbol{u}\cdot\boldsymbol{n}=-\frac{w+\boldsymbol{u}_\mathrm{h}\cdot\nabla_\mathrm{h}h}{\sqrt{1+(\nabla_\mathrm{h}h)^2}} \tag{2.11}$$

In most applications, the square root terms in the above relations can be replaced by 1 because the slopes of the interfaces are very small.

## 2.2 Mass Conservation

As discussed in Section 1.2.1, it is sufficient for almost all dynamical purposes to take the chemical composition of seawater as constant. Thus it suffices to consider the two components salt and pure water, with concentrations $s$ and $w=1-s$ measured in kg salt, respectively kg freshwater, per kg seawater, and densities $\rho_\mathrm{s}=\rho s=\rho S/1{,}000$, $\rho_\mathrm{w}=\rho w$, in kg m$^{-3}$. The total mass of the respective molecules must be conserved, i. e. the general conservation equation must hold for mass (notice that we ignore here freezing of freshwater and melting of sea ice). Formally, it is possible to simply use $\chi=1$ and $\boldsymbol{J}_\chi=0$ and $C_\chi=0$ in (2.4) (the total mass has neither sources/sinks nor a nonadvective flux) to derive a conservation budget for total mass. However, we consider first budgets for the partial masses of salt and freshwater independently. The combination yields the total mass and the salt conservation equation.

### 2.2.1 Total Mass and Salt Conservation Equation

Consider the partial mass $\delta M_s$ of salt in a moving material volume element with volume $\delta V$ which is defined with respect to the mean motion $\boldsymbol{u}_s$ of the salt molecules. This volume element is moving with the salt velocity which may differ from the flow velocity $\boldsymbol{u}$, and thus corresponds to a Lagrangian derivative $\mathrm{D}_s/\mathrm{D}t$ formed with $\boldsymbol{u}_s$. In this Lagrangian frame of reference $\delta M_s$ must be constant, thus $\mathrm{D}_s(\delta M_s)/\mathrm{D}t = 0$. For the density $\rho_s = \delta M_s/\delta V$ of salt we thus have

$$\frac{\mathrm{D}_s}{\mathrm{D}t}\rho_s = -\frac{\delta M_s}{(\delta V)^2}\frac{\mathrm{D}_s}{\mathrm{D}t}\delta V$$

According to (1.9), the volume change is given by $\mathrm{D}_s(\delta V)/\mathrm{D}t = \delta V \nabla \cdot \boldsymbol{u}_s$, and it follows that

$$\frac{\mathrm{D}_s}{\mathrm{D}t}\rho_s = -\rho_s \nabla \cdot \boldsymbol{u}_s \tag{2.12}$$

Both components, salt and freshwater, must independently satisfy a conservation equation of the form (2.12). In the Eulerian form, these become

$$\frac{\partial \rho_s}{\partial t} + \nabla \cdot \rho_s \boldsymbol{u}_s = 0\,, \quad \frac{\partial \rho_w}{\partial t} + \nabla \cdot \rho_w \boldsymbol{u}_w = 0 \tag{2.13}$$

The transport velocities $\boldsymbol{u}_s$ and $\boldsymbol{u}_w$ for both components generally differ. The total mass per volume has the density $\rho = \rho_s + \rho_w$, and the mass-weighted mean velocity – the flow velocity $\boldsymbol{u}$ – is defined as the barycentric mean

$$\rho \boldsymbol{u} = \rho_s \boldsymbol{u}_s + \rho_w \boldsymbol{u}_w = \rho(s\boldsymbol{u}_s + w\boldsymbol{u}_w) \tag{2.14}$$

Addition of the two equations in (2.13) yields the conservation of total mass,

$$\frac{\partial \rho}{\partial t} + \nabla \cdot \rho \boldsymbol{u} = 0 \tag{2.15}$$

which is also called the *continuity equation*.

The velocities of the partial masses are not easily measurable and hence are not suitable to describe the flow. Using $\rho_s = \rho S/1{,}000$ and introducing the advective flux of salinity by the flow velocity $\boldsymbol{u}$, one obtains from (2.13) and (2.14) the salt conservation in the form

$$\frac{\partial \rho S}{\partial t} + \nabla \cdot \rho S \boldsymbol{u} = -\nabla \cdot [(\boldsymbol{u}_s - \boldsymbol{u})\,\rho S] = -\nabla \cdot \boldsymbol{J}_S \tag{2.16}$$

Here, the *diffusive* salinity flux $\boldsymbol{J}_S = (\boldsymbol{u}_s - \boldsymbol{u})\rho S$ is introduced as a consequence of eliminating the salt velocity $\boldsymbol{u}_s$. It results from the difference of the mean velocity $\boldsymbol{u}_s$ of salt and the flow velocity $\boldsymbol{u}$. As outlined in Section 2.5 below, the diffusive flux is empirically related to the salinity gradient (and possibly gradients of other fluid properties).

The continuity equation (2.15) allows to rewrite the general conservation equation (2.4) in the form

$$\rho \frac{\mathrm{D}\chi}{\mathrm{D}t} = -\nabla \cdot \boldsymbol{J}_\chi + C_\chi \tag{2.17}$$

which will be referred to as the *parcel form*, in contrast to the *flux form* (2.4). The flux form (2.4) is preferred for integral budgets, whereas the particle form (2.17) permits to follow the particle properties. Mathematically, both formulations are equivalent. Alternative formulations of mass conservation (2.15) are given by

$$\frac{1}{\rho}\frac{\mathrm{D}\rho}{\mathrm{D}t} = -\nabla \cdot \boldsymbol{u}\,, \quad \rho\frac{\mathrm{D}\upsilon}{\mathrm{D}t} = \nabla \cdot \boldsymbol{u} \tag{2.18}$$

with the *specific volume* $\upsilon = 1/\rho$. Note that the equations are in parcel form; thus the velocity divergence is the volume source of the specific volume.

The total (advective plus diffusive) salt flux (kg salt per $\mathrm{m}^2$ and s) is given by $\rho S\boldsymbol{u} + \boldsymbol{J}_\mathrm{S}$. It enters the *flux form* of salt conservation,

$$\frac{\partial \rho S}{\partial t} = -\nabla \cdot (\rho S \boldsymbol{u} + \boldsymbol{J}_\mathrm{S}) \tag{2.19}$$

corresponding to the general form (2.4), without interior sources/sinks. Alternatively, the *parcel form* of salt conservation reads

$$\rho\frac{\mathrm{D}S}{\mathrm{D}t} = -\nabla \cdot \boldsymbol{J}_\mathrm{S} \tag{2.20}$$

### 2.2.2 Boundary Conditions for the Fluxes of Total Mass and Salt

For $\chi = 1$ and $\boldsymbol{J}_\chi = 0$ the relation (2.7) yields the boundary condition for the total mass at the sea surface,

$$\rho\,(\boldsymbol{u} - \boldsymbol{u}_I) \cdot \boldsymbol{n} = \mathcal{J}_\mathrm{mass} = \mathcal{E} - \mathcal{P} \tag{2.21}$$

The mass flux $\mathcal{J}_\mathrm{mass} = \mathcal{E} - \mathcal{P}$ (in $\mathrm{kg\,m^{-2}\,s^{-1}}$) across the air-sea interface (positive upwards) is given as the difference between evaporation $\mathcal{E}$ and precipitation $\mathcal{P}$ of pure water[2]. With (2.10) one obtains

$$\frac{\partial \zeta}{\partial t} + \boldsymbol{u}_\mathrm{h} \cdot \nabla_\mathrm{h}\zeta - w = -\frac{\mathcal{J}_\mathrm{mass}}{\rho} = \frac{\mathcal{P} - \mathcal{E}}{\rho} \quad \text{at} \quad z = \zeta(\boldsymbol{x}_\mathrm{h}, t) \tag{2.22}$$

which is referred to as *kinematical* boundary condition since no forces are involved. We have assumed $(\nabla_\mathrm{h}\zeta)^2 \ll 1$ to write the condition in the above commonly used form.

At a solid boundary as the bottom, where $\boldsymbol{u}_I \equiv 0$ and $\mathcal{J}_\mathrm{mass} = 0$, (2.7) leads to kinematical boundary condition

$$\boldsymbol{u} \cdot \boldsymbol{n} = \boldsymbol{u}_\mathrm{h} \cdot \nabla h + w = 0 \quad \text{at} \quad z = -h(\boldsymbol{x}_\mathrm{h}) \tag{2.23}$$

so that the normal component of the velocity vector has to vanish.

The processes of evaporation and precipitation exchange pure water with the atmosphere, but there is no salt flux through the interface. With $\chi = S$ and $\boldsymbol{J}_\chi = \boldsymbol{J}_\mathrm{S}$ equation (2.7) yields

$$[\rho S\,(\boldsymbol{u} - \boldsymbol{u}_I) + \boldsymbol{J}_\mathrm{S}] \cdot \boldsymbol{n} = 0 \tag{2.24}$$

[2] The mass flux of other substances is very small and negligible for dynamical purposes.

and with (2.21), this can be written as

$$\boldsymbol{J}_S \cdot \boldsymbol{n} = -S\mathcal{J}_{\text{mass}} = S(\mathcal{P}-\mathcal{E}) = \rho \frac{S}{1-S/1000}(\mathcal{P}_w - \mathcal{E}_w) \tag{2.25}$$

at the sea surface. In the last relation $\mathcal{E}_w = \mathcal{E}/\rho_w, \mathcal{P}_w = \mathcal{P}/\rho_w$ are the corresponding *volume* fluxes (in units of $\mathrm{m^3\,m^{-2}\,s^{-1}} = \mathrm{m\,s^{-1}}$). Notice that although no flux is transported through the surface, (2.25) appears as if a virtual flux of salt enters the ocean from the atmosphere and is carried further there by salt diffusion. Therefore, the right hand side $S(\mathcal{P}-\mathcal{E})$ is called *equivalent* salinity flux. It describes the changes of salinity due to the freshwater flux $\mathcal{P}-\mathcal{E}$. At the bottom, assuming zero exchange of salt with the ocean floor, one has a vanishing diffusive salt flux normal to the bottom, $\boldsymbol{J}_S \cdot \boldsymbol{n} = 0$. Note, however, that the transport of water and salt by rivers can be relevant.

## 2.3 Conservation of Momentum

With respect to spatial scales, two different kinds of forces acting on any part of a fluid can be distinguished. Long-range forces have a *macroscopic* range and affect all parts of a fluid. They are generally proportional to the fluid volume, and for that reason they are called *volume forces*. The principal volume force influencing motions in ocean and atmosphere is the force of gravity. Other relevant volume forces are the centrifugal and Coriolis forces discussed in Section 2.3.3 below. Short-range forces have a *microscopic* range of the order of the distance between molecules. If a volume is deformed, the relative positions of molecules change and the deformation generates internal forces, so-called *stresses*, that attempt to bring the volume shape back into its equilibrium state. These internal stresses cannot exist in a nondeformed volume; they are the result of molecular forces which have a very short range, of the order of the distance between molecules (e. g. VAN DER WAALS[3] forces). These forces hence must be proportional to the surface area and are independent of the volume. They can affect the motion of a fluid parcel only through contact of the parcel with the surrounding fluid (or the boundary of the fluid), and, therefore, they are termed *surface forces*. Examples for surface forces are molecular friction or surface tension. The mathematical treatment of surface forces is attributed to CAUCHY[4].

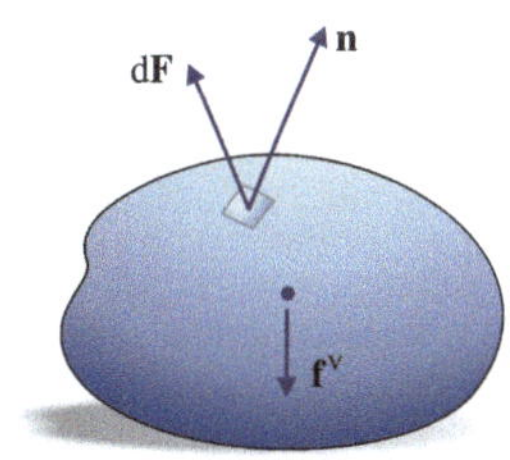


Fig. 2.2 Material fluid parcel experiencing a body (volume) force $\boldsymbol{f}^{\text{v}}$ and a surface force $\mathrm{d}\boldsymbol{F}$ acting on on infinitesimal surface element

Following NEWTON's[5] second law of motion, the conservation of momentum is the balance between the mass times acceleration and the sum of all forces applied to a fluid parcel[6],

$$\rho \frac{\mathrm{D}\boldsymbol{u}}{\mathrm{D}t} = \boldsymbol{f} = \boldsymbol{f}^{\text{v}} + \boldsymbol{f}^{\text{s}} \tag{2.26}$$

[3] JOHANNES DIDERIK VAN DER WAALS, *1837 in Leiden, †1923 in Amsterdam, physicist and thermodynamicist.

[4] AUGUSTIN LOUIS CAUCHY, *1789 in Paris, †1857 in Sceaux, mathematician.

[5] SIR ISAAC NEWTON, *1643 in Woolsthorpe-by-Colsterworth/Lincolnshire, †1727 in Kensington, physicist, philosopher, mathematician and astronomer.

[6] The differential form of Newton's second law (2.26) is actually not found in Newton's *Principia* of 1687 or any of his work. Newton still formulated differential equations with geometrical approaches. It was Euler who first wrote the fundamental law of dynamics in the above form (in 1749).

Here $\boldsymbol{f}$ is the force per volume, separated into those forces arising from volume ($\boldsymbol{f}^{\mathrm{v}}$, long range) and from surface ($\boldsymbol{f}^{\mathrm{s}}$, short range; with infinitesimal contribution $\mathrm{d}\boldsymbol{F}$) related origin (see sketch in Figure 2.2). Note that the form of momentum conservation given by (2.26) is valid only in an inertial frame. Rotating frames of reference, more appropriate to the ocean, are considered in Section 2.3.3. In the following section, we will consider the mathematical representation of the forces proportional to the surface, while in Section 2.3.4 we consider the forces proportional to volume. Boundary conditions for momentum are discussed in Section 2.3.2.

### 2.3.1 Stresses, Pressure and Frictional Forces

A first result is gained by considering (2.26) for a *material* volume $V = V(t)$ where volume forces $\boldsymbol{f}^{\mathrm{v}}$ must appear as an integral over the volume and surface forces $\boldsymbol{f}^{\mathrm{s}}$ as an integral of a vector $\boldsymbol{t}^{\mathrm{s}}$ over the surface

$$\frac{\mathrm{d}}{\mathrm{d}t} \int\limits_{V(t)} \rho \boldsymbol{u} \mathrm{d}V = \int\limits_{V(t)} \boldsymbol{f}^{\mathrm{v}} \mathrm{d}V + \int\limits_{A(t)} \boldsymbol{t}^{\mathrm{s}} \mathrm{d}A \tag{2.27}$$

The vector $\boldsymbol{t}^{\mathrm{s}}$ is thus a force per unit area, called a *stress*. The relation of $\boldsymbol{t}^{\mathrm{s}}$ to a representation in terms of a force $\boldsymbol{f}^{\mathrm{s}}$ per volume is not yet clear, but the integral form allows immediately to point out an important property of surface forces: because the volume terms in (2.27) are proportional to $d^3$ and the surface integral to $d^2$ in terms of a characteristic scale $d$ of the volume, the limit of small $d$ cancels the volume contribution faster than the surface contribution. Hence

$$\lim_{d \to 0} \frac{1}{d^2} \int\limits_{A(t)} \boldsymbol{t}^{\mathrm{s}} \mathrm{d}A = 0 \tag{2.28}$$

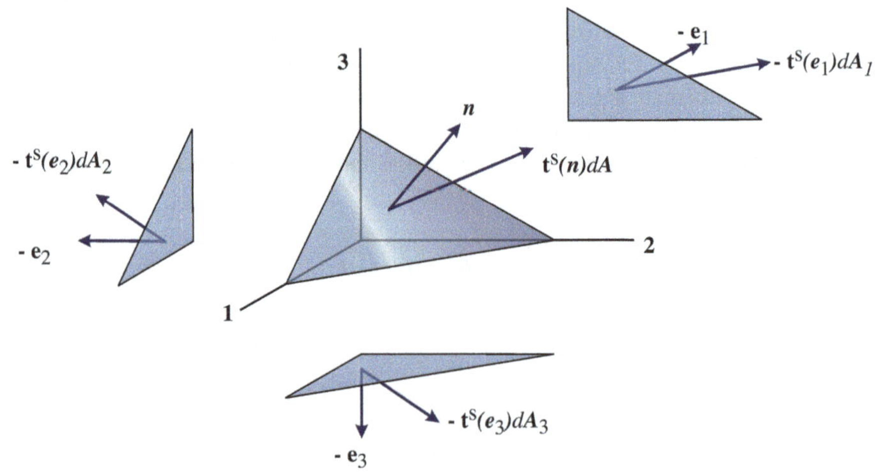


Fig. 2.3 The tetrahedron with side areas, associated units vectors, and stress components. The tetrahedron is bounded by three surfaces which are normal to the coordinate axes and a surface with arbitrary orientation specified by the normal vector $\boldsymbol{n}$. Redrawn from Aris (1989)

stating that the stresses must be locally in equilibrium. This statement allows to express the relation of the surface stress $\boldsymbol{t}^{\mathrm{s}}$ to the volume term $\boldsymbol{f}^{\mathrm{s}}$ in (2.26) in a simple mathematical form. The direction of the resulting force $\mathrm{d}\boldsymbol{F} = \boldsymbol{t}^{\mathrm{s}}\mathrm{d}A$ may have an arbitrary angle with the surface element which is characterized by the normal vector $\mathrm{d}\boldsymbol{A} = \boldsymbol{n}\mathrm{d}A$ (see Figure 2.2). Cauchy postulated that $\mathrm{d}\boldsymbol{F}$ at any given time $t$ and at any position $\boldsymbol{x}$ only depends upon the orientation $\boldsymbol{n}$, i. e. the curvature of the surface is irrelevant for the stress. This postulate is indicated by the notation $\boldsymbol{t}^{\mathrm{s}} = \boldsymbol{t}^{\mathrm{s}}(\boldsymbol{n}; \boldsymbol{x}, t)$. Expressing now the stress equilibrium (2.28) for a small tetrahedron (see Figure 2.3), having three of its faces parallel to the coordinate planes with unit normals $\boldsymbol{e}_j$ and area elements $\mathrm{d}A_j$, $j = 1, 2, 3$, and the fourth with the normal $\boldsymbol{n}$ and area $\mathrm{d}A$ (see Figure 2.3), we find

$$\boldsymbol{t}^{\mathrm{s}}(\boldsymbol{n})\mathrm{d}A - \boldsymbol{t}^{\mathrm{s}}(\boldsymbol{e}_j)\mathrm{d}A_j = 0$$

using $\boldsymbol{t}^{\mathrm{s}}(-\boldsymbol{e}_j) = -\boldsymbol{t}^{\mathrm{s}}(\boldsymbol{e}_j)$, a consequence of the principle of action and reaction, which also follows from Cauchy's postulate. A geometric consideration yields $\mathrm{d}A_j = n_j\mathrm{d}A$ for $\boldsymbol{n} = (n_1, n_2, n_3)$, and hence

$$\boldsymbol{t}^{\mathrm{s}}(\boldsymbol{n}) = n_j\boldsymbol{t}^{\mathrm{s}}(\boldsymbol{e}_j) \quad \text{or} \quad t_i^{\mathrm{s}}(\boldsymbol{n}) = n_j t_i^{\mathrm{s}}(\boldsymbol{e}_j)$$

stating a *linear proportionality* of the stress with the normal of the respective surface element. The latter relation is component-wise, and denoting the $i$th component of the stress vector $\boldsymbol{t}^{\mathrm{s}}(\boldsymbol{e}_j)$, acting on the area element normal to the coordinate axis $j$, by $\Pi_{ji}$, we may write

$$\mathrm{d}F_i = \mathrm{d}An_j\Pi_{ji} \quad \text{or} \quad \mathrm{d}\boldsymbol{F} = \mathrm{d}\boldsymbol{A}\cdot\Pi \tag{2.29}$$

for the stress force $\mathrm{d}\boldsymbol{F}$ acting on the infinitesimal area $\mathrm{d}A$. The proportionality factor $\Pi_{ji}$ stands for the $i$-component of the force per unit area on the area perpendicular to the $j$-axis. The $\Pi_{ji}$ can be combined into a tensor $\Pi = (\Pi_{ij})$ (with $3 \times 3$ components), the *stress tensor*. The components with associated surfaces are displayed in Figure 2.4. Hence $n_j\Pi_{ji}$ is the $i$-component of the force per unit area on the area perpendicular to $\boldsymbol{n}$.

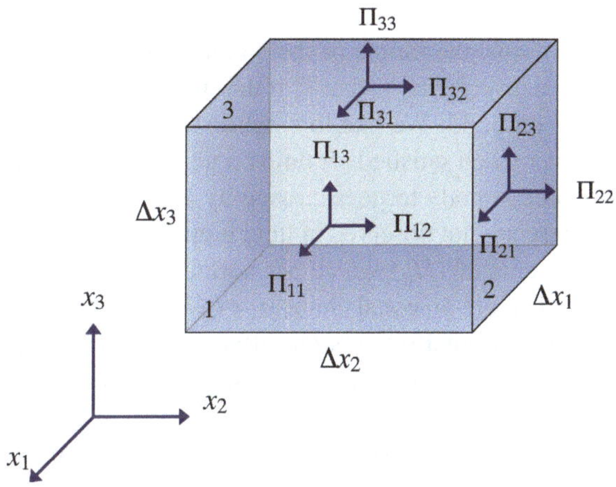


Fig. 2.4 The components of the stress tensor

The resulting force $\boldsymbol{F}$ on a volume $V$ results from all forces $\mathrm{d}\boldsymbol{F}$ on the surface of the volume, and integration yields

$$\boldsymbol{F} = \oint_A \mathrm{d}\boldsymbol{A} \cdot \Pi = \int_V \nabla \cdot \Pi \mathrm{d}V = \int_V \boldsymbol{f}^{\mathrm{s}} \mathrm{d}V \tag{2.30}$$

which clarifies that the volume form $\boldsymbol{f}^{\mathrm{s}}$ of the surface forces is given by the divergence $\nabla \cdot \Pi$ of the stress tensor. The relation follows by application of Gauss' theorem. This mathematical form guarantees that the surface forces, acting between different interior parts of the volume, must sum up to zero as action and reaction must compensate on every interior surface. The $i$-component $\partial \Pi_{ji}/\partial x_j$ constitutes a force per volume in the $i$-direction. Likewise, $\Pi_{ji}$ is the flux of $i$-momentum in the *negative* $j$-direction. Notice that this sign convention used for the molecular flux of momentum, given by $\Pi$ and below also by $\Sigma$, differs from the convention of sign used for molecular fluxes of partial masses and heat (compare to the Sections 2.2 and 2.4.3).

It may be somewhat counterintuitive that the last relation expresses the *surface* forces again by a *volume* term, but now the spatial variations of the stress tensor come into play. In fact, if $\Pi$ is constant, the first expression in (2.30) vanishes as well. Taking (2.30) for an infinitesimal volume $V$, the force per volume becomes $\nabla \cdot \Pi$, and the momentum balance (2.26) is written in the form

$$\rho \frac{\mathrm{D}\boldsymbol{u}}{\mathrm{D}t} = \nabla \cdot \Pi + \boldsymbol{f}^{\mathrm{V}} \tag{2.31}$$

This form of momentum conservation is due to Cauchy. It is valid for any continuum, but the stress tensors of different materials can be rather different.

The stress tensor for fluids is symmetric, i. e. $\Pi_{ij} = \Pi_{ji}$, according to BOLTZMANN's[7] postulate. It may be shown that the symmetry of the stress tensor is equivalent to the postulate that the angular momentum of a material fluid element is only changed by torques $\boldsymbol{x} \times \boldsymbol{f}^{\mathrm{V}}$ associated with the volume forces, and torques $\boldsymbol{x} \times \boldsymbol{t}^{\mathrm{S}}$ arising from the stress forces, expressed by

$$\frac{\mathrm{d}}{\mathrm{d}t} \int_{V(t)} \rho \boldsymbol{x} \times \boldsymbol{u} \mathrm{d}V = \int_{V(t)} \boldsymbol{x} \times \boldsymbol{f}^{\mathrm{V}} \mathrm{d}V + \int_{A(t)} \boldsymbol{x} \times \boldsymbol{t}^{\mathrm{S}} \mathrm{d}A \tag{2.32}$$

Put in mathematical terms, the torque of the volume representation $\boldsymbol{x} \times \nabla \cdot \Pi$ can only be expressed by a divergence $\nabla \cdot \boldsymbol{x} \times \Pi$ if the stress tensor is symmetric. Details are found in the box on p. 35. We like to emphasize that the symmetry of $(\Pi_{ij})$ is a postulate which is very well satisfied for fluids: a nonsymmetric stress tensor would generate stress couples as body torques, arising by a net volume contribution from the stress forces. These are not observed in fluid motions.

The diagonal elements of $(\Pi_{ij})$ which act normal to the corresponding surface are *normal stresses*. The off-diagonal elements of $(\Pi_{ij})$ act tangential; they are the *tangential stresses*. The normal stresses on different surfaces of a volume element need not be identical. The mean normal inward stress is the *pressure*

$$p = -\frac{1}{3}(\Pi_{11} + \Pi_{22} + \Pi_{33}) = -\frac{1}{3}\Pi_{ii} = -\frac{1}{3}\mathrm{tr}\,\Pi \tag{2.33}$$

[7] LUDWIG BOLTZMANN, *1844 in Linz, †1908 in Duino near Triest, physicist.

The relation (2.33) defines the *mechanical pressure* which is now assumed to be identical[8] to the thermodynamical pressure introduced in Section 1.2.

It is useful to decompose the stress tensor according to the isotropic pressure part and the remaining terms,

$$\Pi_{ij} = -p\delta_{ij} + \Sigma_{ij} \tag{2.34}$$

The remainder ($\Sigma_{ij}$) is called the *friction tensor* (or viscous stress tensor or deviatoric stress tensor). It does not transport any normal component of momentum in the mean since by definition the sum of the viscous normal stresses must vanish, $\Sigma_{ii} = 0$. For the surface force per volume $\boldsymbol{f}^{\mathrm{s}}$ acting on a volume element we have then

$$\boldsymbol{f}^{\mathrm{s}} = \nabla \cdot \Pi = -\nabla p + \nabla \cdot \Sigma$$

In a fluid at rest or in uniform motion all tangential stresses must vanish; otherwise fluid elements would start to move relative to neighboring elements. Likewise, all normal stresses must be identical, otherwise the volume would be deformed and also induce relative motion. More generally, an *ideal or perfect fluid* is defined by the vanishing of all viscous stresses, $\Sigma \equiv 0$. The surface forces are then given by the gradient of pressure alone, $\boldsymbol{f}^{\mathrm{s}} = -\nabla p$. In real fluids, the viscous stresses do generally not vanish. The behavior of a solid body and a fluid is fundamentally different. In a deformed body, forces occur directly as result of deformations and hence a relation between stress tensor and deformation tensor is expected. This *constitutive law* or *stress–strain relation* is a material property of the medium. With respect to changes

**7. Balance of Angular Momentum**

The stress tensor for fluids must be symmetric, $\Pi_{ij} = \Pi_{ji}$, as any antisymmetric component would result in net torques internally in the fluid and hence would lead to a rotation of respective volume elements. To prove this property, we write (2.31) in the flux form and proceed to the balance of angular momentum (with respect to the coordinate origin),

$$\frac{\partial}{\partial t}\rho\boldsymbol{x} \times \boldsymbol{u} + \boldsymbol{x} \times \nabla \cdot (\rho\boldsymbol{u}\boldsymbol{u} - \Pi) = \boldsymbol{x} \times \boldsymbol{f}^{\mathrm{v}}$$

obtained by taking the cross-product with the position vector $\boldsymbol{x}$. As $\boldsymbol{u}\boldsymbol{u}$ is a symmetric tensor, we have

$$\boldsymbol{x} \times \nabla \cdot (\rho\boldsymbol{u}\boldsymbol{u}) = \nabla \cdot \rho\boldsymbol{u}(\boldsymbol{x} \times \boldsymbol{u})$$

so that only the divergence of the advective flux of angular momentum is of relevance. Likewise, for the stress tensor we must require

$$\boldsymbol{x} \times \nabla \cdot \Pi = \nabla \cdot \boldsymbol{x} \times \Pi \tag{B7.1}$$

as otherwise torques would arise from the stresses in the interior of any volume. Written in component form, this results in $\epsilon_{ijk}\Pi_{jk} = 0$, which directly shows that the tensor $\Pi$ must be symmetric. The relation (B7.1) or directly $\Pi_{ij} = \Pi_{ji}$ is known as Boltzmann's postulate. Then, in the Lagrangian view, the angular momentum balance becomes

$$\rho\frac{\mathrm{D}(\boldsymbol{x} \times \boldsymbol{u})}{\mathrm{D}t} = \nabla \cdot (\boldsymbol{x} \times \Pi) + \boldsymbol{x} \times \boldsymbol{f}^{\mathrm{v}}$$

[8] The thermodynamic pressure relies on the concept of thermodynamic equilibrium; the mechanical pressure is defined for a moving fluid which is not in equilibrium. The equivalence of the two pressure concepts is nontrivial; we refer the reader to Batchelor (1977).

in volume, most materials react in a similar elastic way. Differences between materials occur, however, when the deformation causes a strain rather than a volume change.

In a fluid, displacements generally do not result in forces. Only a relative motion of a neighboring parcel exerts a stress. It is phenomenologically well known that for a constant velocity field all tangential stresses vanish. A Newtonian fluid is defined by a linear relation between stress and velocity shear, for instance a shear flow $u(y)$ in the $x$-direction the tangential stress $\Sigma_{yx}$ (describing flux of $x$-momentum in the negative $y$-direction) is proportional to the gradient $\partial u/\partial y$, i. e.

$$\Sigma_{yx} = \nu \frac{\partial u}{\partial y}$$

with a down-gradient direction of transport (note here again the above-mentioned sign convention). The (dynamical) viscosity $\nu$ is a property of the fluid. A general *linear* relation between stress and strain must involve both the stress tensor and the deformation tensor. Both tensors are symmetric, but the stress tensor has a vanishing trace. The deviator $D^\star_{ij}$ of the deformation tensor $D_{ij}$, introduced in (1.10), describes shearing motion without a change in volume. It is symmetric and also has a vanishing trace. The simplest relation between stress and strain hence is given by

$$\Sigma_{ij} = 2\nu \mathrm{D}^\star_{ij} = 2\nu \left( \mathrm{D}_{ij} - \frac{1}{3}\mathrm{D}_{\ell\ell}\delta_{ij} \right) \tag{2.35}$$

or explicitly

$$\Sigma_{ij} = \nu \left( \frac{\partial u_i}{\partial x_j} + \frac{\partial u_j}{\partial x_i} - \frac{2}{3}\frac{\partial u_\ell}{\partial x_\ell}\delta_{ij} \right) \tag{2.36}$$

Relation (2.36) defines a general *Newtonian fluid*, and has been confirmed experimentally for gases and most fluids. However, there are fluids which behave differently, e. g. where a nonlinear or more complicated, (e. g. time-dependent) relation between stress and strain is found. Such fluids are called non-Newtonian fluids; prominent examples are suspensions like paint, which gets less viscous when painted and more viscous when it remains at the wall or paper, or ketchup, which also often needs some time-dependent treatment to get less viscous and leave the bottle. The dynamical viscosity of pure water is about $10^{-3}\,\mathrm{N\,s\,m^{-2}}$ at 20°C and decreases (increases) significantly for higher (lower) temperatures. There are not many fluids which are less viscous than water, e. g. acetone has a viscosity of $3 \times 10^{-4}\,\mathrm{N\,s\,m^{-2}}$ at 25°C, but many which are more viscous, e. g. olive oil with a viscosity of $0.08\,\mathrm{N\,s\,m^{-2}}$. Figure 2.5 shows the dynamical viscosity of seawater, which is very similar to pure water, for range of temperature and salinity representative for the ocean. The dependency of $\nu$ on salinity is much weaker compared to the one on temperature.

The divergence of the frictional stress tensor, i. e. the frictional force per volume, is given by

$$\frac{\partial \Sigma_{ji}}{\partial x_j} = \frac{\partial}{\partial x_j}\left( \nu \frac{\partial u_i}{\partial x_j} \right) + \frac{\partial}{\partial x_j}\left( \nu \frac{\partial u_j}{\partial x_i} \right) - \frac{2}{3}\frac{\partial}{\partial x_i}\left( \nu \frac{\partial u_j}{\partial x_j} \right) \tag{2.37}$$

In most applications it is sufficiently accurate to neglect the spatial variations of $\nu$ (which is due to the dependence of $\nu$ on thermodynamic variables as shown in

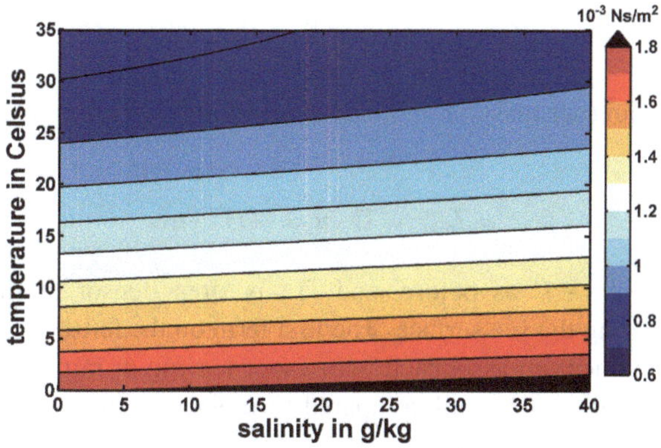


Fig. 2.5 Dynamical viscosity of seawater $\nu$ in $10^{-3}\,\mathrm{N\,s\,m^{-2}}$ as a function of temperature and salinity. After Sharqawy et al. (2010)

Figure 2.5) so that

$$\frac{\partial \Sigma_{ji}}{\partial x_j} = \nu \left( \frac{\partial^2 u_i}{\partial x_j^2} + \frac{1}{3} \frac{\partial^2 u_j}{\partial x_i \partial x_j} \right) \tag{2.38}$$

Incorporating the force of the molecular stresses into Newton's law of motion, the balance of momentum (2.26) thus takes the form

$$\rho \frac{\mathrm{D}\boldsymbol{u}}{\mathrm{D}t} = \nabla \cdot \Pi + \boldsymbol{f}^{\mathrm{V}} = -\nabla p + \nabla \cdot \Sigma + \boldsymbol{f}^{\mathrm{V}} \tag{2.39}$$

where $\boldsymbol{f}^{\mathrm{V}}$ denotes the volume force which still remains to be specified (compare Section 2.3.4). Using the continuity equation (2.15), we obtain the equivalent flux form

$$\frac{\partial}{\partial t} \rho \boldsymbol{u} = -\nabla \cdot (\rho \boldsymbol{u}\boldsymbol{u} - \Pi) + \boldsymbol{f}^{\mathrm{V}} = -\nabla p - \nabla \cdot (\rho \boldsymbol{u}\boldsymbol{u} - \Sigma) + \boldsymbol{f}^{\mathrm{V}} \tag{2.40}$$

with the *momentum flux tensor* $\rho \boldsymbol{u}\boldsymbol{u} - \Pi$ which includes the momentum flux both through advection and through surface forces. Occasionally, only the advective contribution is called momentum flux.

Expressing the divergence of the frictional tensor by (2.38) (i. e. neglecting the spatial dependency of viscosity) leads to the NAVIER–STOKES[9] equations

$$\rho \frac{\mathrm{D}\boldsymbol{u}}{\mathrm{D}t} = -\nabla p + \nu \nabla^2 \boldsymbol{u} + \frac{\nu}{3} \nabla (\nabla \cdot \boldsymbol{u}) + \boldsymbol{f}^{\mathrm{V}} \tag{2.41}$$

written here in the parcel form. These equations form the basis of most applications of hydrodynamics.

[9] CLAUDE-LOUIS NAVIER, *1785 in Dijon, †1836 in Paris, engineer and physicist. He laid down the complete Navier–Stokes equations in 1823 for an incompressible fluid. His work received little contemporary attention, and so the equations were rediscovered at least four times: by Cauchy in 1823, by Poisson in 1829, by Saint-Venant in 1837, and by Stokes in 1845 (Darrigol, 2009). GEORGE GABRIEL STOKES, *1819 in Skreen, †1903 in Cambridge, mathematician and physicist.

### 2.3.2 Boundary Condition for the Momentum Flux

For the momentum (per mass) $\boldsymbol{u}$ and the momentum flux $\Pi$ one obtains from (2.7) and (2.21)

$$\boldsymbol{u}\mathcal{J}_{\text{mass}} - \Pi \cdot \boldsymbol{n} = -\Pi^a \cdot \boldsymbol{n} \tag{2.42}$$

where $\mathcal{J}_{\text{mass}} = \mathcal{E} - \mathcal{P}$ as before, and $\Pi^a$ is stress tensor of the neighboring medium, considering the sea surface. The first term on the left-hand side represents the transfer of momentum associated with vapor leaving or rain drops entering the ocean. It is generally very small compared to molecular stresses exchanging momentum (at the sea surface: $\mathcal{J}_{\text{mass}}/\rho = (\mathcal{E} - \mathcal{P})\rho$ is a very small velocity of order $1\,\text{m/y} \sim 10^{-7}\,\text{m s}^{-1}$, and the second term is of order $\nu/(\rho\Delta) \sim 10^{-6}/\Delta\,\text{m}^2\,\text{s}^{-1}$ where $\Delta$ is the scale of the velocity gradient. The first term in (2.42) is, therefore, of negligible magnitude.

Decomposing (2.42) into normal and tangential components, we obtain the *dynamic* boundary conditions

$$-p + \Sigma_{33} = -p^a + \Sigma_{33}^a \tag{2.43}$$

$$\Sigma_{j3} = \Sigma_{j3}^a = \tau_j^a\,, \quad j = 1, 2 \tag{2.44}$$

where $\boldsymbol{\tau}^a$ is the tangential force (windstress) and $p^a$ the pressure at the lower boundary of the atmosphere, and $\boldsymbol{n}$ has been taken vertically for simplicity. Remember, however, that the stresses considered here are of molecular character, and the implied molecular boundary layers at the interfaces contain extremely large velocity shears to carry e. g. the flux of momentum imparted by the surface wind field. Usually the viscous part in (2.43) is small, so that the pressure field is very nearly continuous at the surface, $p \approx p^a$.

At the bottom of the ocean, the pressure within the solid earth is not constrained by the oceanic pressure, and the relation (2.43) does not apply. In fluid mechanics two possibilities for the dynamic boundary conditions are popular at solid interfaces: $\Sigma \cdot \boldsymbol{n} = 0$, representing the case of no stress or *free slip*, and $\boldsymbol{u} = 0$, requiring a vanishing flow speed, termed *no slip*. Figure 2.6 displays the structure of the near-bottom velocity for the two conditions. These conditions then require a certain stress at the bottom which is determined from the solution of the respective hydrodynamical problem. A parameterization which is intermediate between the limiting forms

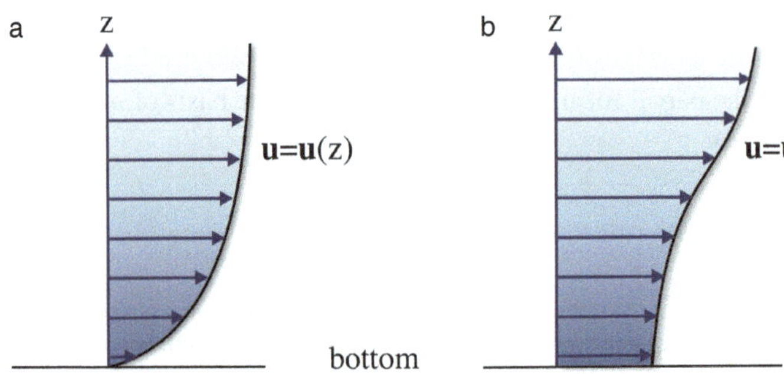


Fig. 2.6 Sketch of no-slip (**a**) and free-slip (**b**) conditions at the ocean bottom

of free slip and no slip is given by

$$\Sigma \cdot \boldsymbol{n} = r\boldsymbol{u} \cdot \boldsymbol{t} \qquad (2.45)$$

where $\boldsymbol{t}$ is the tangential unit vector in the bottom surface. The coefficient $r$ is empirical, frequently expressed as $r = r^*|\boldsymbol{u} \cdot \boldsymbol{t}|$ which makes the parameterization nonlinear. Clearly, for small $r$ we regain the free-slip condition and for large $r$ the no-slip condition, because additionally to $\boldsymbol{u} \cdot \boldsymbol{t} = 0$ we also have $\boldsymbol{u} \cdot \boldsymbol{n} = 0$ from the the kinematic condition at the sea floor.

### 2.3.3 Conservation Equations on the Rotating Earth

So far we have used the field variables without any reference to a particular coordinate system. We have introduced scalar fields $\chi$ and vector fields $\boldsymbol{v}$ (not necessarily velocity) and implicitly used the notion of a Cartesian coordinate system. However, whereas two observers with their orientation in two different coordinate frames would certainly agree on the value of a scalar property $\chi(t)$ of a fluid parcel at all times and its temporal change $\mathrm{D}\chi/\mathrm{D}t$, they would not necessarily agree on vector properties $\boldsymbol{v}(t)$. A vector is characterized by a length and direction, and both depend on a predefined frame of reference. In particular, the two observers might get different values for $\boldsymbol{v}(t)$ and its material rate of change $\mathrm{D}\boldsymbol{v}/\mathrm{D}t$.

#### Rotating Coordinate System

The Earth is turning itself but also rotating around the Sun. It is, therefore, convenient to consider a rotating coordinate system. An *inertial* frame is characterized by constant unit vectors $\boldsymbol{e}_i{}^\star$ in the coordinate direction $i = 1, 2, 3$, while for a *rotating* frame of reference the unit vectors $\boldsymbol{e}_i(t)$ are time dependent and rotate with the angular velocity $\boldsymbol{\Omega}$. Figure 2.7 outlines the situation in a two-dimensional sketch. The magnitude of $\boldsymbol{\Omega}$ denotes the angular rotation speed of $\Omega = \mathrm{d}\lambda/\mathrm{d}t$, where $\lambda(t)$ denotes the angle between the individual unit vectors $\boldsymbol{e}_i(t)$ and $\boldsymbol{e}_i{}^\star$ of the two coordinate systems. The direction of $\boldsymbol{\Omega}$ defines the rotation axis, which points out of the plane in the scheme of Figure 2.7. While the rotating frame has time-dependent unit vectors $\boldsymbol{e}_i(t)$ when viewed from the inertial frame, an observer in the rotating system would, of course, see 'his' $\boldsymbol{e}_i$ as vectors which are constant in time. In the following we assume that the coordinate systems have the same origin; the situation of the Earth is a bit more complicated and discussed in the box on p. 41.

It is worth mentioning that though $\mathrm{D}\chi/\mathrm{D}t$ is the same for the two observers, they obtain different local rates of change and different advective rates of change, i. e. expanding $\mathrm{D}/\mathrm{D}t$ in the inertial and the rotating system into the Eulerian form,

$$\frac{\mathrm{D}\chi}{\mathrm{D}t} = \frac{\partial\chi}{\partial t} + \boldsymbol{u}^\star \cdot \nabla^\star\chi \quad \text{and} \quad \frac{\mathrm{D}\chi}{\mathrm{D}t} = \frac{\partial\chi}{\partial t} + \boldsymbol{u} \cdot \nabla\chi$$

the $\partial\chi/\partial t$ are different in the two expressions because different Eulerian spatial coordinates are held fixed. These are $x_i^\star = \boldsymbol{e}_i^\star \cdot \boldsymbol{x}$ and $x_i = \boldsymbol{e}_i \cdot \boldsymbol{x}$, respectively.

We will now compare the material rate of change $\mathrm{D}\boldsymbol{v}/\mathrm{D}t$ of a vector, measured in the inertial frame, to the rate of change which an observer in the rotating frame

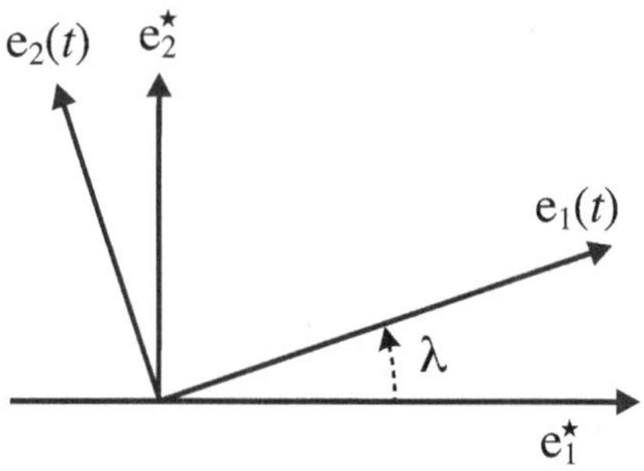


Fig. 2.7 A sketch outlining the rotating coordinate system. The inertial frame is given by $\boldsymbol{e}_1{}^\star$ and $\boldsymbol{e}_2{}^\star$. The frame given by $\boldsymbol{e}_1$ and $\boldsymbol{e}_2$ is rotating with an angular velocity $\Omega = \mathrm{d}\lambda/\mathrm{d}t$

obtains. We will first consider the arbitrary vector $\boldsymbol{v}(t)$, afterward the result will be evaluated for the position vector and the velocity of a fluid parcel. Because $\boldsymbol{e}_1(t + \Delta t) = \boldsymbol{e}_1(t) + \Delta\lambda \boldsymbol{e}_2(t)$ for a unit base vector which is normal to rotation axis (see Figure 2.7), we obtain $\mathrm{d}\boldsymbol{e}_1/\mathrm{d}t = \Omega \boldsymbol{e}_2$ and similarly $\mathrm{d}\boldsymbol{e}_2/\mathrm{d}t = -\Omega \boldsymbol{e}_1$ where $\Omega = \mathrm{d}\lambda/\mathrm{d}t$. Hence, in general

$$\frac{\mathrm{d}\boldsymbol{e}_i}{\mathrm{d}t} = \boldsymbol{\Omega} \times \boldsymbol{e}_i$$

is observed in the inertial frame. Consider now an arbitrary time-dependent vector $\boldsymbol{v}(t)$ with components $v_i^*(t) = \boldsymbol{v}(t) \cdot \boldsymbol{e}_i{}^*$ in the inertial frame and components $v_i(t) = \boldsymbol{v}(t) \cdot \boldsymbol{e}_i(t)$ in the rotating frame. The temporal rate of change of $\boldsymbol{v}(t)$ is thus

$$\frac{\mathrm{d}\boldsymbol{v}}{\mathrm{d}t} = \frac{\mathrm{d}v_i^*}{\mathrm{d}t}\boldsymbol{e}_i{}^* = \frac{\mathrm{d}v_i}{\mathrm{d}t}\boldsymbol{e}_i + v_i\frac{\mathrm{d}\boldsymbol{e}_i}{\mathrm{d}t}$$

The first term in the second expression is the change which a corotating observer measures. We use the notation $(\mathrm{d}/\mathrm{d}t)^{\mathrm{rot}}\boldsymbol{v}$ to write this term in vector form. The second term is just $\boldsymbol{\Omega} \times \boldsymbol{v}$, hence

$$\frac{\mathrm{d}\boldsymbol{v}}{\mathrm{d}t} = \left(\frac{\mathrm{d}}{\mathrm{d}t}\right)^{\mathrm{rot}} \boldsymbol{v} + \boldsymbol{\Omega} \times \boldsymbol{v}$$

Application to the position vector $\boldsymbol{X}(t)$ of a fluid parcel, now a material property, yields

$$\frac{\mathrm{D}\boldsymbol{X}}{\mathrm{D}t} = \boldsymbol{u} = \left(\frac{\mathrm{d}}{\mathrm{d}t}\right)^{\mathrm{rot}} \boldsymbol{X} + \boldsymbol{\Omega} \times \boldsymbol{X} = \boldsymbol{u}^{\mathrm{rot}} + \boldsymbol{\Omega} \times \boldsymbol{X}$$

The first term, $\boldsymbol{u}^{\mathrm{rot}} = (\mathrm{d}/\mathrm{d}t)^{\mathrm{rot}}\boldsymbol{X}$, is the velocity vector measured by the corotating observer. Further, for the acceleration, we find

$$\begin{aligned}
\frac{\mathrm{D}\boldsymbol{u}}{\mathrm{D}t} &= \frac{\mathrm{d}}{\mathrm{d}t}\left(\frac{\mathrm{d}}{\mathrm{d}t}\right)^{\mathrm{rot}} \boldsymbol{X} + \boldsymbol{\Omega} \times \frac{\mathrm{d}\boldsymbol{X}}{\mathrm{d}t} \\
&= \left(\frac{\mathrm{d}^2}{\mathrm{d}t^2}\right)^{\mathrm{rot}} \boldsymbol{X} + \boldsymbol{\Omega} \times \left(\frac{\mathrm{d}}{\mathrm{d}t}\right)^{\mathrm{rot}} \boldsymbol{X} + \boldsymbol{\Omega} \times \left[\left(\frac{\mathrm{d}}{\mathrm{d}t}\right)^{\mathrm{rot}} \boldsymbol{X} + \boldsymbol{\Omega} \times \boldsymbol{X}\right] \\
&= \left(\frac{\mathrm{d}}{\mathrm{d}t}\right)^{\mathrm{rot}} \boldsymbol{u}^{\mathrm{rot}} + 2\boldsymbol{\Omega} \times \boldsymbol{u}^{\mathrm{rot}} + \boldsymbol{\Omega} \times (\boldsymbol{\Omega} \times \boldsymbol{X})
\end{aligned} \tag{2.46}$$

**8. A Rotating Coordinate System for the Earth**

A natural coordinate system attached to the Earth is the geocentric Cartesian frame with its origin in the Earth's center. However, the Earth rotates around the Sun and also around itself. Further, the solar system and the whole galaxy might rotate, which, however, should add only small contributions. We first consider the rotation of the Earth around the Sun using a Cartesian system in the center of the Sun, rotating with a period of $365.24$ days. We assume a constant angular velocity of $\Omega_{\text{sun}} = 2\pi/(365.24 \times 86400\,\text{s}) = 0.02 \times 10^{-5}\,\text{s}^{-1}$ and a circular path of the Earth around the Sun, such that the Earth remains now at a fixed position in the rotating frame. In this coordinate system a corresponding Coriolis and centrifugal force in the momentum budget (2.47) will appear.

Next, we consider the rotation of the Earth itself. First we move the center of the coordinate system into the center of the Earth and we also rotate the system (once) from the ecliptic into the equatorial plane of the Earth (which has an angle with the ecliptic of about $23°$). Note that this simple transformation will have no consequences in the momentum budget, except that $\boldsymbol{\Omega}_{\text{sun}}$ will be rotated by about $-23°$ with respect to the equatorial plane. Since the Earth is still rotating, we finally switch to a rotating coordinate system with rotation vector $\boldsymbol{\Omega}_{\text{earth}}$ pointing towards the North Pole with magnitude $\Omega_{\text{earth}} = 2\pi 86400\,\text{s} = 7.272 \times 10^{-5}\,\text{s}^{-1}$. This last transformation will add another Coriolis and centrifugal force, such that the momentum budget becomes

$$\rho \frac{\mathrm{D}\boldsymbol{u}}{\mathrm{D}t} = -2\rho \boldsymbol{\Omega} \times \boldsymbol{u} - \nabla p + \nabla \cdot \Sigma - \rho \boldsymbol{\Omega}_{\text{earth}} \times (\boldsymbol{\Omega}_{\text{earth}} \times \boldsymbol{X}) \\ - \rho \boldsymbol{\Omega}_{\text{sun}} \times (\boldsymbol{\Omega}_{\text{sun}} \times \boldsymbol{X}) + \boldsymbol{f}^{\text{v}}$$

with $\boldsymbol{\Omega} = \boldsymbol{\Omega}_{\text{earth}} + \boldsymbol{\Omega}_{\text{sun}}$ with magnitude $\Omega \approx 7.292 \times 10^{-5}\,\text{s}^{-1}$ and direction almost parallel to $\boldsymbol{\Omega}_{\text{earth}}$. Note that the centrifugal forces can be written as potentials which then both add to the geopotential discussed in Section 2.3.4. The contribution of the centrifugal potential of the Sun will be much smaller than the one of the Earth.

The left-hand side is the acceleration of the fluid element as seen in the absolute frame, and the first term on the right-hand side is the acceleration which the corotating observer measures for the same parcel. Note that if $\boldsymbol{\Omega}$ is time-dependent, a term $\mathrm{d}\boldsymbol{\Omega}/\mathrm{d}t \times \boldsymbol{X}$ will appear on the right-hand side of (2.46).

### Momentum Budget in a Rotating System

The equations of motion in the preceding sections hold for an absolute coordinate system (*inertial system*). In a *rotating* frame, the left hand side of the momentum equation (2.39) must hence be replaced by the right hand side of (2.46). Dropping the index rot, this yields

$$\rho \frac{\mathrm{D}\boldsymbol{u}}{\mathrm{D}t} = -2\rho \boldsymbol{\Omega} \times \boldsymbol{u} - \nabla p + \nabla \cdot \Sigma - \rho \boldsymbol{\Omega} \times (\boldsymbol{\Omega} \times \boldsymbol{X}) + \boldsymbol{f}^{\text{v}} \tag{2.47}$$

where $\boldsymbol{u}$ now denotes the velocity seen by the corotating observer. Two additional forces appear in the rotating frame which are entirely due to the coordinate transformation and, therefore, have a virtual character:

- The CORIOLIS[10]–force
  $-2\rho\boldsymbol{\Omega} \times \boldsymbol{u}$ (or Coriolis acceleration $2\boldsymbol{\Omega} \times \boldsymbol{u}$) depends on the flow velocity, always trying to bend the parcel's path normal to the direction of $\boldsymbol{\Omega}$ and to $\boldsymbol{u}$. In other

[10] GASPARD GUSTAVE DE CORIOLIS, *1792 in Paris, †1843 in Paris, engineer and mathematician. Coriolis published his work in 1835, but almost 60 years earlier, in 1768, PIERRE-SIMON LAPLACE correctly posed the problem of forces on the rotating Earth in his work on tides.

words: looking down with the North Pole in view, we see a counterclockwise rotation of the Earth and thus a deviation of the motion on the northern hemisphere to the right. With the South Pole in view, we see a clockwise rotation, and the motion on the southern hemisphere is seen to deviate to the left. The Coriolis force is fundamental for all large-scale motions in ocean and atmosphere.

- The *centrifugal force*
  $-\rho\boldsymbol{\Omega} \times (\boldsymbol{\Omega} \times \boldsymbol{X})$ is independent of the flow variables; it will be discussed in the next section.

Both forces are also referred to as *apparent forces* as they are apparent only to a corotating observer.

### 2.3.4 The Force of Gravity on the Earth

According to Newton's law of gravitation a point mass $M$ attracts another point mass $m$ with the force

$$\mathcal{F}(\boldsymbol{r}) = -G_{\mathrm{N}} \frac{mM}{r^2} \frac{\boldsymbol{r}}{r} \tag{2.48}$$

Here $\boldsymbol{r}$ denotes the distance vector from $M$ to $m$, and $G_{\mathrm{N}} = 6.673 \times 10^{-11}\,\mathrm{m^3\,kg^{-1}\,s^{-2}}$ is the universal constant of gravitation. The force $\mathcal{F}/m$ per unit mass can be derived from the gravity potential according to

$$\frac{\mathcal{F}}{m} = -\nabla\Phi^{\star} = \nabla G_{\mathrm{N}} \frac{M}{r} \tag{2.49}$$

The force per mass is $-\nabla\Phi^{\star}$, and the force per volume is $-\rho\nabla\Phi^{\star}$. Since forces are additive, the potentials from different masses add as well, and the attraction of a fluid parcel by different masses can hence be treated separately. The volume force exerted by gravity can thus be represented by a potential $\Phi^{\star}$. In oceanography and meteorology the gravity potentials of the Earth, Moon, and Sun are relevant, i. e. $\Phi^{\star} = \Phi_{\mathrm{E}} + \Phi_{\mathrm{m}} + \Phi_{\mathrm{s}}$. We discuss how to treat the contributions from the Earth and the Moon. The contribution of the Sun is analogous to the Moon.

#### Gravity Potential of the Earth

Newton's law of attraction holds if both $m$ and $M$ can be considered as *mass points* with negligible spatial extent. A result of potential theory is that Newton's law also holds for a body of finite extent, provided that its mass distribution is *spherically symmetric*, i. e. only dependent on the distance from the center of mass. The Earth's surface has, however, no spherical symmetry but is more accurately described as oblate spheroidal ellipsoid. Moreover, the mass distribution in the interior is also not spherically symmetric. Outside the Earth, the gravity potential is, therefore, more accurately given by an expansion of the potential in spherical harmonics with the leading terms

$$\Phi_{\mathrm{E}}(r,\lambda,\varphi) = -G_{\mathrm{N}} \frac{M_{\mathrm{E}}}{r}\left[1 - \varepsilon_1 \frac{a_{\mathrm{e}}^2}{r^2}\left(3\sin^2\varphi - 1\right) + \varepsilon_{\star}(r,\lambda,\varphi)\right] \tag{2.50}$$

(see e.g. Stacey, 1992). Here $r$ denotes the distance from the mass center of the Earth, $\lambda$ is the geographical longitude and $\varphi$ the geographical latitude. The mass of the Earth is $M_E = 5.97 \times 10^{24}$ kg, the equatorial radius of the Earth is $a_e = 6378.1$ km, and $\varepsilon_1 = 0.541 \cdot 10^{-3}$ is a dimensionless coefficient. The residual term $\varepsilon_\star(r, \lambda, \varphi)$ (see Figure 2.9 below) has a magnitude of $\approx 10^{-5}$ and can be neglected for all dynamical purposes. Likewise, corrections to (2.50) resulting from the atmospheric mass distribution (see e. g. Rummel and Rapp, 1976) can be neglected.

## Geopotential

Similar to the gravity force, the centrifugal force in the momentum budget (2.47) can be rewritten in terms of a potential as shown in this section. Note that we only consider here the centrifugal force resulting from the Earth rotation since this is the most important (compare the box on p. 41). Only the component $\boldsymbol{X}_\perp$ of the position vector which is perpendicular to the axis of rotation contributes to the centrifugal force. This is because of $\boldsymbol{\Omega} \times \boldsymbol{X} = \boldsymbol{\Omega} \times \boldsymbol{X}_\perp$. With the vector identity $\boldsymbol{a} \times (\boldsymbol{b} \times \boldsymbol{c}) = (\boldsymbol{a} \cdot \boldsymbol{c})\boldsymbol{b} - (\boldsymbol{a} \cdot \boldsymbol{b})\boldsymbol{c}$ one finds $\boldsymbol{\Omega} \times (\boldsymbol{\Omega} \times \boldsymbol{X}_\perp) = -\Omega^2 \boldsymbol{X}_\perp$. Assuming that the axis of rotation is the $z$–direction, we have $\boldsymbol{X}_\perp = (x, y, 0) = \frac{1}{2}\nabla(x^2+y^2) = \frac{1}{2}\nabla r^2 \cos^2\varphi$ where $\varphi$ is the latitude coordinate and $r$ the length of the vector $\boldsymbol{X}$. The centrifugal acceleration can thus be derived from a potential according to

$$\boldsymbol{\Omega} \times (\boldsymbol{\Omega} \times \boldsymbol{X}) = -\nabla\left(\frac{1}{2}\Omega^2 r^2 \cos^2\varphi\right)$$

and hence be combined with the gravity, arising for the Earth's attraction, to the *geopotential* $\Phi = \Phi_E - \frac{1}{2}\Omega^2 r^2 \cos^2\varphi$ (frequently called 'apparent' gravity potential) so that

$$\Phi(r, \lambda, \varphi) = -G_N \frac{M_E}{r}\left[1 - \varepsilon_1 \frac{a_e^2}{r^2}\left(3\sin^2\varphi - 1\right) + \frac{1}{2}\varepsilon_2\left(\frac{r}{a_e}\right)^3 \cos^2\varphi + \varepsilon_\star(r, \lambda, \varphi)\right] \tag{2.51}$$

where $\varepsilon_2 = \Omega^2 a_e^3/(G_N M_E) \approx 3.47 \times 10^{-3}$ is a dimensionless constant. The combined force in the equation of motion is thus $-\rho\boldsymbol{\Omega} \times (\boldsymbol{\Omega} \times \boldsymbol{X}) - \rho\nabla\Phi_E = -\rho\nabla\Phi$. The geometric condition of the gravitational and geopotential surfaces with associated accelerations is depicted in Figure 2.8.

Surfaces of constant geopotential $r_g = r_g(\lambda, \varphi)$ are defined by $\Phi(r_g, \lambda, \varphi) =$ const. With (2.51), and neglecting the residual $\varepsilon_\star(r, \lambda, \varphi)$, one has

$$-G_N \frac{M_E}{r_g}\left[1 - \varepsilon_1\left(\frac{a_e}{r_g}\right)^2\left(3\sin^2\varphi - 1\right) + \frac{1}{2}\varepsilon_2\left(\frac{r_g}{a_e}\right)^3 \cos^2\varphi\right] = \text{const} \tag{2.52}$$

The geopotential surface corresponding to the mean sea level is called the *geoid*. It can be derived from (2.52) to a first approximation by replacing $r_g \approx a_e$ in the small terms, with the result

$$r_g \approx a_e\left(1 - \varepsilon_3 \sin^2\varphi\right) \tag{2.53}$$

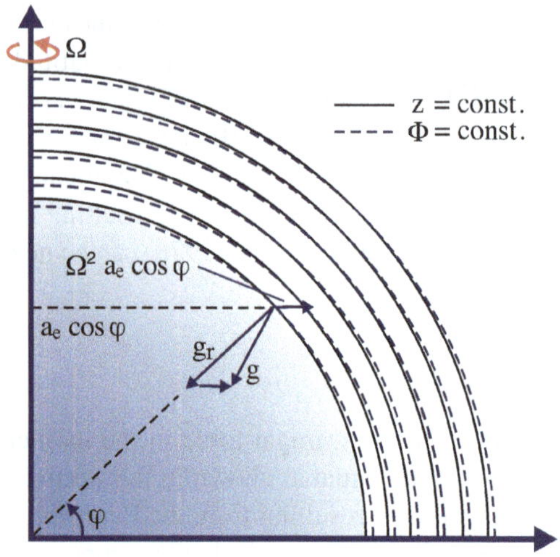


Fig. 2.8 Gravity as sum of the Earth attraction and the centrifugal acceleration. Surfaces of equal height (solid lines) and equal geopotential (dashed lines) are schematically displayed, i. e. the difference between both is exaggerated

with $\varepsilon_3 \approx 3\varepsilon_1 + \varepsilon_2/2 = 3.36 \times 10^{-3}$. In this approximation, the geoid hence has the shape of an oblate ellipsoid: its height increases from the pole (polar radius $a_\mathrm{p} = 6356.8\,\mathrm{km}$) to the equator by $\varepsilon_3 a_\mathrm{e} \approx 21.4\,\mathrm{km}$ as demonstrated in Figure 2.8.

Deviations of the geoid from the reference ellipsoid (shown in Figure 2.9) are related to the residual term $\varepsilon_\star(r, \lambda, \varphi)$ in (2.51). These deviations reflect the non-spherical mass distribution in the Earth, and have the form of irregular undulations with a maximum amplitude of about 100 m. The exact form of the geoid is of great importance in various contexts, in particular in connection with satellite altimetry.

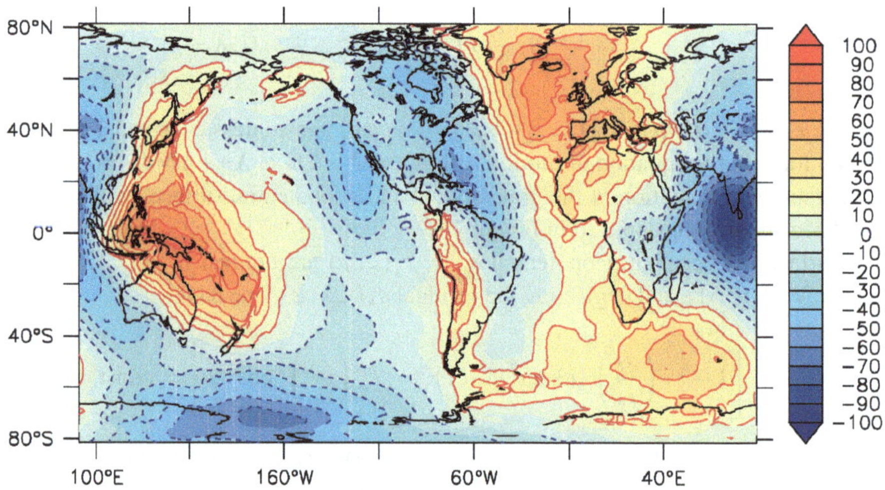


Fig. 2.9 Deviation of the geoid from the reference ellipsoid, related to the residual term $\varepsilon_\star(r, \lambda, \varphi)$ in (2.50) (in m). Data are taken from the official Earth Gravitational Model EGM2008 released by the US National Geospatial-Intelligence Agency. Black lines denote the continental shape

The gravitational acceleration (in short *gravity*) $\boldsymbol{g} = -\nabla\Phi$ is perpendicular to the geopotential surfaces. It follows from (2.52) that $\boldsymbol{g}$ is very nearly parallel to the radius vector $\boldsymbol{r}$ (the ratio of horizontal to vertical component is at most $\varepsilon_3$). With (2.51) and (2.53), the magnitude of gravity $g = |\nabla\Phi| \approx \partial\Phi/\partial r$ at the Earth's surface $r = r_g(\varphi)$ is hence approximately given by

$$g = \frac{G_N M_E}{a_e^2\left(1 - \varepsilon_3 \sin^2\varphi\right)}\left[1 - 3\varepsilon_1\left(3\sin^2\varphi - 1\right) - \varepsilon_2\cos^2\varphi\right] \approx g_e\left(1 + \varepsilon_4\sin^2\varphi\right) \tag{2.54}$$

with $g_e = (G_N M_E/a_e^2)(1 + 3\varepsilon_1 - \varepsilon_2)$ and $\varepsilon_4 = 2\varepsilon_2 - 3\varepsilon_1 \approx 5.3 \times 10^{-3}$. The gravitational acceleration is thus not constant on geopotential surfaces but increases from the equatorial value $g_e \approx 9.78\,\mathrm{m\,s^{-2}}$ to $g_e(1+\varepsilon_4) \approx 9.83\,\mathrm{m\,s^{-2}}$ at the poles. It is customary to ignore these changes and to apply a constant value of $g = 9.81\,\mathrm{m\,s^{-2}}$, causing a relative error of $3 \times 10^{-3}$.

The component of gravity in the horizontal directions (normal to $\boldsymbol{r}$) is at most around $\varepsilon_3 g \approx 0.03\,\mathrm{m\,s^{-2}}$. While this is very small compared to the vertical component, it is not small in comparison to other acceleration terms in the same (horizontal) direction. It is hence very convenient and useful to use a coordinate system which has $\Phi = \mathrm{const}$ as one coordinate surface. For orthogonal coordinates, gravity must thus coincide exactly with one coordinate direction, i. e. $\boldsymbol{g} = (0, 0, -g)$. The geopotential is then dependent on the vertical coordinate $z$. Referring the potential to the mean surface, i. e. $\Phi(z = 0) = 0$, we have

$$\Phi(z) = \int_0^z \mathrm{d}z' g(z') \approx gz \tag{2.55}$$

The geopotential is thus the work which must be applied to lift a unit mass from $z = 0$ to the height $z$.

### Tidal Potential

By far the largest part of the attraction from the Moon and the Sun is compensated by the centrifugal force arising from the orbital motion. The compensation is not complete, however, and a residual *tidal force* remains. As example let us consider the tidal potential of the Moon. The distance between the centers of mass of Earth and Moon is $r_m = 384{,}000\,\mathrm{km}$, approximately 60 times the Earth's radius, and the mass ratio is $M_E/M_m \approx 81.5$. To a good approximation, both bodies rotate on circular orbits around their joint center of mass which is located at a distance $s = r_m M_m/(M_E + M_m) \approx 4{,}600\,\mathrm{km}$ from the Earth's center of mass, still within the Earth. It is sufficient to consider the Earth as a sphere with radius $a = 6{,}371\,\mathrm{km}$.

The motion of the Earth around the joint center of mass is a revolution without rotation rather than a rigid rotation, i. e. each mass point within the Earth follows a circle with the same radius $s$ but a different origin, as shown in Figure 2.10. Hence the centrifugal force $\omega^2 \boldsymbol{s}$ is identical for all points. In order for the Earth–Moon system to remain in equilibrium, mass attraction by the Moon and centrifugal force must compensate exactly in the Earth's center of mass,

$$\omega^2\boldsymbol{s} - G_N\frac{M_m}{r_m^2}\frac{\boldsymbol{r}_m}{r_m} = 0 \tag{2.56}$$

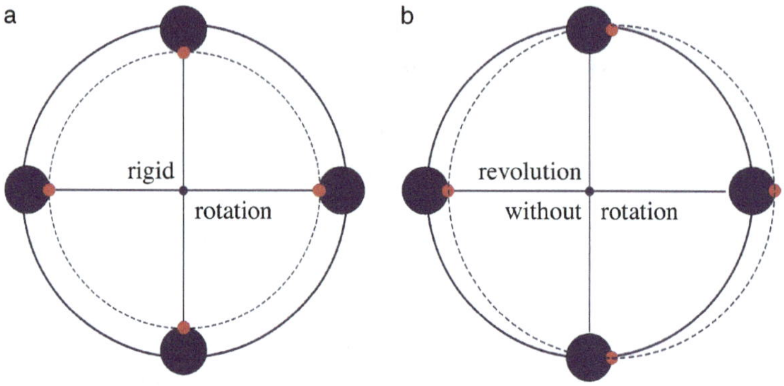


Fig. 2.10 **a** shows a rigid rotation of the large blue body, where all points (red dot) on the body move on circles with a common center but different radii. For a revolution without rotation, shown in **b**, the points rotate on circles with identical radii but different origin

At any point $\boldsymbol{r}$ which is in a distance $r_\star$ from the Moon so that $\boldsymbol{r}_\mathrm{m} + \boldsymbol{r} = \boldsymbol{r}_\star$ (see Figure 2.11), the total force per mass (i. e. the acceleration) is thus

$$\boldsymbol{f} = -G_\mathrm{N}\frac{M_\mathrm{m}}{r_\star^2}\frac{\boldsymbol{r}_\star}{r_\star} + \omega^2\boldsymbol{s} = -G_\mathrm{N}M_\mathrm{m}\left(\frac{\boldsymbol{r}_\star}{r_\star^3} - \frac{\boldsymbol{r}_\mathrm{m}}{r_\mathrm{m}^3}\right)$$

where (2.56) has been used for the second form. The acceleration can be expressed as gradient of a potential, i. e. $\boldsymbol{f} = -\nabla\Phi_\mathrm{m}$, with

$$\Phi_\mathrm{m} = -G_\mathrm{N}M_\mathrm{m}\left(\frac{1}{r_\star} - \frac{1}{r_\mathrm{m}} - \frac{r}{r_\mathrm{m}^2}\cos\vartheta\right) \tag{2.57}$$

For convenience, the constant $G_\mathrm{N}M_\mathrm{m}/r_\mathrm{m}$ has been added here. For the distance $r_\star$ it follows geometrically (law of cosines, see Figure 2.11) that $r_\star^2 = r_\mathrm{m}^2 + r^2 - 2rr_\mathrm{m}\cos\vartheta$ (here $\vartheta$ is the angle between $\boldsymbol{r}$ and $\boldsymbol{r}_\mathrm{m}$). As $r/r_\mathrm{m} \approx 0.017$ at the Earth's surface, a Taylor expansion for $r/r_\mathrm{m} \ll 1$ yields

$$\frac{r_\mathrm{m}}{r_\star} = \left(1 - 2\frac{r}{r_\mathrm{m}}\cos\vartheta + \frac{r^2}{r_\mathrm{m}^2}\right)^{-\frac{1}{2}} = 1 + \frac{r}{r_\mathrm{m}}\cos\vartheta - \frac{1}{2}\frac{r^2}{r_\mathrm{m}^2} + \frac{3}{2}\frac{r^2}{r_\mathrm{m}^2}\cos^2\vartheta + o\left(\frac{r^3}{r_\mathrm{m}^3}\right)$$

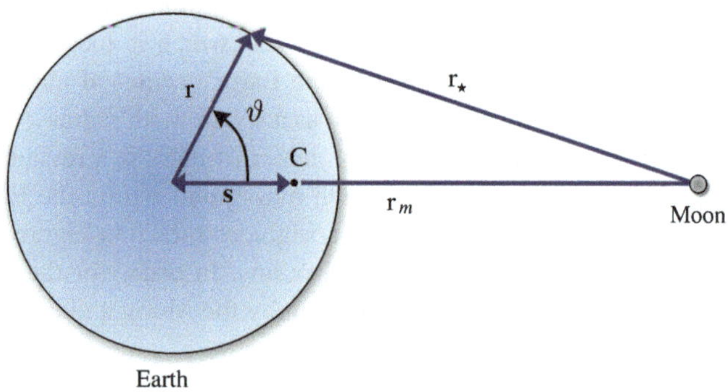


Fig. 2.11 Configuration of the Earth–Moon system. $C$ is the center of mass of the two bodies

Hence the Moon's potential follows from (2.57), up to second order in $r/r_{\mathrm{m}}$, as

$$\Phi_{\mathrm{m}}(r,\vartheta) = -G_{\mathrm{N}} M_{\mathrm{m}} \frac{r^2}{r_{\mathrm{m}}^3} \frac{1}{2} \left(3\cos^2\vartheta - 1\right) \tag{2.58}$$

The magnitude of the tidal force $F^{\mathrm{r}} = \partial\Phi_{\mathrm{m}}/\partial r$ in the radial direction compared to the gravity force at the Earth's surface is very small,

$$\frac{F^{\mathrm{r}}}{g^{\star}} = \frac{G_{\mathrm{N}} M_{\mathrm{m}} a / r_{\mathrm{m}}^3}{G_{\mathrm{N}} M_{\mathrm{E}} / a^2} = \frac{M_{\mathrm{m}}}{M_{\mathrm{E}}} \left(\frac{a}{r_{\mathrm{m}}}\right)^3 \approx 10^{-7}$$

The vertical component of the tidal force hence is always negligible. For the horizontal component it suffices to set $r = a$ in (2.58).

The angle $\vartheta$ depends on the astronomical variables describing the Moon's orbit relative to the position $\boldsymbol{r}$, and can be expressed as a function of geographic longitude and latitude as well as time, $\vartheta = \vartheta(\lambda, \varphi, t)$. The details are not given here, see e. g. Pugh (1987). However, it is easy to see that because of the Earth rotation and the rotation of the Moon, $\vartheta$ will contain a near-daily and a near-monthly period, and that the potential will be dominated by approximately half-daily and half-monthly periods (note that due to the relation $\cos^2\vartheta = (1 + \cos 2\vartheta)/2$ the forcing period in (2.58) is halved). Such periods can in fact be seen in many tidal records.

For the Sun the situation is completely analogous. Both potentials are of nearly comparable magnitude, $\Phi_{\mathrm{s}}/\Phi_{\mathrm{m}} \sim (M_{\mathrm{s}}/M_{\mathrm{m}}) \cdot (r_{\mathrm{m}}/r_{\mathrm{s}})^3 \approx 0.45$. The sum of the potentials of the Moon and the Sun, $\Phi_{\mathrm{tide}}(\lambda, \varphi, t) = \Phi_{\mathrm{m}} + \Phi_{\mathrm{s}}$, is the full tidal potential, and $\Phi + \Phi_{\mathrm{tide}}$ combines the attraction by the Earth, Moon, and Sun and includes the effect of the centrifugal force. With this, the momentum budget finally takes the form

$$\rho \frac{\mathrm{D}\boldsymbol{u}}{\mathrm{D}t} = -2\rho\boldsymbol{\Omega} \times \boldsymbol{u} - \nabla p + \nabla \cdot \Sigma - \rho\nabla\left(\Phi + \Phi_{\mathrm{tide}}\right) \tag{2.59}$$

Remember that the geopotential $\Phi$ can be considered time-independent whereas the tidal potential $\Phi_{\mathrm{tide}}$ is time-varying.

In the remainder of this book, tidal phenomena will not be considered, except for the energy budgets in Section 2.4 where tidal forcing will be included by redefining $\Phi \to \Phi + \Phi_{\mathrm{tide}}$.

## 2.4 Energy Conservation

In this section, we look at various forms of energy in a moving fluid and the corresponding energy budgets. In Section 1.2, the internal energy $E$ has been introduced which contains, among other contributions, the kinetic energy of molecular motion. In a moving fluid, one has to include the kinetic energy of the *macroscopic* hydrodynamic motion, described in terms of the flow velocity $\boldsymbol{u}$, as there may be exchanges between both forms of energy. Motion in the presence of gravity readily leads to the concept of potential energy. Finally, a budget combining all relevant forms of energy is presented. All energy budgets are based on the first law of thermodynamics discussed in Section 1.2 and on the equations discussed in the previous sections of this chapter.

### 2.4.1 Contributions to the Change of Energy in a Material Volume

The kinetic energy of a parcel with mass $m$ moving with the flow velocity $\boldsymbol{u}$ is $m\boldsymbol{u}^2/2$. Hence the sum of both internal and macroscopic kinetic energies per mass is $E + \boldsymbol{u}^2/2$. In analogy to the first law of thermodynamics (1.22) we formulate the conservation for the energy $\int_V \rho(E + \boldsymbol{u}^2/2)\mathrm{d}V$ in a *material* volume moving with the flow velocity. Note that we could also consider a volume fixed in space as done for the general conservation law in Section 2.1 (for which we would have to include an advective flux of energy in addition to the contributions considered below in this section). We will also derive in the following Section 2.4.2 an individual budget for the macroscopic kinetic energy $\boldsymbol{u}^2/2$ from the momentum budget, which we then subtract from the budget for internal and kinetic energy, which finally yields a budget for the internal energy only. Note that we have to take this route, since it appears cumbersome to directly derive a budget for the internal energy from the first law of thermodynamics given by (1.23).

Changes of energy within the material volume can occur through external fluxes, work on the volume which has been discussed in Section 2.3, and change of the mass composition due to the different enthalpy of the constituents. The individual contributions to the energy budget, i. e. the respective change of energy per volume and time, are as follows:

- **External fluxes**
  Energy can be brought into or out of a volume through molecular heat conduction, resulting in a flux $\boldsymbol{J}_\mathrm{T}$ (in W m$^{-2}$), and also through radiation by the flux $\boldsymbol{J}_\mathrm{rad}$. The combined flux $\boldsymbol{J}_\mathrm{T} + \boldsymbol{J}_\mathrm{rad}$ will be referred to as heat flux. The energy flux per unit time through $\mathrm{d}\boldsymbol{A}$ is $(\boldsymbol{J}_\mathrm{T} + \boldsymbol{J}_\mathrm{rad}) \cdot \mathrm{d}\boldsymbol{A}$, and the transport out of the volume is

$$\oint_A (\boldsymbol{J}_\mathrm{T} + \boldsymbol{J}_\mathrm{rad}) \cdot \mathrm{d}\boldsymbol{A} \tag{2.60}$$

  where the surface element $\mathrm{d}\boldsymbol{A}$ is directed outward.
- **Work on the volume I: surface forces**
  As shown in Section 2.3.1, the net surface force on a surface element is $\Pi \cdot \mathrm{d}\boldsymbol{A}$. Within a time $\delta t$, the boundary of a material fluid element is displaced by $\boldsymbol{u}\,\delta t$. Hence the work on the volume per unit time is given by

$$\oint_A \boldsymbol{u} \cdot \Pi \cdot \mathrm{d}\boldsymbol{A} \tag{2.61}$$

- **Work on the volume II: volume forces**
  The force of gravity (per volume) is $-\rho\nabla\Phi$, the work per unit time hence is

$$-\int_V \rho\boldsymbol{u} \cdot \nabla\Phi \mathrm{d}V \tag{2.62}$$

- **Energy exchange through diffusion of partial masses**
  According to (1.23), a change of $\mathrm{d}S$ in salt concentration causes an energy change of $(\partial H/\partial S)\mathrm{d}S$. The diffusive flux $\boldsymbol{J}_\mathrm{S}$ transports the amount $\boldsymbol{J}_\mathrm{S} \cdot \mathrm{d}\boldsymbol{A}$ and hence the energy $(\partial H/\partial S)\boldsymbol{J}_\mathrm{S} \cdot \mathrm{d}\boldsymbol{A}$ through a surface element. The total energy

transport out of the volume is thus

$$\oint_A \frac{\partial H}{\partial S} \boldsymbol{J}_S \cdot \mathrm{d}\boldsymbol{A} = \oint_A \boldsymbol{J}_{\text{chem}} \cdot \mathrm{d}\boldsymbol{A} \tag{2.63}$$

where the flux of 'chemical' energy (also in W m$^{-2}$) is given by

$$\boldsymbol{J}_{\text{chem}} = \frac{\partial H}{\partial S} \boldsymbol{J}_S \tag{2.64}$$

Adding all terms of the energy budget and applying the Gaussian theorem, we obtain the local energy conservation in the form

$$\rho \frac{\mathrm{D}}{\mathrm{D}t}(E + E_k) = -\nabla \cdot \boldsymbol{J}^{\text{tot}} - \rho \boldsymbol{u} \cdot \nabla \Phi \tag{2.65}$$

with the macroscopic kinetic energy (per mass) $E_k = \boldsymbol{u}^2/2$. Here $\boldsymbol{J}^{\text{tot}}$ is given by

$$\boldsymbol{J}^{\text{tot}} = \boldsymbol{J}_T + \boldsymbol{J}_{\text{rad}} + \boldsymbol{J}_{\text{chem}} - \boldsymbol{u} \cdot \Pi \tag{2.66}$$

and denotes the sum of all nonadvective energy fluxes. The last contribution $-\boldsymbol{u} \cdot \Pi$ will be identified below as flux of mechanical energy. The flux form of (2.66) is

$$\frac{\partial}{\partial t} \rho (E + E_k) = -\nabla \cdot \left[\rho (E + E_k) \boldsymbol{u} + \boldsymbol{J}^{\text{tot}}\right] - \rho \boldsymbol{u} \cdot \nabla \Phi \tag{2.67}$$

The sum of internal and kinetic energy hence can be changed locally not only through a divergent flux of total energy, but also through the gravity force. In the ocean phase changes occur mainly at the surface (by freezing and melting of sea ice) and, therefore, are generally considered in boundary conditions only. Otherwise a local volume term, describing the conversion, would enter (2.67).

### 2.4.2 Mechanical Energy

On the rotating Earth we have found the momentum budget (2.59), written as

$$\rho \frac{\mathrm{D}\boldsymbol{u}}{\mathrm{D}t} + 2\rho \boldsymbol{\Omega} \times \boldsymbol{u} = \nabla \cdot \Pi - \rho \nabla \Phi \tag{2.68}$$

Scalar multiplication with $\boldsymbol{u}$ yields

$$\begin{aligned} \rho \frac{\mathrm{D}E_k}{\mathrm{D}t} &= \boldsymbol{u} \cdot (\nabla \cdot \Pi) - \rho \boldsymbol{u} \cdot \nabla \Phi \\ &= -\boldsymbol{u} \cdot \nabla p + \boldsymbol{u} \cdot (\nabla \cdot \Sigma) - \rho \boldsymbol{u} \cdot \nabla \Phi \end{aligned} \tag{2.69}$$

as budget for the macroscopic kinetic energy (per mass) $E_k = \boldsymbol{u}^2/2$. Note that the Coriolis term does not contribute. In the second relation we have used the splitting of the stress according to (2.34), revealing $-\boldsymbol{u} \cdot \nabla p$ as the amount of kinetic energy produced by the work of the pressure force, and $\boldsymbol{u} \cdot \nabla \cdot \Sigma$ as the amount of kinetic energy produced by viscous effects.

The geopotential $\Phi$ satisfies the (trivial) conservation equation

$$\rho \frac{\mathrm{D}\Phi}{\mathrm{D}t} = \rho \frac{\partial \Phi}{\partial t} + \rho \boldsymbol{u} \cdot \nabla \Phi \tag{2.70}$$

Here local changes can only be caused by tidal forces (cf. Section 2.3.4), i.e. $\partial\Phi/\partial t = \partial\Phi_{\text{tide}}/\partial t$. The advection term in (2.70) occurs, with opposite sign, also in (2.69). We identify the geopotential $\Phi$ with a *potential* energy; hence the advection term describes a conversion between potential and kinetic energy. It is common to use a coordinate system with $z$ normal to geopotential surfaces. Then, with $\boldsymbol{u} = (u, v, w)$, we obtain $\rho\boldsymbol{u} \cdot \nabla\Phi \equiv \rho g w$.

The first term in (2.69) can be split as $\boldsymbol{u} \cdot (\nabla \cdot \Pi) = \nabla \cdot (\boldsymbol{u} \cdot \Pi) - (\Pi \cdot \nabla) \cdot \boldsymbol{u}$ and the last term in this relation can be rewritten as

$$(\Pi \cdot \nabla) \cdot \boldsymbol{u} = \Pi_{ij}\frac{\partial u_i}{\partial x_j} = \Pi_{ij}D_{ij} = \left(-p\delta_{ij} + \Sigma_{ij}\right) D_{ij}$$

with the deformation tensor $D_{ij}$ as defined by (1.6). Separating $D_{ij} = D_{\ell\ell}\delta_{ij}/3 + D^*_{ij}$ into the deviator $D^*_{ij}$ and the remainder, we find $\Sigma_{ij}D_{ij} = \Sigma_{ij}D^*_{ij}$ because the frictional stress tensor has zero trace, $\Sigma_{ii} = 0$. With the definition

$$\rho\epsilon = \Sigma_{ij}D^*_{ij} \tag{2.71}$$

we then obtain $\boldsymbol{u}\cdot(\nabla\cdot\Pi) = \nabla\cdot(\boldsymbol{u}\cdot\Pi) + p\nabla\cdot\boldsymbol{u} - \rho\epsilon$. Inserting this result into (2.69) yields the kinetic energy conservation as

$$\rho\frac{\mathrm{D}E_{\mathrm{k}}}{\mathrm{D}t} = -\nabla \cdot (-\boldsymbol{u} \cdot \Pi) + p\nabla \cdot \boldsymbol{u} - \rho\epsilon - \rho\boldsymbol{u} \cdot \nabla\Phi \tag{2.72}$$

Addition of (2.69) and (2.70) yields the conservation of *mechanical* energy, i.e. the sum of kinetic and potential energies, as

$$\rho\frac{\mathrm{D}}{\mathrm{D}t}(E_{\mathrm{k}} + \Phi) = -\nabla \cdot (-\boldsymbol{u} \cdot \Pi) + p\nabla \cdot \boldsymbol{u} - \rho\epsilon + \rho\frac{\partial\Phi_{\text{tide}}}{\partial t} \tag{2.73}$$

The work terms for pressure and viscous forces is separated into a flux divergence and a remainder. The flux of mechanical energy is $-\boldsymbol{u} \cdot \Pi = p\boldsymbol{u} - \boldsymbol{u} \cdot \Sigma$. The role of the remaining terms will be discussed in the next section.

### 2.4.3 Internal Energy and Enthalpy

Subtracting the kinetic energy budget (2.72) from (2.65) leads to the conservation of internal energy,

$$\rho\frac{\mathrm{D}E}{\mathrm{D}t} = -\nabla \cdot \boldsymbol{J}_{\mathrm{H}} - p\nabla \cdot \boldsymbol{u} + \rho\epsilon \tag{2.74}$$

Here the flux $\boldsymbol{J}_{\mathrm{H}}$ of internal energy (and also of enthalpy, see below) is given as

$$\boldsymbol{J}_{\mathrm{H}} = \boldsymbol{J}_{\mathrm{T}} + \boldsymbol{J}_{\text{rad}} + \boldsymbol{J}_{\text{chem}} \tag{2.75}$$

Two source/sink terms in (2.74) appear which constitute a conversion between internal and mechanical energy because they appear with opposite sign in the mechanical energy budget (2.73). The term $\rho\epsilon$ is the rate of *energy dissipation* (also known as Joule heating) (per volume, in $\mathrm{kg\,m^{-1}\,s^{-3}}$). As will be shown in Section 2.5.1, $\epsilon$ is always positive, and hence constitutes an *irreversible* transformation of mechanical into internal energy caused by friction. The term $p\nabla \cdot \boldsymbol{u}$ is the amount of internal energy converted to mechanical energy, which occurs also in reversible systems, and is usually denoted the *reversible exchange*. This interpretation is, however, strictly correct only in the absence of heat conduction and diffusion.

It is often convenient to formulate (2.74) in terms of the enthalpy $H = E + pv$. Using the mass conservation (2.18) one obtains

$$\rho \frac{\mathrm{D}H}{\mathrm{D}t} = \frac{\mathrm{D}p}{\mathrm{D}t} - \nabla \cdot \boldsymbol{J}_{\mathrm{H}} + \rho\epsilon \tag{2.76}$$

We have discussed $\rho\boldsymbol{u} \cdot \nabla\Phi \equiv \rho g w$ as exchange term between kinetic and potential energy. Note that the reversible term in (2.76) is approximately of the same form, $\mathrm{D}p/\mathrm{D}t \approx w\partial p/\partial z \approx -\rho w g$, for hydrostatic conditions.

### 2.4.4 Total Energy and Total Enthalpy

The sum of internal, kinetic and potential energy

$$E^{\mathrm{tot}} = E + E_{\mathrm{k}} + \Phi \tag{2.77}$$

is referred to as *total energy*. Adding (2.65) and (2.70) yields

$$\rho \frac{\mathrm{D}E^{\mathrm{tot}}}{\mathrm{D}t} = -\nabla \cdot \boldsymbol{J}^{\mathrm{tot}} + \rho \frac{\partial \Phi_{\mathrm{tide}}}{\partial t} \tag{2.78}$$

where the total energy flux is given by (2.66). The flux form of (2.78) is

$$\frac{\partial}{\partial t}\rho E^{\mathrm{tot}} = -\nabla \cdot \left(\rho E^{\mathrm{tot}}\boldsymbol{u} + \boldsymbol{J}^{\mathrm{tot}}\right) + \rho \frac{\partial \Phi_{\mathrm{tide}}}{\partial t} \tag{2.79}$$

Only the fluxes $\boldsymbol{J}^{\mathrm{tot}}$ at the boundaries change the integrated total energy, otherwise the fluxes can only redistribute this quantity. In the diagram of Figure 2.12 and the box on p. 52 we summarize the energy compartments with the exchange terms and external fluxes.

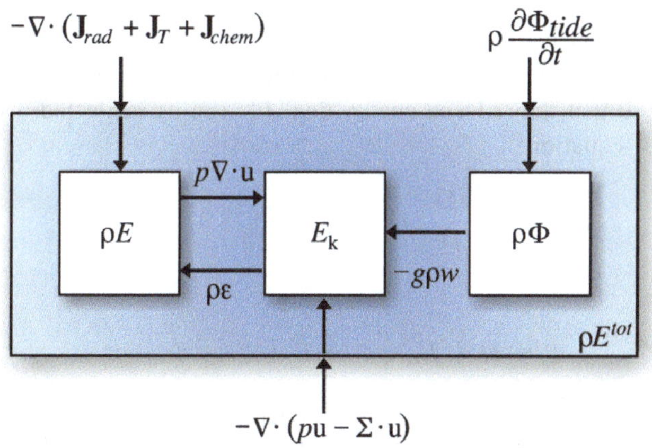


Fig. 2.12 Local source terms for the various energy compartments

**9. Fluxes and Source Terms in Energy Budgets**

| Variable | *Symbol* | Flux | Source/sink (per volume) |
|---|---|---|---|
| Pot. energy | $\Phi$ | 0 | $g\rho w + \rho \frac{\partial \Phi_{\text{tide}}}{\partial t}$ |
| Kin. energy | $E_{\text{k}}$ | $\boldsymbol{J}_{\text{mech}} = p\boldsymbol{u} - \Sigma \cdot \boldsymbol{u}$ | $-g\rho w - \rho\epsilon + p\boldsymbol{\nabla} \cdot \boldsymbol{u}$ |
| Mech. energy | $E_{\text{k}} + \Phi$ | $\boldsymbol{J}_{\text{mech}} = p\boldsymbol{u} - \Sigma \cdot \boldsymbol{u}$ | $-\rho\epsilon + p\boldsymbol{\nabla} \cdot \boldsymbol{u} + \frac{\partial \Phi_{\text{tide}}}{\partial t}$ |
| Int. energy | $E$ | $\boldsymbol{J}_{\text{H}} = \boldsymbol{J}_{\text{rad}} + \boldsymbol{J}_{\text{T}} + \frac{\partial H}{\partial S} \boldsymbol{J}_{\text{S}}$ | $\rho\epsilon - p\boldsymbol{\nabla} \cdot \boldsymbol{u}$ |
| Enthalpy | $H$ | $\boldsymbol{J}_{\text{H}} = \boldsymbol{J}_{\text{rad}} + \boldsymbol{J}_{\text{T}} + \frac{\partial H}{\partial S} \boldsymbol{J}_{\text{S}}$ | $\rho\epsilon + \frac{\text{D}p}{\text{D}t}$ |
| Total energy | $E^{\text{tot}} = E + E_{\text{k}} + \Phi$ | $\boldsymbol{J}^{\text{tot}} = \boldsymbol{J}_{\text{H}} + \boldsymbol{J}_{\text{mech}}$ | $\rho \frac{\partial \Phi_{\text{tide}}}{\partial t}$ |
| Total enthalpy | $H + E_{\text{k}} + \Phi$ | $\boldsymbol{J}^{\text{tot}} = \boldsymbol{J}_{\text{H}} - \Sigma \cdot \boldsymbol{u}$ | $\frac{\partial p}{\partial t} + \rho \frac{\partial \Phi_{\text{tide}}}{\partial t}$ |

Decomposing the mechanical energy flux in (2.66) according to $\boldsymbol{u} \cdot \Pi = -p\boldsymbol{u} + \boldsymbol{u} \cdot \Sigma$, the total energy flux may be represented as

$$\boldsymbol{J}^{\text{tot}} = p\boldsymbol{u} - \boldsymbol{u} \cdot \Sigma + \boldsymbol{J}_{\text{H}} \tag{2.80}$$

Invoking the identity

$$\rho \frac{\text{D}}{\text{D}t}(pv) = \frac{\partial p}{\partial t} + \nabla \cdot p\boldsymbol{u}$$

which can be derived by means of the continuity equation, one obtains from (2.78)

$$\rho \frac{\text{D}}{\text{D}t}(H + E_{\text{k}} + \Phi) = -\boldsymbol{\nabla} \cdot \boldsymbol{J}_{*}^{\text{tot}} + \frac{\partial p}{\partial t} + \rho \frac{\partial \Phi_{\text{tide}}}{\partial t} \tag{2.81}$$

where the reduced energy flux $\boldsymbol{J}_{*}^{\text{tot}} = \boldsymbol{J}^{\text{tot}} - p\boldsymbol{u} = -\boldsymbol{u} \cdot \Sigma + \boldsymbol{J}_{\text{H}}$ contains the fluxes due to molecular friction, diffusion, heat conduction and radiation. In many cases, especially when considering large-scale motions, the pressure is mainly determined by depth, and its local rate of change is negligible. If, in addition, the conditions are adiabatic so that the reduced energy flux $\boldsymbol{J}_{*}^{\text{tot}}$ can be neglected, we obtain the BERNOULLI[11]-equation

$$\frac{\text{D}}{\text{D}t}(H + E_{\text{k}} + \Phi) \approx 0 \tag{2.82}$$

The quantity

$$B = H + E_{\text{k}} + \Phi = E + E_{\text{k}} + \Phi + pv \tag{2.83}$$

[11] JOHANN BERNOULLI, *1667 in Basel, †1748 in Basel, mathematician, and his son DANIEL BERNOULLI, *1700 in Groningen, †1782 in Basel, are the most prominent members of the scientist family Bernoulli. The above theorem is due to Daniel B. whose most famous work 'Hydrodynamics' appeared in 1738.

is the Bernoulli function in its most general form. Under the above-mentioned conditions, the Bernoulli function of a fluid element is approximately constant. In such situations (2.82) can be more convenient than energy conservation. The Bernoulli function can also be viewed as total enthalpy.

### 2.4.5 Boundary Condition for the Enthalpy Flux

A boundary condition is conveniently formulated in terms of the enthalpy $H$ of seawater because boundary processes take place at (approximately) constant pressure. From the conservation equation (2.76) it follows that the appropriate flux of enthalpy is $\boldsymbol{J}_{\mathrm{H}}$ which is defined in (2.75) as sum of (sensible) heat flux, radiative heat flux and a flux related to the chemical potential (the latter will later be shown to be small, cf. Section 2.6). Application of (2.7), with $\chi = H$ and $\boldsymbol{J}_\chi = \boldsymbol{J}_{\mathrm{H}}$, yields for the ocean-atmosphere boundary

$$H \mathcal{J}_{\mathrm{mass}} + \boldsymbol{J}_{\mathrm{H}} \cdot \boldsymbol{n} = \mathcal{J}_{\mathrm{H}}^{\mathrm{atm}} \tag{2.84}$$

**10. Alternative Derivation of the Budget for Internal Energy**

An alternative derivation of the budget for internal energy can be found using the Gibbs relation (1.47), for which the differentials $\mathrm{d}E$, $\mathrm{d}\eta$, etc are written as material derivatives $\mathrm{D}E/\mathrm{D}t$, $\mathrm{D}\eta/\mathrm{D}t$, etc.

$$\frac{\mathrm{D}E}{\mathrm{D}t} = \mu \frac{\mathrm{D}S}{\mathrm{D}t} + T\frac{\mathrm{D}\eta}{\mathrm{D}t} + p/\rho^2 \frac{\mathrm{D}\rho}{\mathrm{D}t}$$

where $-p\mathrm{d}v = p/\rho^2 \mathrm{d}\rho$ was used. Combing this particle form of the Gibbs relation with the budgets for salt and total mass and assuming the standard form for the entropy budget with a nonadvective flux of entropy $\boldsymbol{J}_\eta$ and a source or sink of entropy $C_\eta$ given by $\rho \mathrm{D}\eta/\mathrm{D}t = -\nabla \cdot \boldsymbol{J}_\eta + C_\eta$ yields

$$\rho \frac{\mathrm{D}E}{\mathrm{D}t} = -\mu \nabla \cdot \boldsymbol{J}_{\mathrm{S}} - T \nabla \cdot \boldsymbol{J}_\eta + T C_\eta - p \nabla \cdot \boldsymbol{u}$$

We rewrite this budget for the internal energy again in the standard form including a nonadvective flux and a source or sink term of internal energy, as

$$\rho \frac{\mathrm{D}E}{\mathrm{D}t} = -\nabla \cdot (\mu \boldsymbol{J}_{\mathrm{S}} + T \boldsymbol{J}_\eta) + \boldsymbol{J}_{\mathrm{S}} \cdot \nabla \mu + \boldsymbol{J}_\eta \cdot \nabla T + T C_\eta - p \nabla \cdot \boldsymbol{u}$$

Combining this budget for internal energy with the kinetic and potential energy budgets yields as budget for total energy

$$\rho \frac{\partial}{\partial t}(E + E_{\mathrm{k}} + \Phi) = -\nabla \cdot (\mu \boldsymbol{J}_{\mathrm{S}} + T \boldsymbol{J}_\eta - \boldsymbol{u} \cdot \Pi) + \boldsymbol{J}_{\mathrm{S}} \cdot \nabla \mu + \boldsymbol{J}_\eta \cdot \nabla T + T C_\eta - \rho \epsilon + \rho \frac{\partial \Phi}{\partial t}$$

Now we assume that the total energy is conserved (except for tidal forcing). It follows for the source term of entropy that

$$T C_\eta = -\boldsymbol{J}_{\mathrm{S}} \cdot \nabla \mu - \boldsymbol{J}_\eta \cdot \nabla T + \rho \epsilon$$

must hold, which is identical to (2.91). From the internal energy budget, we also see that the enthalpy flux is given by $\boldsymbol{J}_{\mathrm{H}} = \mu \boldsymbol{J}_{\mathrm{S}} + T \boldsymbol{J}_\eta$. With $\partial H/\partial S = \mu + T \partial \eta/\partial S$ and the entropy flux from (2.90) we recover the form (2.75) for the enthalpy flux.

Here $\mathcal{J}_{\mathrm{H}}^{\mathrm{atm}}$ is the net enthalpy flux and $\mathcal{J}_{\mathrm{mass}} = \mathcal{E} - \mathcal{P}$ the net mass flux, both normal to the surface and pointing into the atmosphere. On the atmospheric side of the boundary, water is transported as water vapor (with enthalpy $H_{\mathrm{q}}$) due to evaporation, and as liquid water (with enthalpy $H_{\mathrm{p}}$) due to precipitation. Therefore, (2.84) can be written as $H(\mathcal{E} - \mathcal{P}) + \boldsymbol{J}_{\mathrm{H}} \cdot \boldsymbol{n} = H_{\mathrm{q}}\mathcal{E} - H_{\mathrm{p}}\mathcal{P} + \boldsymbol{J}_{\mathrm{H}}^{a} \cdot \boldsymbol{n}$ where $\boldsymbol{J}_{\mathrm{H}}^{a}$ is the atmospheric sensible plus radiative heat flux. The *latent heat* of condensation $L_{\mathrm{q}}$ is defined as

$$L_{\mathrm{q}} = H_{\mathrm{q}} - H_{\mathrm{f}} \tag{2.85}$$

and has the value of $L_{\mathrm{q}} \approx 2.5 \times 10^6\,\mathrm{J\,kg^{-1}}$ at 0°C and atmospheric standard pressure. Here $H_{\mathrm{f}}$ is the enthalpy of pure water. Since with (2.85) one has $H_{\mathrm{q}} - H = L_{\mathrm{q}} + H_{\mathrm{f}} - H$, the boundary condition finally takes the form

$$\boldsymbol{J}_{\mathrm{H}} \cdot \boldsymbol{n} = \boldsymbol{J}_{\mathrm{H}}^{a} \cdot \boldsymbol{n} + L_{\mathrm{q}}\mathcal{E} + (H_{\mathrm{f}} - H)\mathcal{E} - (H_{\mathrm{p}} - H)\mathcal{P} \tag{2.86}$$

Sensible and radiative heat fluxes on both sides of the boundary hence differ by three terms associated with the air-sea mass flux, of which the first term (latent heat flux) is by far the most important. The enthalpy differences between pure water and sea water, and between precipitating water and sea water, are much smaller so that the last two terms in (2.86) can usually be ignored.

The exchange of enthalpy between ocean and atmosphere is an important driving mechanism for the ocean circulation. We continue with a detailed discussion in Chapter 13. The exchange at the sea floor is comparatively small, and usually ignored, i. e. $\boldsymbol{J}_{\mathrm{H}} \cdot \boldsymbol{n} = 0$ is assumed. Locally it may however be necessary to include geothermal heating at places such as hydrothermal vents. Furthermore, geothermal heating may play a certain role for the large-scale circulation over long time scales (see e.g. Adcroft et al., 2001; Emile-Geay and Madec, 2009).

## 2.5 Entropy Budget

The entropy $\eta(S, T, p)$ can be computed numerically as a derivative of the known Gibbs function according to (1.49). Therefore, a consideration of the entropy budget is not in itself necessary to obtain a closed system of equations which allows determination of all state variables. The second law, however, allows conclusions on the direction of frictional/diffusive molecular fluxes.

### 2.5.1 Entropy Sources and Flux–Gradient Relations

Starting from the Gibbs relation (1.28) in the form

$$T\frac{\mathrm{D}\eta}{\mathrm{D}t} = \frac{\mathrm{D}H}{\mathrm{D}t} - \upsilon\frac{\mathrm{D}p}{\mathrm{D}t} - \mu\frac{\mathrm{D}S}{\mathrm{D}t} \tag{2.87}$$

one obtains with (2.76) and (2.20)

$$\rho\frac{\mathrm{D}\eta}{\mathrm{D}t} = \frac{1}{T}\left[-\nabla\cdot\left(\boldsymbol{J}_{\mathrm{rad}} + \boldsymbol{J}_{\mathrm{T}} + \frac{\partial H}{\partial S}\boldsymbol{J}_{\mathrm{S}}\right) + \rho\epsilon + \mu\nabla\cdot\boldsymbol{J}_{\mathrm{S}}\right] \tag{2.88}$$

After some algebra (expansion of $\nabla\mu$, and use of (1.50)) one obtains the entropy conservation in the standard form

$$\rho\frac{\mathrm{D}\eta}{\mathrm{D}t} = -\nabla\cdot\boldsymbol{J}_\eta + C_\eta \tag{2.89}$$

where the entropy flux is defined by

$$\boldsymbol{J}_\eta = \frac{\boldsymbol{J}_{\mathrm{rad}} + \boldsymbol{J}_{\mathrm{T}}}{T} + \frac{\partial\eta}{\partial S}\boldsymbol{J}_{\mathrm{S}} \tag{2.90}$$

The entropy source in (2.89) consists of external and internal contributions, $C_\eta = C_\eta^{\mathrm{ext}} + C_\eta^{\mathrm{int}}$. The external source

$$C_\eta^{\mathrm{ext}} = -\frac{\boldsymbol{J}_{\mathrm{rad}}}{T^2}\cdot\nabla T$$

is due to radiation, and the internal sources, summarized by

$$C_\eta^{\mathrm{int}} = \frac{1}{T}\rho\epsilon - \frac{\boldsymbol{J}_{\mathrm{T}}}{T^2}\cdot\nabla T - \frac{\boldsymbol{J}_{\mathrm{S}}}{T}\cdot\left(\frac{\partial\mu}{\partial S}\nabla S + \frac{\partial\mu}{\partial p}\nabla p\right) \tag{2.91}$$

are due to a suite of irreversible processes within the fluid. The second law of thermodynamics requires the internal entropy source to be positive, $C_\eta^{\mathrm{int}} > 0$. This requirement is certainly satisfied when *all* individual contributions in (2.91) are positive, i. e.

$$\epsilon > 0\,, \quad -\boldsymbol{J}_{\mathrm{T}}\cdot\nabla T > 0\,, \quad -\boldsymbol{J}_{\mathrm{S}}\cdot\left(\frac{\partial\mu}{\partial S}\nabla S + \frac{\partial\mu}{\partial p}\nabla p\right) > 0 \tag{2.92}$$

These inequalities lead to the following restrictions for the direction of molecular transports, as discussed in the following.

- **Mechanical dissipation**
  From (2.71), the definition of the viscous stress tensor (2.36), and the stress–strain relation (2.35) one finds

$$\rho\epsilon = \Sigma_{ij}D_{ij}^\star = 2\nu D_{ij}^\star D_{ij}^\star = \frac{1}{2}\nu\sum_{i,j}\left(\frac{\partial u_i}{\partial x_j} + \frac{\partial u_j}{\partial x_i} - \frac{2}{3}\frac{\partial u_\ell}{\partial x_\ell}\delta_{ij}\right)^2 \tag{2.93}$$

  Hence the *energy dissipation rate* $\epsilon$ (per mass, in $\mathrm{m^2\,s^{-3}}$) is positive if the molecular viscosity $\nu$ is positive. Instead of $\nu$ one frequently uses the *kinematical* viscosity $\kappa_{\mathrm{m}} = \nu/\rho$. For water at 10°C temperature, the kinematic viscosity is $\kappa_{\mathrm{m}} \approx 1.3\times10^{-6}\,\mathrm{m^2\,s^{-1}}$. It decreases with temperature and varies by ±25% under oceanic conditions.
- **Heat Flux**
  For molecular heat conduction FOURIER'S[12] law states that the diffusive heat flux is opposite to the gradient of temperature,

$$\boldsymbol{J}_{\mathrm{T}} = -\lambda\nabla T \tag{2.94}$$

  with the heat conductivity $\lambda$. Hence we have $-\boldsymbol{J}_{\mathrm{T}}\cdot\nabla T > 0$ provided that $\lambda$ is positive, i. e. that the heat flux is directed from warm to cold regions. Normally, instead of $\lambda$, the temperature conductivity $\kappa_{\mathrm{T}} = \lambda/\rho c_p$ is preferred. In the ocean, typical values are $\kappa_{\mathrm{T}} \approx 1.4\times10^{-7}\,\mathrm{m^2\,s^{-1}}$.

[12] JEAN BAPTIST JOSEPH FOURIER, *1768 near Auxerre, †1830 in Paris, mathematician and physicist.

- **Salinity Flux**
The FICKIAN[13] law relates the salinity flux to the gradient by

$$\boldsymbol{J}_{\mathrm{S}} \approx -\rho\kappa_{\mathrm{S}}\nabla S \tag{2.95}$$

However, from the last expression in (2.92) we see that (2.95) is *not* sufficient to guarantee positivity of the entropy production. Including the salinity flux due to the pressure gradient according to

$$\boldsymbol{J}_{\mathrm{S}} = -\rho\kappa_{\mathrm{S}}\left(\nabla S + \frac{\partial\mu/\partial p}{\partial\mu/\partial S}\nabla p\right) \tag{2.96}$$

leads, however, to a positive entropy production term provided that the salt diffusivity $\kappa_{\mathrm{S}}$ is positive. Typical values for seawater are $\kappa_{\mathrm{S}} \approx 1.2\times10^{-9}\,\mathrm{m}^2\,\mathrm{s}^{-1}$. A consequence of (2.96) is carried out in the box on p. 57.

### 2.5.2 Onsager Relations

A generalization of the phenomenological relations (2.94) and (2.96) follows from the irreversible thermodynamics developed by ONSAGER[14] in 1931 (Onsager, 1931a,b). The positivity of entropy production (2.91) can also be satisfied by the weaker condition that only the sum, rather than the individual contributions, be positive. We consider here only heat and salt fluxes, and obtain the requirement that

$$C_\eta^\star = -\frac{\boldsymbol{J}_{\mathrm{T}}}{T^2}\cdot\nabla T - \frac{\mu_{\mathrm{S}}\boldsymbol{J}_{\mathrm{S}}}{T}\cdot\left(\nabla S + \frac{\partial\mu/\partial p}{\partial\mu/\partial S}\nabla p\right) > 0$$

A more general linear and isotropic relation between the fluxes $\boldsymbol{J}_{\mathrm{T}}$ and $\boldsymbol{J}_{\mathrm{S}}$ and the vectors on the right-hand side of (2.94) and (2.96) can be written as

$$\boldsymbol{J}_{\mathrm{T}} = -\rho c_p\kappa_{\mathrm{T}}\nabla T - \rho\kappa_{\mathrm{TS}}\left(\nabla S + \frac{\partial\mu/\partial p}{\partial\mu/\partial S}\nabla p\right) \tag{2.97}$$

$$\boldsymbol{J}_{\mathrm{S}} = -\rho\kappa_{\mathrm{S}}\left(\nabla S + \frac{\partial\mu/\partial p}{\partial\mu/\partial S}\nabla p\right) - \rho\kappa_{\mathrm{ST}}\nabla T \tag{2.98}$$

The last term in (2.97) leads to a heat flux in the presence of a combined salinity/pressure gradient (DUFOUR[15]-effect), the last term in (2.98) to a salinity flux due to a temperature gradient (SORET[16]-effect). The entropy production due to heat and salt fluxes then follows as

$$\begin{aligned} C_\eta^\star/\rho = {} & \frac{c_p\kappa_{\mathrm{T}}}{T^2}(\nabla T)^2 + \frac{\kappa_{\mathrm{S}}}{T}\frac{\partial\mu}{\partial S}\left(\nabla S + \frac{\partial\mu/\partial p}{\partial\mu/\partial S}\nabla p\right)^2 \\ & + \left(\frac{\kappa_{\mathrm{TS}}}{T^2} + \frac{\kappa_{\mathrm{ST}}}{T}\frac{\partial\mu}{\partial S}\right)\nabla T\cdot\left(\nabla S + \frac{\partial\mu/\partial p}{\partial\mu/\partial S}\nabla p\right) \end{aligned} \tag{2.99}$$

[13] ADOLF FICK, *1829 in Kassel, †1901 in Blankenberge/Flanders. Interestingly, he was not a physicist but a physiologist. He formulated in 1855 the three Fickian diffusion laws.
[14] LARS ONSAGER, *1903 in Oslo, †1976 in Coral Gables/Florida, chemist and physicist.
[15] LOUIS DUFOUR, *1832 in Veytaux, †1892 in Lausanne, physicist.
[16] CHARLES SORET, *1854 in Geneva, †1904 in Geneva, physicist and chemist.

From arguments based on statistical mechanics, Onsager has deduced that the two off-diagonal coefficients must be related by *reciprocity relations* which for $\kappa_{\mathrm{TS}}$ and $\kappa_{\mathrm{ST}}$ read

$$\frac{\partial \mu}{\partial S}\kappa_{\mathrm{ST}} = \frac{\kappa_{\mathrm{TS}}}{T} \tag{2.100}$$

As a consequence, both additional terms contribute equally to entropy production. Positivity of (2.99) then requires that

$$\kappa_{\mathrm{T}} > 0\,, \quad \kappa_{\mathrm{S}} > 0\,, \quad \text{and} \quad |\kappa_{\mathrm{TS}}|^2 < \kappa_{\mathrm{T}}\kappa_{\mathrm{S}} c_p T \frac{\partial \mu}{\partial S}$$

More details are given in Kamenkovich (1977) where also the general case with inclusion of the mechanical dissipation is discussed. However, since experiments show that both $\kappa_{\mathrm{ST}}$, the Soret-coefficient, and $\kappa_{\mathrm{TS}}$, the Dufour coefficient, are very small for seawater, they are usually neglected.

## 2.6 Temperature Equations

With the conservation of internal energy (2.74) and with the relevant thermodynamic relations considered in Section 1.2, we now have all basic conservation laws which in principle are sufficient to determine the state variables $\boldsymbol{u}$, $\rho$, $S$, $p$, and $E$. In the following we use the enthalpy $H$ instead of $E$. The temperature $T$ which is e. g. needed to evaluate the state equation, could in principle be determined through inversion of the known thermodynamic relationship $H = H(S, T, p)$. The more familiar procedure is, however, to use that relationship to eliminate $H$ from the system of conservation laws, and to derive an equation directly for the temperature. This elimination is useful because (1) temperature is an *independent* state variable and more convenient than $H$ to evaluate thermodynamic expressions, (2) in contrast to internal energy or enthalpy, temperature can be easily observed, and (3) in this way one can achieve a reduction in the number of variables.

In a second step, the temperature will be replaced by a *conservative temperature* variable which is closer to conservation than *in-situ* temperature. The price to be paid is an increase in algebraic complexity in the respective conservation equations. The conservative temperature replaces the historically used *potential temperature*, since

**11. Diffusive Equilibrium**

A pressure gradient can cause a diffusive salt flux. Note that $\partial\mu/\partial p = \partial v/\partial S = -\gamma v < 0$ while $\partial\mu/\partial S > 0$. Therefore, the pressure-related component of salt flux is directed towards higher pressure, i. e. downward in the ocean. As an example, consider a state of rest which is in hydrostatic equilibrium (cf. Section 2.9 below) and assume a vanishing diffusive salt flux $\boldsymbol{J}_{\mathrm{S}} = 0$, i. e. a diffusive equilibrium. For the vertical component of (2.96) one finds

$$\frac{\partial S}{\partial z} = -\frac{\partial\mu/\partial p}{\partial\mu/\partial S}\frac{\partial p}{\partial z} \approx \frac{\partial\mu/\partial p}{\partial\mu/\partial S} g\rho \approx -3\,(\mathrm{g\,kg^{-1}})/\mathrm{km}$$

a value which is far above observed salinity gradients in the ocean. It follows that the observed oceanic state is far from its diffusive equilibrium state. In fact, small-scale turbulent mixing as discussed in Chapter 11 and advection by the large-scale ocean circulation prevents a diffusive equilibrium of the ocean.

the former has the best conservation properties of all variables measuring ocean heat content. It is hence the preferred temperature variable. The potential temperature budget is derived here for comparison and for historical reasons.

### 2.6.1 In-situ Temperature

The enthalpy conservation (2.76) can be written, by expanding the total derivative of $H = H(S, T, p)$, as

$$\rho\left(\frac{\partial H}{\partial S}\frac{\mathrm{D}S}{\mathrm{D}t} + \frac{\partial H}{\partial T}\frac{\mathrm{D}T}{\mathrm{D}t} + \frac{\partial H}{\partial p}\frac{\mathrm{D}p}{\mathrm{D}t}\right) = \frac{\mathrm{D}p}{\mathrm{D}t} - \nabla\cdot\boldsymbol{J}_{\mathrm{H}} + \rho\epsilon$$

The derivatives of the enthalpy, appearing in this relation, are evaluated as

$$\frac{\partial H}{\partial T} = c_p \quad \text{and} \quad \frac{\partial H}{\partial p} = \upsilon + T\frac{\partial\eta}{\partial p} = \upsilon - T\frac{\partial\upsilon}{\partial T} = \upsilon(1-\alpha T)$$

with the help of (1.50) and (1.57). Use of the salt conservation (2.20) then results in the *in-situ* temperature equation

$$\rho c_p\frac{\mathrm{D}T}{\mathrm{D}t} = -\nabla\cdot\boldsymbol{J}_{\mathrm{H}} + \alpha T\frac{\mathrm{D}p}{\mathrm{D}t} + Q_{\mathrm{T}} \tag{2.101}$$

Hence temperature can be changed not only by heat-flux convergence, but also by a change of pressure or by the temperature source $Q_{\mathrm{T}}$ which is given by

$$Q_{\mathrm{T}} = \rho\epsilon + \frac{\partial H}{\partial S}\nabla\cdot\boldsymbol{J}_{\mathrm{S}} \tag{2.102}$$

Note that the contributions from the dissipation $\rho\epsilon$ and from the salt fluxes $\boldsymbol{J}_{\mathrm{S}}$ for the temperature source term $Q_{\mathrm{T}}$ are small and usually neglected, as discussed below. The *in-situ* temperature budget (2.101) can also be written as

$$\rho c_p\left(\frac{\mathrm{D}T}{\mathrm{D}t} - \Gamma\frac{\mathrm{D}p}{\mathrm{D}t}\right) = -\nabla\cdot\boldsymbol{J}_{\mathrm{H}} + Q_{\mathrm{T}} \quad \text{with} \quad \Gamma(S,T,p) = \frac{\alpha T}{\rho c_p} = -\frac{\partial^2 G/\partial T\partial p}{\partial^2 G/\partial T^2} \tag{2.103}$$

using the relations in Table 1.1. Under adiabatic conditions (and without mass exchange) we have $\boldsymbol{J}_{\mathrm{H}} = 0$ and $Q_{\mathrm{T}} = 0$, and then changes in *in-situ* temperature and pressure are related by $\mathrm{d}T = \Gamma\mathrm{d}p$. The quantity $\Gamma$ is, therefore, called *adiabatic temperature gradient* or *lapse rate*. The gradient is meant here with respect to pressure, in fact the adiabatic lapse rate can also be written as $\Gamma = (\partial T/\partial p)_{\mathrm{ad}}$, defining the adiabatic temperature gradient. A typical value for the ocean is $\Gamma \approx 10^{-8}\,\mathrm{K/Pa} = 10^{-4}\,\mathrm{K/dbar}$ which corresponds about to $\sim 0.1\,\mathrm{K/km}$.

### 2.6.2 Conservative Temperature

The *in-situ* temperature equation (2.101) with the source terms (2.102) for the ocean is preferred to the balances of internal energy or enthalpy because $T$ is measurable.

However, the *in-situ* temperature is not conserved even under adiabatic conditions because changes of pressure always lead to changes in temperature of a fluid parcel, manifested by the nonvanishing lapse rate $\Gamma$. Therefore, the budget (2.101) is in practice not very useful. The concept of the *conservative temperature* introduces a new temperature variable which takes care of the pressure effects and which is conserved under adiabatic conditions.

Consider the enthalpy $H$ of a fluid parcel in terms of its canonical independent variables (see Figure 1.4) salinity, entropy, and pressure, $H = H(S, \eta, p)$. The enthalpy conservation (2.76) takes the form

$$\rho\left[\left(\frac{\partial H}{\partial S}\right)_{\eta p}\frac{\mathrm{D}S}{\mathrm{D}t}+\left(\frac{\partial H}{\partial \eta}\right)_{\mathrm{S}p}\frac{\mathrm{D}\eta}{\mathrm{D}t}+\left(\frac{\partial H}{\partial p}\right)_{\mathrm{S}\eta}\frac{\mathrm{D}p}{\mathrm{D}t}\right]=\frac{\mathrm{D}p}{\mathrm{D}t}-\nabla\cdot\boldsymbol{J}_{\mathrm{H}}+\rho\epsilon \tag{2.104}$$

From Maxwell's relations in Section 1.2.5 one has

$$\left(\frac{\partial H}{\partial \eta}\right)_{S,p}=T\,,\quad \left(\frac{\partial H}{\partial p}\right)_{S,\eta}=\upsilon \tag{2.105}$$

and with $\rho\upsilon = 1$ it is found that both terms with $\mathrm{D}p/\mathrm{D}t$ in (2.104) cancel exactly (in contrast to the *in-situ* temperature budget (2.103)). The *potential enthalpy* $H^0$ is defined by

$$H(S,\eta,p_0)\equiv H^0(S,\eta) \tag{2.106}$$

and is the heat content of a parcel which is brought adiabatically from the local pressure $p$ to a reference pressure $p_0$, i. e. without mass (constant $S$) and energy (constant $\eta$) exchange. The reference pressure is usually chosen as the surface pressure, which means that the water parcel is brought adiabatically from its local depth to the surface. Note that the *in-situ* temperature of the water parcel will change because of the change in pressure according to (2.103). Note also that the index 0 means that the quantity is evaluated at the pressure $p_0$. From the definition of potential enthalpy (2.106) we obtain

$$\rho\frac{\mathrm{D}H^0}{\mathrm{D}t}=\rho\left[\left(\frac{\partial H^0}{\partial S}\right)_{\eta,p}\frac{\mathrm{D}S}{\mathrm{D}t}+\left(\frac{\partial H^0}{\partial \eta}\right)_{S,p}\frac{\mathrm{D}\eta}{\mathrm{D}t}\right] \tag{2.107}$$

From the Maxwell relations (2.105) it follows that the quantity

$$\theta=\left(\frac{\partial H^0}{\partial \eta}\right)_{S,p} \tag{2.108}$$

is the temperature of the water parcel brought adiabatically to the reference pressure $p_0$, therefore, $\theta$ is called the *potential* temperature. Using the salt budget $\rho\mathrm{D}S/\mathrm{D}t = -\nabla\cdot\boldsymbol{J}_\mathrm{S}$ to replace $\mathrm{D}S/\mathrm{D}t$ and the entropy budget in the form (2.88) to replace $\mathrm{D}\eta/\mathrm{D}t$ in (2.107) we obtain

$$\rho\frac{\mathrm{D}H^0}{\mathrm{D}t}=\left(\frac{\theta}{T}\frac{\partial H}{\partial S}-\frac{\partial H^0}{\partial S}\right)\nabla\cdot\boldsymbol{J}_\mathrm{S}+\frac{\theta}{T}(-\nabla\cdot\boldsymbol{J}_\mathrm{H}+\rho\epsilon) \tag{2.109}$$

Following McDougall (2003), the *conservative* temperature $\theta^\star$ is introduced as

$$\theta^\star=H^0/c_p^\star \tag{2.110}$$

where the constant reference value of the specific heat $c_p^\star = 3991.868\,\mathrm{J\,kg^{-1}\,K^{-1}}$ was chosen such that the numerical value of conservative temperature $\theta^\star$ and potential temperature $\theta$ become identical at $S = 35\,\mathrm{g\,kg^{-1}}$. This definition for $\theta^\star$ allows to write the heat balance in the form

$$\rho c_p^\star \frac{\mathrm{D}\theta^\star}{\mathrm{D}t} = \left(\frac{\theta}{T}\frac{\partial H}{\partial S} - \frac{\partial H^0}{\partial S}\right)\nabla\cdot\boldsymbol{J}_\mathrm{S} + \frac{\theta}{T}(-\nabla\cdot\boldsymbol{J}_\mathrm{H} + \rho\epsilon) \tag{2.111}$$

For adiabatic conditions, i. e. for $\boldsymbol{J}_\mathrm{H} = 0$, $\boldsymbol{J}_\mathrm{S} = 0$ and $\epsilon = 0$, the conservative temperature is a conserved quantity, in contrast to the *in-situ* temperature which is not conserved under adiabatic conditions. Conservative temperature is thus more convenient as a state variable than *in-situ* temperature, since adiabatic processes can be characterized by $S = \text{const}$ and $\theta^\star = \text{const}$, instead of $S = \text{const}$ and $\eta = \text{const}$. We will, therefore, use $\theta^\star$ instead of $T$ in the following as state variable (see also Section 2.6.4). The difference between $\theta^\star$ and $T$ is, however, small (see Figure 2.13).

The equation (2.111) for the conservative temperature is exact. The local sources and sinks arise not only from the physical heat sources $\rho\epsilon - \nabla\cdot\boldsymbol{J}_\mathrm{H}$ (modified by a factor $\theta/T$) but also from an additional term related to the salinity flux divergence. A substantial simplification is, however, possible by considering the magnitude of various source terms in (2.111):

- A first approximation involves the first term on the right-hand side of (2.111) which describes the change in conservative temperature due to a salinity flux divergence. Near the surface where $\theta/T \to 1$ and $H^0 \to H$ this term vanishes. At depth its magnitude remains small, and a salinity flux that changes the salinity of a fluid parcel by as much as $1\,\mathrm{g\,kg^{-1}}$ leads to a change in conservative temperature of less than 0.05 K. For all practical purposes, the direct influence of the salt flux divergence on conservative temperature can, therefore, be neglected (see McDougall, 2003, for a detailed analysis). Note that the contribution of the flux $\boldsymbol{J}_\text{chem}$ to the heat flux $\boldsymbol{J}_\mathrm{H} = \boldsymbol{J}_\mathrm{T} + \boldsymbol{J}_\text{rad} + \boldsymbol{J}_\text{chem}$ defined in (2.75) is also small and can be neglected.
- A second approximation involves the magnitude of $\theta/T$. In the ocean, the difference between conservative and *in-situ* temperature rarely exceeds 0.5°C. Both

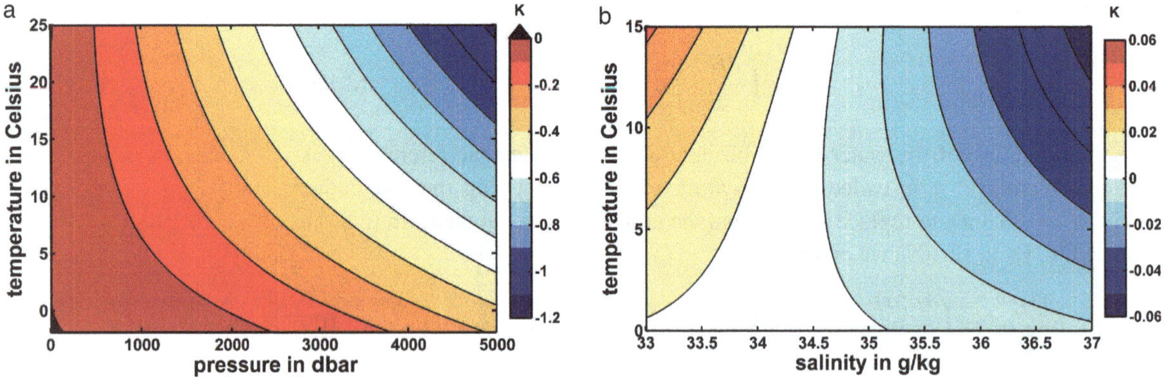


**Fig. 2.13** **a** Differences of conservative temperature $\theta^\star$ and in-situ temperature $T$ as function of $T$ and $p$ at $S = 35\,\mathrm{g\,kg^{-1}}$. **b** The difference of $\theta^\star$ and $\theta$ is shown as function of $S$ and $T$ for $p = 1{,}000\,\mathrm{dbar}$

temperatures are defined in K, and hence the approximation $\theta/T \approx 1$ in (2.111) has a relative error of no more than $2 \times 10^{-3}$, and is even better near the surface where the flux divergence term is most important.

- A third approximation concerns the mechanical dissipation $\epsilon$. As seen from (2.71), the dissipation rate depends on the deformation, respectively shear, of the flow. For a shear of $\partial u/\partial x = 0.1\,\mathrm{s}^{-1}$ (which is already unrealistically high for ocean conditions) one would have $\rho\epsilon \approx 10^{-6}\,\mathrm{m^2s^{-1}} \times 10^3\,\mathrm{kg\,m^{-3}} \times (0.1)^2\,\mathrm{s^{-2}} \approx 10^{-4}\,\mathrm{W\,m^{-3}}$. According to (2.101), such a value corresponds to a temperature increase of $\sim 10^{-6}\,\mathrm{K/d}$. In the deep ocean, typical values of the dissipation are around $10^{-9}\,\mathrm{W\,m^{-3}}$, so that the corresponding dissipative heating is five orders of magnitude smaller. Note, however, that while mechanical dissipation is negligible for the heat budget, its existence is fundamental for the *mechanical* energy budget.

With these approximation, it follows that

$$\rho c_p^\star \frac{\mathrm{D}\theta^\star}{\mathrm{D}t} = -\nabla \cdot \boldsymbol{J}_\mathrm{H} \tag{2.112}$$

Note that this equation is used in most ocean models. Boundary conditions follow from (2.86). For a more detailed discussion of the approximations involved in (2.112) see IOC, SCOR and IAPSO (2010). We show in Figure 2.13 the difference of the conservative temperature $\theta^\star$ with respect to the *in-situ* temperature $T$ (left panel) and to the potential temperature $\theta$ as function of $T$, $S$ and $p$. Significant deviations only occur for pressures exceeding 1,000 dbar.

### 2.6.3 Potential Temperature

We derive in this section a budget for the potential temperature $\theta$. This is done because potential temperature has long been in use in physical oceanography as the 'best' temperature variable, until it became clear that conservative temperature is the superior choice. It is convenient to start from the entropy $\eta = \eta(S, T, p)$ which is a known function of its arguments, as discussed in Section 1.2.5. The potential temperature $\theta$, referred to a fixed reference pressure $p_0$, is implicitly defined through the relation

$$\eta(S, T, p) = \eta(S, \theta, p_0) \equiv \eta^0(S, \theta) \tag{2.113}$$

which states that the potential temperature $\theta$ is the temperature which a particle of temperature $T$ and pressure $p$ will assume when it is brought *adiabatically* (i. e. at constant $\eta$ and $S$) to a location with pressure $p_0$ (usually the surface pressure). Note that the definition (2.113) agrees with (2.108), and also with the more intuitive definition $\theta(S, T, p) = T - \int_{p_0}^{p} \Gamma(S, T(S, \eta, p'), p')\mathrm{d}p'$ as can be seen from (2.118) below. Here $S$ and $\eta$ remain constant during the integration, and the function $T = T(S, \eta, p)$ can be obtained by solving (2.113) for the *in-situ* temperature.

Solving (2.113) for $\theta$ results in

$$\theta = \theta(S, T, p) \tag{2.114}$$

and a prognostic equation for $\theta$ can be derived in terms of the time rates of change of $S$, $T$, and $p$,

$$\frac{\mathrm{D}\theta}{\mathrm{D}t} = \frac{\partial\theta}{\partial S}\frac{\mathrm{D}S}{\mathrm{D}t} + \frac{\partial\theta}{\partial T}\frac{\mathrm{D}T}{\mathrm{D}t} + \frac{\partial\theta}{\partial p}\frac{\mathrm{D}p}{\mathrm{D}t} \tag{2.115}$$

The functional form of (2.114) is not explicitly known, but the partial derivatives can be inferred by considering the total differential of (2.113), i. e.

$$\frac{\partial\eta}{\partial S}\mathrm{d}S + \frac{\partial\eta}{\partial T}\mathrm{d}T + \frac{\partial\eta}{\partial p}\mathrm{d}p = \frac{\partial\eta^0}{\partial S}\mathrm{d}S + \frac{\partial\eta^0}{\partial\theta}\mathrm{d}\theta \tag{2.116}$$

Here the index $^0$ indicates that the function of $(S, T, p)$ is evaluated at temperature $\theta$ and pressure $p_0$. From (1.57) it follows that $\partial\eta/\partial T = c_p(S, T, p)/T$ and $\partial\eta^0/\partial\theta = c_p(S, \theta, p_0)/\theta \equiv c_p^0/\theta$. Using (2.116), the derivatives in (2.115) may be expressed as

$$\frac{\partial\theta}{\partial S} = \frac{\theta}{c_p^0}\left(\frac{\partial\eta}{\partial S} - \frac{\partial\eta^0}{\partial S}\right), \quad \frac{\partial\theta}{\partial T} = \frac{c_p}{c_p^0}\frac{\theta}{T}, \quad \frac{\partial\theta}{\partial p} = \frac{\theta}{c_p^0}\frac{\partial\eta}{\partial p} \tag{2.117}$$

In particular one finds, using (1.50) and (1.54),

$$\frac{\partial\theta/\partial p}{\partial\theta/\partial T} = \frac{T}{c_p}\frac{\partial\eta}{\partial p} = -\frac{\alpha T}{\rho c_p} \equiv -\Gamma(S, T, p) \tag{2.118}$$

From (2.118) we see that the term containing $\mathrm{D}p/\mathrm{D}t$ in (2.115) exactly cancels with the corresponding term from (2.103) when eliminating $\mathrm{D}T/\mathrm{D}t$. Invoking (2.20) to eliminate $\mathrm{D}S/\mathrm{D}t$ from (2.115) as well as (2.102) and the first relation in (1.49), a prognostic equation for potential temperature is obtained as (Bacon and Fofonoff, 1996)

$$\rho c_p^0\frac{\mathrm{D}\theta}{\mathrm{D}t} = \frac{\theta}{T}(\rho\epsilon - \nabla\cdot\boldsymbol{J}_{\mathrm{H}}) + \frac{\theta}{T}\left(\mu - T\frac{\partial\mu^0}{\partial T}\right)\nabla\cdot\boldsymbol{J}_{\mathrm{S}} \tag{2.119}$$

Similar to the conservative temperature $\theta^\star$, given by (2.111), the potential temperature $\theta$ is also conserved for adiabatic processes, i. e. for $\rho\epsilon = 0$, $\boldsymbol{J}_{\mathrm{H}} = 0$, and $\boldsymbol{J}_{\mathrm{S}} = 0$. With the same approximations which lead from (2.111) to (2.112), we obtain for the potential temperature

$$\rho c_p^0\frac{\mathrm{D}\theta}{\mathrm{D}t} = -\nabla\cdot(\boldsymbol{J}_{\mathrm{T}} + \boldsymbol{J}_{\mathrm{rad}}) \quad \text{or} \quad \rho\frac{\mathrm{D}\theta}{\mathrm{D}t} = -\nabla\cdot\boldsymbol{J}_\theta + c_p^0\boldsymbol{J}_\theta\cdot\nabla\frac{1}{c_p^0} \tag{2.120}$$

with $\boldsymbol{J}_\theta = (\boldsymbol{J}_{\mathrm{T}} + \boldsymbol{J}_{\mathrm{rad}})/c_p^0$. It becomes obvious that it is not a reference value $c_p^\star$ for specific heat as for the budget for conservative temperature (2.109), but the specific heat $c_p^0$ at the reference pressure $p_0$ which enters the balance for potential temperature (2.120). Variations of $c_p^0$ hence lead to a source or sink of potential temperature. In the relevant oceanic interval of salinities 34–36 g kg$^{-1}$, values of $c_p^0$ range from 3850 to 4050 J/kg, i. e. within 1% at most (see Figure 1.7). As an (admittedly extreme) example for this effect, the mixture of two water masses with 0°C temperature, 0 g kg$^{-1}$ salinity and 30 °C, 40 g kg$^{-1}$ at surface pressure results in a potential temperature of 14.6°C, and not 15 °C. Neglecting the source term in (2.120), therefore, causes a small error, which is, however, much larger than the approximations discussed above for the conservative temperature. Using conservative temperature completely avoids this error and conservation properties of $\theta^\star$ are, therefore, better by two orders of magnitude.

On the other hand, the differences between $\theta^\star$ and $\theta$ are in general small (see right panel of Figure 2.13). They can reach maximum values of 0.15 K at the surface comparing one ocean basin with another, but there are also comparable differences in regions of strong fronts at the surface, like the Gulf Stream. Differences at depth are much smaller and can in general be neglected, in particular in the light of the various approximations which will be introduced in Part II. Note that we will use conservative temperature as variable, but we will drop the index $^\star$ for convenience in the remainder of the book (except for the next section). Note also that when referring to potential temperature, either $\theta^\star$ is meant, or differences between $\theta^\star$ and $\theta$ are simply neglected.

### 2.6.4 Conservative Temperature as a State Variable

We now introduce the conservative temperature $\theta^\star$ instead of the *in-situ* temperature $T$ as state variable, since $\theta^\star$ is conserved under adiabatic conditions, while $T$ is not conserved. Using $T = T(S, \theta^\star, p)$ for the state equation $\rho = \rho(S, T, p)$, we obtain the density as function of the new set of state variables $S$, $\theta^\star$, and $p$, i. e. by

$$\rho = \rho(S, T(S, \theta^\star, p), p) = F^\star(S, \theta^\star, p) \tag{2.121}$$

An explicit numerical approximation of the function $F^\star(S, \theta^\star, p)$ can be found in IOC, SCOR and IAPSO (2010). For practical applications of the equations of motion (2.121) is the most convenient state equation, whereas for thermodynamic considerations the original form (1.52) of the state equation is preferred.

We will need below the expansion coefficients for density with respect to the new state variables. The coefficients are given by the first derivatives of the relation (2.121) and denoted by

$$\alpha^\star = -\frac{1}{\rho}\left(\frac{\partial \rho}{\partial \theta^\star}\right)_{S,p} \quad \text{modified thermal expansion} \tag{2.122}$$

$$\gamma^\star = \frac{1}{\rho}\left(\frac{\partial \rho}{\partial S}\right)_{\theta^\star,p} \quad \text{modified haline contraction} \tag{2.123}$$

$$\kappa^\star = \frac{1}{\rho}\left(\frac{\partial \rho}{\partial p}\right)_{S,\theta^\star} \quad \text{adiabatic compressibility} \tag{2.124}$$

Note that the coefficients $\alpha^\star$ and $\gamma^\star$ are similar in magnitude but not identical to the corresponding coefficients in (1.54)–(1.56) using *in-situ* temperature $T$ as state variable. To express the modified thermal expansion coefficient $\alpha^\star = -(1/\rho)(\partial\rho/\partial\theta^\star)_{S,p} = \alpha(\partial T/\partial\theta^\star)_{S,p}$ in terms of known quantities, we need to know $\partial T/\partial\theta^\star$. Using $\eta(S, T, p) = \eta^0(S, \theta^\star, p_0)$, thus

$$\frac{\partial \eta}{\partial T} = \frac{\partial \eta^0}{\partial \theta^\star}\frac{\partial \theta^\star}{\partial T}$$

and combining with $\partial\eta/\partial T = c_p/T$ and $\partial\theta^\star/\partial\eta^0 = \theta/c_p^\star$ and with the definition (2.110) for $\theta^\star$, for the modified thermal expansion coefficient

$$\alpha^\star = \alpha \frac{c_p^\star T}{c_p \theta} \tag{2.125}$$

The modified haline contraction coefficient $\gamma^\star$ is related to $\gamma$ by

$$\begin{aligned}\gamma = \frac{1}{\rho}\left(\frac{\partial\rho}{\partial S}\right)_{T,p} &= \frac{1}{\rho}\left[\left(\frac{\partial\rho}{\partial S}\right)_{\theta^\star,p} + \left(\frac{\partial\rho}{\partial\theta}^\star\right)_{S,p}\left(\frac{\partial\theta^\star}{\partial S}\right)_{S,p}\right]\\ &= \gamma^\star - \alpha^\star\left(\frac{\partial\theta^\star}{\partial S}\right)_{S,p}\end{aligned}$$

To find $\partial\theta^\star/\partial S$, use again $\eta(S,T,p) = \eta^0(S,\theta^\star,p_0)$, thus

$$\left.\frac{\partial\eta}{\partial S}\right|_{T,p} = \left.\frac{\partial\eta^0}{\partial S}\right|_{\theta^\star} + \left.\frac{\partial\eta^0}{\partial\theta^\star}\right|_S \frac{\partial\theta^\star}{\partial S} \quad \text{or} \quad \frac{\partial\theta^\star}{\partial S} = \left(\left.\frac{\partial\eta}{\partial S}\right|_{T,p} - \left.\frac{\partial\eta^0}{\partial S}\right|_{\theta^\star}\right)\frac{\theta}{c_p^\star}$$

since $\partial\eta^0/\partial\theta^\star = c_p^\star/\theta$. Thus with (2.125) we find

$$\gamma^\star = \gamma + \alpha^\star\left(\left.\frac{\partial\eta}{\partial S}\right|_{T,p} - \left.\frac{\partial\eta^0}{\partial S}\right|_{\theta^\star}\right)\frac{\theta}{c_p^\star} = \gamma + \alpha\frac{T}{c_p}\left(\left.\frac{\partial\eta}{\partial S}\right|_{T,p} - \left.\frac{\partial\eta^0}{\partial S}\right|_{\theta^\star}\right) \tag{2.126}$$

For the adiabatic compressibility $\kappa^\star$ analogous arguments lead to

$$\kappa^\star = \kappa + \alpha^\star\left(\frac{\partial\theta^\star}{\partial p}\right)_{S,T} = \kappa - \frac{\alpha^2 T}{c_p\rho} \tag{2.127}$$

Since it has a clear physical meaning, we will use in the following the speed of sound $c_s$ instead of the modified adiabatic compressibility $\kappa^\star$. The sound velocity is related to $\kappa^\star$ by

$$\frac{1}{c_s^2} = \kappa^\star\rho = \left(\frac{\partial\rho}{\partial p}\right)_{S,\theta^\star} \tag{2.128}$$

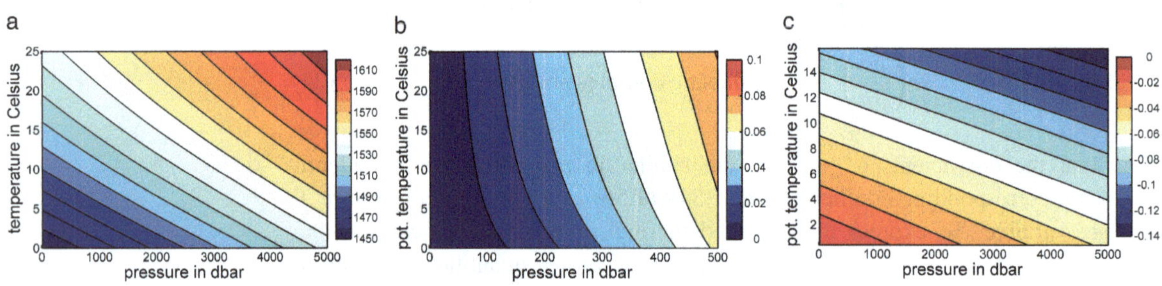


**Fig. 2.14** Velocity of sound in seawater, $c_s$ (**a**, in $\mathrm{m\,s^{-1}}$), difference of the modified thermal expansion coefficient to Figure 1.6, $\alpha^\star - \alpha$ (**b**, in $10^{-4}$ 1/K) as well as difference of the modified haline contraction coefficient to Figure 1.6, $\gamma^\star - \gamma$ (**c**, in $10^{-4}\ (\mathrm{g\,kg^{-1}})^{-1}$), all as a function of $T$ and $p$ at $S = 35\ \mathrm{g\,kg^{-1}}$

The sound speed $c_s$ and the differences in expansion coefficients $\alpha^\star - \alpha$ and $\gamma^\star - \gamma$ for the different state variables are shown in Figure 2.14. Relative differences are for the thermal expansion coefficient significant on the order of 5%, but for the haline contraction coefficient smaller.

Note that from now on, we shall drop the $^\star$ for the coefficients and use the notation $\alpha$ and $\gamma$ for the coefficients defined in (2.122) and (2.123). As mentioned above, we shall also drop the $^\star$ for the conservative temperature (which will be also called occasionally potential temperature, when differences between both are unimportant). Likewise, the notation $F(S, \theta, p)$ will denote the state equation (2.121) whenever conservative temperature is used as independent variable.

## 2.7 Density Variables

The *in-situ* density $\rho$ is not conserved for constant $S$ and $\theta$, i. e. for adiabatic conditions. However, since molecular and turbulent transports divergences (see also next section) acting on the salinity and conservative temperature budgets are often rather small in the interior of the ocean, a "conserved" density is a useful quantity in particular to trace the spreading of water masses in the ocean or the construction of isopycnal models. Similar to the definition of a "conserved" temperature, we introduce the concept of a potential density in this section. On the other hand, it will turn out that water masses spread along potential density surfaces only under certain conditions, which will be clarified by the concept of neutral surface elements.

### 2.7.1 Potential Density

In analogy to the concept of conservative temperature a density variable can be introduced which is conserved for adiabatic conditions. The *potential density* is a useful concept more alike the *in-situ* density than either temperature or salinity alone. It is defined by

$$\rho_\mathrm{p}(S, \theta) = F(S, \theta, p_\star) \tag{2.129}$$

with the *fixed* reference pressure $p_\star = \text{const.}$ For practical use, 1,000 kg m$^{-3}$ is subtracted from the values of potential density. The potential density $\rho_\mathrm{p}$ equals the *in-situ* density of a particle (except for the constant removed in (2.129)), when the particle is moved adiabatically from its *in-situ* pressure $p$ to the reference pressure $p_\star$ as sketched in Figure 2.15. Note that the reference pressure $p_\star$ is in principle unrelated to the reference pressure $p_0$ used in the definition of conservative temperature (2.106). It is, in fact, common to define potential densities not only with reference to the surface but also to deeper levels in the ocean. The reason is that isosurfaces of $\rho_\mathrm{p}$ are close to surfaces of adiabatic spreading of water masses when the *in-situ* pressure is close to $p_\star$, as discussed below. The common notation is as follows: $\sigma_0(\sigma_\theta), \sigma_1, \sigma_4$ for $p_\star = 0$, 1,000, 4,000 dbar etc. The conservation of potential density is described by

$$\frac{\mathrm{D}\rho_\mathrm{p}}{\mathrm{D}t} = \rho_\mathrm{p}\left(\gamma_\star \frac{\mathrm{D}S}{\mathrm{D}t} - \alpha_\star \frac{\mathrm{D}\theta}{\mathrm{D}t}\right) \tag{2.130}$$

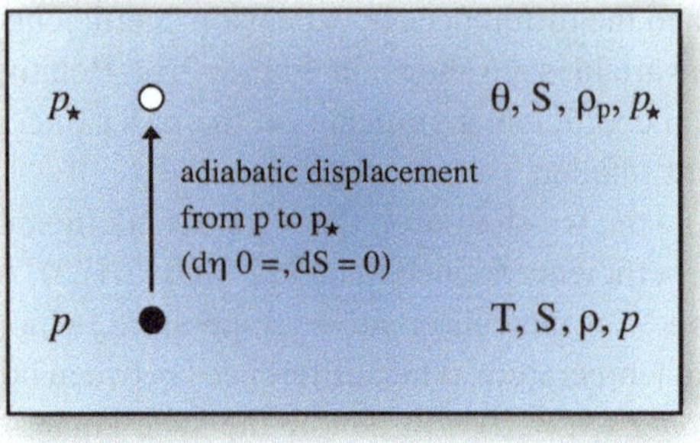


Fig. 2.15 Sketch of adiabatic displacement of a fluid parcel from $p$ to $p_\star$ with respective temperatures, salinities, and densities

where $\gamma_\star, \alpha_\star$ denote the expansion coefficients given by (2.123) and (2.122), respectively, evaluated for the reference pressure $p_\star$. Hence potential density is always conserved for adiabatic motions, for any choice of the reference pressure. Surfaces of constant potential density are called *isopycnal surfaces*. Note, however, that it is the *in-situ* density which appears in the equations of motion. Only near the reference pressure, the potential density is similar to the (dynamically relevant) *in-situ* density. Later in Section 4.1.3 we will derive the balance of *in-situ* density $\rho$ and find that any suitable density variable containing the compressibility is not conserved, even for adiabatic motions.

## 2.7.2 Neutral Surface Elements

Under adiabatic conditions the advection of water masses occurs along isopycnals, however, only if the pressure changes are not substantial, as shown in this section. To derive a more general concept, that of a so-called *neutral surface*, we start with the equation of state (2.121) and consider a small change of density over the distance $\mathrm{d}\boldsymbol{x}$. It is given by

$$\mathrm{d}\rho = \gamma\rho\nabla S \cdot \mathrm{d}\boldsymbol{x} - \alpha\rho\nabla\theta \cdot \mathrm{d}\boldsymbol{x} + \frac{1}{c_\mathrm{s}^2}\nabla p \cdot \mathrm{d}\boldsymbol{x}$$

Likewise, the change due to an adiabatic motion of a parcel (i. e. with constant $S$, $\theta$) over distance $\mathrm{d}\boldsymbol{x}$ is

$$(\mathrm{d}\rho)_\mathrm{ad} = \frac{1}{c_\mathrm{s}^2}\mathrm{d}p = \frac{1}{c_\mathrm{s}^2}\nabla p \cdot \mathrm{d}\boldsymbol{x} \tag{2.131}$$

and we thus obtain

$$\mathrm{d}\rho - (\mathrm{d}\rho)_\mathrm{ad} = \gamma\rho\nabla S \cdot \mathrm{d}\boldsymbol{x} - \alpha\rho\nabla\theta \cdot \mathrm{d}\boldsymbol{x} = \boldsymbol{e} \cdot \mathrm{d}\boldsymbol{x}$$

for the difference of density between the background and the parcel. Here the vector

$$\boldsymbol{e} = \gamma\rho\nabla S - \alpha\rho\nabla\theta \equiv \nabla\rho - \frac{1}{c_\mathrm{s}^2}\nabla p \tag{2.132}$$

is introduced. The difference in density between the parcel and the environment vanishes if

$$e \cdot dx = 0 \tag{2.133}$$

Considering all possible $dx$ that satisfy this condition, a *neutral* surface element is constructed. The vector $e$ is normal to the element and also normal to the isopycnal $\rho_p = \text{const}$ defined for the local pressure $p = p_\star$, i. e. at the reference point we have $e = \nabla \rho_p$ (see Figure 2.16). Note that since the pressure gradient is almost vertical, the vectors $e$ and $\nabla \rho$ differ by an almost vertical vector.

The condition (2.133), however, defines a surface only locally. A continuous surface of global extent is given by a condition $\chi(x) = \text{const}$ with some scalar function $\chi(x)$ and (2.133) is then replaced by $\nabla\chi \cdot dx = 0$. This implies that only if $e = \varphi\nabla\chi$ would hold with some arbitrary function $\varphi$, an entire neutral surface exists. A sufficient condition is the vanishing of the helicity $\mathcal{H} = e \cdot \nabla \times e$. This can be seen by using the condition $e = \varphi\nabla\chi$ for the helicity, which yields $\mathcal{H} = \varphi\nabla\chi \cdot (\varphi\nabla \times \nabla\chi + \nabla\varphi \times \nabla\chi) = 0$. Using the equation of state, the helicity is given by

$$\begin{aligned}\mathcal{H} &= -\alpha\rho\nabla(\gamma\rho) \times \nabla S \cdot \nabla\theta - \gamma\rho\nabla(\alpha\rho) \times \nabla\theta \cdot \nabla S \\ &= \rho^2 (\alpha\nabla\gamma - \gamma\nabla\alpha) \cdot \nabla\theta \times \nabla S = -\rho^2\gamma^2\nabla\frac{\alpha}{\gamma} \cdot \nabla\theta \times \nabla S\end{aligned}$$

Since the function $\alpha/\gamma$ depends not only on $\theta$ and $S$ but also on pressure, the helicity $\mathcal{H}$ does not vanish in general. Note that the $\theta$ or $S$ dependencies of $\alpha/\gamma$ yield corresponding components of $\nabla\alpha/\gamma$ which are orthogonal to $\nabla\theta \times \nabla S$. Since $\mathcal{H} \neq 0$ there are no neutral surfaces in general. Only for an equation of state in which $\partial(\alpha/\gamma)/\partial p = 0$ would hold (which is not the case), neutral surfaces exists in general. On the other hand, for regions of the ocean with constant $\theta$ or $S$, or for a fixed functional relation $\theta(S)$ or $S(\theta)$, i. e. for a fixed water mass relationship which is often found for rather large regions of the ocean, neutral surfaces can be approximately defined.

In any case, however, there are *neutral trajectories*, i. e. pathways $s$ in the ocean for which $e \cdot s = 0$ holds, such that no buoyant forces would act on water parcels moved along those pathways. McDougall (1987) considers a neutral trajectory in a large circle around the subtropical gyre in the North Atlantic, and finds that the

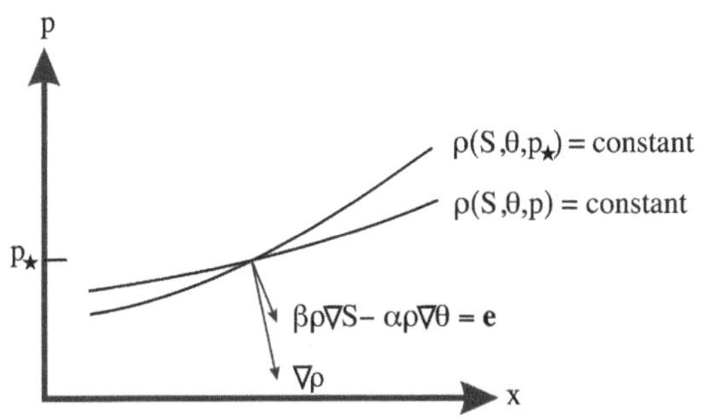


Fig. 2.16 Sketch showing the neutral surface element, the density surface $\rho(S, \theta, p) = \text{const}$, and the local isopycnal $\rho(S, \theta, p_0) = \text{const}$

neutral trajectory of the water parcel would form a large helix, with a change in height in one circle to the other of about 5 m.

**12. Approximate Neutral Density**

A more restrictive constraint on the vector $\boldsymbol{e}$ is given by $\boldsymbol{e} = \nabla \chi$, for which the quantity $\chi$ is called the *neutral density*, denoted by $\gamma_n$. Surfaces of $\chi = $ const would then form neutral surfaces. It is obvious that there is no neutral density in general, but approximate forms can be given, for instance by the condition

$$\nabla^2 \chi = \nabla \cdot \boldsymbol{e}$$

which was used by Eden and Willebrand (1999) to estimate $\chi$ for the North Atlantic. A different method for construction of a neutral density variable $\gamma_n$ has been suggested by Jackett and McDougall (1997), however with an equation of state relating $\gamma_n$ to $S, \theta, p$ which includes an explicit spatial dependence.

## 2.8 Molecular and Turbulent Transports

The conservation equations for mass (2.18), momentum (2.59), salt (2.20), and heat (2.120), in combination with the equation of state (2.121), form a complete system of differential equations for the variables $\rho$, $S$, $\theta$, $p$, and $\boldsymbol{u}$ which we use to represent the oceanic motion. In principle, they can be solved if initial conditions and boundary fluxes of mass, momentum, freshwater, and heat are prescribed.

To understand large-scale properties of the ocean and its interactions with the atmosphere, it is necessary to consider averages over suitable temporal and/or spatial scales. A possible filtering method is discussed below. However, a fundamental property of oceanic dynamics is that the motions at smaller scales influence those at larger scales. That influence occurs through *turbulent* transports which arise from the nonlinear nature of the equations of motion, as shown below. In contrast to molecular transports, which are generally very small, turbulent transports of momentum, heat, and salt are at the heart of ocean dynamics. In this section we review the size of the molecular transports and then derive the mathematical form of the turbulent transports by the commonly used averaging methods. Further treatment of these eddy-driven transports is postponed to Part IV.

### *2.8.1 Magnitude of Molecular Transports*

The nonadvective transports of momentum, salt, and heat are the result of molecular processes. They are usually described by the empirical laws discussed in Section 2.5.1. Ignoring the cross-transports of heat and salt discussed in Section 2.5.2, these laws are (2.36) for the stress in Newtonian fluids, Fourier's law (2.94) for the heat conduction, and the modified Fickian law (2.96) for the salt diffusion.

In order to estimate the role of the molecular fluxes, we assume that typical velocities are of magnitude $U$, and typical length scales of magnitude $L$. Specifically, we consider two limiting cases: (1) small-scale motions with $U = 1\,\mathrm{m\,s^{-1}}$ and

$L = 10\,\mathrm{m}$ which are e. g. typical of ocean surface waves, and (2) large-scale motions with $U = 0.1\,\mathrm{m\,s^{-1}}$ and $L = 1{,}000\,\mathrm{km}$ typical for the ocean circulation. In the equations for momentum, salt, and heat we then have the following balances:

- **Friction**
  The order of magnitude of the momentum advection, the Coriolis and the friction terms in (2.59) are given by

$$\rho \boldsymbol{u} \cdot \nabla \boldsymbol{u} \sim \rho U \frac{U}{L} \tag{2.134}$$

$$2\rho \boldsymbol{\Omega} \times \boldsymbol{u} \sim 2\rho \Omega U \tag{2.135}$$

$$\kappa_{\mathrm{M}} \rho \nabla^2 \boldsymbol{u} \sim \rho \kappa_{\mathrm{M}} \frac{U}{L^2} \tag{2.136}$$

  The ratio of the momentum advection (2.134) to the friction term (2.136) is the REYNOLDS[17]-number $\mathrm{Re} = UL/\kappa_{\mathrm{M}}$. With a kinematic viscosity of $\kappa_{\mathrm{M}} \approx 1.3 \times 10^{-6}\,\mathrm{m^2\,s^{-1}}$ , we obtain for small-scale motions $\mathrm{Re} \approx 10^7$ and for large-scale motions even $\mathrm{Re} \approx 10^{11}$. In both cases it follows that the molecular momentum transport is negligible compared to the momentum advection. The same conclusion arises from the consideration of the ratio of the friction term and the Coriolis term which is the EKMAN[18]-number $\mathrm{Ek} = \kappa_{\mathrm{M}}/(2\Omega L^2)$. With $\Omega \approx 7 \times 10^{-5}\,\mathrm{s^{-1}}$, we obtain $\mathrm{Ek} \approx 10^{-14}$ for $L = 1{,}000\,\mathrm{km}$.
- **Salt diffusion**
  The order of magnitude of the salt advection and diffusion terms in (2.20) is given by

$$\rho \boldsymbol{u} \cdot \nabla S \sim \rho U \frac{\Delta S}{L} \tag{2.137}$$

$$\rho \kappa_{\mathrm{S}} \nabla^2 S \sim \kappa_{\mathrm{S}} \rho \frac{\Delta S}{L^2} \tag{2.138}$$

  The ratio of salt advection (2.137) to salt diffusion (2.138) is the PECLET[19]-number Pe which has a magnitude of $\mathrm{Pe} = UL/\kappa_{\mathrm{S}}$. With $\kappa_{\mathrm{S}} \approx 1.2 \times 10^{-9}\,\mathrm{m^2\,s^{-1}}$, the Peclet-number varies between $\mathrm{Pe} \approx 10^{10}$ (small scales) and $\mathrm{Pe} \approx 10^{14}$ (large scales) for the above examples, and salt diffusion is negligible compared to advection.
- **Heat conduction**
  The situation is analogous to the salt budget. The ratio of salt and temperature coefficients for molecular diffusion is $\kappa_{\mathrm{T}}/\kappa_{\mathrm{S}} \approx 10^2$. It follows that the heat conduction term in (2.120) is of relative order $10^{-8}$–$10^{-12}$ compared to temperature advection and hence negligible.

For almost all oceanographic problems the *direct* influence of molecular transports is thus very small. However, the very existence of molecular transports is of fundamental importance, in particular for turbulent energy cascades but also for specific processes such as double diffusion. Moreover, in boundary layers molecular fluxes are always important. Those boundary layers are, however, very thin and hence usually neglected or – more precisely – included in parameterizations for turbulent fluxes in thicker turbulent boundary layers.

---

[17] OSBORNE REYNOLDS, *1842 in Belfast, †1912 in Watchet/Somerset, mechanical engineer and physicist.

[18] VAGN WALFRID EKMAN, *1874 in Stockholm, †1954 in Gostad, oceanographer and physicist.

[19] JEAN CLAUDE EUGÈNE PÈCLET, *1793 in Besançon, †1857 in Paris, physicist.

### 2.8.2 Reynolds and Hesselberg Averaging

The equations of motion are valid for all scales larger than, say, 1 μm. But we are normally interested only in the oceanic state averaged over much larger temporal and/or spatial scales. Denoting such an average by $\overline{(..)}$, the most simple temporal mean of a variable $\psi$ could e. g. be defined by

$$\overline{\psi}(t) = \frac{1}{\tau} \int_{t-\tau/2}^{t+\tau/2} \psi(t')\mathrm{d}t' \tag{2.139}$$

with an averaging interval of length $\tau$. The average (2.139) is sometimes also called a running mean. Likewise, a spatial or a combined spatial-temporal average could be considered. A disadvantage of any average as (2.139) is, however, that it does *not* satisfy

$$\overline{\overline{\psi}} = \overline{\psi} \tag{2.140}$$

which we will assume nevertheless because it is a rather desirable property, as we see below. Some other averaging operations which satisfy (2.140), but which are more complicated, are discussed in the box on p. 70. For a Boussinesq fluid (which we will introduce in Section 4.1.2), the averaging attributed to Osborn Reynolds is most commonly performed, which is sometimes called *Reynolds averaging*. This procedure will be extensively used and discussed in Chapters 11 and 12. With a non-constant density entering the conservation laws, however, another procedure is more appropriate: the HESSELBERG[20] average (Hesselberg, 1926), which we discuss now.

**13. Weighted and Statistical Averages**

In principle, one could define an average by

$$\overline{\psi} = \int_{-\infty}^{\infty} W(t,t')\psi(t')\mathrm{d}t' \tag{B13.1}$$

which satisfies (2.140) provided that $\int W(t,t')\mathrm{d}t' = 1$ and $\int W(t,t'')W(t'',t')\mathrm{d}t'' = W(t,t')$. The kernel $W(t,t')$, of course, must contain the appropriate scales corresponding to temporal/spatial averaging. If the motions at smaller scales are described statistically, a statistical average arises which would likewise satisfy (2.140).
To simplify the work arising from the noncommutativity of averaging and differential operators, the averaging procedure is frequently formulated in terms of an ensemble of states $\psi(\boldsymbol{x},t;\lambda)$ such that $\overline{\psi}(\boldsymbol{x},t) = \int \mathrm{d}P(\lambda)\psi(\boldsymbol{x},t;\lambda)$ is the expectation with respect to a probability measure $\mathrm{d}P(\lambda)$ and $\psi' = \psi - \overline{\psi}$ is the deviation of a particular realization. We will assume that the averaging operation obeys (2.140) although in practice this may be difficult to achieve.

In an Eulerian framework, it is most straightforward to consider averages of the state variables *per volume*, not per mass. We thus define the mean state variables by mass-weighted averages according to

$$\rho_{\mathrm{m}} = \overline{\rho}\,, \quad S_{\mathrm{m}} = \frac{\overline{\rho S}}{\rho_{\mathrm{m}}}\,, \quad \theta_{\mathrm{m}} = \frac{\overline{\rho\theta}}{\rho_{\mathrm{m}}}\,, \quad \boldsymbol{u}_{\mathrm{m}} = \frac{\overline{\rho\boldsymbol{u}}}{\rho_{\mathrm{m}}} \tag{2.141}$$

[20] THEODOR HESSELBERG, *1885 in Lierne, †1966 in Oslo, meteorologist.

In this way the mean concentrations per volume are preserved since we have e. g. $\rho_\mathrm{m} S_\mathrm{m} = \overline{\rho S}$ etc. As an example, we will explicitly discuss the salt conservation (2.19), the average of which is

$$\frac{\partial \overline{\rho S}}{\partial t} + \nabla \cdot \overline{\rho S \boldsymbol{u}} = -\nabla \cdot \overline{\boldsymbol{J}}_\mathrm{S} \tag{2.142}$$

Decomposing the variables according to

$$\rho = \rho_\mathrm{m} + \rho' , \quad S = S_\mathrm{m} + S' , \quad \boldsymbol{u} = \boldsymbol{u}_\mathrm{m} + \boldsymbol{u}' \quad \text{etc.} \tag{2.143}$$

we obtain

$$\frac{\partial \rho_\mathrm{m} S_\mathrm{m}}{\partial t} + \nabla \cdot \rho_\mathrm{m} S_\mathrm{m} \boldsymbol{u}_\mathrm{m} = -\nabla \cdot \left( \overline{\boldsymbol{J}}_\mathrm{S} + \boldsymbol{J}_\mathrm{S}^\mathrm{turb} \right) \tag{2.144}$$

The mean salinity hence satisfies a conservation equation completely identical to (2.19), except for the addition of a *turbulent* salinity flux $\boldsymbol{J}_\mathrm{S}^\mathrm{turb}$ which is given by

$$\begin{aligned} \boldsymbol{J}_\mathrm{S}^\mathrm{turb} &= \overline{\rho S \boldsymbol{u}} - \rho_\mathrm{m} S_\mathrm{m} \boldsymbol{u}_\mathrm{m} = \overline{(\rho_\mathrm{m} + \rho')(S_\mathrm{m} + S')(\boldsymbol{u}_\mathrm{m} + \boldsymbol{u}')} - \rho_\mathrm{m} S_\mathrm{m} \boldsymbol{u}_\mathrm{m} \\ &= \overline{\rho'} S_\mathrm{m} \boldsymbol{u}_\mathrm{m} + \boldsymbol{u}_\mathrm{m} \left( \rho_\mathrm{m} \overline{S'} + \overline{\rho' S'} \right) + S_\mathrm{m} \left( \rho_\mathrm{m} \overline{\boldsymbol{u}'} + \overline{\rho} \overline{\boldsymbol{u}'} \right) + \rho_\mathrm{m} \overline{S' \boldsymbol{u}'} + \overline{\rho' S' \boldsymbol{u}'} \end{aligned} \tag{2.145}$$

From (2.141) and (2.140) we have $\overline{\rho'} = 0$. Furthermore, from (2.141) it follows that

$$S_\mathrm{m} = \frac{\overline{(\rho_\mathrm{m} + \rho')(S_\mathrm{m} + S')}}{\rho_\mathrm{m}} = \frac{\rho_\mathrm{m} S_\mathrm{m} + \rho_\mathrm{m} \overline{S'} + \overline{\rho' S'}}{\rho_\mathrm{m}} = S_\mathrm{m} + \overline{S'} + \frac{\overline{\rho' S'}}{\rho_\mathrm{m}}$$

and thus

$$\overline{S'} = -\frac{\overline{\rho' S'}}{\rho_\mathrm{m}} , \quad \overline{\boldsymbol{u}'} = -\frac{\overline{\rho' \boldsymbol{u}'}}{\rho_\mathrm{m}} \quad \text{etc.}$$

Hence only the last two terms in (2.145) remain, and we obtain

$$\boldsymbol{J}_\mathrm{S}^\mathrm{turb} = \rho_\mathrm{m} \overline{S' \boldsymbol{u}'} + \overline{\rho' \boldsymbol{u}' S'} \tag{2.146}$$

Even when the turbulent fluctuations are strong, the density fluctuations in the ocean are always small, with a typical magnitude of $\rho'/\rho \sim 10^{-3}$. To a good approximation, the triple-correlations in (2.146) can hence be neglected, and one obtains the more common form

$$\boldsymbol{J}_\mathrm{S}^\mathrm{turb} = \rho_\mathrm{m} \overline{S' \boldsymbol{u}'} \tag{2.147}$$

It is, however, largely irrelevant whether (2.146) or (2.147) is used because, in any case, the turbulent flux ultimately has to be parameterized in terms of the mean state. Analogous considerations can be made for the temperature and momentum equations.

$$\mathrm{J}_\theta^\mathrm{turb} = \rho_\mathrm{m} \overline{\theta' \boldsymbol{u}'} + \overline{\rho' \boldsymbol{u}' \theta'} \approx \rho_\mathrm{m} \overline{\theta' \boldsymbol{u}'} \tag{2.148}$$

$$\Sigma^\mathrm{turb} = -\left( \rho_\mathrm{m} \overline{\boldsymbol{u}' \boldsymbol{u}'} + \overline{\rho' \boldsymbol{u}' \boldsymbol{u}'} \right) \approx -\rho_\mathrm{m} \overline{\boldsymbol{u}' \boldsymbol{u}'} \tag{2.149}$$

Note, however, that the definitions in (2.141) are such that no additional terms appear in the continuity equation (2.15). In fact, a term of turbulent density diffusion would be awkward in view of the absence of molecular density diffusion.

The equation of state (2.121) is nonlinear, and the outcome of averaging as defined in (2.141) is not immediately evident. We perform a Taylor expansion around the mean variables and subsequently average the result to obtain

$$\overline{\rho} = \rho_\mathrm{m} = F(S_\mathrm{m}, \theta_\mathrm{m}, p_\mathrm{m}) + \frac{\partial F}{\partial \theta}\overline{\theta'} + \cdots + \frac{\partial^2 F}{\partial \theta^2}\overline{\theta'^2} + \cdots \tag{2.150}$$

where many further terms of first and second order occur. The derivatives are taken at $(S_\mathrm{m}, \theta_\mathrm{m}, p_\mathrm{m})$. Of all contributions to (2.150), the last term is by far the largest one. With $\partial^2 F/\partial\theta^2 \approx \rho_0 10^{-5}\,\mathrm{K}^{-2}$, its magnitude depends on the strength of the temperature fluctuations. For $(\overline{\theta'^2})^{1/2} = 1\,\mathrm{K}$, the relative contribution of that term is $5 \times 10^{-6}$, which is of same order as the accuracy of the approximations used for the state equation and hence negligible.

It follows that the equations for the mean state variables according to (2.141) are to a good approximation given by (omitting the index $m$ from all variables)

$$\frac{\mathrm{D}\rho}{\mathrm{D}t} = -\rho\nabla\cdot\boldsymbol{u} \tag{2.151}$$

$$\rho\frac{\mathrm{D}\boldsymbol{u}}{\mathrm{D}t} = -2\rho\boldsymbol{\Omega}\times\boldsymbol{u} - \nabla p - \rho\nabla\Phi + \mathcal{F} \tag{2.152}$$

$$\rho\frac{\mathrm{D}S}{\mathrm{D}t} = \mathcal{G}_\mathrm{S} \tag{2.153}$$

$$\rho\frac{\mathrm{D}\theta}{\mathrm{D}t} = \mathcal{G}_\theta \tag{2.154}$$

$$\rho = F(S, \theta, p) \tag{2.155}$$

with the flux divergencies

$$\begin{aligned}
\mathcal{F} &= \nabla\cdot\left(\Sigma^\mathrm{turb} + \overline{\Sigma}\right)\\
\mathcal{G}_\mathrm{S} &= -\nabla\cdot\left(\boldsymbol{J}_\mathrm{S}^\mathrm{turb} + \overline{\boldsymbol{J}}_\mathrm{S}\right)\\
\mathcal{G}_\theta &= -\nabla\cdot\left(\boldsymbol{J}_\theta^\mathrm{turb} + \left(\overline{\boldsymbol{J}}_\mathrm{rad} + \overline{\boldsymbol{J}}_\mathrm{T}\right)/c_p^\star\right)
\end{aligned}$$

as the divergences of the overall (turbulent plus mean molecular) fluxes; the latter are very small as discussed above. All following considerations in this book are based on the filtered system (2.151) to (2.155), if not explicitly stated otherwise. However, for a complete description, expressions, i. e. parameterizations, for $\mathcal{F}$, $\mathcal{G}_\mathrm{S}$ and $\mathcal{G}_\theta$ have to be provided, based on the known averaged quantities. These parameterizations, typically diffusive closures, will be discussed in the Chapters 11 and 12.

## 2.9 The State of Rest

The first application of the governing equations derived in the previous sections is for a state of rest without any forcing except the force of gravity. Due to the presence of gravity, pressure and density variations are closely related. Furthermore, it is discussed under which condition a density stratification caused by temperature and salinity is stable against small perturbations.

### 2.9.1 Hydrostatic Balance

Consider a motionless steady state which is defined by the state variables $\boldsymbol{u}_r \equiv 0$, $p_r$, $\rho_r$, $\theta_r$ and $S_r$ and vanishing turbulent fluxes, i. e. $\mathcal{F} = \mathcal{G}_S = \mathcal{G}_\theta = 0$. The equations of motion (2.152) then reduce to

$$0 = -\nabla p_r - \rho_r \nabla \Phi \tag{2.156}$$

which can only be satisfied if both conditions $\nabla \rho_r \times \nabla \Phi = 0$ and $\nabla \rho_r \times \nabla p_r = 0$ hold. It follows that $\rho_r = \rho_r(\Phi)$ and also $p_r = p_r(\Phi)$, and, therefore, $\mathrm{d}p_r/\mathrm{d}\Phi = -\rho_r$ or with $\Phi = gz$

$$\frac{\mathrm{d}p_r}{\mathrm{d}z} = -g\rho_r \tag{2.157}$$

Equation (2.157) is the *hydrostatic relation*. In the state of rest there are no forces in the surfaces $\Phi$ = const to accelerate the fluid. The equation of state (2.121) takes

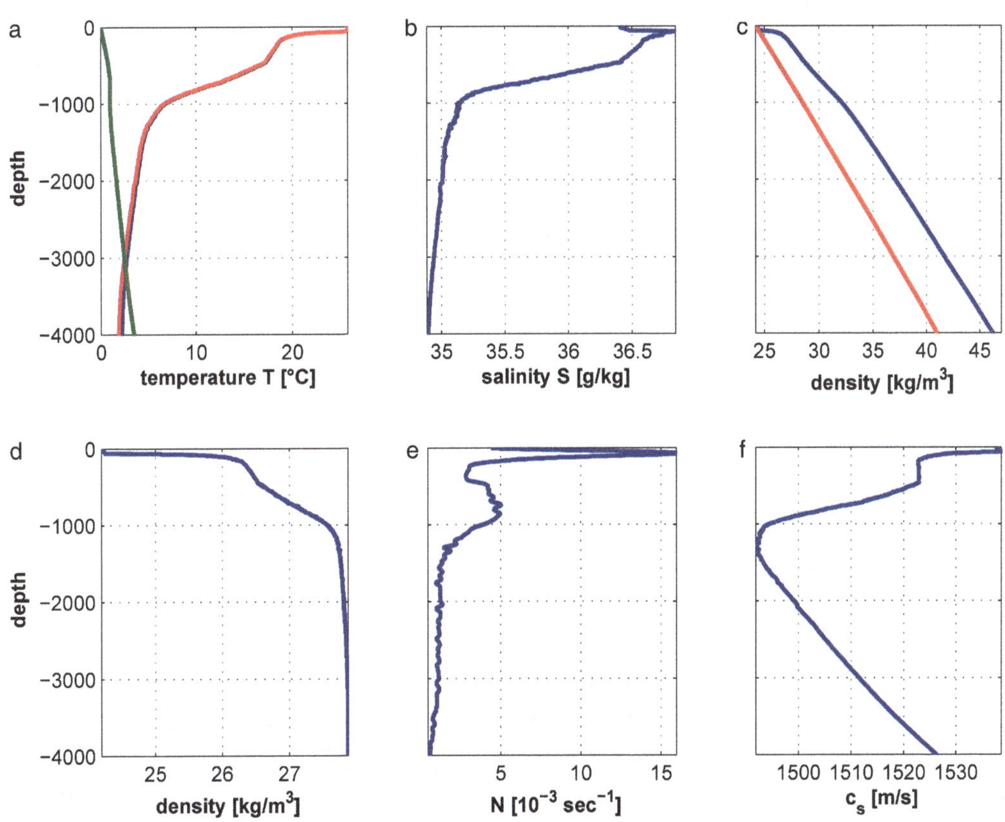


Fig. 2.17 Profiles of *in-situ* temperature (blue) and potential temperature (red) (**a**) and salinity (**b**) in the subtropical North Atlantic near Bermuda. **a** also shows the difference between *in-situ* and conservative temperature (green, amplified by a factor of 10). A profile of density (in $\mathrm{kg\,m^{-3}}$; a reference value of $1{,}000\,\mathrm{kg\,m^{-3}}$ is subtracted), taken at the *in-situ* pressure is shown in the blue curve in **c**. The straight red line in **c** is the *in-situ* density for constant temperature and salinity (the values are those at the surface). It demonstrates the pressure dependence of the density. In **d**, the density is evaluated at the surface pressure. **e** shows the Brunt–Väisälä frequency (in $10^{-3}\,\mathrm{s^{-1}}$) at the station, **f** shows the profile of the sound velocity (in $\mathrm{m\,s^{-1}}$)

the form

$$\rho_r(z) = F(S_r, \theta_r, p_r(z)) \tag{2.158}$$

As the equation of state is nonlinear, it follows that both potential temperature $\theta_r$ and salinity $S_r$ can only depend on $z$ but not on $x, y, t$. For given profiles of $S_r(z)$ and $\theta_r(z)$, the vertical distribution of density and pressure can be calculated by combining (2.157) and the equation of state. By vertical differentiation of (2.158), and invoking (2.125) to (2.128), one obtains

$$\frac{1}{\rho_r}\frac{d\rho_r}{dz} = \gamma\frac{dS_r}{dz} - \alpha\frac{d\theta_r}{dz} - \frac{g}{c_r^2} \tag{2.159}$$

where $c_r(z)$ is the sound velocity for the hydrostatic state[21]. For typical oceanic conditions, the density changes due to the compressibility of sea water (last term in (2.159)) are approximately an order of magnitude larger than those due to the salinity/temperature variations, as demonstrated in Figure 2.17 where profiles of temperature, salinity and appropriate density variables are shown for a CTD cast near Bermuda in the subtropical North Atlantic.

As shown in the box on p. 74, the density variations in the ocean do not exceed 2–3%. Inserting (B14.1) into the hydrostatic balance (2.157), we obtain the depth

**14. Scale Depth**

According to (2.159), the density profile can be calculated in a situation without the stratification of temperature and salinity so that $S_r, \theta_r$ = const. It is convenient to introduce instead of the sound velocity the *scale depth* D by

$$\mathrm{D}(z) = c_r^2(z)/g$$

where $c_r = c_s(S_r, \theta_r, p_r(z))$ is the reference state of the sound speed. It follows from (2.159) that $d\rho_r/dz = -\rho_r/\mathrm{D}$ which has the solution

$$\rho_r(z) = \rho_r(0)\exp\int_z^0 \frac{1}{\mathrm{D}(z')}dz'$$

The scale depth increases with temperature and pressure and is almost independent of salinity. Typical oceanic values are around **210** km in the upper ocean and **240** km at **5,000** m depth. Ignoring the weak dependency of the scale depth on $z$, we may write

$$\rho_r(z) = \rho_r(0)e^{-z/\mathrm{D}} \approx \rho_r(0)(1 - z/\mathrm{D}) \tag{B14.1}$$

The last approximation holds because the water depth is always much smaller than the scale depth. In the state of rest, the relative variations of density hence do not exceed $H/\mathrm{D} \sim 2-3\%$. Most oceanic densities range between $\rho \approx$ **1025–1050** kg m$^{-3}$.
Note the analogy of the oceanic scale depth to the atmospheric *scale height* $H_s$ which for an isothermal atmosphere of mean temperature $T_m$ is given as $H_s = R_m T_c/g \approx$ **7.4** km ($R_m$ is the gas constant). Thus, while in the ocean the scale depth is much larger than the ocean depth, in the atmosphere the scale height is of similar magnitude as the height of the troposphere (see also Section 2.10).

[21] In the following, $\alpha, \gamma$ and $\kappa$ are the modified coefficients (2.122)–(2.124). They are to be evaluated at the local values of $\theta_r$, $S_r$ and $p_r$.

dependence of the reference pressure

$$p_r(z) = p_r(0) + gD\rho_r(0)\left[\exp\left(-z/D\right) - 1\right] \approx p_r(0) - gz\rho_r(0)\left(1 - \frac{z}{2D}\right) \tag{2.160}$$

which can be further approximated (to an accuracy of 2–3%) to the commonly used form

$$p_r(z) \approx p_r(0) - gz\rho_0 \tag{2.161}$$

where $\rho_0 = \rho_r(0)$. Hence the pressure of 1 bar corresponds approximately to a water column of 9.81 m (see also the box on p. 20).

### 2.9.2 Static Stability

The state of rest as described in the previous section constitutes an equilibrium state of the ocean. As we have seen, the stratification of temperature and salinity has only a minor influence on the overall density. Nevertheless, the stratification is fundamental for ocean dynamics as it controls the vertical stability.

Consider a fluid particle which is displaced *adiabatically* by a small amount $\xi$ in the vertical direction from its level $z$ to $z + \xi$. If the displacement is *upwards*, and the particle at its new position is denser than the surrounding fluid, the force of gravity will act to bring the particle back towards its previous level, and the stratification is said to be *stable*. If, on the other hand, the particle at its new position is less dense than the surrounding fluid, gravity will force the particle to continue moving upwards, and the stratification is said to be *unstable*. If, on the other hand, the displacement is *downwards*, the situation is reversed, and a less dense particle will return to its previous position.

Since the particle is displaced adiabatically, its salinity and potential temperature remain constant, and its density can only change due to the different pressure at the new position $z + \xi$. The pressure at the new position is $p_r(z + \xi) \approx p_r(z) - g\rho_r\xi$. The density change $\Delta\rho_{ad}$ of the particle hence follows from (2.158) as

$$\Delta\rho_{ad} = \rho_r\,\kappa\,\Delta p = -\frac{g\rho_r}{c_r^2}\xi \tag{2.162}$$

In the surrounding fluid, not only the pressure but also salinity and potential temperature change, and its density change $\Delta\rho_{env}$ is hence given by

$$\Delta\rho_{env} \approx \frac{\partial\rho_r}{\partial z}\xi = \rho_r\left(-\frac{g}{c_r^2} - \alpha\frac{\partial\theta_r}{\partial z} + \gamma\frac{\partial S_r}{\partial z}\right)\xi$$

The density difference of the particle and the surrounding fluid at the new position is thus

$$\Delta\rho_{ad} - \Delta\rho_{env} = \left(\alpha\frac{\partial\theta_r}{\partial z} - \gamma\frac{\partial S_r}{\partial z}\right)\rho\xi \tag{2.163}$$

For $\xi > 0$, stability requires that the particle is denser than its new environment. Therefore, the stratification is stable provided that $\Delta\rho_{ad} > \Delta\rho_{env}$, and unstable for

$\Delta\rho_{ad} < \Delta\rho_{env}$. Conversely, for negative $\xi$ the stratification is stable for $\Delta\rho_{ad} < \Delta\rho_{env}$, and unstable for $\Delta\rho_{ad} > \Delta\rho_{env}$. In both cases, stability therefore requires that the sign of $\Delta\rho_{ad} - \Delta\rho_{env}$ is the same as the sign of $\xi$. From (2.163) it follows then that a stability criterion can be conveniently formulated in terms of

$$N^2 = g\left(\alpha\frac{\partial\theta_r}{\partial z} - \gamma\frac{\partial S_r}{\partial z}\right) \tag{2.164}$$

The stratification is stable if $N^2 > 0$, unstable if $N^2 < 0$, and neutrally stable if $N^2 = 0$. The variable $N$ has the dimension of an inverse time, and is called BRUNT–VÄISÄLÄ[22]–frequency (also *stability* or *buoyancy* frequency). The momentum balance for the particle is given by $\rho\ddot{\xi} = g(\Delta\rho_{env} - \Delta\rho_{ad})$, or

$$\ddot{\xi} = -N^2\xi \quad \text{with the solution} \quad \xi = \xi_0 e^{iNt}$$

For stable stratification, the particle undergoes *stability oscillations* with a period $2\pi/N$. In the ocean, the stability period varies approximately between 10 min (upper ocean) and 2 h (in the deep ocean).

**15. Other Forms of the Brunt–Väisälä frequency**

For an adiabatic displacement, the *in-situ* temperature is obviously not constant, and its changes are given by

$$(\Delta T)_{ad} = \Gamma\Delta p = -g\rho\Gamma\xi \tag{B15.1}$$

with the lapse-rate $\Gamma$. The stability frequency can be expressed in terms of $T$ and $\Gamma$ as

$$N^2 = g\left[\tilde{\alpha}\left(\frac{\partial T}{\partial z} + g\rho\Gamma\right) - \tilde{\gamma}\frac{\partial S}{\partial z}\right] \tag{B15.2}$$

which is, of course, equivalent to the formulation (2.164). The tilde is used here to indicate the use of the conventional density equation $\rho = \rho(S, T, p)$ using the *in-situ* temperature instead of the conservative temperature as state variable. Another equivalent form for $N^2$ is given by

$$N^2 = -\frac{g}{\rho}\left[\frac{\partial\rho}{\partial z} - \left(\frac{\partial\rho}{\partial z}\right)_{ad}\right] = -\frac{g}{\rho}\frac{\partial\rho}{\partial z} - \frac{g^2}{c_s^2} \tag{B15.3}$$

## 2.10 * Some Differences to Atmospheric Thermodynamics

In this section, we briefly discuss some differences of the previous results of ocean thermodynamics to the atmospheric case, in particular with respect to thermodynamics. The lesson from atmospheric thermodynamics is illuminating since the phase transition between vapor and liquid water (or ice) occurs in principle in seawater in the form of the transition between freshwater and sea ice. Such phase transitions have been neglected for the oceanic case; however, they can easily be incorporated when necessary. The way to do so can be learned from the atmospheric thermodynamics.

[22] SIR DAVID BRUNT, *1886 at Staylittle/Wales, †1965, meteorologist. VILHO VÄISÄLÄ, *1889 in Kontiolahti, †1969 in Helsinki, meteorologist and physicist.

## 2.10.1 Differences in Thermodynamics

Besides phase transitions, which are of great importance for the circulation of the atmosphere, the thermodynamics of the atmosphere is interesting with respect to a further aspect. Some of the thermodynamic properties like the equation of state can be given for the atmospheric case in an (approximate) closed analytical form, since air can be treated approximately as an ideal gas. We briefly review the thermodynamics of the atmosphere in this section.

### Atmospheric Fields

The air of the Earth's atmosphere is here considered simply as a mixture of dry air (basically oxygen and nitrogen) and water vapor with concentration $q$. The partial masses are described by

| | | |
|---|---|---|
| $q$ | specific humidity | in kg water vapor/kg moist air |
| $\rho_\mathrm{q}$ | density of vapor | in kg water vapor/m$^3$ |
| $a$ | content of dry air | in kg dry air/kg moist air |
| $\rho_\mathrm{a}$ | density of dry air | in kg dry air/m$^3$ |
| $f$ | liquid water concentration | kg water/kg moist air |

Particularly for the specification of the water vapor there are various other concepts such as the mixing ratio $r = q/(1-q)$. For the atmospheric partial masses we define the density variables

$$\rho_\mathrm{q} = q\rho\,, \quad \rho_\mathrm{a} = a\rho = (1-q)\rho\,, \quad \rho = \rho_\mathrm{q} + \rho_\mathrm{a}$$

The water vapor is highly variable in the atmosphere. As a total it accounts for about 0.25% of the mass of the atmosphere, but local concentrations range from a few parts per million by volume in the cold high regions of the Earth's atmosphere up to as much as 5% by volume in hot, humid air masses.

### Partial Masses

Air is a mixture of dry air (78% $N_2$, 21% $O_2$, 1% Ar and further trace gases), water vapor, liquid water and ice. It is normally sufficient to consider only water vapor (concentration $m_\mathrm{q} = q$) and dry air (concentration $m_\mathrm{a} = 1-q$) explicitly, and regard liquid water and ice as external components. Formally, counting the number of relevant partial masses, the situation corresponds thus to that in the ocean, with specific humidity $q$ replacing salinity $S$. The independent state variables of the atmospheric flow are thus $q, T, p$. The atmospheric equation of state is thus $\upsilon = \upsilon(q, T, p)$, and we obtain the condition

$$\frac{\partial \upsilon}{\partial q} = \upsilon_\mathrm{q} - \upsilon_\mathrm{a} \tag{2.165}$$

and corresponding relations for all additive variables. As before, $\upsilon = 1/\rho$.

### First Law

The internal energy of moist air – i. e. the mixture of dry air and water vapor – contains thermal energy of dry air and water vapor. For the first law we may hence use an analogous expression to that in the ocean, given in (1.23). In addition, however, we have to account for energy transfer during phase transitions as the enthalpy $H_{\mathrm{f}}$ of liquid water differs from the enthalpy $H_{\mathrm{q}}$ of water vapor. Since we do not account for the enthalpy stored in liquid water, that process has to be considered as an external source/sink. If a net amount $\delta^{\mathrm{e-c}}q$ of water vapor is added (through evaporation of liquid water) to or removed (through condensation) from a fluid element, the energy necessary to vaporize the water is given by $-L_{\mathrm{q}}\delta^{\mathrm{e-c}}q$ (positive when condensation is dominant) where the enthalpy difference liquid water–water vapor $H_{\mathrm{f}} - H_{\mathrm{q}} = L_{\mathrm{q}}$ as defined in (2.85) is the *latent heat* of condensation. Hence the first law has the form

$$\mathrm{d}E = \delta Q - p\mathrm{d}\upsilon + \frac{\partial H}{\partial q}\mathrm{d}q - L_{\mathrm{q}}\delta^{\mathrm{e-c}}q \tag{2.166}$$

where $\delta Q$ is the total energy exchange that is not connected to diffusive exchange of mass and to phase transitions.

### Second Law

In the atmospheric case of the second law we have to include the difference of the entropies of vapor and liquid water if condensation/evaporation occurs. Hence we have from (1.26)

$$\mathrm{d}\eta = \frac{\delta Q}{T} + \frac{\partial \eta}{\partial q}\mathrm{d}q + \left(\eta_{\mathrm{q}} - \eta_{\mathrm{f}}\right)\delta^{\mathrm{e-c}}q \tag{2.167}$$

### Equation of State

For an ideal gas, the state equation is a consequence of the empirical laws of BOYLE–MARIOTTE, GAY–LUSSAC, and AVOGADRO[23]. In retrospect of the corresponding laws and hypotheses, which bear their names, we can define a mole of a substance as the weight in grams of $N_{\mathrm{A}} = 6.022 \times 10^{23}$ of its molecules, $N_{\mathrm{A}}$ being the Avogadro number. For example, a mole of pure water is 18.016 g and the molar weight is $M = 18.016$ (dimensionless). For a mole of any (ideal) gas, the equation of state is $pV = R^{\star}T$ where $V$ is the occupied volume and $R^{\star} = 8314.36\,\mathrm{J\,kmol^{-1}\,K^{-1}}$ is the universal gas constant. For $n$ moles with mass $m = nM$ the volume increases by the factor $n$ (keeping pressure and temperature constant) and the relation becomes

$$pV = nR^{\star}T = \frac{m}{M}R^{\star}T \quad \text{or} \quad \frac{p}{\rho} = \frac{R^{\star}}{M}T = RT$$

where $R = R^{\star}/M$ is now specific for the particular gas.

[23] ROBERT BOYLE, *1627 in Lismore, †1691 in London, physicist and chemist. One of the founders of the Royal Society of London. EDME MARIOTTE, *1620 in Dijon, †1684 in Paris, physicist. LORENZO ROMANO AMEDEO CARLO AVOGADRO, *1776 in Turin, †1858 in Turin, physicist.

We thus define the respective gas constants of water vapor and dry air, $R_\mathrm{q} = R^\star/M_\mathrm{q}^\star$ and $R_\mathrm{a} = R^\star/M_\mathrm{a}^\star$, with the molar weights $M_\mathrm{q}^\star = 18.016$ and $M_\mathrm{a}^\star = 28.966$. The equations of state of the individual constituents are then $p_\mathrm{q}/\rho_\mathrm{q} = R_\mathrm{q}T$ and $p_\mathrm{a}/\rho_\mathrm{a} = R_\mathrm{a}T$, where $p_\mathrm{q}$ and $p_\mathrm{a}$ are the respective partial pressures. The equation of state of the mixture of dry air and water vapor follows from the additivity of the partial pressures to obtain the total pressure, $p = p_\mathrm{q} + p_\mathrm{a}$ (DALTON's[24] law), and the additivity of partial mass or densities to the total density, $\rho = \rho_\mathrm{q} + \rho_\mathrm{a}$. It becomes

$$\rho(q,T,p) = \frac{1}{v(q,T,p)} = \frac{p}{R_\mathrm{m}T} \tag{2.168}$$

which is identical to the state equation for an ideal gas except for the $q$-dependence of the gas constant $R_\mathrm{m}$ of the mixture,

$$R_\mathrm{m} = R_\mathrm{m}(q) = R_\mathrm{a} + \left(R_\mathrm{q} - R_\mathrm{a}\right) q = R_\mathrm{a}\left[1 + \left(M_\mathrm{a}^\star/M_\mathrm{q}^\star - 1\right) q\right] \tag{2.169}$$

### Gibbs Function

An advantage of atmospheric thermodynamics is that moist air can be treated as an ideal gas, at least away from conditions where phase transitions occur. Both constituents, dry air and vapor, behave approximately as ideal gases because intermolecular forces as well as the finite volume of molecules can be neglected, and both components behave as if no other component were present. For an ideal gas, the Gibbs function can be derived in closed form, based on the results of kinetic gas theory, as

$$G^\mathrm{id}(T,p) = RT\left(\ln p - \frac{1}{\kappa}\ln T\right) \tag{2.170}$$

where $R$ is the gas constant and $\kappa = 1/(1+\nu/2)$, with $\nu$ being the number of degrees of freedom for the molecular kinetic energy ($\kappa_\mathrm{a} = 2/7$ for the two-atomic dry air and $\kappa_\mathrm{q} = 2/8$ for the three-atomic water vapor). The Gibbs function for moist air, by additivity, can then be constructed as

$$G(q,T,p) = (1-q)\,G^\mathrm{id}(T,p_\mathrm{a}) + q\,G^\mathrm{id}(T,p_\mathrm{q}) \tag{2.171}$$

We express the partial pressures in terms of the total pressure as $p_\mathrm{a} = p(1-q)R_\mathrm{a}/R_\mathrm{m}$ respectively $p_\mathrm{q} = pqR_\mathrm{q}/R_\mathrm{m}$ with $R_\mathrm{m} = (1-q)R_\mathrm{a} + qR_\mathrm{q}$, and introduce the mass-weighted ratio $\kappa_\mathrm{m}$ for the mixture according to

$$\kappa_\mathrm{m} = \kappa_\mathrm{m}(q) = \frac{R_\mathrm{a} + \left(R_\mathrm{q} - R_\mathrm{a}\right) q}{R_\mathrm{a}/\kappa_\mathrm{a} + \left(R_\mathrm{q}/\kappa_\mathrm{q} - R_\mathrm{a}/\kappa_\mathrm{a}\right) q} \tag{2.172}$$

This equation uses Dalton's law $p = p_\mathrm{q} + p_\mathrm{a}$ and the atmospheric equation of state. The general form (2.171) results in

$$G(q,T,p) = R_\mathrm{m}\,T\left(\ln p - \frac{1}{\kappa_\mathrm{m}}\ln T\right) - T\Delta_\mathrm{m} \tag{2.173}$$

[24] JOHN DALTON, *1766 in Eaglesfield/Cumberland, †1844 in Manchester, chemist, meteorologist and physicist.

The expression for the Gibbs function of humid air hence corresponds to the ideal gas form (2.170), except for the last term in (2.173) which is

$$\Delta_{\mathrm{m}} = \Delta_{\mathrm{m}}(q) = -qR_{\mathrm{q}} \ln\left(qR_{\mathrm{q}}\right) + R_{\mathrm{m}}(q) \ln R_{\mathrm{m}}(q) - (1-q)R_{\mathrm{a}} \ln\left[(1-q)R_{\mathrm{a}}\right] \tag{2.174}$$

It is always positive, $\Delta_{\mathrm{m}}(q) \geq 0$, and specifically $\Delta_{\mathrm{m}}(q) \approx -qR_{\mathrm{q}} \ln q$ for $q \ll 1$. The entropy of moist air follows through differentiation according to (1.49) as

$$\eta(q, T, p) = \frac{R_{\mathrm{m}}}{\kappa_{\mathrm{m}}}(1 + \ln T - \kappa_{\mathrm{m}} \ln p) + \Delta_{\mathrm{m}} \tag{2.175}$$

and corresponds to the entropy of an ideal gas with 'constants' $R_{\mathrm{m}}(q)$ and $\kappa_{\mathrm{m}}(q)$, except for $\Delta_{\mathrm{m}}(q)$ which is the additional entropy due to mixing of the two gases. Other thermodynamic potentials can likewise be derived through differentiation according to (1.49). In particular one obtains the enthalpy, internal energy and the 'chemical potential' as

$$H(q, T, p) = \frac{R_{\mathrm{m}}}{\kappa_{\mathrm{m}}} T \tag{2.176}$$

$$E(q, T, p) = \frac{R_{\mathrm{m}}}{\kappa_{\mathrm{m}}}(1 - \kappa_{\mathrm{m}}) T \tag{2.177}$$

$$\mu(q, T, p) = (R_{\mathrm{q}} - R_{\mathrm{a}}) T \ln p - \left(\frac{R_{\mathrm{q}}}{\kappa_{\mathrm{q}}} - \frac{R_{\mathrm{a}}}{\kappa_{\mathrm{a}}}\right) T \ln T - T \frac{\partial \Delta_{\mathrm{m}}}{\partial q} \tag{2.178}$$

For $q = 0$ one recovers the corresponding relations for dry air. For typical values of $q = O(0.01)$ in the atmosphere, $R_{\mathrm{m}}(q) \approx R_{\mathrm{a}}$ with an error of approximately 1%, and likewise $\kappa_{\mathrm{m}}(q) \approx \kappa_{\mathrm{a}} = 2/7$. Hence for many problems the effect of $q$ on the thermodynamics can be ignored (except when phase transitions are involved where the above consideration of ideal gas behavior is not valid).

### Specific Heat

The specific heat of moist air is evaluated from (1.57) and (1.58),

$$c_p = \frac{\partial H}{\partial T} = \frac{R_{\mathrm{m}}}{\kappa_{\mathrm{m}}} = \frac{7}{2}\left[R_{\mathrm{a}} + \left(\frac{8}{7}R_{\mathrm{q}} - R_{\mathrm{a}}\right) q\right] \tag{2.179}$$

$$c_v = \left(\frac{\partial E}{\partial T}\right)_{\mathrm{v}} = \frac{5}{2}\left[R_{\mathrm{a}} + \left(\frac{6}{5}R_{\mathrm{q}} - R_{\mathrm{a}}\right) q\right] \tag{2.180}$$

Both specific heats are strictly independent of temperature and pressure, as to be expected for a mixture of two ideal gases. Their dependence on specific humidity is very weak, and their ratio is to a very good approximation given as $c_p/c_v \approx 7/5$, i. e. its value for dry air.

## 2.10.2 Differences in Conservation Laws

The differences in the thermodynamics of the atmosphere with respect to the ocean have consequences for the conservation laws for water vapor (replacing salt conservation), energy and the temperature equations. These budgets are discussed in this section.

### Water Vapor Balance

For the atmospheric counterpart, the water vapor $q$ must satisfy a conservation law corresponding to (2.17), with a diffusive water vapor flux $\boldsymbol{J}_\mathrm{q}$ and a source term $C_\mathrm{q} = e - c$, accounting for the phase transition (evaporation and condensation), where $e$ is the rate of evaporation and $c$ the rate of condensation (both in $\mathrm{kg\,m^{-3}\,s^{-1}}$ of liquid water). Hence

$$\rho \frac{\mathrm{D}q}{\mathrm{D}t} = -\nabla \cdot \boldsymbol{J}_\mathrm{q} + C_\mathrm{q} \tag{2.181}$$

The diffusive flux $\boldsymbol{J}_\mathrm{q}$ of water vapor can be expressed in a similar form as outlined above for salt, with a diffusivity $\kappa_\mathrm{q} \approx 2.4 \times 10^{-5}\,\mathrm{m^2\,s^{-1}}$ (at 8°C).

Dry air satisfies a sourceless balance, and strictly speaking, the 'total' mass with density $\rho = \rho_\mathrm{a} + \rho_\mathrm{q}$ would then not be conserved: there is a loss of moist air to the liquid or ice phase. This loss is small and generally ignored in the total mass balance. Strictly, however, two more partial mass compartments must be included, the liquid and the frozen phase of water. Equivalently, an ice phase could be considered for seawater with corresponding source terms in the balances of freshwater and ice.

### Energy Changes by Phase Transitions

Writing $C_\mathrm{q} = e - c$ for the difference between evaporations and condensation (i. e. the source of vapor) per volume and time, we obtain $-L_\mathrm{q} C_\mathrm{q}$ as an additional source term in the enthalpy budget as compared to the oceanic case (2.76) (positive if condensation prevails, i. e. $C_\mathrm{q} < 0$). Similar contributions arise from freezing of melting processes.

### Latent Energy

The conversion term $-L_\mathrm{q} C_\mathrm{q}$, representing the energy exchange due to phase transitions, has no equivalent in the ocean energetics because ice freezing/melting occurs only very localized at the surface and is usually treated in the boundary conditions and not in the interior balances. It is, however, quite important in the atmosphere. With a minor assumption (taking $L_\mathrm{q}$ as constant, i. e. neglecting its temperature dependence), this term can be brought to a flux form as well. The *latent* energy is defined as $E^\mathrm{lat} = L_\mathrm{q} q$, and is that part of internal energy of the liquid phase which is lost in the phase transition and gained by the internal energy of the vapor. From the conservation of water vapor we then obtain

$$\rho \frac{\mathrm{D}E^\mathrm{lat}}{\mathrm{D}t} \approx \rho L_\mathrm{q} \frac{\mathrm{D}q}{\mathrm{D}t} = -L_\mathrm{q} \nabla \cdot \boldsymbol{J}_\mathrm{q} + L_\mathrm{q} C_\mathrm{q} \approx -\nabla \cdot (L_\mathrm{q} \boldsymbol{J}_\mathrm{q}) + L_\mathrm{q} C_\mathrm{q} \tag{2.182}$$

where $L_\mathrm{q} \boldsymbol{J}_\mathrm{v}$ is defined as *latent heat flux*. The combined energy $E^\mathrm{tot} + E^\mathrm{lat}$ is then governed by

$$\rho \frac{\mathrm{D}}{\mathrm{D}t} \left(E^\mathrm{tot} + E^\mathrm{lat}\right) \approx -\nabla \cdot \left(\boldsymbol{J}^\mathrm{tot} + L_\mathrm{q} \boldsymbol{J}_\mathrm{q}\right) + \rho \frac{\partial \Phi_\mathrm{tide}}{\partial t} \tag{2.183}$$

which reveals a total energy flux $\boldsymbol{J}^\mathrm{tot} + L_\mathrm{q} \boldsymbol{J}_\mathrm{q}$ for this energy variable. Only the external tidal forces and the radiative flux at the outer rim of the atmosphere remain to change this form of energy.

### Temperature Equation

The temperature equation for the atmosphere is derived in complete analogy, starting with (2.76) and incorporating the energy transfer expressions for phase transitions and evaporation/condensation terms in the conservation of water vapor. Again, the dissipation of energy and the 'chemical' terms are small and one finds

$$Q_{\mathrm{T}} \approx -\nabla \cdot (\boldsymbol{J}_{\mathrm{T}} + \boldsymbol{J}_{\mathrm{rad}}) - L_{\mathrm{q}} C_{\mathrm{q}} \tag{2.184}$$

### Potential Temperature

Using the form of entropy (2.175) for a moist atmosphere, we obtain an implicit definition of the potential temperature from the entropy relation (2.113), i. e.

$$\begin{aligned}\eta(q, T, p) &= R\left[\frac{1}{\kappa_{\mathrm{m}}}\left(1 + \ln\frac{T}{T_0}\right) - \ln\frac{p}{p_0}\right] + \xi(q)\\ &= R\left[\frac{1}{\kappa_{\mathrm{m}}}\left(1 + \ln\frac{\theta}{T_0}\right)\right] + \xi(q)\end{aligned}$$

where $\xi(q)$ is easily computed from $\Delta_{\mathrm{m}}(q)$. Evaluating for $\theta$ yields

$$\theta = T\left(\frac{p_0}{p}\right)^{\kappa_{\mathrm{m}}} \tag{2.185}$$

which is in accordance with the considerations of the ideal gas physics as discussed above. The only difference is that here $\kappa_{\mathrm{m}} = \kappa_{\mathrm{m}}(q)$ refers to the mixture of dry air and water vapor. For the definition of $\kappa_{\mathrm{m}}(q)$ we refer to (2.172). In fact, the dependence on $q$ is weak. In contrast to the ocean case, however, $\theta$ and $T$ differ quite substantially in the atmosphere. Following (2.179), the specific heat is almost constant, $c_p^0 = c_p$. Neglecting the 'chemical' terms here as well we hence find

$$\rho c_p \frac{\mathrm{D}\theta}{\mathrm{D}t} = -\frac{\theta}{T}\left(\nabla \cdot \boldsymbol{J}_{\mathrm{H}} + L_{\mathrm{q}} C_{\mathrm{q}}\right) \tag{2.186}$$

This form is usually applied. The equivalent form

$$\rho\frac{\mathrm{D}\theta}{\mathrm{D}t} = -\nabla \cdot \left(\frac{\theta}{c_p T}\boldsymbol{J}_{\mathrm{H}}\right) + \frac{\boldsymbol{J}_{\mathrm{H}}}{c_p} \cdot \nabla\frac{\theta}{T} - \frac{\theta}{c_p T} L_{\mathrm{q}} C_{\mathrm{q}} \tag{2.187}$$

oriented at the general form of the conservation equation, is less customary.

The potential temperature is not conserved in the moist atmosphere, not only because of the evaporation/precipitation term, but also because $\theta/T$ is significantly not constant. The use of $\theta$ instead of $T$, therefore, has less advantages than for the ocean. In atmospheric models one thus usually applies the *in-situ* temperature equation (2.101) with the source term (2.184).

### Potential Density

For a mixture of ideal gases, as we treat the atmosphere, the use of $\theta$ instead of $T$ as a variable in the equation of state is straightforward. By combination of (2.185) and (2.168) one obtains

$$\rho(q, T, p) = \frac{p}{R_{\mathrm{m}}(q)\theta}\left(\frac{p_0}{p}\right)^{\kappa_{\mathrm{m}}(q)}$$

### Hesselberg Average

For the atmosphere, the mass-weighted average (see Section 2.8.2) has the advantage of exact validity of the ideal gas law for the averaged fields: indeed, from the equation $p = \rho R T$ we obtain $\overline{p} = \rho_{\mathrm{m}} R T_{\mathrm{m}}$ with the mass-averaged temperature $T_{\mathrm{m}} = \overline{\rho T}/\rho_{\mathrm{m}}$. This can be used for each partial mass (dry air and water vapor) and then transferred to the equation of state.

### State of Rest

It is also interesting to compare the oceanic state of rest with the atmospheric one. With the state equation (2.168), the hydrostatic equation (2.157) takes the form

$$\frac{1}{p}\frac{\mathrm{d}p}{\mathrm{d}z} = -\frac{g}{R_{\mathrm{m}} T} \tag{2.188}$$

which can be used to determine the pressure distribution for a given temperature profile. A simple form is obtained when the atmosphere is assumed isothermal i. e. $T \approx T_{\mathrm{c}} = 250\,\mathrm{K}$ (the error is $< 15\,\%$). It follows that

$$p(z) = p(0)\exp(-z/H_{\mathrm{s}}) \tag{2.189}$$

where $H_{\mathrm{s}} = R_{\mathrm{m}} T_{\mathrm{c}}/g \approx 7.4\,\mathrm{km}$ is the atmospheric *scale height*. In this approximation, the density decreases exponentially with increasing height but more substantially than the oceanic counterpart because $H_{\mathrm{s}}$ is of the order of the tropospheric height.

## 2.11 Vorticity

We have seen in the previous sections that salinity and potential temperature are materially conserved under the restrictive conditions of an adiabatic flow. The quantities keep their value in a moving fluid parcel and the corresponding conservation laws are trivially solved. For momentum such a situation does not occur, even when all external forces vanish and adiabatic conditions apply. The reason is the ubiquitous presence of the pressure force. However, there are materially conserved quantities which relate to momentum.

Forming the curl of the momentum balance, gradient forces cancel out and the balance of vorticity is derived. The concept of vorticity leads to associated quantities which are materially conserved – or very approximately so – in many nontrivial and interesting dynamical situations. In Section 1.1.2 we have shown the similarity of vorticity and angular momentum, and indeed there is some analogy between the vorticity-related conservation theorems introduced in this section and the conservation of angular momentum of a solid body. The name 'vorticity' was introduced by LAMB[25] in 1916.

The vorticity theorems apply to locally defined quantities relating the vorticity vector to integral properties of closed material loops of fluid. In Cartesian coordinates

[25] SIR HORACE LAMB, *1849 in Stockport, †1934 in Cambridge, fluid dynamicist.

the vorticity vector is written as (cf. (1.12))

$$\boldsymbol{\omega} = (\omega_1, \omega_2, \omega_3) = \left( \frac{\partial u_3}{\partial x_2} - \frac{\partial u_2}{\partial x_3}, \ \frac{\partial u_1}{\partial x_3} - \frac{\partial u_3}{\partial x_1}, \ \frac{\partial u_2}{\partial x_1} - \frac{\partial u_1}{\partial x_2} \right) \tag{2.190}$$

In Section 1.1.2 we have related the vorticity vector $\boldsymbol{\omega} = \nabla \times \boldsymbol{u}$ to the angular velocity $(1/2)\boldsymbol{\omega} \cdot \boldsymbol{n}$ of an infinitesimally small disk with normal vector $\boldsymbol{n}$. In fact, for a rigid body rotation where $\boldsymbol{u} = \boldsymbol{\omega}_0 \times \boldsymbol{r}$ with angular velocity $\boldsymbol{\omega}_0$ we find $\boldsymbol{\omega} = 2\boldsymbol{\omega}_0$. Here, parcel trajectories are clearly curved, but this property is not mandatory for nonzero vorticity: in a simple unidirectional shear flow $\boldsymbol{u} = (0, \kappa x, 0)$ the vorticity is nonzero, $\boldsymbol{\omega} = (0, 0, \kappa)$ (see Figure 2.18 for the shape of the shear flow).

The *circulation* $C$ around a closed curve $\Gamma$ in the fluid is defined by the line integral of the tangential velocity vector,

$$C = \oint_{\Gamma} \boldsymbol{u} \cdot \mathrm{d}\boldsymbol{s} \tag{2.191}$$

The closed curve $\Gamma$ can be defined at any initial time anywhere in the fluid. Following the fluid motion of the individual fluid parcels on $\Gamma$ for subsequent times, $C = C(t)$ becomes a material property of the parcel group $\Gamma = \Gamma(t)$. The circulation around a rectangle of dimension $L \times B$ with above specified velocity field is $C = \kappa L B$, in this case equal to vorticity times the enclosed area. In the course of time the rectangle will be deformed to a parallelogram, but the area and the circulation will remain constant. These properties are, of course, more general, as will be investigated in this chapter.

On the other hand, curved trajectories do not imply nonzero vorticity, as can be seen in the two-dimensional point vortex (see right panel of Figure 2.18) described by the stream function (in cylindrical coordinates) $\Psi(r, \varphi) = -(\kappa/2\pi) \ln r$. Here, the velocity is tangential, $u^{\varphi} = \kappa/(2\pi r)$, and the vorticity $\boldsymbol{\omega} = \boldsymbol{k}(1/r)\partial(r u^{\varphi})/\partial r$ is vertical to the two-dimensional motion. It vanishes everywhere except in the center where it is infinite. The circulation around any loop enclosing the center equals $\kappa$, and it vanishes for loops that do not contain the center. Again, $C(t)$ is conserved since material loops cannot cross the center. In these examples we find, with $\kappa > 0$, a counterclockwise local rotation ($\omega_3 > 0$) about the vertical. A vorticity and the corresponding flow is called *cyclonic* (cum sole) if the rotation is the same as the Earth rotation (i. e. positive in the northern hemisphere). An *anticyclone* rotates the other way.

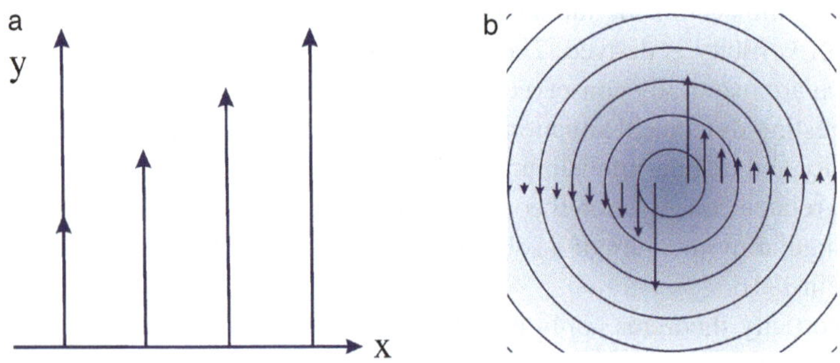


Fig. 2.18 Sketch of linear shear flow with vorticity (**a**) and a point vortex with zero vorticity (**b**)

### 2.11.1 Kinematical Properties

There is a number of simple theorems which relate the circulation $C$ of a loop $\Gamma$ at a certain instant of time to other physical quantities related to the same loop or to other loops. The theorems derived below require that at any instant $\Gamma$ is reducible to a point by subsequent shrinking, without leaving the fluid. Thus there always exists a simply connected surface $A$ in the fluid with the rim $\Gamma$. Then, for any such $A$ we may use Stokes' theorem (see Appendix A) to express $C$ in terms of the vorticity vector,

$$C = \oint_{\Gamma} \boldsymbol{u} \cdot \mathrm{d}\boldsymbol{s} = \int_{A} \mathrm{d}A\boldsymbol{n} \cdot \boldsymbol{\omega} \tag{2.192}$$

The latter integral is referred to as the *vortex flux* through the area $A$. Notice that the vortex flux is a kinematic concept; it should not be confused with the dynamical concept of the flux of vorticity (by advection or diffusion) considered later in this chapter.

The vortex flux has the same value for any surface attached to $\Gamma$. This follows immediately from (2.192), but it also may be inferred from Gauss' theorem applied to the vorticity vector. Since it is nondivergent by construction, i. e. $\nabla \cdot \boldsymbol{\omega} = 0$, there are no sources of vorticity inside the fluid, and we find

$$\int_{V} \mathrm{d}^3x \nabla \cdot \boldsymbol{\omega} = \oint_{A} \mathrm{d}A\boldsymbol{n} \cdot \boldsymbol{\omega} = 0 \tag{2.193}$$

where $A$ is now the surface surrounding the entire volume $V$. Hence circulation and vortex flux through a loop are synonymous concepts.

The quantities $\boldsymbol{\omega}$ and $C$ use the relative velocity $\boldsymbol{u}$ referenced to the rotating Earth whereas the absolute velocity $\boldsymbol{u}_\mathrm{a} = \boldsymbol{u} + \boldsymbol{\Omega} \times \boldsymbol{r}$ is seldom of interest in dynamical considerations. It is for some purposes useful to consider the vorticity and circulation referred to an inertial system. We thus define

$$\boldsymbol{\omega}_\mathrm{a} = \nabla \times \boldsymbol{u}_\mathrm{a} = \nabla \times (\boldsymbol{u} + \boldsymbol{\Omega} \times \boldsymbol{r}) = \boldsymbol{\omega} + 2\boldsymbol{\Omega}$$

and

$$C_\mathrm{a} = C + \oint_{\Gamma} \boldsymbol{\Omega} \times \boldsymbol{r} \cdot \mathrm{d}\boldsymbol{s}$$

correspondingly. We refer to $\boldsymbol{\omega}_\mathrm{a}$ and $\boldsymbol{\omega}$ as absolute and relative vorticity vectors, respectively. As $2\boldsymbol{\Omega}$ is the vorticity of the planetary rotation, it is referred to as planetary vorticity. In the same way we speak of absolute and relative circulation.

There is a simple measure which classifies the importance of the relative and planetary vorticity in the sum, the absolute vorticity. Consider the vertical component of the vorticity vectors,

$$\omega_\mathrm{a}^{(z)} = \eta + f\,, \quad \eta = \boldsymbol{k} \cdot \boldsymbol{\omega} = \frac{\partial v}{\partial x} - \frac{\partial u}{\partial y}\,, \quad f = 2\boldsymbol{k} \cdot \boldsymbol{\Omega} = 2\Omega \sin\varphi$$

where $\varphi$ is the geographic latitude and the vertical $\boldsymbol{k}$ is defined as the radial direction. Notice that with $\eta = O(U/L)$ the ratio of the relative and planetary contribution to the absolute vorticity $\omega_{\mathrm{a}}^{(z)}$,

$$\frac{\eta}{f} = O\left(\frac{U}{fL}\right) = O(\mathrm{Ro})$$

is found to be of order of the ROSSBY[26]-number $\mathrm{Ro} = U/(2\Omega L)$. In a large-scale oceanic flow we have $U = 0.1\,\mathrm{m\,s^{-1}}$, $L = 10^3\,\mathrm{km}$ so that with $f = 10^{-4}\,\mathrm{s^{-1}}$ (corresponding to 30° latitude) we find $\mathrm{Ro} = 10^{-3}$. In the atmosphere larger values occur ($\eta \sim 10^{-5}\,\mathrm{s^{-1}}$, $\mathrm{Ro} \sim 10^{-2} \ldots -10^{-1}$). In any case, in a large-scale flow the relative vorticity is small compared to the planetary vorticity, and thus the sign of the absolute vorticity corresponds to the sign of the Coriolis parameter $f$.

The above kinematic properties of relative vorticity and circulation are carried over to the absolute quantities since $\nabla \cdot \boldsymbol{\omega}_{\mathrm{a}} = 0$ as well. These properties are cast into a similar phrasing by use of the concepts of the *vortex line or filament* and the *vortex tube*. A vortex line is a continuous line of fluid parcels which is everywhere tangent to the instantaneous local vorticity vector. A vortex tube, sketched in Figure 2.19, is the ensemble of vortex lines passing through a given loop $\Gamma$ at a certain instant. These groups of fluid parcels may be defined for the absolute or relative vorticity vectors but, of course, the ensembles differ. Notice, moreover, that neither vortex lines nor vortex tubes are material in general. It is only under rather restricted conditions that a vortex line remains intact in the course of the fluid motion. This will be investigated in the next section.

It is clear, however, that the vortex flux is the same through any intersection of a vortex tube (first Helmholtz theorem), i. e.

$$C_{\mathrm{a}} = \int\limits_{A_1} \mathrm{d}A\boldsymbol{n}_1 \cdot \boldsymbol{\omega}_{\mathrm{a}} = \int\limits_{A_2} \mathrm{d}A\boldsymbol{n}_2 \cdot \boldsymbol{\omega}_{\mathrm{a}} \tag{2.194}$$

where $A_i$ and $\boldsymbol{n}_i$ are two intersecting surfaces with the corresponding normal vectors oriented towards the same side of the tube. The spatial constancy of $C_{\mathrm{a}}$ in a given tube follows directly from (2.193) since the contribution from the tube mantle is

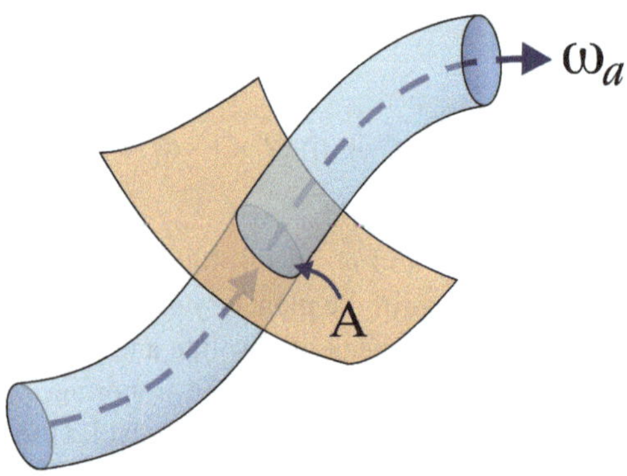


Fig. 2.19 Illustration of a vortex tube. The configuration is restricted to have the form of a simple 'tube', i. e. cross sections shall be simple closed curves. Redrawn after Vallis (2006)

[26] CARL-GUSTAV ARVID ROSSBY, *1898 in Stockholm, †1957 in Stockholm, meteorologist. The dimensionless number named after Rossby was used earlier by the Russian fluid dynamicist I.A. Kibel.

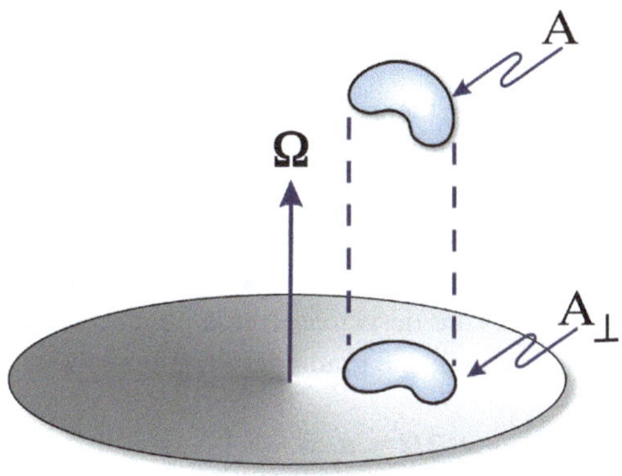


Fig. 2.20 Schematic showing the projection onto the equatorial plane. Redrawn after Vallis (2006)

identical to zero. This theorem may as well be formulated for the relative vortex flux (with a different tube, however). An immediate consequence of the constant strength of the vortex flux is that vortex lines and tubes cannot end somewhere in the fluid; they must close on themselves, extend to infinity or end at boundaries.

The flux of planetary vorticity

$$\int_A \mathrm{d}A 2\boldsymbol{\Omega} \cdot \boldsymbol{n} = 2\boldsymbol{\Omega} \cdot \int_A \mathrm{d}A\boldsymbol{n}$$

through an intersection $A$ of a vortex tube may expressed in terms of the area $A_\perp$ which is the projection of $A$ onto the plane perpendicular to $\boldsymbol{\Omega}$ which is the equatorial plane (see Figure 2.20). We thus may express the absolute vortex flux or circulation in the form

$$C_\mathrm{a} = C + 2\Omega A_\perp$$

Notice that while $C_\mathrm{a}$ is spatially constant along the absolute vortex tube, the individual contributions $C$ and $2\Omega A_\perp$ vary in general. Remember also that all terms may vary in the course of time.

### 2.11.2 Dynamical Properties

The kinematic theorems discussed above are quite general but not very powerful since they do not say anything about the temporal evolution. The vorticity and the circulation (following a material loop) change in the course of time. There is a sequence of vorticity and circulation theorems due to HELMHOLTZ (1858), KELVIN[27] (Thomson, W. (Lord Kelvin), 1869) and BJERKNES[28] (Bjerknes, 1898). For a historical perspective see Thorpe et al. (2003). The physics revealed in these theorems is reflected in the complete evolution equation of the vorticity vector but is not immediately evident from the underlying balance of momentum.

[27] WILLIAM THOMSON, LORD KELVIN, *1824 in Belfast, †1907 in Netherhall near Largs, mathematician and physicist.

[28] VILHELM BJERKNES, *1862 in Kristiania, †1951 in Oslo, physicist and hydrodynamicist.

### Barotropic Fluid

Helmholtz' dynamical vorticity theorem applies to a homogeneous fluid (i. e. constant density $\rho$) and a strictly nondivergent motion or, more general, to a flow in which pressure surfaces and density surfaces coincide. This may be an inherent property of the fluid, as e. g. for an equation of state $\rho = \rho(p)$, or a property of the flow in a certain area of the fluid. This latter case is called a *barotropic* state of flow. If, moreover, the external forces are derivable from a potential (implying that the fluid is frictionless), the acceleration is the gradient of a potential,

$$\frac{\mathrm{D}\boldsymbol{u}}{\mathrm{D}t} + 2\boldsymbol{\Omega} \times \boldsymbol{u} = -\nabla\left(\frac{p}{\rho} + \Phi\right) \tag{2.195}$$

and – after some mathematical manipulations outlined in more detail later in a more general setting (page 91) – the curl of the momentum balance becomes

$$\frac{\mathrm{D}}{\mathrm{D}t}\left(\frac{\boldsymbol{\omega}_{\mathrm{a}}}{\rho}\right) = \left(\frac{\boldsymbol{\omega}_{\mathrm{a}}}{\rho} \cdot \nabla\right)\boldsymbol{u} \tag{2.196}$$

This is the Helmholtz equation, here written for the absolute vorticity $\boldsymbol{\omega}_{\mathrm{a}}$. In this specific set-up the equation is identical to the evolution equation of a material line element $\boldsymbol{\ell}$ derived in (1.7) and repeated here,

$$\frac{\mathrm{D}}{\mathrm{D}t}\boldsymbol{\ell} = (\boldsymbol{\ell} \cdot \nabla)\boldsymbol{u} \tag{2.197}$$

If we choose $\boldsymbol{\ell}$ parallel to a vortex filament at a certain position at some initial time, we conclude from (2.196) and (2.197) that this vortex line coincides with the material line element defined by $\boldsymbol{\ell}$ at all later times (second Helmholtz theorem). Evidently, vortex tubes are material, and the strength of a vortex tube is an integral of the motion (third Helmholtz theorem), i. e. the circulation and vortex flux are not only constant along a specified vortex tube but also constant in time. Notice that (2.196) also implies that a line element which is free of vorticity will never acquire vorticity (Lagrange–Cauchy theorem). The generalization of these Helmholtz' theorems is considered further below.

It is worth mentioning that (2.196) (or (2.197)) may be integrated exactly as shown first by Cauchy in 1815. Using the Lagrangian framework where $\boldsymbol{X}(\boldsymbol{a}, t)$ denotes the particle position with initial value $\boldsymbol{a}$ and $\nabla_{\boldsymbol{a}}\boldsymbol{X}$ is the JACOBIAN[29], it is easily verified that

$$\boldsymbol{\ell} = \boldsymbol{c} \cdot \nabla_{\boldsymbol{a}}\boldsymbol{X}$$

with a materially conserved vector $\boldsymbol{c} = \boldsymbol{c}(\boldsymbol{a})$ is the solution of (2.197) for which $\boldsymbol{\ell}(t = 0) = \boldsymbol{c}$. Hence $\boldsymbol{\omega}_{\mathrm{a}} = (\rho/\rho_0)\boldsymbol{\omega}_0 \cdot \nabla_{\boldsymbol{a}}\boldsymbol{X}$ solves the Helmholtz equation (2.196). Here $\boldsymbol{\omega}_0$ and $\rho_0$ are the initial values.

### Circulation Theorems

The temporal change

$$\frac{\mathrm{D}C}{\mathrm{D}t} = \frac{\mathrm{D}}{\mathrm{D}t}\oint_{\Gamma} \boldsymbol{u} \cdot \mathrm{d}\boldsymbol{s} = \oint_{\Gamma}\left(\frac{\mathrm{D}\boldsymbol{u}}{\mathrm{D}t} \cdot \mathrm{d}\boldsymbol{s} + \boldsymbol{u} \cdot \frac{\mathrm{D}\mathrm{d}\boldsymbol{s}}{\mathrm{D}t}\right) \tag{2.198}$$

[29] CARL JACOBI, *1804 in Potsdam, †1851 in Berlin, mathematician.

of the circulation around a material loop is readily evaluated. Since d$\boldsymbol{s}$ is a material line element, we may use (2.197) to convert the last term on the right hand side of (2.198) as follows,

$$\oint_{\Gamma} \boldsymbol{u}\cdot\frac{\mathrm{D}\mathrm{d}\boldsymbol{s}}{\mathrm{D}t} = \oint_{\Gamma} \boldsymbol{u}\cdot(\mathrm{d}\boldsymbol{s}\cdot\nabla)\,\boldsymbol{u} = \oint_{\Gamma} \mathrm{d}\boldsymbol{s}\cdot\nabla\left(\frac{1}{2}\boldsymbol{u}^2\right) = 0$$

The change of the circulation is thus determined by the line integral of the acceleration,

$$\frac{\mathrm{D}C}{\mathrm{D}t} = \oint_{\Gamma} \mathrm{d}\boldsymbol{s}\cdot\frac{\mathrm{D}\boldsymbol{u}}{\mathrm{D}t} \tag{2.199}$$

and if the acceleration is derivable from a potential as under conditions leading to (2.195), we find that $C$ remains constant; the circulation is materially conserved for frictionless, barotropic flow in case of no rotation (Kelvin's theorem).

In general, however, we have a conservation of momentum of the form (2.47), written here as

$$\rho\frac{\mathrm{D}\boldsymbol{u}}{\mathrm{D}t} = -2\rho\boldsymbol{\Omega}\times\boldsymbol{u} - \nabla p - \rho\nabla\Phi + \mathcal{F} \tag{2.200}$$

where the effect of friction is contained in the force term $\mathcal{F}$. Then

$$\frac{\mathrm{D}C}{\mathrm{D}t} = -\oint_{\Gamma} \mathrm{d}\boldsymbol{s}\cdot\left(2\boldsymbol{\Omega}\times\boldsymbol{u} + \frac{1}{\rho}\nabla p + \nabla\Phi - \frac{1}{\rho}\mathcal{F}\right) \tag{2.201}$$

There are three forces that may lead to a change in the circulation (notice that gravity does not contribute). The Coriolis term may be expressed in terms of the rate of change of the area $A_\perp$ introduced above,

$$-\oint_{\Gamma} \mathrm{d}\boldsymbol{s}\cdot(2\boldsymbol{\Omega}\times\boldsymbol{u}) = -2\Omega\frac{\mathrm{D}A_\perp}{\mathrm{D}t} \tag{2.202}$$

The pressure term is conveniently rewritten in either of the following forms: since $\mathrm{d}\boldsymbol{s}\cdot\nabla p = \mathrm{d}p$, we have

$$-\oint_{\Gamma} \mathrm{d}\boldsymbol{s}\cdot\frac{1}{\rho}\nabla p = -\oint_{\Gamma} \frac{\mathrm{d}p}{\rho} = -\oint_{\Gamma} \upsilon\mathrm{d}p = \oint_{\Gamma} p\mathrm{d}\upsilon \tag{2.203}$$

or, using Stokes' theorem, we find

$$-\oint_{\Gamma} \mathrm{d}\boldsymbol{s}\cdot\frac{1}{\rho}\nabla p = -\int_A \mathrm{d}A\boldsymbol{n}\cdot\nabla\times\left(\frac{\nabla p}{\rho}\right) = \int_A \mathrm{d}A\boldsymbol{n}\cdot\boldsymbol{B} \tag{2.204}$$

where

$$\boldsymbol{B} = \frac{1}{\rho^2}\nabla\rho\times\nabla p$$

is called the *baroclinic vector*, also solenoidal vector, of the flow field. Both formulations are, of course, equivalent. In (2.203) the pressure term is found equal to the expansion work which the ensemble of fluid parcels along the material loop have to perform against the pressure field. In (2.204) it appears that the 'torque' induced by the pressure forces, acting on mass distribution in the surface which is normal to the pressure force, spins up the circulation. It is easy to see that either term vanishes for a barotropic state where the surfaces of constant density and constant pressure coincide and the baroclinic vector $\boldsymbol{B}$ vanishes everywhere. The complementary, more general state with pressure changing on density surfaces is called a *baroclinic* state, sketched in Figure 2.21.

The rate of change of the circulation is then written as

$$\frac{\mathrm{D}C}{\mathrm{D}t} + 2\Omega \frac{\mathrm{D}A_\perp}{\mathrm{D}t} = \frac{\mathrm{D}C_\mathrm{a}}{\mathrm{D}t} = \oint_\Gamma p\mathrm{d}\upsilon + \oint_\Gamma \mathrm{d}s \cdot \frac{\mathcal{F}}{\rho} \tag{2.205}$$

This is the famous circulation theorem of Bjerknes (1898) which shows that the absolute circulation is only changed by the baroclinic torque and friction.

There are various other equivalent forms of the baroclinic torque. As an example, with the first law of thermodynamics, (1.16), in the form $-p\mathrm{d}\upsilon = \mathrm{d}E - T\mathrm{d}\eta$, we have

$$-\oint_\Gamma \upsilon \mathrm{d}p = \oint_\Gamma T\mathrm{d}\eta = \int_A \mathrm{d}A \nabla T \times \nabla \eta \tag{2.206}$$

A perfect (ideal and frictionless) gas where $\eta = \eta(T)$ would thus follow Helmholtz theorems. Plotting the loop $\Gamma$ in the $(\upsilon, p)$-plane, the torque is found equal to the surrounded area (which evidently contracts to a curve for a barotropic state). Plotting in this plane grid lines of $\upsilon = n\Delta\upsilon$, $p = m\Delta p$ with $n, m = 0, \pm 1, \pm 2 \ldots$, we find that the torque is evaluated as the number of elementary $(\upsilon, p)$-cells times the area $\Delta\upsilon\Delta p$. The grid and cells appear and may be counted as well on any surface $A$ spanned by $\Gamma$. Corresponding cells may be defined in the $(T, \eta)$-plane.

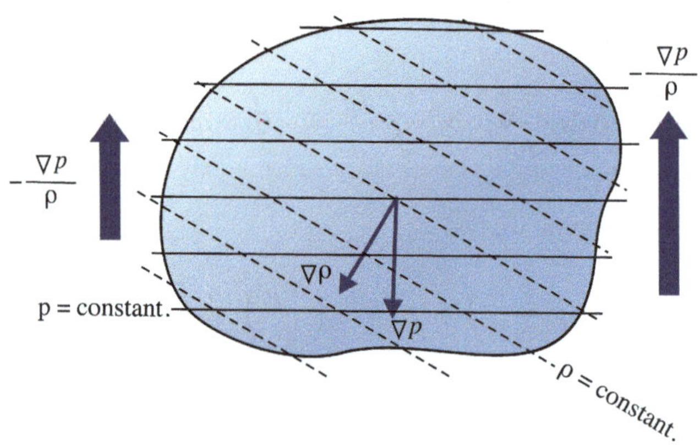


Fig. 2.21 Baroclinic conditions and direction of gradients. Whereas $\nabla\rho$ and $\nabla p$ are constant in this example, the vector $-(1/\rho)\nabla p$ is increasing from left to right

### Vorticity Equation

The local form of the vorticity balance is derived by applying the curl to the momentum balance. This operation is simplified by using the vector identities

$$\boldsymbol{\omega} \times \boldsymbol{u} = (\boldsymbol{u} \cdot \nabla)\boldsymbol{u} - \nabla \boldsymbol{u}^2/2$$
$$\nabla \times (\boldsymbol{\omega} \times \boldsymbol{u}) = \boldsymbol{\omega}(\nabla \cdot \boldsymbol{u}) + (\boldsymbol{u} \cdot \nabla)\boldsymbol{\omega} - (\boldsymbol{\omega} \cdot \nabla)\boldsymbol{u}$$

The first uses $\boldsymbol{\omega} = \nabla \times \boldsymbol{u}$, the second $\nabla \cdot \boldsymbol{\omega} = 0$. The momentum balance thus takes the equivalent form

$$\frac{\partial \boldsymbol{u}}{\partial t} + \boldsymbol{\omega}_{\mathrm{a}} \times \boldsymbol{u} = -\nabla\left(\frac{1}{2}u^2 + \Phi\right) - \frac{1}{\rho}\nabla p + \frac{\mathcal{F}}{\rho} \tag{2.207}$$

and the vorticity balance becomes

$$\frac{\mathrm{D}\boldsymbol{\omega}_{\mathrm{a}}}{\mathrm{D}t} = (\boldsymbol{\omega}_{\mathrm{a}} \cdot \nabla)\boldsymbol{u} - \boldsymbol{\omega}_{\mathrm{a}}(\nabla \cdot \boldsymbol{u}) + \boldsymbol{B} + \nabla \times \frac{\mathcal{F}}{\rho} \tag{2.208}$$

For barotropic steady flows, (2.208) leads to the TAYLOR–PROUDMAN[30] theorem discussed in the box on p. 91.

The divergence term (second on the right-hand side of (2.208)) describes changes of vorticity which arise when the volume and thus density of a parcel changes. Thus,

**16. Taylor–Proudman Theorem**

If the flow is frictionless, barotropic, and steady, the vorticity equation (2.208) takes the simple form

$$\nabla \times (\boldsymbol{\omega}_{\mathrm{a}} \times \boldsymbol{u}) = \nabla \times [(\boldsymbol{\omega} + 2\boldsymbol{\Omega}) \times \boldsymbol{u}] = 0$$

and if, in addition, the flow is slow, in the sense that the relative vorticity is small compared to the planetary part, we find

$$\boldsymbol{\Omega}(\nabla \cdot \boldsymbol{u}) - (\boldsymbol{\Omega} \cdot \nabla)\boldsymbol{u} = 0$$

Evaluating this in a Cartesian coordinate system where $\boldsymbol{\Omega} = (0, 0, \Omega)$ it is found that

$$\frac{\partial u}{\partial z} = \frac{\partial v}{\partial z} = 0\,, \quad \frac{\partial u}{\partial x} + \frac{\partial v}{\partial y} = 0 \tag{B16.1}$$

In a barotropic steady, slow, and frictionless motion, the normal velocity thus cannot vary in the direction of the planetary rotation vector, which is the celebrated Taylor–Proudman theorem. Fluid parcels lying on a line which is parallel to $\boldsymbol{\Omega}$ will always remain in such a state. In addition, the normal velocity vector $(u, v)$ is nondivergent. The fluid moves in a column-like way and we speak of Taylor columns. It is customary to consider the first condition in (B16.1) as a definition or at least a necessary prerequisite of a barotropic flow, though it is a mere consequence in a rather restrictive situation.

Notice that (B16.1) does not imply a two-dimensional flow since the vertical velocity is not constrained at all. If, however, the three-dimensional velocity is nondivergent, $\nabla \cdot \boldsymbol{u} = 0$, this requires a vertically constant $w$, and fluid parcels oriented along $\boldsymbol{\Omega}$ cannot change their distance. Still, $w = w(x, y)$ may vary laterally. An example is the barotropic motion over a fast rotating terrain with topography.

[30] GEOFFREY INGRAM TAYLOR, *1886 in St. John's Wood, †1975 in Cambridge, physicist. JOSEPH PROUDMAN, *1888 in Unsworth near Bury (Lancashire), †1975 in Fordingbridge, Hampshire, mathematician and oceanographer.

if the flow is for instance divergent ($\nabla \cdot \boldsymbol{u} > 0$), the volume expands, the density decreases, and mass is moved away from the center of mass. Consequently the local rate of rotation must diminish. In general, this effect is small in oceanic and even atmospheric flows. It is anyhow an easy matter to eliminate the divergence term by use of the mass conservation, which casts the balance of vorticity into the form of BELTRAMI'S[31] equation

$$\frac{\mathrm{D}}{\mathrm{D}t}\left(\frac{\boldsymbol{\omega}_{\mathrm{a}}}{\rho}\right) = \left(\frac{\boldsymbol{\omega}_{\mathrm{a}}}{\rho}\cdot\nabla\right)\boldsymbol{u} + \frac{1}{\rho}\left(\boldsymbol{B} + \nabla\times\frac{1}{\rho}\boldsymbol{\mathcal{F}}\right) \tag{2.209}$$

We are already familiar with the terms which act here as sources of vorticity, and for $\boldsymbol{B} = 0$ and $\boldsymbol{\mathcal{F}} = 0$, (2.209) is identical to (2.196). The baroclinic vector $\boldsymbol{B}$ and the curl of the frictional force appear in the local balance as they appear in the circulation theorem (2.205) in the integrated form. The first term on the right-hand side is identified with the tilting and stretching of the infinitesimal material line element $\boldsymbol{\ell}$ which is oriented as the local absolute vorticity vector, as explained below. In contrast to the other sources it cannot generate vorticity in a nonrotating state, i. e. only the presence of a nonzero vector $\boldsymbol{\omega}_{\mathrm{a}}$ may lead to a change in the components of $\boldsymbol{\omega}_{\mathrm{a}}$.

Consider a particular component of the vorticity balance (2.209) in a Cartesian frame of reference, say the balance of the vertical component $\omega_{\mathrm{a}}^{(z)} = \partial v/\partial x - \partial u/\partial y + \Omega^{(z)}$,

$$\begin{aligned}\frac{\mathrm{D}}{\mathrm{D}t}\frac{\omega_{\mathrm{a}}^{(z)}}{\rho} &= \frac{\omega_{\mathrm{a}}^{(x)}}{\rho}\frac{\partial w}{\partial x} + \frac{\omega_{\mathrm{a}}^{(y)}}{\rho}\frac{\partial w}{\partial y} + \frac{\omega_{\mathrm{a}}^{(z)}}{\rho}\frac{\partial w}{\partial z}\\ &\quad + \frac{1}{\rho}\left[B^{(z)} + \frac{\partial}{\partial x}(\mathcal{F}^{(y)}/\rho) - \frac{\partial}{\partial y}(\mathcal{F}^{(x)}/\rho)\right]\end{aligned} \tag{2.210}$$

Apparently, if the fluid expands (or shrinks) in the vertical direction, i. e. $\partial w/\partial z \neq 0$, a nonzero $\omega_{\mathrm{a}}^{(z)}$ of a fluid parcel will be changed. This mechanism is accordingly called *vortex stretching*, and the corresponding term in (2.210) is the *stretching term*. In contrast, the first two terms on the right-hand side act even if $\omega_{\mathrm{a}}^{(z)}$ vanishes; they describe the generation of vertical vorticity of the parcel by tilting of vorticity components of other coordinate directions into the vertical direction. This is induced by lateral changes of the vertical velocity. These terms are called the *tilting terms*. This concept is applied accordingly to the other components of the vorticity balance. We will see, however, in later chapters, that for a large-scale geophysical flow on the rotating Earth, the vertical component (i. e. radial with respect to the Earth geometry) of vorticity plays a particularly important role, and the stretching term in turn is the most important term. The three sources of vorticity – stretching, tilting and solenoidal production – are illustrated in Figure 2.22.

### 2.11.3 Ertel's Potential Vorticity

The most complete form of potential vorticity conservation has been found by ERTEL[32]. It is based on the same principle, the vorticity balance (2.209), now written

[31] EUGENIO BELTRAMI, *1835 in Cremona, †1900 in Rome, mathematician.

[32] HANS ERTEL, *1904 in Berlin, †1971 in Berlin, meteorologist and geophysicist.

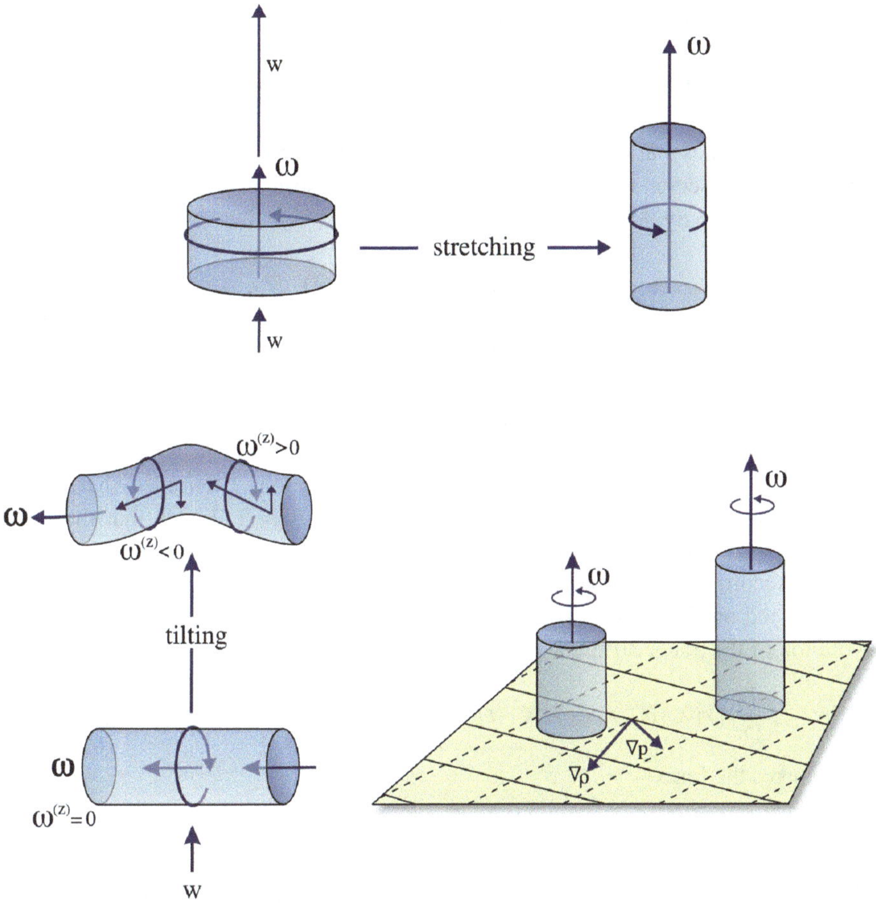


Fig. 2.22 Sketch showing the change of vorticity by stretching, tilting, and solenoidal production by the baroclinic vector. Redrawn after Vallis (2006) and Dutton (1976)

with an expanded baroclinic vector,

$$\frac{\mathrm{D}}{\mathrm{D}t}\left(\frac{\boldsymbol{\omega}_{\mathrm{a}}}{\rho}\right) = \left(\frac{\boldsymbol{\omega}_{\mathrm{a}}}{\rho}\cdot\nabla\right)\boldsymbol{u} + \frac{1}{\rho^3}\nabla\rho\times\nabla p + \frac{1}{\rho}\nabla\times\boldsymbol{\mathcal{F}} \tag{2.211}$$

with the mass conservation being incorporated already. In addition, consider a tracer $\chi$ which satisfies the conservation $\rho\mathrm{D}\chi/\mathrm{D}t = \mathcal{G}_\chi$. In practical applications, $\chi$ is usually taken as potential density, leading to the name 'potential vorticity'. We will, however, retain the diabatic terms $\mathcal{F}$ and $\mathcal{G}_\chi$. Projection of (2.211) onto $\nabla\chi$ is

$$\nabla\chi\cdot\frac{\mathrm{D}}{\mathrm{D}t}\left(\frac{\boldsymbol{\omega}_{\mathrm{a}}}{\rho}\right) = \left[\left(\frac{\boldsymbol{\omega}_{\mathrm{a}}}{\rho}\right)\cdot\nabla\boldsymbol{u}\right]\cdot\nabla\chi + \nabla\chi\cdot\frac{\nabla\rho\times\nabla p}{\rho^3} + \frac{1}{\rho}\nabla\chi\cdot(\nabla\times\boldsymbol{\mathcal{F}})$$

Using the fairly complicated and nonobvious vector identity

$$\frac{\boldsymbol{\omega}_{\mathrm{a}}}{\rho}\cdot\frac{\mathrm{D}}{\mathrm{D}t}\nabla\chi = \left(\frac{\boldsymbol{\omega}_{\mathrm{a}}}{\rho}\cdot\nabla\right)\frac{\mathrm{D}}{\mathrm{D}t}\chi - \left[\left(\frac{\boldsymbol{\omega}_{\mathrm{a}}}{\rho}\cdot\nabla\right)\boldsymbol{u}\right]\cdot\nabla\chi$$

it is then easy to derive Ertel's theorem (Ertel, 1942)

$$\rho \frac{\mathrm{D}}{\mathrm{D}t}\left(\frac{\boldsymbol{\omega}_{\mathrm{a}}}{\rho}\cdot\nabla\chi\right) = \boldsymbol{\omega}_a \cdot \nabla\left(\mathcal{G}_\chi/\rho\right) + \frac{1}{\rho^2}\nabla\chi\cdot\nabla\rho\times\nabla p + \nabla\chi\cdot(\nabla\times\boldsymbol{\mathcal{F}}) \quad (2.212)$$

The theorem states that a frictionless (i. e. $\boldsymbol{\mathcal{F}} = 0$) and barotropic (i. e. $\nabla\rho\times\nabla p = 0$) flow which conserves the tracer $\chi$ (i. e. $\mathcal{G}_\chi = 0$) also conserves Ertel's potential vorticity $Q$ given by

$$Q = \frac{1}{\rho}\boldsymbol{\omega}_{\mathrm{a}}\cdot\nabla\chi = \frac{1}{\rho}(\boldsymbol{\omega} + 2\boldsymbol{\Omega})\cdot\nabla\chi$$

The condition of barotropicity may be relaxed for a tracer which is a function of $\rho$ and $p$ only, i. e. $\chi = \chi(\rho, p)$ because for any such tracer the triple product of gradients in (2.212) vanishes as well.

An illustrative proof of Ertel's theorem can be obtained by direct use of Helmholtz's vorticity equation (2.196) which applies to a frictionless barotropic fluid. Consider two particles with trajectories $\boldsymbol{X}_i(t), i = 1, 2$ such that the $\boldsymbol{\ell} = \boldsymbol{X}_2 - \boldsymbol{X}_1$ is oriented along the local vorticity vector $\boldsymbol{\omega}_{\mathrm{a}}/\rho$ at $\boldsymbol{X}_1$ at the initial time $t = 0$. According to Helmholtz' theorem this continues to be true when the particles proceed on their paths. Thus $\boldsymbol{\ell} = \epsilon\boldsymbol{\omega}_{\mathrm{a}}/\rho$ with a constant $\epsilon$. Consider the time evolution of the tracer difference over the separation $\boldsymbol{\ell}$,

$$\rho\frac{\mathrm{D}}{\mathrm{D}t}\left[\chi(\boldsymbol{X}_2(t), t) - \chi(\boldsymbol{X}_1(t), t)\right] = \mathcal{G}_\chi(\boldsymbol{X}_2(t), t) - \mathcal{G}_\chi(\boldsymbol{X}_1(t), t)$$

which takes the form

$$\rho\frac{\mathrm{D}}{\mathrm{D}t}\boldsymbol{\ell}\cdot\nabla\chi = \boldsymbol{\ell}\cdot\nabla\mathcal{G}_\chi$$

for infinitesimal separation. By this we have indeed derived (2.212) for the case that the baroclinic vector and the friction vanish.

It becomes obvious that the vorticity and circulation theorems are closely connected. In fact, Ertel's theorem is a local formulation of Kelvin's theorem, including the effect of rotation. To see this, consider an infinitesimal closed curve $C$ on a surface where $\chi = \text{const}$. From Kelvin's theorem we find that $\boldsymbol{\omega}_{\mathrm{a}}\cdot\boldsymbol{n}\delta A$ is materially conserved. Because the normal vector is $\boldsymbol{n} = \nabla\chi/|\nabla\chi|$ and the mass in the volume between neighboring $\chi$-surfaces and $\chi + \Delta\chi$ is conserved, $\delta m = \rho\delta A\cdot h = \text{const}$, with $h = \Delta\chi/|\nabla\chi|$, we immediately find

$$\boldsymbol{\omega}\cdot\boldsymbol{n}\delta A = \frac{\boldsymbol{\omega}_{\mathrm{a}}\cdot\nabla\chi}{|\nabla\chi|}\frac{\delta m}{\rho h} = Q\frac{\delta m}{\Delta\chi}$$

Hence $Q$ must be conserved as well.

## 2.12 * Lagrangian Concepts in Fluid Mechanics

So far we have considered the governing equations of motion in Eulerian coordinates. It is instructive to consider the principal form of these equations also in Lagrangian coordinates, which we will do in this section. As well known from classical mechanics, the use of a variational principle greatly simplifies the derivation of conservation

17. Impermeability Theorem for Potential Vorticity

As noted by Obukhov (1962), the terms on the right-hand side of (2.212) may be cast into the form of a divergence and Ertel's theorem takes the form

$$\rho\frac{\mathrm{D}Q}{\mathrm{D}t} = -\nabla\cdot\boldsymbol{J}_{\mathrm{Q}} \tag{B17.1}$$

with flux vector of potential vorticity

$$\boldsymbol{J}_{\mathrm{Q}} = \frac{1}{\rho}\nabla p\times\nabla\chi - \boldsymbol{\omega}_{\mathrm{a}}\mathcal{G}_\chi - \mathcal{F}\times\nabla\chi \tag{B17.2}$$

The proof is easy: $\boldsymbol{\omega}_{\mathrm{a}}\cdot\nabla\mathcal{G}_\chi = \nabla\cdot\boldsymbol{\omega}_{\mathrm{a}}\mathcal{G}_\chi$ because $\boldsymbol{\omega}_{\mathrm{a}}$ is solenoidal, $\nabla\chi\cdot\nabla\times\mathcal{F} = \nabla\cdot(\mathcal{F}\times\nabla\chi)$ because $\nabla\times\nabla\chi = 0$, and $\nabla\chi\cdot\nabla 1/\rho\times\nabla p = \nabla\cdot 1/\rho(\nabla p\times\nabla\chi)$ because $\nabla\cdot(\nabla a\times\nabla b) = 0$ for any $a, b$. Then (B17.1) has the form of the general conservation equation discussed in Chapter 2. In flux form we have

$$\frac{\partial}{\partial t}\rho Q + \nabla\cdot(\boldsymbol{J}_{\mathrm{Q}} + \rho\boldsymbol{u}Q) = 0 \tag{B17.3}$$

The result is a bit surprising because there is no restriction as usual for a sourceless conservation equation. In fact, (B17.3) is valid for a diabatic flow in the presence of friction and baroclinicity. The rate of change of the $Q$-content in an arbitrary volume is obtained by integration and use of Gauss' theorem. We find

$$\frac{\partial}{\partial t}\int_V \rho Q\mathrm{d}V = -\oint\left(\boldsymbol{J}_{\mathrm{Q}} + \rho\boldsymbol{u}Q - \rho Q\frac{\mathrm{d}\boldsymbol{n}}{\mathrm{d}t}\right)\cdot\mathrm{d}\boldsymbol{A} \tag{B17.4}$$

where $\mathrm{d}\boldsymbol{n}/\mathrm{d}t$ denotes the velocity of the boundary (positive for the outward direction), and $\mathrm{d}\boldsymbol{A} = \boldsymbol{n}\mathrm{d}A$ is the surface element. For a volume which is bounded by surfaces $\chi = \chi_1$ (the 'bottom') and $\chi = \chi_2$ (the 'top') this integral budget attains a simpler form because all fluxes through the top and bottom boundaries vanish. The proof uses the fact that the normal vector at the top and the bottom is aligned along the gradient of the tracer, i. e. $\boldsymbol{n} = \nabla\chi/|\nabla\chi|$, and

$$\frac{\mathrm{d}n}{\mathrm{d}t} = -\frac{\partial\chi/\partial t}{|\nabla\chi|}$$

which follows because $\mathrm{d}\chi = (\partial\chi/\partial n)\mathrm{d}n + (\partial\chi/\partial t)\mathrm{d}t$ vanishes on the bounding surfaces $\chi = \text{const}$. Furthermore,

$$\boldsymbol{n}\cdot\boldsymbol{u} = \frac{\boldsymbol{u}\cdot\nabla\chi}{|\nabla\chi|} = \frac{\mathcal{G}_\chi - \partial\chi/\partial t}{|\nabla\chi|} \quad\text{and}\quad \boldsymbol{n}\cdot\boldsymbol{J}_{\mathrm{Q}} = -\frac{\mathcal{G}_\chi\boldsymbol{\omega}_{\mathrm{a}}\cdot\nabla\chi}{|\nabla\chi|} = -\frac{\mathcal{G}_\chi\rho Q}{|\nabla\chi|}$$

by use of the tracer balance equation, the definition of $\boldsymbol{J}_{\mathrm{Q}}$ and the potential vorticity. Hence

$$\boldsymbol{n}\cdot(\boldsymbol{J}_{\mathrm{Q}} + \rho\boldsymbol{u}Q) = -\rho Q\frac{\partial\chi/\partial t}{|\nabla\chi|} = \rho Q\frac{\mathrm{d}n}{\mathrm{d}t}$$

for the top and bottom surfaces in the integral (B17.4). It follows that the content of potential vorticity in a volume bounded by surfaces $\chi = \text{const}$ is conserved, except for fluxes through lateral boundaries (in the oceanic case, with $\chi$ taken as potential density, the potential vorticity is then controlled by mixing at continental boundaries, which is unfortunately an area of major ignorance). This important theorem (*impermeability theorem*) was found by Haynes and McIntyre (1990). Notice that this includes all effects of friction, mixing and baroclinicity. In particular, the surfaces $\chi = \text{const}$ do not need to be material surfaces.

equations and is always possible with Lagrangian coordinates. Furthermore, the use of a variational principle can also be useful in the context of Eulerian coordinates. While not generally possible, it is shown in the following that the complete set of equations of motion – in their adiabatic form – can indeed be derived from a variational principle. Note that we return to the Lagrangian concepts in Chapter 9, where the procedure is used to obtain wave solutions.

We begin with a discussion of the variational principle for discrete systems, which we then extend to continuous systems. Consider a discrete system with $n$ degrees of freedom and which is described by coordinates $q_i(t), i = 1, \ldots, n$ where $t$ is the time. In classical mechanics the action is defined by

$$A[q_i] = \int_{t_1}^{t_2} L\mathrm{d}t \quad \text{with} \quad L = E_\mathrm{k} - E_\mathrm{p}$$

where $L$ is called *Lagrangian* and $E_k$ and $E_p$ are kinetic and potential energy of the system, respectively. The coordinates do not need to be the physical coordinates of the system and, therefore, are usually called generalized coordinates. In general, $L = L(\dot{q}_i, q_i, t)$ is a function of the $\dot{q}_i, q_i$ and the time $t$. The system is Lagrangian if its equations of motion can be derived from HAMILTON'S[33] variational principle

$$\delta A[q_i] = \lim_{\epsilon \to 0} \frac{A[q_i + \epsilon \delta q_i] - A[q_i]}{\epsilon} = 0 \tag{2.213}$$

involving the action $A[q_i]$ which is a functional of the coordinates $q_i$. The variation in (2.213) is performed with $\delta q_i(t_1) = \delta q_i(t_2) = 0$ and yields the *Lagrange equations*

$$\frac{\mathrm{d}}{\mathrm{d}t}\frac{\partial L}{\partial \dot{q}_i} - \frac{\partial L}{\partial q_i} = 0 \tag{2.214}$$

in terms of the Lagrangian. These are the $n$ equations of motion of the system. As a simple example, take the one-dimensional harmonic oscillator where $K = m\dot{q}^2/2$ and $V = kq^2/2$. The Lagrangian formalism then yields via (2.214) the familiar oscillator equation $m\ddot{q} + kq = 0$. When dealing with systems having many degrees of freedom, the handling by the one scalar functional $L$ is clearly of advantage.

An extension of the discrete Lagrange–Hamilton formalism (outlined above and in the box on p. 98), which is important for our purposes of describing fluid systems, concerns the step from discrete to continuous systems, i. e. instead of variables $q_i$ characterized by a discrete index $i$ the system is now described by functions $\phi(x, t)$ of a continuous variable $x$ (usually the position coordinate). The functional $L$ in (2.215) is now a *Lagrangian density* in $x$-space. The appropriate action is

$$A[\phi] = \int_{t_1}^{t_2} \mathrm{d}t \int_D \mathrm{d}x L\left(\frac{\partial \phi}{\partial t}, \frac{\partial \phi}{\partial x}, \phi, x, t\right) \tag{2.215}$$

and variations $\delta\phi(x, t)$ are considered which vanish at limits of the time interval and on the boundary $D$ of the spatial domain. Then, the so-called *Euler–Lagrange equation* results,

$$\frac{\partial}{\partial t}\frac{\partial L}{\partial \phi_t} + \frac{\partial}{\partial x}\frac{\partial L}{\partial \phi_x} - \frac{\partial L}{\partial \phi} = 0 \tag{2.216}$$

[33] SIR WILLIAM ROWAN HAMILTON, *1805 in Dublin, †1865 in Dunsink near Dublin, mathematician and physicist.

with shorthand notation $\phi_\mathrm{t} = \partial\phi/\partial t$ etc. Worth mentioning is that higher order derivatives $\phi_\mathrm{tt}, \phi_\mathrm{tx}, \phi_{xx}, \ldots$ may be considered in the Lagrange function, resulting then in corresponding higher derivatives in the Euler–Lagrange equations. We may actually deal with a vector field $\phi_\alpha(\boldsymbol{x}, t), \alpha = 1, \ldots, m$ of dimension $m$ defined in two- or three-dimensional space which yields

$$\frac{\partial}{\partial t}\frac{\partial L}{\partial \phi_{\alpha,t}} + \frac{\partial}{\partial x_j}\frac{\partial L}{\partial \phi_{\alpha,j}} - \frac{\partial L}{\partial \phi_\alpha} = 0\,, \quad \alpha = 1, \ldots, m \tag{2.217}$$

where the comma notation abbreviates the spatial derivative, and the sum convention is used for the index $j$.

A Hamiltonian theory may be developed for continuous systems as well. In fact, a complete set of balance equations for energy and momentum conservation may be formulated,

$$\begin{aligned} \frac{\partial T_{00}}{\partial t} + \frac{\partial T_{0i}}{\partial x_i} &= -\frac{\partial L}{\partial t} \\ \frac{\partial T_{i0}}{\partial t} + \frac{\partial T_{ij}}{\partial x_j} &= -\frac{\partial L}{\partial x_i} \end{aligned} \tag{2.218}$$

where

$$T_{00} = \sum_\alpha \phi_{\alpha,\mathrm{t}} \frac{\partial L}{\partial \phi_{\alpha,\mathrm{t}}} - L \quad \text{energy density} \tag{2.219}$$

$$T_{0i} = \sum_\alpha \phi_{\alpha,\mathrm{t}} \frac{\partial L}{\partial \phi_{\alpha,i}} \quad \text{energy flux density} \tag{2.220}$$

$$T_{i0} = \sum_\alpha \phi_{\alpha,i} \frac{\partial L}{\partial \phi_{\alpha,\mathrm{t}}} \quad \text{momentum density} \tag{2.221}$$

$$T_{ij} = \sum_\alpha \phi_{\alpha,i} \frac{\partial L}{\partial \phi_{\alpha,j}} - \delta_{ij} L \quad \text{stress tensor} \tag{2.222}$$

Note that the meaning of 'energy' and 'momentum' coincides with the physical energy and momentum only if the Lagrangian is built from physical energies.

Following (2.218), energy is conserved if the Lagrangian does not explicitly depend on time, and momentum is conserved if no explicit spatial dependence occurs. There is a deeper reason for these conservation laws: according to NOETHER's[34] theorem a (sourceless) conservation law exists for each symmetry in the Lagrange function $L$. A symmetry exists if $L$ is invariant to a continuous transformation of its dependent or independent variables. In the above cases (2.218), these are translations in time or space.

In the following we present some examples of fluid mechanical systems which can be treated by a Lagrangian.

### 2.12.1 Incompressible Fluid

In the Lagrangian representation, outlined in Section 1.1, the position of a fluid parcel at time $t$ was denoted by $\boldsymbol{X}(\boldsymbol{a}, t)$ where the initial position $\boldsymbol{a} = \boldsymbol{X}(\boldsymbol{a}, t = 0)$

[34] AMALIE EMMY NOETHER, *1882 in Erlangen, †1935 in Bryn Mawr in Pennsylvania, mathematician.

**18. Discrete Systems – Hamilton's Equations**

From the Lagrangian $L$ we may proceed to the *Hamiltonian* $H$ defined by

$$H = \sum_i p_i \dot{q}_i - L \quad \text{with} \quad p_i = \frac{\partial L}{\partial \dot{q}_i} \tag{B18.1}$$

The $p_i$ is the momentum conjugate to the coordinate $q_i$. To obtain the correct form of the Hamiltonian, the last relation must be inverted to get $\dot{q}_i$ in terms of the $p_i$. Then, $H$ is a function of the $p_i$, $q_i$ and $t$, and inserting $L$ from (B18.1) into (2.213) and varying now $q_i$ and $p_i$ yields $2n$ first-order equations

$$\dot{q}_i = \frac{\partial H}{\partial p_i} \quad \text{and} \quad \dot{p}_i = -\frac{\partial H}{\partial q_i}$$

Two implications are immediately evident: (1) the value of $H$ is conserved if the Hamiltonian is time-independent because the sum in

$$\frac{\mathrm{d}H}{\mathrm{d}t} = \frac{\partial H}{\partial t} + \sum_i \left( \dot{p}_i \frac{\partial H}{\partial p_i} + \dot{q}_i \frac{\partial H}{\partial q_i} \right) = \frac{\partial H}{\partial t} \tag{B18.2}$$

is identical to zero; and (2) for the classical kinetic energy $K = 1/2 \sum m_i \dot{q}_i^2$ we find that $H$ is the total energy because $\dot{q}_i = p_i/m_i$ and

$$H = \sum_i \frac{p_i^2}{2m_i} + V = K + V$$

Considering again the harmonic oscillator, we find from the Hamiltonian formalism $p = \partial L/\partial \dot{q} = m\dot{q}$ and thus $H = p^2/2m + kq^2/2$ and then the two coupled equations $\dot{q} = \partial H/\partial p = p/m, \dot{p} = -\partial H/\partial q = -kq$.

is used as particle label. As $\boldsymbol{u}(\boldsymbol{x}, t) = \partial \boldsymbol{X}/\partial t$ is the Eulerian velocity at the position $\boldsymbol{x} = \boldsymbol{X}(\boldsymbol{a}, t)$, the equations of motion are – in a mixed Lagrangian-Eulerian expression – given by

$$\rho \ddot{X}_i = -\frac{\partial p}{\partial x_i}, \quad i = 1, 2, 3$$

where $\rho(\boldsymbol{x}, t)$ and $p(\boldsymbol{x}, t)$ are Eulerian fields. To simplify the notation, we use the dot for the partial derivative with respect to time. When the initial value $\rho_0(\boldsymbol{a}) = \rho(\boldsymbol{a}, t = 0)$ is taken as Lagrangian variable for the density (which is constant in time), conservation of mass requires $\rho_0 \mathrm{d}^3 a = \rho \mathrm{d}^3 x$ where the volume elements $\mathrm{d}^3 a$ and $\mathrm{d}^3 x$ are mapped onto each other by corresponding parcel trajectories. Write $\mathrm{d}^3 a = \mathrm{d}a_1 \mathrm{d}a_2 \mathrm{d}a_3$ for the volume of a small cube at the initial time. With the transformation $X_i(\boldsymbol{a}, t)$ we find for the small increment $\mathrm{d}a_1$ the displacement $\mathrm{d}X_i(\boldsymbol{a}, t) = (\partial X_i/\partial a_1)\mathrm{d}a_1$ and write this as a vector $\mathrm{d}\boldsymbol{X}^{(1)}$, similarly for $\mathrm{d}\boldsymbol{X}^{(2)}$ and $\mathrm{d}\boldsymbol{X}^{(3)}$. These span the volume at the time $t$

$$\mathrm{d}\boldsymbol{X}^{(1)} \cdot (\mathrm{d}\boldsymbol{X}^{(2)} \times \mathrm{d}\boldsymbol{X}^{(3)}) = \epsilon_{ijk} \frac{\partial X_i}{\partial a_1} \mathrm{d}a_1 \frac{\partial X_j}{\partial a_2} \mathrm{d}a_2 \frac{\partial X_k}{\partial a_3} \mathrm{d}a_3 = J(\boldsymbol{a}, t) \mathrm{d}a_1 \mathrm{d}a_2 \mathrm{d}a_3$$

evaluated by the familiar triple product, with

$$\epsilon_{ijk} = \begin{cases} +1 & \text{if } ijk \text{ is an even permutation of 1,2,3} \\ -1 & \text{if } ijk \text{ is an uneven permutation of 1,2,3} \\ 0 & \text{otherwise} \end{cases}$$

This defines the Jacobian determinant $J$ of the transformation from the initial state $\boldsymbol{a} = \boldsymbol{X}(\boldsymbol{a}, t = 0)$ to the state $\boldsymbol{X}(\boldsymbol{a}, t)$ at time $t$. Hence

$$\rho_0(\boldsymbol{a}) = \rho(\boldsymbol{X}, t) J(\boldsymbol{a}, t) \quad \text{with} \quad J(\boldsymbol{a}, t) = \frac{\partial(X_1, X_2, X_3)}{\partial(a_1, a_2, a_3)}$$

Introducing a Lagrangian pressure variable $\pi(\boldsymbol{a}, t) = p(\boldsymbol{x}, t)$ by use of the inverse trajectory mapping $\boldsymbol{a} = \boldsymbol{a}(\boldsymbol{X}, t) = \boldsymbol{A}(\boldsymbol{x}, t)$ (see (1.2)), we find for the pressure gradient

$$\frac{\partial p}{\partial x_i} = \frac{\partial A_j}{\partial x_i} \frac{\partial \pi}{\partial a_j}$$

and the equations of motion are completely expressed in the Lagrangian variables by

$$\rho_0 \ddot{X}_i = -J \frac{\partial A_j}{\partial x_i} \frac{\partial \pi}{\partial a_j} = -C_{ij} \frac{\partial \pi}{\partial a_j} \tag{2.223}$$

$C_{ij}$ is the cofactor of $\partial X_i / \partial a_j$ in the expansion of the determinant $J = C_{ij} \partial X_i / \partial a_j$. To arrive at the last relation, we have used the identity $\partial A_i / \partial x_j = C_{ij} / J$.

As shown in the box on p. 119 below, the volume is conserved for an *incompressible* flow, $J = 1$, hence $\rho_0 = \rho$. The above equations of motion (2.223) follow for this incompressible state from the Lagrangian density (in $\boldsymbol{a}$-space)

$$L = \frac{1}{2} \rho_0 \dot{X}_i^2 + \pi(J - 1) \tag{2.224}$$

Variation with respect to the Lagrangian multiplier $\pi$ for the incompressibility constraint $J = 1$ trivially recovers this constraint, and variation with respect to $X_i$ yields

$$\rho_0 \ddot{X}_i + \frac{\partial}{\partial a_j} \left( \pi \frac{\partial J}{\partial X_{i,j}} \right) = 0 \tag{2.225}$$

Since each element occurs only once in $J$, we have $\partial J / \partial X_{i,j} = C_{ij}$, and $\partial C_{ij} / \partial a_j = 0$ is used to equate (2.225) with (2.223).

### 2.12.2 Compressible Isentropic Fluid

Note that (2.223) is valid for compressible flow as well, but replacing the constraint $J - 1 = 0$ in the Lagrangian by the compressible version $\rho J - \rho_0 = 0$ does not lead to these equations. An adequate Lagrangian for isentropic (adiabatic) conditions may be constructed by utilization of the internal energy $E = E(\upsilon, \eta)$ where $\upsilon = 1/\rho$ is the specific volume and $\eta$ the specific entropy. In Section 1.2 it was shown that the derivatives of $E$ yield the pressure and the temperature, respectively,

$$\frac{\partial E}{\partial \upsilon} = -p \,, \quad \frac{\partial E}{\partial \eta} = T \tag{2.226}$$

As $\eta$ is conserved, it enters as $\eta(\boldsymbol{x}, t) = \eta_0(\boldsymbol{a})$. This constraint as well as the constraint of mass conservation is implemented by Lagrangian multipliers,

$$L = \frac{1}{2} \rho_0 \dot{X}_i^2 - \rho_0 E(\upsilon, \eta) + \lambda_1(\eta - \eta_0) + \lambda_2(\upsilon - J \upsilon_0) \tag{2.227}$$

and variations of $X_i, \lambda_1, \lambda_2, \upsilon$ and $\eta$ yield

$$\begin{aligned} \rho_0 \ddot{X}_i + \frac{\partial}{\partial a_j}\left(\rho_0 \lambda_2 \frac{\partial J}{\partial X_{i,j}}\right) &= 0 \\ \eta &= \eta_0 \\ \upsilon &= J\upsilon_0 \\ \rho_0 \frac{\partial E}{\partial \upsilon} &= \lambda_2 \\ \rho_0 \frac{\partial E}{\partial \eta} &= \lambda_1 \end{aligned}$$

Use of (2.226) and elimination of $\lambda_2 = -\rho_0 p$ lead to the momentum balance in its proper form. The other multiplier is found to relate to the temperature, $\lambda_1 = \rho_0 T$. With a given thermodynamic potential $E = E(\upsilon, \eta)$, the equation of state follows from (2.226). Note that it is straightforward to include other conserved quantities such as salinity $S$: add $\lambda_3(S - S_0)$ to the Lagrangian and use $E = E(\upsilon, \eta, S)$ with $\mu = -\partial E/\partial S$ as chemical potential (see (1.47)).

### 2.12.3 Rotating Fluid with Gravity

The Lagrangians (2.224) for an incompressible fluid or (2.227) for an isentropic fluid are readily extended to include gravity and rotation. We proceed with (2.224) and take coordinates relative to the rotating frame with a rotation vector $\Omega_i, i = 1, 2, 3$. The coordinate transformation to the rotating frame follows the same route outlined in Section 2.3.3. The appropriate equations of motion are

$$\rho\left(\ddot{X}_i + 2\epsilon_{ijk}\Omega_j \dot{X}_k\right) = -\frac{\partial p}{\partial x_i} - \rho \frac{\partial \Phi}{\partial x_i} \tag{2.228}$$

where $\Phi(x_i)$ is the apparent gravitational potential (including the centrifugal part as described in Section 2.3.4). The Lagrangian form of (2.228) is

$$\rho_0 \left(\ddot{X}_i + 2\epsilon_{ijk}\Omega_j \dot{X}_k + \frac{\partial \Phi}{\partial X_i}\right) = -C_{ij}\frac{\partial \pi}{\partial a_j}$$

It follows from varying the Lagrangian

$$L = \frac{1}{2}\rho_0 \left(\dot{X}_i^2 + 2\epsilon_{ijk}\Omega_i X_j \dot{X}_k\right) - \rho_0 \Phi(X_i) + \pi(J - 1) \tag{2.229}$$

with respect to $X_i$ and $\pi$. The inertial part can be obtained be replacing $\dot{X}_i$ in the Lagrangian (2.224) for the nonrotating frame by $\dot{X}_i + \epsilon_{ijk}\Omega_j X_k$; the term which is quadratic in $\Omega_i$ is contained in the apparent gravity potential.

### 2.12.4 Rotating Stratified Fluid

Next we consider waves residing on a given background stratification on an $f$-plane with $\Omega_i = \delta_{i3} f/2$ with a constant $f$ and gravity acting along the vertical direction,

i. e. $\Phi = gX_3$. We transform from the field variables $x_i, \pi$ to perturbations $\xi$ of displacement and $\varpi$ of pressure about a state of rest,

$$\begin{aligned} \xi_i(\boldsymbol{a}, t) &= X_i(\boldsymbol{a}, t) - a_i \\ \varpi(\boldsymbol{a}, t) &= \pi(\boldsymbol{a}, t) - p_0(a_3) \end{aligned}$$

where $p_0$ is hydrostatically balanced with a stratification $\rho_0(a_3)$,

$$\frac{dp_0}{da_3} = -\rho_0(a_3)$$

The effect of $p_0(a_3)(J - 1)$ in the transformed Lagrangian is identical to a term $p_0(X_3)$ in the Lagrangian, and hence, omitting terms with no effect on the equations of motion, we arrive at

$$L = \frac{1}{2}\rho_0\left[\dot{\xi}_i^2 + f(\xi_1\dot{\xi}_2 - \dot{\xi}_1\xi_2)\right] - U(\xi_3) + \varpi(J - 1) \tag{2.230}$$

with a potential $U$ and Jacobian determinant $J$, given by

$$\begin{aligned} U(\xi_3) &= p_0(\xi_3 + a_3) + \rho_0 g \xi_3 \\ J &= 1 + \frac{\partial \xi_i}{\partial a_i} + \Delta_{ii} + \frac{\partial(\xi_1, \xi_2, \xi_3)}{\partial(a_1, a_2, a_3)} \end{aligned}$$

Here, $\Delta_{ij}$ is the cofactor of $\partial\xi_i/\partial a_j$ in the Jacobian determinant $|\partial\xi/\partial a|$. The equations of motion in the new coordinates are obtained by varying with respect to $\xi_i$ and $\varpi$. The potential may be expanded as

$$U(\xi_3) = p_0(a_3) + \frac{1}{2}\rho_0 N^2(a_3)\xi_3^2 + \frac{1}{3!}\xi_3^3\frac{d^3 p_0}{da_3^3} + \cdots$$

with the squared Brunt-Väisälä frequency $N^2 = -(g/\rho_0)d\rho_0/da_3$. The $p_0$-term can be omitted. This form reveals the ordinary quadratic potential energy $N^2\xi_3^2/2$ resulting from a displacement in stratified environment. Note that $U$ leads to nonlinearities in the equations of motion in addition to those arising from the incompressibility constraint.

### 2.12.5 A Variational Principle for Eulerian Coordinates

Lagrangian variables have the disadvantage that local interactions, as e. g. acceleration by the local pressure gradient force, are not immediately reflected in the governing equations. The Eulerian representation appears, therefore, more useful as it corresponds to a local field theory[35]. Various attempts have been made to construct a Lagrangian theory in Eulerian coordinates (pioneering papers are Lin (1963) and Seliger and Whitham (1968)). A simple-minded translation of the ingredients of the above Lagrangian is given by

$$L_1 = \frac{1}{2}\rho\boldsymbol{u}^2 - \rho E(\upsilon, \eta) + \rho\lambda_1\left(\frac{\partial\eta}{\partial t} + \boldsymbol{u}\cdot\nabla\eta\right) + \lambda_2\left(\frac{\partial\rho}{\partial t} + \nabla\cdot\rho\boldsymbol{u}\right)$$

[35] The transformation from Eulerian to Lagrangian variables is not canonical. Therefore, it is not immediately evident that the Eulerian equations are derivable from a variational principle.

which takes mass conservation and material conservation of entropy as side constraints into account. Because there is no dependence on $\partial \boldsymbol{u}/\partial t$, the evolution equation of $\boldsymbol{u}$ does not follow directly by variation. Instead, variation with respect to $\boldsymbol{u}$ yields a Clebsch representation of the velocity field,

$$\boldsymbol{u} = \nabla\lambda_2 - \lambda_1\nabla\eta \tag{2.231}$$

which seems at first general enough but, in fact, is rather restrictive: the circulation $\oint \boldsymbol{u}\cdot \mathrm{d}\boldsymbol{s}$ for any closed loop in an isentropic surface $\eta = \text{const}$ vanishes according to (2.231). This dilemma is mediated by introduction of additional constraints, so-called Lin constraints. Lin used the initial position $\boldsymbol{A}(\boldsymbol{x},t)$ of a parcel as additional variable: it is materially conserved, and

$$L_2 = L_1 + \rho\boldsymbol{\alpha}\cdot\left(\frac{\partial \boldsymbol{A}}{\partial t} + \boldsymbol{u}\cdot\nabla\boldsymbol{A}\right)$$

yields a representation for the velocity field

$$\boldsymbol{u} = \nabla\lambda_2 - \lambda_1\nabla\eta - \sum_j \alpha_j\nabla A_j \tag{2.232}$$

Actually, using one additional coordinate as constraint is already sufficiently general (Bretherton, 1970) because three scalar fields are only needed to guarantee arbitrary initial conditions for the velocity field. Instead of singling out a particular coordinate it seems thus reasonable to look for an additional scalar field which is conserved.

For a binary fluid like seawater salinity $S$ is such a variable. Consider material salinity conservation as constraint; hence we use

$$L = L_1 + \rho\lambda_3\left(\frac{\partial S}{\partial t} + \boldsymbol{u}\cdot\nabla S\right)$$

Variation of the Lagrangian with respect to all fields yields the equations

$$\delta\boldsymbol{u} : \boldsymbol{u} = \nabla\lambda_2 - \lambda_1\nabla\eta - \lambda_3\nabla S \tag{2.233}$$

$$\delta\lambda_1 : \frac{\partial\eta}{\partial t} + \boldsymbol{u}\cdot\nabla\eta = \frac{\mathrm{D}\eta}{\mathrm{D}t} = 0 \tag{2.234}$$

$$\delta\lambda_3 : \frac{\partial S}{\partial t} + \boldsymbol{u}\cdot\nabla S = \frac{\mathrm{D}S}{\mathrm{D}t} = 0 \tag{2.235}$$

$$\delta\lambda_2 : \frac{\partial\rho}{\partial t} + \nabla\cdot\rho\boldsymbol{u} = 0 \tag{2.236}$$

$$\delta S : \frac{\partial\lambda_3}{\partial t} + \boldsymbol{u}\cdot\nabla\lambda_3 = \frac{\mathrm{D}\lambda_3}{\mathrm{D}t} = \mu \tag{2.237}$$

$$\delta\eta : \frac{\partial\lambda_1}{\partial t} + \boldsymbol{u}\cdot\nabla\lambda_1 = \frac{\mathrm{D}\lambda_1}{\mathrm{D}t} = -T \tag{2.238}$$

$$\delta\rho : \frac{1}{2}\boldsymbol{u}^2 - E - pv + \lambda_1\frac{\mathrm{D}\eta}{\mathrm{D}t} + \lambda_3\frac{\mathrm{D}S}{\mathrm{D}t} - \frac{\mathrm{D}\lambda_2}{\mathrm{D}t} = 0 \tag{2.239}$$

where $\mu = -\partial E/\partial S$ is the chemical potential (see (1.47)). The last equation may be written as

$$\frac{\mathrm{D}\lambda_2}{\mathrm{D}t} = \frac{1}{2}\boldsymbol{u}^2 - H \tag{2.240}$$

where $H = E + pv$ is the enthalpy. It remains to show that in the realm of the above set of equations the conventional momentum balance is valid. After some nontrivial eliminations (see the box on p. 103) we indeed arrive at

$$\frac{\mathrm{D}\boldsymbol{u}}{\mathrm{D}t} = -v\boldsymbol{\nabla} p \tag{2.241}$$

The set of Eulerian balances (2.234)–(2.236) and (2.241) must be supplemented by the equation of state in the form $\rho = F(S, \eta, p)$. It may be derived from (2.226) with salinity as additional variable.

19. Derivation of (2.241)

Applying $\mathrm{D}/\mathrm{D}t$ to (2.233) yields

$$\frac{\mathrm{D}\boldsymbol{u}}{\mathrm{D}t} = \frac{\mathrm{D}\boldsymbol{\nabla}\lambda_2}{\mathrm{D}t} - \lambda_1\frac{\mathrm{D}\boldsymbol{\nabla}\eta}{\mathrm{D}t} - \lambda_3\frac{\mathrm{D}\boldsymbol{\nabla}S}{\mathrm{D}t} + T\boldsymbol{\nabla}\eta - \mu\boldsymbol{\nabla}S \tag{B19.1}$$

where (2.237) and (2.238) has been used. The last two terms will be replaced by the relation for gradient of the enthalpy

$$\boldsymbol{\nabla}H = T\boldsymbol{\nabla}\eta - \mu\boldsymbol{\nabla}S + v\boldsymbol{\nabla}p$$

The first three terms in (B19.1) are evaluated according to the identity

$$\frac{\mathrm{D}\boldsymbol{\nabla}\lambda_2}{\mathrm{D}t} = \boldsymbol{\nabla}\frac{\mathrm{D}\lambda_2}{\mathrm{D}t} - \frac{\partial\lambda_2}{\partial x}\boldsymbol{\nabla}u - \frac{\partial\lambda_2}{\partial y}\boldsymbol{\nabla}v - \frac{\partial\lambda_2}{\partial z}\boldsymbol{\nabla}w$$

and correspondingly for $\eta$ and $S$. Implementing the result into (B19.1), collecting then the term proportional to $\boldsymbol{\nabla}u$, and using the first component of (2.233), and proceeding similarly with $\boldsymbol{\nabla}v$ and $\boldsymbol{\nabla}w$, we find

$$\frac{\mathrm{D}\boldsymbol{u}}{\mathrm{D}t} = -v\boldsymbol{\nabla}p + \boldsymbol{\nabla}\frac{\mathrm{D}\lambda_2}{\mathrm{D}t} + \boldsymbol{\nabla}H - u\boldsymbol{\nabla}u - v\boldsymbol{\nabla}v - w\boldsymbol{\nabla}w \tag{B19.2}$$

where (2.234) and (2.235) was used. Finally, (2.241) results by implementing (2.240).

## Further Reading

The manual *International Thermodynamic Equation of Seawater–2010* (IOC, SCOR and IAPSO, 2010) contains the currently most accurate and comprehensive information on all relevant aspects of the thermodynamics of seawater, including the algorithms for the practical computation of thermodynamic parameters.

The principles of ocean thermodynamics is treated in an accurate though rather condensed way in the first chapters of *Fundamentals of Ocean Dynamics* by Kamenkovich (1977).

*Thermodynamics of Atmospheres and Oceans* by Curry and Webster (1999) covers the subject for both ocean and atmosphere.

Classical textbooks on fluid mechanics are *An Introduction to Fluid Dynamics* by Batchelor (1977) and *Fluid Mechanics* by Landau and Lifshitz (1987). More easy for the beginner is *Fluid Mechanics* by Kundu et al. (2004) which gives an overview on basic conservation laws of fluids and discusses many applications for irrotational, laminar, turbulent and compressible flows.

An overview on the Hamiltonian formulation of fluid dynamics can be found in the *Lectures on Geophysical Fluid Dynamics* by Salmon (1998).

The mathematical tools for the description of hydrodynamic fields are described e. g. in *Mathematical Principles of Classical Fluid Mechanics* by Serrin (1959), *Vectors, Tensors, and the Basic Equations of Fluid Mechanics* by Aris (1989), and at a more elementary level in *Fundamentals of Atmospheric Dynamics and Thermodynamics* by Riegel (1992).

The article *The First Law of Thermodynamics in a Salty Ocean* by Warren (2006) is an excellent review of the first law from an oceanic perspective.

For a historical review of fluid dynamics see *Worlds of Flow: a History of Hydrodynamics from the Bernoullis to Prandtl* by Darrigol (2009).

# Part II Common Approximations

In Part I we combined the Navier–Stokes equations and the first and second law of thermodynamics to find the governing equations of motion for the ocean. In principle, the equations (2.151)–(2.155) describe the full spectrum of oceanic motions, from sound waves with time-scales of milliseconds to the thermohaline circulation with periods of up to thousands of years. Although the governing equations have not been derived rigorously from first principles but more in a heuristic or empirical way, there is no reason to question their validity. However, the broad spectrum of scales involved in oceanic motions leads to difficulties when one attempts to solve the equations. As an example, when the wind sets on at the ocean's surface, it simultaneously generates responses at all time-scales in the ocean, namely planetary waves, gravity waves, but also sound waves and capillary surface waves. Due to the very different temporal and spatial scales of these processes, one is usually only interested in the response of only one kind of waves. The presence of the other kinds of waves complicates the structure of analytical solutions, and the different scales lead to formidable practical difficulties when one applies numerical solution techniques.

In this chapter we will look at a number of common approximations to the system (2.151)–(2.155). The purpose of these approximations is to modify the system in such a way that the manifold of solutions is reduced, in order that the system can be solved or analyzed more easily, and that one can focus on specific processes while ignoring others. In particular, for considering motions with large scales related to the general circulation of the oceans, it is almost mandatory to eliminate from the system those processes occurring at very small spatial and/or temporal scales. Note that these approximations will generally restrict the validity of the resulting relations and may not always reduce (in fact, occasionally even increase) the *algebraic* complexity of the system.

There are different methods to approximate the system and to reduce its complexity. We have already met an important method in Section 2.8, i. e. the Reynolds average of the equation of motions. The simplest Reynolds filter is a time-mean; other possibilities are spatial, ensemble or zonal means or a combination of them. By looking at the Reynolds averaged (filtered) equations one considers only that part of the spectrum of possible motions above the cut-off frequency of the Reynolds filter. However, a drawback of the method is given by the appearance of the turbulent fluxes containing the effect of the filtered motions on the larger (unfiltered) scales.

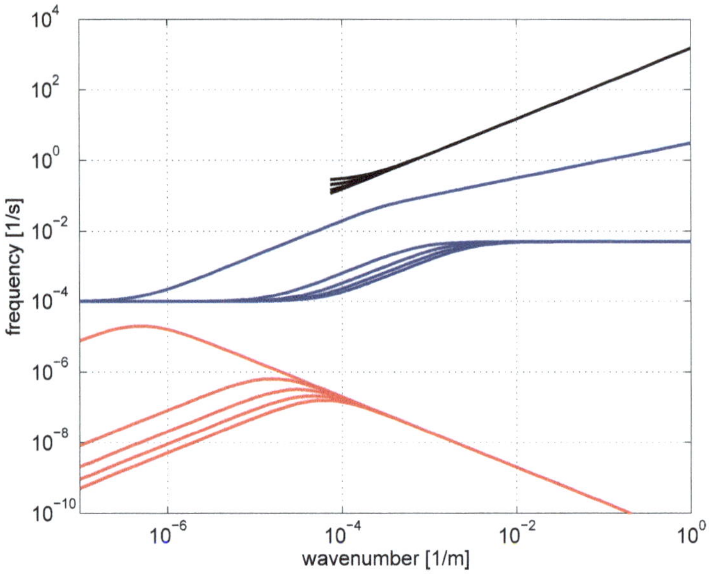


Overview of spatial and time-scales of oceanic waves. The horizontal axis denotes the horizontal wave number and the vertical axis angular frequency $\omega$. The dispersion relation of sound waves is shown by black lines, gravity waves by blue lines and planetary waves by red lines. Several branching dispersion relations for a certain kind of wave regime are related to the impact of vertical boundaries (the figure shows the first four modes) as discussed in Part III. The parameters are: Sound velocity $c_s = 1{,}500\,\mathrm{m\,s^{-1}}$ (see Section 6.1), Brunt–Väisälä frequency $N = 5 \times 10^{-3}\,\mathrm{s^{-1}}$ defined in (5.9), Coriolis frequency $f = 10^{-4}\,\mathrm{s^{-1}}$ and $\beta = 2 \times 10^{-11}\mathrm{m^{-1}\,s^{-1}}$ as defined in (5.18), water depth $h = 4{,}000\,\mathrm{m}$, vertical wave number $k_3 = n\pi/h$, baroclinic Rossby radius $R_i = N/(fk_3)$ and barotropic Rossby radius $R_b = 2 \times 10^6\,\mathrm{m}$ (see Section 8.1.1).

Part IV of this book is devoted to a detailed discussion of these turbulent fluxes and methods how to deal with them.

Another way to approximate the system is given by mode filtering of the system, eliminating certain wave branches from all possible wave solutions of the system. This procedure is described in Chapter 3 and gives a first overview of the possible wave solutions and the possible approximations to the system. However, the most practical way to approximate the system is achieved by neglecting certain terms since they might be identified to be small under certain circumstances, or by other direct manipulations in the equations of motions. Various approximations which are common in oceanography are discussed in detail in Chapters 4 and 5. Chapter 4 deals with basic approximations which will lead to the primitive equations representing the fundamental description of the large-scale oceanic circulation while Chapter 5 deals with more restrictive approximations which will be suited for geostrophically balanced systems. For all approximations we will monitor the important consequences for the energy cycle and potential vorticity.

# Approximations Derived from Mode Filtering

# 3

In this chapter we will give a first preliminary account of the different time-scales that are contained in the system (2.151) to (2.155) of oceanic motions, and of practical ways to eliminate certain modes without substantially modifying those that remain. Note that this chapter is directed to the more advanced reader and not meant to be studied in detail since many of the issues, which are only touched here, are covered in more detail later. References to the more detailed discussions are given in the respective context.

The system (2.151)–(2.155) is *complete* with respect to these seven variables, in the sense that the temporal evolution of the system is completely determined, provided that suitable initial and boundary conditions are specified, and that the forcing terms $\mathcal{F}$, $\mathcal{G}_S$, $\mathcal{G}_\theta$ are prescribed in terms of known quantities. Although complete, the system is however not *immediately* suitable to actually compute the evolution of the system. This can be seen when one attempts to calculate the evolution of all fields from a given initial state. While it is obvious that (2.151)–(2.154) allow to determine the variables $\rho$, $\boldsymbol{u}$, $S$ and $\theta$, respectively, for a small time interval $\delta t$ ahead, there is no equation to directly determine the pressure $p$. Instead, there is in addition to (2.151) another equation, namely the state equation (2.155), which also allows to determine $\rho$, provided that $S, \theta, p$ are known. It might seem obvious to resolve this difficulty by determining the pressure *diagnostically* from the state equation (2.155) by solving for $p = p(S, \theta, \rho)$. However, the very small compressibility of seawater (which, as we shall see, is in fact often neglected altogether) precludes this option. A much better alternative is to use the state equation to derive a *prognostic* differential equation for the pressure. The set of equations, achieved by this reorganization, allows to derive and filter wave modes in a straightforward and consistent way.

## 3.1 A Prognostic Equation for the Pressure

Forming the substantial derivative of the state equation (2.155) and using the continuity equation (2.151), the salt conservation (2.153) and the temperature equa-

D. Olbers, J. Willebrand, C. Eden, *Ocean Dynamics*,
DOI 10.1007/978-3-642-23450-7_3, © Springer-Verlag Berlin Heidelberg 2012

tion (2.154), one obtains

$$\frac{1}{c_s^2}\frac{\mathrm{D}p}{\mathrm{D}t} = -\rho\nabla\cdot\boldsymbol{u} - \mathcal{G}_\rho \tag{3.1}$$

Here $c_s$ denotes sound velocity as given by (2.128), and $\mathcal{G}_\rho$ is defined as

$$\mathcal{G}_\rho = \gamma\mathcal{G}_S - \alpha\mathcal{G}_\theta \tag{3.2}$$

and is associated with *diapycnal* processes resulting from turbulent salt and heat fluxes which affect the density of a water parcel and vanish in case of adiabatic motions. In this relation, $\alpha$ and $\gamma$ denote the modified thermal and haline expansion coefficients, respectively, as introduced in Section 2.6.4. While less common, the use of (3.1) instead of the continuity equation (2.151) in connection with (2.152)–(2.155) is conceptually preferable because the full system now consists of six *prognostic* equations for the variables $u, v, w, p, S$ and $\theta$, with differential equations that are of first order in time. The variable $\rho$ can in principle be eliminated from the system by invoking the *diagnostic* algebraic relation given by the state equation (2.155).

## 3.2 Linear Waves

A simple view of the time-scales and the types of motion exhibited by the system (3.1) and (2.152) to (2.155) can be obtained by a linear wave analysis. Wave solutions of the system will be discussed in detail in Part III while here the essential results from the more complete discussion are used to motivate the separation of the system into different wave regimes. An overview of the different wave branches existing in the ocean is given in the figure on page 106.

Consider small *adiabatic* perturbations from a state of rest, the latter being characterized by a vertical stratification of salinity and temperature as in Section 2.9. Adiabatic in this context means zero turbulent or molecular fluxes, i. e. $\mathcal{G}_S = \mathcal{G}_\theta = 0$. All prognostic variables are expanded according to $\chi = \chi_r + \chi'$ (where $\chi$ can be one of $u, v, w, p, S$ and $\theta$) around a basic reference state which is defined as

$$u_r = v_r = w_r = 0\,, \quad S_r = S_r(z)\,, \quad \theta_r = \theta_r(z)$$

$$\frac{\partial p_r}{\partial z} = -g\rho_r\,, \quad N_\theta^2 = g\alpha\frac{\partial\theta_r}{\partial z}\,, \quad N_S^2 = -g\gamma\frac{\partial S_r}{\partial z}$$

Neglecting for the moment all terms which are quadratic or of higher order in the perturbation quantities (refer to Section 6.1 for a more detailed and mathematically more rigorous derivation of this assumption), one obtains

$$\frac{\partial u'}{\partial t} - fv' + \frac{1}{\rho_r}\frac{\partial p'}{\partial x} = 0 \tag{3.3}$$

$$\frac{\partial v'}{\partial t} + fu' + \frac{1}{\rho_r}\frac{\partial p'}{\partial y} = 0 \tag{3.4}$$

$$\frac{\partial w'}{\partial t} + \frac{1}{\rho_r}\left(\frac{\partial p'}{\partial z} + \frac{g}{c_s^2}p'\right) + g\left(\gamma S' - \alpha\theta'\right) = 0 \tag{3.5}$$

$$\frac{1}{c_s^2\rho_r}\frac{\partial p'}{\partial t} + \nabla\cdot\boldsymbol{u}' = 0 \tag{3.6}$$

$$\frac{\partial S'}{\partial t} - \frac{N_S^2}{\gamma g}w' = 0 \tag{3.7}$$

$$\frac{\partial \theta'}{\partial t} + \frac{N_\theta^2}{\alpha g}w' = 0 \tag{3.8}$$

Instead of the horizontal velocity components, we now introduce the horizontal stream function $\psi$ and velocity potential $\phi$ (see also Appendix A.1.4 which are given by

$$u' = \frac{\partial\phi}{\partial x} - \frac{\partial\psi}{\partial y}\,, \qquad v' = \frac{\partial\phi}{\partial y} + \frac{\partial\psi}{\partial x}$$

so that $\partial v'/\partial x - \partial u'/\partial y = \nabla_h^2\psi$ and $\partial u'/\partial x + \partial v'/\partial y = \nabla_h^2\phi$. Furthermore, instead of $S'$ and $\theta'$, we introduce the scalar variables $\nu'$ and $\rho'$ as

$$\rho' = \gamma S' - \alpha\theta'\,, \qquad \nu' = \gamma N_\theta^2 S' + \alpha N_S^2\theta'$$

From (3.3)–(3.8) one can derive the following equations (dropping all primes from now on):

$$\epsilon_1\nabla_h^2\frac{\partial\psi}{\partial t} + f\nabla_h^2\phi + \beta\psi_x = 0 \tag{3.9}$$

$$\epsilon_2\nabla_h^2\frac{\partial\phi}{\partial t} - f\nabla_h^2\psi + \frac{1}{\rho_r}\nabla_h^2 p = 0 \tag{3.10}$$

$$\epsilon_3\frac{\partial w}{\partial t} + \frac{1}{\rho_r}\left(\frac{\partial p}{\partial z} + g\rho\right) = 0 \tag{3.11}$$

$$\epsilon_4\frac{1}{c_s^2\rho_r}\frac{\partial p}{\partial t} + \left(\nabla_h^2\phi + \frac{\partial w}{\partial z}\right) = 0 \tag{3.12}$$

$$\epsilon_5\frac{\partial\rho}{\partial t} - \frac{N^2}{g}w = 0 \tag{3.13}$$

$$\epsilon_6\frac{\partial\nu}{\partial t} = 0 \tag{3.14}$$

The definition $N^2 = N_\theta^2 + N_S^2$ for the Brunt–Vaisälä frequency (cf. Section 2.9.2) has been used. The dimensionless factors $\epsilon_1 \ldots \epsilon_6$ have been attached to the time derivatives for later use; they all should be set to 1 to represent the correct system. A number of approximations have been made in deriving (3.9)–(3.14). The Earth's rotation has been replaced by $2\boldsymbol{\Omega} \approx (0, 0, f)$, and $\beta = \partial f/\partial y$ describes the lateral variation of the Coriolis parameter. The first two equations are derived by taking the horizontal curl and divergence of (3.3) and (3.4), but a term $\beta u$ in (3.10) was neglected for reasons discussed in Section 5.2. Furthermore, some subtleties regarding the state equation have been ignored. For maximum simplicity, all coefficients in the system (3.3)–(3.8) are now assumed to be constant and positive, an approximation which greatly facilitates the identification of the basic modes of motion. The validity range of these approximations will not be considered at this point because only some qualitative properties of the system are discussed. A more complete treatment of linear wave systems is given in Part III.

As for any linear system with constant coefficients, linear wave solutions can be found. Combining the dependent variables into a vector $\boldsymbol{b}$, a plane wave ansatz with wave number $\boldsymbol{k} = (k_1, k_2, k_3)$ and frequency[1] $\omega$ according to

$$(\psi, \phi, w, p/\rho_r, \rho/\rho_r, \nu)^T = \boldsymbol{b}(x, y, z, t) = \boldsymbol{b}_0 e^{i(k_1 x + k_2 y + k_3 z - \omega t)} \tag{3.15}$$

leads to a standard eigenvalue problem (for eigenvalue problems see the box on p. 9) of the form

$$\boldsymbol{A}\boldsymbol{b}_0 = i\omega \boldsymbol{b}_0 \tag{3.16}$$

with the matrix (here with all $\epsilon_i = 1$)

$$\boldsymbol{A} = \begin{pmatrix} -i\beta k_1/k_h^2 & f & 0 & 0 & 0 & 0 \\ -f & 0 & 0 & 1 & 0 & 0 \\ 0 & 0 & 0 & ik_3 & g & 0 \\ 0 & -c_s^2 k_h^2 & ic_s^2 k_3 & 0 & 0 & 0 \\ 0 & 0 & -N^2/g & 0 & 0 & 0 \\ 0 & 0 & 0 & 0 & 0 & 0 \end{pmatrix}$$

Here the notation $k_h^2 = k_1^2 + k_2^2$, and $k^2 = k_1^2 + k_2^2 + k_3^2$ has been used. The eigenfrequencies $\omega$ are given as roots of the characteristic equation

$$\omega^6 + \omega^5 \frac{\beta k_1}{k_h^2} - \omega^4 \left(f^2 + N^2 + c_s^2 k^2\right) - \omega^3 \frac{\beta k_1}{k_h^2} \left(c_s^2 k^2 + N^2\right)$$
$$+ \omega^2 \left(c_s^2 k_h^2 N^2 + c_s^2 k_3^2 f^2 + f^2 N^2\right) + \omega \beta k_1 c_s^2 N^2 = 0 \tag{3.17}$$

which has six solutions $\omega_i, i = 1, \ldots, 6$. One of these solutions corresponds to zero frequency, say $\omega_6 = 0$, reflecting the fact that the variable $\nu$ is dynamically passive and does not occur in any of the equations (3.9)–(3.13). The other roots of (3.17) refer to more interesting wave branches, as discussed below. Note, however, that in certain less simplified situations the variable $\nu$ may actually become active. This can occur when changes of salinity and temperature have different time-scales, e. g. as a consequence of differences in mixing or in the exchange with the atmosphere, but also as a consequence of nonlinearities in the equation of state.

Hence effectively a polynomial of order 5 remains in (3.17). Note that the form with inclusion of the $\epsilon$-factors is displayed in the box on p. 111. Several (inverse) time-scales appear which under typical oceanic conditions have *very* different magnitudes:

[1] Throughout this book, we refer to $\omega$ defined as in (3.15) as frequency. Strictly speaking, however, $\omega$ is the *circular* frequency, and the frequency is $\omega/2\pi$.

- The planetary frequency scale $\beta/k_\mathrm{h}$, with magnitude of $2\times10^{-11}$–$2\times10^{-6}\,\mathrm{s}^{-1}$ where the former number holds for a wave length of $k^{-1}\sim 1\,\mathrm{m}$ , the latter for $k^{-1}\sim 100\,\mathrm{km}$.
- The Coriolis frequency with a magnitude of $f\sim10^{-4}\,\mathrm{s}^{-1}$ for midlatitudes.
- The buoyancy frequency $N\sim 2\times10^{-3}\,\mathrm{s}^{-1}$ for a typical situation in the interior of the ocean (compare with Figure 2.17).
- The frequency of sound waves $c_\mathrm{s}k$, with magnitude $1.5\times10^{-2}\,\mathrm{s}^{-1}$–$1.5\times10^{3}\,\mathrm{s}^{-1}$ for a wave lengths of $k^{-1}\sim 1$ m to $k^{-1}\sim 100\,\mathrm{km}$.

Under normal oceanic situations one can assume that

$$\frac{\beta}{k_\mathrm{h}} \ll f \lesssim N \ll c_\mathrm{s}k \tag{3.18}$$

where the $\ll$ sign is understood to denote a difference in magnitude by a factor

$$\delta \sim \frac{\beta}{f k_\mathrm{h}} \sim \frac{N}{c_\mathrm{s}k} \approx 10^{-6}\text{–}10^{-1}$$

depending on the scale of motion. Because the relevant time scales are so different, approximate values for the roots of (3.17) can be found. It is straightforward to show that for $\delta \ll 1$, (3.17) can be approximated by

$$\omega^5 - \omega^3 c_\mathrm{s}^2 k^2 + \omega\left(c_\mathrm{s}^2 k_\mathrm{h}^2 N^2 + c_\mathrm{s}^2 k_3^2 f^2\right) + \beta k_1 c_\mathrm{s}^2 N^2 \approx 0 \tag{3.19}$$

For the smallest root, say $\omega_5$, the main contribution must come from the two last terms in (3.19), and we have

$$\omega_5 = -\frac{\beta k_1}{k_\mathrm{h}^2 + k_3^2 f^2/N^2} + O\left(\delta^2\right) \tag{3.20}$$

which is identified as the dispersion relation of planetary or Rossby waves (see Section 8.2). On the other hand, for the bigger roots of order $f, N$ or $c_\mathrm{s}k$, the term in (3.19) containing $\beta k_1/k_\mathrm{h}^2$ must be small, and we obtain

$$\omega^4 - \omega^2 c_\mathrm{s}^2 k^2 + c_\mathrm{s}^2\left(k_\mathrm{h}^2 N^2 + k_3^2 f^2\right) \approx 0$$

which can be readily solved. Of the two pairs of solution, the bigger one is approximated by

$$\omega_{1,2}^2 = c_\mathrm{s}^2 k^2 + O\left(\delta^2\right) \tag{3.21}$$

**20. Dispersion Relation including the $\epsilon_i$**

When all $\epsilon$-factors are included in the eigenvalue problem (3.16), the dispersion relation (3.17) changes to

$$\begin{aligned}
&\omega^6\epsilon_1\epsilon_2\epsilon_3\epsilon_4\epsilon_5\epsilon_6 + \omega^5\frac{\beta k_1}{k_\mathrm{h}^2}\epsilon_2\epsilon_3\epsilon_4\epsilon_5\epsilon_6\\
&\quad - \omega^4\left(f^2\epsilon_3\epsilon_4\epsilon_5\epsilon_6 + N^2\epsilon_1\epsilon_2\epsilon_4\epsilon_6 + c_\mathrm{s}^2k_\mathrm{h}^2\epsilon_1\epsilon_3\epsilon_5\epsilon_6 + c_\mathrm{s}^2k_3^2\epsilon_1\epsilon_2\epsilon_5\epsilon_6\right)\\
&\quad - \omega^3\frac{\beta k_1}{k_\mathrm{h}^2}\left(c_\mathrm{s}^2k_\mathrm{h}^2\epsilon_3\epsilon_5\epsilon_6 + c_\mathrm{s}^2k_3^2\epsilon_2\epsilon_5\epsilon_6 + N^2\epsilon_2\epsilon_4\epsilon_6\right)\\
&\quad + \omega^2\left(c_\mathrm{s}^2k_\mathrm{h}^2N^2\epsilon_1\epsilon_6 + c_\mathrm{s}^2k_3^2f^2\epsilon_5\epsilon_6 + f^2N^2\epsilon_4\epsilon_6\right)\\
&\quad + \omega\beta k_1 c_\mathrm{s}^2 N^2\epsilon_6 = 0
\end{aligned}$$

and constitutes the dispersion relation for sound waves (see Chapter 6). The smaller pair is given by

$$\omega_{3,4}^2 = \frac{k_\mathrm{h}^2 N^2 + k_3^2 f^2}{k^2} + O\left(\delta^2\right) \tag{3.22}$$

and constitutes the dispersion of internal gravity waves (see Chapter 7). Both for sound waves and gravity waves there exist *two* solutions each because along any coordinate, waves can propagate into both positive and negative directions. Rossby waves, on the other hand, have a phase propagation to the west only.

## 3.3 Filtering of Modes

We have seen that the full system (3.3) to (3.8) has five nontrivial modes of motion which correspond to three wave processes with very different time-scales. We will now demonstrate that certain approximations to the full system can change this behavior, and in fact it is possible to single out any of the above modes. In a linear system, such a reduction to a single mode could most effectively be achieved by projecting the equations of motion onto the eigenvector of the particular mode of interest (this procedure is explained in Section 10.1). In a general nonlinear system, a convenient method which is particularly useful to eliminate fast modes is to consider the *diagnostic* limit of certain equations that may be associated with those fast modes. The argument for this is not that the time rate of change of the corresponding variable is generally small, but rather that the (fast) adjustment process has already taken place and has reached its steady state limit. For the system (3.9)–(3.14), this is equivalent to setting some of the $\epsilon_i$ to zero. The consequences for the dispersion relation are then easily found from the box on p. 111.

### Removing sound waves, $\epsilon_4 = 0$

The neglect of the time dependence in (3.12) is the most common of all possible approximations in this context. The dispersion relation is then given by

$$\omega\left[\omega^3 k^2 + \omega^2 \frac{\beta k_1}{k_\mathrm{h}^2} k^2 - \omega\left(k_\mathrm{h}^2 N^2 + k_3^2 f^2\right) - \beta k_1 N^2\right] = 0 \tag{3.23}$$

We note that the order of the polynomial has changed, and though the approximated system (3.9)–(3.14) still has 6 variables, we have only four eigenfrequencies. One is again identified as the trivial root $\omega = 0$ which must always appear. For the smallest of the remaining three solutions, the main balance is between the last two terms in the bracket, yielding the Rossby wave frequency $\omega_5$ from (3.20). For the other two solutions, the first two terms balance, and we obtain the gravity wave frequencies $\omega_{3,4}$ from (3.22). Hence the neglect of the time dependence in (3.12) is indeed sufficient to remove the sound waves completely from the system, without significant distortion of either the gravity or the Rossby modes. This approximation will be discussed in Sections 4.1.1 and 4.1.2.

### Hydrostatic approximation, $\epsilon_3 = 0$

The neglect of the time dependence in (3.11) will be related in Section 4.2 to the *hydrostatic approximation*. The dispersion is found to be

$$\omega\left[\left(\omega^3 + \omega^2\frac{\beta k_1}{k_\mathrm{h}^2}\right)\left(c_\mathrm{s}^2k_3^2 + N^2\right) - \omega\left(c_\mathrm{s}^2k_\mathrm{h}^2N^2 + c_\mathrm{s}^2k_3^2f^2 + f^2N^2\right) - \beta k_1 c_\mathrm{s}^2 N^2\right] = 0$$

Invoking (3.18), this may be approximated as

$$\omega\left[\omega^3\left(c_\mathrm{s}^2k_3^2 + N^2\right) - \omega c_\mathrm{s}^2\left(k_\mathrm{h}^2N^2 + k_3^2f^2\right) - \beta k_1 c_\mathrm{s}^2N^2\right] = 0 \qquad (3.24)$$

Again the trivial root and the Rossby mode from (3.20) remain unchanged. For the fast waves we obtain

$$\omega^2 = \frac{k_\mathrm{h}^2N^2 + k_3^2f^2}{k_3^2 + N^2/c_\mathrm{s}^2} + O\left(\delta^2\right) \qquad (3.25)$$

If $c_\mathrm{s}k_3 \gg N$, we obtain a relation similar to (3.22), with $k^2$ in the denominator replaced by $k_3^2$, which coincides with the gravity wave frequency provided that also $k_3 \gg k_\mathrm{h}$, i. e. for *long* gravity waves (see Section 8.1.2). If, however, $c_\mathrm{s}k_3 \ll N$, we obtain instead

$$\omega^2 = c_\mathrm{s}^2k_\mathrm{h}^2 + O\left(\delta^2\right) \qquad (3.26)$$

and thus obtain the sound wave relation (3.21), provided that the waves propagate horizontally (i. e. $k_3 \ll k_\mathrm{h}$). The waves described by (3.24) are referred to as LAMB[2] waves.

Without further approximations, the hydrostatic approximation hence leaves a combination of long gravity waves and horizontal sound waves in the system and is usually not advisable unless specific precautions are taken (e. g. through suitable boundary conditions which can eliminate the Lamb waves).

### Shallow water approximation, $\epsilon_3 = \epsilon_4 = 0$

The combination of the two previous approximations is straightforward, and is at the heart of the *shallow water approximation* to be discussed in Section 4.2. We find

$$\omega c_\mathrm{s}^2\left[\omega^3k_3^2 + \omega^2\frac{\beta k_1}{k_\mathrm{h}^2}k_3^2 - \omega\left(k_\mathrm{h}^2N^2 + k_3^2f^2\right) - \beta k_1 N^2\right] = 0$$

Hence Rossby waves remain unchanged in the system, but only the long part of the gravity waves as in (3.25) is left.

[2] SIR HORACE LAMB, *1849 in Stockport, †1934 in Cambridge, fluid dynamicist.

### Quasi-geostrophic approximation, $\epsilon_2 = \epsilon_3 = \epsilon_4 = 0$

We find in this limit

$$\omega c_\mathrm{s}^2 \left[\omega \left(k_\mathrm{h}^2 N^2 + k_3^2 f^2\right) + \beta k_1 N^2\right] = 0$$

and hence recover the Rossby wave dispersion (3.20) without distortion but no gravity modes. This is the essential property of the *quasi-geostrophic* equations which will be discussed in Section 5.2 below.

### Planetary geostrophic approximation, $\epsilon_1 = \epsilon_2 = \epsilon_3 = \epsilon_4 = 0$

We find

$$\omega c_\mathrm{s}^2 \left(\omega k_3^2 f^2 + \beta k_1 N^2\right) = 0$$

and thus have

$$\omega = -\frac{\beta k_1 N^2}{k_3^2 f^2} + O\left(\delta^2\right) \tag{3.27}$$

which agrees with the relation (3.20) for Rossby waves provided that $k_\mathrm{h} \ll k_3 f/N$, i. e. long Rossby waves. This is the basis for the (planetary) geostrophic approximation discussed in Section 5.3.

### Short barotropic Rossby wave, $\epsilon_3 = \epsilon_4 = \epsilon_5 = 0$

No vertical motions are allowed in this case, and we find

$$\omega N^2 c_\mathrm{s}^2 \left(\omega k_\mathrm{h}^2 + \beta k_1\right) = 0$$

Hence only one mode exists, with

$$\omega = -\frac{\beta k_1}{k_\mathrm{h}^2} + O\left(\delta^2\right)$$

which is the short barotropic Rossby wave (see Section 8.2).

Many further wave types exist when the existence of vertical or lateral boundaries is taken into account which is, however, not our concern here. On the other hand, we have seen how the most important wave solutions of the linearized system can be found, and how they relate to certain time-scales. The aim of our discussion was to illustrate how these wave solutions are related to different approximations which are detailed in the following sections.

# 4 Approximations Relating to Density Changes and Geometric Conditions

This chapter explains various approximations which are commonly used for the discussion of the large scale ocean circulation. These approximations are based on the smallness of oceanic density variations – which will lead to the inelastic and ultimately to the Boussinesq approximation, the geometry of the Earth, in particular its almost spherical shape, and the scales of ocean circulation – which lead to the hydrostatic approximation. The combination of all approximations form the so-called primitive equations, which represent the basis of current numerical ocean general circulation models.

In most situations, sound waves are not of interest for the discussion of the circulation of the ocean. It appears, therefore, of benefit to filter out sound waves from the governing equations, by using an approximated set of equations. Such approximations are all based on the fact that the density variations in the ocean are very small compared the mean value of density of seawater. However, although small, the density variations are nevertheless dynamically crucial, and the approximation of the equations of motion requires thus some care.

## 4.1 Approximations Involving Density

We first define a *reference* state, which is a state of rest, as introduced in Section 2.9, but now with *constant* values for potential temperature and salinity which may be chosen e. g. as $\theta_c = 0°\mathrm{C}$, $S_c = 35$. All fields are now decomposed into reference value and derivation thereof, i. e.

$$p = p_c(z) + \tilde{p} \,, \quad S = S_c + \tilde{S} \,, \quad \theta = \theta_c + \tilde{\theta} \,, \quad \rho = \rho_c(z) + \tilde{\rho}$$

but $\boldsymbol{u} = \tilde{\boldsymbol{u}}$ since $\boldsymbol{u}_c = 0$. It is emphasized that the reference state (labeled with the index 'c' and used for all the approximations that follow in the present chapter) has no stratification of temperature and salinity. This choice avoids remnants of the reference state gradients in the evolution equations for temperature and salinity so that these equations remain unchanged when formulated with $\tilde{S}, \tilde{\theta}$ instead of $S, \theta$. Note that pressure and density of the reference state are not constant, and that the depth dependence of its density reflects the compressibility of the water. As discussed

D. Olbers, J. Willebrand, C. Eden, *Ocean Dynamics*,
DOI 10.1007/978-3-642-23450-7_4, 

in Section 2.9, any state of rest is hydrostatically balanced, and equations (2.157) and also (2.158) must hold. For that reason we may use the identity

$$-\nabla p - \rho\nabla\Phi \equiv -\nabla\tilde{p} - \tilde{\rho}\nabla\Phi \tag{4.1}$$

in the momentum balance (2.152). In other words, for pressure gradient force and gravity force, only deviations from reference pressure, respectively density, are dynamically relevant. We will, therefore, refer to $\tilde{p}$ and $\tilde{\rho}$ as the *dynamically relevant* parts of pressure, respectively density. Note, however, that the full density still appears as a coefficient in the equations for momentum, salt and heat (2.152)–(2.154), and in the pressure equation (3.1), and likewise the full pressure appears in the state equation (2.155).

### 4.1.1 Anelastic Approximation

The dynamically relevant part of the density in the ocean is very small compared to the complete density. It follows from the state equation (2.155) as the difference

$$\tilde{\rho} = F(S_c + \tilde{S}, \theta_c + \tilde{\theta}, p_c + \tilde{p}) - F(S_c, \theta_c, p_c) \tag{4.2}$$

Variations of salinity, temperature and pressure cause variations of $\tilde{\rho}$ which have a magnitude of $|\tilde{\rho}|/\rho_c \lesssim 2\times 10^{-3}$ (cf. Figure 2.17). Furthermore, the reference density $\rho_c(z) = F(S_c, \theta_c, p_c)$ itself varies by only a few percent. The relative smallness of $\tilde{\rho}$ and the fact that $\rho_c(z)$ is nearly constant are the basis for the approximations considered in the following.

#### Elimination of sound waves

Consider first (3.1) in the form

$$\frac{\partial\tilde{p}}{\partial t} + \boldsymbol{u}\cdot\nabla\tilde{p} + w\frac{\partial p_c}{\partial z} = -c_s^2\left(\rho\nabla\cdot\boldsymbol{u} + \mathcal{G}_\rho\right) \tag{4.3}$$

So far, no approximation has been made. Equation (4.3) describes the response of the dynamically relevant part of the pressure to the flow divergence $\nabla\cdot\boldsymbol{u}$ and the diabatic terms contained in $\mathcal{G}_\rho$. As will be shown in Chapter 6, these changes in pressure propagate through the water with the sound velocity $c_s \approx 1{,}500\ \mathrm{m\,s^{-1}}$ which is much faster than any other velocity in the system, and the time-scale of sound waves is typically a fraction of a second. On the much longer time-scales of interest in dynamical oceanography, after the passage of sound waves, the balance (4.3) effectively has reached its steady-state limit. Hence by dropping the term $\partial\tilde{p}/\partial t$ in (4.3), the sound waves are filtered out of the system.

Assuming further that the gravity force is always important (i. e. not negligible compared to the pressure gradient), the magnitude of $\tilde{p}$ cannot exceed $O(gH\tilde{\rho})$ where $H$ is the maximum length scale of the system in the direction of gravity, i. e. the water depth. It follows that the relative magnitude of the advection term $\boldsymbol{u}\cdot\nabla\tilde{p}$ to the individual contributions of the $\nabla\cdot\boldsymbol{u}$–term on the right hand side in (4.3) is

$$\frac{UgH\tilde{\rho}}{Lc_s^2\rho U/L} \sim \frac{\tilde{\rho}}{\rho}\frac{H}{c_s^2/g} \sim 4\times 10^{-5} \ll 1$$

and, therefore, the advection term can also be neglected. It follows that (4.3) is no longer a prognostic differential equation to calculate $\tilde{p}$ but now a *diagnostic* relation which allows to compute $\nabla \cdot \boldsymbol{u}$. It takes the form

$$\rho \nabla \cdot \boldsymbol{u} = -\frac{w}{c_s^2}\frac{\partial p_c}{\partial z} - \mathcal{G}_\rho \tag{4.4}$$

Together with the momentum budget (2.152) in combination with (4.1), the heat and salt budgets and the state equation (4.2), equation (4.4) constitutes a complete system, with a relative accuracy of $4 \times 10^{-5}$. That system is, however, rarely used.

### Approximation of volume

In a second step where a relative error level of $\sim 10^{-3}$ is accepted, we now use the fact that $\tilde{\rho} \ll \rho_c$ and neglect $\tilde{\rho}$ whenever it is compared to $\rho_c$, which is tantamount to replacing $\rho$ by the reference density $\rho_c(z)$. At this level of accuracy the right-hand side of (4.4) can also be further simplified. Using the state equation (2.158) for the reference state, we obtain for the first term on the right-hand side of (4.4) the identity

$$\frac{w}{c_s^2}\frac{\partial p_c}{\partial z} = w\frac{c_c^2}{c_s^2}\frac{\partial \rho_c}{\partial z} = w\frac{\partial \rho_c}{\partial z} - w\frac{\partial \rho_c}{\partial z}\frac{c_s^2 - c_c^2}{c_s^2} \tag{4.5}$$

where $c_c$ is the sound speed for the reference state. While the relative magnitude of the first term on the right-hand side of (4.5) compared to the left-hand side of (4.4) is $gH/c_c^2 \sim 0.02$, which is not negligible in this approximation, the second term has the relative magnitude is $(gH/c_c^2) \cdot 2\Delta c/c_s \sim 0.001$ which is negligible.

The magnitude of the diapycnal fluxes $\mathcal{G}_\rho$ cannot be given in general without further knowledge about the processes that are involved. However, nearly all interesting situations in the ocean are such that in the heat and the salt balance the advective terms dominate, i. e. that the Peclet-number is $\mathrm{Pe} \gtrsim 1$ (see Section 2.8). It follows that $\mathcal{G}_\rho \lesssim \mathrm{Pe}^{-1}\tilde{\rho}U/L$, and hence one finds for the relative magnitude of the diapycnal flux in (4.4) compared to the left-hand side

$$\frac{\mathcal{G}_\rho}{\rho U/L} \sim \frac{1}{\mathrm{Pe}}\frac{\tilde{\rho}}{\rho} \lesssim 2 \times 10^{-3}$$

Hence the $\mathcal{G}_\rho$-term in (4.4) can be neglected in this approximation, and we finally obtain

$$\rho_c(z)\frac{\mathrm{D}\boldsymbol{u}}{\mathrm{D}t} = -2\rho_c(z)\boldsymbol{\Omega} \times \boldsymbol{u} - \nabla\tilde{p} - \tilde{\rho}\nabla\Phi + \boldsymbol{\mathcal{F}} \tag{4.6}$$

$$\nabla \cdot (\rho_c(z)\boldsymbol{u}) = 0 \tag{4.7}$$

$$\rho_c(z)\frac{\mathrm{D}}{\mathrm{D}t}\begin{pmatrix}\tilde{S}\\ \tilde{\theta}\end{pmatrix} = \begin{pmatrix}\mathcal{G}_S\\ \mathcal{G}_\theta\end{pmatrix} \tag{4.8}$$

which, together with the full equation of state (4.2), constitute the *anelastic approximation* to the equations of motion which was first developed by Ogura and Phillips (1962) for the atmosphere. The main effect of this approximation is that sound waves are filtered out of the system. Moreover, the terms involving $\rho_c$ in (4.6)–(4.8) are now *linear* in the respective field variables (except nonlinearities resulting from a material derivative), though with a variable coefficient. The anelastic approximation has an error of about $2 \times 10^{-3}$ and is mainly used in numerical models of the large-scale atmospheric circulation. In the ocean it is less useful since the density changes are smaller than in the atmosphere.

Note that the neglect in (4.7) of all terms depending on $\tilde{p}$ means that changes in the ocean volume resulting e. g. from thermal expansion cannot be computed from the system (4.6)–(4.8).

### 4.1.2 Boussinesq Approximation

It is customary to introduce a further step of approximation which takes into account that the variations of $\rho_c(z)$ are at most 2–3%, as was shown in Section 2.9. Frequently, one can afford to ignore errors of this magnitude and replace the reference density $\rho_c(z)$ by $\rho_0 = \text{const}$ in (4.6)–(4.8). With the same accuracy, the reference pressure $p_c(z) \approx p_0(z)$ is *linear* with depth, $\partial p_0(z)/\partial z = -g\rho_0$. A further approximation concerns the equation of state (4.2). The magnitude of density changes due to small changes in $\tilde{S}, \tilde{\theta}, \tilde{p}$ is

$$\Delta\tilde{\rho} = \rho\gamma\Delta\tilde{S} - \rho\alpha\Delta\tilde{\theta} + \frac{1}{c_s^2}\Delta\tilde{p} \tag{4.9}$$

It is shown in Section 4.2.4 below that in many situations not only the reference pressure but also the dynamically relevant pressure $\tilde{p}$ follows a hydrostatic scaling, so that $\Delta\tilde{p} \sim gh\Delta\tilde{\rho}$ with a vertical length scale $h$. The relative contribution of the last term in (4.9) is thus of order $gh/c_s^2$ and hence does not exceed 2–3% even for a maximum scale of $h = 5\,\text{km}$. With this accuracy, the dependence of (4.2) on $\tilde{p}$ can be ignored, and we obtain the equations in the BOUSSINESQ[1] approximation

$$\rho_0\frac{\mathrm{D}\boldsymbol{u}}{\mathrm{D}t} = -2\rho_0\boldsymbol{\Omega}\times\boldsymbol{u} - \nabla\tilde{p} - \tilde{\rho}\nabla\Phi + \mathcal{F} \tag{4.10}$$

$$\nabla\cdot\boldsymbol{u} = 0 \tag{4.11}$$

$$\rho_0\frac{\mathrm{D}}{\mathrm{D}t}\begin{pmatrix}\tilde{S}\\ \tilde{\theta}\end{pmatrix} = \begin{pmatrix}\mathcal{G}_S\\ \mathcal{G}_\theta\end{pmatrix} \tag{4.12}$$

$$\tilde{\rho} = F(S_c + \tilde{S}, \theta_c + \tilde{\theta}, p_0(z)) - F(S_c, \theta_c, p_0(z)) \tag{4.13}$$

with a constant $\rho_0$. The system (4.10)–(4.13) is the starting point for most oceanographic applications. Neither the exact form (2.151)–(2.155) nor the inelastic form of the equations of motions are frequently used. The error in the Boussinesq approximation is of order 0.02–0.03, roughly a factor of 10 higher than the error in the inelastic approximation. Many numerical models for the large-scale oceanic circulation are based on the Boussinesq approximation.

#### Determination of pressure in the Boussinesq system

The system (4.10)–(4.13) has an analogous problem as the complete set (2.151)–(2.155). Again there is no equation to determine the pressure (but now there are four

[1] VALENTIN JOSEPH BOUSSINESQ, *1842 in Saint-André-de-Sangonis, †1929 in Paris, mathematician and physicist. His work was published in 1903. Similar approximations were considered earlier by A. OBERBECK in 1879 and 1888.

Equation (4.11) is usually called the *incompressibility condition*, a somewhat unfortunate name because the notion of incompressibility is commonly used in two slightly different ways. A *fluid* is called incompressible when its (adiabatic) compressibility vanishes, i. e. when $c_s \to \infty$ holds. On the other hand, a *flow* is said to be incompressible when it satisfies equation (4.11), i. e. when the continuity equation – which was originally derived from *mass* conservation – formally states *volume* conservation (cf. (1.9)).

As seen from (3.1), both definitions agree only if the fluid motion is also adiabatic which is not generally true for the system (4.10)–(4.13). Note also that the Boussinesq system indeed contains an effect of the finite compressibility of sea water through the pressure dependence in (4.13).

21. Two Definitions of Incompressibility

equations for the three velocity components). Hence in this form, the system is again not directly suitable for the solution of initial value problems. It is possible, however, to derive a diagnostic equation for the pressure. Taking the divergence of (4.10), and using (4.11), leads to

$$\nabla^2 \tilde{p} = \nabla \cdot (-\rho_0 \boldsymbol{u} \cdot \nabla \boldsymbol{u} - 2\rho_0 \boldsymbol{\Omega} \times \boldsymbol{u} - \tilde{\rho} \nabla \Phi + \mathcal{F}) \tag{4.14}$$

which constitutes a POISSON[2] equation for the pressure. Equation (4.14) can be solved with the appropriate boundary conditions which will not be discussed here. Analogous forms can be derived based on (4.7) or (4.4). Equation (4.14) is the basis for all numerical models which explicitly simulate convection processes (nonhydrostatic models). Further approximations, such as e. g. the hydrostatic approximation discussed below, make the use of (4.14) unnecessary.

### 4.1.3 Dynamical Role of Sea Water Compressibility

In the Boussinesq approximation the dynamical influence of finite compressibility is not completely eliminated and occurs through the equation of state (4.13). Using the salt and heat budgets (4.12) and the hydrostatic relation (2.157) from Section 2.9, it is easily shown that the material derivative of the density perturbation $\tilde{\rho} = \rho - \rho_c(z)$ is given by

$$\frac{\mathrm{D}\tilde{\rho}}{\mathrm{D}t} = \mathcal{G}_\rho - g\rho_0 w \left( \frac{1}{c_s^2} - \frac{1}{c_c^2} \right) \tag{4.15}$$

where $\mathcal{G}_\rho$ is the diapycnal density flux as defined as in (3.2).

The last term on the right hand side of (4.15) shows that it is the dependence of the sound velocity (i. e. compressibility) on $\theta$ and $S$ that may cause a dynamically relevant change of density. This can be understood as follows: Two water bodies (marked A and B in Figure 4.1), which have different $\theta, S$ but the same (*in-situ*) density at pressure $p_0$, have different densities at another pressure $p_4$. At higher pressure ($p_4$) the more salty water body (B) is less dense, and a dynamical pressure gradient results.

The relative importance of this last term can be estimated as follows. Let $\Delta c_s = c_c - c_s$ be the change in sound velocity due to differences in $S$ or $\theta$. The magnitude

[2] SIMEON DENIS POISSON, *1781 in Pithiviers, †1840 in Paris, physicist and mathematician.

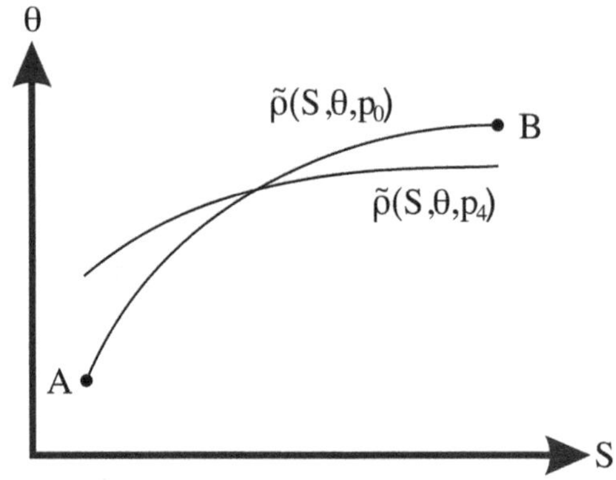


**Fig. 4.1** Density for two different pressures in the $\theta - S$-diagram. Two water masses A,B have the same density at pressure $p_0$ (full line) but different densities at higher pressure $p_4$ (thin line)

of the compressibility term in (4.15) is then of order $(2\Delta c_s/c_s)w\rho_0/c_s^2$. The vertical advection term in (4.15) scales as $w\Delta\rho/H$. Typical values are $\Delta c_s/c_s \lesssim 0.05$, $H/D \lesssim 0.02$, $\rho_0/\Delta\rho \approx 500$. Hence the ratio of both terms is given by

$$\frac{(2\Delta c_s/c_s)w\rho_0/c_s^2}{w\Delta\rho/H} = 2\frac{\Delta c_s}{c_s}\frac{H}{D}\frac{\rho_0}{\Delta\rho} \lesssim 1 \tag{4.16}$$

and may locally reach $O(1)$, although it is usually much smaller. The dynamical effect is related to a vertical motion, more precisely to differences in sound velocity, respectively compressibility in regions of upwelling and downwelling. The compressibility effect is not dominating, but may not always be neglected altogether.

Note also that according to (4.15) only deviations of the sound velocity from the reference state are dynamically relevant. As the state of rest is not uniquely defined, the last term in (4.15) cannot have a unique interpretation. However, when integrating e. g. along a horizontal surface, the contribution of the reference term drops out completely.

### 4.1.4 Energetics in the Boussinesq Approximation

Scalar multiplication of the momentum equation (4.10) with $\boldsymbol{u}$, in combination with (4.11) yields the kinetic energy budget of the Boussinesq approximation,

$$\frac{\mathrm{D}}{\mathrm{D}t}\left(\frac{1}{2}\rho_0\boldsymbol{u}^2\right) = -\nabla\cdot\boldsymbol{u}\tilde{p} - g\tilde{\rho}w + \boldsymbol{u}\cdot\boldsymbol{\mathcal{F}} \tag{4.17}$$

Note that in this approximation the energy per volume is given by $\rho_0\boldsymbol{u}^2/2$ rather than the exact form $\rho\boldsymbol{u}^2/2$. The terms on the right-hand side of (4.17) have an obvious meaning. The term $\boldsymbol{u}\cdot\boldsymbol{\mathcal{F}}$ describes the mechanical work on the system due to Reynolds stresses (or molecular stress in case that we have not applied a Reynolds filter yet). The term $-\nabla\cdot\boldsymbol{u}\tilde{p}$ is the divergence of a mechanical energy flux with respect to the dynamic pressure $\tilde{p}$. The term $-g\tilde{\rho}w$ is the rate of work against the gravity force.

The first two terms contain $\tilde{p}, \tilde{\rho}$ rather than the full variables $p, \rho$ as in (2.69). Due to the identity $\nabla \cdot \boldsymbol{u}\tilde{p} + gw\tilde{\rho} \equiv \nabla \cdot \boldsymbol{u}p + gw\rho$, however, this has no consequences for the kinetic energy but has the advantage that two very large terms which are connected to the reference state cancel exactly.

Previously, in Section 2.4.2, we have interpreted $\Phi = gz$ as a potential energy (per mass) and shown that the work against gravity can be interpreted as an exchange between kinetic and potential energy. In the Boussinesq approximation this interpretation is no longer useful. The reason is that the replacement of *mass* conservation (2.151) by *volume* conservation (4.11) means that mass is no longer strictly conserved. Hence all conservation laws now must be formulated for properties per volume, not per mass. The identification of a potential energy $E_{\mathrm{p}}$ to close (4.17) in a conservation of some total energy variable is nevertheless fairly straightforward. With the definition $E_{\mathrm{p}} = gz\,\tilde{\rho}/\rho_0$ and (4.15) one finds

$$\rho_0 \frac{\mathrm{D}E_{\mathrm{p}}}{\mathrm{D}t} = g\tilde{\rho}w + gz\frac{\mathrm{D}\tilde{\rho}}{\mathrm{D}t} \tag{4.18}$$

Compared to (4.17), the work against gravity appears with the opposite sign, showing that $E_{\mathrm{p}}$ is indeed an appropriate definition for the potential energy relative to the reference state.

However, on the right-hand side of (4.18) a new term $gz\,\mathrm{D}\tilde{\rho}/\mathrm{D}t$ appears which is related to changes of the dynamically relevant density and is not present in the exact potential energy budget (2.70). Its interpretation is facilitated by noting that the energy exchange $-p\rho\mathrm{D}\upsilon/\mathrm{D}t$ between kinetic and internal energy appears in the exact kinetic energy budget (2.69) but is absent from (4.17). Obviously,

$$-p\rho\frac{\mathrm{D}\upsilon}{\mathrm{D}t} = \frac{p}{\rho}\frac{\mathrm{D}\left(\rho_{\mathrm{c}} + \tilde{\rho}\right)}{\mathrm{D}t} \approx -gz\frac{\mathrm{D}\tilde{\rho}}{\mathrm{D}t}$$

(here we have ignored a term $-gz\,\mathrm{D}\rho_{\mathrm{c}}/\mathrm{D}t$ on the right-hand side which applies to the background state). We see that the term $gz\,\mathrm{D}\tilde{\rho}/\mathrm{D}t$ appears with opposite sign in the budget of internal energy and hence can be interpreted as an exchange between mechanical and internal energy. As a consequence of the Boussinesq approximation, the exchange occurs, however, with the potential rather than the kinetic energy.

By expressing $\mathrm{D}\tilde{\rho}/\mathrm{D}t$ by (4.18) with (4.15), this exchange is seen to consist of two contributions. The term $gz\mathcal{G}_\rho$ arises from *diapycnal* processes resulting from turbulent salt and heat fluxes which according to (3.2) affect the density of a water parcel and hence its potential energy. The term $-g^2 z\rho_0 w(c_{\mathrm{s}}^{-2} - c_{\mathrm{c}}^{-2})$ describes the (reversible) exchange with the internal energy due to the compressibility of sea water, more precisely the *dynamically relevant* fraction of the exchange with internal energy. Its relative magnitude, compared to the exchange between kinetic and potential energy in (4.17), is given by the same ratio as in (4.16) and is indeed usually small, although it can reach locally $O(1)$.

### 4.1.5 Potential Vorticity in the Boussinesq Approximation

The derivation of an equation for absolute and potential vorticity from (4.10)–(4.12) follows the same line as in Section 2.11 and will not be repeated here. The potential vorticity budget is found to be

$$\frac{\mathrm{D}}{\mathrm{D}t}\left(\frac{\boldsymbol{\omega}_{\mathrm{a}}}{\rho_0}\cdot\nabla\chi\right)=\frac{\boldsymbol{\omega}_{\mathrm{a}}}{\rho_0}\cdot\nabla\mathcal{G}_\chi+\frac{1}{\rho_0}\nabla\chi\cdot(\nabla\times\boldsymbol{\mathcal{F}})+\nabla\Phi\cdot\nabla\left(\frac{\tilde{\rho}}{\rho_0}\right)\times\nabla\chi \tag{4.19}$$

and is readily identified as the Boussinesq approximation to Ertel's theorem (2.212). It differs from the original mainly because the last term in (4.19) – the baroclinicity term – takes a different form. As $\nabla\Phi$ is a vertical vector (cf. Section 4.2.2), this term may also be written as

$$B=\nabla\Phi\cdot\nabla\left(\frac{\tilde{\rho}}{\rho_0}\right)\times\nabla\chi=\frac{g}{\rho_0}\nabla_{\mathrm{h}}\tilde{\rho}\times\nabla_{\mathrm{h}}\chi \tag{4.20}$$

where the index h denotes the horizontal vector. A special form for this term is obtained when $\chi$ is chosen as potential density $\rho_{\mathrm{p}}$ according to (2.129). Evaluating the cross-product in (4.20) yields

$$B=g\rho_0\left(\gamma_\star\alpha-\gamma\alpha_\star\right)\nabla_{\mathrm{h}}S\times\nabla_{\mathrm{h}}\theta \tag{4.21}$$

where $\alpha_\star$, $\gamma_\star$ denote the expansion coefficients at the reference pressure of the potential density. Hence the baroclinicity term vanishes always at the reference pressure and in addition also if the horizontal gradients of $S$ and $\theta$ are parallel (which is often approximately the case when both follow a close $\theta$–$S$ relation in a region). For those reasons, $\rho_{\mathrm{p}}$ is usually the preferred scalar for defining potential vorticity.

Furthermore, a constant factor $\rho_0$ instead of $\rho$ appears in the denominator of all terms in (4.19). It is hence possible (and often customary) to drop that factor in the definition of potential vorticity.

### 4.1.6 Full Incompressibility and Combination of Salt and Heat Budgets

The ultimate approximation with respect to density is the complete neglect of compressibility effects in (4.15), which is equivalent to the use of a pressure-independent equation of state. In other words, in this approximation the difference between potential density and *in-situ* density is ignored, and the density balance is taken as

$$\frac{\mathrm{D}\tilde{\rho}}{\mathrm{D}t}=\mathcal{G}_\rho \tag{4.22}$$

Frequently, this approximation is combined with a linearized equation of state, i. e.

$$\tilde{\rho}=\rho_0\left(\gamma_0\tilde{S}-\alpha_0\tilde{\theta}\right) \tag{4.23}$$

with constant coefficients $\alpha_0$ and $\gamma_0$, implying that $\tilde{S}$, $\tilde{\theta}$ vary only in a sufficiently small range. If, in addition, the source term $\mathcal{G}_\rho$ can be expressed in terms of $\tilde{\rho}$ instead of $\tilde{S}$ and $\tilde{\theta}$ individually (including the relevant boundary conditions), then (4.22) may replace equations (4.12) and (4.13). Hence the temperature and salinity need not be considered individually, or in other words: salinity on isopycnal surfaces is considered as a *passive* tracer. Note that because the difference between *in-situ* density and potential density is ignored in this approximation, it follows that the baroclinicity term in the potential vorticity budget (4.21) can also be neglected.

Quite frequently, the terminus Boussinesq approximation is understood to include the approximation of full incompressibility. An even further approximation is that of an *ideal fluid* where all diabatic terms are neglected. An *ideal Boussinesq fluid* thus satisfies

$$\rho_0 \frac{\mathrm{D}\boldsymbol{u}}{\mathrm{D}t} = -2\rho_0 \boldsymbol{\Omega} \times \boldsymbol{u} - \nabla \tilde{p} - \tilde{\rho} \nabla \Phi \tag{4.24}$$

$$\nabla \cdot \boldsymbol{u} = 0 \tag{4.25}$$

$$\frac{\mathrm{D}\tilde{\rho}}{\mathrm{D}t} = 0 \tag{4.26}$$

The constant factor $\rho_0$ frequently is eliminated by redefining density and pressure. When specifying the geopotential $\Phi = gz$ (see next chapter), it is convenient to introduce the buoyancy $b = -g\tilde{\rho}/\rho_0$ replacing the density $\tilde{\rho}$.

## 4.2 Shallow Water Approximation

The equations in the Boussinesq approximation, (4.10)–(4.13), are the common starting point for many problems in oceanography. We proceed in this section to introduce a coordinate system for (4.10)–(4.13), which is more useful than Cartesian coordinates for large-scale motions in the ocean. It is obvious that at horizontal scales larger than a few tens of kilometers, it becomes important to use a coordinate system that takes into account the Earth's geometry.

### 4.2.1 Oblate Spheroidal Coordinates

Due to the overriding magnitude of the gravity force in the equations of motion, it is most useful to choose a coordinate system where surfaces of constant geopotential coincide with a constant coordinate, which we will call the vertical coordinate. Without such a choice, the gravity force would not only dominate the vertical momentum balance but would also appear in the horizontal directions, thus making the physical interpretation of the horizontal momentum balance very difficult. As discussed in Section 2.3.4, geopotential surfaces are to a very good approximation given by an oblate ellipsoid. The deviations of geopotential surfaces from a perfect (best fit) ellipsoid vary for instance at maximum only between −110 m in the Indian Ocean to +90 m over southeast Asia (see Figure 2.9).

Therefore, oblate spheroidal coordinates represent the best-suited coordinate system to describe oceanic motions. They are defined as

$$x_1 = \left(\mu^2 + c^2\right)^{\frac{1}{2}} \cos\psi \cos\lambda\ , \quad x_2 = \left(\mu^2 + c^2\right)^{\frac{1}{2}} \cos\psi \sin\lambda\ , \quad x_3 = \mu \sin\psi \tag{4.27}$$

Here $\lambda$ is the eastward longitude, and the geometric meaning of the coordinates $\psi$ and $\mu$ follows from Figure 4.2. The angle $\psi$ and the length $\mu$ are related to the geographical latitude $\varphi$ and the distance from the origin $r = (x_1^2 + x_2^2 + x_3^2)^{\frac{1}{2}}$ by

$$r^2 = \mu^2 + c^2 \cos^2 \psi \quad \text{and} \quad \sin^2 \varphi = \frac{\sin^2 \psi}{1 + (c^2/\mu^2) \cos^2 \psi} \tag{4.28}$$

The constant $c$ is given by $c^2 = a_\mathrm{e}^2 - a_\mathrm{p}^2$ where $a_\mathrm{e} \approx 6{,}378.1$ km and $a_\mathrm{p} \approx 6{,}356.8$ km are the equatorial and polar Earth radii, respectively (cf. Section 2.3.4). The geoid surface is given by $\mu = a_\mathrm{p}$, and the coordinate $\mu$ is strictly along the direction of the local gravity force. Note that other definitions of ellipsoidal coordinates exist which differ from (4.27) by the scaling of the coordinates (see e.g. Morse and Feshbach, 1953). Furthermore, as noted by White et al. (2008), while *confocal* oblate spheroidal coordinates such as defined by (4.27) describe the Earth's shape correctly, the relation $g = |\nabla \Phi|$ based on these coordinates does not imply the correct variation of gravity according to (2.54). To also implement the correct variation, White et al. (2008) propose a different system which is also based on oblate spheroids but is defined by a numerical coordinate transformation. However, this approach is not further discussed here, and the following discussion is restricted to the analytical system defined by (4.27).

The transformation of the equations of motion to oblate spheroidal coordinates is in principle straightforward, and has been discussed e. g. by Veronis (1973) and Gates (2004). As shown in Appendix A.4.2, the momentum equations (4.10) for the velocity components $(u_\lambda, u_\psi, u_\mu) \equiv (u, v, w)$ and the mass conservation (4.11) in

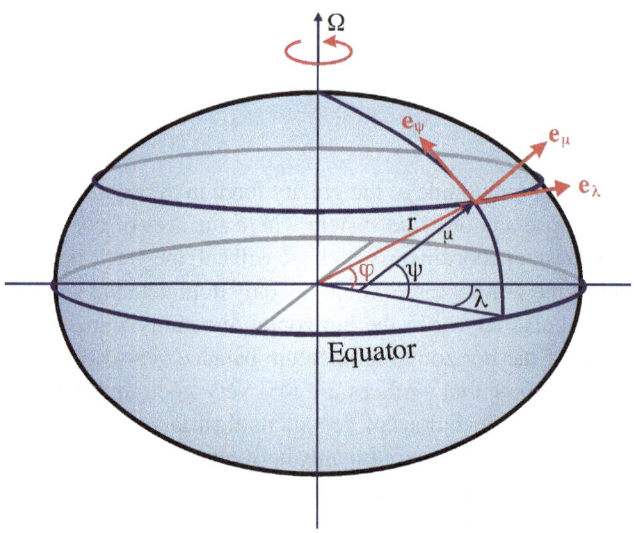


Fig. 4.2 Geometry of oblate spheroidal coordinates

**22. Angular Momentum**

The angular momentum (per mass) of a fluid element at a distance $\boldsymbol{r}$ from the Earth' center is given as $\boldsymbol{r} \times \boldsymbol{u}$ relative to a rotating coordinate system, and as $\boldsymbol{r} \times (\boldsymbol{u} + \boldsymbol{\Omega} \times \boldsymbol{r})$ in an absolute frame. The balance of angular momentum is obtained by vectorial multiplication of the momentum equation (4.10) with $\boldsymbol{r}$. Using the identity $\boldsymbol{u} = \mathrm{D}\boldsymbol{r}/\mathrm{D}t$, one obtains

$$\rho_0 \frac{\mathrm{D}}{\mathrm{D}t}(\boldsymbol{r} \times \boldsymbol{u}) + 2\rho_0 \boldsymbol{r} \times (\boldsymbol{\Omega} \times \boldsymbol{u}) = \boldsymbol{r} \times (-\nabla \tilde{p} - \tilde{\rho}\nabla\Phi + \mathcal{F})$$

The axial component $M$ of the absolute angular momentum is given as $M = \boldsymbol{e}_3 \cdot [\boldsymbol{r} \times (\boldsymbol{u} + \boldsymbol{\Omega} \times \boldsymbol{r})]$ where $\boldsymbol{e}_3$ is the (constant) unit vector parallel to the Earth' rotation axis. Scalar multiplication of the angular momentum balance with $\boldsymbol{e}_3$, using the fact that $\boldsymbol{e}_3 \cdot \boldsymbol{r} \times \nabla\Phi = 0$, yields

$$\rho_0 \frac{\mathrm{D}M}{\mathrm{D}t} = \boldsymbol{e}_3 \cdot [\boldsymbol{r} \times (-\nabla \tilde{p} + \mathcal{F})] \tag{B22.1}$$

Hence the axial angular momentum $M$ is conserved in the absence of pressure gradients and forcing.
For large-scale motions, consideration of the axial angular momentum balance can be useful, even though the balance follows directly from the momentum equations and hence is not independent. Note that further approximations in the momentum equations must also be reflected in the definition of $M$. In principle, the angular momentum conservation places a constraint on possible approximations, since approximations which do not imply a conservation law analogously to (B22.1) lack an important property and can be considered as physically inconsistent (see e. g. Section 4.2.3 below).

the Boussinesq approximation take the form

$$\rho_0\left(\frac{\mathrm{D}}{\mathrm{D}t}u - \frac{uv}{A}\tan\psi + \frac{\mu}{AB}uw - 2\frac{B}{A}\Omega v\sin\psi + 2\frac{\mu}{A}\Omega w\cos\psi\right) = -\frac{1}{B\cos\psi}\frac{\partial\tilde{p}}{\partial\lambda} + \mathcal{F}_u \tag{4.29}$$

$$\rho_0\left(\frac{\mathrm{D}}{\mathrm{D}t}v + \frac{u^2}{A}\tan\psi + \frac{\mu}{A^2}vw - \frac{CBw^2}{A} + 2\frac{B}{A}\Omega u\sin\psi\right) = -\frac{1}{A}\frac{\partial\tilde{p}}{\partial\psi} + \mathcal{F}_v \tag{4.30}$$

$$\rho_0\left(\frac{\mathrm{D}}{\mathrm{D}t}w - \frac{\mu}{AB}u^2 - \frac{\mu}{A^2}v^2 + \frac{CB}{A}vw - 2\frac{\mu}{A}\Omega u\cos\psi\right) = -\frac{B}{A}\frac{\partial\tilde{p}}{\partial\mu} - g\tilde{\rho} + \mathcal{F}_w \tag{4.31}$$

$$\frac{1}{B\cos\psi}\frac{\partial u}{\partial\lambda} + \frac{1}{A^2\cos\psi}\frac{\partial}{\partial\psi}(vA\cos\psi) + \frac{1}{A^2}\frac{\partial}{\partial\mu}(wAB) = 0 \tag{4.32}$$

Here the material advection operator is given as

$$\frac{\mathrm{D}}{\mathrm{D}t} = \frac{\partial}{\partial t} + \frac{u}{B\cos\psi}\frac{\partial}{\partial\lambda} + \frac{v}{A}\frac{\partial}{\partial\psi} + \frac{w}{AB^{-1}}\frac{\partial}{\partial\mu} \tag{4.33}$$

and the coefficients $A, B, C$ are defined as

$$A(\psi,\mu) = (\mu^2 + c^2\sin^2\psi)^{\frac{1}{2}}, \quad B(\mu) = (\mu^2 + c^2)^{\frac{1}{2}},$$
$$C(\psi,\mu) = \frac{c^2\sin\psi\cos\psi}{BA^2} \tag{4.34}$$

Temperature and salinity budgets in the Boussinesq system (4.12) remain formally unchanged if one remembers the transformed material advection operator (4.33). Likewise, the budgets of kinetic and potential energies remain as in (4.17) and (4.18), with the divergence operator defined as in (A.57) in Appendix A.4.2. Note the appearance of metric terms in (4.29)–(4.31) which appear in any curvilinear coordinate system (cf. Appendix A.4.1). Like the Coriolis terms, these terms do not contribute to the budget of kinetic energy.

The conservation of axial angular momentum (cf. the box on p. 125) can be derived directly by multiplication of (4.29) with $B\cos\psi$, using the identities $\mathrm{D}\psi/\mathrm{D}t = v/A$ and $\mathrm{D}\mu/\mathrm{D}t = wB/A$. The results is

$$\rho_0 \frac{\mathrm{D}}{\mathrm{D}t}\left[(u + \Omega B\cos\psi)B\cos\psi\right] = -\frac{\partial\tilde{p}}{\partial\lambda} + \mathcal{F}_u B\cos\psi \tag{4.35}$$

Note that the metric terms do not appear explicitly in (4.35) which also could be derived from the general form discussed in the box on p. 125.

### 4.2.2 Spherical Approximation

The use of ellipsoidal coordinates in (4.29)–(4.32) – while not very common – is straightforward and causes no difficulties e. g. for the formulation of a numerical model. For many purposes, however, such a coordinate system is not a practical choice. In particular, the algebraic complexity connected to the nonconstant coefficients $A, B, C$ is often inconvenient.

Now the shape of the Earth is very close to that of a sphere. The deviation from the spherical shape is measured by the *ellipticity*[3] (or flattening parameter) $e = 1 - a_\mathrm{p}/a_\mathrm{e} \approx 1/298$. It hence follows that $c^2/a_\mathrm{e}^2 \approx c^2/a_\mathrm{p}^2 \approx 2e \ll 1$. Expanding now the coefficients in (4.34) for small $e$, and replacing $\mu$ by $a_\mathrm{p}$ in the small terms, one obtains

$$A \approx \mu(1 + e\sin^2\psi)\,, \quad B \approx \mu(1+e)\,, \quad C \approx \frac{e}{\mu}\sin\psi\cos\psi \tag{4.36}$$

up to terms of quadratic and higher order in $e$. From (4.28) one likewise obtains

$$r \approx \mu(1 + e\cos^2\varphi) \quad \text{and} \quad \sin\psi \approx (1 + e\cos^2\varphi)\sin\varphi \tag{4.37}$$

Specifically, the geoid is given as

$$r_\mathrm{g} = (a_p^2 + c^2\cos^2\psi)^{\frac{1}{2}} = a_\mathrm{e}\left(1 - \frac{c^2}{a_\mathrm{e}^2}\sin^2\psi\right)^{\frac{1}{2}} \approx a_\mathrm{e}(1 - e\sin^2\varphi)$$

which agrees with (2.53), with $e \approx \varepsilon_3 = 3.36 \times 10^{-3}$.

In the *spherical approximation*, the differences between the ellipsoid and a sphere of mean radius $a = (a_\mathrm{e} + a_\mathrm{p})/2 \approx 6{,}367.4\,\mathrm{km}$ are ignored. Formally, the approximation is obtained by taking the limit $e \to 0$ in (4.36) and (4.37), so that $A \approx \mu$, $B \approx \mu$, $C \approx 0$ and also $r \approx \mu$ and $\psi \approx \varphi$. The error incurred in the spherical approximation is indeed of order $e$, as to be expected. The spherical geometry is displayed in Figure 4.3. Introducing for abbreviation the locally vertical component $f$ (referred to as *Coriolis parameter*) and the horizontal component $f_\mathrm{h}$ of the rotation vector $\boldsymbol{\Omega}$ by

$$f = 2\Omega\sin\varphi\,, \quad f_\mathrm{h} = 2\Omega\cos\varphi$$

[3] The ellipticity is related to the eccentricity $\varepsilon$ by $1 - \varepsilon^2 = (1-e)^2$.

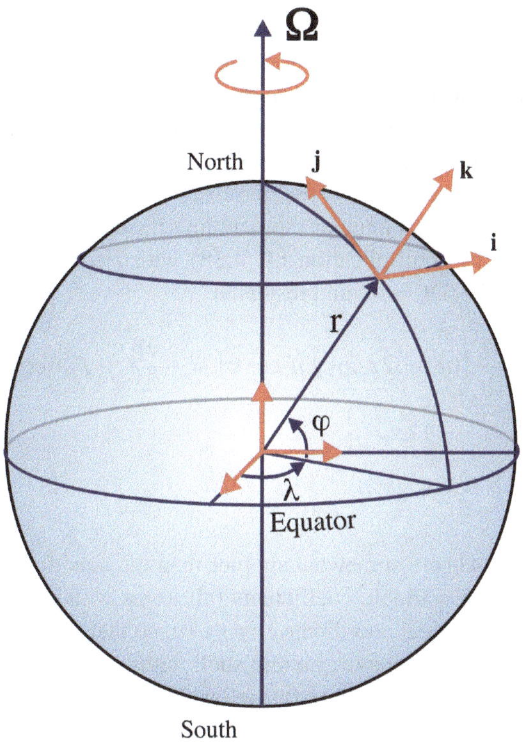


Fig. 4.3 Cartesian and spherical coordinates of the Earth. Redrawn after Vallis (2006)

equations (4.29)–(4.32) then take the form

$$\rho_0\left(\frac{\mathrm{D}u}{\mathrm{D}t}+\frac{uw}{r}-\frac{uv}{r}\tan\varphi-fv+f_{\mathrm{h}}w\right)=-\frac{1}{r\cos\varphi}\frac{\partial\tilde{p}}{\partial\lambda}+\mathcal{F}_u \tag{4.38}$$

$$\rho_0\left(\frac{\mathrm{D}v}{\mathrm{D}t}+\frac{wv}{r}+\frac{u^2}{r}\tan\varphi+fu\right)=-\frac{1}{r}\frac{\partial\tilde{p}}{\partial\varphi}+\mathcal{F}_v \tag{4.39}$$

$$\rho_0\left(\frac{\mathrm{D}w}{\mathrm{D}t}-\frac{u^2+v^2}{r}-f_{\mathrm{h}}u\right)=-\frac{\partial p}{\partial r}-g\tilde{\rho}+\mathcal{F}_w \tag{4.40}$$

$$\frac{1}{r\cos\varphi}\left(\frac{\partial u}{\partial\lambda}+\frac{\partial v\cos\varphi}{\partial\varphi}\right)+\frac{1}{r^2}\frac{\partial r^2 w}{\partial r}=0 \tag{4.41}$$

with the material advection operator

$$\frac{\mathrm{D}}{\mathrm{D}t}=\frac{\partial}{\partial t}+\frac{u}{r\cos\varphi}\frac{\partial}{\partial\lambda}+\frac{v}{r}\frac{\partial}{\partial\varphi}+w\frac{\partial}{\partial r} \tag{4.42}$$

The system (4.38)–(4.41) can be interpreted as still being expressed in ellipsoidal coordinates, in the limit of small $e$. In this view, the geopotential surfaces still form an ellipsoid, so that the coordinate $r \approx \mu$ is still along the direction of the gravity force. Alternatively, (4.38)–(4.42) can be obtained directly by transforming the equations of motion into spherical coordinates (see Appendix A.4.3), provided, however, that the geopotential surfaces are approximated by spherical surfaces, so that again the gravity force does not appear in the locally horizontal directions.

As shown in Section 2.3.4, the gravity $g = |\nabla\Phi|$ at the Earth's surface depends on latitude $\varphi$ (cf. (2.54)). However, the relative changes of $g(\varphi)$ are of order $3 \times 10^{-3}$, and within the accuracy of the spherical approximation we can assume $g = 9.81\,\mathrm{m\,s^{-2}} = \text{const}$. Furthermore, as pointed out by White et al. (2005), keeping the variation of $g$ in this approximation would be inconsistent and lead to spurious vorticity sources.

The conservation of axial angular momentum (cf. the box on p. 125) can again be derived directly by multiplication of (4.38) with $r\cos\varphi$, using the identities $\mathrm{D}\varphi/\mathrm{D}t = v/r$ and $\mathrm{D}r/\mathrm{D}t = w$, and results in

$$\rho_0 \frac{\mathrm{D}}{\mathrm{D}t}\left[(u + \Omega r\cos\varphi)r\cos\varphi\right] = -\frac{\partial\tilde{p}}{\partial\lambda} + \mathcal{F}_u r\cos\varphi$$

## 4.2.3 Thin-Shell Approximation

Equations (4.38)–(4.41) are somewhat simpler than the equations in spheroidal coordinates but still have variable coefficients (all terms with $r$). It is convenient to introduce the locally vertical coordinate $z = r - a$, so that $\partial/\partial r \equiv \partial/\partial z$.

Now the ocean only extends over a thin shell with a vertical range up to a depth $H \approx 5\,\mathrm{km}$, which is a factor of 1,000 smaller than the Earth's mean radius. It is, therefore, very suggestive to approximate $r = a + z \approx a$ in all terms in (4.38)–(4.42) since the error involved is a most $H/a \approx 10^{-3}$. Furthermore, those metric terms containing the vertical velocity $w$ are smaller by a factor $H/a$ than the respective vertical advection terms and can likewise be neglected. This applies to the terms $uw/r$ in (4.38) and $vw/r$ in (4.39). In this *thin-shell approximation*, (4.38)–(4.41) then take the form

$$\frac{\mathrm{D}u}{\mathrm{D}t} - \frac{uv}{a}\tan\varphi - fv + f_\mathrm{h} w = -\frac{1}{\rho_0 a\cos\varphi}\frac{\partial\tilde{p}}{\partial\lambda} + \mathcal{F}_u \tag{4.43}$$

$$\frac{\mathrm{D}v}{\mathrm{D}t} + \frac{u^2}{a}\tan\varphi + fu = -\frac{1}{\rho_0 a}\frac{\partial\tilde{p}}{\partial\varphi} + \mathcal{F}_v \tag{4.44}$$

$$\frac{\mathrm{D}w}{\mathrm{D}t} - \frac{u^2 + v^2}{a} - f_\mathrm{h} u = -\frac{1}{\rho_0}\frac{\partial\tilde{p}}{\partial z} - g\frac{\tilde{\rho}}{\rho_0} + \mathcal{F}_w \tag{4.45}$$

$$\frac{1}{a\cos\varphi}\left(\frac{\partial u}{\partial\lambda} + \frac{\partial v\cos\varphi}{\partial\varphi}\right) + \frac{\partial w}{\partial z} = 0 \tag{4.46}$$

with the substantial derivative

$$\frac{\mathrm{D}}{\mathrm{D}t} = \frac{\partial}{\partial t} + \frac{u}{a\cos\varphi}\frac{\partial}{\partial\lambda} + \frac{v}{a}\frac{\partial}{\partial\varphi} + w\frac{\partial}{\partial z} \tag{4.47}$$

At first sight it appears that we have achieved a useful form with a small error $\lesssim 10^{-3}$ and constant coefficients $a$ instead of $r$. However, it turns out that the system (4.43)–(4.46) does not lead to a consistent angular momentum budget, a fact which was already noted by Phillips (1966). This can e. g. be seen by multiplication of (4.43)

**23. Components of Coriolis Vector**

The rotation vector has the two components $f = 2\Omega \sin\varphi$ and $f_h = 2\Omega \cos\varphi$, as seen from the adjacent figure. The Coriolis force in the momentum equations (4.38)–(4.40) appears in four terms, two related to $f$ and two to $f_h$. Following Gerkema et al. (2008), all four terms can be derived from elementary mechanical principles, alternatively to the straightforward algebra. While the equations hold in a coordinate system rotating with the Earth, it is useful to take the perspective of an observer in a nonrotating system. A fluid parcel is at rest when all forces acting on it are balanced. Viewed from a nonrotating observer, the parcel moves along a latitudinal circle of radius $R(\varphi) = a\cos\varphi$ at an eastward velocity $U = \Omega a \cos\varphi$. Consider now the case when the parcel is not at rest but has a velocity in one of the three coordinate directions (for the Northern Hemisphere, the situation in the Southern Hemisphere is equivalent):

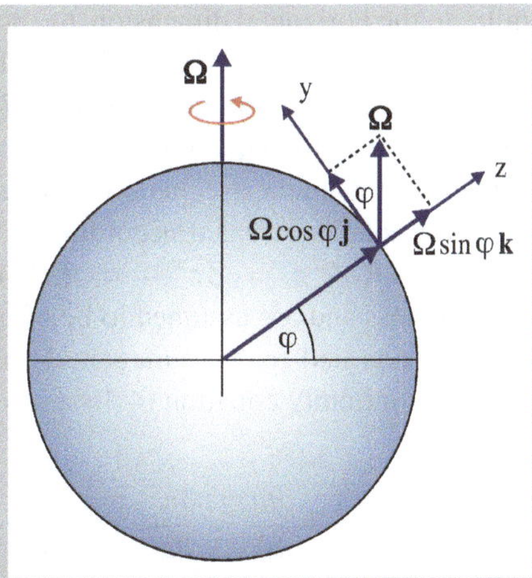


*Components of the Earth rotation vector* $\boldsymbol{\Omega}$*, shown on a section through the Earth.*

**Eastward**: A parcel moving eastward with velocity $u$ has a total velocity $U+u$, where $u \ll U$ can be assumed. Hence the required centripetal force increases by $2Uu/R(\varphi) = 2\Omega u$. This increment tends to push the parcel outward in the latitudinal plane. This outward acceleration can be decomposed into a radial component $2\Omega u \cos\varphi = f_h u$ and a southward one, $-2\Omega u \sin\varphi = -fu$, terms which indeed appear in (4.40) respectively (4.39).
**Northward**: When moving northward to the latitude $\varphi' = \varphi + \delta\varphi$, the parcels new circle of latitude is smaller by approximately $-a\sin\varphi\delta\varphi$. Conservation of angular momentum requires $UR(\varphi) = (U+u)R(\varphi')$, so that the parcel will obtain an eastward velocity $u = \Omega\sin\varphi a\delta\varphi$. Now resting parcels at latitude $\varphi'$ rotate at $U' = \Omega R(\varphi') \approx U - \Omega\sin\varphi a\delta\varphi$. Relative to those ambient parcels, the initially northward moving parcel will thus have an eastward velocity difference of $2\Omega\sin\varphi a\delta\varphi = fa\delta\varphi$. This corresponds to an eastward acceleration $fv$, which indeed appears in (4.38).
**Upward**: For a parcel moving initially upward (i. e. , radially outward) to $a' = a + \delta a$, conservation of angular momentum $Ua\cos\varphi = (U+u)a'\cos\varphi$ requires that the parcels eastward velocity will become smaller by $u = -\Omega\cos\varphi\delta a$. At this higher altitude, parcels at rest have an eastward velocity $U' = \Omega a'\cos\varphi$, differing from $U$ by $\Omega\cos\varphi\delta a$. Relative to those parcels, the initially upward moving parcel thus has an excess of westward velocity of $-2\Omega\cos\varphi\delta a = -f_h\delta a$, corresponding to a westward acceleration $-f_h w$, as appearing in (4.38).

with $a\cos\varphi$, and proceeding as in the previous section. The result is

$$\rho_0 \frac{\mathrm{D}}{\mathrm{D}t}[(u + \Omega a\cos\varphi)a\cos\varphi] + f_h a w\cos\varphi = -\frac{\partial\tilde{p}}{\partial\lambda} + r\cos\varphi\mathcal{F}_u$$

This form does not agree with the general form discussed in the box on p. 125, due to an extra term which originates from the Coriolis acceleration due to the horizontal component of the rotation vector in (4.43). For this reason, the system (4.43)–(4.47) is not fully consistent and should be used with care.

### 4.2.4 Small Aspect Ratio

For a further simplification of the system, we now use the fact that the horizontal dimensions of ocean basins are much larger than the ocean depth. We assume that similar to the horizontal dimension, the horizontal scale $L$ for motions of large scale is also much larger than the vertical scale $H$. It is useful to define the *aspect ratio* $\delta = H/L$. The restriction to scales $H \ll L$, i. e. $\delta \ll 1$, leads now to a further simplification of the equations of motion. To determine the order of magnitude of individual terms in these equations, scales are introduced as described in Table 4.1.

All variables are scaled according to $(u,v) = U(u^*,v^*)$, $w = Ww^*$, $\tilde{p} = \tilde{P}p^*$, $\tilde{\rho} = \tilde{R}\rho^*$, $z = Zz^*$, $t = Tt^*$, etc. where the variables with asterisk are dimensionless and are assumed to be of order one. Only the scales $L$, $H$, etc. carry the dimension and relevant magnitude for the motions we want to consider now. From the continuity equation (4.46) one obtains as relative orders of magnitude (or scaling)

$$\frac{1}{a\cos\varphi}\frac{\partial u}{\partial\lambda} \sim \frac{\partial w}{\partial z} \quad \text{or} \quad \frac{U}{L} \sim \frac{W}{H} \quad \text{or} \quad W \sim \frac{H}{L}U = \delta U \tag{4.48}$$

We see that the aspect ratio $\delta$ equals the ratio of vertical to horizontal velocity and describes the slope of particle trajectories in a vertical plane. It should be noted that this scaling is revised later in Section 5.1 since although $\delta$ is small, the relation (4.48) still overestimates the magnitude of the vertical velocity in many situations.

Under oceanic conditions, the mechanical forcing in the horizontal momentum budget (4.43) and (4.44) is rarely the dominant term in the horizontal momentum balance. Hence the pressure gradient has to be balanced by either the Coriolis force or by the acceleration term. This leads to the scaling relation

$$\tilde{P} \sim \rho_0 UL\Omega \max\left(1, \frac{1}{\Omega T}\right) \tag{4.49}$$

For $\Omega T \gg 1$ the acceleration terms are small, which is the case for the large-scale ocean circulation. On the other hand, for $\Omega T \ll 1$ the acceleration terms dominate, which occurs e. g. for tidal currents.

We now consider the individual terms in the horizontal momentum balance (4.43). In Table 4.2 the magnitude relative to the horizontal pressure gradient is evaluated,

**Table 4.1** Characteristic scales for large-scale oceanic motions and some resulting dimensionless (small) numbers

| | | | |
|---|---|---|---|
| $L$ | Horizontal length scale | $\mathrm{Ro} = U/2\Omega L$ | Rossby number |
| $H$ | Vertical length scale | $\mathrm{Ek} = \mathcal{F}_u/\Omega U$ | Ekman number |
| $U$ | Horizontal velocity scale | $\delta = H/L$ | Aspect ratio |
| $W$ | Vertical velocity scale | $\ell = L/a$ | Metric ratio |
| $T$ | Time ($L/U$ or $H/W$) scale | | |
| $\tilde{P}$ | Scale of pressure perturbation | | |
| $\tilde{R}$ | Scale of density perturbation | | |
| $a$ | Earth radius | | |

**Table 4.2** Scaling of the terms in the horizontal momentum balance, with $\mu = \min(1, \Omega T)$

| Term | Scaling | Relative to pressure gradient |
|---|---|---|
| $\partial u/\partial t$ | $U/T$ | $\mu$ |
| $\boldsymbol{u}\cdot\boldsymbol{\nabla} u$ | $U^2/L$ | $(U/2\Omega L)\mu \le \mathrm{Ro}$ |
| $2\Omega v \sin\varphi$ | $\Omega U$ | $\mu$ |
| $2\Omega w \cos\varphi$ | $\Omega W$ | $\delta\mu \le \delta$ |
| $(uv/a)\tan\varphi$ | $U^2/a$ | $(U/L\Omega)L/a\mu \le \mathrm{Ro}\ell$ |
| $(\rho_0 a\cos\varphi)^{-1}\partial\tilde{p}/\partial\lambda$ | $\tilde{P}/\rho_0 L$ | 1 |
| $\mathcal{F}_u$ | $\mathrm{Ek}\Omega U$ | $\mathrm{Ek}\mu$ |

with the help of (4.48) and (4.49). A corresponding analysis can be applied to the vertical momentum balance (4.40) where the magnitudes relative to the *vertical* pressure gradient are computed. This is shown in Table 4.3. In addition to the aspect ratio, three additional dimensionless numbers appear. The Rossby number $\mathrm{Ro} = U/\Omega L$ is the ratio of the inertial (momentum advection) terms to Coriolis terms in the horizontal momentum equation. The Ekman number Ek is the ratio of mechanical forcing (friction) to the Coriolis term in the horizontal momentum equation. Both numbers are usually small but may approach $O(1)$ in certain situations. We will assume that $\mathrm{Ro} \lesssim 1$ and $\mathrm{Ek} \lesssim 1$. Furthermore, $\ell = L/a$ denotes the ratio of the length scale and Earth radius. It is usually $O(0.1)$ or smaller and measures the importance of the metric terms compared to the advection. The scaling assumes $\sin\varphi$, $\tan\varphi \sim 1$ and must thus be modified when approaching the equator or the poles.

The assumption $\delta \ll 1$ now allows to identify small and thus unimportant terms in the momentum budget. As seen in Table 4.3, in the vertical momentum budget (4.45), all inertial terms and the mechanical friction term are by a factor $\delta^2$ smaller, and the term containing $f_\mathrm{h}$ in (4.45) is by a factor of $\delta$ smaller than the vertical pressure gradient. It follows that only the buoyancy term can balance the pressure gradient, and that flows of small aspect ratio are hydrostatically balanced. The neglect of all terms except pressure gradient and gravity force in the vertical momentum equation is called the *hydrostatic approximation*. Accordingly, the density must scale as

$$\tilde{R} = \frac{\tilde{P}}{gH}$$

The term containing $f_\mathrm{h} = \Omega\cos\varphi$ in the horizontal momentum balance (4.43) is of order $\delta$. Formally, neglecting both terms with $f_\mathrm{h}$ is equivalent to ignoring the locally horizontal component of the vector $\boldsymbol{\Omega}$, an approach which is called *traditional approximation* (see e. g. Gerkema et al. (2008) for an extensive discussion on the validity of this approximation). Note that this conclusion is valid both for $\Omega T \gg 1$ as well as for $\Omega T \ll 1$.

Note that once the hydrostatic approximation is made, the conservation of kinetic energy applies to the horizontal part $(u^2 + v^2)/2$ only as shown below in Section 4.2.6. Consequently, neglection of the Coriolis term in (4.45) requires to neglect the $\Omega\cos\varphi$-term in (4.43) as well (which may introduce work in the energy budget otherwise). We note that dropping this term may lead to problems close to the equator where $\sin\varphi \ll 1$.

Table 4.3 Scaling of the terms in the vertical momentum balance, with $\mu = \min(1, \Omega T)$

| Term | Scaling | Relative to Pressure Gradient |
|---|---|---|
| $\partial w/\partial t$ | $W/T$ | $(H^2/L^2)\mu \leq \delta^2$ |
| $\boldsymbol{u}\cdot\boldsymbol{\nabla} w$ | $UW/L$ | $(U/L\Omega)(H^2/L^2)\mu \leq \mathrm{Ro}\delta^2$ |
| $(u^2 + v^2)/a$ | $U^2/a$ | $(U/L\Omega)(H/a)\mu \leq \mathrm{Ro}\ell\delta$ |
| $2\Omega u\cos\varphi$ | $\Omega U$ | $(H/L)\mu \leq \delta$ |
| $(1/\rho_0)\partial\tilde{p}/\partial z$ | $\tilde{P}/\rho_0 H$ | 1 |
| $g\tilde{\rho}/\rho_0$ | $g\tilde{R}/\rho_0$ | $gH\tilde{R}/\tilde{P}$ |
| $\mathcal{F}_w$ | $\mathrm{Ek}\Omega W$ | $\mathrm{Ek}(H^2/L^2)\mu \leq \mathrm{Ek}\delta^2$ |

### 4.2.5 Primitive Equations

We now have completed all steps for the *shallow water approximation* which involves transformation to spherical coordinates, thin-shell approximation and assuming a small aspect ratio for relevant motions. The combination with the Boussinesq approximation yields the set of equations most frequently used for ocean circulation models, which are often called the *primitive equations*. One obtains

$$\rho_0\left(\frac{\mathrm{D}u}{\mathrm{D}t}-\frac{uv}{a}\tan\varphi-fv\right)=-\frac{1}{a\cos\varphi}\frac{\partial\tilde{p}}{\partial\lambda}+\mathcal{F}_u \tag{4.50}$$

$$\rho_0\left(\frac{\mathrm{D}v}{\mathrm{D}t}+\frac{u^2}{a}\tan\varphi+fu\right)=-\frac{1}{a}\frac{\partial\tilde{p}}{\partial\varphi}+\mathcal{F}_v \tag{4.51}$$

$$\frac{\partial\tilde{p}}{\partial z}=-g\tilde{\rho} \tag{4.52}$$

$$\frac{\partial w}{\partial z}=-\frac{1}{a\cos\varphi}\left(\frac{\partial u}{\partial\lambda}+\frac{\partial v\cos\varphi}{\partial\varphi}\right) \tag{4.53}$$

$$\rho_0\frac{\mathrm{D}\tilde{S}}{\mathrm{D}t}=\mathcal{G}_{\mathrm{S}} \tag{4.54}$$

$$\rho_0\frac{\mathrm{D}\tilde{\theta}}{\mathrm{D}t}=\mathcal{G}_{\theta} \tag{4.55}$$

$$\tilde{\rho}=F(\tilde{S},\tilde{\theta},p_{\mathrm{c}}) \tag{4.56}$$

where the operator $\mathrm{D}/\mathrm{D}t$ is defined as in (4.47). The primitive equations describe many problems in large-scale oceanography. They are the preferred set for ocean general circulation models. However, they also can be used to describe tidal motions when the appropriate forcing potential is included.

Note that the neglect of the $\mathrm{D}w/\mathrm{D}t$-term in the vertical momentum equation also changes the character of the system, and the pressure can directly be obtained by integration of (4.52), instead of solving the Poisson-equation (4.14).

### 4.2.6 Energetics and Potential Vorticity in the Shallow Water Approximation

Scalar multiplication of (4.50)–(4.52) with the respective velocity components and subsequent addition, in combination with (4.53), yields the kinetic energy budget

$$\frac{\mathrm{D}}{\mathrm{D}t}\frac{1}{2}\rho_0\boldsymbol{u}_{\mathrm{h}}^2=-\left(\nabla_{\mathrm{h}}\cdot\tilde{p}\boldsymbol{u}_{\mathrm{h}}+\frac{\partial\tilde{p}w}{\partial z}\right)-g\tilde{\rho}w+\boldsymbol{u}_{\mathrm{h}}\cdot\boldsymbol{\mathcal{F}}_{\mathrm{h}} \tag{4.57}$$

Compared to the general Boussinesq form of kinetic energy (4.17), we see that only the horizontal velocity components contribute to the kinetic energy and also to the mechanical forcing term, i. e. the contribution of the respective vertical terms is small and neglected. The general Boussinesq form of potential energy (4.18) remains unchanged.

A theorem of conservation for potential vorticity may again be derived in the usual way outlined in Section 4.1.5, by considering the budget of absolute vorticity based on (4.50)–(4.52) and combining with a passive tracer budget. The somewhat lengthy derivation is omitted here, and we just give the result

$$\frac{\mathrm{D}}{\mathrm{D}t}Q^{\mathrm{SW}} = \frac{\boldsymbol{\omega}_{\mathrm{a}}^{\mathrm{SW}}}{\rho_0} \cdot \nabla \mathcal{G}_\chi + \frac{1}{\rho_0}\frac{\partial \chi}{\partial z}\nabla_{\mathrm{h}} \times \boldsymbol{\mathcal{F}}_{\mathrm{h}} + B \tag{4.58}$$

where $\boldsymbol{\omega}_{\mathrm{a}}^{\mathrm{SW}}$ and $Q^{\mathrm{SW}}$ are the shallow water forms of absolute vorticity and potential vorticity

$$\boldsymbol{\omega}_{\mathrm{a}}^{\mathrm{SW}} = \left[-\frac{\partial v}{\partial z}, \frac{\partial u}{\partial z}, \frac{1}{a\cos\varphi}\left(\frac{\partial v}{\partial \lambda} - \frac{\partial u \cos\varphi}{\partial \varphi}\right) + f\right]$$

$$Q^{\mathrm{SW}} = \frac{1}{a\rho_0 \cos\varphi}\left[-\frac{\partial \chi}{\partial \lambda}\frac{\partial v}{\partial z} + \cos\varphi \frac{\partial \chi}{\partial \varphi}\frac{\partial u}{\partial z} + \frac{\partial \chi}{\partial z}\left(\frac{\partial v}{\partial \lambda} - \frac{\partial u\cos\varphi}{\partial \varphi} + f\right)\right]$$

respectively, and $B$ is the baroclinicity term as in (4.20).

### Rossby's Potential Vorticity

Earlier than Ertel's concept of potential vorticity a simpler version was found by Rossby. Consider a homogeneous layer of fluid with a height $h(\boldsymbol{x}_{\mathrm{h}}, t)$ along the vertical direction $z$, where $\boldsymbol{x}_{\mathrm{h}}$ denoting the horizontal coordinates. Assume a frictionless flow so that the vorticity is governed by Helmholtz' equation (2.196). Further assume that the horizontal velocity is independent of depth. The motion occurs thus in the form of Taylor columns (see also Section 2.11.2 and Figure 4.4). Using the notation $\eta$ and $f$ for the vertical components of relative and planetary vorticity, as in Section 2.11.1, the vertical component of (2.196) becomes

$$\frac{\mathrm{D}}{\mathrm{D}t}(\eta + f) = (\eta + f)\frac{\partial w}{\partial z} \tag{4.59}$$

Mass conservation requires a nondivergent velocity field,

$$\frac{\partial w}{\partial z} = -\left(\frac{\partial u}{\partial x} + \frac{\partial v}{\partial y}\right) \tag{4.60}$$

which integrates to

$$\frac{\mathrm{D}h}{\mathrm{D}t} = w_{\mathrm{top}} - w_{\mathrm{bottom}} = -h\left(\frac{\partial u}{\partial x} + \frac{\partial v}{\partial y}\right) \tag{4.61}$$

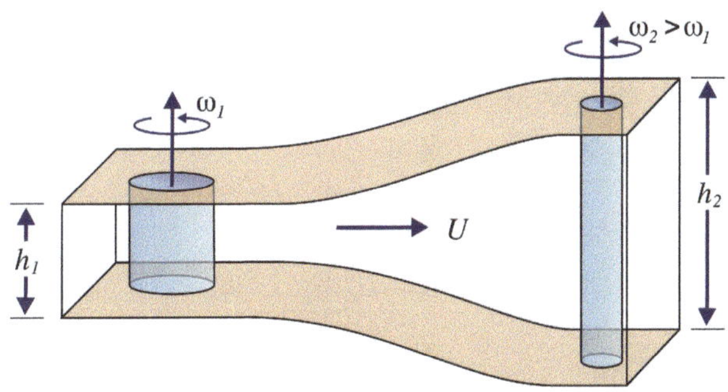


Fig. 4.4 Sketch illustrating the change of relative vorticity by stretching of the water column

Eliminating the divergence between (4.59) and (4.61), we obtain a conservation equation for Rossby's form of the potential vorticity,

$$\frac{\mathrm{D}}{\mathrm{D}t}\left(\frac{\eta + f}{h}\right) = 0 \tag{4.62}$$

In a frictionless, barotropic fluid the quantity $(\eta + f)/h$ is conserved following the motion of the Taylor columns. For instance, if $h$ increases, the column must increase its counterclockwise spin or move to a place where the planetary vorticity is larger, i. e. to higher latitudes. This statement is easily understood if the Taylor column is considered as a cylinder of height $h$ and radius $R$ in solid rotation. Then, because pressure forces cannot affect any torque on an infinitesimally thin cylinder, conservation of its angular momentum is simply $I\omega = \text{const}$ where $I = \pi R^2 h\rho/2$ is the moment of inertia and $\omega$ the angular velocity. If the cylinder changes its height $h$ while conserving mass, $\pi R^2 h\rho = \text{const}$, the quantity $\omega/h$ is thus conserved. With $\omega = (\eta + f)/2$, we find indeed (4.62).

# Geostrophic and Quasi-Geostrophic Motions

# 5

In this chapter, further approximations to the primitive equations derived in Section 4.2.5 are considered, which are valid for situations characterized by a small Rossby number. Geostrophic balance is assumed to dominate for lateral scales comparable to the Rossby radius – leading to the quasi-geostrophic approximation – or on scales comparable to Earth's radius – the planetary geostrophic approximation. The latter set of equations can be reformulated for an explicit expression for the absolute velocity.

From the scaling introduced in Section 4.2.4, it became evident that large-scale motions with small Rossby number $\mathrm{Ro} = U/\Omega L$, small aspect ratio $\delta = H/L$ and small Ekman number Ek are *geostrophic*, i. e. the horizontal pressure force is nearly in balance with the Coriolis force. Employing, for a moment, Cartesian coordinates, the horizontal momentum budget in geostrophic balance becomes

$$-f\rho v = -\partial p/\partial x + O(U^2/L, \Omega L\mathrm{Ek}) \,, \quad f\rho u = -\partial p/\partial y + O(U^2/L, \Omega L\mathrm{Ek}) \tag{5.1}$$

So far we have used the continuity equation to infer a scale for the vertical velocity, with the result $W = U\delta$. However, as we shall see in the following, the geostrophic balance implies a different and much smaller size of $W$.

## 5.1 Geostrophic Scaling

More information on the scale of the vertical velocity can be obtained by eliminating the pressure by cross-differentiating and subtracting both equations (5.1). With the continuity equation, and $\beta = \partial f/\partial y$, one obtains

$$f\frac{\partial w}{\partial z} - \beta v = O(U^2/L^2, \Omega \mathrm{Ek}) \tag{5.2}$$

which is referred to as the planetary vorticity equation. It is now in a form that can be used to infer the magnitude $W$. If the balance occurs between the terms of the left-hand side of (5.2), one finds a scale $W = \beta U/\Omega = (\beta L/\Omega)U\delta$ which is smaller by the factor $\beta L/\Omega \sim L/a$ (assuming that $f$ varies on the Earth radius $a$,

D. Olbers, J. Willebrand, C. Eden, *Ocean Dynamics*,
DOI 10.1007/978-3-642-23450-7_5, © Springer-Verlag Berlin Heidelberg 2012

i. e. $\beta \sim \Omega/a$) than the scaling inferred from the continuity equation only. Other regimes result if the stretching term $(f\,\partial w/\partial z)$ in (5.2) is in balance with one or several of the terms on the right-hand side as detailed below. However, the fact that for geostrophically balanced motions the magnitude of the vertical velocity appears to be smaller than the scaling used in Chapter 4.2, implies the possibility of further approximations which will be explained below. Such approximations include the quasi-geostrophic approximation (cf. Section 5.2) and the planetary-geostrophic approximation (cf. Section 5.3).

A more detailed classification of geostrophically balanced motions can be obtained by considering a scale analysis of the complete vorticity equation which was discussed in Section 2.11.2. The vertical component of Beltrami's equation (2.209), resembling the complete version of (5.2), reads

$$\underbrace{\rho\frac{\partial}{\partial t}\left(\frac{\eta}{\rho}\right)}_{\mathrm{I}} + \underbrace{\rho\boldsymbol{u}\cdot\nabla\left(\frac{\eta}{\rho}\right)}_{\mathrm{II}} + \underbrace{\beta v}_{\mathrm{III}} = \underbrace{(\eta+f)\frac{\partial w}{\partial z}}_{\mathrm{IV}} + \underbrace{\left(\omega_{\mathrm{a}}^{(x)}\frac{\partial w}{\partial x} + \omega_{\mathrm{a}}^{(y)}\frac{\partial w}{\partial y}\right)}_{\mathrm{V}} + \boldsymbol{B}^{(z)} + \mathrm{fric} \tag{5.3}$$

where $\eta = \omega_{\mathrm{a}}^{(z)} = \partial v/\partial x - \partial u/\partial y$ denotes the vertical component of the relative vorticity, and fric denotes frictional contributions. Only the terms labeled I to V matter in the following scaling analysis since the vertical component of the baroclinicity vector $\boldsymbol{B}$ and of the frictional term remain small in (5.3). The dominant terms are given by

I: the tendency term, with magnitude $U/LT$,
II: the advection of relative vorticity, with magnitude $U^2/L^2$,
III: the advection of planetary vorticity, with magnitude $\beta U$,
IV: the stretching term, with magnitude $\Omega W/H$ since $\eta/f \sim$ Ro, and
V: the tilting term, with magnitude $UW/HL$.

The vertical velocity scale $W$ can be inferred from this balance, provided that the stretching term plays a dominant role compared with some of the other terms. Since V/IV $\sim$ Ro, the tilting term V is always small compared to the stretching term and can be discarded. Thus we can differentiate between three different regimes: IV $\sim$ III, which yields the planetary geostrophic scaling, IV $\sim$ II which yields the quasi-geostrophic scaling, and finally IV $\sim$ I, which we will call the Rossby wave regime.

For the following discussion it is useful to introduce a length scale and a dimensionless number, the RHINES[1] scale $L_\beta$ (Rhines, 1975) and the $\beta$-Rossby number or Rhines number $\mathrm{R}_\beta$ as

$$L_\beta = \left(\frac{U}{\beta}\right)^{\frac{1}{2}} \quad \text{and} \quad \mathrm{R}_\beta = \frac{U}{\beta L^2} = \frac{L_\beta^2}{L^2} = \frac{a}{L}\mathrm{Ro} \tag{5.4}$$

Note that the ratio III/II divides the range of possible length scales: for II $\geq$ III we find $L \geq L_\beta$. A typical oceanic value for $L_\beta$ is 100 km. The Rhines scale separates motions with respect to their transport of relative or planetary vorticity: if $L \geq L_\beta$ the transport of planetary vorticity exceeds that of relative vorticity, and vice versa. In terms of the $\beta$-Rossby number, the threshold dividing the regimes of predominant

[1] PETER RHINES, *1942 in Hartford, Connecticut, fluid dynamicist.

transport of planetary or relative vorticity is defined by $R_\beta = 1$. For $R_\beta > 1$ the advection of relative vorticity dominates and for $R_\beta < 1$ the transport of planetary vorticity dominates, where the former regime might be called the mesoscale flow and the latter the gyre-scale flow. Typical values of the relevant scales for mesoscale and gyre-scale flow are given in Table 5.1.

Next consider the role of the time tendency term in (5.3). We first note that in addition to the Rhines scale $L_\beta$, two more length scales can be formed in the scaling analysis of (5.3) by comparing the relative importance of the time tendency term (I) with the advection of relative (II) and planetary vorticity (III): II $\geq$ I yields $L \leq UT = L_{ad}$, the advective length scale (which we leave out of the discussion), and III $\geq$ I yields $L \geq 1/\beta T = L_w$ which can be viewed as the wave length scale.

In the Rossby wave regime where the stretching term IV in (5.3) is balanced by the time tendency term (I), it follows that $W \sim UH/L\Omega T$. The magnitude of the appropriate time-scale $T$ must however be estimated from further relations. Restricting the analysis to the hydrostatic and geostrophic regime leads to a scaling of pressure as $P \sim UL\Omega$ and of density as $R \sim P/gH$, respectively, as introduced in Section 4.2. Anticipating the equation for the density perturbation (5.15) (with zero source term) which is derived in Section 5.2 below (see also Section 3.2), one finds the scaling $T \sim LU\Omega/(HN^2W)$, and hence $L = NH/\Omega = R_i$ which will later be identified as the (first) internal Rossby radius for constant $N$ (cf. Section 5.2.1). Finally, setting $L_w \sim R_i$ yields the time-scale $T = 1/\beta R_i$ of the (baroclinic) Rossby waves.

To summarize, the scale analysis of Beltrami's equation (5.3) has identified three important dynamical regimes. The *quasi-geostrophic regime* is characterized by IV $\sim$ II $\geq$ III, i.e. $W\Omega/H \approx U^2/L^2 \geq \beta U$. Hence

$$W \approx \mathrm{Ro}UH/L \quad \text{and} \quad L \leq L_\beta \tag{5.5}$$

For $L = L_\beta$ we find $W = UH/a$. The quasi-geostrophic regime is relevant for mesoscale flow and is discussed in more detail in Section 5.2. The *planetary geostrophic regime* is characterized by IV $\sim$ III $\geq$ II, i.e. $W\Omega/H \approx \beta U \geq (U/L)^2$. Hence

$$W \approx \frac{\beta}{\Omega}UH \approx (L/a)UH/L = UH/a \quad \text{and} \quad L \geq L_\beta \tag{5.6}$$

The planetary-geostrophic regime is relevant for gyre-scale flow and is discussed in more detail in Section 5.3. The *Rossby wave regime* is characterized by IV $\sim$ I, hence

$$W \approx UH/a \quad \text{and} \quad L \sim R_i \quad \text{and} \quad T = 1/\beta R_i$$

**Table 5.1** Typical magnitudes of velocity and length scales and dimensionless numbers for mesoscale and gyre-scale oceanic flows (for $H = 4{,}000$ m, $\beta = 2 \times 10^{-11}\,\mathrm{m^{-1}\,s^{-1}}$, $\Omega = 7.3 \times 10^{-5}\,\mathrm{s^{-1}}$ and $a = 6{,}371$ km). For mesoscale $W$ is estimated from (5.5), for gyre-scale from (5.6)

| | Mesoscale | Gyre-scale |
|---|---|---|
| $U$ [m s$^{-1}$] | 0.2 | 0.05 |
| $W$ [m s$^{-1}$] | $2 \times 10^{-4}$ | $3 \times 10^{-5}$ |
| $L$ [m] | $10^5$ | $10^6$ |
| $\ell = L/a$ | 0.015 | 0.15 |
| $\mathrm{Ro} = U/(\Omega L)$ | 0.03 | $7 \times 10^{-4}$ |
| $R_\beta = U/(\beta L^2)$ | 1 | 0.003 |

The wave regime is relevant for the adjustment of the circulation to changes in the forcing, and is discussed in more detail in Section 8.2.

The tilting term (V) in the scale analysis of (5.3) has been neglected from the beginning since its size is small, both in the planetary geostrophic regime where $\mathrm{V}/\mathrm{IV} = L_\beta^2/aL \ll 1$ and in the quasi-geostrophic regime where $\mathrm{V}/\mathrm{III} = \mathrm{Ro} \ll 1$. This means that in either case, the vertical vorticity component $\eta$ cannot be altered with any significance by tilting, i. e. the motion tends to occur in columnar form, expressing a vertical stiffness of the flow. For this reason, the vertical vorticity balance and the potential vorticity are dominant in geophysical flows.

We finally note that the corresponding scaling of the equations of the horizontal vorticity components does not reveal new insight into the different regimes. Advection, stretching and tilting are all of order $U^2/LH$, except the planetary part of the tilting, $fu_z = O(\Omega U/H)$. As their ratio is $O(\mathrm{Ro})$, the planetary term must be balanced by the baroclinic vector, i. e.

$$f\rho\frac{\partial u}{\partial z} = -\frac{1}{\rho}\left(\frac{\partial\rho}{\partial y}\frac{\partial p}{\partial z} - \frac{\partial\rho}{\partial z}\frac{\partial p}{\partial y}\right) + O(U^2/LH) + \mathrm{fric}$$

and using the hydrostatic equation $\partial p/\partial z = -g\rho$ we find

$$f\rho\frac{\partial u}{\partial z} = g\frac{\partial\rho}{\partial y} - \frac{1}{\rho}\frac{\partial\rho}{\partial z}\frac{\partial p}{\partial y} + O(U^2/LH) + \mathrm{fric}$$

The term on the left-hand side and the first term on the right-hand side are of the same order in our scaling, but the second term on the right-hand side scales as $\mathrm{Ro}(L\Omega)^2/(gH) \ll 1$ compared to both terms. Thus, to order $O(\mathrm{Ro})$, the horizontal vorticity equations result in the thermal wind balance (see Section 5.3).

## 5.2 Quasi-Geostrophic Approximation

The quasi-geostrophic regime is characterized by a set of small dimensionless numbers: as we have worked out in the previous section, the conditions are summarized by $\mathrm{Ro} = U/\Omega L \ll 1, \ell = L/a \ll 1$ and $\mathrm{Ek} = \mathcal{F}/\Omega U \ll 1$. All these numbers are assumed to be roughly of the same order; in particular we have $\mathrm{Ro}/\ell = U/\beta L^2 = \mathrm{R}_\beta = O(1)$ or $L \sim L_\beta$. A further important restriction is made with respect to the thermohaline state of the fluid: temperature and salinity are combined into a density variable perturbed about a stratified background state while all effects of compressibility are neglected. The resulting equations describe small perturbations of this stratified background state, like Rossby waves and other quasi-geostrophic mesoscale motions.

To start with a proper definition of the background stratification, we consider the thermohaline equation (4.15) of the Boussinesq approximation. Remember that this balance is written for the perturbation density $\tilde{\rho} = \rho - \rho_c(z)$, defined relative to the reference state $\rho_c$ with *constant* temperature and salinity (cf. Section 4.1, see List of Symbols for a list of all density variables). For the quasi-geostrophic approximation we need to consider perturbations with respect to a specified background state which reflects the basic stratification of temperature, salinity and density in the area of interest. The background state, denoted by the index $b$, could be defined as the areal average $S_b(z), \theta_b(z)$ of the stratification of temperature and salinity, depending only on the vertical coordinate. The background density is defined by

$$\rho_b(z) = F(S_b(z), \theta_b(z), p_c) \tag{5.7}$$

with the hydrostatic pressure counterpart $p_b(z)$ governed by $\partial p_b/\partial z = -g\rho_b$, and the Boussinesq reference pressure $p_c = -g\rho_0 z$. Hence we define a density perturbation $\hat{\rho}$ with associated pressure perturbation $\hat{p}$ relative to the stratified background state by

$$\hat{\rho} = \rho - \rho_b(z) \quad \text{and} \quad \hat{p} = p - p_b(z)$$

With the equation of state in the form $\hat{\rho} = F(S, \theta, p_c) - F(S_b(z), \theta_b(z), p_c)$ we find

$$\frac{D\hat{\rho}}{Dt} - w\frac{\rho_b N^2}{g} = \mathcal{G}_\rho - g\rho_0 w\left(\frac{1}{c_s^2} - \frac{1}{c_b^2}\right) \tag{5.8}$$

where the density source $\mathcal{G}_\rho$ is defined by (3.2), and the (squared) Brunt–Väisälä frequency associated with the background state is

$$N^2(z) = g\left(\alpha\frac{\partial\theta_b}{\partial z} - \gamma\frac{\partial S_b}{\partial z}\right) \tag{5.9}$$

The difference in compressibility between the *in-situ* state and the background state, given by the last terms in (5.8), will be neglected, so that *in-situ* density and potential density become identical. Furthermore, the density factor in the second term on the left-hand side will be replaced by the constant $\rho_0$. The Brunt–Väisälä frequency $N(z)$ is then the only manifestation of the background state in the equation.

In quasi-geostrophic theory the equations of motion are expanded about a fixed latitude $\varphi_0$ which could be in principle anywhere except at the equator where $f \to 0$ so that the geostrophic balance breaks down. To facilitate the notation, we introduce new coordinates $x, y$ defined by

$$x = \lambda a \cos\varphi_0 , \quad y = a(\varphi - \varphi_0)$$

Note that these coordinates should *not* be considered as Cartesian coordinates although they resemble Cartesian coordinates near the latitude $\varphi_0$. Furthermore, to simplify notation, the equations will be written in this section in terms of scaled pressure and density variables,

$$p = \frac{\hat{p}}{\rho_0} , \quad \rho = \frac{\hat{\rho}}{\rho_0} \tag{5.10}$$

so that the constant factor $\rho_0$ is eliminated, $\rho$ is dimensionless, and the pressure $p$ is measured in $m^2\,s^{-2}$. The friction term $\mathcal{F}$ is rescaled appropriately as well. The primitive equations then take the form

$$\frac{\mathrm{D}u}{\mathrm{D}t} - \frac{uv}{a}\tan\varphi - fv = -\frac{\cos\varphi_0}{\cos\varphi}\frac{\partial p}{\partial x} + \mathcal{F}_u \tag{5.11}$$

$$\frac{\mathrm{D}v}{\mathrm{D}t} + \frac{u^2}{a}\tan\varphi + fu = -\frac{\partial p}{\partial y} + \mathcal{F}_v \tag{5.12}$$

$$\frac{\partial p}{\partial z} = -g\rho \tag{5.13}$$

$$\frac{\partial w}{\partial z} = -\frac{\cos\varphi_0}{\cos\varphi}\left(\frac{\partial u}{\partial x} + \frac{\partial}{\partial y}v\frac{\cos\varphi}{\cos\varphi_0}\right) \tag{5.14}$$

$$\frac{\mathrm{D}\rho}{\mathrm{D}t} - \frac{N^2}{g}w = \mathcal{G}_\rho \tag{5.15}$$

where

$$\frac{\mathrm{D}}{\mathrm{D}t} = \frac{\partial}{\partial t} + u\frac{\cos\varphi_0}{\cos\varphi}\frac{\partial}{\partial x} + v\frac{\partial}{\partial y} + w\frac{\partial}{\partial z} \tag{5.16}$$

The approximations in the density equation (5.15) with respect to effects of compressibility are the only approximations made so far.

### 5.2.1 Expansion for Small Parameters

Remember from Section 4.2.4 that the magnitudes of the terms in the momentum balances (5.11) and (5.12) are $\mathrm{Ro} : \ell\mathrm{Ro} : 1 : 1 : \mathrm{Ek}$, in the order as they appear. The scaling of the density equation (5.15) has been considered previously in Section 5.1. With the geostrophic and hydrodynamic scaling introduced in Section 4.2.4, i. e. $p \sim P = \Omega UL$ and $\rho \sim R = \Omega UL/gH$, respectively, and the magnitude of the vertical velocity for the quasi-geostrophic regime, $W = \mathrm{Ro}\, UH/L$, the ratio of the second to the first term on the left-hand side of (5.15) becomes[2]

$$\frac{WN^2/g}{UR/L} = \frac{N^2H^2}{\Omega^2L^2} = \mathrm{Bu} = (\mathrm{R_i}/L)^2 \tag{5.17}$$

which defines the BURGER[3] number Bu (Burger, 1958). It is the squared ratio of two length scales, the internal Rossby radius $\mathrm{R_i} = NH/\Omega$ (for constant $N$) and the length scale $L$. Typical oceanic magnitudes for $\mathrm{R_i}$ range between 10 km (in polar latitudes) to about 30–50 km in midlatitudes. As long as $\mathrm{Bu} \geq O(1)$, the time rate of change of density is of similar size as the heaving of mass associated with the background stratification. Usually the source of density $\mathcal{G}_\rho$ is small compared to the other terms, say of order Ek. Then, the terms in the density balance scale as $\mathrm{Ro} : \mathrm{RoBu} : \mathrm{Ek}$. Notice that as long as $\mathrm{Bu} \sim O(1)$, the entire density equation is of first order in the small parameters; it will not contribute to zero order. A proper expansion, however, must also consider the smallness of $\ell$ in the trigonometric expressions involving $\varphi - \varphi_0 = y/a = O(\ell) \ll 1$. Taylor expansion of these expressions

[2] In this scaling analysis $N$ is a typical constant magnitude of $N(z)$.

[3] ALEWYN P. BURGER, *1927 in Middelburg, †2003 in Pretoria, mathematician and fluid dynamicist.

about $\varphi = \varphi_0$ becomes

$$\sin\varphi = \sin\varphi_0 + \frac{y}{a}\cos\varphi_0 + O(\ell^2)\,,$$
$$\frac{\cos\varphi}{\cos\varphi_0} = 1 - \frac{y}{a}\tan\varphi_0 + O(\ell^2)\,,$$
$$\frac{\cos\varphi_0}{\cos\varphi} = 1 + \frac{y}{a}\tan\varphi_0 + O(\ell^2)$$

which implies a linear dependence of the Coriolis parameter on the meridional coordinate to order $\ell^2$,

$$f = 2\Omega\sin\varphi = 2\Omega\sin\varphi_0 + \beta y \equiv f_0 + \beta y \tag{5.18}$$

with $f_0 = 2\Omega\sin\varphi_0$ and $\beta = (2\Omega/a)\cos\varphi_0$.

Now a rigorous expansion of all fields can be performed. Assuming for simplicity that all small parameters are of the same order $\epsilon \ll 1$, all variables are now written as a series in powers of the small parameter $\epsilon$, starting with a zero order contribution indicated by the subscript 0, etc.

$$u = u^{(0)} + \epsilon u^{(1)} + \cdots\,, \quad v = v^{(0)} + \epsilon v^{(1)} + \cdots\,, \quad w = \epsilon w^{(1)} + \cdots$$
$$\rho = \rho^{(0)} + \epsilon\rho^{(1)} + \cdots\,, \quad p = p^{(0)} + \epsilon p^{(1)} + \cdots$$

Note that the zero order contribution of $w$ vanishes since it has a leading term of order Ro. Indeed, we will see below that the zero order horizontal velocities are free of divergence. Inserting this expansion into (5.11)–(5.16), we collect the terms of same order in $\epsilon$ and require that the equations are satisfied for each order. Fortunately, only the two lowest orders will be of interest here.

### Equations of 0-order

The 0-order momentum equations

$$-f_0 v^{(0)} = -\frac{\partial p^{(0)}}{\partial x} \tag{5.19}$$
$$f_0 u^{(0)} = -\frac{\partial p^{(0)}}{\partial y} \tag{5.20}$$
$$0 = -\frac{\partial p^{(0)}}{\partial z} - g\rho^{(0)} \tag{5.21}$$

express the geostrophic and hydrostatic balance of the basic state. Since $f_0$, the value of the Coriolis parameter at the reference latitude, is constant, a geostrophic stream function $\psi = p^{(0)}/f_0$ can be defined such that

$$u^{(0)} = -\frac{\partial\psi}{\partial y}\,, \quad v^{(0)} = \frac{\partial\psi}{\partial x}$$

The horizontal geostrophic velocity is obviously free of divergence

$$\frac{\partial u^{(0)}}{\partial x} + \frac{\partial v^{(0)}}{\partial y} = 0 \tag{5.22}$$

which is in agreement with the 0-order continuity equation, i. e. the continuity equation (5.14) does not constrain the lowest order at all. The density balance does not have a 0-order contribution, and thus, counting the number of zero-order field variables (there are four: $u^{(0)}$, $v^{(0)}$, $\rho^{(0)}$ and $p^{(0)}$) and the number of equations to determine them (there are only three: (5.19)–(5.21)), it becomes obvious that the 0-order problem is incomplete, and it is necessary to consider higher orders of the governing equations.

### Equations of 1-order

The 1-order equations read

$$\frac{\mathrm{D}^0}{\mathrm{D}t}u^{(0)} - \beta y v^{(0)} - f_0 v^{(1)} = -\frac{\partial}{\partial x}\left(p^{(1)} + p^{(0)}\frac{y}{a}\tan\varphi_0\right) + \mathcal{F}_u \tag{5.23}$$

$$\frac{\mathrm{D}^0}{\mathrm{D}t}v^{(0)} + \beta y u^{(0)} + f_0 u^{(1)} = -\frac{\partial}{\partial y}p^{(1)} + \mathcal{F}_v \tag{5.24}$$

$$0 = -\frac{\partial p^{(1)}}{\partial z} - g\rho^{(1)} \tag{5.25}$$

$$\frac{\partial u^{(1)}}{\partial x} + \frac{\partial v^{(1)}}{\partial y} + \frac{\partial w^{(1)}}{\partial z} + \frac{y}{a}\tan\varphi_0\frac{\partial u^{(0)}}{\partial x} - \frac{v^{(0)}}{a}\tan\varphi_0 = 0 \tag{5.26}$$

$$\frac{\mathrm{D}^0}{\mathrm{D}t}\rho^{(0)} - \frac{N^2}{g}w^{(1)} = \mathcal{G}_\rho \tag{5.27}$$

The substantial derivative $\mathrm{D}^0/\mathrm{D}t = \partial/\partial t + u^{(0)}\partial/\partial x + v^{(0)}\partial/\partial y$ is based on the 0-order geostrophic velocities, there is thus no vertical advection. The first-order momentum balances are clearly not geostrophic since now all forces and the inertial term are present. The above first order balances and the 1-order fields are, therefore, called ageostrophic.

Of course, also the 1-order problem yields only five equations and brings in five new unknown fields[4], so it seems hopeless to close the problem at this – or any higher – order. But a closer inspection of (5.23)–(5.27) shows some peculiarities of the expansion.

The 1-order density equation (5.27) does not predict the 1-order density, instead it determines the 1-order vertical velocity $w^{(1)}$ in terms of 0-order fields ($\rho^{(0)}$, $u^{(0)}$ and $v^{(0)}$). The 1-order continuity equation (5.14) can then be used to evaluate the divergence of the horizontal 1-order velocities, and, finally, forming the vorticity of the 1-order momentum balances, the 1-order pressure field can be eliminated, and the only other 1-order contribution comes from the divergence of the ageostrophic velocity. The ageostrophic vorticity balance thus yields a condition on the flow which involves only 0-order fields, which we shall show in the next section.

[4] Some derivations of quasi-geostrophic equations do not make a rigorous asymptotic expansion of all field variables. They do not expand the pressure and density to first order. Then the equations are trivially closed at the first order: $u^{(1)}, v^{(1)}, w^{(1)}$ follow as functional of $p^{(0)}$ from (5.23), (5.24) and (5.27). They have to satisfy continuity (5.26) which yields the quasi-geostrophic potential vorticity equation in the identical form as given below.

### 5.2.2 Quasi-Geostrophic Vorticity Equation

The derivation of the ageostrophic vorticity equation from (5.23)–(5.26) is straightforward but somewhat lengthy, and we drop the mathematics here (for further details we refer to Pedlosky, 1987). The result is

$$\frac{\mathrm{D}^0}{\mathrm{D}t}(\eta^{(0)} + \beta y) = f_0 \frac{\partial w^{(1)}}{\partial z} + \operatorname{curl} \mathcal{F} \tag{5.28}$$

Here the vertical vorticity $\eta^{(0)}$ of the geostrophic flow is defined as

$$\eta^{(0)} = \frac{\partial v^{(0)}}{\partial x} - \frac{\partial u^{(0)}}{\partial y} = \nabla_{\mathrm{h}}^2 \psi$$

and the frictional vorticity source is

$$\operatorname{curl} \mathcal{F} = \frac{\partial \mathcal{F}_v}{\partial x} - \frac{\partial \mathcal{F}_u}{\partial y} \tag{5.29}$$

(remember that $\mathcal{F}$ is horizontal in the shallow water approximation). The balance (5.28) considers advection of relative and planetary vorticity, the stretching term and the curl of the frictional forces. There is no tilting effect in this order, consistent with the scaling we have discussed in Section 5.1.

With $w^{(1)}$ determined by the 0-order density equation (5.27),

$$w^{(1)} = \frac{g}{N^2}\left(\frac{\mathrm{D}^0}{\mathrm{D}t}\rho^{(0)} - \mathcal{G}_\rho\right) = \frac{g}{N^2}\left[\frac{\mathrm{D}^0}{\mathrm{D}t}\left(-\frac{1}{g}\frac{\partial p^{(0)}}{\partial z}\right) - \mathcal{G}_\rho\right]$$

we express the stretching term $f_0 \partial w/\partial z$ in (5.28) by the 0-order pressure field and obtain

$$\frac{\mathrm{D}^0}{\mathrm{D}t}(\eta^{(0)} + \beta y) + f_0 \frac{\partial}{\partial z}\frac{\mathrm{D}^0}{\mathrm{D}t}\frac{1}{N^2}\frac{\partial p^{(0)}}{\partial z} = \operatorname{curl} \mathcal{F} - f_0 \frac{\partial}{\partial z}\frac{g\mathcal{G}_\rho}{N^2} \tag{5.30}$$

The vertical derivative and the substantial advection operator in fact commute here so that

$$\frac{\partial}{\partial z}\frac{\mathrm{D}^0}{\mathrm{D}T}\frac{1}{N^2}\frac{\partial p^{(0)}}{\partial z} \equiv \frac{\mathrm{D}^0}{\mathrm{D}t}\left(\frac{\partial}{\partial z}\frac{1}{N^2}\frac{\partial p^{(0)}}{\partial z}\right) \tag{5.31}$$

The proof of (5.31) is, however, not entirely trivial, it uses the 0-order balances. Then, finally, with $\eta^{(0)} = \nabla_{\mathrm{h}}^2 \psi$ and $p^{(0)} = f_0 \psi$, we arrive at the quasi-geostrophic vorticity balance in the form

$$\frac{\mathrm{D}}{\mathrm{D}t}\left(\nabla_{\mathrm{h}}^2 \psi + \beta y + \frac{\partial}{\partial z}\frac{f_0^2}{N^2}\frac{\partial \psi}{\partial z}\right) = \operatorname{curl} \mathcal{F} - f_0 g \frac{\partial}{\partial z}\frac{\mathcal{G}_\rho}{N^2} \tag{5.32}$$

The index 0 of the advective operator is omitted from now on, but it should be remembered that it involves only the two-dimensional advection by the geostrophic velocities,

$$\frac{\mathrm{D}}{\mathrm{D}t} = \frac{\partial}{\partial t} + u\frac{\partial}{\partial x} + v\frac{\partial}{\partial y} = \frac{\partial}{\partial t} + \frac{\partial \psi}{\partial x}\frac{\partial}{\partial y} - \frac{\partial \psi}{\partial y}\frac{\partial}{\partial x} = \frac{\partial}{\partial t} + \mathcal{J}(\psi, \ldots)$$

where $\mathcal{J}$ is the Jacobian differential operator with respect to $x$ and $y$. The stream function $\psi$ is the only prognostic variable which determines the entire flow at any instant, as seen in the relations

$$u = -\frac{\partial \psi}{\partial y} \tag{5.33}$$

$$v = \frac{\partial \psi}{\partial x} \tag{5.34}$$

$$w = -\frac{f_0}{N^2}\frac{\mathrm{D}}{\mathrm{D}t}\frac{\partial \psi}{\partial z} - \frac{g\mathcal{G}_\rho}{N^2} \tag{5.35}$$

$$\rho = -\frac{f_0}{g}\frac{\partial \psi}{\partial z} \tag{5.36}$$

Notice that $w$ is here first order but the other fields are zero order. In fact, the vertical velocity is the only first order field which is uniquely determined by quasi-geostrophic theory, in particular the ageostrophic horizontal velocities and the ageostrophic density and pressure fields remain unknown at this level of approximation.

The quasi-geostrophic vorticity balance (5.32) is valid for $\mathrm{Ro} = U/\Omega L \ll 1, \ell = L/a \ll 1$ and $\mathrm{Ek} = \mathcal{F}/\Omega U \ll 1$ and $\mathrm{Bu} = (\mathrm{R_i}/L)^2 = O(1)$ and is of fundamental importance in geophysical fluid dynamics. It describes the kinematic and dynamic properties of planetary waves, barotropic and baroclinic instabilities, the dynamics of geophysical turbulence and mesoscale eddies, and – with some straining of the scaling which we have applied – even most features of the large-scale wind-driven circulation.

### 5.2.3 Quasi-Geostrophic Potential Vorticity

We have supplemented the approximations of the complete equations of motions, described previously in this chapter, with the appropriate theorems of potential vorticity and energy conservation. The derivation of (5.32) made it clear already that this basic prognostic equation of quasi-geostrophic theory is a balance of potential vorticity. In the quasi-geostrophic form of potential vorticity,

$$Q_{\mathrm{qg}} = \nabla_{\mathrm{h}}^2 \psi + \frac{\partial}{\partial z}\frac{f_0^2}{N^2}\frac{\partial \psi}{\partial z} + f$$

relative and planetary vorticity and the stretching part appear additive (note that $f = f_0 + \beta y$, and that the constant value $f_0$ of planetary vorticity has been added since it has no effect in (5.32)). The quantity $Q_{\mathrm{qg}}$ is materially conserved in adiabatic condition, following however the projection of the fluid parcel's path onto the horizontal plane, as outlined by the geostrophic velocities, and not the three-dimensional path.

The derivation of this theorem from the complete set of equations of motion is somewhat elaborate (as demonstrated above where we even have left out much of the mathematics). On the other hand, with the aim known and the scaling and expansion in mind it is, in fact, easy to derive the conservation of $Q_{\mathrm{qg}}$ from Ertel's theorem (2.212) in a few steps. The appropriate tracer in Ertel's potential vorticity

$$Q = \frac{(2\boldsymbol{\Omega} + \boldsymbol{\omega}) \cdot \nabla \chi}{\rho}$$

is the density $\chi = \tilde{\rho} = \rho_b + \hat{\rho}$ which is conserved according to (5.15) if the diabatic source $\mathcal{G}_\rho$ is absent. Implementing the Boussinesq and shallow water approximations, Ertel's theorem becomes

$$\left(\frac{\mathrm{D}}{\mathrm{D}t}\right)_{3-\mathrm{d}}\left[(f+\omega_3)\left(\frac{\partial\rho_b}{\partial z}+\frac{\partial\rho}{\partial z}\right)\right] = \omega_3\frac{\partial\mathcal{G}_\rho}{\partial z}+\left(\frac{\partial\rho_b}{\partial z}+\frac{\partial\rho}{\partial z}\right)\operatorname{curl}\mathcal{F}$$

because the baroclinicity term vanishes identically for $\chi = \rho$. Expanding this equation with respect to all small parameters so that $f = f_0 + \beta y$, $\beta y \ll f_0$, $\omega_3 \ll f_0$, $\rho \ll \rho_b$ and taking $w$ small as well, we find, retaining all terms up to the second order,

$$\frac{\mathrm{D}}{\mathrm{D}t}(f+\omega_3)\frac{\partial\rho_b}{\partial z}+\frac{\mathrm{D}}{\mathrm{D}t}f_0\frac{\partial\rho}{\partial z}+w\frac{\partial}{\partial z}f_0\frac{\partial\rho_b}{\partial z}\approx f_0\frac{\partial\mathcal{G}_\rho}{\partial z}+\frac{\partial\rho_b}{\partial z}\operatorname{curl}\mathcal{F}$$

With $\omega_3 = \nabla_h^2\psi$, $\partial\rho_b/\partial z = -N^2/g$, using the hydrostatic balance (5.13) and eliminating the vertical velocity using (5.15), this becomes

$$\frac{\mathrm{D}}{\mathrm{D}t}Q_{\mathrm{qg}} = \operatorname{curl}\mathcal{F} - f_0 g\frac{\partial}{\partial z}\frac{\mathcal{G}_\rho}{N^2} \tag{5.37}$$

i. e. the same result as in (5.32). Notice that a purely geostrophic approximation of Ertel's potential vorticity does not yield the correct form of the quasi-geostrophic potential vorticity.

### 5.2.4 Boundary Conditions

The quasi-geostrophic potential vorticity equation (5.32) is second order in the vertical derivative and thus needs two boundary conditions. They naturally arise by considering the kinematic conditions of zero mass flux through the surface $z = \zeta(\boldsymbol{x}, t)$ and the bottom $z = -h(\boldsymbol{x})$. These are (see Chapter 2)

$$\frac{\mathrm{D}\zeta}{\mathrm{D}t} - w = 0 \quad \text{at} \quad z = \zeta\,, \qquad \boldsymbol{u}_h\cdot\nabla h + w = 0 \quad \text{at} \quad z = -h \tag{5.38}$$

The bottom boundary condition is easily translated into the quasi-geostrophic framework by use of (5.33) to (5.35) which yields

$$\frac{\mathrm{D}}{\mathrm{D}t}\frac{\partial\psi}{\partial z} = \frac{N^2}{f_0}\mathcal{J}(\psi, h) \quad \text{at} \quad z = -h$$

Note that the density source $\mathcal{G}_\rho$ is assumed to vanish at the bottom. To exploit the upper kinematic boundary condition, we must relate the sea surface elevation to the geostrophic pressure. This is done by expanding (5.38) about the mean sea surface height $z = 0$ (then to lowest order in $\zeta$ all quantities in (5.38) are taken at $z = 0$), and also expanding the total pressure[5] $p^{(0)} + p_b + p$ about $z = 0$, we find to lowest order in $\zeta$

$$p^{(0)}(0) + p_b(0) + p(\boldsymbol{x}, 0) + \zeta\left(\frac{\partial p^{(0)}}{\partial z}\right)_{z=0} + \cdots = p_{\mathrm{atm}}$$

[5] For simplicity of notation, all pressures are scaled according to (5.10).

because the reference state gives the largest contribution to the vertical pressure gradient. Then, with $p^{(0)}(0) = f_0\psi(0)$, we obtain

$$\frac{\mathrm{D}\zeta}{\mathrm{D}t} = \frac{f_0}{g}\frac{\mathrm{D}\psi(0)}{\mathrm{D}t} - \frac{1}{g}\frac{\mathrm{D}p_{\mathrm{atm}}}{\mathrm{D}t}$$

and use this to write the boundary condition at the upper surface

$$\frac{g}{N^2}\frac{\mathrm{D}}{\mathrm{D}t}\frac{\partial\psi}{\partial z} + \frac{\mathrm{D}\psi}{\mathrm{D}t} = -\frac{g^2\mathcal{G}_\rho}{f_0 N^2} \quad \text{at} \quad z = 0 \tag{5.39}$$

Here we have neglected the contribution from the atmospheric pressure, which is generally small compared with the frictional effect of the surface wind (in approximate geostrophic balance with the surface pressure) on the flow.

Both boundary conditions must be in accordance with the quasi-geostrophic scaling, i. e. $W = O(\mathrm{Ro}UH/L)$. While this is no problem for the surface velocity, there may occur too strong vertical velocities at the bottom. If a horizontal current crosses a ridge with slope $\Delta h/L$, it produces at maximum vertical velocities of the size $W \sim U\Delta h/L = (\Delta h/h)Uh/L$. We thus must require that the relative height does not exceed the Rossby number, $\Delta h/h = O(\mathrm{Ro})$. This condition represents a severe limit for the applicability of the quasi-geostrophic approach since it inhibits strong (realistic) variations in bottom topography.

### 5.2.5 Energetics of Quasi-Geostrophic Motions

The conservation theorem of kinetic energy is derived by multiplying the first order momentum balances (5.23) and (5.24) by $u$ and $v$, respectively, and adding the results, which yields

$$\frac{\partial}{\partial t}\frac{1}{2}(u^2+v^2) + \nabla_{\mathrm{h}}\cdot\left[\boldsymbol{u}_{\mathrm{h}}\frac{1}{2}(u^2+v^2)\right] + f_0\left(u^{(1)}v - v^{(1)}u\right)$$
$$= -\nabla_{\mathrm{h}}\cdot\left(\boldsymbol{u}_{\mathrm{h}}p^{(1)}\right) - \frac{y\tan\varphi_0}{a}u\frac{\partial p}{\partial x} + \boldsymbol{u}_{\mathrm{h}}\cdot\boldsymbol{\mathcal{F}}$$

We have abandoned the index 0 such that $\boldsymbol{u}_{\mathrm{h}} = (u, v)$ denotes here the zero-order geostrophic velocity, $p$ the zero-order geostrophic pressure and $\rho$ the zero-order density in hydrostatic balance with $p$, but we keep the index 1 for ageostrophic components. Some tedious mathematical manipulations cast this balance into a form which separates between flux and exchange terms,

$$\frac{\partial}{\partial t}E_{\mathrm{k}} + \nabla_{\mathrm{h}}\cdot\boldsymbol{J}_{\mathrm{h}} + \frac{\partial}{\partial z}(pw^{(1)}) = w^{(1)}\frac{\partial p}{\partial z} + \boldsymbol{u}_{\mathrm{h}}\cdot\boldsymbol{\mathcal{F}} \tag{5.40}$$

where $E_{\mathrm{k}} = (u^2+v^2)/2$ is the quasi-geostrophic kinetic energy. The horizontal flux of kinetic energy is found as the complicated expression

$$\boldsymbol{J}_{\mathrm{h}} = \boldsymbol{u}_{\mathrm{h}}\left(E_{\mathrm{k}} + p^{(1)}\right) + \boldsymbol{u}_{1\mathrm{h}}p - \boldsymbol{j}\frac{y\tan\varphi_0}{a}vp$$

and the vertical flux is simply $w^{(1)}p$. The term $w^{(1)}\partial p/\partial z = -g\rho w^{(1)}$ describes the production of kinetic energy by lifting mass. As we will see below, it is exchanging this amount per unit time with the potential energy. The last term of the right-hand side of (5.40) is the sink/source due to friction and external stresses. A further conservation theorem is obtained from the first order density balance (5.27). Multiplying by $\rho$, we find

$$\frac{\partial}{\partial t}\frac{1}{2}\rho^2 + \nabla_{\mathrm{h}} \cdot \left(\boldsymbol{u}_{\mathrm{h}}\frac{1}{2}\rho^2\right) - \frac{N^2}{g}\rho w^{(1)} = \rho \mathcal{G}_\rho$$

We identify

$$E_{\mathrm{p}}^{\mathrm{a}} = \frac{1}{2}\frac{g^2\rho^2}{N^2} \tag{5.41}$$

as a potential energy (per mass) of the quasi-geostrophic system. The above equation is then written in the form

$$\frac{\partial}{\partial t}E_{\mathrm{p}}^{\mathrm{a}} + \nabla_{\mathrm{h}} \cdot \left(\boldsymbol{u}_{\mathrm{h}} E_{\mathrm{p}}^{\mathrm{a}}\right) = -w^{(1)}\frac{\partial p}{\partial z} + \frac{g^2\rho\mathcal{G}_\rho}{N^2}$$

and adding the conservation equation of kinetic energy and the conservation theorem for the total quasi-geostrophic energy, we obtain

$$\frac{\partial}{\partial t}\left(E_{\mathrm{k}} + E_{\mathrm{p}}^{\mathrm{a}}\right) + \nabla \cdot \boldsymbol{J} = \mathcal{S}$$

with an energy flux vector and sink/source term

$$\begin{aligned}
\boldsymbol{J} &= \boldsymbol{i}\left[u\left(E_{\mathrm{k}} + E_{\mathrm{p}}^{\mathrm{a}}\right) + up^{(1)} + u^{(1)}p\right] \\
&\quad + \boldsymbol{j}\left[v\left(E_{\mathrm{k}} + E_{\mathrm{p}}^{\mathrm{a}}\right) + vp^{(1)} + v^{(1)}p - \frac{y\tan\varphi_0}{a}vp\right] + \boldsymbol{k}w^{(1)}p \\
\mathcal{S} &= \boldsymbol{u}_{\mathrm{h}} \cdot \boldsymbol{\mathcal{F}} + \frac{g^2\rho\mathcal{G}_\rho}{N^2}
\end{aligned}$$

Apart from flux divergences only the diabatic forcing terms remain as source/sink of total energy, which proves that we have identified the correct form for the quasi-geostrophic potential energy. A few words of caution and interpretation are, however, needed to put the results into perspective. An alternative derivation of the energy balance is given in the box on p. 148.

An examination of the quasi-geostrophic potential energy $E_{\mathrm{p}}^{\mathrm{a}}$ is required. The quadratic form emanating from the above analysis is certainly not the conventional potential energy introduced in Section 2.4. We postpone this question to the next Section 5.2.6.

### 5.2.6 Available Potential Energy

In the quasi-geostrophic framework a potential energy variable was defined by (5.41). Another form of this quantity is found by considering the vertical displacement $\xi$ of an isopycnal[6] represented by $\varrho = \rho_{\mathrm{b}} + \rho$ from its unperturbed

[6] Remember that densities are here potential densities.

**24. Quasi-Geostrophic Energy Balance**

As we have pointed out above, there is no access to the ageostrophic velocities $\boldsymbol{u}_{1\mathrm{h}}$ and the ageostrophic pressure $p^{(1)}$ (unless higher order equations are considered). Hence the flux $\boldsymbol{J}_\mathrm{h}$ cannot be evaluated: Although the conservation theorems are physically meaningful and closed, they are not of much value because the quasi-geostrophic potential vorticity equation only yields the geostrophic flow field and not the ageostrophic horizontal velocities and pressures.
There is, however, a form of energy conservation equation which is free of these shortcomings. In fact, if the potential vorticity balance (5.32) is multiplied by $\psi$, we obtain after some trivial mathematical transformations

$$\frac{\partial E_\mathrm{qg}}{\partial t} + \boldsymbol{\nabla} \cdot \boldsymbol{J}_\mathrm{qg} = \mathcal{S}_\mathrm{qg}$$

where

$$E_\mathrm{qg} = \frac{1}{2}\left[\left(\frac{\partial \psi}{\partial x}\right)^2 + \left(\frac{\partial \psi}{\partial y}\right)^2 + \frac{f_0^2}{N^2}\left(\frac{\partial \psi}{\partial z}\right)^2\right]$$

$$\boldsymbol{J}_\mathrm{qg} = -\boldsymbol{i}\left[\psi\frac{\partial^2\psi}{\partial t\partial x} + u\psi(Q_\mathrm{qg} - f) + \frac{1}{2}\beta\psi^2\right] - \boldsymbol{j}\left[\psi\frac{\partial^2\psi}{\partial t\partial y} + v\psi(Q_\mathrm{qg} - f)\right] - \boldsymbol{k}\left[\frac{f_0^2}{N^2}\psi\frac{\partial^2\psi}{\partial t\partial z}\right]$$

$$\mathcal{S}_\mathrm{qg} = -\psi\left(\mathrm{curl}\,\mathcal{F} - f_0 g\frac{\partial}{\partial z}\frac{\mathcal{G}_\rho}{N^2}\right)$$

Apparently, $E_\mathrm{qg}$ is a quadratic invariant with its associated flux vector $\boldsymbol{J}_\mathrm{qg}$ and adiabatic source term $\mathcal{S}_\mathrm{qg}$. While it is easy to see, using (5.33)–(5.36), that the quadratic invariant is the total energy of the previous conservation theorem, $E_\mathrm{qg} = E_\mathrm{k} + E_\mathrm{p}^\mathrm{a}$, the fluxes and source terms do not agree individually, i. e. $\boldsymbol{J} \neq \boldsymbol{J}_\mathrm{qg}, \mathcal{S} \neq \mathcal{S}_\mathrm{qg}$. But evidently, we must have $\boldsymbol{\nabla} \cdot \boldsymbol{J} - \mathcal{S} = \boldsymbol{\nabla} \cdot \boldsymbol{J}_\mathrm{qg} - \mathcal{S}_\mathrm{qg}$ which, in fact, can be proven (see e. g. Pedlosky, 1987).

background state $\rho_\mathrm{b}$ (the constant contribution $\rho_\mathrm{c}$ is ignored). From a truncated Taylor expansion of $\rho$ one finds $\xi \approx -\rho/(\partial\rho_\mathrm{b}/\partial z) = g\rho/N^2$ and

$$E_\mathrm{p}^\mathrm{a} = \frac{1}{2}\frac{g^2\rho^2}{N^2} = \frac{1}{2}\xi^2 N^2 \tag{5.42}$$

showing a similarity to the potential energy of a harmonic oscillator with frequency $N$ in a Lagrangian view of the fluid displacement. Though (5.42) is a well-defined local variable of the state of the flow, its relation to the total potential energy $g\varrho z$ can only be understood when considering the entire system in an integral sense. We integrate $g\varrho z$ over a piece of the fluid which is bounded at the bottom and the top by two isopycnals $\varrho_1$ and $\varrho_2$ and obtain by partial integration

$$\int_{z(\varrho_1)}^{z(\varrho_2)} g\varrho z \mathrm{d}z = g\int_{\varrho_1}^{\varrho_2} z(\boldsymbol{x}_\mathrm{h}, \varrho, t)\varrho\frac{\partial z}{\partial \varrho}\mathrm{d}\varrho = g\left.\frac{1}{2}z^2\varrho\right|_{\varrho_1}^{\varrho_2} - g\int_{\varrho_1}^{\varrho_2}\frac{1}{2}z^2\mathrm{d}\varrho \tag{5.43}$$

The integral has been transformed from the vertical coordinate $z$ to the density coordinate by inverting $\varrho = \varrho(\boldsymbol{x}_\mathrm{h}, z, t)$ to $z = z(\boldsymbol{x}_\mathrm{h}, \varrho, t)$ which gives the height of the isopycnal $\varrho$. Note that we have assumed stable stratification for the coordinate transformation to work. Now we define the height of the unperturbed isopycnals in the same manner, $z_\mathrm{b} = z(\rho_\mathrm{b})$. Then we use $z = z_\mathrm{b} + \xi$, or

$$\xi(\rho) = z(\rho_\mathrm{b} + \hat{\rho}) - z_\mathrm{b}(\rho_\mathrm{b}) \approx \frac{g\rho}{N^2}$$

Suppose that the reference state $\rho_b$ has been defined such that the positive and negative displacements of each isopycnal averaged out in an integral over the whole system. We consider the horizontal mean of (5.43) and concentrate on the part representing the interior contribution. The mean of $z^2$ becomes

$$\overline{z^2} = z_b^2 + 2z_b\bar{\xi} + \overline{\xi^2} = z_b^2 + \overline{\xi^2}$$

and thus, apart from boundary contributions and constants deriving from the background state, we have

$$\int \mathrm{d}\boldsymbol{x}_h \int_{z(\varrho_1)}^{z(\varrho_2)} g\varrho z \mathrm{d}z = g \int \mathrm{d}\boldsymbol{x}_h \int_{\varrho_1}^{\varrho_2} \frac{1}{2}\xi^2 \mathrm{d}\varrho + \text{const} = \int \mathrm{d}\boldsymbol{x}_h \int_{z(\varrho_1)}^{z(\varrho_2)} \frac{1}{2}\xi^2 N^2 \mathrm{d}z + \text{const}$$

The rephrasing of the conventional potential energy in the above indicated integral sense sorts out the quasi-geostrophic potential energy as a part associated with adiabatic displacements of the fluid elements. The part stored in the background stratification is not touched. Lorenz (1955) has introduced the concept of available potential energy (APE) as the part of total potential energy which can be converted to kinetic energy solely by adiabatic processes. More precisely, APE is the difference between the total potential energy and the minimum potential energy one can obtain from an adiabatic rearrangement of the fluid particles. Defined in this way, APE cannot be associated with a single fluid element or position in the fluid. Nonetheless, in the quasi-geostrophic framework it is the integral of what we have called quasi-geostrophic potential energy which is a well-defined local quantity.

To further elucidate the concept of APE in a simple setting, we consider a box of height $H$ and length $L$ filled by two fluids of different density, a lighter fluid with density $\varrho_1$ on top of a heavier fluid with density $\varrho_2$, the interface being tilted such that the depth of the heavier fluid is $h = H/2 + \alpha x$, with coordinates $(x, z)$ such that $x = 0$ at the center and $z = 0$ at the bottom (see the sketch in Figure 5.1). The total potential energy (per unit width) is

$$E_p = g \int_{-L/2}^{L/2} \int_0^H \varrho z \mathrm{d}z \mathrm{d}x = \frac{1}{8} g H^2 L (\varrho_2 + 3\varrho_1) + \frac{1}{24} g (\varrho_2 - \varrho_1)\alpha^2 L^3 \quad (5.44)$$

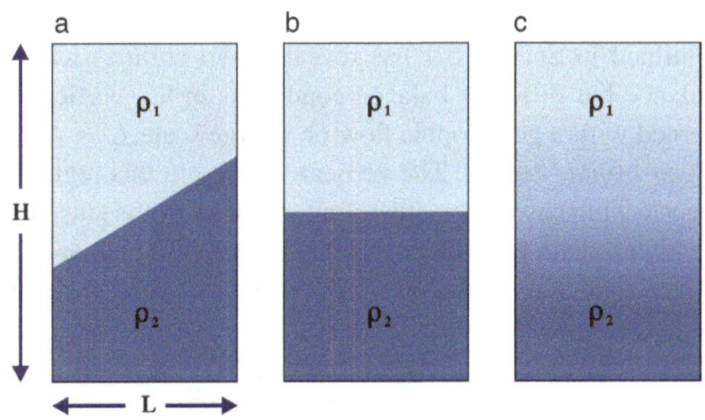


Fig. 5.1 Illustrating the redistribution of water to define APE. The state in **b** is obtained by adiabatic rearrangements, the state in **c** by mixing the state in **a** horizontally

The minimum is apparently achieved for zero displacement of the interface, i. e. $\alpha = 0$. It is easy to show that the displacement $\xi = \alpha(x - L/2)$ of the interface has a mean square value $\overline{\xi^2} = (\alpha L)^2/12$, and thus the potential energy is

$$E_\mathrm{p} = E_\mathrm{p,min} + \frac{1}{2} g(\varrho_2 - \varrho_1) L\overline{\xi^2}$$

Dividing by the mass $(\varrho_1 + \varrho_2)LH/2$ per unit width, the amount of available potential energy per unit mass is found as

$$E_\mathrm{p}^\mathrm{a} = \frac{1}{2} g \frac{\Delta\varrho}{\varrho H} \overline{\xi^2}$$

with $\Delta\varrho = \varrho_2 - \varrho_1$, $\varrho = (\varrho_1 + \varrho_2)/2$. This expression is in close agreement with the definition (5.42).

We have interpreted the available potential energy $E_\mathrm{p}^\mathrm{a}$ as the difference of the actual potential energy $E_\mathrm{p}$ to a state of minimum total potential energy $E_\mathrm{p,min}$ with respect to adiabatic rearrangements of the fluid particles. In contrast, density mixing, by molecular or subgrid scale (turbulent) processes, is a diabatic process. Note that mixing in the depth range of the tilted interface of the above example (instead of flattening the layer interface by adiabatic rearrangements) results in a higher potential energy (the second term in (5.44) is increased by a factor 3). This additional amount of energy must be transferred from some unspecified process (subgrid scale turbulence) into the mean stratification to achieve the mixing.

We finally note in this section that in a quasi-geostrophic flow the available potential energy $E_\mathrm{p}^\mathrm{a}$ is in a relatively fixed proportion to the kinetic energy. Using the scaling $\Omega UL/(gH)$ for density perturbations, we can give an estimate for the ratio of the kinetic and potential energies

$$\frac{E_\mathrm{k}}{E_\mathrm{p}^\mathrm{a}} \sim \frac{N^2(u^2 + v^2)}{(g\rho)^2} \sim \frac{N^2H^2}{\Omega^2 L^2} = \frac{R_i^2}{L^2} = \mathrm{Bu}$$

Available potential energy thus exceeds kinetic energy when the scale of the flow is larger than the internal Rossby radius and vice versa. Since $\mathrm{Bu} = O(1)$ within the quasi-geostrophic approximation, both forms of energy give similar contributions to the total mechanical energy.

## 5.3 Planetary-Scale Geostrophic Motions

The scaling outlined in Section 5.1 has revealed two regimes with a geostrophic momentum balance but different balance conditions in the vorticity equation. We are now concerned with a geostrophic flow on a large scale $L \geq (U/\beta)^{1/2}$, named the *planetary geostrophic regime*. The vertical velocity in this regime is larger than in the quasi-geostrophic case and is, therefore, retained in the substantial derivative operator $\mathrm{D}/\mathrm{D}t$ (here again the full three-dimensional). The equations are obtained from the primitive equations in Section 4.2.5 by ignoring the terms of order Ro in the horizontal momentum equations. They read

$$-fv = -\frac{1}{\rho_0 a \cos\varphi} \frac{\partial p}{\partial \lambda} + \mathcal{F}_u \tag{5.45}$$

$$fu = -\frac{1}{\rho_0 a} \frac{\partial p}{\partial \varphi} + \mathcal{F}_v \tag{5.46}$$

Although the forcing terms are small, they have been retained so that external stresses can act through boundary layers at the top and the bottom. Together with the remaining equations (4.52)–(4.56), this set of (viscous) *planetary-geostrophic equations* forms the basis of some simplified ocean circulation models, with forcing by wind stress and buoyancy flux included.

It is instructive to compute the horizontal divergence $\nabla_h \cdot \boldsymbol{u}_h$ and the vertical component of the relative vorticity $\omega^{(z)}$ of a geostrophic flow of planetary scale. One finds

$$\nabla_h \cdot \boldsymbol{u}_h = \frac{1}{a\cos\varphi}\left(\frac{\partial u}{\partial\lambda} + \frac{\partial v\cos\varphi}{\partial\varphi}\right) = -\frac{\beta}{f}v \tag{5.47}$$

$$\begin{aligned}\omega^{(z)} &= \frac{1}{a\cos\varphi}\left(\frac{\partial v}{\partial\lambda} - \frac{\partial u\cos\varphi}{\partial\varphi}\right) = \frac{\beta}{f}u \\ &+ \frac{1}{a^2 f\cos^2\varphi}\left(\frac{\partial^2 p}{\partial\lambda^2} + \cos\varphi\frac{\partial}{\partial\varphi}\cos\varphi\frac{\partial p}{\partial\varphi}\right)\end{aligned} \tag{5.48}$$

which reveals that $\nabla_h \cdot \boldsymbol{u}_h$ is controlled by variations of the Coriolis parameter whereas $\omega^{(z)}$ is controlled by variations of the pressure field because the pressure terms exceed the $\beta$-term by $a/L$.

From the horizontal divergence equation (5.47), with friction neglected, and the full continuity equation $\nabla_h \cdot \boldsymbol{u}_h + \partial w/\partial z = 0$ we immediately find the planetary vorticity balance in the form

$$f\frac{\partial w}{\partial z} = \beta v \tag{5.49}$$

which is used at various places throughout this book and frequently referred to as the local SVERDRUP[7] balance. As external sources of momentum can only be applied through frictional terms, the frictionless form of the planetary-geostrophic equations can be valid only in the interior of the fluid away from boundaries. A prominent example is the oceanic thermo- or pycnocline where it is customary to combine (4.54) and (4.55) to a density balance

$$\frac{\mathrm{D}\rho}{\mathrm{D}t} = \mathcal{G}_\rho = \frac{\partial}{\partial z}K_\mathrm{v}\frac{\partial\rho}{\partial z} \tag{5.50}$$

where the diabatic term is written here for simplicity as vertical diffusion of density, and compressibility effects are ignored (see Section 4.1.3).

Two further important equations follow from (5.45), (5.46), and (5.50) for the frictionless case. The potential vorticity of the planetary-geostrophic equations is $Q_{pg} = f\,\partial\rho/\partial z$. One finds

$$\frac{\mathrm{D}Q_{pg}}{\mathrm{D}t} = f\frac{\partial\mathcal{G}_\rho}{\partial z} \tag{5.51}$$

A further useful quantity is the appropriate Bernoulli function $B = p + g\rho z$ which is governed by

$$\frac{\mathrm{D}B}{\mathrm{D}t} = \frac{\partial p}{\partial t} + gz\mathcal{G}_\rho \tag{5.52}$$

It is conserved in the stationary and adiabatic case. Note the resemblance of $B/\rho = pv + gz = pv + \Phi$ (in these expressions $v$ is the specific volume) with the Bernoulli function (2.83) in its general form.

[7] HARALD ULRIK SVERDRUP, *1888 in Sogndal, †1957 in Oslo, oceanographer and meteorologist.

### 5.3.1 The M-Representation

As noted by Needler (1967), these equations may be reduced to a single nonlinear differential equation for the pressure. A simpler form was found by Welander (1971), defining the $M$-function

$$M(\lambda, \varphi, z) = \int_{z_0}^{z} p(\lambda, \varphi, z')\mathrm{d}z' + M_0(\lambda, \varphi)$$

which allows to express all fields as partial derivatives of the one variable $M$,

$$p = \frac{\partial M}{\partial z}, \quad \rho = -\frac{1}{g}\frac{\partial^2 M}{\partial z^2}, \quad u = -\frac{1}{fa}\frac{\partial^2 M}{\partial \varphi \partial z} \quad \text{and} \quad v = \frac{1}{fa\cos\varphi}\frac{\partial^2 M}{\partial \lambda \partial z} \tag{5.53}$$

The vertical velocity may as well be expressed by $M$: from (5.49), we find by integration

$$w = \frac{\beta}{f}\int_{z_0}^{z} v\mathrm{d}z' + w_0 = \frac{1}{2\Omega a^2 \sin^2\varphi}\frac{\partial}{\partial \lambda}(M - M_0) + w_0 = \frac{1}{2\Omega a^2 \sin^2\varphi}\frac{\partial M}{\partial \lambda} \tag{5.54}$$

where $w_0(\lambda, \varphi) = w(\lambda, \varphi, z_0)$, and the last relation holds for the choice $M_0 = 2\Omega a^2 \sin^2\varphi \int_0^\lambda w_0 \mathrm{d}\lambda'$. Obviously, $M$ is the state variable of the system, in a similar way as $\psi$ for the quasi-geostrophic case. With the above density equation, one obtains a single nonlinear partial differential equation of second degree and fourth order for $M$,

$$\begin{aligned} & a^2\Omega \sin 2\varphi \frac{\partial^3 M}{\partial z^2 \partial t} + \frac{\partial(\partial M/\partial z, \partial^2 M/\partial z^2)}{\partial(\lambda, \varphi)} + \cot\varphi \frac{\partial M}{\partial \lambda}\frac{\partial^3 M}{\partial z^3} \\ & = a^2\Omega \sin 2\varphi \frac{\partial}{\partial z}\left(K_\mathrm{v}\frac{\partial^3 M}{\partial z^3}\right) \end{aligned} \tag{5.55}$$

This so-called $M$-equation has been a starting point for finding certain analytic solutions describing the oceanic pycnocline, which will be discussed below in Section 14.3.2.

### 5.3.2 Thermal Wind-Equations

A consequence of the frictionless geostrophic momentum equations (with the hydrostatic equation (4.52)) are the *thermal-wind equations*

$$-f\frac{\partial v}{\partial z} = g\frac{\partial \rho}{\partial x} \quad \text{and} \quad f\frac{\partial u}{\partial z} = g\frac{\partial \rho}{\partial y} \tag{5.56}$$

using for simplicity now Cartesian coordinates. The name originates from meteorology where density depends mostly on temperature. The thermal-wind equations relate the vertical shear of the horizontal current to the lateral density gradients. They are the basis of the *geostrophic method* (or dynamical method) which allows to determine the horizontal geostrophic current relative to a reference level $z_0$ by the observed gradients of temperature and salinity: the hydrographic data $\theta(y,z)$, $S(y,z)$ along a section with coordinate $y$ yields the relative velocity $u(y,z)$ normal to the section, involving, however, the reference velocity $u_0 = u(y,z_0)$. Finding $u_0$ has for long been a classical problem of oceanography.

### 5.3.3 Planetary Ideal Fluid Equations

The *ideal fluid equations* for motions of planetary scale are the adiabatic form of (4.24)–(4.26), given there for a Boussinesq fluid. In steady state, they have some remarkable properties which we will discuss in the following. Let us consider these equations with usual approximations of the planetary-scale motion,

$$f\boldsymbol{k}\times\boldsymbol{u} = -\nabla p - g\rho\boldsymbol{k} \tag{5.57}$$

$$\boldsymbol{u}\cdot\nabla\rho = 0 \tag{5.58}$$

$$\nabla\cdot\boldsymbol{u} = 0 \tag{5.59}$$

where $\boldsymbol{u} = (u,v,w)$ is the three-dimensional velocity and $\boldsymbol{k}$ the vertical unit vector. As stated above, the equation imply the validity of the planetary vorticity balance (5.49). Furthermore, in addition to the density, two more conserved quantities exist: the potential vorticity $Q = f\partial\rho/\partial z$ and the Bernoulli function $B = p + g\rho z$ (see also previous section). Hence

$$\boldsymbol{u}\cdot\nabla Q = 0 \quad \text{and} \quad \boldsymbol{u}\cdot\nabla B = 0 \tag{5.60}$$

The three-dimensional velocity vector $\boldsymbol{u}$ must lie in all three surfaces $\rho =$ const, $Q =$ const and $B =$ const. Therefore, the iso-surfaces of density, potential vorticity, and Bernoulli function must intersect in lines which are streamlines. Hence there is a functional relation $\Phi(\rho,Q,B) = 0$ for each ideal fluid system. The function $\Phi$ is usually determined by conditions on the boundaries. It can be quite complicated and solutions for one of the variables, e. g. $Q = F(\rho,B)$, might be multivalued and not continuous. Solving $\Phi(\rho,Q,B) = 0$ for $B = G(\rho,Q)$ a second order differential equation for density is achieved,

$$\left(gz - \frac{\partial G}{\partial\rho}\right)\frac{\partial\rho}{\partial z} - \frac{\partial G}{\partial Q}f\frac{\partial^2\rho}{\partial z^2} = 0 \tag{5.61}$$

provided that the derivatives of $G$ exist. Solutions for the case of the oceanic pycnocline will be discussed in Section 14.3.3. Here, our concern is to investigate the potential of the ideal fluid approach to represent the velocity in terms of the density field.

As a first step, we use $B = p + g\rho z = G(\rho,Q)$ to eliminate the pressure from the geostrophic balances in (5.57). This determines $u$ and $v$ as functional of the density field. Computing the vertical velocity from the density equation (5.58) yields

$$u = -\frac{1}{f}\left[\left(\frac{\partial G}{\partial\rho} - gz\right)\frac{\partial\rho}{\partial y} + \frac{\partial G}{\partial Q}\frac{\partial Q}{\partial y}\right], \quad v = \frac{1}{f}\left[\left(\frac{\partial G}{\partial\rho} - gz\right)\frac{\partial\rho}{\partial x} + \frac{\partial G}{\partial Q}\frac{\partial Q}{\partial x}\right]$$

$$w = \frac{\partial G/\partial Q}{Q}\left(\frac{\partial\rho}{\partial x}\frac{\partial Q}{\partial y} - \frac{\partial\rho}{\partial y}\frac{\partial Q}{\partial x}\right) \tag{5.62}$$

Note that the continuity equation (5.59) is satisfied by these relations. Combining (5.62) with (5.61), the first of the NEEDLER's[8] formulae (Needler, 1985) is obtained,

$$\boldsymbol{u} = \alpha \nabla \rho \times \nabla Q \quad \text{with} \quad \alpha = (\partial G / \partial Q)/Q \tag{5.63}$$

It involves derivatives of the density field up to the second order and suffers, of course, from the dependence of $\boldsymbol{u}$ on the unknown function $G$. Note that the factor $\alpha$ is a conserved quantity as well, $\boldsymbol{u} \cdot \nabla \alpha = 0$.

Needler's second formula avoids to fall back on the unknown $G$. It may be obtained in two steps. Differentiate the conservation equation of the potential vorticity with respect to $z$ to find

$$\boldsymbol{u} \cdot \nabla f \frac{\partial Q}{\partial z} = g \left( \frac{\partial \rho}{\partial x} \frac{\partial Q}{\partial y} - \frac{\partial \rho}{\partial y} \frac{\partial Q}{\partial x} \right) = g \boldsymbol{k} \cdot (\nabla \rho \times \nabla Q) \tag{5.64}$$

Here, the thermal-wind equations (5.56) and the planetary vorticity equation (5.49) have been used. Multiplication of (5.63) with $\nabla f \partial Q / \partial z$ and insertion of (5.64) then yields

$$\alpha = g \frac{\boldsymbol{k} \cdot (\nabla \rho \times \nabla Q)}{\nabla f \partial Q / \partial z \cdot (\nabla \rho \times \nabla Q)} \tag{5.65}$$

which involves now up to third-order derivatives of the density field. With this representation, Needler (1985) was the first who formally solved the classical problem of oceanography, the determination of the absolute velocity from the density field. The relation (5.65) is exact for ideal fluids, but unfortunately it is practically useless for the following reasons:

1. The derivatives of the density field are of high order, and severe noise problems may occur if the density field is taken from observations.
2. $\boldsymbol{u} = \alpha \nabla \rho \times \nabla Q$ from (5.65) does only fulfill all equations (especially continuity) if the density satisfies the equation $\boldsymbol{u} \cdot \nabla \rho = 0$. This is generally not the case if $\rho$ is taken from data.
3. The expression degenerates if $\nabla \rho \times \nabla Q = 0$, i. e. when the potential vorticity is constant on isopycnals. This state is actually found in many ocean areas. It is a consequence of isopycnal mixing which is ignored in the ideal fluid approach.

Further representations (in fact infinitely many) with increasingly higher derivations are possible (see Olbers et al., 1985). Note also the derivation by McDougall (1995) of a related representation which includes diffusion.

Eliminating $(\partial G / \partial Q)/Q$ between the horizontal components of (5.63) or the corresponding factor in (5.65), the $\beta$-spiral equation

$$u \left( \frac{\partial \rho}{\partial z} \frac{\partial Q}{\partial x} - \frac{\partial \rho}{\partial x} \frac{\partial Q}{\partial z} \right) + v \left( \frac{\partial \rho}{\partial z} \frac{\partial Q}{\partial y} - \frac{\partial \rho}{\partial y} \frac{\partial Q}{\partial z} \right) = 0 \tag{5.66}$$

is derived for an ideal fluid. It is usual rewritten as

$$u \frac{\partial}{\partial z} \left( \frac{\partial \rho / \partial x}{\partial \rho / \partial z} \right) + v \left[ \frac{\partial}{\partial z} \left( \frac{\partial \rho / \partial y}{\partial \rho / \partial z} \right) + \frac{\beta}{f} \right] = 0 \tag{5.67}$$

[8] GEORGE TREGLOHAN NEEDLER, *1935 in Summerside, Prince Edward Island, †2002 Dartmouth, Nova Scotia, oceanographer.

in terms of the density gradients alone. The coefficients of $u$ and $v$ in this equation can be determined from the density field at the cross point of two hydrographic sections; hence (5.67) has been involved in many applications to estimate the absolute horizontal velocity from hydrographic data (Olbers et al., 1985; Stommel and Schott, 1977). Note that only the slope of the isopycnals appears in (5.67). The slope and the velocity turns with depth on account of a nonzero $\beta$ (therefore, the name '$\beta$-spiral'), as immediately can be inferred from

$$\frac{\partial}{\partial z}\frac{u}{v} = \frac{v\partial u/\partial z - u\partial v/\partial z}{v^2} = \frac{g}{fv^2}\left(u\frac{\partial \rho}{\partial x} + v\frac{\partial \rho}{\partial y}\right) = -\frac{g\partial \rho/\partial z}{fv^2}w \tag{5.68}$$

and $w \sim \beta$ from (5.49). If friction and diabatic processes are included, these terms also contribute to the spiraling of the current.

## Further Reading

The textbooks by Pedlosky (1987) and Gill (1982) have defined the standard for a description of the dynamics of the large-scale ocean circulation. *Geophysical Fluid Dynamics* by Pedlosky provides a detailed derivation of the relevant equations for both ocean and atmosphere, with great mathematical rigor. *Atmosphere-Ocean Dynamics* by Gill contains an excellent description of the physics of large-scale circulation, with a particular focus on geostrophic adjustment and equatorial dynamics.

Shorter and easily accessible is the *Introduction to Geophysical Fluid Dynamics* by Cushman-Roisin (1994).

To be recommended is also *Atmospheric and Oceanic Fluid Dynamics* by Vallis (2006) which treats the large-scale atmospheric and oceanic dynamics in a profound and uniform way.

The equations of geophysical motions are derived and discussed in many text books. A recent, very concise view of the various approximations is found in White (2002), however, in an atmospheric context.

# Part III
# Ocean Waves

Ocean waves can be considered as the most basic form of oceanic motions. They cover temporal scales from fractions of a second up to many years, and spatial scales from millimeters up to thousands of kilometers. Ocean waves can be grouped into three types which have very different physics. Sound waves cause the adjustment of the pressure field to velocity divergences, with compression as the main restoring force, and propagate at speeds around 1,500 m $s^{-1}$. Inertial-gravity waves bring about the adjustment towards geostrophic momentum balance, with gravity as the main restoring force. They propagate at speeds around 1 m $s^{-1}$ (baroclinic) and up to 200 m $s^{-1}$ (barotropic). Rossby waves mediate changes in ocean circulation due to changes of the forcing, and propagate at speeds of a few cm/s. The spatial and temporal scales of free unbounded ocean waves are displayed in the figure on page 106. Note that wave types that depend on specific aspects of geometry (e. g. coastlines, equator) or topography (e. g. seamounts, continental shelf) are not included in the figure.

In the following chapters an overview over the various wave motions is given, and the physical processes governing their motion are discussed. The focus is on the characteristics of free waves which – once excited – can propagate over large regions and for a long time. Finally, in Chapter 10 we outline a conceptual framework for how to treat forced wave problems with a few applications.

The theoretical description of ocean waves starts with the general equations that were derived earlier, e. g. in (2.151)–(2.155). It is however convenient to replace the continuity equation (2.151) with the equivalent equation (3.1) for the pressure, and use (3.2). One obtains the system

$$\frac{\partial p}{\partial t} + \boldsymbol{u} \cdot \nabla p = -c_s^2 \left( \rho \nabla \cdot \boldsymbol{u} + \gamma \mathcal{G}_S - \alpha \mathcal{G}_\theta \right) \tag{5.69}$$

$$\frac{\partial \boldsymbol{u}}{\partial t} + \boldsymbol{u} \cdot \nabla u = -\frac{1}{\rho} \nabla p - \nabla \Phi - 2\boldsymbol{\Omega} \times \boldsymbol{u} + \mathcal{F} \tag{5.70}$$

$$\frac{\partial \theta}{\partial t} + \boldsymbol{u} \cdot \nabla \theta = \frac{1}{\rho} \mathcal{G}_\theta \tag{5.71}$$

$$\frac{\partial S}{\partial t} + \boldsymbol{u} \cdot \nabla S = \frac{1}{\rho} \mathcal{G}_S \tag{5.72}$$

$$\rho = F(S, \theta, p) \tag{5.73}$$

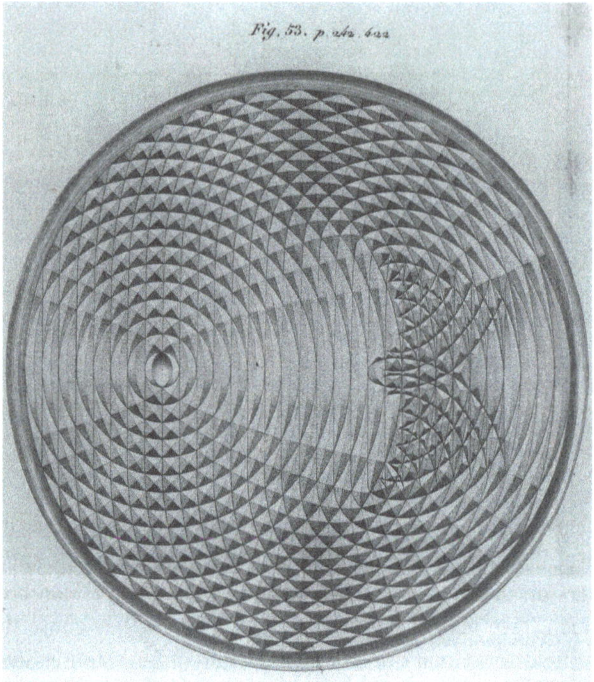

Wave generation, reflection and interference, exemplified by an illustration by the Weber brothers in their book of 1825 (Weber and Weber, 1825, page 583). Mercury was arranged dropping through a paper funnel on the surface of a circular disk of mercury. By controlling the rate of the falling drops, the Webers produced this picture by hand drawing.

where $\gamma$ and $\alpha$ are the usual haline contraction and thermal expansion coefficients (see Section 2.6.4). To a very good approximation, these equations are valid for any form of oceanic motions, including all waves. The turbulent fluxes contained in the divergence terms $\mathcal{F}, \mathcal{G}_\theta$ and $\mathcal{G}_S$ are important for the generation and dissipation of waves; they do not play any significant role for the balance and propagation of free waves. We will, therefore, assume in most of this part that $\mathcal{F}, \mathcal{G}_S, \mathcal{G}_\theta = 0$, and only include the fluxes where necessary (e. g. when considering forced waves in Section 10.1).

# 6 Sound Waves

Sound waves play an important role for the transmission of information in the ocean. This property is used in various measurement methods, such as, e. g. echo-sounding, acoustic floats, ADCP's, and acoustic tomography. As will become clear below, the mathematical description of sound waves is relatively simple in comparison with other types of waves. Therefore, some of the concepts which are useful for describing *all* wave processes, such as, e. g. plane wave propagation, dispersion and group velocity, propagation in waveguides and in inhomogeneous media, can be introduced most easily for sound waves.

Sound waves spread over a wide range of time-scales, about $10^{-6}$–$10^{-1}$ s, i. e. over many decades. Compared to the period of the Earth's rotation these time-scales are all very short, so that the Earth's rotation in (5.70) can be neglected to a very good approximation. Likewise, the gravity force which fundamentally influences oceanic motions is unimportant for sound waves.

## 6.1 Approximations and Perturbation Expansion

There are no general methods of solution for the equations (5.69)–(5.73). As for all wave processes, the most useful approximation results from a perturbation expansion, where small oscillations around an arbitrarily chosen and often idealized background state are considered. The aim is to obtain a set of *linear* differential equations, if possible even with constant coefficients because such equations can be solved easily. The solutions of the linearized equations, of course, deviate from the exact solutions, but they usually form a good starting point for learning about the system's physics.

The system (5.69)–(5.73) has several nonlinearities. The advective nonlinearity, such as the term $\boldsymbol{u} \cdot \nabla\theta$ in the temperature equation, is by far the most important form of nonlinearity. Products of state variables involving density, such as $\rho\nabla \cdot \boldsymbol{u}$ in the pressure equation, are usually less important because $\rho$ only varies within 2–3% in the ocean. The nonlinearity of the equation of state $\rho = F(S, \theta, p)$ is important for some aspects of the circulation but usually less so for waves. The related nonlinearity in the sound velocity $c_s = c_s(S, \theta, p)$ is, however, important for the propagation of

D. Olbers, J. Willebrand, C. Eden, *Ocean Dynamics*,
DOI 10.1007/978-3-642-23450-7_6, © Springer-Verlag Berlin Heidelberg 2012

sound waves in inhomogeneous environments, especially through the dependence on temperature and pressure.

To derive the simplest case of sound waves, we choose a rather idealized background state defined by $\rho_0, p_0, \theta_0, S_0 = \text{const}$ and $\boldsymbol{u}_0 = 0$. In analogy to the previously discussed perturbation approach (see Section 5.2.1) we expand all variables as

$$\rho = \rho_0 + \epsilon\rho_1 + \dots\,, \qquad p = p_0 + \epsilon p_1 + \dots\,, \qquad \boldsymbol{u} = 0 + \epsilon\boldsymbol{u}_1 + \dots \tag{6.1}$$

The background state (with index 0) automatically satisfies all equations of motion (5.69)–(5.72) exactly, provided that $\rho_0 = F(S_0, \theta_0, p_0)$. After insertion of the above expansion, we obtain from (5.69)–(5.73)

$$\epsilon\frac{\partial p_1}{\partial t} + \epsilon\boldsymbol{u}_1\cdot\nabla(p_0+\cdots)+\cdots = -(\rho_0+\cdots)c_s^2(S,\theta,p)\nabla\cdot\epsilon\boldsymbol{u}_1 \tag{6.2}$$

$$(\rho_0+\epsilon\rho_1+\cdots)\left[\epsilon\frac{\partial\boldsymbol{u}_1}{\partial t}+\epsilon\boldsymbol{u}_1\cdot\nabla(\epsilon\boldsymbol{u}_1)\right]+\cdots = -\nabla(p_0+\epsilon p_1+\cdots) \tag{6.3}$$

$$\epsilon\frac{\partial S_1}{\partial t} + \epsilon\boldsymbol{u}_1\cdot\nabla(S_0+\cdots)+\cdots = 0 \tag{6.4}$$

$$\epsilon\frac{\partial \theta_1}{\partial t} + \epsilon\boldsymbol{u}_1\cdot\nabla(\theta_0+\cdots)+\cdots = 0 \tag{6.5}$$

$$\rho_0+\epsilon\rho_1+\cdots = F(S,\theta,p) \tag{6.6}$$

The perturbation parameter is small, i. e. $\epsilon \ll 1$. The equation of state and the sound velocity are expanded in a Taylor series which yields

$$F(S,\theta,p) = F(S_0,\theta_0,p_0) + \epsilon\frac{\partial F}{\partial p}p_1 + \dots \quad \text{and}$$
$$c_s(S,\theta,p) = c_s(S_0,\theta_0,p_0) + \dots \tag{6.7}$$

The terms $\sim S_1$ and $\sim \theta_1$ are irrelevant as they become constant, as shown below. Equation (6.6) then results in

$$\rho_0 = F(S_0,\theta_0,p_0) \tag{6.8}$$

and we shall use the abbreviation $c_{s,0} = c_s(S_0,\theta_0,p_0)$, which is a constant. To the lowest order ($\epsilon^0$) the equations (6.2)–(6.5) are identically satisfied. The equations of first order then become

$$\frac{\partial\epsilon p_1}{\partial t} = -\rho_0 c_{s,0}^2\nabla\cdot\epsilon\boldsymbol{u}_1 \tag{6.9}$$

$$\frac{\partial\epsilon\boldsymbol{u}_1}{\partial t} = -\frac{1}{\rho_0}\nabla\epsilon p_1 \tag{6.10}$$

$$\frac{\partial\epsilon S_1}{\partial t} = 0 \tag{6.11}$$

$$\frac{\partial\epsilon\theta_1}{\partial t} = 0 \tag{6.12}$$

$$\epsilon\rho_1 = \frac{1}{c_{s,0}^2}\epsilon p_1 \tag{6.13}$$

These are the equations for linear sound waves in a homogeneous medium ($c_{s,0}$ is constant). The equations are linear with constant coefficients and can, therefore, be solved quite easily. The perturbation density $\rho_1$ does not appear in the first two equations and can be determined *diagnostically* from $p_1$. Furthermore, $S_1, \theta_1 = \text{const}$ is found. Therefore, it suffices to consider (6.9) and (6.10).

To simplify the notation, from now $\epsilon$ is incorporated in the first order fields and the index 1 omitted so that $\epsilon p_1 \to p$ etc. , and the index 0 of $c_{s,0}$ is dropped as well.

## 6.2 Plane Waves

Forming the time derivative $\partial/\partial t$ of equation (6.9) and the divergence $\nabla\cdot$ of (6.10), and the elimination of $\boldsymbol{u}$ leads to the *wave equation*

$$\frac{\partial^2 p}{\partial t^2} - c_s^2 \nabla^2 p = 0 \quad \text{respectively} \quad \frac{\partial^2 p}{\partial t^2} - c_s^2 \left( \frac{\partial^2 p}{\partial x^2} + \frac{\partial^2 p}{\partial y^2} + \frac{\partial^2 p}{\partial z^2} \right) = 0 \tag{6.14}$$

This linear partial differential equation of second order in space and time is one of the best-studied equations in physics. It occurs in many different fields and has a multitude of solutions. For one-dimensional situations, the solutions must be of the form

$$p(x,t) = F(x - c_s t) + G(x + c_s t) \tag{6.15}$$

with arbitrary functions $F$ and $G$. This solution can be found by transforming to two new independent variables according to

$$\xi = x - c_s t \ , \quad \frac{\partial}{\partial t} = -c_s \frac{\partial}{\partial \xi} + c_s \frac{\partial}{\partial \eta}$$

$$\eta = x + c_s t \ , \quad \frac{\partial}{\partial x} = \frac{\partial}{\partial \xi} + \frac{\partial}{\partial \eta}$$

with derivatives as indicated. Insertion into (6.14) results, after a brief calculation, in $\partial^2 p(\xi,\eta)/\partial\xi\partial\eta = 0$ with the integrals $\partial p(\xi,\eta)/\partial\eta = g(\eta)$ and hence $p(\xi,\eta) = G(\eta) + F(\xi)$. In particular, for $G \equiv 0$, we obtain the solution

$$p(x,t) = F(x - c_s t) \tag{6.16}$$

which corresponds to a signal propagating in positive $x$-direction with the speed $c_s$ while preserving its shape (see the sketch in Figure 6.1). For $F(\xi) = \cos(k\xi)$ the normal harmonic wave results. Analogously, $F \equiv 0$ in (6.15) corresponds to a wave proceeding in *negative* $x$-direction.

In three dimensions there is no general solution of (6.14). As can be confirmed by insertion, however, a solution analogous to (6.15) is given by

$$p(x,y,z,t) = F(\boldsymbol{e} \cdot \boldsymbol{x} - c_s t) \tag{6.17}$$

with an arbitrary unit vector $\boldsymbol{e}$, and the coordinate vector $\boldsymbol{x} = (x, y, z)$. For a harmonic wave, $F(\eta) \sim \cos(\eta)$, we thus obtain for the pressure

$$p(x,y,z,t) = P\cos(\boldsymbol{k} \cdot \boldsymbol{x} - \omega t) = P\cos(k_1 x + k_2 y + k_3 z - \omega t) \tag{6.18}$$

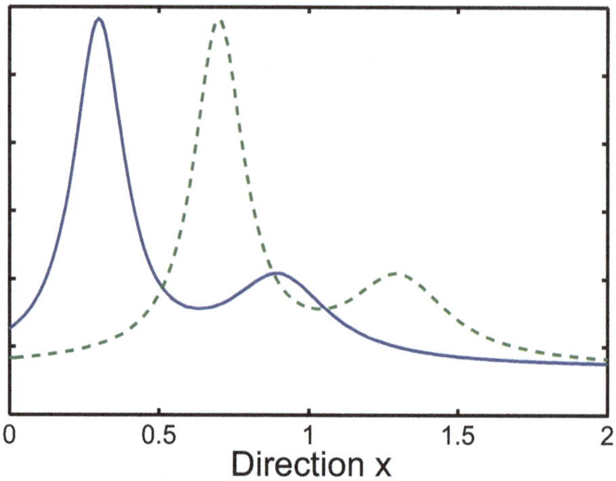


Fig. 6.1 Preservation of form during the propagation of plane sound waves in positive $x$-direction. Full line: $t = 0$, dashed line: $t = 1$

Here, the wave number vector (henceforth wave vector) $\boldsymbol{k} = (k_1, k_2, k_3) = k\boldsymbol{e}$ and the angular frequency $\omega = kc_s$ have been introduced as well as the pressure amplitude $P$. Inserting (6.18) in the wave equation (6.14) yields the *dispersion relation*

$$\omega^2 = c_s^2 k^2 \Rightarrow \omega = \pm c_s |\boldsymbol{k}| = \pm c_s \left(k_1^2 + k_2^2 + k_3^2\right)^{\frac{1}{2}} \tag{6.19}$$

which associates a frequency to each wave number, or equivalently the wave period $T = 2\pi/\omega$ with the wavelength $\lambda = 2\pi/|\boldsymbol{k}|$ according to $T = \lambda/c_s$. For a typical sound speed of 1,500 m s$^{-1}$, a wavelength $\lambda = 1$ m corresponds to a period $T = 0.7 \times 10^{-3}$ s.

The existence of a dispersion relation is a characteristic property of all wave processes. For a given $\boldsymbol{k}$, the two roots in (6.19) are physically not different since the properties of the solution (6.18) do not change when $\omega \rightarrow -\omega$ and $\boldsymbol{k} \rightarrow -\boldsymbol{k}$. Therefore, we restrict ourselves to the case $\omega > 0$ and write

$$\omega = \Omega(\boldsymbol{k}) \tag{6.20}$$

with a positive function $\Omega$ as the general form of a dispersion relation. Note, however, that the wave vector $\boldsymbol{k}$ can take all possible values.

Equation (6.18) represents a *plane wave* in space and time (the situation is displayed in Figure 6.2). It describes propagation in the positive $\boldsymbol{k}$-direction, with the constant *phase speed* $c = \omega/|\boldsymbol{k}| = c_s$. As the phase speed is the same for all wave numbers, the waves are called *nondispersive*; this very property is the reason why the wave form is always preserved.

Note that the phase velocity is always a scalar. The propagation velocity in the direction of $\boldsymbol{k}$ is $c = \omega/k$, with $k = |\boldsymbol{k}|$. The propagation velocity, e. g. along the $x_1$-axis, however, is $ck/k_1$ and not $ck_1/k$. Therefore, $\boldsymbol{c} = c\boldsymbol{k}/k$ would not serve as a meaningful vector[1].

With (6.10), the solution for $\boldsymbol{u}$ corresponding to (6.18) is found to be

$$\boldsymbol{u} = \boldsymbol{U} \cos(\boldsymbol{k} \cdot \boldsymbol{x} - \omega t) \tag{6.21}$$

[1] The *slowness* $\boldsymbol{s} = \boldsymbol{k}/\omega$ is a sensibly defined vector but will not be used here.

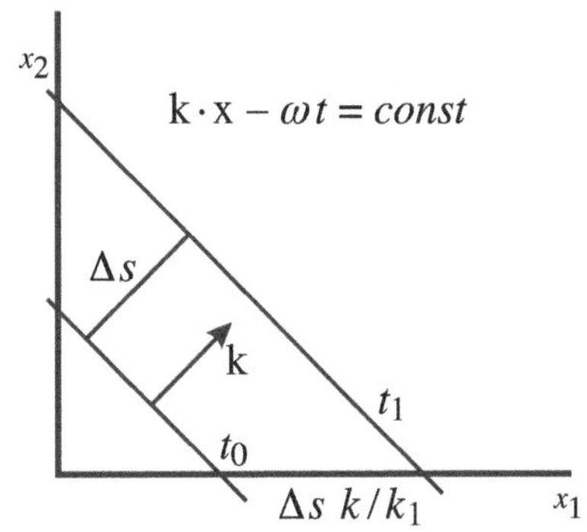


Fig. 6.2 Propagation of a wave front and direction of the phase velocity

where $\boldsymbol{U} = P\boldsymbol{k}/\rho_0\omega$. As $\boldsymbol{U}$ is parallel to $\boldsymbol{k}$, the velocity and thus the particle displacement are always in the direction of the phase propagation. Therefore, sound waves are always *longitudinal* waves. The particle displacement is governed by

$$\frac{\mathrm{d}\boldsymbol{x}(t)}{\mathrm{d}t} = \boldsymbol{u}(\boldsymbol{x},t) = \boldsymbol{U}\cos(\boldsymbol{k}\cdot\boldsymbol{x} - \omega t)\,, \quad \boldsymbol{x}(t_0) = \boldsymbol{x}_0 \tag{6.22}$$

For small excursions around the initial point $\boldsymbol{x}_0$, i. e. $k(x - x_0) \ll 1$, the cosine can be developed in a Taylor expansion $\mathrm{d}\boldsymbol{x}(t)/\mathrm{d}t = \boldsymbol{U}\cos(\boldsymbol{k}\cdot\boldsymbol{x}_0 - \omega t) + \ldots$ with the solution

$$\boldsymbol{x}(t) = \boldsymbol{x}_0 - \frac{\boldsymbol{U}}{\omega}\sin(\boldsymbol{k}\cdot\boldsymbol{x}_0 - \omega t) \tag{6.23}$$

Note that $k(x - x_0) \ll 1$ is equivalent to the condition $U/\omega \ll \lambda$ or $kU \ll \omega$ for the velocity amplitude, or likewise $U/c_\mathrm{s} \ll 1$, i. e. the particle velocity is much smaller than the phase speed.

Further remarks (without proof):

1. The ratio $U/c_\mathrm{s} = kU/\omega \ll 1$ corresponds to $|\boldsymbol{u}\cdot\nabla\boldsymbol{u}|/|\boldsymbol{u}_t| \ll 1$ in the momentum balance (6.3) and can, therefore, be considered as a dimensionless measure for the amplitude, respectively the expansion parameter $\epsilon$ used in Section 6.1.
2. Gravity can be neglected as long as the wavelength $\lambda$ satisfies the inequality $\lambda \ll c_\mathrm{s}^2/g \approx 150$–$200$ km.
3. The Earth's rotation can be neglected as long as the wavelength $\lambda$ satisfies the inequality $\lambda \ll c_\mathrm{s}/\Omega \approx 15{,}000$ km (i. e. always).

### 6.2.1 Group Velocity I: Interference of Waves

The sum of two or more solutions of a linear homogeneous system is itself a solution. Specifically, the superposition of two waves of identical amplitude $P$ but with different wave numbers results in the solution $p = P\cos(\boldsymbol{k}\cdot\boldsymbol{x} - \omega t) + P\cos(\boldsymbol{k}'\cdot\boldsymbol{x} - \omega' t)$ which by trigonometric identities is equivalent to

$$p = 2P\cos\left(\frac{\boldsymbol{k}' - \boldsymbol{k}}{2}\cdot\boldsymbol{x} - \frac{\omega' - \omega}{2}t\right)\cos\left(\frac{\boldsymbol{k}' + \boldsymbol{k}}{2}\cdot\boldsymbol{x} - \frac{\omega' + \omega}{2}t\right)$$

Note that the dispersion relation must be satisfied for both waves, i. e. $\omega = \Omega(\boldsymbol{k})$ and $\omega' = \Omega(\boldsymbol{k}')$. If both wave numbers are not very different, i. e. if $\boldsymbol{k}' = \boldsymbol{k} + \Delta\boldsymbol{k}$ with small $\Delta\boldsymbol{k}$, then a Taylor expansion yields

$$\omega' = \Omega(\boldsymbol{k}') = \Omega(\boldsymbol{k} + \Delta\boldsymbol{k}) = \Omega(\boldsymbol{k}) + \sum_{i=1}^{3} \frac{\partial \Omega}{\partial k_i} \cdot \Delta k_i + \cdots = \omega + \boldsymbol{c}_{\mathrm{g}} \cdot \Delta\boldsymbol{k} + \cdots$$

with the vector

$$\boldsymbol{c}_{\mathrm{g}} = \left( \frac{\partial \Omega}{\partial k_1}, \frac{\partial \Omega}{\partial k_2}, \frac{\partial \Omega}{\partial k_3} \right) = \frac{\partial \Omega}{\partial \boldsymbol{k}} \tag{6.24}$$

The solution, therefore, has the approximate form, with $\boldsymbol{k}_{\mathrm{g}} = \Delta\boldsymbol{k}/2$,

$$p = 2P \cos\left[\boldsymbol{k}_{\mathrm{g}} \cdot \left(\boldsymbol{x} - \boldsymbol{c}_{\mathrm{g}} t\right)\right] \cos\left(\boldsymbol{k} \cdot \boldsymbol{x} - \omega t\right)$$

and consists of a product of two plane waves with rather different wavelengths (as depicted in Figure 6.3). The short wave (wavelength $2\pi/|\boldsymbol{k}|$) has the usual propagation speed $\omega/|\boldsymbol{k}| = c_{\mathrm{s}}$. The long wave (wavelength $2\pi/|\boldsymbol{k}_{\mathrm{g}}|$) describes the envelope (that is the shape of the wave group) and propagates with the velocity $\boldsymbol{c}_{\mathrm{g}}$ which, therefore, is called *group velocity*. From (6.20) it follows

$$\boldsymbol{c}_{\mathrm{g}} = c_{\mathrm{s}} \frac{\boldsymbol{k}}{k} \tag{6.25}$$

for sound waves. In contrast to the scalar (and positive) phase speed, the group velocity $\boldsymbol{c}_{\mathrm{g}}$ is always a vector. For sound waves it has the magnitude of the phase speed and the direction of the wave-number vector. Note that the latter is always the case when the dispersion relation only depends on the magnitude $|\boldsymbol{k}|$ but not on the direction of $\boldsymbol{k}$. Other important waves in the ocean do not share this property.

The above description refers to free waves far from boundaries. The vertical boundary of the ocean is generally not far away, but it often only plays a minor role (due to the sound channel, see below). It is not in principle difficult to take the boundaries into account, at least in case of simple geometry, as long as the boundary conditions are known (e. g. zero normal velocity; at the seafloor, however, one has to take into account that sound can also propagate into the solid earth). We will not deal with this issue further.

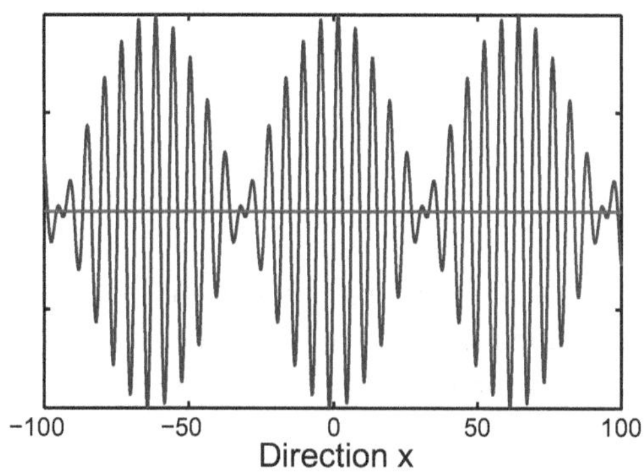


Fig. 6.3 Superposition of 2 waves with similar wave numbers

### 6.2.2 Energy Conservation I: Kinetic and Elastic Energy

The conservation of energy can be directly derived from the linearized form of the sound wave equations. Multiplying (6.9) with $p/\rho_0 c_s^2$, and (6.10) with $\rho_0 \boldsymbol{u}$ (note that $p_1 \to p$, etc. ), addition of the resulting equations yields

$$\rho_0 \frac{\partial}{\partial t}\left(\frac{1}{2}u^2 + \frac{1}{2}\frac{p^2}{\rho_0^2 c_s^2}\right) = -p\nabla\cdot\boldsymbol{u} - \boldsymbol{u}\cdot\nabla p \equiv -\nabla\cdot\boldsymbol{u}p \tag{6.26}$$

The first term on the left-hand side of (6.26) corresponds to the change of kinetic energy $E_k = \boldsymbol{u}^2/2 \equiv u^2/2$ (energy per mass). In the linear approximation, $\rho_0 E_k$ is, therefore, the kinetic energy per volume.

According to the first law of thermodynamics (1.22), under adiabatic conditions a small change $\mathrm{d}\upsilon$ in specific volume $\upsilon = 1/\rho$ at pressure $p$ corresponds to work done on the volume and causes a change $\mathrm{d}E$ in internal energy $E$ (per mass) of

$$\mathrm{d}E = -p\mathrm{d}\upsilon = -p\mathrm{d}\frac{1}{\rho} = +\frac{p}{\rho^2}\mathrm{d}\rho = \frac{p}{\rho^2 c_s^2}\mathrm{d}p \approx \mathrm{d}\left(\frac{1}{2}\frac{p^2}{\rho_0^2 c_s^2}\right) \tag{6.27}$$

Here we have used (2.128), and for the last relation assumed that $\rho \approx \rho_0$ and $c_s \approx$ const. It is, therefore, plausible that the change of

$$E_e = \frac{1}{2}\frac{p^2}{\rho_0^2 c_s^2} \tag{6.28}$$

is the contribution of the (reversible) volume work to the internal energy in the linear approximation. Note that the pressure in (6.27) and (6.28) has to be replaced by $p_0 + p$ according to (6.1), however since $p_0^2 =$ const and $p_0 p$ vanishes on average, (6.28) remains valid as well for the perturbation pressure $p$. The quantity $E_e$ is called *elastic* energy. From (6.26) we obtain the energy conservation for sound waves in the form

$$\frac{\partial}{\partial t}\rho_0 (E_k + E_e) = -\nabla\cdot p\boldsymbol{u} \tag{6.29}$$

with the (mechanical) energy flux $p\boldsymbol{u}$ (in W m$^{-2}$) as discussed in (2.73). The sum of elastic and kinetic energy is conserved, except for the divergence of the energy flux term which vanishes in the integral over a domain when there is no flux of energy into the volume. With the wave solution (6.18), and using $\overline{\cos^2(\ldots)} = 1/2$ we obtain as average over a wavelength/period

$$\overline{E_e} = \frac{1}{2}\frac{1}{\rho_0^2 c_s^2}\overline{P^2\cos^2(\ldots)} = \frac{1}{4}\frac{1}{\rho_0^2 c_s^2}P^2 = \overline{E_k} \tag{6.30}$$

and thus an equal distribution between both energy forms. As mean total energy *per volume* we can define

$$E_w = \rho_0\left(\overline{E_e} + \overline{E_k}\right) = \frac{1}{2}\frac{P^2}{\rho_0 c_s^2} \tag{6.31}$$

To give an idea of the orders of magnitude involved: a pressure amplitude $P = 10^{-6}$ Pa corresponds to an energy $E_w \sim 10^{-18}$ J/kg, and an rms particle velocity $U \sim 10^{-9}$ m s$^{-1}$. An energy flux of $J_0 = 10^{-12}$ W m$^{-2}$ is the smallest intensity that the human ear can perceive (at 1000 Hz in the air). Intensities are frequently given in dB (decibel), defined as $10 \log J/J_0$.

### 6.2.3 Sound Waves in a Mean Current

The analysis in Section 6.1 can easily be generalized for a reference state which is moving with the constant velocity $\boldsymbol{u}_0$. The only change is in the substantial derivative $\mathrm{D}/\mathrm{D}t$ which is now linearized as

$$\frac{\mathrm{D}}{\mathrm{D}t} = \frac{\partial}{\partial t} + \boldsymbol{u}_0 \cdot \nabla$$

Therefore, one only has to replace $\partial/\partial t$ by $\partial/\partial t + \boldsymbol{u}_0 \cdot \nabla$ in the equations. In case of a wave ansatz of the form $\mathrm{e}^{i(\boldsymbol{k} \cdot \boldsymbol{x} - \omega t)}$ it follows that in all previous formulae one has to make the replacement of $\omega$ by $\omega^i = \omega - \boldsymbol{u}_0 \cdot \boldsymbol{k}$. The dispersion relation (6.20) thus becomes $(\omega - \boldsymbol{u}_0 \cdot \boldsymbol{k})^2 = c_s^2 k^2$ or

$$\omega = \boldsymbol{u}_0 \cdot \boldsymbol{k} + \omega^i = \Omega(\boldsymbol{k}) \tag{6.32}$$

The part $\omega^i = \Omega^i(\boldsymbol{k}) = c_s k$ is referred to as *intrinsic frequency*, i. e. the frequency seen by an observer who is moving along with the current $\boldsymbol{u}_0$. The frequency $\omega$, on the other hand, is noted by a stationary observer. The frequency shifts by the amount $\boldsymbol{u}_0 \cdot \boldsymbol{k}$, called DOPPLER[2] shift. The group velocity

$$\boldsymbol{c}_g = \frac{\partial \Omega}{\partial \boldsymbol{k}} = \boldsymbol{u}_0 + c_s \frac{\boldsymbol{k}}{k} \tag{6.33}$$

is a vectorial addition of the mean current $\boldsymbol{u}_0$ and the (intrinsic) group velocity $\boldsymbol{c}_g^i = c_s \boldsymbol{k}/k$ seen by a moving observer.

## 6.3 Propagation in a Variable Environment: WKBJ Approximation

So far we have assumed that $\theta, S, p_0, \rho_0 = \text{const}$ and also $c_s = \text{const}$. As can be seen from Figure 2.14, the sound velocity varies with temperature and pressure in a range of about 1,460–1,540 m s$^{-1}$ (the dependence on the salinity is negligible). This variation, while small, is crucial for the propagation of sound waves in the ocean. Sound velocity increases with pressure, and it generally increases towards the seafloor. Often it also increases toward the surface because of higher temperature, and frequently there is a zone of minimum sound velocity in the upper 1,000 m. A typical profile of $c_s$ (at Bermuda) with the above described features is shown in Figure 2.17. Horizontal temperature gradients also may cause corresponding gradients of the sound velocity.

In the linear approximation, the sound wave equations (6.9)–(6.13) remain valid for a variable sound velocity $c_s(\boldsymbol{x}, t)$ and lead to the modified wave equation

$$\frac{\partial}{\partial t}\left(\frac{1}{c_s^2(\boldsymbol{x}, t)} \frac{\partial p}{\partial t}\right) = \nabla^2 p \tag{6.34}$$

In practice, the dependence of sound velocity on location rather than time is of particular importance, but it is no problem to deal with the general case. Analytical

[2] CHRISTIAN ANDREAS DOPPLER, *1803 in Salzburg, †1853 in Venice, mathematician and physicist.

solutions of (6.34), however, are no longer possible offhand, and we have to search for methods of approach. The WKBJ method, named after WENTZEL, KRAMERS, BRILLOUIN and JEFFREYS[3] is the best-known approximation in case that wave lengths and periods are short compared to the scales of the mean environmental fields. The method is widespread for many wave processes and often useful for ordinary as well as partial differential equations. We will derive the approximation for the sound wave equation (6.34); its validity and applicability, however, extend much further.

For $c_s = \text{const}$, the solution (6.18) to the wave equation can be written as $p(\boldsymbol{x},t) = a\cos\phi = \Re\{a\mathrm{e}^{i\phi}\}$, with constant amplitude $a$, and the phase $\phi = \boldsymbol{k}\cdot\boldsymbol{x} - \omega t$ which is linear in $\boldsymbol{x}$ and $t$. In the following we will use the complex representation; when necessary, squares can be formed with the complex conjugate. To allow for variable sound velocity, an approximate solution to (6.34) is written as

$$p(\boldsymbol{x},t) = a(\boldsymbol{x},t)\mathrm{e}^{i\phi(\boldsymbol{x},t)} \tag{6.35}$$

For the derivatives of $\phi$ we introduce the abbreviations

$$\omega(\boldsymbol{x},t) = -\frac{\partial\phi}{\partial t}\,, \quad \boldsymbol{k}(\boldsymbol{x},t) = \nabla\phi \tag{6.36}$$

and $\omega$ and $\boldsymbol{k}$ are interpreted as *local* frequency, respectively wave number. The derivatives of $p$ thus can be expressed as

$$\begin{aligned}
\frac{\partial p}{\partial t} &= \left(\frac{\partial a}{\partial t} - i\omega a\right)\mathrm{e}^{i\phi}\,, \quad \nabla p = (\nabla a + i\boldsymbol{k}a)\mathrm{e}^{i\phi} \\
\nabla^2 p &= \left(\nabla^2 a + 2i\boldsymbol{k}\cdot\nabla a - k^2 a + ia\nabla\cdot\boldsymbol{k}\right)\mathrm{e}^{i\phi} \\
\frac{\partial}{\partial t}\left(\frac{1}{c_s^2}\frac{\partial p}{\partial t}\right) &= \left[\frac{1}{c_s^2}\left(\frac{\partial^2 a}{\partial t^2} - 2i\omega\frac{\partial a}{\partial t} - \omega^2 a - ia\frac{\partial\omega}{\partial t}\right) + \left(\frac{\partial a}{\partial t} - i\omega a\right)\frac{\partial c_s^{-2}}{\partial t}\right]\mathrm{e}^{i\phi}
\end{aligned} \tag{6.37}$$

So far, nothing has been gained, and instead of one unknown function $p$ we now have two, namely $a$ and $\phi$. To proceed, we now require that the amplitude, frequency, and wave number are approximately constant over a wavelength and period, and hence that the phase $\phi(\boldsymbol{x},t)$ is approximately linear in $\boldsymbol{x}$ and $t$. These requirements can be expressed with the inequalities

$$\frac{\partial a}{\partial t} \ll \omega a\,, \quad |\nabla a| \ll |\boldsymbol{k}|a\,, \quad \frac{\partial\omega}{\partial t} \ll \omega\cdot\omega\,, \quad \frac{\partial k}{\partial t} \ll \omega k \quad \text{etc.} \tag{6.38}$$

or abbreviated $\partial/\partial t \ll \omega$, $|\nabla| \ll |\boldsymbol{k}|$. Physically, the request (6.38) means that the solution (6.35) *locally* has the character of a plane wave. Formally, we introduce a parameter $\epsilon$ which measures the deviation from constancy of the amplitude over a wavelength/period, i. e.

$$\epsilon = o\left(\frac{1}{\omega a}\frac{\partial a}{\partial t}\right) \quad \text{or likewise} \quad \epsilon = o\left(\frac{1}{|\boldsymbol{k}|a}|\nabla a|\right)$$

[3] GREGOR WENTZEL, *1898 in Düsseldorf, †1978 in Ascona, physicist; HENDRIK KRAMERS, *1894 in Rotterdam, †1952 in Oegstgeest, physicist; LEON BRILLOUIN, *1889 in Sèvres, †1969 in New York, physicist; HAROLD JEFFREYS, *1891 in Fatfield/Durham, †1989 in Cambridge, mathematician, statistician, geophysicist and astronomer.

It follows from (6.38) that $\epsilon \ll 1$, and this allows a perturbation expansion in powers of $\epsilon$. The expansion is not based on the smallness of amplitudes but on the smallness of the scale ratio, and it is often referred to as *multiple-scale expansion*. Note that one cannot in general expect convergence of the series. Usually, however, *asymptotic* series are obtained which are such that for $\epsilon \to 0$ a finite number of terms in the series (in practice the lowest order term) is a very good approximation to the true solution. A brief introduction to asymptotic series can be found e. g. in Gradshteyn and Ryzhik (2000).

Inserting (6.37)–(6.36) into the wave equation (6.34) and rearranging for orders of $\epsilon$, one obtains

$$-\epsilon^0 \left(\omega^2 - c_s^2 k^2\right) a + i\epsilon^1 \left(-2\omega \frac{\partial a}{\partial t} - a\frac{\partial \omega}{\partial t} - \omega a c_s^2 \frac{\partial c_s^{-2}}{\partial t} - 2c_s^2 \boldsymbol{k} \cdot \nabla a - a c_s^2 \nabla \cdot \boldsymbol{k}\right)$$
$$+ \epsilon^2 \left(\frac{\partial^2 a}{\partial t^2} + c_s^2 \frac{\partial c_s^{-2}}{\partial t}\frac{\partial a}{\partial t} - c_s^2 \nabla^2 a\right) = 0 \qquad (6.39)$$

Here the common factor $e^{i\phi}$ has been dropped.

### 6.3.1 General Wave Kinematics

To the order $\epsilon^0$ one finds

$$\omega^2(\boldsymbol{x}, t) = c_s^2(\boldsymbol{x}, t) k^2(\boldsymbol{x}, t) \qquad (6.40)$$

from (6.39). This is the dispersion relation (6.20), but now (6.40) is *locally* valid. Accordingly, $\omega(\boldsymbol{x}, t)$ and $\boldsymbol{k}(\boldsymbol{x}, t)$ represent the local frequency and wave number. From (6.40) it follows that the derivatives of the phase $\phi(\boldsymbol{x}, t)$ are algebraically coupled according to

$$\left(\frac{\partial \phi}{\partial t}\right)^2 = c_s^2(\boldsymbol{x}, t)\,(\nabla\phi)^2 \qquad (6.41)$$

This equation is called *Eikonal* equation. It is a partial differential equation for the phase $\phi(\boldsymbol{x}, t)$. Due to its strongly nonlinear structure, a direct solution of (6.41) is, however, very difficult and will not be attempted. As will be seen, an alternative way to proceed which is based on the determination of $\boldsymbol{k}(\boldsymbol{x}, t)$ is mathematically easier and also physically more illustrative.

Note that from the definition of $\boldsymbol{k}$ in (6.36) it follows that

$$\frac{\partial k_j}{\partial x_\ell} \equiv \frac{\partial k_\ell}{\partial x_j} \qquad (6.42)$$

a relation which may also be written as $\nabla \times \boldsymbol{k} = 0$. The integral $\int \boldsymbol{k} \cdot d\boldsymbol{s}$ along some path from point $A$ to $B$ equals the $2\pi$ times the number crests. Because the vector field $\boldsymbol{k}$ is free of rotation, this number is independent of the connecting path (a consequence of the Stokes theorem, see Appendix A.1). Furthermore, from (6.36) we find the relation

$$\frac{\partial \boldsymbol{k}}{\partial t} + \nabla\omega = 0 \qquad (6.43)$$

In the direction of $\boldsymbol{k}$, the magnitude $|\boldsymbol{k}|$ is the number of wave crests per unit length at a fixed time, and $\omega$ is the number of wave crests passing a fixed location per unit time (wave crest flux). Therefore, (6.43) can be interpreted as the conservation of the number of wave crests.

According to the dispersion relation (6.40), the dependence of frequency on space and time results from the wave number $\boldsymbol{k}(\boldsymbol{x},t)$, and also explicitly from the varying sound velocity, i. e.

$$\omega(\boldsymbol{x},t) = c_s(\boldsymbol{x},t)|\boldsymbol{k}(\boldsymbol{x},t)| = \Omega(\boldsymbol{k};\boldsymbol{x},t) \tag{6.44}$$

In the last relation only the $(\boldsymbol{x},t)$-dependence of $c_s$ is written as argument of the function $\Omega$. In the following, only the existence of the dispersion relation (6.44) but not its specific form for sound waves will be used. Therefore, the conclusions will be valid for *all* wave processes that have a dispersion relation. From (6.44) we obtain (in tensor notation) the relation

$$\frac{\partial \omega}{\partial x_\ell} = \frac{\partial \Omega}{\partial k_j}\frac{\partial k_j}{\partial x_\ell} + \frac{\partial \Omega}{\partial x_\ell} \tag{6.45}$$

Note that although $\omega = \Omega$ according to (6.44), one has to distinguish between spatial derivatives of $\omega$ and $\Omega$ because in the first case only $t$ is kept constant while in the second both $\boldsymbol{k}(\boldsymbol{x},t)$ and $t$ are kept constant. This is analogously valid for temporal derivatives. With the group velocity $\boldsymbol{c}_g$ as defined in (6.25), (6.43) finally results in

$$\frac{\partial \boldsymbol{k}}{\partial t} + \boldsymbol{c}_g \cdot \nabla \boldsymbol{k} = -\frac{\partial \Omega}{\partial \boldsymbol{x}} \tag{6.46}$$

The derivative $\partial/\partial \boldsymbol{x}$ on the right-hand side refers to the *explicit* dependence of frequency $\Omega$ on the spatial coordinate via the sound velocity according to (6.44) (therefore, we are not using the gradient symbol). Equation (6.46) is a *quasi-linear* partial differential equation for $\boldsymbol{k}(\boldsymbol{x},t)$. Its physical significance will be discussed, along with a practical solution method, in the next section. With the solution $\boldsymbol{k}(\boldsymbol{x},t)$ of (6.46), the local frequency $\omega(\boldsymbol{x},t)$ follows from the dispersion (6.40). If needed, the phase $\phi(\boldsymbol{x},t)$ can be obtained by simple integration of (6.36). It is often useful to consider the relation corresponding to (6.46) for the frequency $\omega$. In analogy to (6.45) one has

$$\frac{\partial \omega}{\partial t} = \frac{\partial \Omega}{\partial k_j}\frac{\partial k_j}{\partial t} + \frac{\partial \Omega}{\partial t}$$

and with (6.43) it follows that

$$\frac{\partial \omega}{\partial t} + \boldsymbol{c}_g \cdot \nabla \omega = \frac{\partial \Omega}{\partial t} \tag{6.47}$$

As the dispersion relation (6.40) relates wave number and frequency, (6.46) and (6.47) are not independent of each other; however, there can be practical advantages to use both simultaneously (cf. Section 6.3.4).

### 6.3.2 Group Velocity II: Rays and Wave Packages

Equation (6.46) is a *quasi-linear* partial differential equation of first order for $\boldsymbol{k}(\boldsymbol{x},t)$ (*linear*, because the derivatives of $\boldsymbol{k}$ appear linearly, and *quasi*, because $\boldsymbol{c}_g$ and $\Omega$

*algebraically* depend on $\boldsymbol{k}$). A straightforward solution can be obtained from the method of characteristics, a mathematical procedure which is conceptually analogous to using Lagrangian rather than Eulerian coordinates in the equations of motion (see Section 2.12). The equations (6.46) for the wave number, respectively (6.47) for the frequency, have the general form

$$\frac{\partial \psi}{\partial t} + \boldsymbol{c}_{\mathrm{g}}(\boldsymbol{k}, \boldsymbol{x}, t) \cdot \nabla \psi = Q_{\psi}$$

where $\boldsymbol{c}_{\mathrm{g}} = \boldsymbol{c}_{\mathrm{g}}(\boldsymbol{k}, \boldsymbol{x}, t)$. As an analogy, the propagation of a substance $\chi$ is governed by

$$\frac{\partial \chi}{\partial t} + \boldsymbol{u} \cdot \nabla \chi = Q \tag{6.48}$$

Here all sources and sinks (including diffusion) are contained in $Q$. Instead of solving the partial differential equation (6.48), one may consider in the Lagrangian description the movement of material elements along the particle paths which are given by

$$\frac{\mathrm{d}\boldsymbol{x}}{\mathrm{d}t} = \boldsymbol{u}(\boldsymbol{x}, t) \tag{6.49}$$

Along the particle path one then has

$$\frac{\mathrm{d}\chi}{\mathrm{d}t} = Q \tag{6.50}$$

Note that equations (6.49) and (6.50) are *ordinary* differential equations. From each particle's concentration $\chi(\boldsymbol{x}, 0) = \chi_0(\boldsymbol{x})$ at an initial time $t = 0$, the value $\chi(\boldsymbol{x}, t)$ along each trajectory can be determined, given $Q$ (assuming that $Q$ only depends on the values of $\chi$ on the respective trajectory). In principle, this allows construction of the solution for every point in the region that is reached by a trajectory (see the sketch in Figure 6.4).

Similar considerations can be used for (6.46) and (6.47) since mathematically it does not matter that $\boldsymbol{c}_{\mathrm{g}}$ is a group velocity rather than a particle velocity. One

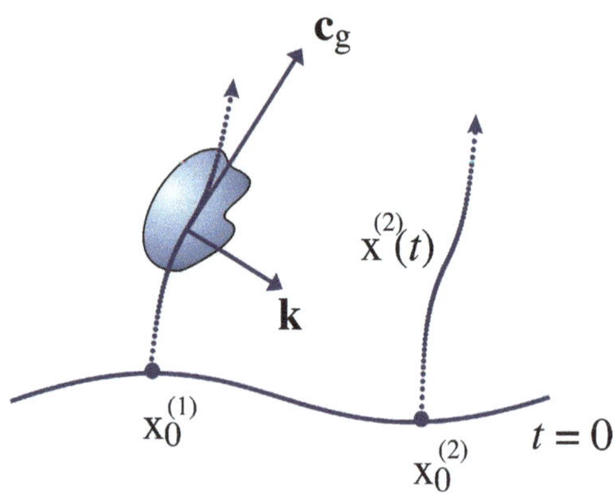


Fig. 6.4 Rays starting from initial points $\boldsymbol{x}_0^{(i)}$ at time $t = 0$. For sound waves $\boldsymbol{k}$ and $\boldsymbol{c}_{\mathrm{g}}$ are aligned parallel. Most other waves do not have this property. A perpendicular orientation, as depicted here, is realized in internal gravity waves (see Chapter 7)

can define a *wave group* that is described by its location $\boldsymbol{x}$ and its wave number $\boldsymbol{k}$ and moves with its specific group velocity $\boldsymbol{c}_g(\boldsymbol{x}, t)$. The wave group is analogous to a substance particle which propagates along a particle path defined by the particle velocity. The 'group path' is defined by the group velocity here, i. e.

$$\dot{\boldsymbol{x}} = \boldsymbol{c}_g = \frac{\partial \Omega}{\partial \boldsymbol{k}} \tag{6.51}$$

Changes in time of a property $\psi$ along the motion of a wave packet are governed by

$$\dot{\psi} = \frac{\partial \psi}{\partial t} + \boldsymbol{c}_g \cdot \nabla \psi = Q_\psi \tag{6.52}$$

The left side of the equation (6.46) is the 'substantial' derivative of $\boldsymbol{k}$, hence

$$\dot{\boldsymbol{k}} = -\frac{\partial \Omega}{\partial \boldsymbol{x}} \tag{6.53}$$

If the dispersion relation is dependent on the position, the wave number $\boldsymbol{k}$ for the propagating wave group will change, generally both in magnitude and in direction. This is called *refraction*. It is often useful to also consider the changes in frequency $\omega$ along the characteristics. According to (6.47)

$$\dot{\omega} = \frac{\partial \Omega}{\partial t} \tag{6.54}$$

In the frequent case that the dispersion relation does not explicitly depend on time, the frequency of a wave group remains constant.

### 6.3.3 Energy Conservation II: Energy Flux and Group Velocity

To complete the (approximate) solution of (6.34), the amplitude $a(\boldsymbol{x}, t)$ still needs to be determined. The terms of the order $\epsilon^1$ in (6.39) are

$$\frac{\partial a}{\partial t} + \frac{c_s^2 \boldsymbol{k}}{\omega} \cdot \nabla a + \frac{a}{2\omega} \frac{\partial \omega}{\partial t} + \frac{a c_s^2}{2} \frac{\partial c_s^{-2}}{\partial t} + \frac{a}{2\omega} c_s^2 \nabla \cdot \boldsymbol{k} = 0 \tag{6.55}$$

With the relation $\boldsymbol{c}_g = c_s^2 \boldsymbol{k}/\omega$ for sound waves, which follows from (6.25) and the dispersion relation, one can replace the term containing $\nabla \cdot \boldsymbol{k}$ in (6.55) with the identity

$$\nabla \cdot \boldsymbol{k} = \nabla \cdot \left( \frac{\omega}{c_s^2} \boldsymbol{c}_g \right) = c_s^{-2} \boldsymbol{c}_g \cdot \nabla \omega + \omega \boldsymbol{c}_g \cdot \nabla c_s^{-2} + \frac{\omega}{c_s^2} \nabla \cdot \boldsymbol{c}_g$$

and obtains, after some reordering,

$$\frac{\partial a}{\partial t} + \boldsymbol{c}_g \cdot \nabla a + \frac{a c_s^2}{2} \left( \frac{\partial c_s^{-2}}{\partial t} + \boldsymbol{c}_g \cdot \nabla c_s^{-2} \right) + \frac{a}{2\omega} \left( \frac{\partial \omega}{\partial t} + \boldsymbol{c}_g \cdot \nabla \omega \right) = -\frac{a}{2} \nabla \cdot \boldsymbol{c}_g$$

This complicated equation can be rewritten as

$$\frac{\partial}{\partial t} \frac{a}{c_s} + \boldsymbol{c}_g \cdot \nabla \frac{a}{c_s} + \frac{1}{2} \frac{a}{c_s} \frac{1}{\omega} \left( \frac{\partial \omega}{\partial t} + \boldsymbol{c}_g \cdot \nabla \omega \right) = -\frac{a}{2 c_s} \nabla \cdot \boldsymbol{c}_g \tag{6.56}$$

As the period of sound waves is very short compared to time-scales of changes in the environment, the time dependence of the sound velocity is of little importance, and we can assume $c_s = c_s(\boldsymbol{x})$, i. e. time-constant sound velocity. With this assumption, it follows from (6.54) that the term in brackets in (6.56) vanishes. After multiplication by $a^*/\rho_0 c_s$, adding the complex conjugate equation, averaging over a wave period and identification of the mean total energy according to (6.31), i. e.

$$E_w = \frac{1}{2} \frac{\overline{aa^*}}{\rho_0 c_s^2}$$

one finally obtains the energy conservation in the common form

$$\frac{\partial E_w}{\partial t} = -\nabla \cdot \boldsymbol{c}_g E_w \tag{6.57}$$

The total energy is thus conserved in the spatial integral, as can be seen by integration of (6.57) over the total volume and using the Gaussian theorem (A.6). We identify

$$\boldsymbol{J}_E = \boldsymbol{c}_g E_w \tag{6.58}$$

as total energy flux (in W m$^{-2}$). The relation (6.58) is valid for most wave processes under general conditions, as remarked earlier.

As discussed in Section 6.3.2, the most convenient solution of (6.57) is obtained by using the characteristics. The change in total energy along the wave group path is given by

$$\dot{E}_w = -E_w \nabla \cdot \boldsymbol{c}_g \tag{6.59}$$

and can be solved together with (6.51), (6.53) and (6.54). Equation (6.59) indicates that the energy of a wave group can change *locally* through convergence/divergence of the groups.

As mentioned above, the time dependence of the sound velocity is not very important, but the problem is interesting for educational purposes. Obviously, if we retain the term in the brackets of equation (6.56), we find that the energy $E_w$ is not conserved, and (6.57) must be replaced by

$$\frac{\partial E_w}{\partial t} + \nabla \cdot \boldsymbol{c}_g E_w = -\frac{E_w}{\omega} \frac{\partial \Omega}{\partial t} = -E_w \frac{\partial c_s/\partial t}{c_s} \tag{6.60}$$

which is found using (6.54). We note, that while a spatial dependence of the background (in a WKBJ sense) does not spoil the energy conservation, the time dependence leads to a source/sink term associated with change of frequency along the wave path. In Chapter 9 we will show that the relation (6.60) has a more general validity. Here we just mention that a conserved quantity still can be found: it is the *wave action* which for this particular system reads $\mathcal{A} = \omega E_w$. In fact, (6.60) transforms to

$$\frac{\partial \mathcal{A}}{\partial t} + \nabla \cdot \boldsymbol{c}_g \mathcal{A} = 0 \tag{6.61}$$

A summary of the WKBJ analysis and the resulting ray equations is given in the box on p. 173.

**25. Ray Equations**

In summary of the WKBJ analysis, we have obtained the relations

$$\dot{\boldsymbol{x}} = \frac{\partial \Omega}{\partial \boldsymbol{k}} \tag{B25.1}$$

$$\dot{\boldsymbol{k}} = -\frac{\partial \Omega}{\partial \boldsymbol{x}} \tag{B25.2}$$

$$\dot{\omega} = \frac{\partial \Omega}{\partial t} \tag{B25.3}$$

$$\dot{\mathcal{A}} = -\mathcal{A}\nabla \cdot \boldsymbol{c}_g \tag{B25.4}$$

The *ray equations* (B25.1)–(B25.4) are *ordinary* differential equations, and the first three are the *characteristic equations* of the partial differential equation (6.46). Equation (B25.4) determines the wave action. Their solution is straightforward. The solution of the Eikonal equation (6.41) can then be constructed from all possible characteristics. The equations (B25.1) to (B25.3) resemble the Hamiltonian equations of the canonical variables $\boldsymbol{x}$ and $\boldsymbol{k}$ with Hamiltonian and energy $\Omega$ (see Section 2.12 and the box on p. 98). We thus can view the rays either as trajectories in the physical space, with $\boldsymbol{k}$ as property changing along the path, or in the phase space spanned by the six (or correspondingly less in a less dimensional physical space) variables $\boldsymbol{x}$ and $\boldsymbol{k}$. The latter has the advantage to use the Liouville theorem, stating that the phase space volume $\mathrm{d}x^3 \times \mathrm{d}k^3$ is conserved in the phase space flow.

The equation (B25.3) for $\omega$ also follows directly from the dispersion relation (6.44) together with (B25.1) and (B25.2) and is, therefore, redundant. Nevertheless, it can often conveniently be used to replace one of the equations (B25.2).

One might ask why the Lagrangian representation is not preferred to the Eulerian form in fluid dynamics. The reason is that through the equation of motion, specifically by the pressure gradient, different particle paths are coupled and cannot be calculated independently. A further coupling may arise through frictional or diffusive processes. This can be physically understood since different particles cannot occupy the same volume element. In contrast, ray paths may pass the same volume at the same time. It may, however, happen that the underlying assumption of the WKBJ approach – the high wave-number asymptotic – breaks down for crossing rays, e. g. in focal points or more complicated caustic lines.

Note that in derivation of (B25.1)–(B25.3), only the *existence* of a dispersion relation but not the specific dispersion relation of the sound waves has been used. Therefore, the relations are valid for *all* wave processes with a dispersion relation.

### 6.3.4 Pathways of Sound Wave Propagation in the Ocean

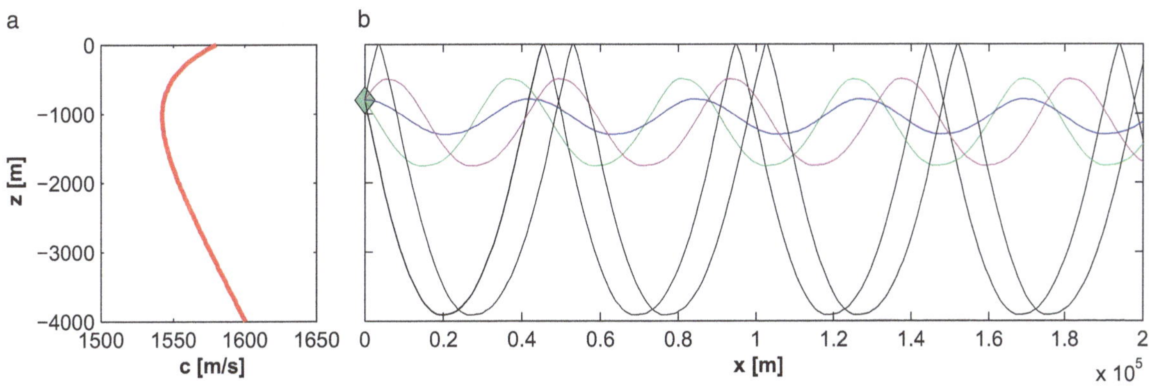


**Fig. 6.5** Sound wave rays along a sound channel given by the $c$-profile in **a**. All rays (**b**) start at a depth of 800 m but have different initial angles $\varphi_\star = -15, -5, 0, 5, 15°$

In the ocean, the speed of sound varies mainly with depth, due to the dependence of pressure and temperature on depth. For an approximate determination of the sound wave pathways, we thus consider the specific case $c_s = c_s(z)$. The dispersion relation (6.44) then is $\omega = c_s(z)(k_1^2 + k_2^2 + k_3^2)^{1/2}$, and the characteristic pathway of wave groups is given by

$$\dot{x}_\ell = c_s \frac{k_\ell}{|\boldsymbol{k}|}, \quad \ell = 1, 2, 3 \tag{6.62}$$

i. e. the propagation always proceeds in the direction of $\boldsymbol{k}$. Equations (B25.2) and (B25.3) now read

$$\dot{k}_1 = -\frac{\partial \Omega}{\partial x_1} \equiv 0 \Rightarrow k_1 = \text{const} = k_\text{h} \tag{6.63}$$

$$\dot{k}_2 = -\frac{\partial \Omega}{\partial x_2} \equiv 0 \Rightarrow k_2 = \text{const} = 0 \tag{6.64}$$

$$\dot{k}_3 = -\frac{\partial \Omega}{\partial x_3} = -\frac{\partial c_s}{\partial z}|\boldsymbol{k}| \tag{6.65}$$

$$\dot{\omega} = \frac{\partial \Omega}{\partial t} = 0 \Rightarrow \omega = c_s \left(k_h^2 + k_3^2\right)^{1/2} = \text{const} \tag{6.66}$$

with some immediate consequences indicated. Let the minimum of the sound velocity $c_0 = c_s(z_0)$ be at $z = z_0$. The vertical wave number $k_3$ decreases with time above the minimum of $c_s$, and increases below the minimum. A wave group starting at the level $z_0$ with initially upwards propagation hence undergoes refraction, and unless it reaches the surface where it will be reflected, eventually it will propagate downward. Hence the path effectively oscillates around the level $z = z_0$ and remains within a certain range of depth, the maximum/minimum depth $z_\text{max/min}$ of which is reached at $k_3 = 0$ respectively $c_s(z_\text{max/min})k_\text{h} = \omega$. In between, we find the SOFAR[4] channel. Because of the constancy of $k_\text{h}$ and with $c_s(z_\text{max/min}) = c_\text{m}$, (6.66) is equivalent to

$$\frac{c_\text{m}}{c_s(z)} \left(\frac{k_\text{h}}{k_\text{h}^2 + k_3^2(z)}\right)^{1/2} = n \cos\varphi = \text{const} \tag{6.67}$$

Here we have introduced the angle $\varphi$ between $\boldsymbol{k}$ and the horizontal, and the refraction index $n(z) = c_\text{m}/c_s(z)$ as ratio of the maximum propagation velocity to its actual value. In its second form, (6.67) directly corresponds to SNELLIUS'[5] law of refraction in optics. Figure 6.5 displays the propagation of sound rays in a realistic sound channel.

[4] **SO**und **F**ixing **A**nd **R**anging

[5] WILLEBRORD SNELLIUS, *1580 and †1626 in Leiden, astronomer and mathematician.

**26. Analytical solution for sound propagation**

An approximative analytical solution of the equation of sound rays can be obtained for small vertical excursions around the sound velocity minimum (i. e. $\varphi \ll 1$, or $|k_3| \ll k_h$), whereas the 3-d wave-number varies only slightly. With the parabolic sound velocity profile

$$c(z) = c_0 + \frac{1}{2}\gamma(z - z_0)^2$$

we find the simple equation for the horizontal component of the ray

$$\dot{x}_h = c_s(z)\frac{k_h}{k(z)} \approx c_s(z_0)\frac{k_h}{k(z_0)} \approx c_0$$

from (6.62) (remember that $k_h = \text{const}$). For the vertical component of the wave path

$$\dot{z} = c_s(z)\frac{k_3}{k(z)} \approx c_0\frac{k_3}{k(z_0)} \quad \text{(B26.1)}$$

also follows from (6.62). From the refraction equation (6.65)

$$\dot{k}_3 = -\frac{\partial c_s}{\partial z}k(z) \approx -\frac{\partial c_s}{\partial z}k(z_0) \approx -\gamma k(z_0)(z - z_0) \quad \text{(B26.2)}$$

results. A further differentiation of (B26.1) and insertion of (B26.2), using $\zeta = z - z_0$, then leads to

$$\ddot{\zeta} + \gamma c_0 \zeta = 0$$

We thus obtain an oscillatory path with the oscillation period $T = 2\pi/(\gamma c_0)^{1/2}$. As the horizontal propagation velocity is approximately given by $c_0$, this corresponds to a horizontal length scale

$$L = c_0 T = 2\pi\sqrt{c_0/\gamma}$$

With the value $\gamma \approx 2 \times 30\,\text{m}\,\text{s}^{-1}/(1{,}000\,\text{m})^2 \approx 0.6 \times 10^{-4}\,\text{m}^{-1}\,\text{s}^{-1}$, one obtains $L \approx 30\,\text{km}$ for the distance between e. g. two maxima of the ray curve which has the right order of magnitude compared to Figure 6.5.

# 7 Gravity Waves

Gravity waves play an important role in the adjustment of ocean currents towards geostrophic equilibrium. Internal gravity waves contribute by breaking to mixing processes in the ocean interior, especially to diapycnal mixing. Surface gravity waves are effective in the exchange of momentum, heat, water, and gases between the ocean and the atmosphere, in addition to their practical significance for shipping.

The force of gravity is fundamental for the existence of gravity waves and indeed for nearly all oceanic motions with the exception of sound waves. Earth's rotation is also important and leads to significant modifications. For internal gravity waves, the existence of a density stratification caused by temperature and salinity variations with depth is crucial.

The density stratification is described by a background density $\rho_b(z)$ or equivalently the Brunt–Väisälä frequency $N(z)$ defined in Equation (5.9). The role of $N$ for the stability of the water column and frequency of particle oscillations about a reference level was discussed in Section 2.9.2. In the ocean, the period of stability oscillations normally is between 10 and 20 min (upper ocean) and a few hours at great depths. Close to the surface very large values of $N$ may occur in the seasonal thermocline. A typical profile of $N(z)$ in midlatitudes is shown in Figure 2.17.

## 7.1 Governing Equations

The equations of motion (5.69)–(5.73) remain valid for gravity waves. Equation (5.69) describes the adjustment of the pressure field as a consequence of a divergent velocity field. As shown in Chapter 6, the adjustment occurs with the time-scales of sound waves which are much shorter than for all other ocean processes. On the much longer time-scale of gravity waves, it is, therefore, possible to approximate (5.69) by assuming that the sound waves are infinitely fast so that in effect the adjustment has already been accomplished. Formally, this approximation can be achieved by taking the limit $c_s \to \infty$ in (5.69), thereby eliminating sound waves from the system.

Equivalently, as the starting point one can use the Boussinesq system (4.10)–(4.11) and (4.22) derived in Section 4.1. It is, however, preferable to formulate these

D. Olbers, J. Willebrand, C. Eden, *Ocean Dynamics*,
DOI 10.1007/978-3-642-23450-7_7, 

equations for the density variable $\hat{\rho} = \rho - \rho_b(z)$ which is the deviation of the *in-situ* density from the stationary background state. The background state is characterized by the Brunt–Väisälä frequency $N(z)$ as in the quasi-geostrophic theory discussed in Section 5.2. Hence

$$\rho_0 \left( \frac{\mathrm{D}\boldsymbol{u}}{\mathrm{D}t} + 2\boldsymbol{\Omega} \times \boldsymbol{u} \right) = -\nabla \hat{p} - \hat{\rho} \nabla \Phi + \mathcal{F} \tag{7.1}$$

$$\nabla \cdot \boldsymbol{u} = 0 \tag{7.2}$$

$$\frac{\mathrm{D}\hat{\rho}}{\mathrm{D}t} - \frac{\rho_0 N^2}{g} w = \mathcal{G}_\rho \, . \tag{7.3}$$

The treatment of the forcing terms $\mathcal{F}$ and $\mathcal{G}_\rho$ is postponed to Chapter 10.1. Here we will assume that both are zero.

In the traditional approximation for the Coriolis force, the vector $\boldsymbol{\Omega}$ is approximated by its *locally* vertical component, i. e. $2\boldsymbol{\Omega} \approx \boldsymbol{f}$ with $\boldsymbol{f} = (0, 0, 2\Omega \sin\varphi) = (0, 0, f)$. As shown in Section 4.2.4, this approximation holds if the aspect ratio (ratio of vertical to horizontal length scales of motion) is small. It will become apparent that for short gravity waves where the aspect ratio is not necessarily small, rotation is altogether insignificant. Therefore, this approximation is used in the following; for a detailed discussion on the consequences of the traditional approximation for gravity waves see Gerkema and Shrira (2005). Furthermore, in the present chapter the Coriolis frequency $f$ will be treated as constant.

Consider now small deviations from the background state. Instead of a formal expansion of the variables in terms of a small expansion parameter such as in Section 6.1, it suffices for the linear approximation to neglect all products of the field variables (the validity of this approach can be considered *a posteriori*). One obtains the linear system

$$\frac{\partial \boldsymbol{u}_\mathrm{h}}{\partial t} + \boldsymbol{f} \times \boldsymbol{u}_\mathrm{h} = -\frac{1}{\rho_0} \nabla_\mathrm{h} p \tag{7.4}$$

$$\frac{\partial w}{\partial t} = -\frac{1}{\rho_0} \frac{\partial p}{\partial z} - g \frac{\rho}{\rho_0} \tag{7.5}$$

$$\frac{\partial \rho}{\partial t} - \frac{\rho_0 N^2}{g} w = 0 \tag{7.6}$$

$$\nabla_\mathrm{h} \cdot \boldsymbol{u}_\mathrm{h} + \frac{\partial w}{\partial z} = 0 \, . \tag{7.7}$$

Note that we have explicitly distinguished between horizontal and vertical vector components, and simplified the notation for the density variable, replacing $\hat{\rho}$ by $\rho$.

### Boundary conditions:

The condition that no mass flux crosses the boundary (*kinematic* boundary condition) follows from (2.22) and (2.23). At the bottom $z = -h$ it takes the form $w = -\boldsymbol{u}_\mathrm{h} \cdot \nabla_\mathrm{h} h$. For a free surface located at $z = \zeta(\boldsymbol{x}_\mathrm{h}, t)$, and ignoring the term $\boldsymbol{u}_\mathrm{h} \cdot \nabla_\mathrm{h} \zeta$ which is small in accordance with linear wave theory, that condition is expressed as $w = \partial\zeta/\partial t$ at $z = \zeta$. In the absence of atmospheric pressure changes, the surface elevation $\zeta$ is related to pressure by $\zeta = p/g\rho_0$ at $z = 0$ (cf. Section 5.2.4).

Therefore, the boundary condition at the surface is given as

$$w = \frac{1}{g\rho_0}\frac{\partial p}{\partial t} \quad \text{at} \quad z = 0\,. \tag{7.8}$$

In many situations, (7.8) is approximated by the simpler condition

$$w = 0 \quad \text{at} \quad z = 0\,. \tag{7.9}$$

For gravity waves, the accuracy of the *rigid-lid condition* (7.9) is discussed in Section 7.4.4 below.

## 7.2 Plane Gravity Waves

The set of equations (7.4)–(7.7) is the starting point for all wave problems discussed in this chapter. Since all coefficients are independent of $\boldsymbol{x}_\text{h}$ and $t$, a solution describing plane waves in the horizontal directions can be found by a separation ansatz of the form

$$\begin{Bmatrix} \boldsymbol{u} \\ p \\ w \\ \rho \end{Bmatrix} = \begin{Bmatrix} \boldsymbol{U}_0 \mathrm{d}W/\mathrm{d}z \\ P_0 \mathrm{d}W/\mathrm{d}z \\ W(z) \\ R_0 N^2(z) W(z) \end{Bmatrix} \mathrm{e}^{i(\boldsymbol{k}_\text{h}\cdot\boldsymbol{x}_\text{h} - \omega t)} \tag{7.10}$$

with an educated guess of the vertical dependence. Here $\boldsymbol{U}_0, P_0, R_0$ are constants, and $W(z)$ is the vertical structure function which will be determined below. Inserting (7.10) in equations (7.4), (7.6) and (7.7) yields the algebraic relations

$$\boldsymbol{U}_0 = i\,(\omega \boldsymbol{k}_\text{h} - i\,\boldsymbol{f} \times \boldsymbol{k}_\text{h})\,/\omega k_\text{h}^2 \tag{7.11}$$

$$P_0 = i\rho_0(\omega^2 - f^2)/\omega k_\text{h}^2 \tag{7.12}$$

$$R_0 = i\rho_0/g\omega \tag{7.13}$$

for the coefficients $\boldsymbol{U}_0, P_0, R_0$. Note that all physical fields in equations (7.4)–(7.7) have the proper physical dimensions, however, the dimensions of the factors $\boldsymbol{U}_0, P_0, R_0$ differ from those of the fields $\boldsymbol{u}, p, \rho$. Inserting of (7.10) in (7.5), using (7.12) and (7.13), then results in

$$\frac{\mathrm{d}^2 W}{\mathrm{d}z^2} + k_\text{h}^2 \frac{N^2(z) - \omega^2}{\omega^2 - f^2} W = 0 \tag{7.14}$$

as a differential equation for the function $W(z)$ representing the vertical dependence which will be extensively used in the following.

### *7.2.1 Propagation Characteristics*

In this section, the density changes are idealized to be linear with depth so that $N =$ const. While this idealization does not well reflect the typical stratification in the ocean, it facilitates to obtain a first view of internal wave propagation characteristics. For constant $N$, (7.14) is solved by

$$W(z) = W_0 \mathrm{e}^{ik_3 z} \quad \text{with} \quad k_3^2 = k_\mathrm{h}^2 \frac{N^2 - \omega^2}{\omega^2 - f^2} \tag{7.15}$$

and leads to the dispersion relation

$$\omega^2 = \frac{N^2 k_\mathrm{h}^2 + f^2 k_3^2}{k_\mathrm{h}^2 + k_3^2} \equiv f^2 + \frac{N^2 - f^2}{1 + k_3^2/k_\mathrm{h}^2} \tag{7.16}$$

The period of the internal gravity waves is limited by the stability period $2\pi/N \sim$ 20 min in the upper ocean, increasing to a few hours in the deep water column, and by the inertial period $2\pi/f \geq 12\,\mathrm{h}$. In general, $N > f$, hence $f < \omega < N$ holds. With (7.15), the solution describes plane waves with a phase proceeding in the direction of the three-dimensional wave vector $(k_1, k_2, k_3)$, with phase speed $c = \omega/k$ where $k = (k_\mathrm{h}^2 + k_3^2)^{1/2}$ is the total wave number. Other useful forms of the dispersion relation are

$$k_3^2 = k^2 \frac{N^2 - \omega^2}{N^2 - f^2} \quad \text{and} \quad k_\mathrm{h}^2 = k^2 \frac{\omega^2 - f^2}{N^2 - f^2}\,.$$

Let $\varphi$ be the angle of the wave vector to the horizontal plane so that $\cos^2\varphi = k_\mathrm{h}^2/k^2, \sin^2\varphi = k_3^2/k^2$. Then, the dispersion relation (7.16) can be written in the useful form

$$\omega^2 = N^2 \cos^2\varphi + f^2 \sin^2\varphi \tag{7.17}$$

which shows that the frequency $\omega$ only depends on the angle $\varphi$ but not on the magnitude $|\boldsymbol{k}|$ of the three-dimensional wave vector. This property of internal gravity waves is very different from e. g. sound waves where the frequency only depends on the magnitude but not on the direction of the wave vector.

In a $(k_\mathrm{h}, k_3)$ plane, the lines of a constant $\omega$ are straight lines through the origin. As $N \gg f$ the frequency is almost a cosine of $\varphi$. We have $\omega \simeq N$ if the wave vector is close to horizontal ($\varphi \simeq 0$), and $\omega \simeq f$ if it is almost vertical ($\varphi \simeq \pm\pi/2$). The frequency $\omega$ is shown in Figure 7.1 (left panels) as function of $k_\mathrm{h}$ and $k_3$. In the logarithmic view the dependence of $\omega$ on $k_\mathrm{h}$ or $k_3$ is expressed as a mere shift of the profile along the respective wave number. This is so because $\omega$ is a function of $k_3/k_\mathrm{h}$ and $\log k_\mathrm{h}/k_3 = \log k_\mathrm{h} - \log k_3$.

Since the group velocity $\boldsymbol{c}_\mathrm{g}$ is given as $\boldsymbol{c}_\mathrm{g} = \partial\omega/\partial\boldsymbol{k}$, i. e. as the gradient of $\omega$ in $\boldsymbol{k}$-space, and $\omega$ only depends on the direction of $\boldsymbol{k}$, the group velocity must be orthogonal to the wave-number vector, $\boldsymbol{c}_\mathrm{g} \perp \boldsymbol{k}$ or $\boldsymbol{c}_\mathrm{g} \cdot \boldsymbol{k} = 0$. For $k_3 > 0$ it follows that $c_{\mathrm{g}3} < 0$, and vice versa. Hence the case $k_3 > 0$ ($k_3 < 0$) corresponds to the upward (downward) phase propagation and downward (upward) propagation of wave groups and thus also of energy. The situation is visualized in Figure 7.2 where the energy propagation is directed to the upper right whereas the phase propagates to the lower right. This behavior can be algebraically verified by direct calculation of

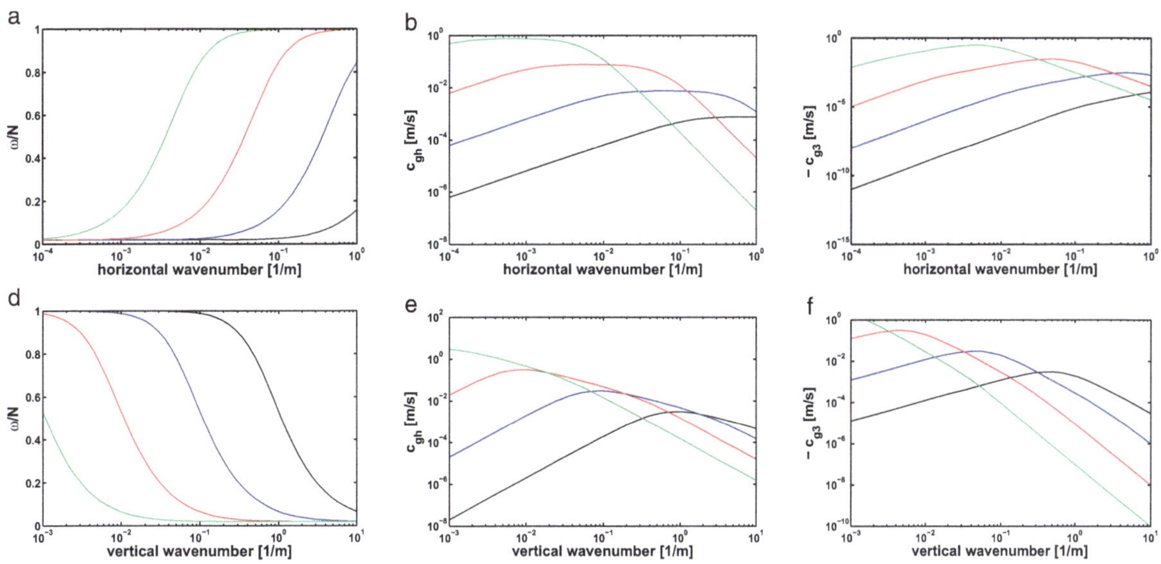


Fig. 7.1 Dispersion relation (**a**,**d**, in the the scaled form $\omega/N$), horizontal (**b**,**e**) and vertical (**c**,**f**) components of the group velocity of internal waves, for $f = 10^{-4}\,\mathrm{s}^{-1}$ and $N = 5 \times 10^{-3}\,\mathrm{s}^{-1}$. **a–c** show these quantities as function of the horizontal wave number $k_\mathrm{h}$ for 4 vertical wavelengths $2\pi/k_3 = 1, 10, 100, 1{,}000\,\mathrm{m}$, **d–f** as function of the vertical wave number $k_3$ for 4 horizontal wavelengths $2\pi/kh = 10, 10^2, 10^3, 10^4\,\mathrm{m}$ (colors are for both in the order black, blue, red, green). Note that the wave-number axes and the scales for group velocity are logarithmic (the negative of $c_3$ is displayed, because $c_3 < 0$ for $k_3 > 0$)

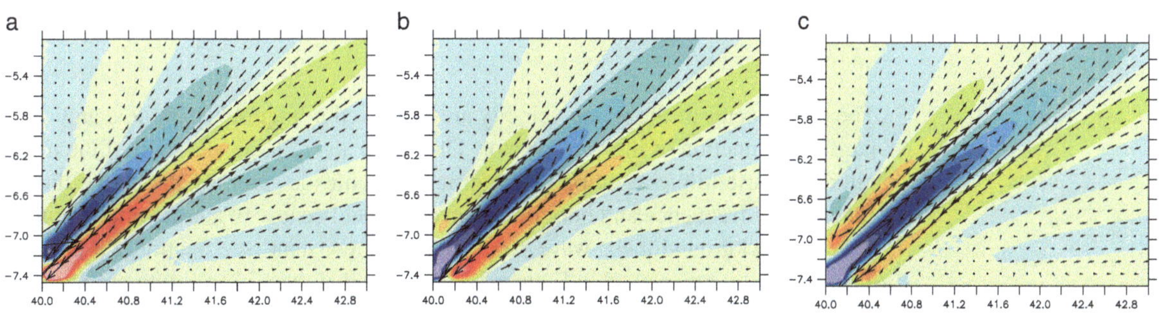


Fig. 7.2 Propagation of internal waves in the $x$–$z$-plane, as simulated by a numerical model. At time $t = 0$, an oscillatory forcing with period $T$ is switched on which is concentrated in the bottom left corner. Shown are the density anomaly (in color) and the velocity vectors, at times $4T$ (**a**), $4.25T$ (**b**) and $4.5\,T$ (**c**). The numerical experiment is a replica of the classical laboratory experiment by Mowbray and Rarity (1967) in which an internal wave is generated by a small disk oscillating in a stratified fluid

the group velocities, resulting in

$$\boldsymbol{c}_{\mathrm{gh}} = \frac{\partial \omega}{\partial \boldsymbol{k}_\mathrm{h}} = \frac{N^2 - f^2}{\omega} \frac{k_3^2}{\left(k_\mathrm{h}^2 + k_3^2\right)^2} \boldsymbol{k}_\mathrm{h} = \frac{(N^2 - \omega^2)(\omega^2 - f^2)}{\omega(N^2 - f^2)} \frac{\boldsymbol{k}_\mathrm{h}}{k_\mathrm{h}^2} \tag{7.18}$$

$$c_{\mathrm{g}3} = \frac{\partial \omega}{\partial k_3} = -\frac{N^2 - f^2}{\omega} \frac{k_\mathrm{h}^2}{\left(k_\mathrm{h}^2 + k_3^2\right)^2} k_3 = \mp \frac{(\omega^2 - f^2)^{3/2}(N^2 - \omega^2)^{1/2}}{\omega(N^2 - f^2)} \frac{1}{k_\mathrm{h}} \tag{7.19}$$

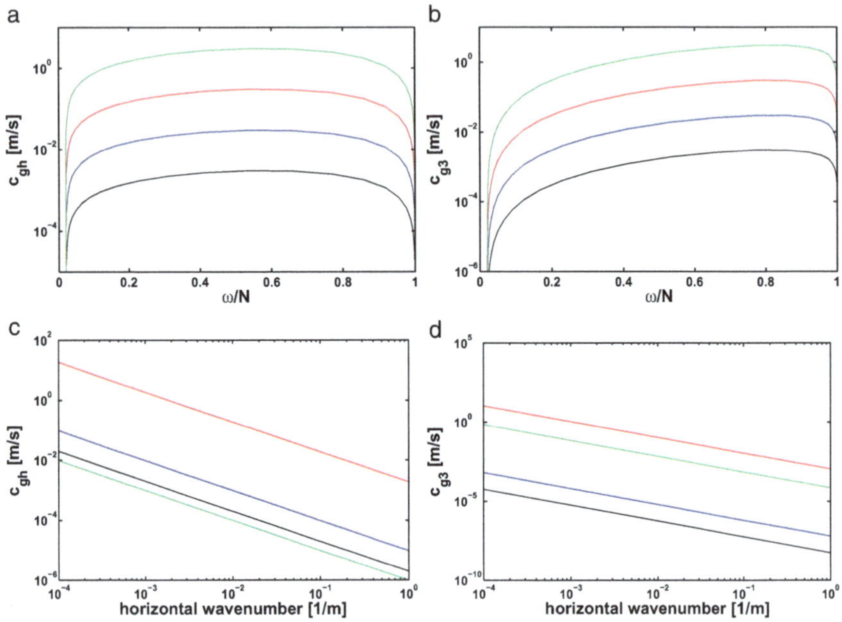


Fig. 7.3 Horizontal (**ac**) and vertical (**bd**) components of the group velocity of internal waves for $f = 10^{-4}\,\mathrm{s}^{-1}$ and $N = 5 \times 10^{-3}\,\mathrm{s}^{-1}$. **a**,**b** show these quantities as function of the scaled frequency $\omega/N$ for 4 horizontal wavelengths $2\pi/k_{\mathrm{h}} = 10, 10^2, 10^3, 10^4$ m, **c**,**d** as function of horizontal wave number $k_{\mathrm{h}}$ for 4 frequencies $\omega = 1.01f, 1.05f, 0.5N, 0.99N$ (colors are for both in the order black, blue, red, green). Note that the wave-number axes and the scales for group velocity are logarithmic ($c_3$ is displayed for $k_3 < 0$)

The components of the group velocity are displayed in Figure 7.1 (middle and right panels) as function of the wave vector. They have a very large variation of the magnitudes. A useful presentation of the group velocities is given by the above shown functions of frequency and horizontal wave number because in a horizontally homogeneous background these wave parameters stay constant (see Section 7.3 below). These forms are displayed in Figure 7.3. We note that there is no group propagation for the limiting frequencies $\omega = f$ and $\omega = N$. Here internal waves degenerate to nonpropagating oscillations, called inertial and buoyancy oscillations, respectively.

### 7.2.2 Energy Conservation

The mechanical energy conservation is derived in the usual way (cf. Section 2.4.2). Multiplying (7.4) with $\rho_0 \boldsymbol{u}_{\mathrm{h}}$, (7.5) with $\rho_0 w$, and the density conservation (7.6) with $g^2 \rho/\rho_0 N^2$, one obtains after addition

$$\rho_0 \frac{\partial}{\partial t}\left(\frac{1}{2}\boldsymbol{u}_{\mathrm{h}}^2 + \frac{1}{2}w^2 + \frac{1}{2}\frac{g^2}{\rho_0^2 N^2}\rho^2\right) = -\boldsymbol{u}_{\mathrm{h}} \cdot \nabla_{\mathrm{h}} p - w\frac{\partial p}{\partial z} \tag{7.20}$$

which describes the energy conservation for free linear gravity waves. The kinetic energy (per volume) is

$$E_{\mathrm{k}} = \frac{1}{2}\rho_0 \boldsymbol{u}_{\mathrm{h}}^2 + \frac{1}{2}\rho_0 w^2 = E_{\mathrm{k}}^{\mathrm{h}} + E_{\mathrm{k}}^{\mathrm{v}}$$

27. Particle Orbits

The displacement of particles by the wave motion follows from the vector $\boldsymbol{X}(t) = (X(t), Y(t), Z(t))$, determined by $\dot{\boldsymbol{X}} = \boldsymbol{u}$. To derive their shape, it is useful to transform from the coordinates $(x, y, z)$ to a particular coordinate system $(x', y', z')$ in which the wave vector is oriented along the $z'$-axis and has zero component in the $x'$-direction. We thus express the wave vector as $\boldsymbol{k} = (0, -\cos\varphi, \sin\varphi)$, which implies the transformation $x' = x$, $y' = y\sin\varphi + z\cos\varphi$, $z' = -y\cos\varphi + z\sin\varphi$, as sketched in the figure below. The orbit $\boldsymbol{X}(t) = (X'(t), Y'(t), Z'(t))$ in the new coordinates is transformed accordingly. Note, however, that the incompressibility constraint (7.7) yields $-\dot{Y}\cos\varphi + \dot{Z}\sin\varphi = 0$ and hence $\dot{Z}' \equiv 0$ and consequently $Z' \equiv 0$. The orbits is thus in the $(x, y')$ plane perpendicular to $\boldsymbol{k}$.

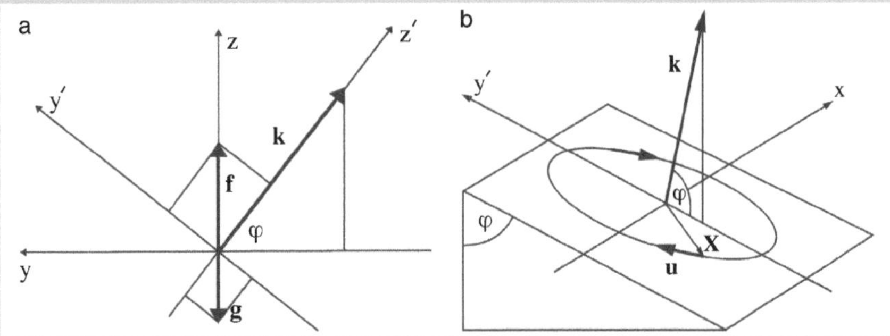


*a Sketch of the coordinate systems $(x, y, z)$ and $(x, y', z')$ ($x$ is pointing into the page). b A three-dimension view: the $(x, y')$-plane has the inclination $\varphi$ with respect to the $(x, y)$-plane and the wave vector $\boldsymbol{k} = (0 - \cos\varphi, \sin\varphi)$ is normal to the $(x, y')$-plane. The orbit is elliptical with the major axis along the line of maximum slope. The displacement vector $\boldsymbol{X}$ and the particle velocity $\boldsymbol{u}$ is indicated, the latter for the northern hemisphere. Note that the group velocity (not shown) is pointing down-slope in this configuration (along the negative $y'$-axis).*

Transforming (7.4)–(7.6) to the new coordinates yields

$$\ddot{X} = \dot{Y}' f \sin\varphi \quad \text{and} \quad \ddot{Y}' = -\dot{X} f \sin\varphi - Y' N^2 \cos^2\varphi$$

In fact, the component of the Coriolis vector perpendicular to the orbit plane is $f \sin\varphi$, which enters in the Coriolis force. Furthermore, the motion feels the density change along the plane, leading to $\cos\varphi$ times the vertical density gradient, and $g\cos\varphi$ is the component of the gravity vector in the plane. Their product enters in the buoyancy force.

The system is solved by

$$X(t) = \frac{f \sin\varphi}{\omega} Y_0 \sin\omega t \qquad \text{and} \qquad Y'(t) = Y_0 \cos\omega t$$

and $\omega$ given by the dispersion relation (7.17). The orbit is thus an ellipse with the ratio of along-slope to up-slope axes of $(f/\omega)\sin\varphi$. It is sketched in the above figure. For near-inertial frequencies we have $\boldsymbol{k}$ almost vertical ($\sin\varphi \approx 1$) and an almost circular orbit and almost horizontal motion. At higher frequencies, the wave vector tends to the horizontal, the ellipse towards the vertical, and it becomes more and more elongated in the $y'$-direction, pointing almost in the $z$-direction. Hence, an almost vertical motion is found approaching $\omega = N$ (where $\sin\varphi = 0$). The motion along the orbit is anticyclonic, i. e. clockwise in the northern hemisphere.

where $E_{\mathrm{k}}^{\mathrm{h}}$ and $E_{\mathrm{k}}^{\mathrm{v}}$ are horizontal and vertical kinetic energy, respectively. The last term in the bracket on the left-hand side of (7.20) is identified as *potential* energy per volume,

$$E_{\mathrm{p}} = \frac{1}{2}\rho_0 \frac{g^2}{\rho_0^2 N^2}\rho^2$$

(cf. Section 5.2.6). With the definitions $E_{\text{mech}} = E_{\text{k}} + E_{\text{p}}$ for the total mechanical energy and $\boldsymbol{J}_{\text{h}} = \boldsymbol{u}_{\text{h}} p$ respectively $J_3 = wp$ for the horizontal and vertical components of the mechanical energy flux vector, and with the continuity equation (7.7) one obtains the energy conservation in the standard form

$$\frac{\partial E_{\text{mech}}}{\partial t} = -\boldsymbol{\nabla}_{\text{h}} \cdot \boldsymbol{J}_{\text{h}} - \frac{\partial J_3}{\partial z} \tag{7.21}$$

It is illuminating to evaluate energy and energy flux with the harmonic wave solution (7.10) and (7.15). The ratio of vertical and horizontal kinetic energies is

$$\frac{E_{\text{k}}^{\text{v}}}{E_{\text{k}}^{\text{h}}} = \frac{\omega^2}{\omega^2 + f^2} \frac{\omega^2 - f^2}{N^2 - \omega^2}$$

The horizontal kinetic energy dominates near the inertial frequency $\omega \to f$, the vertical at $\omega \to N$. The ratio between potential and kinetic energy is given by

$$\frac{E_{\text{p}}}{E_{\text{k}}} = \frac{N^2(\omega^2 - f^2)}{\omega^2(N^2 - f^2) + f^2(N^2 - \omega^2)} \leq 1$$

The potential energy is always smaller than the kinetic energy and becomes very small near the inertial frequency. The Earth's rotation prevents an equi-distribution between both energy forms which prevailed for the sound waves. Only for $\omega \to N$, or for $f \to 0$, both energy forms have similar size. For the total energy one obtains

$$E_{\text{mech}} = \frac{1}{2}\rho_0 |W_0|^2 2 \frac{N^2 - f^2}{\omega^2 - f^2} \tag{7.22}$$

With the harmonic wave solution, the energy flux components $p\boldsymbol{u}$ and $pw$ in (7.21) can also be expressed[1] in terms of $W_0$ as

$$\boldsymbol{J}_{\text{h}} = (p\boldsymbol{u}_{\text{h}}^* + p^*\boldsymbol{u}_{\text{h}})/2 = \frac{1}{2}\rho_0 |W_0|^2 \frac{N^2 - \omega^2}{\omega^2 k_{\text{h}}^2} 2\omega \boldsymbol{k}_{\text{h}} = \frac{N^2 - \omega^2}{N^2 - f^2} \frac{\omega^2 - f^2}{\omega k_{\text{h}}} \frac{\boldsymbol{k}_{\text{h}}}{k_{\text{h}}} E_{\text{mech}} \tag{7.23}$$

For the last identity the relation (7.22) has been used. From the dispersion relation (7.16), we obtain after a brief calculation $\boldsymbol{J}_{\text{h}} = \boldsymbol{c}_{\text{g}_{\text{h}}} E_{\text{mech}}$ and $J_3 = c_{\text{g}_3} E_{\text{mech}}$, and the energy conservation (7.20) has again the canonical form

$$\frac{\partial E_{\text{mech}}}{\partial t} = -\boldsymbol{\nabla} \cdot \boldsymbol{c}_{\text{g}} E_{\text{mech}} \tag{7.24}$$

which is formally identical with equation (6.57) for sound waves, with potential energy replacing elastic energy.

## 7.3 Propagation in Variable Stratification

The wave solutions derived in the previous sections are valid for the idealized case of constant environmental conditions, in particular constant buoyancy frequency. For a variable stratification, $N = N(z)$, one can expect refraction and a corresponding modification of the propagation characteristics. Variations on the meridional direction of the Coriolis parameter can also lead to refraction; these are mainly relevant for long gravity waves and will be dealt with in Section 8.3.1.

[1] Since the field variables are complex but physical quantities are real, the real part of products is to be taken, e. g. for the horizontal energy flux $\Re(\boldsymbol{u}_{\text{h}} p) = (p\boldsymbol{u}_{\text{h}}^* + p^*\boldsymbol{u}_{\text{h}})/2$.

### 7.3.1 WKBJ Approximation for Internal Waves

Approximate solutions can be found with the WKBJ technique discussed in Section 6.3, provided that the vertical wavelength is short compared with the scale of variation of $N(z)$. Analogous to (6.35), the approximate solution is written as $w = W(\boldsymbol{x}_\mathrm{h}, z, t)\mathrm{e}^{\mathrm{i}\phi(\boldsymbol{x}_\mathrm{h},z,t)}$. Wave number and frequency are related to the phase $\phi$ by

$$\boldsymbol{k}_\mathrm{h} = \nabla_\mathrm{h}\phi\,, \quad k_3 = \frac{\partial\phi}{\partial z}\,, \quad \omega = -\frac{\partial\phi}{\partial t}$$

The kinematic treatment of internal gravity waves is completely analogous to the sound waves in Chapter 6.3.1 and yields the *local* dispersion relation

$$\omega = \Omega(\boldsymbol{k}_\mathrm{h}, k_3, \boldsymbol{x}_\mathrm{h}, z, t) = \left(\frac{N^2(z)k_\mathrm{h}^2 + f^2k_3^2}{k_\mathrm{h}^2 + k_3^2}\right)^{\frac{1}{2}} \tag{7.25}$$

In the dispersion relation the environmental parameters $f$ (constant on the $f$-plane) and $N = N(z)$ appear. Therefore, only the explicit dependence on $z$ needs to be considered. The evaluation of the characteristic equations leads to

$$\dot{\boldsymbol{x}}_\mathrm{h} = \frac{\partial\Omega}{\partial\boldsymbol{k}_\mathrm{h}} = \boldsymbol{c}_\mathrm{gh} \tag{7.26}$$

$$\dot{z} = \frac{\partial\Omega}{\partial k_3} = c_{\mathrm{g}3} \tag{7.27}$$

$$\dot{\boldsymbol{k}}_\mathrm{h} = -\frac{\partial\Omega}{\partial\boldsymbol{x}_\mathrm{h}} = 0 \tag{7.28}$$

$$\dot{k}_3 = -\frac{\partial\Omega}{\partial z} = -\frac{\partial\Omega}{\partial N}\frac{\partial N}{\partial z} = -\frac{k_\mathrm{h}^2}{\left(k_\mathrm{h}^2 + k_3^2\right)}\frac{N}{\omega}\frac{\partial N}{\partial z} \tag{7.29}$$

$$\dot{\omega} = \frac{\partial\Omega}{\partial t} = 0 \tag{7.30}$$

Note that (7.30) is redundant, but nevertheless often quite convenient. According to (7.29), the sign of $\dot{k}_3$ is always opposite to that of $\partial N/\partial z$. The stratification $N(z)$ has a maximum in the main thermocline. Below the thermocline $\dot{k}_3 < 0$, above $\dot{k}_3 > 0$. Therefore, the path of the wave groups must have a form as outlined in Figure 7.4. It is, however, not necessary to explicitly solve (7.29) since from (7.28) and (7.30) the integrals $\boldsymbol{k}_\mathrm{h} = $ const and $\omega = $ const follow, and, therefore, $k_3$ can be determined algebraically from (7.25) as

$$k_3(z) = \pm\left(\frac{N^2(z) - \omega^2}{\omega^2 - f^2}\right)^{\frac{1}{2}} k_\mathrm{h} \tag{7.31}$$

Equations (7.26) and (7.27) reflect the fact that wave groups propagate with the group velocity which is orthogonal to the wave-number vector. With (7.18) and (7.19), they can now be solved since the wave numbers are known.

At a rigid boundary, we can expect reflection of the wave group so that frequency and the wave-number component parallel to the boundary are unchanged. To the extent that the boundaries (air-sea interface and bottom) can be idealized as a plane surfaces, it follows that the horizontal wave number $\boldsymbol{k}_\mathrm{h}$ as well as the frequency $\omega$ remain constant while the vertical component $k_3$ from (7.31) changes its sign. This occurs for two of the waves in Figure 7.4.

**28. WKBJ Solution**

In order to find an approximate solution for the amplitude $W(z)$, equation (7.14) is rewritten as

$$\frac{\mathrm{d}^2 W}{\mathrm{d}z^2} + q(z)W = 0 \quad \text{with} \quad q(z) = \frac{N^2(z) - \omega^2}{\omega^2 - f^2} k_\mathrm{h}^2 \tag{B28.1}$$

We transform the independent and dependent variables according to $\xi = \gamma(z)$ and $H = \gamma'^{\frac{1}{2}} W(z)$ with a yet unspecified, monotonous function $\gamma(z)$, and $\gamma' = \mathrm{d}\gamma/\mathrm{d}z$. It follows that

$$\frac{\mathrm{d}^2 H}{\mathrm{d}\xi^2} + \frac{q}{\gamma'^2} H = \left( \frac{1}{2} \frac{\gamma'''}{\gamma'^3} - \frac{3}{4} \frac{\gamma''^2}{\gamma'^4} \right) H \tag{B28.2}$$

The specific choice $\gamma'^2 \equiv q$, or $\gamma(z) = \int_{z_0}^{z} q^{1/2}(z')\mathrm{d}z'$ with arbitrary $z_0$ yields

$$\frac{\mathrm{d}^2 H}{\mathrm{d}\xi^2} + H = \left( \frac{1}{4} \frac{q''}{q^2} - \frac{5}{16} \frac{q'^2}{q^3} \right) H = r(\xi) H \tag{B28.3}$$

So far no approximations have been made. For $q = \text{const}$ the right-hand side of (B28.3) would vanish. If $q(z)$ changes slowly, the right side is small and can be neglected, yielding the approximate solution $H(\xi) \approx \text{const}\ \mathrm{e}^{\pm \mathrm{i}\xi}$, or in the original variables

$$W(z) \approx \frac{\text{const}}{[q(z)]^{1/4}} \exp\left( \pm i \int_{z_0}^{z} q^{1/2}(z')\mathrm{d}z' \right) \tag{B28.4}$$

This is the standard WKBJ solution of (B28.1) which in many situations is an excellent approximation.

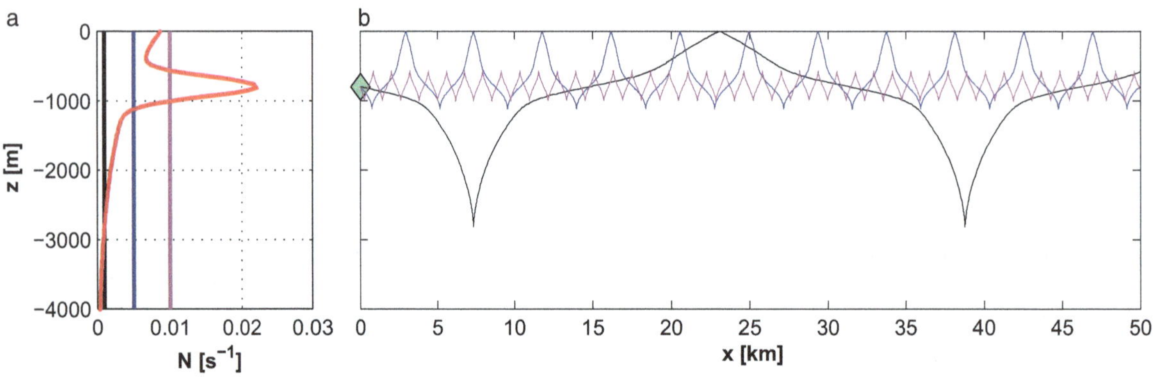


Fig. 7.4 **b** Example of ray propagation in a variable $N$-profile for three frequencies with initially positive vertical wave numbers. All the waves, therefore, propagate downward at the initial time. For each case the horizontal wave length is 1,000 m, the source in 800 m depth. **a** shows the Brunt–Väisälä frequency (red) and the three frequencies marked by straight lines

## 7.3.2 Turning Points

Analytical solutions of the general wave equation (7.14) along a WKBJ approximation are discussed in the box on p. 186. The solution breaks down when $k_3(z)$ becomes very small, thus violating the WKBJ condition of a small wavelength in comparison with the background variations. It is interesting to consider what happens at a depth level $z_\star$ where $k_3(z_\star) = 0$. From (7.14) it follows that in regions where $N(z) < \omega$ the exact solution $W(z)$ is no longer oscillatory but exponentially decaying, and has a turning point at $N(z_\star) = \omega$. From (7.31) it follows that $k_3 \sim (N^2 - \omega^2)^{1/2} \to 0$. According to (7.18), respectively (7.19), $\boldsymbol{c}_g \to 0$ is valid for all components of the group velocity. Note that although the vertical group velocity becomes small, the wave group still reaches the level $z_\star$ in a finite time as $c_{g3} \sim (N^2 - \omega^2)^{1/2} \sim |z - z_\star|^{1/2}$ implies $t_\star - t_0 \sim |z_0 - z_\star|^{1/2}$ for the time to proceed from $z_0$ to $z_\star$.

The failure of the WKBJ solution (B28.4) at turning points can also be seen from the energy equation (7.24) which for $\partial/\partial t = 0$ (stationarity) and $\nabla_h = 0$ (horizontal homogeneity) leads to

$$\frac{\partial c_{g3} E_{\text{mech}}}{\partial z} = 0 \quad \text{or} \quad c_{g3} E_{\text{mech}} = \text{const} \quad \text{for} \quad z \to z_\star$$

i. e. the energy density would become singular because $c_{g3} \to 0$. The reason for the failure of the WKBJ solution is that the precondition $\partial/\partial z \ll |k_3|$ is no longer satisfied because $k_3$ vanishes at $z = z_\star$. Physically it is clear that waves cannot propagate into the region where $N(z) < \omega$. As dissipation is unlikely to play any role because the vertical scale is large ($k_3 \to 0$), the only plausible continuation of the characteristic at the turning point is, therefore, obtained by reflection at $z = z_\star$. All three waves in Figure 7.4 have turning points on their way downward and the one with the highest frequency also on the way up. The latter is trapped in the channel by the depth-varying $N$-profile.

The wave form in the vicinity of a turning point can be determined analytically and expressed in terms of AIRY[2]-functions, as shown in the box on p. 188.

## 7.4 The Influence of Boundaries

Typical length scales of internal waves are 10 m to 10 km in the horizontal and 10 m to 1 km in the vertical. Therefore, it is of interest to consider what happens when the waves approach the boundaries at the sea floor and/or the air-sea interface.

### 7.4.1 Reflection at a Plane Interface

Consider a wave of the form (7.15) which propagates toward the sea floor at $z = -H = \text{const}$. A solution can be found by superposition of two solutions in the form

$$W(z) = A_+ \mathrm{e}^{ik_3 z} + A_- \mathrm{e}^{-ik_3 z}$$

One of the two solutions is the *incident* wave, the other one the *reflected* wave. Physically, only the wave with a group velocity directed toward the reflecting interface can be the incident wave. For reflection at the bottom boundary this is the wave with negative $c_{g3}$ and hence positive $k_3$. The other wave with negative $k_3$ is the

[2] GEORGE B. AIRY, *1801 in Alnwick, †1892 in Greenwich, mathematician and astronomer.

**29. Turning Point Solution**

It is possible to find an approximate solution for the amplitude $W(z)$ which remains valid near a turning point. In (B28.2) we now choose the transformation

$$\xi = \gamma(z) = \frac{3}{2}i\left[\int_{z_\star}^{z} q^{1/2}(z')\mathrm{d}z'\right]^{2/3}$$

and obtain

$$\frac{\mathrm{d}^2H}{\mathrm{d}\xi^2} - \xi H = r(\xi)H$$

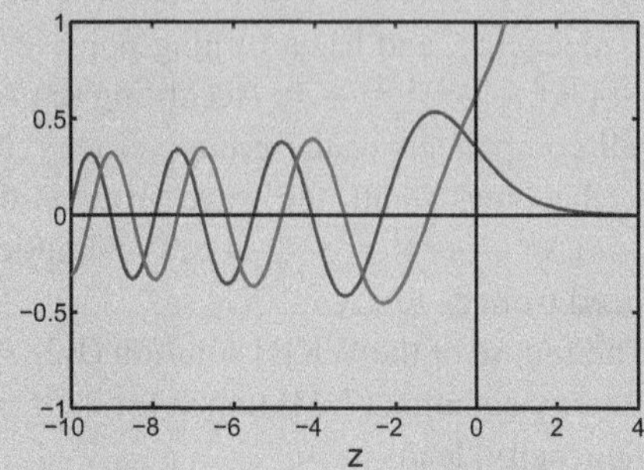


*Airy functions* $\mathrm{Ai}(z)$ *[blue] and* $\mathrm{Bi}(z)$ *[green].*

with $r(\xi)$ as in (B28.3). If the right side can again be neglected, we obtain $H(\xi) \approx c_1\mathrm{Ai}(\xi) + c_2\mathrm{Bi}(\xi)$ with $\mathrm{Ai}(\xi)$ and $\mathrm{Bi}(\xi)$ being the Airy functions which exponentially decrease for a positive argument (the Ai term), respectively increase (the Bi term, this term is thus discarded), and oscillate for a negative argument. In the original variables, the solution then is

$$W(z) \approx \mathrm{const}\left(\frac{\gamma(z)}{q(z)}\right)^{1/4} \mathrm{Ai}\left(\gamma(z)\right)$$

This approximation which was derived by Langer (1949) is regular at $z = z_\star$ and has a turning point like the exact solution. At a large distance from the turning point, it merges into the WKBJ solution.

reflected wave. With the choice $A_- = -A_+\mathrm{e}^{-2ik_3H}$, the boundary condition $w = 0$ at $z = -H$ is satisfied. The solution is then of the form

$$W(z) = A_+\left(\mathrm{e}^{ik_3z} - \mathrm{e}^{-ik_3z}\mathrm{e}^{-2ik_3H}\right) \equiv A\sin k_3(z+H) \tag{7.32}$$

with $A = 2iA_+\mathrm{e}^{-ik_3H}$, and has the form of a *standing wave* in the vertical direction. It can be interpreted as the superposition of two propagating waves which have the same amplitude and a phase difference of 180°.

### *7.4.2 Reflection at a Sloping Bottom*

The reflection problem can be generalized to a plane bottom sloping by an angle $\alpha$ against the horizontal, i. e. $h = H_0 - x_1\tan\alpha$. For simplicity, consider the case $k_2 = 0$ and assume again a constant buoyancy frequency $N(z) = \mathrm{const} = N_\mathrm{b}$, representative for the stratification at the bottom. It is convenient to introduce new coordinates $x_\parallel$, $x_\perp$ which are parallel, respectively vertical, to the bottom (see Fig. 7.5). In the new coordinates, a harmonic wave solution can be written as $\psi = \psi_0\exp i(k^\mathrm{i}_\parallel x_\parallel + k^\mathrm{i}_\perp x_\perp - \omega^\mathrm{i}t)$ for all field variables including the velocity components. Consider an incoming wave (index $i$) propagating in positive $x_1$-direction toward shallower waters, so that its group velocity has a downward component. The solution for the velocity component $u_\perp$ is written as a sum of incident and reflected wave,

$$u_\perp = u^\mathrm{i}_\perp \exp i\left(k^\mathrm{i}_\parallel x_\parallel + k^\mathrm{i}_\perp x_\perp - \omega^\mathrm{i}t\right) + u^\mathrm{r}_\perp \exp i\left(k^\mathrm{r}_\parallel x_\parallel + k^\mathrm{r}_\perp x_\perp - \omega^\mathrm{r}t\right)$$

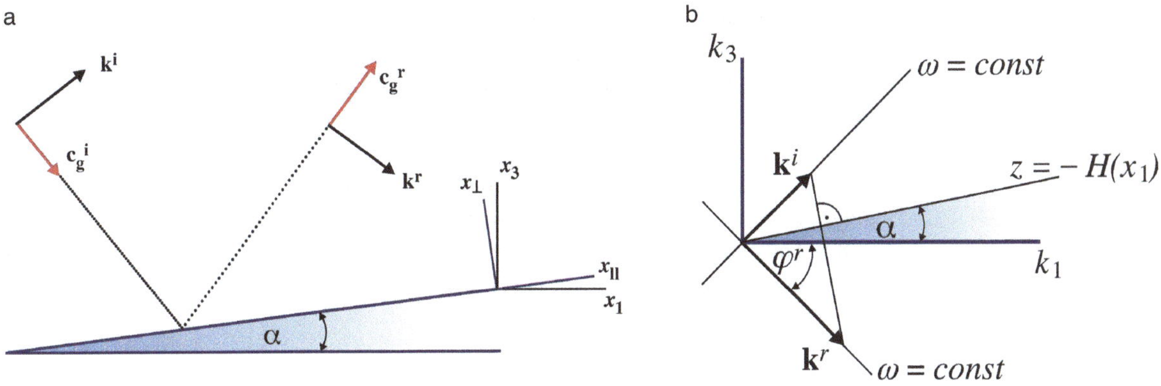


Fig. 7.5 **a** Reflection of an internal gravity wave in the wavenumber plane. **b** Sketch of the incident and reflected waves in the wavenumber plane. See text

Mass conservation at the bottom requires $w = -\boldsymbol{u}_h \cdot \nabla h$ or $u_\perp = 0$ at $x_\perp = 0$ at all $x_\parallel$ and $t$. This condition can be satisfied only if $k^r_\parallel = k^i_\parallel$ and $\omega^r = \omega^i$. Therefore, the wave number $k^r_\perp$ of the reflected wave can be determined algebraically by rewriting the dispersion relation (7.16) in terms of $k_\parallel$ and $k_\perp$. It is, however, illuminating to determine the reflected wave number graphically by considering the situation in a $(k_1, k_3)$ respectively $(k_\parallel, k_\perp)$ diagram (see Figure 7.5). The wave number of the incident wave has a magnitude $k^i$ and an angle $\varphi^i = \varphi$ against the horizontal. The reflected wave must have the same frequency, and, therefore, $\varphi^r = -\varphi$ must hold. Geometrically, we see that $k^i_\parallel / k^i = \cos(\varphi - \alpha)$ and $k^r_\parallel / k^r = \cos(\varphi + \alpha)$. With $k^r_\parallel = k^i_\parallel$, one obtains for the reflected wave number

$$\frac{k^r}{k^i} = \frac{\cos(\alpha - \varphi)}{\cos(\alpha + \varphi)} \quad \text{and also} \quad \frac{k^r_\perp}{k^i_\perp} = \frac{\tan(\alpha + \varphi)}{\tan(\alpha - \varphi)} \tag{7.33}$$

Since normally $\alpha \ll 1$ in the ocean, the case $\varphi + \alpha < \pi/2$ is most common for wave vectors which are not nearly vertical. Consider a situation where the group velocity of the incoming wave is directed downward and to regions of shallower water, such as sketched in the left panel of Figure 7.5, so that $k^i_3 > 0$ and $k^i_1 > 0$. For the reflected wave one has $k^r_3 < 0$ and $k^r_1 > 0$ so that the wave continues to propagate toward shallower water. For the magnitude one finds $k^r > k^i$. Qualitatively, the situation resembles the flat-bottom reflection.

Of special interest is the situation when $\varphi + \alpha \to \pi/2$. In this case the reflected wave number becomes orthogonal to the bottom and very large, $k^r \to \infty$, and its group velocity vanishes. For a given $\omega$, respectively $\varphi$, the critical angle $\alpha_c$ for the bottom slope (*critical slope*) is given by

$$\omega^2 = N^2 \sin^2 \alpha_c + f^2 \cos^2 \alpha_c \Leftrightarrow \alpha_c(\omega) = \arctan\left(\frac{\omega^2 - f^2}{N^2 - \omega^2}\right)^{\frac{1}{2}} \tag{7.34}$$

(note that $\cos \alpha_c = \sin \varphi$). In the vicinity of the critical angle $k^r \gg k^i$ is valid. Hence the process of near-critical reflection moves wave energy toward small scales. Here one can expect enhanced nonlinearity and enhanced dissipation since the group velocity of the reflected wave is very slow. This process is thought to play a role for near-bottom diapycnal mixing in the ocean (St. Laurent and Garrett, 2002).

If the incoming wave number is nearly vertical so that $\varphi + \alpha > \pi/2$, one has $k_3^{\mathrm{r}} > 0$ but $k_1^{\mathrm{r}} < 0$ for the reflected wave. The reflected wave now propagates back toward deeper water, and qualitatively the situation resembles reflection at a vertical boundary.

In the general case, with $k_2 \neq 0$, the right diagram in Figure 7.5 and also equation (7.33) have to be modified because the lines of constant frequency in the $(k_1, k_3)$ diagram are hyperbolae rather than straight lines (for a full analysis see e. g. Eriksen, 1982). However, the main conclusion regarding the transport of wave energy toward small scales remains valid, and the critical slope is still given by (7.34).

### 7.4.3 Vertical Modes

The solution (7.32) describes, for constant $N$, a wave reflected at the bottom boundary. The inclusion of a second boundary will lead to a further reflection. For the sea surface, consider for now the simplified condition (7.9). Obviously, the only way to satisfy $W(z) = 0$ both at $z = 0$ and $z = -H$ with (7.32) is to permit only discrete vertical wave numbers, i. e. $k_3 = n\pi/H$ with the solutions

$$W_n(z) = a_n \sin\frac{n\pi z}{H}\,, \quad n = 1, 2, \ldots \tag{7.35}$$

which describes standing modes in the vertical direction. When the length scale

$$R_n = NH/(fn\pi) \tag{7.36}$$

is introduced, the frequency is given by

$$\omega_n^2 = f^2\frac{1 + k_{\mathrm{h}}^2 R_n^2}{1 + k_{\mathrm{h}}^2 H^2/(n\pi)^2} \tag{7.37}$$

which relates, for each value of $n$, the frequency to the horizontal wave number $k_{\mathrm{h}}$ (see Figure 7.6). The length scale $R_n$ can be interpreted as Rossby radius for $N =$ const (see Section 8.1.1).

Since usually $N \gg f$ and thus also $R_n \gg H/(n\pi)$, one can identify three ranges in the dispersion diagram in Figure 7.6 with different dynamical balances. In the long wavelength range with $k \lesssim 1/R_n$ and thus also $k_h \ll n\pi/H$, the dispersion relation can be approximated as

$$\omega_n^2 \approx f^2 + \frac{N^2 k_{\mathrm{h}}^2}{(n\pi/H)^2} = f^2\left(1 + k_{\mathrm{h}}^2 R_n^2\right) \tag{7.38}$$

In this range, the waves are dispersive, rotation is crucial, and the motion is in hydrostatic balance (compare with Chapter 8). In the intermediate range $1/R_n \ll k_{\mathrm{h}} \ll n\pi/H$, the motion still is in hydrostatic balance but rotation is now insignificant. The dispersion relation is approximately

$$\omega_n^2 \approx \frac{N^2 k_{\mathrm{h}}^2}{(n\pi/H)^2} \tag{7.39}$$

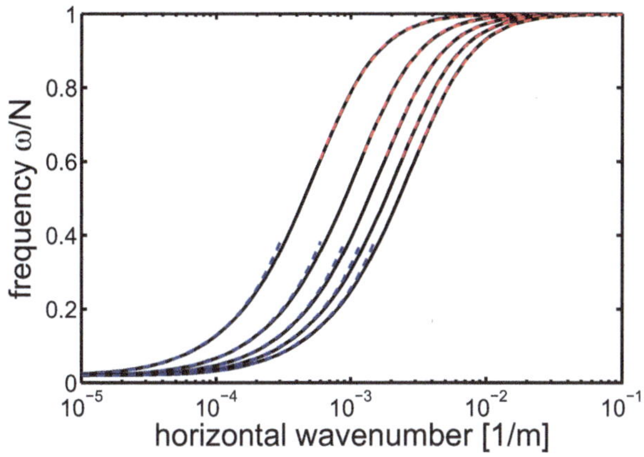


Fig. 7.6 Dispersion relation of internal waves in the wave guide for the first 5 modes, as function of the horizontal wave number $k_{\rm h}$ for $f = 10^{-4}\,{\rm s}^{-1}$ and constant $N = 5\times10^{-3}\,{\rm s}^{-1}$. The complete dispersion relation (black) reflects the approximations for long (blue dashed), intermediate and short (red dashed) wavelengths. The frequency is displayed in the scaled form $\omega/N$

so that wave propagation in this range is nondispersive. For short wavelengths $k_{\rm h} \gtrsim n\pi/H$ and hence $k_{\rm h} \gg 1/R_n$, the dispersion is approximated by

$$\omega_n^2 \approx \frac{N^2 k_{\rm h}^2}{k_{\rm h}^2 + (n\pi/H)^2} \tag{7.40}$$

The vertical momentum balance in this range is *not* hydrostatic. Rotation is unimportant, and wave propagation is again dispersive.

The sinusoidal vertical structure (7.35) and the dispersion relation (7.37) are only valid for $N = \text{const}$ which does not well represent oceanic conditions. For arbitrary $N = N(z)$, an analytical solution cannot be obtained; however, it is possible to derive important properties of the solution. With the definition

$$\lambda = \frac{k_{\rm h}^2}{\omega^2 - f^2} \quad \text{and} \quad r(z) = N^2(z) - \omega^2 \tag{7.41}$$

one obtains from (7.42)

$$\frac{{\rm d}^2 W}{{\rm d}z^2} + \lambda r(z) W = 0 \quad \text{with} \quad W = 0 \quad \text{at} \quad z = 0\,, \quad -H \tag{7.42}$$

Note that $\lambda > 0$. Equation (7.42) is an eigenvalue problem of the STURM–LIOUVILLE[3] type (see the box on p. 192) which can only be solved for certain discrete values of $\lambda$, the *eigenvalues* $\lambda_n$, $n = 1, 2, \ldots$, which belong to the eigenfunctions (*modes*) $W_n(z)$ so that

$$\frac{{\rm d}^2 W_n}{{\rm d}z^2} = -\lambda_n r(z) W_n\,, \quad n = 1, 2, \ldots$$

[3] CHARLES FRANÇOIS STURM, *1803 in Geneva, †1900 in Paris, mathematician. JOSEPH LIOUVILLE, *1809 in Saint-Omer, †1882 in Paris, mathematician.

**30. The Sturm–Liouville Eigenvalue Problem**

The Sturm–Liouville problem is of the form

$$\frac{\mathrm{d}}{\mathrm{d}z}\left(p(z)\frac{\mathrm{d}\phi}{\mathrm{d}z}\right)+q(z)\phi=-\lambda r(z)\phi$$

with homogeneous boundary conditions

$$\phi+B\frac{\mathrm{d}\phi}{\mathrm{d}z}=0 \quad \text{at} \quad z=a \quad \text{and} \quad C\phi+D\frac{\mathrm{d}\phi}{\mathrm{d}z}=0 \quad \text{at} \quad z=b$$

Here $p(z), q(z)$ and $r(z)$ are given functions, $A, B$ and $C, D$ are given constants, $\phi(z)$ is the eigenfunction and $\lambda$ the eigenvalue. In general, $p(z)>0$ is assumed in order that the term with the higher derivative does not become singular in the interval. If, in addition $r(z)>0, q(z)\leq 0$, the solutions have the following properties:

1. there is an infinite countable number of solutions $\phi_n(z), \lambda_n, n=0,1,\ldots$ with $0<\lambda_0<\lambda_1<\ldots$. The eigenfunction $\phi_n(z)$ has $n$ zeros in the open interval $(a,b)$.
2. the set $\{\phi_n\}$ is orthogonal, expressed by

$$\int_a^b r(z)\phi_n(z)\phi_m(z)\mathrm{d}z=\delta_{nm}$$

3. the set $\{\phi_n\}$ is complete, expressed by

$$\sum_{n=0}^{\infty} r(z)\phi_n(z)\phi_n(z')\mathrm{d}z=\delta(z-z')$$

so that any function $g(z)$ can be represented as an infinite sum $g(z)=\sum_{n=0}^{\infty} g_n\phi_n(z)$ where $g_n=\int_a^b r(z)\phi_n(z)g(z)\mathrm{d}z$.

As discussed in Section 7.3.2, the solutions have an oscillating character where $N(z)>\omega$, i. e. $r(z)>0$, an exponential character where $r(z)<0$ and a turning point for $r(z)=0$.

Since $r(z)=N^2(z)-\omega^2$, the eigenfunctions and eigenvalues depend on the value of $\omega$, and hence a more accurate notation would be $\lambda_n=\lambda_n(\omega)$ for the eigenvalues and $W_n(z,\omega)$ for the eigenfunctions. Only if everywhere $N^2 \gg \omega^2$, that dependence can be ignored (see Section 8.1 below). For a given $\omega$, the horizontal wave number follows from (7.41) as $k_\mathrm{h}^2=k_\mathrm{h}^2(\omega)=(\omega^2-f^2)\lambda_n(\omega)$ which can be inverted to yield the dispersion relation

$$\omega=\Omega_n(k_\mathrm{h}) , \quad n=1,2,\ldots \tag{7.43}$$

which depends on the index $n$. Relation (7.43) is generally not available in a closed analytical form and has to be constructed by numerical solution of (7.42). The form of the dispersion curve depends on the details of $N(z)$ and can show a complex behavior. Examples for a typical $N(z)$ profile of the subtropical North Atlantic are presented in Figure 7.7. For $N=$ const we obtain the previous relations (7.35) and (7.37).

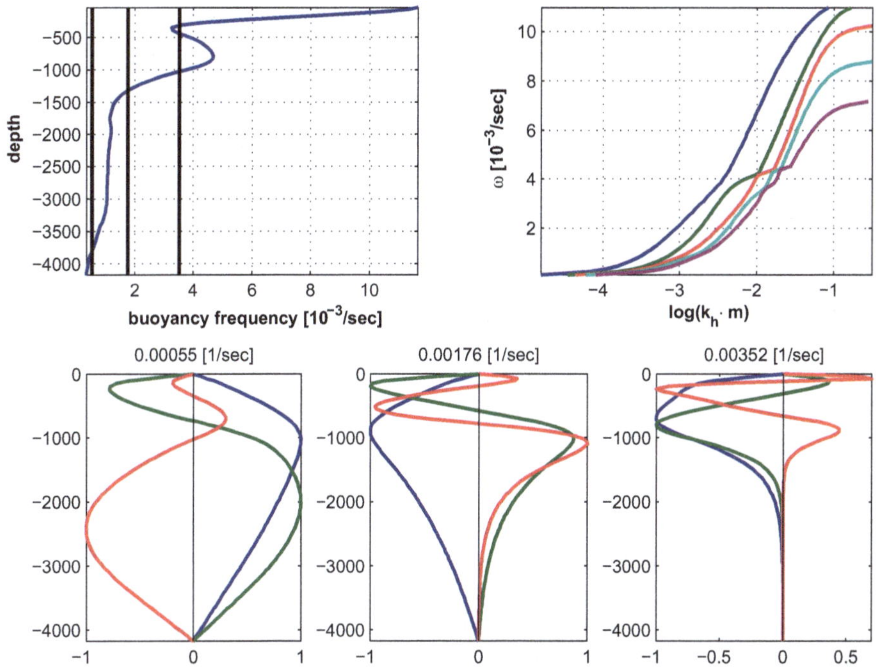


Fig. 7.7 Dispersion relation and vertical eigenmodes for an example of the Brunt-Väisälä frequency profile and $f = 10^{-4}\,\mathrm{s}^{-1}$. The profile of the Brunt–Väisälä frequency, the dispersion relation of the first 5 modes and the first 3 eigenfunctions at three selected frequencies are displayed

### 7.4.4 Accuracy of the Rigid-Lid Condition

So far, the boundary condition at the sea surface has been approximated by the rigid-lid condition (7.9). The accuracy of this approximation is investigated in the following. With the plane wave solution (7.10), the exact surface condition (7.8) takes the form

$$W = \frac{\omega^2 - f^2}{g k_\mathrm{h}^2} \frac{\mathrm{d}W}{\mathrm{d}z} \quad \text{at} \quad z = 0 \tag{7.44}$$

It is sufficient to consider only the case $N = \text{const}$, so we insert expression (7.32) which already satisfies the boundary condition at $z = -H$ into (7.44). It follows that the upper boundary condition can be satisfied only if

$$\omega^2 = f^2 + \frac{k_\mathrm{h}^2}{k_3^2} g k_3 \tan k_3 H \tag{7.45}$$

Using (7.45) to replace $\omega$ in (7.31) leads to

$$\frac{H\left(N^2 - f^2\right)}{g} = \left[\left(k_\mathrm{h}^2 + k_3^2\right) H^2\right] \frac{\tan k_3 H}{k_3 H} \tag{7.46}$$

which can be used to determine the vertical wave number $k_3$ for any given $k_\mathrm{h}$. The left side of (7.46) has the magnitude

$$\epsilon = H \frac{N^2 - f^2}{g} \approx \frac{N^2 H}{g} \sim \frac{H}{g} \frac{g}{\rho_0} \frac{\Delta\rho}{H} \sim \frac{\Delta\rho}{\rho_0} \approx 2 \times 10^{-3} \ll 1$$

Therefore, the right side of (7.46) also must be $\ll 1$, and thus at least one of the two dimensionless factors in (7.46) must be $\ll 1$. The case $\tan(k_3 H)/k_3 H \ll 1$ requires

$$k_3 H = n\pi + O(\epsilon)\,, \quad n = 1, 2, \ldots \quad \text{(but not } n = 0!\text{)} \tag{7.47}$$

Up to terms of order $\epsilon$, this corresponds exactly to the solutions (7.35) obtained with the rigid-lid condition. Therefore, the error caused by the *rigid-lid approximation* is small, approximately 0.2% for internal gravity waves, and the solutions derived in the previous sections remain valid.

## 7.5 Surface Waves

As stated before, the restoring force of gravity can be effective only in the presence of vertical density gradients. The largest possible density gradient occurs at the air-sea interface where the density changes by three orders of magnitude. Therefore, one can expect a wave mode associated with excursions of the free surface.

As starting point consider (7.46) which implicitly relates vertical to horizontal wave number, and is based on the exact surface boundary condition (7.8). We already have found (7.47) as one possibility to satisfy (7.46). The only alternative possibility is that the other dimensionless factor in (7.46) is small, i. e. $(k_\mathrm{h}^2 + k_3^2)H^2 \ll 1$. For arbitrary values of $k_\mathrm{h}$ this cannot be satisfied with a real $k_3$. Allowing, however, imaginary values for $k_3$, one has

$$k_3 = i k_\mathrm{h} + O(\epsilon) \tag{7.48}$$

The vertical structure becomes now exponentially decreasing from the surface. With (7.48) and the identity $ix \tan ix = -x \tanh x$, from (7.45) one obtains

$$\omega^2 = f^2 + g k_\mathrm{h} \tanh k_\mathrm{h} H + O(\epsilon) \tag{7.49}$$

which is dispersion relation of surface gravity waves, shown in Figure 7.8 for various values of the water depth $H$.With (7.48) and (7.32), the vertical profiles of all field variables are given as hyperbolic functions. With (7.11), the solution for the velocity field is given in real form as

$$w = a\omega \frac{\sinh k_\mathrm{h}(z+H)}{\sinh k_\mathrm{h} H} \sin\theta = W(z)\sin\theta \tag{7.50}$$

$$\boldsymbol{u} = a\omega \frac{\cosh k_\mathrm{h}(z+H)}{\sinh k_\mathrm{h} H} \left( \frac{\boldsymbol{k}_\mathrm{h}}{k_\mathrm{h}} \cos\theta - \frac{\boldsymbol{f} \times \boldsymbol{k}_\mathrm{h}}{\omega k_\mathrm{h}} \sin\theta \right) = \boldsymbol{U}_\mathrm{c}(z)\cos\theta + \boldsymbol{U}_\mathrm{s}(z)\sin\theta \tag{7.51}$$

with $\theta = \boldsymbol{k}_\mathrm{h} \cdot \boldsymbol{x}_\mathrm{h} - \omega t$. Here the amplitude $a$ is chosen such that the surface elevation is $\zeta = a\cos\theta$. With (7.12) and (7.13), solutions for the other field variables can be obtained. The solutions correspond to those of internal gravity waves, except for the imaginary vertical wave number. Note that to order $\epsilon$, both the solution (7.50) and (7.51) and the dispersion relation (7.49) are independent of stratification.

Two limiting cases can be distinguished. For long waves with $k_\mathrm{h} H \ll 1$, one has $\tanh k_\mathrm{h} H \approx k_\mathrm{h} H$, and pressure as well as horizontal velocity do no longer depend

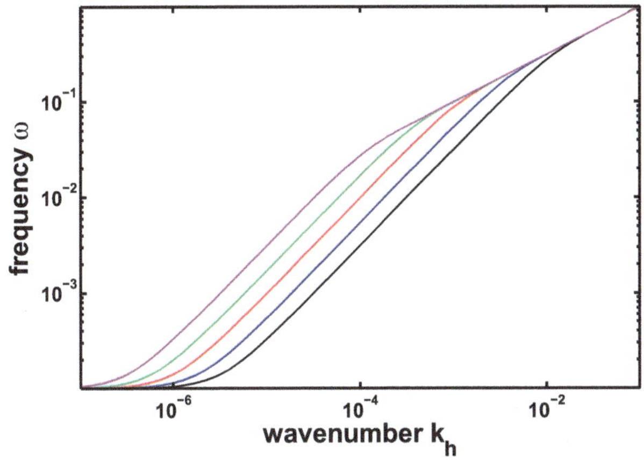


Fig. 7.8 Dispersion relation (7.49) of surface gravity waves for $f = 10^{-4}\,\mathrm{s}^{-1}$ and various depths $H = 100, 300, 1{,}000, 3{,}000, 10{,}000\,\mathrm{m}$ (colors in the order black to magenta), as function of the horizontal wave number $k_\mathrm{h}$. Units: frequency in $\mathrm{s}^{-1}$, wave number in $\mathrm{m}^{-1}$

on depth while $w$ decreases linearly from surface to bottom. The dispersion relation becomes $\omega^2 = f^2 + gHk_\mathrm{h}^2$, and rotation is only significant on scales of order $\sqrt{gH}/f \sim 2{,}000\,\mathrm{km}$, such as e. g. for tides. The phase speed $c = \sqrt{f^2/k_\mathrm{h}^2 + gH}$ approaches $c \approx \sqrt{gH}$, and exceeds this value only for very long waves. The amplitudes of the horizontal velocity are vertically constant, and the vertical velocity decreases linearly to the bottom. On the other hand, for short waves with $k_\mathrm{h}H \gg 1$, $\tanh k_\mathrm{h}H \rightarrow 1$ applies. Since $gk_\mathrm{h} \gg g/H \gg f^2$, the dispersion relation is $\omega^2 = gk_\mathrm{h}$ so that in the dispersion diagram Figure 7.8 eventually (on the far right side) all curves fall together. From (7.50) it further follows that pressure and thus all other fields decrease exponentially in the vertical direction.

### Particle motion

The motion of particles can be determined from the velocity field in (7.50)–(7.51). Consider specifically the case that $\omega \gg f$ so that in (7.51) $|\boldsymbol{U}_s| \ll |\boldsymbol{U}_c|$ and hence $\boldsymbol{u}_\mathrm{h}||\boldsymbol{k}_\mathrm{h}$. For convenience we choose the orientation of the coordinate system such that wave propagation is in the $x_1$-direction, i. e. $k_2 = 0$.

With $x = x_1$, the particle path is described by the ordinary differential equations $\dot{x} = u(x, z, t)$ and $\dot{z} = w(x, z, t)$. Since the velocity amplitude is small, the particle excursions are also small, and one can attempt the iterative solution

$$\dot{x}_{n+1} = u(x_n, z_n, t) \tag{7.52}$$

and likewise for $z$, with $x_0, z_0$ as constant initial values. With (7.50) and (7.51) and $\theta_0 = k_\mathrm{h}x_0 - \omega t$, the first approximation is obtained as

$$\dot{x}_1 = U_c(z_0)\,\cos\theta_0 \quad \text{and} \quad \dot{z}_1 = W(z_0)\sin\theta_0$$

which can immediately be integrated to

$$x_1 = x_0 - \frac{U_c(z_0)}{\omega}\sin\theta_0 \quad \text{and} \quad z_1 = z_0 + \frac{W(z_0)}{\omega}\cos\theta_0$$

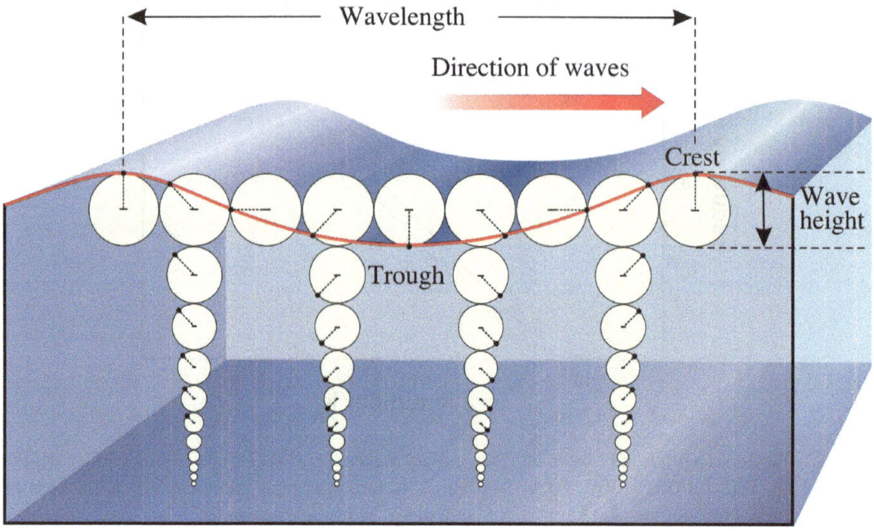


Fig. 7.9 Sketch of particle paths in a surface wave in deep water

and describes elliptical orbits in the vertical plane, centered around $(x_0, z_0)$ and with half axes

$$A = \frac{U_c(z_0)}{\omega} = a\frac{\cosh k_h(z_0 + H)}{\sinh k_h H} \quad \text{and} \quad B = \frac{W(z_0)}{\omega} = a\frac{\sinh k_h(z_0 + H)}{\sinh k_h H} \tag{7.53}$$

For $k_h H \ll 1$, it follows from (7.53) that $A = a/k_h H$, $B = a(1 + z_0/H)$, and the particle paths are ellipses with the axis ratio $B/A = k_h H(1+z_0/H) \ll 1$, i. e. nearly horizontal. On the other hand, for $k_h H \gg 1$ (deep water) one has $A = ae^{k_h z_0}$, $B = ae^{k_h z_0}$, and the particle paths are circles with the radius $A = B = ae^{k_h z_0}$ which vanishes exponentially for $|k_h z_0| \gg 1$. This case is sketched in Figure 7.9.

In the general case when rotation cannot be neglected, the orientation of the elliptical path is no longer vertical but tilted in the direction orthogonal to $\boldsymbol{k}_h$, and the half-axes in (7.53) describe the projection to the vertical plane. The next iteration of (7.52) is discussed in the box on p. 197 on the Stokes drift.

## 7.6 Group Velocity III: Initial Value Problems and Stationary Phase Method

In this section we will look from a different angle at the group velocity in a dispersive wave system. Consider a one-dimensional wave solution

$$\psi(x, t) = ae^{i(kx-\omega t)} \quad \text{with} \quad \omega = \Omega(k)$$

for any dependent variable $\psi$. According to the principle of superposition, which is valid for all linear wave processes, a finite or infinite sum of solutions is also a solution, i. e.

$$\psi(x,t) = \sum_n a_n e^{i[k_n x - \Omega(k_n)t]}$$

or likewise

$$\psi(x,t) = \int_{-\infty}^{\infty} a(k) e^{i[kx - \Omega(k)t]} dk \tag{7.54}$$

for an arbitrary amplitude distribution $a(k)$. The representation (7.54) can now be used to find a wave solution to a given initial condition $\psi(x,0) = \psi_0(x)$ at the time $t = 0$. The initial pattern $\psi_0(x)$ must satisfy

$$\psi(x,0) = \psi_0(x) = \int_{-\infty}^{\infty} a(k) e^{ikx} dk \tag{7.55}$$

so that from the inversion of the Fourier transformation the amplitude function can be determined according to

$$a(k) = \frac{1}{2\pi} \int_{-\infty}^{\infty} \psi_0(x') e^{-ikx'} dx' \tag{7.56}$$

Insertion into (7.54) results in the solution of the initial value problem in the form

$$\psi(x,t) = \int_{-\infty}^{\infty} e^{-i\Omega(k)t} \left( \frac{1}{2\pi} \int_{-\infty}^{\infty} \psi_0(x') e^{ik(x-x')} dx' \right) dk \tag{7.57}$$

**31. Stokes Drift**

It is interesting to consider the next approximation of (7.52) which is

$$\begin{aligned}\dot{x}_2 &= u(x_1, z_0, t) \approx u(x_0, z_0, t) + \frac{\partial u}{\partial x} \cdot (x_1 - x_0) + \dots \\ &= U \cos(k_h \cdot x_0 - \omega t) + U^2 \frac{k_h}{\omega} \sin^2(k_h \cdot x_0 - \omega t) + \dots\end{aligned}$$

Averaging over the phase, a nonvanishing part results

$$u_{st} = \langle \dot{x}_2 \rangle = \frac{k_h}{2\omega} U^2 = \frac{U}{2} \frac{U}{c}$$

Due to

$$\frac{U}{c} = \begin{cases} ak_h, & k_h H \gg 1 \\ a/H, & k_h H \ll 1 \end{cases}$$

we find $u_{st} : U : c = (k_h a)^2 : k_h a : 1$. The velocity $u_{st}$ is called *Stokes drift* and causes a transport into the direction of $\boldsymbol{k}_h$. Although its amplitude is proportional to the square of the wave amplitude and thus is rather small, it is important for momentum and mass transport of waves. Note, however, that the Stokes term might not be the only second order contribution in an expansion of a specific problem.

Of particular interest is the specific case $a(k) = \text{const} = 1$ which corresponds to a very localized initial distribution at $x = 0$, i. e. $\psi_0(x) = 2\pi\delta(x)$ (see the box on p. 199). The solution (7.54) is then

$$\psi(x,t) = \int_{-\infty}^{\infty} e^{i\phi(k,x,t)} dk \tag{7.58}$$

with the phase $\phi(k,x,t) = kx - \omega t = x[k - \Omega(k)t/x]$. Consider now very large values of $x$ and $t$ but a fixed ratio $t/x$. The phase $\phi(k,x,t)$ is very large, and the integrand will generally oscillate quite rapidly as a function of $k$ and thus have very small contributions to the integral. The situation changes when there is a wave number $k_0$ where the phase becomes *stationary*, i. e. when

$$\frac{\partial\phi(k,x,t)}{\partial k} = 0 = x - t\frac{\partial\Omega}{\partial k} \tag{7.59}$$

for $k = k_0 = k_0(x,t)$. In the vicinity of $k_0$, the exponent is still large but does not lead to strong oscillations of the integrand because near-independence of $k$. Hence we can expect considerable contributions to the integral which can be determined as follows. Expansion of the phase into a Taylor series at $k = k_0$, with

$$\left.\frac{\partial\Omega}{\partial k}\right|_{k=k_0} = c_g = \frac{x}{t}$$

leads to

$$\phi = \phi_0 + \left.\frac{\partial\phi}{\partial k}\right|_0 (k-k_0) + \frac{1}{2}\left.\frac{\partial^2\phi}{\partial k^2}\right|_0 (k-k_0)^2 = \phi_0 - t\frac{\partial c_g}{\partial k_0}(k-k_0)^2 + \dots$$

because the linear term vanishes by (7.59).

With the new integration variable $k' = k - k_0$ and the abbreviation $\mathrm{D}(k) = \partial c_g/\partial k$, we obtain for the integral (7.57)

$$\psi(x,t) \approx e^{i\phi_0} \int_{-\infty}^{\infty} e^{-i\frac{t\mathrm{D}(k_0)}{2}k'^2} dk' + \dots = \sqrt{\frac{2\pi}{|t\mathrm{D}(k_0)|}} e^{i[\phi_0 \pm \pi/4]} + \dots \tag{7.60}$$

The $\pm$ sign in the exponent corresponds to the sign of $-t\mathrm{D}(k_0)$. For an arbitrary amplitude spectrum $a(k)$ which is weakly dependent on $k$, the wave amplitude is thus proportional to

$$\psi(x,t) \sim \frac{a(k_0)}{|t\mathrm{D}(k_0)|^{\frac{1}{2}}} e^{i[k_0 x - \omega_0 t]} \tag{7.61}$$

The originally broad spectrum is now constricted to a wave of a single frequency $\omega_0(x,t)$ and wave number $k_0(x,t)$. The ratio $x/t$ exactly matches the corresponding group velocity, a fact that shows directly that the group velocity is indeed governing the propagation of wave energy.

As the denominator of (7.61) shows, the stationary phase method only works as long as $\mathrm{D}(k_0) = \partial c_g/\partial k|k_0 \neq 0$. This is physically evident because only in the presence of dispersion, where different wave components have different velocities, one component can be expected to dominate at a large distance/time. The stationary phase method can, therefore, be used in all wave processes with dispersion, also in the case of internal gravity waves but not in the case of e. g. long surface gravity waves or sound waves. The generalization to more than one dimension is straightforward (Lighthill, 1978).

32. The $\delta$-function

The delta-function $\delta(x)$ is an often useful idealization which can be viewed as a limiting case of a property distribution concentrated at a single point $x = 0$ with total "mass" 1, e. g.

$$\delta(x) \cong \lim_{\epsilon \to 0} \frac{1}{\epsilon\sqrt{2\pi}} e^{-\frac{1}{2}(x/\epsilon)^2}$$

(cf. the figure in the box on p. 349). The delta-function is not a function in the usual sense but rather a "generalized function" which is only suitable for use under an integral (Lighthill, 1958). It can be implicitly defined by

$$f(x) = \int_{-\infty}^{\infty} \delta(x - x') f(x') \mathrm{d}x'$$

for *any* suitable function $f(x)$. Another equivalent definition is

$$\delta(x) = 0 \quad \text{if} \quad x \neq 0 \quad \text{and} \quad \int_{-\infty}^{\infty} \delta(x)\mathrm{d}x = 1$$

## 7.7 Influence of a Mean Flow

So far, our treatment of internal oceanic waves has considered a motionless background, manifesting itself in the Brunt-Väisälä frequency $N(z)$. A more realistic scenario is the inclusion of a mean current on which the waves are superimposed, as discussed for sound waves in Section 6.2.3. Such a current is mainly horizontal and usually has large spatial and time-scales compared to the wavelength and period of the wave. To examine the impact of a mean current on the waves, consider a background state with a mean horizontal velocity vector

$$\boldsymbol{U} = (U_1(z), U_2(z), 0) \tag{7.62}$$

Instead of (7.4)–(7.7), the following linearized equations for the perturbation values (leave out the tilde) result from the perturbation expansion of (7.1)–(7.3),

$$\frac{\partial \boldsymbol{u}_\mathrm{h}}{\partial t} + \boldsymbol{U} \cdot \nabla_\mathrm{h} \boldsymbol{u}_\mathrm{h} + w \cdot \frac{\partial \boldsymbol{U}}{\partial z} + \boldsymbol{f} \times \boldsymbol{u}_\mathrm{h} = -\frac{1}{\rho_0} \nabla_h p \tag{7.63}$$

$$\frac{\partial w}{\partial t} + \boldsymbol{U} \cdot \nabla_\mathrm{h} w = -\frac{1}{\rho_0}\frac{\partial p}{\partial z} - g\frac{\rho}{\rho_0} \tag{7.64}$$

$$\frac{\partial \rho}{\partial t} + \boldsymbol{U} \cdot \nabla_\mathrm{h} \rho - \frac{\rho_0 N^2}{g} w = 0 \tag{7.65}$$

$$\nabla_\mathrm{h} \cdot \boldsymbol{u}_\mathrm{h} + \frac{\partial w}{\partial z} = 0 \tag{7.66}$$

For the case $\boldsymbol{U} = \text{const}$, the wave ansatz (7.15) remains valid to solve (7.63)–(7.66), but one has to replace $\omega$ by $\omega - \boldsymbol{U} \cdot \boldsymbol{k}_h$ in the dispersion relation (7.16), as discussed in Section 6.2.3. Hence

$$(\omega - \boldsymbol{U} \cdot \boldsymbol{k}_\mathrm{h})^2 = \frac{N^2 k_\mathrm{h}^2 + f^2 k_3^2}{k_\mathrm{h}^2 + k_3^2} = \left(\Omega^\mathrm{i}(\boldsymbol{k})\right)^2 \tag{7.67}$$

for the intrinsic frequency, with frequency $\omega = \boldsymbol{U} \cdot \boldsymbol{k}_\mathrm{h} + \Omega^\mathrm{i}(\boldsymbol{k})$ and group velocity

$$\boldsymbol{c}_\mathrm{g} = \frac{\partial \omega}{\partial \boldsymbol{k}} = \boldsymbol{U} + \boldsymbol{c}_\mathrm{g}^\mathrm{i} \tag{7.68}$$

### 7.7.1 Critical Layer Absorption

If $\boldsymbol{U}(z)$ and $N(z)$ are changing *slowly* with regard to $z$, the propagation in a vertical plane can be examined with the help of the WKBJ approximation. From the ray equations (7.26)–(7.30) we obtain

$$\dot{\boldsymbol{x}}_h = \boldsymbol{U} + \frac{\partial \Omega^{\mathrm{i}}}{\partial \boldsymbol{k}_{\mathrm{h}}} \tag{7.69}$$

$$\dot{z} = \frac{\partial \Omega^{\mathrm{i}}}{\partial k_3} \tag{7.70}$$

$$\dot{k_3} = -\boldsymbol{k}_{\mathrm{h}} \cdot \frac{\partial \boldsymbol{U}}{\partial z} - \frac{\partial \Omega^{\mathrm{i}}}{\partial z} \tag{7.71}$$

The equations (7.69) and (7.70) state that the paths move in the direction of the group velocity which is, however, no longer necessarily vertical to the wave-number vector. Since the environmental variables $\boldsymbol{U}$ and $N$ are independent of $\boldsymbol{x}_{\mathrm{h}}$ and $t$, it follows

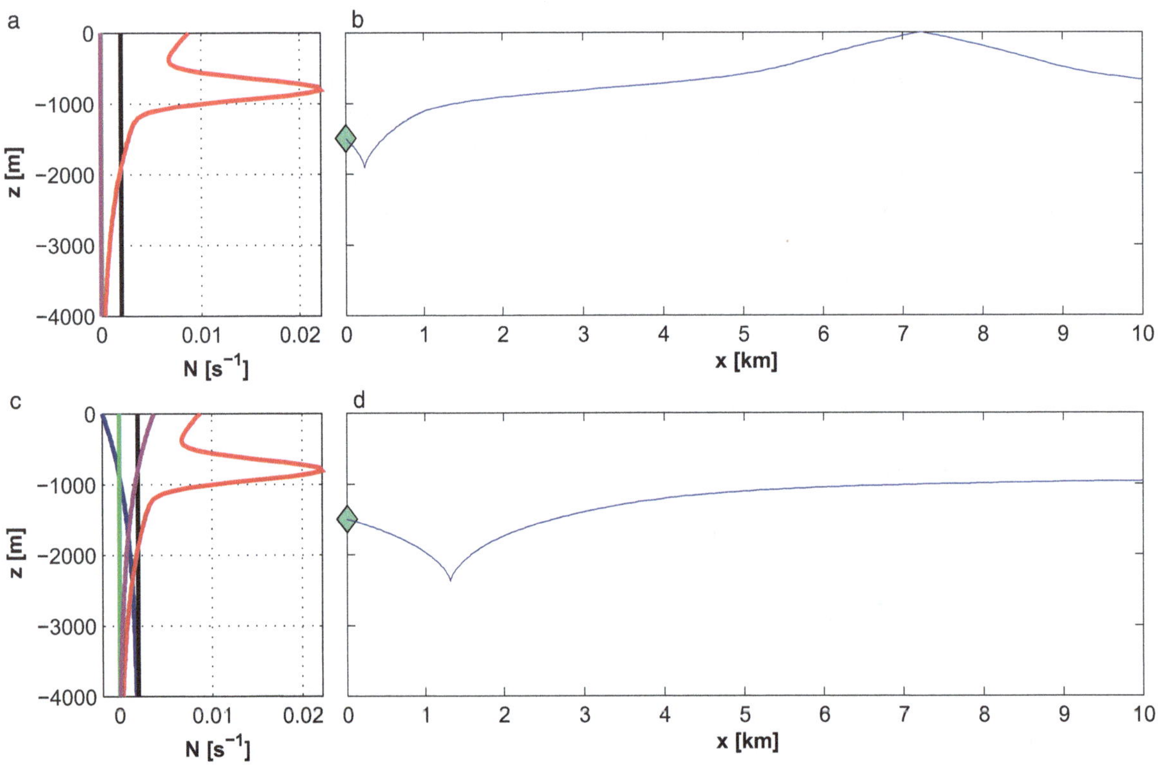


Fig. 7.10 The two row contrast the situations of the ray propagation for zero mean flow (**a**, **b**) and the critical layer case **c**, **d**, occurring with mean flow present. In **a**, **c**, $\boldsymbol{k}_{\mathrm{h}} \cdot \boldsymbol{U}$ is the magenta curve, the Brunt–Väisälä frequency is red, $\omega$ is black, $\omega^{\mathrm{i}}$ is blue, and $f$ is green). For both cases the wave starts with identical initial conditions (horizontal wavelength 1,000 m) and proceeds initially downward. For zero mean flow the wave experiences a turning point and then propagates to the surface where it is reflected. In the case with mean flow the wave also runs into a turning point and then proceeds into a critical layer well below the surface (at the depth where the blue curve ($\omega^{\mathrm{i}}$) and the green curve ($f$) in **a** and **c** intersect)

that $\boldsymbol{k}_\mathrm{h} = \mathrm{const}$ and $\omega = \mathrm{const}$. Solving the dispersion relation (7.67) for $k_3$, one obtains

$$k_3^2 = k_\mathrm{h}^2 \frac{N^2(z) - (\omega - \boldsymbol{U} \cdot \boldsymbol{k}_\mathrm{h})^2}{(\omega - \boldsymbol{U} \cdot \boldsymbol{k}_\mathrm{h})^2 - f^2} \tag{7.72}$$

At a depth where the numerator of (7.72) vanishes, i. e. where $[\omega - \boldsymbol{U}(z) \cdot \boldsymbol{k}_\mathrm{h}]^2 = N^2(z)$, we expect a turning point. The situation is analogous to that of a turning point without mean flow which has been analyzed in Section 7.3.2 and needs no further discussion. However, the denominator of (7.72) can also vanish, this occurs at a depth where

$$[\omega - \boldsymbol{U}(z) \cdot \boldsymbol{k}_\mathrm{h}]^2 = f^2 \tag{7.73}$$

The depth $z = z_\mathrm{c}$ at which this occurs is called *critical layer*. Here, $k_3 \to \infty$ and $c_{g_3} \to 0$ are valid (after (7.19) $c_{g_3}$ is proportional to $1/k_3^2$). Because of the mean flow, however, the horizontal group velocity remains finite with the value $\boldsymbol{c}_\mathrm{gh} = \boldsymbol{U}(z_c)$, i. e. the waves have the same propagation velocity as the mean flow so that the path of a wave packet becomes horizontal near the critical layer.

The wave packet asymptotically approaches the critical layer but cannot leave it anymore. Because of $k_3 \to \infty$ the vertical scale near the critical layer will become small, and it is plausible that eventually dissipation may occur. If the waves steepen, there may also be nonlinear effects and an energy exchange with the mean flow. The process is called *critical layer absorption*.

In contrast to turning points (cf. Section 7.3.2) where the WKBJ approximation is singular but the actual solution is regular, the singularity at the critical layer at $z = z_\mathrm{c}$ is real and indicates that the waves can exchange energy with the background state. The behavior of a ray entering a critical layer is shown in Figure 7.10.

### 7.7.2 Propagation in a Geostrophic Current

The mean flow (7.62) satisfies the continuity equation. The conservation of momentum, however, is only satisfied if the thermal-wind relations (5.56) govern the vertical shear, i. e. $\boldsymbol{f} \times \boldsymbol{U}_z = g\nabla\rho_b/\rho_0$. Hence horizontal gradients of the mean density field $\rho_b$ must exist. The simplest system with such a geostrophic balance is one in which all mean fields are independent of one coordinate, say $y$, and vary in the $(x, z)$-plane only in the direction of a constant vector $(a, b)$ (dimension $\mathrm{m}^{-1}$). Thus we assume $\rho_b = \rho_0 + \Delta\rho\chi(ax + bz)$ with an arbitrary function $\chi(\eta)$, implying

$$N^2(x, z) = -gb\frac{\Delta\rho}{\rho_0}\chi'(ax + bz)\,, \quad V(x, z) = -\frac{ga}{fb}\frac{\Delta\rho}{\rho_0}\chi(ax + bz) + \mathrm{const} \tag{7.74}$$

for the Brunt–Väisälä frequency and the mean current $V$, respectively. For realistic conditions we must have $\chi' < 0$ if $b > 0$. Furthermore, the vector $(a, b)$ should be almost vertical, i. e. $a \ll b$.

The propagation of rays in this system can be treated completely in analytical form (cf. Olbers, 1981). We consider the simplified case where $\chi(\eta)$ is linear (i. e. we take $\chi(\eta) = -\eta$). Then $N = gb\Delta\rho/\rho_0$ is constant and $V(x,z) = V_0(ax+bz)$ with $V_0 = aN^2/(fb^2)$ and shear components $V_x = aV_0$ and $V_z = bV_0$ which are constant. The corresponding ray equations are

$$\dot{x} = \frac{k}{k_\mathrm{h}} \frac{(N^2-\omega_0^2)(\omega_0^2-f^2)}{k_\mathrm{h}\omega_0(N^2-f^2)} \tag{7.75}$$

$$\dot{y} = \frac{\ell}{k_\mathrm{h}} \frac{(N^2-\omega_0^2)(\omega_0^2-f^2)}{k_\mathrm{h}\omega_0(N^2-f^2)} + V_0(ax+bz) \tag{7.76}$$

$$\dot{z} = -\frac{m}{k_\mathrm{h}} \frac{(\omega_0^2-f^2)^2}{k_\mathrm{h}\omega_0(N^2-f^2)} \tag{7.77}$$

$$\dot{k} = -a\ell V_0 \tag{7.78}$$

$$\dot{m} = -b\ell V_0 \tag{7.79}$$

where $\boldsymbol{k} = (k,\ell,m)$ is the wave vector, $k_\mathrm{h}^2 = k^2+\ell^2$, and $\omega_0 = \omega - \ell V$ is the intrinsic frequency. The intrinsic part of the group velocity is identical to our previous problem. Note that $\omega$ and $\ell$ remain constant, and thus, for prescribed $\omega$ and $\ell$,

$$\omega_0 = \omega_0(x,z) = \omega - \ell V_0(ax+bz) \tag{7.80}$$

is a given linear field in the $(x,z)$-plane. Furthermore, the component $k_\perp$ of $(k,m)$ normal to the orientation vector $(a,b)$ obviously stays constant as well. Only the parallel component $k_\parallel$ is varying on account of the current shear. We define the unit vector $(\alpha,\beta) = (a,b)/\sqrt{a^2+b^2}$ so that $k_\perp = -\beta k + \alpha m$, $k_\parallel = \alpha k + \beta m$, $k = \alpha k_\parallel - \beta k_\perp$. Indeed, we find $\dot{k}_\perp = 0$, $\dot{k}_\parallel = -\ell V_0\sqrt{a^2+b^2}$. The dispersion relation may be expressed in the wave numbers $k_\perp$, $k_\parallel$ and the intrinsic frequency $\omega_0$ as

$$\left(L^2-\omega_0^2\right)k_\parallel^2 + \left(M^2-\omega_0^2\right)k_\perp^2 + \left(N^2-\omega_0^2\right)\ell^2 + 2ab(N^2-f^2)k_\perp k_\parallel = 0 \tag{7.81}$$

where we have introduced the frequencies $L$ and $M$ by

$$L^2 = \alpha^2N^2 + \beta^2 f^2\,, \quad M^2 = \beta^2 N^2 + \alpha^2 f^2 \tag{7.82}$$

for convenience, which are constant for our set-up. Note that $f^2 \le (L^2, M^2) \le N^2$. A third constant frequency, defined by

$$\omega_C^2 = \frac{L^2\ell^2 + f^2k_\perp^2}{\ell^2+k_\perp^2} \tag{7.83}$$

enters the problem to characterize the wave guide in the $(x,z)$-plane. Waves can only exist (i. e. a real $k_\parallel$ results from the above dispersion relation (7.81)) within the strip given by $\omega_C^2 \le \omega_0^2(x,z) \le N^2$. There may be two such strips, associated with the $\pm$-solutions. The lines $\omega_0 = \pm L$, lying inside the strips, are of particular interest, as explained below, while the corresponding lines for $\omega_0 = \pm M$, also lying inside, have no particular impact.

The analysis of the dispersion relation and the group velocity, not given here in detail, reveals that a ray approaching $\omega_0 = \pm N$ gets $m \to 0$ and zero intrinsic

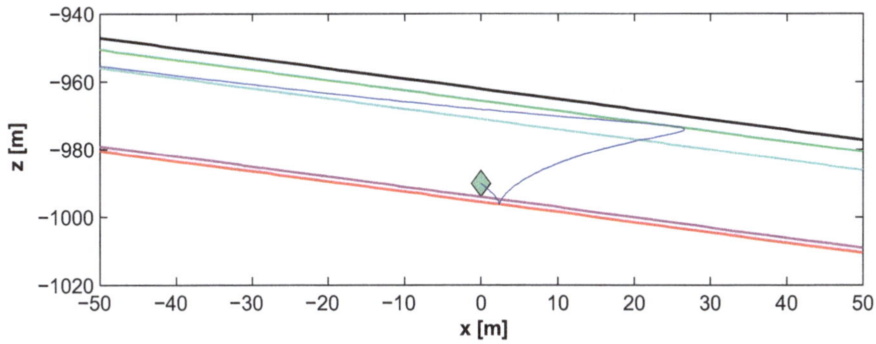


Fig. 7.11 An example of ray progression in a geostrophic shear current. The line $\omega_0 = N$ is red, $\omega_0 = f$ is black, $\omega_0 = L$ is cyan, $\omega_0 = M$ is magenta, and $\omega_0 = \omega_C$ is green. The ray (blue) starts at $x = 0$, $z = -990$ m, first approaches a turning point at $\omega_0 = N$, and then proceeds (from below) into a critical layer. The wave is shown to pass the line $\omega_0 = L$ and is reflected at the wave guide boundary $\omega_0 = \omega_C$. Then it gets captured in a critical layer at $\omega_0 = L$, approaching now from above. The values of the Coriolis and Brunt–Väisälä frequency are $f = 10^{-4}\,\mathrm{s}^{-1}$, $N = 3.1 \times 10^{-3}\,\mathrm{s}^{-1}$. In order to display this valve-type behavior the parameters of shear flow $a = 3 \times 10^{-5}\,\mathrm{m}^{-1}$, $b = 10^{-4}\,\mathrm{m}^{-1}$ are chosen unrealistically such that $V_0 = 285\,\mathrm{m\,s}^{-1}$ is extremely large

group velocity (all three components) if $\beta \neq 0$. A turning point results. At the other boundary, $\omega_0 = \pm\omega_C$, internal reflection occurs. At the lines $\omega_0 = \pm L$ inside the wave guide, waves approaching from one side run into a critical layer while those approaching from the other side just pass through (note that there are always two solutions for $k_\parallel$ from (7.81)).

The behavior of rays in the discussed regimes is illustrated in Figure 7.11. Note that the cusp-like reflection at $\omega_0 = \pm N$ does not appear so in three dimensions because the wave generally progresses in the $y$-direction at this layer with the mean flow speed $V(x, z)$. This behavior applies as well to the critical layer at $\omega_0 = L$. In the above described figure we use $\alpha = 0.29$, $\beta = 0.95$. It should be emphasized that this mathematically interesting valve effect is of little importance in realistic situations because then $\alpha \ll \beta$, and the domain between $\omega_0^2 = \omega_C^2$ and $\omega_0^2 = L^2$ collapses to very thin strips of ocean close to $\omega_0^2 = f^2$.

### 7.7.3 Stability of Shear Flows

The WKBJ approximation is not suitable for a more exact treatment of the stability of shear currents, and we proceed from (7.63)–(7.66). We will ignore the Earth's rotation in this section; its inclusion does not bring completely new aspects. With the wave ansatz

$$\{\boldsymbol{u}, w, p, \rho\} = \{\hat{\boldsymbol{u}}(z), \hat{w}(z), \hat{p}(z), \hat{\rho}(z)\} \exp \mathrm{i}\,(\boldsymbol{k}_h \cdot \boldsymbol{x} - \omega t)$$

with the horizontal wave number $\boldsymbol{k}_\text{h}$, we obtain

$$-\mathrm{i}\,(\omega - \boldsymbol{U}\cdot\boldsymbol{k}_\text{h})\,\hat{\boldsymbol{u}} + \hat{w}\boldsymbol{U}_z = -\frac{\mathrm{i}}{\rho_0}\boldsymbol{k}_\text{h}\hat{p} \tag{7.84}$$

$$-\mathrm{i}\,(\omega - \boldsymbol{U}\cdot\boldsymbol{k}_\text{h})\,\hat{w} = -\frac{1}{\rho_0}\hat{p}_z - g\frac{\hat{\rho}}{\rho_0} \tag{7.85}$$

$$-\mathrm{i}\,(\omega - \boldsymbol{U}\cdot\boldsymbol{k}_\text{h})\,\hat{\rho} - \frac{\rho_0 N^2}{g}\hat{w} = 0 \tag{7.86}$$

$$\mathrm{i}\boldsymbol{k}_\text{h}\cdot\hat{\boldsymbol{u}} + \hat{w}_z = 0 \tag{7.87}$$

In this section $\hat{w}_z = \partial\hat{w}/\partial z$ etc. is used as a shorthand notation for the derivative. The mean flow $\boldsymbol{U} = \boldsymbol{U}(z)$ does not necessarily have the same direction everywhere. It is convenient to introduce its component parallel to the wave number,

$$U_\parallel(z) = \frac{\boldsymbol{U}(z)\cdot\boldsymbol{k}_\text{h}}{k_\text{h}}$$

In the following, for convenience we write $U$ instead of $U_\parallel$. Scalar multiplication of (7.84) with $\boldsymbol{k}$ and taking the vertical derivative leads to

$$(\omega - Uk)\,\hat{w}_{zz} + \hat{w}kU_{zz} = -\frac{\mathrm{i}}{\rho_0}k^2\hat{p}_z \tag{7.88}$$

Elimination of $\hat{\rho}$ from (7.85) and (7.86) results in

$$-(\omega - Uk)^2\,\hat{w} = \mathrm{i}\,(\omega - Uk)\,\frac{1}{\rho_0}\hat{p}_z - N^2\hat{w} \tag{7.89}$$

and elimination of $\hat{p}_z$ from (7.88) and (7.89) yields

$$\hat{w}_{zz} + \left[k^2\frac{N^2 - (\omega - Uk)^2}{(\omega - Uk)^2} + \frac{kU_{zz}}{\omega - Uk}\right]\hat{w} = 0 \tag{7.90}$$

For $U \equiv 0$ this corresponds to the previously derived relation (7.14) (with $f = 0$). With the phase velocity $c = \omega/k$, the above equation can be rewritten as

$$\hat{w}_{zz} + \left[\frac{N^2}{(U - c)^2} - \frac{U_{zz}}{U - c} - k^2\right]\hat{w} = 0 \tag{7.91}$$

which is often called TAYLOR–GOLDSTEIN[4] equation (Goldstein, 1931; Taylor, 1931). Note that this name is sometimes reserved for the inelastic form of (7.91).

With the boundary conditions $\hat{w}(0) = \hat{w}(-H) = 0$, we obtain an eigenvalue problem with the eigenvalue $c$. However, since the eigenvalue appears nonlinearly in (7.91), the problem is not of the Sturm-Liouville type (see the box on p. 192). Therefore, it is not certain that the solutions for $\omega$, respectively $c$ and $\hat{w}(z)$, are real. Note that (7.91) can have a singularity only for real $c$, i. e. $c_\text{i} = 0$ since otherwise $U - c \neq 0$ is valid everywhere. A necessary condition for instability can be derived from the Taylor–Goldstein equation (7.91). With the definition $a(z) = U(z) - c$ (note that $a_z \equiv U_z$), (7.91) can be written as

$$(a\hat{w}_z)_z - (a_z\hat{w})_z + \left(\frac{N^2}{a} - k^2 a\right)\hat{w} = 0 \tag{7.92}$$

[4] SYDNEY GOLDSTEIN, *1903 in Hull, †1989 in Belmont (Mass.), mathematician.

after multiplication with $a = U - c$, and introducing the new dependent variable $G = \hat{w}a^{-\frac{1}{2}}$, we find

$$(aG_z)_z - \left(\frac{1}{2}U_{zz} + k^2 a + \frac{\frac{1}{4}U_z^2 - N^2}{a}\right) G = 0 \qquad (7.93)$$

Multiplication of with $G^*$ and integration leads to the integral constraint

$$\int\limits_{-H}^{0} \left[ a\left(|G_z|^2 + k^2|G|^2\right) + \frac{1}{2}U_{zz}|G|^2 + a^* \left|\frac{G}{a}\right|^2 \left(\frac{1}{4}U_z^2 - N^2\right)\right] \mathrm{d}z = 0 \quad (7.94)$$

Both the real and the imaginary part of (7.94) must vanish. The imaginary part is

$$c_i \int\limits_{-H}^{0} \left[ |G_z|^2 + k^2|G|^2 + \left|\frac{G}{a}\right|^2 \left(N^2 - \frac{1}{4}U_z^2\right)\right] \mathrm{d}z = 0 \qquad (7.95)$$

which shows that a solution with $c_i \neq 0$ is possible only if the integral vanishes. In this case the integrand must be negative over some depth range, i. e. $N^2 - \frac{1}{4}U_z^2 < 0$. This condition can be expressed in terms of the local RICHARDSON[5] number $\mathrm{Ri} = \mathrm{Ri}(z)$ defined as

$$\mathrm{Ri} = \frac{N^2}{U_z^2} \qquad (7.96)$$

Instability can occur only if at least in certain regions $\mathrm{Ri} < 1/4$ is valid. We distinguish between 2 cases:

1. If $\mathrm{Ri} > 1/4$ *everywhere*, the mean flow is stable.
2. If $\mathrm{Ri} < 1/4$ in certain areas (or everywhere), the current does not necessarily need to be unstable, but in practice this is often the case.

Consider a situation with two layers of different densities, rather than a continuous stratification, e. g. $\rho$ and $\rho + \Delta\rho$, and different mean flow $U$, respectively $U + \Delta U$. A corresponding investigation leads to the stability condition (without proof)

$$\frac{g\Delta\rho/\rho_0}{(\Delta U)^2} > \frac{k_\mathrm{h}}{2}$$

For perturbations with sufficiently large $k_\mathrm{h}$, this condition is always violated. In contrast to the continuous case leading to (7.95), the situation in this two-layer case is always unstable to perturbations at a small scale. This is the Kelvin–Helmholtz instability. The occurrence is not surprising because both layers can be viewed as separated by a thin thermocline (depth $d$) where the shear is very large and, therefore, the Richardson number becomes very small,

$$\mathrm{Ri} = \frac{g\Delta\rho}{\rho_0 d}\frac{d^2}{(\Delta U)^2} = \frac{g\Delta\rho}{\rho_0\,(\Delta U)^2}\,d \to 0 \quad \text{for} \quad d \to 0$$

[5] LEWIS F. RICHARDSON, *1881 in Newcastle upon Tyne, †1953 in Kilmun, Argyll, physicist and meteorologist.

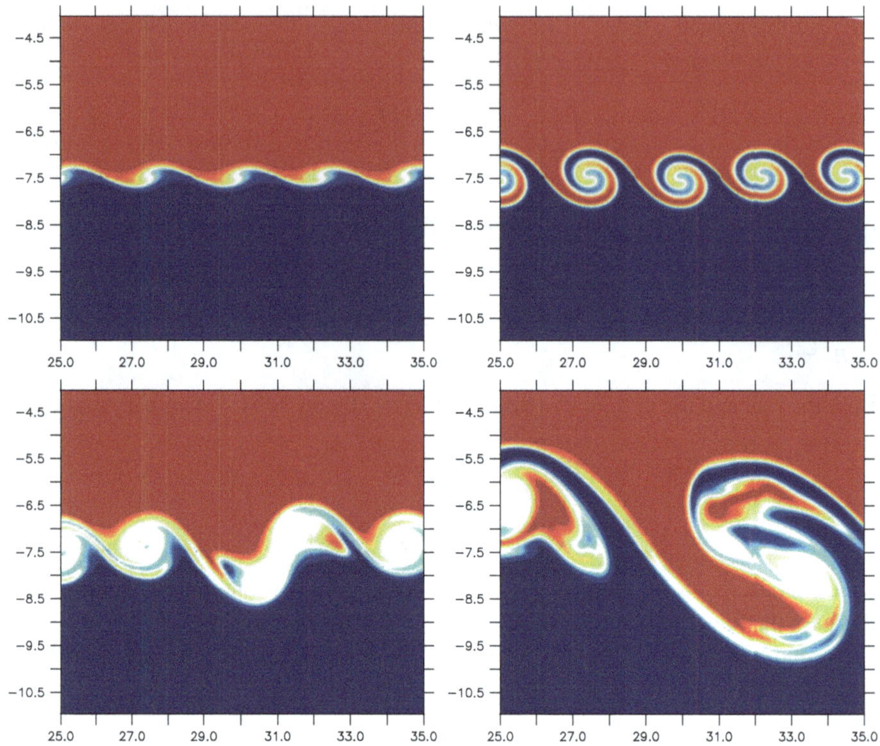

Fig. 7.12 Numerical simulation of Kelvin–Helmholtz instability in a two-layer system with vertically sheared flow. The upper layer is moving to the right while the lower layer is moving to the left. The density is shown for four different times, with an increasing stage of the instability. Red colors denote lower density. At the layer interface a rapidly growing instability is observed, with subsequent wave breaking

A numerical solution of the Kelvin–Helmholtz instability in a two-layer fluid is shown in Figure 7.12.

A geometric condition for unstable solutions can be derived from (7.92). With $F = \hat{w}/a$, this equation changes into

$$[a(aF)_z]_z - (a_z a\ F)_z + \left(\frac{N^2}{a} - k^2 a\right) aF = 0$$

or with $[a(aF)_z]_z \equiv (a_z aF + a^2 F_z)_z$, we obtain

$$\left(a^2 F_z\right)_z + \left(N^2 - k^2 a^2\right) F = 0 \tag{7.97}$$

Multiplication with $F^*$ (complex conjugate) and integration over the depth results in

$$0 = a^2 F_z F^* \Big|_{-H}^{0} - \int_{-H}^{0} \left\{a^2 |F_z|^2 - \left(N^2 - k^2 a^2\right) |F|^2\right\} \mathrm{d}z \tag{7.98}$$

The boundary terms vanish. Furthermore, $a^2 = (U-c)^2 = (U-c_\mathrm{r})^2 - c_i^2 - 2\mathrm{i}c_i(U-c_\mathrm{r})$. The real and the imaginary part of the integrals in (7.98) both must vanish, so that

$$\int_{-H}^{0} \{[U^2 - 2Uc_r + c_r^2 - c_i^2]\, Q^2 - N^2|F|^2\}\, dz = 0 \tag{7.99}$$

$$-2ic_i \int_{-H}^{0} (U - c_r)\, Q^2 dz = 0 \tag{7.100}$$

with $Q^2 = Q^2(z) = |F_z|^2 + k^2|F|^2$. If $c_i \neq 0$, it follows from (7.100) that $U(z) - c_r$ must be zero somewhere: otherwise the integrand would have the same sign everywhere. Therefore, $U_{min} < c_r < U_{max}$ is a *necessary* condition for instability. Furthermore, it follows from (7.100) that

$$c_r \int_{-H}^{0} Q^2 dz = \int_{-H}^{0} UQ^2 dz \tag{7.101}$$

Using this result to replace $c_r U$ in the corresponding term in (7.99), we obtain

$$\int_{-H}^{0} \left[U^2 - (c_r^2 + c_i^2)\right] Q^2 dz = \int_{-H}^{0} N^2|F|^2 dz \tag{7.102}$$

Now, $(U - U_{min})(U - U_{max}) = U^2 - [U_{min} + U_{max}]U + U_{min}U_{max} < 0$, and thus with (7.102) and using once more (7.101), we finally obtain

$$\left[c_r^2 + c_i^2 - (U_{min} + U_{max})\, c_r + U_{min}U_{max}\right] \int_{-H}^{0} Q^2 dz < 0$$

Hence one has $c_r^2 + c_i^2 - c_r(U_{min} + U_{max}) + U_{min}U_{max} < 0$ or

$$\left[c_r - \frac{1}{2}(U_{min} + U_{max})\right]^2 + c_i^2 < \left(\frac{U_{min} - U_{max}}{2}\right)^2 \tag{7.103}$$

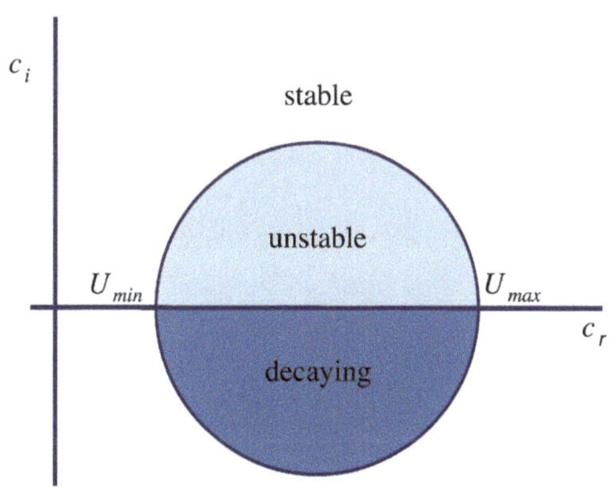


Fig. 7.13 Sketch of the semicircle theorem in the plane formed by real and complex parts of the phase velocity $c$

This condition, which limits the possible eigenvalues, can easily be interpreted geometrically. All solutions with $c_i \neq 0$ lie within a circle in the $c_r$-$c_i$-plane, as sketched in Figure 7.13. The eigenvalues with $c_i > 0$ corresponding to unstable mean flow lie in a semicircle above the abscissa. The stability condition (7.103) is the famous *semicircle theorem*, first derived by HOWARD[6] (Howard, 1961).

[6] LOUIS NORBERG HOWARD, *1929, applied mathematician and fluid dynamicist.

# 8 Long Waves

As shown in Section 4.2.4, motions with horizontal scales much larger than the water depth are in hydrostatic balance. Waves in hydrostatic balance (generally called *long* waves) are of particular importance in the ocean. We have already encountered the hydrostatic limit for gravity waves in Sections 7.4.1 and 7.5. In addition to long gravity waves, planetary and topographic waves are in hydrostatic balance and hence also are long waves. Near the equator, both long gravity and planetary waves have a specific form. In the present chapter, some aspects of long waves will be discussed. We also present the influence of a mean flow on long waves including the important forms of instabilities.

As discussed in Chapter 7, the hydrostatic limit is achieved for $\omega^2 \ll N^2$. From (7.14) it follows that in this limit, the vertical vertical structure of the eigenfunctions $W_n(z)$ no longer depends on frequency. Furthermore, for long waves it is more convenient to use the vertical structure of the horizontal velocities rather than that of the vertical velocity. Therefore, the vertical structure is rederived in the following, starting from the linearized equations of motion (7.4)–(7.7) (neglecting the time derivative in the vertical momentum equation (7.5)), i. e.

$$\frac{\partial u}{\partial t} - fv = -\frac{1}{\rho_0}\frac{\partial p}{\partial x} + \mathcal{F}_u \tag{8.1}$$

$$\frac{\partial v}{\partial t} + fu = -\frac{1}{\rho_0}\frac{\partial p}{\partial y} + \mathcal{F}_v \tag{8.2}$$

$$0 = -\frac{\partial p}{\partial z} - g\rho \tag{8.3}$$

$$\frac{\partial \rho}{\partial t} - \frac{\rho_0 N^2}{g} w = \mathcal{G}_\rho \tag{8.4}$$

$$\frac{\partial u}{\partial x} + \frac{\partial v}{\partial y} + \frac{\partial w}{\partial z} = 0 \tag{8.5}$$

For later reference we have included forcing terms, which are now abandoned but will be used in the treatment of forced waves in Section 10.3. The Coriolis parameter $f$ is assumed constant. Later, for the equatorial wave guide, this assumption will be abandoned. Likewise, the ocean is assumed constant, except when topographic waves are studied. All equations are derived in Cartesian form to show the basic

D. Olbers, J. Willebrand, C. Eden, *Ocean Dynamics*, 
DOI 10.1007/978-3-642-23450-7_8, 

concepts in the most simple conditions. Note, however, that for very long waves (such as e. g. tides) the use of spherical coordinates is required.

## 8.1 Long Gravity Waves

The variables $w$ and $\rho$ can be eliminated from (8.3)–(8.5), either by multiple differentiations or by multiple integrations. The latter way is preferred since solutions may be lost during differentiation, and also the boundary conditions are immediately accounted for. Vertical integration of (8.5) from $-H$ to $z$, using the boundary condition $w = 0$ at $z = -H$, results in

$$w = -\int_{-H}^{z} \nabla_h \cdot \boldsymbol{u}(z'')\mathrm{d}z'' \tag{8.6}$$

Integration of (8.3), with the linearized dynamical boundary condition $p|_{z=0} = g\rho_0\zeta$ (cf. Section 7.1), yields

$$p = \rho_0 g\zeta + g\int_{z}^{0} \rho(z')\mathrm{d}z' \tag{8.7}$$

Temporal differentiation of (8.7), using $\partial\zeta/\partial t = w|_{z=0}$ and (8.4), results in

$$\frac{\partial p}{\partial t} = \rho_0 g w|_{z=0} + \rho_0 \int_{z}^{0} N^2(z')w(z')\mathrm{d}z' \tag{8.8}$$

Inserting now the expression (8.6) for the vertical velocity, one obtains

$$\begin{aligned}\frac{\partial p}{\partial t} &= -\rho_0 g\int_{-H}^{0} \nabla_h \cdot \boldsymbol{u}(z'')\mathrm{d}z'' - \rho_0\int_{z}^{0} N^2(z')\int_{-H}^{z'} \nabla_h \cdot \boldsymbol{u}(z'')\mathrm{d}z''\mathrm{d}z' \\ &= -\rho_0 M\,\nabla_h \cdot \boldsymbol{u}\end{aligned} \tag{8.9}$$

where the integral operator $M$ is defined as

$$M = g\int_{-H}^{0} \mathrm{d}z'' + \int_{z}^{0} \mathrm{d}z' N^2(z')\int_{-H}^{z'} \mathrm{d}z'' \tag{8.10}$$

In the momentum equations (8.1)–(8.3), the rotation vector is assumed to be parallel to the direction of gravity. Since also the boundaries are perpendicular to gravity, it is possible to separate the vertical structure from horizontal and time dependence with the ansatz

$$\begin{pmatrix} u, v, p(x,y,z,t) \\ w(x,y,z,t) \\ \rho(x,y,z,t) \end{pmatrix} = \begin{pmatrix} \tilde{u}, \tilde{v}, \rho_0\tilde{p}(x,y,t) & \Phi(z) \\ \tilde{w}(x,y,t) & N^{-2}(z)\partial\Phi(z)/\partial z \\ \rho_0\tilde{\rho}(x,y,t) & \partial\Phi(z)/\partial z \end{pmatrix} \tag{8.11}$$

**33. Rayleigh Approximation for the Eigenvalue**

The Rayleigh quotient, obtained by multiplication of (8.15) with $\Phi_n$ and integration, represents the eigenvalues $\lambda_n$ in terms of the eigenfunctions $\Phi_n(z)$ in the form

$$\lambda_n = \frac{\Phi_n^2(0) + \int_{-H}^{0} \frac{g}{N^2}(\frac{d\Phi_n}{dz})^2 \mathrm{d}z}{g\int_{-H}^{0} \Phi_n^2 \mathrm{d}z} \tag{B33.1}$$

The Rayleigh quotient is particularly useful in situations when the eigenfunctions are only approximately known since the eigenvalue computed from (B33.1) may still be fairly accurate.

Note that the physical dimensions of the tilde variables $\tilde{u}, \tilde{v}\ldots$ depend on the dimension of $\Phi$ which has not yet been specified, and are generally different from those of the corresponding variables $u, v\ldots$ which do have the proper physical dimensions. Insertion of (8.11) into (8.9), and division by $\Phi\nabla_h\cdot\tilde{\boldsymbol{u}}$, results in

$$\frac{\partial\tilde{p}/\partial t}{\nabla_h\cdot\tilde{\boldsymbol{u}}} = -\frac{M\Phi}{\Phi} \tag{8.12}$$

Here the left-hand side is a function of $x, y, t$, the right-hand side a function of only $z$. To be equal, both sides can neither depend on $x, y, t$ nor on $z$ but must both be equal to a constant, say $-1/\lambda$. Hence (8.9) results in *two* relations, namely

$$\frac{\partial\tilde{p}}{\partial t} = -\frac{1}{\lambda}\left(\frac{\partial\tilde{u}}{\partial x} + \frac{\partial\tilde{v}}{\partial y}\right) \tag{8.13}$$

$$M\Phi = \frac{1}{\lambda}\Phi \tag{8.14}$$

with the operator $M$ given by (8.10), and a yet unknown constant $\lambda$. Equation (8.13) will be further discussed in Section 8.1.2. Equation (8.14) is an integral equation governing the vertical structure $\Phi(z)$ of horizontal velocity and pressure. Note that the more common differential formulation is obtained by application of $\partial/\partial z(N^{-2}\partial/\partial z)$ to (8.14), and leads to

$$\frac{\mathrm{d}}{\mathrm{d}z}\left(\frac{1}{N^2}\frac{\mathrm{d}\Phi}{\mathrm{d}z}\right) = -\lambda\Phi \tag{8.15}$$

$$\frac{\mathrm{d}\Phi}{\mathrm{d}z} + (N^2/g)\Phi = 0\,, \quad z = 0 \tag{8.16}$$

$$\frac{\mathrm{d}\Phi}{\mathrm{d}z} = 0\,, \quad z = -H \tag{8.17}$$

Equation (8.14), or equivalently (8.15)–(8.17), constitute a Sturm–Liouville problem (see the box on p. 192). A solution $\Phi = \Phi_n(z)$ exists only for discrete eigenvalues $\lambda_n, n = 0, 1, 2, \ldots$ which are real and positive and can be sorted in ascending order. Note that $\Phi_n(z)$ and $\lambda_n$ are independent of $\omega$ which contrasts the long-wave eigenvalue problem with the one for internal gravity waves discussed in Section 7.4.3.

### 8.1.1 Barotropic and Baroclinic Modes

As shown in Section 7.4.4, $N^2H/g \sim \Delta\rho/\rho_0 \approx 2\times10^{-3} = \epsilon \ll 1$ is valid in the ocean. The second term in (8.16) is of the order $O(\epsilon)$ and can be neglected in the lowest approximation. We then distinguish between two types of solutions:

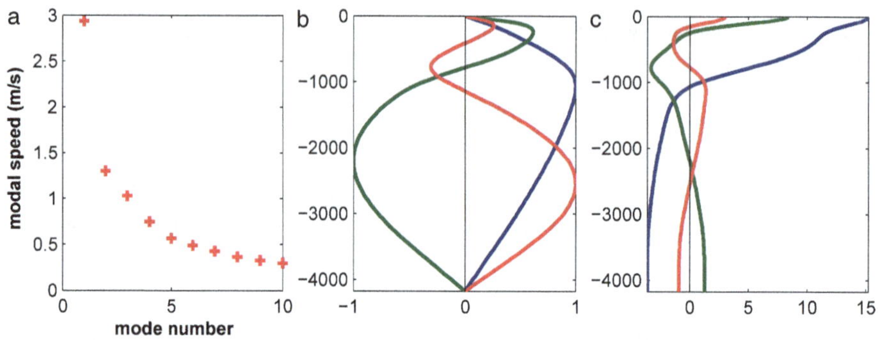


Fig. 8.1 Eigenmodes for long waves for the Brunt-Väisälä frequency profile shown in Figure 7.7. The modal speed (**a**) of the first ten modes in m s$^{-1}$ and the first three eigenfunctions of the vertical (**b**) and horizontal (**c**) velocity are displayed

## Barotropic solution

With the approximate boundary condition $\Phi_z \approx 0$ at $z = 0$, a nontrivial solution of (B33.1) is given by $\Phi_0(z) \approx$ const, with the corresponding eigenvalue $\lambda_0 \approx 0$. Here the index $n = 0$ is set arbitrarily. The approximation $\lambda_0 = 0$ is formally correct to order $\epsilon$ but nevertheless insufficient since the true eigenvalues are always positive. An approximation accurate to order $\epsilon^2$ can be found with the help of the RAYLEIGH[1] coefficient in the box on p. 211. With $\Phi_0 =$ const, one obtains

$$\lambda_0 = \frac{1}{gH} \tag{8.18}$$

## Baroclinic solutions

Neglecting the term $O(\epsilon)$ in the boundary condition for baroclinic modes leads to $\Phi_z = 0$ at $z = 0, -H$, which corresponds to $w = 0$. For arbitrary profiles of $N(z)$, the eigenfunctions have to be calculated numerically. Some analytical and numerical solutions of the eigenvalue problem of long waves are shown below (see Figure 8.1 and the box on p. 214). With the normalization $\int \Phi_n^2 \mathrm{d}z = 1$, one obtains from (B33.1) for the baroclinic eigenvalues

$$\lambda_n \approx \int \frac{1}{N^2} \left( \frac{d\Phi_n}{dz} \right)^2 \mathrm{d}z \,, \quad n = 1, 2 \ldots \tag{8.19}$$

The separation parameter $\lambda_n$ characterizes the stratification in the baroclinic case and the water depth in the barotropic case. In both cases, it has the dimension [s$^2$ m$^{-2}$]. It is convenient to replace $\lambda_n$ by one of the following parameters which are physically more intuitive:

1. *The equivalent depth*

$$h_n = \frac{1}{g\lambda_n}$$

[1] JOHN WILLIAM STRUTT, LORD RAYLEIGH, *1842 in Maldon, Essex, †1919 in Witham, Essex, physicist.

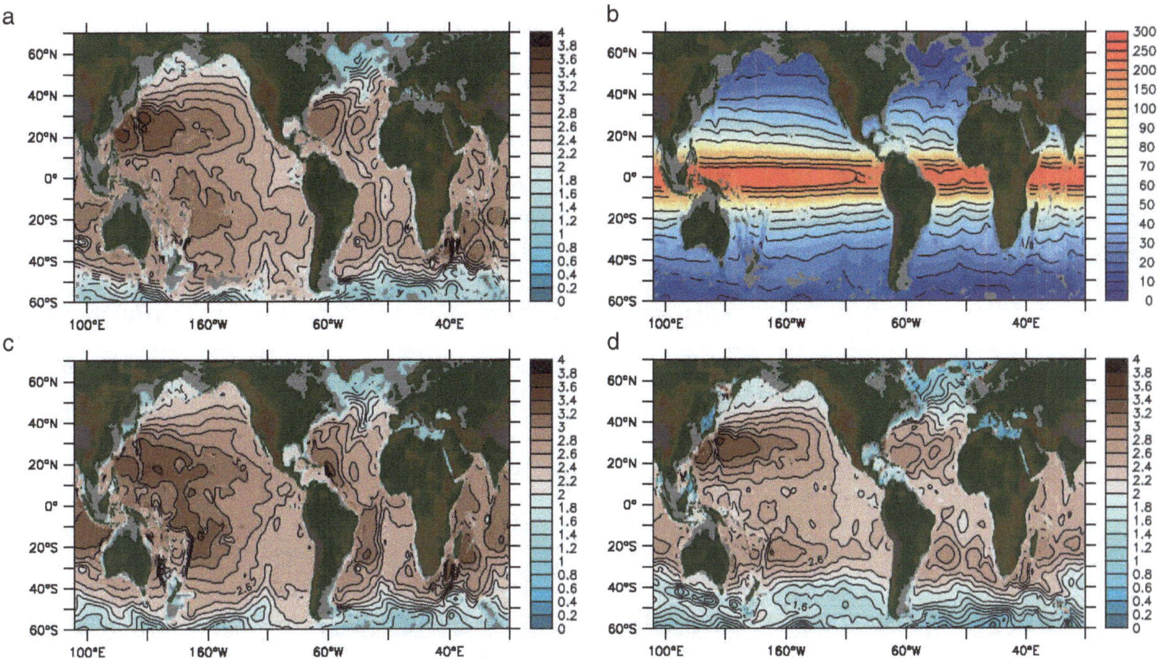


Fig. 8.2 Global distribution of **a**, the baroclinic eigenvalue $c_1$, and **b**, the associated Rossby radius $R_1$, according to the analysis of Chelton et al. (1998). In **c**, **d** we show the resulting $c_1$ from the WKBJ in **c** and the Rayleigh approximation in **d** (see text), using the WOA05 climatology. All data have been smoothed over three grid points of the $1^\circ \times 1^\circ$ resolution

has the dimension of a length. The suggestive name is due to the fact that $h_0 = 1/g\lambda_0 \approx H$. For $n > 1$, however, $h_n$ has no immediate physical significance.

2. *The gravity wave speed*

$$c_n = \frac{1}{\lambda_n^{1/2}} \equiv \sqrt{g h_n}$$

has the dimension $[\mathrm{m\,s^{-1}}]$. For both barotropic and baroclinic modes, $c_n$ is the phase speed of long gravity waves *without* rotation. Note, however, that $c_n$ is *not* identical to the phase speed of waves in a rotating system.

3. *The Rossby radius*

$$R_n = \frac{1}{f\lambda_n^{1/2}} = \frac{c_n}{f}$$

again has the dimension of a length. For gravity waves it is the length scale above which rotation effects are significant.

For $f = 10^{-4}\,\mathrm{s}^{-1}$ we roughly obtain the values in Table 8.1; they depend on the stratification.

Table 8.1 Typical oceanic ranges of equivalent depth $h_n$, phase speed $c_n$ and Rossby radius $R_n$ for the barotropic and the first two baroclinic modes

| $n$ | $h_n$ [m] | $c_n$ [m s$^{-1}$] | $R_n$ [km] |
|---|---|---|---|
| 0 | 4,000 | 200 | 2,000 |
| 1 | 0.1–1 | 1.0–3.0 | 10–30 |
| 2 | 0.004–0.04 | 0.2–0.6 | 2–6 |

**34. Eigenmodes for Constant and Exponential Brunt-Väisälä Frequency**

We introduce two frequently used models of the Brunt–Väisälä frequency and the associated long-wave eigenmodes determined by (8.15)–(8.17). The eigenfunctions are normalized as

$$\int_{-H_0}^{0} dz \Phi_n(z)\Phi_m(z) = \delta_{nm}$$

1. **Constant** $N$: For $N = N_0 = \text{const}$ the eigenvalue problem is solved by

$$\Phi_0(z) = \sqrt{1/H_0} \qquad \lambda_0 = 0$$
$$\Phi_m(z) = \sqrt{2/H_0}\cos(m\pi z/H_0) \qquad \lambda_m = (m\pi)^2/(N_0 H_0)^2$$

2. **Exponential** $N$: For an exponential Brunt–Väisälä frequency $N(z) = N_0 e^{z/b}$ with $b > 0$ we find an analytical solution for the baroclinic modes in the terms of Bessel functions (the box on p. 247, see also e.g. Abramowitz and Stegun, 1984)

$$\mathcal{J}_n(\zeta) = J_n(\zeta)Y_0(\alpha_m) - Y_n(\zeta)J_0(\alpha_m)\,, \quad n = 0, 1$$

with $\alpha_m = N_0 b\sqrt{\lambda_m}$. The eigenfunctions are

$$\Phi_m(z) = D_m e^{z/b}\mathcal{J}_1(\alpha_m e^{z/b})$$

with the normalization constant $D_m = \sqrt{2/b}\{\mathcal{J}_1^2(\alpha_m) - \mathcal{J}_1^2(\alpha_m e^{-H/b})\}^{-1/2}$, and the $\alpha_m$ follows from the dispersion relation

$$\mathcal{J}_0(\alpha_m e^{-H/b}) = 0$$

Some estimates of the wave speed $c_1$ and the Rossby radius $R_1$ of the first baroclinic mode are displayed in Figure 8.2. The upper panel uses the data of analysis of Chelton et al. (1998) who have solved the eigenvalue problem locally for the Brunt–Väisälä frequency derived from the WOA05 hydrographic climatology[2]. The Rossby radius $R_1 = c_1/|f|$ becomes singular at the equator. Here, $R_1$ is replaced by the equatorial value $R_e = \sqrt{c_1/\beta}$ (see Section 8.3.2). The result for $c_1$ is compared in the lower panels with WKBJ estimate $c_1 \sim \int N(z)dz/\pi$ and the estimate obtained from the Rayleigh formula in the box on p. 211. For the latter we have approximated the eigenfunction with the eigenfunction of an exponential Brunt–Väisälä frequency, see the box on p. 214. The better accuracy of the Rayleigh approximation becomes obvious.

## 8.1.2 Dispersion Relation and Group Velocity

From (8.1), (8.2), (8.9) and (8.14), equations for the $\tilde{u}, \tilde{v}, \tilde{p}(x, y, t)$ follow. With $c_n^2 = 1/\lambda_n$, one obtains

$$\frac{\partial \tilde{u}}{\partial t} - f\tilde{v} + \frac{\partial \tilde{p}}{\partial x} = 0 \tag{8.20}$$

[2] Chelton et al. (1998) solve the eigenvalue problem for vertical velocity. This is $W_{zz} + \lambda N^2 W = 0$ with boundary conditions $W = 0$ at $z = 0, -H$. Note that $W = N^{-2}\Phi_z$.

$$\frac{\partial \tilde{v}}{\partial t} + f\tilde{u} + \frac{\partial \tilde{p}}{\partial y} = 0 \tag{8.21}$$

$$\frac{\partial \tilde{p}}{\partial t} + c_n^2 \left( \frac{\partial \tilde{u}}{\partial x} + \frac{\partial \tilde{v}}{\partial y} \right) = 0 \tag{8.22}$$

These equations are often referred to as long-wave equations. All variables refer to the respective vertical eigenfunction with the index $n$ which from now on is suppressed for convenience. The wave ansatz $(\tilde{u}, \tilde{v}, \tilde{p}) = (\hat{U}, \hat{V}, \hat{P})e^{\mathrm{i}(k_1 x + k_2 y - \omega t)}$ leads us (with $f = \text{const}$) directly to the dispersion relation

$$\omega^2 = f^2 + c_n^2 k^2 = f^2(1 + R_n^2 k^2) \tag{8.23}$$

which is the long-wave limit of the gravity waves, previously considered in Section 7.4.3.

Long gravity waves in the presence of rotation are also called POINCARÉ[3] waves. From the dispersion relation (8.23) we obtain their phase speed as

$$c = \frac{\omega}{k} = \frac{(f^2 + c_n^2 k^2)^{\frac{1}{2}}}{k} \equiv \left( \frac{f^2}{k^2} + c_n^2 \right)^{\frac{1}{2}} \equiv c_n \left( 1 + \frac{1}{k^2 R_n^2} \right)^{\frac{1}{2}}$$

It follows that always $c > c_n$, and in the limit $kR_n \gg 1$ one finds $c \approx c_n$. As the frequency depends only on the magnitude of the wave-number vector, the group velocity has the direction of $\boldsymbol{k}$; its magnitude $c_\mathrm{g}$ is obtained by differentiation of (8.23) as $2\omega \partial\omega/\partial k \equiv 2\omega c_\mathrm{g} = 2c_n^2 k$ or

$$\boldsymbol{c}_\mathrm{g} = \frac{c_n^2}{\omega}\boldsymbol{k} = \frac{c_n}{(k^2 + R_n^{-2})^{\frac{1}{2}}}\boldsymbol{k} \tag{8.24}$$

in vectorial form. Note that the relation $cc_\mathrm{g} \equiv c_n^2$ holds at all frequencies.

The amplitudes $\hat{U}, \hat{V}$ can be expressed by $\hat{P}$ according to

$$\begin{pmatrix} \hat{U} \\ \hat{V} \end{pmatrix} = \frac{1}{\omega^2 - f^2} \begin{pmatrix} \omega k_1 + \mathrm{i} f k_2 \\ \omega k_2 - \mathrm{i} f k_1 \end{pmatrix} \hat{P} \tag{8.25}$$

In particular, the ratio of velocity amplitudes $\hat{V}$ and $\hat{U}$ is given as

$$\frac{\hat{V}}{\hat{U}} = \frac{\omega k_2 - \mathrm{i} f k_1}{\omega k_1 + \mathrm{i} f k_2} \tag{8.26}$$

a relation that will be needed below (cf. Section 8.1.4).

### 8.1.3 Geostrophic Adjustment

In this section, the role of long gravity waves for the adjustment from a non-geostrophic initial distribution to the geostrophic balance will be considered.

[3] JULES HENRI POINCARÉ, *1854 in Nancy, †1912 in Paris, mathematician, theoretical physicist and philosopher.

### Conservation of potential vorticity

The system (8.20)–(8.22) can be written as

$$\frac{\partial \tilde{u}}{\partial t} - fv + g\frac{\partial \zeta}{\partial x} = 0 \tag{8.27}$$

$$\frac{\partial \tilde{v}}{\partial t} + fu + g\frac{\partial \zeta}{\partial y} = 0 \tag{8.28}$$

$$\frac{\partial \zeta}{\partial t} + h_n\left(\frac{\partial \tilde{u}}{\partial x} + \frac{\partial \tilde{v}}{\partial y}\right) = 0 \tag{8.29}$$

dropping the tilde and introducing the elevation $\zeta = p/g$. Here $h_n = c_n^2/g$ is the equivalent depth (in the following the index $n$ will be dropped). Forming the curl of (8.27) and (8.28) yields (note that $f = \text{const}$)

$$\frac{\partial}{\partial t}\left(\frac{\partial v}{\partial x} - \frac{\partial u}{\partial y}\right) + f\left(\frac{\partial u}{\partial x} + \frac{\partial v}{\partial y}\right) = 0 \tag{8.30}$$

With (8.29), this leads to $\partial(\partial v/\partial x - \partial u/\partial y - (f/h)\zeta)/\partial t = 0$, and we note that

$$Q = \frac{\partial v}{\partial x} - \frac{\partial u}{\partial y} - \frac{f}{h}\zeta = Q_0 \tag{8.31}$$

is always constant in time. The equation $\partial Q/\partial t = 0$ corresponds to the eigenfrequency $\omega = 0$ of the system (8.27)–(8.29). Here $Q$ is an approximated form of the *potential vorticity*.

To obtain an equation for $\zeta$ alone, the divergence of (8.27) and (8.28) is taken; in combination with (8.29) this leads to

$$-\frac{1}{h}\frac{\partial^2 \zeta}{\partial t^2} - f\left(\frac{\partial v}{\partial x} - \frac{\partial u}{\partial y}\right) = -g\nabla^2\zeta \tag{8.32}$$

The relative vorticity $\partial v/\partial x - \partial u/\partial y$ in (8.32) can be eliminated using (8.31) with the aid of the potential vorticity, and we obtain an equation governing the evolution of the displacement $\zeta$

$$-\frac{\partial^2 \zeta}{\partial t^2} + gh\nabla^2\zeta - f^2\zeta = fhQ \equiv fhQ_0 \tag{8.33}$$

The inhomogeneity of this equation comes about by integration of the homogeneous system (8.27)–(8.29), and one constant of integration ($Q_0$) has appeared.

35. Green's Function

The Green's function is a very useful concept for the solution of linear forced boundary value problems. Consider the problem

$$\mathcal{L}\{\psi\} = q(z)$$

where $\mathcal{L}\{\psi\}$ is a linear differential operator acting on the scalar function $\psi(z)$, with linear homogeneous boundary conditions at, say, $z = a$ and $z = b$, and a forcing $q(z)$. The associated Green's function $G(z, z')$ satisfies identical boundary conditions but solves a somewhat different problem which – instead of $q(z)$ – has a point source of infinite strength (see box on p. 199) located at $z = z'$. In other words, as function of $z$ the Green's function is a solution of the unforced problem everywhere except at $z = z'$ where it is discontinuous. With proper normalization it hence satisfies

$$\mathcal{L}\{G(z, z')\} = \delta(z - z')$$

For an arbitrary source $q(z)$ all the different contributions from the point source have to be summed with the appropriate response $G(z, z')$, and the function

$$\psi(z) = \int_a^b G(z, z')q(z')\mathrm{d}z'$$

constitutes the solution to the forced problem, as can be shown easily. The solution can be extended to nonhomogeneous boundary conditions.

### Adjustment of an initial disturbance

The solution of the initial value problem and depends on the initial distribution of the displacement field and is not pursued here. The final state after a long time can, however, be obtained in a rather simple way. In equilibrium, the time derivative term in (8.33) vanishes, and for the equilibrium solution $\zeta_{\mathrm{eq}}$ we obtain

$$gh\nabla^2\zeta_{\mathrm{eq}} - f^2\zeta_{\mathrm{eq}} = fhQ_0 \tag{8.34}$$

Consider an initially motionless field, $u_0 = 0$, $v_0 = 0$ with an initial elevation $\zeta_0(y)$ and hence an initial potential vorticity distribution $Q_0(y) = -(f/h)\zeta_0(y)$. For symmetry reasons, the final distribution also depends only on $y$, and with the Rossby radius $R = \sqrt{gh}/f$ it follows that

$$-R^2\frac{\partial^2\zeta_{\mathrm{eq}}}{\partial y^2} + \zeta_{\mathrm{eq}} = \zeta_0(y) \tag{8.35}$$

A solution of (8.35) is given by

$$\zeta_{\mathrm{eq}}(y) = \int_{-\infty}^{\infty} \zeta_0(y')G(y - y')\mathrm{d}y' \tag{8.36}$$

with the Green's function (see the box on p. 217)

$$G(y - y') = \frac{1}{2R}\mathrm{e}^{-|y-y'|/R} \tag{8.37}$$

The behavior of the equilibrium solution $\zeta_{\mathrm{eq}}(y)$ depends on the scale, relative to the Rossby radius, of variation of the initial distribution. For a qualitative discussion it is sufficient to consider the limiting cases that the scale of variation is either very small or very large compared to the Rossby radius.

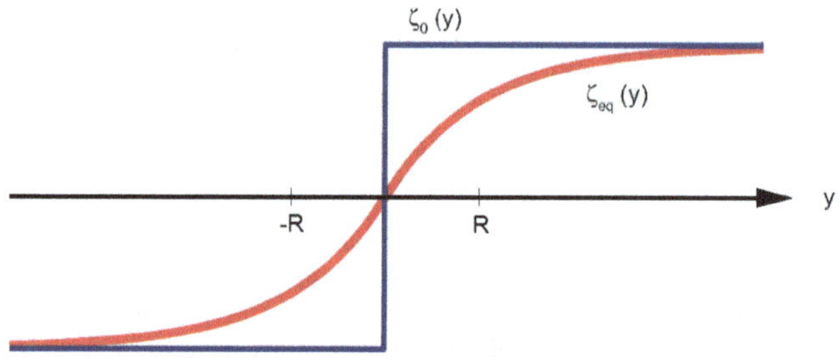


Fig. 8.3 Adjustment of an initial height disturbance with a discontinuity (*blue*) into geostrophic equilibrium (*red*)

### Small-scale initial distribution

In case that the initial distribution varies over scales much smaller than $R$, i. e. $|\partial\zeta_0/\partial y| \gg \zeta_0/R$, it follows from (8.36) and (8.37) that those rapid variations will effectively be smoothed over the distance $R$. For example, if the initial elevation has a jump with amplitude $2a_0$, i. e. $\zeta_0 = a_0 \,\mathrm{sign}(y)$, the equilibrium distribution can be evaluated with (8.36) to be

$$\zeta_{\mathrm{eq}} = a_0(1 - \mathrm{e}^{-|y|/R})\,\mathrm{sign}(y) \tag{8.38}$$

(cf. Figure 8.3). The corresponding zonal velocity follows from (8.28). According to (8.28), the geostrophic relation

$$f u_{\mathrm{eq}} = -g\frac{\partial\zeta_{\mathrm{eq}}}{\partial y} \tag{8.39}$$

must hold in equilibrium. For the solution (8.38) corresponding to an initial jump this leads to

$$u_{\mathrm{eq}} = a_0\,\frac{g}{fR}\mathrm{e}^{-|y|/R} \tag{8.40}$$

Displacement and current only differ from their initial condition within a range of a Rossby radius from the initial discontinuity; the adjustment only takes place here.

From (8.27)–(8.29) one obtains, in the usual way, the energy conservation as

$$\frac{\partial}{\partial t}\left[\frac{1}{2}h(u^2+v^2)+\frac{1}{2}g\zeta^2\right] = -\nabla\cdot(g\zeta h\boldsymbol{u}) \tag{8.41}$$

The potential energy per length unit is initially $\frac{1}{2}ga_0^2$, i. e. infinitely large in the total area. Initially, there is no kinetic energy. For the difference of the respective energies, minus the initial values, integrated over all $y$ we find

$$\Delta E_{\mathrm{k}} = \int_{-\infty}^{\infty}\frac{1}{2}h[u_{\mathrm{eq}}^2(y) - 0]\mathrm{d}y, \quad \text{and} \quad \Delta E_{\mathrm{p}} = \int_{-\infty}^{\infty}\frac{1}{2}g[\zeta_{\mathrm{eq}}^2(y) - a_0^2]\mathrm{d}y$$

Evaluation of the integrals with (8.38) and (8.40) results in $\Delta E_{\mathrm{k}} = gRa_0^2/2$ and $\Delta E_{\mathrm{p}} = -3gRa_0^2/2$. The increase in kinetic energy related to the adjustment is smaller than the loss of potential energy,

$$\Delta E_{\mathrm{k}} = -\frac{1}{3}\Delta E_{\mathrm{p}}$$

The remaining potential energy has been radiated away by gravity waves. The factor $1/3$ is, however, not universally valid but specific to this example and depends on how far the initial distribution deviates from geostrophy. If that is already in a geostrophic balance, no energy is radiated (as long as $f = \text{const}$). Only a finite part of the potential energy (which initially is infinitely large) is lost. We conclude that the presence of rotation makes it difficult to change potential energy into kinetic energy.

#### Large-scale initial distribution

On the other hand, if the initial distribution varies over scales much larger than $R$, i. e. $|\partial\zeta_0/\partial y| \gg \zeta_0/R$, then under the integral in (8.36), $\zeta_0(y')$ is effectively constant over the range $|y - y'| < R$ where the Green's function significantly differs from zero. Therefore, under the integral the approximation $\zeta_0(y') \approx \zeta_0(y)$ is valid. With $\int_{-\infty}^{\infty} G(y - y')\mathrm{d}y' = 1$ it then follows that

$$\zeta_{\text{eq}}(y) \approx \zeta_0(y) \tag{8.42}$$

The corresponding velocity field can again be found from (8.39). Since the elevation is virtually unchanged, the potential energy is also unchanged. As will be shown later (Section 8.2.2), the kinetic energy of near-geostrophic motions with a scale of motion much larger than $R$ is always small compared to potential energy. It follows that only a small fraction of potential energy is transformed into kinetic energy during the adjustment.

### 8.1.4 Influence of Horizontal Boundaries

When a horizontal boundary crosses the propagation path of a long gravity wave, reflection can be expected. The details of reflection are, however, considerably modified by the Earth's rotation. Furthermore, one finds a new type of waves which can only exist near the boundary.

#### Reflection at a lateral wall

The situation is more complicated in the general case with rotation. Consider a wave with wave number $\boldsymbol{k}^i = (k_1^i < 0, k_2^i > 0)$ (incoming wave) propagating towards a boundary at $x = 0$ where $\tilde{u}(0, y, t) = 0$ must be valid. We write

$$\tilde{u}(x, y, t) = U^i \mathrm{e}^{\mathrm{i}(k_1^i x + k_2^i y - \omega^i t)} + U^r \mathrm{e}^{\mathrm{i}(k_1^r x + k_2^r y - \omega^r t)}$$

and likewise for $\tilde{v}$ and $\tilde{p}$. Again, the reflection condition requires

$$k_1^r = -k_1^i = k_1 \,, \quad k_2^r = k_2^i = k_2 \,, \quad \omega^r = \omega^i \quad \text{and} \quad U^r = -U^i$$

**36. Eigenoscillations in a nonrotating basin**

If the dimensions of a basin are considerably smaller than the Rossby radius, rotation does not play a major role. The results mentioned above in Section 8.1.2 remain valid, with $f = 0$. The reflection at plane boundaries does not cause any difficulties in this case. For a rectangular basin of the dimension $L_1, L_2$, the boundary condition of no flow normal to the boundaries can be satisfies only for discrete horizontal wave numbers according to

$$k_1 = \frac{\ell\pi}{L_1}, \quad k_2 = \frac{m\pi}{L_2}$$

with $\ell, m = 0, 1, 2\ldots$ Pressure field and eigenfrequencies are given by

$$p(x, y, t) = P_{\ell m} \cos\frac{\ell\pi x_1}{L_1} \cos\frac{m\pi x_2}{L_2} \cos\omega^n_{\ell m} t$$

$$\omega^n_{\ell m} = c_n \left[\frac{\ell^2\pi^2}{L_1^2} + \frac{m^2\pi^2}{L_2^2}\right]^{\frac{1}{2}}$$

The basic oscillation (longitudinal oscillation) $\ell = 1$, $m = 0$, $n = 0$ in a basin with the length $L_1$ then has a period

$$T = \frac{2\pi}{\omega^0_{1,0}} = 2\frac{L_1}{\sqrt{gH}}$$

a formula attributed to Merian (1828).

The solution for $\tilde{u}$ is thus given by $\tilde{u} = U_0 \sin(k_1 x) \exp\left(\mathrm{i}(k_2 y - \omega t)\right)$ with $U_0 = 2\mathrm{i}U^r$. A second boundary condition in $x$-direction (channel geometry) leads, as usual, to a discretization of the wave number $k_1$.

Reflection at a further boundary in $y$-direction causes, however, greater difficulties for the analytical treatment. From (8.26), it follows that

$$V = \gamma \mathrm{e}^{\mathrm{i}\chi} U \quad \text{with} \quad \gamma \mathrm{e}^{\mathrm{i}\chi(k_1,k_2)} = \frac{\omega k_2 - \mathrm{i} f k_1}{\omega k_1 + \mathrm{i} f k_2}$$

for the meridional velocity component. This relation must hold for both incident and reflected waves, i. e.

$$V^i = \gamma \mathrm{e}^{\mathrm{i}\chi(k_1,k_2)} U^i, \quad V^r = \gamma \mathrm{e}^{\mathrm{i}\chi(-k_1,k_2)} U^r$$

For $\tilde{v}$ we thus obtain the solution

$$\tilde{v} = \gamma U^i \left(\mathrm{e}^{\mathrm{i}[k_1 x + \chi(k_1,k_2)]} - \mathrm{e}^{\mathrm{i}[-k_1 x + \chi(-k_1,k_2)]}\right) \mathrm{e}^{\mathrm{i}(k_2 y - \omega t)}$$

Except for the particular case $f = 0$ (where $\chi = 0$ or $\pi$; see the box on p. 220), the dependence of the phase $\chi$ on the wave number $(k_1, k_2)$ is neither symmetric nor antisymmetric, i. e. $\chi(k_1, -k_2) \neq \pm\chi(k_1, k_2)$. For this reason, it is no longer possible to satisfy the boundary condition $\tilde{v} = 0$ at, e.g., $y = 0$ by superposition of $k_2$ and $-k_2$ in the $y$-direction. The gravity waves in a rotating rectangular basin can, therefore, no longer be represented by a simple superposition of standing waves. Nevertheless, an analytical solution can be obtained by series expansion.

### 8.1.5 Kelvin Waves

Consider a particular solution of (8.20)–(8.22) with $\tilde{u}(x, y, t) \equiv 0$ everywhere. Such a solution automatically satisfies the boundary condition in the $x$-direction. With

$$\begin{pmatrix} \tilde{p} \\ \tilde{v} \end{pmatrix} = \begin{pmatrix} \hat{P}(x) \\ \hat{V}(x) \end{pmatrix} \mathrm{e}^{\mathrm{i}(k_2 y - \omega t)} \tag{8.43}$$

one obtains

$$-f\hat{V} = -\frac{\partial \hat{P}}{\partial x} \tag{8.44}$$

$$-\mathrm{i}\omega\hat{V} = -\mathrm{i}k_2\hat{P} \tag{8.45}$$

$$-\frac{\mathrm{i}\omega}{c_n^2}\hat{P} + \mathrm{i}k_2\hat{V} = 0 \tag{8.46}$$

From (8.44) and (8.45) it follows that $\hat{V} = (k_2/\omega)\hat{P}$ and $(fk_2/\omega)\hat{P} = \partial\hat{P}/\partial x$ with the solution

$$\hat{P}(x) = P_0 \mathrm{e}^{(fk_2/\omega)x} \quad \text{and} \quad \hat{V}(x) = \frac{k_2}{\omega} P_0 \mathrm{e}^{(fk_2/\omega)x}$$

Since by convention always $\omega > 0$, $k_2$ must have a sign that causes the solution to decline for $x \to \infty$. For this reason, $k_2 < 0$ applies in the above-mentioned example. Equation (8.46) leads to

$$\omega^2 = c_n^2 k_2^2 \tag{8.47}$$

With the Rossby radius $R_n = c_n/f = \omega/|k_2|f$, we obtain

$$(\tilde{p}(x, y, t), \tilde{v}(x, y, t)) = (P_0, V_0)\mathrm{e}^{-x/R_n}\mathrm{e}^{\mathrm{i}(k_2 y - \omega t)} \tag{8.48}$$

which describes a wave propagating in $-y$ direction (in the northern hemisphere, $k_2 < 0$, and the coast remains to the right with respect to the direction of propagation). It is called Kelvin wave. The Kelvin wave is *trapped* to the coast and declines within a distance of a Rossby radius. In the southern hemisphere $f < 0$ applies, and the propagation is in opposite direction.

Kelvin waves are free of dispersion. Their phase velocity $c = \omega/|k_2| \equiv c_n$ is identical to that of long gravity waves without rotation. The Earth's rotation does not occur in the dispersion relation (8.47) but is fundamental for the existence of Kelvin waves (the scale of decay in the $x$-direction depends on $f$). Note that the frequencies can be very low and are no longer bounded by the inertial frequency. For the first baroclinic mode ($c_n \sim 2\,\mathrm{m\,s^{-1}}$) the period is about 1 month if the wavelength is 6,000 km. Kelvin waves thus constitute a fast mechanism for propagating information over large distances along coastlines (cf. Section 8.4).

### 8.1.6 Hydraulic Control: Wave Propagation and Nonlinearity

The characteristics of linear long gravity waves can be decisive for the dynamics of stationary and strongly nonlinear currents. Hydraulically controlled flows constitute a good example for this fact.

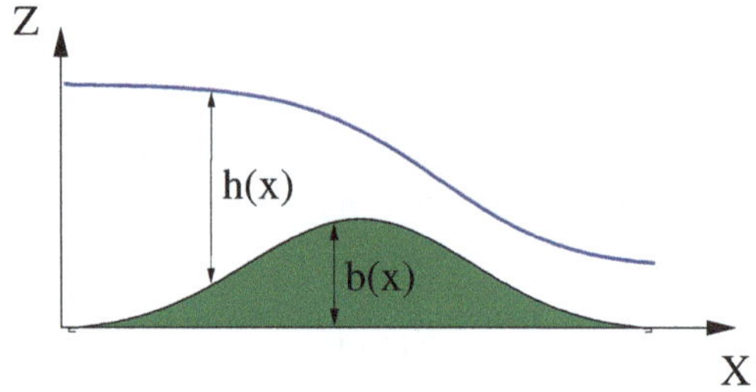


Fig. 8.4 Sketch of the configuration assumed for the hydraulic control flow: $h(x)$ is the depth of the fluid above the topographic barrier of height $b(x)$

The simplest case of hydraulic control occurs in an unforced flow of a homogeneous fluid in a nonrotating channel of constant width. When friction is small, the current is independent of depth (cf. Section 2.11.2); if the width is small in proportion to the length, the dependence on the transverse coordinate can be neglected. In hydrostatic balance, the vertically integrated equations of motion result as follows

$$\frac{\partial u}{\partial t} + u\frac{\partial u}{\partial x} + g\frac{\partial}{\partial x}(h + b) = 0 \tag{8.49}$$

$$\frac{\partial h}{\partial t} + \frac{\partial uh}{\partial x} = 0 \tag{8.50}$$

Here $x$ is the coordinate in longitudinal direction, $u(x,t)$, $h(x,t)$ are velocity and height of water column, respectively; and $b(x)$ is the height of the bottom topography above a reference depth (see the sketch in Figure 8.4). The linearized flat-bottom solution would consist of long gravity waves with phase velocity $c = \sqrt{gH}$ with a mean water depth $H$ (propagation in both positive and negative $x$ direction).

The equilibrium solutions of (8.49) and (8.50) obey the relations

$$u\frac{\partial u}{\partial x} + g\frac{\partial h}{\partial x} = -g\frac{\partial b}{\partial x} \tag{8.51}$$

$$h\frac{\partial u}{\partial x} + u\frac{\partial h}{\partial x} = 0 \tag{8.52}$$

with the integrals

$$\frac{1}{2}u^2 + gh + gb = B = \text{const} \tag{8.53}$$

$$hu = U = \text{const} \tag{8.54}$$

Hence in a steady frictionless balance, both the Bernoulli function $B$ as well as the transport (per unit width) $U$ are constant, i. e. they are independent of $x$. Therefore, the solution $h(x)$ and $u(x)$ can be determined once $B$ and $U$ are known.

An important dimensionless parameter is the FROUDE[4] number $F^2 = u^2/gh$, or equivalently $F = u/\sqrt{gh} = u/c$ which is the ratio of the flow velocity to the speed of long gravity waves. Equations (8.51) and (8.52) can be interpreted as a linear system for $\partial h/\partial x$ and $\partial u/\partial x$. For arbitrary $\partial b/\partial x \neq 0$, the system has a unique

[4] WILLIAM FROUDE, *1810 in Devon, †1879 in Simonstown/South Africa, engineer and hydrodynamicist.

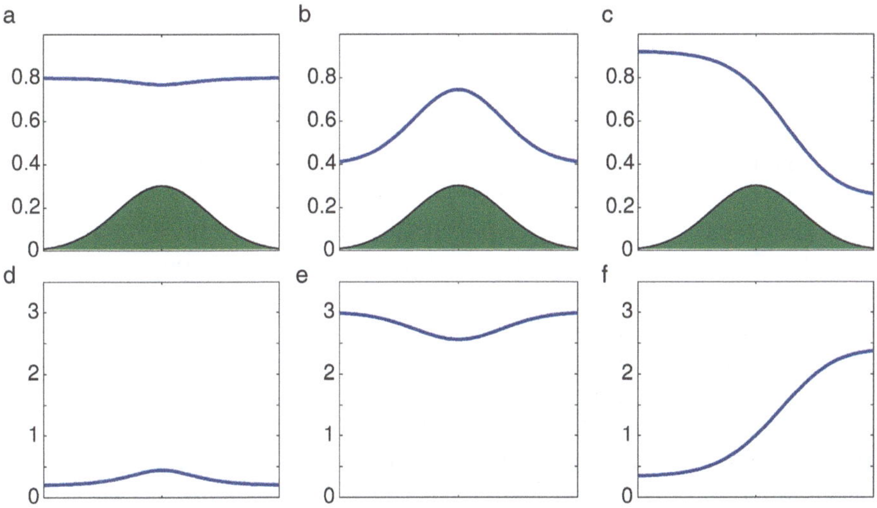


Fig. 8.5 Steady solutions of the system (8.53)–(8.54). **a–c** show the free surface $h(x) + b(x)$, together with the topography $b(x)$, and **d–f** show the corresponding Froude number. **a, d** Flow is subcritical everywhere. **b, e** Flow is supercritical everywhere. **c, f** Flow (from left to right) is hydraulically controlled

solution only if the determinant $u^2 - gh \neq 0$, i. e. $F \neq 1$. Specifically, at the sill where $\partial b/\partial x = 0$, for $F \neq 1$ it follows that $\partial h/\partial x = 0$ and $\partial u/\partial x = 0$. Therefore, the character of the equilibrium solutions depends on the magnitude of $F$.

If $F < 1$ (subcritical flow, see left panel in Figure 8.5), the wave speed is larger than the flow velocity, and wave characteristics are important for the solution. For $F \ll 1$, in (8.53) one has $u^2/2 \ll gh$, hence to first approximation $h + b \approx$ const so that in this limit the free surface is nearly level. For finite $F < 1$ we note that $h + b$ is somewhat smaller above the topography.

If $F > 1$ (supercritical flow, see middle panel in Figure 8.5), the flow velocity is larger than the wave propagation speed. For this reason, no information can be transported by waves upstream against the current. For $F \gg 1$, $u^2/2 \gg gh$ applies in (8.53), and hence approximately $u \approx$ const. It also follows that $h \approx$ const. For finite $F > 1$, $u$ is somewhat smaller above the topography.

The case $F = 1$ at the sill is of particular interest. Here the solution changes its character, from $F < 1$ to $F > 1$. Note also that $\partial h/\partial x \neq 0$ and $\partial u/\partial x \neq 0$. Physically a transition from subcritical to supercritical can be expected in the direction of flow when an initially subcritical flow is accelerated. The location where $F = 1$ is called *hydraulic control* point, and the flow is hydraulically controlled (see right panel in Figure 8.5). Note that hydraulic control can also occur at a lateral constriction in a channel, completely analogous to a sill.

Height and velocity at the critical point $x_c$ can be related to the upstream conditions. At a location $x_0$ far upstream of the control point where the bottom is flat (i. e., $b_0 = 0$) and the Froude number is small ($F \ll 1$), it follows from the constancy of the Bernoulli function (8.53) that

$$\frac{1}{2}u_c^2 + g(h_c + b_c) = \frac{1}{2}u_0^2 + g(h_0 + b_0) \approx gh_0$$

With $u_c^2 = gh_c$, one obtains the geometrical relation

$$h_c = \frac{2}{3}(h_0 - b_c) \quad \text{and thus} \quad u_c = \left[\frac{2}{3}g(h_0 - b_c)\right]^{\frac{1}{2}}$$

Hence the flow properties at the control point only depend on the upstream height and the topography at the control point.

In the presence of weak friction, a supercritical flow must eventually become subcritical again, somewhere downstream of the control point. The enforced transition to the downstream conditions occurs in a *hydraulic jump*. Dissipation can be expected to be important at the location of the jump.

The previous considerations remain valid when considering the transport of a bottom layer (thickness $h$, density $\rho + \Delta\rho$) which lies under a motionless layer (density $\rho$). Then, $g$ must be replaced by the *reduced gravity* $g' = g\Delta\rho/\rho \sim 2\times 10^{-3}$ g. Hence the transport in a hydraulically controlled channel flow can be obtained from thickness measurements at the control point, $M = uhD = u_c h_c D$, where $D$ is the channel width. With $u_c^2 = g' h_c$, this is equivalent to $M = g'^{\frac{1}{2}} h_c^{3/2} D$. In the presence of rotation, one can expect that the transport depends on the relation between the basin width $D$ and the Rossby radius $R \sim \sqrt{g'h}/f$. If the channel is wide ($D \gg R$), the current essentially remains within a Rossby radius of the basin's boundary. In that case, a simple approximation is to replace $D \to O(R)$. The calculation by Whitehead (1998) results in $D \to R/2$ and leads to

$$M \approx \frac{1}{2}\frac{g' h_c^2}{f} = \frac{1}{2}\frac{\Delta\rho}{\rho}\frac{g}{f}h_c^2$$

Further aspects occur in the case of a baroclinic current in two or more layers which are not considered here.

Known examples for currents which are – at least temporarily – hydraulically controlled are the outflow of Mediterranean Water and the overflow through the Denmark Strait between Iceland and Greenland.

## 8.2 Planetary Waves in Midlatitudes

Planetary or Rossby waves are central for the adjustment of the oceanic circulation to changes in forcing. Their governing equations can in principle be derived from the long wave equations (8.1)–(8.5), now including variations of the Coriolis parameter $f$ with latitude. A more convenient starting point is, however, the quasi-geostrophic vorticity equation (5.32). Omission of the forcing terms in (5.32) and linearization yields the linear potential vorticity equation for the stream function $\psi$ as

$$\frac{\partial}{\partial t}\left[\nabla_h^2\psi + \frac{\partial}{\partial z}\left(\frac{f_0^2}{N^2}\frac{\partial\psi}{\partial z}\right)\right] + \beta\frac{\partial\psi}{\partial x} = 0 \tag{8.55}$$

The vertical structure is that of long waves given by (8.14), and with $\psi(x,y,z,t) = \Phi_n(z)\tilde{\psi}(x,y,t)$ one obtains for the horizontal structure

$$\nabla_\mathrm{h}^2\frac{\partial\tilde{\psi}}{\partial t} - \frac{1}{R_n^2}\frac{\partial\tilde{\psi}}{\partial t} + \beta\frac{\partial\tilde{\psi}}{\partial x} = 0 \tag{8.56}$$

with the Rossby radius $R_n = c_n/f$ (the index $n$ will be dropped in the following).

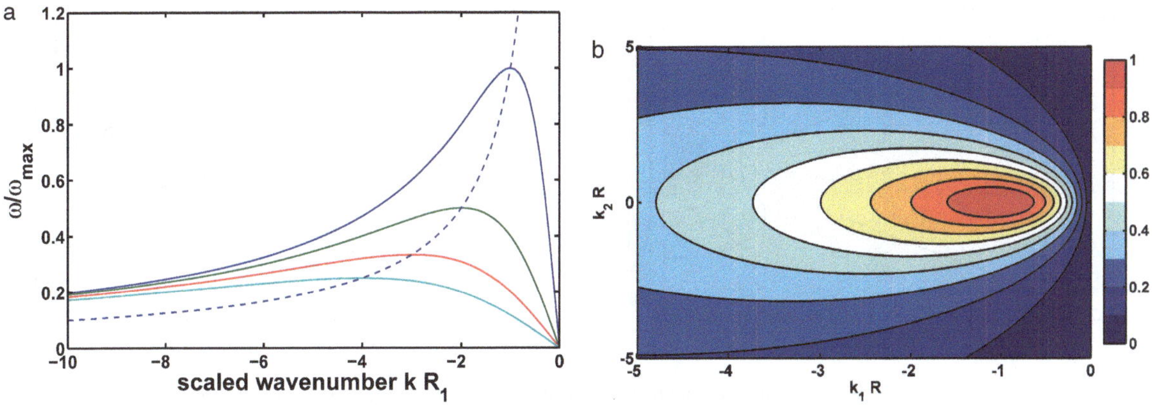


Fig. 8.6 Dispersion relation of baroclinic Rossby waves. **a** Frequency as function of scaled zonal wave number, for $k_2 = 0$ and modes $n = 1, 2, 3, 4$ (*colors blue to cyan*). The dashed line is along the locations of maximum frequency. **b** Frequency $\omega/\omega_{max}$ as function of the scaled wave numbers $(k_1 R, k_2 R)$

## 8.2.1 Propagation Characteristics

Since (8.56) has constant coefficients, the exponential ansatz $\psi = \psi_0 e^{i(k_1 x + k_2 y - \omega t)}$ immediately yields the dispersion relation

$$\omega = -\frac{\beta k_1}{k_1^2 + k_2^2 + R^{-2}} \tag{8.57}$$

which is displayed in Figure 8.6. For positive $\omega$ the zonal wave number must be negative, $k_1 < 0$, i.e. the phase always propagates westward. For fixed $k_2$, the maximum value of $\omega$ lies at $k_1 = -(k_2^2 + R^{-2})^{1/2}$. The maximum frequency for $k_2 = 0$ is $\omega_{max} = \beta R/2$; it is reached at $k_1 = -1/R$. For $k_2^2 \ll k_1^2$ and $\beta = 2 \times 10^{-11}\,\mathrm{m^{-1}\,s^{-1}}$ we obtain for the barotropic case ($R_0 = 2{,}000$ km) a maximum frequency $\omega_{max} = 2\pi/(3\ \mathrm{d})$, and for the baroclinic case ($R_1 = 30$ km) the value $\omega_{max} = 2\pi/(200\ \mathrm{d})$.

For waves with scales that are much larger than the Rossby radius (i.e. $k_1^2, k_2^2 \ll R^{-2}$) the dispersion relation is approximately

$$\omega = -\beta R^2 k_1 \tag{8.58}$$

Hence in this limiting case the waves are *nondispersive*.

Wave groups are propagating with the group velocity

$$\boldsymbol{c}_g = \begin{pmatrix} \partial\omega/\partial k_1 \\ \partial\omega/\partial k_2 \end{pmatrix} = \begin{pmatrix} c_g^x \\ c_g^y \end{pmatrix} \tag{8.59}$$

Using the dispersion relation (8.57), this implies for the zonal component

$$c_g^x = \beta \frac{k_1^2 - k_2^2 - R^{-2}}{(k_1^2 + k_2^2 + R^{-2})^2} \tag{8.60}$$

The group velocity vanishes at $k_1 = -k_\star = -\sqrt{k_2^2 + R^{-2}}$. For *long* waves ($k_1 > -k_\star$) $c_g^x < 0$, and group and phase both propagate westward. For *short* waves ($k_1 <$

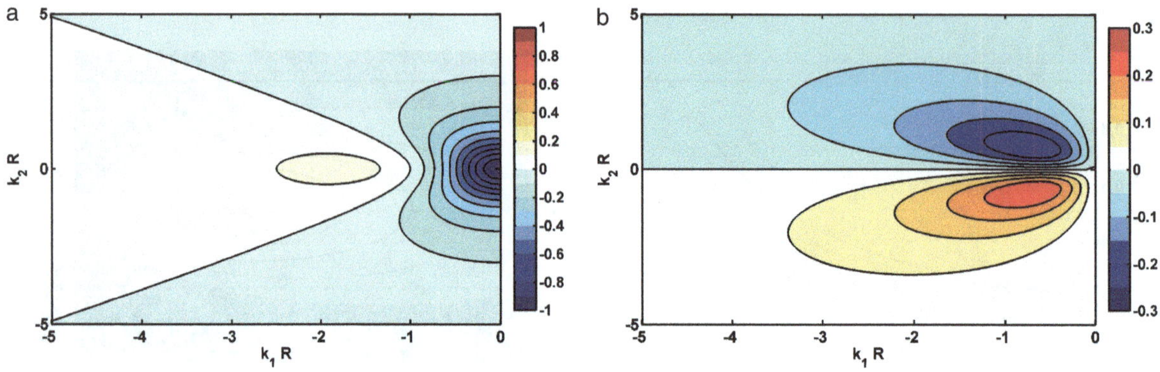


Fig. 8.7 Zonal (**a**) and meridional (**b**) components of the group velocity Rossby waves, according to (8.60) and (8.61), as function of the scaled wave numbers $(k_1 R, k_2 R)$. Unit is m s$^{-1}$

$-k_\star)\ c_g^x > 0$, and group and phase propagate in opposite directions. Maximum values of the magnitude of $c_g^x$ lie at

$$k_1 \to 0 \qquad \text{where} \quad c_g^x = -\frac{\beta}{k_\star^2}$$

$$k_1 = -\sqrt{3}k_\star \qquad \text{where} \quad c_g^x = \frac{1}{8}\frac{\beta}{k_\star^2}$$

Short waves are thus at least by factor 8 slower in $x$-direction than long waves. The group velocities are displayed in Figure 8.7. The group velocity in meridional direction is

$$c_g^y = \frac{2\beta k_1 k_2}{(k_1^2 + k_2^2 + R^{-2})^2} \tag{8.61}$$

As $k_1 < 0$, the sign of $c_g^y$ is always opposed to $k_2$, i.e group and phase propagate in opposite directions.

The existence of the planetary waves obviously hinges on the $\beta$-effect, but how the waves come about is not easily understood. In the box on p. 227 an elementary explanation of the restoring mechanism is presented.

### 8.2.2 Energy of Planetary Waves

The energy balance of free Rossby waves can be derived from the general form shown in the box on p. 148, with the appropriate approximations and linearization. Alternatively, multiplication of (8.56) by $\psi$ yields

$$\nabla\cdot\left(\psi\frac{\partial}{\partial t}\nabla\psi\right) - \nabla\psi\cdot\frac{\partial}{\partial t}\nabla\psi - \psi\frac{\partial}{\partial t}\frac{\psi}{R^2} + \beta\frac{\partial}{\partial x}\frac{1}{2}\psi^2 = 0$$

After reordering we obtain

$$\frac{\partial}{\partial t}\left[\frac{1}{2}(\nabla\psi)^2 + \frac{1}{2}\frac{\psi^2}{R^2}\right] = \nabla\cdot\left(\psi\frac{\partial}{\partial t}\nabla\psi\right) + \frac{\partial}{\partial x}\frac{\beta}{2}\psi^2 \tag{8.62}$$

**37. Restoring Mechanism of Planetary Waves**

To understand the restoring mechanism, it is useful to consider the vorticity balance. For simplicity, we restrict the discussion to the case $k_2 = 0$, i. e. there is no meridional pressure gradient and hence $u \equiv 0$, and, furthermore, to long waves, $k_1 \ll R^{-1}$, so that the stretching plays no role. If stretching as well as mechanic force and baroclinic vector are neglected, the vertical component $f + \eta$ of the absolute vorticity is governed by

$$\frac{D}{Dt}(f + \eta) = 0 \quad \text{or linearized} \quad \frac{\partial \eta}{\partial t} \approx -\beta v$$

where $D/Dt = \partial/\partial t + v\partial/\partial y$ has been used. Consider now a parcel that moves northwards, i. e. $v > 0$. Then $f = f_0 + \beta y$ increases and thus the planetary vorticity increases as well. Hence the relative vorticity $\eta$ must decrease. Consider the initial distribution $v_0(x) = V \sin(x/L)$, with a relative vorticity $\eta_0(x) = \partial v_0/\partial x = V/L \cos(x/L)$. This results in

$$\frac{\partial \eta_0}{\partial t} = -\beta v_0 = -\beta V \sin \frac{x}{L}$$

and integration over a short period yields

$$\eta = \eta_0(x) - t\,\beta V \sin \frac{x}{L} = \frac{V}{L} \cos \frac{x}{L} - t\,\beta V \sin \frac{x}{L} \approx \frac{V}{L} \cos\left(\frac{x}{L} + \beta L t\right)$$

As discussed in the treatment of planetary waves, westward propagation is the result.

Here $E_{\text{kin}} = (1/2)(\nabla\psi)^2$ is the *horizontal* kinetic energy (per unit mass). With similar arguments as in Section 5.2.5, $E_{\text{pot}} = (1/2)(\psi^2/R^2)$ can be identified as available potential energy of the waves. Note that integration of (8.62) over all $x, y$ gives no contribution from the right side as long as the normal velocity vanishes at the boundary, or equivalently $\psi = 0$ holds at the boundary.

For a wave solution $\psi = a\cos(k_1 x + k_2 y - \omega t)$ the relation between kinetic and potential energies is found as

$$\frac{E_{\text{kin}}}{E_{\text{pot}}} = \left(k_1^2 + k_2^2\right) R^2 \simeq \frac{R^2}{L^2} \tag{8.63}$$

where an average over a wave period/wave length is implied. Hence for short waves the kinetic energy dominates, for long waves the potential energy. Furthermore, the total energy is found to be

$$E = \frac{1}{2}\left(k_1^2 + k_2^2 + \frac{1}{R^2}\right) a^2$$

The energy conservation (8.62) can be rewritten, in the WKBJ sense (cf. Section 6.3), as

$$\begin{aligned} \frac{\partial E}{\partial t} &= \frac{\partial}{\partial x}\left(\frac{2\omega k + \beta}{k_1^2 + k_2^2 + R^{-2}} E\right) + \frac{\partial}{\partial y}\left(\frac{2\omega l}{k_1^2 + k_2^2 + R^{-2}} E\right) \\ &= -\frac{\partial}{\partial x} c_{\text{g}}^x E - \frac{\partial}{\partial y} c_{\text{g}}^y E = -\nabla \cdot \boldsymbol{c}_{\text{g}} E \end{aligned} \tag{8.64}$$

with the group velocity $\boldsymbol{c}_{\text{g}} = (c_{\text{g}}^x, c_{\text{g}}^y)$.

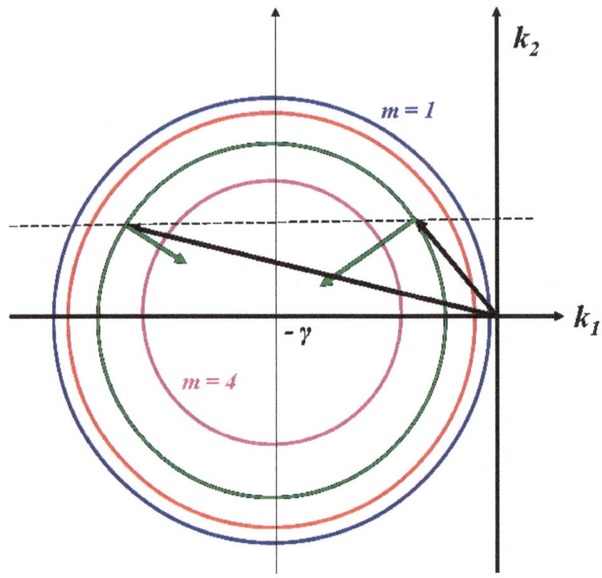


Fig. 8.8 The slowness circle with 4 modes is displayed. For mode $n = 3$ an eastward (short) and a westward propagating (long) wave is shown with wave vector and group velocity. The center of the circle is at $(-\gamma, 0)$ with $\gamma = \beta/2\omega$, the radius is $\gamma^2 - 1/R^2$. The wave vector is shown in black, the group velocity in green

### 8.2.3 Reflection at Meridional Boundaries

Reflection at lateral boundaries can significantly change the characteristics of planetary waves. A convenient way to discuss reflection is to consider the dispersion relation (8.57) in the $(k_1, k_2)$-plane, i. e.

$$k_1^2 + k_2^2 + \frac{1}{R^2} + \frac{\beta k_1}{\omega} = 0 \quad \Rightarrow \quad \left(k_1 + \frac{1}{2}\frac{\beta}{\omega}\right)^2 + k_2^2 = \left(\frac{1}{2}\frac{\beta}{\omega}\right)^2 - \frac{1}{R^2} \tag{8.65}$$

Hence $\omega = \text{const}$ corresponds to a circle depicted in Figure 8.8. For given $k_2$ and $\omega$, equation (8.65) has two solutions $k_1^i$ and $k_1^r$. With $1/R_\star^2 = k_2^2 + 1/R^2$, these can be written as

$$k_1^{r,i} = -\frac{1}{2}\frac{\beta}{\omega} \pm \left(\left(\frac{1}{2}\frac{\beta}{\omega}\right)^2 - \frac{1}{R_\star^2}\right)^{\frac{1}{2}} = -\frac{1}{2}\frac{\beta}{\omega}\left[1 \pm \left(1 - \frac{4\omega^2}{\beta^2 R_\star^2}\right)^{\frac{1}{2}}\right] \tag{8.66}$$

For small $\omega$ (i. e. $\omega/\beta R_\star \ll 1$) both wave numbers have very different magnitudes,

$$k_1^r \approx -\frac{\beta}{\omega}\,, \quad k_1^i \approx -\frac{\omega}{\beta R_\star^2}$$

with ratio $|k_1^i|/|k_1^r| = \omega^2/\beta^2 R_\star^2 \ll 1$.

Consider now the reflexion at a western boundary, situated at $x = 0$. To satisfy the kinematic boundary condition $u = 0$ or equivalently $\psi = \text{const} = 0$, set

$$\psi = \psi_i + \psi_r = a_i \cos\left(k_1^i x + k_2^i y - \omega^i t\right) + a_r \cos\left(k_1^r x + k_2^r y - \omega^r t\right)$$

The incoming wave $k_1^i, k_2^i$ is given, and $\omega^i$ follows from (8.57). Physically, this only makes sense if $c_g^i < 0$, i.e. the incoming wave must be a long wave. Note that $k_1^r, k_2^r, \omega^r$ also have to satisfy the dispersion relation. From $\psi = 0$ at $x = 0$ we obtain

$$a_i \cos\left(k_2^i y - \omega^i t\right) + a_r \cos\left(k_2^r y - \omega^r t\right) = 0$$

for all $y, t$, and it follows that $a_r = -a_i, k_2^r = k_2^i, \omega^r = \omega^i$. Based on the dispersion relation, $k_r$ is the short wave mentioned above, with an eastward (but small) group velocity. As the reflected wave is generally short, it is more affected by (scale-dependent) dissipation. Furthermore, since its group velocity is small, it is unlikely to propagate eastward over very large distances.

For the same reason, the midlatitude reflection problem at the eastern boundary is of less interest, since eastward traveling waves are short, have a slow group velocity, and are hence more prone to dissipation. Reflexion at the northern/southern boundary does not result in any new aspects because of the symmetry of the dispersion relation in view of $k_2$; it is $a_r = -a_i$, $k_r = k_i, \omega^r = \omega^i$ and thus $k_2^r = -k_2^i$. A diagram like Figure 8.8 can generally be used to obtain a graphical solution, even in cases when the boundary is along other directions.

### 8.2.4 Topographic-Planetary Waves

The topography (called orography in meteorology) plays an important role for Rossby waves and, more generally, for quasigeostrophic motions. The linear vorticity equation (8.55) form remains a convenient starting point. The planetary gradient $\beta$ has been central for planetary waves, but for topographic waves, there is the gradient $\nabla h = (h_x, h_y)$ of ocean depth entering the problem as well. The topography appears through the kinematic boundary condition at the bottom. The boundary conditions are

$$\frac{N^2}{g}\frac{\partial \psi}{\partial t} + \frac{\partial^2 \psi}{\partial z \partial t} = 0 \quad \text{at } z = 0 \tag{8.67}$$

$$\frac{N^2}{f_0}\nabla \psi \cdot \underset{\neg}{\nabla} h + \frac{\partial^2 \psi}{\partial z \partial t} = 0 \quad \text{at } z = -h \tag{8.68}$$

Here the notation $\underset{\neg}{\nabla} h = (-h_y, h_x)$ for the depth gradient $\nabla h$ rotated 90° in the counterclockwise direction has been used (compare also the box on p. 444). Note that the rigid-lid approximation (for the condition at the surface) has not yet been made. Consider the case of a weak bottom slope, $h_x/h, h_y/h \ll 1$, but at the same time we take $h \approx H_0 = \text{const}$ (this approach is analogous to the $\beta$-plane approximation $f = f_0 + \beta y$ with $\mathrm{d}f/\mathrm{d}y = \beta$ but $f \approx f_0$ and $\beta L/f_0 = L/a \ll 1$). Then $\beta$ as well as $\nabla h$ may be treated as constant in the equations. We assume solutions of the form

$$\psi = \Phi(z) \exp \mathrm{i}(\boldsymbol{k} \cdot \boldsymbol{x} - \omega t) \tag{8.69}$$

with a wave vector $\boldsymbol{k} = (k_1, k_2)$ and a vertical eigenfunction $\Phi(z)$ which must satisfy

$$\frac{\mathrm{d}}{\mathrm{d}z}\left[\frac{f_0^2}{N^2}\frac{\mathrm{d}\Phi}{\mathrm{d}z}\right] + b^2\Phi = 0 \tag{8.70}$$

$$\Phi\frac{N^2}{g} + \frac{\mathrm{d}\Phi}{\mathrm{d}z} = 0 \quad \text{at } z = 0 \tag{8.71}$$

$$\Phi\frac{N^2}{\omega f_0}\boldsymbol{k}\cdot\underset{\neg}{\nabla} h - \frac{\mathrm{d}\Phi}{\mathrm{d}z} = 0 \quad \text{at } z = -h \tag{8.72}$$

The parameters $k, b$, and $\omega$ are found to be related by $b^2 = -(k^2 + \beta k_1/\omega)$ which for $\beta \neq 0$ is equivalent to the dispersion relation

$$\omega = -\frac{\beta k_1}{k^2 + b^2} \tag{8.73}$$

where $k^2 = k_1^2 + k_2^2$. The cases $\nabla h = 0$ or $\beta = 0$ are generally considered. For $\nabla h = 0$ we find the flat-bottom planetary waves with $b = 1/R_n$, as discussed previously. For $\beta = 0$ (with $b^2 = -k^2$ and $\omega$ as eigenvalue) we obtain the pure topographic Rossby waves (see below). Note that $\boldsymbol{k}\cdot\underset{\neg}{\nabla} h$ relates to the component of $\boldsymbol{k}$ which is normal to $\nabla h$, i. e. along the isolines of equal depth.

### Pure Form of Topographic Waves

We will first discuss the waves existing for nonzero bottom slope without the planetary effect. For $\beta = 0$, only topography can provide a restoring mechanism, and we expect to obtain 'pure' topographic waves. Assuming for simplicity a bottom having only a meridional slope, $h = H_0 + \alpha y$, we obtain

$$\frac{\mathrm{d}\Phi}{\mathrm{d}z} = 0 \quad \text{at } z = 0 \tag{8.74}$$

$$\omega\frac{\mathrm{d}\Phi}{\mathrm{d}z} + \frac{N^2}{f_0}\alpha k_1\Phi = 0 \quad \text{at } z = -H_0 \tag{8.75}$$

with a simplified boundary condition at the bottom, now taken at the constant depth $H_0$. The rigid-lid approximation is assumed at the surface in (8.74). The wave ansatz results in (8.70) with $b^2 = -k^2$, or

$$\frac{\mathrm{d}}{\mathrm{d}z}\left[\frac{f_0^2}{N^2}\frac{\mathrm{d}\Phi}{\mathrm{d}z}\right] - k^2\Phi = 0 \tag{8.76}$$

An analytical solution can be obtained for a density profile with constant $N(z) = N_0$. With the vertical scale

$$d = \frac{|f_0|}{N_0 k} \tag{8.77}$$

the solution of (8.76) is given by

$$\Phi = A\cosh\frac{z}{d} + B\sinh\frac{z}{d} \tag{8.78}$$

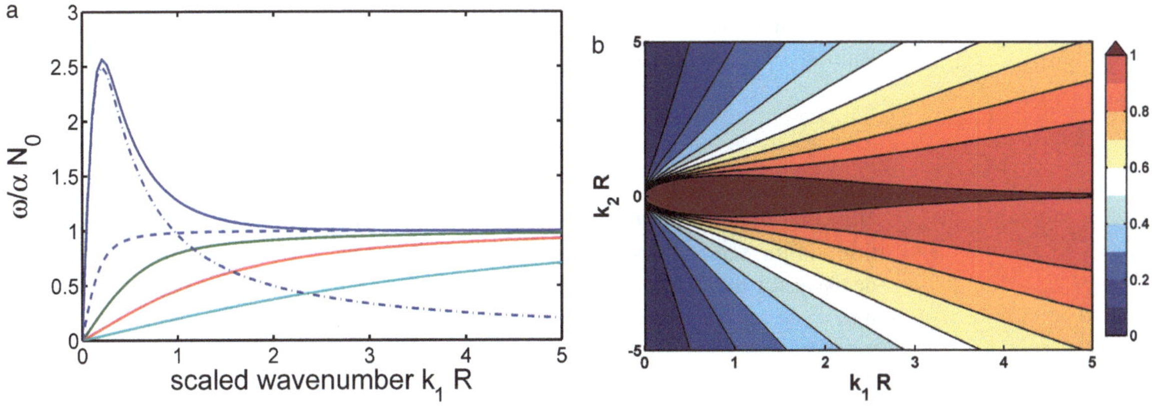


Fig. 8.9 **a** Dispersion relation for topographic waves as function of the scaled wave number $k_1 R$ for $k_2 R = 0.2, 1, 2, 5$ (*solid lines from blue to cyan*), showing for $k_2 R = 0.2$ (*blue*) also the short and long wave limits by the dashed curves. The frequency is displayed in scaled form as $\omega/(\alpha N_0)$. In **b** the dispersion relation is displayed as function of both components of the scaled wave vector $\boldsymbol{k} R$

The condition (8.74) leads to $B = 0$ while condition (8.75) results in

$$\frac{\omega}{d} \sinh \frac{H_0}{d} = \alpha \frac{N_0^2}{f} k_1 \cosh \frac{H_0}{d} \tag{8.79}$$

and with $d$ from (8.77) we obtain the dispersion relation in the form

$$\omega = \omega(k_1, k_2) = \operatorname{sign}(f) \frac{\alpha N_0 k_1}{k \tanh Rk} \tag{8.80}$$

with the baroclinic Rossby radius $R = N_0 H_0 / |f|$. Dispersion curves are displayed in Figure 8.9. For $f > 0$ and $k_1 > 0$ we must have $\alpha > 0$ to obtain a positive $\omega$. Therefore, in the northern hemisphere the shallower water always is to the right from the direction of phase propagation. The group velocity may apparently attain large values close to the origin of the wave-number space. For large wave numbers, however, the frequency flattens to a plateau and waves do not propagate (see below).

The ratio $H_0/d$ can be expressed in terms of the Rossby radius $R$. With (8.77), one obtains

$$\frac{H_0}{d} = \frac{H_0 N_0 k}{f} = kR \approx \frac{R}{L}$$

where $L = 1/k$. Note that because of the intensification of the eigenfunction towards the bottom the Rossby radius can now be defined with the near-bottom value $N = N_0$ of the Brunt–Väisälä frequency. With $N_0 = 10^{-3}\,\mathrm{s}^{-1}$ and $H_0 = 4\,\mathrm{km}$, one obtains $R \sim 40\,\mathrm{km}$. Depending on the horizontal scale $L$, there are 2 limiting cases:

### Short waves

where $L \ll R$. Hence the vertical scale is small compared to the water depth, i. e. $d \ll H_0$ and

$$\Phi(z) = A \cosh \frac{z}{d} = \frac{1}{2} A\, \mathrm{e}^{H_0/d} \left(\mathrm{e}^{(z-H_0)/d} + \mathrm{e}^{-(z+H_0)/d}\right) \sim \mathrm{e}^{-(z+H_0)/d}$$

These waves are *bottom-trapped* and are also called *fast baroclinic* waves (see Rhines, 1977). With tanh $H_0/d \approx 1$, we obtain the dispersion relation

$$\omega = \operatorname{sign}(f)\frac{\alpha N_0 k_1}{k} \tag{8.81}$$

Note that in the short wave limit $\omega$ does not depend on the value of $f$ but the solution via the scale $d$ does, in analogy to Kelvin waves. Obviously, rotation remains decisive for the existence of topographic waves. For $k_1^2 \gg k_2^2$, the dispersion turns into

$$\omega = \alpha N_0 \tag{8.82}$$

For these pure topographic waves $\omega$ does not depend on the wave number at all (but the limitation to $k_1\alpha > 0$ remains). Typical numerical values are $N_0 \approx 10^{-3}\,\mathrm{s}^{-1}$, $\alpha = 10^{-2}$ (continental shelf) so that $\omega = 10^{-5}\,\mathrm{s}^{-1}$, $2\pi/\omega \approx 6\,\mathrm{d}$. The group velocity of the limiting case (8.82) vanishes identically, i. e. pure topographic waves cannot transport energy.

### Long waves

where $L \gg R$. For $H_0 \ll d$ we obtain tanh $H_0/d \approx H_0/d$. The dispersion relation for this case

$$\omega = \frac{\alpha f}{H_0}\frac{k_1}{k^2} \tag{8.83}$$

is independent of $N_0$. Comparison with barotropic planetary waves in the rigid-lid limit shows that $\beta$ is replaced by the 'topographic $\beta$' given by $\beta_* = -\alpha f/H_0$. The vertical structure $\Phi(z)$ is approximately constant, i. e. the wave is barotropic.

## Mixed Topographic-Planetary Waves

Now we solve (8.70)–(8.73) for $\beta \neq 0$, again for constant $N^2 = N_0^2$ but with an arbitrary direction of $\nabla h$. The equations may be written in the form

$$\Phi'' + m^2\Phi = 0 \tag{8.84}$$

$$\Phi' + r^2\Phi = 0 \quad \text{at } z/H_0 = 0 \tag{8.85}$$

$$\Phi' + \eta[(kR)^2 + m^2]\Phi = 0 \quad \text{at } z/H_0 = -1 \tag{8.86}$$

where the dash denotes differentiation with respect to $z/H_0$ and $\Phi$ is now considered as function of $z/H_0$. Furthermore, we use the abbreviations

$$R = \frac{N_0 H_0}{|f_0|} \quad m^2 = (bR)^2 \quad r = \frac{R}{R_0} \quad \eta = \frac{f_0}{H_0}\frac{\boldsymbol{k}\cdot\underset{\neg}{\nabla} h}{\beta k_1} \tag{8.87}$$

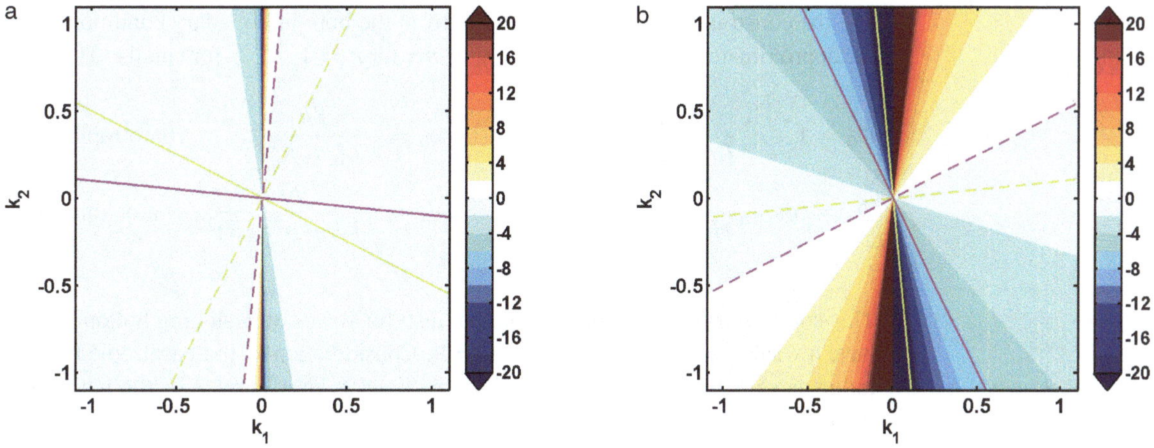


Fig. 8.10 The dimensionless parameter $\eta$ as function of $\boldsymbol{k}R$. The purple cross is oriented at $\nabla h$, the yellow cross at $\nabla(f/h)$ (the dashed lines are along the gradients, the full lines are along $h = \text{const}$ and $f/h = \text{const}$, respectively). The bottom in the **a** is close to a northward slope ($h_x = 1 \times 10^{-4}$, $h_y = 1 \times 10^{-3}$) where $\eta$ becomes almost constant. The bottom in **b** is sloping upward towards the north-east ($h_x = 2 \times 10^{-3}$, $h_y = 1 \times 10^{-3}$). For both examples we have $f_0/\beta H_0 = 1250$. The dashed lines are: $\eta = 0$ magenta ($\nabla h$ is directed along $\eta = 0$ and upward in the examples), $\eta = -1$ yellow ($\nabla f/h$ is directed along $\eta = -1$ and downward in the examples)

Here $R_0 = \sqrt{gH_0}/f_0$ is the barotropic Rossby radius. The squared ratio of the baroclinic and barotropic Rossby radii is small, $r^2 \ll 1$. The last parameter, $\eta = (f_0/\beta H_0)[-h_y + h_x k_2/k_1]$ depends only on the direction of the wave vector: $\eta = 0$ is the line in $(k_1, k_2)$-space along $\nabla h$ (magenta dashed in Figure 8.10), and $\eta = -1$ is along $\nabla f/h = (\beta/H_0)[-(f_0/\beta H_0)\nabla h + (0,1)]$ (yellow dashed in Figure 8.10). In the region close to the $k_2$-axis the parameter $\eta$ becomes very large (it approaches $\pm\infty$): here we find the regime with small effect of the $\beta$-term, which was discussed above.

The dispersion relation becomes

$$\omega = -\frac{\beta R^2 k_1}{(kR)^2 + m^2} \tag{8.88}$$

when $m^2$ is considered as the eigenvalue. There are sinusoidal solutions (for $m^2 > 0$) and exponential solutions (for $\mu^2 = -m^2 > 0$). We have to find $m^2$ as function of $r^2$, $(kR)^2$ and $\eta$. Only the solutions with positive $\omega$ are of interest. The following cases can be distinguished:

### Case $\eta = 0$, planetary Rossby waves

These have been considered previously in Section 8.2. The eigenfunctions are

$$\Phi = -\frac{r^2}{m}\sin m\frac{z}{H_0} + \cos m\frac{z}{H_0} \tag{8.89}$$

and the constraint $m \tan m = r^2$ results from the bottom boundary condition, with the approximate solutions $m_0 = r$, $m_n = n\pi$ for $n = 1, 2, \ldots$ for small $r$. Hence

$$\begin{aligned}
\Phi_0 &= 1 - r^2\frac{z}{H_0} - \frac{1}{2}\left(\frac{rz}{H_0}\right)^2 + \cdots & \omega_0 &= -\frac{\beta k_1}{k^2 + R_0^{-2}} & &\text{barotropic mode} \\
\Phi_n &= \cos m_n \frac{z}{H_0} + \cdots & \omega_n &= -\frac{\beta k_1}{k^2 + (n\pi/R)^2} & &\text{baroclinic mode}
\end{aligned} \tag{8.90}$$

This solution applies to a flat bottom but also for waves on a sloping bottom which have a wave vector $\boldsymbol{k} \| \nabla h$ and thus $\eta = 0$. Consideration of the barotropic Rossby radius is seen to generate only small corrections, with exception of the barotropic mode at very low wave numbers.

### Case $\eta \neq 0$, $m^2 > 0$, planetary-topographic Rossby waves

The above discussed planetary waves become deformed in the presence of a sloping bottom. The eigenfunction cannot remain zero at the bottom any longer, they must satisfy the kinematic condition of zero *normal* velocity. Eigenfunctions and the constraint on $m$ become

$$\begin{aligned}
\Phi &= -\frac{r^2}{m}\sin m\frac{z}{H_0} + \cos m\frac{z}{H_0} \\
m \tan m &= -m^2\frac{\eta[(kR)^2 + m^2] - r^2}{r^2\eta[(kR)^2 + m^2] + m^2} \approx -\eta[(kR)^2 + m^2] + r^2
\end{aligned} \tag{8.91}$$

The $r^2$-terms are only relevant for solutions with very small $m$. We will put $r = 0$ (rigid-lid approximation) in some parts of the following analysis. There are solutions $m_n$ close to $\pi/2, 3\pi/2, \ldots, (2n+1)\pi/2, \ldots$, just between the modes of the flat-bottom problem. More specific, one finds $n\pi \le m_n \le (2n+1)\pi/2$ for $\eta > 0$ and $(2n-1)\pi/2 \le m_n \le n\pi$ for $\eta < 0$ with $n = 1, 2, 3, \ldots$ These are familiar topographically modified baroclinic Rossby waves, the *slow baroclinic waves* in Rhines' terminology (Rhines, 1977), having a sinusoidal profile. In addition, there is a $m_0$-solution which exists only for $\eta < 0$ (see left panel of Figure 8.11) and hence requires a negative $k_1$ because we may restrict the discussion to positive $\omega$. It is discussed below in more detail.

### Case $\eta \neq 0$, $\mu = -m^2 > 0$, fast baroclinic and barotropic topographic waves

While the waves of the above case can be considered as modifications of the familiar flat-bottom Rossby waves, the present case waves only exist due to a sloping bottom. Eigenfunctions and the constraint on $\mu$ become for this case

$$\begin{aligned}
\Phi &= -\frac{r^2}{\mu}\sinh \mu\frac{z}{H_0} + \cosh \mu\frac{z}{H_0} \\
\mu \tanh \mu &= -\mu^2\frac{\eta[(kR)^2 - \mu^2] - r^2}{r^2\eta[(kR)^2 - \mu^2] - \mu^2} \approx \eta[(kR)^2 - \mu^2] - r^2
\end{aligned} \tag{8.92}$$

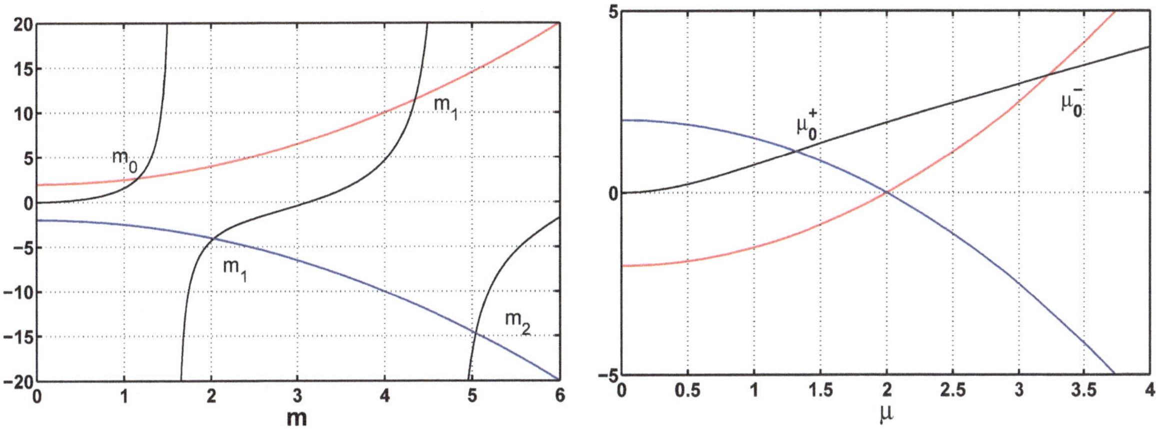


Fig. 8.11 Sketch of the constraints (8.91) and (8.92) for $r^2 = 0$ and two values of $\eta = \pm 0.5$. The left-hand side of each equation is plotted as black curves, and the right-hand side as red ($\eta < 0$) or blue ($\eta > 0$) curves

Relating the present form to the case of pure topographic waves, discussed above, we must set $\mu = H_0/d$. The constraint allows for only one solution $\mu_0$ which is found below $kR$ if $\eta > 0$, $k_1 < 0$ (denoted by $\mu_0^+$, see right panel of Figure 8.11), and above $kR$ if $\eta < 0$, $k_1 > 0$ (denoted by $\mu_0^-$). There is no solution in the remaining sector $\eta > 0$, $k_1 > 0$. The four sectors of the $(k_1, k_2)$-plane, where solutions $m_0$ of the previous case, the solutions $\mu_0^\pm$, and the 'no waves' case are valid, are thus separate and cover the whole plane (see Figures 8.10, 8.12 and 8.13).

Inserting (8.92) into (8.88) as well as assuming $\mu \approx kR$, we recover the pure topographic case from the previous section,

$$\begin{aligned} \omega &= -\frac{\beta R^2 k_1}{(kR)^2 - \mu^2} \approx -\operatorname{sign}(f_0) N_0 \frac{\boldsymbol{k} \cdot \underset{\neg}{\nabla} h}{k \tanh kR} \\ &= \operatorname{sign}(f_0) \frac{\alpha N_0 k_1}{k \tanh kR} \end{aligned} \tag{8.93}$$

where $\nabla h = (0, \alpha)$ has been assumed in the latter relation. This branch thus yields the *fast baroclinic (bottom trapped) mode* and the *fast barotropic topographic mode* of Rhines (1977), as discussed previously for $\beta = 0$. Obviously, they exist in identical form (with the current approximations) for $\beta \neq 0$.

## The lowest mode

The $m_0$-solution, which is possible only for $\eta < 0$, is sketched in the left panel of Figure 8.11. For short waves, $(kR)^2 \gg 1$, one finds $m_0 \lesssim \pi/2$. In this regime the mode has a half-sinusoid as vertical profile. For long waves, $(kR)^2 \ll 1$, an analytical approximation is obtained by expanding the tangent by its linear form at small $m$ to yield the condition $m_0^2 + \cdots = -\eta m_0^2 - \eta (kR)^2 + r^2$ where the approximated form of (8.91) has been used. One finds $m_0^2 = (-\eta (kR)^2 + r^2)/(\eta + 1)$

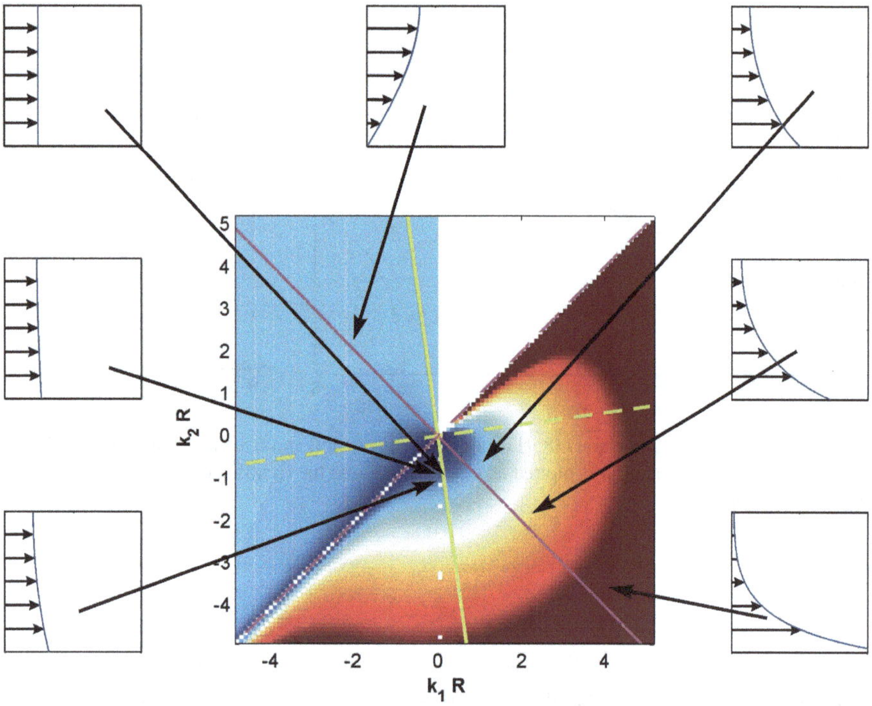


**Fig. 8.12** Eigenvalues $m_0$ and $\mu_0^\pm$ as function of $\boldsymbol{k}R$ and the different regimes of the lowest planetary-topographic mode. The bottom in this example is sloping upward towards north-east. The $m_0$-solution applies to the sector between the positive $k_2$-axis and the $\eta = 0$ line [*dashed magenta*] with $k_1 < 0$. The $\mu_0^+$-solution applies to the sector between the $\eta = 0$ line and the negative $k_2$-axis, again with $k_1 < 0$. The $\mu_0^-$-solution applies to the sector between the $\eta = 0$ line and the negative $k_2$-axis, now with $k_1 > 0$. There are no solutions (with positive $\omega$) between the $\eta = 0$ line and the positive $k_2$-axis with $k_1 > 0$. The vertical profile of the horizontal velocity of the solutions is indicated. The color scale of the eigenvalues ranges linearly from 0 (*dark blue*) to 5 (*dark brown*)

and hence

$$\omega_0 = -H_0 \frac{\boldsymbol{k} \cdot \underline{\nabla} f/h}{k^2 + R_0^{-2}} \tag{8.94}$$

This describes the topographical modification of the *barotropic Rossby wave*, the *fast barotropic mode* of Rhines's terminology. The small $m_0$ of these long-wave solutions yields an almost constant profile of $\Phi$, i. e. a barotropic mode. The eigenvalues of the lowest mode and the vertical profiles are displayed in Figure 8.12.

The $\mu_0^\pm$-solutions are sketched in the right panel of Figure 8.11. The assumption $\mu \approx kR$, made above, does, in fact, not apply everywhere in wave-number space. It is valid in the limit $\eta \to \pm\infty$, occurring for very small $\beta$ or in the vicinity of the $k_2$-axis (see Figure 8.10). Away from this regime the solutions $\mu_0^\pm$ may be much smaller or much larger than $kR$, respectively. Approximate analytical solutions for $\mu_0^\pm$ may be easily found for the various regions of the wave-number space. For short waves we find an analytical form replacing $\tanh \mu$ by its value at infinity and hence

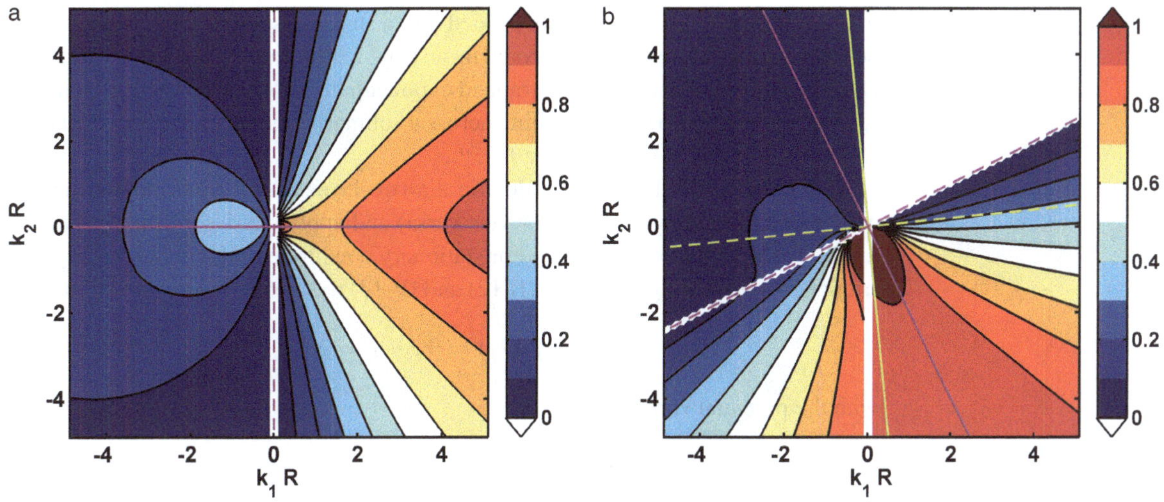


Fig. 8.13 Dispersion relation of the lowest mode for topographic-planetary waves as function of $\boldsymbol{k}R$. The frequency is shown in the scaled form as $\omega/(|\nabla h|N_0)$. **a** refers to the conditions $h_x = 0$, i. e. about case in Figure 8.10b, **b** refers to the case of a north-eastward slope, shown in Figure 8.10b. The purple cross is oriented at $\nabla h$, the yellow cross at $\nabla(f/h)$, as in Figure 8.10. The solution is patched from the three parts $m_0$, $\mu_0^+$ and $\mu_0^-$ described in the text (they are separated by white gaps)

$\mu_0 + \cdots = -\eta\mu_0^2 + \eta(kR)^2 - r^2$ which yields after some computations

$$\omega_0 = N_0 \frac{\boldsymbol{k} \cdot \underset{\neg}{\nabla} h}{k} \tag{8.95}$$

in agreement with (8.81). Waves associated with this plateau are bottom trapped (see Figure 8.12). Long waves are found by expanding $\tanh\mu \approx \mu$ which yields $\mu_0^2 = (\eta(kR)^2 + r^2)/(\eta + 1)$ and then again (8.94).

In Figure 8.13 we present numerical solutions of (8.91) and (8.92), using in each case the simplified relation with $r^2 = 0$. Then the $m_0$-solution is separated from the $\mu_0^+$-solution by a zero line along $\eta = 0$ (magenta dashed in the figures). To the left of the $\eta = 0$ line we find the half-sinusoid solution for $m_0$ in the large wave-number region and the barotropic solutions, including the particular form (8.94), at small wave numbers. The truly topographic waves are found on the right side. We find here the extension of the barotropic topographic wave (8.94) at low wave numbers and the bottom trapped mode at high wave numbers. The left panel, appropriate for small $h_x$, are the planetary version of the pure topographic waves, discussed above and shown in Figure 8.9.

### 8.2.5 Stationary Rossby Waves in a Baroclinic Flow over a Ridge

Flow over varying topography is the topic in many chapters of this book, particularly in the Chapters 14 and 16. Here we address a fairly elementary situation: the perturbation of a zonal current by a topographic feature which is a meridionally oriented step in the ocean depth. In an initial value problem with a prescribed upstream flow, waves are generated at the step and propagate up- and downstream. After this initial response has radiated away or died out, a steady response remains which can be classified as Rossby waves in zonal mean flow with zero frequency (including the Doppler shift) and perturbations trapped exponentially at the step. The problem has relevance to zonal flows propagating across a submarine ridge, as e. g. the Antarctic Circumpolar Current (see Chapter 16).

We consider zonal baroclinic flow in a a two-layer ocean from a higher terrain over a step into a lower terrain (or vice versa) and apply quasi-geostrophic dynamics, presented in Appendix B.1.3. For stationary conditions and without forcing and friction, the governing equations (B.12) and (B.13) reduce to

$$\mathcal{J}(\psi_j, Q_j) = 0\,, \quad j = 1, 2 \tag{8.96}$$

with the potential vorticities

$$\begin{aligned} Q_1 &= \nabla^2\psi_1 - F_1(\psi_1 - \psi_2) + f_0 + \beta y \\ Q_2 &= \nabla^2\psi_2 + F_2(\psi_1 - \psi_2) + f_0 + \beta y + f_0 b/H_2 \end{aligned}$$

where $F_j = f_0^2/(g^* H_j)$ and $g^* = \Delta\rho/\rho$. The $f_0$ contribution to the potential vorticities is irrelevant and will abandoned in the following. Note that the ocean is assumed infinite and thus there are no boundary conditions. Far upstream, $x \to -\infty$, the flow is assumed zonal and eastward.

The step is assumed at $x = 0$, and to the west we have a depth $H_1$, and to the east we have $H_2 - b$. In the examples, we shall consider a flow in the southern hemisphere, i. e. $f_0 < 0$. For a downward step $b$ is negative. Note further that only the product $f_0 b$ is of relevance for the flow pattern: an upward step in the northern hemisphere produces the same flow as downward step in the southern hemisphere. Consider now a downward step. It is expected that the eastward flow, after entering the deeper domain, is deflected towards the south due to the stretching of the fluid columns: $\beta v = f_0 \partial w/\partial z$ implies negative $v$ for $f_0 < 0$ and stretching. However, this argument only applies to the barotropic component of the flow; the flow in the individual layers may behave differently (see e. g. the cases displayed in Figure 8.16). The $\beta$-effect will then act as a restoring force (see the box on p. 227) and a stationary Rossby wave pattern will evolve eastward of the step. The barotropic problem, including the initial value case, is solved by McIntyre (1968). The baroclinic problem is discussed to some extent in Salmon (1998).

The solution of (8.96) may be expressed by a functional relation $Q_j = G_j(\psi_j)$ between the stream function and the potential vorticity, applying to both sides of the step. The functions $G_j(\psi_j)$ can be evaluated from the condition on the far upstream western part, $x \to -\infty$, where we have a strictly zonal flow,

$$\psi_1 \to -u_1 y\,, \quad \psi_2 \to -u_2 y$$

with constant velocities $u_j$, assumed positive. A vertical shear $U = u_1 - u_2$ in the upstream profile is thus allowed. As the relative vorticity is zero in the far upstream region, we obtain

$$\begin{aligned} G_1(\psi_1) &= \Lambda_1 U y + \beta y = -\frac{\beta + \Lambda_1 U}{u_1}\psi_1 = -c_1\psi_1 \\ G_2(\psi_2) &= -\Lambda_2 U y + \beta y = -\frac{\beta - \Lambda_2 U}{u_2}\psi_2 = -c_2\psi_2 \end{aligned}$$

On the eastern part we have nonzero vorticity and have to solve the coupled system $Q_j = G_j(\psi_j)$ of differential equations, given by

$$\begin{aligned} \nabla^2\psi_1 - \Lambda_1(\psi_1 - \psi_2) + \beta y &= -c_1\psi_1 \\ \nabla^2\psi_2 + \Lambda_2(\psi_1 - \psi_2) + \beta y + f_0 b/H_2 &= -c_2\psi_2 \end{aligned} \tag{8.97}$$

Note that the problem has become linear though the problem is originally governed by the fully nonlinear conservation equation of potential vorticity. The $c_j$ are functions of the upstream velocity. With $m_j = c_j - \Lambda_j$ and defining the matrix

$$\mathcal{M} = \begin{pmatrix} m_1 & \Lambda_1 \\ \Lambda_2 & m_2 \end{pmatrix} \tag{8.98}$$

the set of equations (8.97) may be written in vector form,

$$\nabla^2\boldsymbol{\psi} + \mathcal{M}\boldsymbol{\psi} = -\begin{pmatrix} \beta y \\ \beta y + f_0 b/H_2 \end{pmatrix} = -\boldsymbol{\beta} \tag{8.99}$$

with $\boldsymbol{\psi} = (\psi_1, \psi_2)^{\mathrm{T}}$. The solution to (8.99) may then be expressed in terms of a particular solution of the inhomogeneous system, e. g. one which independent of $x$,

$$\boldsymbol{\psi}^{\mathrm{inh}} = -\mathcal{M}^{-1}\boldsymbol{\beta} = -\det(\mathcal{M})^{-1}\begin{pmatrix} m_2 & -\Lambda_1 \\ -\Lambda_2 & m_1 \end{pmatrix}\boldsymbol{\beta}$$

and the general solution of the homogeneous problem. The inhomogeneous part is thus easily solved but the homogeneous part requires some consideration.

The solution of the homogeneous part

$$\nabla^2\boldsymbol{\psi} + \mathcal{M}\boldsymbol{\psi} = 0$$

has the form $\psi_j = (1 + \hat{b}_j y)\chi_j(x)$ with arbitrary constant $\hat{b}_j$. For the $x$-dependence we consider the ansatz $\chi_j(x) = E_j \exp(\mathrm{i}\lambda x)$ which leads to the eigenvalue problem

$$-\lambda^2 E_1 + m_1 E_1 + \Lambda_1 E_2 = 0 \quad \text{and} \quad -\lambda^2 E_2 + \Lambda_2 E_1 + m_2 E_2 = 0$$

The squared eigenvalues and the corresponding eigenvectors are easily found,

$$\begin{aligned} \lambda_\pm^2 &= \frac{1}{2}(m_1 + m_2) \pm \frac{1}{2}\sqrt{4\Lambda_1\Lambda_2 + (m_1 - m_2)^2} \quad \text{and} \\ E_2^\pm &= E_1^\pm(\lambda_\pm^2 - m_1)/\Lambda_1 \end{aligned}$$

The eigenvalues are displayed in Figure 8.14 as function of the upstream velocities $u_1$ and $u_2$. The black line in the figures is $\lambda_-^2 = 0$, or $u_2 = (\beta - \Lambda_2 u_1)/\Lambda_1$ and the blue line marks the limit $U = u_1 - u_2 = \beta/\Lambda_2$ for the condition of baroclinic instability for the two-layer system (see Section 8.5.3). The two-layer flow is unstable to the right of the blue line; the solutions given below are irrelevant in this sector. The lines intersect at $u_1 = \beta/\Lambda_2$. We distinguish three cases: the condition $\lambda_+^2 > 0, \lambda_-^2 < 0$ (case 1) seems most interesting because oscillatory and decaying solutions are possible; the condition $\lambda_+^2 > 0, \lambda_-^2 > 0$ (case 2) has only oscillatory solutions; and a further case, $\lambda_+^2 < 0$, $\lambda_-^2 < 0$ (case 3), remains which is entirely of decaying nature.

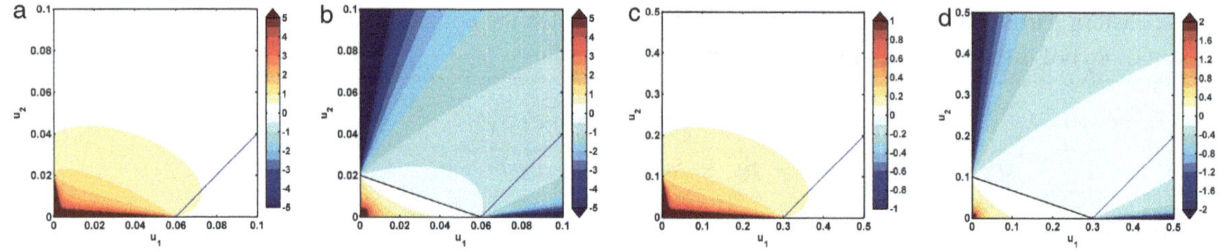

Fig. 8.14 **a** $\lambda_+^2$ and **b** $\lambda_-^2$ as function of $u_1$ and $u_2$ for $H_1 = 1{,}000\,\mathrm{m}$, $H_2 = 3{,}000\,\mathrm{m}$, $f_0 = -10^{-4}\,\mathrm{s}^{-1}$, $\beta = 2\times10^{-11}\,\mathrm{m}^{-1}\,\mathrm{s}^{-1}$ and $g^* = 0.01\,\mathrm{m\,s}^{-2}$. **c** and **d** same but for $g^* = 0.05\,\mathrm{m\,s}^{-2}$. The black is $\lambda_-^2 = 0$ and the blue line is $U = \beta/\Lambda_2$. Note the different axes and color scales

**38. A simple case: $u_1 = u_2$**

It is illuminating to consider the flow over the step in terms of the barotropic and baroclinic stream functions $\psi_e = (H_1\psi_1 + H_2\psi_2)/(H_1 + H_2)$, $\psi_i = \psi_1 - \psi_2$, as defined in Appendix B.1.3. To simplify further, we assume the shear to be zero, i. e. $u_1 = u_2 = u$, because the components $\psi_e$ and $\psi_i$ decouple for this particular condition. From (8.97) we find indeed

$$\begin{aligned} \nabla^2\psi_e + \frac{\beta}{u}\psi_e &= -\beta y - f_0 b/(H_1 + H_2) \\ \nabla^2\psi_i + \left(\frac{\beta}{u} - R^{-2}\right)\psi_i &= f_0 b/H_2 \end{aligned} \tag{B38.1}$$

where $R^{-2} = \Lambda_1 + \Lambda_2$ defines the internal Rossby radius $R$ (see Appendix B.1.3). The $\beta y$ term in the barotropic equation can be eliminated by writing the equations for the deviation $\psi_j' = \psi_j + uy$, $j = 1, 2$ from the upstream flow. The barotropic and baroclinic components are then seen to be forced by the step with different signs of $f_0 b$.
The eigenvalues of the homogeneous part of the solution (see text) now take a simple form: $\lambda_+^2 = \beta/u$ and $\lambda_-^2 = \beta/u - R^{-2}$. For eastward flow, $u > 0$, on the upstream side the barotropic solution is thus always oscillatory: steady Rossby waves with a wave number $\sqrt{\beta/u}$, excited by the step in the topography, appear downstream. The baroclinic response, on the other hand, may be decaying or oscillatory, depending on the sign of $u - \beta R^2$, i. e. the sum of the upstream flow velocity and the (westward) group velocity of the fastest long baroclinic Rossby waves (see Section 8.2.1). If $u < \beta R^2$ there is a wake of Rossby waves behind the step; if $u > \beta R^2$ the response is decaying, i. e. trapped at the step.
In the above qualitative arguments we have placed the wave response only on the downstream side of the step. This is because steady Rossby waves have no upstream propagation of energy. For a vanishing shear, as discussed in this box, the proof is easy. The group velocity $c_g^x + u$ comes into play, with the intrinsic component $c_g^x$ given by (8.60). Take $k_2 = 0$ for simplicity. For steady waves one has $\omega = -\beta k_1/(k_1^2 + R^{-2}) + k_1 u = 0$ which yields $k_1^2 = \beta/u - R^{-2}$. Insertion into the expression for the group velocity yields $c_g^x + u = 2u(1 - u/(\beta R^2))$. Hence $c_g^x + u > 0$ if $u < \beta R^2$: the group velocity is eastward and downstream Rossby waves appear. Furthermore, $c_g^x + u < 0$ if $u > \beta R^2$: the group velocity is now westward but the response is decaying because $k_1^2 < 0$. We will use this argument in the solution of the general case as well.

### Case 1: $\lambda_+^2 > 0, \lambda_-^2 < 0$

Here we have oscillatory and exponential solutions. We make the ansatz

$$\begin{aligned} \psi_{ju} &= -u_j y + (1 + d_j y) C_j \mathrm{e}^{|\lambda_-|x} \\ \psi_{jd} &= (1 + b_j y)[A_j \cos\lambda_+ x + B_j \mathrm{e}^{-|\lambda_-|x}] + \psi_j^{\mathrm{inh}}(y) \end{aligned}$$

for the up- and downstream stream functions. The form of the ansatz follows from the discussion of energy propagation in the box on p. 240. These are solutions if the eigenvector relations

$$A_2 = A_1(\lambda_+^2 - m_1)/\Lambda_1\,, \quad B_2 = B_1(\lambda_-^2 - m_1)/\Lambda_1\,, \quad C_2 = C_1(\lambda_-^2 - m_1)/\Lambda_1 \tag{8.100}$$

are satisfied. Continuity of $\psi_j$ at $x = 0$ requires

$$-u_j y + (1 + d_j y)C_j = (1 + b_j y)[A_j + B_j] + \psi_j^{\text{inh}}(y)$$

In the following we write the inhomogeneous solution as sum, $\psi_j^{\text{inh}}(y) = \psi_j^0 - u_j^\beta y$ with $\psi_j^0 = \psi_j^{\text{inh}}(y = 0)$. Take $y = 0$ and find

$$C_j = A_j + B_j + \psi_j^0 \quad \text{and thus} \quad -u_j + d_j C_j = b_j(A_j + B_j) - u_j^\beta \tag{8.101}$$

Note that continuity of the $\psi_j$ implies continuity of the $Q_j$ because of the functional relation. Continuity of $\partial\psi_j/\partial x$ at $x = 0$ implies

$$(1 + d_j y)C_j = -(1 + b_j y)B_j$$

hence

$$C_j = -B_j \quad \text{and} \quad b_j = d_j \tag{8.102}$$

Combining (8.101) and (8.102) we find

$$A_j = -2B_j - \psi_j^0 \quad \text{and} \quad -u_j = -d_j\psi_j^0 - u_j^\beta$$

The last relation determines $d_j$ and thus also $b_j$ which are found to be zero. Finally, the $A$'s and $B$'s follow from

$$\begin{aligned}
A_1 + 2B_1 &= -\psi_1^0 \\
A_2 + 2B_2 &= -\psi_2^0 \\
\Lambda_1 B_2 - B_1(\lambda_-^2 - m_1) &= 0 \\
\Lambda_1 A_2 - A_1(\lambda_+^2 - m_1) &= 0
\end{aligned}$$

The solution is exemplified in Figure 8.15 for two cases of stratification.

### Case 2: $\lambda_+^2 > 0, \lambda_-^2 > 0$

Here we have only oscillatory solutions. We make the ansatz

$$\begin{aligned}
\psi_{ju} &= -u_j y \\
\psi_{jd} &= (1 + b_j y)[A_j \cos\lambda_+ x + B_j \cos\lambda_- x] + \psi_j^{\text{inh}}(y)
\end{aligned}$$

again with (8.100) being valid. Continuity of $\psi$ at $x = 0$ yields

$$-u_j y = (1 + b_j y)[A_j + B_j] + \psi_j^{\text{inh}}(y)$$

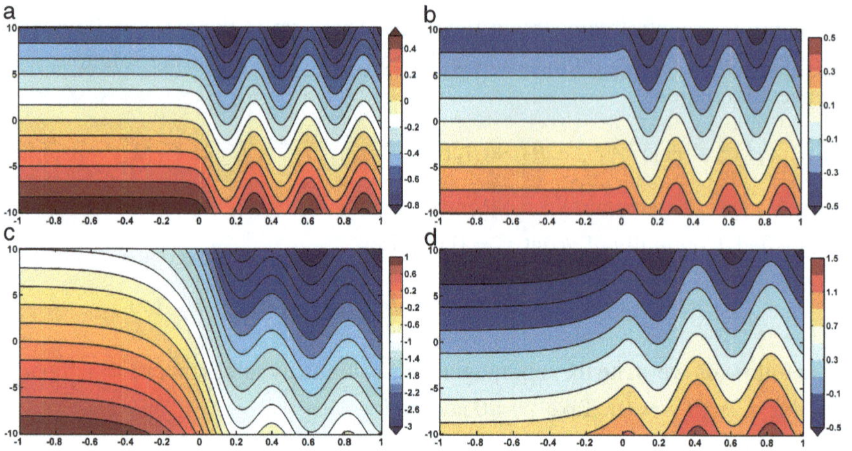


Fig. 8.15 Case 1. Stream functions (in $m^2 s^{-1}$) for $H_1 = 1{,}000\,m$, $H_2 = 3{,}000\,m$, $b = -2{,}000\,m$, $f_0 = -10^{-4}\,s^{-1}$ and $g^* = 0.01\,m\,s^{-2}$, $u_1 = 0.06\,m\,s^{-1}$, $u_2 = 0.04\,m\,s^{-1}$ **a**, **b** and $g^* = 0.05\,m\,s^{-2}$, $u_1 = 0.10\,m\,s^{-1}$, $u_2 = 0.08\,m\,s^{-1}$ **c**, **d**. Note that $u_1$ and $u_2$ are different for the two cases of stratification. **a** and **c** refer to the upper layer, **b** and **d** to the lower layer. The coordinates are in units of 1,000 km

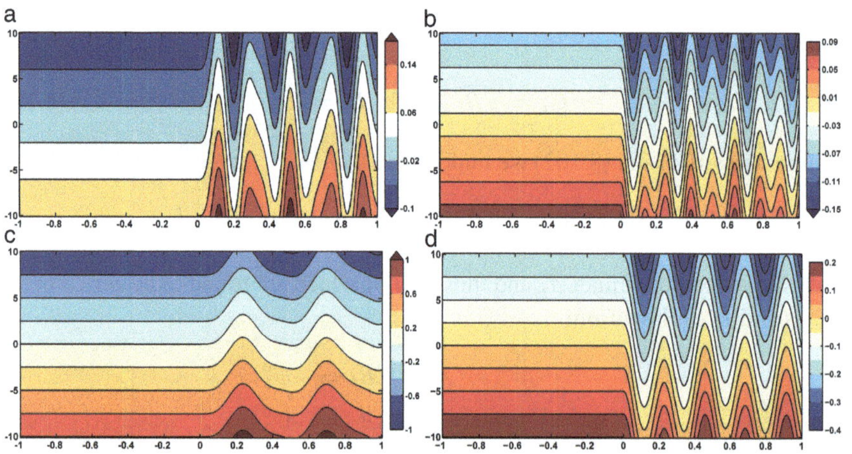


Fig. 8.16 Case 2. Stream functions (in $m^2 s^{-1}$) for $H_1 = 1{,}000\,m$, $H_2 = 3{,}000\,m$, $b = -2{,}000\,m$, $f_0 = -10^{-4}\,s^{-1}$ and $g^* = 0.01\,m\,s^{-2}$, $u_1 = 0.01\,m\,s^{-1}$, $u_2 = 0.008\,m\,s^{-1}$ **a**, **b** and $g^* = 0.05\,m\,s^{-2}$, $u_1 = 0.08\,m\,s^{-1}$, $u_2 = 0.02\,m\,s^{-1}$ **c**, **d**. Note that $u_1$ and $u_2$ are different for the two cases of stratification. **a** and **c** refer to the upper layer, **b** and **d** to the lower layer. The coordinates are in units of 1,000 km

Take $y = 0$ and find

$$0 = A_j + B_j + \psi_j^0 \quad \text{and thus} \quad -u_j = b_j(A_j + B_j) - u_j^\beta \tag{8.103}$$

Continuity of $\partial\psi/\partial x$ at $x = 0$ is satisfied by the above ansatz. We find

$$A_j = -B_j - \psi_j^0\,, \quad -u_j = -b_j\psi_j^0 - u_j^\beta$$

and hence, the $A$'s and $B$'s follow from

$$
\begin{aligned}
A_1 + B_1 &= -\psi_1^0 \\
A_2 + B_2 &= -\psi_2^0 \\
\Lambda_1 B_2 - B_1(\lambda_-^2 - m_1) &= 0 \\
\Lambda_1 A_2 - A_1(\lambda_+^2 - m_1) &= 0
\end{aligned}
$$

The $b_j$ are again zero. The solution is exemplified in Figure 8.16.

#### Case 3: $\lambda_+^2 < 0, \lambda_-^2 < 0$

Here we have only exponential solutions. We may proceed as before but the case is less interesting because there are no wavy patterns.

### 8.2.6 Spin-up of the Wind-Driven Basin Circulation

Planetary waves are responsible for the adjustment of large-scale disturbances, after the flow has passed the phase to a local geostrophic balance by radiation of gravity waves (see Section 8.1.3). Of particular interest is the response of an initially quiet ocean to a suddenly applied forcing and the subsequent spin-up of a basin-wide circulation. Forcing and boundary effects play an important role, but most of the processes focus on wave propagation and reflection. For this reason, we discuss the spin-up of the wind-driven ocean circulation at this point and not in Chapter 10, which is about forced wave problems. Most of the analysis follows the work of Anderson and Gill (1975) who have first studied the spin-up problem.

We consider an ocean in a simple rectangular configuration $0 < x < B, 0 < y < L$ with a flat bottom and forcing by a zonal windstress $\boldsymbol{\tau} = (\tau^{(x)}, 0)$ with a sinusoidal profile

$$\tau^{(x)}(y) = -\tau_0 \cos \frac{\pi y}{L} \tag{8.104}$$

The set-up is identical to the one used later in Chapter 14 on the analysis of the steady wind-driven ocean circulation. We add the appropriate forcing term to the equation in Section 8.2. The potential vorticity balance (8.56) for the vertical mode $n$ is augmented by the forcing term due to a stress divergence $\mathcal{F} = (\mathcal{F}_u, \mathcal{F}_v)$ in the corresponding momentum balances (8.1) and (8.2) in the form

$$\frac{\partial}{\partial t}\left(\nabla^2 \tilde{\psi}_n - \frac{1}{R_n^2}\tilde{\psi}_n\right) + \beta \frac{\partial \tilde{\psi}_n}{\partial x} = \frac{\partial \mathcal{F}_{vn}}{\partial x} - \frac{\partial \mathcal{F}_{un}}{\partial y} \tag{8.105}$$

Here, $\tilde{\psi}_n$ and $\mathcal{F}_{un}, \mathcal{F}_{vn}$ are modal amplitudes. We assume a simple body-force model for the windstress coupling: the corresponding forcing terms in (8.105) are written as $\mathcal{F} = (\mathcal{F}_u, \mathcal{F}_v) = \boldsymbol{\tau}\mathrm{d}S(z)/\mathrm{d}z$, i.e. the stress $\boldsymbol{\tau}$ has a vertical structure given by $S(z)$, which may be taken linearly decreasing from the surface value 1 to 0 at the mixed layer base at $z = -d$. Then $\mathrm{d}S(z)/\mathrm{d}z = 1/\mathrm{d}$ for $-d < z < 0$ and zero elsewhere. The only important property is that $\mathrm{d}S(z)/\mathrm{d}z$ is confined to the surface layer. The specific form, however, will be irrelevant, but note that $\mathrm{d}S(z)/\mathrm{d}z \to \delta(z)$ approaches a $\delta$-function for $d \to 0$ (see the box on p. 199). Expanding this limit of the structure function into the vertical normal modes, $\mathrm{d}S(z)/\mathrm{d}z = \sum_n s_n \Phi_n(z)$, one finds $s_n = \Phi_n(0)$.

The equation governing long planetary waves of the vertical mode $n$ in the presence of the above described forcing becomes

$$\frac{\partial}{\partial t}\left(\nabla^2 \tilde{\psi}_n - \frac{1}{R_n^2}\tilde{\psi}_n\right) + \beta \frac{\partial \tilde{\psi}_n}{\partial x} = -s_n \frac{\partial \tau^{(x)}}{\partial y}\Theta(t) \tag{8.106}$$

The windstress appears as curl of the stress, and the sudden turning on of the forcing at time $t = 0$ is formally expressed by the HEAVISIDE[5] function $\Theta(t)$. The ansatz $\tilde{\psi}_n(x, y, t) = -\phi_n(x, t) s_n \tau_0 \ell \sin \ell y$ with $\ell = \pi/L$ leads to

$$\frac{\partial}{\partial t}\left(\frac{\partial^2 \phi}{\partial x^2} - \left(\ell^2 + \frac{1}{R^2}\right)\phi\right) + \beta \frac{\partial \phi}{\partial x} = \Theta(t) \tag{8.107}$$

The mode index $n$ is omitted for the moment. The equation governs the amplitude function $\phi(x, t)$ of a wave with meridional wave number $\ell$ to a unit force which is suddenly switched on at time $t = 0$. For $t \le 0$ we thus have $\phi \equiv 0$. At the eastern and western boundaries, $x = 0$ and $x = B$, the amplitude function $\phi$ must vanish at all times. Because the forcing is homogeneous in $x$, an $x$-independent (so-called 'local') solution

$$\phi_{\mathrm{loc}}(t) = -\frac{t}{\ell^2 + R^{-2}} = \frac{t c_{\mathrm{g}}}{\beta} \qquad \text{with} \qquad c_{\mathrm{g}} = -\frac{\beta}{\ell^2 + R^{-2}} \tag{8.108}$$

is easily found, but it cannot prevail. The nonvanishing of $\phi_{\mathrm{loc}}$ at the boundaries requires that waves are excited which propagate from the specific boundary into the interior. We thus write $\phi = \phi_{\mathrm{loc}} + \phi_{\mathrm{wave}}$ where $\phi_{\mathrm{wave}}$ is the wave response. Waves starting at the eastern boundary $x = B$ must possess a westward group velocity and must hence be long planetary waves (see Section 8.2.1). For this part of the wave response, we may approximate (8.107) by the long-wave limit

$$\frac{\partial \phi_{\mathrm{wave}}^E}{\partial t} + c_{\mathrm{g}} \frac{\partial \phi_{\mathrm{wave}}^E}{\partial x} = 0 \tag{8.109}$$

The long-wave limit is nondispersive. The general solution of (8.109) is of the form $\phi_{\mathrm{wave}}^E(x, t) = F(x - c_{\mathrm{g}} t)$ with an arbitrary function $F(\xi)$ (see Section 6.2), but the boundary condition $\phi_{\mathrm{wave}}(x = B, t) = -\phi_{\mathrm{loc}}(t)$ at the eastern rim determines this function to $F(B - c_{\mathrm{g}} t) = -t c_{\mathrm{g}}/\beta$ or $F(\xi) = -(B - \xi)/\beta$. The response is a front which propagates with the speed $c_{\mathrm{g}}$, and the position $x_{\mathrm{f}}(t)$ of the front at time $t$ is given by $B - x_{\mathrm{f}} = -c_{\mathrm{g}} t$, likewise the time $t_{\mathrm{f}}(x)$ when the front reaches a given position $x$ in the interior is $t_{\mathrm{f}}(x) = -(B - x)/c_{\mathrm{g}}$ (note that $c_{\mathrm{g}}$ is negative). Then, for $t < t_{\mathrm{f}}$, the response at $x$ is given by the local solution (8.108), and if $t > t_{\mathrm{f}}$, the wave response has to be added.

Disregarding for the moment the contribution from the western boundary, the solution is given by

$$\phi^E(x, t) = \phi_{\mathrm{loc}}(t) + \phi_{\mathrm{wave}}^E(x, t) = \begin{cases} t c_{\mathrm{g}}/\beta & \text{for } t < t_{\mathrm{f}} \\ -(B - x)/\beta & \text{for } t > t_{\mathrm{f}} \end{cases} \tag{8.110}$$

[5] OLIVER HEAVISIDE, *1850 in London, †1925 in Torquay, mathematician and physicist. The Heaviside function is the step function with unit amplitude: $\Theta(t) = 1$ for $t > 0$ and $\Theta(t) = 0$ for $t \le 0$.

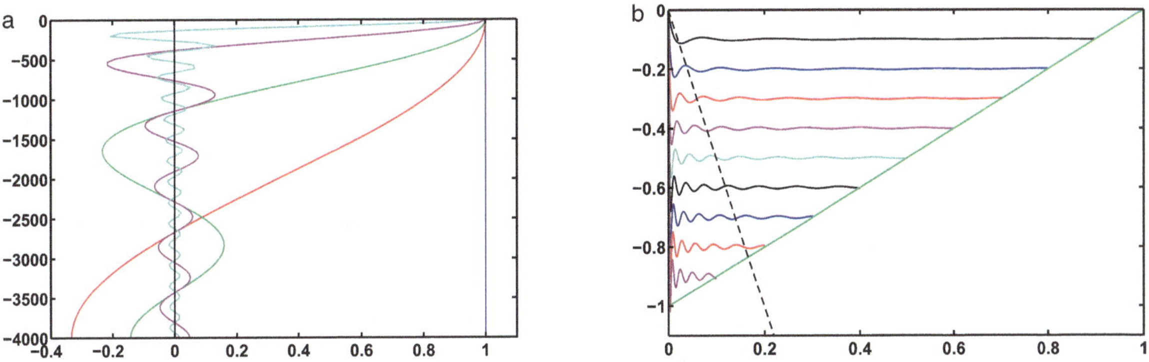


Fig. 8.17 **a** Demonstration of the Sverdrup catastrophe. The figure displays the modal sum $\sum_0^N \Phi_n(0)\Phi_n(z)$ over $N$ modes as function of $z$: $N = 0$ [*blue*], $N = 1$ [*red*], $N = 3$ [*green*], $N = 10$ [*magenta*], and $N = 30$ [*cyan*]. The curves are normalized by their respective value at $z = 0$. **b** Total response $\phi_{\text{loc}} + \phi^E_{\text{wave}} + \phi^W_{\text{wave}}$, scaled by $B/\beta$, as function of $x/B$ for various times $(0.1, 0.2, \ldots, 0.9) \times (-B/c_g)$. The dashed black line indicates the maximum distance that a short wave can travel from the western coast at the respective time

After the wave front has passed the point $x$, the field is stationary from that position to the eastern boundary. It seems that the wave response switches off the time-dependent local response in the wake of its front. In fact, as $-(B-x)/\beta = t_f c_g/\beta$, the local response is locked at the current value, when the front passes at $t = t_f$. Note that the steady response $-(B-x)/\beta$ is identical for all vertical modes $n$, but the time $t = t_f$, when it is reached, depends on $n$ via the group velocity $c_g$. With $c_g \approx \beta R^2$ and the Rossby radii from the table in Section 8.1.1, we find $c_g \approx 10^2\,\text{m s}^{-1}$ for the barotropic mode $n = 0$ and $c_g \approx 0.01\,\text{m s}^{-1}$ for the first baroclinic mode $n = 1$. The barotropic front thus passes a basin of width 5,000 km in about a day; the baroclinic front needs some years.

The passage of the barotropic wave leads to the vertically constant solution $\psi = \tilde{\psi}_0 \Phi_0$ after some days of forcing. When $n = N$ modes have passed the basin, they contribute the steady response

$$\sum_{n=0}^{N} \tilde{\psi}_n(x,y)\Phi_n(z) = \frac{B-x}{\beta}\tau_0 \ell \sin \ell y \sum_{n=0}^{N} s_n \Phi_n(z) = -\frac{B-x}{\beta} \operatorname{curl} \boldsymbol{\tau} \sum_{n=0}^{N} s_n \Phi_n(z) \tag{8.111}$$

to the complete solution, and there is still the time-dependent local response from the modes $n > N$ to be added. With increasing time and hence increasing $N$, the terms of the local response diminish, and the modal sum in the steady response approaches more and more the derivative of the structure function $\mathrm{d}S(z)/\mathrm{d}z$ of the windstress body force (see left panel of Figure 8.17). In other words: each passing baroclinic mode diminishes the interior response and leads to an increasing confinement of the vertical structure to the near-surface layer, which is directly influenced by the wind. Remember that the baroclinic response is slow – it takes some years – but ultimately, the ocean below the surface layer is motionless. This feature of the ocean spin-up has been named 'Sverdrup catastrophe' for reason to be explained immediately (see also Section 14.2.3).

To put the above results in a physical frame, we must anticipate some ingredients of the wind-driven circulation which will be discussed in detail later in Chapter 14.

We do this in brief. The steady response in (8.110) for $t > t_f$ is identical the stationary Sverdrup response $-(B - x)/\beta$ for the stream function (see the later discussion in Chapter 14; specifically the box on p. 454). It is the solution of the steady version of (8.106), $\beta\partial\psi/\partial x = \text{curl}\,\boldsymbol{\tau}$, the so-called Sverdrup balance, with the boundary condition $\psi = 0$ at the eastern coast. Note that $\psi$ is the barotropic stream function ($\psi_0\Phi_0$ of the mode $n = 0$) in the Sverdrup model. From the above analysis, we learn that its steady state is reached when the barotropic wave has passed across the basin. A look, however, into the baroclinic support, appearing in a stratified ocean, discloses the 'Sverdrup catastrophe'.

We have so far ignored the response that comes from the waves excited at the western boundary. These waves must propagate eastward, and hence the waves are in the short-wave regime. As discussed in Section 8.2.1, their group velocity is much slower (at least a factor of 8 for each vertical mode) than the long westward propagating waves. Moreover, if friction is present, waves of smaller wavelength are more likely to be affected than long waves. It may hence be assumed that the wave response from the western boundary will not propagate far into the interior in a realistic viscous ocean. Nevertheless, it is of interest to find solutions for the short-wave response at the western boundary for the frictionless case. The western response is governed by the short-wave approximation of (8.107)

$$\frac{\partial}{\partial t}\frac{\partial^2\phi^W_{\text{wave}}}{\partial x^2} + \beta\frac{\partial\phi^W_{\text{wave}}}{\partial x} = 0 \tag{8.112}$$

and boundary conditions are $\phi^W_{\text{wave}}(x = 0, t) = -\phi^E(x = 0, t)$ and $\phi^W_{\text{wave}}(x = B, t) = 0$. Here $\phi$ is given by (8.110). A trivial scaling of the coordinates $\xi = x/B, \theta = t\beta B$ cancels the $\beta$-factor if $\phi^W_{\text{wave}}(\xi, \theta)$ is expressed in the new variables. The equation may be integrated once to the form

$$\frac{\partial}{\partial\theta}\frac{\partial\phi^W_{\text{wave}}}{\partial\xi} + \phi^W_{\text{wave}} = a(\theta) \tag{8.113}$$

where $a(\theta)$ is a yet unknown function of time. We may assume, however, that the response is virtually confined to a boundary layer at the western side, so that $\phi^W_{\text{wave}}$ and all of its derivatives approach zero very fast outside the layer (the boundary condition at $x = B$ is then reformulated accordingly). This leads to $a \equiv 0$. A solution of (8.113) for that case is generated by writing $\phi^W_{\text{wave}}$ in a similarity form

$$\phi^W_{\text{wave}}(\xi, \theta) = \left(\frac{\theta}{\xi}\right)^{m/2} G(2\sqrt{\xi\theta}) \tag{8.114}$$

It is found from (8.113) that $G(z)$ satisfies BESSEL[6]'s differential equation (see the box on p. 247)

$$z^2G'' + zG' + (z^2 - m^2)G = 0 \tag{8.115}$$

Hence $G(z) = A\,J_m(z)$ equals the Bessel function $J_m(z)$ of degree $m$, which decays for large $z$. Here $A$ is an arbitrary amplitude factor that is fixed by considering the condition at the boundary $x = 0$. We rescale the variables,

$$\phi^W_{\text{wave}}(x, t) = A\left(\frac{t\beta B^2}{x}\right)^{m/2} J_m(2\sqrt{xt\beta}) \tag{8.116}$$

[6] FRIEDRICH WILHELM BESSEL *1784 in Minden-Ravensberg, †1846 in Königsberg, mathematician and astronomer.

39. Bessel Differential Equation and Functions

Bessel functions are defined as solutions of the second-order differential equation

$$x^2y'' + xy' + (x^2 - \nu^2)y = 0 \quad \text{(B39.1)}$$

for $y = y(x)$ with an arbitrary constant $\nu$. The equation is considered on the interval $0 \leq x \leq \infty$, and then $x = 0$ is a singular point. There are two types of Bessel functions, $J_\nu(x)$ (Bessel function of the first kind) and $Y_\nu(x)$ (Bessel function of the second kind), which are linearly independent. They occur as solutions in many physical problems, frequently with integer index $\nu = m = 0, \pm 1, \pm 2, \ldots$ They are studied extensively and tabulated (see Abramowitz and Stegun, 1984) but also implemented in many mathematical tools like MATLAB or MAPLE. For $m = 0$ and positive $m$, the $J_m$ behave regular at the singular point, while the $Y_m$ become singular,

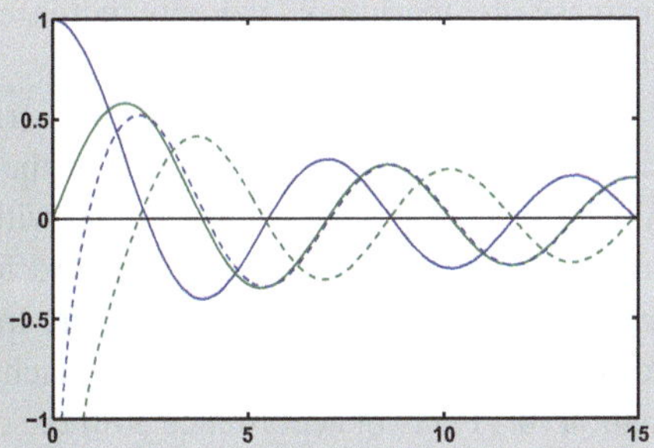


*Bessel functions of the first kind $J_0(x)$ [blue] and $J_1(x)$ [green]. The corresponding functions of the second kind $Y_m(x)$, $m = 0, 1$ as shown as dashed curves.*

$$J_m(x) \sim (x/2)^m/m! \quad \text{and} \quad Y_m(x) \sim \frac{(m-1)!}{\pi}(x/2)^{-m} \quad \text{as} \quad x \to 0$$

Both are of oscillatory nature and decay with increasing $x$,

$$J_m(x) = \sqrt{2/(\pi x)}\cos(x - m\pi/2 - \pi/4) \quad \text{and}$$

$$Y_m(x) = \sqrt{2/(\pi x)}\sin(x - m\pi/2 - \pi/4) \quad \text{as} \quad x \to \infty$$

and use the behavior of $J_m(z) = (z/2)^m/m!$ for small arguments. The choice for the index $m$ is different for the case $t < t_f$, when the boundary value $\phi^E(x = 0, t)$ is linearly increasing in time according to (8.110), and $t > t_f$, where $\phi^E(x = 0, t)$ is a constant. A finite nonzero value and a linear response in time at the western boundary is obviously obtained by setting $m = 1$. Then $\phi^W_{\text{wave}}(x \to 0, t) = At\beta B$ and matching the boundary condition yields $A = -c_g/(\beta^2 B)$. To obtain a constant value at $x = 0$, the only choice is $m = 0$, then $\phi^W_{\text{wave}}(x \to 0, t) = A$, which fits the corresponding boundary condition with $A = B/\beta$.

The complete solution is the sum of the local part $\phi_{\text{loc}}$ and the two wave parts $\phi^E_{\text{wave}}$ and $\phi^W_{\text{wave}}$, respectively. It is displayed in the right panel of Figure 8.17 for various times. The green curve refers to the time $-B/c_g$ and is the response after the front of long eastward moving waves has just crossed the entire basin. The other curves refer to times before this has taken place: the solution is then pieced together at the collision point of the wavy part and the green part. These short westward moving waves, occurring during the 'local' spin-up phase, are made by the Bessel function part with $m = 1$. Note that the amplitude of the oscillations near the western coast increases in this phase and the spatial scale decreases, as obvious from the mathematical expression (8.116). For times longer than $-B/c_g$, the $m = 0$ part comes into play and leads to a wavy western boundary layer in the 'green' solution (not shown for reasons discussed below). While the amplitude of the oscillation now becomes locked at finite values (because $|J_0| < 1$), the spatial scales continue to decrease (the meridional velocity increases!), and eventually friction must come into play to stop the march into a singularity.

The dashed black line in Figure 8.17 (right panel) displays the maximum distance that information can propagate from the western coast by short planetary waves. Apparently, there is rough agreement with $J_1$-solution at the respective times. This is not the case for the $J_0$-solution, and for this reason we have not plotted the $J_0$ response. Note that there is no feature in these solutions which could intrinsically constrain the response appropriately such that a solution of the short-wave approximation (8.112) is in accordance with the complete wave equation. In fact, Anderson and Gill (1975) show mostly numerical solutions of the complete equation.

## 8.3 Equatorial Waves

As we have seen, the velocity relevant for group propagation depends, besides on wave number, on the wave speed parameter $c$ which reflects stratification, and on the Coriolis parameter $f$. For very long waves in the limit $k \ll f/c$, that dependency is given by $c_g \sim kc^2/f$ (gravity waves) and $c_g \sim \beta c^2/f^2$ (planetary waves). Therefore, for a given stratification, the propagation velocity for both wave types increases toward lower latitudes.

In deriving equations for long gravity and planetary waves in midlatitudes, it has been assumed that either the Coriolis parameter $f$ is constant (gravity waves), or that both $f$ and $\beta = \mathrm{d}f/\mathrm{d}y$ are constant (planetary waves). In the tropics, these approximations fail since $f$ changes its sign at the equator. A better approximation for $f$ is obtained by expansion of $f = 2\Omega \sin\varphi = 2\Omega \sin(y/a)$ at $y = 0$ (here $y = \varphi a$ is the latitudinal distance from the equator, and $a$ is Earth radius), yielding

$$f(y) \approx \beta y \quad \text{with} \quad \beta = \frac{2\Omega}{a} \tag{8.117}$$

which is referred to as equatorial $\beta$-plane approximation. With (8.117), we can expect to obtain a different view for both gravity and planetary waves valid in the tropics which will be discussed in Section 8.3.2 and those following.

### *8.3.1 Refraction due to Variations of the Coriolis Parameter*

Before investigating wave solutions near the equator, it is useful to consider how the propagation of long gravity and planetary waves is modified by a varying Coriolis parameter. As long as that variation is gentle compared with the changes over a wavelength, one can apply the WKBJ-technique from Section 6.3 to follow the wave pathway, even though that technique is stretched to its limits near the equator where $f(y) \to 0$. Starting from the appropriate dispersion relations, i. e. (8.23) for long gravity waves and (8.57) for planetary waves, the local dispersion can be written as

$$\omega = \Omega(k_1, k_2, y) = \begin{cases} [f^2(y) + c^2k_1^2 + c^2k_2^2]^{1/2} \\ -\beta k_1 [k_1^2 + k_2^2 + f^2(y)/c^2]^{-1} \end{cases} \tag{8.118}$$

where $c \equiv c_n$ is the gravity wave speed parameter (here, as in the following expressions, the upper form applies to gravity waves, the lower to planetary waves). It is obvious that for both wave types, both $\omega$ and $k_1$ will remain constant along the wave path. The evolution of $k_2$ along the path is governed by the refraction equation (cf. Section 6.3.2)

$$\dot{k}_2 = -\frac{\partial \Omega}{\partial y} = \begin{cases} -\beta f/\omega \\ 2\omega\beta f \left[c^2\left(k_1^2 + k_2^2\right) + f^2\right]^{-1} \end{cases} \tag{8.119}$$

Hence for the northern hemisphere it follows that $\dot{k}_2 < 0$ for gravity waves, and $\dot{k}_2 > 0$ for planetary waves. As the meridional propagation of wave groups is in the direction of $+k_2$ for gravity waves, but $-k_2$ for planetary waves (cf. (8.61)), for both wave types the sign of $\dot{k}_2$ indicates refraction of the ray path such that any wave group will eventually propagate toward the tropics, as first shown by Anderson and Gill (1979).

Note that the refraction equation (8.119) is not explicitly needed for a quantitative analysis of the wave path, since with (8.118) $k_2(y)$ can be algebraically expressed in terms of the constants $\omega$ and $k_1$ as

$$k_2(y) = \begin{cases} \left[\omega^2 - c^2k_1^2 - f^2(y)\right]^{\frac{1}{2}}/c \\ \left[-\beta k_1/\omega - k_1^2 - f^2(y)/c^2\right]^{\frac{1}{2}} \end{cases} \tag{8.120}$$

The propagation of wave groups is governed by $(\dot{x}, \dot{y}) = (c_g^x, c_g^y)$, with the group velocity from (8.24) (gravity waves) and (8.60) and (8.61) (planetary waves), respectively. With this approximation and (8.120) one finds for the path of the wave groups

$$\frac{\dot{y}}{\dot{x}} = \frac{dy}{dx} = \begin{cases} \left(\omega^2 - c^2k_1^2 - f^2\right)^{\frac{1}{2}}/ck_1 \\ (2\omega/\beta c)\left[-\frac{\beta k_1 c^2}{\omega}(1 + \omega k_1/\beta) - f^2\right]^{\frac{1}{2}}/(1 + 2\omega k_1/\beta) \end{cases} \tag{8.121}$$

With the additional approximation (8.117), in both cases the solution is given by

$$y(x) = y_0 \cos b(x - x_0) \tag{8.122}$$

where $b$ and $y_0$ are given by

$$b = \begin{cases} \beta/ck_1 \\ 2(\omega/c)/(1 + 2\omega k_1/\beta) \end{cases} \quad \text{and} \quad y_0 = \begin{cases} \left(\omega^2 - c^2k_1^2\right)^{\frac{1}{2}}/\beta \\ \left[-k_1c^2/\beta\omega(1 + \omega k_1/\beta)\right]^{\frac{1}{2}} \end{cases} \tag{8.123}$$

and $x_0$ is an integration constant. From (8.122) it follows that $b$ describes the scale of zonal variation of the wave path, and $y_0$ is the maximum distance of the wave group from the equator which is obtained where the wave number is purely zonal, i. e. $k_2(y_0) = 0$. Hence the pathway of long waves oscillates between $\pm y_0$ around the equator. For planetary waves the solutions are displayed in Figure 8.18.

The long wave limit in (8.121) and (8.123) is obtained for $c|k_1| \ll \omega$ (gravity waves) and $\omega k_1/\beta \ll 1$ (planetary waves). In this limit, the intersection of all curves in Figure 8.18 occurs along the equator at the same longitude.

Note that these results do not hold in the presence of strong mean flows.

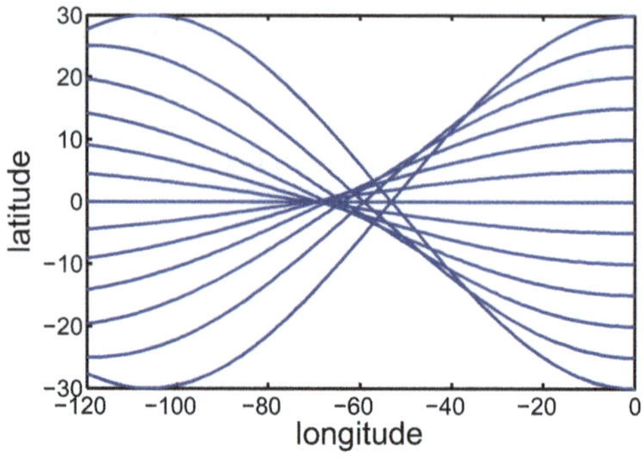


Fig. 8.18 Ray pathways for planetary waves with a period of one year, starting at the eastern boundary with an initially zero meridional wave number. Vertical and horizontal axes are latitude and longitude (in degrees)

## 8.3.2 Equation for the Meridional Velocity

The results from the preceding section suggest that long waves in the tropics have a tendency to remain within a latitude range centered at the equator. To find solutions which are valid near the equator, one has to start from equations (8.20)–(8.22). With the approximation (8.117), and dropping the tilde, one obtains

$$\frac{\partial u}{\partial t} - \beta y v + \frac{\partial p}{\partial x} = \mathcal{F}_u \tag{8.124}$$

$$\frac{\partial v}{\partial t} + \beta y u + \frac{\partial p}{\partial y} = \mathcal{F}_v \tag{8.125}$$

$$\frac{1}{c_n^2}\frac{\partial p}{\partial t} + \frac{\partial u}{\partial x} + \frac{\partial v}{\partial y} = Q \tag{8.126}$$

For later reference we have included forcing terms (which are here projected on the respective vertical mode). They are abandoned until the treatment of forced waves in Section 10.3.

The above system has $\beta$ and $c$ as the only dimensional constants. From these constants, the time-scale $T_e = 1/\sqrt{c_n\beta}$ and length scale $R_e = \sqrt{c_n/\beta}$ can be defined. The length scale is called *equatorial Rossby radius*, and it will be shown that it is indeed analogous to the Rossby radius in midlatitudes. In the following, the index $n$ is omitted.

It is convenient to introduce new dependent variables according to

$$q = \frac{p}{c} + u \quad \text{and} \quad r = \frac{p}{c} - u \tag{8.127}$$

In the new variables, the system (8.124)–(8.126) reads

$$\frac{\partial q}{\partial t} + c\frac{\partial q}{\partial x} + c\frac{\partial v}{\partial y} - \beta y v = 0 \tag{8.128}$$

$$\frac{\partial v}{\partial t} + \frac{1}{2}\left(c\frac{\partial q}{\partial y} + \beta y q + c\frac{\partial r}{\partial y} - \beta y r\right) = 0 \tag{8.129}$$

$$\frac{\partial r}{\partial t} - c\frac{\partial r}{\partial x} + c\frac{\partial v}{\partial y} + \beta y v = 0 \tag{8.130}$$

To shorten the notation, the differential operators

$$D_\pm = \frac{\partial}{\partial t} \pm c\frac{\partial}{\partial x} \quad \text{and} \quad L_\pm = \frac{\partial}{\partial y} \pm \frac{y}{R_\mathrm{e}^2} \tag{8.131}$$

are introduced, so that the system (8.128) to (8.129) becomes

$$D_+ q + cL_- v = 0 \tag{8.132}$$

$$\frac{\partial v}{\partial t} + \frac{c}{2}(L_+ q + L_- r) = 0 \tag{8.133}$$

$$D_- r + cL_+ v = 0 \tag{8.134}$$

A single equation for the meridional velocity can be derived by applying the operator $D_+D_-$ to (8.133), and eliminating $D_+q$ with (8.132) and $D_-r$ with (8.134). Note that $D_-$ commutes with $D_+$, and both commute with the $L_\pm$ operations. The result is

$$D_+D_-\frac{\partial v}{\partial t} - \frac{c^2}{2}(D_-L_+L_-v + D_+L_-L_+v) = 0 \tag{8.135}$$

It is useful to further introduce the operator

$$L = \frac{\partial^2}{\partial y^2} - \frac{y^2}{R_\mathrm{e}^4} \tag{8.136}$$

because of the convenient identities

$$L_+L_- = L - \frac{1}{R_\mathrm{e}^2} \quad \text{and} \quad L_-L_+ = L + \frac{1}{R_\mathrm{e}^2} \tag{8.137}$$

following from (8.131). With these relations, and $D_+D_-v_t = v_{ttt} - c^2 v_{xxt}$, (8.135) takes the explicit form

$$\frac{\partial^3 v}{\partial t^3} - c^2\frac{\partial^3 v}{\partial x^2 \partial t} - c^2 L\frac{\partial v}{\partial t} - \beta c^2\frac{\partial v}{\partial x} = 0 \tag{8.138}$$

Equation (8.138) is a convenient starting point for discussing the meridional structure. However, since the derivation of (8.138) has been obtained by differentiation, it is likely that not all solutions can be found from (8.138) since solutions vanishing under the operation $D_+D_-$ are lost. We will later present an approach that guarantees the identification of *all* solutions.

### 8.3.3 Meridional Eigenfunctions

With the ansatz $v = \hat{v}(y)\mathrm{e}^{\mathrm{i}(kx-\omega t)}$, one finds from (8.138) after division by $\omega c^2$ and reordering

$$L\hat{v} = -\alpha^2 \hat{v} \tag{8.139}$$

where for brevity the parameter

$$\alpha^2 = \frac{\omega^2}{c^2} - k^2 - \frac{\beta k}{\omega} \tag{8.140}$$

has been introduced. The appropriate boundary condition is that $\hat{v}$ vanishes at meridional boundaries. If these boundaries are located many Rossby radii away from the equator, it suffices to require that $\hat{v} \to 0$ for $y \to \pm\infty$. With this boundary condition, (8.139) is an eigenvalue problem, and one expects that solutions (eigenfunctions of the operator $L$) can only exist for certain discrete values of the parameter $\alpha^2$. This is shown in an elementary way in the following.

With $y_\star = y/R_e$ as dimensionless independent variable, (8.139) can be written as

$$\hat{v}'' - y_\star^2 \hat{v} = -\alpha_\star^2 \hat{v} \tag{8.141}$$

with $\alpha_\star = \alpha R_e$. The prime denotes differentiation with respect to $y_\star$. Note that the solutions of (8.141) are oscillatory for $|y_\star| < \alpha_\star$, and exponential outside. It follows that equatorial waves are trapped within a distance $R_e$ from the equator.

Introducing a new dependent variable $g(y_\star)$ in (8.141) according to

$$\begin{aligned} \hat{v} &= g(y_\star)\mathrm{e}^{-\frac{1}{2}y_\star^2}, \quad \hat{v}' = (g' - y_\star g)\mathrm{e}^{-\frac{1}{2}y_\star^2} \\ \hat{v}'' &= [g'' - 2y_\star g' + (y_\star^2 - 1)g]\mathrm{e}^{-\frac{1}{2}y_\star^2} \end{aligned} \tag{8.142}$$

one obtains

$$g'' - 2y_\star g' + (\alpha_\star^2 - 1)g = 0 \tag{8.143}$$

The solution of (8.143) is found with a power series approach

$$g(y_\star) = \sum_{j=0}^{\infty} a_j y_\star^j, \quad g' = \sum_{j=1}^{\infty} j a_j y_\star^{j-1}$$

$$g' = \sum_{j=2}^{\infty} j(j-1)a_j y_\star^{j-2} \equiv \sum_{j=0}^{\infty} (j+2)(j+1)a_{j+2} y_\star^j$$

where, in the last form, the index $j$ was replaced by $j+2$. Insertion into (8.143) and comparison of the coefficients yields

$$a_{j+2} = \frac{2j + 1 - \alpha_\star^2}{(j+2)(j+1)} a_j \tag{8.144}$$

The even and the odd $a_j$ are independent of each other, and (8.144) has two solutions corresponding to the fundamental solutions of the differential equation (8.143), respectively (8.141), a solution which is symmetric in $y_\star$ (starting with $a_0 = 1$, $a_1 = 0$) and an antisymmetric solution (starting with $a_0 = 0$, $a_1 = 1$). The corresponding $a_j$ can be calculated recursively from (8.144). All linear combinations of the two solutions are solutions as well.

The asymptotic behavior of $g(y_\star)$ for large $y_\star$ follows from the behavior of the coefficients $a_j$ for large $j$ which is found to be

$$\frac{a_{j+2}}{a_j} \to \frac{2}{j} \tag{8.145}$$

Since for all $j > (\alpha_\star^2 - 1)/2$ all $a_j$ have the same sign, it follows that $g(y_\star) \to \infty$ for large $y_\star$. The behavior of $g(y_\star)$ for large $y_\star$ can be compared with that of the exponential function

$$e^{y_\star^2} = \sum_{m=0}^{\infty} \frac{y_\star^{2m}}{m!} = \sum_{j=0}^{\infty} c_j y_\star^j \quad \text{with} \quad c_j = \begin{cases} 0 & \text{for } j \text{ uneven} \\ \frac{1}{(j/2)!} & \text{for } j \text{ even} \end{cases}$$

This leads to $c_{j+2}/c_j \to 2/j$, so that the $c_j$ and $a_j$ have the same asymptotic behavior. The inverse conclusion is also valid: if there is asymptotic behavior of the form (8.145), then for large $y_\star$ asymptotically $g(y_\star) \sim e^{y_\star^2}$ is valid. Hence $\hat{v}(y_\star) = g(y_\star)e^{-\frac{1}{2}y_\star^2} \to \infty$, and we have to conclude that in general (i. e. for arbitrary values of the eigenvalue $\alpha_\star^2$) there is no solution satisfying the boundary conditions.

If, however,

$$\alpha_\star^2 = 2m + 1 \tag{8.146}$$

holds for any *integer* $m$, then $a_j = 0$ is valid for all $j > m$, i. e. the series is terminated and $g(y_\star)$ is a polynomial of order $m$, written as $g(y_\star) = H_m(y_\star)$. For even $m$ one finds $a_0, a_2, \ldots, a_m \neq 0$, all other $a_j$ vanish. Furthermore, $a_1, a_3, \ldots, a_m \neq 0$ applies for odd $m$, and all other $a_j$ vanish. The $H_m(y_\star)$ are now determined except for a constant. The usual normalization is

$$H_m(y_\star) = (2y_\star)^m + \cdots y_\star^{m-2} + \cdots$$

which defines the HERMITE[7] polynomials (see e. g. Abramowitz and Stegun, 1984, p. 773). In particular, the first polynomials are

$$H_0(y_\star) = 1\,, \quad H_1(y_\star) = 2y_\star\,, \quad H_2(y_\star) = 4y_\star^2 - 2\,, \quad \ldots$$

Hermite *functions*, defined as

$$\psi_m(y_\star) = H_m(y_\star)e^{-\frac{1}{2}y_\star^2} \tag{8.147}$$

are displayed in Figure 8.19, and the connection to parabolic cylinder functions $D_m(x)$ is given by $\psi_m(y_\star) = 2^{m/2} D_m(\sqrt{2}y_\star)$. An orthogonality relation can be derived from (8.139),

$$\int_{-\infty}^{\infty} \psi_m(y_\star)\psi_n(y_\star)\, dy_\star = 2^m m! \sqrt{\pi}\delta_{mn} \tag{8.148}$$

The following recurrence relations will be used below

$$\begin{aligned} L_+\psi_m(y/R_e) &= \frac{2m}{R_e}\psi_{m-1}(y/R_e) \\ L_-\psi_m(y/R_e) &= -\frac{1}{R_e}\psi_{m+1}(y/R_e) \end{aligned} \tag{8.149}$$

Hence the operator $L_+$ decreases and $L_-$ increases the order of the eigenfunctions by 1, in addition to a factor $\sim 1/R_e$. Specifically, for $m = 0$ it follows that $L_+\psi_0 \equiv 0$.

[7] CHARLES HERMITE, *1822 in Dieuze/Lorraine, †1901 in Paris, mathematician.

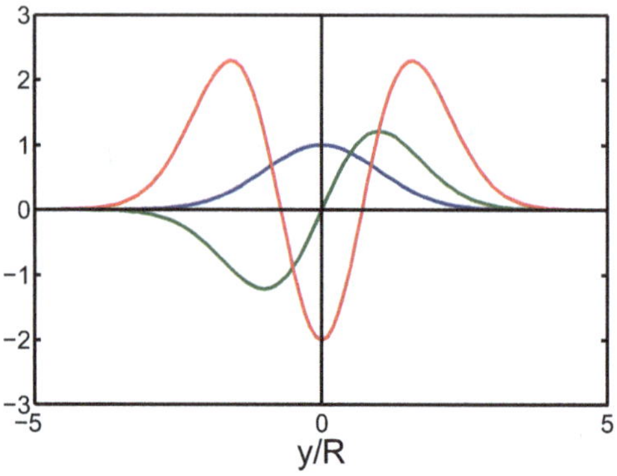


Fig. 8.19 The first three Hermite functions $m = 0, 1, 2$

The solution of (8.141) is then (now again in dimensioned coordinates, i. e. $y_\star = y/R_e$)

$$v(x, y, t) = \psi_m(y/R_e)e^{i(kx-\omega t)} \tag{8.150}$$

The solution decays exponentially for $|y| \gtrsim \sqrt{(2m+1)}R_e$, thus defining an *equatorial wave guide* (cf. Section 8.4 below). Hermite functions $\psi_m(y/R_e)$ are eigenfunctions of the operator $L$ with eigenvalue $-(2m+1)/R_e^2$.

### 8.3.4 Wave Solutions

With the meridional eigenfunctions $\psi_m(y/R_e)$ derived above, it is now possible to find a solution of the system (8.132)–(8.134) without differentiation and the associated danger of losing solutions. From (8.132) and (8.134) it follows that $q \sim L_- v$ and $r \sim L_+ v$. With $v \sim \psi_m$ and the properties shown in (8.149), it follows that $q \sim \psi_{m+1}$ and $r \sim \psi_{m-1}$. In order to have variables which all have the same index, we introduce new dependent variables $(\theta_1, \theta_2, \theta_3)$ according to

$$\begin{aligned} q &= \theta_1 \\ v &= L_+\theta_2 \\ r &= L_+L_+\theta_3 \end{aligned} \tag{8.151}$$

instead of $q, v, r$. Here $\theta_2$ and $\theta_3$ have the role of potentials, only the original parameters are physically relevant.

For the following, it is important that the differential operator $L_+$ has a unique inverse (the inverse of $L_-$ is not needed). The actual form of the inverse which is important only for inhomogeneous problems is given by

$$L_+^{-1}\chi = e^{-\frac{1}{2}y^2/R_e^2} \int_{-\infty}^{y} e^{\frac{1}{2}y'^2/R_e^2}\chi(y')dy' \tag{8.152}$$

for an arbitrary function $\chi(y)$ which satisfies the same boundary conditions as $\hat{v}$ in the previous section, i. e. which vanishes exponentially for $|y| \to \infty$. It is left to the reader to show, by partial integration, that indeed always $L_+^{-1}L_+\chi = \chi$ and $L_+L_+^{-1}\chi = \chi$ are valid. Insertion of (8.151) in (8.132)–(8.134) yields

$$D_+\theta_1 + cL_-L_+\theta_2 = 0 \tag{8.153}$$

$$\frac{\partial}{\partial t}L_+\theta_2 + \frac{c}{2}L_+\theta_1 + \frac{c}{2}L_-L_+L_+\theta_3 = 0 \tag{8.154}$$

$$D_-L_+L_+\theta_3 + cL_+L_+\theta_2 = 0 \tag{8.155}$$

With (8.137), equation (8.153) can be written as

$$D_+\theta_1 + c(L + 1/R_e^2)\theta_2 = 0 \tag{8.156}$$

Application of the operator $L_+^{-1}$ to (8.154) yields

$$\frac{\partial}{\partial t}\theta_2 + \frac{c}{2}\theta_1 + \frac{c}{2}L_+^{-1}L_-L_+L_+\theta_3 = 0 \tag{8.157}$$

With repeated use of (8.137), the fourfold operator product in (8.157) can be simplified to

$$L_+^{-1}(L_-L_+)L_+ = L_+^{-1}\left(L_+L_- + \frac{2}{R_e^2}\right)L_+ = L_-L_+ + \frac{2}{R_e^2} \tag{8.158}$$

From (8.156) and (8.157) with (8.158) and application of the operator $L_+^{-1}L_+^{-1}$ to (8.155) one obtains the system

$$D_+\theta_1 + c\left(L + \frac{1}{R_e^2}\right)\theta_2 = 0 \tag{8.159}$$

$$\frac{\partial}{\partial t}\theta_2 + \frac{c}{2}\theta_1 + \frac{c}{2}\left(L + \frac{3}{R_e^2}\right)\theta_3 = 0 \tag{8.160}$$

$$D_-\theta_3 + c\theta_2 = 0 \tag{8.161}$$

which contains only the operator $L$, the eigenfunctions of which are known. A solution is found with the ansatz

$$(\theta_1, \theta_2, \theta_3) = (a_1, a_2, a_3)\psi_\ell(y/R_e)e^{i(kx-\omega t)}$$

with $\ell = 0, 1, 2, \ldots$. With $D_\pm \to -i(\omega \mp ck)$ and $L \to -(2\ell + 1)/R_e^2$, it follows that

$$-i(\omega - ck)a_1 - \frac{2c\ell}{R_e^2}a_2 = 0 \tag{8.162}$$

$$-i\omega a_2 + \frac{c}{2}a_1 + c\frac{-\ell + 1}{R_e^2}a_3 = 0 \tag{8.163}$$

$$-i(\omega + ck)a_3 + ca_2 = 0 \tag{8.164}$$

which is a homogeneous system for the amplitudes $a_1, a_2, a_3$. Hence the determinant must vanish, and one obtains the characteristic equation, the dispersion relation,

$$\frac{\omega^2}{c^2} - k^2 - \frac{\beta k}{\omega} - \frac{2\ell - 1}{R_e^2} = 0 \tag{8.165}$$

which is a third-order polynomial in terms of the frequency and quadratic in terms of the wave number. The dispersion relation (8.165) also follows from (8.140) and (8.146), with $\ell - 1 \to m$. Note that (8.165) holds for $\ell = 0, 1, 2, \ldots$, corresponding to $m = -1, 0, 1, \ldots$

To determine the amplitudes $a_1, a_2, a_3$ in a homogeneous system, assume $a_3 = a \neq 0$. From (8.163) and (8.164) we find

$$a_1 = -2\left[\frac{\omega}{c}\left(\frac{\omega}{c} + k\right) + \frac{1-\ell}{R_e^2}\right] a \tag{8.166}$$

$$a_2 = \mathrm{i}\left(\frac{\omega}{c} + k\right) a \tag{8.167}$$

Hence for all $\ell \geq 0$ the full solution is given by

$$\hat{q} = -2a\left[\frac{\omega}{c}\left(\frac{\omega}{c} + k\right) + \frac{1-\ell}{R_e^2}\right] \psi_\ell\left(\frac{y}{R_e}\right) \tag{8.168}$$

$$\hat{v} = \mathrm{i}a\left(\frac{\omega}{c} + k\right) L_+ \psi_\ell\left(\frac{y}{R_e}\right) \tag{8.169}$$

$$\hat{r} = aL_+L_+\psi_\ell\left(\frac{y}{R_e}\right) \tag{8.170}$$

### 8.3.5 Equatorial Kelvin Waves

Consider the solution for $\ell = 0$. Since $L_+\psi_0 \equiv 0$, it follows from (8.169) and (8.170) that $v \equiv 0$ and $r \equiv 0$, independent of the nonzero amplitudes $a_2$ and $a_3 = a$. For the amplitude $a_1$ one has from (8.166)

$$a_1 = -2\left[\frac{\omega}{c}\left(\frac{\omega}{c} + k\right) + \frac{1}{R_e^2}\right] a \tag{8.171}$$

For $\ell = 0$, the dispersion relation (8.165), after multiplication by $\omega$, is

$$\frac{\omega^3}{c^2} - \omega k^2 - \beta k + \frac{\omega}{R_e^2} = (\omega - ck)\left[\frac{\omega}{c}\left(\frac{\omega}{c} + k\right) + \frac{1}{R_e^2}\right] = 0 \tag{8.172}$$

One of the factors in (8.172) must vanish. The case that the second bracket vanishes is, however, only of academic interest since comparison with (8.171) shows that in this case $a_1 = 0$; hence $q = v = r = 0$ so that the corresponding solutions of the system (8.168) to (8.170) are physically meaningless. Hence the first bracket in (8.172) has to vanish,

$$\omega = ck \tag{8.173}$$

This is the dispersion curve for equatorial Kelvin waves which are nondispersive (see Figure 8.20). The wave number $k > 0$ is always positive, corresponding to an eastward propagation with constant phase speed $c$.

From (8.171) it follows that $a_1 \neq 0$. Hence the solution (8.168)–(8.170) can be simplified to $\hat{q} = a_1\psi_0(y/R_e)$ or in the original variables

$$\frac{p}{c} = u = \frac{q}{2} = a_K e^{-\frac{1}{2}y^2/R_e^2}\cos(kx - \omega t)$$

and $v \equiv 0$, where $a_K$ is the amplitude of the Kelvin wave.

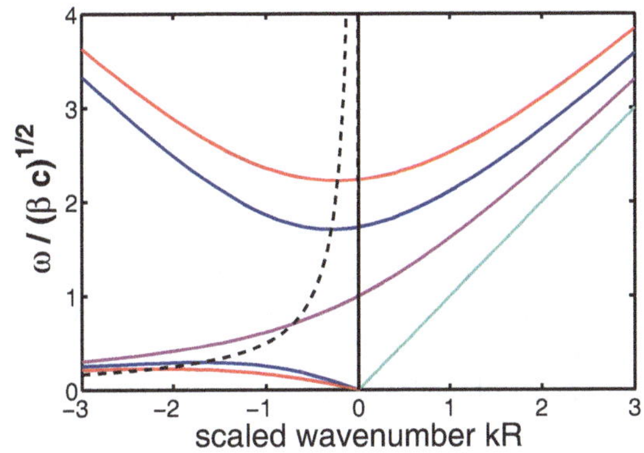


Fig. 8.20 Dispersion curves for equatorial waves. Kelvin wave (*cyan*, $\ell = 0$), Yanai-wave (*magenta*, $\ell = 1$), equatorial Rossby and gravity waves for $\ell = 2, 3$ (*blue* and *red*). The dashed line connects minima of the gravity curves with maxima of the Rossby wave curves

### 8.3.6 Yanai Waves

Next consider the solution for $\ell = 1$. Since $L_+ L_+ \psi_1 \sim L_+ \psi_0 \equiv 0$, it follows that $r \equiv 0$ (although $a_3 \neq 0$). The amplitudes $a_1, a_2$ follow from (8.166) and (8.167),

$$a_1 = -\frac{2\omega}{c}\left(\frac{\omega}{c} + k\right) a$$
$$a_2 = \mathrm{i}\left(\frac{\omega}{c} + k\right) a$$

For $\ell = 1$ the dispersion relation (8.165) can be factorized as

$$\frac{\omega^3}{c^2} - \omega k^2 - \beta k - \frac{\omega}{R_\mathrm{e}^2} = \left(\frac{\omega^2}{c^2} - \frac{k\omega}{c} - \frac{1}{R_\mathrm{e}^2}\right)(\omega + ck) = 0 \tag{8.174}$$

Again one of the factors must vanish. The case that $\omega = -ck$ is not of interest since from (8.167) and (8.166) it follows that $a_1 = 0$ und $a_2 = 0$ and hence a trivial solution in the physical variables. Vanishing of the first factor

$$\frac{\omega^2}{c^2} - \frac{k\omega}{c} - \frac{1}{R_\mathrm{e}^2} = 0 \tag{8.175}$$

results in the relation $k(\omega)$,

$$k = \frac{\omega}{c} - \frac{\beta}{\omega} \tag{8.176}$$

which is algebraically more convenient than $\omega(k)$. This is the dispersion curve for YANAI[8] waves (also called *mixed planetary–gravity waves*, see Figure 8.20). The group velocity of the Yanai waves

$$\frac{\partial \omega}{\partial k} = \left(\frac{\partial k}{\partial \omega}\right)^{-1} = \left(\frac{1}{c} + \frac{\beta}{\omega^2}\right)^{-1} = \frac{\omega^2 c}{\omega^2 + \beta c} = \frac{c}{1 + \beta c/\omega^2} > 0$$

[8] MICHIO YANAI, *1934 in Chigasaki, †2010 in Santa Monica, meteorologist.

is always directed eastwards and less than $c$. Two limiting cases can be identified:

$$k \gg (\beta/c)^{1/2}: \quad \omega \approx ck\,, \quad \text{and} \quad c_g \approx c$$
$$k \ll -(\beta/c)^{1/2}: \quad \omega \approx -\frac{\beta}{k} \quad \text{and} \quad c_g \approx \frac{\beta}{k^2} \ll c$$

Hence for large positive wave numbers, Yanai waves are approximately nondispersive with the velocity of the gravity waves, whereas for large negative wave numbers the Yanai waves behave like short Rossby waves, with westward phase speed and slow eastward group propagation. At $k = 0$ we find $\omega = (\beta c)^{1/2}$ and $c_g = c/2$.

The meridional structure of the Yanai waves is given by

$$\hat{q} = -2a\left[\frac{\omega}{c}\left(\frac{\omega}{c} + k\right)\right]\psi_1(y/R_e)$$
$$\hat{v} = \mathrm{i}a\left(\frac{\omega}{c} + k\right)\frac{2}{R_e}\psi_0(y/R_e)$$

and the solutions in the original variables can be written as

$$u = \frac{p}{c} = a_Y\frac{\omega R_e}{c}y\mathrm{e}^{-\frac{1}{2}y^2/R_e^2}\cos(kx - \omega t)$$
$$v = a_Y\mathrm{e}^{-\frac{1}{2}y^2/R_e^2}\sin(kx - \omega t)$$

### 8.3.7 Equatorial Rossby and Gravity Waves

Before considering the solutions for $\ell \geq 2$, it is useful to look at the dispersion relation (8.165) which is of third order in $\omega$. Algebraically, it is more convenient to solve for $k$,

$$k_\pm = -\frac{\beta}{2\omega} \pm \left[\frac{\omega^2}{c^2} + \left(\frac{\beta}{2\omega}\right)^2 - \frac{\beta}{c}(2\ell - 1)\right]^{1/2} \tag{8.177}$$

Whether or not this equation has real solutions depends on the expression in the square bracket which is positive for very large as well as for very small values of $\omega$. The frequencies $\omega_0$ where the bracket vanishes are given by

$$\frac{\omega_0^2}{c^2} + \left(\frac{\beta}{2\omega_0}\right)^2 - \frac{\beta}{c}(2\ell - 1) = 0 \quad \text{or} \quad \frac{\omega_0^4}{\beta^2 c^2} - (2\ell - 1)\frac{\omega_0^2}{\beta c} + \frac{1}{4} = 0 \tag{8.178}$$

which is a quadratic equation in $\omega_0^2$ and has two zeroes,

$$\frac{\omega_0^2}{\beta c} = \frac{2\ell - 1}{2} \pm \left[\frac{(2\ell - 1)^2}{4} - \frac{1}{4}\right]^{\frac{1}{2}} = \ell - \frac{1}{2} \pm \sqrt{(\ell - 1)\ell} = \left(\sqrt{\frac{\ell}{2}} \pm \sqrt{\frac{\ell - 1}{2}}\right)^2$$

Hence in the interval

$$\sqrt{\frac{\ell}{2}} - \sqrt{\frac{\ell - 1}{2}} < \frac{\omega}{\sqrt{c\beta}} < \sqrt{\frac{\ell}{2}} + \sqrt{\frac{\ell - 1}{2}} \tag{8.179}$$

$k_\pm$ is not real, and no wave solutions can exist in this interval which is nontrivial for $\ell \geq 2$. Hence one has to distinguish two wave types, depending on $\omega$ being above or below the interval given by (8.179).

### Gravity waves

If $\omega$ is above the 'prohibited' area, then

$$\frac{\omega}{(\beta c)^{1/2}} > \left(\frac{\ell}{2}\right)^{1/2} + \left(\frac{\ell-1}{2}\right)^{1/2} \approx \begin{Bmatrix} 1.7, \ \ell = 2 \\ 2.2, \ \ell = 3 \\ \vdots \end{Bmatrix}$$

These numbers can be used to estimate the maximum of two terms in the dispersion relation

$$\frac{\omega^2}{\beta c} - (kR_e)^2 - \frac{ck}{\omega} = 2\ell - 1 \tag{8.180}$$

The first term is always larger than $1.7^2$, the third term always less than $kR_e/1.7$. Hence the term $ck/\omega$ in this expression is always small, either (if $kR_e \ll 1$) against the first one or (if $kR_e \gg 1$) against the second one. Therefore,

$$\omega^2 \approx (ck)^2 + c\beta(2\ell - 1) = c^2\left[k^2 + \frac{2(\ell-1)}{R_e^2} + \frac{1}{R_e^2}\right] \tag{8.181}$$

is a good approximation to the exact dispersion curve. The similarity with long gravity waves in midlatitudes follows because (8.23) can be rewritten as

$$\omega^2 = f^2 + (c_\star k)^2 = c_\star^2\left[k_1^2 + k_2^2 + \frac{1}{R_\star^2}\right] \tag{8.182}$$

with $R_\star = c_\star/f$, the Rossby radius in midlatitudes. A complete analogy results by identifying the meridional wave number with $\sqrt{2(\ell-1)}/R_e$, and $R_e$ with the midlatitude Rossby radius $R_\star$.

### Rossby waves

Below the prohibited area the relation

$$\frac{\omega}{(\beta c)^{1/2}} < \left(\frac{\ell}{2}\right)^{1/2} - \left(\frac{\ell-1}{2}\right)^{1/2} = \begin{Bmatrix} 0.29, \ \ell = 2 \\ 0.22, \ \ell = 3 \\ \vdots \end{Bmatrix}$$

applies, and in the dispersion relation

$$\frac{\omega^2}{\beta c} - (kR_e)^2 - \frac{ck}{\omega} = 2\ell - 1 \tag{8.183}$$

the first term is always small compared to the right side. Therefore,

$$\omega \approx -\frac{ck}{R_e^2 k^2 + 2\ell - 1} = -\frac{\beta k}{k^2 + 2(\ell-1)/R_e^2 + 1/R_e^2} \tag{8.184}$$

holds to a good approximation. The frequency has a maximum for $k = k_{max} = \sqrt{2\ell - 1}/R_e$. Comparison with the dispersion relation (8.57) for Rossby waves in midlatitudes shows again the analogy between equatorial and midlatitude waves (this also holds for the group velocity). The phase propagation is always directed westwards. For the group velocity in east-west direction we obtain

$$c_g = \frac{\partial \omega}{\partial k} = \beta \frac{k^2 - (2\ell - 1)/R_e^2}{\left[k^2 + (2\ell - 1)/R_e^2\right]^2}$$

It vanishes at $k = k_{max}$. As in midlatitudes, this permits a division into *long* ($k \ll k_{max}$) and *short* ($k \gg k_{max}$) planetary waves. For long planetary waves

$$\omega \approx -\frac{\beta k}{(2\ell - 1)/R_e^2} = -\frac{ck}{2\ell - 1} \tag{8.185}$$

is valid. They are not dispersive; the group velocity is

$$c_g = -\frac{c}{2\ell - 1}$$

which is e. g. $-c/3$ for $\ell = 2$. The short waves' group velocity has a maximum at $k = \sqrt{3}k_{max}$, and the short waves are at least by a factor 8 slower than long waves, as for midlatitude Rossby waves.

Unlike the approximation (8.181), the exact dispersion (8.165) is not fully symmetric for positive and negative $k$. Hence for $k > 0$ the frequency is somewhat larger and for $k < 0$ somewhat smaller than expected from (8.181), and the minimum frequency is attained at a non-zero $k$.

The locations of the minima of $\omega(k)$ (i. e. $c_g = 0$) for gravity and maxima for Rossby waves are found by differentiation of (8.165),

$$\frac{2\omega}{c^2} c_g - 2k - \frac{\beta}{\omega} + \frac{\beta k}{\omega^2} c_g = 0 \tag{8.186}$$

Hence $c_g = 0$ occurs at $k = -\beta/2\omega$. The maximum frequency $\omega_{max}$ and the smallest frequency $\omega_{min}$ of the gravity waves are obtained to

$$\frac{\omega_{max}}{(c\beta)^{1/2}} = \frac{1}{2(2\ell - 1)^{1/2}} \quad \text{and} \quad \frac{\omega_{min}}{(c\beta)^{1/2}} \approx (2\ell - 1)^{1/2} \tag{8.187}$$

For baroclinic waves with $c = 2.8\,\mathrm{m\,s^{-1}}$ and $\ell = 2, 3, 4$, we find the minimum period for planetary waves as $T_{min} = 2\pi/\omega_{max} = 31,\ 41,\ 74$ days, and the maximum period for gravity waves as $T_{max} = 2\pi/\omega_{min} = 5.5,\ 4,\ 3$ days.

The meridional structure for both gravity and Rossby waves is given by

$$\hat{v} = -\mathrm{i}a2R_e \left(\frac{\omega}{c} - k\right) \psi_{\ell-1} \tag{8.188}$$

$$\hat{u} = a \left[\psi_\ell + 2(\ell - 1)\frac{\omega - kc}{\omega + kc}\, \psi_{\ell-2}\right] \tag{8.189}$$

$$\frac{\hat{p}}{c} = a \left[\psi_\ell - 2(\ell - 1)\frac{\omega - kc}{\omega + kc}\, \psi_{\ell-2}\right] \tag{8.190}$$

Thus a complete system of solutions has been obtained. Note that if $\ell$ is even, then $p$ an $u$ are symmetric to the equator, but $v$ is antisymmetric. If $\ell$ is odd, then $p$ and $u$ are antisymmetric, but $v$ is symmetric.

### 8.3.8 Reflection at Meridional Boundaries

We look at an equatorial wave which meets a meridional western or eastern boundary, and we limit ourselves to planetary and Kelvin waves[9]. According to (8.189), one has

$$u(x, y, z, t) = a[\psi_\ell + C_\ell(k)\,\psi_{\ell-2}]\phi_n(z)\mathrm{e}^{\mathrm{i}(kx-\omega t)} \tag{8.191}$$

for the zonal velocity where we have introduced the abbreviation

$$C_\ell(k) = 2(\ell-1)\frac{\omega - kc}{\omega + kc}$$

The incoming wave (denoted $i$) is characterized by the parameters $k_i, \omega_i, a_i$, and the meridional order $\ell_i$ (and, by the way, also the vertical order $n$ which has been implicit in all derivations). The incoming wave must have a group velocity towards the boundary. At the boundary (at $x = 0$) the condition $u(0, y, z, t) \equiv 0$ (by mass conservation) is to be achieved, and as usual we try to satisfy this condition by superposition with another wave (reflected wave, index $r$), and obtain immediately $\omega_r = \omega_i = \omega$ for the frequencies as well as $n_r = n_i = n$ for the vertical index. The further procedure depends on the fact whether the reflexion takes place at a western or at an eastern boundary.

#### Western boundary

Planetary waves only have a westward group velocity if $k > -\beta/2\omega$. Hence only those waves can come in at the western boundary. The index of the incoming wave must, therefore, be $\ell_i \equiv \ell \geq 2$. Following (8.191), we obtain

$$u_i \sim a_i[\psi_\ell + C_\ell(k_i)\psi_{\ell-2}]$$

The attempt to take a reflecting wave with the same index $\ell$ and a wave number $k_r$ according to the dispersion relation fails because of the meridional structure, in fact

$$a_i[\psi_\ell + C_\ell(k_i)\psi_{\ell-2}] + a_r[\psi_\ell + C_\ell(k_r)\psi_{\ell-2}] \neq 0 \tag{8.192}$$

always applies: with $a_r = -a_i$ the coefficient of $\psi_\ell$ vanishes, but then, because of $k_i \neq k_r$, the coefficient of $\psi_{\ell-2}$ does not vanish. The solution is to add another reflected wave with the index $\ell - 2$. Its coefficient must be chosen in a way that the term is balanced with $\psi_{\ell-2}$ in (8.192). Then, however, the term with $\psi_{\ell-4}$ is unbalanced, and the procedure must be repeated by adding another wave with the index $\ell - 4$. However, the series is terminated. If $\ell$ is even, the last wave ($\ell = 0$) is the Kelvin wave. If $\ell$ is uneven, the last wave ($\ell = 1$) is the Yanai wave.

To be more specific, consider an incoming long planetary wave with index $\ell = 2$. Then, the reflected wave consists of a short planetary wave with the wave number $k_r$ and $\ell = 2$ as well as a Kelvin wave with the wave number $k_{\text{Kelvin}} = \omega/c$ and $\ell = 0$. Their amplitudes are determined from

$$a_i[\psi_2 + C_2(k_i)\psi_0] + a_r[\psi_2 + C_2(k_r)\psi_0] + a_{\text{Kelvin}}\,\psi_0 = 0$$

[9] The reflexion of gravity waves at the edge proceeds in a mathematically analogous way but does not offer any new aspects compared to the reflexion in midlatitudes.

with the result $a_r = -a_i$ as well as

$$a_{\text{Kelvin}} = [C_2(k_r) - C_2(k_i)]a_i$$

Usually, the reflected planetary wave is not overly important because of its large wave number and its small group velocity. The main contrast to reflexion in midlatitudes, however, is the reflected Kelvin wave which has a wave number comparable to the incoming wave and by a factor of 3 (generally $2\ell - 1$) faster eastward propagation velocity.

### Eastern boundary

Kelvin and Yanai waves as well as planetary waves can come in here with an eastward group velocity. Formally, the problem is similar to that at the western boundary, but that solution does not work here, because there is no westward group velocity in the case of the smallest values $\ell = 0$ and $\ell = 1$. We can try reflecting waves with higher indices instead.

Look at the case which is most important in practice: a Kelvin wave (index $\ell = 0$) comes in at the eastern boundary. The reflected solution consists of Rossby waves as far as they exist at that frequency, i. e.

$$u_r = \sum_{\ell=2,4,6\ldots} a_{r,\ell}[\psi_\ell + C_\ell(k_r)\psi_{\ell-2}]$$

as before, except for the factor $\phi_n(z)\mathrm{e}^{\mathrm{i}(kx-\omega t)}$. The amplitudes can now be determined successively:

$$\begin{aligned} a_{r,2}C_2(k_r) &= -a_{\text{Kelvin}} \\ a_{r,4}C_4(k_r) &= -a_{r,2} \\ &\vdots \end{aligned}$$

While the series does not come to an end, convergence can be expected as $C_\ell \sim 2(\ell - 1)$ so that the amplitudes decrease with increasing index.

However, another complication occurs. The wave number of the reflected wave is determined from the dispersion relation by (8.177),

$$k_r = -\frac{\beta}{2\omega} \pm \left[\frac{\omega^2}{c^2} + \left(\frac{\beta}{2\omega}\right)^2 - \frac{\beta}{c}(2\ell - 1)\right]^{1/2} \tag{8.193}$$

Only for sufficiently small $\ell$, the bracket in (8.193) is positive so that the reflected waves have a real wave number $k_r$. For larger values of $\ell$, eventually all further $k_r$, therefore, are complex. With the notation $k = \Re(k) + i\Im(k)$ for real and complex parts of a complex variable, it then follows that

$$k_r = -\frac{\beta}{2\omega} - i\Im(k_r)$$

If the incoming Kelvin wave has a frequency in the interval between Rossby and gravity waves, *all* reflected waves have a complex wave number. The imaginary part

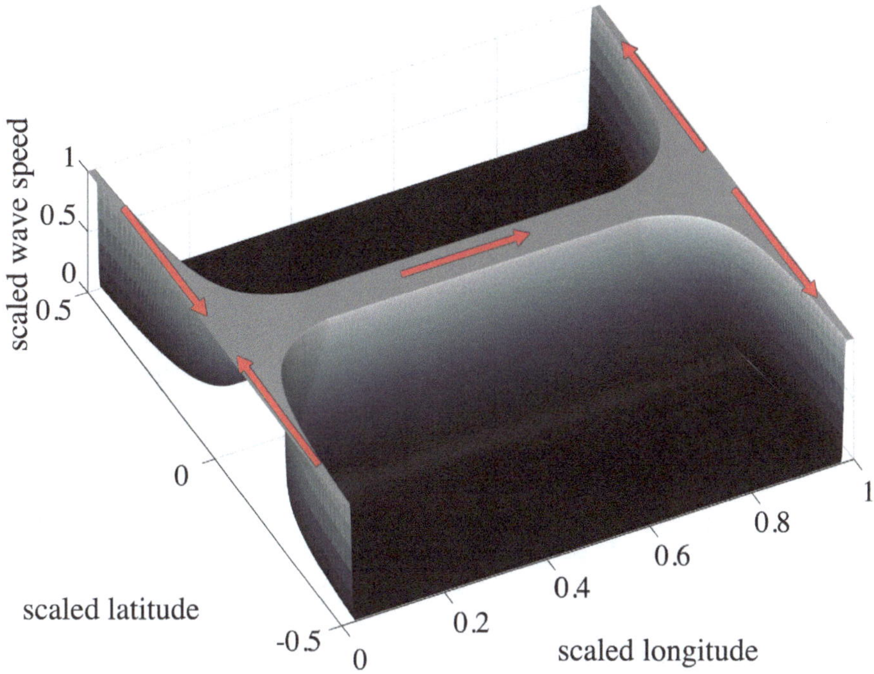


Fig. 8.21 Maximum propagation speed (for the first baroclinic mode; scaled by $c$) of oceanic waves in an ocean with an idealized box shape. The vertical axis indicates the speed of the fastest waves at the respective location. The arrows show the direction of Kelvin wave propagation which has a scaled speed of 1. Away from the boundary regions (east and west coasts and the equatorial band with a width of order of the Rossby radius) the surface depicts the speed of propagation of long Rossby waves. The equatorial Rossby radius is set to 0.04 of the basin width, the midlatitude Rossby radius is 0.008 of the basin width at $y = \pm 0.5$

follows from

$$[\Im(k_r)]^2 + \frac{\omega^2}{c^2} + \left(\frac{\beta}{2\omega}\right)^2 = (2\ell - 1)\frac{\beta}{c} \quad \text{or} \quad [\Im(k_r)]^2 + \left(\frac{\omega}{c} + \frac{\beta}{2\omega}\right)^2 = 2\ell\frac{\beta}{c}$$

The sign of the imaginary part is to be chosen so that there is exponential decrease towards the west. The waves with a complex wave number are neither Rossby nor gravity waves. Collectively, the waves declining in westward direction correspond to Kelvin waves attached to the eastern boundary. This is plausible when considering energy conservation: As the energy propagating towards the eastern boundary cannot be reflected westwards, it has to be transported away from the equator along the eastern boundary, and this is performed by coastal Kelvin waves.

## 8.4 The Oceanic Waveguide

The long waves discussed in the previous sections determine (in the linear approximation) the ocean's reaction to a perturbation. The wave velocity in the interior ocean is very small compared to the speed of equatorial waves and that of Kelvin waves at the basins' boundaries. This fact is the basis for the concept of the oceanic waveguide which is sketched in Figure 8.21.

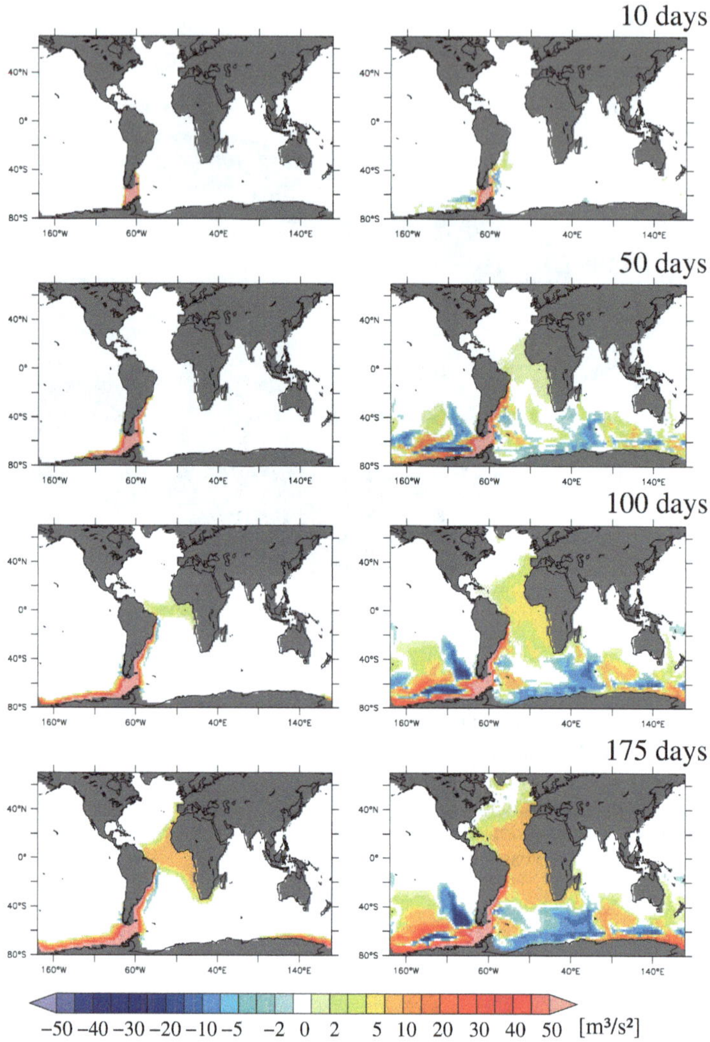


Fig. 8.22 Aspects of oceanic wave propagation, as simulated with the BARBI model (see Appendix B.2). The wave is initiated in Drake Passage by a baroclinic perturbation and propagates in the ocean wave guide around the globe. Left without topography, right with realistic topography. Both simulations are for a mean state with prescribed Brunt–Väisälä frequency without any mean flow

In an idealized ocean basin, outside the equatorial zone at the eastern and western boundaries, there is always a boundary region of the width of the latitude-dependent Rossby radius $R = c/f = c/(2\Omega \sin\varphi)$, where Kelvin waves can propagate equatorward (in the Northern Hemisphere at the western side of the basin) or poleward (in the Northern Hemisphere at the eastern side). Furthermore, there is the region of the equatorial Kelvin waves within a distance $R_e = \sqrt{c/\beta}$ from the equator. The Kelvin waves propagate with the phase velocity $c \sim 1\,\mathrm{m\,s^{-1}}$. In the interior ocean, gravity waves propagate with the same velocity, but due to the geostrophic adjustment (see Section 8.1.3) they are ineffective for signal propagation for scales which are considerably larger than the Rossby radius. The long planetary waves in the interior ocean propagate with the velocity

$$\frac{\omega}{|k|} = \beta R^2 = \beta \frac{c^2}{f^2} = c\frac{c}{f}\frac{\beta}{f} = c\frac{c}{fa} \qquad (a = \text{Earth radius})$$

Outside the equatorial zone, $c \ll |f|a$ and hence $\omega/|k| \ll c$ holds. In the regions with a high wave velocity, signals can propagate over large distances within a comparatively short time.

The waveguide is mainly relevant for the baroclinic reactions. Note however that impacts like small-scale topography and strong mean currents can considerably influence the reaction. This is exemplified in Figure 8.22, which displays the ocean wave response to a sudden change of a baroclinic property in the Drake Passage area (the simulation is done with the BARBI model, see Section B.2). For a flat-bottom ocean (left set of figures) we see the perturbation proceeding as Kelvin waves along the continents and into the equatorial wave guide, followed by Rossby waves in the later stage. All response is entirely baroclinic. With topography implemented, the baroclinic impulse also generates barotropic Rossby waves when baroclinic waves meet topography, and vice versa (the modes are linearly coupled, see Section B.2.6). The evolution becomes highly complex, as shown in the right set of figures. Other aspects are also important for the barotropic reaction: it is essentially influenced by bottom topography, and the barotropic Rossby radius nearly has the size of the ocean basins.

## 8.5 Influence of a Mean Flow on Planetary Waves

So far planetary waves have been considered as linear perturbations of a state of rest. In the following, some aspects of planetary wave propagation on a mean shear flow will be discussed. Consider the vorticity equation (5.32) in the form

$$\left(\frac{\partial}{\partial t} + \boldsymbol{u}\cdot\nabla\right)\left[\nabla_h^2\psi + \frac{\partial}{\partial z}\frac{f_0^2}{N^2}\frac{\partial\psi}{\partial z}\right] + \beta\frac{\partial\psi}{\partial x} = 0 \tag{8.194}$$

Defining the quasi-geostrophic potential vorticity $q = \nabla^2\psi + \partial(f^2/N^2\partial\psi/\partial z)/\partial z + \beta y$ and using $u = -\partial\psi/\partial y$ and $v = \partial\psi/\partial x$, one can rewrite (8.194) in the more convenient form

$$\frac{\partial q}{\partial t} + \frac{\partial\psi}{\partial x}\frac{\partial q}{\partial y} - \frac{\partial\psi}{\partial y}\frac{\partial q}{\partial x} \equiv \frac{\partial q}{\partial t} + \mathcal{J}(\psi, q) = 0 \tag{8.195}$$

where the Jacobian determinant $\mathcal{J}(\psi, q) = (\partial\psi/\partial x)(\partial q/\partial y) - (\partial\psi/\partial y)(\partial q/\partial x)$ is used for abbreviation.

We now define a background state which is described by a mean stream function $\bar{\psi}$ and which must satisfy the (time independent) equation of motion (8.194), respectively (8.195), i. e.

$$\frac{\partial\bar{\psi}}{\partial x}\frac{\partial Q}{\partial y} - \frac{\partial\bar{\psi}}{\partial y}\frac{\partial Q}{\partial x} = 0 \tag{8.196}$$

with $Q = \bar{q}$. This condition is always satisfied with the choice $\bar{\psi} = \Psi(y,z)$ and thus also $\bar{q} = Q(y,z)$. Hence we limit the discussion to a *zonal* shear flow as a geostrophically balanced background state, i. e.

$$\bar{\psi} = \Psi(y,z)\,, \quad \bar{u} = -\frac{\partial \Psi}{\partial y} = U(y,z)\,, \quad \bar{v} = \frac{\partial \Psi}{\partial x} \equiv 0\,, \quad \bar{\rho}(y,z) = -\frac{f}{g}\frac{\partial \Psi}{\partial z}$$

$$\bar{q} = \frac{\partial^2 \bar{\psi}}{\partial y^2} + \frac{\partial}{\partial z}\left(\frac{f}{N^2}\frac{\partial \bar{\psi}}{\partial z}\right) + \beta y = -\frac{\partial U}{\partial y} - \frac{\partial}{\partial z}\left(\frac{gf}{N^2}\bar{\rho}\right) + \beta y = Q(y,z)$$

Expanding (8.195) about the background condition, $\psi = \Psi + \psi'$, $q = Q + q'$ etc., results in

$$q' = \frac{\partial^2 \psi'}{\partial x^2} + \frac{\partial^2 \psi'}{\partial y^2} + \frac{\partial}{\partial z}\left(\frac{f^2}{N^2}\frac{\partial \psi'}{\partial z}\right) \tag{8.197}$$

and

$$\frac{\partial}{\partial t}(Q + q') + \mathcal{J}(\bar{\psi}, Q) + \mathcal{J}(\bar{\psi}, q') + \mathcal{J}(\psi', Q) + \mathcal{J}(\psi', q') = 0 \tag{8.198}$$

Some of the terms trivially vanish, and the last term is of second order in the perturbation field. Linearization, i. e. neglecting terms which are products of two perturbation quantities, thus results in

$$\frac{\partial q'}{\partial t} + U\frac{\partial q'}{\partial x} + \frac{\partial Q}{\partial y}\frac{\partial \psi'}{\partial x} = 0 \tag{8.199}$$

$$\text{with} \quad \frac{\partial Q}{\partial y} = \beta - \frac{\partial^2 U}{\partial y^2} - \frac{\partial}{\partial z}\left(\frac{f^2}{N^2}\frac{\partial U}{\partial z}\right) \tag{8.200}$$

We note that the mean flow has two effects, i) a Doppler shift through advection of $q'$ (the $U\partial q'/\partial x$ term in (8.199)), and ii) a change in the mean potential vorticity gradient in ((8.200) by the terms related to mean shear). The latter may also influence the phase velocity of the waves.

### 8.5.1 Modification of the Doppler Shift

Consider a mean flow $U = U(z)$ only depending on $z$, so that $\bar{\psi} = -yU(z)$. With (8.197) and (8.200), one finds from (8.199)

$$\begin{aligned}
&\frac{\partial}{\partial t}\left[\nabla_h^2 \psi' + \frac{\partial}{\partial z}\left(\frac{f^2}{N^2}\frac{\partial \psi'}{\partial z}\right)\right] + \beta\frac{\partial \psi'}{\partial x} \\
&\quad = -U\nabla_h^2\frac{\partial \psi'}{\partial x} - U\frac{\partial}{\partial z}\left(\frac{f^2}{N^2}\frac{\partial^2 \psi'}{\partial z \partial x}\right) + \frac{\partial \psi'}{\partial x}\frac{\partial}{\partial z}\left(\frac{f^2}{N^2}\frac{\partial U}{\partial z}\right)
\end{aligned} \tag{8.201}$$

The right-hand side of (8.201) would vanish without a mean flow. The terms related to the mean flow correspond to i) mean flow advection of relative vorticity, ii) advection of the vorticity stretching and iii) change in background potential vorticity through the mean current. The eigenfunctions $\Phi_n$ of the system without mean flow defined by (cf. Section 8.1)

$$\frac{\partial}{\partial z}\left(\frac{f^2}{N^2}\frac{\partial \Phi_n}{\partial z}\right) = -\frac{1}{R_n^2}\Phi_n \tag{8.202}$$

and the appropriate boundary conditions form a complete system. Therefore, both the solution of (8.201) and the mean velocity profile can be expanded in this base, resulting in

$$\psi'(x, y, z, t) = \sum_i \psi_i(x, y, t)\Phi_i(z) \tag{8.203}$$

$$U(z) = \sum_j U_j \Phi_j(z) \tag{8.204}$$

Insertion of (8.203) and (8.204) into (8.201)[10], multiplication with $\Phi_n(z)$, integration over $z$, and use of (8.202) result in

$$L^{(n)}\psi_n = -\sum_{i,j} C_{ijn}\left[U_j \frac{\partial}{\partial x}\nabla_\mathrm{h}^2\psi_i - \frac{U_j}{R_i^2}\frac{\partial\psi_i}{\partial x} + \frac{U_j}{R_j^2}\frac{\partial\psi_i}{\partial x}\right] \tag{8.205}$$

where the terms on the right-hand side directly correspond to those in (8.201). Here, $L^{(n)} = \partial/\partial t(\nabla_\mathrm{h}^2 - 1/R_n^2) + \beta\partial/\partial x$ is the linear operator of the free system, and $C_{ijn} = \int_{-H}^{0} \Phi_i(z)\Phi_j(z)\Phi_n(z)\mathrm{d}z$ is a coefficient that only depends on the indices and implicitly on the form of the eigenfunctions.

According to (8.205), for an arbitrary profile $U$ the propagation characteristics of the mode with index $n$ depend on all other modes. Compensation can occur between advection of stretching and the change in background potential vorticity, provided both have the same vertical scale. As an example, consider the lowest mode $n = 0$ and the mean flow component $U_0$. Note that if one of the indices in the coefficient $C_{ijn}$ is 0 (corresponding to a barotropic component), using orthonormality of the $\Phi_n$ and $\Phi_0(z) \approx \text{const} = 1/\sqrt{H}$ one finds $C_{0jn} = 1/\sqrt{H}\delta_{jn}$. It follows that

$$L^{(0)}\psi_0 = -\frac{U_0}{\sqrt{H}}\frac{\partial}{\partial x}\nabla_\mathrm{h}^2\psi_0$$

corresponding to a dispersion relation of the barotropic Rossby waves

$$\omega = Uk - \frac{\beta + UR_0^{-2}}{k^2 + \ell^2 + R_0^{-2}}k$$

(see e.g. Rhines, 1993). While according to (8.199) the Doppler shift acts on both the relative vorticity and the stretching parts of $q'$ in (8.197), for the latter part a compensation occurs through the change in the mean potential vorticity gradient. For long waves, $k, \ell \ll 1/R_0$, the compensation is exact, $\omega \approx -\beta R_0^2 k$ (*non-Doppler effect*). In general, compensation of the Doppler shift occurs when mean flow and wave have a similar vertical structure. The compensation applies only to the stretching part of potential vorticity, not to the relative vorticity, and is hence more important for long planetary waves. An analogous compensation is also found in layer models.

If the vertical profiles of wave and mean flow are different, other effects can occur. For instance, a mean zonal flow with a vertical structure resembling the 2nd vertical mode *increases* the phase speed of long first-mode planetary waves (Killworth et al., 1997).

[10] Note that the individual terms in the sum in (8.203) are not a solution of (8.201), but the sum is.

### 8.5.2 Energy Transfer Between Waves and Mean Flow

From the quasi-geostrophic vorticity equation (8.199), an energy balance can be derived which provides information on the energy transfer between waves and mean flow. The derivation is completely analogous to Section 8.2.2. Multiplication of (8.199) with $\psi'$ results in the energy conservation

$$\frac{\partial}{\partial t}(E_\mathrm{k} + E_\mathrm{p}^i) = \nabla \cdot \left(\psi' \frac{\partial}{\partial t} \nabla \psi'\right) + \frac{\partial}{\partial z}\left(\psi' \frac{f^2}{N^2} \frac{\partial}{\partial t} \frac{\partial \psi'}{\partial z}\right) + \psi' \left(U \frac{\partial q'}{\partial x} + \frac{\partial Q}{\partial y} \frac{\partial \psi'}{\partial x}\right) \tag{8.206}$$

Here $E_\mathrm{k} = (\nabla\psi')^2/2$ denotes the kinetic energy of the perturbations, and $E_\mathrm{p}^i = (f^2/N^2)(\partial\psi'/\partial z)^2/2$ the available potential energy associated with the elevation of density surfaces (the index $i$ stands for the ocean interior).

Consider the energy balance in a finite volume. For simplicity, the geometry of a channel that is periodic in the $x$-direction and bounded in the $y$-direction is chosen. Then, an integration is performed over $y$ and $z$, and averaging over $x$ is denoted as $\langle \cdots \rangle$. Note that always $\langle \partial/\partial x \cdots \rangle = 0$. The first term on the right side of (8.206) is a divergence term which does not contribute to the integral since at the boundaries $\psi' = \text{const}$ or the periodicity condition hold. Vertical integration of the second term in (8.206) results in a contribution from the surface (at the bottom there is no contribution because of the boundary condition $\partial\psi'/\partial z = 0$). The surface displacement is $\zeta = (f/g)\psi'(z = 0)$, and the boundary condition at the free surface according to (8.16), is $\partial\psi'/\partial z + (N^2/g)\psi' = 0$. We find

$$\int_{-H}^{0} \frac{\partial}{\partial z}\left(\psi' \frac{f^2}{N^2} \frac{\partial^2 \psi'}{\partial t \partial z}\right) \mathrm{d}z = \psi' \frac{f^2}{N^2} \frac{\partial^2 \psi'}{\partial t \partial z}\bigg|_0 = -\psi' \frac{f^2}{g} \frac{\partial \psi'}{\partial t} = -g\zeta \frac{\partial \zeta}{\partial t} = -g \frac{1}{2} \frac{\partial}{\partial t} \zeta^2 = -\frac{\partial E_\mathrm{p}^0}{\partial t}$$

where $E_\mathrm{p}^0$ is the potential energy associated with free surface elevations. When we define the total energy as $\mathcal{E} = \int \langle E_\mathrm{k} + E_\mathrm{p}^i \rangle \mathrm{d}y\mathrm{d}z + \int \langle E_\mathrm{p}^0 \rangle \mathrm{d}y$, the energy budget

$$\frac{\partial \mathcal{E}}{\partial t} = \int \left[ U \left\langle \frac{\partial q'}{\partial x} \psi' \right\rangle + \frac{\partial Q}{\partial y} \left\langle \psi' \frac{\partial \psi'}{\partial x} \right\rangle \right] \mathrm{d}y\mathrm{d}z \tag{8.207}$$

is obtained. As $\langle \psi' \partial\psi'/\partial x \rangle = \langle \partial/\partial x \psi'^2/2 \rangle = 0$, the second term in (8.207) does not contribute to the integral. According to (8.197) the first term under the integral in (8.207) consists of three parts,

$$U\left\langle \frac{\partial q'}{\partial x} \psi' \right\rangle = U \left\langle \psi' \frac{\partial^3 \psi'}{\partial x^3} \right\rangle + U \left\langle \psi' \frac{\partial^3 \psi'}{\partial y^2 \partial x} \right\rangle + U \left\langle \psi' \frac{\partial^2}{\partial x \partial z} \left( \frac{f^2}{N^2} \frac{\partial \psi'}{\partial z} \right) \right\rangle \tag{8.208}$$

The first term on the right-hand side of (8.208) vanishes since $\langle \psi' \partial^3 \psi'/\partial x^3 \rangle = \langle \partial/\partial x \psi' \partial^2 \psi'/\partial x^2 \rangle - \langle \partial/\partial x (\partial\psi'/\partial x)^2/2 \rangle = 0$. The second term in (8.208) can be written as

$$U\left\langle\psi'\frac{\partial^3\psi'}{\partial y^2\partial x}\right\rangle=\frac{\partial}{\partial y}\left(U\left\langle\psi'\frac{\partial^2\psi'}{\partial x\partial y}\right\rangle\right)-\frac{\partial U}{\partial y}\left\langle\psi'\frac{\partial^2\psi'}{\partial x\partial y}\right\rangle-U\left\langle\frac{\partial\psi'}{\partial y}\frac{\partial^2\psi'}{\partial x\partial y}\right\rangle \tag{8.209}$$

The first term in (8.209) vanishes in the integral over $y$ since $\psi'=0$ at the boundary. The last term vanishes because $\langle\partial\psi'/\partial y\partial^2\psi'/\partial x\partial y\rangle=\langle\partial/\partial x(\partial\psi'/\partial y)^2\rangle=0$. For the remaining term in (8.209) we use $\langle\psi'\partial^2\psi'/\partial x\partial y\rangle=\langle\partial/\partial x\psi'\partial\psi'/\partial y\rangle-\langle\partial\psi'/\partial x\partial\psi'/\partial y\rangle$. The last term in (8.208) can be written, with $\langle\psi'\partial^2\psi'/\partial x\partial z\rangle=-\langle\partial\psi'/\partial x\partial\psi'/\partial z\rangle$, as

$$\begin{aligned}U\left\langle\psi'\frac{\partial}{\partial z}\frac{f^2}{N^2}\frac{\partial}{\partial x}\frac{\partial\psi'}{\partial z}\right\rangle=&-\frac{\partial}{\partial z}\left(U\frac{f^2}{N^2}\left\langle\frac{\partial\psi'}{\partial x}\frac{\partial\psi'}{\partial z}\right\rangle\right)\\&+\frac{\partial U}{\partial z}\frac{f^2}{N^2}\left\langle\frac{\partial\psi'}{\partial x}\frac{\partial\psi'}{\partial z}\right\rangle-U\frac{f^2}{N^2}\left\langle\frac{\partial\psi'}{\partial z}\frac{\partial^2\psi'}{\partial x\partial z}\right\rangle\end{aligned} \tag{8.210}$$

The last term in (8.210) vanishes since it can be written as $\langle\partial/\partial x\cdots\rangle$. The first term on the right-hand side of (8.210) vanishes in the integral over $z$ when the boundary conditions are used.

With the above results, (8.207) takes the form

$$\frac{\partial\mathcal{E}}{\partial t}=\int\left[\frac{\partial U}{\partial y}\left\langle\frac{\partial\psi'}{\partial x}\frac{\partial\psi'}{\partial y}\right\rangle+\frac{\partial U}{\partial z}\frac{f^2}{N^2}\left\langle\frac{\partial\psi'}{\partial x}\frac{\partial\psi'}{\partial z}\right\rangle\right]\mathrm{d}y\mathrm{d}z \tag{8.211}$$

As $\partial\psi'/\partial x=v'$ and $\partial\psi'/\partial y=-u'$ and $\partial\psi'/\partial z=-(g/f)\rho'$, it follows that the energy transformation is described by the 'Reynolds' terms $\langle v'u'\rangle\partial U/\partial y$ and $\langle v'\rho'\rangle\partial U/\partial z$ which correspond to the transfer of kinetic respectively potential energy, as discussed in the energy cycle of Section 12.2.1.

An alternate form of the transfer terms can be derived from (8.211) by a further partial integration to the form

$$\frac{\partial\mathcal{E}}{\partial t}=-\int U\left[\frac{\partial}{\partial y}\left\langle\frac{\partial\psi'}{\partial x}\frac{\partial\psi'}{\partial y}\right\rangle+\frac{\partial}{\partial z}\left(\frac{f^2}{N^2}\left\langle\frac{\partial\psi'}{\partial x}\frac{\partial\psi'}{\partial z}\right\rangle\right)\right]\mathrm{d}y\mathrm{d}z \tag{8.212}$$

again except for the boundary terms. From (8.197) it follows that

$$\begin{aligned}\langle v'q'\rangle&=\left\langle\frac{\partial\psi'}{\partial x}\left[\frac{\partial^2\psi'}{\partial x^2}+\frac{\partial^2\psi'}{\partial y^2}+\frac{\partial}{\partial z}\left(\frac{f^2}{N^2}\frac{\partial\psi'}{\partial z}\right)\right]\right\rangle\\&=\frac{\partial}{\partial y}\left\langle\frac{\partial\psi'}{\partial x}\frac{\partial\psi'}{\partial y}\right\rangle+\frac{\partial}{\partial z}\left(\frac{f^2}{N^2}\left\langle\frac{\partial\psi'}{\partial x}\frac{\partial\psi'}{\partial z}\right\rangle\right)\end{aligned}$$

The proof is analogous to the one above. Thus, apart from the boundary terms,

$$\frac{\partial\mathcal{E}}{\partial t}=-\int U\langle v'q'\rangle\mathrm{d}y\mathrm{d}z \tag{8.213}$$

The energy exchange with the mean current is hence proportional to the meridional flux of potential vorticity, which is a consequence of the quasi-geostrophic scaling considered here.

### 8.5.3 Conditions for Instability

In principle, energy exchange between waves and mean flow can take place in both directions, and the direction of the exchange cannot be inferred from (8.211). In Section 12.2.1, numerical estimates of the energy exchange in the Lorenz energy cycle are shown. In that estimate, the energy exchange related to the Reynolds stress $\langle u'v'\rangle$ takes both signs (in the general situation of nonzonal mean flow there are further contributions), whereas the term related to the horizontal eddy density flux $\langle v'\rho'\rangle$ is predominantly positive. Of particular interest is the latter case, i. e. where the right side of (8.211) is positive so that energy is transferred from the mean flow to the waves, which means that the mean flow is unstable to small perturbations. Two physically different mechanisms can be distinguished, corresponding to energy transfer by the two terms in (8.211), namely barotropic and baroclinic instability.

#### Barotropic Instability

Barotropic instability is associated with the transport term $-(\partial U/\partial y)\langle u'v'\rangle$ in (8.211) and can only occur if there is a horizontal shear of the background velocity $\partial U/\partial y \neq 0$. Since no vertical shear of $U$ is needed (in contrast to baroclinic instability, see next section) a barotropic background current can produce the instability, which explains the name. An analysis of the energetics of the mean current shows that the energy is exchanged with the *kinetic* energy of the mean flow, analogous to the corresponding term in the Lorenz energy cycle in Section 12.2.1.

We can obtain a simple criterion for instability in the completely barotropic case where both the mean flow $U$ as well as the disturbance $\psi'$ are independent of $z$. With $\psi' = \phi(y)\mathrm{e}^{\mathrm{i}(kx-\omega t)}$ one obtains from (8.200)

$$\frac{\partial^2\phi}{\partial y^2} - k^2\phi + \frac{\partial Q}{\partial y}\frac{1}{U-c}\phi = 0 \tag{8.214}$$

where the phase velocity $c = \omega/k$ has been used. Note that with the transformation $N^2 \to 0$, $-z \to y$, $H \to L$, (8.214) corresponds to the Taylor-Goldstein equation (7.91), except for the $\beta$-term in $\partial Q/\partial y$. The examination of barotropic stability is, therefore, mathematically very similar to the instability of gravity waves considered in Section 7.7.3, even though both processes have quite different physics (i.e., instability of vertically sheared flow to gravity waves vs. instability of horizontally sheared flow to planetary waves).

Consider again a current in a zonal channel with the boundaries $y = 0$ and $y = L$. With these boundary conditions, (8.214) poses an eigenvalue problem where the eigenvalue $c$ is not necessarily real. Instability (exponential growth of the waves) occurs for a positive imaginary part of $c$. To derive a condition for $c_i$, multiplication of (8.214) with $\phi^*$, and integration over the area according to partial integration with use of the boundary conditions yields

$$\int_0^L \left(\left|\frac{\partial\phi}{\partial y}\right|^2 + k^2|\phi|^2\right)\mathrm{d}y - \int_0^L (U-c^*)\frac{\partial Q}{\partial y}\frac{1}{|U-c|^2}|\phi|^2\mathrm{d}y = 0 \tag{8.215}$$

The real and the imaginary part of (8.215) must vanish individually. Therefore,

$$c_i \int_0^L \frac{\partial Q}{\partial y} \frac{|\phi^2|}{|U-c|^2} \mathrm{d}y = 0 \tag{8.216}$$

Instability ($c_i > 0$) can only occur if the integral in (8.216) vanishes. As the second factor is always positive, it is a necessary condition for instability that $\partial Q/\partial y$ changes its sign in the interval. According to (8.200), the gradient of the mean potential vorticity is $\partial Q/\partial y = \beta - \partial^2 U/\partial y^2$. Therefore, a necessary condition for barotropic instability is $\partial^2 U/\partial y^2 = \beta$ somewhere in the domain, say at $y = y_0$. With the lateral scale $L$ so that $\partial^2 U/\partial y^2 \sim \Delta U/L^2$, it follows that instability can only occur if $L < L_\beta = \sqrt{\Delta U/\beta}$. The length scale $L_\beta$ was introduced in Section 5.1 as the Rhines scale. For $\Delta U \approx 10\,\mathrm{cm\,s^{-1}}$, we obtain $L_\beta \approx 70\,\mathrm{km}$, so that instability can occur only on scales $\lesssim 70\,\mathrm{km}$.

One can even prove the slightly stricter condition

$$\int_0^L [U(y) - U(y_0)] \left(\beta - \frac{\partial^2 U}{\partial y^2}\right) \frac{|\phi|^2}{|U-c|^2} \mathrm{d}y > 0$$

In analogy to the derivation in Section 7.7.3, a *semicircle theorem* can be proven, modified through presence of the $\beta$-term. The corresponding analogue to (7.103) is

$$\left(c_r - \frac{U_{\min} + U_{\max}}{2}\right)^2 + c_i^2 \le \left(\frac{U_{\min} - U_{\max}}{2}\right)^2 + \frac{\beta L^2 (U_{\max} - U_{\min})}{2(\pi^2 + k^2 L^2)} \tag{8.217}$$

One of the important consequences of barotropic instability is that flows with small length scales are likely to be unstable. In fact, it can be shown that interacting short Rossby waves are unstable (Gill, 1974), which also explains the turbulent nature of mesoscale motions in the ocean, as discussed in Chapter 12.

## Baroclinic Instability

Baroclinic instability is related to the transport term $fg/N^2(\partial U/\partial z)\langle v'\rho'\rangle$ in (8.211). The energy is exchanged with the (available) *potential* energy of the basic state, analogous to the corresponding term in the Lorenz energy cycle in Section 12.2.1. A necessary condition for baroclinic instability can be found in a similar way as above. With the wave ansatz

$$\psi'(x,y,z,t) = \phi(y,z)\mathrm{e}^{\mathrm{i}(kx-\omega t)} \tag{8.218}$$

one obtains from (8.199), analogous to (8.216), the equation

$$\frac{\partial^2 \phi}{\partial y^2} + \frac{\partial}{\partial z}\left(\frac{f^2}{N^2}\frac{\partial \phi}{\partial z}\right) + \frac{\phi}{U-c}\left[\frac{\partial Q}{\partial y} - k^2(U-c)\right] = 0 \tag{8.219}$$

The boundary conditions at the top and the bottom are crucial. They are simplified here as $w(0) = w(-H) = 0$. According to (5.35), in the quasi-geostrophic approximation the vertical velocity is given as

$$\begin{aligned} w &= -\frac{f_0}{N^2}\frac{\mathrm{d}}{\mathrm{d}t}\frac{\partial \psi}{\partial z} = -\frac{f_0}{N^2}\frac{\mathrm{D}}{\mathrm{D}t}\left(\frac{\partial \Psi}{\partial z} + \frac{\partial \psi'}{\partial z}\right) \\ &= -\frac{f_0}{N^2}\left(\frac{\partial}{\partial t} + U\frac{\partial}{\partial x} + u'\frac{\partial}{\partial x} + v'\frac{\partial}{\partial y}\right)\left(\frac{\partial \psi'}{\partial z} + \frac{\partial \Psi}{\partial z}\right) \end{aligned}$$

Linearization results in

$$w \approx -\frac{f_0}{N^2}\left(\frac{\partial^2 \psi'}{\partial t \partial z} + U\frac{\partial^2 \psi'}{\partial z \partial x} - \frac{\partial U}{\partial z}\frac{\partial \psi'}{\partial x}\right)$$

With (8.218) the boundary conditions $w = 0$ are

$$(U - c)\frac{\partial \phi}{\partial z} - \frac{\partial U}{\partial z}\phi = 0 \quad \text{at} \quad z = 0\,, \quad -H \tag{8.220}$$

with $c = \omega/k$. Multiplication of (8.219) with $\phi^*$ and subsequent integration over $y$ and $z$, with (8.220) and $\phi = 0$ at $y = 0, L$, results (after partial integration) in a complex expression, the imaginary part of which is given by

$$c_i \int_0^L \left\{ \int_{-H}^0 \frac{|\phi^2|}{|U - c|^2}\frac{\partial Q}{\partial y}\mathrm{d}z + \left[\frac{f^2}{N^2}\frac{|\phi^2|}{|U - c|^2}\frac{\partial U}{\partial z}\right]_{-H}^0 \right\} \mathrm{d}y = 0 \tag{8.221}$$

For instability ($c_i > 0$) it is necessary that the integral vanishes. As the factors of $\partial Q/\partial y$ and $\partial U/\partial z$ are always positive, necessary conditions for instability are that either

1. $\partial Q/\partial y = \beta - \partial^2 U/\partial y^2 - \partial(f_0^2/N^2 \partial U/\partial z)/\partial z$ changes its sign somewhere in the domain, or
2. that the sign of $\partial Q/\partial y$ is opposite to that of $\partial U/\partial z$ at surface or bottom.

Note that for $\partial U/\partial z \equiv 0$ the condition (8.216) is recovered. When we consider the Ertel potential vorticity (given by (4.58)) for the mean flow instead of the quasi-geostrophic form for $Q$, and neglect the horizontal terms related to $\partial u/\partial z$, which are small, i. e. $Q \sim (\partial \rho/\partial z)(f - \partial u/\partial y)$, and further neglect the relative vorticity compared to the planetary vorticity, i. e. $Q \sim f(\partial \rho/\partial z)$, it becomes possible to assess the necessary condition for baroclinic instability from hydrographic observations of density. It turns out that the condition $\partial(f\,\partial\rho/\partial z)/\partial y = 0$ approximately holds over some areas of the world's oceans, particularly at mid-depth in the bowl of the subtropical gyre. Rhines and Young (1982) have argued that a uniform distribution of potential vorticity might result from a systematic stirring by the oceanic eddy field (see Section 14.3.5). Note also that baroclinic instability is the major source of mesoscale eddy energy in the ocean and the atmosphere, as discussed in Section 12.2.1.

### Eady's Solution

In certain particularly simple situations, statements exceeding the general condition (8.221) can be made in analytical form (numerical solutions are discussed in the box on p. 275). Eady (1949) considered the case $\beta = 0$, $N = \text{const}$, $U = U_0(1 + z/H)$ and hence $\Psi = -U_0 y(1 + z/H)$ (see also Figure 8.23a). The first assumption eliminates the propagation of planetary waves but allows to solve the problem analytically. It follows that $\partial Q/\partial y \equiv 0$. Note that with vanishing $\partial Q/\partial y$ and constant vertical shear $\partial U/\partial z = U_0/H$, the integral in (8.221) is always zero, and no general conditions for instability result. In fact, it will become clear that perturbations of the Eady problem become unstable by implementing boundary conditions.

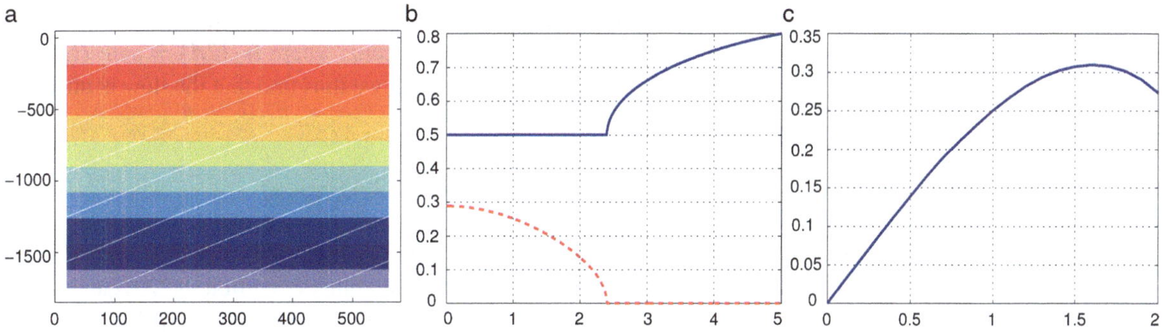


Fig. 8.23 **a** The mean flow $U = U_0(1 + z/H)$ with $U_0 = 0.5\,\mathrm{m\,s^{-1}}$ (red denotes positive and blue negative values) and mean (total) buoyancy $B = f\partial\Psi/\partial z + N^2 z = fU_0 y/H + N^2 z$ (white lines) with $N = 0.004\,\mathrm{s^{-1}}$. **b** $c/U_0$ from equation (8.228) as a function of $h/d$. The dashed red line denotes the imaginary part of $c$ for $c_i > 0$, the solid blue line the real part of $c$. **c** The function $\xi F(\xi)$ from equation (8.231)

With constant coefficients in $x$ and $y$, (8.199) is solved with $\psi'(x, y, z, t) = \phi(z)e^{i(k_1 x + k_2 y - \omega t)}$ and leads to

$$\frac{\partial^2\phi}{\partial z^2} - \frac{N^2}{f^2}(k_1^2 + k_2^2)\phi = 0 \tag{8.222}$$

for the vertical dependence (actually the same equation as the one governing topographic waves, see (8.76)), with the solution

$$\phi = A\cosh(z/d) + B\sinh(z/d) \tag{8.223}$$

Here, the vertical scale

$$d = \frac{|f|}{N\sqrt{k_1^2 + k_2^2}} \tag{8.224}$$

is introduced. According to (8.220) the boundary conditions are

$$(U_0 - c)\frac{\partial\phi}{\partial z} - \frac{U_0}{H}\phi = 0 \quad \text{at} \quad z = 0 \tag{8.225}$$

$$-c\frac{\partial\phi}{\partial z} - \frac{U_0}{H}\phi = 0 \quad \text{at} \quad z = -H \tag{8.226}$$

with $c = \omega/k_1$. From (8.223) it follows with (8.225) that

$$A = \frac{U_0 - c}{U_0}\frac{H}{d}B \tag{8.227}$$

From (8.226) and (8.227) one obtains, after a brief calculation, an equation for the phase speed $c$ given by

$$\begin{aligned}\frac{c}{U_0} &= \frac{1}{2} \pm \frac{d}{H}\left[1 + \frac{1}{4}\frac{H^2}{d^2} - \frac{H}{d}\coth\frac{H}{d}\right]^{\frac{1}{2}} \\ &= \frac{1}{2} \pm \frac{d}{H}\left[\left(\frac{H}{2d} - \coth\frac{H}{2d}\right)\left(\frac{H}{2d} - \tanh\frac{H}{2d}\right)\right]^{\frac{1}{2}}\end{aligned} \tag{8.228}$$

A real solution for $c$ can only exist if the expression under the square root is positive. As $\xi \geq \tanh \xi$, the sign change of the bracket occurs where $H/(2d) = \coth H/(2d)$ or $H/d \approx 2.4$ (see Figure 8.23). With (8.224), we thus find that the flow is unstable if

$$k = \sqrt{k_1^2 + k_2^2} < 2.4\frac{f}{NH} \tag{8.229}$$

For a constant $N$ it was shown earlier (cf. Section 7.4.3) that $R = NH/(\pi f)$ is the first baroclinic Rossby radius. Therefore, the instability condition (8.229) can also be written as $kR < 0.76$. Hence the flow is unstable to perturbations with sufficiently long wavelengths, and stable for perturbations with short wavelengths. For $kR < 0.76$, the imaginary part is given by

$$c_i = U_0\frac{d}{H}\sqrt{\left(\coth\frac{H}{2d} - \frac{H}{2d}\right)\left(\frac{H}{2d} - \tanh\frac{H}{2d}\right)} \equiv U_0 F(H/d) \tag{8.230}$$

For the growth rate $\gamma = k_1 c_i$ and the special case $k_2 = 0$ it follows from (8.230) and (8.224) that

$$\gamma = k_1 c_i = \frac{fU_0}{Nd}F(H/d) = \frac{fU_0}{NH} \times \frac{H}{d}F(H/d) \tag{8.231}$$

The function $\xi F(\xi)$ has a maximum value of $\approx 0.3$ at $\xi \approx 1.6$ (see Figure 8.23). Hence the maximum growth rate becomes

$$\gamma_{\max} \approx 0.3U_0\frac{f}{NH} \approx 0.1\frac{U_0}{R} \quad \text{at} \quad k_{\max} \approx 1.6\frac{f}{NH} \approx \frac{0.5}{R}$$

The scale of the most unstable perturbations is, therefore, comparable to the Rossby radius, and the maximum perturbation energy can be expected at this scale. Note that for $k_2 \neq 0$ the growth rates are somewhat smaller. As the mean shear $\partial U/\partial z = U_0/H$ is related with the meridional density gradient via the thermal wind relation, i. e. $U_0/H = (g/f)\partial\bar{\rho}/\partial y$, the maximum growth rate can be expressed as

$$\gamma_{\max} \approx 0.3\left|\frac{\partial \rho}{\partial y}\right|\frac{g}{N\rho_0} = 0.3\frac{M^2}{N} \tag{8.232}$$

with $M^2 = g|\partial\bar{\rho}/\partial y|/\rho_0$, and thus can directly be determined by hydrographic observations. It turns out that the maximum Eady growth rates are largest in regions of maximum mesoscale eddy kinetic energy, i. e. in western boundary currents, along the equator and in the Antarctic Circumpolar Current, where also the largest eddy energy production terms related to baroclinic instability show up, as discussed in Section 12.2.1. Note that also that the introduction of nonzero $\beta$ does not change the Eady solution much, as discussed in the box on p. 275.

For unstable flows with $c = c_r + ic_i$, it follows from (8.228) that $c_r = U_0/2$, i. e. the increasing perturbations propagate in $x$ direction with half the *mean* zonal particle velocity. The complete solution for the perturbation stream function $\psi'$ for unstable flow is given by

$$\Re(\psi') \sim |\phi| \sin(k_1 x + k_2 y - \omega t + \alpha) \tag{8.233}$$

40. Numerical Solution of the Instability Problem

Solutions of (8.219) and (8.220) of the form $\psi' = \phi(y,z)e^{i(kx-\omega t)}$ for general $U(y,z)$, $N(z)$, and $Q(y,z)$ can be found numerically: discretizing the vertical and meridional derivatives in (8.219) and (8.220) leads to a linear system of algebraic equations. The eigenvalues ($\omega$) and eigenvectors ($\phi$) of the corresponding matrix of the linear system can then be found numerically and calculated as a function of the wave number. The figure below compares the solution of the standard Eady problem and the unstable modes in the numerical solution for the modified Eady problem with $\partial Q/\partial y = \beta$.

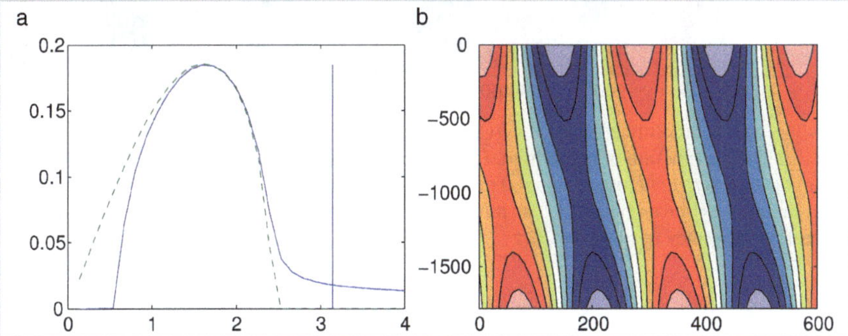


*a shows the growth rate $\gamma = k_1 c_i$ as a function of the scaled wave number $k_1 Nh/f$ in* $1/\mathrm{d}$ *for the standard Eady problem ($\beta = 0$, dashed) and the growth rates calculated numerically for $\beta = 2\times 10^{-11}\,(\mathrm{ms})^{-1}$ (solid). Also shown is the inverse of the scaled Rossby radius $Rf/(Nh) = 1/\pi$ as the vertical solid line. All parameters are the same as in Figure 8.24, and Figure 8.27. b shows the complete solution $\psi'$ of the most unstable mode for $\beta = 2\times 10^{-11}\,1/(\mathrm{ms})$ as a function of depth in* m *and longitude in* km.

It becomes clear that the introduction of the planetary vorticity does not modify the solution much, i. e. maximum growth rates are similar and at a similar wave length, and the vertical structure of the most unstable mode is also similar to Figure 8.24. However, in contrast to the standard Eady problem, shorter wave lengths than the most unstable one become now all unstable while wave lengths larger as about four times the Rossby radius become stable.

with the depth-dependent phase $\alpha$ and amplitude $|\phi|$

$$\tan\alpha(z) = \pm\frac{U_0}{c_i}\frac{(1/2)\cosh(z/d) + (d/H)\sinh(z/d)}{\cosh(z/d)}$$

$$|\phi(z)| = \sqrt{\left(\frac{H}{2d}\cosh(z/d) + \sinh(z/d)\right)^2 + \left(\frac{H}{d}\frac{c_i}{U_0}\cosh(z/d)\right)^2}$$

Figure 8.24 shows phase $\alpha$, amplitude $\phi$ and stream function $\psi'$ for the most unstable mode and parameters listed in the caption.

## Phillips' Model

Another model of a baroclinically unstable system, that is frequently studied and solved analytically, is the model of Phillips (1954). It studies the unstable modes in a two-layer channel flow, governed by the conservation of the quasi-geostrophic potential vorticity. The model is described in Appendix B.1.3 by (B.12) to (B.13) with the definition (B.14) of the potential vorticities. The coordinates are $x$ along the channel and $y$ across with a channel width $B$; the mean layer depths are $H_1$ and $H_2$ with a total depth $H = H_1 + H_2$, and the reduced gravity is denoted by

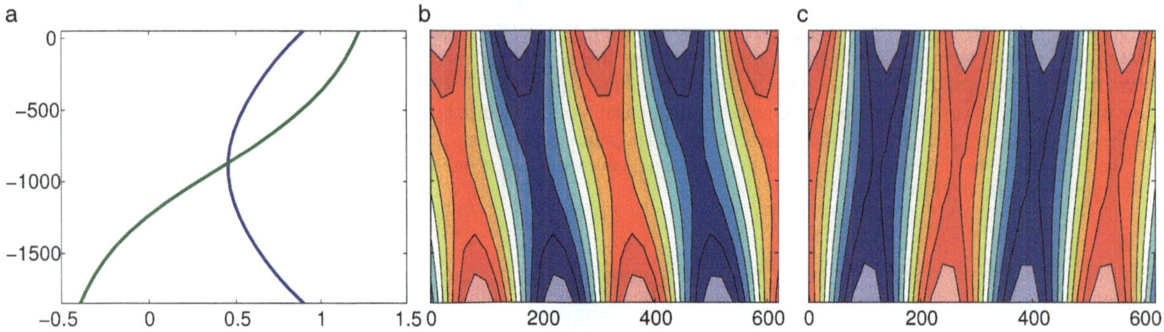


**Fig. 8.24** **a** The amplitude function $|\phi(z)|$ (*blue*) and the phase $\alpha(z)$ (*green*) of the unstable mode (8.233) as a function of depth in m, for mean flow and stratification as shown in Figure 8.23a and for $k_1 = 1.6f/(NH)$ and $k_2 = 0\,\mathrm{m}^{-1}$, $H = 1800\,\mathrm{m}$ and $f = 10^{-4}\,\mathrm{s}^{-1}$. **b** Perturbation stream function $\psi'$ given by (8.233) as a function of depth in m and $x$ in km. **c** Same as **b** but for perturbation buoyancy $b' = f\partial\psi'/\partial z$

$g^*$. Compared with the Eady model, the flow has a simpler (less resolved) vertical structure, but there is advantage that now $\beta$ can be taken nonzero.

Perturbations, denoted by primed quantities, residing on a mean current $U_j(y)$, $j = 1, 2$ in the two layers, satisfy (see (8.199))

$$\left(\frac{\partial}{\partial t} + U_j \frac{\partial}{\partial x}\right) q'_j + \frac{\partial \psi'_j}{\partial x}\frac{\partial Q_j}{\partial y} = 0 \tag{8.234}$$

with

$$q'_j = \nabla^2 \psi'_j \pm \Lambda_j(\psi'_2 - \psi'_1) \quad \text{and} \quad \frac{\partial Q_j}{\partial y} = \beta - \frac{\partial^2 U_j}{\partial y^2} \pm \Lambda_j(U_1 - U_2) \tag{8.235}$$

and $\Lambda_j = f_0^2/(g^* H_j)$. The upper sign refers to the upper layer $i = 1$, the lower sign to the lower layer $i = 2$. The ansatz $\psi'_j = \phi_j(y)\mathrm{e}^{\mathrm{i}k(x-ct)}$ yields the eigenvalue problem

$$(U_j - c)\left(\frac{\partial^2 \phi_j}{\partial y^2} - k^2\phi_j \pm \Lambda_j(\phi_2 - \phi_1)\right) + \phi_j \frac{\partial Q_j}{\partial y} = 0 \quad \text{for} \quad j = 1, 2 \tag{8.236}$$

with boundary conditions $\phi_j = 0$ on both channel walls. It is, in fact, the discrete form of (8.219), and also the condition (8.221), leading to the instability constraint, is transferred correspondingly with a sum over the two layers replacing the vertical integral.

We proceed now to the simplest case that the mean flow $U_j$ is independent of $y$ such that $U_s = U_1 - U_2$ and $\mathrm{d}Q_j/\mathrm{d}y = \beta \pm \Lambda_j U_s$ are constants. The condition necessary for instability, $c_i \neq 0$, is that the gradients of the mean potential vorticity must have different signs. They read as follows:

- if $U_s > 0$, then $\mathrm{d}Q_1/\mathrm{d}y = \beta + \Lambda_1 U_s > 0$, and hence $\mathrm{d}Q_2/\mathrm{d}y = \beta - \Lambda_2 U_s$ must be less than zero, or

$$U_s = U_1 - U_2 > \beta/\Lambda_2 \tag{8.237}$$

- if $U_s < 0$, then $\mathrm{d}Q_2/\mathrm{d}y = \beta - \Lambda_2 U_s > 0$, and hence $\mathrm{d}Q_1/\mathrm{d}y = \beta + \Lambda_1 U_s$ must be less than zero, or

$$U_s = U_1 - U_2 < -\beta/\Lambda_1 \tag{8.238}$$

Note that the shear $U_s$ relates to the tilt of the mean layer interface $\bar{\eta}$ by the thermal-wind equation $U_s = (g^*/f_0) d\bar{\eta}/dy$. The above conditions are thus constraints on a minimum of the available potential energy of the mean state in case of an unstable flow.

The conditions (8.237) and (8.238) are sufficient for instability as well, as can be proven by a solution of (8.236). The ansatz $\phi_j(y) = A_j \sin \ell y$ with constant amplitudes $A_j$ and the discrete meridional wave number $\ell = n\pi/B, n = 1, 2, \ldots$ satisfies the boundary conditions. The second order linear system for the $A_j$, resulting from the equations (8.236), has a solution if the determinant of the coefficients vanishes, leading to

$$\begin{aligned} c &= U_2 + \frac{K^2 U_s (K^2 + 2\Lambda_2) - \beta(2K^2 + \Lambda_1 + \Lambda_2) \pm D^{1/2}}{2K^2(K^2 + \Lambda_1 + \Lambda_2)} \\ &= \frac{U_1(K^2 + 2\Lambda_2) + U_2(K^2 + 2\Lambda_1)}{2(K^2 + \Lambda_1 + \Lambda_2)} - \frac{\beta(2K^2 + \Lambda_1 + \Lambda_2) \mp D^{1/2}}{2K^2(K^2 + \Lambda_1 + \Lambda_2)} \end{aligned} \tag{8.239}$$

with $K^2 = k^2 + \ell^2$ and the discriminant

$$D(U_s, K) = \beta^2(\Lambda_1 + \Lambda_2)^2 + 2\beta U_s K^4(\Lambda_1 - \Lambda_2) + K^4 U_s^2 (K^4 - 4\Lambda_1\Lambda_2) \tag{8.240}$$

Let us first consider some limiting cases:

- $U_s = 0$, i. e. a barotropic mean flow $U_1 = U_2 = U$. The two solutions for the phase speed may be written as

$$c = U - \frac{\beta}{K^2} \quad \text{and} \quad c = U - \frac{\beta}{K^2 + \Lambda_1 + \Lambda_2} \tag{8.241}$$

  and coincide with the barotropic and baroclinic Rossby wave speed for the two-layer system, including the Doppler shift by the barotropic current (see Section 8.2). The flow is stable because the barotropic instability cannot operate for a constant $U$.
- $\beta = 0$. The solution becomes

$$c = \frac{U_1(K^2 + 2\Lambda_2) + U_2(K^2 + 2\Lambda_1) \pm U_s(K^4 - 4\Lambda_1\Lambda_2)^{1/2}}{2(K^2 + \Lambda_1 + \Lambda_2)} \tag{8.242}$$

  An imaginary part of the phase speed develops, and the system becomes unstable for all $U_s \neq 0$ if $K^4 < 4\Lambda_1\Lambda_2$, or

$$(KR)^2 < 2R^2\sqrt{\Lambda_1\Lambda_2} = 2R^2 \frac{f_0^2}{g^*\sqrt{H_1 H_2}} = 2\frac{\sqrt{H_1 H_2}}{H} \tag{8.243}$$

  where $R = (g^* H_1 H_2 / f_0^2 H)^{1/2}$ is the baroclinic Rossby radius of the channel configuration (see Appendix B.1.3). Perturbations with wavelengths roughly exceeding the baroclinic Rossby radius become unstable. More precisely, there is an interval $(0, k_{\max})$ of wave numbers for which the perturbations $\sim e^{ikx} \sin n\pi y/B$ are unstable. The maximum is attained for $n = 1$, hence $(k_{\max} R)^2 = 2\sqrt{H_1 H_2}/H - (R\pi/B)^2$. The overall result is similar to the condition (8.229) of the Eady model.

- $U_1 = -U_2, H_1 = H_2$. This symmetric case is most easily analyzed. With $\Lambda = \Lambda_1 = \Lambda_2$, we find

$$c = \frac{-\beta(K^2 + \Lambda^2) \pm \sqrt{4\beta^2 \Lambda^2 + K^4 U_s^2 (K^4 - 4\Lambda^2)}}{K^2(K^2 + 2\Lambda^2)} \tag{8.244}$$

Instability occurs if

$$U_s^2 > U_{\text{crit}}^2 = \frac{4\beta^2 \Lambda^2}{K^4(4\Lambda^2 - K^4)} \tag{8.245}$$

which requires that $K^2 < 2\Lambda$. Positive and negative shears are thus affected in the same way. The minimum of $U_{\text{crit}}^2$ follows from $dU_{\text{crit}}^2/dK^4 = 0$; it occurs at $K^4 = 2\Lambda^2$ and has the value $U_{\text{crit}}^2 = (\beta/\Lambda)^2$ in agreement with the results of integral conditions discussed above. Note that, in contrast to the case $\beta = 0$, there is now a range of shear values $-\beta/\Lambda < U_s < \beta/\Lambda$ which is stable. Hence the presence of $\beta$ stabilizes the flow. A further property is that (8.244) implies for unstable modes the ordering

$$-\frac{\beta}{K^2} < \Re(c) = -\frac{\beta(K^2 + \Lambda^2)}{K^2(K^2 + 2\Lambda^2)} < -\frac{\beta}{K^2 + 2\Lambda^2} \tag{8.246}$$

i. e. unstable modes propagate westward with a speed which lies between the fast barotropic and the slow baroclinic phase speeds of ordinary Rossby waves with the same wave number $K$.

In the general case, considering now the complete discriminant $D$ as function of $U_s$ and $K$ with nonzero $\beta$, we notice that the curve $D(U_s, K) = 0$ separates two regions in the $(U_s, K)$-plane which are no longer symmetric. The curve may be expressed as two different branches $U^{\pm}(K)$ given by

$$U^{\pm} = \frac{\beta}{K^2(4\Lambda_1\Lambda_2 - K^4)} \left( K^2(\Lambda_1 - \Lambda_2) \pm 2\sqrt{\Lambda_1\Lambda_2}\sqrt{((\Lambda_1 + \Lambda_2)^2 - K^4)} \right) \tag{8.247}$$

and displayed in Figure 8.25 where the stable and unstable regimes are laid out. Unstable perturbations can only occur where $K^4 < (\Lambda_1 + \Lambda_2)^2$, which transfers to $KR < 1$. The minimum of the branch $U_+$ of positive shears is $\min(U^+) = \beta/\Lambda_2$; it occurs at $K^2 = K_+^2 = \sqrt{\Lambda_2^2 + \Lambda_1\Lambda_2}$. The maximum of the branch $U^-$ of negative shears is $\max(U^-) = -\beta/\Lambda_1$, it occurs at $K^2 = K_-^2 = \sqrt{\Lambda_1^2 + \Lambda_1\Lambda_2}$. These statements fully agree with the above derived results of the integral constraints and show that the necessary conditions are indeed sufficient. Note that

$$(K_+ R)^2 = \sqrt{\frac{H_1}{H}} \quad \text{and} \quad (K_- R)^2 = \sqrt{\frac{H_2}{H}} \tag{8.248}$$

For given positive shear above the critical one, perturbations in a whole range of wave numbers, below the inverse Rossby radius, are unstable. A corresponding statement holds for a negative shear. Hence long waves are stable and very short waves as well.

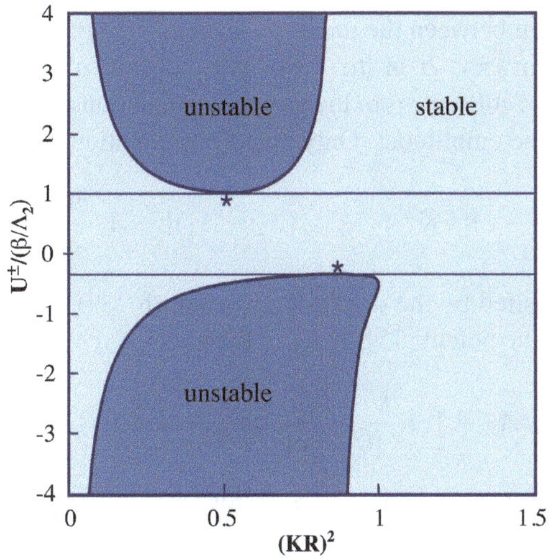


Fig. 8.25 Regime diagram showing the critical shear $U^{\pm}$, scaled by $\beta/\Lambda_2$, as function of $(KR)^2$. The parameters are $H_1 = 1{,}000$ m, $H_2 = 3{,}000$ m, $\beta = 2\times10^{-11}$ m$^{-1}$ s$^{-1}$, $f_0 = 10^{-4}$ s$^{-1}$, $g^* = 0.02$, the Rossby radius for these values is $R = 39$ km

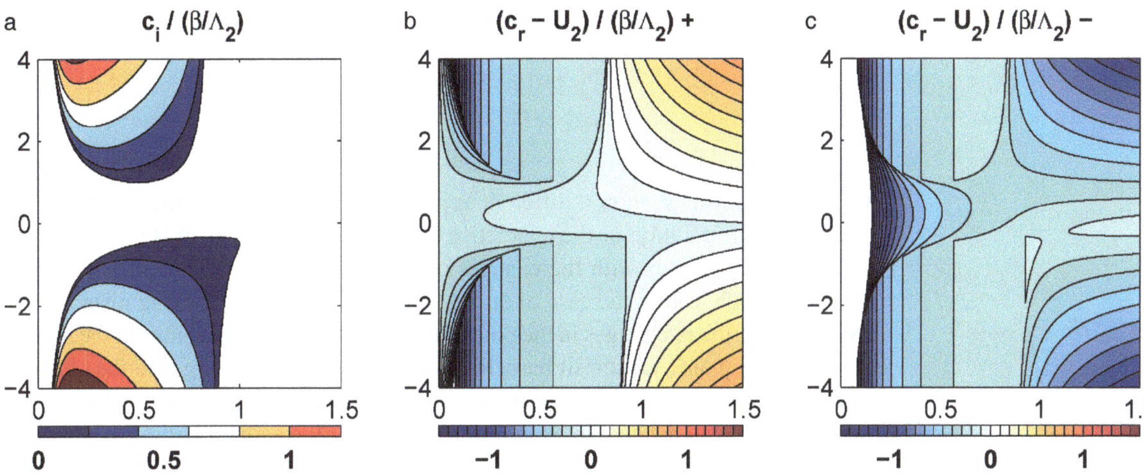


Fig. 8.26 The complex phase speed $c = c_r + ic_i$, scaled by $\beta/\Lambda_2$, as function of $(KR)^2$ and $U_s/(\beta/\Lambda_2)$. **a** shows the imaginary part $c_i$; **b** and **c** show the two solutions for the real part $c_r - U_2$ relative to the deep velocity $U_2$. The parameters are as in Figure 8.25. The values of the $-$ solution at very small $K$ are blended out

As $H_1 < H_2$ is usually taken in applications of the layer model, the marginally unstable perturbations have a larger wavelength for positive shear compared to the situation with negative shear.

The general dispersion relation (8.239) is shown in Figure 8.26. The figure displays the positive imaginary part of $c = c_r + \mathrm{i}c_i$ and the two real parts of $c_r - U_2$. In the region of unstable modes, the two parts coincide, but note that the solution $c_r - U_2$ is here strictly negative. Unstable modes thus propagate westward relative to $U_2$. This is not generally true for stable modes. The stable modes on the axis $U_s = 0$ represent the ordinary barotropic (+ solution) and baroclinic ($-$ solution) Rossby waves.

The phase relation between the mode $\psi_1 = A_1 e^{ikx} \sin n\pi y/B$ in the upper layer and $\psi_2 = A_2 e^{ikx} \sin n\pi y/B$ in the lower layer is similar to that observed in the Eady model. A proof follows from the governing equations (8.236) which becomes a relation between the amplitudes. Only one of the equations is needed,

$$(U_2 - c)[-K^2 A_2 - \Lambda_2(A_2 - A_1)] + A_2 \frac{\partial Q_2}{\partial y} = 0 \tag{8.249}$$

the other one is satisfied by the dispersion relation (8.239). Inserting $c = c_r + ic_i$ and the gradient of the potential vorticity in the lower layer, we find

$$A_1 = A_2 \left[ K^2/\Lambda_2 + 1 + \frac{U - \beta/\Lambda_2}{|U_2 - c|^2}(U_2 - c_r + \mathrm{i}c_i) \right] = A_2 \alpha \mathrm{e}^{\mathrm{i}\theta} \tag{8.250}$$

We consider an unstable mode, i. e. $c_i > 0$, for an eastward shear, i. e. $U_s > 0$ and hence $U_s > \beta/\Lambda_2$ by the instability constraint. The real part of the factor in brackets is positive because $U_2 - c_r > 0$. The imaginary part is positive as well. Hence the phase $\theta$ of $A_1$ relative to $A_2$ is in the first quadrant, i. e. $0 < \theta < \pi/2$. Thus, the wave in the upper layer leads the wave in the lower layer by less than half a wavelength, a pattern which looks in the continuous case like the figure in the box on p. 275. The unstable modes are thus tilted upstream.

## 8.5.4 Energetics of Parcel Exchanges

We discuss in this section a different approach to look at instabilities of the mean flow by a consideration of the energetics of parcel exchanges as in Haine and Marshall (1998). We start with the changes of potential energy by vertical parcels exchanges, which lead in the case of an unstable stratification to growing instabilities. By consideration of changes in the potential energy of parcel excursions in the meridional/vertical plane, we revisit baroclinic instability, and by taking also the changes in zonal kinetic energy into account we derive the condition for inertial or symmetric instability, which can play an important role near the equator and for (slantwise) convection in the weakly stratified mixed layer.

### Static Instability

Consider the potential energy change of a vertical exchange of a parcel with density $\rho_1$ initially at $z_1$ and a parcel with density $\rho_2$ initially at $z_2$ which is given by

$$\Delta P = g(\rho_1 z_2 + \rho_2 z_1) - g(\rho_1 z_1 + \rho_2 z_2) = -g(\rho_2 - \rho_1)(z_2 - z_1) \equiv -g\Delta\rho\Delta z \tag{8.251}$$

We assume that the density of the two parcels does not change during their displacement, i.e. we consider adiabatic rearrangements, similar to the discussion of available potential energy in Section 5.2.6. While they are moved vertically, the kinetic energy of the two parcels is given by $K = (1/2)\rho_1 w_1^2 + (1/2)\rho_2 w_2^2 \approx \rho_0 w^2$ where $\rho_0$ is the mean density (equivalent to the reference density in the Boussinesq approximation) of the two parcels with velocities $w_1$ and $w_2$ of opposite

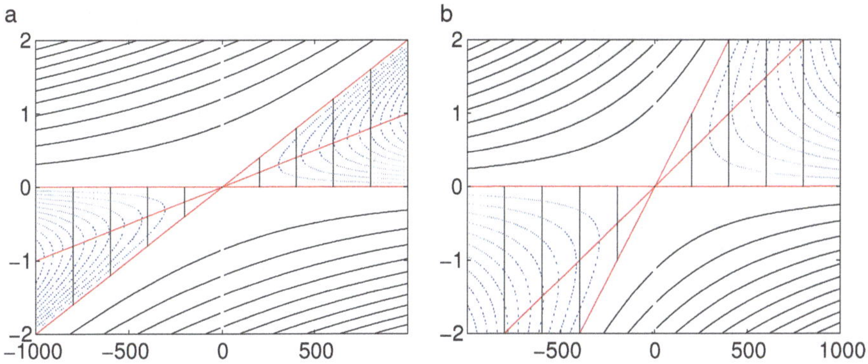


Fig. 8.27 The change in potential energy $\Delta P$ by adiabatic displacements of two particles as a function of $\Delta y$ (horizontal axis) and $\Delta z$ in m. Solid black lines denote positive values of $\Delta P$, while the blue dotted lines denote negative values where potential energy can be transferred to the kinetic energy of the particles, i.e. where an instability shows up. The red line with the largest slope denotes the isopycnal, with a slope of $s_b = 2\times10^{-3}$ in **a** and $s_b = 5\times10^{-3}$ in **b**. Also shown as red lines are the geopotential as the horizontal line and the surface with half the slope of the isopycnal, $s_b/2$, for which the minimum potential energy $\Delta P_{\min}$ is obtained. The vertical black lines connect values on the isopycnal and the line for slope $s_b/2$ with identical $\Delta y$, to allow an easy identification of the minima in $\Delta P$

sign ($w = w_1 = -w_2$). Adding both energies yields $(\mathrm{d}\Delta z/\mathrm{d}t)^2 + \Delta b \Delta z = 0$ with $w = \mathrm{d}\Delta z/\mathrm{d}t$ and the buoyancy difference $\Delta b = -g\Delta\rho/\rho_0$. Rewriting the equation as $-(\Delta b/\Delta z)\Delta z^2 = (\mathrm{d}\Delta z/\mathrm{d}t)^2$ yields for an unstable stratification $\Delta b/\Delta z = N^2 < 0$ a first order differential equation in $\Delta z$

$$\sqrt{|N^2|}\Delta z = \frac{\mathrm{d}}{\mathrm{d}t}\Delta z$$

The solution is an unstable oscillation $\Delta z \sim \exp \omega t$ with the growth rate $\omega = \sqrt{|N^2|}$. For $N^2 > 0$ we find $\mathrm{d}\Delta z/\mathrm{d}t = \mathrm{i}N\Delta z$ which yields a vertical oscillation with the Brunt–Väisälä frequency (compare Section 7.2).

### Baroclinic Instability

Now consider the change in potential energy when we also allow for a meridional exchange of two parcels, which now have their initial position at $(y_1, z_1)$ and $(y_2, z_2)$. Since only vertical excursions matter for the change, the potential energy is still identical to (8.251). The density or buoyancy difference has now vertical and lateral components given by $\Delta b = M^2\Delta y + N^2\Delta z$ with $N^2 = \partial b/\partial z$ and $M^2 = \partial b/\partial y$. Note that the lateral and vertical buoyancy gradients are assumed to be constant, and we now assume gravitationally stable conditions, i.e $N^2 > 0$. We consider the isopycnal slope $s_b = -M^2/N^2$ and the slope of the exchange direction given by $s_e = \Delta z/\Delta y$. Writing $\Delta b = -N^2 s_b \Delta y + N^2 s_e \Delta y$, for the potential energy change becomes

$$\Delta P = \rho_0 N^2 \Delta y^2 s_e (s_e - s_b) \tag{8.252}$$

Figure 8.27 shows $\Delta P$ as a function of $\Delta y$ and $\Delta z$ for two different values for the isopycnal slope $s_b$, assuming stable stratification. The sign of $\Delta P$ depends on the factor $s_e(s_e - s_b)$. Figure 8.27 shows that for most displacements, $\Delta P$ is positive, i. e. energy is needed for the displacements. For horizontal displacements $s_e = 0$

or for displacements in the isopycnal plane $s_e = s_b$, there will be no change in potential energy, i. e. those displacements are possible without doing work against the gravity force. For slopes $s_e$ between $s_b$ and zero, $\Delta P$ is negative and energy is released by those displacements and can be transferred to the kinetic energy of the parcels, which leads to a (baroclinic) instability. Since the change in potential energy is always positive (or zero) for $s_b = 0$, we need inclined isopycnals to release (available) potential energy to kinetic energy. $\Delta P$ is at minimum for $s_e = s_b/2$ and is given by $\Delta P_{\min} = -\rho_0 N^2 \Delta y^2 s_b^2/4$.

The solution for the lateral displacement is given by $\Delta y \sim \exp \omega t$ and the kinetic energy, related to the displacement, is $K = \rho_0 v^2$ (we count the meridional velocities of both parcels) where $v = \mathrm{d}(\Delta y)/\mathrm{d}t$. Adding again $\Delta P_{\min}$ and the kinetic energy $K$ yields

$$\omega^2 = \frac{1}{4} N^2 s_b^2 = \left( \frac{1}{2} \frac{M^2}{N} \right)^2 \tag{8.253}$$

which is, except for a numerical factor of order one, the growth rate of the fastest growing wave in the Eady problem discussed in 8.5.3. The fastest growing wave thus extracts the maximum available potential energy from the system. The parcel excursions for the fastest growing wave are along half of the isopycnal slope of the background state. However, note that this holds for the onset of the baroclinic instability only. In the balanced state, mean parcel excursions will be along isopycnals, in agreement with our expectation (compare also Sections 12.2.2 and 12.3.5).

### Inertial Instability

If the isopycnal slopes are inclined in the meridional direction, geostrophic balance applies to the zonal velocity, and the parcels, therefore, also have zonal velocities. If this zonal velocity becomes large, we also have to take it into account for the energetics of parcel exchanges. The parcels must obey the zonal momentum budget, which we write for $f = \text{const}$ in the form

$$\frac{\mathrm{d}}{\mathrm{d}t}(u - fy) = 0 \tag{8.254}$$

where we have assumed that $\partial p/\partial x = 0$. Note that the quantity $u - fy$ is the (absolute) zonal momentum in the fixed (nonrotating) frame of reference, and (8.254) becomes the angular momentum conservation equation. As a consequence of (8.254), $u_2 - u_1 = \Delta u = f \Delta y$ must hold for the meridional displacements. Two parcels, which have their initial position again at $(y_1, z_1)$ and $(y_2, z_2)$, have initially the kinetic energy $(1/2)\rho_0 u_1^2 + (1/2)\rho_0 u_2^2$ and after the exchange the zonal velocities change according (8.254) to $u_1 \to u_1 + f\Delta y$ and $u_2 \to u_2 - f\Delta y$ with corresponding kinetic energies. The change in kinetic energy is given by

$$\Delta K = \frac{1}{2}\rho_0 \left( (u_1 + f\Delta y)^2 + (u_2 - f\Delta y)^2 - u_1^2 - u_2^2 \right) = \rho_0 \Delta y^2 f (f - \Delta u/\Delta y) \tag{8.255}$$

With $\Delta u = (\partial u/\partial y)\Delta y + (\partial u/\partial z)\Delta z$ and using the geostrophic balance $f(\partial u/\partial z) = -M^2 = N^2 s_b$, the total change in mechanical energy $\Delta E = \Delta P + \Delta K$ becomes

$$\Delta E = \rho_0 \Delta y^2 \left( N^2 s_e (s_e - 2 s_b) + f \omega_a \right)$$

with the vertical component of the absolute vorticity $\omega_a = f - \partial u/\partial y$ as defined in Section 2.11 (for $\partial/\partial x = 0$). If $\Delta E < 0$, this change in the mean mechanical energy can be used to accelerate the parcels in the meridional direction. For given $f\omega_a$, the function $\Delta E$ has a minimum $\Delta E_{\min}$ at $s_e = s_b$. Adding again the meridional kinetic energy $\rho_0 v^2$ (both parcels are counted) and $\Delta E_{\min}$, we find a growth rate of the meridional displacement (for negative $\Delta E_{\min}$) of

$$\omega^2 = \left(\frac{M^2}{N}\right)^2 - f\omega_a = \frac{f}{N^2}\left(-\omega_a N^2 - \frac{\partial u}{\partial z}\frac{\partial b}{\partial y}\right) = -\frac{fQ}{N^2} \tag{8.256}$$

Here $Q$ is the Ertel potential vorticity in the Boussinesq and hydrostatic approximations, as given in (4.58), which in Cartesian coordinates takes the form

$$Q = \omega_a N^2 + \frac{\partial u}{\partial z}\frac{\partial b}{\partial y} \tag{8.257}$$

for $\partial/\partial x = 0$ and $\chi = b$, ignoring the constant factor $\rho_0$. For $fQ < 0$, we find $\Delta E_{\min} < 0$ and, therefore, exponential growth with a rate equal to $\omega$. The process is called *inertial* or *symmetric instability*. Note that the term $\omega_a N^2$ often dominates the Ertel potential vorticity. In that situation or for vanishing vertical shear, the condition for inertial instability becomes $f(f - \partial u/\partial y) < 0$. However, the condition for inertial instability is hardly satisfied under normal conditions; inertial instability plays a role in the ocean only close to the equator where $|f|$ is small enough such that $|f| \sim |\partial u/\partial y|$, or for weak stratification and strong vertical shear in $u$, as in the mixed layer in boundary currents. We will consider the latter issue in the following section. Note that the condition for inertial instability $fQ < 0$ holds for the Southern Hemisphere as well.

### Slantwise Convection

From (8.257), the potential vorticity can be written as $Q = -\partial\,(u - fy)/\partial y \partial b/\partial z + \partial(u - fy)/\partial z \partial b/\partial y = -\mathcal{J}(u - fy, \rho)$ where the Jacobian $\mathcal{J}$ is applied in the $y$-$z$-plane. The angle between surfaces of constant absolute momentum $u - fy$ and the isopycnals determines the sign of the potential vorticity $Q$. If $Q < 0$, there will be inertial instability (for $f > 0$) which will force the system towards a state with at least $Q = 0$, where isopycnals and isosurfaces of $u - fy$ become parallel. When the mean flow is constant we have $fQ = f^2 \partial b/\partial z$, which is always positive for $\partial b/\partial z > 0$ and negative for static instability. In this case, static and inertial instability become identical. With a vertical shear in $u$, however, the isosurfaces of $u - fy$ are tilted. If $fQ < 0$, convection and inertial instability will lead to a state where isosurfaces of $u - fy$ and isopycnals become parallel, which is called *slantwise convection*. Convection then does not occur in vertical direction but along isosurfaces of $u - fy$. This case applies sometimes to the atmosphere where there is deep convection at the equator and where large shears in the zonal flow and small $f$ lead to slantwise convection. In the ocean, there is no deep convection at the equator, but the process can be noticed sometimes in the mixed layer in case of large lateral density gradients and related geostrophic flow.

Figure 8.28 shows a numerical simulation of slantwise convection in the oceanic mixed layer in three configurations with different vertically sheared background

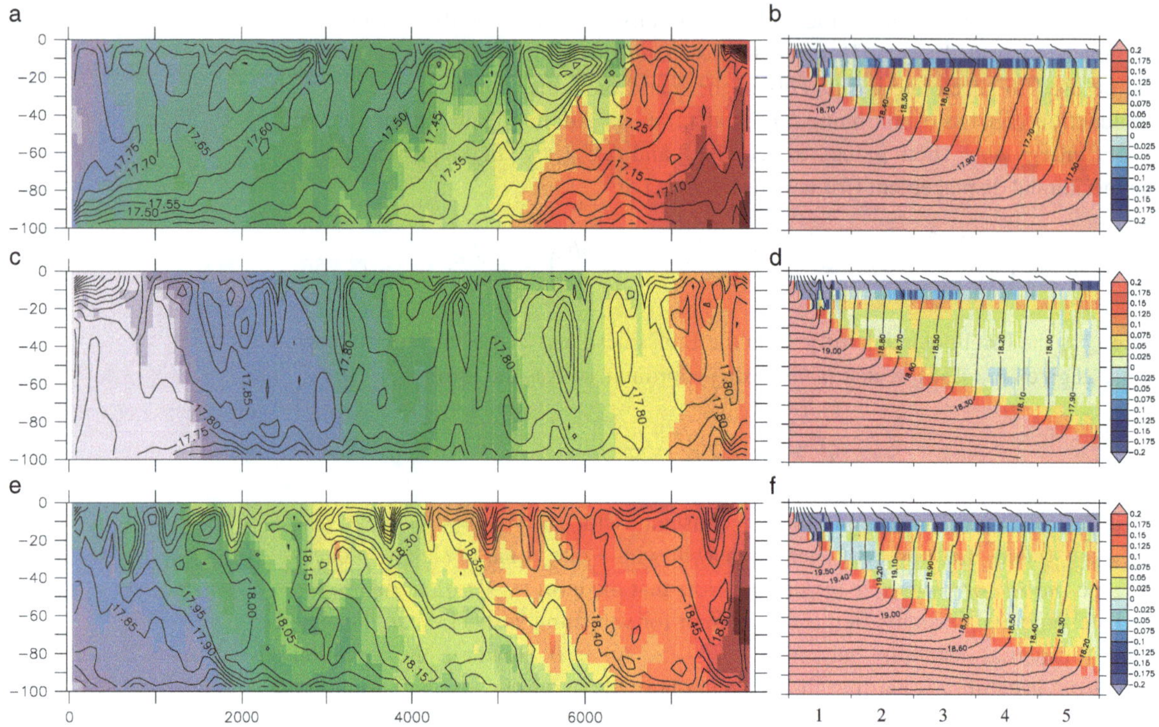

Fig. 8.28 Slantwise convection in the mixed layer: **a** Snapshot of absolute momentum $u - fy$ (color, ranging from 0 to $1\,\mathrm{m\,s^{-1}}$) and temperature (lines in °C) in a numerical nonhydrostatic simulation of convection in the presence of large vertical shears in the geostrophically balanced background flow $u$. The vertical axis is depth in m and the horizontal axis latitude in m. The model is two-dimensional in the $y$-$z$ plane, with a constant Coriolis frequency $f = 10^{-4}\,\mathrm{s^{-1}}$, initialized with a meridional and vertical buoyancy gradient ($N_0 = 7\times10^{-3}\,\mathrm{s^{-1}}$) and is forced with a surface heat loss of $1{,}000\,\mathrm{W\,m^{-2}}$. Salinity is constant and temperature proportional to buoyancy. The snapshot shows the simulation after 5 d. **b** Time series of zonally averaged temperature (lines in °C) and zonally averaged $N^2(z,t)/N_0^2$ (color) as function of depth and time in d. **c** and **d** Same as a) and b) but without background flow $u = 0$. **e** and **f** Same as **a** and **b** but for reversed geostrophic background flow $u$

flow $u$. A clear tendency of parallel isosurfaces of $u - fy$ and isopycnals can be noticed. In the case without vertically sheared mean flow (Figure 8.28c), the isopycnals are predominantly vertical, but in the cases with vertically sheared $u$, the isopycnals indeed clearly follow the isosurfaces of $u - fy$ (Figure 8.28a and e). Note that while there is no stratification in the mixed layer in the case of upright convection (Figure 8.28d), slantwise convection leads to a nonvanishing stratification in the mixed layer (Figure 8.28b and f).

# * Lagrangian Theory of Ocean Waves

# 9

In Section 2.12, we have introduced the Lagrangian formalism for fluid mechanical equations and discussed some examples of Lagrangians, in particular for incompressible flow, for isentropic compressible flow and for a rotating stratified fluid system. Some of the wave equations, derived and analyzed in the previous chapters, may be transferred into this Lagrangian framework, including Rossby waves and gravity waves. The present chapter is devoted to the description of slowly varying wave trains and resonant wave-wave interaction, both based on a Lagrangian formalism.

It might be questioned what use there is in a Lagrangian treatment if only the governing equations of the respective system are reproduced by the variational algorithm. As one advantage we mentioned in Section 2.12 the immediate access to general conservation theorems of energy and momentum in a condensed form, as introduced by (2.218). A related issue is the conservation of 'almost conserved' quantities, so-called *adiabatic invariants* which are energy-type variables obeying conservation theorems in a slowly varying state of the wave-carrying background. The general and beautiful treatment of this topic is due to WHITHAM[1]; further references on this subject can be found in his monograph (Whitham, 1974). Especially the use of Lagrangian theory leads to very powerful recipes for studying wave propagation in a WKBJ-type environment.

## 9.1 Sound Waves as Example

We start with a resumption of the WKBJ analysis of the wave systems in the previous chapters, in particular with the specific example of sound waves in (6.61), as this is a scalar problem and a Lagrangian exists. The wave equation for sound is given in Section 6.3),

$$\frac{\partial}{\partial t}\left[\frac{1}{c_{\mathrm{s}}^2(\boldsymbol{x},t)}\frac{\partial p}{\partial t}\right] - \nabla^2 p = 0 \tag{9.1}$$

[1] GERALD BERESFORD WHITHAM, *1927 in Halifax, West Yorkshire, applied mathematician.


DOI 10.1007/978-3-642-23450-7_9, 

and written here with a variable sound velocity $c_s(\boldsymbol{x}, t)$. It may indeed be derived by variation of the Lagrangian

$$L\left(p, \frac{\partial p}{\partial t}, \nabla p, \boldsymbol{x}, t\right) = \frac{1}{2}\left[\frac{1}{c_s^2(\boldsymbol{x}, t)}\left(\frac{\partial p}{\partial t}\right)^2 - (\nabla p)^2\right] \tag{9.2}$$

with respect to the pressure field $p(\boldsymbol{x}, t)$. Evidently, the two terms comprising $L$ are not kinetic and elastic energy, found in Section 6.2.2 to be the appropriate compartments of the energy of sound waves. Strictly speaking, the terms in $L$ are not energies at all, and will be referred to as *pseudoenergies*. In fact, the form of the Lagrangian (9.2) is found by an educated guess, which is a procedure that is often applied in Lagrangian theory of fluid mechanics, as shown in the examples of this section and in Section 2.12. The 'energy' $T_{00}$ and its flux $T_{0i}$, promoted by the Lagrangian framework according to (2.219) and (2.220), are found from (9.2) as

$$T_{00} = \frac{1}{2}\left[\frac{(\partial p/\partial t)^2}{c_s^2} + (\nabla p)^2\right] \quad \text{and} \quad T_{0i} = \frac{\partial p}{\partial t}\nabla p \tag{9.3}$$

We insert the plane-wave solution $p = a\sin(\boldsymbol{k}\cdot\boldsymbol{x} - \omega t)$ and average over a wave period or wave length (or over the phase; indicated by a bar), as done for the WKBJ approach in Section 6.3.3, and find $\bar{T}_{00} = (\omega a/c_s)^2/2$ and $\bar{T}_{0i} = \omega \boldsymbol{k} a^2/2$, or

$$\bar{T}_{00} = \omega^2 E_w \quad \text{and} \quad \bar{T}_{0i} = \boldsymbol{c}_g \omega^2 E_w \tag{9.4}$$

related to the physical energy $E_w$ of sound waves, defined by (6.31) (apart from the $\rho_0$-factor). This true energy satisfies (6.60) whereas the Lagrangian form of 'energy', in its phase-averaged state $\bar{T}_{00}$, is governed by the phase averaged form of (2.218), found to be

$$\frac{\partial}{\partial t}\omega^2 E_w + \nabla\cdot\boldsymbol{c}_g\omega^2 E_w = -\frac{\partial \bar{L}}{\partial t} = \omega^2 E_w \frac{\partial c_s/\partial t}{c_s} = \omega E_w \frac{\partial \Omega}{\partial t} \tag{9.5}$$

The last but one relation is obtained by inserting the plane-wave solution into the Lagrangian (9.2), performing the phase average, and taking the time derivative with respect to the explicit time dependence of the sound velocity. The last relation follows from the dispersion relation $\omega = \Omega(\boldsymbol{k}, \boldsymbol{x}, t) = c_s(\boldsymbol{x}, t)k$ of sound waves. As expected, we are not gaining a completely new theorem of energy conservation: it is easily verified that (9.5) is equivalent to (6.60). Note, however, that the Lagrangian treatment avoids the cumbersome expansion work of the WKBJ approach of Section 6.3 (and the following ones). As explained in the following sections, the use of Whitham's theory makes the work even simpler than presented above. Remember for the following that the action $\mathcal{A} = \omega E_w$, found in Section 6.3.3 to govern the specific sound wave system in terms of the simple conservation equation (6.61), relates to Lagrangian 'energy' by $\mathcal{A} = \bar{T}_{00}/\omega$. This latter relation and the source/sinkless conservation (6.61) of wave action will be found to be of general validity.

## 9.2 Adiabatic Invariants

We briefly introduce the concept of an adiabatic invariant a for discrete system Landau and Lifshitz (see Section 2.12, and e.g. 1982). Given is a Hamilton function

$H[p, q, \lambda(t)]$ of a dynamical system[2] which depends on a slowly varying parameter $\lambda(t)$ and has a periodic solution $p_0(t), q_0(t)$ for constant $\lambda = \lambda_0$. Slowly varying means that $\lambda$ does not change much over the period $T$ of $p_0(t)$ and $q_0(t)$, i. e. $T\mathrm{d}\lambda/\mathrm{d}t \ll 1$. According to (B18.2) the energy of the system with varying $\lambda$ is not constant,

$$\frac{\mathrm{d}E}{\mathrm{d}t} = \frac{\mathrm{d}H}{\mathrm{d}t} = \frac{\partial H}{\partial \lambda}\frac{\mathrm{d}\lambda}{\mathrm{d}t} \tag{9.6}$$

Averaging over a period yields

$$\overline{\frac{\mathrm{d}E}{\mathrm{d}t}} = \frac{1}{T}\int_0^T \frac{\partial H}{\partial \lambda}\frac{\mathrm{d}\lambda}{\mathrm{d}t}\mathrm{d}t \approx \overline{\frac{\mathrm{d}\lambda}{\mathrm{d}t}}\frac{1}{T}\int_0^T \frac{\partial H}{\partial \lambda}\mathrm{d}t = \overline{\frac{\mathrm{d}\lambda}{\mathrm{d}t}}\frac{1}{T}\oint \frac{\partial H/\partial \lambda}{\partial H/\partial p}\mathrm{d}q \tag{9.7}$$

where the approximated form bears on the slow variation of $\lambda$. The crucial assumption is a further approximating step in which the remaining integral is not evaluated for the actual path $[p(t), q(t)]$ in the phase space $(p, q)$ but rather for the 'unperturbed' periodic solution $[p_0(t), q_0(t)]$. In the integration, we take the Hamiltonian $H = H_0(p_0, q_0, \lambda_0)$ evaluated from the periodic solution, but let $\lambda_0$ take its varying trajectory $\lambda(t)$. The above relation becomes

$$\oint \frac{\partial H_0[p_0, q, \lambda(t)]/\partial \lambda_0}{\partial H_0[p_0, q, \lambda(t)]/\partial p_0}\mathrm{d}q = -\oint \frac{\partial p_0}{\partial \lambda_0}\mathrm{d}q \tag{9.8}$$

because $H_0 = E_0 = \text{const}$, leading to $(\partial H_0/\partial p_0)\partial p_0/\partial \lambda_0 + \partial H_0/\partial \lambda_0 = 0$. Combing now (9.7) and (9.8) into

$$T\overline{\frac{\mathrm{d}E}{\mathrm{d}t}} + \overline{\frac{\mathrm{d}\lambda}{\mathrm{d}t}}\oint \frac{\partial p_0}{\partial \lambda_0}\mathrm{d}q = 0 \tag{9.9}$$

and using

$$T = \int_0^T \mathrm{d}t = \oint \frac{\mathrm{d}q}{\dot{q}_0} = \oint \frac{\partial p_0}{\partial H_0}\mathrm{d}q \tag{9.10}$$

we finally arrive at

$$\overline{\frac{\mathrm{d}E}{\mathrm{d}t}}\oint \frac{\partial p_0}{\partial E_0}\mathrm{d}q + \oint \overline{\frac{\mathrm{d}\lambda}{\mathrm{d}t}}\frac{\partial p_0}{\partial \lambda_0}\mathrm{d}q = \frac{\mathrm{d}}{\mathrm{d}t}\oint p_0[q, E(t), \lambda(t)]\mathrm{d}q = 0 \tag{9.11}$$

proving that $A = \oint p_0\mathrm{d}q$ is an adiabatic invariant, called wave 'action'. For linear systems, one finds $A(t) = E(t)/\omega(t)$, as exemplified by the harmonic oscillator with slowly changing frequency, $H = p^2/2m + m\omega^2(t)q^2/2 = E(t)$. Using $H = p\dot{q} - L(\dot{q}, q, \lambda) = E$ and expressing $q_0(t)$ by a periodic function $Q(\theta)$ of the phase $\theta$, i. e. $q_0(t) = Q(\theta)$ with $\theta = \omega t + \vartheta$, we arrive at

$$A = \oint p_0\mathrm{d}q = \frac{1}{2\pi}\int_0^{2\pi} \{L[\omega Q_\theta, Q, \lambda(t)] + E\}\mathrm{d}\theta \tag{9.12}$$

[2] In this section, $q$ and $p$ denote the generalized coordinate and the conjugate momentum, see the box on p. 98.

Hence $A$ relates to the averaged Lagrangian,

$$\mathcal{L}(\omega, E, \lambda) = \frac{1}{2\pi}\int_0^{2\pi} L[\omega Q_\theta, Q, \lambda(t)]\mathrm{d}\theta = \frac{A}{T} - E = \frac{\omega}{2\pi}A - E \tag{9.13}$$

The trick is now to remember that $\omega$ is the time derivative of $\theta$ and to allow variations of $\mathcal{L}$ with respect to $\theta$ and $E$, resulting in the Euler–Lagrange equations (see Section 2.12)

$$\frac{\mathrm{d}}{\mathrm{d}t}\frac{\partial \mathcal{L}}{\partial(\partial\theta/\partial t)} - \frac{\partial \mathcal{L}}{\partial \theta} = \frac{\mathrm{d}}{\mathrm{d}t}\frac{\partial \mathcal{L}}{\partial \omega} = 0 \quad \text{and} \quad \frac{\partial \mathcal{L}}{\partial E} = 0 \tag{9.14}$$

The first relation corresponds to the conservation of action, (9.11), and we have $\mathcal{L}_\omega = A/2\pi$. The second relation simply becomes

$$\frac{\omega}{2\pi}\oint \frac{\partial p}{\partial E}\mathrm{d}q - 1 = 0 \tag{9.15}$$

or, using (9.10), one finds $\omega = 2\pi/T$ which corresponds to a dispersion relation.

## 9.3 Variational Approach to Wave Trains

To begin with Whitham's theory, we describe the recipe in the light of the sound wave equation (9.1) with the associated Lagrangian (9.2). The solution of the wave equation for constant background conditions $c_\mathrm{s}$ = const is a sinusoidal pressure field

$$p(\boldsymbol{x}, t) = a \sin\theta \quad \text{with the phase} \quad \theta = \boldsymbol{k}\cdot\boldsymbol{x} - \omega t \tag{9.16}$$

and a constant amplitude $a$ and a dispersion relation $\omega = \pm c_\mathrm{s} k$; relating the prescribed constant wave vector $\boldsymbol{k}$ to the frequency $\omega$. If the sound speed is slowly changing on scales that are large compared to the wavelength $2\pi/k$ and period $2\pi/\omega$, a solution is sought in the above sinusoidal form but with slowly varying $a, \boldsymbol{k}$ and $\omega$. Following Whitham's recipe of the averaged Lagrangian, the periodic form (9.16) is inserted into the Lagrangian, which is then averaged over the fast-changing phase, resulting in the 'averaged Lagrangian'

$$\mathcal{L} = \frac{1}{4}|a|^2\left(\frac{\omega^2}{c_\mathrm{s}^2} - k^2\right) \tag{9.17}$$

The adiabatic invariant and its governing equation follow by variation of $\mathcal{L}$ with respect to $\theta$, as done for the discrete system in the previous section. Since $\mathcal{L}$ does not depend anymore on the phase $\theta$ itself but only on its derivatives $\theta_\mathrm{t} = -\omega$ and $\nabla\theta = \boldsymbol{k}$, we obtain

$$\frac{\partial}{\partial t}\mathcal{L}_\omega - \nabla\cdot\mathcal{L}_{\boldsymbol{k}} = 0 \tag{9.18}$$

with the abbreviations $\mathcal{L}_\omega = \partial\mathcal{L}/\partial\omega$ and $\mathcal{L}_{\boldsymbol{k}} = (\partial\mathcal{L}/\partial k_1, \partial\mathcal{L}/\partial k_2, \partial\mathcal{L}/\partial k_3) = \nabla_{\boldsymbol{k}}\mathcal{L}$. The temporal and spatial derivatives in (9.18) act on the slow variations of $c_\mathrm{s}(\boldsymbol{x}, t)$. We find

$$\mathcal{A} = \mathcal{L}_\omega = \frac{1}{2}|a|^2\frac{\omega}{c_\mathrm{s}^2} \quad \text{and} \quad \boldsymbol{J}_{\mathcal{A}} = -\mathcal{L}_{\boldsymbol{k}} = \frac{1}{2}|a|^2\boldsymbol{k} \tag{9.19}$$

for the 'wave action density' $\mathcal{A}$ and 'action flux density' $\boldsymbol{J}_{\mathcal{A}}$ of sound waves. Varying $\mathcal{L}$ with respect to the amplitude, we obtain the dispersion relation, i. e. $\mathcal{L}_{\mathrm{a}} = 0$ yields $\omega^2 = c_{\mathrm{s}}^2 k^2$.

Hence for this linear system $\mathcal{L}$ has the value zero, implying two features. We define, as usual, the 'energy' of the system by $E = \theta_{\mathrm{t}} \partial L / \partial \theta_{\mathrm{t}} - L$. Averaging yields the wave energy $\mathcal{E}$ and its important relation to the wave action $\mathcal{A}$,

$$\mathcal{E} = \omega \mathcal{L}_{\omega} - \mathcal{L} = \omega \mathcal{L}_{\omega} = \omega \mathcal{A} = \frac{1}{2} |a|^2 \frac{\omega^2}{c_{\mathrm{s}}^2} \tag{9.20}$$

Furthermore, as $\mathrm{d}\mathcal{L} = \mathcal{L}_{\omega} \mathrm{d}\omega + \mathcal{L}_{\boldsymbol{k}} \mathrm{d}\boldsymbol{k} = 0$ because of $\mathcal{L} = 0$, we obtain the expression

$$\boldsymbol{c}_{\mathrm{g}} = \frac{\mathrm{d}\omega}{\mathrm{d}\boldsymbol{k}} = -\frac{\mathcal{L}_{\boldsymbol{k}}}{\mathcal{L}_{\omega}} = \frac{\boldsymbol{J}_{\mathcal{A}}}{\mathcal{A}} = \frac{\omega}{c_{\mathrm{s}}^2} \boldsymbol{k} \tag{9.21}$$

for the group velocity. This relation enables us to rewrite the action conservation (9.18) in the 'standard' form, already derived in (6.61),

$$\frac{\partial \mathcal{A}}{\partial t} + \nabla \cdot \boldsymbol{c}_{\mathrm{g}} \mathcal{A} = 0 \tag{9.22}$$

which governs the evolution of the squared wave amplitude on the large scale variations of $c_{\mathrm{s}}$. Indeed, we have regained the results from the elementary analysis in the preceding sections. Compared to the lengthy derivation of (6.61) in the WKBJ framework of Section 6.3.3, the Lagrangian approach is rather elegant and algebraically simpler. The difference is even more drastic when a cumbersome multiple field case as in the internal gravity wave problem or the Rossby wave problem (see below) is considered. Note that except for the special expressions of the sound wave problem, the above equations are valid for *any* linear wave problem.

## 9.4 A Rigorous Derivation

The above described recipe is elegant and works well for the particular example of sound waves. Nevertheless, why are we allowed to consider variations of the averaged Lagrangian with respect to phase and amplitude? Remember that the original Lagrangian is a functional of the field (pressure $p$ in the above example), and the Lagrangian framework allows variation with respect to this variable only. A proof of the method is given here by use of the method of multiple time scales (the name of the method is used not only for the temporal but also for the spatial variables). For the sake of simple notation, we consider only one spatial dimension and a problem with a single dynamical field $\phi$.

While changes of $x$ and $t$ are first felt on fast scales of the wavelength and period, there are much slower variations on the scales of the background parameters (the sound speed $c_{\mathrm{s}}$ in the above example). Formally, we may introduce two time-scales $t$ and $T = \epsilon t$ where $\epsilon$ is a small parameter (do not confuse $T$ with the previous notation used for the period). Likewise, for the spatial domain we have $x$ and $X = \epsilon x$. When the fast variable $t$ runs through a period (an $O(1)$ variation) of the wave, the slow variable $T$ has not changed much (there is only a $O(\epsilon)$ variation). The background hence depends on $X, T$, and the wave amplitude frequency and wave number as well. An appropriate ansatz for the field is then

$$\phi = \phi(x, t, X, T) = \Phi(\theta, X, T; \epsilon) \quad \text{with} \quad \theta = \Theta(X, T)/\epsilon \tag{9.23}$$

where $\Phi$ is assumed periodic in $\theta$ (we assume a period $2\pi$ without restriction). Frequency and wave number are given as derivatives of the phase function,

$$\begin{aligned} \omega(X, T) &= -\frac{\partial \Theta(X, T)/\epsilon}{\partial t} = -\Theta_{\mathrm{T}} \\ k(X, T) &= \frac{\partial \Theta(X, T)/\epsilon}{\partial x} = \Theta_{\mathrm{X}} \end{aligned} \tag{9.24}$$

so that $\theta \approx k(X, T)x - \omega(X, T)t$.

The Lagrangian of a slowly varying wave field is of the form $L = L(\phi_{\mathrm{t}}, \phi_x, \phi, X, T)$ where field gradients have the two-timing form

$$\begin{aligned} \frac{\partial \phi}{\partial t} &= -\omega \Phi_\theta + \epsilon \Phi_{\mathrm{T}} \\ \frac{\partial \phi}{\partial x} &= k \Phi_\theta + \epsilon \Phi_{\mathrm{X}} \end{aligned} \tag{9.25}$$

reflecting the changes in both contributions due to the fast oscillations and the slow modulations. Likewise, the Euler–Lagrange equations become

$$\left(-\omega \frac{\partial}{\partial \theta} + \epsilon \frac{\partial}{\partial T}\right) \frac{\partial L}{\partial \alpha_1} + \left(k \frac{\partial}{\partial \theta} + \epsilon \frac{\partial}{\partial X}\right) \frac{\partial L}{\partial \alpha_2} - \frac{\partial L}{\partial \alpha_3} = 0 \tag{9.26}$$

where

$$L = L(-\omega \Phi_\theta + \epsilon \Phi_{\mathrm{T}}, k \Phi_\theta + \epsilon \Phi_{\mathrm{X}}, \Phi, X, T) \tag{9.27}$$

and the $\alpha_n, n = 1, 2, 3$ are the first three arguments of this function. Note that (9.26) is the equation determining the function $\Phi(\theta, X, T)$ with its three variables $\theta$, $X$ and $T$, now treated as independent. We cast (9.26) into the equivalent conservative form

$$\frac{\partial}{\partial \theta}\left[\left(-\omega \frac{\partial L}{\partial \alpha_1} + k \frac{\partial L}{\partial \alpha_2}\right) \Phi_\theta - L\right] + \epsilon \frac{\partial}{\partial T}\left(\Phi_\theta \frac{\partial L}{\partial \alpha_1}\right) + \epsilon \frac{\partial}{\partial X}\left(\Phi_\theta \frac{\partial L}{\partial \alpha_2}\right) = 0 \tag{9.28}$$

This is obtained by multiplying (9.26) with $\Phi_\theta$ and adding the identity

$$\frac{\partial L}{\partial \theta} = \frac{\partial L}{\partial \alpha_1}(-\omega \Phi_{\theta\theta} + \epsilon \Phi_{T\theta}) + \frac{\partial L}{\partial \alpha_2}(k \Phi_{\theta\theta} + \epsilon \Phi_{X\theta}) + \frac{\partial L}{\partial \alpha_3} \Phi_\theta \tag{9.29}$$

Now, we average (9.28) over the phase $\theta$. Making use of the above assumed periodicity in the $\theta$-variable, the first part vanishes, and we obtain

$$\frac{\partial}{\partial T} \frac{1}{2\pi} \int_0^{2\pi} \Phi_\theta \frac{\partial L}{\partial \alpha_1} \mathrm{d}\theta + \frac{\partial}{\partial X} \frac{1}{2\pi} \int_0^{2\pi} \Phi_\theta \frac{\partial L}{\partial \alpha_2} \mathrm{d}\theta = 0 \tag{9.30}$$

The most important and surprising result of this step is the fact that (9.28) and (9.30) are just the Euler–Lagrange equations of the variational principle

$$\delta \int \frac{1}{2\pi} \int_0^{2\pi} L(-\omega\Phi_\theta + \epsilon\Phi_\mathrm{T}, k\Phi_\theta + \epsilon\Phi_\mathrm{X}, \Phi, X, T)\mathrm{d}\theta \mathrm{d}X \mathrm{d}T = 0 \tag{9.31}$$

if variations of the Lagrangian with respect to the carrier function $\Phi$ *and* the phase $\Theta$ are considered. The first yields (9.28), using the particular form (9.27) of $L$, and the second yields (9.30), using (9.24). The latter result may be expressed as

$$\frac{\partial \mathcal{L}_\omega}{\partial T} - \frac{\partial \mathcal{L}_k}{\partial X} = 0 \tag{9.32}$$

with averaged Lagrangian

$$\mathcal{L}(\omega, k, X, T) = \frac{1}{2\pi} \int_0^{2\pi} L(-\omega\Phi_\theta + \epsilon\Phi_\mathrm{T}, k\Phi_\theta + \epsilon\Phi_\mathrm{X}, \Phi, X, T)\mathrm{d}\theta \tag{9.33}$$

Equation (9.32) is the conservation of action density $\mathcal{L}_\omega$, which is a slowly varying quantity. Note that we have not assumed linearity of the system (or any particular wave system, as in the previous section). In addition, the result is correct to all orders in $\epsilon$.

We summarize the results that we got so far. After the transformation (9.23) of the field variable, the Euler–Lagrange equation (9.26) determines $\Phi$ as function of $\theta, T$ and $X$. The conservative form (9.28) is a statement about the conservation of wave energy: the lowest order of (9.28) in $\epsilon$ indeed yields a first integral for the fast dependence of carrier function $\Phi^{(0)}$,

$$\left(-\omega \frac{\partial L^{(0)}}{\partial \alpha_1} + k \frac{\partial L^{(0)}}{\partial \alpha_2}\right) \Phi_\theta^{(0)} - L^{(0)} = \Phi_\theta^{(0)} \frac{\partial L^{(0)}}{\partial \Phi_\theta^{(0)}} - L^{(0)} = \mathcal{E}^{(0)}(X, T) \tag{9.34}$$

with $L^{(0)} = L(-\omega\Phi_\theta^{(0)}, k\Phi_\theta^{(0)}, \Phi^{(0)}, X, T)$. Averaging over the phase reveals that $\mathcal{E}^{(0)}(X, T)$ is the average pseudo wave energy density to lowest order,

$$\mathcal{E}^{(0)}(X, T) = \omega\mathcal{L}_\omega^{(0)} - \mathcal{L}^{(0)} = \omega\mathcal{A}^{(0)} - \mathcal{L}^{(0)} \tag{9.35}$$

For the general case of a nonlinear system, we find

$$\frac{\partial}{\partial T}\left(\omega\mathcal{L}_\omega^{(0)} - \mathcal{L}^{(0)}\right) - \frac{\partial}{\partial X}\omega\mathcal{L}_k^{(0)} = -\frac{\partial \mathcal{L}^{(0)}}{\partial T} \tag{9.36}$$

using (9.32), (9.35), and $k_\mathrm{T} + \omega_\mathrm{X} = 0$. The time derivative on the left-hand side is taken with respect to the explicit slow time dependence of the Lagrangian. Note that only for linear systems it can be shown that the Lagrangian $\mathcal{L}^{(0)}$ has the value zero, and hence $\mathcal{A}^{(0)} = \mathcal{E}^{(0)}/\omega$. Then, $-\mathcal{L}_k^{(0)}/\mathcal{L}_\omega^{(0)}$ can be identified with the group velocity.

The conservation (9.36) of pseudo energy $\mathcal{E}^{(0)} = \omega\mathcal{A}$ can be augmented by a corresponding conservation theorem for a pseudo momentum $\mathcal{P}^{(0)} = k\mathcal{A}$. The conservation theorems (2.218), discussed in Section 2.12, carry over in the averaged

**41. Waves in a Mean Current**

Whitham's theory is easily extended to the case that a slowly varying mean flow is imposed upon the wave motion. The frequency of encounter (see Section 6.2.3) then contains the Doppler shift, $\omega = \Omega(k, X, T) = \Omega'(k, X, T) + k(X, T)U(X, T)$ where $\omega' = \Omega'(k, X, T)$ is the dispersion relation that an observer finds who is moving with the mean current $U(X, T)$ (notice that $k' = k$). The ray equations summarized in the box on p. 173 remain valid (the group velocity contains the intrinsic part and the mean current, $c_g = c'_g + U$).

The observer in the frame moving with $U$ finds the averaged Lagrangian $\mathcal{L}'^{(0)} = \mathcal{L}'^{(0)}(\omega', k', X', T)$ with pseudo energy $\mathcal{E}'^{(0)} = \omega' \mathcal{L}'^{(0)}_{\omega'} = \omega' \mathcal{A}'$, and the action $\mathcal{A}'$ satisfies

$$\frac{\partial \mathcal{A}'}{\partial T} + \frac{\partial}{\partial X} c'_g \mathcal{A}' = 0$$

The observer in the nonmoving frame finds a Lagrangian $\mathcal{L}^{(0)} = \mathcal{L}^{(0)}(\omega, k, X, T) = \mathcal{L}'^{(0)}(\omega - kU, k, X - UT, T)$. Because of $\mathcal{L}^{(0)}_{\omega} = \mathcal{L}'^{(0)}_{\omega'}$ the action in both frames is identical, $\mathcal{A} = \mathcal{A}'$, but the group velocity differs by the Doppler shift which derives from $\mathcal{L}^{(0)}_{k} = \mathcal{L}'^{(0)}_{k'} - U\mathcal{L}'^{(0)}_{\omega'}$. Hence written in terms of the energy $\mathcal{E}'(0)$ found in the moving system and the frequency of encounter (which is constant if $\Omega_T = 0$), the action conservation becomes

$$\frac{\partial}{\partial T} \frac{\mathcal{E}'^{(0)}}{\omega - kU} + \frac{\partial}{\partial X}\left(c'_g + U\right) \frac{\mathcal{E}'^{(0)}}{\omega - kU} = 0$$

Lagrangian theory to a pseudo energy-momentum tensor which may be expressed in terms of the wave action as

$$\begin{pmatrix} \mathcal{E}^{(0)} & \mathcal{F}^{(0)}_{e} \\ \mathcal{P}^{(0)} & \mathcal{F}^{(0)}_{p} \end{pmatrix} = \begin{pmatrix} \omega & c_g \omega \\ k & c_g k \end{pmatrix} \mathcal{A} \tag{9.37}$$

given here in a one-dimensional space. The conservation theorems are

$$\begin{aligned} \frac{\partial}{\partial T} \omega \mathcal{A} + \frac{\partial}{\partial X} c_g \omega \mathcal{A} &= -\frac{\partial \mathcal{L}^{(0)}}{\partial T} \\ \frac{\partial}{\partial T} k \mathcal{A} + \frac{\partial}{\partial X} c_g k \mathcal{A} &= -\frac{\partial \mathcal{L}^{(0)}}{\partial X} \end{aligned} \tag{9.38}$$

These equations follow as well from the action conservation and the ray equations. The generalization to the presence of a slowly varying mean flow is given in the box on p. 292.

## 9.5 Rossby Waves and Internal Gravity Waves as Examples

Two applications of Whitham's theory are considered where a Lagrangian is explicitly averaged and the action function of the respective wave train solution is derived.

### Action Conservation for Rossby Waves

The case of sound waves, treated as example in the previous sections, is singular in so far as sound waves are nondispersive, and equation (9.26) leaves the carrier function $\Phi$ undetermined, in agreement with (6.15). To bring Whitham's method to life for a more complicated problem, we consider the Rossby wave case of the box on p. 293.

The fields $\psi, \chi$ and $\mu$, entering the Lagrangian (B42.2), are expressed in terms of the slow and fast independent variables according to $\psi = P(\theta, X, T), \chi = C(\theta, X, T)$ and $\mu = M(\theta, X, T)$. The transformed Lagrangian becomes

$$L^{(0)} = \frac{1}{2}\left(k^2 P_\theta^2 + \frac{P^2}{R^2}\right) - \frac{1}{2}\beta k_1 P C_\theta + M\left(P + \omega C_\theta\right) \tag{9.39}$$

to lowest order in $\epsilon$, and varying with respect to $P, C$ and $M$ yields

$$\begin{aligned} k^2 P_{\theta\theta} - \frac{P}{R^2} + \frac{1}{2}\beta k_1 C_\theta - M &= 0 \\ -\frac{1}{2}\beta k_1 P_\theta + \omega M_\theta &= 0 \\ P + \omega C_\theta &= 0 \end{aligned} \tag{9.40}$$

We find $C_\theta = -P/\omega, M_\theta = (\beta k_1/2\omega) P_\theta$ from the last two equations, and thus

$$P_{\theta\theta} - \lambda P = 0 \quad \text{with} \quad \lambda = \frac{\omega + \beta R^2 k_1}{(kR)^2 \omega} \tag{9.41}$$

from the first equation. Solutions are required to be periodic with period $2\pi$. Hence $\lambda$ must be equal to $-1$, and we obtain $P = a \sin\theta$ and $\omega = -\beta k_1/(k^2 + R^{-2})$, which is indeed the dispersion relation (8.57) of linear Rossby waves. Inserting the solution for the fast time behavior of $P, C$ and $M$ into (9.39) and averaging over the phase, we find

$$\mathcal{L}^{(0)} = \frac{1}{4}a^2\left(k^2 + R^{-2} + \frac{\beta k_1}{\omega}\right) \tag{9.42}$$

**42. A Lagrangian for Linear Rossby Waves**

Linear Rossby waves are governed by the vorticity equation

$$\frac{\partial}{\partial t}\left(\nabla^2\psi - \frac{1}{R^2}\psi\right) + \beta\frac{\partial\psi}{\partial x} = 0 \tag{B42.1}$$

where $\psi$ is the stream function and $R$ the Rossby radius. A Lagrangian for this equation is

$$L = \frac{1}{2}\left[(\nabla\psi)^2 + \frac{\psi^2}{R^2}\right] - \frac{1}{2}\beta\psi\frac{\partial\chi}{\partial x} + \mu\left(\psi - \frac{\partial\chi}{\partial t}\right) \tag{B42.2}$$

where $\chi$ is an auxiliary field and $\mu$ a Lagrangian multiplier. Varying with respect to $\psi, \mu$ and $\chi$ yields

$$\begin{aligned} \nabla^2\psi - \frac{\psi}{R^2} + \frac{1}{2}\beta\frac{\partial\chi}{\partial x} - \mu &= 0 \\ \psi - \frac{\partial\chi}{\partial t} &= 0 \\ -\frac{1}{2}\beta\frac{\partial\psi}{\partial x} - \frac{\partial\mu}{\partial t} &= 0 \end{aligned}$$

and elimination of the auxiliary variables $\mu$ and $\chi$ leads to (B42.1).

and hence $\mathcal{A} = \mathcal{L}_\omega = -(a^2/4)\beta k_1/\omega^2 = (a^2/4)(k^2 + R^{-2})/\omega$ for the action density of Rossby waves. Furthermore, the group velocity components (8.60) and (8.61) result from $\boldsymbol{c}_{\mathrm{g}} = -\mathcal{L}_{\boldsymbol{k}}/\mathcal{L}_\omega$, and variation of $\mathcal{L}^{(0)}$ with respect to the amplitude $a$ yields the dispersion relation (8.57).

## Action Conservation for Internal Gravity Waves

As a second example, we derive the action conservation for internal gravity waves, starting with Lagrangian 2.230 from which we separate the quadratic part

$$L_2 = \frac{1}{2}\rho_0 \left[\dot{\xi}_i^2 + f(\xi_1\dot{\xi}_2 - \dot{\xi}_1\xi_2) - N^2\xi_3^2\right] + \varpi\frac{\partial \xi_i}{\partial a_i} \tag{9.43}$$

to begin with linear wave motion. Higher orders in the field amplitudes and nonlinear wave processes are considered below. Remember that Lagrangian coordinates are used: the fields are functions of $\boldsymbol{a}$ and $t$ where the vector $\boldsymbol{a}$ is the Lagrangian spatial coordinate and $t$ is time.

We consider wave propagation in a slowly varying medium; in the present case the Brunt–Väisälä frequency would be space and time dependent with the now familiar WKBJ conditions. The displacement $\xi_j$, $j = 1, 2, 3$ and the pressure variable $\varpi$ are written as a slowly varying wave train in the form

$$\begin{Bmatrix} \xi_j(\boldsymbol{a},t) \\ \varpi(\boldsymbol{a},t) \end{Bmatrix} = \begin{Bmatrix} A_j(\boldsymbol{X},T) \\ B(\boldsymbol{X},T) \end{Bmatrix} \mathrm{e}^{\mathrm{i}\Theta(\boldsymbol{X},T)/\epsilon} + \text{c.c.} \tag{9.44}$$

as used before with $\boldsymbol{X} = \epsilon\boldsymbol{a}$, $T = \epsilon t$ but assuming now a sinusoidal form from the beginning. The time rate of change is

$$\dot{\xi}_j = \left(-\mathrm{i}\omega A_j + \epsilon\frac{\partial A_j}{\partial T}\right)\mathrm{e}^{\mathrm{i}\Theta(\boldsymbol{X},T)/\epsilon} + \text{c.c.} \tag{9.45}$$

and correspondingly for $\varpi$ and the spatial derivatives. Inserting this form into (9.43) and averaging over the phase yields the averaged Lagrangian. To lowest order in $\epsilon$, the quadratic part $L_2$ yields (summation over $j$ is implied)

$$\begin{aligned} \bar{L}_2^{(0)} = \mathcal{L} = \rho_0 \big[&\omega^2 A_j A_j^* - N^2 A_3 A_3^* + \mathrm{i}\omega f\left(A_1 A_2^* - A_1^* A_2\right) \\ &-ik_j\left(BA_j^* - B^* A_j\right)\big] \end{aligned} \tag{9.46}$$

Variation with respect to $A_j^*$ and $B^*$ yields a homogeneous set of equations for the amplitudes which are determined as

$$(A_j, B) = A(Z_j, P) \tag{9.47}$$

with an arbitrary amplitude $A(\boldsymbol{X}, T)$ and an eigenvector part given by

$$\begin{aligned} Z_1 &= \left(-k_1 - \mathrm{i}\frac{f}{\omega}k_2\right)\frac{k_3}{k_\mathrm{h}^2}, \quad Z_2 = \left(-k_2 + \mathrm{i}\frac{f}{\omega}k_1\right)\frac{k_3}{k_\mathrm{h}^2} \\ Z_3 &= 1, \quad P = \mathrm{i}\rho_0\left(\omega^2 - f^2\right)\frac{k_3}{k_\mathrm{h}^2} \end{aligned} \tag{9.48}$$

provided the dispersion relation (7.16) is valid. Note that not only $A$ is slowly varying but also the wave vector and frequency, as described in the previous sections. Up to a common factor the above eigenvector is identical to (7.11) and (7.12) (written there for velocity instead of displacement). Inserting the eigenvector representation (9.47) into the averaged Lagrangian yields

$$\mathcal{L} = \rho_0 |A|^2 \left[ \omega^2 - N^2 + \left(\omega^2 - f^2\right) \frac{k_3^2}{k_h^2} \right] \tag{9.49}$$

The action and energy are found to be[3]

$$\mathcal{A} = \mathcal{L}_\omega = 2\rho_0 |A|^2 \omega \frac{k^2}{k_h^2} \quad \text{and} \quad \mathcal{E} = \omega \mathcal{A} = 2\rho_0 |A|^2 \left( \frac{\omega k}{k_h} \right)^2 \tag{9.50}$$

They are governed by (9.32), or more specifically by the vectorial form (9.22). Comparison with (7.22) yields $2\omega^2|A|^2 = |W_0|^2$ in terms of the amplitude $W_0$ of the vertical velocity. The group velocity components (7.18) and (7.19) result from $\boldsymbol{c}_g = -\mathcal{L}_{\boldsymbol{k}}/\mathcal{L}_\omega$, and variation of $\mathcal{L}$ with respect to the amplitude $A$ again yields the dispersion relation (7.16).

What is to learn from the next order in $\epsilon$ of the Lagrangian (9.43) which derives from the slow space and time dependence of the parameters of the wave train? The phase averaged Lagrangian of first order in $\epsilon$, derived from $L_2$, is given by

$$\begin{aligned} \bar{L}_2^{(1)} &= \rho_0 \left( \alpha A \frac{\partial A^*}{\partial T} + \beta_j A \frac{\partial A^*}{\partial X_j} + \alpha^* A^* \frac{\partial A}{\partial T} + \beta_j^* A^* \frac{\partial A}{\partial X_j} \right) \\ \text{with} \quad \alpha &= \mathrm{i}\left[-\omega|Z_j|^2 + f\Im(Z_1 Z_2^*)\right] \quad \text{and} \quad \beta_j = P Z_j^* \end{aligned} \tag{9.51}$$

where the representation (9.47) was used. Variation with respect to $A^*$ yields the equation governing the slow dependence of the amplitude,

$$\frac{\partial \alpha A}{\partial T} + \frac{\partial \beta_j A}{\partial X_j} - \alpha^* \frac{\partial A}{\partial T} - \beta_j^* \frac{\partial A}{\partial X_j} = 0 \tag{9.52}$$

Because $\alpha$ is imaginary, we introduce $\gamma = -\mathrm{i}\alpha$ and write the above equation as

$$2\mathrm{i}\gamma \frac{\partial A}{\partial T} + 2\mathrm{i}\Im(\beta_j) \frac{\partial A}{\partial X_j} + A \left\{ \mathrm{i} \frac{\partial \gamma}{\partial T} + \frac{\partial}{\partial X_j} \left[ \Re(\beta_j) + \mathrm{i}\Im(\beta_j) \right] \right\} = 0 \tag{9.53}$$

Evaluation of the coefficients leads to

$$\gamma = -\omega \frac{k^2}{k_h^2} \quad \text{and} \quad \Im(\beta_j) = \gamma c_{gj}\,, \quad j = 1, 2, 3 \tag{9.54}$$

The imaginary part of $\boldsymbol{\beta} = (\beta_j)$ is thus expressed by the group velocity $\boldsymbol{c}_g$ of internal gravity waves (cf. (7.18) and (7.19)). If $\omega$ remains constant (i. e. no slow time dependence according to the box on p. 173), the real part of $\boldsymbol{\beta}$, entering as $\nabla \cdot \Re(\boldsymbol{\beta})$, may be expressed by the curl of $\boldsymbol{c}_g$ (curl is here, as usually, the vertical component of the $\nabla\times$ operator). Hence, the change of the amplitude along the path of the wave train (the ray) is determined from

$$\frac{\partial A}{\partial T} + \boldsymbol{c}_g \cdot \nabla A = -\frac{1}{2} \frac{A}{\gamma} \left( \frac{\partial \gamma}{\partial T} + \boldsymbol{c}_g \cdot \nabla \gamma \right) - \frac{1}{2} A \left( \nabla \cdot \boldsymbol{c}_g + \mathrm{i} \frac{f k_h^2}{\omega k^2} \operatorname{curl} \frac{k^2}{k_h^2} \boldsymbol{c}_g \right) \tag{9.55}$$

[3] In later applications we will use amplitudes $a = A\omega k/k_h$ which are renormalized to directly describe the energy $\mathcal{E} = 2\rho_0|a|^2$.

The vector gradient and divergence operators are meant to act on the slow spatial dependence. The equation is consistent with the action conservation: the action is $\mathcal{A} = -\rho_0 \gamma A A^*$, and indeed, after multiplication with $A^*$ and addition of the conjugate equation, it is found that the right-hand side can be absorbed and (9.22) results.

## 9.6 Wave–Wave Interactions

In the previous sections, a Lagrangian theory was developed to understand the deformation of a wave train due to propagation in a nonuniform medium where, as a consequence of the nonhomogeneity, the amplitude, wave vector, and frequency are slowly changing. The perturbation analysis was oriented at the smallness of the WKBJ parameter $\epsilon \simeq k\ell$, the ratio of the length scale $\ell$ of variation of the background medium and the wavelength $2\pi/k$ (correspondingly for the temporal conditions of change). We have, however, disregarded nonlinearities so far: we have neglected all terms in the Lagrangian which are of higher order than quadratic. To be more specific, the Lagrangian (2.230) for internal gravity waves, treated as example in the previous section, contains besides the quadratic part (9.43) terms which are of third and higher order in the field amplitudes, i. e. $L = \delta^2 L_2 + \delta^3 L_3 + \cdots$, with $L_2$ given by (9.43) and

$$L_3 = \varpi \left(\Delta_{jj} + \Delta\right) + \frac{1}{3!}\xi_3^3 \frac{\mathrm{d}^3 \rho_0}{da_3^3} + \cdots \tag{9.56}$$

Here $\Delta_{ij}$ is the cofactor of $\partial\xi_i/\partial a_j$ in the Jacobian determinant $\Delta = |\partial\xi/\partial a|$. The parameter $\delta$ measures the size of the fields amplitudes, and for the internal wave case one obtains $\delta \simeq \varpi\Delta_{jj}/\dot{\xi}_j^2 \simeq k\xi_j$ as ratio of the third order nonlinear term and the quadratic inertial term. The other cubic term in (9.56) yields $\delta \simeq \xi_3/\ell$ by comparing to $\xi_3^2 N^2/2$ in $L_2$. A generally valid definition of $\delta$ is not possible. For systems in which the nonlinearities arise from advection one has $\delta = k\xi = v/c$ where $v$ is a typical particle velocity and $c$ the phase speed.

To conclude: we have to deal with two effects and two parameters, $\epsilon$ controlling the effect of nonhomogeneity of the medium and $\delta$ governing nonlinearity of the dynamics. Both effects lead to slow (provided the parameters are sufficiently small) changes of the wave amplitudes. Though a consistent perturbation analysis in $\epsilon$ and $\delta$ is, of course, possible, we will assume from now on that the background medium is homogeneous and will concentrate exclusively on nonlinear effects on the wave motion. We consider an ensemble of waves where the amplitudes – apart from the intrinsic wave oscillation – are slowly changing due to interactions among them, and again we shall develop a Lagrangian theory because it reveals the symmetry properties of wave interactions to a greater deal than an approach starting directly from the equations of motion.

We continue with the internal gravity wave case as example and go on from the linear representation (9.44) to (9.48), written now for an ensemble of waves which are numbered by the counter $\lambda$,

$$\begin{Bmatrix} \xi_j(\boldsymbol{a},t) \\ \varpi(\boldsymbol{a},t) \end{Bmatrix} = \sum_\lambda A^\lambda(T) \begin{Bmatrix} Z_j(\boldsymbol{k}^\lambda,\omega^\lambda) \\ P(\boldsymbol{k}^\lambda,\omega^\lambda) \end{Bmatrix} \mathrm{e}^{\mathrm{i}(\boldsymbol{k}^\lambda\cdot\boldsymbol{a}\,-\omega^\lambda t)} \tag{9.57}$$

The amplitude has a slow time dependence with $T = \delta t, \delta \ll 1$. While the contributions to the sum in (9.57) are complex, the result is always real by taking the sum over $\lambda \gtrless 0$ and assuming $\boldsymbol{k}^{-\lambda} = -\boldsymbol{k}^{\lambda}, \omega^{-\lambda} = -\omega^{\lambda}, A^{-\lambda} = (A^{\lambda})^*$. Note that $Z_j(\boldsymbol{k}^{-\lambda}, \omega^{-\lambda}) = Z_j^*(\boldsymbol{k}^{\lambda}, \omega^{\lambda})$ and $P(\boldsymbol{k}^{-\lambda}, \omega^{-\lambda}) = P^*(\boldsymbol{k}^{\lambda}, \omega^{\lambda})$ are satisfied in accordance with (9.48). Inserting the representation into the Lagrangian yields[4]

$$
\begin{aligned}
\bar{L} &= \int (\delta^2 L_2 + \delta^3 L_3) \mathrm{d}^3 a \\
&= \delta^3 \sum_{\lambda} \frac{1}{2} \mathrm{i} \gamma^{\lambda} \left( A^{\lambda} \frac{\partial A^{-\lambda}}{\partial T} - A^{-\lambda} \frac{\partial A^{\lambda}}{\partial T} \right) \\
&\quad + \delta^3 \sum_{\lambda\mu\nu} A^{\lambda} A^{\mu} A^{\nu} D^{\lambda\mu\nu} \mathrm{e}^{-\mathrm{i}(\omega^{\lambda} + \omega^{\mu} + \omega^{\nu})t} + O(\delta^4)
\end{aligned}
\tag{9.58}
$$

for the third order term in $\delta$, arising from $L_2$ due to the slow time dependence of the amplitudes and from $L_3$ from the cubic field terms. The coefficient $\gamma^{\lambda}$ is given by (9.54). The spatial integral of Hamilton's variational principle (2.213) has been performed already; the temporal integration still has to be done. The first term is the same as (9.51) (without the contributions from the spatial derivatives), the second term follows from (9.56). In the coupling coefficient $D^{\lambda\mu\nu}$, we have collected all third order contributions from respective products of the eigenvectors.

As a major advantage of the derivation from a Lagrangian, $D^{\lambda\mu\nu}$ can be taken symmetric in the index triple $\lambda\mu\nu$ because the amplitude triple is symmetric: other terms than symmetric ones drop out of the sum. Because of the spatial integral the first is only over one index, and in the second sum only terms have survived that satisfy $\boldsymbol{k}^{\lambda} + \boldsymbol{k}^{\mu} + \boldsymbol{k}^{\nu} = 0$. The constraint is taken as implemented in the definition of $D^{\lambda\mu\nu}$ (it is zero for all other combinations of the wave vectors). A further property is $D^{-\lambda-\mu-\nu} = \left(D^{\lambda\mu\nu}\right)^*$ (reality condition for the Lagrangian). The actual form of $D^{\lambda\mu\nu}$ will be irrelevant for the following discussion, but for completeness we give its form,

$$
D^{\lambda\mu\nu} = \frac{\mathrm{i}k_3^{\lambda}}{6\left(k_{\mathrm{h}}^{\lambda}\right)^2} \left[ \left(\omega^{\lambda}\right)^2 - f^2 \right] (\boldsymbol{k}^{\mu} \cdot \boldsymbol{Z}^{\nu})(\boldsymbol{k}^{\nu} \cdot \boldsymbol{Z}^{\mu}) + (\lambda \leftrightarrow \mu) + (\lambda \leftrightarrow \nu) \tag{9.59}
$$

which is valid on the resonance surface $\omega^{\lambda} + \omega^{\mu} + \omega^{\nu} = 0$. Only such interactions will gain importance, as shown in the following.

The $O(\delta^4)$ contribution in (9.58) contains terms $O(A^n), n > 3$ with similar coupling coefficients of higher order, involving the interaction of quartets of waves and higher orders. We shall confine the analysis to triplet interactions. There are also quadratic products of $\dot{A}$ which, due to the slow time dependence, are of fourth order in $\delta$.

Variation of $\bar{L}$ with respect to the amplitude $A^{-\lambda}$ yields

$$
2\gamma^{\lambda} \dot{A}^{\lambda} = -3\mathrm{i} \sum_{\mu\nu} A^{\mu} A^{\nu} D^{-\lambda\mu\nu} \mathrm{e}^{\mathrm{i}(\omega^{\lambda} - \omega^{\mu} - \omega^{\nu})t} + O(\delta) \tag{9.60}
$$

[4] The irrelevant factor $\rho_0$ will from now on be dropped.

The dot time derivation of $A^\lambda$ is now with respect to slow time $T = \delta t$. The sum over $\mu$ and $\nu$ is over all wave vectors $\boldsymbol{k}^\mu, \boldsymbol{k}^\nu$ satisfying $\boldsymbol{k}^\mu + \boldsymbol{k}^\nu = \boldsymbol{k}^\lambda$. The resolved term on the right-hand side presents the interaction of waves $\mu$ and waves $\nu$, changing the amplitude of wave $\lambda$.

For later purposes, it will be convenient to renormalize the amplitude such that $a = A\omega k/k_\mathrm{h}$ is directly related to the energy $\mathcal{E} = 2|A|^2(\omega k/k_\mathrm{h})^2 = 2|a|^2$ of each wave (cf. previous section). In terms of these new amplitudes $a$, the equations of motion (9.60) become

$$\dot{a}^\lambda = -3\mathrm{i}\omega^\lambda \sum_{\mu\nu} a^\mu a^\nu C^{-\lambda\mu\nu} \mathrm{e}^{\mathrm{i}(\omega^\lambda - \omega^\mu - \omega^\nu)t} + O(\delta) \tag{9.61}$$

with $C^{\lambda\mu\nu} = \frac{1}{2} D^{\lambda\mu\nu} / \left[(\omega^\lambda k^\lambda / k_\mathrm{h}^\lambda)(\omega^\mu k^\mu / k_\mathrm{h}^\mu)(\omega^\nu k^\nu / k_\mathrm{h}^\nu)\right]$ which shares all essential properties with the $D$-coefficient mentioned above.

## Resonance

Averaging (9.61) over the fast phase dependence leads to annihilation of the factor $\exp \mathrm{i}(\omega^\lambda - \omega^\mu - \omega^\nu)t$ in a two-time-scale approach if $\Delta\omega = \omega^\lambda - \omega^\mu - \omega^\nu$ is nonzero. Then the response of wave $\lambda$ becomes negligible. This does, however, not hold if $\Delta\omega$ is very small of order $\delta$ or less. In this case, the phase averaging leaves the exponential term essentially unchanged. The initial value problem of three single waves, with $a_j(t = 0), j = \lambda, \mu, \nu$, given leads to

$$a^\lambda(t) = a^\lambda(0) - 3\mathrm{i}\omega^\lambda C^{-\lambda\mu\nu} a^\mu(0) a^\nu(0) \int_0^t \mathrm{e}^{(\omega^\lambda - \omega^\mu - \omega^\nu)t'} \mathrm{d}t' \tag{9.62}$$

For resonant triads, for which

$$\omega^\lambda - \omega^\mu - \omega^\nu = 0 \qquad \text{and also} \qquad \boldsymbol{k}^\lambda - \boldsymbol{k}^\mu - \boldsymbol{k}^\nu = 0 \tag{9.63}$$

a secular behavior is found in the limit $\Delta\omega \to 0$ because

$$\lim_{\Delta\omega \to 0} \int_0^t \mathrm{e}^{\mathrm{i}\Delta\omega t'} \mathrm{d}t = \lim_{\Delta\omega \to 0} \frac{\mathrm{e}^{\mathrm{i}\Delta\omega t} - 1}{\mathrm{i}\Delta\omega} = t \tag{9.64}$$

Wave $\lambda$ thus grows linearly on the slow time-scale on account of interacting with the duo $\mu$ and $\nu$ satisfying the resonance condition (9.63). Even if this condition is satisfied only approximately, i. e. $\Delta\omega \neq 0$ but small compared to the natural wave frequencies, the growth will be disastrous.

Evidently, the resonant response of wave $\lambda$ has the same frequency as the original wave and may thus be interpreted as a slow variation of the normal mode $a^\lambda \exp(-\mathrm{i}\omega^\lambda t)$. Note that nonresonant terms do not carry the intrinsic wave period. It should be emphasized that the resonance condition (9.63) for triad interaction cannot be satisfied for all kinds of dispersion relations. For internal gravity waves, nontrivial solutions exist; for surface gravity waves, on the other hand, resonant behavior is only possible in quartet interactions. In such a case the Lagrangian must be expanded to the fourth order in the amplitudes.

### 9.6.1 Resonant Wave Triads

We consider a single set of three specific waves $\lambda = 1, \mu = 2, \nu = 3$ (a triad) satisfying the resonance condition (9.63). Their amplitudes are governed by

$$\begin{aligned} \dot{a}_1 &= -\mathrm{i}\omega_1 \eta a_2 a_3 \\ \dot{a}_2 &= -\mathrm{i}\omega_2 \eta^* a_1 a_3^* \\ \dot{a}_3 &= -\mathrm{i}\omega_3 \eta^* a_1 a_2^* \end{aligned} \tag{9.65}$$

where $\eta = 3C^{-123} = 3(C^{-21-3})^* = 3(C^{-31-2})^*$. These equalities arise because of the reality condition of the Lagrangian and the symmetry of the coupling coefficient. There are a number of theorems following from (9.65), allowing an intuitive interpretation in terms of conservation of a wave 'energy', 'action' and 'momentum'.

Multiplying each equation by the respective conjugate amplitude and adding the complex conjugate of the result, we find

$$\frac{1}{\omega_1}\frac{\mathrm{d}}{dt}|a_1|^2 = -\frac{1}{\omega_2}\frac{\mathrm{d}}{\mathrm{d}t}|a_2|^2 = -\frac{1}{\omega_3}\frac{\mathrm{d}}{\mathrm{d}t}|a_3|^2 = \mathrm{i}\eta^* a_1 a_2^* a_3^* + \text{c.c.} = -2\Im(h_{\text{int}}) \tag{9.66}$$

with $h_{\text{int}} = \eta^* a_1 a_2^* a_3^*$. The first two equalities are known as the *Manley–Rowe relations* (Manley and Rowe, 1956). They are independent of the coupling coefficient[5]. We may interpret $|a_j|^2/\omega_j$ as the 'action' of the wave $j$. Equation (9.66) then states that if the waves 2 and 3 each lose a quantum of action, wave 1 gains one quantum: action is thus not conserved in the triplet interaction.

By suitable addition of the equations of (9.66), we find a form of energy conservation

$$\frac{\mathrm{d}}{\mathrm{d}t}\left(|a_1|^2 + |a_2|^2 + |a_3|^2\right) = -2\Im(h_{\text{int}})\,(\omega_1 - \omega_2 - \omega_3) = 0 \tag{9.67}$$

Hence $|a_j|^2$ is proportional to the 'energy' of the wave $j$, in agreement with the definition of the amplitudes $a$ given above. The total energy of the triad is constant because of the resonance condition. Likewise,

$$\frac{\mathrm{d}}{\mathrm{d}t}\left(\frac{|a_1|^2}{\omega_1}\boldsymbol{k}_1 + \frac{|a_2|^2}{\omega_2}\boldsymbol{k}_2 + \frac{|a_3|^2}{\omega_3}\boldsymbol{k}_3\right) = -2\Im(h_{\text{int}})\,(\boldsymbol{k}_1 - \boldsymbol{k}_2 - \boldsymbol{k}_3) = 0 \tag{9.68}$$

and thus the interacting triad conserves the total 'momentum' if $|a_j|^2\boldsymbol{k}_j/\omega_j$ is attributed as momentum to a single wave. A further conserved quantity is a cubic expression in the amplitudes. Using (9.65), it is easily confirmed that

$$\frac{\mathrm{d}}{\mathrm{d}t}\Re(h_{\text{int}}) = \frac{\mathrm{d}}{\mathrm{d}t}\left(\eta^* a_1 a_2^* a_3^* + \eta a_1^* a_2 a_3\right) = 0 \tag{9.69}$$

The two Manley–Rowe relations in (9.66) and the constancy of $\Re(h_{\text{int}})$ are three independent integrals of the triad problem (9.65), which, therefore, is completely integrable. In fact, the general solution may be expressed in terms of elliptic functions.

[5] If we do not use the Lagrangian framework but start from the differential equations, the coefficients in (9.65) generally differ. A suitable renormalization of the amplitudes then leads to the Manley–Rowe relations as well. This, however, can only be done for one particular triplet of waves; it is not possible if a wave from a triplet interacts with another set of waves that should be renormalized as well.

The behavior depends critically on the sign of the frequencies and the interaction coefficient (for details see e. g. Craik (1988)). A great variety of types of temporal behavior includes cases with periodic exchanges of the energies, but also explosive solutions exist under specific conditions.

## 9.6.2 Interaction Theory for Random Wave Fields

The weak interaction framework was developed so far for plane waves having an infinite extent and small amplitudes. In a realistic geophysical situation, the wave field is more likely described by a superposition of a great number of wave trains, each localized in the physical space and having a dominant wave vector, frequency, and amplitude, which slowly change as a consequence of propagation and refraction (see the WKBJ framework in the previous sections). When two wave trains occupy the same region, they might interact resonantly for a short finite time and build up a third component. The dominant wave numbers and frequencies of the interacting wave are unaffected if the interaction region is small in the WKBJ sense, but their energies and phases change. The triad members then separate again and propagate out of the interaction region. The triad interaction (or higher order resonant interaction) thus still serves as the elementary coupling process in the wave field, much like particle collisions in a dilute gas.

We describe such a wave field by an ensemble of waves, such that we find a great number of waves with different wave vectors $\boldsymbol{k}$ and amplitudes $a(\boldsymbol{k}, \boldsymbol{x}, t)$ at the position $\boldsymbol{x}$ and time $t$. A particular wave will be identified by $\boldsymbol{k}$ instead of the counter $\lambda$ used before, and $\boldsymbol{k}$ will be taken continuous. Following (9.44), we represent the field by a Fourier integral,

$$\boldsymbol{\xi}(\boldsymbol{x}, t) = \sum_{s=\pm} \int \mathrm{d}^3k\, a^s_{\boldsymbol{k}}(\boldsymbol{x}, t) \boldsymbol{Z}^s_{\boldsymbol{k}} \mathrm{e}^{\mathrm{i}(\boldsymbol{k}\cdot\boldsymbol{x} - \omega^s_{\boldsymbol{k}} t)} \tag{9.70}$$

and may regard this representation as fitting for the internal gravity wave field. However, it is appropriate for an arbitrary wave field that is described by the state vector $\boldsymbol{\xi}(\boldsymbol{x}, t)$ and the associated eigenvector $\boldsymbol{Z}^s_{\boldsymbol{k}}$ of its linear appearance. To make $\boldsymbol{\xi}$ real, the index $s = \pm$ is introduced with the convention explained in the context of (9.44) where $\lambda$ appeared with two signs. We assume that the amplitude $a^s_{\boldsymbol{k}}(\boldsymbol{x}, t)$ is governed by the generic equation (9.61).

The wave amplitude $a^s_{\boldsymbol{k}}(\boldsymbol{x}, t)$ is regarded as a random variable (for details see Appendix A.3), such that for a particular set of amplitudes (appropriate for a particular position and time), the field (9.70) is a particular realization taken from a statistical ensemble. We are interested in average quantities, denoted by cornered brackets, such as second-order correlations $\langle a^s_{\boldsymbol{k}}(\boldsymbol{x}, t) a^{s'}_{\boldsymbol{k}'}(\boldsymbol{x}, t)\rangle$, describing the statistical dependence of a wave with wave vector $\boldsymbol{k}$ to another one with wave vector $\boldsymbol{k}'$ at the same position and time. The evolution equation for second-order correlations is readily obtained from the equation governing the amplitude, e. g. (9.61) yields

$$\begin{aligned} \frac{\partial}{\partial t}\left\langle a^s_{\boldsymbol{k}} a^{s'}_{\boldsymbol{k}'}\right\rangle = -3\mathrm{i}\omega^s_{\boldsymbol{k}} \int \mathrm{d}^3k'' \int \mathrm{d}^3k'' C^{-ss''s'''}_{-\boldsymbol{k}\boldsymbol{k}''\boldsymbol{k}'''} \langle a^{s''}_{\boldsymbol{k}''} a^{s'''}_{\boldsymbol{k}'''} a^{s'}_{\boldsymbol{k}'}\rangle \mathrm{e}^{\mathrm{i}\Delta\omega t} \\ + \left(\begin{smallmatrix} s \\ \boldsymbol{k} \end{smallmatrix} \leftrightarrow \begin{smallmatrix} s' \\ \boldsymbol{k}' \end{smallmatrix}\right) \end{aligned} \tag{9.71}$$

where $\Delta\omega = \omega_{\boldsymbol{k}}^{s} - \omega_{\boldsymbol{k}''}^{s''} - \omega_{\boldsymbol{k}'''}^{s'''}$ is the frequency misfit of the interacting triad. The notation of the indices obviously becomes a bit awkward. We use from now on the abbreviation with a lower index $s$ to stand for ${}^{s}_{\boldsymbol{k}}$ as e. g. $a_s = a_{\boldsymbol{k}}^{s}$ for the amplitude and likewise for frequencies and coupling coefficients.

For quadratic nonlinearities, the equation thus involves triple correlations, and we are facing the familiar problem that correlation equations of nonlinear systems are not closed but form an infinite hierarchy: the evolution equation for triple correlations needs in turn correlations of fourth order, and so on. As in turbulence theory (see Part IV), a closure hypothesis is needed. There is no generally accepted scheme by which hierarchies describing strongly nonlinear systems can be cut short, but weakly nonlinear wave fields never depart much from a Gaussian state (see Appendix A.3). Indeed, it has been shown by Prigogine (1962) that the nonlinear forcing terms on the right-hand side of (9.71) can be determined under the assumption that the lowest order amplitudes $a_s^{(1)}$ – in a perturbation expansion with respect to the nonlinearity parameter – are elements of a Gaussian ensemble. This implies that a correlation $\langle a_{s_1}^{(1)} \cdots a_{s_n}^{(1)} \rangle$ of this lowest order state vanishes for odd $n$ and becomes a sum of all possible products $\langle a_{s_i}^{(1)} a_{s_j}^{(1)} \rangle \langle a_{s_\ell}^{(1)} a_{s_m}^{(1)} \rangle$ for even $n$. We will identify the state $a_s^{(1)}$ with an arbitrary initial state at some time $t = 0$ and further assume that

$$\left\langle a_s^{(1)} a_{s'}^{(1)} \right\rangle = \frac{1}{2}\mathcal{E}(\boldsymbol{k})\delta_{ss'}\delta(\boldsymbol{k} - \boldsymbol{k}') \tag{9.72}$$

This means that waves with different wave vectors are statistically independent. There is only a correlation between $a_s^{(1)}$ and its conjugate complex $a_{-s}^{(1)}$ which actually describe the same physical wave component. If the eigenvector $Z_{\boldsymbol{k}}^{s}$ is suitably normalized, the function $\mathcal{E}(\boldsymbol{k}) = \mathcal{E}(\boldsymbol{k}, \boldsymbol{x}, t)$ is the local energy spectrum[6], now taking the spatial and time dependence explicitly into consideration. Hence

$$E(\boldsymbol{x}, t) = \int \mathrm{d}^3 k \mathcal{E}(\boldsymbol{k}, \boldsymbol{x}, t) \tag{9.73}$$

is the total energy of the wave field at the position $\boldsymbol{x}$ and time $t$. Note that $E$ is still an energy density with respect to the spatial dependence. Due to the Gaussian property, the energy spectrum $\mathcal{E}(\boldsymbol{k}, \boldsymbol{x}, t)$ gives a complete description of a Gaussian random wave field.

Our aim is thus to derive the evolution equation for $\mathcal{E}(\boldsymbol{k}, \boldsymbol{x}, t)$. Remembering the discussion of propagation and refraction for waves in a slowly changing background medium (see Section 6.3 and the previous sections of this chapter) and the consequent definition of wave action, it seems reasonable to use the action spectrum $\mathcal{N}(\boldsymbol{k}, \boldsymbol{x}, t) = \mathcal{E}(\boldsymbol{k}, \boldsymbol{x}, t)/\omega(\boldsymbol{k}, \boldsymbol{x}, t)$ instead of the energy spectrum. Because the action is now written as function of $\boldsymbol{k}$ in addition to previous independent variables $\boldsymbol{x}$ and $t$, and the wave vector is also slowly changing, the action conservation (9.22) reads for the random case

$$\frac{\partial \mathcal{N}}{\partial t} + \nabla_{\boldsymbol{x}} \cdot (\dot{\boldsymbol{x}} \mathcal{N}) + \nabla_{\boldsymbol{k}} \cdot (\dot{\boldsymbol{k}} \mathcal{N}) = \mathcal{S} \tag{9.74}$$

Here, propagation and refraction are given by the ray equations $\dot{\boldsymbol{x}} = \nabla_{\boldsymbol{k}} \Omega$ (the group velocity) and $\dot{\boldsymbol{k}} = -\nabla_{\boldsymbol{x}} \Omega$ with the dispersion relation $\omega = \Omega(\boldsymbol{k}, \boldsymbol{x}, t)$, and $\mathcal{S}$

[6] Note that the eigenvector of the internal wave example in the previous section is not normalized in this way.

is a source representing all processes that may lead to a change of the action spectrum, except for the slow propagation and refraction processes which are explicitly accounted for on the left-hand side. Note that $\nabla_{\boldsymbol{x}} \cdot \dot{\boldsymbol{x}} + \nabla_{\boldsymbol{k}} \cdot \dot{\boldsymbol{k}} = 0$, so that (9.74) may be rephrased as

$$\left(\frac{\partial}{\partial t} + \dot{\boldsymbol{x}} \cdot \nabla_{\boldsymbol{x}} + \dot{\boldsymbol{k}} \cdot \nabla_{\boldsymbol{k}}\right) \mathcal{N}(\boldsymbol{k}, \boldsymbol{x}, t) = \mathcal{S}(\boldsymbol{k}, \boldsymbol{x}, t) \tag{9.75}$$

This equation (or the flux form (9.74)) is called 'radiative transfer equation' of the specific wave field. The resemblance between wave groups and interacting particles becomes obvious: the ray equations are the Hamiltonian equations (see the box on p. 98) with $\Omega(\boldsymbol{k}, \boldsymbol{x}, t)$ as Hamiltonian for a 'particle' with the generalized coordinate $\boldsymbol{x}$, momentum $\boldsymbol{k}$, and energy $\omega$. The radiative transfer equation is the analogue of the transport equation governing the particle distribution function (see below).

The similarity becomes even closer when the source term $\mathcal{S}_{\text{triad}}$ for resonant triad interactions (particle collisions) in the wave field is evaluated. We briefly indicate how this can be done. The wave amplitude is expanded in terms of the nonlinearity parameter,

$$a_s = a_s^{(1)} + a_s^{(2)} + \cdots \tag{9.76}$$

and inserted into the triple moment on the right-hand side of (9.71). Because the basic state $a_s^{(1)}$ is Gaussian, the lowest (third) order of the rate of change of $\langle a_s a_{-s} \rangle$ vanishes because a triple moment occurs on the right-hand side, and the first nonzero contribution is of fourth order. It takes the form

$$\begin{aligned} \frac{\partial}{\partial t} \langle a_s a_{-s} \rangle &= -6\omega_s \Im \sum_{s's''} C_{-ss's''} \mathrm{e}^{\mathrm{i}(\omega_s - \omega_{s'} - \omega_{s''})t} \\ &\quad \times \left[ 2\langle a_{-s}^{(1)} a_{s'}^{(1)} a_{s''}^{(2)} \rangle + \langle a_{-s}^{(2)} a_{s'}^{(1)} a_{s''}^{(1)} \rangle \right] \end{aligned} \tag{9.77}$$

Here, $a_s^{(1)} = a_s(t = 0)$ is the value at some initial time, and $a_s^{(2)}$ is given by (9.62) in the corresponding form

$$a_s^{(2)}(t) = -3\mathrm{i}\omega_s \sum_{s's''} C_{-ssS''} a_{s'}^{(1)} a_{s''}^{(1)} \Delta(\omega_s - \omega_{s'} - \omega_{s''}) \tag{9.78}$$

with the previously defined resonance function

$$\Delta(\omega) = \frac{\mathrm{e}^{\mathrm{i}\omega t} - 1}{\mathrm{i}\omega} \tag{9.79}$$

Inserting (9.78) into (9.77), we note that fourth-order moments of $a_s^{(1)}$ arise which can be broken by the Gaussian assumption to three products of second-order moments. Using (9.72), we find

$$\begin{aligned} \frac{\partial}{\partial t} \langle a_s a_{-s} \rangle &= -36\omega_s \Re \sum_{s's''} |C_{-ss's''}|^2 \mathrm{e}^{\mathrm{i}(\omega_s - \omega_{s'} - \omega_{s''})t} \\ &\quad \times [2\omega_{s''} \langle a_s a_{-s} \rangle \langle a_{s'} a_{-s'} \rangle \Delta(\omega_{s''} - \omega_s - \omega_{s'}) \\ &\quad - \omega_s \langle a_{s'} a_{-s'} \rangle \langle a_{s''} a_{-s''} \rangle \Delta(\omega_s - \omega_{s'} - \omega_{s''})] \end{aligned} \tag{9.80}$$

after some lengthy operations during which integration over some $\delta$ functions of the wave vectors have to be evaluated. Here, properties of the coupling coefficient come into play, as e. g. the symmetry in the indices and that it vanishes unless the sum of the corresponding wave vector triple is zero (see previous section). Note the behavior of the resonance function

$$\Re e^{-i\omega t}\Delta(\omega) = \Re\frac{1-e^{-i\omega t}}{i\omega} = \frac{\sin\omega t}{\omega} \to \pi\delta(\omega) \quad \text{as} \quad t\to\infty \tag{9.81}$$

As explained previously for the amplitudes, in the long run only resonant triads make a noticeable contribution to the spectral development. Using the above relation and implementing the action spectrum $\mathcal{N}(\boldsymbol{k}) = 2\langle a_+a_-\rangle/\omega(\boldsymbol{k})$, we arrive after some elementary manipulations at the final form

$$\begin{aligned}\mathcal{S}_{\text{triad}} = \int d^3k' \int d^3k'' \{&T^+\delta(\boldsymbol{k}-\boldsymbol{k}'-\boldsymbol{k}'')\delta(\omega-\omega'-\omega'')\\ &\times\left[\mathcal{N}(\boldsymbol{k}')\mathcal{N}(\boldsymbol{k}'')-\mathcal{N}(\boldsymbol{k})\mathcal{N}(\boldsymbol{k}')-\mathcal{N}(\boldsymbol{k})\mathcal{N}(\boldsymbol{k}'')\right] + 2T^-\delta(\boldsymbol{k}-\boldsymbol{k}+\boldsymbol{k}'')\\ &\times\delta(\omega-\omega'+\omega'')\left[\mathcal{N}(\boldsymbol{k}')\mathcal{N}(\boldsymbol{k}'')+\mathcal{N}(\boldsymbol{k})\mathcal{N}(\boldsymbol{k}')-\mathcal{N}(\boldsymbol{k})\mathcal{N}(\boldsymbol{k}'')\right]\}\end{aligned} \tag{9.82}$$

where the transfer function (or cross section of the scattering process for triad interactions)

$$T^\mu = 18\pi\omega\omega'\omega''|C^{-+\mu}_{-\boldsymbol{k}\boldsymbol{k}'\mu\boldsymbol{k}''}|^2 \quad \text{for} \quad \mu = \pm \tag{9.83}$$

controls the efficiency of sum and difference interactions (compare the arguments of the $\delta$ function in the two contributions to the scattering integral (9.82)). The general form of the scattering integral $\mathcal{S}_{\text{triad}}$ for triad interaction and also for higher order resonant interactions has been worked out by Hasselmann (1966, 1967b), who also pointed out the formal similarity with Boltzmann's collision integral (see e. g. Huang, 1987) for interacting particles in a dilute gas. Hasselmann also derived the scattering integrals for interaction between waves of different types, e. g. surface and internal waves, and even for scattering of waves at a stationary random field, as e. g. random topography. The formal procedure is, of course, always similar to that given above.

The above described procedure to derive the scattering integral as a functional of the second moment – the action or energy spectrum – seems at first inconsistent: the initial state has Gaussian property by assumption but it develops correlations by the interaction of the wave groups which are certainly non-Gaussian, as can be seen in the nonvanishing triple moment in (9.77). After evaluation in terms of quadruple moments – in (9.80) – a Gaussian state is again assumed. Hence the wave field cannot be Gaussian in the rigorous sense that the wave groups are statistically independent. The resonant interaction between two wave groups leads to a correlation which literally would hold for an indefinite time. But leaving the interaction volume the group propagate away from one another and eventually interact with other members. The correlation between the groups disperses then into a fine-structure of the distribution function. The Gaussian hypothesis assumes that this fine-structure can be ignored in the further development. The same assumption hold for the derivation of the Boltzmann collision integral for interacting particles (Huang, 1987).

The internal resonant interactions, described by the scattering integral $\mathcal{S}_{\text{triad}}$, conserve energy and momentum (but not so for action) of the wave field,

$$\int d^3k \begin{pmatrix}\omega\\ \boldsymbol{k}\end{pmatrix} \mathcal{S}_{\text{triad}} = 0 \tag{9.84}$$

so that the wave interactions only redistribute energy and momentum within the wave field. Hence evaluating (9.82) for any given spectrum shows a transfer of energy from a certain part of the $\boldsymbol{k}$-space to another part, but sources and sinks must balance. If the wave field is in equilibrium, the transfer induced by resonant interactions must be balanced by external generation processes and dissipation of wave energy. These processes generally affect the wave spectrum on vastly different scales. The transfer by resonant interactions establishes the spectral transfer which is necessary for a balance, but unlike the situation in turbulent fields where the spectral transfer couples neighboring regions in the wave-number space in a spectral cascade, the resonant transfer in wave fields can strongly deviate from a local or diffusive behavior in wave-number space.

# Forced Waves 10

The equations of oceanic motions are nonlinear and contain forcing terms, either in the boundary conditions or in the dynamical equations, all of which are neglected in the previous chapters on wave kinematics. The present chapter discusses the effect of prescribed forcing functions, such as the atmospheric pressure or the wind stress which may excite waves from a state of rest. Specifically, the generation of long waves both in midlatitudes and in the tropics will be discussed. Furthermore, the forcing of internal gravity waves will be considered where the nonlinear interactions play a crucial role. Nonlinearities lead to coupling of the waves branches and modes by which one wave may be excited through the presence of others – waves interact and can be created by other waves. We have discussed a general treatment in Section 9.6, using a Lagrangian framework. We expand the treatment to a situation where a random ensemble of waves is an adequate representation for the wave field.

The description of waves in the preceding chapters is focussed on linear, unforced wave systems and free-wave properties (except in Section 8.2.6 and in Section 9.6): the emphasis is on the relation of the wave frequency and the wave number and on dispersion and refraction and changes of the wave parameters in inhomogeneous media during propagation. The cause of the wave motion, what process has set them into being, is left from the discussion. Any initial push on a later freely propagating wave is a generation process involving external 'forces', indicated in the governing equations (5.69)–(5.73) by the $\mathcal{G}$ and $\mathcal{F}$ terms and similarly in the later versions (7.1)–(7.3) and (8.1)–(8.5). In fact, waves may be forced not only at an initial time but modified continuously during their life time.

The generation of a wave by some forcing is as simple as the excitement of a linear oscillator by an external force with periodic time behavior. What makes the wave system complicated from a mathematical point of view is the complicated structure of the oceanic wave guide and the resonance in wave-number-frequency space, deriving from the dispersion relation. What make it physically attractive is the wave propagation in concert with the forcing process.

D. Olbers, J. Willebrand, C. Eden, *Ocean Dynamics*,
DOI 10.1007/978-3-642-23450-7_10, 

## 10.1 The Forcing Functions of Long Waves

In the next sections, we derive a formalism allowing to treat forced wave problems. The relevant nonlinear terms and forcing function, abandoned in the previous sections, are restored in the governing equations, and their response on the wave system is determined. We restrict the treatment, however, to long waves, starting with the forced version of (8.1) to (8.5). Redefining the pressure as $p^\star = (p - p_\mathrm{a})/\rho_0$ where $p_\mathrm{a}$ is the atmospheric pressure, and then dropping the star, we write these equations as

$$\frac{\partial u}{\partial t} - fv + \frac{\partial p}{\partial x} = \mathcal{F}_u \tag{10.1}$$
$$\frac{\partial v}{\partial t} + fu + \frac{\partial p}{\partial y} = \mathcal{F}_v \tag{10.2}$$
$$\frac{\partial b}{\partial t} + wN^2 = \mathcal{G}_\mathrm{b} \tag{10.3}$$
$$\frac{\partial p}{\partial z} - b = 0 \tag{10.4}$$
$$\nabla \cdot \boldsymbol{u} + \frac{\partial w}{\partial z} = 0 \tag{10.5}$$

Here the buoyancy $b = -g\rho/\rho_0$ is introduced to replace density. The horizontal momentum balances and the buoyancy balance now contain the nonlinear advection terms and other nonspecified body forces or sources or frictional/diffusive terms $\boldsymbol{\mathcal{F}}^\star, \mathcal{G}_\mathrm{b}^\star$, i. e.

$$\boldsymbol{\mathcal{F}} = (\mathcal{F}_u, \mathcal{F}_v) = -\nabla p_\mathrm{a} - \boldsymbol{u} \cdot \nabla \boldsymbol{u} - w\frac{\partial \boldsymbol{u}}{\partial z} + \boldsymbol{\mathcal{F}}^\star \tag{10.6}$$
$$\mathcal{G}_\mathrm{b} = -\boldsymbol{u} \cdot \nabla b - w\frac{\partial b}{\partial z} + g(\alpha\mathcal{G}_\theta - \gamma\mathcal{G}_\mathrm{S}) + \mathcal{G}_\mathrm{b}^\star \tag{10.7}$$

where $\alpha$ and $\gamma$ are the coefficients of thermal and haline expansion (see Section 2.6.4).

The ocean is a wave guide because of the vertical boundaries and the mean stratification entering the theory via the Brunt–Väisälä frequency $N(z)$. The kinematic boundary conditions $D\zeta/Dt - w = (\mathcal{P} - \mathcal{E})/\rho_0$ at the sea surface $z = \zeta$, and $w = -\boldsymbol{u} \cdot \nabla h$ at the bottom $z = -h$, as well as the dynamic boundary condition $p = 0$ (actual pressure equal to the atmospheric pressure) at $z = \zeta$ were introduced before (see Sections 2.2 and 2.3). Here, the rate of precipitation minus evaporation, $\mathcal{P} - \mathcal{E}$ enters as external forcing function. We expand these conditions about the mean sea surface $z = 0$ and the mean bottom $z = -H$ and obtain

$$\frac{\partial \zeta}{\partial t} - w = \mathcal{Z} \quad \text{and} \quad p - g\zeta = \mathcal{T} \quad \text{at} \quad z = 0 \tag{10.8}$$
$$w = \mathcal{W} \quad \text{at} \quad z = -H \tag{10.9}$$

where $\mathcal{Z}, \mathcal{T}$ and $\mathcal{W}$ contain the forcing terms and the nonlinear terms arising in this expansion,

$$\mathcal{T} = -\left(\frac{1}{2}N^2\zeta^2 + \cdots\right)_{z=0} \tag{10.10}$$

$$\mathcal{Z} = \frac{1}{\rho_0}(\mathcal{P} - \mathcal{E}) + \left(\zeta\frac{\partial w}{\partial z} + \cdots\right)_{z=0} \tag{10.11}$$

$$\mathcal{W} = \left[-\boldsymbol{u}\cdot\nabla h - (h - H)\frac{\partial w}{\partial z} - \cdots\right]_{z=-H} \tag{10.12}$$

Notice that $h$ is the actual depth and $H$ a mean constant depth.

Putting the right-hand side of the equations (10.1)–(10.3) and (10.8)–(10.9) to zero, we regain the familiar linear wave equations, analyzed in the previous chapters. However, not all of the terms on the right-hand side are nonlinear: we also find the diffusive terms for momentum and buoyancy which are attached to boundary conditions, introducing the surface flux of horizontal momentum (windstress) and the surface flux of buoyancy, combining the surface heat flux and the flux of freshwater (see e. g. Section 3.1).

Though the above system of equations contains four time derivatives, the wave state is described by a 3-dimensional state vector. If we take $(u, v, p)$ as state vector, the remaining fields follow from diagnostic equations: (10.4) determines $b$; (10.5) together with the kinematic bottom boundary condition determines $w$; and (10.8) determines $\zeta$ as functionals of $(u, v, p)$. A prognostic equation for the pressure to supplement (10.1) and (10.2) is obtained from (10.3)–(10.5), and (10.8)–(10.9). Analogously to the homogeneous equation (8.9), one obtains

$$\frac{\partial p}{\partial t} + M\nabla\cdot\boldsymbol{u} = \mathcal{Q} \tag{10.13}$$

where

$$\mathcal{Q} = \frac{\partial\mathcal{T}}{\partial t} - g\mathcal{Z} - \int_z^0 \mathcal{G}_b \mathrm{d}z + \mathcal{W}\int_z^0 N^2 \mathrm{d}z' \tag{10.14}$$

collects forcing and nonlinear terms. The integral operator

$$M = g\int_{-H}^0 \mathrm{d}z'' + \int_z^0 \mathrm{d}z' N^2(z') \int_{-H}^{z'} \mathrm{d}z'' \tag{10.15}$$

is defined as in (8.10) and acts only on the vertical structure. The properties of $M$, and of the homogeneous system (10.1)–(10.5), are discussed in Section 8.1.

## 10.2 Forced Midlatitude Waves

For midlatitude waves, it is preferable to consider the budgets of horizontal divergence $\delta$ and the vorticity $\eta$ which are defined as

$$\delta = \frac{\partial u}{\partial x} + \frac{\partial v}{\partial y} \quad \text{and} \quad \eta = \frac{\partial v}{\partial x} - \frac{\partial u}{\partial y} \tag{10.16}$$

These are used to replace[1] $u = \nabla^{-2}(\partial\delta/\partial x - \partial\eta/\partial y), v = \nabla^{-2}(\partial\eta/\partial x + \partial\delta/\partial y)$. With (10.13), the evolution equations become

$$\frac{\partial\delta}{\partial t} - f\eta + \beta u + \nabla^2 p = \nabla\cdot\mathcal{F} \tag{10.17}$$

$$\frac{\partial\eta}{\partial t} + f\delta + \beta v = \operatorname{curl}\mathcal{F} \tag{10.18}$$

$$\frac{\partial p}{\partial t} + M\delta = \mathcal{Q} \tag{10.19}$$

in terms of the new state vector $(\delta, \eta, p)$. Here $\beta = \mathrm{d}f/\mathrm{d}y$ arises from the differential rotation due to the change of the Coriolis parameter $f(y)$ with latitude. From (10.17)–(10.19), a scalar wave equation is easily derived for the $f$-plane case with uniform rotation $f = f_0 = \text{const}$ and $\beta = 0$. Then, the pressure equation

$$\frac{\partial}{\partial t}\left[\frac{\partial^2}{\partial t^2} + (f_0^2 - M\nabla^2)\right] p = \frac{\partial^2\mathcal{Q}}{\partial t^2} + f_0^2\mathcal{Q} - M\nabla\cdot\frac{\partial\mathcal{F}}{\partial t} - f_0 M \operatorname{curl}\mathcal{F} \tag{10.20}$$

is obtained. Plane wave solutions $p \sim \Phi(z)\exp\mathrm{i}(\boldsymbol{k}\cdot\boldsymbol{x} - \omega t)$ with $\boldsymbol{k} = (k_1, k_2)$ are found for the unforced equation with three branches (i. e. different wave types). The dispersion relations are $\omega_n^{(1,2)} = \pm f_0[1 + (kR_n)^2]^{1/2}$ (the gravity branches) and $\omega_n^{(3)} = 0$ (the geostrophic branch). The latter is degenerated on the $f$-plane: it describes a steady geostrophically balanced current where Coriolis and pressure forces balance, $-f_0 v = -\partial p/\partial x$ and $f_0 u = -\partial p/\partial y$.

The degeneracy is relieved on the $\beta$-plane where both $f$ and $\beta$ are retained and considered constant in the above equations (10.17)–(10.19). This approximation can be justified by a proper expansion into various small parameters (see QG approximation in Section 5.2). Here we simply assume $\beta L/f \ll 1$ where $L$ is a typical horizontal scale, and $f/\beta$ is of the order of the Earth radius, and $\omega \ll f$. The first order correction of the geostrophic branch can then be derived by extracting the geostrophic balances from (10.1) and (10.2), evaluating the geostrophic vorticity as $\eta = (1/f_0)\nabla^2 p$ and using (10.17) and (10.19) to obtain

$$\frac{\partial}{\partial t}\left(f_0^2 - M\nabla^2\right) p - \beta M\frac{\partial p}{\partial x} = f_0^2\mathcal{Q} - f_0 M \operatorname{curl}\mathcal{F} \tag{10.21}$$

We deduce $\omega_n^{(3)} = -\beta k_1/\left[k_1^2 + k_2^2 + R_n^{-2}\right]$ which is the dispersion relation of linear planetary Rossby waves. Another way to filter out gravity waves is to neglect the tendency term in the divergence equation (10.17).

The separation of the wave spectrum of motions is however not completed with the above analysis: (10.20) describes all wave branches on the $f$-plane, and one of them is degenerate while (10.21) yields the correct form of the geostrophic wave response. A complete separation into the wave branches requires a proper diagonalization of the linear matrix operator appearing in the system (10.17)–(10.19). With the same approximations used above to obtain (10.21), this goal can strictly be achieved. Consider the wave evolution equations (10.17)–(10.19), written in the form

$$\frac{\partial\boldsymbol{\psi}}{\partial t} + \mathrm{i}(\boldsymbol{H} + \boldsymbol{B})\cdot\boldsymbol{\psi} = \boldsymbol{q} \tag{10.22}$$

[1] The use of the inverse of the Laplace operator $\nabla^2$ and of other operators involving $\nabla^2$ in this section is a sloppy shorthand notation. The symbolic form $f = \nabla^{-2}g$ is equivalent to solving $\nabla^2 f = g$ with the associated boundary conditions.

Here the state vector $\boldsymbol{\psi}$ and forcing function $\boldsymbol{q}$ are given by

$$\boldsymbol{\psi} = \left(\delta, \eta, \frac{1}{f_0}\nabla^2 p\right)^{\mathrm{T}} \quad \text{and} \quad \boldsymbol{q} = \left(\nabla \cdot \mathcal{F}, \operatorname{curl} \mathcal{F}, \frac{1}{f_0}\nabla^2 Q\right)^{\mathrm{T}} \tag{10.23}$$

and the matrix operators $\boldsymbol{H}$ and $\boldsymbol{B}$ are defined as

$$\boldsymbol{H} = -\mathrm{i} f_0 \begin{pmatrix} 0 & -1 & 1 \\ 1 & 0 & 0 \\ \mathcal{M}\nabla^2 & 0 & 0 \end{pmatrix} \quad \boldsymbol{B} = -\mathrm{i}\beta\nabla^{-2} \begin{pmatrix} \partial/\partial x & -\partial/\partial y & 0 \\ \partial/\partial y & \partial/\partial x & 0 \\ 0 & 0 & 0 \end{pmatrix} \tag{10.24}$$

Note that the modified operator $\mathcal{M} = M/f_0^2$ for the vertical eigenfunctions has been used, and the imaginary unit $i$ is introduced in (10.22) for convenience in the notation. Because $||\boldsymbol{B}||/||\boldsymbol{H}|| = O(\beta L/f_0) \ll 1$ and $\boldsymbol{H}$ is easily diagonalized, we approach the problem by expansion in terms of $\beta L/f_0$. To lowest order in $\beta L/f_0$, the eigenvalue problem of the evolution operator is $\boldsymbol{H} \cdot \boldsymbol{R}^s = \Omega^{(s)} \boldsymbol{R}^s$ where $s = \pm, 0$ counts the three eigensolutions, and no summation over $s$ is implied. This is solved by

$$\Omega^{(s)} = s\Omega = s f_0 \left(1 - \mathcal{M}\nabla^2\right)^{1/2} \quad \text{and} \quad \boldsymbol{R}^s = \left[\mathrm{i}s\Omega/f_0, 1, 1 - (s\Omega/f_0)^2\right]^{\mathrm{T}} \tag{10.25}$$

for $s = \pm, 0$. Note that this eigenvalue problem concerns only the algebraic aspects in the space of dependent variables of the full problem (10.22), and both eigenvectors and eigenvalues still contain integral/differential operators related to $\mathcal{M}$ and $\nabla^2$. The left eigenvectors, obtained from $\boldsymbol{P}^s \cdot \boldsymbol{H} = \Omega^{(s)} \boldsymbol{P}^s$, are given by

$$\boldsymbol{P}^{\pm} = \frac{1}{2}(\Omega/f_0)^{-2}(\mp\mathrm{i}\Omega/f_0, 1, -1) \quad \text{and} \quad \boldsymbol{P}^0 = (\Omega/f_0)^{-2}\left[0, (\Omega/f_0)^2 - 1, 1\right] \tag{10.26}$$

We will call $\boldsymbol{R}^s$ *representation vector* and $\boldsymbol{P}^s$ *projection vector* for reasons which will immediately become evident. Both vectors are mutually orthogonal and normalized, $\boldsymbol{P}^s \cdot \boldsymbol{R}^{s'} = \delta^{ss'}$ and $\sum_s \boldsymbol{R}^s \boldsymbol{P}^s = \boldsymbol{I}$ (3 × 3 unit matrix). The branches $s = \pm$ with the corresponding field functions $\psi^{\pm}$, describing gravity waves, and the branch $s = 0$ with field function $\psi^0$, representing the steady geostrophic flow, are defined as projections of the fields $\boldsymbol{\psi}$ in the form

$$\psi^s = \boldsymbol{P}^s \cdot \boldsymbol{\psi} \quad (s = +, -, 0) \quad \text{and} \quad \boldsymbol{\psi} = \sum_s \boldsymbol{R}^s \psi^s \tag{10.27}$$

The vectors $\boldsymbol{R}^s$ achieve thus the representation of the state vector $\boldsymbol{\psi}$ in terms of the wave field function $\psi^s$, whereas the $\boldsymbol{P}^s$ projects the state $\boldsymbol{\psi}$ on the field function $\psi^s$. The first relation in (10.27) is explicitly given as

$$\begin{aligned} \psi^0 &= \left(\frac{f_0}{\Omega}\right)^2 \nabla^2 \left[-\mathcal{M}\eta + p/f_0\right] \quad \text{and} \\ \psi^{\pm} &= \frac{1}{2}\left(\frac{f_0}{\Omega}\right)^2 \left[\mp\mathrm{i}\frac{\Omega}{f_0}\delta + \eta - \frac{1}{f_0}\nabla^2 p\right] \end{aligned} \tag{10.28}$$

For gravity waves, (10.17)–(10.19) yield $p = M\delta/(\mathrm{i}\omega)$, $\eta = f_0\delta/(\mathrm{i}\omega)$, hence $\psi^0 = 0$. For planetary waves, we have $\delta = 0$, $\eta = \nabla^2 p/f_0$ and hence $\psi^{\pm} = 0$. In general, however, the operation leading to $\psi^0$ extracts the geostrophically balanced part from the general state $(\delta, \eta, \nabla^2 p/f_0)$, and $\psi^\pm$ extracts the gravity wave part.

Finally, the first order correction of the eigenvalue problem is $\boldsymbol{B} \cdot \boldsymbol{R}^s + \boldsymbol{H} \cdot \tilde{\boldsymbol{R}}^s = \Omega^{(s)} \tilde{\boldsymbol{R}}^s + \tilde{\Omega}^{(s)} \boldsymbol{R}^s$ where the operator $\boldsymbol{B}$ and the tilde quantities are of first order in $\beta L/f_0$. Since the eigenvectors are normalized, only changes in direction are relevant, and we require that $\boldsymbol{P}^s \cdot \tilde{\boldsymbol{R}}^s = 0$. It yields corrections to the representation vector (which are not needed because the lowest order of these quantities is nonzero) and of the eigenvalue operator $\tilde{\Omega}^{(s)} = \boldsymbol{P}^s \cdot \boldsymbol{B} \cdot \boldsymbol{R}^s$. For the geostrophic branch we thus obtain

$$\tilde{\Omega}^{(0)} = \mathrm{i}\beta \mathcal{M} \left(1 - \mathcal{M}\nabla^2\right)^{-1} \frac{\partial}{\partial x} \tag{10.29}$$

as correction to the geostrophic evolution operator $\Omega^{(0)}$ (which is zero to lowest order, and thus the correction is needed). We denote from now $\tilde{\Omega}^{(0)}$ by $\Omega^{(0)}$.

The wave evolution is then governed by the three independent problems

$$\frac{\partial \psi^s}{\partial t} + \mathrm{i}\Omega^{(s)} \psi^s = q^s = \boldsymbol{P}^s \cdot \boldsymbol{q}\,, \quad s = \pm, 0 \tag{10.30}$$

These abstract equations of wave evolution can be cast into a more familiar form by expressing $\Omega^{(s)}$ with (10.25) and $\boldsymbol{P}^s$ with (10.26). Writing (10.30) separately for the gravity branch ($s = \pm$) and the Rossby wave branch ($s = 0$), one obtains

$$\frac{\partial \psi^\pm}{\partial t} \pm \mathrm{i} f_0 (1 - \mathcal{M}\nabla^2)^{1/2} \psi^\pm = \frac{1}{2}\left(\frac{f_0}{\Omega}\right)^2 \left(\mp \mathrm{i}\frac{\Omega}{f_0} \nabla \cdot \mathcal{F} + \operatorname{curl} \mathcal{F} - \frac{1}{f_0}\nabla^2 \mathcal{Q}\right) \tag{10.31}$$

$$\frac{\partial}{\partial t}\left(1 - \mathcal{M}\nabla^2\right)\psi^0 - \beta\mathcal{M}\frac{\partial \psi^0}{\partial x} = -\mathcal{M}\nabla^2 \operatorname{curl}\mathcal{F} + \frac{1}{f_0}\nabla^2 \mathcal{Q} \tag{10.32}$$

The gravity wave branch fields are conjugate to each other, $\psi^+ = (\psi^-)^*$. Note that (10.32) seems to be equivalent to (10.21) which was derived by a much simpler procedure from the governing equations. There is, however, an important difference: the pressure $p$ in (10.21) is the total pressure without concern of a geostrophic balance, while $\psi^0$, defined by (10.28), is the amplitude of the geostrophically balanced branch of the motion. A similar remark applies to the equations (10.31) and (10.20).

For practical applications, the wave field functions $\psi^s$ should be expanded in the eigenfunctions of $\mathcal{M}$, which simply replaces $\mathcal{M}$ by $c_n^2/f_0^2 = R_n^2$ and, e.g., the Rossby wave field $\psi^0(x, y, z, t)$ by the modal amplitudes $\psi_n^0(x, y, t)$. For laterally homogeneous problems, further simplification is achieved by expansion into the eigenfunctions $\exp(\mathrm{i}\boldsymbol{k} \cdot \boldsymbol{x})$ of the $\nabla^2$-operator which replaces $\nabla^2$ by $-k^2$ and modal amplitudes $\psi_n^0(\boldsymbol{k}, t)$ arise.

### Generation of Long Gravity Waves by Pressure Fluctuations

As a first example of the above procedure, consider the generation of long gravity waves by a variety of mechanisms, such as variations of atmospheric surface fluxes (i. e. windstress, air pressure, and mass fluxes) as well as tidal forcing. To be specific, we will consider only the barotropic component of long waves. Similar considerations, though different in detail, apply to baroclinic waves. For the barotropic mode, we have $c_0 = \sqrt{gH}$ and $R_0^2 = gH/f_0$ where $R_0 = c_0/f_0 \sim 2{,}000$ km is the barotropic Rossby radius. The gravity wave equation (10.31) becomes

$$(-\mathrm{i}\omega \pm \mathrm{i}\Omega)\,\psi^{\pm} = \frac{1}{2}\left(\frac{f_0}{\Omega}\right)^2 \left(\mp \mathrm{i}\frac{\Omega}{f_0}\nabla \cdot \boldsymbol{\mathcal{F}} + \operatorname{curl} \boldsymbol{\mathcal{F}} - \frac{1}{f_0}\nabla^2 \mathcal{Q}\right) \qquad (10.33)$$

for disturbances $\sim \exp(-\mathrm{i}\omega t)$. We replace $\Omega/f_0$ by $(1 + (R_0 k)^2)^{1/2}$, so that $\psi^{\pm}$ is the response in the barotropic mode at the wave number $\boldsymbol{k}$ and frequency $\omega$. Forcing can be easily represented in terms of atmospheric pressure variations $p_{\mathrm{a}}$ and windstress $\boldsymbol{\tau}$. All these forcing functions are contained in $\boldsymbol{\mathcal{F}}$ which then takes the form

$$\boldsymbol{\mathcal{F}} = -\nabla p_{\mathrm{a}} + \boldsymbol{\tau}/H \qquad (10.34)$$

The forcing terms in (10.33) allow a direct comparison of the magnitude of the different driving forces. For the windstress, the divergence term dominates the curl term at time-scales of order $f_0^{-1}$ or shorter, whereas at longer time-scales the windstress curl (last term) dominates the wind forcing. We continue without wind forcing because it is interesting and particularly simple to consider the barotropic response to forcing by air pressure and tidal forcing. With the definition $\zeta_{\mathrm{e}} = -p_{\mathrm{a}}/g$ where $\zeta_{\mathrm{e}}$ is the equilibrium response to pressure in a motionless state, one obtains from (10.33)

$$(\Omega \mp \omega)\,\psi^{\pm} = -\frac{1}{2}\left(\frac{f_0}{\Omega}\right)\nabla^2 g\zeta_{\mathrm{e}} \qquad (10.35)$$

and the pressure field $p$ and displacement field $\zeta = p/g$ of the ocean response in the gravity wave branch follow from the representation $\psi_3 = \nabla^2 p/f_0 = R_3^+\psi^+ + R_3^-\psi^-$, hence

$$\zeta(\boldsymbol{k},\omega) = \frac{k^2 R_0^2}{1 + k^2 R_0^2 - \omega^2/f_0^2}\zeta_{\mathrm{e}}(\boldsymbol{k},\omega) = r^{\mathrm{grav}}(k,\omega)\,\zeta_{\mathrm{e}}(\boldsymbol{k},\omega) \qquad (10.36)$$

The resonance function $r^{\mathrm{grav}}(k,\omega)$ is displayed in Figure 10.1. Depending on the frequency, the following limiting cases can be identified:

1. $\omega^2/f_0^2 \gg 1 + k^2 R_0^2$, i. e. forcing period $\ll 1$ day. One finds $r^{\mathrm{grav}}(k,\omega) < 0$ and $|r^{\mathrm{grav}}(k,\omega)| \ll 1$; hence the ocean's reaction is small, with a phase opposite to that of the forcing.
2. $\omega^2/f_0^2 = 1 + k^2 R_0^2$, i. e. forcing period and scale correspond exactly to those of long gravity waves. Then, $r^{\mathrm{grav}}(k,\omega) \to \infty$ applies, and *resonance* is present. In this case, the amplitude has to be limited by nonlinear effects and/or dissipation.
3. $\omega = f_0$, i. e. forcing at inertial period. Here $r^{\mathrm{grav}}(k,\omega) \equiv 1$ applies, independent of $k$, and $\zeta = \zeta_{\mathrm{e}}$ follows (equilibrium response). This case is, however, less interesting because inertial waves are mainly generated by the wind.
4. $\omega^2/f_0^2 \ll 1 + k^2 R_0^2$, and thus $r^{\mathrm{grav}}(k,\omega) = k^2 R_0^2/(1 + k^2 R_0^2) < 1$. Note that for sufficiently small wavelengths ($kR_0 \gg 1$) it follows that $r^{\mathrm{grav}}(k,\omega) \to 1$, and thus $\zeta \approx \zeta_{\mathrm{e}}$ (equilibrium response). For air-pressure forcing ($\zeta_{\mathrm{e}} = -p_{\mathrm{a}}/g$), this situation is referred to as *inverse barometer* response. For $kR_0 \gtrsim 2$, which is typical for many synoptic variations in air pressure, one finds $r^{\mathrm{grav}}(k,\omega) \gtrsim 0.8$. Hence at these scales the inverse barometer is a good approximation.

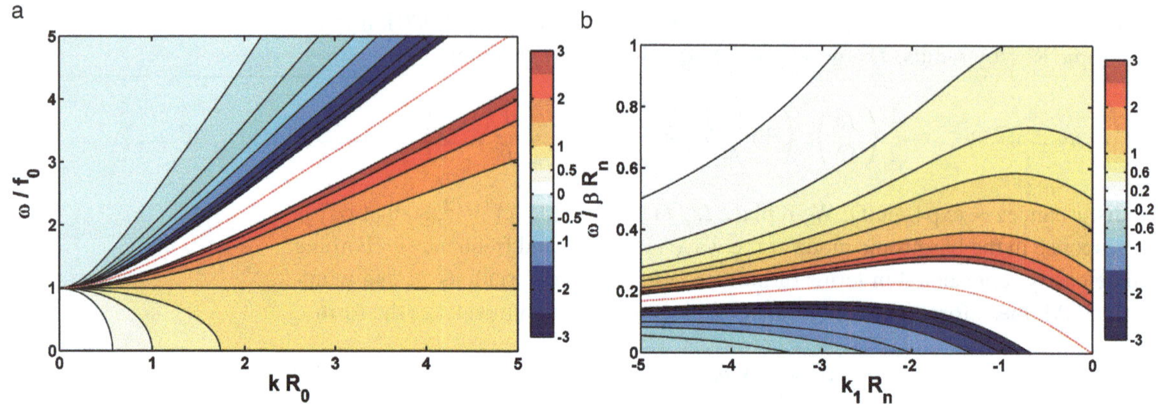


**Fig. 10.1** The resonance functions $r^{\text{grav}}$ (**a**) and $r_n^{\text{ross}}$ (**b**) as function of wave number and frequency, defined by (10.36) and (10.39), respectively . The *dashed red curve* is the respective resonance line. Values above 3 and below $-3$ are wiped out. For the Rossby wave case, we have taken $k_2 R_n = 2$

### Generation of Baroclinic Rossby Waves by Windstress Fluctuations

We assume a simple body-force model for the windstress coupling, $\mathcal{F} = S'(z)\boldsymbol{\tau}_0$ (the stress has a vertical structure given by $S(z)$). Expanding the structure function into the vertical normal modes, $S'(z) = \sum_n s_n \Phi_n(z)$, one finds $s_n \approx \Phi_n(0) \sim 1/\sqrt{H}$ (see also Section 8.2.6 on the use of the body-force model). Looking for the response of the Rossby wave amplitude $\psi_n^0(\boldsymbol{k}, \omega)$ of the mode $n$ at the wave number $\boldsymbol{k}$ and frequency $\omega$, we immediately obtain from (10.32) the result

$$\psi_n^0(\boldsymbol{k}, \omega) = -\frac{s_n (R_n k)^2}{1 + (R_n k)^2} \frac{k_1 \tau_0^y(\boldsymbol{k}, \omega) - k_2 \tau_0^x(\boldsymbol{k}, \omega)}{\omega - \omega_n} \tag{10.37}$$

To avoid the resonant singularity when the frequency $\omega$ matches the Rossby wave frequency $\omega_n = -\beta k_1 R_n^2/(1 + (k R_n)^2)$, friction must be added. The amplitude $\psi_n^0$ relates to the vorticity and pressure via $\psi_n^0 = [(R_n k)^2 \eta_n - k^2 p_n / f_0]/[1 + (R_n k)^2]$, which becomes $\psi_n^0 = -k^2 p_n / f_0$ for a geostrophically balanced state. We then find

$$p_n(\boldsymbol{k}, \omega) = \frac{s_n R_n^2 f_0}{1 + (R_n k)^2} \frac{k_1 \tau_0^y(\boldsymbol{k}, \omega) - k_2 \tau_0^x(\boldsymbol{k}, \omega)}{\omega - \omega_n} \tag{10.38}$$

for the pressure variations. Restricting the analysis to a zonal windstress, the pressure amplitude becomes

$$\begin{aligned} s_n p_n(\boldsymbol{k}, \omega) &= \frac{k_2 R_n}{k_1 R_n + [1 + (R_n k)^2]\omega/\bar{\omega}_n} \frac{s_n^2 f_0}{\beta} \tau_0^x(\boldsymbol{k}, \omega) \\ &= r_n^{\text{ross}}(\boldsymbol{k}, \omega) \frac{s_n^2 f_0}{\beta} \tau_0^x(\boldsymbol{k}, \omega) \end{aligned} \tag{10.39}$$

where $\bar{\omega}_n = 2\omega_n^{\max} = \beta R_n$ is twice the maximum Rossby wave frequency of the mode $n$ (see Section 8.2.1). Note that $s_n p_n$ as well $\tau_0^x$ have the dimension of $\text{m}^2\,\text{s}^{-2}$ (after scaling the pressure and stress by a reference density). Furthermore, $s_n^2 f_0/\beta \approx a/H$ where $a$ is the Earth radius. Here are some numbers to characterize the response: for an amplitude $\delta\tau = 10^{-5}\,\text{m}^2\,\text{s}^{-2}$ of windstress fluctuations,

$a/H = 10^3$ and $r_n^{\text{ross}} = 1$, we find a pressure amplitude $\delta p = 10^{-3}$ dbar or a displacement of a density interface of roughly 1 m. One has to be quite close to the resonance (i. e. large $r_n^{\text{ross}}$) to obtain a significant response. The simple structure of the resonance function $r_n^{\text{ross}}(\boldsymbol{k}, \omega)$ is displayed in Figure 10.1. There is a strong decrease of $|r_n^{\text{ross}}|$ away from the resonance at $\omega = \omega_n$. Disturbances of the pressure are in phase with the forcing for $\omega > \omega_n$ and out of phase for $\omega < \omega_n$.

## 10.3 Forced Equatorial Waves

The Coriolis parameter vanishes at the equator and, as presented in Section 8.3, a special wave theory must be developed. The equatorial $\beta$-plane uses $f = \beta y$ with constant $\beta = 2\Omega/a$ in (10.1), (10.2) and (10.13). As discussed in Section 8.3, the system supports gravity and Rossby type waves which are trapped vertically (as in midlatitudes) but also meridionally. Equations for the wave branches are obtained by vertical decomposition (replacement of $M$ by $c_n^2$; we will omit the index $n$). We use the state vector $(\theta_1, \theta_2, \theta_3)$ as defined in Section 8.3.4, $\theta_1 = p/c + u$, $\theta_2 = L_+^{-1} v$ and $\theta_3 = L_+^{-2}(p/c - u)$ and write the forced version of (8.159)–(8.161) as

$$\left(\frac{\partial}{\partial t} + c\frac{\partial}{\partial x}\right)\theta_1 + c\left(L + \frac{1}{R_\text{e}^2}\right)\theta_2 = \mathcal{Q}/c + \mathcal{F}_u = \mathcal{T}_1 \tag{10.40}$$

$$\frac{\partial \theta_2}{\partial t} + \frac{c}{2}\theta_1 + \frac{c}{2}\left(L + \frac{3}{R_\text{e}^2}\right)\theta_3 = L_+^{-1}\mathcal{F}_v = \mathcal{T}_2 \tag{10.41}$$

$$\left(\frac{\partial}{\partial t} - c\frac{\partial}{\partial x}\right)\theta_3 + c\theta_2 = L_+^{-2}\left(\mathcal{Q}/c - \mathcal{F}_u\right) = \mathcal{T}_3 \tag{10.42}$$

and for convenience repeat the definitions (8.136) and (8.152) of the operators appearing in this representation,

$$L = \frac{\partial^2}{\partial y^2} - \frac{y^2}{R_\text{e}^4} \qquad L_+^{-1}\chi = \mathrm{e}^{-\frac{1}{2}y^2/R_\text{e}^2}\int_{-\infty}^{y} \mathrm{e}^{\frac{1}{2}y'^2/R_\text{e}^2}\chi(y')\mathrm{d}y' \tag{10.43}$$

As shown in Section 8.3, eigenfunctions of $L$ are the Hermite functions $n_\ell \psi_\ell(y/R_\text{e}), \ell = 0, 1, 2, \ldots$ with the normalization factors $n_\ell = (2^\ell \ell! \sqrt{\pi})^{(-1/2)}$ and eigenvalues $-(2\ell+1)/R_\text{e}^2$. We note that the equatorial wave problem, expressed in the potentials $\theta_i$ (in contrast to $(q, v, r)$ or $(u, v, p)$), is diagonal with respect to an expansion of the variables in Hermite functions, i. e. with $L$ replaced by $-(2\ell+1)/R_\text{e}^2$ and $\mathcal{T}_i$ by their amplitudes $\mathcal{T}_{i\ell}$ of mode $\ell$, we may consider (10.40) to (10.42) as equations determining the amplitudes $\theta_{i\ell}$ of the meridional mode $\ell$. We find

$$\left(\frac{\partial}{\partial t} + c\frac{\partial}{\partial x}\right)\theta_{1\ell} - 2\beta\ell\theta_{2\ell} = \mathcal{T}_{1\ell} \tag{10.44}$$

$$\frac{\partial \theta_{2\ell}}{\partial t} + \frac{c}{2}\theta_{1\ell} - (\ell - 1)\beta\theta_{3\ell} = \mathcal{T}_{2\ell} \tag{10.45}$$

$$\left(\frac{\partial}{\partial t} - c\frac{\partial}{\partial x}\right)\theta_{3\ell} + c\theta_{2\ell} = \mathcal{T}_{3\ell} \tag{10.46}$$

which may be simplified further by scaling time $t$ by $\tau = (2\beta c)^{-1/2}$, the zonal coordinate $x$ by $\xi = (c/2\beta)^{1/2}$, the potentials as $\theta_{i\ell} = (\xi^i/\tau)\theta'_{i\ell}$, and the forcing terms as $\mathcal{T}_{i\ell} = (\xi^i/\tau^2)\mathcal{T}'_{i\ell}, i = 1, 2, 3$. This yields, dropping the primes,

$$\left(\frac{\partial}{\partial t} + \frac{\partial}{\partial x}\right)\theta_{1\ell} - \ell\theta_{2\ell} = \mathcal{T}_{1\ell} \tag{10.47}$$

$$\frac{\partial \theta_{2\ell}}{\partial t} + \frac{1}{2}\theta_{1\ell} - \frac{1}{2}(\ell - 1)\theta_{3\ell} = \mathcal{T}_{2\ell} \tag{10.48}$$

$$\left(\frac{\partial}{\partial t} - \frac{\partial}{\partial x}\right)\theta_{3\ell} + \theta_{2\ell} = \mathcal{T}_{3\ell} \tag{10.49}$$

The component $\ell = 0$ is the Kelvin wave branch, $\ell = 1$ the Yanai wave branch, and for $\ell \geq 2$ we have gravity and Rossby waves.

There is a simple procedure to filter out the gravity and Yanai waves from the system: this is performed by omitting the time derivative in (10.48) leading to the diagnostic relation $\theta_{1\ell} - (\ell - 1)\theta_{3\ell} = 2\mathcal{T}_{2\ell}$. Introducing the variable (an equatorial potential vorticity)

$$\varphi_\ell = \theta_{1\ell} + (\ell - 1)\theta_{3\ell} \quad \ell = 2, 3, \ldots \tag{10.50}$$

one finds

$$\left(\frac{\partial}{\partial t} - \frac{1}{2\ell - 1}\frac{\partial}{\partial x}\right)\varphi_\ell = 2\frac{\ell - 1}{2\ell - 1}\mathcal{T}_{1\ell} + 2\left(\frac{\partial}{\partial t} - \frac{1}{2\ell - 1}\frac{\partial}{\partial x}\right)\mathcal{T}_{2\ell} + 2\frac{\ell(\ell + 1)}{2\ell - 1}\mathcal{T}_{3\ell} \tag{10.51}$$

Only long Rossby waves are retained as free solutions of (10.51) (with $\omega = -k/(2\ell - 1)$, compare with (8.185)).

By projection onto the meridional modes, various problems of trapped equatorial wave motion can be formulated. If the system is zonally unbounded or periodic, the waves propagate independently. Interesting problems arise in a zonally bounded wave guide (e. g. the Pacific Ocean) since wave reflection at zonal boundaries couples the wave branches. The reflection process is complicated, as discussed in Section 8.3.8.

### Example

As example, we consider the system of Kelvin and long Rossby waves excited by a localized wind patch as only forcing function in (10.40)–(10.42), i. e. $\mathcal{F}_v = \mathcal{Q} = 0$ and $\mathcal{F}_u = F(x)G(y)$ where $F(x)$ and $G(y)$ have a finite support of a size which is large compared to the Rossby radius. Hence $\mathcal{T}_1 = F(x), \mathcal{T}_2 = 0, \mathcal{T}_3 = -F(x)L_+^{-2}G(y)$. So we consider

$$\left(\frac{\partial}{\partial t} + \frac{\partial}{\partial x}\right)\theta_{10} = \alpha_0 F(x) \tag{10.52}$$

$$\left(\frac{\partial}{\partial t} - \frac{1}{2\ell - 1}\frac{\partial}{\partial x}\right)\varphi_\ell = \frac{1}{2\ell - 1}\left[(\ell - 1)\alpha_\ell - \ell(\ell + 1)\beta_\ell\right]F(x) \tag{10.53}$$

The coefficients $\alpha_\ell$ and $\beta_\ell$ derive from projecting the $G(y)$-terms. The first equation determines the Kelvin wave amplitude, the second is for the Rossby wave of mode

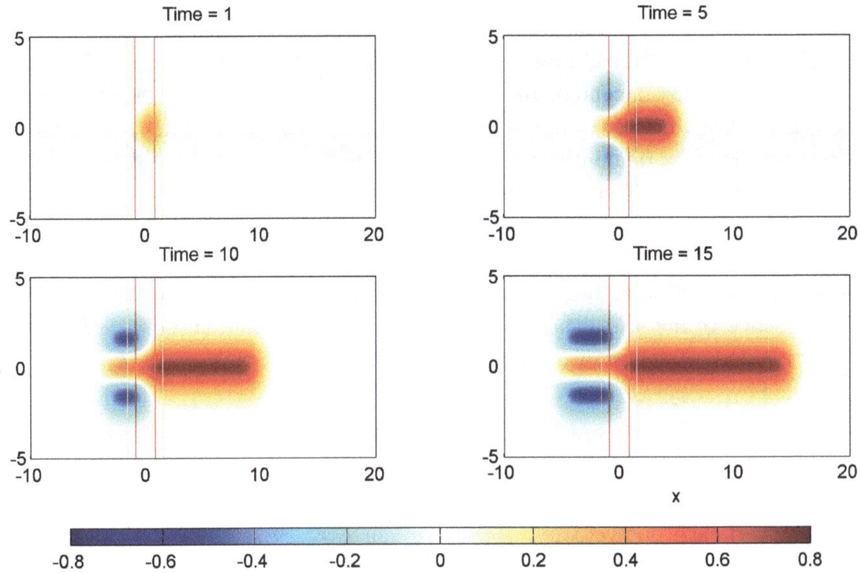


Fig. 10.2 Kelvin wave and Rossby wave $\ell = 2$, generated by a steady wind patch localized as indicated by the red lines. The Kelvin wave propagates to the east out of the forcing region, the Rossby wave to the west

$\ell = 2, 3, \ldots$ The solution is easily constructed because the waves have no dispersion. For instance, the Kelvin wave solution is

$$\theta_{10}(x,t) = \alpha_0 \left[ \int_0^x F(x')\mathrm{d}x' - \int_0^{x-t} F(x')\mathrm{d}x' \right] \tag{10.54}$$

for the initial condition $\theta_{10}(x, t = 0) = 0$. The forced wave system consisting of the Kelvin wave and the long Rossby wave $\ell = 2$ is shown for different times in Figure 10.2.

## 10.4 * Energetics of a Random Gravity Wave Field

In Section 9.6.2 we have laid the fundament for the description of statistical wave fields, and the weak-interaction theory was developed for the resonant energy transfer a the wave spectrum as one contribution to the source function $\mathcal{S}$ in the radiative transfer equation (9.74). We have noticed the highly complicated nature of the nonlinear transfer, leading to scattering integrals like (9.82) for the case of triad interactions. In addition to wave-wave interactions, an oceanic wave field, however, is subject to very specific sources and sinks of energy in the wave-number space, and thus in general, all terms of the radiative transfer equation (RTE)

$$\frac{\partial \mathcal{N}}{\partial t} + \nabla_{\boldsymbol{x}} \cdot (\dot{\boldsymbol{x}}\mathcal{N}) + \nabla_{\boldsymbol{k}} \cdot (\dot{\boldsymbol{k}}\mathcal{N}) = \mathcal{S} = \mathcal{S}_{\text{gen}} + \mathcal{S}_{\text{wwi}} + \mathcal{S}_{\text{diss}} \tag{10.55}$$

must be considered. We have separated the source term $\mathcal{S}$ into three contributions: $\mathcal{S}_{\text{gen}}$ describes the generation of waves by external processes, $\mathcal{S}_{\text{wwi}}$ represents the energy in the spectrum due to resonant wave-wave interactions, and $\mathcal{S}_{\text{diss}}$ stands for dissipation terms which eliminate wave energy. All sources and sinks, in particular $\mathcal{S}_{\text{wwi}}$ and $\mathcal{S}_{\text{diss}}$, may depend on the action spectrum $\mathcal{N}(\boldsymbol{k},\boldsymbol{x},t)$, and the prototype for $\mathcal{S}_{\text{wwi}}$ is the triad term $\mathcal{S}_{\text{triad}}$ in (9.82). In the present section we intend to exemplify the energetics of a random wave field, having internal gravity waves in mind, and we choose the three-dimensionally propagating form of internal waves developed in Section 7.3. Hence $\boldsymbol{x} = (\boldsymbol{x}_{\text{h}}, x_3)$ and $\boldsymbol{k} = (\boldsymbol{k}_{\text{h}}, k_3)$ are three-dimensional.

To simplify, we assume that the wave field is stationary and horizontally homogeneous, so that only a vertically varying Brunt–Väisälä frequency profile $N(x_3)$ and the presence of the sea surface at $x_3 = 0$ and (flat) bottom at $x_3 = -H$ lead to reflection and refraction of the waves. The action spectrum $\mathcal{N} = \mathcal{N}(\boldsymbol{k}, x_3)$ is then governed by

$$\frac{\partial}{\partial x_3}(\dot{x}_3\mathcal{N}) + \frac{\text{d}}{\text{d}k_3}(\dot{k}_3\mathcal{N}) = \mathcal{S} = \mathcal{S}_{\text{gen}} + \mathcal{S}_{\text{wwi}} + \mathcal{S}_{\text{diss}} \tag{10.56}$$

with the vertical group velocity $\dot{x}_3 = c_{g3}$, given by (7.19), and the vertical refraction $\dot{k}_3$, given by (7.29). Note that $\dot{k}_3 \equiv 0$ for a constant Brunt–Väisälä frequency $N$. We proceed with this simpler condition in the following.

To use the radiative transfer equation (10.56), the action flux $\mathcal{F}_3(\boldsymbol{k}, x_3) = c_{g3}(\boldsymbol{k})\mathcal{N}(\boldsymbol{k}, x_3)$ must be specified at the boundaries $x_3 = 0, -H$. Consider for example the sea surface. A wave impinging from the interior has $k_3 < 0, c_{g3} > 0$ and is reflected into one with the same modulus of $k_3$ but reversed signs for the vertical wave number and group velocity, $k_3 > 0, c_{g3} < 0$ (see Section 7.4). If there is no source of action (or equivalently energy) at the surface, the spectrum must be symmetric with respect to $k_3$ and the flux antisymmetric: $\mathcal{N}(\boldsymbol{k}_{\text{h}}, k_3, x_3 = 0) = \mathcal{N}(\boldsymbol{k}_{\text{h}}, -k_3, x_3 = 0)$ and $\mathcal{F}_3(\boldsymbol{k}_{\text{h}}, k_3, x_3 = 0) = -\mathcal{F}_3(\boldsymbol{k}_{\text{h}}, -k_3, x_3 = 0)$, respectively. Suppose, however, that there is a wave-maker (as e. g. windstress fluctuations) at the surface, radiating waves with horizontal wave vectors $\boldsymbol{k}_{\text{h}}$ and frequencies $\omega$, with the corresponding downward action flux $\tilde{\Phi}(\boldsymbol{k}_{\text{h}}, \omega)$ (the unit is action per area and time and per wave vector increment $\text{d}\boldsymbol{k}_{\text{h}}$ and frequency increment $\text{d}\omega$). We shall use the convention that $\omega > 0$. The downward flux of action, associated with the wave-maker, as density in $\boldsymbol{k}$-space is thus $\Phi(\boldsymbol{k}) = \Phi(\boldsymbol{k}_{\text{h}}, k_3) = \tilde{\Phi}(\boldsymbol{k}_{\text{h}}, \omega)\text{d}\omega/\text{d}k_3$, where $\omega = \omega(\boldsymbol{k})$ is given by the dispersion relation and the flux is placed at $k_3 > 0$. Note that $\Phi(\boldsymbol{k})$ is negative (downward). Then

$$\mathcal{F}_3(\boldsymbol{k}_{\text{h}}, k_3) + \mathcal{F}_3(\boldsymbol{k}_{\text{h}}, -k_3) = \Phi(\boldsymbol{k}) \quad \text{at} \quad x_3 = 0 \tag{10.57}$$

must hold. We shall demonstrate below how to obtain $\Phi(\boldsymbol{k})$ for windstress forcing of internal gravity waves.

The RTE framework has been successfully used to build a prediction model for surface gravity waves (Komen et al., 1994). The source functions for generation and dissipation have been parameterized and an efficient code was developed for the scattering integral of the resonant transfer (which are of quadruplet nature for surface waves), building on the work by Klaus Hasselmann. As surface waves are on the fringe of dynamical oceanography and not in the center of this book, we will elucidate the above described concepts of the energy balance of a wave field and the use of (10.55) for the case of internal gravity waves. The energy balance of internal gravity waves was formulated by Olbers (1974, 1976) and Müller and Olbers (1975), and an overall assessment of the source, sink and transfer terms was given in Munk

(1981) and Olbers (1983, 1986). Many questions have been clarified since then and details worked out, but a breakthrough for a predictive use of the RTE has still to come. Problems exist in manifold ways. First, the scattering integral, being responsi-

**43. The Garrett–Munk Spectrum**

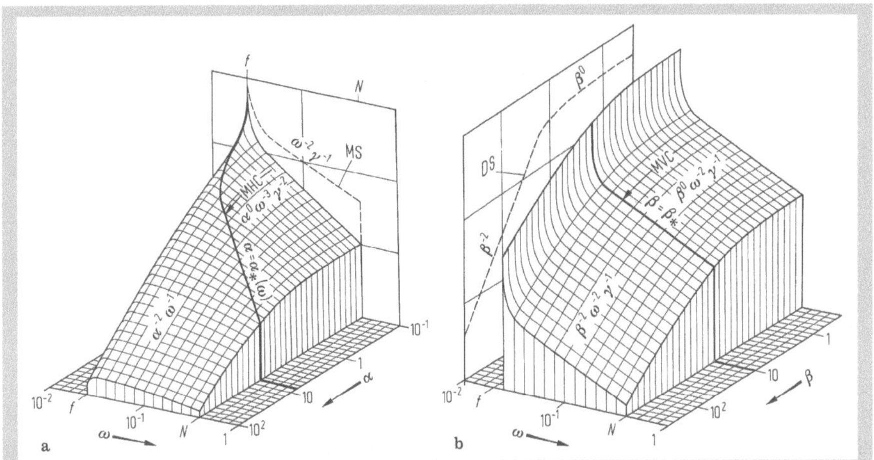


*The GM model, displayed in* ***a*** *as* $\mathcal{E}(k_h, \omega)$, *and in* ***b*** *as* $\mathcal{E}(\beta, \omega)$, *where* $\beta$ *is the modulus of* $k_3$. *The coordinates are plotted logarithmically so that plane surfaces represent powers laws, some of which are indicated in the graph. The partially integrated forms MS and DS of the moored and dropped spectra, respectively, are displayed as respective projections, the moored coherences MHC and MVC are related to the corresponding bandwidths, as indicated. We use* $\alpha = k_h$ *and* $\gamma = (1 - f^2/\omega^2)^{1/2}$ *in the figure. After Garrett and Munk (1975).*

A unified picture of the oceanic internal wave field was developed by Garrett and Munk (GM), first in 1972 and later extended in refined form Garrett and Munk (1975). The basic features of the GM spectral model are: horizontal isotropy and vertical symmetry, a −2 slope as well in the frequency continuum and the (horizontal and vertical) wave-number domain, and a bandwidth of equivalently 10 modes at each frequency. These properties are in accordance with observations, basically the moored spectrum MS (the frequency spectrum of the horizontal current), the towed spectrum TS (the spectrum obtained from horizontally towed thermistors), the dropped spectrum DS (spectrum obtained from vertically dropped thermistors), and the moored coherence spectra for horizontal (MHC) and vertical (MVC) separation. Later more refined measurements were used to confirm and improve the model.
The spectrum is expressed in the factorized form

$$\begin{aligned}\mathcal{E}(k_h, \omega, z) &= E(z)A(k_h, \omega)B(\omega) \quad \text{with} \\ A(k_h, \omega) &= \tilde{A}[k_h/k_h^*(\omega)]/k_h^*(\omega) \quad \text{and} \\ B(\omega) &= \frac{2}{\pi}\frac{f}{\omega}(\omega^2 - f^2)^{-1/2}\end{aligned}$$

where $E(z)$ is the total (kinetic plus potential) energy density in physical space (see Section 7.2.2) and the factors $\tilde{A}(\lambda)$ and $B(\omega)$ are normalized, integrated over their respective arguments. The frequency spectrum is thus given by $E(z)B(\omega)$ which has a peak at the inertial frequency, followed by a continuum $\sim \omega^{-2}$, consistent with the moored spectrum MS. The wave-number-dependent part $A(k_h, \omega)$ has the same shape at each frequency and

$$k_h^*(\omega) = \frac{n^* \pi}{b N_0}(\omega^2 - f^2)^{1/2}$$

characterizes its width at each frequency. A corresponding bandwidth in $(k_3, \omega)$-space follows from the dispersion relation (7.15). It is adjusted such that a constant number $n^*$ of equivalent WKBJ modes is present at low frequencies (the bandwidth in the mode-number domain is then

43. (continued)

$n^*\pi$). To appreciate this statement consider the vertical integral of (7.31) over the ocean depth,

$$\int k_3(z)\mathrm{d}z = k_\mathrm{h}\int\left[\frac{N^2(z)-\omega^2}{\omega^2-f^2}\right]^{1/2}\mathrm{d}z \simeq \frac{bN_0}{(\omega^2-f^2)^{1/2}}$$

The left-hand side is approximately $n\pi$ where $n$ is the mode number, and $bN_0$ represents the integral of $N(z)$. The wave-number dependence of the GM spectrum is specified as

$$A(\lambda) = \frac{2}{\pi}(1+\lambda^2)^{-1}$$

which is consistent with the towed and dropped spectra TS $\sim k_\mathrm{h}^{-2}$ and DS $\sim k_3^{-2}$, respectively. The dependence on the vertical coordinate is assumed as $E(z) = E_0 N(z)/N_0$ in agreement with a WKBJ scaling, as giving by (7.22) and the box on p. 186. The standard parameters of GM are: $E_0 = 4\times 10^3\,\mathrm{J\,m^{-2}}$, $n^*\pi = 10$, $bN_0 = 6.5\,\mathrm{m\,s^{-1}}$ for typical conditions.
The general form of the spectral density is $\mathcal{E}^{\pm}(k_\mathrm{h},\phi,\omega)$, where the index $\pm$ stands for the up- and downward parts of spectral energy ($k_3 < 0$ and $k_3 > 0$, respectively) and $\phi$ is the angle of the horizontal wave vector $\boldsymbol{k}_\mathrm{h}$. The symmetric and isotropic GM spectrum is embedded into this form as $\mathcal{E}^{\pm}(k_\mathrm{h},\phi,\omega) = (1/2)\mathcal{E}(k_\mathrm{h},\omega)/(2\pi)$.

ble for the transfer of energy in the wave-number domain, must face a huge range of situations in a prediction model, from the smooth transfer in an almost equilibrium spectrum (like the Garrett and Munk type, see the box on p. 317 and Garrett and Munk (1975)) to the spectral adjustment of localized energy outbursts and spatially varying monochromatic sources, occurring e. g. for the generation of internal waves by tidal forcing. There is a lack of efficient numerical algorithms to compute the scattering integral in such conditions far from equilibrium. Second, there is quite a number of potential sources affecting the wave field, some of which are briefly described in the next section. Finally, in contrast to the two-dimensional surface wave problem, the internal wave problem is three-dimensional in space because the most dominating sources are located at the sea surface and the ocean bottom.

Instead of showing observations we elucidate the typical internal wave activity by presenting in the box on p. 317 a condensed model, namely the Garrett and Munk model of the spectral energy distribution (the so-called GM spectrum). It is based, as described in the box, on various types of measurements from current meter and thermistors (as a proxy of the vertical displacement, as e. g. described in Section 2.9.2). It represents within a factor about 3 most internal wave properties observed in the World Ocean: the spectral form in wave-number-frequency space, the spatial coherence of the wave field, and the rms velocities (about 5 cm/s) and the rms vertical displacement (about 7 m), but also the vertical dependence of these quantities. Note that the energy in GM is found mainly at low frequencies and large vertical scales while the shear spectrum is concentrated in small vertical scales. Larger deviations from GM are found in singular places such as in the vicinity of extreme topographic features (seamounts, canyons). Moreover, the GM spectrum does not represent the spectral peaks at the inertial and the tidal frequencies adequately. They have a high time and space variability and generally also show a vertical asymmetry. Near-inertial waves mostly propagate downwards, hinting at a likely generation by the wind. In 1973, the experiment IWEX (Internal Wave Experiment) was performed in the Sargasso Sea, using a very stable three-legged mooring, to test the GM model. IWEX was found in good agreement with GM (Müller et al., 1978) though small but significant deviations due to layered fine-structure in the temperature field was revealed.

### 10.4.1 Generation Processes

The most prominent generation processes occur very localized in the frequency domain, mainly close to the inertial frequency $\omega = f$ and at the tidal frequency $\omega = M_2$. Near-inertial waves are excited by windstress fluctuations and propagate long distances horizontally and down from the surface layer. The second, even more monochromatic source comes from the scattering of the barotropic tide at topography, predominantly the continental shelf, mid-ocean ridges but also the more random-type small-scale roughness of the seafloor. The energy spectrum, on the other hand, is observed to be quite continuous in the wave-number-frequency domain, and the energy is also not overwhelmingly concentrated in space, which would be the case if one or the other of the above prominent, spatially and temporally very localized processes would dominate. More specific, the energy in the near-inertial and in the tidal frequency bands are observed to be highly variable in space, but they reside on a spectral continuum with a more universal strength and shape (see the box on p. 317). There are thus very likely sources of energy in the wave-number continuum, part of which may be attributed to nonlinear transfer contained in the scattering integral $\mathcal{S}_{\mathrm{wwi}}$, but other candidates are discussed. These are generation of internal waves by geostrophic adjustment of large-scale disturbances and by interaction with mesoscale eddies. Internal waves arise as lee-waves by large-scale currents flowing over topography (see e.g. Bell Jr., 1975), and by resonantly interacting surface gravity waves (see e.g. Olbers and Herterich (1979)). The associated source terms in the RTE are quite complicated, and we elucidate the generation process by deriving in some detail the source terms for windstress and tidal forcing of internal waves.

#### Windstress Forcing

A straightforward way to derive the surface action flux $\Phi^w(\boldsymbol{k})$ due to windstress fluctuations, prescribed at the sea surface, starts by implementing the stress divergence $(\mathcal{F}_u, \mathcal{F}_v) = \partial(\tau^x, \tau^y)/\partial z$ into the horizontal momentum equations (10.1) and (10.2) and integrate vertically over the surface layer influenced by the wind (the Ekman layer, see Section 14.1). At the surface $\boldsymbol{\tau}$ equals the windstress $\boldsymbol{\tau}_0(\boldsymbol{x}_{\mathrm{h}}, t)$ (scaled as usually by a reference density; the unit is $\mathrm{m^2\,s^{-2}}$), and at the base of the layer and below the stress vanishes. After Fourier transformation with respect to space and time, e. g. for the velocity in the $x$-direction, integrated vertically over the surface layer,

$$U(\boldsymbol{x}_{\mathrm{h}}, t) = \int \mathrm{d}_{\mathrm{h}}^{k} \int \mathrm{d}\omega\, U(\boldsymbol{k}_{\mathrm{h}}, \omega) \mathrm{e}^{\mathrm{i}(\boldsymbol{k}_{\mathrm{h}} \cdot \boldsymbol{x}_{\mathrm{h}} - \omega t)}$$

we find

$$U = \frac{\mathrm{i}\omega\tau_0^x - f\tau_0^y}{\omega^2 - f^2} \quad \text{and} \quad V = \frac{\mathrm{i}\omega\tau_0^y + f\tau_0^x}{\omega^2 - f^2} \tag{10.58}$$

for the windstress-driven $(U, V)$, and

$$w(-d) = \mathrm{i}\boldsymbol{k}_{\mathrm{h}} \cdot (U, V) = -\frac{\omega \boldsymbol{k}_{\mathrm{h}} + \mathrm{i} f \underline{\mathbf{k}}_{\mathrm{h}}}{\omega^2 - f^2} \cdot \boldsymbol{\tau}_0 \tag{10.59}$$

for the vertical velocity at the base $z = -d$ of the directly wind-driven layer. Note that $w(-d)$ is the Ekman pumping for time-dependent conditions (compare with Section 14.1). All field quantities in (10.58) and (10.59) are Fourier components, i. e. they are function of $(\boldsymbol{k}_\mathrm{h}, \omega)$. The pumping induced by $w(-d)$ establishes the above mentioned wave-maker, exciting the waves which radiate downward from the surface layer into the interior. The appropriate vertical velocity $\dot{\xi}_3$ of the interior wave field follows from the representation (9.44) (and the following relations in that section),

$$\dot{\xi}_3 = -\mathrm{i}\frac{k_\mathrm{h}}{k} a \mathrm{e}^{\mathrm{i}(\boldsymbol{k}\cdot\boldsymbol{x}-\omega t)} + \text{c.c.} \tag{10.60}$$

where the renormalization to energy amplitudes $a$, described in Section 9.6, was assumed. We equate $w(\boldsymbol{k}_\mathrm{h}, \omega, -d)$ with $\dot{\xi}_3$ at $x_3 = -d$, where $k_3 = k_3(\omega) > 0$ has to be taken for a downward radiating wave, and obtain

$$a(\boldsymbol{k}_\mathrm{h}, \omega) = \mathrm{i}\frac{k}{k_\mathrm{h}} w(\boldsymbol{k}_\mathrm{h}, \omega, -d) \mathrm{e}^{\mathrm{i}k_3(\omega)d} = -\mathrm{i}\frac{k}{k_\mathrm{h}} \frac{\omega \boldsymbol{k}_\mathrm{h} + \mathrm{i} f \underline{\boldsymbol{k}}_\mathrm{h}}{\omega^2 - f^2} \cdot \boldsymbol{\tau}_0(\boldsymbol{k}_\mathrm{h}, \omega) \mathrm{e}^{\mathrm{i}k_3(\omega)d} \tag{10.61}$$

which is the required relation of the wave field to the windstress fluctuations. Only frequencies in the range $f < \omega < N$ are allowed. The remaining frequency range in the forcing does not force internal waves. It is important to note that $a(\boldsymbol{k}_\mathrm{h}, \omega)$ is the wave amplitude in $(\boldsymbol{k}_\mathrm{h}, \omega)$-space (not in $(\boldsymbol{k}_\mathrm{h}, k_3(\omega))$-space!). Hence

$$\langle a(\boldsymbol{k}_\mathrm{h}, \omega) a^*(\boldsymbol{k}'_\mathrm{h}, \omega') \rangle = \frac{1}{2} \delta(\boldsymbol{k}_\mathrm{h} - \boldsymbol{k}'_\mathrm{h}) \delta(\omega - \omega') \mathcal{E}(\boldsymbol{k}_\mathrm{h}, \omega) \tag{10.62}$$

defines the appropriate energy spectrum of the downward radiating waves, and

$$\langle \dot{\xi}_3 \varpi \rangle = \int \mathrm{d}^2 k_\mathrm{h} \int_f^N \mathrm{d}\omega c_{g3} \mathcal{E}(\boldsymbol{k}_\mathrm{h}, \omega) \tag{10.63}$$

is the vertical energy flux. Inserting the above relations we find

$$\begin{aligned} \langle \dot{\xi}_3 \varpi \rangle|_{x_3=-d} &= \rho_0 \int \mathrm{d}^2 k_\mathrm{h} \int_f^N \mathrm{d}\omega c_{g3} \frac{k^2 \omega^2}{(\omega^2 - f^2)^2} \frac{k_\alpha k_\gamma}{k_\mathrm{h}^2} D_{\alpha\beta} D^*_{\gamma\delta} F^\tau_{\beta\delta}(\boldsymbol{k}_\mathrm{h}, \omega) \\ &= -\rho_0 \int \mathrm{d}^2 k_\mathrm{h} \int_f^N \mathrm{d}\omega \frac{\omega k_\mathrm{h} (N^2 - \omega^2)^{1/2}}{(\omega^2 - f^2)^{3/2}} \frac{k_\alpha k_\gamma}{k_\mathrm{h}^2} D_{\alpha\beta} D^*_{\gamma\delta} F^\tau_{\beta\delta}(\boldsymbol{k}_\mathrm{h}, \omega) \end{aligned} \tag{10.64}$$

The scalar products $\boldsymbol{k}_\mathrm{h}$ and $\boldsymbol{\tau}_0$ in (10.61) have been converted to a matrix notation ($\alpha, \beta, \ldots$ run over 1, 2), using

$$(D_{\alpha\beta}) = \begin{pmatrix} 1 & \mathrm{i}f/\omega \\ -\mathrm{i}f/\omega & 1 \end{pmatrix}$$

and the proper form for the vertical group velocity was inserted. The windstress cospectrum $F^\tau_{\alpha\beta}(\boldsymbol{k}_\mathrm{h}, \omega)$ is defined such that its integral over $\boldsymbol{k}_\mathrm{h}$ and $\omega$ yields the covariance $\langle \tau_{0\alpha}\tau_{0\beta}\rangle$. Simplifying to a spatially isotropic spectrum (see Section 11.1) for the windstress fluctuations,

$$F^{\tau}_{\alpha\beta}(\boldsymbol{k}_{\mathrm{h}},\omega) = \left(\delta_{\alpha\beta} - \frac{k_\alpha k_\beta}{k_{\mathrm{h}}^2}\right)\frac{F^{\tau}(k_{\mathrm{h}},\omega)}{2\pi k_{\mathrm{h}}} \tag{10.65}$$

we finally arrive at

$$\langle \dot{\xi}_3 \varpi\rangle|_{x_3=-d} = -\rho_0 \int \mathrm{d}k_{\mathrm{h}} \int_f^N \mathrm{d}\omega \frac{\omega k_{\mathrm{h}}(N^2-\omega^2)^{1/2}}{(\omega^2-f^2)^{3/2}}\frac{f^2}{\omega^2} F^{\tau}(k_{\mathrm{h}},\omega) \tag{10.66}$$

The appropriate energy flux, which enters the boundary condition (10.57), can easily be identified, however, here as density in $(k_{\mathrm{h}},\omega)$-space; the angular dependence has been integrated. Note that there appears a nonintegrable singularity at the inertial frequency. It derives from the pumping velocity (10.59) because the momentum equations (10.1) and (10.2) for the upper layer show a resonance at $\omega = \pm f$. This can be avoided by the convenient implementation of (horizontal) friction. Adding $(-ru, rv)$ to $(\mathcal{F}_u, \mathcal{F}_v)$, the singular denominator is then replaced according to

$$\frac{1}{(\omega^2-f^2)^{3/2}} \rightarrow \frac{(\omega^2-f^2)^{1/2}}{(\omega^2-f^2)^2 + 2r^2(\omega^2+f^2) + r^4} \tag{10.67}$$

The frequency band close to $f$ now assumes a vanishing response, as expected for the vanishing of the vertical group velocity for inertial waves. The maximum of this expression is roughly at $\omega_m^2 - f^2 \simeq (2/\sqrt{3})rf$, and the function is sharply peaked at $\omega_{\mathrm{m}}$ for small $r/f$.

Nevertheless, the response comes predominantly from the near-inertial range of the windstress fluctuations. Alford (2001) reports values up to $30\,\mathrm{mW\,m^{-2}}$ for the global range of the flux from the wind to mixed-layer near-inertial motions, with zonal means $\sim 3\,\mathrm{mW\,m^{-2}}$ in the strong west-wind zone and a global mean of $0.98\,\mathrm{mW\,m^{-2}}$. Some is dissipated and used to erode the mixed-layer base, a significant fraction must penetrate into the deeper layers. A rough estimate of the flux from (10.66), using $\omega \simeq \omega_{\mathrm{m}}$ for the frequency dependence, is $\rho_0(N/f)(f/r)^{3/2}(2\pi/L_{\mathrm{h}})F^{\tau}(f)$ where $L_{\mathrm{h}}$ is the length scale and $F^{\tau}(f)$ is the spectral density of the stress fluctuations at low synoptic frequencies. For typical conditions, values of order 1–10 $\mathrm{mW\,m^{-2}}$ are found for $\langle\dot{\xi}_3\varpi\rangle$: take $N/f =$ 3–50, $f/r = 10^2$, $L_{\mathrm{h}} =$ 10–100 km, $fF^{\tau}(f) = 10^{-10}\,\mathrm{m^2\,s^{-2}}$. The energy content of the Garrett–Munk spectrum is $4\times10^3\,\mathrm{J\,m^{-2}}$, leading to an 'overturning' time scale of 10–100 d.

### Conversion of Barotropic to Baroclinic Tides at Topography

Without proof we refer to the result of Bell Jr. (1975) who derived the vertical flux due to scattering of the barotropic tide at topography,

$$\begin{aligned}\langle \dot{\xi}_3\varpi\rangle|_{x_3=-H} = \rho_0 &\int \mathrm{d}^2k k_{\mathrm{h}}^{-1} F^{\mathrm{top}}(\boldsymbol{k}_{\mathrm{h}}) \\ &\times \sum_n \frac{n\omega_{\mathrm{T}}}{2\pi^2}\left[\left(N^2-n^2\omega_{\mathrm{T}}^2\right)\left(n^2\omega_{\mathrm{T}}^2-f^2\right)\right]^{1/2} J_n^2(\beta)\end{aligned} \tag{10.68}$$

radiating upward from the bottom. Here $\omega_T$ is the fundamental frequency of the tide $M_2$, and $(U_T, V_T)$ are the amplitudes. Furthermore, $\beta^2 = (k_1^2 U_T^2 + k_2^2 V_T^2)/\omega_T^2$. The sum over the integer $n$ is restricted to $n\omega_T < N$, and $J_n$ is the Bessel function of order $n$ (see the box on p. 247). Most important, the spectral density $F^{\mathrm{top}}(\boldsymbol{k}_\mathrm{h})$ of the random topography enters the expression (the integral over the two-dimensional wave vector yields the variance of the height fluctuations) and largely determines the dependence on the wave number. The frequency dependence of the flux, on the other hand, is localized in a $\delta$-function comb at the harmonics $n\omega_T < N$ of the tide.

The following estimate restricts (10.68) to $n = 1$ and assumes $k_\mathrm{h} U_T \ll \omega_T$. Then, using $J_1(x) \approx x/2$ for small $x$ and $\omega_T^2 \ll N^2$ we find

$$\langle \dot{\xi}_3 \varpi \rangle|_{x_3=-H} = \rho_0 \frac{U_T^2}{8\pi^2} \frac{N}{\omega_T} \left(\omega_T^2 - f^2\right)^{1/2} \int \mathrm{d}^2 k k_\mathrm{h} F^{\mathrm{top}}(\boldsymbol{k}_\mathrm{h}) \qquad (10.69)$$

for an isotropic tide where $U_T = V_T$. Typical values to estimate the factor in front of the integral are $N = 10^{-5}\,\mathrm{s}^{-1}$ (for the near bottom part of the water column), $U_T = 5 \times 10^{-2}\,\mathrm{m\,s}^{-1}$ and $\omega_T = 1.4 \times 10^{-4}\,\mathrm{s}^{-1}$ for the tidal parameters. The spectral integral is bit critical: we replace it by a variance $\langle (\delta h)^2 \rangle$ of topographic fluctuations and a spatial scale $2\pi/L$. With $\langle (\delta h)^2 \rangle = (50\,\mathrm{m})^2$ and $L = 100\,\mathrm{m}$ the flux becomes $4\,\mathrm{mW\,m}^{-2}$. This magnitude is confirmed by Polzin (2009) in a more detailed evaluation of (10.68).

### 10.4.2 Dissipation Mechanisms

A survey of the many mechanisms, by which internal wave energy can dissipate, has been given by Thorpe (2005). They are manifold: gravitational instability occurs when the fluid velocity exceeds the phase speed; shear instability requires that the local Richardson number is less that $1/4$ in parts of the region (see Section 7.7.3). This condition may be caused by the superposition of the waves themselves or, for instance, at critical layers where the wave shear becomes large (see Section 7.7.1). The most likely candidate for the oceanic wave field is wave breaking by shear instability, handing the wave energy over to turbulence. The physics is described in Section 11.3.3 (the mean flow in that framework is now the internal wave).

We make use of a parametrization of wave breaking for the spectral domain, proposed by Natarov and Müller (2005). They argue for a quasi-linear form of the dissipation source function

$$\mathcal{S}_{\mathrm{diss}}(\boldsymbol{k}) = -\gamma(\boldsymbol{k}, \mathrm{Ri}) \mathcal{N}(\boldsymbol{k}) \qquad (10.70)$$

for wave action, where $\gamma(\boldsymbol{k}, \mathrm{Ri})$ is a coefficient, depending on the wave vector and the overall Richardson number Ri supported by the wave field itself. It may be computed from the spectrum of the vertical shear, related to the energy spectrum by

$$S(\boldsymbol{k}) = (\omega^2 + f^2) k_3^2 \frac{\mathcal{N}(\boldsymbol{k})}{\omega} = (\omega^2 + f^2) k_3^2 \mathcal{E}(\boldsymbol{k}) \qquad (10.71)$$

The Richardson number follows from

$$\mathrm{Ri}^{-1} = \int \mathrm{d}^3 k \frac{S(\boldsymbol{k})}{N^2} = \int \mathrm{d}^3 k \frac{\omega^2 + f^2}{N^2} k_3^2 \mathcal{E}(\boldsymbol{k}) \qquad (10.72)$$

All quantities may, of course, depend on space and time. The form of (10.71) follows from the eigenvector (9.48), and the integral of the shear spectrum yields the variance of the wave-induced vertical shear, $\int \mathrm{d}^3 k S(\boldsymbol{k}) = \langle (\partial u_1/\partial x_3)^2 + (\partial u_2/\partial x_3)^2 \rangle$. Note that (10.72) is an implicit relation for Ri because the wave spectrum depends on Ri via $\gamma(\boldsymbol{k}, \mathrm{Ri})$ in the RTE balance.

We have shown in Section 7.7.3 that shear flows may become unstable if Ri $< 1/4$, hence it is reasonable to assume that the coefficient $\gamma(\boldsymbol{k}, \mathrm{Ri})$ increases with decreasing Ri. Equally important is the dependence on the wave number. Long waves should be less affected than short waves, i. e. $\gamma(\boldsymbol{k}, \mathrm{Ri})$ should be increasing with the (vertical) wave number. Natarov and Müller (2005) give specific forms for $\gamma(k_3, \mathrm{Ri})$. Except for the above general features we will make no use of them when we use the dissipation model in the following section.

### 10.4.3 Some Prototype Balances

We will illustrate the use of the radiative transfer equation in the reduced form (10.56) and the associated boundary condition (10.57) with some simple examples. Of course, we cannot treat the wave-wave-interaction in a simple way, and thus we will ignore it in the present section. Some information on the working of this process in the spectral balance will be given in the next section.

It is useful to split the RTE explicitly in the up- and downward propagating parts. We write it for energy instead of action, and we define upward propagating part of the spectrum as $\mathcal{E}^+(\boldsymbol{k}) = \mathcal{E}(\boldsymbol{k}_\mathrm{h}, k_3 < 0)$ and zero for $k_3 > 0$, because the vertical group velocity $c_{g3}$ is upward for negative $k_3$. The downward part $\mathcal{E}^-(\boldsymbol{k})$ is defined accordingly. The source terms are split in the same way. The RTE becomes

$$\pm c \frac{\partial \mathcal{E}^\pm}{\partial x_3} = \omega \mathcal{S}^\pm \quad \text{with} \quad c(\mathcal{E}^+ - \mathcal{E}^-) = F \quad \text{at the respective boundary} \quad (10.73)$$

where $c = |c_{g3}|$ is the modulus of the group velocity. Here $F = \omega \Phi$ is the flux through the boundary, and we will use $F^{\mathrm{surf}} < 0$ and $F^{\mathrm{bot}} > 0$ for the fluxes at the sea surface and the bottom, respectively. Remember that we have assumed a constant Brunt–Väisälä frequency so that the refraction term is absent in (10.73) (in addition, the group velocity is spatially constant). Of course, the above equation is easily solved if the sources $\omega \mathcal{S}^\pm$ and $F$ are given. In the following we will use the dissipation term (10.70) and prescribed boundary sources.

#### No Dissipation, no Fluxes

If there are no forcing and dissipation, we find $\mathcal{E}^\pm(\boldsymbol{k}, x_3) = \mathrm{const} = \mathcal{E}_0(\boldsymbol{k})$. The spectrum is thus symmetric with respect to $k_3$ and vertically constant; the function $\mathcal{E}_0(\boldsymbol{k})$ is arbitrary. The situation, envisioned here, is one in which a wave field has been excited and then left alone without dissipation. Reflection at the surface and bottom lead to symmetric conditions.

### With Dissipation, no Fluxes

Here we face the balance (10.73) with $\omega\mathcal{S}^{\pm} = -\gamma\mathcal{E}^{\pm}$ so that the up/down spectra are exponentially in the vertical with the spatial scale $c/\gamma$. The boundary conditions, however, lead to $\mathcal{E}^{\pm} \equiv 0$, as expected: with zero input of energy the dissipation eliminates all waves.

### With Dissipation, with Fluxes

Solving (10.73) with $\omega\mathcal{S}^{\pm} = -\gamma\mathcal{E}^{\pm}$ subject to forcing at the surface and the bottom, we obtain the relations

$$\mathcal{E}^{\pm}(\boldsymbol{k}, x_3) = \mathcal{E}_0^{\pm}(\boldsymbol{k})\epsilon^{\pm}(x_3) \quad \text{with} \quad \epsilon^{\pm}(x_3) = \left[\epsilon^{\mp}(x_3)\right]^{-1} = \exp\left(\pm\int_{x_3}^{0} \gamma/c\,\mathrm{d}x_3'\right)$$
$$\mathcal{E}_0^{+} - \mathcal{E}_0^{-} = F^{\mathrm{surf}}/c \quad \text{and} \quad \mathcal{E}_0^{+}\epsilon^{+}(-H) - \mathcal{E}_0^{-}\epsilon^{-}(-H) = F^{\mathrm{bot}}/c \tag{10.74}$$

and hence

$$c\mathcal{E}_0^{\pm}(\boldsymbol{k}) = \frac{F^{\mathrm{bot}}(\boldsymbol{k}) - \epsilon^{\mp}(-H)F^{\mathrm{surf}}(\boldsymbol{k})}{\epsilon^{+}(-H) - \epsilon^{-}(-H)} \tag{10.75}$$

The form and amplitude of the wave spectrum is uniquely determined by the fluxes at the surface and the bottom. The upward spectrum $\mathcal{E}^{+}$ decreases and the downward spectrum $\mathcal{E}^{-}$ increases with depth with the scale $c/\gamma$, which is generally wave-number-dependent. If $\gamma$ is, as described above, a function of the gross Richardson number Ri supported by the wave field, the solution must be completed by determining Ri. This is achieved by the implicit condition (10.72).

Prescribed or linear interior sources/sinks can easily be incorporated into the above solution, but it unlikely at all that it mirrors any observed state, as e. g. the GM spectrum from the box on p. 317. In this linear model, generation and dissipation must occur at the same wave number. If the dissipation coefficient $\gamma$ does not change much over the range of the horizontal wave numbers of the generated waves, the vertical depth scale $c/\gamma$ becomes extremely wave-number-dependent: referring to Figures 7.1 and 7.3, showing the vertical group velocity $c$ as function of frequency and horizontal wavelength, we note that $c$ changes drastically as function of the horizontal wavelength. Longer waves of say 10 km horizontal wavelengths need about 10–100 days to reach the bottom, shorter ones of 1 km wavelength need about 100–1,000 days (for near-inertial frequencies; see Figure 7.3). Hence the exponential decrease in the linear solution (10.74) diversifies the spectrum in the vertical: we will find considerable asymmetry in the linear solution, actually at all frequencies and wavelengths. This is in contrast to the observations.

## 10.4.4 Resonant Transfer

The cross-spectral transfer by wave-wave interaction is missing in the previous linear model. This process is a main actor in the interior of the ocean. In the above described state, where wave energy is injected either at the sea surface or the bottom leading to the asymmetric energy spectrum $\mathcal{E}^{\pm}(\boldsymbol{k})$, the resonant coupling of up- and downward propagating waves could lead to a rapid redistribution of the spectral shape. In the present section we will discuss the spectral transfer by wave-wave interactions on a rather qualitative level. For details we refer to the literature already cited in Section 9.6.2 and in the present section. A review is given by Müller et al. (1986).

The spectral transfer due to interactions of wave triads is governed by the scattering integral (9.82). Two wave with wave vectors $\boldsymbol{k}', \boldsymbol{k}''$ and associated frequencies $\omega', \omega''$ transfer action (energy) to a third one, characterized by $\omega, \boldsymbol{k}$ by either sum or difference interaction,

$$\omega = \omega' \pm \omega'' \quad \text{and} \quad \boldsymbol{k} = \boldsymbol{k}' \pm \boldsymbol{k}'' \tag{10.76}$$

The essential point is that not any wave triad may interact resonantly because the frequencies have to satisfy the dispersion relation and the above frequency condition becomes $\omega(\boldsymbol{k}) = \omega(\boldsymbol{k}') \pm \omega(\boldsymbol{k} - \boldsymbol{k}')$ (eliminating $\boldsymbol{k}''$; in writing the last term we have used that the frequency of internal wave only depends on the modulus of the wave vector). For given output wave vector $\boldsymbol{k}$, this condition defines a surface in the three-dimensional $\boldsymbol{k}'$-space, which turns out to be highly complicated and multiconnected. By evaluating the $\delta$-functions, the scattering integral (9.82) for triad interactions can then be reduced to a two-dimensional integration over this domain. Note that the efficiency of the transfer – the magnitude of $\mathcal{S}_{\text{triad}}(\boldsymbol{k})$ – not only depends on the range of the resonance but also on the size of the transfer function (9.83) and the spectral power of the interacting triplets.

We discuss two examples in a qualitative way. First consider the model from the previous section where near-inertial waves are generated at the sea surface and reflected at the bottom. The difference interaction in (10.76) is excluded for such waves because the difference of two near-inertial frequencies falls below the range of the internal wave frequency band. The sum interaction of a downward propagating wave and its bottom-reflected upward propagating partner should have a response at about twice the inertial frequency and double the horizontal wave number. The restriction is, however, that the opposing vertical wave numbers sum to a low value, and it is unlikely that a resonant triad results: with the resulting $\boldsymbol{k}$ being almost horizontal the wave frequency is rather close to $N$ than to $2f$. The near inertial band is thus likely more stable than the wave continuum at higher frequencies.

For waves of higher frequencies (larger than $2f$) the difference interaction is allowed, and hence two partners with opposing vertical wave numbers may interact with a wave in the inertial band which has about double the vertical wave number (see Figure 10.3a). Likewise we may describe the same situation as the interaction of (say) a downward propagating wave (with $k_3' > 0$) with near-inertial motion (with $k_3 \sim 2k_3'$) to excite an upward moving wave (with $k_3'' = k_3' - k_3 \sim -k_3' < 0$). This is condition of elastic BRAGG[2] scattering. It leads to vertically symmetric wave field at all frequencies (with exclusion of the very low near-inertial ones) in a fairly rapid transition process (of the order of the wave period, see the work of McComas (1977)). Note that the GM model (see the box on p. 317) of the basic state of the oceanic internal wave field is vertically symmetric. On the other hand, the near-

[2] SIR WILLIAM LAWRENCE BRAGG, *1880 in North Adelaide, Australia, †1971 in Waldringfield near Ipswich, Suffolk, physicist. He received the Noble Price for physics in 1915 jointly with his father for their work X-ray diffraction.

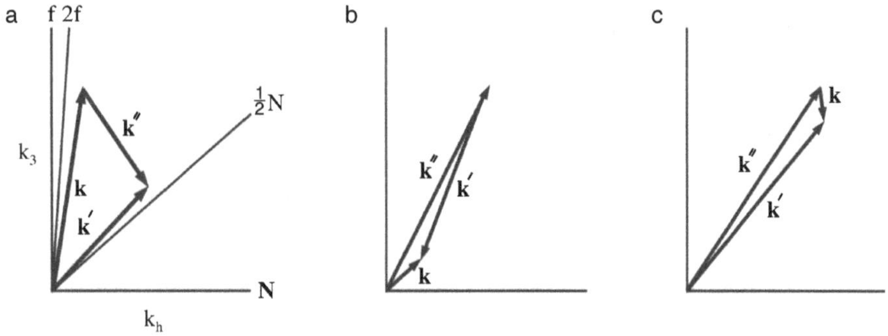


Fig. 10.3 Resonant triads for the **a** elastic Bragg scattering, **b** parametric subharmonic instability, and **c** induced diffusion mechanisms, explained in the text. The wave vectors are sketched in the $(k_h, k_3)$-space but as the frequency of internal waves is only dependent on the direction of the wave vector, the sketches also include lines of constant frequency. Note that $\omega = f$ is along $k_h = 0$ and $\omega = N$ is along $k_3 = 0$. Note that **a** and **c** are difference interactions according to (10.76), and **b** is a sum interaction

inertial band (which is assumed symmetric in GM as well) is found asymmetric in most observations.

The spectral transfer in the GM model spectrum (and slightly perturbed versions) has been investigated in detail by evaluating the scattering integral (9.82) by a numerical code. Olbers (1974, 1976) has found that wave energy is systematically transferred in the GM spectrum from the intermediate frequency range to $f < \omega < 2f$ and towards high vertical wave numbers with a delivery time scale of a few days. McComas and Bretherton (1977) have identified parametric subharmonic instability as the main contributor for this transfer. In this process a low-vertical-wave-number component decays into two components of high vertical wave number and half the frequency of the primary wave (see Figure 10.3b). In successive stages of having the frequency the energy is ultimately pushed to the near-inertial band at high vertical wave numbers.

The instability mechanism is readily understood from the triad equations (9.65): if the amplitudes $a_2$ and $a_3$ in the triplet are much smaller than $a_1$, we may treat the latter as approximately constant, leading to

$$\ddot{a}_2 = \omega_2 \omega_3 |a_1|^2 a_2$$

by elimination of $a_3$ (a corresponding equation is valid for component 3). Remember that the frequencies in the above notation were allowed to carry both signs (see Section 9.6) so we conclude that it is only the sum interaction, where both frequencies $\omega_2$ and $\omega_3$ are positive, which leads to an unstable growth of the components 2 and 3. The growth is exponential with a growth rate $\sqrt{\omega_2 \omega_3}|a_1|$. Then $|\omega_1| = |\omega_2| + |\omega_3|$. The wave with the largest frequency is thus exponentially unstable if the other two waves of a triad which have the small amplitudes. The condition is known as HASSELMANN's[3] criterion (Hasselmann, 1967a). In the realm of resonant triads the condition that all waves have to satisfy the dispersion relation is most easily met if $|\omega_2| = |\omega_3| = |\omega_1|/2$, i. e. we are facing the subharmonic situation described above. In plasma physics the wave 1 is referred to as 'pump wave' because it pumps energy into the subharmonic components 2 and 3.

[3] KLAUS HASSELMANN, *1931 in Hamburg, physicist. Work on wave dynamics, climate change and particle physics.

Another prominent process among a resonant triad, also identified by McComas and Bretherton (1977), is the induced diffusion mechanism. Here a high-frequency, high-wave-number component interacts with a wave of much lower frequency and wave number to generate another high-frequency, high-wave-number component (see Figure 10.3c). The process has some similarity with Brownian motion leading to a random walk of a test particle in the macroscopic physical space. It occurs here, however, in the wave-number space and expresses a diffusion of action in the high-wave-number region. Small-scale waves with short periods are randomly perturbed with small excursions in the wave-number-frequency domain by interaction with larger-scale more persistent wave motion. McComas and Bretherton (1977) show that the scattering integral (9.82) for triads interacting in the induced diffusion limit is approximated by a diffusion term, in the simplest case in the vertical wave-number space,

$$\frac{\partial \mathcal{N}(\boldsymbol{k}_\mathrm{h}, k_3)}{\partial t} = \mathcal{S}_\mathrm{triad}(\boldsymbol{k}_\mathrm{h}, k_3) \approx \frac{\partial}{\partial k_3}\left[ D(\boldsymbol{k}_\mathrm{h}, k_3) \frac{\partial \mathcal{N}(\boldsymbol{k}_\mathrm{h}, k_3)}{\partial k_3} \right]$$

The diffusion coefficient $D$ is given by an integral over the shear spectrum of the large-scale low-frequency wave motion. Note that action $\mathcal{N}$, but not energy $\mathcal{E} = \omega \mathcal{N}$, is diffusing through the vertical wave-number space. However, both quantities are in equilibrium if the action flux $\mathcal{J}_\mathrm{a} = D \partial \mathcal{N} / \partial k_3$ is constant (as function of $k_3$; it may still depend on $\boldsymbol{k}_\mathrm{h}$) and no other sources or sinks occur in the respective wave-number domain. Any flux of action across a wave number $k_3^-$ is transferred undiminished across any other wave number $k_3^+$ in that domain. Knowing $\mathcal{J}_\mathrm{a}$, say from boundary condition at $k_3^-$, and evaluating $D$ as function of $k_3$, we may easily determine the spectral form of $\mathcal{N}$ and $\mathcal{E}$ in the constant flux region by a simple integration (McComas and Bretherton (1977) give examples but note also that $\mathcal{J}_\mathrm{a}$ is generally unknown).

The basic physics of the induced diffusion mechanism is disclosed by the above statements. We may, however, make matters more complicate by considering the behavior of energy in the constant action flux regime. Associated with the action flux $\mathcal{J}_\mathrm{a}$ there is an energy flux $\mathcal{J}_\mathrm{e} = \omega \mathcal{J}_\mathrm{a}$ but as $\omega$ is not constant during the diffusive march through the domain ($\boldsymbol{k}_\mathrm{h}$ remain constant but $k_3$ and thus $\omega$ changes), the energy flux has a nonzero divergence. In other words, though we have $\partial \mathcal{N} / \partial t = 0$ and $\partial \mathcal{E} / \partial t = 0$ for $\mathcal{J}_\mathrm{a} = \mathrm{const}$ in the induced diffusion domain, we find that – in terms of energy – the divergence of $\mathcal{J}_\mathrm{e}$ is locally balanced by a source/sink term $\mathcal{J}_\mathrm{a} \partial \omega / \partial k_3$. Note, however, that energy is lost/gained by the divergence term at corresponding rate.

### 10.4.5 The Link to Mixing

Finally, we would like to point out that investigations of the RTE for internal gravity waves are by no means an academic problem. The march of energy through the spectrum from large to small scales ends in the dissipation range of small-scale turbulence, and the aim is to relate the wave energy balance to the turbulent mixing processes associated with the wave dissipation (see Section 11.3.3).

It was shown in Section 9.6.2 that wave-wave interactions conserve the total energy. Hence the rest of the terms in (10.55) must balance in the overall integral[4],

$$\int \mathrm{d}^3k \left[ \frac{\partial \mathcal{E}}{\partial t} + \omega \nabla_{\boldsymbol{x}} \cdot (\dot{\boldsymbol{x}} \mathcal{N}) + \omega \nabla_{\boldsymbol{k}} \cdot (\dot{\boldsymbol{k}} \mathcal{N}) \right] = \int \mathrm{d}^3k \omega \left( \mathcal{S}_{\mathrm{gen}} + \mathcal{S}_{\mathrm{diss}} \right) \quad (10.77)$$

after conversion to the energy spectrum $\mathcal{E} = \omega \mathcal{N}$. The propagation and refraction terms on the left-hand side may be converted to the divergence (in physical space) of the energy flux carried by the wave field, and we obtain

$$\frac{\partial E}{\partial t} + \nabla_{\boldsymbol{x}} \cdot \int \mathrm{d}^3k \dot{\boldsymbol{x}} \mathcal{E} = \int \mathrm{d}^3k \omega \left( \mathcal{S}_{\mathrm{gen}} + \mathcal{S}_{\mathrm{diss}} \right) \quad (10.78)$$

where $E$ is the total energy (spatial) density. Note that surface and bottom processes enter via boundary conditions. If (10.78) is integrated over the ocean depth, they appear explicitly. In steady state and under spatially homogeneous conditions, the generation and dissipation of wave energy must balance in the overall integral. There is likely a large scale difference between them as function of the wave number because generation is likely at large scales and dissipation at small scales. It is bridged by the transfer in wave-number space due to resonant interactions, however, this process does no longer appear in the integral balance. Other processes might contribute to this transfer from large to small scales, such refection at sloping bottom profiles (see Section 7.4.2).

The budget of the kinetic energy TKE, residing in the turbulence field, will be derived in Section 11.3.1. It is of the form

$$\frac{\partial \mathrm{TKE}}{\partial t} = \frac{\partial F_{\mathrm{TKE}}}{\partial z} - \overline{\boldsymbol{u}_{\mathrm{h}}' w'} \cdot \frac{\partial \overline{\boldsymbol{u}}_{\mathrm{h}}}{\partial z} + \overline{b' w'} - \epsilon \quad (10.79)$$

where primed quantities refer to the turbulent fields and mean quantities refer to the wave field (only the mean shear $\partial \overline{\boldsymbol{u}}_{\mathrm{h}} / \partial z$ appears). Furthermore, $F_{\mathrm{TKE}}$ is the vertical flux of turbulent kinetic energy and $\overline{b'w'}$ is the vertical buoyancy flux, supported by the turbulence, and $\epsilon$ is the dissipation rate of TKE. The exchange of energy with the wave field is found in the second term on the right-hand side, the so-called 'shear production term' of the turbulence. Assuming that the shear production term equates to the transfer of wave energy to high wave numbers by the resonant interactions and neglecting the turbulent flux divergence and the tendency term in (10.79), we find

$$-\overline{\boldsymbol{u}_{\mathrm{h}}' w'} \cdot \frac{\partial \overline{\boldsymbol{u}}_{\mathrm{h}}}{\partial z} \simeq -\frac{1}{\rho_0} \int_{\mathrm{high}\, \boldsymbol{k}} \mathrm{d}^3k \omega \mathcal{S}_{\mathrm{wwi}} \simeq \epsilon - \overline{b' w'} = \epsilon + K_{\mathrm{b}} N^2 \quad (10.80)$$

In the last relation we have implemented the diffusive parametrization of the turbulent buoyancy flux (see Section 11.3.3). Assuming further that the dissipation of wave energy is dominated by the high wave-number domain and using then (10.78), the thread from wave generation to the diapycnal diffusivity is revealed. It should be mentioned, however, that breaking internal waves are not the only mechanism to generate small-scale mixing. For instance, large-scale geostrophic flow may be unstable and produce small-scale turbulence directly (by Kelvin–Helmholtz instability, see Section 7.7.3) without passing any energy through the internal wave compartment. For a comprehensive determination of the diffusivity from (10.80), these contributions have to added on the left-hand side of (10.80).

[4] Note that for simplicity the frequency $\omega$ was assumed here time-independent. Otherwise an additional source term would come up (see the discussion in Section 9.3).

A rough estimate of $K_b$ is obtained by equating the transfer rate from wave-wave interactions to a typical generation rate, e. g. by the windstress. Because this process enters via a flux through ocean surface, a vertical integral must applied to (10.78) to find $\int dz \int d^3\omega S_{wwi} \simeq F^{surf}$. Taking $F^{surf} = 1–10\,mW\,m^{-2}$ as estimated above for the surface flux due to forcing by windstress, $N = 3 \times 10^{-3}\,s^{-1}$, and an ocean depth of $H = 5{,}000\,m$, we find $K_b \approx 0.2–2 \times 10^{-4}\,m^2\,s^{-2}$ if $\epsilon$ is neglected in (10.80). However, it is generally accepted that only a small fraction of the energy transfer to high wave numbers is used to erode the mean density gradient. Anticipating the results from Section 11.3.3 we may use

$$\frac{\epsilon}{K_b N^2} = \frac{1 - \mathrm{Ri_f}}{\mathrm{Ri_f}} \approx 5$$

where $\mathrm{Ri_f} = 0.15$ has been used for the flux Richardson number. In fact, observations yield values for $\epsilon$ as high as $10^{-9}\,m^2\,s^{-3}$ in the ocean interior (see Section 11.3.3) which is of the same magnitude as $F^{surf}/(\rho_0 H)$. The above range of values for $K_b$ is thus overestimated and values of a sixth of the range result if the above ratio is implemented. Note that the reasoning in this section differs from that in Section 11.3.3, where $K_b$ is inferred from $\epsilon$ measurements and the flux Richardson number.

## Further Reading

A comprehensive account of ocean waves is given in *Waves in the Ocean* by LeBlond and Mysak (1980).

An introduction in the dynamics of both oceanic and atmospheric waves is provided in *Waves in the Ocean and Atmosphere* by Pedlosky (2003).

More on sound waves in the ocean can be found e. g. in *Fundamentals of Ocean Acoustics* by Brekhovskikh and Lysanov (2003), and in Chapter 7 of *Principles of Ocean Physics* by Apel (1987).

*Ocean Acoustic Tomography* by Munk et al. (1995) provides an exhaustive treatment of the use of sound waves to infer ocean temperature and other fields.

*The Dynamics of the Upper Ocean* by Philipps (1977) has for long been a standard text on small-scale waves and turbulent motions in the upper ocean.

An in-depth discussion of the interactions between waves and mean flow can be found in the monograph *Waves and Mean Flows* by Bühler (2009).

*Atmosphere-Ocean Dynamics* by Gill (1982) contains many aspects of large-scale dynamics in ocean and atmosphere. In particular the forcing of equatorial motions is discussed here extensively, and also in *El Niño, La Niña*, and the Southern Oscillation by Philander (1990).

The theory of linear and nonlinear internal gravity waves is presented in *Dynamics of Internal Gravity Waves in the Ocean* by Miropolsky (2001), including a Hamiltonian description.

In *The Turbulent Ocean*, Thorpe (2005) gives an extensive account of internal waves, instability of stratified shear flows and the transition to turbulence. Furthermore, the physics and observations of mixing and turbulence in different ocean environments is discussed.

The linear forcing of various wave types in terms of the corresponding Green's function is discussed in *Analytical Theory of Forced Oceanic Waves* by Fennel and Lass (1989).

# Part IV
# Oceanic Turbulence and Eddies

Many flows in the ocean and the atmosphere are turbulent. Turbulent flows are characterized by large fluctuations in space and time and occur on many space and time-scales, from the large-scale oceanic and atmospheric circulation down to small-scale processes such as in the planetary boundary layer of the atmosphere or the surface mixed layer of the ocean. The figure on p. 333 shows the population of oceanic energy of the dominating motion in a space-time scale diagram, namely the general circulation, the wave compartments, and the turbulence (two-dimensional mesoscale eddies and three-dimensional isotropic turbulence). Turbulent flows are governed by the Navier-Stokes equations derived in Chapter 2. The ultimate cause of turbulence lies in the instability of flows, which tends to occur when inertial forces become large compared to other forces in the momentum balance. This circumstance can often be expressed by the Reynolds number, which compares the magnitude of inertial to frictional forces and is always very large in turbulent flows. Associated is a high sensitivity to small changes in initial and boundary conditions. Therefore, the evolution of turbulent flows is not predictable in detail, although certain average features of the flow are well predictable. Note that, to some extent, the characterization of a flow as "turbulent" can depend on which scale the observer is interested in. For example, on time-scales of a few days, the evolution of mesoscale eddies in the ocean is of a fairly deterministic nature and can be predicted with considerable success. However, on time-scales relevant to ocean climate change, only certain mean features of the eddying system are relevant and eventually predictable.

An important feature of turbulence is its diffusive nature. In a turbulent flow, the separation of two particles that are initially close together will on average increase with time. Although turbulent mixing is almost always much larger than molecular mixing, both are linked together, and the existence of the latter is a prerequisite for the former. To describe this linkage is one task of turbulence theory.

The starting point of any theory of turbulence is to accept the impossibility of predicting turbulent flows. A consequence is that all flow variables have to be considered as stochastic, so that only statistical parameters of the flow are meaningful. A complete theory of turbulence does not exist so far, and only under rather restrictive assumptions, is it possible to derive certain aspects of turbulent flows based on the governing equations. The prime example for this approach is Kolmogorov's theory of homogeneous turbulence discussed in Section 11.1 below. On the other hand,

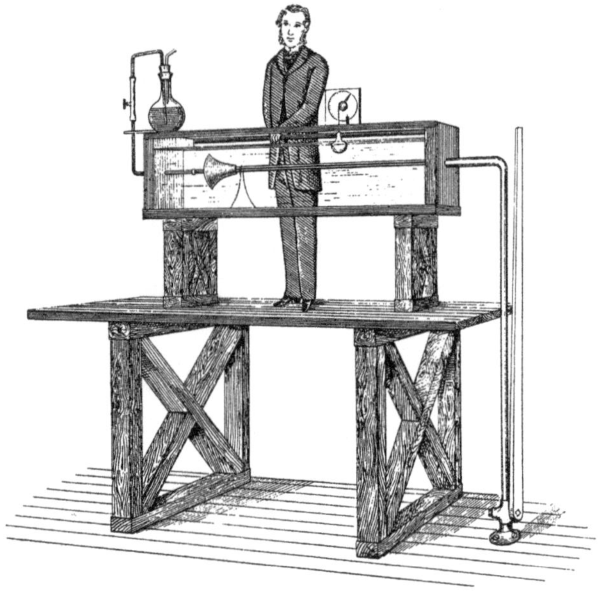

Drawing of the apparatus, which still can be seen at the University of Manchester, used by Osborne Reynolds to study the transition from laminar to turbulent motion. In the tank, water enters the horizontal glass tube through a conical funnel. To visualize the flow, colored water drawn into the tube from a container seen on the left top of the tank. By controlling the flux through the tube, the transition to turbulent flow could be observed and related to the dimensionless number which bears Reynolds' name. Reynolds states in his 1883 paper: "The internal motion of water assumes one or other of two broadly distinguishable forms – either the elements of the fluid follow one another along lines of motion which lead in the most direct manner to their destination, or they eddy about in sinuous paths the most indirect possible" From Reynolds (1883).

it is not surprising that the lack of a comprehensive theory of turbulence has led to a large variety of concepts, often based on heuristic arguments, to describe certain aspects of turbulent flows. These concepts and some basic parameterizations which rely on those concepts are discussed in Sections 11.2 and 11.3 respectively. They are oriented toward practical applications and can have considerable skill in simulating the effects of turbulence, although they do not necessarily provide much physical insight into the nature of turbulent flows.

While Chapter 11 considers the basic aspects of turbulent flow for the case of small-scale, three-dimensional turbulence, with or without effects of gravity, in Chapter 12 some aspects of large-scale and (quasi) two-dimensional turbulence are discussed. Section 12.1 refers to the application of Kolmogorov's theory to the isotropic two-dimensional case, and Section 12.2 discusses concepts to understand mesoscale turbulence and its mixing effect in the ocean and discusses some basic attempts for parameterizations. An important aspect of mesoscale eddy mixing is its advective nature. Section 12.3 discusses consistent differentiations between advective and diffusive effects of turbulent mixing.

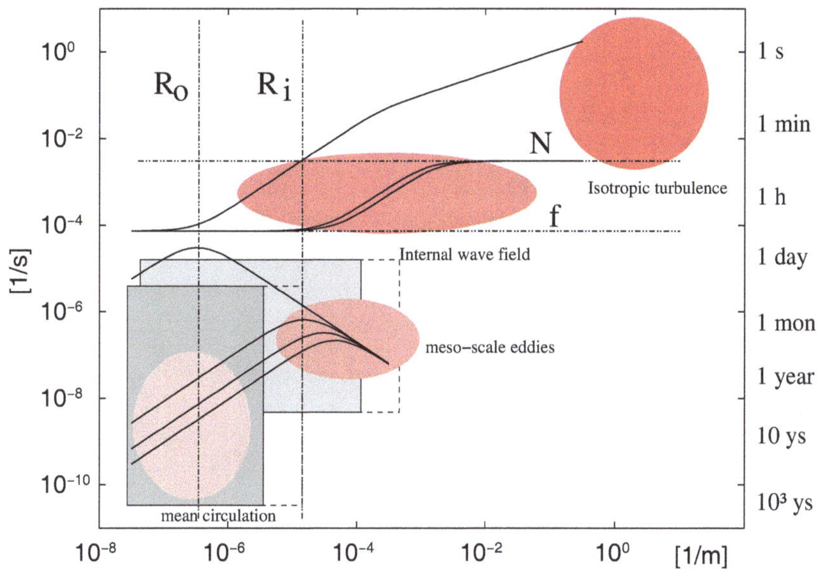


Space-time scales of important oceanic processes (pink areas) and scales explicitly resolved by ocean models (grey rectangular areas). The lower left rectangle represents modern global ocean climate models and the upper right rectangle eddy resolving basin-scale models. Also shown are dispersion curves (solid lines) for linear gravity waves (upper set) and planetary waves (lower set) (compare with the figure on page 106). Vertical dotted lines indicate the external ($R_o$) and first internal ($R_i$) Rossby radii and horizontal dotted lines indicate stability frequency ($N$) and Earth rotation rate ($f$). The large-scale mean circulation involves spatial scales ranging from the circumference of the Earth down to a couple of 100 kilometers, given e. g. by the width of western boundary currents, and involves time scales of seasons to about 1,000 years, given by the time scale of a water parcel to circulate along the entire thermohaline circulation loop. The relevant wave processes of the large-scale mean circulation are characterized by the nondispersive long wave branch of the baroclinic planetary waves. The mesoscale eddy field, on the other hand, involves much smaller spatial scales of several 100 kilometers down to kilometers and time scales of days to months. The mesoscale eddy field is characterized by the short wave branch of the baroclinic planetary waves. The time scales of baroclinic inertio-gravity waves are rather sharply defined as being in between the stability frequency $N$ and Earth rotation frequency $f$, while spatial scales can range from global scale in case of long barotropic gravity wave down to a couple of 10 m for the baroclinic gravity wave branch. On even smaller time and space scales the internal wave regime approaches isotropic turbulence which then connects to the regime of ultimate dissipation of energy by molecular processes.

# 11 Small-Scale Turbulence

In this chapter, we will discuss turbulence on small spatial and temporal scales. For small scales, Earth's rotation is not important, and the length scales in horizontal and vertical directions are comparable, so that the turbulence can often be idealized as isotropic in three dimensions. Furthermore, we expect that the inertial forces are larger than viscous or gravity forces. Small-scale turbulence dominates the motion in the oceanic surface layer and is also important in other boundary layers, e. g. near the ocean floor. The theory of homogeneous turbulence and some important results are discussed in Section 11.1, while basic concepts of turbulent mixing are described in Section 11.2. The application to the real, i. e. inhomogeneous, ocean in the presence of density stratification and large gravity force and the development of basic turbulence closure schemes can be found in Section 11.3.

The theory of homogeneous turbulence by KOLMOGOROV[1] (Kolmogorov, 1941) gives important insight into the nature of turbulent flows even though it is not of immediate help in obtaining parameterizations for practical problems. Kolmogorov's statistical theory was developed more than 70 years ago and is still considered to be robust in many respects. However, the theory has been found less successful in predicting higher order structure functions and cannot quantitatively address the pervasive intermittency of turbulence, which is due to the spatial concentration of shear in physical space. Beside this complication, Kolmogorov's theory yields a successful description of important properties of turbulent flows.

Note that the Boussinesq approximation is employed throughout this chapter, so that $\partial u_i/\partial x_i = 0$, and that we will use the sum convention according to which summation is implied if an index occurs twice within one term.

## 11.1 Kolmogorov's Theory of Homogeneous Turbulence

In this section, we consider the idealized case in that all statistical properties of the turbulent flow are independent of location, a situation referred to as *homogeneous* turbulence. Consequently, all physical variables are considered as homogeneous and stationary random variables, as detailed in Appendix A.3 which gives a description

[1] ANDREY NIKOLAEVICH KOLMOGOROV, *1903 in Tambov, †1987 Moscow, mathematician.

D. Olbers, J. Willebrand, C. Eden, *Ocean Dynamics*,
DOI 10.1007/978-3-642-23450-7_11, 

of basic concepts associated with stochastic variables. A further consequence of homogeneity is the absence of a mean density stratification, therefore, it is assumed that density variations are unimportant, i. e. $\rho \approx$ const. Furthermore, the Coriolis force is assumed to be dominated by the inertial force and is, therefore, neglected.

Considered as a random function, the turbulent velocity $\boldsymbol{u}(\boldsymbol{x},t)$ can be described by its mean and its covariance tensor $R_{ij}(\boldsymbol{r},t) = \overline{u_i(\boldsymbol{x},t)u_j(\boldsymbol{x}+\boldsymbol{r},t)}$. The mean velocity is however not of interest in homogeneous turbulence and is assumed to be zero, since one might choose a coordinate system moving with the mean flow. The covariance, or equivalently the spectral energy tensor $E_{ij}(\boldsymbol{k},t)$, given by the Fourier transform

$$R_{ij}(\boldsymbol{r},t) = \int E_{ij}(\boldsymbol{k},t)\mathrm{e}^{\mathrm{i}\boldsymbol{k}\cdot\boldsymbol{r}}\mathrm{d}\boldsymbol{k} \tag{11.1}$$

contains all relevant information about the turbulent field. In particular, the *scalar* energy spectrum

$$E(k,t) = \frac{1}{2}\int E_{ii}(\boldsymbol{k},t)\mathrm{d}f(\boldsymbol{k}) \tag{11.2}$$

will be of interest. Here $\mathrm{d}f(\boldsymbol{k})$ is the surface element of a sphere with radius $k$ in wave-number space, e. g. in polar coordinates $\mathrm{d}f = k^2 \sin\varphi\, \mathrm{d}\varphi\, \mathrm{d}\lambda$ with $k = |\boldsymbol{k}|$. The kinetic energy[2] of turbulent motion $\mathcal{E}$ is given by

$$\mathcal{E} \equiv \frac{1}{2}\overline{u_i u_i} = \int_0^\infty E(k)\mathrm{d}k \tag{11.3}$$

hence $E$ is the spectral density of turbulent kinetic energy (per mass).

The turbulent velocity field has to satisfy the Navier–Stokes equations. The continuity equation $\partial u_i/\partial x_i = 0$ leads to constraints for $R_{ij}$ and $E_{ij}$:

$$\frac{\partial R_{ij}(\boldsymbol{r})}{\partial r_j} = \overline{u_i(\boldsymbol{x})\frac{\partial}{\partial r_j}u_j(\boldsymbol{x}+\boldsymbol{r})} = 0 \quad \text{and likewise} \quad \frac{\partial R_{ij}}{\partial r_i} = 0 \tag{11.4}$$

Note that the time dependence of $R_{ij}$ and $E_{ij}$ is not explicitly stated in the following.

### 11.1.1 Isotropy

So far only homogeneity (i. e. invariance against translation) and continuity have been invoked. It is often useful to consider the special situation that the turbulent flow is also isotropic. Formally, isotropy means that the statistical variables are invariant against rotation around an arbitrary axis, against reflection at an arbitrary plane and against translation in an arbitrary direction. Isotropic turbulence, therefore, implies that all statistical properties, including the mean and the covariance tensor, are also invariant against any possible combination of rotation, mirroring, and translation.

[2] In this chapter we will the symbol $\mathcal{E}$ for the turbulent kinetic energy and $\mathcal{V}$ for the turbulent potential energy to avoid additional subscripts.

An isotropic scalar statistical property $\lambda$ cannot depend on the direction of $\boldsymbol{r}$, hence $\lambda = \lambda(r)$. Isotropy of a vector, as for example the mean of $\boldsymbol{u}$, implies that $\overline{\boldsymbol{u}} = 0$. To derive the isotropic form of the covariance tensor $R_{ij}(\boldsymbol{r})$, consider the covariance $Q$ of velocity components in the direction of two fixed vectors of unit length, $\boldsymbol{a}$ and $\boldsymbol{b}$, given as

$$Q(\boldsymbol{r},\boldsymbol{a},\boldsymbol{b}) = \overline{(\boldsymbol{a}\cdot\boldsymbol{u}(\boldsymbol{x},t))(\boldsymbol{b}\cdot\boldsymbol{u}(\boldsymbol{x}+\boldsymbol{r},t))} \equiv a_i b_j R_{ij} \tag{11.5}$$

To be invariant, the covariance $Q$ must be a function of the invariants which can be formed from its arguments, hence $Q(\boldsymbol{r},\boldsymbol{a},\boldsymbol{b}) = Q(\boldsymbol{a}\cdot\boldsymbol{b},\boldsymbol{a}\cdot\boldsymbol{r},\boldsymbol{b}\cdot\boldsymbol{r},r^2)$. According to (11.5), $Q$ must also be bilinear in $\boldsymbol{a}$ and $\boldsymbol{b}$, and hence must have the form

$$Q = \boldsymbol{a}\cdot\boldsymbol{b}G(r) + \boldsymbol{a}\cdot\boldsymbol{r}\boldsymbol{b}\cdot\boldsymbol{r}F(r) \equiv a_i b_j \left(G(r)\delta_{ij} + F(r)r_i r_j\right)$$

where $F(r)$ and $G(r)$ are arbitrary functions. Comparing with (11.5), the form of an isotropic velocity covariance tensor is thus given by

$$R_{ij}(\boldsymbol{r}) = F(r)r_i r_j + G(r)\delta_{ij} \tag{11.6}$$

A consequence of continuity in (11.4) is that

$$\frac{\partial}{\partial r_j}(F(r)r_i r_j + G(r)\delta_{ij}) = \frac{\partial F}{\partial r}\frac{r_j}{r}r_i r_j + 3F(r)r_i + F(r)r_j\delta_{ij} + \frac{\partial G}{\partial r}\frac{r_j}{r}\delta_{ij} = 0$$

holds. Separating the common factor $r_i$, one obtains that the functions $F(r)$ and $G(r)$ are related by

$$\frac{\partial G}{\partial r} = -\left(4rF(r) + r^2\frac{\partial F}{\partial r}\right) \tag{11.7}$$

Specifying $F$, (11.7) determines $G$; the integration constant is chosen such that $G \to 0$ for $r \to \infty$. For isotropic turbulence, a single scalar function $F$ is, therefore, sufficient to describe the velocity covariance. Corresponding to (11.6), the isotropic form of the energy spectral tensor has the general form

$$E_{ij}(\boldsymbol{k}) = A(k)k_i k_j + B(k)\delta_{ij} = A(k)\left[k_i k_j - k^2\delta_{ij}\right] \tag{11.8}$$

The second form is valid since the continuity (11.4) requires that $k_i E_{ij} = 0$ which implies $B(k) = -k^2 A(k)$. With (11.2), it then follows that

$$E(k) = \frac{1}{2}4\pi k^2 E_{ii}(k) = -4\pi k^4 A(k)$$

and with (11.8) the energy spectral tensor can be written in terms of the function $E(k)$ as

$$E_{ij}(\boldsymbol{k}) = \frac{E(k)}{4\pi k^4}\left(k^2\delta_{ij} - k_i k_j\right) \tag{11.9}$$

The scalar energy spectrum $E(k)$ is the preferred variable for a discussion of isotropic turbulent flows.

**44. Longitudinal Velocity Correlation and Velocity Variance**

Of interest is often the covariance of velocity components in the direction of the separation vector, e. g. in $r_1$ direction, which from (11.6) follows as (with $r = r_1$)

$$R_{11}(r, 0, 0) = r^2 F(r) + G(r) \equiv \overline{u_1^2} f(r)$$

The longitudinal velocity correlation $f(r)$ is the covariance of the velocity component in the direction of the separation vector $\boldsymbol{r}$ and is normalized as $f(0) = 1$. Using (11.1), $f(r)$ can be related to the energy spectrum with $R_{11} = \int E_{11} \cos(k_1 r) \mathrm{d}\boldsymbol{k}$ as

$$\begin{aligned} \overline{u_1^2} f(r) &= \int \frac{E}{4\pi k^4}(k^2 - k_1^2) \cos(k_1 r) k^2 \sin\phi \, \mathrm{d}\phi \, \mathrm{d}\lambda \, \mathrm{d}k \\ &= 2 \int_0^\infty E(k) \frac{1}{k^2 r^2} \left( \frac{\sin kr}{kr} - \cos kr \right) \mathrm{d}k \end{aligned} \tag{B44.1}$$

with $k_1 = k \cos\phi$. Letting $r \to 0$ within the integral in (B44.1) yields again (11.3),

$$\begin{aligned} \lim_{r \to 0} \overline{u_1^2} f(r) &\approx 2 \int_0^\infty E(k) \lim_{r \to 0} \frac{1}{k^2 r^2} \left( \frac{kr - (kr)^3/6}{kr} - \left( 1 - \frac{(kr)^2}{2} \right) \right) \mathrm{d}k \\ &= \frac{2}{3} \int E(k) \mathrm{d}k \end{aligned} \tag{B44.2}$$

which will be needed below.

### 11.1.2 Momentum and Kinetic Energy in Homogeneous Turbulence

In addition to the continuity equation, the dynamics of turbulent motion is governed by the momentum equation. For small-scale motions with time-scales much shorter than one day, the Earth's rotation can be neglected. For constant $\rho$, the momentum equation (4.10) can be written in tensor notation as

$$\frac{\partial u_i}{\partial t} = -\frac{\partial u_l u_i}{\partial x_l} - \frac{\partial p^\star}{\partial x_i} + \kappa_\mathrm{m} \frac{\partial^2 u_i}{\partial x_l^2} \tag{11.10}$$

where $\mathcal{F} = \kappa_\mathrm{m} \nabla^2 \boldsymbol{u}$ was used and $\kappa_\mathrm{m}$ denotes the kinematic viscosity of sea water. Furthermore, $p^\star = p/\rho + \Phi$ is the combined scaled pressure, and gravity potential (the star is omitted in the following). According to (11.10), changes in momentum are caused by inertial, pressure and frictional forces. The statistical mean of (11.10) is given by

$$\frac{\partial \overline{u_i}}{\partial t} = -\frac{\partial \overline{u_l u_i}}{\partial x_l} - \frac{\partial \bar{p}}{\partial x_i} + \kappa_\mathrm{m} \frac{\partial^2 \overline{u_i}}{\partial x_l^2} = 0$$

because of homogeneity of all mean values (compare Appendix A.3). Hence the mean velocity is constant in time and space and can be ignored (if necessary by choosing a coordinate system moving with the mean velocity).

### Equations for Velocity Covariance

To derive an equation for the evolution of $R_{ij}(\boldsymbol{r}, t)$, consider the momentum equation (11.10) at two points, $\boldsymbol{x}$ and $\boldsymbol{x}' = \boldsymbol{x} + \boldsymbol{r}$,

$$\boldsymbol{x}: \frac{\partial u_i}{\partial t} = -\frac{\partial u_l u_i}{\partial x_l} - \frac{\partial p}{\partial x_i} + \kappa_m \frac{\partial^2 u_i}{\partial x_l^2} \tag{11.11}$$

$$\boldsymbol{x}': \frac{\partial u'_j}{\partial t} = -\frac{\partial u'_l u'_j}{\partial x'_l} - \frac{\partial p'}{\partial x'_j} + \kappa_m \frac{\partial^2 u'_j}{\partial x'^2_l} \tag{11.12}$$

where $u'_i = u_i(\boldsymbol{x}', t)$ etc. Note that the covariance tensor $R_{ij}$ depends on the separation $\boldsymbol{r}$ but not on the location $\boldsymbol{x}$. Multiplication of (11.11) with $u'_j$ and of (11.12) with $u_i$, adding both results and averaging yields

$$\overline{u'_j \frac{\partial}{\partial t} u_i} + \overline{u_i \frac{\partial}{\partial t} u'_j} = -\overline{u'_j \frac{\partial}{\partial x_l} u_l u_i} - \overline{u_i \frac{\partial}{\partial x'_l} u'_l u'_j} - \overline{u'_j \frac{\partial}{\partial x_i} p} - \overline{u_i \frac{\partial}{\partial x'_j} p'}$$
$$+ \kappa_m \overline{u'_j \frac{\partial^2}{\partial x_l^2} u_i} + \kappa_m \overline{u_i \frac{\partial^2}{\partial x'^2_l} u'_j} \tag{11.13}$$

The terms on the left-hand side of (11.13) can be combined to the rate of change of the covariance tensor $\partial R_{ij}/\partial t = \partial \overline{u'_j u_i}/\partial t$. Equation (11.13) is thus an evolution equation for $R_{ij}$ and can be written as

$$\frac{\partial R_{ij}(\boldsymbol{r})}{\partial t} = T_{ij}(\boldsymbol{r}) + P_{ij}(\boldsymbol{r}) + 2\kappa_m \frac{\partial^2 R_{ij}(\boldsymbol{r})}{\partial r_\ell \partial r_\ell} \tag{11.14}$$

The three terms on the right-hand side of (11.14) correspond directly to the inertial, pressure, and frictional terms in the momentum equation (11.11) and (11.12), respectively. Two new tensors have been introduced, the inertial tensor $T_{ij}$ and the pressure tensor $P_{ij}$, which can be brought into the form

$$T_{ij}(\boldsymbol{r}) \equiv -\overline{u'_j \frac{\partial}{\partial x_l} u_l u_i} - \overline{u_i \frac{\partial}{\partial x'_l} u'_l u'_j} = \frac{\partial}{\partial r_\ell}\left(\overline{u_i u'_j u_\ell} - \overline{u_i u'_j u'_\ell}\right) \tag{11.15}$$

$$P_{ij}(\boldsymbol{r}) \equiv -\overline{u'_j \frac{\partial}{\partial x_i} p} - \overline{u_i \frac{\partial}{\partial x'_j} p'} = \frac{\partial}{\partial r_i}\overline{p u'_j} - \frac{\partial}{\partial r_j}\overline{p' u_i} \tag{11.16}$$

since $u'_j$ does not depend on $x_l$ and $u_i$ does not depend on $x'_l$ and $\partial/\partial x_l = -\partial/\partial r_l$ and $\partial/\partial x'_l = \partial/\partial r_l$. Note that for the inertial tensor $T_{ij}(\boldsymbol{r} = \boldsymbol{0}) = 0$ holds for all $i, j$, and that by summing over $i = j$ the trace of the pressure tensor vanishes for all $\boldsymbol{r}$ since

$$P_{ii}(\boldsymbol{r}, t) = \frac{\partial}{\partial r_i}\overline{u_i(\boldsymbol{x} + \boldsymbol{r}) p(\boldsymbol{x})} - \frac{\partial}{\partial r_i}\overline{u_i(\boldsymbol{x} - \boldsymbol{r}) p(\boldsymbol{x})}$$
$$= \overline{p(\boldsymbol{x}) \frac{\partial}{\partial r_i} u_i(\boldsymbol{x} + \boldsymbol{r})} - \overline{p(\boldsymbol{x}) \frac{\partial}{\partial r_i} u_i(\boldsymbol{x} - \boldsymbol{r})} = 0$$

as a consequence of the continuity equation $\partial u_i/\partial r_i = 0$. The results concerning $T_{ij}(0)$ and $P_{ii}$ will be used below.

A direct solution of (11.14) is not possible since the inertial tensor $T_{ij}$ is a third-order moment, and $P_{ij}$ can also be expressed as a third-order moment (not shown).

One approach is to write down equations for the necessary third-order moments needed to predict $T_{ij}$ and $P_{ij}$. While this is in principle possible, the approach fails because fourth-order moments will appear in the respective equations, and so forth. Therefore, one always will have more variables than equations. This fact – which is referred to as 'closure problem' – precludes a direct solution.

### Spectral Description

All tensors in (11.14) can be represented by the spectral description

$$\left\{R_{ij}, T_{ij}, P_{ij}, 2\kappa_{\mathrm{m}}\frac{\partial^2 R_{ij}}{\partial r_l^2}\right\} = \int \left\{E_{ij}(\boldsymbol{k}), \Gamma_{ij}(\boldsymbol{k}), \Pi_{ij}(\boldsymbol{k}), -2\kappa_{\mathrm{m}}k^2 E_{ij}(\boldsymbol{k})\right\} \mathrm{e}^{\mathrm{i}\boldsymbol{k}\cdot\boldsymbol{r}}\,\mathrm{d}\boldsymbol{k}$$

with the inversions

$$E_{ij}(\boldsymbol{k}) = \frac{1}{(2\pi)^3}\int R_{ij}(\boldsymbol{r})\mathrm{e}^{-\mathrm{i}\boldsymbol{k}\cdot\boldsymbol{r}}\,\mathrm{d}\boldsymbol{r}$$

and correspondingly for inertial spectral tensor $\Gamma_{ij}$ and pressure spectral tensor $\Pi_{ij}$. The spectral balance is obtained by multiplication of (11.14) with $\mathrm{e}^{-\mathrm{i}\boldsymbol{k}\cdot\boldsymbol{r}}$ and integration over $\boldsymbol{r}$, yielding

$$\frac{\partial E_{ij}(\boldsymbol{k})}{\partial t} = \Gamma_{ij}(\boldsymbol{k}) + \Pi_{ij}(\boldsymbol{k}) - 2\kappa_{\mathrm{m}}k^2 E_{ij}(\boldsymbol{k}) \tag{11.17}$$

Since $\int E_{ij}(\boldsymbol{k})\mathrm{d}\boldsymbol{k} = \overline{u_i u_j}$, the tensor $E_{ij}(\boldsymbol{k})$ denotes the spectral energy density of $\overline{u_i u_j}$ in the (three-dimensional) wave-number space. The last term in (11.17) represents viscous dissipation. The spectral inertial tensor $\Gamma_{ij}(\boldsymbol{k})$ in (11.17) describes the effect of inertial forces on the energy tensor $E_{ij}(\boldsymbol{k})$ and it holds that

$$T_{ij}(\boldsymbol{0}, t) \equiv 0 \Leftrightarrow \int \Gamma_{ij}(\boldsymbol{k})\mathrm{d}\boldsymbol{k} = 0 \tag{11.18}$$

as shown above. Inertial forces, therefore, cannot change the moment $\overline{u_i u_j}$ as well as the total energy, but can redistribute energy between different wave numbers. It is plausible for three-dimensional turbulence that the inertial forces transport energy from regions with low wave numbers to high ones (where dissipation is mostly active), corresponding to a down-gradient transport in wave-number space.

For the pressure tensor, it holds that

$$P_{ii}(\boldsymbol{r}) \equiv 0 \Leftrightarrow \Pi_{ii}(\boldsymbol{k}) = 0 \tag{11.19}$$

as shown above. The pressure forces $\Pi_{ij}(\boldsymbol{k})$ in (11.17), therefore, cannot change the kinetic energy at any wave number, but can redistribute kinetic energy between different velocity components. It is plausible that the transport will be directed from components with high to components with low energy. Hence pressure forces reduce anisotropy, and in the absence of external forces one can assume that the turbulent flow is nearly isotropic.

### Scalar Energy Balance

The scalar form of the spectral balance can be obtained by integration of (11.17) over a sphere with radius $k$ in wave-number space. The scalar projection $\Gamma(k)$ of the inertial tensor is given, in analogy to (11.2), as

$$\Gamma(k) = \frac{1}{2} \int \Gamma_{ii}(\boldsymbol{k})\mathrm{d}f(\boldsymbol{k}) \tag{11.20}$$

and satisfies $\int \Gamma(k)\mathrm{d}k = 0$ as in (11.18). It is convenient to define the spectral energy transport in wave-number space as convergence of an energy flux in wave-number space,

$$F(k) = -\int_0^k \Gamma(k')\mathrm{d}k' \tag{11.21}$$

which satisfies $F(0) = 0$ and $F(\infty) = 0$. With the energy spectrum $E(k)$ according to (11.2) and (11.21), the energy balance can be written as

$$\frac{\partial}{\partial t} E(k) = -\frac{\partial F(k)}{\partial k} - 2\kappa_{\mathrm{m}} k^2 E(k) \tag{11.22}$$

The flux $F(k)$ redistributes energy among different wave numbers. Since dissipation $2\kappa_{\mathrm{m}} k^2 E(k)$ is more effective at small scales, it is plausible that energy has to be transported by $F(k)$ from large scales (small $k$) to small scales (large $k$). Integration of (11.22) over all wave numbers $k$ yields

$$\frac{\mathrm{d}\mathcal{E}}{\mathrm{d}t} = -\epsilon \tag{11.23}$$

for the budget of the total kinetic energy $\mathcal{E} = \int E(k)\mathrm{d}k = \frac{1}{2}\overline{u_i u_i}$ from (11.3). In (11.23), $\epsilon = 2\kappa_{\mathrm{m}} \int k^2 E(k)\mathrm{d}k$ denotes the energy dissipation and is a central parameter in the following discussion. Using the definition for mechanical dissipation of Section 2.4.2 in the Boussinesq approximation and for isotropic turbulence, it can be written as (compare also the box on p. 354)

$$\epsilon = 2\kappa_{\mathrm{m}} \overline{D_{ij}^2} = \kappa_{\mathrm{m}} \overline{\frac{\partial u_j}{\partial x_i}\frac{\partial u_j}{\partial x_i}} + \kappa_{\mathrm{m}} \frac{\partial}{\partial x_i}\frac{\partial}{\partial x_j}\overline{u_j u_i} = \kappa_{\mathrm{m}} \overline{\frac{\partial u_j}{\partial x_i}\frac{\partial u_j}{\partial x_i}} \tag{11.24}$$

with the deformation tensor $D_{ij} = (1/2)(\partial u_i/\partial x_j + \partial u_j/\partial x_i)$. Since $\epsilon$ is always positive, the total kinetic energy $\mathcal{E}$ will decrease in time as long as no energy is injected into the system.

## *11.1.3 Large and Small Length Scales*

We have assumed in the proceeding section that the flux $F(k)$ in wave-number space transports energy from large (small $k$) to small scales (large $k$), since dissipation is more effective at small scales. To further discuss the concept of such an energy flux (or cascade) we need to specify the definitions of large and small scales, which will become the integral length scale and the micro scale, respectively.

### Integral Length Scale

The longitudinal correlation defined in the box on p. 338 is suitable to discuss characteristic length scales of turbulence. The *integral length scale* $L$ (also called *macro scale* or *longitudinal scale*) is defined as

$$L = \int_0^\infty f(r)\mathrm{d}r \tag{11.25}$$

and describes the distance over which the correlation is essentially nonzero (cf. Figure 11.1). Hence $L$ must be roughly equivalent to the scale of the largest coherent fluctuations (i. e. to the largest eddies). Integrating (B44.1) over all $r$ (exchanging the integrations over $k$ and $r$) and using (B44.2), the length scale can be expressed in terms of the energy spectrum as

$$L = \frac{3\pi}{4} \frac{\int k^{-1} E(k)\mathrm{d}k}{\int E(k)\mathrm{d}k} \tag{11.26}$$

Large eddies with length scale $L$ will predominantly contribute to the total energy $\mathcal{E}$. On the other hand, the total energy (i. e. its rate of decrease if there is no energy input, or its forcing if there is) and thus the properties of the large eddies are directly related to the dissipation $\epsilon$ by (11.23). Therefore, one can assume that the dissipation is determined by the energy and scale of the large eddies, i. e. that $\epsilon = f(\mathcal{E}, L)$. A simple dimensional argument shows that the only possible combination between $\mathcal{E}$ and $L$ to form the dimension of $\epsilon$ is given by

$$\epsilon = c_\epsilon \frac{\mathcal{E}^{3/2}}{L} \tag{11.27}$$

where $c_\epsilon$ denotes a dimensionless constant which is experimentally found as $c_\epsilon = O(1)$. The form (11.27) can be used with (11.23) to determine the time-scale for the decrease of kinetic energy as

$$T_0 = \frac{\mathcal{E}}{\mathrm{d}\mathcal{E}/\mathrm{d}t} = \frac{\mathcal{E}}{c_\epsilon \mathcal{E}^{\frac{3}{2}}/L} = \frac{1}{c_\epsilon}\frac{L}{\mathcal{E}^{\frac{1}{2}}} = \frac{1}{c_\epsilon}\frac{L}{U} \tag{11.28}$$

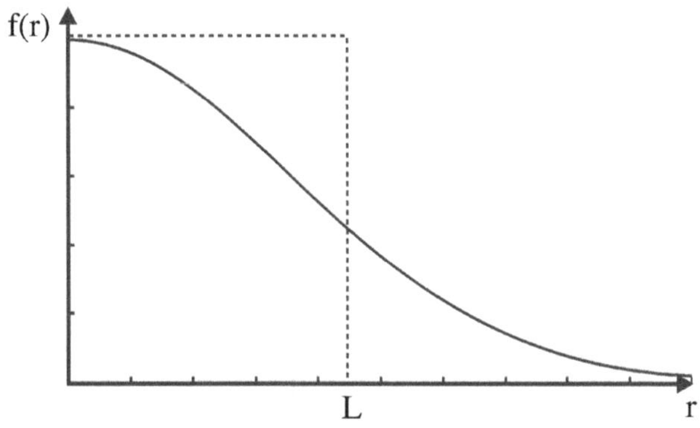


Fig. 11.1 Sketch of the longitudinal correlation and the integral length scale $L$, which is a measure of the scale of the largest coherent fluctuations, i. e. to the largest eddies

where $U$ is a velocity scale of the large eddies and where $L/U = T_{\text{adv}}$ denotes an advective time-scale of the large eddies;, therefore, the energy content is changing on the time-scale $T_{\text{adv}}$. As a consequence, there must be energy input on the scale $L$ to sustain the turbulent flow.

### Micro Scale

For small $r$, the longitudinal correlation $f(r)$ (cf. the box on p. 338) can be expanded in a Taylor series

$$f(r) = 1 + rf'(0) + \frac{1}{2}r^2 f''(0) + \ldots = 1 - \frac{1}{2}\frac{r^2}{\lambda^2} + \cdots$$

since $f'(0) = 0$ for any reasonable symmetric correlation function (it is also a direct consequence of (11.7)) and with

$$\lambda^2 = -\frac{1}{f''(0)} \tag{11.29}$$

The length scale $\lambda$ reflects the curvature of $f(r)$ at $r = 0$ and is referred to as TAYLOR's[3] *micro scale*. The micro scale can also be expressed in terms of the energy spectrum. Differentiating (B44.1) twice with respect to $r$ and then letting $r \to 0$ leads to

$$\overline{u_1^2} f''(0) = -\frac{2}{15}\int k^2 E(k)\mathrm{d}k$$

With (11.29) and using (B44.2), one obtains for the micro scale

$$\frac{1}{\lambda^2} = \frac{1}{5}\frac{\int k^2 E(k)\mathrm{d}k}{\int E(k)\mathrm{d}k} \tag{11.30}$$

Replacing the integrals in (11.30) with $\mathcal{E}$ and $\epsilon$ from (11.23), one finds

$$\lambda^2 = 10\kappa_{\text{m}}\mathcal{E}\epsilon^{-1} \tag{11.31}$$

It follows that the micro scale $\lambda$ is related to dissipation. However, it will turn out that $\lambda$ is not the scale at which the viscous forcing dominates inertial forces; it is still much larger.

### Relation between Integral Length Scale and Micro Scale

To relate the integral length scale with the micro scale, we use (11.26) and (11.31) and express the dissipation via (11.27) by the total energy and the integral length scale. It follows that $10\kappa_{\text{m}}\mathcal{E}\lambda^{-2} = c_\epsilon \mathcal{E}^{3/2}L^{-1}$ or

$$\frac{L^2}{\lambda^2} = c_\epsilon \frac{\mathcal{E}^{1/2}L}{10\kappa_{\text{m}}} = c_\epsilon \frac{\text{Re}_L}{10} \tag{11.32}$$

where the Reynolds number of large eddies $\text{Re}_L = UL\kappa_{\text{m}}^{-1} = \mathcal{E}^{1/2}L\kappa_{\text{m}}^{-1}$ has been introduced. If $\text{Re}_L \gg 1$, which is the case for turbulent flows, it follows that the magnitude of both length scales is very different, with $L \gg \lambda$. Note that for an energy spectrum with strong (exponential) decay for $k > 1/L$, the length scales $L$ and $\lambda$ would have the same order of magnitude. We conclude that the spectral decay for a typical turbulence spectrum is weaker than exponential, as will be seen below.

[3] GEOFFREY INGRAM TAYLOR, *1886 in St. John's Wood, †1975 in Cambridge, physicist.

### 11.1.4 Equilibrium Range and Inertial Subrange

As we have seen, there has to be input of energy into the turbulent flow on the scale $L$ to sustain the turbulence. At smaller scales, however, this is not necessary, and it can be expected that in a range of wave numbers $k \gg 1/L$ the spectrum $E(k)$ will not change much over time. In this *equilibrium range* the energy balance (11.22) reduces to

$$\frac{\partial F(k)}{\partial k} = -2\kappa_\mathrm{m} k^2 E(k) \quad \text{or in integral form} \quad F(k) = 2\kappa_\mathrm{m} \int_k^\infty k'^2 E(k')\mathrm{d}k' \tag{11.33}$$

for any wave number $k$ in the equilibrium range. Here $F(k) > 0$, i. e. energy, is indeed transported through the spectrum toward higher wave numbers and is finally dissipated at small scales.

We further assume that the energy is transported between wave numbers which are close together (turbulent cascade), so that in the equilibrium range the details of the flow at the large scale $L$ are irrelevant. It is then clear that the spectrum $E(k)$ must depend on $k$, $\epsilon$ and $\kappa_\mathrm{m}$ in the equilibrium range. In the absence of other, e. g. geometric, constraints, that dependence must be universal, i. e. $E = \Phi(k, \epsilon, \kappa_\mathrm{m})$. The variables in this universal relation have the dimensions $[E] = \mathrm{m^2\,s^{-2}}$, $[k] = \mathrm{m^{-1}}$, $[\kappa_\mathrm{m}] = \mathrm{m^2\,s^{-1}}$, $[\epsilon] = \mathrm{m^2\,s^{-3}}$. While an infinite number of dimensionless ratios can be built from these four variables, only two of these are independent, e. g. $k\kappa_\mathrm{m}^{3/4}\epsilon^{-1/4}$ and $E\epsilon^{-2/3}k^{5/3}$. From dimensional analysis it follows that the relation

$$E\epsilon^{-2/3}k^{5/3} = \phi(k\kappa_\mathrm{m}^{3/4}\epsilon^{-1/4}) = \phi(k\lambda_\mathrm{d}) \tag{11.34}$$

must hold, where $\phi(x)$ is an arbitrary dimensionless function, and

$$\lambda_\mathrm{d} = (\kappa_\mathrm{m}^3/\epsilon)^{1/4} \tag{11.35}$$

The length scale $\lambda_\mathrm{d}$ is given by dissipation and viscosity and is called the dissipation length scale or Kolmogorov's microscale. The energy spectrum has the form

$$E(k) = \phi(k\lambda_\mathrm{d})\epsilon^{2/3}k^{-5/3} \tag{11.36}$$

One can show that the dissipation spectrum $k^2E(k)$ has a maximum near the wave number $1/\lambda_\mathrm{d}$. Defining a velocity scale in the dissipation range as $U_\mathrm{d} = (\epsilon\kappa_\mathrm{m})^{1/4}$, which is the only combination of $\epsilon$ and $\kappa_\mathrm{m}$ with the correct dimension, the Reynolds number at the dissipation length scale $\lambda_\mathrm{d}$ follows as $\mathrm{Re_d} = U_\mathrm{d}\lambda_\mathrm{d}/\kappa_\mathrm{m} = 1$. Hence at the scale $\lambda_\mathrm{d}$ inertial forces no longer dominate, and the turbulent transfer of energy ends here. Typical values of $\lambda_\mathrm{d}$ are given in the box on p. 345.

To determine the relation between Taylor's microscale $\lambda$ and the dissipation length $\lambda_\mathrm{d}$, we use (11.31) and $\lambda_\mathrm{d} = (\kappa_\mathrm{m}^3/\epsilon)^{1/4}$. It follows that

$$\begin{aligned}\frac{\lambda^2}{\lambda_\mathrm{d}^2} &= \frac{10\kappa_\mathrm{m}\mathcal{E}\epsilon^{-1}}{(\kappa_\mathrm{m}^3/\epsilon)^{1/2}} = 10\frac{\mathcal{E}}{\kappa_\mathrm{m}^{1/2}\epsilon^{1/2}} = 10c_\epsilon^{-1/2}\frac{\mathcal{E}^{1/4}L^{1/2}}{\kappa_\mathrm{m}^{1/2}} \\ &= 10c_\epsilon^{-1/2}\frac{(UL)^{1/2}}{\kappa_\mathrm{m}^{1/2}} \approx 10\mathrm{Re}_L^{1/2}\end{aligned}$$

**45. Dissipation in a Kitchen Mixer**

The magnitude of $\lambda_d$ can be estimated for a simple kitchen experiment, where a mixer with 100 W power is used to mix 1 kg water, thereby generating turbulent flow. Assuming that the energy input into the mixer is dissipated mechanically only in the water, the dissipation is $\epsilon = 100\,\mathrm{W}/1\,\mathrm{kg} = 100\,\mathrm{m^2\,s^{-3}}$. With $\kappa_m = 10^{-6}\,\mathrm{m^2\,s^{-1}}$ as kinematic molecular viscosity of water, it follows that $\lambda_d = (\kappa_m^3/\epsilon)^{1/4} = 0.01\,\mathrm{mm}$. In the ocean, the dissipation $\epsilon$ is much smaller, e. g. in the interior ocean measurements give typically values of $\epsilon \approx 10^{-9}\,\mathrm{m^2\,s^{-3}}$, thus $\lambda_d \approx 5\,\mathrm{mm}$. However, in the mixed layer of the ocean dissipation can reach values of $10^{-4}\,\mathrm{m^2\,s^{-3}}$, thus $\lambda_d \approx 0.3\,\mathrm{mm}$. In the atmosphere, a typical value is $\lambda_d \approx 1\,\mathrm{mm}$.

Here (11.27) has been used to eliminate the dissipation, and $\mathrm{Re}_L$ again denotes the Reynolds number of large eddies. With (11.31), the orders of magnitude are related as

$$\lambda_d : \lambda : L = 1 : \sqrt{10}\mathrm{Re}_L^{1/4} : \mathrm{Re}_L^{3/4} \approx 1 : 60 : 6{,}000$$

where the numbers are given for the case $\mathrm{Re}_L = 10^5$. The three length scales $L$, $\lambda$ and $\lambda_d$ are, therefore, well separated in the normal case where the Reynolds number is large. As noted above, the micro scale $\lambda$ is (for $\mathrm{Re} \gg 1$) much smaller than the integral scale $L$, but also much larger than the dissipation scale $\lambda_d$ at which viscous and inertial forces become comparable. The microscale $\lambda$ thus represents a "geometric mean" between $L$ and $\lambda_d$ somewhere in the equilibrium range.

## Inertial Subrange

For wave numbers in the range

$$1/L \ll k \ll 1/\lambda_d$$

dissipation must be small, i. e. only the inertial forces are important for the flow in the inertial subrange (see display in Figure 11.2). According to (11.33), the energy

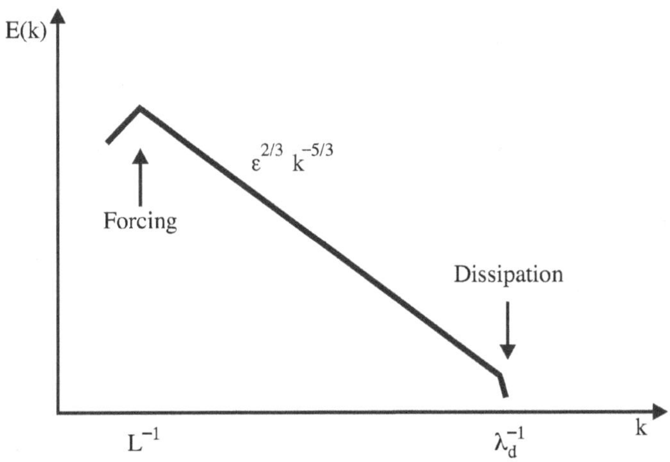


Fig. 11.2 Conceptual sketch of the turbulent energy spectrum in the inertial subrange (logarithmic scaling), which extents from the integral length scale of the largest eddies $L$ given by (11.25) to the dissipation length scale $\lambda_d$ given by (11.35). The micro scale $\lambda$ given by (11.31) represented a "geometric mean" between $L$ and $\lambda_d$

flux $F(k)$ is constant and equals the dissipation $\epsilon$, as seen from (11.33). Therefore, one can assume that the characteristic features of the turbulence depend only on $\epsilon$ but not on the viscosity $\kappa_m$. For the energy spectrum (11.34) this means that $\phi(k\lambda_d) =$ const $= c_k$, hence

$$E(k) = c_k \epsilon^{2/3} k^{-5/3} \tag{11.37}$$

Empirically one finds that $c_k \approx 1.5$. This result, in particular the spectral power law $k^{-5/3}$, is well supported by many oceanographic and meteorological observations. Prerequisite for the existence of an inertial subrange is a very large Reynolds number on the large scale $L$, an assumption which often holds in the ocean and the atmosphere. Note that this spectral law holds only for small-scale three-dimensional and isotropic turbulence, an important restriction as shown in the next chapter.

## 11.2 Turbulent Mixing

In this section, some aspects of tracer diffusion in a turbulent flow will be discussed. Consider the concentration of a substance or tracer $\psi$ in a turbulent velocity field which in the Boussinesq approximation is governed by

$$\frac{\partial \psi}{\partial t} + \boldsymbol{u} \cdot \nabla \psi = \nabla \cdot \kappa \nabla \psi \tag{11.38}$$

The Fickian law has been used for the molecular tracer flux (i. e. a simplified version of (2.96)). As discussed in Section 2.8.2, both $\boldsymbol{u}$ and $\psi$ can be decomposed into a statistical mean[4] and deviation thereof, i. e.

$$\boldsymbol{u} = \bar{\boldsymbol{u}} + \boldsymbol{u}' \quad \text{and} \quad \psi = \bar{\psi} + \psi' \tag{11.39}$$

where $\overline{\psi'} = 0$ and $\overline{\overline{\psi}} = \overline{\psi}$ (and equivalently for $\boldsymbol{u}$). Further useful properties of the statistical mean are $\overline{(\psi + \lambda\phi)} = \overline{\psi} + \lambda\overline{\phi}$ and $\overline{\phi\overline{\psi}} = \overline{\phi}\,\overline{\psi}$ for any constant $\lambda$ and random functions $\phi$ and $\psi$ and that it commute with differential operators. With (11.39), statistical averaging of (11.38) yields

$$\frac{\partial \bar{\psi}}{\partial t} + \bar{\boldsymbol{u}} \cdot \nabla \bar{\psi} = \nabla \cdot \left(\kappa \nabla \bar{\psi} - \overline{\boldsymbol{u}'\psi'}\right) \tag{11.40}$$

The averaged equation (11.40) is similar to the instantaneous equation (11.38), except for the additional turbulent flux $\boldsymbol{J}_\psi^{\text{turb}} = \overline{\boldsymbol{u}'\psi'}$, which originates from the advective term in (11.38) and which is generally unknown but almost always much larger than the (mean) molecular tracer flux $\kappa \nabla \bar{\psi}$.

In analogy to the Fickian law, it is customary to formulate the turbulent flux with an Austauschansatz in terms of the mean gradient as

$$\overline{\boldsymbol{u}'\psi'} = -K \nabla \bar{\psi} \tag{11.41}$$

with the turbulent diffusivity $K$. Since for $K > 0$ relation (11.41) corresponds to down-gradient tracer transport, it is also referred to as down-gradient parameterization. The remainder of this section will be used to present some empirical mixing formulations, to discuss the plausibility of the parameterization (11.41), and to derive constraints for the choice of the turbulent diffusivity $K$.

[4] The statistical mean is defined in Appendix A.3. However, often a time mean as in Section 2.8.2 is used instead for simplicity, for which the relation $\overline{\overline{\psi}} = \overline{\psi}$ does not hold in general.

### 11.2.1 Heuristic Approaches

Heuristic considerations lead to specific forms for the parameterization (11.41). In the following, we will discuss two classical approaches which have proven useful, based on mixing by energy-containing eddies and the mixing-length concept. We further discuss a practical approach to represent the effect of partially resolved eddies.

#### Mixing by Energy-Containing Eddies

As discussed in Section 11.1, a turbulent flow is characterized by fluctuations over a range of different length scales. A simple form of (11.41) can be obtained by assuming that mixing is mainly caused by eddies with maximum energy, with a length scale $L$ as discussed in Section 11.1.3.

Consider a tracer with the mean concentration $\overline{\psi}(x)$ (one-dimensional for simplicity). Let the turbulent flow be characterized by eddies with characteristic swirl velocity $U$ and diameter $L$. To estimate the transport of $\psi$ by the turbulent flow, we note that the tracer flux from $x_0 - L/2$ towards $x_0$ is approximately $U\overline{\psi}(x_0 - L/2)$, and likewise the flux from $x_0 + L/2$ towards $x_0$ is approximately $-U\overline{\psi}(x_0 + L/2)$ (cf. Figure 11.3). The total tracer flux $J$ summed over both cases is approximately

$$J = U\left[\overline{\psi}(x_0 - L/2) - \overline{\psi}(x_0 + L/2)\right] \equiv -UL\frac{\overline{\psi}(x_0 + L/2) - \overline{\psi}(x_0 - L/2)}{L}$$

Provided that the scale over which the mean tracer varies is larger than $L$, this can be approximated as $J \approx -UL\partial\overline{\psi}/\partial x$. Identifying $J$ with the turbulent tracer flux $J = \overline{u'\psi'}$, it follows that (11.41) is recovered with the turbulent diffusivity

$$K = UL \sim \mathcal{E}^{\frac{1}{2}} L \qquad (11.42)$$

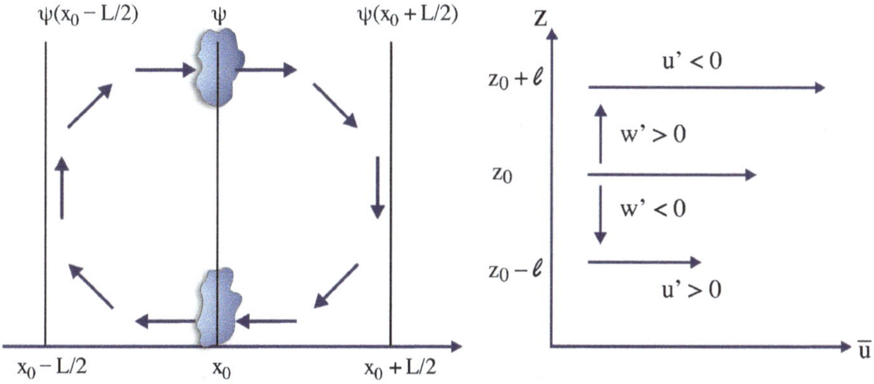


Fig. 11.3 *Left*: Schematic of tracer transport by large eddies with diameter $L$ and swirl velocity $U$. The resulting turbulent flux of the tracer concentration $\psi$ is given by $\overline{u'\psi'} \approx -UL\partial\overline{\psi}/\partial x$, where $K = UL$ can be interpreted as a turbulent diffusivity. *Right*: Schematic of the mixing-length concept in which a sheared horizontal mean flow $\overline{u}(z)$ is subject to turbulent vertical exchange of momentum. The vertical turbulent flux of horizontal momentum can be written as $\overline{u'w'} = -K_u\partial\overline{u}/\partial z$, where the turbulent viscosity is given by $K_u \sim \ell^2|\partial\overline{u}/\partial z|$ with the mixing length $\ell$

where $\mathcal{E}$ is the turbulent kinetic energy. The result (11.42) is very useful because often the magnitudes of the characteristic velocity $U$ and length scale $L$ of eddies are approximately known so that the magnitude of $K$ can be estimated. It is, therefore, not surprising that (11.42) is the basis of many parameterizations.

### Mixing Length

A concept which is particularly useful to describe mixing of momentum was introduced by PRANDTL[5] (Prandtl, 1925). Consider a horizontally homogeneous mean shear flow $\overline{u}(z)$ with $\partial\overline{u}/\partial z > 0$ (cf. Figure 11.3). A turbulence element with a vertical velocity perturbation $w' > 0$ will carry a fluid element (and its mean momentum) from the initial level $z_0$ upward to a new level $z_0 + \ell'$ before it is mixed. Here, the velocity of the fluid element deviates from the ambient velocity by $u' = \overline{u}(z_0) - \overline{u}(z_0 + \ell') < 0$. Likewise, if $w' < 0$, then $u' > 0$. In both cases, it is clear that $\overline{u'w'} < 0$. Expansion in a Taylor series leads to

$$u' = -\overline{u}(z_0 + \ell') + \overline{u}(z_0) \approx -\ell' \left.\frac{\partial\overline{u}}{\partial z}\right|_{z_0} + \ldots \tag{11.43}$$

It follows that

$$\overline{u'w'} = -\overline{w'\ell'}\frac{\partial\overline{u}}{\partial z} = -K_\mathrm{u}\frac{\partial\overline{u}}{\partial z} \tag{11.44}$$

which corresponds to the form (11.41) for a tracer. One can expect $w'$ and $\ell'$ to be positively correlated, $\overline{w'\ell'} > 0$ or $\overline{w'\ell'} = c(\overline{w'^2}\,\overline{\ell'^2})^{1/2}$ with a positive $c \leq 1$. Assuming isotropy so that $\overline{w'^2} \approx \overline{u'^2}$, this result is equivalent to $\overline{w'\ell'} = c(\overline{u'^2}\,\overline{\ell'^2})^{1/2} = c(\overline{\ell'^2}(\partial\overline{u}/\partial z)^2\overline{\ell'^2})^{1/2}$, where (11.43) has been used. It follows that the diffusivity $K_\mathrm{u} = \overline{w'\ell'}$ can be expressed as

$$K_\mathrm{u} = c\ell^2\left|\frac{\partial\overline{u}}{\partial z}\right| \tag{11.45}$$

The length scale $\ell = (\overline{\ell'^2})^{1/2}$ is the (mean) *mixing-length*. Under certain conditions it can be comparable to the scale of the energy-containing eddies (see Section 11.3.2 below).

Note that termination of the Taylor series in (11.43) after the first term requires that the length scale $\ell$ is small compared to the scales of the mean flow. Therefore, the mixing-length concept (11.45) can be valid only if $\ell \ll |\partial\overline{u}/\partial z|/|\partial^2\overline{u}/\partial z^2|$.

### Diffusion in the Inertial Subrange

The concept (11.42) can also be modified for situations where the large eddies are explicitly resolved, and turbulent diffusion is caused by velocity fluctuations on smaller scales. Specifically, assume that only wave numbers $k \geq k_\star$ in the inertial subrange (cf. Section 11.1.4) contribute to velocity. The energy $\mathcal{E}_\star$ of these fluctuations is

$$\mathcal{E}_\star = \int_{k_\star}^{k_\mathrm{d}} c_\mathrm{K}\epsilon^{2/3}k^{-5/3}\mathrm{d}k \approx \int_{k_\star}^{\infty} c_\mathrm{K}\epsilon^{2/3}k^{-5/3}\mathrm{d}k = \frac{3c_\mathrm{K}}{5}\epsilon^{2/3}k_\star^{-2/3}$$

[5] LUDWIG PRANDTL, *1875 Freising, † 1953 Göttingen, fluid dynamicist.

**46. An Aspect of Molecular Diffusion**

To discuss the plausibility of relation (11.41) for the turbulent transports, it is useful to briefly reconsider molecular diffusion as the most familiar form of mixing. In its pure form, molecular diffusion of tracer concentration $\psi$ is governed by (11.38) without mean flow, and restricting for simplicity to one spatial dimension with constant $\kappa$, one has

$$\frac{\partial \psi}{\partial t} = \kappa \frac{\partial^2 \psi}{\partial x^2} \tag{B46.1}$$

For a given initial tracer distribution $\psi_0(x) = \psi(x, t_0)$ at time $t_0$, the solution of the diffusion equation (B46.1) at later times can be expressed with the help of a Green's function as

$$\psi(x,t) = \int_{-\infty}^{\infty} P(x,x',t,t_0)\psi_0(x')\mathrm{d}x' \tag{B46.2}$$

In an unbounded domain, the Green's function $P(x,x',t,t_0)$ is given by

$$P(x,x',t,t_0) = \frac{1}{[4\pi\kappa(t-t_0)]^{1/2}} \mathrm{e}^{-\frac{(x-x')^2}{4\kappa(t-t_0)}} \tag{B46.3}$$

With (B46.2) and (B46.3) (or integrating (B46.1) in an bounded domain), it is straightforward to show that the following three integral relations hold:

$$M = \int \psi(x,t)\mathrm{d}x = \text{const} \tag{B46.4}$$

$$x_c = \frac{1}{M}\int x\psi(x,t)\mathrm{d}x = \text{const} \tag{B46.5}$$

$$\sigma^2(t) = \frac{1}{M}\int (x-x_c)^2\,\psi(x,t)\mathrm{d}x = 2\kappa t \tag{B46.6}$$

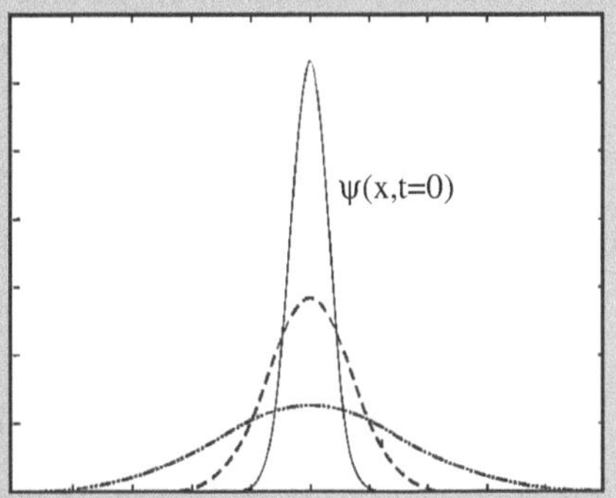


*Molecular diffusion of a tracer with concentration $\psi(x,t)$ which initially is concentrated at $x = 0$. $\psi$ is shown at three subsequent times (solid, dashed, dotted).*

Hence molecular diffusion leaves both the total amount $M$ of tracer and the center $x_c$ of tracer mass constant. Most importantly, the tracer "variance" $\sigma^2$, which is a measure of the spread of the concentration around $x_c$, increases linearly with time, a basic property of diffusion processes.

We assume that the diffusivity $K_s$ associated with these fluctuations is given as $K_s = \mathcal{E}_\star^{1/2}\ell_\star$ with $\ell_\star = 1/k_\star$ (in analogy to (11.42)). One obtains (up to a factor $O(1)$)

$$K_s \approx \epsilon^{1/3}\ell_\star^{4/3} = \mathcal{E}^{\frac{1}{2}} L\,(\ell_\star/L)^{4/3} \tag{11.46}$$

where the relation (11.27) has been used for the last equality. In numerical circulation models, (11.46) is occasionally used to parameterize subgrid scale motions. It is possible to relate the scale $\ell_\star$ with the resolution, i. e. $\ell_\star \approx \Delta x$, which is particularly useful when the resolution is variable.

### 11.2.2 Turbulent Diffusion in the Lagrangian Reference System

In Lagrangian coordinates, it is possible to relate the mean rate of spreading with statistical properties of the flow field. The following description was given by Taylor (1921).

A particle is advected by a turbulent flow $u(x,t)$, which for simplicity is assumed to be one-dimensional and stationary, with $\overline{u}(x,t) = 0$. The particle starts from the location $x = 0$ at time $t = 0$, and is at the time $t$ at the location $x(t)$. In the Lagrangian system one has

$$\frac{\mathrm{d}x(t)}{\mathrm{d}t} = v(t) \quad \text{or} \quad x(t) = \int_0^t v(t')\mathrm{d}t'$$

where $v(t) = u(x(t),t)$ denotes the Lagrangian velocity. Since $\overline{v} = 0$, it follows that $\overline{x} = 0$, i. e. on average the particle does not move away from its origin. The mean quadratic excursion of the particle $\overline{x^2}$ is given by

$$\frac{\mathrm{d}\overline{x^2}}{\mathrm{d}t} = \overline{2x(t)\frac{\mathrm{d}x(t)}{\mathrm{d}t}} = \overline{2v(t)\int_0^t v(t')\mathrm{d}t'} = 2\int_0^t \overline{v(t)v(t')}\mathrm{d}t'$$

and thus leads to Taylor's diffusion equation

$$\frac{\mathrm{d}\overline{x^2}}{\mathrm{d}t} = 2\int_0^t R^L(\tau)\mathrm{d}\tau \tag{11.47}$$

Here $R^L(\tau) = R^L(t-t')$ denotes the Lagrangian velocity covariance which is characterized by its value at origin, $R^L(0) = \overline{v^2}$, and the integral time scale

$$T_{\text{int}} = \frac{1}{R^L(0)}\int_0^\infty R^L(\tau)\mathrm{d}\tau$$

Of particular interest is the limiting case for which the diffusion time $t$ is much larger than the integral time-scale, $t \gg T_{\text{int}}$. In this case, $\int_0^t R^L(\tau)\mathrm{d}\tau \approx \int_0^\infty R^L(\tau)\mathrm{d}\tau = \overline{v^2}T_{\text{int}}$, and it follows that

$$\overline{x^2}(t) = 2Kt \quad \text{with} \quad K = \overline{v^2}T_{\text{int}} \tag{11.48}$$

In analogy to (B46.6) in the box on p. 349, $K > 0$ is referred to as turbulent diffusivity. The result (11.48) can be used to infer the diffusivity in situations where $\overline{v^2}$ and $T_{\text{int}}$ are known. In three dimensions, the appropriate generalization is $K_{ij} \approx \int_0^\infty \overline{v_i(t)v_j(t+\tau)}\mathrm{d}\tau$ for $t \gg T_{\text{int}}$, where $K_{ij}$ is symmetric and positive definite.

### 11.2.3 Eulerian Diffusion by Small-Scale Turbulence

In this section, we investigate under which conditions we can expect the analogy between molecular and turbulent diffusion as expressed by (11.41). In a one dimensional situation with a tracer distribution $\overline{\psi}(x)$, one can view $\overline{\psi}(x)\Delta x$ as the number of particles in an interval $\Delta x$ around $x$ (up to a constant factor). Turbulent diffusion corresponds to a random transport process. Specifically, assume that a particle which initially (at time $t'$) is located at $x'$ is found at location $x$ at a later time $t$ with a probability $\Phi(x, x', t, t')$. Note that if the turbulence is homogeneous and stationary, then the probability depend only on $r = x - x'$ and $\tau = t - t'$, i. e. $\Phi = \Phi(x - x', t - t')$. It follows that the tracer distributions at both times are related by

$$\overline{\psi}(x,t) = \int_{-\infty}^{\infty} \Phi(x - x', t - t')\overline{\psi}(x', t')\mathrm{d}x' \tag{11.49}$$

While the probability distribution $\Phi$ is not known for turbulent flows, it has to satisfy the following physical constraints:

$$\text{mass conservation} \qquad \int_{-\infty}^{\infty} \Phi(x - x', t - t')\mathrm{d}x' = 1 \tag{11.50}$$

$$\text{center of mass} \qquad \int_{-\infty}^{\infty} (x - x')\Phi(x - x', t - t')\mathrm{d}x' = 0 \tag{11.51}$$

Mass conservation requires (11.50), as seen by integration of (11.49) over all $x$. If the turbulent motion has no preferred direction, i. e. is isotropic, then $\Phi(x - x', t - t') = \Phi(x' - x, t - t')$, and (11.51) follows. This also means that there is no movement of the center of mass related to the random transport process. If the particle excursion $x - x'$ is small compared to the scales of variation of $\overline{\psi}(x)$, then $\overline{\psi}(x', t')$ under the integral in (11.49) can be expanded into a Taylor series. With $r = x - x'$, one obtains

$$\begin{aligned}\overline{\psi}(x,t) &= \int_{-\infty}^{\infty} \Phi(r, t - t')\left[\overline{\psi}(x,t') - r\frac{\partial\overline{\psi}(x,t')}{\partial x} + \frac{1}{2}r^2\frac{\partial^2\overline{\psi}(x,t')}{\partial x^2} + O(r^3)\right]\mathrm{d}r \\ &= \overline{\psi}(x,t')\int_{-\infty}^{\infty} \Phi(r, t - t')\mathrm{d}r - \frac{\partial\overline{\psi}(x,t')}{\partial x}\int_{-\infty}^{\infty} r\Phi(r, t - t')\mathrm{d}r \\ &\quad + \frac{\partial^2\overline{\psi}(x,t')}{\partial x^2}\int_{-\infty}^{\infty}\frac{1}{2}r^2\Phi(r, t - t')\mathrm{d}r + \int_{-\infty}^{\infty}\Phi(r, t - t')O(r^3)\mathrm{d}r\end{aligned} \tag{11.52}$$

Using (11.50) and (11.51), the first two integrals in (11.52) can be evaluated. With $\tau = t - t'$, one finds

$$\overline{\psi}(x,t) = \overline{\psi}(x, t - \tau) + K\tau\frac{\partial^2\overline{\psi}(x, t - \tau)}{\partial x^2} + O(r^3) \tag{11.53}$$

with the definition

$$\int_{-\infty}^{\infty} r^2\Phi(r,\tau)\mathrm{d}r = 2K\tau \tag{11.54}$$

in analogy to the case of molecular diffusion in (B46.6). Note that the second moment $\int r^2\Phi(r,\tau)\mathrm{d}r$ becomes also equivalent to the Lagrangian mean quadratic particle excursion which is given by (11.48). However, this holds only for time intervals $\tau \gg T_{\mathrm{int}}$, as discussed in Section 11.2.2. Provided that the tracer concentration changes little over the time $\tau \gg T_{\text{int}}$ and the distance $r$, then (11.53) can be approximated by

$$\frac{\partial \overline{\psi}}{\partial t} = K \frac{\partial^2 \overline{\psi}}{\partial x^2} \tag{11.55}$$

Hence one obtains the standard diffusion equation, with a down-gradient diffusivity $K$ defined in (11.48). It is, however, important to note that (11.55) only holds under the restrictive conditions that the tracer $\psi$ varies little over the temporal and spatial scales of the eddy field, i. e. $|\partial\psi/\partial t| \ll |\psi/\tau|$ and $|\partial\psi/\partial x| \ll |\psi/r|$ such that the Taylor expansion in (11.52) rapidly converges.

## 11.3 Inhomogeneous Three-Dimensional Turbulence

In the presence of mean flow, stratification, forcing, and boundary conditions, turbulence in the ocean cannot be expected to be homogeneous. The most important aspect of inhomogeneity is given by stratification. However, numerical simulations confirm that many aspects from Kolmogorov's theory for isotropic turbulence can be transferred to the inhomogeneous case: In particular the predicted shape of the spectrum in horizontal direction and the forward energy cascade in the inertial subrange can also be found for nonisotropic turbulence in the presence of strong stratification (e. g. Brethouwer et al., 2007). On the other hand, a closed analytical theory like the one for isotropic turbulence is missing. In the following sections, energetic constraints for inhomogeneous turbulence are derived, and applications to the dynamics of the ocean mixed layer are discussed.

Starting point are the equations of motion in the Boussinesq approximation (4.10)–(4.13), with the molecular form for the diabatic terms. In the upper ocean, the state equation (4.13) can be approximated[6] by the linear form (4.23). With these changes, and introducing the buoyancy $b = -g\tilde{\rho}/\rho_0$ the system

$$\frac{\partial u_j}{\partial t} + u_i \frac{\partial u_j}{\partial x_i} + 2\Omega_l u_m \epsilon_{lmj} = -\frac{\partial p^*}{\partial x_j} + b\delta_{j3} + \kappa_{\text{m}} \frac{\partial^2 u_j}{\partial x_i^2} \tag{11.56}$$

$$\frac{\partial u_i}{\partial x_i} = 0 \tag{11.57}$$

$$\frac{\partial b}{\partial t} + u_i \frac{\partial b}{\partial x_i} = -\frac{\partial J_i^{\text{b}}}{\partial x_i} + \kappa \frac{\partial^2 b}{\partial x_i^2} \tag{11.58}$$

is obtained from (4.10)–(4.13), written here in tensor notation. Note that we use the sum convention as before, according to which summation is implied if an index occurs twice within one term. The variable $p^* = p/\rho_0$ denotes the scaled pressure (the star will be dropped in the following), and $\boldsymbol{J}^{\text{b}}$ is the buoyancy flux arising from radiation. Note that (11.58) is a shorthand notation for both the heat and salt budgets, and is strictly valid only to the extent that variations of the expansion coefficients and the difference between the coefficients of heat conductivity and salt diffusivity can be ignored. Introducing the Reynolds decomposition (cf. Section 2.8) $u_j = \overline{u}_j + u'_j$,

[6] This is only needed for the diffusion term.

$b = \overline{b} + b'$ etc. and averaging yields

$$\frac{\partial \overline{u}_j}{\partial t} + \overline{u}_i \frac{\partial \overline{u}_j}{\partial x_i} + 2\Omega_l \overline{u}_m \epsilon_{lmj} = -\frac{\partial \overline{p}}{\partial x_j} + \overline{b}\delta_{j3} - \frac{\partial}{\partial x_i}\left(\overline{u'_i u'_j}\right) - \kappa_{\mathrm{m}} \frac{\partial^2 \overline{u}_j}{\partial x_i^2} \tag{11.59}$$

$$\frac{\partial \overline{u}_i}{\partial x_i} = 0 \tag{11.60}$$

$$\frac{\partial \overline{b}}{\partial t} + \overline{u}_i \frac{\partial \overline{b}}{\partial x_i} = -\frac{\partial}{\partial x_i}\left(J_i^{\mathrm{b}} + \overline{u'_i b'} - \kappa \frac{\partial \overline{b}}{\partial x_i}\right) \tag{11.61}$$

The equations for the mean fields hence are identical to (11.56)–(11.58), except for the additional turbulent fluxes $\overline{u'_i b'}$ and $\overline{u'_i u'_j}$ which need to be known to predict the mean variables.

### 11.3.1 Energetic Constraints

Parameterization of the turbulent flux terms in (11.59) and (11.61) involves issues similar to the parameterization of turbulent tracer fluxes discussed in Section 11.2. However, for the turbulent momentum and buoyancy fluxes there are additional constraints arising from consideration of the energy budget.

#### Turbulent Kinetic Energy

Analogous to Section 4.1.4, a conservation equation for turbulent kinetic energy (TKE) is obtained by multiplication of (11.56) with $u'_j$, averaging and summation over all $j$. The derivation is straightforward and not given in detail. With the turbulent kinetic energy $\mathcal{E} = \frac{1}{2}\overline{u'_i u'_i}$, it follows that

$$\frac{\partial \mathcal{E}}{\partial t} + \overline{u}_i \frac{\partial \mathcal{E}}{\partial x_i} = -\frac{\partial}{\partial x_i}\left(\overline{u'_i p'} + \frac{1}{2}\overline{u'_i u'_j u'_j} - \kappa_{\mathrm{m}} \frac{\partial \mathcal{E}}{\partial x_i}\right)$$
$$- \overline{u'_j u'_i} \frac{\partial \overline{u}_j}{\partial x_i} + \overline{b' w'} - \kappa_{\mathrm{m}} \overline{\left(\frac{\partial u'_i}{\partial x_j}\right)^2} \tag{11.62}$$

The physical interpretation of the individual terms in the TKE equation (11.62) is the following: the terms on the left-hand side describe the local change in time plus the advection by the mean flow, i. e. the material derivative of $\mathcal{E}$ along the mean flow $\overline{\boldsymbol{u}}$. The last term of the right-hand side is always negative and represents dissipation of turbulent kinetic energy and is identical to the definition (11.24) (compare also the box on p. 354). Three terms in (11.62) contribute to a flux divergence which redistributes $\mathcal{E}$. These are the turbulent flux of mechanical energy $\overline{u'_i p'}$, the turbulent flux of turbulent kinetic energy $\overline{u'_i u'_j u'_j}/2$, which has the form of a triple correlation, and a diffusive flux $\kappa_{\mathrm{m}} \partial \mathcal{E}/\partial x_i$. It follows that $\boldsymbol{F} = \overline{\boldsymbol{u}' p'} + \overline{\boldsymbol{u}' \boldsymbol{u}'^2}/2 - \kappa_{\mathrm{m}} \nabla \mathcal{E}$ is the total flux of turbulent kinetic energy. The remaining production terms $\overline{u'_j u'_i} \partial \overline{u}_j / \partial x_i$ and $\overline{b' w'}$ in (11.62) will be discussed below. Note that for isotropic turbulence, i. e. when all mean fields are independent of space, (11.62) reduces to the form (11.23).

**47. Alternative Formulation of the TKE Equation**

In Boussinesq approximation, the mechanical dissipation of Section 2.4.2, takes the form

$$\epsilon = 2\kappa_{\mathrm{m}} D_{ij}^2$$

with the deformation tensor $D_{ij} = (1/2)(\partial u_i/\partial x_j + \partial u_j/\partial x_i)$ and the kinematic viscosity $\kappa_{\mathrm{m}}$. Using the Reynolds decomposition of $\boldsymbol{u}$ and taking the mean yields

$$\overline{\epsilon} = \kappa_{\mathrm{m}} \overline{\left(\frac{\partial u_i'}{\partial x_j}\right)^2} + \kappa_{\mathrm{m}} \frac{\partial}{\partial x_i}\frac{\partial}{\partial x_j}\overline{u_i' u_j'} + \kappa_{\mathrm{m}} \left(\frac{\partial \bar{u}_i}{\partial x_j}\right)^2 + \kappa_{\mathrm{m}} \frac{\partial}{\partial x_i}\frac{\partial}{\partial x_j}\bar{u}_i \bar{u}_j$$

The part of $\overline{\epsilon}$ which is related to velocity fluctuations differs from (11.24) and the dissipation in (11.62) by an additional molecular flux $\kappa_{\mathrm{m}}\partial \overline{u_j' u_i'}/\partial x_j$, which vanishes in the case of isotropic turbulence. The same holds for the part of $\overline{\epsilon}$ related to the mean velocity gradients. To be consistent with the mean dissipation rate $\overline{\epsilon}$, the TKE equation is sometimes rewritten as

$$\begin{aligned} \frac{\partial \mathcal{E}}{\partial t} + \overline{u}_i \frac{\partial \mathcal{E}}{\partial x_i} &= -\frac{\partial}{\partial x_i}\left(\overline{u_i' p'} + \frac{1}{2}\overline{u_i' u_j' u_j'} - 2\kappa_{\mathrm{m}} \overline{D_{ij}' u_j'}\right) \\ &\quad - \overline{u_j' u_i'}\frac{\partial \overline{u}_j}{\partial x_i} + \overline{b' w'} - 2\kappa_{\mathrm{m}} \overline{(D_{ij}')^2} \end{aligned}$$

with $D_{ij}' = (1/2)(\partial u_i'/\partial x_j + \partial u_j'/\partial x_i)$, and similar for the mean kinetic energy equation. This modified form and (11.62) differ only by the interpretation of a molecular flux term, which does not figure in an integral budget over a closed domain.

### Exchange with Mean Kinetic Energy

The term $\overline{u_j' u_i'}\partial \overline{u}_j/\partial x_i$ in (11.62) describes the interaction of turbulence with a mean shear. Further insight into its meaning can be gained by considering the conservation equation for mean kinetic energy (MKE) which is obtained by multiplication of (11.59) with $\overline{u}_j$. With the mean kinetic energy $\frac{1}{2}\overline{u}_i\overline{u}_i$ one finds

$$\begin{aligned} \left(\frac{\partial}{\partial t} + \overline{u}_i \frac{\partial}{\partial x_i}\right)\frac{1}{2}\overline{u}_i\overline{u}_i &= -\frac{\partial}{\partial x_i}\left(\overline{u}_i\overline{p} + \frac{1}{2}\overline{u}_j\overline{u_i' u_j'} - \kappa_{\mathrm{m}}\frac{\partial}{\partial x_i}\frac{1}{2}\overline{u}_i\overline{u}_i\right) + \\ &\quad + \overline{u_i' u_j'}\frac{\partial \overline{u}_j}{\partial x_i} + \overline{b}\overline{w} - \kappa_{\mathrm{m}}\left(\frac{\partial \overline{u}_j}{\partial x_i}\right)^2 \end{aligned} \tag{11.63}$$

All terms in (11.63) are analogous to those in (11.62) and will not be discussed in detail. The term $\overline{u_j' u_i'}\partial \overline{u}_j/\partial x_i$, however, appears both in (11.62) and (11.63), but with opposite signs, and hence constitutes an exchange between mean and turbulent kinetic energies. The direction of this exchange is normally such that energy flows from mean to turbulent kinetic energy. This is consistent with a parametrization $\overline{u_i' u_j'} = -K_u \partial \overline{u}_j/\partial x_i$ which for $K_u > 0$ leads to

$$\overline{u_i' u_j'}\frac{\partial \overline{u}_j}{\partial x_i} = -K_u\left(\frac{\partial \overline{u}_j}{\partial x_i}\right)^2 < 0$$

For small-amplitude fluctuations, this energy transfer is related to the process of Kelvin-Helmholtz instability discussed in Section 7.7.3. Note, however, that the perturbation analysis from that section does not apply to fully developed turbulence.

### Exchange with Mean Potential Energy

A term of the form $bw = -g\rho w/\rho_0$ describes work against gravity and can be interpreted as exchange between kinetic and potential energy, as discussed in Section 4.1.4. In the Boussinesq approximation employed here, a useful potential energy is given by $\mathcal{V} = gz\rho/\rho_0 = -zb$. From (4.18), with the simplified state equation (4.23), the conservation equation for mean potential energy $\overline{\mathcal{V}}$ follows as

$$\frac{\mathrm{D}\overline{\mathcal{V}}}{\mathrm{D}t} = -\overline{b}\overline{w} - \overline{b'w'} - \frac{\partial \overline{u'_j\mathcal{V}'}}{\partial x_j} + z\left(\frac{\partial \overline{J_i^{\mathrm{b}}}}{\partial x_i} - \kappa\frac{\partial \overline{b}}{\partial x_i}\right) \tag{11.64}$$

The vertical buoyancy fluxes $\overline{b}\overline{w}$ and $\overline{b'w'}$ also occur in the kinetic energy equations (11.63) and (11.62), respectively, and hence constitute exchanges of mean and turbulent kinetic energy with mean potential energy. Note that the definition of $\mathcal{V}$ agrees with the definition (4.18), but differs from the quadratic form (5.41) for available potential energy. The exchange term $\overline{bw}$ with kinetic energy and its physical interpretation, however, remains identical for both formulations.

It is easy to see that in a stably stratified environment, i. e. with $\partial\overline{b}/\partial z > 0$, the term $\overline{b'w'}$ will lead to a transfer from TKE to potential energy. When turbulent velocity moves a particle upward or downward from its mean position, buoyancy forces will tend to bring the particle back to its mean position. Hence the particle will decelerate, and its turbulent kinetic energy will be reduced. For the turbulent buoyancy flux it follows that $\overline{b'w'} < 0$, hence this leads to a decreasing tendency $\partial\mathcal{E}/\partial t < 0$. Conversely, in case of an unstable stratification, the particle will be further accelerated, and the energy transfer is in the opposite direction. Note that in both cases, an Austauschansatz $\overline{b'w'} = -K_{\mathrm{b}}\partial\overline{b}/\partial z$ with $K_{\mathrm{b}} > 0$ leads to the correct sign for the energy transfer.

## 11.3.2 Turbulence Models for the Surface Boundary Layer

Models of ocean mixed layer turbulence are based on (11.59) and (11.61) and face the need to express the fluxes $\overline{u'_i u'_j}$ and $\overline{b'u'_j}$ in terms of the mean fields. For a qualitative discussion, it is sufficient to consider a simplification based on the fact that the horizontal scales in the mixed layer are much larger than the vertical scales. The approximation of horizontal homogeneity neglects horizontal gradients of all mean quantities. From the mean continuity equation (11.60), it then follows that $\partial\overline{w}/\partial z = 0$ and hence also $\overline{w} = 0$; and, therefore, in this approximation the advection of mean fields through the mean flow is neglected. Accordingly, the mean field equations (11.59) and (11.61) simplify to

$$\frac{\partial \overline{\boldsymbol{u}}_{\mathrm{h}}}{\partial t} - \boldsymbol{f} \times \overline{\boldsymbol{u}}_{\mathrm{h}} = -\frac{\partial}{\partial z}\overline{\boldsymbol{u}'_{\mathrm{h}}w'} \tag{11.65}$$

$$\frac{\partial \overline{b}}{\partial t} = -\frac{\partial}{\partial z}\left(\overline{J_3^{\mathrm{b}}} + \overline{b'w'}\right) \tag{11.66}$$

Closures for $\overline{b'w'}$ and $\overline{\boldsymbol{u}'_{\mathrm{h}}w'}$ are needed. Furthermore, the TKE equation (11.62) is – in various forms – a central element of nearly all models of the oceanic mixed layer.

**48. Richardson Flux Number**

As discussed above, of the two production terms in the TKE equation (11.62) the buoyancy flux normally leads to a decrease in TKE (for stable mean stratification) whereas the interaction with mean shear leads to an increase in TKE. Hence the ratio of both terms

$$\mathrm{Ri_f} = \frac{\overline{b'w'}}{\overline{u'_i u'_j}\,\partial \overline{u}_j/\partial x_i} \approx \frac{\overline{b'w'}}{\overline{w'u'_j}\,\partial \overline{u}_j/\partial z} \tag{B48.1}$$

is a dimensionless measure for the net production of TKE and referred to as flux Richardson number $\mathrm{Ri_f}$ (the second expression holds for the typical situation of a horizontal mean flow that is vertically sheared). For $\mathrm{Ri_f} > 1$, the buoyancy flux dominates and the net production of TKE is negative, whereas for $\mathrm{Ri_f} < 1$ the shear term dominates and the net production (except dissipation) is positive. If both $\overline{b'w'}$ and $\overline{w'u'_j}$ can be described by exchange laws with coefficients $K_b$ respectively $K_u$, one has

$$\mathrm{Ri_f} = \frac{K_b \partial \overline{b}/\partial z}{K_m \left(\partial \overline{u}_j/\partial z\right)^2} = \frac{K_b}{K_m}\frac{N^2}{\left(\partial \overline{u}_j/\partial z\right)^2} = \frac{K_b}{K_u}\mathrm{Ri}$$

Thus under these assumptions the flux Richardson number is proportional to the gradient Richardson number Ri as defined in (7.96). The ratio of turbulent viscosity and diffusivity is called the turbulent Prandtl number

$$\mathrm{Pr} = \frac{K_u}{K_b} = \frac{\mathrm{Ri}}{\mathrm{Ri_f}}$$

The Prandtl number is frequently expressed as a function of the Richardson number Ri. Experience suggests that $\mathrm{Pr} \approx 1$ for $\mathrm{Ri} \ll 1$, i. e. for weak stratification and/or strong shear, and that the Prandtl number increases for larger values of Ri (Anderson, 2009; Webster, 1964).

With horizontal homogeneity, (11.62) takes the form

$$\frac{\partial \mathcal{E}}{\partial t} = \frac{\partial F}{\partial z} - \overline{\boldsymbol{u}'_h w} \cdot \frac{\partial \overline{\boldsymbol{u}}_h}{\partial z} + \overline{b'w'} - \epsilon \tag{11.67}$$

Here the vertical transport $F = \overline{w'(p' + u'_j u'_j/2)}$ (defined as positive into the ocean) and the dissipation $\epsilon$ are two more variables for which closures are needed. We have neglected the molecular TKE flux which is very small. In the following, several approaches for closing the system (11.65)–(11.67) will be briefly discussed.

### Bulk Models

In bulk mixed layer models, which were pioneered by Kraus and Turner (1966, 1967), the observed buoyancy structure within the mixed layer is taken as given, and no attempt is made to model the causes of this structure. Specifically, it is assumed that the buoyancy has a constant value within a well-mixed layer of depth $h$, so that the buoyancy profile is given by $b = b_0(t)$ for $0 \geq z \geq -h(t)$, and by a linear function of depth from $b_0$ to $b_\star$ between $z = -h(t)$ and $z = -h_\star$ representing a thin pycnocline (see Figure 11.4). The deep buoyancy $b_\star$ has to be specified in this model. The parameter $\Delta = h_\star - h$ characterizes the pycnocline thickness and is assumed to be constant and small, $\Delta \ll h$, so that effectively the pycnocline is modeled as a discontinuity at the mixed layer base.

Only the integral budgets of buoyancy and TKE over the mixed layer are considered. Integration of the momentum balance (11.65) leads to an equation for

**49. Seasonal Cycle of the Oceanic Mixed Layer**

The structure of the oceanic near-surface layer is characterized by a layer of nearly constant density, residing over a thin sharp pycnocline (a layer with a strong vertical gradient of density), followed by a more gradual increase of density down to the abyss. The turbulence in the mixed layer which leads to almost complete mixing is caused by a combination of wind and buoyancy forcing at the surface. In a typical summer mixed layer (30–50 m thick), wind-induced turbulence normally dominates; in winter, the mixed layer can deepen to several hundred meters through buoyancy loss.

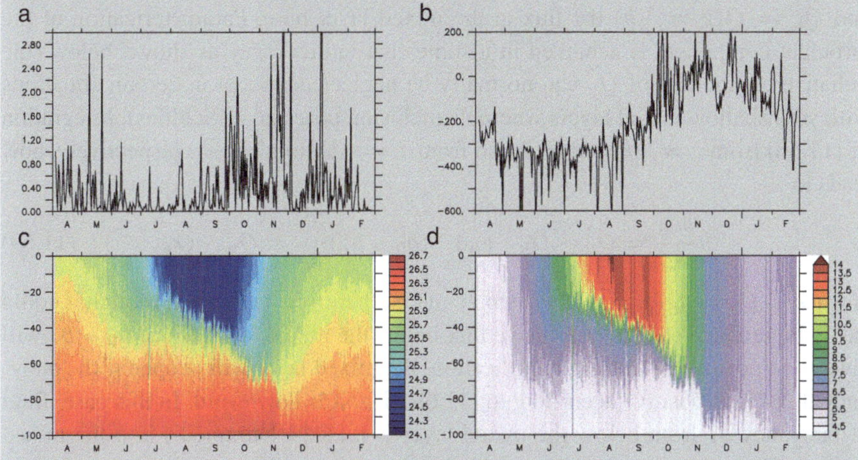


*Time series during 1961 at Ocean Weather Ship PAPA* (145° W, 50° N). ***a*** *Turbulent kinetic energy input at the sea surface* $F(z = 0)$ *in (11.67) in* $10^{-5}\,\mathrm{m^3\,s^{-3}}$, ***b*** *surface heat flux* $Q(z = 0) = -(J_3^b + b'w')|_{z=0}$ *in (11.66) in* $\mathrm{W\,m^{-2}}$, ***c*** *potential density (anomaly) in* $\mathrm{kg\,m^{-3}}$ *and* ***d*** *potential temperature in* °C *at station PAPA. Data was taken from GOTM (www.gotm.net). Note that since station PAPA is located in a region where the horizontal advection of heat and salt is assumed to be small, the seasonal cycle of the mean buoyancy is given to a good approximation by (11.66).*

the Ekman transport which is not discussed further (see Section 14.1). The physics of the mixed layer model should determine both $b_0(t)$ and $h(t)$ from the specified surface buoyancy flux and the characteristics of the turbulence in that layer.

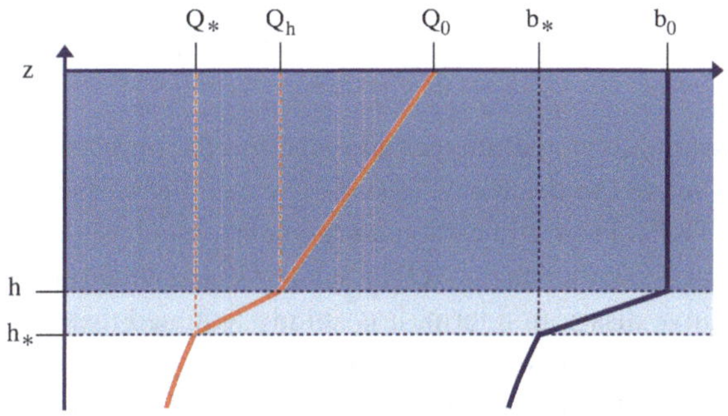


**Fig. 11.4** Sketch of profiles of buoyancy (*blue line*) and buoyancy flux (*red line*) in bulk mixed layer models. The buoyancy $b$ is constant ($b_0$) in the mixed layer which extends from $z = -h$ to the surface. The buoyancy decreases linearly from $b_0$ to the prescribed value $b_\star$ in the thin pycnocline between $z = -h$ and $z = -h_\star$. The vertical (radiative plus turbulent) buoyancy flux $Q$ enters at the surface (with the prescribed value $Q_0$) and decreases linearly towards its value $Q_\mathrm{h}$ at the mixed layer base at $z = -h$. The flux $Q_\star$ at the base of the pycnocline at $z = -h_\star$ is given by the small interior diffusive flux and might be set to zero

It is useful to define $Q = -\overline{(J_3^{\mathrm{b}} + b'w')}$ as the vertical (radiative plus turbulent) buoyancy flux (positive downwards into the ocean). As evident from (11.66), for $b$ to be constant in the mixed layer, $Q(z)$ must be a linear function of depth,

$$Q(z) = Q_0 + \frac{z}{h}(Q_0 - Q_{\mathrm{h}}) \tag{11.68}$$

Here $Q_0$ is the net surface buoyancy flux which is considered as a prescribed forcing, and $Q_{\mathrm{h}} = Q(z = -h)$ the flux at the mixed layer base. Parameterization of the turbulent component is achieved in a somewhat indirect way as shown below (the radiative component of $Q_{\mathrm{h}}$ can normally be neglected, except in certain situations with very shallow mixed layers where its inclusion poses no difficulties). Integration of (11.66) from $z = -h$ to $z = 0$ and from $z = -h_\star$ to $z = -h$, respectively, now leads to

$$h\frac{\partial b_0}{\partial t} = Q_0 - Q_{\mathrm{h}} \quad \text{and} \quad (b_0 - b_\star)\frac{\partial h}{\partial t} = Q_{\mathrm{h}} - Q_\star \tag{11.69}$$

Normally, $Q_\star = Q(z = -h_\star)$ is small and is, therefore, neglected, though it could easily be retained e. g. as a diffusive flux below the mixed layer. Since $b_0 - b_\star$ will always be positive in a stable stratification, the mixed layer will deepen if the buoyancy flux $Q_{\mathrm{h}}$ is positive according to (11.69), i. e. it is downward. In this case, fluid of buoyancy $b_\star$ from below the mixed layer is mixed ('entrained') into the mixed layer and brought to $b_0$. As discussed below, this mixing consumes turbulent energy and hence can only take place if sufficient turbulent kinetic energy is available for mixing fluid from below into the mixed layer.

The basic assumption of mixed layer physics is that $Q_{\mathrm{h}}$ originates from turbulent processes within the mixed layer and, therefore, cannot become negative since there is no 'unmixing' of fluid from the mixed layer. Instead, when $Q_{\mathrm{h}}$ drops to zero, there is insufficient turbulent energy to continue mixing at the mixed layer base. Mixing and deepening of the mixed layer then stops, and a new mixed layer is established at a shallower depth.

The mixed layer depth is determined from the budget of turbulent kinetic energy, which usually equilibrates within a few minutes. Ignoring, for simplicity, the shear term in (11.67), integration from $-h$ to 0 with (11.68) then yields

$$0 = F_0 - F_{\mathrm{h}} - \frac{h}{2}(Q_0 + Q_{\mathrm{h}}) - \int_{-h}^{0} \epsilon \mathrm{d}z \tag{11.70}$$

A few further assumptions (parameterizations) close the problem. The turbulent energy input $F_0$ is related to the wind stress $\tau_0$ (by exciting surface waves, which by breaking create TKE). From dimensional arguments, one postulates $F_0 = c|\tau_0|^{3/2}$ with a dimensionless coefficient $c = O(1)$, while $F_{\mathrm{h}}$ is assumed small and neglected. To close the positive dissipation term, it is simply assumed that a fraction $r_1$ of the (positive) wind input $F_0$ is dissipated. In situations when buoyancy is lost at the surface, i. e. when $Q_0 < 0$, which occurs mainly by cooling, the potential energy of the heavier fluid is converted to TKE, and it is likewise assumed that a fraction $r_2$ of this TKE gain is dissipated. Hence the dissipation term is expressed as

$$\int_{-h}^{0} \epsilon \mathrm{d}z = r_1 F_0 + r_2 \frac{h}{4}(|Q_0| - Q_0)$$

with $0 < r_1, r_2 < 1$. This relation, together with (11.70), finally leads to the required parameterization of $Q_h$ in the form[7]

$$\frac{h}{2}Q_h = (1-r_1)c|\tau_0|^{3/2} - \frac{h}{2}\left[\left(1-\frac{r_2}{2}\right)Q_0 + \frac{r_2}{2}|Q_0|\right] \tag{11.71}$$

With specified forcing terms $\tau_0$ and $Q_0$, the recipe to calculate the mixed layer depth $h(t)$ is now as follows. If $Q_h$ from (11.71) is positive, it is used to compute $h(t)$ from the *prognostic* equation (11.69). If $Q_h$ from (11.71) becomes negative (which can happen only when $Q_0 > 0$), then $h$ is determined from the *diagnostic* equation (11.71) as balance between mechanical forcing and buoyancy gain by setting $Q_h = 0$, i. e.

$$h = 2(1-r_1)c|\tau_0|^{3/2}/Q_0 \tag{11.72}$$

The algorithm reflects the fundamental difference between the processes of deepening and shallowing of the mixed layer. The physics of the model is straightforward, and in this approximation there are only two adjustable parameters, $(1-r_1)c$ and $r_2$. Typical parameter values are $r_1 \approx 0.1$, $r_2 \approx 0.9$, and $c \approx 1$.

Bulk mixed layer models have been commonly used as part of ocean circulation models. On the positive side is their conceptual (and computational) simplicity. In particular, the wind-induced deepening of the mixed layer is usually well described by these models. On the other hand, bulk models perform less well in convective situations and are unable to resolve any structures within the boundary layer. Furthermore, a discontinuity at the mixed layer base – an essential part of the models – is not generally observed.

### Models based on Parameterization of Vertical Buoyancy and Momentum Fluxes

The standard closure for the turbulent buoyancy and momentum fluxes in (11.65) and (11.66) is the down-gradient parameterization

$$\overline{w'\psi'} = -K_\psi \frac{\partial \overline{\psi}}{\partial z} \tag{11.73}$$

where $\psi$ stands for $\boldsymbol{u}$ or $b$, and the coefficient $K_\psi$ can be different for buoyancy and momentum. It follows from (11.66) that this specification of constant or slowly varying values of $K_\psi$ will completely fail to reproduce the observed property structure in the mixed layer.

Various parameterizations have been proposed which express $K_\psi$ as function of the local Richardson number Ri. Specifically, Pacanowski and Philander (1981) have proposed functional relations for $K_u(\mathrm{Ri})$ and $K_b(\mathrm{Ri})$, which permit variation of the coefficients over two (for $K_u$) or even three (for $K_b$) orders of magnitude, with the highest values obtained for $\mathrm{Ri} \to 0$. The parameterization works well in regions of strong shear, such as in the tropical oceans. However, in a situation where the buoyancy is vertically well mixed so that everywhere in the mixed layer $\mathrm{Ri} = 0$, these methods are less suited since effectively they imply constant coefficients.

[7] Note that we used here a scaled stress variable $\tau_0$ (divided by the reference density) in units $\mathrm{m^2\,s^{-2}}$.

The parameterization (11.73) fails in situations where an unstable stratification induces convective transport independent of the local gradient. The generalization

$$\overline{w'\psi'} = -K_\psi \left( \frac{\partial \overline{\psi}}{\partial z} - \gamma_\psi \right) \tag{11.74}$$

includes a nonlocal contribution $\gamma_\psi$ which is modeled as an empirical function of stratification and forcing, and applies only in convective situations. The parameterization (11.74) was originally developed for convection in the planetary boundary layer of the atmosphere (Deardorff, 1966). In the KPP (K-profile parameterization) formulation of ocean mixed layer turbulence, (11.74) is used together with a specification of mixed layer depth $h$ by a bulk Richardson number criterion and with specified profiles of both the coefficients $K_\mathrm{u}$, $K_\mathrm{b}$ and the nonlocal parameters $\gamma_\mathrm{u}, \gamma_\mathrm{b}$ based on empirical results (Large et al., 1994).

### Models Based on the TKE Equation

TKE models are based on the parameterization (11.73) and link the diffusivity with TKE and a characteristic length scale with the heuristic relation (11.42) as

$$K_\psi = c_\psi \mathcal{E}^{1/2} L \tag{11.75}$$

Here $c_\psi$ is an empirical dimensionless coefficient (or 'stability function'), either chosen as constant or as function of other variables such as stability, normally with $c_\psi = O(1)$. The turbulent kinetic energy $\mathcal{E}$ can in principle be determined from the TKE equation (11.62). The dissipation is expressed as $\epsilon = c_\epsilon \mathcal{E}^{3/2}/L$ according to (11.27), again with $c_\epsilon = O(1)$. The vertical transport term in (11.67) has been modeled (Gaspar et al., 1990) as a diffusive flux of TKE, i. e.

$$F = c_\mathrm{E} \mathcal{E}^{1/2} L \frac{\partial \mathcal{E}}{\partial z} \tag{11.76}$$

with $c_\mathrm{E} = O(1)$, but note that the flux of TKE can take complicated forms in particular for convective situations. With (11.73), (11.75), (11.76), and (11.27), the TKE equation (11.67) can be written as

$$\frac{\partial \mathcal{E}}{\partial t} = \frac{\partial}{\partial z}\left( c_\mathrm{E} \mathcal{E}^{1/2} L \frac{\partial \mathcal{E}}{\partial z} \right) + c_\mathrm{u} \mathcal{E}^{1/2} L \left( \frac{\partial \overline{u}}{\partial z} \right)^2 - c_\mathrm{b} \mathcal{E}^{1/2} L \frac{\partial \overline{b}}{\partial z} - c_\epsilon \frac{\mathcal{E}^{3/2}}{L} \tag{11.77}$$

For a qualitative discussion of (11.77), the dimensionless coefficients will be ignored in the following, i. e. $c_\mathrm{E}, c_\mathrm{u}, c_\mathrm{b}, c_\epsilon \to 1$. Note, however, that modifications may occur in practical applications, due to different choices for the dimensionless coefficients.

As discussed earlier in (11.28), the time-scale of adjustment of TKE to changes in the forcing is $T = L/\sqrt{\mathcal{E}}$, which typically is much shorter than 1 h. On longer timescales, the energy equation, therefore, is effectively in equilibrium so that the terms on the right-hand side of (11.77) must balance. Since dissipation is a central aspect of turbulence, any term balances not involving dissipation are not very plausible.

Note that a principal difficulty remains, namely the parameterization of the length scale $L$, which often can be interpreted as scale of the large eddies (see Section 11.2.1). Several plausible limits for $L$ can, however, be derived from considerations of stratification, shear, and geometrical factors:

- In the presence of a stable mean stratification $\partial\overline{b}/\partial z = N^2 > 0$, consider a particle that is vertically displaced from its mean position by a turbulent eddy with the length scale $L$. In this way, the particle gains potential energy (per mass) $E_p = L^2N^2/2$ (see e. g. Section 7.2.2). The potential energy which is gained cannot be larger than the eddy TKE, i. e. $E_p \leq \mathcal{E}$. It follows that

$$L \lesssim L_b \approx \sqrt{2\mathcal{E}/N^2} \tag{11.78}$$

In an unstable mean stratification, the buoyancy term in (11.77) is positive, and TKE is produced. However, the convective TKE production cannot be larger than dissipation, i. e. $\mathcal{E}^{1/2}L|\partial\overline{b}/\partial z| \lesssim \mathcal{E}^{3/2}/L$, if one assumes for simplicity that the first two terms on the right-hand side of (11.77) are also positive (note that the flux of TKE in (11.77) takes complicated form in convective situations). It follows for that case that

$$L \lesssim L_b^* = \sqrt{\mathcal{E}/|\partial\overline{b}/\partial z|} \tag{11.79}$$

which is formally similar to (11.78).
- In the presence of mean shear, consider variations of the mean flow $\overline{\boldsymbol{u}}$ over the eddy length scale $L$. For the concept of the parameterization (11.73) with (11.75) to be valid, $\overline{\boldsymbol{u}}$ should not change over the eddy length scale by more than $u_{rms} = \mathcal{E}^{1/2}$, i. e. $|\overline{\boldsymbol{u}}(z+L) - \overline{\boldsymbol{u}}(z)| \approx |L\partial\overline{\boldsymbol{u}}/\partial z| \lesssim \mathcal{E}^{1/2}$, and hence

$$L \lesssim L_u = \mathcal{E}^{1/2}/|\partial\overline{\boldsymbol{u}}/\partial z| \tag{11.80}$$

Another interpretation of the condition (11.80) is that the shear term in (11.77) cannot be much larger than the dissipation term. Note that for $L = L_u$, with (11.75) it follows that $K = |\partial\overline{\boldsymbol{u}}/\partial z|\, L^2$, a relation which was derived in Section 11.2.1 based on the mixing length concept.
- In the absence of stratification and mean shear, observations suggest that turbulent mixing rarely extends deeper than $L_{max} \approx 50–100$ m. This situation is described by a balance between diffusion of energy and dissipation in (11.77), i. e.

$$\frac{\partial}{\partial z}\left(\mathcal{E}^{1/2}L\frac{\partial\mathcal{E}}{\partial z}\right) = \frac{\mathcal{E}^{3/2}}{L}$$

which can be solved analytically with the approximation of constant $L$. The solution is given as $\mathcal{E} = \mathcal{E}_0 e^{z/d}$ with $d = \sqrt{3/2}\,L \sim L$. Hence the TKE decreases exponentially with $z$ in the oceanic mixed layer, which is to be expected since only dissipation is acting in the interior. Since $L$ is not very different from the mixed layer depth scale $d$ in this case, one finds the further condition $L \lesssim L_{max}$. Note that the magnitude of TKE is determined by the surface flux $F_0$ of TKE into the ocean which appeared in (11.70) as $\mathcal{E}_0 = ((3/2)F_0)^{2/3}$.
- For geometrical reasons, eddies cannot be larger than their distance from a boundary. In particular, near the surface $z = 0$ it follows that $L \lesssim L_r = |z|$.

In summary, from the preceding considerations it follows that $L \lesssim L_{min} = \min(L_b, L_b^*, L_u, L_{max}, L_r)$. A very simple choice for $L$ would be to assume $L = L_{min}$. Other models have used $L \approx L_b$ or an integral variant thereof as a closure for $L$, but there is no consensus which model is best. This variety of possibilities also includes the choice for the stability functions and dimensionless parameters in the closure. Turbulence models based on the TKE equation (11.77) and a diagnostic specification of $L$ succeed in general in simulating the large variations in diffusivity/viscosity required for a simulation of a well-mixed surface layer. However, the parameterization of the combined TKE flux and pressure correlation as TKE diffusion in (11.76), which is central for TKE models, is not supported by numerical simulations of boundary layer turbulence; it appears instead that buoyancy effects dominate that flux (Mironov et al., 2000).

### k-$\epsilon$ and Second Order Turbulence Models

To avoid a diagnostic specification of the length scale $L$, prognostic equations have been used for the dissipation $\epsilon$. From (11.56), an equation for $\epsilon = \kappa_{\mathrm{m}}\overline{(\partial u_i/\partial x_j)(\partial u_i/\partial x_j)}$ can be derived in a straightforward way. The terms in the resulting equation somewhat resemble those in the energy equation (11.62), but are of higher order of differentiation and have a less direct physical interpretation. These terms have been modeled in analogy to the TKE equation, and the resulting equation is often of the form

$$\frac{\partial \epsilon}{\partial t} = \frac{\partial}{\partial z} c_4 \mathcal{E}^{1/2} L \frac{\partial \epsilon}{\partial z} + \frac{\epsilon}{\mathcal{E}} \left[ c_1 \mathcal{E}^{1/2} L \left( \frac{\partial \overline{\boldsymbol{u}}}{\partial z} \right)^2 - c_3 \mathcal{E}^{1/2} L \frac{\partial \overline{b}}{\partial z} - c_2 \frac{\mathcal{E}^{3/2}}{L} \right] \tag{11.81}$$

which can in principle be solved together with the TKE equation (11.77), by expressing the length scale $L$ through (11.27). Here $c_1 \ldots c_4 = O(1)$ denote additional dimensionless stability functions which need to be specified. It can be shown that $c_2$ determines the free turbulence decay time scale and $c_3$ the steady-state Richardson number and mixing efficiency for homogeneous shear layers (Umlauf and Burchard, 2005). Using (11.81) together with the TKE equation is often referred to as $k$-$\epsilon$-model ($k$ stands for TKE). Mellor and Yamada (1982) have used a similar equation for the variable $\mathcal{E}L$ instead of $\epsilon$.

Another possibility for the construction of a turbulence model is to consider equations for *all* second order moments. These equations can be obtained in a straightforward way, by multiplication of (11.56) and (11.58) with $u_i'$ respectively $b'$, adding and averaging. With further assumptions (assuming that the turbulent length scale is smaller than a mean length scale and that isotropy prevails at high wave numbers, neglecting the radiative component in (11.58) and assuming horizontal homogeneity), the result is given by

$$\begin{aligned} \frac{\partial \overline{u_i' u_j'}}{\partial t} &= -\overline{u_i' w'} \frac{\partial \overline{u}_j}{\partial z} - \overline{u_j' w'} \frac{\partial \overline{u}_i}{\partial z} - \frac{\partial \overline{u_i' u_j' w'}}{\partial z} - f(\overline{u_k' u_j'} \epsilon_{3,k,i} + \overline{u_k' u_i'} \epsilon_{3,k,j}) \\ &\quad + \overline{p' \left( \frac{\partial u_j'}{\partial x_i} + \frac{\partial u_i'}{\partial x_j} \right)} + (\overline{b' u_i'} \delta_{j,3} + \overline{b' u_j'} \delta_{i,3}) - \frac{2}{3} \epsilon \delta_{ij} \end{aligned} \tag{11.82}$$

$$\frac{\partial \overline{b' u_i'}}{\partial t} = \overline{b' w'} \frac{\partial \overline{u}_i}{\partial z} + \overline{w' u_i'} \frac{\partial \overline{b}}{\partial z} - f \overline{b' u_k'} \epsilon_{3,k,i} + \overline{p' \frac{\partial b'}{\partial x_i}} + \overline{b'^2} \delta_{i,3} \frac{\partial \overline{b' u_i' w'}}{\partial z} \tag{11.83}$$

$$\frac{\partial \overline{b'^2}/2}{\partial t} = -\overline{b' w'} \frac{\partial \overline{b}}{\partial z} - \frac{\partial \overline{b'^2 w/2}}{\partial z} - \epsilon_{\mathrm{b}} \tag{11.84}$$

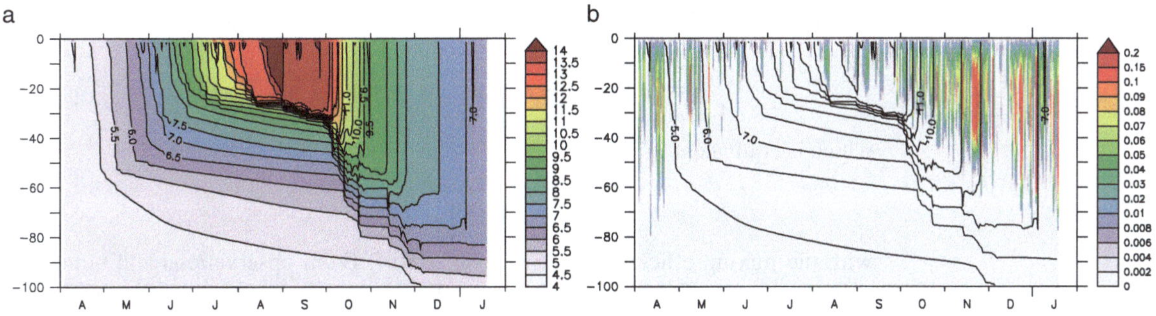


**Fig. 11.5** Mixed layer simulation using a standard $k$-$\epsilon$-model at Ocean Weather Ship PAPA for the same year as in the box on p. 357. **a** Temperature in $^\circ$C, **b** vertical turbulent diffusivity in $m^2\,s^{-1}$ (*colors*) and temperature (*contours*). Model and forcing was taken from GOTM (www.gotm.net). Since station PAPA is located in a region where the horizontal advection of heat and salt is assumed to be small, it is often used for validating turbulence closure schemes. Comparing the simulation with the observations in the box on p. 357 shows indeed a good agreement

where $\epsilon_b = \kappa\overline{(\partial b/\partial x_i)^2}$. Note that summation of (11.82) over the diagonal terms $i = j$ yields the TKE equation (11.67). A main advantage of (11.82)–(11.84) is that parameterizations of the type (11.73) are no longer needed since the fluxes are now modeled explicitly. However, parameterizations are now needed for the pressure correlation terms $\overline{p'\partial b'/\partial x_i}$ and $\overline{p'(\partial u_j'/\partial x_i + \partial u_i'/\partial x_j)}$ and, as before, for the dissipation and the moments of third order. Equations (11.82)–(11.84) are often used in their equilibrium version (equating right hand sides to zero).

Both $k - \epsilon$ and second order turbulence models are frequently used in practical engineering and have also successfully been applied to the ocean boundary layer modeling; an example is given in Figure 11.5. It is probably fair to say that the development of these models is guided more by the results than by the physical insight. The number of adjustable parameters in these models is fairly large, and it remains to be shown that the results justify the increase in complexity and computational efforts.

### 11.3.3 Turbulence in the Ocean Interior

Turbulent mixing in the stratified interior of the ocean is small compared to the boundary layers, but nevertheless of high interest. Among other aspects, interior mixing can influence the global meridional overturning circulation (see Chapter 15 and the box on p. 364) in important ways, in particular its upwelling branch, by mixing of the cold dense water with warmer and lighter water masses above them.

Mixing through surfaces of constant density (isopycnals) is called diapycnal mixing (in contrast to mixing along isopycnals, referred to as isopycnal mixing). Since the slopes of isopycnals in the ocean are very small (a consequence of $|\partial\rho/\partial z| \gg \max(|\partial\rho/\partial x|, |\partial\rho/\partial y|)$) the diapycnal mixing is almost identical to vertical mixing.

A relation between diffusivity and dissipation can be obtained by consideration of the TKE equation (11.67). Neglecting the transport term and assuming stationarity leads to a local balance between the production terms and the dissipation. With the definition of the flux Richardson number $\mathrm{Ri_f}$ from (B48.1), one obtains

$$0 = -\overline{w'u'_j}\frac{\partial \bar{u}_j}{\partial z} + \overline{b'w'} - \epsilon = \overline{b'w'}(1 - 1/\mathrm{Ri_f}) - \epsilon$$

which is equivalent to

$$-\overline{b'w'} = \gamma\epsilon \tag{11.85}$$

with the mixing efficiency $\gamma = \mathrm{Ri_f}/(1 - \mathrm{Ri_f})$. From observations and numerical simulations, one finds for stratified flows like the ocean that $\mathrm{Ri_f} \approx 0.15$, thus for the mixing efficiency $\gamma \approx 0.2$ (Osborn, 1980). Using the Austauschansatz (11.75) for $\overline{b'w'}$ yields

$$K_\mathrm{b} = \gamma\epsilon N^{-2}$$

Therefore, the turbulent diffusivity is related to the local dissipation, a consequence of the assumption of a local balance between production and dissipation of TKE. Hence the turbulent diffusivity can be inferred from measurements of the dissipation rate $\epsilon$. From microstructure observations in the interior ocean, one typically estimates $\epsilon \approx 10^{-10}$–$10^{-9}\,\mathrm{m^2\,s^{-3}}$ and thus $K_\mathrm{b} \approx 10^{-5}\,\mathrm{m^2\,s^{-1}}$ for the typical stratification in the interior of the ocean. Note that there is an order of magnitude discrepancy between this local estimate of $K_\mathrm{b}$ and the global mean value inferred from (B50.1) to balance the meridional overturning (cf. also Section 15.3). Further, observations indicate that mixing in the ocean is not uniformly distributed but rather concentrated in certain regions associated with rough bottom topography (Polzin et al., 1997).

The TKE, which is dissipated by $\epsilon$ and used to change the potential energy by the transfer term $\overline{b'w'}$, is supplied by the shear instability transfer term $\overline{w'u'_j}\partial\bar{u}_j/\partial x_j$. The mean velocity $\bar{u}_j$, whose shear is responsible for the mixing, needs some consideration. The vertical shear of the large-scale flow, e. g. related to the meridional overturning, is much weaker than the shear related to the internal wave field, since the latter has vertical scales much smaller than the large-scale or mesoscale ocean circulation. The way how the internal waves are generated and how they transfer energy from source to the dissipation, which is then related to mixing described by (11.85), is discussed in Section 10.4.5.

**50. Munk's Abyssal Recipe**

In equilibrium, the diffusive vertical flux of tracers has to be balanced by an advective vertical flux. Munk (1966) assumed a balance

$$\overline{w}\frac{\partial \overline{b}}{\partial z} = K_\mathrm{b}\frac{\partial^2 \overline{b}}{\partial z^2} \tag{B50.1}$$

which shall hold everywhere below the permanent thermocline of the oceans. Additionally, he assumed that $\overline{w}$ and $\overline{b}$ are horizontally constant to first order. As a solution one obtains $\overline{b} = b_0 e^{z/H}$ with $H = K_b/\overline{w}$. Actually, the observed vertical profiles of $\overline{b}$ are often almost exponential, such that $H$ can be easily determined from data ($\approx$ 800 m). Assuming a value of 25–30 Sv for the global overturning in the ocean, one obtains $w \approx 1$ cm/d as vertical velocity (assuming equally distributed upwelling) and $K_\mathrm{b} \approx 10^{-4}\,\mathrm{m^2\,s^{-1}}$. This simple discussion shows that global overturning and the vertical diffusivity are roughly proportional to each other. Increasing $K_\mathrm{b}$ by an order of magnitude, the global overturning also has to be increased by one order of magnitude, i. e. from 25 to 250 Sv. The order of magnitude of $K_\mathrm{b}$ appears, therefore, important. It should be kept in mind that Munk's value for $K_\mathrm{b}$ is applicable to the large-scale global mean. Locally $K_\mathrm{b}$ may be much smaller.

Another possibility to infer $K_\mathrm{b}$ from data follows from the buoyancy variance equation (11.84). Again assuming stationarity and neglecting the transport terms yields

$$\overline{b'w'}\frac{\partial \overline{b}}{\partial z} = -\epsilon_\mathrm{b} \tag{11.86}$$

With $\overline{b'w'} = -K_\mathrm{b}N^2$ and $\epsilon_\mathrm{b} = \kappa\overline{(\partial b/\partial x_i)^2}$, the Osborn–Cox relation

$$\frac{K_\mathrm{b}}{\kappa} = \mathrm{C} = \frac{\overline{(\partial b'/\partial x_i)^2}}{\left(\partial \overline{b}/\partial z\right)^2} \tag{11.87}$$

is obtained where the Cox number C denotes the ratio between the variance of the vertical buoyancy (or temperature) gradient and the square of the mean vertical gradient (Osborn and Cox, 1972). For a more detailed discussion of the Osborn–Cox relation and its underlying assumptions we refer to Section 12.3.3. The Cox number also gives the ratio between turbulent and molecular diffusivity and is normally much larger than 1. The relation (11.87) demonstrates how the turbulence enhances the effect of molecular diffusivity by increasing the instantaneous gradients relative to the gradients of the mean buoyancy.

As shown e. g. in Section 7.2.2, the buoyancy variance is related to the turbulent potential energy $E_\mathrm{p}$ by $E_\mathrm{p} = \frac{1}{2}\overline{b'^2}/N^2$. Therefore, the dissipation of buoyancy variance $\epsilon_\mathrm{b}$ in (11.84) must be related to the dissipation of potential energy $\epsilon_\mathrm{p}$ by $\epsilon_\mathrm{p} = \epsilon_\mathrm{b}/N^2$. Eliminating the flux $\overline{b'w'}$ from (11.86) and (11.85) leads to

$$\epsilon_\mathrm{p} = \gamma\epsilon$$

Hence the dissipation of turbulent potential energy is a fraction (around 25%) compared to the dissipation of TKE.

# 12 Geostrophic Turbulence

In this chapter, we will discuss aspects of turbulent flows on spatial scales around and larger than the (first baroclinic) Rossby radius, which are ubiquitous and important for many aspects of the large-scale ocean circulation and often are referred to as mesoscale ocean eddies. As discussed in Section 5.1, motions on these scales are strongly constrained by rotation and stratification and have a very small aspect ratio and hence are nearly two-dimensional. Turbulent motions in two dimensions differ in important aspects from the three-dimensional motions considered in Chapter 11. Therefore, this chapter could also be called two-dimensional turbulence. Idealized two-dimensional turbulence will be discussed in the following section, while Section 12.2 is devoted to the more realistic inhomogeneous situation of the real ocean. In Section 12.3 we discuss alternative averaging frameworks to differentiate between advective and diffusive effects of turbulent mixing.

Mesoscale fluctuations are almost always and anywhere present in the ocean and can be conceptually distinguished from the large-scale circulation by their smaller time and space scales. Mesoscale fluctuations (in the following referred to as mesoscale eddies, although all kinds of fluctuations in space and time and not just the coherent eddies are meant here) have spatial scales around (i. e. not more than several times and not less than a substantial fraction of) the first internal Rossby radius and time-scales ranging from the inertial period to several weeks or at most months and are, therefore, dynamically well described by the quasi-geostrophic approximation (compare Section 5.2). The large-scale circulation of the ocean, on the other hand, refers in general to larger space and time-scales, in fact often referring to the time-mean and basin-scale circulation. Since the flow related to the mesoscale eddies is often of turbulent nature, it is therefore sometimes also convenient to refer to mesoscale eddies as deviations from the time mean. On the other hand, it is clear that the large-scale circulation may also change and includes – for instance in the western boundary current – also spatial scales comparable to the mesoscale eddies. Therefore, it should be kept in mind that the distinction between large-scale and mesoscale (and small-scale) circulation of the ocean is in practice often not very sharp.

D. Olbers, J. Willebrand, C. Eden, *Ocean Dynamics*,

DOI 10.1007/978-3-642-23450-7_12, 

## 12.1 Homogeneous Turbulence in Two Dimensions

We start the discussion of geostrophic turbulence considering the idealized case of homogeneous two-dimensional turbulence similar to the discussion in Section 11.1 for three dimensions. The results obtained there, in particular the scalar energy balance (11.22), remain valid, with modifications where appropriate.

As discussed in Section 5.2, oceanic flows at the scale of the internal Rossby radius are almost two-dimensional and nearly in geostrophic balance and, therefore, are well represented by the quasi-geostrophic approximation, which is expressed by the quasi-geostrophic potential vorticity equation (5.37). A most simplified form of (5.37) is obtained by neglecting both the $\beta$-effect and effects of the stratification. While this seems not a good approximation for oceanic mesoscale eddies, it suffices for the moment to discuss some basic aspects of homogeneous two-dimensional turbulence. Neglecting, furthermore, the diapycnal forcing and specifying the horizontal mechanical forcing as $\mathcal{F} = A_\mathrm{h}\nabla^2 \boldsymbol{u}$ results in

$$\frac{\partial \eta}{\partial t} + \boldsymbol{u}\cdot\nabla\eta = A_\mathrm{h}\nabla^2\eta \tag{12.1}$$

for the relative vorticity $\eta = \partial v/\partial x - \partial u/\partial y$. Note that (12.1) could also be derived directly, i. e. without reference to the QG approximation, from the momentum equation (4.10) in two dimensions. As will be shown below, the existence of the vorticity equation (12.1) leads to substantial differences in the physics of two-dimensional turbulence, compared to turbulence on three dimensions.

Multiplication of (12.1) by $\eta$ and averaging, noting that spatial derivatives of mean quantities vanish due to homogeneity, yields

$$\frac{\partial}{\partial t}\frac{1}{2}\overline{\eta^2} = -A_\mathrm{h}\overline{(\nabla\eta)^2} \tag{12.2}$$

as an equation for the "enstrophy" $\overline{\eta^2}/2$. Spectral decomposition of $u$, $v$ and $\eta$ is now analogous to Section 11.1.2, and the details are not repeated here. It is again sufficient to consider the scalar energy spectrum $E(k)$ and the enstrophy spectrum $G(k)$ which satisfy

$$\frac{1}{2}\overline{(u^2+v^2)} = E_0 = \int E(k)\mathrm{d}k \quad \text{and} \quad \frac{1}{2}\overline{\eta^2} = G_0 = \int G(k)\mathrm{d}k$$

The derivation of the scalar energy balance from the momentum equation in Section 11.1.2 remains valid in two dimensions. Additionally, from the vorticity equation (12.1) a conservation equation for the scalar enstrophy spectrum $G(k)$ can be derived in a similar way, resulting in

$$\frac{\partial}{\partial t}E(k) = \Gamma(k) - 2A_\mathrm{h}k^2E(k) \quad \text{and} \quad \frac{\partial}{\partial t}G(k) = \Gamma_\mathrm{G}(k) - 2A_\mathrm{h}k^2G(k) \tag{12.3}$$

where the term $\Gamma(k)$ is the scalar projection of the inertial forces defined in analogy to (11.20) (but for two dimensions), and $\Gamma_\mathrm{G}(k)$ denotes the spectral form of the inertial term $\nabla\cdot(\overline{\boldsymbol{u}\eta\eta'} + \overline{\boldsymbol{u}'\eta\eta'})$ following from (12.1). Since $\eta = \partial v/\partial x - \partial u/\partial y$, both spectra are related by $G(k) = k^2E(k)$. Multiplication of (12.3) with $k^2$ then shows that also $\Gamma_\mathrm{G}(k) = k^2\Gamma(k)$. The inertial forces can change neither the total energy $E_0$ nor the total enstrophy $G_0$, i. e.

$$\int_0^\infty \Gamma(k)\mathrm{d}k = 0 \quad \text{and} \quad \int_0^\infty k^2\Gamma(k)\mathrm{d}k = 0$$

In analogy to (11.21), the spectral transports of energy and entropy are given as

$$F(k) = -\int_0^k \Gamma(k')\mathrm{d}k' \quad \text{and} \quad F_\mathrm{G}(k) = -\int_0^k k'^2\Gamma(k')\mathrm{d}k'$$

Compared to the three-dimensional case, there are now two constraints given by (12.3) for the energy transport $F(k)$ by the inertial forces in wave-number space. The consequence of the additional constraint is that in two dimensions the situation changes completely with respect to the direction of the energy transport, as shown below.

### 12.1.1 Inverse Energy Cascade

In analogy to the statistical moments, the "moment" of order $n$ in wave-number space can be defined as

$$\overline{k^n} = E_0^{-1}\int_0^\infty k^n E(k)\mathrm{d}k\,, \quad n = 0, 1, \ldots \tag{12.4}$$

In this section, the symbol $\overline{()}$ is used to indicate the spectral moment according to (12.4) (thereafter the usual definition as statistical or time average will hold again). Of particular interest are the "mean" wave number $\overline{k}$, which indicates the region where most energy is concentrated, and the "variance" $\overline{(k-\overline{k})^2}$ as a measure of the spread of the energy spectrum around $\overline{k}$. Note that all moments depend on time, as a consequence of the time dependence of the energy spectrum. From definition (12.4), it follows that

$$\overline{(k-\overline{k})^2} = E_0^{-1}\int (k-\overline{k})^2 E(k)\mathrm{d}k = \overline{k^2} - (\overline{k})^2 \tag{12.5}$$

If the spectrum is initially concentrated around a certain wave number, it is plausible to assume that the inertial forces will tend to widen the maximum in analogy to molecular diffusion, as indicated in Figure 12.1. From $\partial\overline{(k-\overline{k})^2}/\partial t > 0$ and (12.5), one finds

$$\frac{\partial\overline{k^2}}{\partial t} - \frac{\partial(\overline{k})^2}{\partial t} = \frac{\partial\overline{k^2}}{\partial t} - 2\overline{k}\frac{\partial\overline{k}}{\partial t} > 0 \tag{12.6}$$

To infer the sign of $\partial\overline{k}/\partial t$, we have to know the sign of $\partial\overline{k^2}/\partial t$. Integration of (12.3) over all wave numbers, together with the definition (12.4), yields the integral budgets of kinetic energy and enstrophy in the form

$$\frac{\partial E_0}{\partial t} = -2A_\mathrm{h}\overline{k^2}E_0 \tag{12.7}$$

$$\frac{\partial G_0}{\partial t} = \frac{\partial(\overline{k^2}E_0)}{\partial t} = \overline{k^2}\frac{\partial E_0}{\partial t} + E_0\frac{\partial\overline{k^2}}{\partial t} = -2A_\mathrm{h}\overline{k^4}E_0 \tag{12.8}$$

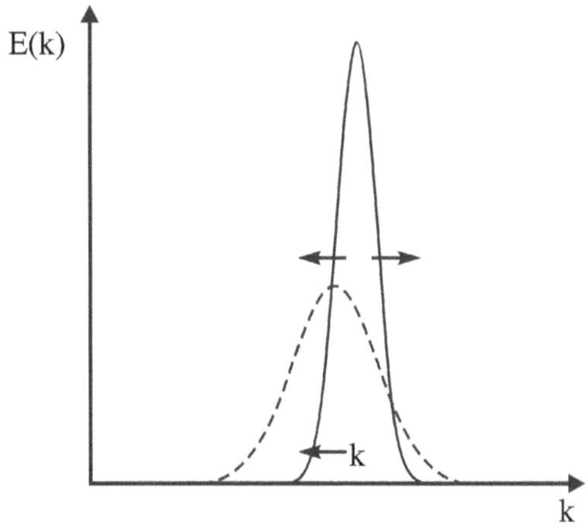


Fig. 12.1 An initially narrow spectrum (*full line*) with the energy centering around $\overline{k}$ will widen in time, and the center will move toward lower wave numbers. The *dashed line* shows the spectrum at a later time

Multiplying (12.7) with $\overline{k^2}$, and subtracting the result from (12.8), results in

$$E_0 \frac{\partial \overline{k^2}}{\partial t} = -2A_\mathrm{h} E_0 \left[ \overline{k^4} - (\overline{k^2})^2 \right]$$

Since $\overline{k^4} - (\overline{k^2})^2 > 0$, as shown from

$$0 < \int \left( k^2 - \overline{k^2} \right)^2 E(k) \mathrm{d}k = E_0 \left[ \overline{k^4} - 2\overline{k^2 k^2} + (\overline{k^2})^2 \right] = E_0 \left[ \overline{k^4} - (\overline{k^2})^2 \right]$$

it follows that $\partial \overline{k^2}/\partial t < 0$. Finally, with (12.6), we have $\partial \overline{k}/\partial t < 0$. Therefore, the main energy concentration in time will move toward smaller wave numbers, as indicated in Figure 12.1. This characteristic behavior of two-dimensional turbulence is referred to as red or inverse energy cascade because energy moves to larger scales, in contrast to three dimensions where the energy is transferred to smaller scales.

### 12.1.2 A Numerical Example of Two-Dimensional Turbulence

As an example of two-dimensional turbulence, we consider a numerical solution of the vorticity equation (12.1) on a periodic domain. The temporal evolution of the stream function $\psi$ from the (specified) initial condition is shown in Figure 12.2. Note that $u = -\partial \psi/\partial y$ and $v = \partial \psi/\partial x$ and thus $\eta = \nabla^2 \psi$. The initial conditions for $\psi$ and $\eta$ (Figure 12.2 upper and lower left panel) are dominated by a prescribed high wave number on which a small background noise with randomly chosen wave numbers is superimposed. After several days of integration, the flow becomes unstable due to shear instability (Figure 12.2 middle and right panels). The temporal evolution of the stream function illustrates the red cascade discussed in the previous section, i. e. the shift of kinetic energy from initially small scales, as prescribed in the initial conditions, toward larger scales. This red kinetic energy cascade is responsible for the fact that the stream function becomes dominated by lower and lower wave numbers in the subsequent integration. The final result is that the computational domain is filled with only a few large eddies.

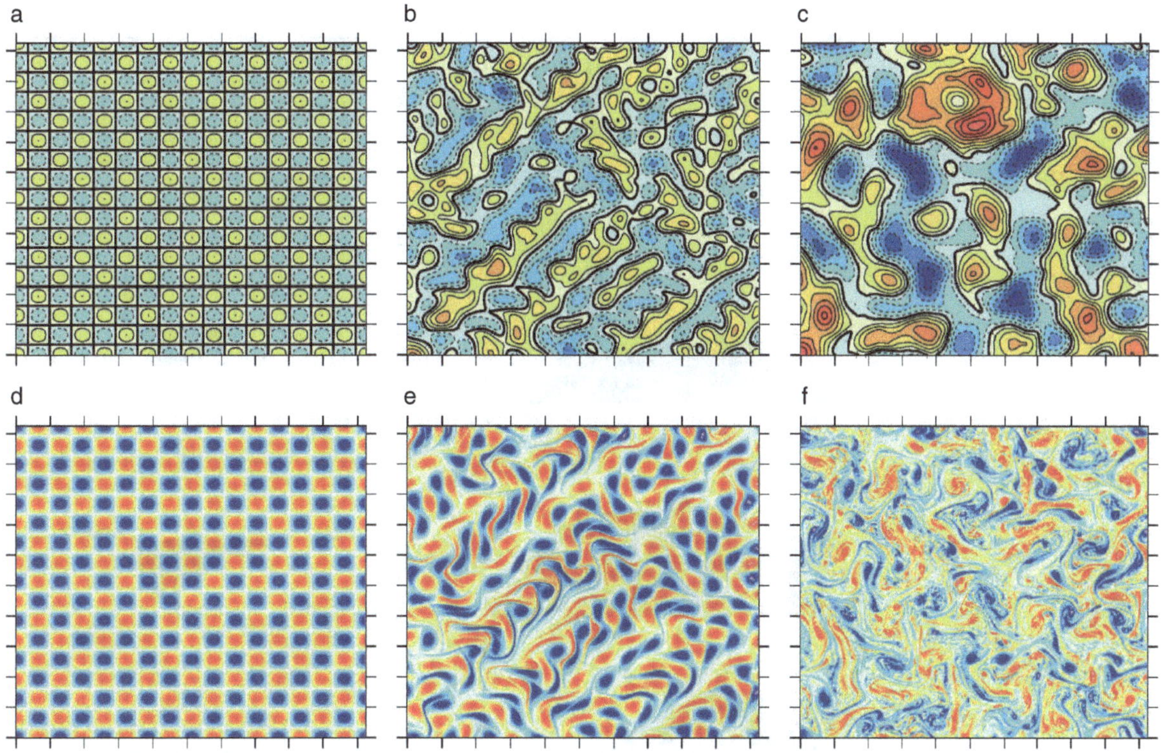


Fig. 12.2 Illustration of two-dimensional turbulence, as described by numerical solutions of the vorticity equation (12.1). The stream function $\psi$ (**a–c**, with $u = -\partial\psi/\partial y$ and $v = \partial\psi/\partial x$) is shown at the initial time (**a**) and for two later times (**b** and **c**) after the shear instability sets in. Also shown is the vorticity $\eta$ (**d–f**). The domain shown here is $5{,}000 \times 5{,}000$ km, but the actual computational domain is larger with periodic boundary conditions. The grid resolution is 10 km, and we have used hyperviscosity instead of lateral friction term in (12.1), i. e. we have added the term $-A_{hy}\nabla^4\eta$ to the right-hand side of (12.1) with hyperviscosity of $A_{hy} = 3.9 \times 10^9\ \mathrm{m^4\,s^{-1}}$ while we have set $A_h = 0$

Figure 12.2 also shows the vorticity $\eta$. In contrast to the stream function, it does not show a dominance of lower wave numbers in the subsequent integration. In fact, $\eta$ is dominated by small-scale features with high wave numbers in the subsequent integration, i. e. it is much noisier than in the prescribed initial conditions. While the kinetic energy shows a red (or inverse) cascade in wave-number space, the enstrophy features a direct cascade, i. e. enstrophy is transported towards smaller scales and higher wave numbers.

Figure 12.3 shows the spectral characteristics of the solutions in terms of the energy spectrum $E(k)$ and the energy flux $F(k)$ in wave-number space. Since the mechanical forcing in (12.1) only consists of a frictional term, the total kinetic energy $E_0$ associated with the initial conditions decreases with time. The energy spectrum $E(k)$ is initially dominated by the high wave number $k_i$ associated with the initial distribution (Figure 12.3a). At the time when the instability indicated in Figure 12.2 sets in, an energy flux shows up, which continues at later times, growing in time, and rapidly changes the energy spectrum $E(k)$. As seen in Figure 12.3d the energy flux $F(k)$ is initially directed in both directions, i. e. it is positive and directed towards larger $k$ for $k > k_i$ and negative towards smaller $k$ for the part of the spec-

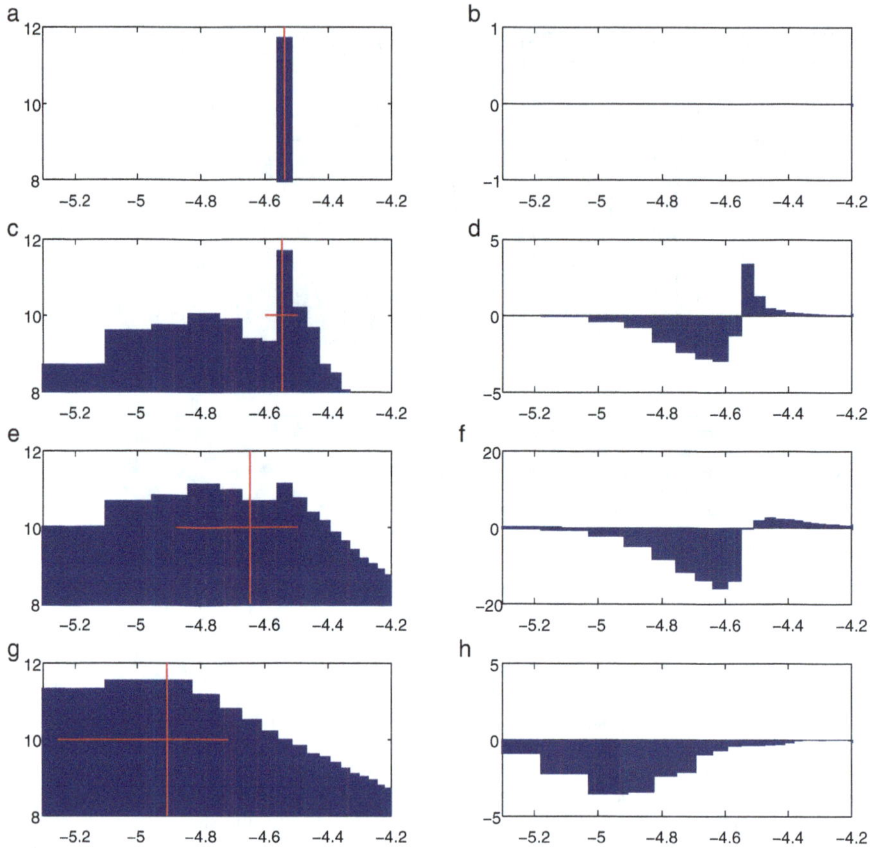


**Fig. 12.3** Spectral properties of numerical solutions of equation (12.1). The *left column* shows the evolution of the wave-number spectrum of the kinetic energy, $\log E(k)$, at subsequent times as a function of wave number $\log k$. The *vertical and horizontal red lines* denotes the mean wave number $\overline{k}$ and its standard deviation $\overline{k} \pm (\overline{k^2} - \overline{k}^2)^{1/2}$ respectively. The $\overline{k^n}$ are defined by (12.4). The *right column* shows the evolution of the energy flux $F(k)$ as a function of $\log k$ at the corresponding times. The *upper row* shows the initial time, where the spectrum is dominated by a sharp peak in wave-number space also seen in Figure 12.2. *Lower rows* correspond to later times when the instability seen in Figure 12.2 sets in and progresses

trum with $k < k_i$. Note that this flux is induced by the inertial term in (12.1) related to the growing instabilities. As a manifestation of the red (inverse) energy cascade, the negative flux towards smaller $k$ is much larger than the positive one such that the mean wave number $\overline{k}$, also shown in Figure 12.3, decreases in the subsequent integration, while the wave-number variance $\overline{k^2} - \overline{k}^2$ increases due to the nonlinear interactions. At later times, the positive part of energy flux $F(k)$ becomes small (Figure 12.3d), such that a red energy cascade is found. In contrast to the energy flux $F(k)$, the enstrophy flux $F_G(k)$ (not shown) is positive at large scales, i. e. towards higher $k$ and smaller scales.

**51. The $\beta$-Effect and the Rhines Scale**

In the vorticity equation (12.1), we have neglected so far variations in the planetary vorticity. If we consider instead

$$\frac{\partial \eta}{\partial t} + \boldsymbol{u} \cdot \nabla \eta + \beta v = A_\mathrm{h} \nabla^2 \eta \qquad \text{(B51.1)}$$

with $\beta = \partial f / \partial y$, the flow will feature (short barotropic) Rossby waves (cf. Section 8.2). In the numerical solutions shown in Figures 12.2 and 12.3, using the vorticity equation (B51.1) instead of (12.1), the $\beta$-effect manifests itself in a distinct anisotropy, i. e. in the appearance of preferred zonal flow in the simulation with $\beta \neq 0$ which features, in fact, quasi-stationary zonal jets.

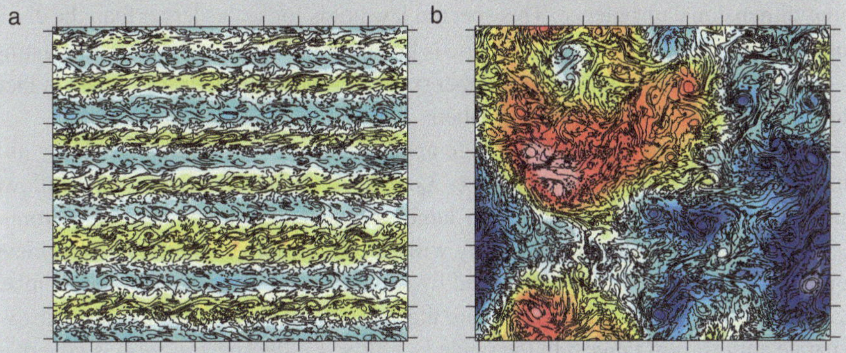

The above figure shows a simulation using (B51.1) (**a**) and (12.1) (**b**). The difference to the simulation shown in Figure 12.2 is that we have also included a forcing term on the right-hand side of (B51.1) and (12.1) and a linear damping term (in addition to the hyperviscosity to allow for a more effective dissipation on larger scales). The figure shows a snapshot of stream function $\psi$ and vorticity $\eta$ (contours) after the system is equilibrated for the energy input by the forcing and removal by dissipation by hyperviscosity and linear damping.

In both experiments, kinetic energy is injected on a small scale $k_i$, identical to the initial conditions in the simulation shown in Figure 12.2. The red energy cascade leads to a transport of that energy to larger scales as seen before, where it is then dissipated by the linear damping term. However, it turns out that when the turbulent scales become larger, Rossby wave dynamics – present for the case $\beta \neq 0$ – become important for the turbulent flow. The dispersion relation for Rossby waves of the form $\psi \sim \exp i(kx - \omega t)$ from (a linearized version of) (B51.1) is given by $\omega = -\beta/k$. Suppose that an eddy of scale $L$ transfers its energy to larger scales during the time $T = L/U$, where $U$ is the (turnover) velocity scale of the eddy which is usually taken simply as the rms velocity of the turbulent flow. When $T$ is smaller than the Rossby wave period $T_\mathrm{r} = 1/|\omega|$, the eddy will not be much affected by the wave propagation, but at the scale $L \sim \sqrt{U/\beta}$ where $T \sim T_\mathrm{r} = \beta L$, the red cascade will be effected, i. e. the flow will become anisotropic for scales $\geq L$ as seen in the simulation. Note that the red cascade will eventually be stopped by friction as in the simulation with $\beta = 0$ (Danilov and Gurarie, 2002).

The scale $\sqrt{U/\beta}$ is called the Rhines scale (Rhines, 1977) (see also Section 5.1) and can be related with the meridional spacing of the zonal jets appearing in the simulation for $\beta \neq 0$. Zonal jets related to the $\beta$-effect in geostrophic turbulence are often met in the ocean and the atmosphere (the jet stream), but also in the atmosphere of giant planets like Jupiter. Compare also the numerical model solution in the box on p. 380.

### 12.1.3 Equilibrium Range

In contrast to energy, the spectral transport of enstrophy (and passive tracers) towards larger scales is expected to be in the "normal" direction, i. e. from large to small scales. In analogy to Section 11.1.4, it is possible to define again an equilibrium range in the enstrophy conservation equation (12.3) where inertial and frictional terms are in balance. The dissipation of enstrophy is given as $\epsilon_G = 2A_h \int_0^\infty k^4 E(k)dk$. In the equilibrium range, both enstrophy and energy spectra will depend on $\epsilon_G$, $k$ and $A_h$. If an inertial subrange exists where the inertial terms are dominating over dissipation, then $E = E(k, \epsilon_G)$ is expected to be again a universal function. On dimensional arguments, the only possible functional form is given by

$$E(k) = \text{const}\,\epsilon_G^{2/3} k^{-3} \tag{12.9}$$

and describes a somewhat steeper decay of $E(k)$ with wave number compared to three-dimensional turbulence. This spectral law holds for scales larger than the dissipation length scale $\lambda_d$, which in this case is given by $\lambda_d = A_h^{3/6} \epsilon_G^{-1/6}$. It is interesting to note that the form of the wave-number spectrum shown in Figure 12.3g is, in fact, close to a $k^{-3}$ law for large wavenumbers.

However, another equilibrium range appears possible. Assume that energy and enstrophy are injected at a scale $k_i^{-1} \gg \lambda_d$. From the preceding argument it follows that enstrophy will be transported to larger $k$ where it will be dissipated around $k_d = 1/\lambda_d$. On the other hand, energy will be transported to smaller $k$. To achieve equilibrium, energy has to be removed by some process at a smaller wave number $k_*$, at a rate of, say, $\epsilon_*$. Provided now that $k_* \ll k_i \ll k_d$, one can expect two separate equilibrium ranges: In the range $k_i < k < k_d$, the spectral form is governed by enstrophy transport and follows the $k^{-3}$–law from (12.9). In the range $k_* < k < k_i$, a similar dimensional argument as in the three-dimensional case can be made, although the direction of the energy transport is now reversed. In analogy to (11.37), the spectral form is, therefore, given by $E = \text{const}\,\epsilon_*^{2/3} k^{-5/3}$.

The resulting conceptual spectrum is sketched in Figure 12.4, which might help to understand energy fluxes in ocean and atmosphere: At the scale $k_i$ which can be identified as the internal Rossby radius, mean kinetic and potential energy is converted to fluctuating energy by barotropic and baroclinic instability. Kinetic energy is then transported to larger scales where it has to be dissipated. The energy dissipation at large scales by linear damping is often interpreted as bottom friction (cf. Section 14.1). Note that on very large scales comparable to the Rhines scale, the $\beta$-effect may become important, as discussed in the box on p. 373.

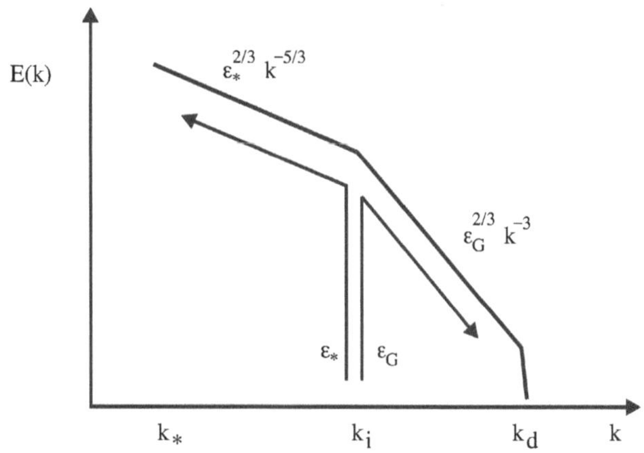


Fig. 12.4 Conceptual diagram showing the form of the energy spectrum for 2-d turbulence when energy and enstrophy are injected at a scale $k_i$. See text for further explanation

However, in numerical experiments of two-dimensional turbulence (observations are sparse in this respect), the spectral laws discussed above are often not found. This is because the scales are not very different (for instance the wave numbers in Figure 12.3 extend only over a single decade), the energy input is spread over a wide vicinity of $k_i$, and dissipation (by linear damping, whose effect is independent of wave number) is present in both equilibrium ranges, which is inconsistent with our assumptions. Furthermore, it seems that wave interaction can extend over large distances in wave-number space such that both equilibrium ranges can be coupled.

## 12.2 Mesoscale Eddies and Their Impact on the Mean Flow

Although some aspects of mesoscale ocean eddies seem consistent with the results of homogeneous two-dimensional turbulence as discussed in the preceding section, the complex situation in the real ocean including all other interacting dynamical regimes often hampers the application of the idealized homogeneous theory. This section is, therefore, devoted to the description of some important aspect of mesoscale eddies, and in particular to the interpretation of mixing effects by mesoscale eddies in the ocean. This mixing is strongly anisotropic and, therefore, differs in interesting ways from the mixing concepts met in the previous sections.

Oceanic eddies derive their energy mainly through the (barotropic and baroclinic) instability processes considered in Section 8.5.3. This energy transport has an important influence on the large-scale circulation. It needs to be parameterized in circulation models when the mesoscale eddies are not explicitly resolved. In the following, some aspects of the energy exchange between large-scale circulation and mesoscale eddies are discussed, and concepts for the parameterization of eddies are introduced.

### 12.2.1 Energetics of Mesoscale Eddies and the Lorenz Cycle

The equations of motion in the Boussinesq approximation (4.50) can be written as

$$\frac{\partial \boldsymbol{u}_\mathrm{h}}{\partial t} + \boldsymbol{u} \cdot \nabla \boldsymbol{u}_\mathrm{h} + \boldsymbol{f} \times \boldsymbol{u}_\mathrm{h} = -\nabla_\mathrm{h} p + \mathcal{F}_\mathrm{h} \tag{12.10}$$

together with the hydrostatic relation $b = \partial p/\partial z$ and the continuity equation $\nabla \cdot \boldsymbol{u} = 0$. Here $p$ denotes pressure divided by the reference density $\rho_0$, and $b = -g\tilde{\rho}/\rho_0$ denotes the buoyancy, where $\tilde{\rho}$ denotes the density perturbation with respect to the constant reference density $\rho_0$. The symbols $\boldsymbol{u}_\mathrm{h}$ and $\nabla_\mathrm{h}$ denote the horizontal velocity and the horizontal gradient operator, respectively (note that both are simultaneously used also in their three-dimensional forms, $\boldsymbol{u}$ and $\nabla$).

For small excursions of the isopycnals around a background stratification $b_\mathrm{b}(z)$, we can use quasi-geostrophic theory as discussed in Section 5.2. According to (5.8), small perturbations $b(x, y, z, t)$ of the background state are governed by

$$\frac{\partial b}{\partial t} + \boldsymbol{u}_\mathrm{h} \cdot \nabla_\mathrm{h} b + wN^2 = \mathcal{G}_\mathrm{b} \tag{12.11}$$

where $N^2(z) = \partial b_\mathrm{b}/\partial z$ is the buoyancy frequency. Note that the vertical advection is ignored since it is small compared to the term $wN^2$, and that the small compressibility terms in (5.8) have been neglected as well. When averaging (12.10) and

(12.11) to obtain equations for $\overline{\boldsymbol{u}}$ and $\overline{b}$, turbulent eddy fluxes $\overline{\boldsymbol{u}'\boldsymbol{u}'}$ and $\overline{\boldsymbol{u}'b'}$ appear again, which have to be known in order to describe the mean variables.

For the energy cycle considered in the following, it will be important to specify the viscous and diffusive closures in the underlying equations, i. e. the terms $\mathcal{F}_{\mathrm{h}}$ in (12.10) and $\mathcal{G}_{\mathrm{b}}$ in (12.11). Conceptually, these equations are based on a Reynolds average of the full equations of motions. This Reynolds filter, implicit to the primitive equations, averages over the temporal and/or spatial range of valid motions within the most restrictive approximation, which has led to the primitive equations. The most restrictive approximation is given by the hydrostatic approximation. Therefore, we assume that the terms $\mathcal{F}_{\mathrm{h}}$ and $\mathcal{G}_{\mathrm{b}}$ contain the divergence of turbulent fluxes given by small-scale nonhydrostatic motions of the type considered in Chapter 11.3.

The effect of the small-scale turbulence in the stratified ocean and atmosphere, with its small aspect ratio, is predominantly given by vertical processes. We, therefore, specify the mechanical forcing $\mathcal{F}_{\mathrm{h}} = \partial\boldsymbol{\tau}/\partial z$ in (12.10) as a divergence of a vertical turbulent flux $\boldsymbol{\tau}$ of horizontal momentum, and the diabatic buoyancy source term in (12.11) as $\mathcal{G}_{\mathrm{b}} = -\partial J/\partial z$, i. e. as a divergence of a vertical turbulent flux of buoyancy.

### Mean and Eddy Kinetic Energy

As discussed in Section 4.2.6, only the horizontal velocity components contribute to kinetic energy in the hydrostatic approximation. With this difference, conservation equations for the mean kinetic energy (MKE, $E_{\mathrm{mk}} = \frac{1}{2}[\overline{u}^2 + \overline{v}^2]$) and the eddy kinetic energy (EKE, $E_{\mathrm{ek}} = \frac{1}{2}[\overline{u'^2} + \overline{v'^2}]$) can be derived as in Section 11.3.1, i. e. by taking the mean of (12.10) and multiplication with $\overline{\boldsymbol{u}}_{\mathrm{h}}$ and multiplication of (12.10) with $\boldsymbol{u}'_{\mathrm{h}}$ and taking the mean, respectively, resulting in

$$\left(\frac{\partial}{\partial t} + \overline{\boldsymbol{u}}\cdot\nabla\right)E_{\mathrm{mk}} + \nabla\cdot\boldsymbol{F}_{\mathrm{mk}} = -S + \overline{b}\overline{w} + \overline{\boldsymbol{u}}_{\mathrm{h}}\cdot\frac{\partial\overline{\boldsymbol{\tau}}}{\partial z} \tag{12.12}$$

$$\left(\frac{\partial}{\partial t} + \overline{\boldsymbol{u}}\cdot\nabla\right)E_{\mathrm{ek}} + \nabla\cdot\boldsymbol{F}_{\mathrm{ek}} = +S + \overline{b'w'} + \overline{\boldsymbol{u}'_{\mathrm{h}}\cdot\frac{\partial\boldsymbol{\tau}'}{\partial z}} \tag{12.13}$$

Here $\boldsymbol{F}_{\mathrm{mk}} = \overline{\boldsymbol{u}}_{\mathrm{h}}\cdot\overline{\boldsymbol{u}'\boldsymbol{u}'_{\mathrm{h}}} + \overline{\boldsymbol{u}}\,\overline{p}$ and $\boldsymbol{F}_{\mathrm{ek}} = \overline{\boldsymbol{u}'|\boldsymbol{u}'_{\mathrm{h}}|^2/2} + \overline{\boldsymbol{u}'p'}$ are the fluxes of mean and eddy kinetic energy, respectively. The mean might be a statistical or a time mean. The physical interpretation of the individual terms is analogous to the conservation equations considered in Section 11.3.1. The terms $\overline{\boldsymbol{u}}_{\mathrm{h}}\cdot\partial\overline{\boldsymbol{\tau}}/\partial z$ and $\overline{\boldsymbol{u}'_{\mathrm{h}}\cdot\partial\boldsymbol{\tau}'/\partial z}$ constitute the input of mean and eddy energy by mechanical forcing. The production term $S = -\overline{\boldsymbol{u}'_{\mathrm{h}}\boldsymbol{u}'} : \nabla\overline{\boldsymbol{u}}_{\mathrm{h}} = -\overline{u'\boldsymbol{u}'}\cdot\nabla\overline{u} - \overline{v'\boldsymbol{u}'}\cdot\nabla\overline{v}$ (here the colon-symbol denotes the double-scalar product, see Appendix A.1.6) is the exchange between EKE and MKE, which is related to barotropic instability considered in Section 8.5.3.

### Mean and Eddy Available Potential Energy

In Section 11.3.1, we have shown that the terms $\overline{b}\overline{w}$ in (12.12) and $\overline{b'w'}$ in (12.13) can be interpreted as exchange between mean or turbulent kinetic energy with potential energy. However, the conservation equation of potential energy defined as $E_{\mathrm{p}} = -zb$, i. e. relative to a completely mixed state with constant density, is not very useful when considering ocean eddies. Most of $E_\mathrm{p}$ describes the energy necessary to homogenize the strongly stratified state. However, such a complete homogenization never occurs since motions associated with ocean eddies are nearly adiabatic, and potential energy is changed mainly by *reversible* vertical excursions of the isopycnals.

As shown in Section 5.2.5, a useful measure of the adiabatic exchange of potential energy is the *available potential energy* (per mass) $E^a_\mathrm{p} = b^2/(2N^2)$. With $b = \overline{b} + b'$, it is now possible to define mean available potential energy (MPE, $E^a_\mathrm{mp} = \overline{b}^2/(2N^2)$) and eddy available potential energy (EPE, $E^a_\mathrm{ep} = \overline{b'^2}/(2N^2)$). Conservation equations for $E^a_\mathrm{mp}$ and $E^a_\mathrm{ep}$ are obtained in the usual way. Multiplication of (12.11) with $\overline{b}$ and $b'$, respectively, and division by $N^2$ results in

$$\left(\frac{\partial}{\partial t} + \overline{\boldsymbol{u}}_\mathrm{h} \cdot \nabla_\mathrm{h}\right) E^a_\mathrm{mp} + \nabla_\mathrm{h} \cdot \boldsymbol{F}_\mathrm{mp} = -P - \bar{b}\overline{w} - \bar{b}\frac{\partial \bar{J}}{\partial z} \Big/ N^2 \tag{12.14}$$

$$\left(\frac{\partial}{\partial t} + \overline{\boldsymbol{u}}_\mathrm{h} \cdot \nabla_\mathrm{h}\right) E^a_\mathrm{ep} + \nabla_\mathrm{h} \cdot \boldsymbol{F}_\mathrm{ep} = +P - \overline{b'w'} - \overline{b'\frac{\partial J'}{\partial z}} \Big/ N^2 \tag{12.15}$$

The interpretation of the individual terms is as before, with $\boldsymbol{F}_\mathrm{mp} = \bar{b}\overline{\boldsymbol{u}'_\mathrm{h} b'}/N^2$ and $\boldsymbol{F}_\mathrm{ep} = \frac{1}{2}\overline{\boldsymbol{u}'_\mathrm{h} b'^2}/N^2$ describing transports of MPE and EPE, respectively, by the eddies. The term $\bar{b}\overline{w}$ which appears with opposite sign in (12.12) is an exchange between MPE and MKE, and likewise $\overline{b'w'}$ is an exchange between EPE and EKE. The production term $P = -\overline{\boldsymbol{u}'_\mathrm{h} b'} \cdot \nabla_\mathrm{h}\overline{b}/N^2$, which appears with opposite sign in both equations, describes the exchange between MPE and EPE, and is related to baroclinic instability.

### Lorenz Energy Cycle

In the previous section, the conservation equations (12.12)–(12.15) for the four variables MKE, EKE, MPE, and EPE have been derived. We have seen that the different forms of mean and eddy mechanical energy of the large-scale circulation are ultimately linked by exchange terms appearing with opposite signs in the respective conservation equations. These exchanges describe a cycle of energy, which is often called the LORENZ[1] energy cycle (Lorenz, 1955).

When integrating the conservation equations of EKE, MKE, EPE, and MPE over a volume $V$, assumed closed, all advective flux terms in the equations vanish since the normal component of the velocity vanishes at the boundary. The volume integral can be split in integrals over the horizontal area $\Omega$ and the vertical direction, $\int_V \mathrm{d}V = \int_\Omega \mathrm{d}\Omega \int_{-h}^{0} \mathrm{d}z$. When the standard down-gradient parameterization $\boldsymbol{\tau} = A_\mathrm{v}\partial \boldsymbol{u}_\mathrm{h}/\partial z$ is used, with the turbulent vertical viscosity $A_\mathrm{v}$, integration of the viscous term in (12.12) from surface $z = 0$ to bottom at $z = -h$ (assuming a rigid lid) yields

$$\int_{-h}^{0} \overline{\boldsymbol{u}}_\mathrm{h} \cdot \frac{\partial \overline{\boldsymbol{\tau}}}{\partial z} \mathrm{d}z = \overline{\boldsymbol{u}}_\mathrm{h} \cdot \overline{\boldsymbol{\tau}}\big|_{-h}^{0} - \int_{-h}^{0} A_\mathrm{v} \left|\frac{\partial \overline{\boldsymbol{u}}_\mathrm{h}}{\partial z}\right|^2 \mathrm{d}z \tag{12.16}$$

[1] EDWARD NORTON LORENZ, *1917 in West Hartford (USA) †2008 in Cambridge (USA), mathematician and meteorologist.

Relations similar to (12.16) are obtained for the EKE conservation equation and for the MPE and EPE conservation equation for $J = -K_v \partial b/\partial z$ with the turbulent vertical diffusivity $K_v$. Note that the surface windstress $\boldsymbol{\tau}|^0$, bottom stress $\boldsymbol{\tau}|_{-h}$, and the surface buoyancy flux $J|^0$ (the buoyancy flux across the bottom is assumed to be zero here) enter the integral balances. Note also that the second term in (12.16) is sign definite and resembles dissipation. The integral budgets become

$$\frac{\partial}{\partial t}\int_V E_{\mathrm{mk}}\mathrm{d}V = \int_V (-S + \bar{b}\bar{w} - \epsilon_{\mathrm{mk}})\mathrm{d}V - \int_\Omega \bar{\boldsymbol{u}}_{\mathrm{h}} \cdot \bar{\boldsymbol{\tau}}|_{-h}^0 \mathrm{d}\Omega \tag{12.17}$$

$$\frac{\partial}{\partial t}\int_V E_{\mathrm{ek}}\mathrm{d}V = \int_V (+S + \overline{b'w'} - \epsilon_{\mathrm{ek}})\mathrm{d}V - \int_\Omega \overline{\boldsymbol{u}'_{\mathrm{h}} \cdot \boldsymbol{\tau}'}|_{-h}^0 \mathrm{d}\Omega \tag{12.18}$$

$$\frac{\partial}{\partial t}\int_V E^a_{\mathrm{mp}}\mathrm{d}V = \int_V (-P - \bar{b}\bar{w} - \epsilon_{\mathrm{mp}})\mathrm{d}V - \int_\Omega N^{-2}\bar{b}\bar{J}|^0 \mathrm{d}\Omega \tag{12.19}$$

$$\frac{\partial}{\partial t}\int_V E^a_{\mathrm{ep}}\mathrm{d}V = \int_V (+P - \overline{b'w'} - \epsilon_{\mathrm{ep}})\mathrm{d}V - \int_\Omega N^{-2}\overline{b'J'}|^0 \mathrm{d}\Omega \tag{12.20}$$

Here $\epsilon_{\mathrm{mk}} = A_v|\partial\bar{\boldsymbol{u}}_{\mathrm{h}}/\partial z|^2$, $\epsilon_{\mathrm{ek}} = A_v\overline{|\partial\boldsymbol{u}'_{\mathrm{h}}/\partial z|^2}$, $\epsilon_{\mathrm{mp}} = K_v N^{-2}(\partial\bar{b}/\partial z)^2$, and $\epsilon_{\mathrm{ep}} = K_v N^{-2}\overline{(\partial b'/\partial z)^2}$ denote the dissipation of MKE, EKE, MPE, and EPE, respectively, by small-scale turbulence. The relations (12.17)–(12.20) are the integral form of the Lorenz energy cycle. A schematic representation of the energy cycle is shown in Figure 12.5 (left panel).

### Lorenz Cycle in the Atmosphere

Figure 12.5 (right panel) shows a recent estimate of the Lorenz energy cycle of the atmosphere based on a standard reanalysis dataset (ERA40) of the atmosphere. There is a large reservoir of MPE which is almost an order of magnitude larger than the other forms of mechanical energy, which are similar in magnitude. There is a large energy transfer from MPE to EPE and an even larger one from EPE to EKE. Notice also the energy exchange directed from EKE to MKE, which is in both hemispheres larger than the transfer from MPE to MKE, which means that the mean flow is predominantly driven by the eddy activity. One can infer from the residuals that the differential heating by radiative heat fluxes generates a large reservoir of MPE from which EPE and EKE is generated by baroclinic instability. Mechanical energy is apparently dissipated predominantly by dissipation of EKE. the box on p. 380 illustrates a numerical example similar to the atmospheric situation.

### Lorenz Cycle in the Ocean

In the ocean, such detailed reanalysis data sets as for the atmosphere are not available, which inhibits the estimation of the energy cycle based on observations. However, there are some bulk estimates of the energy cycle given by different authors which have been summarized by Wunsch and Ferrari (2004). On the other hand, a different way to infer information about the energy cycle in the ocean is offered by ocean models. Here, however, one should keep in mind that dissipation and thus

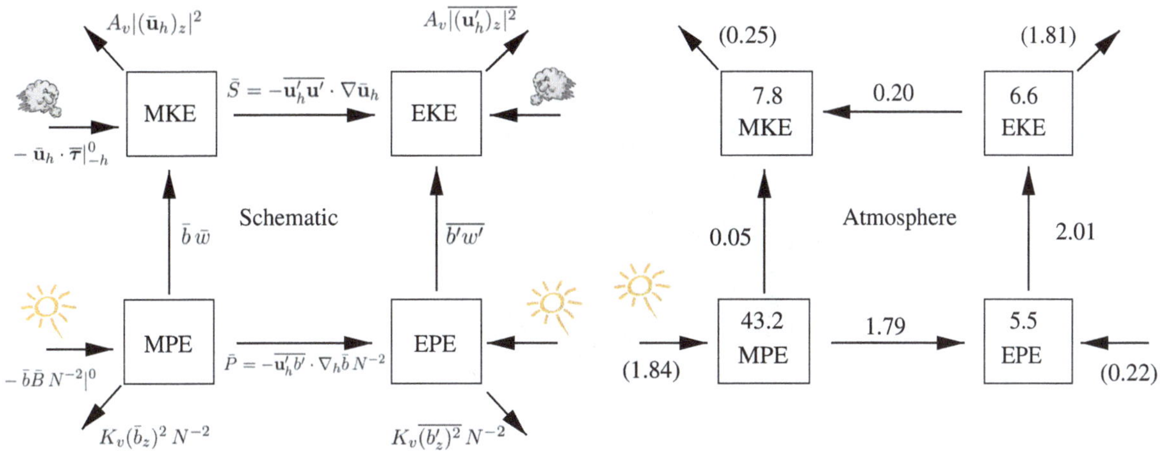


Fig. 12.5 *Left*: Schematic of the Lorenz energy cycle (12.17)–(12.20). Note that the local conservation equations also contain flux terms. It is assumed that dissipation is due to vertical mixing and friction only. *Right*: Lorenz energy cycle estimates from the ECMWF reanalysis dataset ERA40 by Li et al. (2007) for the atmosphere. Energies are given in $10^5\,\mathrm{J\,m^{-2}}$ and exchange terms in $\mathrm{W\,m^{-2}}$, and the integrals are taken over the entire globe. Numbers in brackets are deduced from the residuals; they contain forcing and dissipative terms

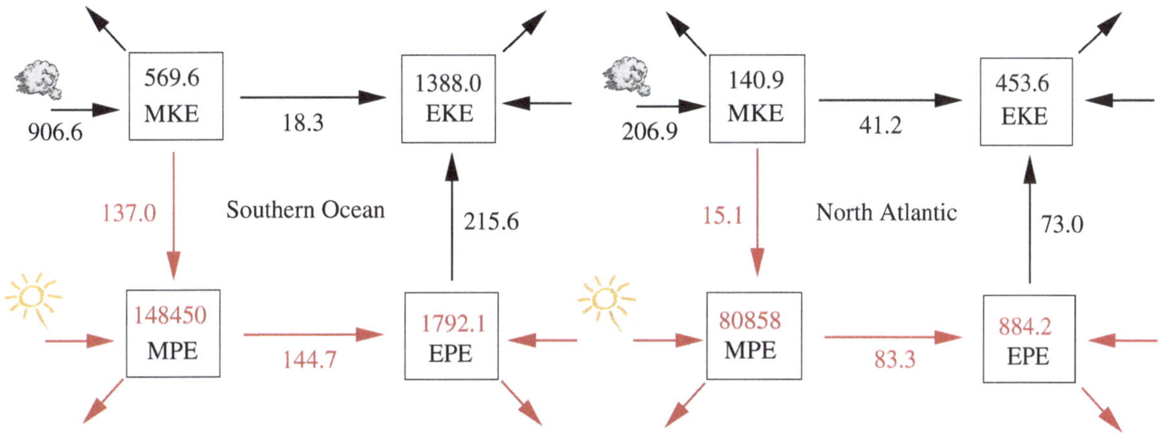


Fig. 12.6 Lorenz energy cycle estimates from regional eddy-permitting models of the North Atlantic and of the Southern Ocean. Energies are given in $10^{12}\,\mathrm{m^5\,s^{-2}} \sim 10^{15}\,\mathrm{J}$ and exchange terms in $10^6\,\mathrm{m^5\,s^{-3}} \sim 10^9\,\mathrm{W}$ (remind that we write energy usually as energy density, correct physical unit is J, however). The budget for the North Atlantic is taken from $10°$S to $68°$N and $100°$W to $0°$ and for the Southern Ocean from $78°$S to $32°$S and $180°$W to $180°$E and from surface to bottom in both cases. The background buoyancy profile is the mean over both regions weighted by the area of the domains. Numbers and arrows in red depend on the definition of the background buoyancy profiles; numbers and arrows in black are independent

the whole energy cycle in models depends on their subgrid-scale parameterizations, which are known to have deficits. Figure 12.6 shows estimates of the Lorenz energy cycle based on regional eddy-permitting models of the North Atlantic Ocean and the Southern Ocean. The models have a lateral resolution of $1/12°$ in the North Atlantic and $1/10°$ in the Southern Ocean and 45 and 43 vertical levels, respectively, and are both based on the same $z$-level finite difference numerical representation of the

**52. Lorenz Cycle in an Idealized Ocean Model**

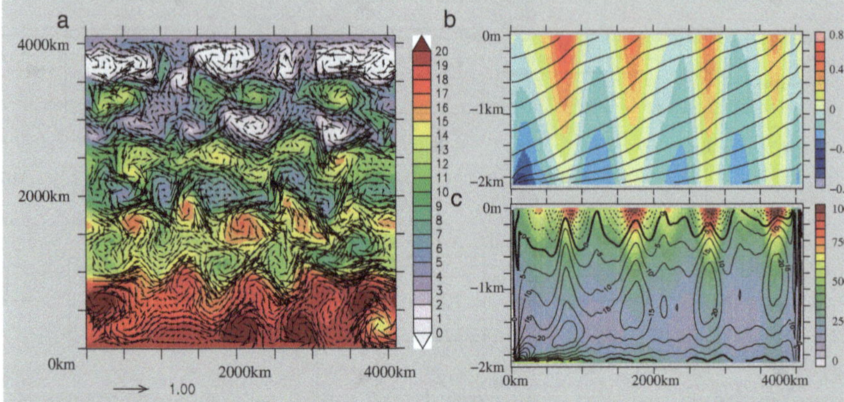


The above figure shows in **a** a snapshot of instantaneous pressure (divided by reference density, in $m^2\,s^{-2}$) and velocity in $m\,s^{-1}$ (arrows) at 500 m depth in an idealized primitive equation model simulation. **b** shows the zonally (and time) averaged zonal velocity $\bar{u}$ in $m\,s^{-1}$ (colors) and buoyancy $\bar{b}$ (contours). The simulation features four energetic zonal jets which can be seen in the snapshots but are most prominent in the zonal averages. The jets are driven by strong baroclinic forcing realized by relaxation zones at the northern and southern boundaries of the zonally periodic domain. The relaxation, resembling large-scale forcing like the radiative forcing in the atmosphere, generates and sustains the domain-wide meridional buoyancy gradient that can be seen in $\bar{b}$ and drives the system. Note that the zonal jets and their dynamics are analogous to the ones found in the barotropic case in the box on p. 373.
Integral mechanical energies (here for the zonally averaged quantities) in the model in $10^9\,m^4\,s^{-2}$ and energy transfer terms in $m^4\,s^{-3}$ (bold in the table) are given by

| $E_{mk}$ | $\rightarrow$ | $E^a_{mp}$ | $\rightarrow$ | $E^a_{ep}$ | $\rightarrow$ | $E_{ek}$ | $\rightarrow$ | $E_{mk}$ |
|---|---|---|---|---|---|---|---|---|
| 0.14 | **−6.70** | 30.6 | **108** | 0.27 | **88.6** | 0.26 | **23.0** | |

As seen in Figures 12.5b and 12.6, $E_{mp}$ dominates the other energy reservoirs by far. There is a large supply of $E^a_{mp}$ by the relaxation zones at the northern and southern boundaries. This energy is then transferred predominantly to $E^a_{ep}$ where a fraction is dissipated and further transferred to $E_{ek}$ where most of that energy is lost to dissipation. However, a significant amount (about 20%) of the energy originally supplied by the large-scale forcing is transferred from $E_{ek}$ back to the mean flow $E_{mk}$, which is, to a lesser extent, also fed directly from $E_{mp}$.
The transfer of eddy energy to the mean flow takes place inside the zonal jets. The above figure, **c**, shows in colors $E_{ek}$ in $10^{-4}\,m^2\,s^{-2}$ while the energy transfer to $E_{ek}$, i. e. $S+\overline{b'w'}$, is denoted by lines in $10^{-9}\,m^2\,s^{-3}$ where the contour spacing is $5\times10^{-9}\,m^2\,s^{-3}$. Note that positive values correspond to solid lines and denote production of $E_{ek}$. Note also that in the upper 200–400 m, $S+\overline{b'w'}$ is negative since here it is dominated by (negative) $S$ related to the effect of barotropic instability. Correspondingly the mean flow is energetic in this upper layer. Below that surface layer, $S+\overline{b'w'}$ is positive and dominated by (positive) $\overline{b'w'}$ related to the effect of baroclinic instability. Note that the term $\overline{p'w'}$ in the energy flux $\boldsymbol{F}_{ek}$ predominantly fluxes the energy from regions with positive $\overline{b'w'}$ to regions with negative $S$.

primitive equations. Note also that the energy budgets for both ocean basins are not closed, i. e. there is a transport of energy in and out of the domains.

As for the atmosphere, there is a large reservoir of MPE, which is roughly two orders of magnitude larger than the other forms of mechanical energy, which are of similar magnitude. Note, however, that the value of MPE and the associated energy transfer terms depend to a large extent on the definition of the background stratifica-

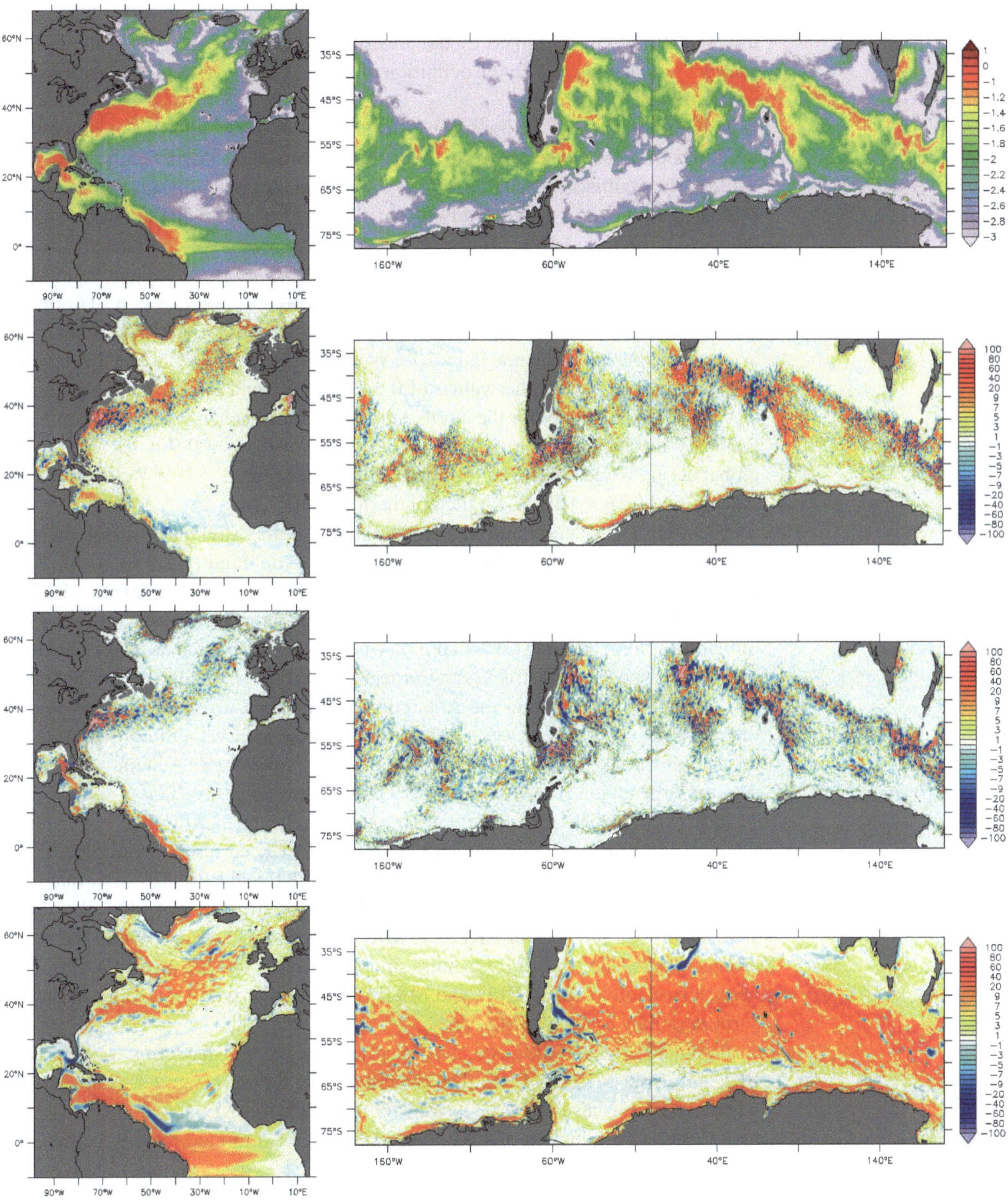


Fig. 12.7 *Upper row*: Eddy kinetic energy (EKE) in regional models of the North Atlantic and the Southern Ocean in $\log_{10}(\mathrm{EKE}/[\mathrm{m}^2\,\mathrm{s}^{-2}])$ at 100 m depth. *Second row*: Energy transfer from EPE to EKE, $\int_{-h}^{0} \overline{b'w'}\mathrm{d}z$ in $10^{-6}\,\mathrm{m}^3\,\mathrm{s}^{-3} \sim 10^{-3}\,\mathrm{W\,m}^{-2}$. Positive values denote production of EKE. *Third row*: Energy transfer from MKE to EKE, $\int_{-h}^{0} S\mathrm{d}z$ in $10^{-6}\,\mathrm{m}^3\,\mathrm{s}^{-3} \sim 10^{-3}\,\mathrm{W\,m}^{-2}$. Positive values denote production of EKE. *Lowest row*: Wind forcing of MKE, $-\overline{\boldsymbol{u}} \cdot \overline{\boldsymbol{\tau}}$, in $10^{-6}\,\mathrm{m}^3\,\mathrm{s}^{-3} \sim 10^{-3}\,\mathrm{W\,m}^{-2}$. Positive values denote production of MKE

tion, which was chosen as the mean over the mean profile of the North Atlantic and the Southern Ocean weighted by the area of the domains. All terms of the Lorenz energy cycle that depend on the definition of the background stratification are shown in red in Figure 12.6.

In both oceans, EKE is more than twice as large as MKE, and EPE is larger than EKE in both cases. In the North Atlantic, the ratio between EPE and EKE is around 2 while in the Southern Ocean this ratio is slightly smaller. Wunsch and Ferrari (2004) estimated $13 \times 10^{18}$ J for the eddy mechanical energy of the global ocean, i.e. for the sum of EKE and EPE. The regional model estimates are consistent with this number, i.e. they yield $3.2 \times 10^{18}$ J eddy mechanical energy in the Southern Ocean and $1.3 \times 10^{18}$ J in the North Atlantic, leaving enough room for contributions to global eddy mechanical energy from the Pacific and Indian Ocean. On the other hand, Wunsch and Ferrari estimated $0.8 \times 10^{12}$ W wind forcing of MKE while the regional model estimates yield higher values of $0.9 \times 10^{12}$ W wind forcing for the Southern Ocean and $0.2 \times 10^{12}$ W for the North Atlantic. For the exchange rate between mean and eddy mechanical energy, Wunsch and Ferrari estimated $0.9 \times 10^{12}$ W for the global ocean, while the model estimates are lower, i.e. $0.12 \times 10^{12}$ W for the North Atlantic and $0.16 \times 10^{12}$ W for the Southern Ocean.

There are large lateral and vertical variations in the different forms of energies and the exchange terms of the energy cycle in the ocean. Figure 12.7 shows the EKE simulated by the two regional models at 100 m depth. In the North Atlantic, EKE is large at the equator and in the western boundary currents, in particular in the Gulf Stream and North Atlantic Current region where the EKE peaks at around 1000 J m$^{-3}$ while in the interior EKE is several orders of magnitude smaller. In the Southern Ocean, EKE is large within the Antarctic Circumpolar Current (ACC) and also at the east coasts of South America and Australia where western boundary currents are located with peak values of EKE similar to those in the North Atlantic while there are also large regions in the Southern Ocean with low values of EKE.

EPE shows a lateral and vertical structure similar to that of EKE with values which are in general slightly higher than those of EKE, as already indicated in the integral budgets. MKE shows large values where EKE and EPE also show maxima, with magnitudes slightly less than EKE. The lateral structures of MKE, however, are sharper than the ones of EKE, i. e. MKE is more concentrated within the mean currents while EKE and EPE are more smoothed out around the mean currents. The vertical structure of EKE, EPE, and MKE is also similar with large values near the surface, rapidly decaying with depth.

Figure 12.7 shows the vertically integrated energy transfer terms from EPE to EKE and from MKE to EKE. The magnitudes of the exchange terms are largest where the EKE is also largest. The transfer from EPE to EKE is predominantly positive, denoting production of EKE while there are also limited regions in the Gulf Stream, in the ACC, and at the equator where $\int_{-h}^{0} \overline{b'w'}\mathrm{d}z$ is getting negative. The lateral and vertical structure of the transfer $P$ from MPE to EPE is similar to $\overline{b'w'}$ and thus not shown. The transfer $S$ from MKE to EKE, on the other hand, shows a lateral pattern of large magnitudes but with rapidly fluctuating signs. Note that in the atmosphere, the global integral of $S$ is negative, i. e. representing a transfer from EKE to MKE, while for the basin integrals of the North Atlantic and the Southern Ocean, there is a (small) transfer from MKE to EKE. However, integrals for certain regions in the regional models also can show transfer from EKE to MKE similar to the atmosphere. This is in particular the case for regions of free zonal flow such as the Gulf Stream/North Atlantic Current system and the Antarctic Circumpolar Current. Figure 12.7 also shows the wind forcing of the energy cycle which is in general positive and large over the ACC and the northern and southern edges of the subtropical gyre of the North Atlantic. The wind forcing of EKE (not shown) is rather small because of the lack of correlation between mesoscale eddies and wind fluctuations.

### 12.2.2 Isopycnal Mixing Tensor

Mesoscale eddies in the ocean are characterized by a small aspect ratio, i. e. their horizontal scales are much larger than their vertical ones, and they have much stronger horizontal than vertical velocities. When the mixing by eddies is represented by an Austauschansatz, one has to expect that the magnitude of mixing will depend on the direction in space and, in particular, that it will be different for the vertical and horizontal directions. With the appropriate anisotropic form of down-gradient diffusion, the diffusive tracer flux is given as $\overline{\boldsymbol{u}'\psi'} = -\mathsf{K} \cdot \nabla \overline{\psi}$ with a $3 \times 3$ diffusion tensor $\mathsf{K} = (K_{mn})$.

The simplest form of $\mathsf{K}$ which accounts for the small aspect ratio is given by $K_{11} = K_{22} = K_{\mathrm{h}}$, $K_{33} = K_{\mathrm{v}}$ and $K_{mn} = 0$ for $m \neq n$, with horizontal and vertical coefficients $K_{\mathrm{h}} \gg K_{\mathrm{v}}$. While commonly used, this simple form is, however, not well founded since the transports by mesoscale eddies in the ocean are mainly along the local isopycnal direction. Due to the stable stratification, mixing of water masses across isopycnal surfaces decreases the available potential energy (compare also Section 5.2.6), whereas mixing along isopycnals does not effect the potential energy (besides small thermodynamic effects in the presence of gradients in temperature and salinity on isopycnals, as discussed in the box on p. 384). Therefore, a parametrization is preferable that explicitly distinguishes between mixing in the directions along and across isopycnals.

As discussed in Section 2.7.2, the vector $\boldsymbol{e} = \gamma\rho\nabla S - \alpha\rho\nabla\theta$ defined in (2.132) is normal to the local neutral surface element. One of the principal directions of the mixing tensor should be in the direction of $\boldsymbol{e}$. To determine the form of the mixing tensor, we note that an arbitrary vector $\boldsymbol{a}$ can be decomposed in its components normal to the surface (*diapycnal*, index d) and along the surface (*isopycnal*, index i) by

$$\boldsymbol{a}_{\mathrm{d}} = \left(\boldsymbol{a} \cdot \frac{\boldsymbol{e}}{|\boldsymbol{e}|}\right) \frac{\boldsymbol{e}}{|\boldsymbol{e}|} = \frac{\boldsymbol{e}\boldsymbol{e}}{e^2} \cdot \boldsymbol{a} \,, \quad \boldsymbol{a}_{\mathrm{i}} = \boldsymbol{a} - \boldsymbol{a}_{\mathrm{d}} = \left(\mathsf{I} - \frac{\boldsymbol{e}\boldsymbol{e}}{e^2}\right) \cdot \boldsymbol{a}$$

with the unit tensor $\mathsf{I}$. Note that $\boldsymbol{e}\boldsymbol{e}$ denotes the dyadic or tensor product defined in (A.15). Therefore, the simplest form[2] of a mixing tensor with different diffusivities in isopycnal and diapycnal directions is given as

$$\mathsf{K} = \mathsf{K}_{\mathrm{d}} + \mathsf{K}_{\mathrm{i}} \quad \text{with} \quad \mathsf{K}_{\mathrm{d}} = K_{\mathrm{d}} \frac{\boldsymbol{e}\boldsymbol{e}}{e^2} \quad \text{and} \quad \mathsf{K}_{\mathrm{i}} = K_{\mathrm{i}} \left(\mathsf{I} - \frac{\boldsymbol{e}\boldsymbol{e}}{e^2}\right) \tag{12.21}$$

[2] The form (12.21) of the mixing tensor assumes isotropic mixing in the along isopycnal direction. Other forms, i. e. isopycnally anisotropic (but symmetric) tensors, are possible but yield more complicated representations.

Note that the mixing tensor as defined in (12.21) is *symmetric*, i. e. $K_{mn} = K_{nm}$. According to (12.21), the (negative of the) diapycnal flux of a tracer $\psi$ is hence given as

$$\mathsf{K}_\mathrm{d} \cdot \nabla\psi = K_\mathrm{d}\frac{\boldsymbol{e}\boldsymbol{e}}{e^2} \cdot \nabla\psi = K_\mathrm{d}\frac{\boldsymbol{e}_\mathrm{h} \cdot \nabla_\mathrm{h}\psi + e_3\partial\psi/\partial z}{e_\mathrm{h}^2 + e_3^2}\boldsymbol{e} \approx K_\mathrm{d}\left(0, 0, \frac{\partial\psi}{\partial z}\right) \quad (12.22)$$

Here the last expression holds for small aspect ratio, so that $|e_1|, |e_2| \ll |e_3|$, and corresponds to the flux divergence $\nabla \cdot (\mathsf{K}_\mathrm{d} \cdot \nabla\psi) \approx \partial\left(K_\mathrm{d}\partial\psi/\partial z\right)/\partial z$. In this approximation, diapycnal and vertical mixing are, therefore, equivalent.

**53. Density Effects due to Isopycnal Mixing of Temperature and Salinity**

The form (12.21) applies to passive tracers but also to turbulent mixing of temperature and salinity. For temperature it leads to a flux divergence $\mathcal{G}_\theta = \nabla_\rho^\dagger \cdot [(\mathsf{K}_\mathrm{d}^\theta + \mathsf{K}_\mathrm{i}^\theta) \cdot \nabla_\rho\theta]$, and likewise for salinity. According to (3.2), these terms are linked to a source term $\mathcal{G}_\rho = \gamma\mathcal{G}_S - \alpha\mathcal{G}_\theta$ in the density conservation equation.
Isopycnal mixing of temperature and salinity by eddies has no direct effect on density since for both the exact and the approximate forms in (12.23) the identity $\mathsf{K}_\mathrm{i} \cdot \boldsymbol{e} \equiv 0$ holds. However, there can be indirect effects through nonlinearities in the state equation. With (12.24), the density forcing $\mathcal{G}_\rho^i$ is

$$\begin{aligned}\mathcal{G}_\rho^i &= \gamma\rho\mathcal{G}_S^i - \alpha\rho\mathcal{G}_\theta^i = \gamma\rho\nabla_\rho^\dagger \cdot (K_\mathrm{i}\nabla_\rho S) - \alpha\rho\nabla_\rho^\dagger \cdot (K_\mathrm{i}\nabla_\rho\theta) \\ &= -K_\mathrm{i}\left[\nabla_\rho(\gamma\rho) \cdot \nabla_\rho S - \nabla_\rho(\alpha\rho) \cdot \nabla_\rho\theta\right] = K_\mathrm{i}\gamma\rho\nabla_\rho(\alpha/\gamma) \cdot \nabla_\rho\theta\end{aligned}$$

Here the identity $\gamma\nabla_\rho S \equiv \alpha\nabla_\rho\theta$ has been used. Expanding $\alpha(S, T, p)$ and $\gamma(S, T, p)$ with the state equation and using $\partial\gamma/\partial\theta = -\partial\alpha/\partial S$, one obtains the form

$$\mathcal{G}_\rho^i = K_\mathrm{i}\rho\left[\frac{\partial\alpha}{\partial\theta} + 2\frac{\alpha}{\gamma}\frac{\partial\alpha}{\partial S} - \frac{\alpha^2}{\gamma^2}\frac{\partial\gamma}{\partial S}\right](\nabla_\rho\theta)^2 + K_\mathrm{i}\rho\left[\alpha_p - \frac{\alpha}{\gamma}\frac{\partial\gamma}{\partial p}\right]\nabla_\rho\theta \cdot \nabla_\rho p \quad \text{(B53.1)}$$

which has been discussed in detail by McDougall (1987). The first term (*cabbeling*) in (B53.1) is caused by the dependence of $\alpha/\gamma$ on temperature and salinity along the isopycnal surface. It is always positive and corresponds to the fact that mixing of two fluid elements with different temperatures and salinities but initially the same potential density leads to a mixing product with higher density, due to the nonlinearity in the state equation (cf. Figure 1.5). The second term (*thermobaricity*) in (B53.1) is caused by the dependence of $\alpha/\gamma$ on pressure along the isopycnal surface. It can have either sign and is normally somewhat smaller in magnitude than the cabbeling term. To balance both terms, an upward vertical velocity of a few $10^{-8}\,\mathrm{m\,s^{-1}}$ would be necessary. This indicates that the effect of isopycnal mixing on density may be small but not altogether negligible.

The isopycnal mixing tensor follows from (12.21) as

$$\mathsf{K}_\mathrm{i} = \frac{K_\mathrm{i}}{e^2}\begin{pmatrix} e_2^2 + e_3^2 & -e_1e_2 & -e_1e_3 \\ -e_2e_1 & e_1^2 + e_3^2 & -e_2e_3 \\ -e_3e_1 & -e_3e_2 & e_1^2 + e_2^2 \end{pmatrix} \approx K_\mathrm{i}\begin{pmatrix} 1 & 0 & s_1 \\ 0 & 1 & s_2 \\ s_1 & s_2 & s_1^2 + s_2^2 \end{pmatrix} \quad (12.23)$$

where $s_1, s_2$ are the components of a horizontal vector $\boldsymbol{s} = -\boldsymbol{e}_\mathrm{h}/e_3$, which describes the slope of the local isopycnal surface against the horizontal plane. The last approximation holds again for small aspect ratio. With this approximation the isopycnal tracer flux is given as

$$\mathsf{K}_\mathrm{i} \cdot \nabla\psi = K_\mathrm{i}\begin{pmatrix} \nabla_\mathrm{h}\psi + \boldsymbol{s}\partial\psi/\partial z \\ \boldsymbol{s} \cdot \nabla_\mathrm{h}\psi + s^2\partial\psi/\partial z \end{pmatrix} = K_\mathrm{i}\begin{pmatrix} \nabla_\rho\psi \\ \boldsymbol{s} \cdot \nabla_\rho\psi \end{pmatrix}$$

**54. Rotated Vectors in the Zonal Mean Case**

In Section 12.2.3 and following, zonally averaged buoyancy and tracer conservation equation of the form (12.25) are discussed. The following notation is used for convenience: All vectors and gradients are two-dimensional in the meridional-vertical plane, e. g. $\overline{\boldsymbol{v}} = (\bar{v}, \bar{w})$ in (12.25). Furthermore, the following definition of rotation for vectors (by $\pi/2$ in the counterclockwise sense) is used

$$\underline{\boldsymbol{v}} = (-w, v) \quad \text{if} \quad \boldsymbol{v} = (v, w)$$

This rotation will also be applied to other vectors; in particular, the vector operator $\underline{\boldsymbol{\nabla}} = (-\partial/\partial z, \partial/\partial y)$ denotes the rotated gradient in the meridional-vertical plane, $\boldsymbol{\nabla} = (\partial/\partial y, \partial/\partial z)$. Note that $\underline{\boldsymbol{\nabla}} \cdot \boldsymbol{\nabla}\psi = \boldsymbol{\nabla} \cdot \underline{\boldsymbol{\nabla}}\psi = 0$. Compare also with the box on p. 444.

where for brevity the operator $\boldsymbol{\nabla}_\rho = \boldsymbol{\nabla}_\mathrm{h} + s\partial/\partial z$ has been introduced which is the gradient along the local isopycnal surface. For the divergence of the isopycnal tracer transport it follows that

$$\boldsymbol{\nabla} \cdot (\mathsf{K}_\mathrm{i} \cdot \boldsymbol{\nabla}\psi) = \boldsymbol{\nabla}_\mathrm{h} \cdot (K_\mathrm{i}\boldsymbol{\nabla}_\rho\psi) + \frac{\partial}{\partial z}(K_\mathrm{i}\boldsymbol{s} \cdot \boldsymbol{\nabla}_\rho\psi) = \boldsymbol{\nabla}_\rho^\dagger \cdot (K_\mathrm{i}\boldsymbol{\nabla}_\rho\psi) \tag{12.24}$$

with the local isopycnal divergence operator $\boldsymbol{\nabla}_\rho^\dagger = \boldsymbol{\nabla}_\mathrm{h} + \partial/\partial z\boldsymbol{s}$ (note that the relation $\boldsymbol{\nabla}_\rho^\dagger \cdot (\lambda\boldsymbol{a}) = \lambda\boldsymbol{\nabla}_\rho^\dagger \cdot \boldsymbol{a} + \boldsymbol{a} \cdot \boldsymbol{\nabla}_\rho\lambda$ holds for arbitrary $\lambda$ and $\boldsymbol{a}$). The isopycnal mixing tensor (12.23) is defined in $z$-coordinates. In Section 12.3.5 and the box on p. 422 it is demonstrated how isopycnal mixing and the form (12.24) are related to mixing of a tracer in isopycnal coordinates.

### 12.2.3 Transformed Eulerian Mean

The mixing tensor $\mathsf{K}$ as defined in (12.21) is anisotropic, reflecting the preferred direction along the isopycnal surface. When applied to density or buoyancy, the isopycnal component essentially vanishes, except for the (small) nonlinear terms considered in the box on p. 384, i. e. $(\mathsf{K}_\mathrm{d} + \mathsf{K}_\mathrm{i}) \cdot \boldsymbol{\nabla}b \approx \mathsf{K}_\mathrm{d} \cdot \boldsymbol{\nabla}b = K_\mathrm{d}\boldsymbol{\nabla}b$, so that only the diapycnal component remains. Now eddies can influence the mean density structure even in adiabatic systems, i. e. in the absence of any diapycnal mixing, which can be better described as an advective effect as detailed below. A prominent example of such an adiabatic process is that of baroclinic instability discussed in Section 8.5.3 and below in Section 12.2.4. To parameterize such effects, the form (12.21) is not sufficient, and a representation beyond an isopycnal mixing tensor is needed.

To introduce the concepts, it is convenient to start the discussion with a two-dimensional flow. The three-dimensional case is discussed in the box on p. 389. To be specific, consider a zonally averaged flow, with zonal averaging denoted as $\overline{(\ldots)}$. The zonally averaged buoyancy conservation equation is given by

$$\frac{\partial \bar{b}}{\partial t} + \boldsymbol{\nabla} \cdot (\overline{\boldsymbol{v}}\bar{b}) + \boldsymbol{\nabla} \cdot \boldsymbol{F}_\mathrm{b} = \overline{\mathcal{G}}_\mathrm{b} \tag{12.25}$$

and is similar to the temporally (or statistically) averaged equation (here $\boldsymbol{F}_\mathrm{b} = \overline{\boldsymbol{v}'b'}$ is the turbulent buoyancy flux). In the remainder of this section, all vectors and vector operators are two-dimensional in the meridional plane, i. e. the zonal component vanishes. The notation follows the box on p. 385. In particular, the two-dimensional current vector $(v, w)$ in the $(y, z)$-plane is denoted by $\boldsymbol{v}$.

Andrews et al. (1987) have suggested to decompose the eddy buoyancy flux into its components parallel and perpendicular to $\nabla\overline{b}$, i. e.

$$\boldsymbol{F}_{\mathrm{b}} = B\underset{\neg}{\nabla}\overline{b} - K_{\mathrm{d}}\nabla\overline{b} \tag{12.26}$$

The coefficients $B$ and $K_{\mathrm{d}}$ are given by

$$K_{\mathrm{d}} = -|\nabla\overline{b}|^{-2}\boldsymbol{F}_{\mathrm{b}} \cdot \nabla\overline{b} \quad \text{and} \quad B = |\nabla\overline{b}|^{-2}\boldsymbol{F}_{\mathrm{b}} \cdot \underset{\neg}{\nabla}\overline{b} \tag{12.27}$$

The approximation $B \approx -\overline{v'b'}/(\partial\overline{b}/\partial z)$ holds for small aspect ratio[3]. With (12.26), the buoyancy equation (12.25) can be written as

$$\frac{\partial}{\partial t}\overline{b} + (\overline{\boldsymbol{v}} - \underset{\neg}{\nabla}B) \cdot \nabla\overline{b} = \overline{\mathcal{G}}_{\mathrm{b}} + \nabla \cdot K_{\mathrm{d}}\nabla\overline{b} \tag{12.28}$$

The flux component $-K_{\mathrm{d}}\nabla\overline{b}$ in (12.26) corresponds to a diffusive flux, and, therefore, the coefficient $K_{\mathrm{d}}$ can be identified with the diapycnal diffusivity $K_{\mathrm{d}}$ in the diapycnal diffusivity tensor given by (12.22).

According to (12.28), the flux component along the mean gradient is equivalent to an advection of $\overline{b}$ with an additional velocity $-\underset{\neg}{\nabla}B$, so that $B$ can be viewed as stream function for that velocity, which might be called an *eddy-driven velocity*. Combining $\underset{\neg}{\nabla}B$ and $\overline{\boldsymbol{v}}$ to the *residual velocity*[4] $\boldsymbol{v}^* = \overline{\boldsymbol{v}} - \underset{\neg}{\nabla}B$, the equation (12.28) can be written as

$$\frac{\partial\overline{b}}{\partial t} + \boldsymbol{v}^* \cdot \nabla\overline{b} = \overline{\mathcal{G}}_{\mathrm{b}} + \nabla \cdot K_{\mathrm{d}}\nabla\overline{b} \tag{12.29}$$

It can be expected that the flux $\boldsymbol{F}_{\mathrm{b}}$ is mainly directed along the lines of constant $\overline{b}$, and that the diapycnal diffusion term is small. It is shown below in Section 12.3.1 that $K_{\mathrm{d}}$ can be linked to $\mathcal{G}_{\mathrm{b}}$, i. e. it is possible to show that $K_{\mathrm{d}} = 0$ follows from $\mathcal{G}_{\mathrm{b}} = 0$. Furthermore, it follows that if $\mathcal{G}_{\mathrm{b}}$ is small and can be neglected, then $K_{\mathrm{d}}$ shall also be small and neglected, such that the mean buoyancy equation becomes

$$\frac{\partial\overline{b}}{\partial t} + \boldsymbol{v}^* \cdot \nabla\overline{b} = 0 \tag{12.30}$$

which has a form very similar to the instantaneous buoyancy equation, however with an advection velocity $\boldsymbol{v}^*$ which is different from $\boldsymbol{v}$. In the stationary case $(\partial\overline{b}/\partial t = 0)$, the residual velocity is directed along lines of $\overline{b} = \text{const}$. In other words, it is the residual velocity that corresponds to the mean particle pathways, not the Eulerian mean velocity $\overline{\boldsymbol{v}}$. The residual velocity $\boldsymbol{v}^*$ is, therefore, also called *Transformed Eulerian Mean* (TEM) velocity.

[3] Note that in well-mixed boundary layers of the ocean and the atmosphere the approximation $|\partial\overline{b}/\partial y| \gg |\partial\overline{b}/\partial z|$ is often used, such that $B \approx \overline{w'b'}/(\partial\overline{b}/\partial y)$.

[4] The name residual velocity derives from the fact that often $\underset{\neg}{\nabla}B$ and $\overline{\boldsymbol{v}}$ are large and of opposite sign, such that their sum is only small. This is for instance the case for the meridional overturning circulation in the Southern Ocean discussed in Section 16.3.

55. Antisymmetric Diffusivity Tensor and Skew Diffusivity

Note that the decomposition (12.26) can also be written as

$$\boldsymbol{F}_{\mathrm{b}} = -\mathsf{K} \cdot \nabla \bar{b} \quad \text{with} \quad \mathsf{K} = \begin{pmatrix} K_{\mathrm{d}} & -B \\ B & K_{\mathrm{d}} \end{pmatrix} \tag{B55.1}$$

The tensor K in (B55.1) is not symmetric, unlike the mixing tensor considered in the previous section. Decomposition into its symmetric and antisymmetric parts, $\mathsf{K} = \mathsf{K}_{\mathrm{symm}} + \mathsf{K}_{\mathrm{anti}}$, yields

$$\mathsf{K}_{\mathrm{symm}} = \begin{pmatrix} K_{\mathrm{d}} & 0 \\ 0 & K_{\mathrm{d}} \end{pmatrix} \quad \text{and} \quad \mathsf{K}_{\mathrm{anti}} = \begin{pmatrix} 0 & -B \\ B & 0 \end{pmatrix}$$

As discussed above, the symmetric tensor in (12.26) is associated with diapycnal diffusion and usually small. The extra advection is formally associated with an antisymmetric tensor and is also referred to as *skew* diffusion.

### 12.2.4 Gent and McWilliams Parameterization and the Bolus Velocity

The TEM form of the buoyancy equation (12.28) states that the zonally averaged buoyancy is advected by the sum of the zonally averaged mean flow $\bar{\boldsymbol{v}}$ and the eddy-driven velocity $-\underset{\neg}{\nabla} B$. The latter is often called the *bolus* velocity, but note that eddy-driven and bolus velocity are not identical, as discussed below in Section 12.3.5 and the box on p. 417. However, for the sake of finding a simple parameterization for the effect, the difference is not of practical importance.

To achieve a parametrization for the eddy-driven velocity, one has to relate the stream function $B$, or equivalently the eddy flux $\overline{v'b'}$ in (12.27), with the mean field. The standard down-gradient formulation for the horizontal component of the isopycnal eddy flux is

$$\overline{v'b'} = -K_\ell \frac{\partial \bar{b}}{\partial y} \tag{12.31}$$

Assuming $K_{\mathrm{d}} = 0$ in (12.27) (compare the discussion in the box on p. 390 concerning this assumption) leads to $\overline{b'w'} = s\overline{v'b'}$ and yields for the stream function

$$B = |\nabla \bar{b}|^{-2} \left( -\overline{v'b'} \frac{\partial \bar{b}}{\partial z} + \overline{b'w'} \frac{\partial \bar{b}}{\partial y} \right) = -\overline{v'b'} \Big/ \frac{\partial \bar{b}}{\partial z} = -s K_\ell \tag{12.32}$$

where $s = -(\partial \bar{b}/\partial y)/(\partial \bar{b}/\partial z)$ denotes the mean isopycnal slope and $K_\ell$ a lateral diffusivity. This parameterization for $B$ is equivalent to the one proposed by Gent and McWilliams (1990). Note that with $K_{\mathrm{d}} = 0$, it also follows $B = \overline{b'w'}/(\partial \bar{b}/\partial y)$ such that the boundary condition $B = 0$ at $z = 0, -h$ is satisfied. In general, however, one has to insure that $B = 0$ at all boundaries. In the parameterization, this condition has to be applied to the diffusivity $K_\ell$, because the isopycnal slopes might be nonzero at the boundaries.

The effect of the parameterization (12.32) for ocean dynamics can be illustrated by a simple example of a sloping front. Figure 12.8 shows the zonally averaged buoyancy $\bar{b}$ and the stream function $B$ for the eddy-driven velocity $-\underset{\neg}{\nabla} B$ in a simulation

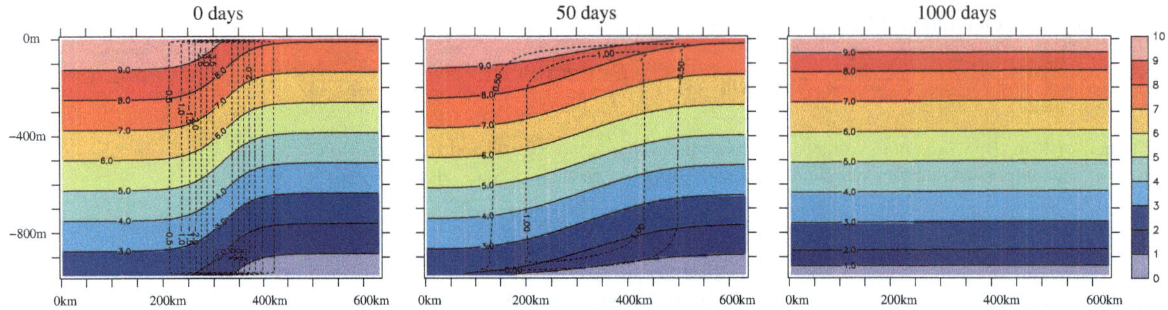


Fig. 12.8 Idealized model simulation of the effect of a simple parameterization of the eddy-driven velocity on a narrow front. Shown is the zonally averaged buoyancy $\bar{b}$ (here scaled to a temperature in °C) in colors and the stream function $\boldsymbol{B}$ for the eddy-driven velocity $\boldsymbol{v}$ as contour lines (contour spacing of $0.5\,\mathrm{m^2\,s^{-1}}$) for three subsequent times as indicated in the panels

with an idealized (two-dimensional) primitive equation model, using the parameterization (12.32) in a zonally periodic domain for three subsequent times. There is no external forcing, all fields are zonally constant, and $K_\ell = 2{,}000\,\mathrm{m^2\,s^{-1}}$ was chosen. The left panel of Figure 12.8 shows the initial condition, resembling a narrow front with a meridional scale of about 100 km. In the ocean, we expect this front to become baroclinically unstable such that mesoscale eddies will grow, i. e. the mean available potential energy stored in the sloping front will be released to eddy energy.

In the model, there are actually no mesoscale eddies (because of the zonal homogeneity), but the effect of the parameterization is indeed what we expect: At the surface above the front, a northward eddy-driven velocity of $O(10\,\mathrm{cm\,s^{-1}})$ develops, compensated by a southward velocity at the bottom and by up- and downward velocities at the southern and northern flanks of the front. Note that $B = 0$ at the upper and lower boundary was assumed. The effect is a flattening of the sloping isopycnals, and the final stage is motionless without any available potential energy. Using the parameterization (12.31) in the conservation equation of mean potential energy (12.15) yields for the production term $P = -\overline{v'b'}(\partial\bar{b}/\partial y)/N^2$ (equivalent for the zonally averaged case), indeed a sign definite energy sink $P = K_\ell(\partial\bar{b}/\partial y)^2/N^2 > 0$, which reduces the mean available potential energy.

Since there is no mixing of density involved in the parameterization ($K_\mathrm{d} = 0$ was assumed), the area between adjacent isolines of $\bar{b}$ in the entire model domain stays constant. But at the surface and the bottom, the thickness of the uppermost and lowermost layers is meridionally homogenized, i. e. mixed. Therefore, the effect of the parameterization is sometimes called isopycnal thickness mixing, and $K_\ell$ is accordingly called the *isopycnal thickness diffusivity* (cf. Section 12.3.5). Note that a value of $K_\ell$ of $O(1{,}000\,\mathrm{m^2\,s^{-1}})$ fits the time-scale of the flattening of isopycnals by baroclinic instability quite well and is, therefore, chosen in numerical models.

## 12.2.5 Isopycnal Mixing and Transformed Eulerian Mean

A tracer with gradients on isopycnals will be mixed along isopycnals, which differs from the turbulent mixing of buoyancy where this isopycnal mixing effect vanishes (except for the small diapycnal effects discussed in the box on p. 384). Therefore, the

**56. Transformed Eulerian Mean in Three Dimensions**

While the zonal mean case is somewhat easier to handle algebraically, all results in the previous section can be generalized to the three-dimensional case. For a three-dimensional eddy buoyancy flux $\boldsymbol{F}_\mathrm{b} = \overline{\boldsymbol{u}'b'}$ the decomposition (12.26) becomes

$$\boldsymbol{F}_\mathrm{b} = -K_\mathrm{d}\boldsymbol{\nabla}\overline{b} + \boldsymbol{B} \times \boldsymbol{\nabla}\overline{b} \tag{B56.1}$$

Note that the overbar denotes in this box a temporal mean and that all vectors are three-dimensional. The diapycnal flux component is again neglected, with the same arguments as in the two-dimensional case. In the mean buoyancy equation, the divergence of the eddy buoyancy flux is given as

$$\boldsymbol{\nabla} \cdot \boldsymbol{F}_\mathrm{b} = \boldsymbol{\nabla} \cdot (\boldsymbol{B} \times \boldsymbol{\nabla}\overline{b}) = (\boldsymbol{\nabla} \times \boldsymbol{B}) \cdot \boldsymbol{\nabla}\overline{b} \equiv \boldsymbol{v} \cdot \boldsymbol{\nabla}\overline{b} \tag{B56.2}$$

introducing the three-dimensional eddy-driven velocity $\boldsymbol{v} = \boldsymbol{\nabla} \times \boldsymbol{B}$. The eddy effect is again of advective nature, this time expressed by the three-dimensional eddy-driven velocity which is given in terms of a vector stream function $\boldsymbol{B}$ for which

$$\boldsymbol{F}_\mathrm{b} \times \boldsymbol{\nabla}\overline{b} = (\boldsymbol{B} \times \boldsymbol{\nabla}\overline{b}) \times \boldsymbol{\nabla}\overline{b} = -\boldsymbol{\nabla}\overline{b} \times (\boldsymbol{B} \times \boldsymbol{\nabla}\overline{b}) = -\boldsymbol{B}(\boldsymbol{\nabla}\overline{b} \cdot \boldsymbol{\nabla}\overline{b}) + \boldsymbol{\nabla}\overline{b}(\boldsymbol{B} \cdot \boldsymbol{\nabla}\overline{b})$$

holds and which can be determined using the gauge condition $\boldsymbol{B} \cdot \boldsymbol{\nabla}\overline{b} = 0$, thus

$$\begin{aligned} \boldsymbol{B} = -|\boldsymbol{\nabla}\overline{b}|^{-2}\boldsymbol{F}_\mathrm{b} \times \boldsymbol{\nabla}\overline{b} &= -|\boldsymbol{\nabla}\overline{b}|^{-2} \begin{pmatrix} \overline{v'b'}\partial\overline{b}/\partial z - \overline{w'b'}\partial\overline{b}/\partial y \\ \overline{w'b'}\,\partial\overline{b}/\partial x - \overline{u'b'}\partial\overline{b}/\partial z \\ \overline{u'b'}\partial\overline{b}/\partial y - \overline{v'b'}\partial\overline{b}/\partial x \end{pmatrix} \\ &\approx -\left(\frac{\partial\overline{b}}{\partial z}\right)^{-2} \begin{pmatrix} \overline{v'b'}\partial\overline{b}/\partial z \\ -\overline{u'b'}\partial\overline{b}/\partial z \\ \overline{u'b'}\partial\overline{b}/\partial y - \overline{v'b'}\partial\overline{b}/\partial x \end{pmatrix} \end{aligned}$$

For the last step, the assumption $|\partial\overline{b}/\partial z| >> |\boldsymbol{\nabla}_\mathrm{h}\overline{b}|$ (small aspect ratio) was used. The standard downgradient parameterization for the horizontal eddy flux $\overline{\boldsymbol{u}_\mathrm{h}'b'} = -K_\ell\boldsymbol{\nabla}_\mathrm{h}\overline{b}$ leads to the three-dimensional version of the parameterization by Gent and McWilliams (1990) for the eddy-driven velocity

$$\boldsymbol{\nabla} \times \boldsymbol{B} = \begin{pmatrix} -\partial(K_\ell s_1)/\partial z \\ -\partial(K_\ell s_2)/\partial z \\ \boldsymbol{\nabla}_\mathrm{h} \cdot (K_\ell \boldsymbol{s}) \end{pmatrix} \tag{B56.3}$$

where $\boldsymbol{s} = (s_1, s_2)$ denotes the isopycnal slope vector as in (12.23). The eddy-driven velocity is added to the Eulerian mean velocity $\overline{\boldsymbol{u}}$ to obtain the residual velocity $\boldsymbol{u}^* = \overline{\boldsymbol{u}} + \boldsymbol{\nabla} \times \boldsymbol{B}$, which then advects the mean buoyancy $\overline{b}$ in the three-dimensional mean buoyancy equation analogous to (12.29).

TEM concept for buoyancy is slightly extended here to account for isopycnal mixing. Consider the zonally averaged conservation equation for a tracer concentration $T$,

$$\frac{\partial\overline{T}}{\partial t} + \overline{\boldsymbol{v}} \cdot \nabla\overline{T} + \nabla \cdot \overline{\boldsymbol{v}'T'} = \overline{\mathcal{G}}_\mathrm{T}$$

where sources and sinks of the tracer are denoted by $\mathcal{G}_\mathrm{T}$. Following the TEM framework, the tracer eddy flux $(\overline{\boldsymbol{v}'T'})$ is decomposed into components along and across isolines of the mean tracer $\overline{T}$, which yields analogous to (12.28)

$$\frac{\partial\overline{T}}{\partial t} + (\overline{\boldsymbol{v}} - \underset{\neg}{\nabla} B_\mathrm{T}) \cdot \nabla\overline{T} = \nabla \cdot K_\mathrm{T}\nabla\overline{T} + \overline{\mathcal{G}}_\mathrm{T} \tag{12.33}$$

where $B_\mathrm{T}$ and $K_\mathrm{T}$ take analogous physical meaning and definitions as $B$ and $K_\mathrm{d}$ in (12.28). It is clear that the eddy-driven velocity $-\underset{\neg}{\nabla} B_\mathrm{T}$ will not be identical to

**57. Diapycnal Mixing by Mesoscale Eddies**

In the Gent and McWilliams parameterization, discussed in Section 12.2.4, the diapycnal diffusivity $K_\mathrm{d}$ in (12.29) is set to zero, i. e. it is assumed that the eddy buoyancy fluxes $\boldsymbol{F}_\mathrm{b}$ are always parallel to the local isopycnal plane. Note that only the component of $\boldsymbol{F}_\mathrm{b}$ perpendicular to the isopycnal leads to a nonzero $K_\mathrm{d}$. It will be shown in Section 12.3.1 that only vanishing diapycnal terms, summarized in $\bar{\mathcal{G}}_\mathrm{b}$ in the mean buoyancy equation (12.25), will lead to $K_\mathrm{d} = 0$ in steady state. Therefore, there is no real justification to set $K_\mathrm{d} = 0$ in the ocean since it is known that $\bar{\mathcal{G}}_\mathrm{b} \neq 0$ (even if one neglects the nonlinearity equation of state leading to diapycnal mixing as discussed in the box on p. 384). In the interior of the ocean, $\bar{\mathcal{G}}_\mathrm{b}$ is in general rather small such that even small diapycnal diffusivities related to mesoscale eddy mixing might be of importance for the large-scale circulation.

It was discussed in Section 12.2.4 that the Gent and McWilliams parameterization represents a sink of potential energy of the mean flow. This energy is transferred by baroclinic instability to eddy energy if the system (or the model) under consideration contains mesoscale eddy dynamics. If it does not contain this dynamical regime, the released energy of the parameterization is simply lost. Tandon and Garrett (1996) noted that the further fate of the energy released by the parameterization might be involved in mixing. In the ocean, the kinetic and potential mesoscale eddy energy is either transferred to the mean flow, as discussed for instance in the box on p. 380, or it is dissipated. The latter means that the energy is again transferred to the smaller-scale dynamical regime, i. e. that of small-scale, stratified turbulence discussed in Chapter 11.3 and the internal wave field discussed in Section 10.4.5. When this turbulent kinetic energy (TKE) related to small-scale turbulence is finally dissipated, there will be density (diapycnal) mixing involved. However, the mechanism of the energetic transfer from the geostrophically balanced large-scale and mesoscale ocean circulation to smaller scale motions, are unclear at the moment (Klein et al., 2008).

The discussion in Section 11.3.3 shows that in the interior of the ocean about 25% of the dissipated TKE is used for density mixing (this ratio is also called the mixing efficiency). Tandon and Garrett (1996) speculated that this energy might represent, in fact, a substantial fraction of the energy needed to sustain the global overturning circulation. Assuming an even lower mixing efficiency and local dissipation of the mesoscale energy, Eden and Greatbatch (2008) demonstrated that in energetic regions of the ocean, e. g. in western boundary currents, the dissipated mesoscale energy is related to diffusivities of a greater or equal order of magnitude as the vertical diffusivities in other small-scale mixing processes.

the eddy-driven velocity $-\underset{\neg}{\nabla} B$ in the buoyancy equation (12.28). The same holds for the diffusivities $K_\mathrm{T}$ and $K_\mathrm{d}$. On the other hand, it is more convenient to have only a single residual velocity in all tracer and buoyancy equations, namely that for buoyancy $\boldsymbol{v}^* = \overline{\boldsymbol{v}} - \underset{\neg}{\nabla} B$. Rewriting (12.33) accordingly yields

$$\frac{\partial \overline{T}}{\partial t} + \boldsymbol{v}^* \cdot \nabla \overline{T} = \nabla \cdot K_\mathrm{T} \nabla \overline{T} - \underset{\neg}{\nabla}(B - B_\mathrm{T}) \cdot \nabla \overline{T} + \overline{\mathcal{G}}_\mathrm{T} \tag{12.34}$$

In order to make the connection of the TEM framework with isopycnal mixing of a tracer, the right-hand side of (12.34) should now be written as isopycnal and diapycnal diffusion, i. e. as

$$\frac{\partial \overline{T}}{\partial t} + \boldsymbol{v}^* \cdot \nabla \overline{T} = \nabla \cdot \left(\mathsf{K}_\mathrm{i} \nabla \overline{T}\right) + \nabla \cdot \left(\mathsf{K}_\mathrm{d} \nabla \overline{T}\right) + \overline{\mathcal{G}}_\mathrm{T} \tag{12.35}$$

where $\mathsf{K}_\mathrm{i}$ and $\mathsf{K}_\mathrm{d}$ are the two-dimensional versions of the tensors defined in (12.21). Comparing the right-hand sides of (12.34) and (12.35) leads to the system

$$\begin{pmatrix} K_\mathrm{T} & -(B - B_\mathrm{T}) \\ B - B_\mathrm{T} & K_\mathrm{T} \end{pmatrix} \nabla \overline{T} = \frac{K_\mathrm{i}}{1 + s^2} \begin{pmatrix} 1 & s \\ s & s^2 \end{pmatrix} \nabla \overline{T} + \frac{K_\mathrm{d}}{1 + s^2} \begin{pmatrix} s^2 & -s \\ -s & 1 \end{pmatrix} \nabla \overline{T}$$

where $K_\mathrm{i}$ denotes isopycnal diffusivity, $K_\mathrm{d}$ diapycnal diffusivity, and $s = -e_2/e_3$ the slope of the mean isopycnals. Solving for $K_\mathrm{i}$ and $K_\mathrm{d}$ yields after some manipulation

$$K_\mathrm{i} = K_\mathrm{T} - (B - B_\mathrm{T})\frac{1+st}{s-t} = K_\mathrm{T} - \frac{B - B_\mathrm{T}}{\tan\phi} \tag{12.36}$$

$$K_\mathrm{d} = K_\mathrm{T} + (B - B_\mathrm{T})\frac{s-t}{1+st} = K_\mathrm{T} + (B - B_\mathrm{T})\tan\phi \tag{12.37}$$

where $t = -(\partial\overline{T}/\partial y)/(\partial\overline{T}/\partial z)$ denotes the slope of mean tracer contours and where $\phi$ is the angle between the gradients of $\bar{T}$ and $\bar{b}$ (or the angle between isopycnals and isolines of the mean tracer). Note that there is a singularity for $t = s$ or $\phi = 0$, but in that case isopycnals and tracer isolines coincide, and isopycnal diffusion is then meaningless, i.e. the value of $K_\mathrm{I}$ is not relevant anymore. Note also that the result (12.36) and (12.37) carries over to the three-dimensional case as discussed by Eden and Greatbatch (2009) although one has to account for anisotropic mixing in the isopycnal plane, which complicates the algebra considerably.

If one assumes that slopes of tracers and buoyancy are small in the ocean interior, specifically that $|st| \ll 1$, and that $B - B_\mathrm{T}$ is larger than or at least of the same order of magnitude as $K_\mathrm{T}$, the following expression will be a good approximation to (12.36)

$$K_\mathrm{i} \approx \frac{B - B_\mathrm{T}}{t - s} \tag{12.38}$$

In other words, in the interior of the ocean, the isopycnal diffusivity is approximately given by the difference in the stream functions for eddy-induced velocities of tracer and buoyancy divided by the difference in their slopes. In numerical models that do not resolve mesoscale eddy activity, the isopycnal diffusivity $K_\mathrm{i}$ has to be parameterized, and a value identical to the thickness diffusivity $K_\ell$ is often used. When the Gent and McWilliams parameterization as in Section 12.2.4 is used for $B = -\overline{v'b'}/(\partial\bar{b}/\partial z) = -sK_\ell$ and also for $B_\mathrm{T} = -tK_\ell$, using identical thickness diffusivities for buoyancy and tracer, (12.38) indeed yields $K_\mathrm{i} = K_\ell$.

### 12.2.6 * Mesoscale Eddy Effects in the Momentum Equation

In this section, we discuss the effects of mesoscale eddies in the momentum equation and possibilities for parameterizations. The equations of motion in the Boussinesq approximation (4.50) can be written as

$$\frac{\partial \boldsymbol{u}_\mathrm{h}}{\partial t} + \boldsymbol{u}_\mathrm{h}\cdot\nabla_\mathrm{h}\boldsymbol{u}_\mathrm{h} + f\,\underline{\boldsymbol{u}}_\mathrm{h} = -\nabla_\mathrm{h}p + \mathcal{F}_\mathrm{h} \tag{12.39}$$

The vertical advection of horizontal momentum is assumed to be small and was neglected in (12.39), and $p$ denotes pressure divided by the reference density $\rho_0$. In this section, we return from the zonal mean case (compare the box on p. 385) to a three-dimensional configuration; the symbols $\boldsymbol{u}_\mathrm{h} = (u, v)$ and $\nabla_\mathrm{h} = (\partial/\partial x, \partial/\partial y)$ denote the horizontal velocity and the horizontal gradient operator, respectively, and the subscript $\underline{\;\cdot\cdot\;}$ denote in this section rotation by 90° in the horizontal plane, i. e. $\underline{\boldsymbol{u}}_\mathrm{h} = (-v, u)$. Compare also with the box on p. 444.

Taking the mean of the momentum equation as in Section 12.2.1 yields

$$\frac{\partial \overline{\boldsymbol{u}}_\mathrm{h}}{\partial t} + \overline{\boldsymbol{u}}_\mathrm{h}\cdot\nabla_\mathrm{h}\overline{\boldsymbol{u}}_\mathrm{h} + f\,\underline{\overline{\boldsymbol{u}}}_\mathrm{h} = -\nabla_\mathrm{h}\overline{p} - \nabla_\mathrm{h}\cdot\overline{\boldsymbol{u}'_\mathrm{h}\boldsymbol{u}'_\mathrm{h}} + \overline{\mathcal{F}}_\mathrm{h} \tag{12.40}$$

The overbar $\overline{(\ldots)}$ denotes a statistical or time mean (in contrast to the zonal mean as used in Section 12.2.3). Note the appearance of the mesoscale eddy momentum fluxes $\overline{\boldsymbol{u}_h'\boldsymbol{u}_h'}$, which have to be parameterized when mesoscale eddies are not resolved by a model. It turns out that the straightforward parameterization of the form $\overline{\boldsymbol{u}_h'\boldsymbol{u}_h'} = -K_m \nabla_h \overline{\boldsymbol{u}}_h$ fails in the context of geostrophic turbulence, since eddy momentum fluxes do not have a clear relation to the mean momentum gradient and are even often upgradient, which would imply a negative lateral viscosity $K_m$ (Starr, 1968). It was often suggested (Gent and McWilliams, 1996; Marshall, 1981; Treguier et al., 1997; Welander, 1973) that a better means to parameterize the eddy momentum fluxes is given by considering eddy potential vorticity fluxes, which are excepted to show a relation to the mean potential vorticity gradient.

Consider the quasi-geostrophic form of potential vorticity, which is, as discussed in Section 5.2, given by $Q = \beta y + \underline{\nabla} \cdot \boldsymbol{u}_h + \partial/\partial z(b f_0/N^2)$, with the relative vorticity $\underline{\nabla} \cdot \boldsymbol{u}_h$ and the perturbation buoyancy $b$ and background stratification $N^2$ as defined in Section 12.2.1. The eddy potential vorticity flux $\overline{\boldsymbol{u}_h' Q'}$ can be written as

$$\overline{\boldsymbol{u}_h' Q'} = \overline{\boldsymbol{u}_h' \underline{\nabla} \cdot \boldsymbol{u}_h'} + \overline{\boldsymbol{u}_h' \frac{\partial}{\partial z} \frac{b' f_0}{N^2}} = \underline{\nabla} \cdot \overline{\boldsymbol{u}_h' \boldsymbol{u}_h'} + f_0 \frac{\partial}{\partial z}\left(\frac{\overline{\boldsymbol{u}_h' b'}}{N^2}\right) - \underline{\nabla}\left(\frac{\overline{b'^2}}{2N^2}\right) \tag{12.41}$$

where the geostrophic relation $f_0 \partial \boldsymbol{u}_h'/\partial z = \underline{\nabla} b'$ was used in the second step – valid within the quasi-geostrophic approximation – and where the identity $\underline{\nabla} \cdot \overline{\boldsymbol{u}_h' \boldsymbol{u}_h'} = \overline{\boldsymbol{u}_h' \underline{\nabla} \cdot \boldsymbol{u}_h'}$ was used. The relation (12.41) between the fluxes of eddy momentum, buoyancy and potential vorticity is sometimes called the *Taylor identity* (Taylor, 1915). The last term in the Taylor identity vanishes when taking the horizontal divergence of the eddy potential vorticity flux, i. e. it has no effect for the mean potential vorticity equation of the form (5.32) written here for vanishing $\mathcal{F}_h$ and $\mathcal{G}_\rho$ as

$$\frac{\partial \overline{Q}}{\partial t} + \overline{\boldsymbol{u}}_h \cdot \nabla_h \overline{Q} = -\nabla_h \cdot \overline{\boldsymbol{u}_h' Q'} \tag{12.42}$$

With $\overline{\underline{\boldsymbol{u}}_h' \underline{\nabla} \cdot \boldsymbol{u}_h'} = \nabla_h \cdot \overline{\boldsymbol{u}'_h \boldsymbol{u}_h'} - \nabla_h (\overline{|\boldsymbol{u}_h'|^2})/2$ the Taylor identity (12.41) can also be written as

$$\overline{\underline{\boldsymbol{u}}_h' Q'} = \nabla_h \cdot \overline{\boldsymbol{u}_h' \boldsymbol{u}_h'} - f_0 \underline{\boldsymbol{u}}_h^e + \nabla_h \left(\frac{\overline{b'^2}}{2N^2} - \frac{\overline{|\boldsymbol{u}_h'|^2}}{2}\right) \tag{12.43}$$

where we have replaced the second term on the right-hand side of (12.41), related to the eddy buoyancy fluxes, as the Coriolis force of the eddy-driven advection velocity $\boldsymbol{u}^e = \nabla \times \boldsymbol{B}$ given by the box on p. 389. Using (12.43) to replace the divergence of the eddy momentum flux, the mean momentum equation (12.40) becomes

$$\frac{\partial \overline{\boldsymbol{u}}_h}{\partial t} + \overline{\boldsymbol{u}}_h \cdot \nabla_h \overline{\boldsymbol{u}}_h + f \underline{\overline{\boldsymbol{u}}}_h^* = -\nabla_h \overline{p} + \nabla_h \left(\frac{\overline{b'^2}}{2N^2} - \frac{\overline{|\boldsymbol{u}_h'|^2}}{2}\right) - \overline{\underline{\boldsymbol{u}}_h' Q'} + \overline{\mathcal{F}}_h \tag{12.44}$$

Note that the Coriolis force in the momentum equation (12.44) is related to the residual momentum $\overline{\boldsymbol{u}}^* = \overline{\boldsymbol{u}} + \nabla \times \boldsymbol{B}$. It is often convenient to rewrite the momentum equation in this form, which will be further discussed for instance in Section 16.4 of the dynamics of the Southern Ocean for an example.

**58. The Eliassen–Palm Flux and the Non-acceleration Theorem**

Taking a zonal average of (12.44), indicated in this box by $\langle .. \rangle$, and assuming zonally periodic boundary conditions, which is the case for the atmosphere or the zonally unbounded part of the Southern Ocean, the zonal component of (12.44) becomes

$$\frac{\partial}{\partial t}\langle \bar{u} \rangle + \frac{\partial}{\partial y}\langle \overline{v u} \rangle - f\langle \overline{v}^* \rangle = \langle \overline{v' Q'} \rangle + \langle \overline{\mathcal{F}_u} \rangle \quad \text{(B58.1)}$$

since the zonal average of all zonal derivatives vanish, e. g. $\langle \partial \overline{p}/\partial x \rangle = 0$, because of the zonal periodicity. Since there is no zonal pressure gradient, the zonally averaged meridional geostrophic velocity vanishes, and consequently, the advective term on the left-hand side of (B58.1) vanishes in quasi-geostrophic approximation (in fact, only $\langle \overline{v} \rangle \langle \overline{u} \rangle$ vanishes and it is possible that there is a so-called standing eddy momentum flux, given by $\langle \overline{v u} \rangle - \langle \overline{v} \rangle \langle \overline{u} \rangle$; this complication is further discussed in the box on p. 563).
In steady state, the zonally averaged meridional residual velocity $\langle \overline{v}^* \rangle$ is thus balanced to leading order only by the meridional eddy potential vorticity flux $\langle \overline{v' Q'} \rangle$ or the small-scale forcing $\langle \overline{\mathcal{F}_u} \rangle$. When the former vanishes, the residual velocity $\langle \overline{v}^* \rangle$ is only driven by $\langle \overline{\mathcal{F}_u} \rangle$, which is also often very small in the interior of the ocean. This situation, i. e. a vanishing residual circulation is sometimes called the non-acceleration condition. The mesoscale eddy forcing in the zonally averaged zonal momentum equation is given by $\langle \overline{v' Q'} \rangle = -\partial \langle \overline{v' u'} \rangle / \partial y + \partial \langle f_0 \overline{v' b'}/N^2 \rangle / \partial z$, which is the divergence of the Eliassen–Palm (EP) flux vector $(-\langle \overline{v' u'} \rangle, \langle f_0 \overline{v' b'}/N^2 \rangle)$ (Andrews and McIntyre, 1976).
The zonal average of the quasi-geostrophic potential vorticity equation (12.42) becomes

$$\frac{\partial \langle \overline{Q} \rangle}{\partial t} = -\frac{\partial}{\partial y}\langle \overline{v' Q'} \rangle$$

Thus for steady state and vanishing small-scale forcing, the EP flux divergence is zero and in consequence, the residual meridional circulation vanishes if the small-scale forcing $\langle \overline{\mathcal{F}_u} \rangle$ is absent. This statement is called the *non-acceleration theorem* (Andrews and McIntyre, 1976).

In the eddy potential vorticity flux, the term related to the eddy momentum flux is often much smaller than the term related to the eddy buoyancy flux. Neglecting, therefore, the former and using for the eddy buoyancy flux a downgradient parameterization as in the Gent and McWilliams (1990) parameterization, yields in (12.44)

$$-\overline{\underline{u}'_h Q'} \approx f_0 \frac{\partial}{\partial z}\left( \frac{K_\ell \underline{\nabla} \overline{b}}{N^2} \right) \approx \frac{\partial}{\partial z}\left( K_v \frac{\partial \overline{\underline{u}}_h}{\partial z} \right) \quad (12.45)$$

where the geostrophic relation $f_0 \partial \underline{u}_h / \partial z = \underline{\nabla} b$ was used in the second step and with $K_v = K_\ell f^2/N^2$. The eddy term is converted to an apparent vertical friction term with the apparent vertical viscosity $K_v$ which is usually much larger than the vertical viscosities related to small-scale mixing in the interior of the ocean. Note that the same result as (12.46) can be obtained by adding the Coriolis force related to the eddy-driven velocity on both sides of (12.40), neglecting the eddy momentum fluxes and making the same replacement using the Gent and McWilliams (1990) parameterization for the eddy-driven velocity (Greatbatch and Lamb, 1990).

For small Rossby number (or within the limits of the quasi-geostrophic approximation) it is possible to replace in (12.44) the Eulerian Mean velocity $\overline{\underline{u}}_h$ with the residual mean velocity $\overline{\underline{u}}^*_h$. Equation (12.46) with (12.45) implemented (and the gradient force ignored),

$$\begin{aligned}\frac{\partial \overline{\boldsymbol{u}}_{\mathrm{h}}^{*}}{\partial t}+\overline{\boldsymbol{u}}_{\mathrm{h}}^{*}\cdot\nabla_{\mathrm{h}}\overline{\boldsymbol{u}}_{\mathrm{h}}^{*}+f\underline{\overline{\boldsymbol{u}}}_{\mathrm{h}}^{*}&=-\nabla_{\mathrm{h}}\overline{p}+f_0\frac{\partial}{\partial z}\left(\frac{K_\ell\underline{\nabla}\overline{b}}{N^2}\right)+\overline{\mathcal{F}}_{\mathrm{h}}\\&=-\nabla_{\mathrm{h}}\overline{p}+\frac{\partial}{\partial z}\left(K_{\mathrm{v}}\frac{\partial\overline{\boldsymbol{u}}_{\mathrm{h}}^{*}}{\partial z}\right)+\overline{\mathcal{F}}_{\mathrm{h}}\end{aligned}\tag{12.46}$$

is then called the *residual mean momentum equation*. This formulation is convenient for a model, since it predicts the velocity variable which is needed in the mean buoyancy equation, i. e. the residual velocity $\overline{\boldsymbol{u}}^{*}$. It is then only necessary to implement the parameterization for the effect of mesoscale eddy activity in the momentum equation, where it appears in this version simply as a vertical momentum flux divergence. This form of the momentum equation will be further discussed in Section 16.4 to construct simple dynamical models of the Southern Ocean.

In the form (12.44) of the momentum equation, the effect of mesoscale eddy activity is given by the horizontal eddy potential vorticity flux $\overline{\underline{\boldsymbol{u}}_{\mathrm{h}}' Q'}$ and a force related to the horizontal gradient of eddy available potential energy $\overline{b'^2}/(2N^2)$ and eddy kinetic energy $\overline{|\boldsymbol{u}_{\mathrm{h}}'|^2}/2$. Formally, this gradient stands on the same footing as the pressure gradient and consequently, it can be replaced a Coriolis-type force, defining an additional horizontal velocity $\boldsymbol{u}_{\mathrm{h}}^{(\mathrm{rot})}$ by $\nabla_{\mathrm{h}}\left(\overline{b'^2}/(2N^2)-\overline{|\boldsymbol{u}_{\mathrm{h}}'|^2}/2\right)=-f_0\underline{\boldsymbol{u}}_{\mathrm{h}}^{(\mathrm{rot})}$. Note that $\boldsymbol{u}_{\mathrm{h}}^{(\mathrm{rot})}$ is divergence-free and thus has no vertical counterpart. Therefore, it has no relevance for the quasi-geostrophic form of the buoyancy equation (12.11), given in averaged form as

$$\frac{\partial\overline{b}}{\partial t}+\overline{\boldsymbol{u}}_{\mathrm{h}}\cdot\nabla_{\mathrm{h}}\overline{b}+(\overline{w}+w^e)N^2=\overline{\mathcal{G}_{\mathrm{b}}}\tag{12.47}$$

with the vertical component of the eddy-driven velocity $w^e=\underline{\nabla}\cdot\boldsymbol{B}_{\mathrm{h}}=\nabla_{\mathrm{h}}\cdot\overline{\boldsymbol{u}_{\mathrm{h}}'b'}/N^2$ (compare also the box on p. 403). The gradient force and the related velocity $\boldsymbol{u}_{\mathrm{h}}^{(\mathrm{rot})}$ is also irrelevant for the mean potential vorticity equation. One might, therefore, argue that the gradient force in the momentum equation (12.44) is irrelevant and can be ignored. In fact we will show next, that the gradient force can be canceled by introducing a properly chosen rotational component in the eddy potential vorticity flux.

Let us consider the consequences of the following decomposition of the eddy potential vorticity flux,

$$\overline{\boldsymbol{u}_{\mathrm{h}}'Q'}=-K_q\nabla_{\mathrm{h}}\overline{Q}+\nu\underline{\nabla}\overline{Q}+\underline{\nabla}\theta\tag{12.48}$$

that has a downgradient term, a component directed perpendicular to the gradient of the mean potential vorticity, and a purely rotational flux component. It is immediately evident that the choice $\theta=\overline{|\boldsymbol{u}_{\mathrm{h}}'|^2}/2-\overline{b'^2}/(2N^2)$ cancels the above discussed gradient force from the momentum balance (12.44). We consider two limiting cases. The first case, $\nu=0$, is presented in the box on p. 395. The second case, $K_q=0$, follows.

When the small-scale forcing $\mathcal{F}_{\mathrm{h}}$ and $\mathcal{G}_\rho$ in the potential vorticity equation (12.42) vanishes, there is good reason to assume that in steady state the component of the

**59. A Simple Parameterization for Eddy Momentum Fluxes**

The relevant case for the ocean is where the small-scale forcing $\mathcal{F}_\mathrm{h}$ and $\mathcal{G}_\rho$ does not vanish in the vorticity equation (12.42). For that case, we expect a significant downgradient component of the eddy potential vorticity flux (compare also Section 12.3.3). We, therefore, use for the eddy potential vorticity flux a downgradient parameterizations of the form $\overline{\boldsymbol{u}_\mathrm{h}' Q'} = -K_q \boldsymbol{\nabla}_\mathrm{h}\bar{Q} + \underline{\boldsymbol{\nabla}}\theta$. We have accounted for a rotational flux component $\underline{\boldsymbol{\nabla}}\theta$, which has no effect in (12.42) which becomes

$$\frac{\partial \overline{Q}}{\partial t} + \overline{\boldsymbol{u}}_\mathrm{h} \cdot \boldsymbol{\nabla}_\mathrm{h}\overline{Q} = \boldsymbol{\nabla}_\mathrm{h} \cdot K_q \boldsymbol{\nabla}_\mathrm{h}\overline{Q}$$

Therefore, based only on the mean potential vorticity equation, we have freedom to choose $\theta$. However, $\theta$ does figure in the momentum equation (12.40) which becomes

$$\frac{\partial \overline{\boldsymbol{u}}_\mathrm{h}}{\partial t} + \overline{\boldsymbol{u}}_\mathrm{h} \cdot \boldsymbol{\nabla}_\mathrm{h}\overline{\boldsymbol{u}}_\mathrm{h} + f\underline{\overline{\boldsymbol{u}}}_\mathrm{h}^* = -\boldsymbol{\nabla}_\mathrm{h}\overline{p} + K_q \underline{\boldsymbol{\nabla}}\bar{Q} + \boldsymbol{\nabla}_\mathrm{h}\left(\theta + \frac{\overline{b'^2}}{2N^2} - \frac{\overline{|\boldsymbol{u}_\mathrm{h}'|^2}}{2}\right) + \overline{\mathcal{F}}_\mathrm{h} \tag{B59.1}$$

In (B59.1) several new forces related to the mean potential vorticity gradient

$$K_q \underline{\boldsymbol{\nabla}}\overline{Q} = K_q \left[-\beta \boldsymbol{i} + \nabla_\mathrm{h}^2 \overline{\boldsymbol{u}}_\mathrm{h} + \frac{\partial}{\partial z}\left(\frac{f_0^2}{N^2}\frac{\partial \overline{\boldsymbol{u}}_\mathrm{h}}{\partial z}\right)\right]$$

appear which need interpretation. For constant lateral diffusivity $K_q$ and $\beta$, the first term $-K_q \beta \boldsymbol{i}$ yields a constant westward force. This force was introduced by the parameterization of the eddy momentum fluxes. However, in a global integral of (B59.1), the effect of the eddy momentum fluxes vanish, and the same must hold for any parameterization of the eddy momentum fluxes (Bretherton, 1966; Eden, 2010). Therefore, we have to adjust the free variable $\theta$ in the parameterization such that the integral of the force $-K_q \beta \boldsymbol{i}$ is balanced, and the same for all other forces in (B59.1) introduced by the parameterization. For constant $K_q$ and $\beta$, the force $-K_q \beta \boldsymbol{i}$ thus vanishes completely, and the momentum equation (B59.1) can be formulated in the more familiar form

$$\frac{\partial \overline{\boldsymbol{u}}_\mathrm{h}^*}{\partial t} + \overline{\boldsymbol{u}}_\mathrm{h}^* \cdot \boldsymbol{\nabla}_\mathrm{h}\overline{\boldsymbol{u}}_\mathrm{h}^* + f\underline{\overline{\boldsymbol{u}}}_\mathrm{h}^* = -\boldsymbol{\nabla}_\mathrm{h}\overline{p} + \boldsymbol{\nabla} \cdot K_q \boldsymbol{\nabla}\overline{\boldsymbol{u}}_\mathrm{h}^* + \frac{\partial}{\partial z}\left(K_\mathrm{v}\frac{\partial \overline{\boldsymbol{u}}_\mathrm{h}^*}{\partial z}\right) + \overline{\mathcal{F}}_\mathrm{h} \tag{B59.2}$$

with $K_\mathrm{v} = K_q f_0^2/N^2$. In (B59.2) the mesoscale eddy effect is given by horizontal and vertical friction, predominantly balanced by the mean pressure gradient and the Coriolis force related to the residual velocity $\boldsymbol{u}^*$ (assuming that the mean advection term $\overline{\boldsymbol{u}}_\mathrm{h} \cdot \boldsymbol{\nabla}_\mathrm{h}\overline{\boldsymbol{u}}_\mathrm{h}$ remains small). Note that the global integral of the frictional terms (except for the small-scale friction $\overline{\mathcal{F}}_\mathrm{h}$) has to vanish by the choice of $\theta$, which is not explicitly indicated in (B59.2).

eddy potential vorticity flux across isolines of mean potential vorticity, i. e. the downgradient component, vanishes as well, except for a purely rotational flux. This general property of eddy fluxes will be shown analytically in Section 12.3.1. Therefore, we assume that for the case of vanishing small-scale forcing $\mathcal{F}_\mathrm{h}$ and $\mathcal{G}_\rho$, the eddy potential vorticity fluxes can be described by (12.48) with $K_q = 0$. The steady mean potential vorticity equation (12.42) then becomes

$$\underline{\nabla}(\overline{\psi} - \nu) \cdot \nabla\overline{Q} = 0$$

with the quasi-geostrophic stream function $\psi$ with $\boldsymbol{u}_\mathrm{h} = \underline{\nabla}\psi$. Thus $\nu = \bar{\psi} + \Xi(\bar{Q})$ with some function $\Xi(x)$. The steady momentum equation becomes after some manipulations

$$f_0\left(1-\frac{\partial}{\partial z}\frac{\overline{b}}{N^2}\right)\underline{\overline{u}}_{\mathrm{h}}^{*}=-\nabla_{\mathrm{h}}\left(\overline{p}+\frac{|\overline{u}^{*}|^{2}}{2}\right) \tag{12.49}$$

with $f = f_0 + \beta y$ and setting $\theta = \overline{|u_{\mathrm{h}}'|^2}/2 - \overline{b'^2}/(2N^2) - \overline{Q}\,\overline{\psi} - F(\bar{Q})$ with $F'(x) = \Xi(x)$. Eddy terms have disappeared from (12.49) and a nonlinear balanced residual momentum equation results, but with a different Coriolis term. Equation (12.49) resembles nonacceleration condition for three-dimensional flow, in which the residual circulation does not vanish. Several other forms are also possible, but a statement that the residual circulation takes a special form for steady state and vanishing small-scale forcing as in the zonal mean case has not been found (Plumb, 1990).

## 12.3 * Alternative Averaging Frameworks

The general conservation equation for a tracer, $\mathrm{D}\chi/\mathrm{D}t = \mathcal{G}_\chi$, describes the changes $\mathcal{G}_\chi = \mathcal{G}$ of a property $\chi$ of a fluid particle during its way in a given flow[5]. If this flow is turbulent, a description of the average evolution of the property is often more useful than the instantaneous one. We have considered so far only the Eulerian Mean $\bar{\chi}$ of the Eulerian quantity $\chi(\boldsymbol{x}, t)$, which performs the average at a fixed location. In this section, we will discuss and compare also other averaging frameworks.

Using the Eulerian Mean framework tends to mix effects of advection and irreversible changes of the particle's properties given by $\mathcal{G}$, which then complicates or even sometimes inhibits a useful physical interpretation of the averaged conservation equation. This undesired mixing of physically different processes manifests itself in the occurrence of the turbulent (Eulerian Mean) eddy fluxes, as seen e. g. in (12.25) with $\chi$ as buoyancy. The eddy fluxes are usually difficult to interpret. In the TEM framework of Section 12.2.3, we have seen that the dominant part of the eddy buoyancy flux was related to the eddy-driven velocity, i. e. to an advective effect, instead to the diapycnal diffusivity $K_{\mathrm{d}}$. It will be shown below that in the TEM framework part of the eddy flux related to $K_{\mathrm{d}}$ is still unrelated to irreversible changes of properties (i. e. to mixing of properties) but to advective processes. The aim of the discussion in the present section is to connect only the irreversible changes of $\chi$ – which are related to $\mathcal{G}$ – to a turbulent diffusivity like $K_{\mathrm{d}}$. The differentiation between advective effects of turbulent mixing and irreversible changes becomes particularly important when the amount of diapycnal mixing by mesoscale eddies (cf. the box on p. 390) is to be determined. On the other hand, it is also of conceptual importance to separate both effects.

We will discuss in Section 12.3.1 how to decipher advective and diffusive processes in the Eulerian Mean eddy fluxes, by revisiting and modifying the TEM framework. This will be done by introducing and defining rotational eddy fluxes, which do not figure in the divergence of the eddy fluxes in the Eulerian Mean conservation equation, but which influence the estimates of turbulent diffusivity and the eddy-driven velocity. The physical meaning of the rotational eddy fluxes will be discussed in Section 12.3.2. The concept is called the Temporal Residual Mean (TRM) framework, and it relates the turbulent diffusivity to irreversible and temporal changes of

[5] We omit the index of the source term $\mathcal{G}_\chi$ in the following discussion.

60. Summary of Alternative Averaging Frameworks

- **Temporal Residual Mean**: Transport velocity is the Residual Mean velocity $\overline{\boldsymbol{u}}^*$ given by the sum of the Eulerian Mean $\overline{\boldsymbol{u}}$ plus the eddy-driven velocity $\overline{\boldsymbol{u}}^e$. The Residual velocity transports the Eulerian Mean property $\overline{\chi}$. The eddy-driven velocity is given by a stream function and differs for different $\overline{\chi}$.
- **Generalized Lagrangian Mean**: Transport velocity is the Lagrangian Mean velocity $\overline{\boldsymbol{u}}^{\mathrm{L}}$ given by the sum of the Eulerian Mean $\overline{\boldsymbol{u}}$ plus the Stokes velocity $\overline{\boldsymbol{u}}^{\mathrm{S}}$. The Lagrangian Mean velocity transports the Lagrangian Mean property $\overline{\chi}^{\mathrm{L}}$, given by the sum of the Eulerian Mean property $\overline{\chi}$ and the Stokes correction $\overline{\chi}^{\mathrm{S}}$. The Stokes velocity is in general divergent, but it is independent of the particular property $\overline{\chi}^{\mathrm{L}}$ under consideration. The Lagrangian Mean might not exists for specific cases.
- **Semi-Lagrangian (Isopycnal) Mean**: Transport velocity is $\overline{\boldsymbol{u}}^{\#}$ given by the sum of the Eulerian Mean $\overline{\boldsymbol{u}}$ plus the Quasi-Stokes velocity $\overline{\boldsymbol{u}}^{+}$. The velocity $\overline{\boldsymbol{u}}^{\#}$ transports the semi-Lagrangian Mean (or modified) property $\overline{\chi}^{\mathrm{I}}$. The Quasi-Stokes velocity is given by a stream function and differs for different $\overline{\chi}^{\mathrm{I}}$. If $\chi$ is specified as density in the definition of the semi-Lagrangian Mean, the procedure is also called Isopycnal Mean.

$\overline{\chi}$ and its statistical moments in a consistent way. In fact, it will be discussed in Section 12.3.3 how the concept leads to a generalized Osborn–Cox relation, which we met before in an approximate form in Section 11.3.3.

However, there are alternative methods for a consistent differentiation of the different physical processes, using a Lagrangian approach to redefine both the mean quantity and the transport velocity. We discuss the Generalized Lagrangian Mean in Section 12.3.4 and the semi-Lagrangian Mean in Section 12.3.5, which are compared with each other and the Temporal Residual Mean in Section 12.3.6. In the remainder of the present section, we briefly introduce the alternative averaging frameworks and summarize their most important aspects.

### Temporal Residual Mean

The Temporal Residual Mean (TRM) framework of Eden et al. (2007) is based on the Eulerian Mean, i. e. taking the average at a fixed location and represents a generalization of the TEM framework by Andrews et al. (1987) and the TRM[6] framework of McDougall and McIntosh (1996). The averaging operator can be a zonal average, a temporal average, or an statistical average. The TRM framework represents a modification of the TEM framework, in which rotational fluxes are introduced in the decomposition of the eddy fluxes which do not appear in the divergence of the eddy fluxes in the mean conservation equation, but which affect the eddy-driven velocity and the turbulent diffusivity. Furthermore, the rotational eddy fluxes show up in the conservation equations for the variance of the property $\chi$, which offers an opportunity for the choice of the rotational eddy fluxes as discussed in Section 12.3.1. It turns out that for an consistent choice, fluxes of higher order moments have to be

[6] Note that the name TRM was first introduced by McDougall and McIntosh (1996), where a framework for the Eulerian mean density is considered. Here, we call TRM the generalization of the concept by McDougall and McIntosh (1996) as discussed by Eden et al. (2007). In McDougall and McIntosh (2001), however, the name TRM is also used for a framework in which the semi-Lagrangian (isopycnal) mean density is considered. Here, we call this concept the semi-Lagrangian Mean to differentiate it from TRM.

considered as well, which then leads to an infinite sum of corrections to the eddy-driven velocity and diffusivity of the TEM framework, given for the zonal average case in (12.58) and (12.57), respectively.

Consistency implies that the resulting diffusivity is only related to dissipation (related to $\mathcal{G}$) or temporal changes of variance and higher order moments, which was already anticipated in the approximate form in the Osborn–Cox relation of Section 11.3.3. The TRM framework will be shown to lead to an Generalized Osborn–Cox relation as discussed in Section 12.3.3, given by (12.71) when specifying $\mathcal{G}$ as molecular diffusion. Note that the TEM framework, contrary to TRM, does not lead to a consistent turbulent diffusivity in this respect. The eddy-driven velocity in the TRM framework is different for each property $\chi$.

### Lagrangian Mean

The Lagrangian Mean denotes the average property following the fluid particles. Specifying the average operator as an ensemble mean, one could imagine an ensemble of particles over which the mean of a certain property is taken. This Lagrangian Mean property applies to the mean position of the particles. The Generalized Lagrangian Mean Theory of Andrews and McIntyre (1978), discussed below in Section 12.3.4, demonstrates that all those Lagrangian Mean properties are transported by the same *Lagrangian Mean velocity*, and that the Lagrangian Mean properties are changed or forced only by the Lagrangian Mean of the instantaneous changes or forcing $\mathcal{G}$. This result is expressed in the mean conservation equation (12.77). The difference between the Eulerian Mean velocity $\overline{\boldsymbol{u}}$ and the Lagrangian Mean velocity is called the *Stokes velocity* $\overline{\boldsymbol{u}}^{\mathrm{S}}$ and, likewise, the difference between the Eulerian Mean property $\bar{\chi}$ and the Lagrangian Mean property $\overline{\chi}^{\mathrm{L}}$ is called the *Stokes correction* $\bar{\chi}^{\mathrm{S}}$ to the property.

It is important to note that the same Lagrangian Mean velocity advects all Lagrangian Mean properties, which is different to the TRM framework of Section 12.3.1. However, since $\overline{\boldsymbol{u}}^{\mathrm{S}}$ is in general divergent, there is no stream function for the Lagrangian Mean velocity. Further, it is not guaranteed that the Lagrangian Mean is always defined as discussed in Section 12.3.1. On the other hand, advective and other processes in the Lagrangian Mean conservation equation are consistently separated by the approach[7].

### Semi-Lagrangian (Isopycnal) Mean

A further alternative averaging framework, which is often discussed in particular in the oceanographic community, follows by applying the average to a variable evaluated at the depth of a specific material surface instead at constant geopotential height $z$. This material surface is usually taken as an isopycnal, but it could also be the iso-surface of any general property $\chi$. Note that instead of the vertical coordinate, one could also use any other coordinate following a material surface, which is, however, not discussed here. Since only a single coordinate is replaced by particle

[7] It is worth mentioning that only the advective processes remain in the Generalized Lagrangian Mean conservation equation (12.77) for $\mathcal{G} = \mathbf{0}$, similar to the TRM conservation equation discussed above.

excursions in that direction, the procedure can be called semi-Lagrangian Mean. If density is used as material surface, the procedure is also called Isopycnal Mean.

The semi-Lagrangian (Isopycnal) Mean with respect to the density conservation equation is readily taken in isopycnal coordinates (compare Appendix B.1.1), in which the vertical coordinate is replaced by (potential) density. Note that analogous to isopycnal coordinates, one might replace the vertical coordinate with the tracer $\chi$ instead of density, with identical results in the thickness equation (B.6) (but differences in the pressure gradient force in the momentum equation (B.5)). Nonmonotonic functions $\chi(z)$ or $\rho(z)$ are also suitable for the coordinate transformation as shown by Nurser and Lee (2004). In isopycnal (tracer) coordinates, the *bolus velocity* corresponds to the eddy-driven or Stokes velocity in the other averaging frameworks.

However, similar to the Lagrangian compared to the Eulerian framework, the semi-Lagrangian (isopycnal) coordinates are more difficult to work with, such that it is desirable to perform the semi-Lagrangian Mean in an Eulerian coordinate system. The transformation from isopycnal coordinates back to $z$-coordinates yields a conservation equation for the semi-Lagrangian Mean density $\overline{\rho}^{\mathrm{I}}$. It is transported by a velocity given in turn by a stream function, defined by the total lateral transport below an instantaneous isopycnal, and which is changed only by the semi-Lagrangian Mean of the forcing $\mathcal{G}$ (McDougall and McIntosh, 2001). Using the general property $\chi$ instead of the density for the definition of the material surfaces in the semi-Lagrangian Mean yields a conservation equation for the semi-Lagrangian Mean tracer $\overline{\chi}^{\mathrm{I}}$. It is transported by a flow described by a stream function given by the lateral transport below instantaneous isosurfaces of $\chi$, and which is changed only by the semi-Lagrangian Mean of the forcing $\mathcal{G}$. The transport velocity is given by the sum of the Eulerian Mean velocity $\overline{\boldsymbol{u}}$ and the *Quasi-Stokes velocity* $\overline{\boldsymbol{u}}^{+}$. The latter takes the role of the eddy-driven velocity in the TRM framework, or the Stokes velocity in the Generalized Lagrangian Mean, and it differs in general from the bolus velocity.

The semi-Lagrangian concept will be detailed in Section 12.3.5. Note that the Quasi-Stokes velocity is nondivergent but in general different for each tracer, as in the TRM framework, and thus in general different from the Stokes velocity. However, advective and other processes are also consistently separated by the semi-Lagrangian Mean. the box on p. 397 gives an overview of the three different averaging framework, their transport velocities and mean quantities.

### 12.3.1 Temporal Residual Mean

In this section, the TEM framework of Section 12.2.3 is extended, and the connection of the turbulent diffusivity to small-scale mixing, or growth/decay of buoyancy variance, is discussed. For algebraic simplicity, the zonal mean case is considered again (compare the box on p. 385). However, all results carry over to three dimensions, with somewhat higher algebraic complexity, as detailed in the box on p. 403.

#### Rotational Eddy Flux

In Section 12.2.3, the decomposition (12.26) for the eddy buoyancy flux into advective and diffusive parts has been introduced. A constant flux $\boldsymbol{F}_{\star}$ can be added to the eddy flux $\boldsymbol{F}_\mathrm{b}$ since it does not affect the mean conservation equation. It is even possible to add any divergence-free flux given by $\boldsymbol{F}_\star = \underset{\neg}{\nabla}\theta$ since $\boldsymbol{F}_\mathrm{b}$ appears in the mean buoyancy equation inside the divergence operator and $\nabla \cdot \underset{\neg}{\nabla}\theta = 0$. Note that the symbol $\underset{\neg}{\cdot\cdot}$ denotes here as in Section 12.2.3 a counterclockwise rotation by 90° in the meridional/vertical plane since we are working with the zonal mean case as defined in the box on p. 385. It follows that the decomposition of the eddy flux $\boldsymbol{F}_\mathrm{b}$ is defined only up to an arbitrary rotational flux given by the *gauge potential* $\theta$. In the TEM decomposition, this gauge potential is set to zero. In the general case, instead of (12.26) one has

$$\boldsymbol{F}_\mathrm{b} \equiv \boldsymbol{F}_1 = B_1 \underset{\neg}{\nabla}\overline{b} - K_1 \nabla \overline{b} + \underset{\neg}{\nabla}\theta_1 \tag{12.50}$$

with arbitrary $\theta_1$ (the index is introduced for later use). While the choice of $\theta_1$ has no direct influence on the mean buoyancy equation, it is indirectly important since it can affect $B_1$ and $K_1$, which will become different from $B$ and $K_\mathrm{d}$ in (12.26). Further, the choice of $\theta_1$ affects the variance equation, which we consider next.

### Eddy Variance Equation

The conservation equation of eddy variance $\overline{\phi_2} = \overline{b'^2}/2$ is derived in the usual way, by multiplying the instantaneous buoyancy equation with the buoyancy perturbation $b'$ and taking the average, resulting in

$$\frac{\partial \overline{\phi_2}}{\partial t} + \nabla \cdot \boldsymbol{F}_2 = -\boldsymbol{F}_1 \cdot \nabla \overline{b} + \overline{b' \mathcal{G}'_\mathrm{b}} \tag{12.51}$$

The interpretation of the individual terms is the following: $\partial \overline{\phi_2}/\partial t$ denotes growth/decay of variance, and $\boldsymbol{F}_2 = \overline{\boldsymbol{v}}\overline{\phi_2} + \overline{\boldsymbol{v}' b'^2}/2$ the total variance flux, consisting of mean and turbulent variance advection. The term $\overline{b'\mathcal{G}'_\mathrm{b}}$ denotes the dissipation of variance and, depending on the form of $\mathcal{G}_\mathrm{b}$, may contain an additional flux divergence. The term $-\boldsymbol{F}_1 \cdot \nabla \overline{b}$ in (12.51) is a variance production term, arising from the interaction of mean and turbulent flows. With the general flux decomposition (12.50), the term becomes $-\boldsymbol{F}_1 \cdot \nabla \overline{b} = K_1 |\nabla \overline{b}|^2 - \underset{\neg}{\nabla}\theta_1 \cdot \nabla \overline{b}$. The first term is positive for $K_1 > 0$ and hence a source of variance while the second one can have both signs. Note that the gauge potential $\theta_1$ appears in (12.51) as part of the eddy flux $\boldsymbol{F}_1$, a fact which is used now to obtain guidance for the choice of $\theta_1$.

### Setting Rotational Eddy Fluxes

By choosing $\theta_1$ appropriately, the advective variance fluxes in (12.51) can be balanced with the part of the variance production term related to the rotational gauge potential $\theta_1$. To do so, the total advective variance flux is again decomposed into a flux along and a flux across contours of mean buoyancy

$$\boldsymbol{F}_2 = B_2 \underset{\neg}{\nabla}\overline{b} - K_2 \nabla \overline{b} + \underset{\neg}{\nabla}\theta_2 \tag{12.52}$$

With (12.52), the variance equation (12.51) becomes

$$\frac{\partial \overline{\phi_2}}{\partial t} + \underset{\neg}{\nabla}(\theta_1 - B_2) \cdot \nabla \overline{b} = \overline{b'\mathcal{G}'_\mathrm{b}} + K_1 |\nabla \overline{b}|^2 + \nabla \cdot K_2 \nabla \overline{b} \tag{12.53}$$

Note that the rotational flux $\underset{\neg}{\nabla}\theta_2$ does not appear in the variance equation but was introduced here for later use. With the choice $\theta_1 = B_2$, the advective fluxes in the variance equation cancel. In this setting, the rotational flux potential is determined by the flux of variance circulating along contours of mean buoyancy. The "localized" variance equation can then be solved for the turbulent diffusivity. It follows that the $K_1$–term can be balanced by i) growth (or decay) of variance, ii) dissipation of variance and iii) a flux of variance across contours of mean buoyancy (diapycnal flux of variance related to $K_2$).

Neglecting for a moment the diapycnal flux of variance related to $K_2$ (it will be discussed in the next section), the Osborn–Cox relation is recovered, as discussed in Section 11.3.3, i. e. a local balance between production ($K_1|\nabla\overline{b}|^2$) and dissipation of variance ($\overline{b'\mathcal{G}_b'}$) or growth of variance ($\partial\overline{\phi}_2/\partial t$), but here without assuming horizontal homogeneity as in Section 11.3.3, but by setting $\theta_1 = B_2$. For the adiabatic case, $\mathcal{G}_b = 0$, the growth (decay) of variance is proportional to positive (negative) turbulent diffusivity. For the steady case, $\partial\overline{\phi}_2/\partial t = 0$, the turbulent diffusivity is proportional to the dissipation of variance.

## Setting Rotational Eddy and Variance Fluxes

The physical interpretation of the turbulent diffusivity in the eddy variance equation (12.53) is hampered by the presence of the (diapycnal) flux $K_2\nabla\overline{b}$ of variance across contours of mean buoyancy. The meaning of this term is not obvious from the discussion so far, but it can also be related to growth of variance and dissipation as shown in this section. Further insight is gained by considering the full hierarchy of buoyancy moments of order $n$, defined as $\overline{\phi_n} = \overline{b'^n}/n$. Multiplying the instantaneous buoyancy equation with $b'^n$ and averaging results in

$$\frac{\partial\overline{\phi_{n+1}}}{\partial t} + \nabla\cdot\boldsymbol{F}_{n+1} = n\overline{\phi_n\mathcal{G}_b} - n\overline{\phi}_n\frac{\partial\overline{b}}{\partial t} - n\boldsymbol{F}_n\cdot\nabla\overline{b} \tag{12.54}$$

with the flux $\boldsymbol{F}_n = \overline{\boldsymbol{v}}\overline{\phi_n} + \overline{\boldsymbol{v}'b'^n}/n$. For $n = 1$, the variance equation (12.53) is recovered. The interpretation of the terms is as before for $\phi_2$: growth or decay ($\partial\overline{\phi_{n+1}}/\partial t + n\overline{\phi}_n\partial\overline{b}/\partial t$), advection ($\nabla\cdot\boldsymbol{F}_{n+1}$), dissipation ($n\overline{\phi_n\mathcal{G}_b}$), and production ($n\boldsymbol{F}_n\cdot\nabla\overline{b}$) of the $(n+1)$th moment. Consistent with (12.52), the decomposition $\boldsymbol{F}_n = B_n\underset{\neg}{\nabla}\overline{b} - K_n\nabla\overline{b} + \underset{\neg}{\nabla}\theta_n$ leads to

$$\frac{\partial\overline{\phi_{n+1}}}{\partial t} + \underset{\neg}{\nabla}(n\theta_n - B_{n+1})\cdot\nabla\overline{b} = n\overline{\phi_n\mathcal{G}_b} - n\overline{\phi}_n\frac{\partial\overline{b}}{\partial t} + nK_n|\nabla\overline{b}|^2 + \nabla\cdot K_{n+1}\nabla\overline{b} \tag{12.55}$$

Again, the choice $n\theta_n = B_{n+1}$ leads to a local balance, except for the term containing $K_{n+1}$. Evaluating the hierarchy for successive orders $n$, an expression for $K_n$ is obtained given by growth or decay and dissipation of $\phi_{n+1}$ and a term related to $K_{n+1}$. Starting with $n = 1$, one can use the next higher order moment to subsequently eliminate all $K_n$ with the effect that one can relate $K_1$ with terms involving only growth or decay or dissipation of variance and higher order moments, i. e. all flux terms can be eliminated from the conservation equations.

For a simpler derivation, it is convenient to consider two cases, the adiabatic case with $\mathcal{G}_b = 0$ and the steady case $\partial/\partial t = 0$. Writing the complete hierarchy for the adiabatic case yields

$$K_1|\nabla\overline{b}|^2 = \frac{\partial\overline{\phi_2}}{\partial t} - \nabla\cdot K_2\nabla\overline{b}$$
$$K_2|\nabla\overline{b}|^2 = \frac{1}{2}\frac{\partial\overline{\phi_3}}{\partial t} + \overline{\phi}_2\frac{\partial\overline{b}}{\partial t} - \frac{1}{2}\nabla\cdot K_3\nabla\overline{b}$$
$$K_3|\nabla\overline{b}|^2 = \frac{1}{3}\frac{\partial\overline{\phi_4}}{\partial t} + \overline{\phi}_n\frac{\partial\overline{b}}{\partial t} - \frac{1}{3}\nabla\cdot K_4\nabla\overline{b}$$
$$\vdots$$

Combining the full hierarchy to obtain a single equation for the turbulent diffusivity yields

$$\begin{aligned}K_1|\nabla\overline{b}|^2 &= \frac{\partial}{\partial t}\overline{\phi_2} - \frac{1}{2}\mathcal{D}\left(\frac{\partial}{\partial t}\overline{\phi_3}\right) + \frac{1}{3!}\mathcal{D}^2\left(\frac{\partial}{\partial t}\overline{\phi_4}\right) + \cdots \\ &\quad - \mathcal{D}\left(\overline{\phi}_2\frac{\partial}{\partial t}\overline{b}\right) + \frac{1}{2}\mathcal{D}^2\left(\overline{\phi}_3\frac{\partial}{\partial t}\overline{b}\right) - \frac{1}{3!}\mathcal{D}^3\left(\overline{\phi}_4\frac{\partial}{\partial t}\overline{b}\right) + \cdots\end{aligned} \tag{12.56}$$

introducing the operator $\mathcal{D}() = \nabla\cdot\nabla\overline{b}|\nabla\overline{b}|^{-2}()$. The steady case $(\partial\overline{()}/\partial t = 0)$ of (12.55) yields

$$K_1|\nabla\overline{b}|^2 = -\overline{b'\mathcal{G}_b'} + \mathcal{D}(\overline{\phi_2\mathcal{G}_b}) - \frac{1}{2}\mathcal{D}^2(\overline{\phi_3\mathcal{G}_b}) + \frac{1}{3!}\mathcal{D}^3(\overline{\phi_4\mathcal{G}_b}) - \frac{1}{4!}\mathcal{D}^4(\overline{\phi_4\mathcal{G}_b}) + \cdots \tag{12.57}$$

For the general case, both forms just add. In Section 12.3.3, the steady variance equation (12.57) will be further discussed specifying $\mathcal{G}_b$ as molecular diffusion, i. e. $\mathcal{G}_b = \mu\nabla^2 b$. It is also shown, how the infinite sum in (12.57) becomes a Taylor expansion of an exponential function.

Using $B_n|\nabla\overline{b}|^2 = (\boldsymbol{F}_n - \underset{\neg}{\nabla}\theta_n)\cdot\underset{\neg}{\nabla}\overline{b}$ and the choice $n\theta_n = B_{n+1}$, the stream function for eddy-driven flow, $B_1$, is in both cases given by

$$B_1|\nabla\overline{b}| = J_1 - \frac{\partial}{\partial m}J_2 + \frac{1}{2}\left(\frac{\partial}{\partial m}\right)^2 J_3 - \frac{1}{3!}\left(\frac{\partial}{\partial m}\right)^3 J_4 + \cdots \tag{12.58}$$

introducing the operator $\partial()/\partial m = |\nabla\overline{b}|^{-1}(\nabla\overline{b})\cdot\nabla|\nabla\overline{b}|^{-1}()$ and the along-isopycnal variance fluxes $J_n = \overline{\boldsymbol{v}\phi_n}\cdot(\underset{\neg}{\nabla}\overline{b})|\nabla\overline{b}|^{-1}$. Note that the first term in the expansion for $B_1$ is identical to the stream function in the TEM formulation, i. e. $B = J_1$ (compare with (12.27)). The remainder of the expansion is due to the introduction of the rotational flux $\theta_1$ given by

$$\theta_1|\nabla\overline{b}| = J_2 - \frac{1}{2}\frac{\partial}{\partial m}J_3 + \frac{1}{3!}\left(\frac{\partial}{\partial m}\right)^2 J_4 - \frac{1}{4!}\left(\frac{\partial}{\partial m}\right)^3 J_5 - \ldots \tag{12.59}$$

A truncated version of the definition (12.59) was called by McDougall and McIntosh (1996) the *Temporal Residual Mean*, and the concept cumulating in the full version of (12.59) was called accordingly *Generalized Temporal Residual Mean* by Eden et al. (2007). Note that the acronym TRM refers here to the generalized version only. Note also that the name, TRM, was used by McDougall and McIntosh (2001) later also for a framework applied to the isopycnally averaged density instead of the Eulerian Mean density. The concept by McDougall and McIntosh (2001) is discussed in Section 12.3.5 and is called semi-Lagrangian Mean to differentiate it from TRM.

In TRM, from (12.56) and (12.57), it is now possible to relate the turbulent diffusivity $K_1$ to either growth or decay of buoyancy variance and higher order moments, and/or covariances between the small-scale forcing or mixing and buoyancy fluctuations. In the steady and adiabatic case, the turbulent diffusivity will be zero. For the steady case, the powerful result follows: there is no diapycnal turbulent mixing if there is no molecular mixing (or some other form of buoyancy sink $\mathcal{G}_\mathrm{b}$). This result applies to buoyancy and to any other tracer as well and is easily generalized to the three-dimensional case as discussed in the box on p. 403.

**61. The Temporal Residual Mean in Three Dimensions**

TRM for the three-dimensional eddy flux $\boldsymbol{F}_\mathrm{b} = \overline{\boldsymbol{u}'b'}$ is based on the flux decomposition similar to (B56.1) and given by

$$\boldsymbol{F}_\mathrm{b} = -K_\mathrm{d}\nabla\overline{b} + \boldsymbol{B}\times\nabla\overline{b} - \nabla\times\boldsymbol{\theta}$$

Note that within this box the overbar denotes a temporal mean and that all vectors are three-dimensional. The rotational eddy flux is given by the vector potential $\boldsymbol{\theta}$ which is given by an infinite series of along-isopycnal fluxes of variance and higher order moments

$$\boldsymbol{\theta}\,|\nabla\overline{b}| = \boldsymbol{n}\times\overline{\boldsymbol{u}\phi_2} - \frac{1}{2}\frac{\partial}{\partial n}\left(\boldsymbol{n}\times\overline{\boldsymbol{u}\phi_3}/|\nabla\overline{b}|\right) + \frac{1}{3!}\frac{\partial}{\partial n}\left(\frac{\partial}{\partial n}\left(\boldsymbol{n}\times\overline{\boldsymbol{u}\phi_4}/|\nabla\overline{b}|\right)\Big/|\nabla\overline{b}|\right) - \ldots$$

with the cross-isopycnal unit vector $\boldsymbol{n} = (\nabla\overline{b})/|\nabla\overline{b}| = \boldsymbol{e}\,/|\boldsymbol{e}\,|$ and the derivative normal to an isopycnal $\partial/\partial n() = -\boldsymbol{n}\times(\nabla\times())$. Note that $\overline{\boldsymbol{u}\phi_n} = \overline{\boldsymbol{u}}\overline{\phi_n} + \overline{\boldsymbol{u}'\phi_n}$ and that for the vector stream functions $\boldsymbol{B}$ and the diapycnal diffusivity $K_\mathrm{d}$ the relations

$$\boldsymbol{B}|\nabla\overline{b}| = \boldsymbol{n}\times\boldsymbol{F}_\mathrm{b} - \frac{\partial}{\partial n}\boldsymbol{\theta} \quad\text{and}\quad K_\mathrm{d}|\nabla\overline{b}| = -\boldsymbol{F}_\mathrm{b}\cdot\boldsymbol{n} - \boldsymbol{n}\cdot\nabla\times\boldsymbol{\theta} \tag{B61.1}$$

hold. For small aspect ratio ($|e_1|, |e_2| \ll |e_3|$), the vector potential for the rotational eddy flux becomes

$$\boldsymbol{\theta}_\mathrm{h}\frac{\partial\overline{b}}{\partial z} \approx \overline{\underset{\neg}{\boldsymbol{u}_\mathrm{h}}\phi_2} - \frac{1}{2}\frac{\partial}{\partial z}\left(\overline{\underset{\neg}{\boldsymbol{u}_\mathrm{h}}\phi_3}\left(\frac{\partial\overline{b}}{\partial z}\right)^{-1}\right) + \cdots, \quad \theta_3 \approx \boldsymbol{s}\cdot\boldsymbol{\theta}_\mathrm{h} \tag{B61.2}$$

with the isopycnal slope vector $\boldsymbol{s} = (s_1, s_2)$. Note that the symbol $\underset{\neg}{..}$ denotes here counter-clockwise rotation by 90° in the horizontal plane, i.e. $\underset{\neg}{\boldsymbol{u}_\mathrm{h}} = (-v, u)$ for $\boldsymbol{u}_\mathrm{h} = (u, v)$ (compare also the box on p. 385 and the box on p. 444). The vector stream function and the diapycnal diffusivity for the three-dimensional TRM for small aspect ratio become

$$\begin{aligned}\boldsymbol{B}_\mathrm{h}\frac{\partial\overline{b}}{\partial z} &\approx \overline{\underset{\neg}{\boldsymbol{u}'_\mathrm{h}}b'} - \frac{\partial}{\partial z}\boldsymbol{\theta}_\mathrm{h}, \quad B_3 \approx \boldsymbol{s}\cdot\boldsymbol{B}_\mathrm{h} \quad\text{and}\\ K_\mathrm{d}\frac{\partial\overline{b}}{\partial z} &\approx -\overline{b'w'} - \underset{\neg}{\nabla_\mathrm{h}}\cdot\boldsymbol{\theta}_\mathrm{h} + \boldsymbol{s}\cdot\left(\overline{\boldsymbol{u}'_\mathrm{h}b'} + \frac{\partial}{\partial z}\underset{\neg}{\boldsymbol{\theta}_\mathrm{h}}\right)\end{aligned} \tag{B61.3}$$

The horizontal and vertical components of the eddy-driven velocity $\boldsymbol{v}$ are given by $\boldsymbol{v}_\mathrm{h} \approx \partial\underset{\neg}{\boldsymbol{B}_\mathrm{h}}/\partial z$ and $v_3 \approx \underset{\neg}{\nabla_\mathrm{h}}\cdot\boldsymbol{B}_\mathrm{h}$, respectively, and become $\boldsymbol{v}_\mathrm{h} = -\partial(K_\ell\boldsymbol{s})/\partial z$ and $v_3 = \nabla_\mathrm{h}\cdot(K_\ell\boldsymbol{s})$ as in the box on p. 389 with the isopycnal thickness diffusivity $K_\ell = O(1{,}000\,\mathrm{m^2\,s^{-1}})$, employing the standard downgradient closure for the horizontal eddy buoyancy fluxes (while setting $\boldsymbol{\theta}_\mathrm{h} = 0$).

### 12.3.2 Rotational Eddy Fluxes

In this section, it is discussed how rotational eddy fluxes, identified by the TRM framework, generate up- and down-gradient diapycnal turbulent fluxes. When the rotational components are not subtracted from the eddy fluxes and the diapycnal diffusivities then estimated from a flux-gradient relationship, the diapycnal rotational fluxes can lead to spurious diapycnal diffusivities unrelated to irreversible changes of the property under consideration. Furthermore, it is shown how rotational eddy fluxes on isopycnals are able to generate spurious lateral diffusivities appropriate to the Gent and McWilliams (1990) parameterization.

#### Diapycnal Rotational Eddy Fluxes

In the TEM framework, (12.27) states that the diapycnal projection of the eddy flux $\boldsymbol{F}_\mathrm{b}$ is proportional to the diapycnal turbulent diffusivity $K_\mathrm{d}$. In the TRM framework, a rotational eddy flux $\boldsymbol{F}_\star = \underset{\neg}{\nabla}\theta$ was introduced in (12.50), which has to be removed from $\boldsymbol{F}_\mathrm{b}$ to obtain the diapycnal diffusivity $K_\mathrm{d}$, i. e.

$$K_\mathrm{d}|\nabla\bar{b}|^2 = -(\boldsymbol{F}_\mathrm{b} - \boldsymbol{F}_\star)\cdot\nabla\bar{b} \tag{12.60}$$

If the rotational eddy flux is significant, the diapycnal diffusivity estimated from the flux-gradient relation (12.60) will be different when setting $\boldsymbol{F}_\star$ to zero (as in TEM) or when using the TRM form of $\boldsymbol{F}_\star$. In fact, it is often the case that $\boldsymbol{F}_\star$ dominates the total flux $\boldsymbol{F}_\mathrm{b}$ such that very different estimates for the diffusivity result.

If the diffusivity $K_\mathrm{d}$ depends on the choice of the rotational flux potential $\theta$, how can one decide which of the estimates of $K_\mathrm{d}$ (which may differ considerably) is the "real" one? In the TRM framework, the diffusivity $K_\mathrm{d}$ is related either to growth or decay of variance (and higher order moments) as stated in (12.56) or to irreversible dissipation of variance (and high order moments) (12.57) by the diabatic sources and sinks $\mathcal{G}_\mathrm{b}$. Since these features are regarded as essential for a diffusivity, $K_\mathrm{d}$ in TRM may be regarded as "the" physically plausible one. It is, however, not possible to proof that this feature is guaranteed only by the TRM framework. There might be other choices for $\theta$ which lead to a proper result.

Figure 12.9 shows an example for different estimates of the diapycnal diffusivity $K_\mathrm{d}$ in the idealized primitive equation model introduced in the box on p. 380. Figure 12.9 a) shows the diapycnal diffusivity $K_\mathrm{tem}$ which results in setting the rotational eddy flux to zero, i. e. the diapycnal diffusivity appropriate to the original TEM framework given by

$$K_\mathrm{tem}\frac{\partial\bar{b}}{\partial z} = -\overline{b'w'} - \boldsymbol{s}\cdot\overline{\boldsymbol{u}_\mathrm{h}'b'} \tag{12.61}$$

where $\boldsymbol{s} = (-\partial\bar{b}/\partial x, -\partial\bar{b}/\partial y)/(\partial\bar{b}/\partial z)$. Note that small isopycnal slopes $|\boldsymbol{s}| \ll 1$ are assumed in (12.61) and that the three-dimensional form of the TEM/TRM formalism is used in the model analysis as detailed in the box on p. 389 and the box on p. 403. Note also that $\overline{\boldsymbol{u}_\mathrm{h}b'}$ denotes the horizontal, two-dimensional eddy buoyancy flux. Within the energetic zonal jets of the model, the diapycnal diffusivity $K_\mathrm{tem}$ is much larger in magnitude than the prescribed vertical diffusivity used in the numerical model. Further, $K_\mathrm{tem}$ also often takes negative values, as a result of

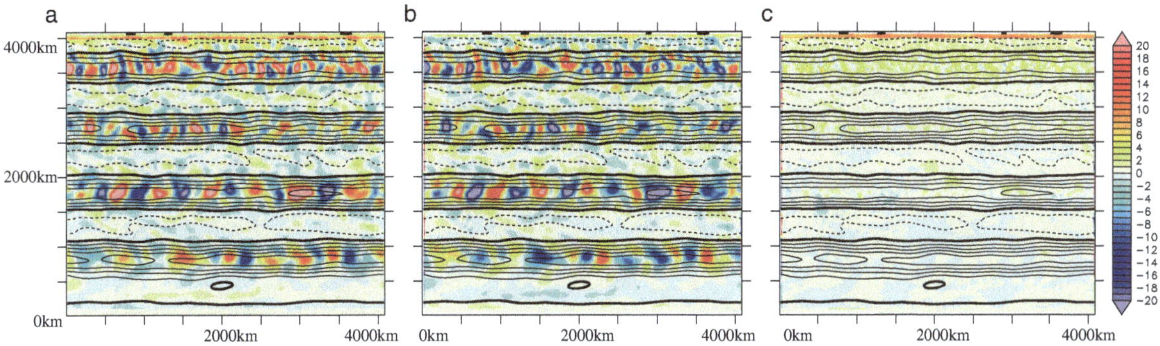


Fig. 12.9 Diapycnal rotational eddy fluxes and diffusivities in the idealized model introduced in the box on p. 380. **a** Diffusivity $K_{\text{tem}}$ from the diapycnal projection of the eddy buoyancy flux given by (12.61) in $\text{cm}^2\,\text{s}^{-1}$ at 500 m depth. Also shown are contours of time-mean zonal velocity. **b** Leading order contribution to the diapycnal diffusivity by the rotational eddy flux, i.e. $K_{\theta_1}$ in $\text{cm}^2\,\text{s}^{-1}$ given by (12.62). **c** The sum $K_{\text{d}} = K_{\text{tem}} + K_{\theta_1}$

upgradient diapycnal eddy buoyancy fluxes. It is clear that the estimated values of $K_{\text{tem}}$ contradict our belief that a diapycnal diffusivity should be small in the interior ocean and that eddy buoyancy fluxes should be downgradient.

Accounting for rotational eddy fluxes, however, it becomes clear that most of the large and sometimes upgradient eddy buoyancy fluxes, generating large and negative $K_{\text{tem}}$, are rotational. Figure 12.9 b) shows the leading order contribution of the diapycnal projection of the rotational eddy fluxes appropriate to TRM (compare (B61.2) and (B61.3)) in terms of a diapycnal diffusivity $K_{\theta_1}$ given by

$$K_{\theta_1}\frac{\partial \bar{b}}{\partial z} = -\nabla_{\text{h}} \cdot \left(\overline{\boldsymbol{u}_{\text{h}}\phi_2} \Big/ \frac{\partial \bar{b}}{\partial z}\right) + \boldsymbol{s} \cdot \frac{\partial}{\partial z}\left(\overline{\boldsymbol{u}_{\text{h}}\phi_2} \Big/ \frac{\partial \bar{b}}{\partial z}\right) \tag{12.62}$$

The leading order term of the rotational eddy flux dominates in the model results while the second and higher order terms in (B61.2) are much smaller (this might not always be the case, as discussed below). Note also that $\overline{\boldsymbol{u}_{\text{h}}\phi_2}$ denotes the horizontal, two-dimensional flux of buoyancy variance. When $K_{\text{tem}}$ and $K_{\theta_1}$ are added, many of the large and sometimes negative values in $K_{\text{tem}}$ cancel, and much smaller and predominantly positive values remain in the leading order diapycnal diffusivity $K_{\text{d}} = K_{\text{tem}} + K_{\theta_1}$ of the TRM framework. This demonstrates the need to account for rotational eddy fluxes when estimating diapycnal diffusivities from flux-gradient relations as (12.60).

Equation (B61.1) in the box on p. 403 shows for the three-dimensional version of TRM that the rotational eddy flux can be related for small aspect ratio to horizontal fluxes of variance and higher order moments. Further, the diapycnal projection of the rotational eddy flux is to first order given in terms of a diapycnal diffusivity by (12.62). It turns out that in the idealized model of Figure 12.9, the vertical component of the rotational eddy flux, given by $\nabla_{\text{h}} \cdot (\overline{\boldsymbol{u}_{\text{h}}\phi_2}/(\partial \bar{b}/\partial z))$, dominates the contribution of the second term in (12.62). Further, the vertical component is dominated by the zonal advection of variance $\phi_2$ by the mean flow, i. e.

$$K_{\theta_1} \approx -\nabla_{\text{h}} \cdot \left(\overline{\boldsymbol{u}_{\text{h}}\phi_2} \Big/ \frac{\partial \bar{b}}{\partial z}\right) \Big/ \frac{\partial \bar{b}}{\partial z} \approx -\bar{u}\frac{\partial \bar{\phi}_2}{\partial x} \Big/ \left(\frac{\partial \bar{b}}{\partial z}\right)^2$$

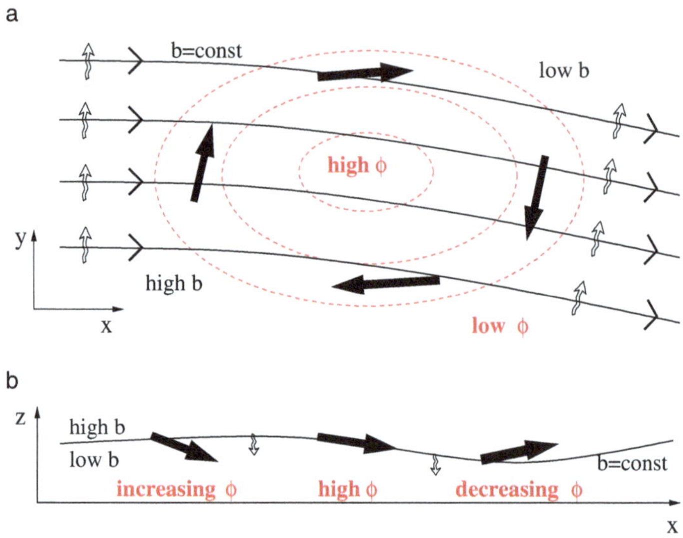


**Fig. 12.10** The concept of diapycnal rotational eddy fluxes similar to a schematic by McDougall and McIntosh (1996). Shown is a situation in which the mean flow enters a region of high variance (**a**). Downgradient (upgradient) diapycnal rotational eddy fluxes, indicated by *bold arrows* in (**b**), develop in the region with increasing (decreasing) variance, which dominate the divergent part of the diapycnal downgradient eddy flux (*small arrows* in **b**). In addition to the diapycnal rotational eddy fluxes (**b**), clockwise rotating isopycnal rotational eddy fluxes (*bold arrows* in **a**) develop in the region of enhanced variance, dominating the divergent isopycnal eddy flux (*small arrows* in **a**)

The variance $\phi_2$ is at maximum inside the zonal jets (compare Figure 12.11 below), such that the mean flow across the (small) zonal variations in the variance $\phi_2$ predominantly generates the diapycnal rotational eddy fluxes in the idealized model. The schematic in Figure 12.10 generalizes this result: when the mean flow enters a region of increased variance, downgradient diapycnal rotational eddy fluxes and large $K_{\theta_1}$ are generated. When the flow leaves the region, upgradient rotational diapycnal eddy fluxes and negative $K_{\theta_1}$ are generated. The rotational eddy fluxes dominate the small residual divergent eddy fluxes, related to $K_\mathrm{d} = K_\mathrm{tem} + K_{\theta_1}$. In the schematic of Figure 12.10 it is assumed that variations in the stratification remain small and that the advection of variance by perturbations $\overline{\boldsymbol{u}'_\mathrm{h}\phi_2}$ is small compared to the advection of $\bar{\phi}_2$ by the mean flow.

### Isopycnal Rotational Eddy Fluxes

Although rotational eddy fluxes are important for estimates of the diapycnal diffusivity $K_\mathrm{d}$, it turns out in the example of the idealized primitive equation model of the box on p. 380 that the rotational eddy flux defined in the box on p. 403 does not much effect the vector stream function $\boldsymbol{B}$. On the other hand, since the vector stream function is given by the eddy buoyancy flux parallel to the isopycnal plane (compare (B61.1)), there might also be isopycnal rotational eddy fluxes. This is indeed the case in the example as discussed in this section.

For the concept of isopycnal rotational eddy fluxes, it is necessary to define a background stratification representing a mean state for which the variance of deviations in buoyancy is related to an available eddy potential energy, as in (12.15).

The isopycnal rotational eddy flux will be defined such that the residual flux is then related to the release of mean available potential energy to eddy energy (or temporal changes of variance). This definition is analogous to the TRM framework in the previous section in which the definition of the rotational flux guarantees that the residual diapycnal eddy flux is related to irreversible mixing of buoyancy (or temporal changes of variance). It is possible to choose an arbitrary mean state, but for convenience a background appropriate to the quasi-geostrophic approximation as discussed in Section 5.2 is chosen. Note that within this approximation, isopycnals related to the background state are simply horizontal planes such that the discussion focuses on horizontal eddy fluxes; the generalization to an arbitrary background state is straightforward.

Consider the mean buoyancy equation (12.11)

$$\frac{\partial \bar{b}}{\partial t} + \bar{\boldsymbol{u}}_\mathrm{h} \cdot \nabla_\mathrm{h} \bar{b} = -N^2 \bar{w} - \nabla_\mathrm{h} \cdot \overline{\boldsymbol{u}'_\mathrm{h} b'} \tag{12.63}$$

in the quasi-geostrophic form. Note that here the buoyancy was decomposed into a background $b_\mathrm{b}(z)$ with $N^2(z) = \partial b_\mathrm{b}/\partial z$ and a deviation $b$. The horizontal velocity $\boldsymbol{u}_\mathrm{h}$ denotes the zero-order geostrophic velocity with $\nabla_\mathrm{h} \cdot \boldsymbol{u}_\mathrm{h} = 0$ while the vertical velocity $w$ is of first order. The diabatic forcing $\mathcal{G}_\mathrm{b}$ was neglected, and Reynolds averaging in time was applied. For the lateral eddy buoyancy flux $\overline{\boldsymbol{u}'_\mathrm{h} b'}$, it is now possible to apply the two-dimensional TRM framework as before for the zonal mean case discussed in Section 12.3.1. Note that the vertical advection of the background stratification $(N^2\bar{w})$ in (12.63) is considered now as the small-scale forcing. The following decomposition is chosen

$$\overline{\boldsymbol{u}'_\mathrm{h} b'} = -\kappa \nabla_\mathrm{h} \bar{b} + \nu \underline{\nabla}_\mathrm{h} \bar{b} + \underline{\nabla}_\mathrm{h} \mathcal{T} \tag{12.64}$$

where $\kappa = -(\overline{\boldsymbol{u}'_\mathrm{h} b'} - \underline{\nabla}_\mathrm{h} \mathcal{T}) \cdot \nabla_\mathrm{h} \bar{b} |\nabla_\mathrm{h} \bar{b}|^{-2}$ denotes a lateral diffusivity and $\nu = (\overline{\boldsymbol{u}'_\mathrm{h} b'} - \underline{\nabla}_\mathrm{h} \mathcal{T}) \cdot \underline{\nabla}_\mathrm{h} \bar{b} |\nabla_\mathrm{h} \bar{b}|^{-2}$ a lateral skew diffusivity (as in (B55.1)), and where $\mathcal{T}$ denotes a stream function for a horizontal rotational eddy buoyancy flux (which will later be added to the three-dimensional rotational flux $\nabla \times \boldsymbol{\theta}$ from the previous section). Note that the subscript $\underline{..}$ denote in this section rotation by 90° in the horizontal plane, i. e. $\underline{\nabla}_\mathrm{h} = (-\partial/\partial y, \partial/\partial x)$ (compare also with the box on p. 444). Using the decomposition (12.64) in the equation for the buoyancy variance $\bar{\phi}_2 = \overline{b'^2}/2$ yields

$$\frac{\partial \bar{\phi}_2}{\partial t} + \nabla_\mathrm{h} \cdot \overline{\boldsymbol{u}_\mathrm{h} \phi_2} + \underline{\nabla}_\mathrm{h} \mathcal{T} \cdot \nabla_\mathrm{h} \bar{b} = \kappa |\nabla_\mathrm{h} \bar{b}|^2 - N^2 \overline{b' w'} \tag{12.65}$$

which is identical to the available eddy potential energy conservation equation (12.15). The lateral diffusivity $\kappa$ is thus related to a sink of eddy potential energy. Using the Gent and McWilliams parameterization for the horizontal eddy buoyancy flux $\overline{\boldsymbol{u}'_\mathrm{h} b'} = -K_\ell \nabla_\mathrm{h} \bar{b}$ (compare with the box on p. 403), we note that $\kappa$ becomes identical to the thickness diffusivity $K_\ell$. However, the decomposition (12.64) accounts not only for the downgradient part of the eddy flux related to $\kappa$, but also for the component of $\overline{\boldsymbol{u}'_\mathrm{h} b'}$ directed along isolines of $\bar{b}$ in the lateral plane related to the skew diffusivity $\nu$, and for a possible rotational eddy flux component given by the stream function $\mathcal{T}$.

Following the TRM formalism, the flux of variance $\overline{\boldsymbol{u}_\mathrm{h} \phi_2}$ in (12.65) is also decomposed into a component across isolines of $\bar{b}$ in the horizontal plane, a component along those isolines and a rotational flux component. Consideration of the complete hierarchy of moments $\overline{\phi}_n$ as in Section 12.3.1 then yields the following definition of the horizontal rotational eddy fluxes

$$\mathcal{T}|\nabla_\mathrm{h}\bar{b}| = \tilde{s}\cdot\overline{\boldsymbol{u}_\mathrm{h}\phi_2} - \frac{1}{2}\frac{\partial}{\partial\tilde{n}}\left(\tilde{s}\cdot\overline{\boldsymbol{u}_\mathrm{h}\phi_3}/|\nabla_\mathrm{h}\bar{b}|\right) + \dots \tag{12.66}$$

with the derivative normal to isolines of $\bar{b}$ in the horizontal plane $\partial/\partial\tilde{n}() = \tilde{\boldsymbol{n}}\cdot\nabla_\mathrm{h}()$ and with the lateral unit vectors $\tilde{s} = |\nabla_\mathrm{h}\bar{b}|^{-1}\underline{\nabla}_\mathrm{h}\bar{b}$ and $\tilde{\boldsymbol{n}} = |\nabla_\mathrm{h}\bar{b}|^{-1}\nabla_\mathrm{h}\bar{b}$ pointing along and across isolines of $\bar{b}$ in the horizontal plane, respectively.

Marshall and Shutts (1981) consider a simplified version of the variance equation (12.65), which yields some insight about the direction of the horizontal rotational eddy flux. They assumed that the flux of variance in the stationary variance equation is dominated by the mean advection, i. e. that $\overline{\boldsymbol{u}_\mathrm{h}\phi_2} \approx \overline{\boldsymbol{u}}_\mathrm{h}\overline{\phi}_2$. Further, they assume that the quasi-geostrophic stream function $\psi$, which determines the geostrophic velocity $\boldsymbol{u}_\mathrm{h} = \underline{\nabla}_\mathrm{h}\psi$, is a function of $\bar{b}$, such that $\underline{\nabla}_\mathrm{h}\psi = (\partial\psi/\partial\bar{b})\underline{\nabla}_\mathrm{h}\bar{b}$ and the approximate variance equation (12.65) becomes

$$\underline{\nabla}_\mathrm{h}\left[\mathcal{T} - (\mathrm{d}\psi/\mathrm{d}\bar{b})\bar{\phi}_2\right]\cdot\nabla_\mathrm{h}\bar{b} \approx \kappa|\nabla_\mathrm{h}\bar{b}|^2 - N^2\overline{b'w'} \tag{12.67}$$

Choosing $\mathcal{T} = (\mathrm{d}\psi/\mathrm{d}\bar{b})\bar{\phi}_2$, the thickness diffusivity becomes $\kappa \approx N^2\overline{b'w'}|\nabla_\mathrm{h}\bar{b}|^{-2}$, i. e. is related to the production of eddy energy by the release of mean available potential energy. Using the TRM framework instead of the approximations chosen by Marshall and Shutts, the thickness diffusivity can be written, using the (steady) hierarchy of variance equations, as

$$\kappa|\nabla_\mathrm{h}\bar{b}|^2/N^2 = \overline{b'w'} - \tilde{D}(\overline{\phi_2 w}) + \frac{1}{2}\tilde{D}^2(\overline{\phi_3 w}) - \frac{1}{3!}\tilde{D}^3(\overline{\phi_4 w}) + \dots$$

with the operator $\tilde{D}() = \nabla_\mathrm{h}\cdot[\tilde{\boldsymbol{n}}()/|\nabla_\mathrm{h}\bar{b}|]$. Thus a very similar form is obtained as within the approximation by Marshall and Shutts, but here with the occurrence of higher order terms. In TRM, the definition of rotational eddy fluxes given by (12.66) guarantees that the diffusivity $\kappa$ is related (in steady state) to the production of variance and higher order moments of perturbation buoyancy on the horizontal plane. This variance production corresponds to leading order with release of mean available potential energy and production of eddy energy, as in the approximative form by Marshall and Shutts. This property of (12.66) contrasts the definition of the three-dimensional rotational eddy flux, which led to the feature that the diapycnal turbulent diffusivity $K_\mathrm{d}$ is related only to irreversible density changes (or temporal changes). In other words, the diapycnal eddy rotational flux in TRM relates the diapycnal diffusivity $K_\mathrm{d}$ to irreversible density mixing, while the isopycnal eddy rotational flux in TRM relates the thickness diffusivity $\kappa$ to the release of mean available potential energy.

Since the left-hand side of (12.67) can also be written as $[\underline{\nabla}_h\mathcal{T} - (\mathrm{d}\psi/\mathrm{d}\bar{b})\underline{\nabla}_\mathrm{h}\bar{\phi}_2]\cdot\nabla_\mathrm{h}\bar{b}$, it becomes clear that within the approximations chosen by Marshall and Shutts, the rotational eddy flux rotates along lines of constant $\bar{\phi}_2$, as indicated in the schematic Figure 12.10. Marshall and Shutts assume that the flux of variance is entirely directed along the isolines of $\bar{b}$ in the horizontal plane. (12.66) shows that in TRM, the leading order term of the potential for the rotational eddy flux is also related to the component of flux of variance directed along isolines of $\bar{b}$ in the

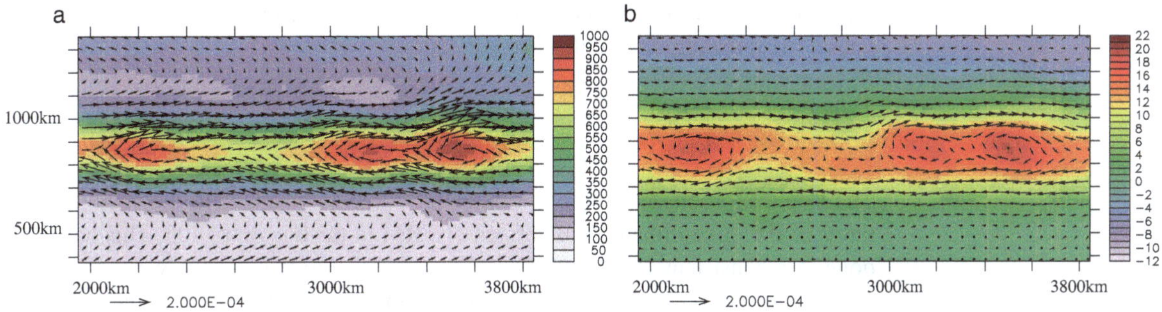


Fig. 12.11 **a** Buoyancy variance, $\bar{\phi}_2$ at 500 m in the eastern part of the southernmost zonal jet in $10^9$ m$^2$ s$^{-4}$ and horizontal eddy buoyancy flux, $\overline{\boldsymbol{u}_{\rm h}'b'}$, in m$^2$ s$^{-3}$ (*arrows*) in the idealized primitive equation model introduced in the box on p. 380. **b** The leading term of the horizontal rotational flux potential $\mathcal{T}_1$ in m$^3$ s$^{-3}$ and the related rotational eddy flux (*arrows*)

horizontal plane, but in TRM it is also possible to account for the perpendicular component of the flux of variance. While in TRM, the direction of the rotational flux in general may, therefore, deviate from isolines of constant $\bar{\phi}_2$, it is often the case that most of the rotational eddy flux in the horizontal plane indeed rotates along lines of constant $\bar{\phi}_2$.

Figure 12.11 shows the variance $\bar{\phi}_2$ and the horizontal eddy flux $\overline{\boldsymbol{u}_{\rm h}'b'}$ in the southernmost jet of the idealized primitive equation model introduced in the box on p. 380. The variance is large within the zonal jets since here eddy energy is also at maximum while $\bar{\phi}_2$ remains low in-between the jets. The same is true for the magnitude of the horizontal eddy fluxes. However, a large fraction of the eddy flux is directed along isolines of $\bar{\phi}_2$, indicating that it is of rotational nature. The figure also shows the leading term $\mathcal{T}_1 = \tilde{s} \cdot \overline{\boldsymbol{u}_{\rm h}\phi_2}/|\nabla_{\rm h}\bar{b}|$ of the rotational eddy flux of (12.66). It has indeed a direction and a magnitude similar to that of the eddy flux, i. e. it rotates along isolines of buoyancy variance. Note that both the eddy flux $\overline{\boldsymbol{u}_{\rm h}'b'}$ and the rotational eddy flux also tend to follow the small zonal variations of $\bar{\phi}_2$.

The leading order term $\mathcal{T}_1$ of the isopycnal rotational flux component is dominated by the zonal mean advection of $\phi_2$, i. e. $\mathcal{T}_1 \approx -\bar{u}\bar{\phi}_2/(\partial\bar{b}/\partial y)$ and is, therefore, large and positive ($\partial\bar{b}/\partial y$ is negative) within the eastward jets and small and slightly negative outside (because of the small westward mean flow between the eastward jets). The related leading order rotational eddy flux, $\underline{\nabla}_{\rm h}\mathcal{T}_1$, therefore, shares the similar direction and magnitude of the horizontal eddy fluxes, i.e. eastward (westward) at the northern (southern) side of the jet. Thus a large part of the zonal eddy buoyancy flux is indeed rotational. In contrast to the diapycnal rotational eddy fluxes, the higher order terms play a larger role for the isopycnal rotational potential flux $\mathcal{T}$ but are not shown here. The residual divergent isopycnal eddy flux $\overline{\boldsymbol{u}_{\rm h}'b'} - \underline{\nabla}_{\rm h}\mathcal{T}$, is much smaller than the total eddy flux $\overline{\boldsymbol{u}_{\rm h}'b'}$ in the jets and downgradient, i. e. northward to a large extent. Note that this implies isotropic isopycnal thickness mixing as anticipated by the Gent and McWilliams parameterization while the skew diffusivity $\nu$ remains small.

The schematic in Figure 12.10 summarizes the situation: when the mean flow enters a region of increased variance, an upgradient isopycnal rotational flux develops while a downgradient isopycnal rotational flux develops when the mean flow leaves the region of enhanced variance. The effect is that the isopycnal rotational flux rotates around the region of enhanced variance.

### Combining Diapycnal and Isopycnal Rotational Eddy Fluxes

The isopycnal rotational eddy flux potential $\mathcal{T}$ has to be added to the vector potential $\boldsymbol{\theta}$ of the three-dimensional version of TRM in the box on p. 403. This has to be done in a consistent way, i. e. it is important to insure that no diapycnal eddy flux and thus spurious diapycnal diffusivities are introduced by the additional isopycnal rotational flux. Since $\boldsymbol{\theta}$ is directed along isopycnals, it is clear that the additional rotational vector flux potential should be directed across isopycnals. A consistent definition for small aspect ratio is given by the modified vector potential for the rotational eddy buoyancy fluxes $\boldsymbol{\zeta} = \boldsymbol{\theta} + \mathcal{T}\boldsymbol{n}$ with the cross-isopycnal unit vector $\boldsymbol{n} = \boldsymbol{e}/|\boldsymbol{e}|$ as defined in Section 12.2.2. The definition for $\boldsymbol{\zeta}$ guarantees that the diapycnal diffusivity $K_\text{d}$ is unaffected by $\mathcal{T}$, while the vector stream function can be written as

$$\boldsymbol{B}_\text{h} N^2 \approx \overline{\underline{\boldsymbol{u}_\text{h}}' b'} + \nabla_\text{h} \mathcal{T} \, , \qquad B_3 \approx -\boldsymbol{s} \cdot \boldsymbol{B}_\text{h}$$

Using the decomposition (12.64), the horizontal eddy-driven velocity is then (also for $|\boldsymbol{s}| \ll 1$) given by $\boldsymbol{v}_\text{h} \approx \partial \underline{\boldsymbol{B}_\text{h}}/\partial z = -\partial(\kappa \boldsymbol{s})/\partial z + \partial(\nu \underline{\boldsymbol{s}})/\partial z$. For vanishing skew diffusivity $\nu$, as in the example of the idealized primitive equation model, this corresponds to the Gent and McWilliams parameterization and implies a northward (southward) directed eddy-driven velocity at the surface (bottom) of the channel flow in the idealized primitive equation model, very similar to the situation shown in Figure 12.8.

Note that without the introduction of an isopycnal rotational eddy flux, the eddy-driven velocity would be given as $\boldsymbol{v}_\text{h} \approx -\partial(\overline{\underline{\boldsymbol{u}'_\text{h}} b'}/N^2)/\partial z$, i. e. directed along the front associated with the zonal jets. Only by the introduction of the isopycnal rotational eddy fluxes, the eddy-driven velocity tends to flatten the isopycnals associated with the jets and reduces the mean available potential energy, in agreement with the physical interpretation of the Gent and McWilliams parameterization.

## 12.3.3 Generalized Osborn–Cox Relation

It is possible to utilize the TRM framework for small-scale mixing (compare Section 11.3). Specifying the buoyancy forcing $\mathcal{G}_\text{b}$ as molecular diffusion, i. e. $\mathcal{G}_\text{b} = \mu \nabla^2 b$, with the molecular[8] diffusivity $\mu \leq O(10^{-7}\,\text{m}^2\,\text{s}^{-1})$ yields for the buoyancy variance equation

$$\frac{\partial \overline{\phi_2}}{\partial t} + \nabla \cdot (\overline{\boldsymbol{v}\phi_2} - \mu \nabla \overline{\phi_2}) = -\mu \overline{|\nabla b'|^2} - \overline{\boldsymbol{v}' b'} \cdot \nabla \bar{b} \tag{12.68}$$

For simplicity, the two-dimensional framework as outlined in the box on p. 385 is used in this section again (although the zonal mean could be replaced with a time-mean when considering a three-dimensional flow).

[8] For simplicity a linear equation of state and identical molecular diffusivities for heat and salt are assumed.

The first term on the right-hand side of (12.68) describes dissipation of variance while the last term describes production of variance. In addition to the advective flux of variance on the left-hand side of (12.68) there is another molecular flux of variance. In the original Osborn–Cox relation (Osborn and Cox, 1972), a steady state $(\partial\overline{()}/\partial t = 0)$ and vanishing advective and molecular fluxes have been assumed, such that production and dissipation balance locally in the variance equation (11.86), i. e. $\mu\overline{|\nabla b'|^2} = -\overline{\boldsymbol{v}'b'}\cdot\nabla\bar{b}$. Assuming a downgradient closure for the turbulent eddy flux $\overline{\boldsymbol{v}'b'}$ or using the TEM decomposition (12.26) then leads to the frequently used Osborn–Cox relation (11.87), which relates the turbulent diapycnal diffusivity $K_{\text{tem}}$ to dissipation and to molecular diffusivity, i. e. $K_{\text{tem}} = \mu\overline{|\nabla b'|^2}/|\nabla\bar{b}|^2 \equiv \mu\bar{C}$, where $C$ denotes the Cox number which becomes very large for turbulent flows.

However, we have seen in Section 12.3.2 that large and sometimes upgradient rotational eddy fluxes can show up in the variance equation such that the TRM decomposition (12.50) is more appropriate than (12.26). It is, therefore, necessary to consider the complete hierarchy of buoyancy moments (12.54), which is given (for $n > 1$) in the case for molecular diffusion

$$\begin{aligned}\frac{\partial\overline{\phi_{n+1}}}{\partial t} + n\bar{\phi}_n\frac{\partial}{\partial t}\bar{b} + \nabla\cdot\boldsymbol{F}_{n+1} = &-\mu n\nabla\overline{\phi_n}\cdot\nabla\bar{b} \\ &- \mu n(n-1)\overline{\phi_{n-1}|\nabla b'|^2} - n\overline{\phi_n\boldsymbol{v}}\cdot\nabla\bar{b}\end{aligned} \tag{12.69}$$

All advective and molecular fluxes in the conservation equation for $\overline{\phi_n}$ are combined here in the flux vector $\boldsymbol{F}_n$, which is decomposed in turn as before into isopycnal and diapycnal components plus rotational fluxes, i. e.

$$\boldsymbol{F}_n = \overline{\boldsymbol{v}\phi_n} - \mu\nabla\overline{\phi_n} - \mu(n-1)\overline{\phi_{n-1}}\nabla\bar{b} = \underset{\neg}{\nabla}\theta_n + B_n\underset{\neg}{\nabla}\bar{b} - K_n\nabla\bar{b}$$

For buoyancy moments $n > 2$, two parts of molecular fluxes appear in $\boldsymbol{F}_n$. Both are small compared to the advective flux in the numerical simulation. As before, all advective and molecular fluxes can be eliminated from all conservation equations of buoyancy moments (12.69) by the choice $n\theta_n = B_{n+1}$. After some manipulations, the steady equations of the buoyancy moments (12.68) and (12.69) become

$$K_1 = \mu\overline{C} - \mathcal{D}(K_2)\,, \quad K_n = \mu(n-1)\overline{\phi_{n-1}(1+C)} - \frac{1}{n}\mathcal{D}(K_{n+1}) + 2\mu\mathcal{F}(\bar{\phi}_n) \tag{12.70}$$

with the operator $\mathcal{F}() = |\nabla\bar{b}|^{-2}\nabla\bar{b}\cdot\nabla()$ and $\mathcal{D}() = |\nabla\bar{b}|^{-2}\nabla\cdot()\nabla\bar{b}$. Solving now subsequently for the turbulent diapycnal diffusivity $K_1$, we obtain the following generalized form of the Osborn–Cox relation

$$\begin{aligned}K_1 + \mu = \mu\Big[&(1+\overline{C}) - \mathcal{D}\overline{\phi_1(1+C)} + \mathcal{D}^2\overline{\phi_2(1+C)} - \frac{1}{2}\mathcal{D}^3\overline{\phi_3(1+C)} + \dots \\ &+\frac{1}{3!}\mathcal{D}^4\overline{\phi_4(1+C)} - 2\mathcal{D}\mathcal{F}\bar{\phi}_2 + 2\frac{1}{2}\mathcal{D}^2\mathcal{F}\bar{\phi}_3 - 2\frac{1}{3!}\mathcal{D}^3\mathcal{F}\bar{\phi}_4 + \dots\Big]\end{aligned} \tag{12.71}$$

In contrast to the original Osborn–Cox relation, $K_{\text{tem}} + \mu = \mu(1+\overline{C})$, the generalized form (12.71) is exact using the setting $n\theta_n = B_{n+1}$ (for steady flow; a corresponding relation can be written for the unsteady case). All advective and molecular fluxes in (12.68) and (12.69) have disappeared by the appropriate choice of the rotational fluxes $n\theta_n = B_{n+1}$. The isopycnal flux of variance in (12.68) is balanced with the rotational eddy flux $\underset{\neg}{\nabla}\theta_1$ while the diapycnal flux of variance has been converted to dissipative terms which are all proportional to the subgrid-scale diffusivity $\mu$. The generalized Osborn–Cox relation (12.71) has now a clear physical meaning: The turbulent diffusivity $K_1$ is locally related to dissipation of buoyancy moments in subsequent order; $K_1$ is zero if there is no subgrid-scale mixing, i. e. if $\mu = 0$.

The first term on the right-hand side of (12.71) corresponds to the original Osborn–Cox relation, $K_{\text{tem}} + \mu = \mu(1 + \overline{C})$, while the remainder shows some resemblance with an expansion of an exponential function, although it is an operator. Formally, we might, therefore, interpret the series as an expansion of an operator $\mathcal{E}()$

$$K_1 + \mu = \mu\overline{\mathcal{E}(b', \bar{b})}$$

The analytical form of the operator $\mathcal{E}()$ remains unknown, but the first term of an expansion of $\mathcal{E}()$ is identical to the original Osborn–Cox relation, which means that a first order truncation to $\mathcal{E}()$ is given by $\mathcal{E}(b', \bar{b}) \approx 1 + C$.

However, it is possible to come closer to the unknown operator $\mathcal{E}$, using the following slightly different expression for the conservation equation of the buoyancy moments. The operator $\mathcal{D}() = |\nabla\bar{b}|^{-2}\nabla \cdot ()\nabla\bar{b}$ in (12.70) is split into two components, i. e. $\mathcal{D} = D + \mathcal{F}$ with $D = (\nabla^2\bar{b})/|\nabla\bar{b}|^2$ denoting an inverse $\bar{b}$-scale related to the curvature of the mean isopycnals. It is then possible to absorb the components related to the operator $\mathcal{F}$ (including the term $2\mu\mathcal{F}(\phi_n)$ in (12.70)) by a modification of the rotational flux $\underset{\neg}{\nabla}\theta_n$. Therefore, the choice for the rotational flux is modified according to $n\theta_n = B_{n+1} + \chi_n$ where the additional rotational gauge potential $\chi_n$ is given by the condition $\underset{\neg}{\nabla}\chi_n \cdot \nabla\bar{b} = \nabla(K_{n+1} - 2n\mu\phi_n) \cdot \nabla\bar{b}$. Here variations in $K_{n+1}$, which can be formally interpreted as a flux, and the corresponding molecular flux are balanced by the new rotational flux related to $\chi_n$. However, it is important to stress that the introduction of $\chi_n$ may lead to an unphysical rotational eddy flux across the boundaries and that the definition of $\chi_n$ becomes ill-posed for the case of closed isopycnals.

After some manipulations, the steady hierarchy of buoyancy moments becomes for the modified choice of $\theta_n$

$$K_1 = \mu\bar{C} - DK_2\,, \quad K_n = \mu(n-1)\overline{\phi_{n-1}(1+C)} - \frac{1}{n}DK_{n+1}$$

The equation is now of algebraic form, i. e. the operator $\mathcal{D}$ has been turned into the inverse buoyancy curvature scale $D$, and the operator $\mathcal{F}$ was absorbed by rotational fluxes. Thus solving for $K_1$ and summing up is now possible and yields an exact expression for the turbulent diffusivity

$$K_1 + \mu = \mu\overline{(1+C)\mathrm{e}^{-Db'}}$$

in terms of the Cox number $C$ and the dimensionless ratio $Db'$ relating the tracer perturbation with the mean curvature scale $D$. With the modified rotational fluxes, the operator $\mathcal{E}$ can be expressed in analytical form as $\mathcal{E}(b', \bar{b}) = (1 + C)\mathrm{e}^{-Db'}$. For $D \to 0$, we recover the first order approximation of the operator $\mathcal{E}$ and thus the original Osborn–Cox relation. The scale $D = (\nabla^2\bar{b})/|\nabla\bar{b}|^2$ in relation to the scale of $b'$, given by $\overline{\phi_2}^{1/2}$, thus yields a good measure for the validity of the original Osborn–Cox relation, i. e. for $D \gg (\overline{\phi_2})^{-1/2}$, we obtain $K_1 \to \mu\overline{C}$.

### 12.3.4 Generalized Lagrangian Mean

We now consider in more detail the Generalized Lagrangian Mean of Andrews and McIntyre (1978), which was introduced above. In the Lagrangian framework of Section 1.1, we have considered a particle with position $\boldsymbol{X}(\boldsymbol{a},t)$ at time $t$ and the initial position $\boldsymbol{a} = \boldsymbol{X}(\boldsymbol{a}, t = 0)$ which was used to label the particle. This concept is now used to define an average of a particle's property during its movement.

We have defined the Eulerian mean $\bar{\chi}$ of an Eulerian quantity $\chi(\boldsymbol{x},t)$ in Section 2.8.2, introducing an averaging operator with the fundamental property $\bar{\bar{\chi}} = \bar{\chi}$. In practice, averaging procedures hardly meet this requirement. The average is usually a time (or space) mean in a system with separated time (or space) scales. A hypothetical ensemble mean, of course, satisfies the requirement exactly. In contrast to the Eulerian Mean, for which we average at a fixed position, we now introduce the Lagrangian Mean $\bar{\chi}^{\mathrm{L}}$ as the average property following the particle. To make the connection of Lagrangian and Eulerian variables, it is useful to introduce the displacement vector $\boldsymbol{\xi}$ by

$$\boldsymbol{x}^{\xi}(t) = \boldsymbol{y} + \boldsymbol{\xi}(\boldsymbol{y},t) = \boldsymbol{x}^{\xi}(\boldsymbol{y},t) \tag{12.72}$$

The displacement $\boldsymbol{\xi}$ is a vector function of space and time and is evaluated at the point $\boldsymbol{y}$. The latter requirement allows to assume that the Eulerian Mean of $\boldsymbol{\xi}(\boldsymbol{y},t)$ will vanish, i. e. $\bar{\boldsymbol{\xi}} = 0$. The mean position of the particle is then given by $\overline{\boldsymbol{x}^{\xi}} = \boldsymbol{y}$ and its actual position by $\boldsymbol{x}^{\xi}$. The relation (12.72) can also be considered as a transformation rule, translating the mean position $\boldsymbol{y}$ to the actual position $\boldsymbol{x}^{\xi}$. It is a necessary assumption for the following analysis that this transformation is invertible, i. e. that we can uniquely relate actual particle positions to their mean position and vice versa. It is important to note that this might not always be the case, indicating a possible breakdown of the theory.

The Lagrangian mean operator can now be defined by

$$\bar{\chi}^{\mathrm{L}} = \overline{\chi(\boldsymbol{x}^{\xi},t)} \equiv \overline{\chi^{\xi}} \tag{12.73}$$

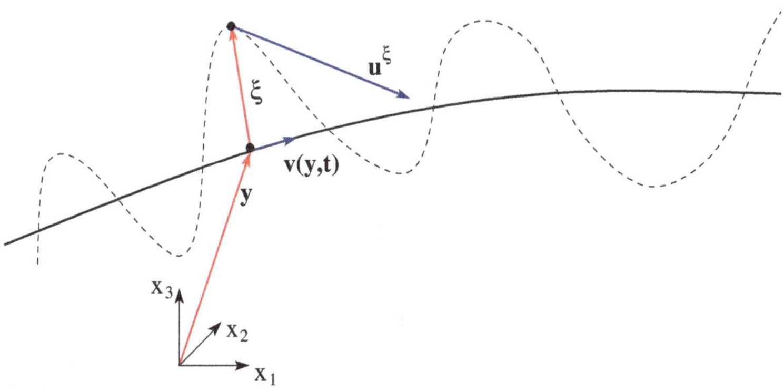


Fig. 12.12 Lagrangian Mean in a flow with two separated (slow and fast) time scales and time mean as averaging operator. The *solid black line* denotes the mean trajectory $\boldsymbol{y}$ and the *dashed line* the actual particle trajectory $\boldsymbol{y} + \boldsymbol{\xi}$. The mean particle position $\boldsymbol{y}$ moves (slowly) with the Lagrangian Mean velocity $\boldsymbol{v} = \overline{\boldsymbol{u}}^{\mathrm{L}}$, while the instantaneous particle position $\boldsymbol{y} + \boldsymbol{\xi}$ moves (fast) with the actual velocity $\boldsymbol{u}^{\xi}$

where $\overline{\chi(\boldsymbol{x}^\xi, t)}$ denotes the averaging operator as defined before, here applied to the property $\chi$ of all particles of the ensemble at their position $\boldsymbol{x}^\xi$. The notation $\chi(\boldsymbol{x}^\xi, t) = \chi^\xi$ is introduced to indicate that $\chi$ is evaluated at the actual position $\boldsymbol{x}^\xi = \boldsymbol{y} + \boldsymbol{\xi}$ during the averaging. The Lagrangian Mean property $\overline{\chi}^{\mathrm{L}} = \overline{\chi}^{\mathrm{L}}(\boldsymbol{y}, t)$ is defined at the mean particle position $\boldsymbol{y}$. Figure 12.12 illustrates the Lagrangian Mean for a time mean in a flow with two separated times scales.

We are interested in spatial and temporal changes of the particle's property $\chi^\xi$. We find using the chain rule for the space and time derivatives

$$\frac{\partial \chi^\xi}{\partial t} = \frac{\partial}{\partial t}\chi(\boldsymbol{x}^\xi, t) = \left.\frac{\partial \chi}{\partial t}\right|_{\boldsymbol{x}^\xi = \mathrm{const}} + \left.\frac{\partial \chi}{\partial x_i'}\right|_{\boldsymbol{x}' = \boldsymbol{x}^\xi}\frac{\partial x_i^\xi}{\partial t} = \left(\frac{\partial \chi}{\partial t}\right)^\xi + \left(\frac{\partial \chi}{\partial x_i}\right)^\xi \frac{\partial \xi_i}{\partial t}$$

$$\frac{\partial \chi^\xi}{\partial y_i} = \frac{\partial}{\partial y_i}\chi(\boldsymbol{x}^\xi, t) = \left.\frac{\partial \chi}{\partial x_j'}\right|_{\boldsymbol{x}' = \boldsymbol{x}^\xi}\frac{\partial x_j^\xi}{\partial y_i} = \left(\frac{\partial \chi}{\partial x_j}\right)^\xi \left(\delta_{ij} + \frac{\partial \xi_j}{\partial y_i}\right)$$

It follows that unlike the Eulerian Mean, the Lagrangian Mean operator $\overline{()}^{\mathrm{L}}$ does not commute with time and space derivatives, i. e. in general $\partial \overline{\chi}^{\mathrm{L}}/\partial x \neq \overline{\partial \chi/\partial x}^{\mathrm{L}}$. This makes the algebra more complicated but it will turn out that the Lagrangian Mean framework yields an averaged tracer conservation equation, where, unlike in the Eulerian Mean tracer conservation equation, eddy fluxes do not appear anymore, i. e. similar to what we found for the TRM framework after considerable effort.

To do so, we combine the rules for time and space differentiation to obtain the material derivative with the velocity $\boldsymbol{v} = \boldsymbol{v}(\boldsymbol{y}, t) = \mathrm{D}\boldsymbol{y}/\mathrm{D}t$ of the mean position $\boldsymbol{y}$

$$\left(\frac{\partial}{\partial t} + v_i \frac{\partial}{\partial y_i}\right)\chi^\xi = \left(\frac{\partial \chi}{\partial t}\right)^\xi + \left(\frac{\partial \chi}{\partial x_j}\right)^\xi \left[\frac{\partial x_j^\xi}{\partial t} + v_i \frac{\partial x_j^\xi}{\partial y_i}\right] \tag{12.74}$$

Remembering that $\boldsymbol{\xi}$ is defined at the mean position, it becomes clear that the last term in brackets of (12.74) is identical to the actual velocity at $\boldsymbol{x}^\xi$, i. e.

$$u_j^\xi = \frac{\partial x_j^\xi}{\partial t} + v_i \frac{\partial x_j^\xi}{\partial y_i} \tag{12.75}$$

Now remember also that $\boldsymbol{\xi}$ is a perturbation quantity with vanishing mean and that the mean position $\overline{\boldsymbol{x}^\xi} = \boldsymbol{y}$ and thus $\boldsymbol{v} = \mathrm{D}\boldsymbol{y}/\mathrm{D}t$ are mean quantities. Taking the mean of (12.75) yields therefore

$$\overline{u_j^\xi} = \overline{\frac{\partial x_j^\xi}{\partial t}} + \overline{v_i \frac{\partial x_j^\xi}{\partial y_i}} = \frac{\partial \overline{\xi_j}}{\partial t} + \overline{v_i \left(\delta_{ij} + \frac{\partial \xi_j}{\partial y_i}\right)} = v_j \equiv \overline{u_j(\boldsymbol{y}, t)}^{\mathrm{L}}$$

Since on the left-hand side $\overline{\xi_j} = 0$ and since $\mathrm{D}y_i/\mathrm{D}t = v_i$ is a mean quantity, we find $v_j = \overline{u_j}^{\mathrm{L}}$. The velocity of the mean point equals the Lagrangian Mean velocity which is equivalent to saying that $\boldsymbol{y}$ denotes the mean particle trajectory. The point $\boldsymbol{y}$ moves with $\overline{\boldsymbol{u}}^{\mathrm{L}}$ on mean trajectories, while $\boldsymbol{x}^\xi$ moves with $\boldsymbol{u}^\xi$ on the actual trajectories.

This allows to define the Lagrangian Mean total derivative $\overline{\mathrm{D}/\mathrm{D}t}^{\mathrm{L}} = \partial/\partial t + v_i \partial/\partial y_i = \partial/\partial t + \overline{\boldsymbol{u}}^{\mathrm{L}} \cdot \nabla$ with advection by the Lagrangian Mean velocity $\overline{\boldsymbol{u}}^{\mathrm{L}}$. Using the Lagrangian Mean total derivative and taking the mean of (12.74) yields

$$\overline{\left(\frac{\partial}{\partial t}+v_i\frac{\partial}{\partial y_i}\right)\chi^\xi}=\overline{\frac{\mathrm{D}}{\mathrm{D}t}}^{\mathrm{L}}\chi^\xi=\frac{\overline{\mathrm{D}}^{\mathrm{L}}}{\mathrm{D}t}\overline{\chi}^{\mathrm{L}}=\overline{\left(\frac{\partial\chi}{\partial t}\right)^\xi+u_i^\xi\left(\frac{\partial\chi}{\partial x_i}\right)^\xi}=\overline{\frac{\mathrm{D}\chi}{\mathrm{D}t}}^{\mathrm{L}} \tag{12.76}$$

This relation reveals the desired property of Lagrangian Mean variables when we consider now the Lagrangian Mean of the evolution equation for a general tracer (Andrews and McIntyre, 1978; Bühler, 2009)

$$\frac{\mathrm{D}\chi}{\mathrm{D}t}=\mathcal{G}_\chi\rightarrow\overline{\frac{\mathrm{D}\chi}{\mathrm{D}t}}^{\mathrm{L}}=\frac{\overline{\mathrm{D}}^{\mathrm{L}}}{\mathrm{D}t}\overline{\chi}^{\mathrm{L}}=\overline{\mathcal{G}_\chi}^{\mathrm{L}} \tag{12.77}$$

If $\mathcal{G}_\chi=0$, the Lagrangian Mean property $\overline{\chi}^{\mathrm{L}}$ is conserved along mean trajectories related to the mean position $\boldsymbol{y}$ and the Lagrangian Mean velocity $\overline{\boldsymbol{u}}^{\mathrm{L}}$. Only the Lagrangian Mean forcing $\overline{\mathcal{G}_\chi}^{\mathrm{L}}$ changes $\overline{\chi}^{\mathrm{L}}$ on the mean trajectories. This result is similar to what we obtained previously in Section 12.3.3 for the TRM formulation in steady state, where we found conservation of the Eulerian Mean property $\overline{\chi}$ on residual mean trajectories given by the residual velocity $\boldsymbol{u}^*$. However, note that since in general $\overline{\chi}\neq\overline{\chi}^{\mathrm{L}}$, residual velocity and Lagrangian Mean velocity will differ. Note also that the Lagrangian Mean velocity will generally be divergent. The difference between the Eulerian Mean velocity $\overline{\boldsymbol{u}}$ and the Lagrangian Mean velocity is the Stokes velocity $\overline{\boldsymbol{u}}^{\mathrm{S}}=\overline{\boldsymbol{u}}^{\mathrm{L}}-\overline{\boldsymbol{u}}$. Since $\overline{\boldsymbol{u}}^{\mathrm{L}}$ is divergent, the Stokes velocity is also divergent. Note that the Stokes velocity $\overline{\boldsymbol{u}}^{\mathrm{S}}$ is in general different from the Stokes drift discussed in the box on p. 197.

### 12.3.5 Semi-Lagrangian (Isopycnal) Mean

A further alternative averaging framework is given by applying the average to a variable evaluated at the depth of a material surface of some tracer instead at constant geopotential height $z$. In principle, any tracer can be used for this reference, but in practice only the potential density is of relevance. In analogy to the definition of the Lagrangian Mean (12.72), the semi-Lagrangian Mean of a variable $\chi$ is given by

$$\overline{\chi}^I=\overline{\chi(\boldsymbol{x}_\mathrm{h},\overline{z}+\xi_3,t)} \tag{12.78}$$

where the vertical displacement $\xi_3$ is defined as the deviation of the instantaneous depth of the above mentioned material surface $z(\boldsymbol{x}_\mathrm{h},t)$ from its mean, i. e. $\xi_3=z-\overline{z}$.

As an example of the averaging procedure involved in the definition (12.78) consider a temporal average. Hence there is a time filter defined, centered at time $t$. Consider an isopycnal with density value $\rho$, which at the position $\boldsymbol{x}_\mathrm{h}$ is moving up and down with an average vertical position $\overline{z}$, and define $\overline{\rho}^{\mathrm{I}}(\boldsymbol{x}_\mathrm{h},\overline{z},t)=\rho(\boldsymbol{x}_\mathrm{h},\overline{z},t)$. During the vertical excursions with amplitude $\xi_3=z-\overline{z}$ of the isopycnal, a range of values of the tracer $\chi$ is met and the isopycnal average $\overline{\chi}^{\mathrm{I}}(\boldsymbol{x}_\mathrm{h},\overline{z},t)$ of $\chi$ is the time mean of these values and attributed to the position $(\boldsymbol{x}_\mathrm{h},\overline{z},t)$. Likewise we may attributed $\overline{\chi}^{\mathrm{I}}$ to $(\boldsymbol{x}_\mathrm{h},\rho,t)$ (in an isopycnal framework; see below). When compared to the Lagrangian Mean (12.73), the semi-Lagrangian Mean resembles a Lagrangian Mean in which only one space coordinate is varied during averaging.

If potential density, i. e. isopycnals are used as the respective material surfaces, the mean defined by (12.78) is called *Isopycnal Mean*. The choice of potential density as semi-Lagrangian coordinate is motivated by the fact that density is conserved in the conservation equation $\mathrm{D}\rho/\mathrm{D}t = \mathcal{G}_\rho$ for vanishing diapycnal forcing $\mathcal{G}_\rho$ so that particles stay on the material surfaces or isopycnals during their movements; only effects included in a nonzero $\mathcal{G}_\rho$ lead to excursions from the isopycnals or material surfaces. Note that it is assumed here for simplicity that the function $\rho(z)$ is monotonic, such that its inverse exists. A generalization of the concept for nonmonotonic functions can be found in Nurser and Lee (2004). In contrast to the Lagrangian Mean, the semi-Lagrangian Mean is, therefore, always well defined, since it does not necessarily rely on a transformation rule as the Lagrangian Mean, which might, therefore, not be invertible in some special cases. Note that, in general, the special treatment of depth $z$ can be replaced by $y$ or $x$ in the definition of the semi-Lagrangian Mean (12.78), which is, however, not considered here.

As mentioned above, any conserved tracer may be used to define the reference material surfaces. Applying (12.78) to $\chi$ and using material surfaces $z(\boldsymbol{x}_\mathrm{h}, \chi, t)$ of $\chi$ itself for the definition of the semi-Lagrangian Mean, yields an average of identical values of $\chi$ and the result is simply $\overline{\chi}^\mathrm{I} = \chi$. This appears trivial, but it is important to note that this average $\overline{\chi}^\mathrm{I}$ applies to the mean height $\overline{z}(\boldsymbol{x}_\mathrm{h}, \chi, t)$, i. e. $\overline{\chi}^\mathrm{I} = \overline{\chi}^\mathrm{I}(\overline{z}, t)$. In other words, $\overline{\chi}^\mathrm{I}$ yields the value of the property $\chi$, whose mean iso-surface height $\overline{z}$ at the horizontal position $\boldsymbol{x}_\mathrm{h}$ equals the actual depth $z$ at which $\overline{\chi}^\mathrm{I}$ is evaluated. Specifying $\chi$ as density, and using isopycnals $z(\boldsymbol{x}_\mathrm{h}, \rho, t)$ as material surfaces in the semi-Lagrangian Mean (12.78), yields the density whose mean isopycnal height corresponds to the actual depth $z$ at which $\overline{\rho}^\mathrm{I}$ is evaluated. This 'isopycnally averaged density' $\overline{\rho}^\mathrm{I}$ was named *modified density* by McDougall and McIntosh (2001). Using isosurfaces $z(\boldsymbol{x}_\mathrm{h}, \chi, t)$ of the general tracer $\chi$ and also $\chi$ itself in (12.78) yields accordingly the *modified property* $\overline{\chi}^\mathrm{I}$. Note that this modified property differs in general from the *isopycnally averaged property*, i. e. using $z(\boldsymbol{x}_\mathrm{h}, \rho, t)$ and $\chi$ in (12.78), and that the modified density differs from $\overline{\rho}^\mathrm{I}$, using $z(\boldsymbol{x}_\mathrm{h}, \chi, t)$ in (12.78). However, note that in many oceanographic applications, isopycnal surfaces are used instead of isosurfaces $z(\boldsymbol{x}_\mathrm{h}, \chi, t)$ of the general tracer $\chi$ in (12.78) and we will follow this convention here.

We will consider the isopycnally averaged tracer (property) $\chi$ in detail below, first, however, we discuss the conservation equation for modified density, or, equivalently, the modified property.

### Isopycnal Coordinates and the Bolus Velocity

A convenient way to apply the Isopycnal Mean is the coordinate transformation of the relevant equations to isopycnal coordinates as described in Appendix B.1.1. In isopycnal coordinates, the (potential) density equation $\mathrm{D}\rho/\mathrm{D}t = \mathcal{G}_\rho$ and the continuity equation $\nabla \cdot \boldsymbol{u}$ are combined to the isopycnal thickness equation (see Appendix B.1)

$$\left.\frac{\partial h_\rho}{\partial t}\right|_{\rho=\mathrm{const}} + \nabla_\rho \cdot (\boldsymbol{u}_\mathrm{h} h_\rho) + \frac{\partial}{\partial \rho}\mathcal{G}_\rho h_\rho = 0$$

with the (infinitesimal) isopycnal thickness $h_\rho = -\partial z(\boldsymbol{x}_\mathrm{h}, \rho, t)/\partial\rho$ (note that $h_\rho \neq z$) and with $\nabla_\rho \chi = \nabla|_{\rho=\mathrm{const}}\chi = \nabla_\mathrm{h}\chi + \partial\chi/\partial z\, \nabla_\rho z(\boldsymbol{x}_\mathrm{h}, \rho, t)$. Note that all variables are function of $(\boldsymbol{x}_\mathrm{h}, \rho, t)$.

**62. The Bolus Velocity in $z$-Coordinates**

To use the parameterization (12.79) for the bolus velocity in $z$-coordinates, it is necessary to consider the modified density $\overline{\rho}^{\mathrm{I}} = \overline{\rho(\boldsymbol{x}_{\mathrm{h}}, z, t)}^{\mathrm{I}}$. This 'isopycnally averaged density' $\overline{\rho}^{\mathrm{I}}$ is a function of depth, and is the density whose mean isopycnal height corresponds to the actual depth $z$ at which $\overline{\rho}^{\mathrm{I}}$ is evaluated. The mean isopycnal thickness $\overline{h_\rho} = -\overline{\partial z/\partial\rho} = -\partial\overline{z}/\partial\rho$ is related to the vertical derivative of the modified density by $\partial\rho/\partial\overline{z} = (\overline{h_\rho})^{-1}$. It becomes now possible to rewrite the bolus velocity at $z = \overline{z}$ as

$$\boldsymbol{u}_{\mathrm{b}} = -K_\ell(\overline{h_\rho})^{-1}\boldsymbol{\nabla}_\rho\overline{h_\rho} = K_\ell\overline{h_\rho}\boldsymbol{\nabla}_\rho(\overline{h_\rho})^{-1} = K_\ell\left(\frac{\partial\overline{\rho}^{\mathrm{I}}}{\partial z}\right)^{-1}\left(\boldsymbol{\nabla}_{\mathrm{h}}\frac{\partial\overline{\rho}^{\mathrm{I}}}{\partial z} + \frac{\partial^2\overline{\rho}^{\mathrm{I}}}{\partial z^2}\boldsymbol{\nabla}_\rho z\right)$$

With $\boldsymbol{\nabla}_\rho z(\boldsymbol{x}_{\mathrm{h}}, \rho, t) = -(\partial\rho/\partial z)^{-1}\boldsymbol{\nabla}_{\mathrm{h}}\rho$ we find

$$\begin{aligned}\boldsymbol{u}_{\mathrm{b}} &= K_\ell\left[\left(\frac{\partial\overline{\rho}^{\mathrm{I}}}{\partial z}\right)^{-1}\boldsymbol{\nabla}_{\mathrm{h}}\frac{\partial\overline{\rho}^{\mathrm{I}}}{\partial z} - \left(\frac{\partial\overline{\rho}^{\mathrm{I}}}{\partial z}\right)^{-2}\frac{\partial^2\overline{\rho}^{\mathrm{I}}}{\partial z^2}\boldsymbol{\nabla}_{\mathrm{h}}\overline{\rho}^{\mathrm{I}}\right] \\ &= K_\ell\frac{\partial}{\partial z}\left[\boldsymbol{\nabla}_{\mathrm{h}}\overline{\rho}^{\mathrm{I}}\left(\frac{\partial\overline{\rho}^{\mathrm{I}}}{\partial z}\right)^{-1}\right] = -K_\ell\frac{\partial\boldsymbol{s}^I}{\partial z}\end{aligned}$$

Here the isopycnal slope vector $\boldsymbol{s}^I$ of the modified density $\overline{\rho}^{\mathrm{I}}$ is introduced. Note that only for vertically constant $K_\ell$, the parameterized bolus velocity $\boldsymbol{u}_{\mathrm{b}}$ becomes identical to the eddy-driven velocity of the parameterization of Gent and McWilliams (1990) as given e. g. by (B56.3). Note also that the isopycnal slopes differ, i. e. given by the modified vs. Eulerian mean density.

We take the mean of the thickness equation in isopycnal coordinates (which is equivalent to the Isopycnal Mean in Eulerian coordinates) and decompose the horizontal velocity and thickness into mean $\overline{\boldsymbol{u}}_{\mathrm{h}}^{\mathrm{I}}$ and $\overline{h_\rho}$ and deviations $\boldsymbol{u}_{\mathrm{h}}'(\boldsymbol{x}_{\mathrm{h}}, \rho, t) \equiv \boldsymbol{u}_{\mathrm{h}}{}'$ and $h_\rho'$ with vanishing means $\overline{\boldsymbol{u}'} = 0$ and $\overline{h_\rho'} = 0$, which yields

$$\left.\frac{\partial\overline{h_\rho}}{\partial t}\right|_{\rho=\mathrm{const}} + \boldsymbol{\nabla}_\rho\cdot\left(\overline{\boldsymbol{u}_{\mathrm{h}}}^{\mathrm{I}}\overline{h_\rho} + \overline{\boldsymbol{u}_{\mathrm{h}}'h_\rho'}^{\mathrm{I}}\right) + \frac{\partial}{\partial\rho}\overline{\mathcal{G}_\rho h_\rho}^{\mathrm{I}} = 0$$

It is now possible to define the thickness weighted mean velocity as

$$\boldsymbol{v}^{\mathrm{thk}} = \overline{\boldsymbol{u}_{\mathrm{h}}h_\rho}^{\mathrm{I}}/\overline{h_\rho} = \overline{\boldsymbol{u}}_{\mathrm{h}}^{\mathrm{I}} + \boldsymbol{u}_{\mathrm{b}} \quad \text{with} \quad \boldsymbol{u}_{\mathrm{b}} = \overline{\boldsymbol{u}_{\mathrm{h}}'h_\rho'}^{\mathrm{I}}\Big/\overline{h_\rho}$$

and the mean thickness equation becomes

$$\left.\frac{\partial\overline{h_\rho}}{\partial t}\right|_{\rho=\mathrm{const}} + \boldsymbol{\nabla}_\rho\cdot\left(\boldsymbol{v}^{\mathrm{thk}}\overline{h_\rho}\right) + \frac{\partial}{\partial\rho}\overline{\mathcal{G}_\rho h_\rho}^{\mathrm{I}} = 0$$

By the definition of the thickness weighted velocity in isopycnal coordinates, the eddy effect in the mean thickness equation is transferred to a redefinition of the advection velocity – here a simple algebraic replacement – analogous to the Generalized Lagrangian Mean tracer conservation equation (12.77) or the TRM tracer conservation equation. The thickness weighted velocity $\boldsymbol{v}^{\mathrm{thk}}$ is the sum of the isopycnally averaged velocity $\overline{\boldsymbol{u}_{\mathrm{h}}}^{\mathrm{I}}$ and the thickness weighted correlation between thickness and velocity. The latter, $\boldsymbol{u}_{\mathrm{b}}$, was named the 'bolus velocity' by Rhines (1982).

It was proposed by Gent and McWilliams (1990) and Gent et al. (1995) to parameterize the bolus velocity $\boldsymbol{u}_{\mathrm{b}} = \overline{\boldsymbol{u}_{\mathrm{h}}'h_\rho'}^{\mathrm{I}}/\overline{h_\rho}$ using the simple down-gradient closure

$$\boldsymbol{u}_{\mathrm{b}} = \frac{\overline{\boldsymbol{u}_{\mathrm{h}}'h_\rho'}^{\mathrm{I}}}{\overline{h_\rho}} = -K_\ell\boldsymbol{\nabla}_\rho\overline{h_\rho} \tag{12.79}$$

The coefficient $K_\ell$ has units of a diffusivity and is accordingly called the thickness diffusivity (compare also Section 12.2.4). Note that $K_\ell$ could also be a diffusivity tensor, even with antisymmetric components. However, $K_\ell$ is chosen in current applications of the Gent and McWilliams (1990) parameterization as a scalar with values of $O(1{,}000\,\mathrm{m^2\,s^{-1}})$. The way to use the parameterization (12.79) in $z$-coordinates is discussed in the box on p. 417.

In isopycnal coordinates, the bolus velocity occurs in the mean isopycnal thickness equation as an 'eddy-advection velocity' of the mean isopycnal thickness, which adds to the (isopycnal) mean velocity similar to the Stokes velocity in $z$-coordinates. On the other hand, it is unclear which role the bolus velocity plays after transformation to $z$-coordinates (shown in the box on p. 417), i. e. which tracer – the Eulerian Mean, the Lagrangian Mean, or some other kind of average – the bolus velocity advects. It is also unclear to which mean velocity the bolus velocity adds. These points are getting clearer in a framework using the isopycnally averaged tracer $\overline{\rho}^{\mathrm{I}}$ evaluated in $z$-coordinates, similar to the Generalized Lagrangian Mean framework, as discussed in the next section.

### The Quasi-Stokes Stream Function

The Quasi-Stokes velocity of McDougall and McIntosh (2001) adds to the Eulerian Mean velocity in $z$-coordinates. It will be shown below in the conservation equation (12.83) that the Quasi-Stokes velocity transports the modified density $\overline{\rho}^{\mathrm{I}}$, where advective processes are consistently separated from the forcing $\mathcal{G}_\rho$, similar to the TRM framework and the Generalized Lagrangian Mean. Note that it is possible to replace the density $\rho$ with the general property $\chi$ in the definition for the material surfaces in the semi-Lagrangian Mean (12.78), which yields a mean conservation equation identical to (12.83) but for the modified property $\overline{\chi}^{\mathrm{I}}$ instead of the modified density $\overline{\rho}^{\mathrm{I}}$. Note that the Quasi-Stokes velocity which transports $\overline{\rho}^{\mathrm{I}}$ using $z(\rho)$ in (12.78) will in general be different from the Quasi-Stokes velocity transporting $\overline{\chi}^{\mathrm{I}}$ using $z(\chi)$ in (12.78). Here, we follow again the oceanographic convention and discuss the isopycnal case, but note that results are identical for modified property and modified density.

Consider the volume flux $\boldsymbol{V}$ between the bottom $z = -h$ and a material surface given by the isopycnal $z = z(\boldsymbol{x}_\mathrm{h}, \rho, t)$, corresponding to the modified density $\overline{\rho}^{\mathrm{I}}$, and take the mean, which yields

$$\overline{\boldsymbol{V}}(\boldsymbol{x}_\mathrm{h}, \rho, t) = \overline{\int_{-h}^{z(\boldsymbol{x}_\mathrm{h},\rho,t)} \boldsymbol{u}_\mathrm{h}(\boldsymbol{x}_\mathrm{h}, \tilde{z}, t)\mathrm{d}\tilde{z}} = \int_{-h}^{\overline{z}} \overline{\boldsymbol{u}}_\mathrm{h}\mathrm{d}\tilde{z} + \overline{\int_{\overline{z}}^{\overline{z}+z'} \boldsymbol{u}_\mathrm{h}\mathrm{d}\tilde{z}} \tag{12.80}$$

with the mean isopycnal depth $\overline{z}$ and its deviation $z' = z - \overline{z}$. Note that the rightmost part of the relation (12.80) can be considered as a function of the mean depth $\overline{z}$, while the left-hand side is a function of modified density $\rho = \overline{\rho}^{\mathrm{I}}$. The very last term on the right-hand side of (12.80) is related to a possible correlation between isopycnal depth and velocity fluctuations and can in fact be shown to be related to an 'eddy-driven velocity', such as Stokes velocity (in $z$-coordinates) or bolus velocity (in isopycnal coordinates).

It is possible to express the mean component $\int \overline{\boldsymbol{u}}_\mathrm{h}\mathrm{d}\tilde{z}$ on the right-hand side of (12.80) by the mean stream function $\overline{\boldsymbol{\Psi}}(\boldsymbol{x}_\mathrm{h}, \overline{z}, t) = -\int_{-h}^{\overline{z}} \underline{\overline{\mathbf{u}}}_\mathrm{h}\mathrm{d}\tilde{z}$ with $\overline{\boldsymbol{u}}_\mathrm{h} = \partial\underline{\overline{\boldsymbol{\Psi}}}/\partial z$.

Note that $\overline{\boldsymbol{\Psi}}$ is here a horizontal vector, i. e. $\overline{\boldsymbol{e}_3 \cdot \boldsymbol{\Psi}} = 0$. In analogy, we may define a stream function for the eddy component of the total transport $\overline{\boldsymbol{V}}$ as

$$\boldsymbol{\Psi}^+(\boldsymbol{x}_\mathrm{h}, \overline{z}, t) = -\overline{\int_{\overline{z}}^{\overline{z}+z'} \underline{\boldsymbol{u}}_\mathrm{h} \mathrm{d}\tilde{z}} \tag{12.81}$$

which is the exact definition for the Quasi-Stokes stream function of McDougall and McIntosh (2001). Note that also $\boldsymbol{e}_3 \cdot \boldsymbol{\Psi}^+ = 0$. We will discuss below an approximate form of the stream function as a function of Eulerian Mean variables.

We first discuss the physical meaning of the transport $\overline{\boldsymbol{V}}$. Consider two infinitesimal close isopycnals $\rho_1 = \mathrm{const}$ and $\rho_2 = \mathrm{const}$ with depths $z_1$ and $z_2$, respectively, and with $z_1 < z_2$ and $\rho_1 > \rho_2$ with $\Delta\rho = \rho_1 - \rho_2 > 0$. The transport below the isopycnals is given by $\overline{\boldsymbol{V}}_1$ and $\overline{\boldsymbol{V}}_2$, respectively. The difference in the transports yields the transport between both isopycnals and can be written as

$$\overline{\boldsymbol{V}}_2 - \overline{\boldsymbol{V}}_1 = \overline{\int_{z_1}^{z_2} \boldsymbol{u}_\mathrm{h} \mathrm{d}\tilde{z}} = \overline{\int_{\rho_1}^{\rho_2} \boldsymbol{u}_\mathrm{h} \frac{\partial z}{\partial \rho} \mathrm{d}\rho} = \int_{\rho_2}^{\rho_1} \overline{\boldsymbol{u}_\mathrm{h} h_\rho}^\mathrm{I} \mathrm{d}\rho \overset{\Delta\rho \to 0}{=} \overline{\boldsymbol{u}_\mathrm{h} h_\rho}^\mathrm{I} \Delta\rho \tag{12.82}$$

Dividing by $\Delta z = z_2 - z_1 \to 0$, it becomes clear that the velocity on the left-hand side of (12.82) becomes for $\Delta z/\Delta\rho \to \overline{h_\rho}$ the thickness weighted velocity introduced above

$$\boldsymbol{v}^\mathrm{thk} = \frac{\partial \overline{\boldsymbol{V}}}{\partial z} = \frac{\overline{\boldsymbol{u}_\mathrm{h} h_\rho}^\mathrm{I}}{\overline{h_\rho}}$$

While the thickness weighted velocity $\boldsymbol{v}^\mathrm{thk}(\boldsymbol{x}_\mathrm{h}, \rho, t)$ can be decomposed into $\overline{\boldsymbol{u}_\mathrm{h}}^\mathrm{I} + \overline{\boldsymbol{u}'_\mathrm{h} h'_\rho}^\mathrm{I} / \overline{h_\rho}$, i. e. the isopycnally averaged velocity $\overline{\boldsymbol{u}_\mathrm{h}}^\mathrm{I}$ and the bolus velocity $\boldsymbol{u}_\mathrm{b}$ as discussed above, we have here decomposed the total transport $\overline{\boldsymbol{V}}$ into Eulerian mean transport and an eddy component, given by the stream functions $\overline{\boldsymbol{\Psi}}$ and $\boldsymbol{\Psi}^+$, respectively, which are both function of the mean depth $\overline{z}$. The latter takes, therefore, the meaning of a *Quasi-Stokes velocity*. It also becomes clear that it is the modified density $\overline{\rho}^\mathrm{I}$ which is advected by the total advection velocity $\boldsymbol{u}^\# = \nabla \times (\overline{\boldsymbol{\Psi}} + \boldsymbol{\Psi}^+)$. Note that $\boldsymbol{u}^\#$ is a function of depth and $\boldsymbol{v}^\mathrm{thk}$ a function of density, but that both are identical.

To proof that $\boldsymbol{u}^\#$ transports the modified density $\overline{\rho}^\mathrm{I}$, we first vertically integrate the mean thickness equation from the density $\rho$ to the largest density at the bottom which yields

$$\int_{\rho}^{\rho(-h)} \frac{\partial}{\partial t}\Big|_{\rho=\mathrm{const}} \overline{h_\rho} \mathrm{d}\tilde{\rho} + \int_{\rho}^{\rho(-h)} \nabla_\rho \cdot \overline{\boldsymbol{u}_\mathrm{h} h_\rho}^\mathrm{I} \mathrm{d}\tilde{\rho} + \overline{\mathcal{G}_\rho h_\rho}^\mathrm{I}\Big|_{\rho(-h)} - \overline{\mathcal{G}_\rho h_\rho}^\mathrm{I}\Big|_{\rho}$$

$$= \frac{\partial \overline{z}}{\partial t} + \nabla_\rho \cdot \overline{\boldsymbol{V}} - \overline{\mathcal{G}_\rho h_\rho}^\mathrm{I} = 0$$

where we have used the bottom boundary density equation (the upper boundary of the integrals depends on time and space in isopycnal coordinates) and (12.82). Now remember that $\nabla_\rho \chi = \nabla_\mathrm{h} \chi + \partial\chi/\partial z \nabla_\mathrm{h} z$ and use[9] $\partial\overline{z}/\partial t = \overline{h_\rho} \partial\overline{\rho}^\mathrm{I}/\partial t$ and $\nabla_\mathrm{h}\overline{z} =$

[9] The total differential of the iso-surface $\rho(\boldsymbol{x}_\mathrm{h}, z, t) = \mathrm{const}$ is given for $\boldsymbol{x}_\mathrm{h} = \mathrm{const}$ by $\mathrm{d}\rho = \partial\rho/\partial z \mathrm{d}z + \partial\rho/\partial t \mathrm{d}t = 0$, which yields $\partial z/\partial t = -(\partial\rho/\partial z)^{-1}(\partial\rho/\partial t) = h_\rho \partial\rho/\partial t$.

$h_\rho \nabla_h \overline{\rho}^I$ which yields

$$\frac{\partial \overline{\rho}^I}{\partial t} - (\nabla_h \cdot \overline{V})\frac{\partial \overline{\rho}^I}{\partial z} + \frac{\partial \overline{V}}{\partial z} \cdot \nabla_h \overline{\rho}^I = \frac{\overline{\mathcal{G}_\rho h_\rho}^I}{\overline{h_\rho}} \tag{12.83}$$

With $w^\# = -\nabla_h \cdot \overline{V} = \underline{\nabla} \cdot (\overline{\boldsymbol{\Psi}} + \boldsymbol{\Psi}^+)$ and $\boldsymbol{u}_h^\# = \partial(\underline{\overline{\boldsymbol{\Psi}}} + \underline{\boldsymbol{\Psi}}^+)/\partial z = \partial \overline{V}/\partial z$, this is indeed a conservation equation for the modified density $\overline{\rho}^I$, where the transport velocity $\boldsymbol{u}^\# = \nabla \times (\overline{\boldsymbol{\Psi}} + \boldsymbol{\Psi}^+)$, i. e. the sum of the Eulerian Mean and the Quasi-Stokes velocity, transports $\overline{\rho}^I$. Note that the forcing in (12.83) is given by the thickness weighted forcing $\overline{\mathcal{G}h_\rho}^I/\overline{h_\rho}$.

The definition of the Quasi-Stokes stream function (12.81) involves an integral which can be difficult to evaluate. A closed analytical form as a function of Eulerian Mean variables would be more helpful. McDougall and McIntosh (2001) suggest to use the following approximation to $\boldsymbol{\Psi}^+$. Using a truncated Taylor expansion of $\boldsymbol{u}_h$ at the depth $\overline{z}$ yields for $\boldsymbol{\Psi}^+(\boldsymbol{x}_h, \overline{z}, t)$

$$\begin{aligned} \boldsymbol{\Psi}^+ &= -\overline{\int_{\overline{z}}^{\overline{z}+z'} \left( \underline{\mathbf{u}}_h(\boldsymbol{x}_h, \overline{z}, t) + \frac{\partial}{\partial z}\underline{\mathbf{u}}_h \Big|_{\boldsymbol{x}_h, \overline{z}} (z^* - \overline{z}) + O(a^2) \right) \mathrm{d}z^*} \\ &= -\overline{\underline{\mathbf{u}}_h z'} - \frac{1}{2}\overline{z'^2 \frac{\partial}{\partial z}\underline{\mathbf{u}}_h} + O(a^3) \end{aligned} \tag{12.84}$$

where $a$ denotes a perturbation quantity of $z$, $\rho$ or $\boldsymbol{u}$. Note that on the left-hand side of (12.84), velocity and shear are evaluated at $\overline{z}$ such that the integrals can be be readily evaluated. We now use the expansion at $\overline{z}$ for the density at $z'$

$$\rho(\overline{z} + z') = \rho(\overline{z}) + z' \frac{\partial \overline{\rho}}{\partial z} + O(a^2)$$

We take the mean of this expansion and subtract from the instantaneous expansion which yields to first order approximation $z' = -\rho'/\partial\overline{\rho}/\partial z$. Replacing $z'$ in (12.84) yields

$$\boldsymbol{\Psi}^+(\boldsymbol{x}_h, \overline{z}, t)\frac{\partial \overline{\rho}}{\partial z} = \overline{\underline{\mathbf{u}}'_h \rho'} - \left(\frac{\partial}{\partial z}\underline{\overline{\mathbf{u}}}_h\right)\overline{\phi} \Big/ \frac{\partial \overline{\rho}}{\partial z} + O(a^3) \tag{12.85}$$

with $\phi = \rho'^2/2$. The vertical Quasi-Stokes velocity is given by $w^+ = \underline{\nabla} \cdot \boldsymbol{\Psi}^+$. Note that the first term of the approximate form (12.85) is identical to the TRM stream

**63. Third Order Correction to the Quasi-Stokes Stream Function**

The third order correction for the Quasi-Stokes stream function can be derived in a similar way as the second order, and is given by

$$\begin{aligned} \boldsymbol{\Psi}^+ \frac{\partial \overline{\rho}}{\partial z} &= \overline{\underline{\mathbf{u}}'_h \rho'} - \left(\frac{\partial}{\partial z}\underline{\overline{\mathbf{u}}}_h\right)\overline{\phi} \Big/ \frac{\partial \overline{\rho}}{\partial z} - \frac{\partial}{\partial z}\left(\overline{\underline{\mathbf{u}}'_h \phi} \Big/ \frac{\partial \overline{\rho}}{\partial z}\right) \\ &\quad + \left(\overline{\rho' \frac{\partial}{\partial z}\left(\phi \Big/ \frac{\partial \overline{\rho}}{\partial z}\right)}\frac{\partial}{\partial z}\underline{\overline{\mathbf{u}}}_h + \frac{1}{2}\left(\frac{\partial^2}{\partial z^2}\underline{\overline{\mathbf{u}}}_h\right)\overline{\phi_3} \Big/ \frac{\partial \overline{\rho}}{\partial z}\right) \Big/ \frac{\partial \overline{\rho}}{\partial z} + O(a^4) \end{aligned}$$

with $\phi_3 = \rho'^3/3$. It is, however, hard to see how this forms a systematic series in the moments of the density as for the TRM stream function.

function for the small isopycnal slope approximation (B61.3), but that the second term differs. Note also that while the Quasi-Stokes stream function transports the modified density $\overline{\rho}^{\mathrm{I}}$, the TRM stream function transports the Eulerian Mean density. Both stream functions should, therefore, in general be different.

### Isopycnal Mixing in the Isopycnal Framework

Considering the conservation equation for a general passive tracer $\chi$ in isopycnal coordinates, it is possible to rewrite the tracer equation such that the bolus velocity (in isopycnal coordinates) or the Quasi-Stokes velocity (in $z$-coordinates) transports the mean tracer. However, additional correlations between tracer and velocity remain in such a conservation equation, which are usually interpreted as isopycnal and diapycnal diffusion. The instantaneous conservation equation for the property $\chi$ in isopycnal coordinates is given by

$$\frac{\partial}{\partial t}(h_\rho \chi)\Big|_{\rho=\mathrm{const}} + \nabla_\rho \cdot (\boldsymbol{u}_{\mathrm{h}} \chi h_\rho) + \frac{\partial}{\partial \rho}(\chi \mathcal{G}_\rho h_\rho) = \mathcal{G}_\chi h_\rho$$

where $\mathcal{G}_\chi = \mathrm{D}\chi/\mathrm{D}t$ and $\mathcal{G}_\rho = \mathrm{D}\rho/\mathrm{D}t$ denote the forcing of density and tracer conservation equation, respectively. Taking the mean yields

$$\frac{\partial}{\partial t}(\overline{h_\rho \chi}^{\mathrm{I}})\Big|_{\rho=\mathrm{const}} + \nabla_\rho \cdot (\overline{\boldsymbol{u}_{\mathrm{h}} \chi h_\rho}^{\mathrm{I}}) + \frac{\partial}{\partial \rho}\overline{\chi \mathcal{G}_\rho h_\rho}^{\mathrm{I}} = \overline{\mathcal{G}_\chi h_\rho}^{\mathrm{I}}$$

Now we decompose as before the horizontal velocity and the isopycnal thickness into mean $\overline{\boldsymbol{u}}_{\mathrm{h}}^{\mathrm{I}}$ and $\overline{h_\rho}$ and deviations $\boldsymbol{u}'_{\mathrm{h}}(\boldsymbol{x}_{\mathrm{h}}, \rho, t)$ and $h'_\rho$ with vanishing means $\overline{\boldsymbol{u}'_{\mathrm{h}}}^{\mathrm{I}} = 0$ and $\overline{h'_\rho} = 0$. We also introduce the thickness weighted mean tracer $\tilde{\chi} = \overline{\chi h_\rho}^{\mathrm{I}} / \overline{h_\rho}$ and deviation from that mean $\tilde{\chi}' = \chi - \tilde{\chi}$ with $\overline{\tilde{\chi}' h_\rho} = 0$. This decomposition yields

$$\begin{aligned}
&\frac{\partial}{\partial t}(\overline{h_\rho}\tilde{\chi})\Big|_{\rho=\mathrm{const}} + \nabla_\rho \cdot \left(\boldsymbol{v}^{\mathrm{thk}} \tilde{\chi} \overline{h_\rho}\right) + \frac{\partial}{\partial \rho}\overline{\tilde{\chi}\mathcal{G}_\rho h_\rho}^{\mathrm{I}} \\
&= -\nabla_\rho \cdot \overline{\boldsymbol{u}'_{\mathrm{h}} \tilde{\chi}' h_\rho}^{\mathrm{I}} - \frac{\partial}{\partial \rho}\overline{\tilde{\chi}' \mathcal{G}_\rho h_\rho}^{\mathrm{I}} + \overline{\mathcal{G}_\chi h_\rho}^{\mathrm{I}}
\end{aligned}$$

or, by removing the mean thickness equation from the left-hand side

$$\frac{\partial}{\partial t}\tilde{\chi}\Big|_{\rho=\mathrm{const}} + \boldsymbol{u}^{\#}_{\mathrm{h}} \cdot \nabla_\rho \tilde{\chi} + \tilde{\mathcal{G}}_\rho \frac{\partial}{\partial \rho}\tilde{\chi} = -\frac{1}{\overline{h_\rho}}\nabla_\rho \cdot \overline{\boldsymbol{u}'_{\mathrm{h}} \tilde{\chi}' h_\rho}^{\mathrm{I}} - \frac{1}{\overline{h_\rho}}\frac{\partial}{\partial \rho}\overline{\tilde{\mathcal{G}}'_\rho \tilde{\chi}' h_\rho}^{\mathrm{I}} + \tilde{\mathcal{G}}_\chi \tag{12.86}$$

with the isopycnal thickness weighted forcings $\tilde{\mathcal{G}}_\rho = \overline{\mathcal{G}_\rho h_\rho}^{\mathrm{I}} / \overline{h_\rho}$ and $\tilde{\mathcal{G}}_\chi = \overline{\mathcal{G}_\chi h_\rho}^{\mathrm{I}} / \overline{h_\rho}$ and corresponding deviations. In the mean tracer conservation equation in isopycnal coordinates (12.86), the isopycnal thickness weighted tracer $\tilde{\chi}$ is laterally advected by the velocity $\boldsymbol{u}^{\#}$, which is identical to the thickness weighted velocity $\boldsymbol{v}^{\mathrm{thk}}$, and which can be expressed as the sum of the Eulerian Mean velocity and the Quasi-Stokes velocity, as shown above. Note that $\tilde{\mathcal{G}}_\rho$ plays the role of a mean diapycnal advection velocity for $\tilde{\chi}$ in isopycnal coordinates. The first term on the right-hand side of (12.86) is akin to an isopycnal eddy flux and usually interpreted as isopycnal diffusion. The latter two terms on the right-hand side of (12.86) are related to direct forcing of density and the tracer and might be interpreted as diapycnal diffusion.

**64. A Simple Parameterization for Isopycnal Mixing**

The Isopycnal Mean tracer conservation equation (12.87) it is convenient to use in ocean models, since it offers a simple way for parameterization of the remaining eddy flux. Parameterizing the 'isopycnal diffusive flux' as $\overline{\boldsymbol{u}_{\mathrm{h}}'\tilde{\chi}'h_\rho}^{\mathrm{I}} = -K_{\mathrm{i}}\overline{h_\rho}\boldsymbol{\nabla}_\rho\tilde{\chi}$ with the isopycnal diffusivity $K_{\mathrm{i}}$ and expressing the last two terms in (12.87) as diapycnal diffusion $\partial(K_{\mathrm{d}}\partial\tilde{\chi}/\partial z)/\partial z$ with the diapycnal diffusivity $K_{\mathrm{d}}$ yields

$$\frac{\partial\tilde{\chi}}{\partial t} + \boldsymbol{\nabla}\cdot\boldsymbol{u}^{\#}\tilde{\chi} = \frac{1}{\overline{h_\rho}}\boldsymbol{\nabla}_\rho\cdot\overline{h_\rho}K_{\mathrm{i}}\boldsymbol{\nabla}_\rho\tilde{\chi} + \frac{\partial}{\partial z}K_{\mathrm{d}}\frac{\partial\tilde{\chi}}{\partial z}$$

where it is usually assumed that $K_{\mathrm{i}} \gg K_{\mathrm{d}}$. The parameterized isopycnal diffusion becomes in fact identical to (12.24) of Section 12.2.2, except for the weighting factor $\overline{h_\rho}$. The same holds for the diapycnal diffusion term. Note, however, that in Section 12.2.2 the tracer was the Eulerian Mean $\overline{\chi}$, while here we have considered the isopycnal thickness weighted tracer $\overline{\chi}^{\mathrm{I}}$.

We now transform the left-hand side of (12.86) from isopycnal coordinates to $z$-coordinates, remembering that $\boldsymbol{\nabla}_\rho\chi = \boldsymbol{\nabla}_{\mathrm{h}}\chi + \partial\chi/\partial z\boldsymbol{\nabla}_{\mathrm{h}}z$, $\partial\overline{z}/\partial t = \overline{h_\rho}\partial\overline{\rho}^{\mathrm{I}}/\partial t$ and $\boldsymbol{\nabla}_{\mathrm{h}}\overline{z} = h_\rho\boldsymbol{\nabla}_{\mathrm{h}}\overline{\rho}^{\mathrm{I}}$ and using (12.83). We find

$$\frac{\partial\tilde{\chi}}{\partial t} + \boldsymbol{u}_{\mathrm{h}}^{\#}\cdot\boldsymbol{\nabla}_{\mathrm{h}}\tilde{\chi} + w^{\#}\frac{\partial\tilde{\chi}}{\partial z} = -\frac{1}{\overline{h_\rho}}\boldsymbol{\nabla}_\rho\cdot\overline{\boldsymbol{u}_{\mathrm{h}}'\tilde{\chi}'h_\rho}^{\mathrm{I}} - \frac{1}{\overline{h_\rho}}\frac{\partial}{\partial\rho}\overline{\tilde{\mathcal{G}}_\rho'\tilde{\chi}'h_\rho}^{\mathrm{I}} + \tilde{\mathcal{G}}_\chi \quad (12.87)$$

It becomes clear that the transport velocity of the modified density, $\boldsymbol{u}^{\#}$, also transports $\tilde{\chi}$ in the Isopycnal Mean tracer conservation equation (12.87). However, it also becomes clear that an eddy flux term remains on the right-hand side of (12.87). Further, there is an effect of the density forcing $\mathcal{G}_\rho$ on the Isopycnal Mean conservation equation of the general property $\chi$.

The usual interpretation of the remaining eddy flux, i. e. the first term on the right-hand side of (12.87) is mixing along isopycnals, but such isopycnal mixing of $\tilde{\chi}$ cannot be directly related to the instantaneous forcing $\mathcal{G}_\chi$. In fact, we have shown in Section 12.2.5 using the TEM framework, that most of the isopycnal mixing, when formulated for the Eulerian Mean tracer $\overline{\chi}$, is actually related to advection, i. e. related to the differences in the stream functions for eddy-driven advection in the density and the tracer conservation equation. We conclude that part of the remaining eddy flux in (12.87), could be of advective nature and not related to irreversible mixing of the mean tracer. In this respect, (12.87) might not be a consistent mean conservation equation for $\chi$, but it is still useful with respect to a simple parameterization as detailed in the box on p. 422.

Alternatively, we could formulate a semi-Lagrangian Mean conservation equation for $\overline{\chi}^{\mathrm{I}}$ by using $z(\chi)$ as material surfaces in (12.78), in which a (different) Quasi-Stokes velocity occurs, but no remaining eddy flux related to isopycnal mixing, i. e. in which the first two terms on the right-hand side of (12.87) will drop. Corresponding consistent mean conservation equations can be formulated using the TRM framework or the Generalized Lagrangian Mean, i. e. conservation equations where advective effects are consistently separated from the irreversible changes of $\chi$ related to $\mathcal{G}$. We will compare the three averaging frameworks in the next section.

### 12.3.6 Relating Lagrangian, Eulerian, and Semi-Lagrangian Mean

The steady conservation equation for vanishing $\mathcal{G}$'s formulated for the Lagrangian, Eulerian, and semi-Lagrangian Mean framework is given by

$$(\overline{\boldsymbol{u}} + \overline{\boldsymbol{u}}^{\mathrm{S}}) \cdot \nabla \overline{\chi}^{\mathrm{L}} = 0\,, \quad (\overline{\boldsymbol{u}} + \overline{\boldsymbol{u}}^{e}) \cdot \nabla \overline{\chi} = 0\,, \quad (\overline{\boldsymbol{u}} + \overline{\boldsymbol{u}}^{+}) \cdot \nabla \overline{\chi}^{\mathrm{I}} = 0 \tag{12.88}$$

with the Stokes velocity $\overline{\boldsymbol{u}}^{\mathrm{S}} = \overline{\boldsymbol{u}}^{\mathrm{L}} - \overline{\boldsymbol{u}}$ (compare Section 12.3.4), the Quasi-Stokes velocity $\overline{\boldsymbol{u}}^{+}$ given by the stream function (12.81), and the eddy-driven velocity $\overline{\boldsymbol{u}}^{e}$ given by the stream function (B61.1). Each averaging framework defines different eddy-related velocities, different mean tracers and thus different quasi-material trajectories[10]. It is only the Lagrangian Mean in which the quasi-material trajectories are independent of the tracer under consideration. This feature is unique to the Lagrangian Mean. On the other hand, no closed analytical form exists for the Lagrangian Mean velocity. It is, therefore, useful to connect $\overline{\boldsymbol{u}}^{\mathrm{S}}$ to the other velocities for which (approximate) closed forms exist.

Subtracting the Lagrangian Mean and the Eulerian Mean conservation equation in (12.88) yields

$$(\overline{\boldsymbol{u}}^{\mathrm{S}} - \overline{\boldsymbol{u}}^{e}) \cdot \nabla \overline{\chi} = -\overline{\boldsymbol{u}}^{\mathrm{L}} \cdot \nabla \overline{\chi}^{\mathrm{S}} \approx -\overline{\boldsymbol{u}} \cdot \nabla \overline{\chi}^{\mathrm{S}} \tag{12.89}$$

with the Stokes correction $\overline{\chi}^{\mathrm{S}} = \overline{\chi}^{\mathrm{L}} - \overline{\chi}$. The last step in (12.89) approximates the Lagrangian advection by Eulerian Mean advection, which is valid to second order in perturbation quantities, as detailed below. The difference between Stokes and eddy-driven velocity is thus related to the advection of the Stokes correction $\overline{\chi}^{\mathrm{S}}$. A similar relation as (12.89) holds for the difference between Stokes and Quasi-Stokes velocities, which is related to the advection of the difference in Lagrangian and semi-Lagrangian Mean tracer. One might argue that this difference is smaller than the difference between Eulerian and Lagrangian Mean tracer, i. e. Stokes correction in (12.89). However, since $\overline{\chi}^{\mathrm{L}} \neq \overline{\chi}^{\mathrm{I}}$ holds, the Stokes and Quasi-Stokes velocities are not identical. This is also reflected by the fact that the Stokes velocity is divergent, while the Quasi-Stokes velocity is given by a stream function.

To relate the Stokes velocity $\overline{\boldsymbol{u}}^{\mathrm{S}}$, the eddy-driven velocity $\overline{\boldsymbol{u}}^{e}$, and the Quasi-Stokes velocity $\overline{\boldsymbol{u}}^{+}$, a truncated expansion of the Stokes correction $\overline{\chi}^{\mathrm{S}}$ is considered. The results will, therefore, only be valid for small amplitudes of the fluctuating quantities. Note that although (12.89) applies only to the adiabatic and steady limit, small deviations from that limit are possible, as long as the effect of these deviations do not exceed the order of truncation of the expansion. Consider a Taylor expansion at the mean position $\boldsymbol{x}$ up to second order in perturbation quantities of the instantaneous value of the particle property $\chi^{\xi}$

$$\chi^{\xi} = \chi + \xi_i \frac{\partial \overline{\chi}}{\partial x_i} + \xi_i \frac{\partial \chi'}{\partial x_i} + \frac{1}{2} \xi_i \xi_j \frac{\partial^2 \overline{\chi}}{\partial x_i \partial x_j} + O(a^3) \tag{12.90}$$

Taking the mean of (12.90) yields $\overline{\chi}^{\mathrm{L}} = \overline{\chi} + O(a^2)$ since the first order term $\overline{\xi_i \partial \overline{\chi} / \partial x_i}$ vanishes. It follows that the Stokes correction $\overline{\chi}^{\mathrm{S}}$ is of second order. It will turn out below that the Stokes velocity (as the TRM eddy-driven and the Quasi-Stokes velocity) is also of second order in perturbation quantities.

[10] Since actual trajectories are different from the mean, quasi-material trajectories are hypothetical, not material trajectories.

We now assume small irreversible changes in the conservation equation for $\chi$ due to a nonzero $\mathcal{G}_\chi$, such that $\overline{\chi}^{\mathrm{L}} = \chi^\xi + O(a^2) = \bar{\chi} + O(a^2)$ and thus $\xi_i \partial \bar{\chi}/\partial x_i = -\chi' + O(a^2)$ or, formulated in two-dimensions for convenience (the three-dimensional case is discussed below)

$$\boldsymbol{\xi} = -\chi' |\nabla_{\mathrm{h}}\bar{\chi}|^{-2} \nabla_{\mathrm{h}}\bar{\chi} + \beta |\nabla_{\mathrm{h}}\bar{\chi}|^{-2} \underset{\neg}{\nabla}\bar{\chi} + O(a^2)$$

with $\beta = \boldsymbol{\xi} \cdot \underset{\neg}{\nabla}\bar{\chi}$. It is now also assumed that $\beta$ is much smaller than $\chi'$, i.e. a perturbation quantity of second order $\beta = O(a^2)$. We then have $\boldsymbol{\xi} = -\chi' |\nabla_{\mathrm{h}}\bar{\chi}|^{-2} \nabla_{\mathrm{h}}\bar{\chi} + O(a^2)$ and we can calculate the second order Stokes correction in (12.90) after some manipulations as

$$\overline{\xi_i \frac{\partial \chi'}{\partial x_i}} + \frac{1}{2}\overline{\xi_i \xi_j} \frac{\partial^2 \bar{\chi}}{\partial x_i \partial x_j} = -|\nabla_{\mathrm{h}}\bar{\chi}|^{-1} \nabla_{\mathrm{h}}\bar{\chi} \cdot \nabla_{\mathrm{h}}(\overline{\phi}/|\nabla_{\mathrm{h}}\chi|) \tag{12.91}$$

with the Eulerian Mean variance $\overline{\phi} = \overline{\chi'^2}/2$. Decomposing the Eulerian Mean velocity $\bar{\boldsymbol{u}}$ into $u_n = \boldsymbol{u} \cdot \tilde{\boldsymbol{n}}$ and $u_s = \boldsymbol{u} \cdot \tilde{\boldsymbol{s}}$ with the unit vectors $\tilde{\boldsymbol{n}} = |\nabla_{\mathrm{h}}\bar{\chi}|^{-1} \nabla_{\mathrm{h}}\bar{\chi}$ and $\tilde{\boldsymbol{s}} = |\nabla_{\mathrm{h}}\bar{\chi}|^{-1} \underset{\neg}{\nabla}\bar{\chi}$ yields in (12.89) for the Stokes velocity

$$\overline{\boldsymbol{u}}^{\mathrm{S}} \cdot \tilde{\boldsymbol{n}} = \left( \overline{\boldsymbol{u}}^e + |\nabla_{\mathrm{h}}\chi|^{-1} (u_n \nabla_{\mathrm{h}} - u_s \underset{\neg}{\nabla}) \tilde{\boldsymbol{n}} \cdot \nabla_{\mathrm{h}} \frac{\overline{\phi}}{|\nabla_{\mathrm{h}}\chi|} \right) \cdot \tilde{\boldsymbol{n}} + O(a^3) \tag{12.92}$$

Now we can infer from the Eulerian Mean conservation equation for small forcing term $\mathcal{G}$ that $\bar{\boldsymbol{u}} \cdot \nabla_{\mathrm{h}}\bar{\chi} = -\overline{\boldsymbol{u}' \cdot \nabla_{\mathrm{h}}\chi'} = O(a^2)$ and that $\bar{\boldsymbol{u}} = \underset{\neg}{\nabla}\bar{\psi}(\bar{\chi}) + O(a^2) = \bar{\psi}' \underset{\neg}{\nabla}\bar{\chi} + O(a^2)$, i.e. $u_n = O(a^2)$ and $u_s = \bar{\psi}' |\nabla_{\mathrm{h}}\bar{\chi}| + O(a^2)$. This means that we can neglect the term related to $u_n$ in (12.92).

With the TRM form for the eddy-driven velocity $\overline{\boldsymbol{u}}^e = -\underset{\neg}{\nabla}(\boldsymbol{s} \cdot \overline{\boldsymbol{u}'\chi'} |\nabla_{\mathrm{h}}\bar{\chi}|^{-1}) + \underset{\neg}{\nabla}(\boldsymbol{s} \cdot (\underset{\neg}{\nabla}\mathcal{T}) |\nabla_{\mathrm{h}}\bar{\chi}|^{-1})$ and with the definition of the rotational eddy fluxes $\mathcal{T} = u_s |\nabla_{\mathrm{h}}\chi|^{-1} \overline{\phi} + O(a^3)$ from (12.66) we find

$$\begin{aligned} \overline{\boldsymbol{u}}^{\mathrm{S}} \cdot \tilde{\boldsymbol{n}} = \Bigg[ &-\underset{\neg}{\nabla} \frac{\overline{u_s'\chi'}}{|\nabla_{\mathrm{h}}\bar{\chi}|} + \underset{\neg}{\nabla} \frac{\tilde{\boldsymbol{n}} \cdot \nabla_{\mathrm{h}}(u_s \overline{\phi} |\nabla_{\mathrm{h}}\chi|^{-1})}{|\nabla_{\mathrm{h}}\bar{\chi}|} \\ &- \frac{u_s}{|\nabla_{\mathrm{h}}\chi|} \underset{\neg}{\nabla}\tilde{\boldsymbol{n}} \cdot \nabla_{\mathrm{h}} \frac{\overline{\phi}}{|\nabla_{\mathrm{h}}\chi|} \Bigg] \cdot \tilde{\boldsymbol{n}} + O(a^3) \end{aligned} \tag{12.93}$$

It becomes clear that the rotational component in TRM, i. e. the second term in (12.93), is modified while the first component is identical for the Stokes velocity and the eddy-driven velocity. It also becomes clear that the Stokes velocity might have a divergent component given by the last term in brackets in (12.93). A divergent transport velocity for incompressible flow is an inherent feature of the Generalized Lagrangian Mean formulation. However, it will turn out that at second order, the divergent part on the right-hand side of (12.93) vanishes.

By combining a part of the divergent (third) term with the second term in (12.93), where parts cancel, the following alternative form of (12.93) is derived

$$\begin{aligned} \overline{\boldsymbol{u}}^{\mathrm{S}} \cdot \tilde{\boldsymbol{n}} = \Bigg[ &-\underset{\neg}{\nabla} \frac{\overline{u_s'\chi'}}{|\nabla_{\mathrm{h}}\bar{\chi}|} + \underset{\neg}{\nabla} \frac{\tilde{\boldsymbol{n}} \cdot (\nabla u_s) \overline{\phi} |\nabla_{\mathrm{h}}\chi|^{-1}}{|\nabla_{\mathrm{h}}\bar{\chi}|} \\ &+ \tilde{\boldsymbol{n}} \cdot \nabla_{\mathrm{h}} \left( \frac{\overline{\phi}}{|\nabla_{\mathrm{h}}\chi|} \right) \underset{\neg}{\nabla} \frac{u_s}{|\nabla_{\mathrm{h}}\chi|} \Bigg] \cdot \tilde{\boldsymbol{n}} + O(a^3) \end{aligned} \tag{12.94}$$

**65. Generalized Quasi-Stokes Velocity in Three Dimensions**

For three dimensions the Stokes correction (12.91) remains valid (the nabla operator becomes three-dimensional, i. e. $\boldsymbol{\nabla}_{\mathrm{h}} \to \boldsymbol{\nabla}$) given that $\boldsymbol{\xi} = -\chi'|\bar{\chi}|^{-1}\boldsymbol{\nabla}\bar{\chi} + O(a^2)$ still holds. The Eulerian Mean velocity is decomposed using the three-dimensional unit vector $\boldsymbol{n} = |\boldsymbol{\nabla}\bar{\chi}|^{-1}\boldsymbol{\nabla}\bar{\chi}$ as $\overline{\boldsymbol{u}} = \bar{u}_n \boldsymbol{n} + \overline{\boldsymbol{u}}_s \times \boldsymbol{n}$ with $\bar{u}_n = \overline{\boldsymbol{u}} \cdot \boldsymbol{n}$ and with $\overline{\boldsymbol{u}}_s \cdot \boldsymbol{n} = 0$ since then $\overline{\boldsymbol{u}}_s = \boldsymbol{n} \times \overline{\boldsymbol{u}}$. This decomposition yields in (12.89)

$$\overline{\boldsymbol{u}}^{\mathrm{S}} \cdot \boldsymbol{n} = \left[\overline{\boldsymbol{u}}^{e} + |\boldsymbol{\nabla}\bar{\chi}|^{-1}\left(\bar{u}_n - \overline{\boldsymbol{u}}_s \times\right)\boldsymbol{\nabla}\boldsymbol{n} \cdot \boldsymbol{\nabla}\left(\bar{\phi}|\boldsymbol{\nabla}\bar{\chi}|^{-1}\right)\right] \cdot \boldsymbol{n} + O(a^3)$$

With the same argument as above, we can neglect the term related to $\bar{u}_n$. The last term in brackets can be decomposed into a nondivergent and divergent part,

$$-|\boldsymbol{\nabla}\bar{\chi}|^{-1}\overline{\boldsymbol{u}}_s \times \boldsymbol{\nabla}\bar{\chi}^{\mathrm{S}} = \boldsymbol{\nabla} \times \bar{\chi}^{\mathrm{S}}\overline{\boldsymbol{u}}_s|\boldsymbol{\nabla}\bar{\chi}|^{-1} - \bar{\chi}^{\mathrm{S}}\boldsymbol{\nabla} \times \overline{\boldsymbol{u}}_s|\boldsymbol{\nabla}\bar{\chi}|^{-1}$$

for which it turns out (after some algebra) that the latter vanishes in the mean conservation equation to $O(a^3)$. Thus we can rewrite (12.89) with $\overline{\boldsymbol{u}}^{e} = \boldsymbol{\nabla} \times \boldsymbol{\Psi}^{e}$, the TRM vector stream function $\boldsymbol{\Psi}^{e}|\boldsymbol{\nabla}\bar{\chi}| = \boldsymbol{n} \times (\overline{\boldsymbol{u}'\chi'} + \boldsymbol{\nabla} \times \boldsymbol{\theta})$, and the rotational eddy flux $\boldsymbol{\theta}|\boldsymbol{\nabla}\bar{\chi}| = \boldsymbol{n} \times \overline{\boldsymbol{u}\phi} + O(a^3)$ from (B61.2) after some algebra as

$$\overline{\boldsymbol{u}}^{\mathrm{S}} \cdot \boldsymbol{n} = \left[\boldsymbol{\nabla} \times \left(\boldsymbol{\Psi}^{e} + \chi^{\mathrm{S}}\overline{\boldsymbol{u}}_s|\boldsymbol{\nabla}\bar{\chi}|^{-1}\right)\right] \cdot \boldsymbol{n} + O(a^3) = \left(\boldsymbol{\nabla} \times \boldsymbol{\Psi}^{+}\right) \cdot \boldsymbol{n} + O(a^3)$$

$$\boldsymbol{\Psi}^{+}|\boldsymbol{\nabla}\bar{\chi}| = \boldsymbol{n} \times \left(\overline{\boldsymbol{u}'\chi'} + \bar{\phi}|\boldsymbol{\nabla}\bar{\chi}|^{-1}\boldsymbol{\nabla} \times \overline{\boldsymbol{u}}_s\right)$$

$\boldsymbol{\Psi}^{+}$ is the generalized three-dimensional version of the approximate Quasi-Stokes stream function (12.85). To show this we specify $\chi$ as density and consider the strongly stratified case. It follows within the small slope limit for the operator $(\boldsymbol{n} \times \boldsymbol{a})_{\mathrm{h}} \approx \underline{\mathrm{a}}_{\mathrm{h}}$ and thus $(\boldsymbol{\nabla} \times \overline{\boldsymbol{u}}_s)_{\mathrm{h}} \approx -\partial\overline{\boldsymbol{u}}_{\mathrm{h}}/\partial z$, and for the stream functions

$$\boldsymbol{\Psi}^{+}_{\mathrm{h}}\frac{\partial\bar{\chi}}{\partial z} \approx \overline{\underline{\mathrm{u}}'\chi'} - \frac{\partial\underline{\overline{\mathrm{u}}}_{\mathrm{h}}}{\partial z}\bar{\phi}\Big/\frac{\partial\bar{\chi}}{\partial z}, \quad \boldsymbol{\Psi}^{e}_{\mathrm{h}}\frac{\partial\bar{\chi}}{\partial z} \approx \overline{\underline{\mathrm{u}}'\chi'} - \frac{\partial}{\partial z}\left(\underline{\overline{\mathrm{u}}}_{\mathrm{h}}\bar{\phi}\Big/\frac{\partial\bar{\chi}}{\partial z}\right) + O(a^3),$$

$$\overline{\boldsymbol{u}}^{\mathrm{S}}_{\mathrm{h}} \approx \frac{\partial}{\partial z}\underline{\boldsymbol{\Psi}}^{+}$$

This form for $\boldsymbol{\Psi}^{+}$ is identical to (12.85).

The nondivergent part of the form (12.94), i. e. the first and second part in brackets, is a generalized version of the approximate Quasi-Stokes stream function introduced in Section 12.3.5. This becomes more clear in the box on p. 425, where the three-dimensional version of the relation (12.94) within the small isopycnal slope limit is considered, which can be directly compared to the approximate Quasi-Stokes stream function (12.85).

It remains to be shown that the third term in brackets in (12.94) vanishes. Using again $u_s = \bar{\psi}'|\nabla_{\mathrm{h}}\bar{\chi}| + O(a^2)$ for $\bar{\psi} = \bar{\psi}(\bar{\chi}) + O(a^2)$ we can rewrite (12.93) and (12.94) with the generalized Quasi-Stokes stream function $\psi^{+}|\nabla_{\mathrm{h}}\bar{\chi}| = -\overline{u'_s\chi'} + \tilde{\boldsymbol{n}} \cdot (\nabla_{\mathrm{h}}u_s)\bar{\phi}|\nabla_{\mathrm{h}}\chi|^{-1}$ and the TRM stream function $\psi^{e}|\nabla_{\mathrm{h}}\bar{\chi}| = -\overline{u'_s\chi'} + \tilde{\boldsymbol{n}} \cdot \nabla_{\mathrm{h}}(u_s\bar{\phi}|\nabla_{\mathrm{h}}\chi|^{-1}) + O(a^3)$ as

$$\begin{aligned}\overline{\boldsymbol{u}}^{\mathrm{S}} \cdot \tilde{\boldsymbol{n}} &= \left(\underline{\nabla}\psi^{e} + \bar{\psi}'\underline{\nabla}\bar{\chi}^{\mathrm{S}}\right) \cdot \tilde{\boldsymbol{n}} + O(a^3)\\ &= \left(\underline{\nabla}\psi^{+} - \bar{\chi}^{\mathrm{S}}\bar{\psi}''\underline{\nabla}\bar{\chi}\right) \cdot \tilde{\boldsymbol{n}} + O(a^3)\end{aligned}$$

Since $\underline{\nabla}\bar{\chi} \cdot \tilde{\boldsymbol{n}} = 0$, it becomes clear that in the approximate steady and adiabatic Generalized Lagrangian Mean conservation equation $(\overline{\boldsymbol{u}} + \overline{\boldsymbol{u}}^{\mathrm{S}}) \cdot \nabla_{\mathrm{h}}\bar{\chi} + \overline{\boldsymbol{u}} \cdot \nabla_{\mathrm{h}}\bar{\chi}^{\mathrm{S}} = O(a^3)$, the replacement of $\overline{\boldsymbol{u}}^{\mathrm{S}}$ by $\underline{\nabla}\psi^{+}$ would not introduce any error in the second order.

Using in contrast the TRM stream function $\psi^e$, would introduce an error related to the divergent part of the Stokes velocity.

Therefore, the Quasi-Stokes stream function $\boldsymbol{\psi}^+$ resembles a second order approximation to the divergent Stokes velocity, which is valid for weakly diabatic flow. $\boldsymbol{\psi}^+$ is better suited (has smaller error) for advecting the Lagrangian Mean tracer (density) $\overline{\chi}^{\mathrm{L}}$ than the TRM stream function $\boldsymbol{\psi}^e$. On the other hand, for the steady and adiabatic case, $\boldsymbol{\psi}^e$ yields a conservation equation for the Eulerian Mean density $\bar{\chi}$ valid to any order, similar to the Lagrangian Mean conservation equation. The Quasi-Stokes stream function is only valid to second order, if used in the conservation equation for $\overline{\chi}^{\mathrm{L}}$ instead of $\overline{\boldsymbol{u}}^{\mathrm{L}}$, but yields a conservation equation for the modified tracer, as shown in Section 12.3.5. Note that the exact Stokes velocity will be divergent, and thus no stream function exists in general. On the other hand, no closed general form for the exact Stokes velocity exists which may make the approximate Quasi-Stokes velocity useful.

## Further Reading

The theory of homogeneous turbulence is based Kolmogorov's (1941) paper, and the classic *Theory of Homogeneous Turbulence* by Batchelor (1990).

Aspects of that theory are reviewed in a collection edited by Hunt et al. (1991), and in *Turbulence* by Frisch (1995).

*Turbulent Flows* by Pope (2000) gives an account of modeling and simulation of turbulent flows of constant density.

A wealth of material on ocean surface layer turbulence can be found in *Small Scale Processes in Geophysical Fluid Flows* by Kantha and Clayson (2000).

Thorpe (2005) reviews in *The Turbulent Ocean* the measurement of oceanic turbulence and the processes leading to turbulent motions, such as breaking internal waves and instabilities of oceanic currents.

Modeling of turbulence in the oceanic boundary layer has initially been based on bulk models, such as described by Niiler and Kraus (1977).

A review on the parameterization of turbulent mixing in the boundary layer was given by Large et al. (1994), see also Large (1998).

Estimates of diapycnal mixing rates derived from observations are discussed e. g. in Toole (1998).

Some current approaches to the theory and modeling of oceanic turbulence are described in *Marine Turbulence: Theories, Observations, and Models* by Baumert et al. (2005).

The article *The Dynamics of Unsteady Currents* by Rhines (1977) has provided an early overview on ocean waves and turbulence on the beta-plane.

The *Lectures on Geophysical Fluid Dynamics* by Salmon (1998) contain an account of geostrophic turbulence.

The parameterization of eddies in large scale flow is discussed in an article by Killworth (1998). In the same book, McDougall (1998) discusses the three-dimensional residual-mean theory.

A profound discussion of the Generalized Lagrangian Mean theory is found in *Waves and Mean Flows* by Bühler (2009).

Aspects of eddy dynamics in the context of a two-layer isopycnal model are discussed in the book *Fundamentals of Geophysical Fluid Dynamics* by McWilliams (2006).

# Part V
# Aspects of Ocean Circulation Theory

The increasing interest in climate change has focussed the research in physical oceanography during recent years more and more on the role of the ocean for climate variability. Feedback processes between ocean and atmosphere and the resulting modes of coupled variability are among the topics, as well as the potential of the ocean circulation to contribute to past or future changes in the climate system and the ocean's role to extent predictability to time scales longer than those of the atmospheric synoptic variability. Although the eminent importance of these topics for society clearly justifies this research, the understanding of the physics of the ocean should not move to the background. Due to the increasing performance of modern computers, there is the danger that ocean circulation becomes a tool box to be switched on in a numerical experiment, in which the understanding of the results and its underlying physics becomes superfluous. In fact, numerical models can simulate today the ocean circulation in surprisingly rich details, if the resolution is high enough. Numerical modeling is not the topic of this book; we offer instead classical and new simplified analytical models to understand and to study the large-scale ocean circulation for given atmospheric forcing functions. We discuss the basic physical ingredients of ocean circulation physics, starting with the classical suite of circulation models complemented with new extensions in most cases. Although we restrict the presentation to the present steady state conditions, one might use the same (or slightly modified) simplified models for the description and understanding of the variability of the ocean and the role of the ocean in climate change as well.

Chapter 13 introduces the boundary conditions at the sea surface for practical use by bulk formulas and presents the forcing functions, basically the components of the surface fluxes of energy, heat and freshwater and the flux of momentum. The wind-driven circulation is discussed in Chapter 14. Based on the simple concept of an 'Elementary Current System', the classical Stommel–Munk model is derived. The BARBI concept and the associated numerical model are developed to elucidate the wind-driven circulation in the presence of stratification and topography. Furthermore, the theory of thermocline ventilation is reviewed. The physics of the oceanic meridional overturning, specifically the one forced by thermohaline processes, is discussed in Chapter 15. The Stommel-Arons overturning model and the classical and more recent box models are presented, and an extended discussion follows on zonally averaged models and their closure problem. Finally, Chapter 16 is devoted

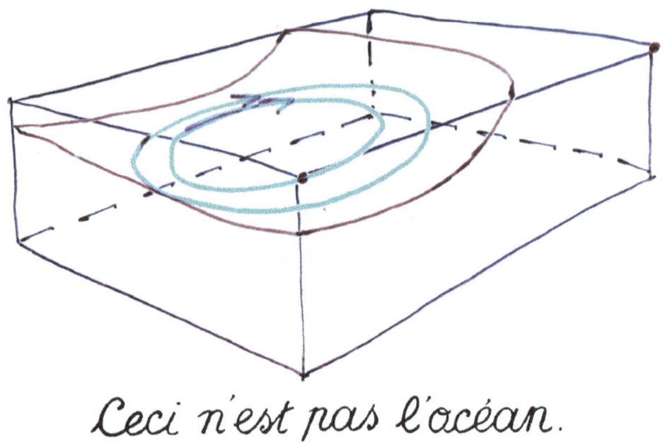


Physics is about models which try to make images of specific features of the real world. Ocean circulation models of analytical kind are often very remote from reality. It is always the hope of a model builder that the few coarse ingredients of his simple model mirror some processes in the real world. The situation is remedied by today's high-resolution numerical models of the ocean circulation. Simple models, however, helped to construct these comprehensive models. They are also valuable to train the physical intuition about the behavior of the system and guide the interpretation of the results of numerical models

to the circulation in the Southern Ocean, in particular its two interleaving branches: the predominantly wind-driven Antarctic Circumpolar Current and the wind- and thermohaline-driven overturning circulation. The dynamics are elucidated by a number of analytical and numerical models which cope with the interaction of stratification, topography and mesoscale eddy effects.

We like to emphasize that the low resolution of many of the simple models, introduced in the Chapters 14–16, is far from being adequate to represent the real ocean circulation. In fact, it is hard to judge whether a two-box system as Stommel's thermohaline model discussed in Section 15.5 is anything more than a dynamical system loaded with interesting nonlinear mathematics, or if its bifurcation structure is to some degree embedded as a 'center manifold' (see Appendix A.2) in more complete circulation models. At least, its bifurcations have been reproduced by more complex three-dimensional general circulation models. In fact, the simplified models and their implications, as derived in the following chapters, are in most cases supported by more advanced models. On the other hand, more complex models like numerical models based on the primitive equations, also rely on turbulence closures which are know to often perform poor in the ocean, as discussed in the previous chapters. The different simplified model types – isopycnal models of various configurations, the two-layer quasi-geostrophic model, the BARBI model of a wind-driven stratified ocean with topography, and the mathematics of spectral and box models – are summarized in Appendix B.

# 13 Forcing of the Ocean

In this chapter, we discuss the surface forcing functions of the large-scale ocean circulation. The exact boundary conditions derived in Chapter 2 are approximated and replaced by the so-called 'bulk formulae' for practical use in the ocean circulation theory. These approximated boundary conditions are discussed on the basis of observational estimates. We also present some simplified boundary conditions which are used to construct the models of the large-scale circulation in the following chapters.

The boundary conditions for those surfaces where the ocean touches the other components of the Earth system, i. e. the atmosphere, the sea ice, and the land, were derived and discussed in detail in Chapter 2. The boundary conditions express the physical requirement of continuity of the fluxes of momentum, partial masses, and internal energy across (and normal to) the boundaries. They are briefly repeated here:

1. The exact kinematic boundary conditions at the sea surface and the bottom of the ocean are given by (2.10) and (2.11), respectively. With respect to the large-scale flow in the ocean, it is reasonable to assume small slopes of the sea surface or the bottom, i. e. to assume that $(\nabla_h \xi)^2 \ll 1$, where $z = \xi(x, y, t)$ denotes either the vertical position of the sea surface ($\xi = \zeta$) or the ocean bottom ($\xi = -h$). Then the kinematic boundary conditions (2.10) and (2.11) simplify to

$$\begin{aligned} \partial\zeta/\partial t + \boldsymbol{u}_h \cdot \nabla_h \zeta - w &= -\mathcal{J}_{mass}/\rho \quad \text{at the sea surface} \quad z = \zeta \\ \boldsymbol{u}_h \cdot \nabla_h h + w &= 0 \quad \text{at the ocean bottom} \quad z = -h \end{aligned} \tag{13.1}$$

   Here, $\mathcal{J}_{mass} = \mathcal{E} - \mathcal{P}$ denotes the air-sea mass flux in kg m$^{-2}$ s$^{-1}$ which is given by the difference between evaporation $\mathcal{E}$ and precipitation $\mathcal{P}$. Note that the melting or freezing of sea ice and continental freshwater run-off might also contribute to $\mathcal{J}_{mass}$, which is, however, not considered here. Furthermore, a zero mass flux through the ocean bottom was assumed.
2. To a good approximation evaporation and precipitation exchange only pure water such that there is zero salt flux through the surface and also through the ocean bottom (although on very long time-scales there might be a nonzero salt flux due to erosion on land and exchange with the Earth's crust). Using the air-sea freshwater flux $\mathcal{E} - \mathcal{P}$ (positive upwards), the boundary condition for the salinity

D. Olbers, J. Willebrand, C. Eden, *Ocean Dynamics*,
DOI 10.1007/978-3-642-23450-7_13, © Springer-Verlag Berlin Heidelberg 2012

budget, (2.25), can be written in form of an equivalent salt flux

$$\begin{aligned} \boldsymbol{J}_S \cdot \boldsymbol{n} &= -S(\mathcal{E} - \mathcal{P}) \quad \text{at} \quad z = \zeta \\ \boldsymbol{J}_S \cdot \boldsymbol{n} &= 0 \quad \text{at} \quad z = -h \end{aligned} \tag{13.2}$$

where $\boldsymbol{n}$ denotes the unit vector normal to the boundary.

3. In Section 2.3, the dynamic boundary condition for the sea surface for the force tangential to the interface was approximated as

$$\boldsymbol{\Sigma} \cdot \boldsymbol{n} = \boldsymbol{\tau}^{\mathrm{a}} \quad \text{at} \quad z = \zeta \tag{13.3}$$

where $\boldsymbol{\tau}^{\mathrm{a}}$ denotes the windstress tangential to the sea surface, and where $\Sigma_{ij}$ represents the frictional stress tensor. The windstress $\boldsymbol{\tau}^{\mathrm{a}}$ represents a viscous *vertical* flux of the horizontal momentum components, if, for simplicity, the sea surface is taken as horizontal (and $\boldsymbol{n}$ as the vertical unit vector). The dynamic boundary condition for the vertical momentum component was approximated as continuity in pressures of ocean and atmosphere, i. e. $p|_{z=\zeta} = p^{\mathrm{a}}$.
At the bottom of the ocean, it is less clear how to proceed. Both free-slip ($\boldsymbol{\Sigma} \cdot \boldsymbol{n} = 0$) and no-slip ($\boldsymbol{u} = 0$) boundary conditions are possible as discussed in Section 2.3. The bottom boundary conditions for momentum are not further discussed here but left to the specific handling in the models under investigation.

4. The flux $\boldsymbol{J}_{\mathrm{H}} = \boldsymbol{J}_{\mathrm{T}} + \boldsymbol{J}_{\mathrm{rad}}$ of enthalpy or internal energy has been defined in (2.75) as the sum of sensible and radiative heat flux, up to a negligible component related to the salinity variations. We follow the convention that the energy exchange at the bottom of the ocean is also usually neglected, but note that such energy fluxes, although small compared to those at the surface, have the potential to generate substantial large-scale flow. The boundary condition (2.86) for the enthalpy flux becomes under this assumption

$$\begin{aligned} \boldsymbol{J}_{\mathrm{H}} \cdot \boldsymbol{n} &= \boldsymbol{J}_{\mathrm{H}}^{\mathrm{a}} \cdot \boldsymbol{n} + L_{\mathrm{q}} \mathcal{E} \quad \text{at} \quad z = \zeta \\ \boldsymbol{J}_{\mathrm{H}} \cdot \boldsymbol{n} &= 0 \quad \text{at} \quad z = -h \end{aligned} \tag{13.4}$$

Here $\boldsymbol{J}_{\mathrm{H}}^{\mathrm{a}} = \boldsymbol{J}_{\mathrm{T}}^{\mathrm{a}} + \boldsymbol{J}_{\mathrm{rad}}^{\mathrm{a}}$ denotes the sum of atmospheric sensible and radiative heat fluxes, and $L_{\mathrm{q}}$ is the latent heat of condensation as defined in (2.85). Note that the last two terms in (2.86), related to enthalpy differences of seawater to freshwater respectively to precipitating water, are small and have been neglected. For the normal flux components in the atmosphere the following notation is used:

$$L_{\mathrm{q}} \mathcal{E} = \mathcal{J}_{\mathrm{L}} \,, \quad \boldsymbol{J}_{\mathrm{T}}^{\mathrm{a}} \cdot \boldsymbol{n} = \mathcal{J}_{\mathrm{T}} \,, \quad \boldsymbol{J}_{\mathrm{rad}}^{\mathrm{a}} \cdot \boldsymbol{n} = \mathcal{J}_{\mathrm{SW}} + \mathcal{J}_{\mathrm{LW}} \tag{13.5}$$

Here $\mathcal{J}_{\mathrm{L}}$, $\mathcal{J}_{\mathrm{T}}$ denote the latent and sensible heat fluxes, and the short-wave ($\mathcal{J}_{\mathrm{SW}}$) and long-wave ($\mathcal{J}_{\mathrm{LW}}$) radiation components have been introduced. These will be discussed further in the following section. The sum $\mathcal{J}_{\mathrm{heat}}$ of these fluxes is the normal component of the net air-sea heat flux[1] given as

$$\mathcal{J}_{\mathrm{heat}} = \mathcal{J}_{\mathrm{SW}} + \mathcal{J}_{\mathrm{LW}} + \mathcal{J}_{\mathrm{T}} + \mathcal{J}_{\mathrm{L}} \tag{13.6}$$

For the radiation inside the ocean, it is usually sufficient to explicitly consider only the short-wave flux divergence. Then the boundary condition (13.4) takes the form

$$\boldsymbol{J}_{\mathrm{T}} \cdot \boldsymbol{n} = \mathcal{J}_{\mathrm{heat}} - \mathcal{J}_{\mathrm{SW}} \quad \text{and} \quad \boldsymbol{J}_{\mathrm{rad}} \cdot \boldsymbol{n} = \mathcal{J}_{\mathrm{SW}} \quad \text{at} \quad z = \zeta \tag{13.7}$$

[1] The quantity $\mathcal{J}_{\mathrm{heat}}$ is referred to as net *energy* flux in the meteorological literature.

Alternatively, under conditions where the radiative flux in the ocean is completely absorbed within the well-mixed layer, it is not necessary to consider the radiative component at all. In this case, (13.7) can be approximated by $\boldsymbol{J}_{\mathrm{T}} \cdot \boldsymbol{n} = \mathcal{J}_{\mathrm{heat}}$.

## 13.1 Bulk Formulae as Boundary Conditions

For practical use in analytical and numerical models of the ocean, the boundary fluxes $\mathcal{J}_{\mathrm{mass}} = \mathcal{E} - \mathcal{P}$, $\boldsymbol{\tau}^{\mathrm{a}}$ and the contributions to the normal heat flux components on the right-hand side of (13.6) have to be specified. It is important to acknowledge first that with respect to the large-scale (Reynolds-averaged) flow in the ocean, the fluxes across the boundaries are in almost any case the result of processes in turbulent boundary layers, i. e. the boundary fluxes are often of turbulent nature and need to be parameterized in terms of resolved (large-scale) quantities of the model. The upper

**66. The Global Wind System**

The figure below (redrawn after Open University, 1989) shows the prevailing surface winds for the situation in the northern summer (left panel) and the northern winter (right panel). Unlike in the ocean, the circulation in the atmosphere is relatively unconstrained in zonal (east-west) direction. The large mountain ranges deflect the winds somewhat, but to a first approximation they are independent of longitude. There is a clear pattern of westward (easterly) blowing trade winds on both sides of the equator, eastward blowing (westerly) winds in somewhat higher latitudes, and also a belt of easterly winds circling the polar regions.

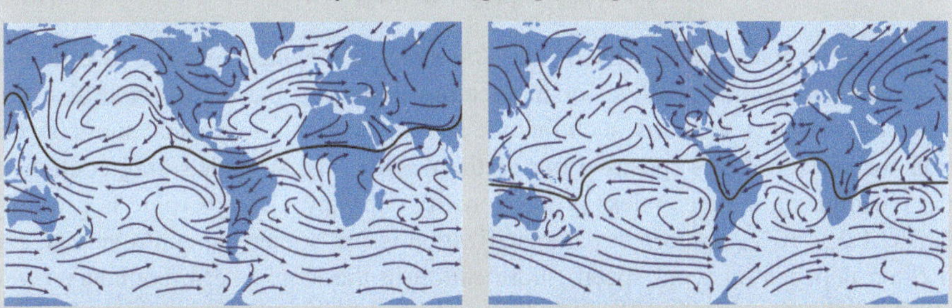

A convergence of the surface winds can be noticed in the 'intertropical convergence zone' (ITCZ) near the equator (indicated by a solid black line in the figure). At the ITCZ, moist air is warmed by the strong solar radiation near the surface; the moist and warm air rises, condensates when the pressure gets lower, and rises even further. This air moves polewards in the upper atmosphere, where it eventually sinks, establishing a 'meridional overturning circulation' in each hemisphere. The equatorward branch of this global circulation is often called the 'Hadley cell'. High pressure regions are located over each of the subtropical gyres, where the air is circulating anticyclonic around those highs in near geostrophic balance. Here, a fraction of the air in the upper atmosphere is sinking and is accordingly dry such that evaporation is high and exceeds precipitation. As a consequence, the salinity in the surface ocean is higher in the subtropics than in tropical or high-latitude regions, where precipitation exceeds evaporation. Low pressure regions can be seen in the North Pacific and Atlantic Ocean, with cyclonic surface circulation, but no such clear indication in the southern hemisphere.
The ITCZ – and with it the upwelling branch of the whole global MOC in the atmosphere – migrates seasonally towards the summer hemisphere. This migration is most pronounced in the Indian Ocean. In the northern part of the Indian Ocean, a seasonal reversal of the surface winds can even be seen. This reversal is related to the Indian monsoon, with south-westerly winds during summer extending from the Indian subcontinent to and even crossing the equator, and north-easterlies during winter. In the tropical Pacific and Atlantic Ocean, similar seasonal changes in the surface winds can be seen, but in general with smaller seasonal signals at higher latitudes.

turbulent boundary layer in the ocean is called the mixed layer (see Section 11.3) and typically of 50–200 m vertical extent, while the lower turbulent layer in the atmosphere is often called the planetary boundary layer with a typical extent of about 1,000 m. Parameterizations of the turbulent air-sea fluxes have been developed over the recent decades in terms of 'bulk formulae', which we briefly describe in this section.

Observations have shown that vertical fluxes of momentum, matter, and energy are constant within a shallow layer of a few meters above the surface, and empirical laws, i. e. the bulk formulae, have been elaborated to relate these fluxes to the values of velocity, partial masses, and temperature at the upper boundary of this 'constant flux layer' and the corresponding sea surface properties. The standard level of this constant flux layer is 10 m height where any sea surface elevation is ignored.

### Windstress

The air-sea boundary flux of horizontal momentum is related to the activity of the sea surface gravity waves – a complex and turbulent process which has to be parameterized by the large-scale variables (in Reynolds-averaged form). The air-sea momentum flux is parameterized by a drag law relating the tangential surface stress, i. e. the windstress $\boldsymbol{\tau}^{\mathrm{a}}$, to the difference between the 10 m wind, $\boldsymbol{u}_{\mathrm{air}}$, in the atmosphere and the surface velocity, $\boldsymbol{u}_{\mathrm{s}}$, in the ocean in the form

$$\boldsymbol{\tau}^{\mathrm{a}} = \rho_{\mathrm{air}} C_{\mathrm{D}} |\boldsymbol{u}_{\mathrm{air}} - \boldsymbol{u}_{\mathrm{s}}| (\boldsymbol{u}_{\mathrm{air}} - \boldsymbol{u}_{\mathrm{s}}) \tag{13.8}$$

where $\rho_{\mathrm{air}}$ denotes the density of air. The stress is oriented into the direction of $\boldsymbol{u}_{\mathrm{air}} - \boldsymbol{u}_{\mathrm{s}}$ and is thus positive for momentum input into the ocean. Note that this sign convention is standard, such that we deviate here from our usual sign convention for surface fluxes (positive out of the ocean). The so-called drag coefficient $C_{\mathrm{D}}$ resembles a dimensionless proportionality constant; a rough value is given by $C_{\mathrm{D}} \approx 1.2 \times 10^{-3}$. Its correct parameterization is still somewhat controversial, and it often depends on wind speed, stratification in the atmospheric boundary layer, etc. Note that since $|\boldsymbol{u}_{\mathrm{s}}| \ll |\boldsymbol{u}_{\mathrm{air}}|$ holds over many regions of the ocean, and the $\boldsymbol{u}_{\mathrm{s}}$ part is often neglected for the bulk formulae forcing ocean models, although the correct formulation might become important in regions of strong ocean currents, e. g. the equator, and for estimates of the work done by the winds on the ocean.

The windstress reflects the large-scale surface wind pattern as shown and discussed in the box on p. 431 and Figure 13.1 shows the long-term and annual mean components of $\boldsymbol{\tau}^{\mathrm{a}}$ based on the bulk parameterization in the NCEP/NCAR reanalysis data-set (Kalnay et al., 1996). Typical magnitudes of the zonal component of $\boldsymbol{\tau}^{\mathrm{a}}$ are $0.1\,\mathrm{N\,m^{-2}}$, positive in the midlatitude west wind region and negative in the subtropical and tropical regions of prevailing trade winds. Maximum zonal windstress shows up in the stormy Southern Ocean with values of $0.2\,\mathrm{N\,m^{-2}}$, exceeding this size specifically in the Indian Ocean part. The meridional component of $\boldsymbol{\tau}^{\mathrm{a}}$ is in general smaller in magnitude and displays the equatorward direction of the trade winds and the poleward deflection of the west winds in the eastern parts of the major ocean basins. Note that the large magnitudes in particular of the meridional component of $\boldsymbol{\tau}^{\mathrm{a}}$ near Greenland and Antarctica are difficult to interpret and might be too large and biased in the reanalysis data-set due to missing observations. A similar caveat has to be mentioned regarding the quality of the windstress from reanalysis products in

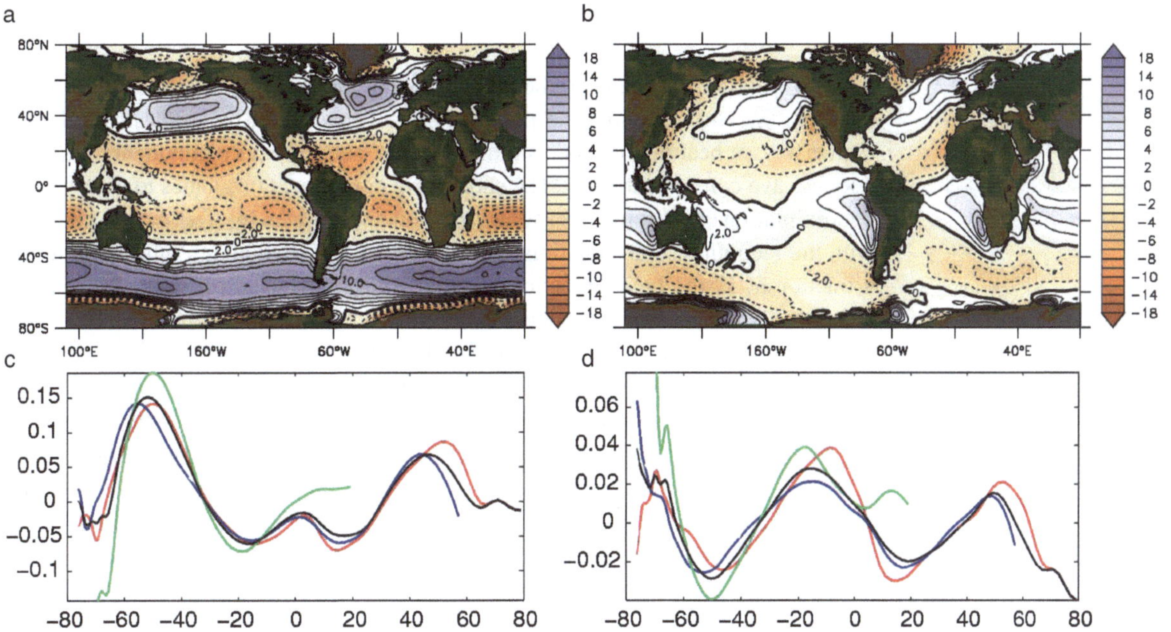

Fig. 13.1 Long-term mean of the zonal (**a**) and the meridional component (**b**) of the windstress $\boldsymbol{\tau}^{a}$ in $10^{-2}\,\mathrm{N\,m^{-2}}$. Data source is the NCEP/NCAR reanalysis (Kalnay et al., 1996). Both stress components have been smoothed over several grid boxes to remove the grid-scale noise. The stress is positive for input of the respective momentum component into the ocean. Also shown are the zonal averages of the zonal (**c**) and meridional component (**d**) of the windstress $\boldsymbol{\tau}^{a}$ in $\mathrm{N\,m^{-2}}$ as a function of latitude. Note the different vertical axes in **c** and **d**. The global mean is denoted by the black line, the zonally averaged components over the Atlantic, Pacific and Indian Ocean are denoted by red, blue and green lines, respectively

equatorial regions, since here atmospheric general circulation models tend to show also large biases.

Figure 13.1 also shows the zonally averaged annual mean zonal and the meridional windstress derived from the NCEP/NCAR data-set. The globally averaged windstress is very similar in both components compared to the windstress averaged over individual ocean basins. The exception is the Indian Ocean which deviates in its northern part from the global mean due to the monsoon (and also near Antarctica). The zonal mean meridional component is smaller than the zonal component, and from the equator polewards around 60°, the zonal mean zonal windstress resembles very much a (negative) cosine function in both hemispheres. Note that such a pattern, i. e. a purely zonal windstress as negative cosine of a scaled latitude, will be used in the following chapters to obtain an idealized (but apparently realistic to a certain extent) forcing in the classical models of the wind-driven circulation.

Note that within the Boussinesq approximation, a factor $\rho_0$ is sometimes absorbed in the windstress for convenience. This is done for instance in Chapter 14. The units of $\boldsymbol{\tau} = \boldsymbol{\tau}^{a}/\rho_0$ are then $\mathrm{m^2\,s^{-2}}$ instead of $\mathrm{N\,m^{-2}}$ for $\boldsymbol{\tau}^{a}$.

### Radiative Surface Heat Flux

Another important forcing for the large-scale circulation of the ocean is the surface heat flux. We first consider the radiative part of this flux. The spectrum of the radiative air-sea heat flux can be summarized into a short-wave component, $\mathcal{J}_{\mathrm{SW}}$, and a long-wave component, $\mathcal{J}_{\mathrm{LW}}$. The short-wave component originates to a very good approximation only from direct solar radiation and is given by $\mathcal{J}_{\mathrm{SW}} = -S_0(1-\alpha_s)$, with the incoming solar radiation $S_0$ at the sea surface and taking into account that a fraction $\alpha_s$ of the incident short-wave radiation is reflected ($\alpha_s$ is the sea surface albedo). The long-wave component is dominated by infrared thermal radiation and, for the upward part, given by the Stefan–Boltzmann law $\mathcal{J}_{\mathrm{LW}}^{\uparrow} = \epsilon\sigma T_s^4$, where $T_s$ denotes surface temperature, $\sigma$ the Stefan–Boltzmann constant and $\epsilon$ an emissivity coefficient. The net long-wave component $\mathcal{J}_{\mathrm{LW}}$ is made up by the sum of $\mathcal{J}_{\mathrm{LW}}^{\uparrow}$ and the downward part $\mathcal{J}_{\mathrm{LW}}^{\downarrow}$ arising by reflection and radiation within the atmosphere.

Figure 13.2 shows the long-term mean of $\mathcal{J}_{\mathrm{SW}}$ and $\mathcal{J}_{\mathrm{LW}}$ as given by the NCEP/NCAR reanalysis data-set. Both fluxes are positive when directed out of the ocean. Clearly, the short-wave component, dominated by the incident solar radiation, is negative anywhere, i. e. it is a heating of the ocean. As expected, it is largest in magnitude towards the tropics with maximum values of 250 W m$^{-2}$. Within the ITCZ and the upwelling regions on the eastern margins of the continents, relative minima in $\mathcal{J}_{\mathrm{SW}}$ show up, related to the almost persistent cloud cover over these regions, which reduces $\mathcal{J}_{\mathrm{LW}}$ sightly.

In contrast to $\mathcal{J}_{\mathrm{SW}}$, the long-wave component $\mathcal{J}_{\mathrm{LW}}$ is positive anywhere, i. e. it is cooling the ocean, but smaller in magnitude with maximum values of 80 W m$^{-2}$. The maxima show up in mid- to high latitudes, while minima of long-wave radiation out of the ocean are located in the tropics – e. g. at the ITCZ and upwelling regions where the cloud cover yields larger $\mathcal{J}_{\mathrm{LW}}^{\downarrow}$ – and in ice-covered regions. Due to the strong solar radiation, however, the total radiative heat flux at the air-sea surface is negative almost everywhere, only during winter at high-latitudes in ice-covered regions, radiative heat fluxes out of the ocean can show up.

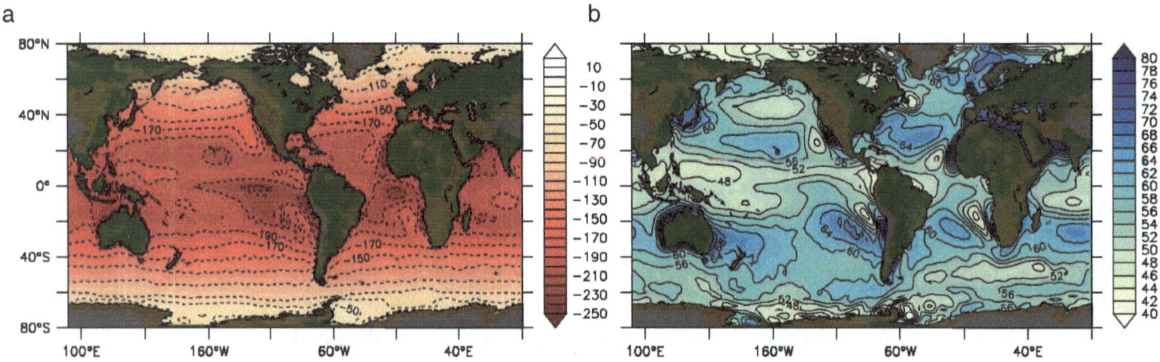


Fig. 13.2 Long-term mean short-wave ($\mathcal{J}_{\mathrm{SW}}$, shown in **a**) and long-wave ($\mathcal{J}_{\mathrm{LW}}$, shown in **b**) radiative heat flux in W m$^{-2}$. Data source is the NCEP/NCAR reanalysis (Kalnay et al., 1996). $\mathcal{J}_{\mathrm{LW}}$ has been smoothed over several grid boxes to remove the grid-scale noise. Fluxes are positive when directed into the atmosphere. Red colors denote heating and blue colors cooling of the ocean

### Turbulent Surface Heat Flux

In addition to the radiative heat fluxes, there are fluxes related to the turbulent transports in the constant flux layer. There is a heat flux associated with the turbulent flux of water vapor in the constant flux layer related to evaporation at the sea surface, the 'latent' heat flux $\mathcal{J}_\mathrm{L}$ defined in (13.5). Further, there is a direct turbulent heat flux, the 'sensible' heat flux $\mathcal{J}_\mathrm{T}$. The sensible heat flux $\mathcal{J}_\mathrm{T}$ is parameterized by the difference of surface air and water temperature, and a similar relation is taken for the rate of evaporation $\mathcal{E}$ for the latent heat flux $\mathcal{J}_\mathrm{L}$, i. e.

$$\mathcal{J}_\mathrm{T} = \rho_\mathrm{air} c_p C_\mathrm{H} |\boldsymbol{u}_\mathrm{air}| (T_\mathrm{s} - T_\mathrm{air}) \,, \quad \mathcal{J}_\mathrm{L} = L_\mathrm{q}\mathcal{E} = L_\mathrm{q}\rho_\mathrm{air} C_\mathrm{E} |\boldsymbol{u}_\mathrm{air}| (q_\mathrm{s} - q_\mathrm{air}) \tag{13.9}$$

with dimensionless coefficients $C_\mathrm{H}$ and $C_\mathrm{E}$ of order $10^{-3}$ (which might be chosen as functions of stratification, wind speed etc.) as before for the windstress in (13.8). Furthermore, $c_p$ denotes the specific heat capacity, $\rho_\mathrm{air}$ the density, and $\boldsymbol{u}_\mathrm{air}$ the velocity of the air (the wind speed). The variables $T_\mathrm{air}$ and $q_\mathrm{air}$ are the air temperature and specific humidity taken at the standard level, and $q_\mathrm{s}(T_\mathrm{s})$ is the saturation value of humidity at the sea surface which is a function of the surface temperature $T_\mathrm{s}$. The sign convention is such that positive $\mathcal{J}_\mathrm{T}$, $\mathcal{J}_\mathrm{L}$ and $\mathcal{E}$ express a loss of the ocean, i. e. a flux from the sea surface to the atmosphere.

Figure 13.3 shows the long-term mean latent heat flux $\mathcal{J}_\mathrm{L}$ and sensible heat flux $\mathcal{J}_\mathrm{T}$ from the NCEP/NCAR reanalysis data-set. The long-term mean evaporation at the sea surface $\mathcal{E}$ from the same data-set is also shown in Figure 13.5. The latent heat flux is positive almost anywhere, i. e. cooling the ocean. There are only very restricted regions and times with $\mathcal{E} < 0$, i. e. where water vapor condensates at the sea surface. $\mathcal{J}_\mathrm{L}$ and $\mathcal{E}$ are largest over the midlatitude subtropical gyres, where evaporation is strongest due to the dry sinking air related to the Hadley circulation. Towards the equator, $\mathcal{J}_\mathrm{L}$ and $\mathcal{E}$ are smaller than their midlatitudinal maxima, since here specific humidity is in general larger and wind speeds are smaller than in the subtropics. Evaporation and $\mathcal{J}_\mathrm{L}$ is in particular large over the western boundary currents with maximum values of up to $200\,\mathrm{W\,m^{-2}}$, where warm water is advected within the western boundary currents polewards and underlies during winter time dry cold air blown from the continents with the midlatitude westerlies. At higher latitudes, the latent heat flux becomes smaller, typically on the order of $20\,\mathrm{W\,m^{-2}}$.

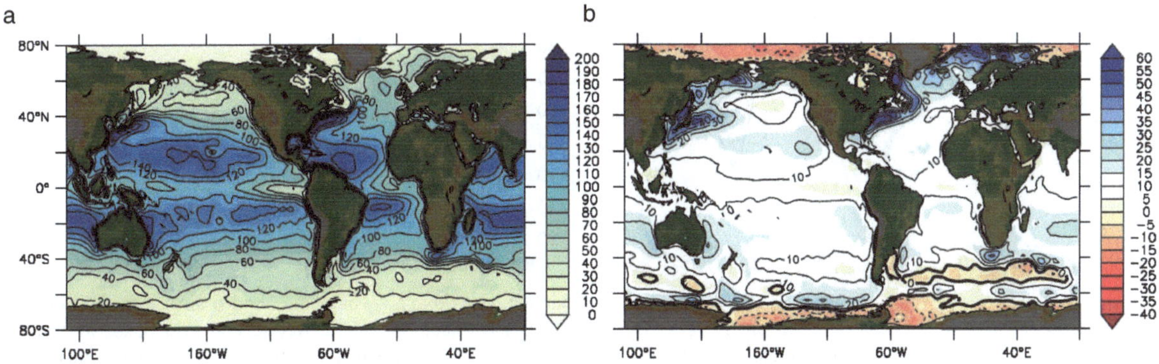


Fig. 13.3 Long-term mean latent ($\mathcal{J}_\mathrm{L}$, shown in **a**) and sensible ($\mathcal{J}_\mathrm{T}$, shown in **b**) turbulent heat flux in $\mathrm{W\,m^{-2}}$. Data source is the NCEP/NCAR reanalysis (Kalnay et al., 1996). Both data sets have been smoothed over several grid boxes to remove the grid-scale noise. Fluxes are positive when directed into the atmosphere. Red colors denote heating and blue colors cooling of the ocean

The smallest contribution to the net air-sea heat flux comes from the sensible heat flux $\mathcal{J}_{\mathrm{T}}$, which is also shown in Figure 13.3 as a long-term mean from the NCEP/NCAR reanalysis data-set. $\mathcal{J}_{\mathrm{T}}$ is positive, i.e. cooling the ocean almost everywhere with magnitudes of about 10–20 W m$^{-2}$ in midlatitudes, corresponding to a 1–2 K air-sea temperature difference. It becomes much larger in the western boundary currents, where, during winter, cold air from the continents and warm water, sustained by the poleward western boundary currents, lead to a much larger air-sea temperature difference and magnitudes of $\mathcal{J}_{\mathrm{T}}$ as large as $\mathcal{J}_{\mathrm{L}}$. The sensible heat flux $\mathcal{J}_{\mathrm{T}}$ can also become important near the sea-ice margin in polar regions of the North Atlantic and Pacific Ocean, where the air-sea difference stays large, with maximum values of 100 W m$^{-2}$.

### Net Surface Heat Flux

All parameterized energy fluxes can be summed up to the net air-sea heat flux $\mathcal{J}_{\mathrm{heat}}$ according to (13.6). Figure 13.4 shows the long-term mean net air-sea heat flux $\mathcal{J}_{\mathrm{heat}}$ from the NCEP/NCAR reanalysis data-set as a global map and as a zonal average. The global pattern of $\mathcal{J}_{\mathrm{heat}}$ is dominated by regions of large heat loss by the ocean within western boundary currents and in the polar regions of the Atlantic Ocean with maximum values of up to 200 W m$^{-2}$ in the Gulf stream region. The ocean takes up heat predominantly in the tropics with maximum magnitudes of about 100 W m$^{-2}$

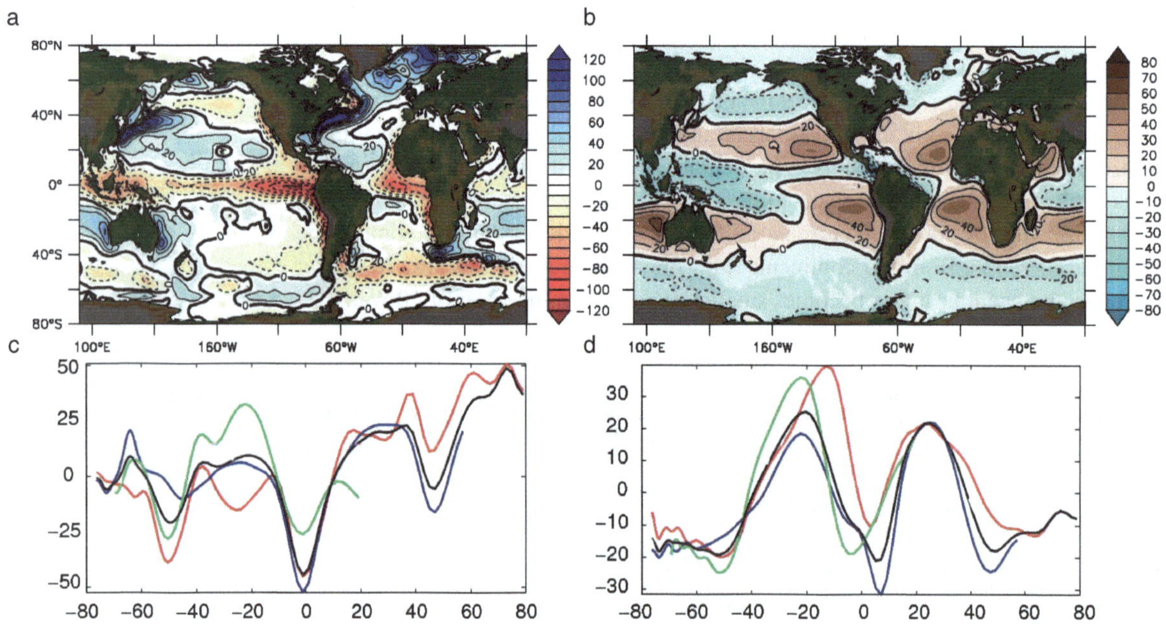


**Fig. 13.4** Long-term mean **a** net air-sea heat flux $\mathcal{J}_{\mathrm{heat}}$, in W m$^{-2}$ and **b** air-sea freshwater flux $\mathcal{P}-\mathcal{E}$, in mg m$^{-2}$s$^{-1}$. Both data sets have been smoothed over several grid boxes to remove the grid-scale noise. Data source is the NCEP/NCAR reanalysis (Kalnay et al., 1996). Fluxes are positive when directed out of the ocean, i.e. heating of the ocean is denoted by red colors and freshening of the ocean by blue colors and vice versa. Also shown are zonal means of $\mathcal{J}_{\mathrm{heat}}$ (**c**) and $\mathcal{P}-\mathcal{E}$ (**d**) in the same units as a function of latitude for the global zonal mean (black line) and zonal averages taken over the Atlantic (red), Pacific (blue), and Indian Ocean (green)

in the eastern tropical Pacific ocean. Clearly, the long-term mean of the global pattern of $\mathcal{J}_{\text{heat}}$ mirrors the lateral heat transport within the ocean, as discussed in the box on p. 438. This lateral heat transport is dominated on the global scale by the meridional overturning circulation (see Chapter 15). The overturning circulation is in particular effective in transporting heat because the temperature difference from the sea surface to the deep ocean is in general one order of magnitude larger than the zonal temperature differences, while the zonally integrated northward mass transport associated with the overturning circulation is of similar strength as the vertically integrated northward transport related to the horizontal circulation.

## Surface Freshwater Flux

Figure 13.4 also shows the long-term mean air-sea freshwater flux $\mathcal{P} - \mathcal{E}$ from the NCEP/NCAR data-set as a global map and zonal averages in individual basins, while Figure 13.5 shows global maps of the individual contributions $\mathcal{P}$ and $\mathcal{E}$. There is freshwater loss by the ocean or an equivalent salt flux into the ocean in subtropical regions, where the sea surface salinity is correspondingly large. This freshwater loss is driven by the large evaporation and small precipitation over the subtropical gyres, which are dominated by the sinking dry air related to the Hadley circulation, inhibiting cloud formation. Towards the equator in the rising branch of the Hadley circulation, however, and in particular within the ITCZ, precipitation is very high, while $\mathcal{E}$ is lower such that the ocean gains freshwater. At mid to high latitudes, the ocean also receives freshwater, since evaporation becomes low but precipitation is slightly increased, in particular within the atmospheric storm tracks.

Note that precipitation $\mathcal{P}$ is difficult to measure on a global scale. Further, atmospheric general circulation models, like the one used for the NCEP/NCAR reanalysis data-set, also have problems in the correct representation of precipitation, since the complex processes involved in cloud formation and precipitation are not resolved but parameterized in those models. Therefore, the observational estimates of $\mathcal{P}$ shown and discussed here are biased. Note that the checker board type structure in $\mathcal{E}$ of Figure 13.5 in the Southern Ocean part appears to be such a model bias.

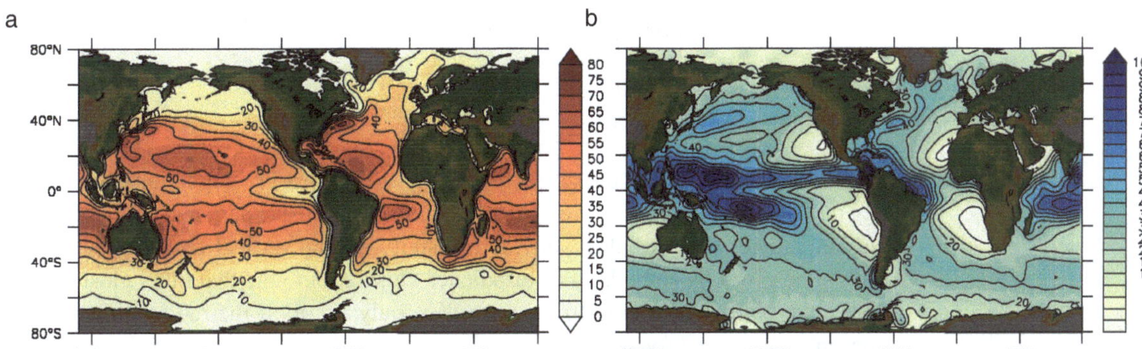


Fig. 13.5 Long-term mean **a** net evaporation $\mathcal{E}$, and **b** precipitation $\mathcal{P}$, both in mg $m^{-2}s^{-1}$. Both data sets have been smoothed over several grid boxes to remove the grid-scale noise. Data source is the NCEP/NCAR reanalysis (Kalnay et al., 1996). Freshening of the ocean is denoted by blue colors, freshwater flux out of the ocean by red colors

**67. The Global Ocean Heat and Freshwater Balance**

The most important role of the ocean in climate are storage and transport of energy. Concerning storage, the ocean has a much larger heat capacity than the atmosphere: 10 m of water weigh roughly as much as the entire atmosphere. Furthermore, the heat capacity of sea water is about 4 times that of air. Hence 2.5 m of water hold as much heat as the entire atmosphere (at a corresponding temperature). In the long run, the net air-sea heat exchange $\mathcal{J}_{\text{heat}}$ (shown in Figure 13.4) is nearly balanced. Local surplus, mostly received during summer, is mixed into the near-surface upper layer of the ocean by wind generated turbulence, is stored there and transported by currents, and later, at other times and places, mostly in the midlatitudes in winter, it is available again for the exchange with the atmosphere. At higher latitudes, the ocean has a negative energy balance at the surface, i. e. in the sum of all exchange processes energy is given into the atmosphere. The deficit is balanced by the meridional transport of the heat gained by the ocean in tropical areas. Large-scale imbalances, which are maintained by the ocean for a long time, represent climate variations. Processes concerning heat storage and transport are, therefore, basic components of the ocean's role in climate.

The three major ocean basins have rather different heat transports. The Pacific and Indian Ocean are similar to the global zonal mean: the heat transport is nearly antisymmetric about the equator, and heat is transported poleward in both hemispheres. The most remarkable pattern of heat transport occurs in the Atlantic: it is northward everywhere. It is up-gradient, from cold to warm, in the South Atlantic and down-gradient in the North Atlantic. The size is typically $1\,\text{PW} = 10^{15}\,\text{W}$, occurring at the latitude of Florida. This large northward heat transport is related to the Meridional Overturning Circulation (see Chapter 15), which also shares the same direction in the South and North Atlantic and provides a significant cross-equatorial heat transport.

Like the heat transport being balanced by the air-sea heat flux, the oceanic freshwater transport balances in the long run the air-sea freshwater flux $\mathcal{E} - \mathcal{P}$ at the sea surface shown in Figure 13.4. In the observational estimate, it is dominated by a large circumpolar transport, which is, however, to a large extent of rotational nature, i. e. the dominant part of it just circles around Antarctica – the Antarctic Circumpolar Current with a transport of $140\,\text{Sv} = 140 \times 10^6\,\text{m}^3\text{s}^{-1}$ (see Chapter 16) – and features (approximately) no zonal divergence which would in steady state correspond to a zero net air-sea freshwater flux. The values of the net meridional transport in the oceanic basins, like the Atlantic, are considerably lower, typically 1 Sv, which is roughly the amount entering the Atlantic from the Arctic Ocean.

## Heat and Freshwater Fluxes below the Surface

Above the constant flux layer in the atmospheric planetary boundary layer and below it in the ocean, all fluxes (except for the short-wave radiation which needs special treatment) are carried further as parameterized by the diffusive approximations. From (13.2), (13.3) and (13.7) one then has

$$-\left[\rho K_{\text{v}} \frac{\partial S}{\partial z}\right]_{z=\zeta} = S_{\text{s}}(\mathcal{P} - \mathcal{E}) \,, \quad \left[\rho A_{\text{v}} \frac{\partial \boldsymbol{u}}{\partial z}\right]_{z=\zeta} = \boldsymbol{\tau}^{\text{a}}$$

$$-\left(\rho c_p K_{\text{v}} \frac{\partial T}{\partial z}\right)_{z=\zeta} = \mathcal{J}_{\text{heat}} - \mathcal{J}_{\text{SW}} \tag{13.10}$$

Here, $S_{\text{s}}$ is the salinity at the sea surface and $\mathcal{P}$ is the rate of precipitation (positive; in $\text{kg}\,\text{m}^{-2}\text{s}^{-1}$). For simplicity, we have taken here the diffusive parameterization for the oceanic fluxes with turbulent vertical diffusivity $K_{\text{v}}$ and vertical viscosity $A_{\text{v}}$, which might be given by some kind of turbulence closure, but note that other closures are possible. Note that in (13.10) the term $\mathcal{J}_{\text{SW}}$ only occurs if the radiation inside the ocean is accounted for separately.

Since the short-wave radiative heat flux $\mathcal{J}_{\text{SW}}$ is able to penetrate into the ocean before it is absorbed, it is necessary to implement this flux separately from the other fluxes, which enter the ocean directly at the sea surface. $\mathcal{J}_{\text{SW}}$ then affects the temperature budget in the form of an interior source term; the penetration depth of the short-wave radiation depends on the turbidity of the sea water, caused by the presence of phytoplankton and suspended matter, and also on wavelength: for perfectly clear water, blue light is to 99% absorbed within the upper 200 m while red light is almost completely absorbed within the upper 10 m. However, for models in which the details of the processes related to surface heating in the first 10–50 m are unimportant, the solar radiation $\mathcal{J}_{\mathrm{SW}}$ is often simply added to the diffusive boundary flux condition at the surface.

## 13.2 Simplified Boundary Conditions

A common way to force an ocean model with realistic surface boundary fluxes is to use prescribed atmospheric parameters, $\boldsymbol{u}_{\mathrm{air}}$, $q_{\mathrm{air}}$, $T_{\mathrm{air}}$, etc. as given e. g. by a data-set or an atmospheric model, and to use the predicted surface values of the ocean model in the bulk formulae, i. e. $\boldsymbol{u}_{\mathrm{s}}$, $T_{\mathrm{s}}$ and $S_{\mathrm{s}}$, in order to obtain the surface boundary fluxes. However, for the net surface heat flux $\mathcal{J}_{\mathrm{heat}}$, a simplified alternative can be derived using the fact that the surface heat flux components depend only on the sea surface temperature $T_{\mathrm{s}}$ and (given) atmospheric parameters. A truncated Taylor expansion of $\mathcal{J}_{\mathrm{heat}}$ around a climatological mean value $T_{\mathrm{s}}^{\mathrm{clim}}$ yields

$$\mathcal{J}_{\mathrm{heat}} = \mathcal{J}_{\mathrm{heat}}(T_{\mathrm{s}}^{\mathrm{clim}}) + \left.\frac{\partial \mathcal{J}_{\mathrm{heat}}}{\partial T_{\mathrm{s}}}\right|_{T_{\mathrm{s}}^{\mathrm{clim}}} (T_{\mathrm{s}}^{\mathrm{clim}} - T_{\mathrm{s}}) + \dots \tag{13.11}$$

The value of $\partial \mathcal{J}_{\mathrm{heat}}/\partial T_{\mathrm{s}} = \partial \mathcal{J}_{\mathrm{LW}}/\partial T_{\mathrm{s}} + \partial \mathcal{J}_{\mathrm{H}}/\partial T_{\mathrm{s}} + \partial \mathcal{J}_{\mathrm{L}}/\partial T_{\mathrm{s}}$ can be analytically derived from the respective bulk formulae using the prescribed atmospheric parameters. When the model simulates a $T_{\mathrm{s}}$ equal to the observed climatology $T_{\mathrm{s}}^{\mathrm{clim}}$, the surface heat flux will be given by the observed climatological heat flux $\mathcal{J}_{\mathrm{heat}}(T_{\mathrm{s}}^{\mathrm{clim}})$. When the simulated $T_{\mathrm{s}}$ deviates from $T_{\mathrm{s}}^{\mathrm{clim}}$, e. g. due to a model bias, the flux will change but assuming unchanged atmospheric conditions.

The surface heat flux boundary condition (13.11) can also be written as

$$\mathcal{J}_{\mathrm{heat}} = \lambda (T_{\mathrm{app}} - T_{\mathrm{s}}) \tag{13.12}$$

with $\lambda = \partial \mathcal{J}_{\mathrm{heat}}/\partial T_{\mathrm{s}}$, which then takes the form of a relaxation condition towards the target temperature $T_{\mathrm{app}} = T_{\mathrm{s}}^{\mathrm{clim}} + \mathcal{J}_{\mathrm{heat}}(T_{\mathrm{s}}^{\mathrm{clim}})/\lambda$. The coefficient $\lambda$ is called the relaxation coefficient and has values of about 50 W m$^{-2}$ K$^{-1}$ in midlatitudes; or, with respect to temperature instead of heat, the corresponding coefficient $\lambda/(\rho c_p)$ is on the order of 1 m/d. Assuming that the flux $\mathcal{J}_{\mathrm{heat}}$ heats a 50 m deep mixed layer with no other processes affecting it, the temperature in the mixed layer will be relaxed to the target temperature $T_{\mathrm{app}}$ on a time-scale of 50 d.

The surface freshwater flux $\mathcal{E} - \mathcal{P}$ is an important forcing for the meridional overturning circulation as discussed in Chapter 15. While evaporation is related to the latent heat flux $\mathcal{J}_{\mathrm{L}}$ and thus on depends ocean-atmosphere parameters similar to the surface heat flux, precipitation is not related to ocean-atmosphere parameters at the surface. In particular, the surface freshwater flux does not at all depend on surface salinity (but on $T_{\mathrm{s}}$), and the (frequently employed) replacement by a relaxation condition for surface salinity similar to the one for surface heat flux (13.12),

$$S_{\mathrm{s}}(\mathcal{P} - \mathcal{E})/\rho = \lambda_{\mathrm{S}}(S_{\mathrm{s}}^{\mathrm{clim}} - S_{\mathrm{s}}) \tag{13.13}$$

has, therefore, no physical justification. On the other hand, the accuracy of observed estimates of the freshwater flux has been far from satisfactory for a long time: as the overall salt content of the ocean is conserved on usual oceanic time-scales, systematic errors in the freshwater budget add up, leading to a drift in the salinity distribution and often to a large drift of the meridional overturning circulation. Therefore, a relaxation boundary condition has often been used in many ocean models, despite its missing physical motivation. A consequence of the relaxation condition (13.13) is that in case of a perfect simulation, i. e. $S^{\mathrm{clim}} = S_{\mathrm{s}}$, the freshwater flux vanishes, which might not be the case. There are ways to treat this problem, but we refrain from a further discussion here.

As formulated by the kinematic surface boundary condition (13.1), the net mass flux $\mathcal{P} - \mathcal{E}$ from the atmosphere to the ocean also leads to a rise of the local sea level by $(\mathcal{P} - \mathcal{E})/\rho$ (adding to the change induced by currents). It is customary to ignore this effect so that the total ocean volume remains constant. Moreover, the *rigid lid approximation* $w = 0$ at $z = 0$ is usually applied as simplified kinematic boundary condition, which also filters out surface gravity waves and long barotropic Rossby waves.

# 14 The Wind-Driven Circulation

In this chapter we discuss the effects of forcing the ocean by a prescribed windstress. We will start the discussion for an ocean in steady state and constant density, followed by a consideration of the effects of stratification and topography. Important concepts and results are the Ekman layers, Ekman spiral and Ekman transport, the Sverdrup transport and the Stommel/Munk gyre models of the large-scale circulation. The impact of topography and stratification on the depth-averaged flow is introduced using the concept of the JEBAR-term and the bottom pressure torque, which are both further investigated using the model BARBI. This model represents the interaction of barotropic and baroclinic fields in a minimal way. The spin-up of the wind-driven circulation by wave processes was presented in Section 8.2.6.

It was discussed in Chapter 11.3 that the momentum, given to the ocean by the surface wind friction, is vertically distributed by small-scale isotropic turbulent mixing. This mixing takes place predominantly from the surface to a depth of 30–50 m – which defines the so-called EKMAN[1] layer. It will be discussed in this chapter how this mixing generates in steady state a directly wind-driven, near-surface current, which is in the vertical integral over the Ekman layer orthogonal to the wind direction and proportional to the windstress magnitude. This volume transport – the so-called the Ekman transport – can be easily calculated when the windstress is given. In the midlatitude west-wind belt, shown for instance in Figure 13.1, the Ekman transport is directed predominantly equatorward, while the subtropical easterly or trade winds force a poleward Ekman transport. Therefore, the Ekman transport produces a convergence in the center of the large-scale subtropical gyre system, i. e. an upwelling of the sea surface and a downward motion of the converging water masses at the lower boundary of the surface Ekman layer. Subtropical gyre systems can be found in each of the major ocean basins; they are indicated in the sketch of the global near surface circulation in Figure 14.1 as red streamlines.

As a consequence of the convergent Ekman transport and the associated downwelling – the so-called Ekman pumping – the density layers, lying stratified upon each other, buckle downwards and form a bowl in the subtropical gyres. The resulting horizontal density gradients lead to pressure forces and thus to a geostrophic

[1] VAGN WALFRID EKMAN, *1874 in Stockholm, †1954 in Gostad, oceanographer and physicist.

D. Olbers, J. Willebrand, C. Eden, *Ocean Dynamics*,
DOI 10.1007/978-3-642-23450-7_14, © Springer-Verlag Berlin Heidelberg 2012

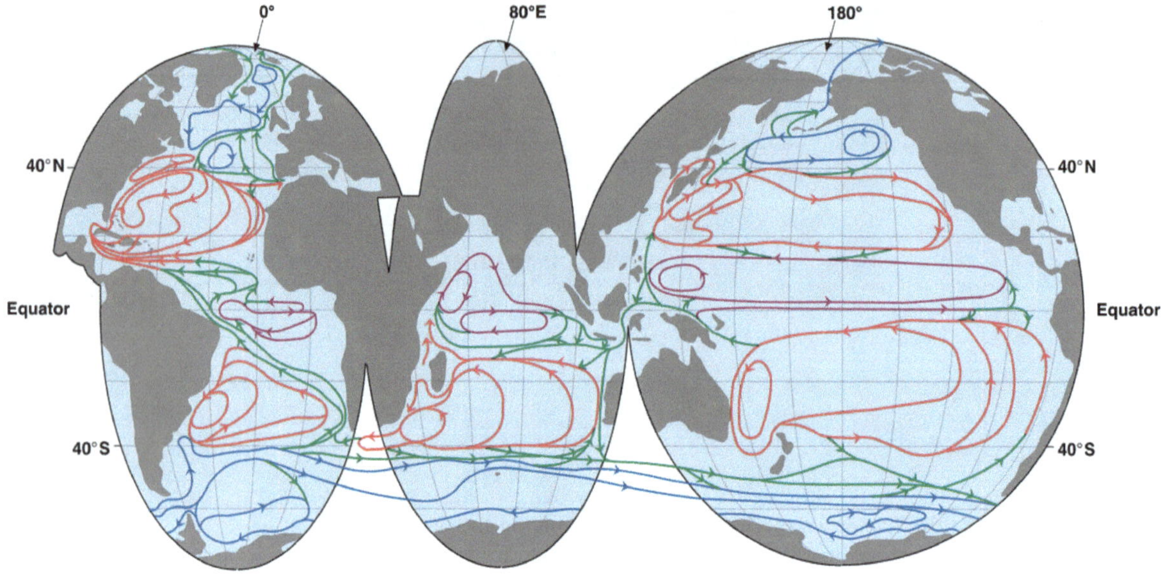


Fig. 14.1 A schematic view of the near-surface ocean circulation (Schmitz Jr., 1996). Subtropical gyres are represented with red, subpolar and polar gyres with *blue*, the equatorial gyres with *magenta lines*. The Antarctic Circumpolar Current is also blue. The *green lines* represent exchange between basins and gyres

circulation where pressure and Coriolis forces balance. This geostrophic transport below the Ekman layer has to compensate the wind-driven Ekman pumping in steady state. It will be discussed in detail in this chapter that in deeper layers, the pressure gradients caused by the density field and the pressure gradients caused by the sloping of the ocean surface mostly compensate, so that the currents strongly decrease from the surface towards the bottom, such that the deep sea in the ocean basins is nearly without currents compared to the surface circulation. The same wind-driven mechanism taking place in the subtropical gyres is also valid for the subpolar gyres – indicated by blue streamlines in Figure 14.1 – with, however, a divergence of the Ekman transport in the center of the subpolar gyres, and, correspondingly, downwelling of the sea surface and upwelling at the lower boundary of the surface Ekman layer.

The existence of the wind-driven gyres in steady state is bound to the blocking effect of the lateral land masses in the east and west of the ocean basins. By accumulating the water, a zonal pressure gradient is built up between these barriers at each side of the basin, which mostly balances the wind input of zonal momentum into the basin. This balance facilitates the weak interior southward branch of the subtropical circulation cell, the so-called Sverdrup circulation, nearly frictionless via the geostrophic balance. At the western boundary, however, other processes come into play, to close the circulation, as discussed in this chapter.

The linear theory of the wind-driven ocean circulation in steady state, which was sketched above, was developed in the first half of the last century in several important studies by Ekman (1905, 1923), Sverdrup (1947), Stommel (1948), and Munk (1950). The importance of the theory manifests itself by its success in describing many features of the steady state ocean circulation. At the same time it is rather simple to derive and to understand since it is accessible to a large extent by analytical means. In particular, many aspects of the large-scale structure of the near-surface ocean currents system, as shown in Figure 14.1, can be well described by the linear theory. It reveals, for instance, how the interior large-scale steady state gyre circulation is related to the wind forcing by Ekman transport and the Sverdrup circulation, and why boundary currents show up at the western side of ocean basins but not at the eastern boundary.

As detailed below in Section 14.1, the linear theory of the homogeneous wind-driven ocean circulation will culminate in the vorticity equation (14.23) for the barotropic stream function $\psi$ of the vertically integrated current velocity. Equation (14.23) is valid on a rotating Earth with meridionally varying Coriolis parameter $f$, and it shows how the stream function $\psi$ is related to the windstress forcing. Although this simple form of the theory is in principle valid for an ocean with a flat bottom only, the large-scale circulation is often surprisingly well described by this simple theory even in regions with large topographic features. The reason for this agreement will be discussed in Section 14.2 focussing on the effects of topography and stratification.

While the first sections in this chapter discuss the depth-averaged flow, the last section focusses on the three-dimensional flow in the interior of the ocean. The flow in the main thermocline of the ocean will be explained by a simple analytical model excluding friction and any diabatic effects in the interior, but incorporating wind-stress forcing at the surface and the observed density structure. This model will demonstrate how and where the thermocline is ventilated with water masses from the surface.

## 14.1 The Flat-Bottom Wind-Driven Circulation

We start with a set of approximations forming the planetary geostrophic equations, given in Section 5.3, which are relevant for the oceanic circulation on lateral scales much larger than the Rossby radius as discussed in Chapter 5.1. However, here we will specify the frictional terms in the horizontal momentum equations (5.45) and (5.46) as divergence of a vertical viscous flux of zonal and meridional momentum. Furthermore, we consider first a homogeneous ocean with $\rho = \rho_0 =$ const such that the salinity balance and the temperature balance are not needed in this section. We also consider a flat-bottom ocean with kinematic boundary conditions $w = 0$ at the top $z = \zeta$ and the bottom $z = -h$. However, for simplicity, we consider the top boundary condition to be taken at $z = 0$ instead of the real surface $z = \zeta$. Since $\zeta \ll h$ and moreover, since we have already linearized the equations, this approximation is acceptable. The pressure field $p$ may be calculated from the hydrostatic relation and expressed in terms of the surface displacement $\zeta$ and the atmospheric pressure; for $\rho =$ const it becomes $p = p_{\text{atm}} - g\rho(z - \zeta)$. For simplicity, we assume $p_{\text{atm}}$ to be constant such that it will be dynamically inactive. Note that, as a consequence, only the horizontal gradients of the surface displacement $\zeta$ will occur in the horizontal momentum equations.

Furthermore, the notation is changed for convenience as detailed in the box on p. 444, and the reference density $\rho_0$ will now be absorbed into the pressure and also in the turbulent stress vector $\boldsymbol{\tau}$ which then both have the dimension $\text{m}^2\,\text{s}^{-2}$. After

**68. Notation**

The following notation is used throughout this chapter: local Cartesian-like coordinates $\boldsymbol{x} = (x, y)$ are introduced by $\mathrm{d}x = a\cos\varphi\,\mathrm{d}\lambda$, $\mathrm{d}y = a\mathrm{d}\varphi$, and the definition of rotation for horizontal vectors (by $\pi/2$ in the counterclockwise sense) is used,

$$\underline{\boldsymbol{u}} = (-v, u) \quad \text{if} \quad \boldsymbol{u} = (u, v)$$

This will also be applied to other vectors and operators as $\boldsymbol{\nabla} = (\partial/\partial x, \partial/\partial y)$. Note that 'double-hooking' a vector $\boldsymbol{a}$ leads to $-\boldsymbol{a}$. A little care has to be taken with this operator since we are working on the sphere. The gradient is

$$\boldsymbol{\nabla}\psi = \mathrm{grad}\,\psi = \left(\frac{\partial}{\partial x}, \frac{\partial}{\partial y}\right)\psi = \left(\frac{1}{a\cos\varphi}\frac{\partial}{\partial\lambda}, \frac{1}{a}\frac{\partial}{\partial\varphi}\right)\psi$$

whereas divergence and curl are given by

$$\boldsymbol{\nabla}\cdot\boldsymbol{u} = \frac{1}{a\cos\varphi}\left[\frac{\partial u}{\partial\lambda} + \frac{\partial}{\partial\varphi}v\cos\varphi\right] \quad \text{and} \quad \underline{\boldsymbol{\nabla}}\cdot\boldsymbol{u} = \frac{1}{a\cos\varphi}\left[\frac{\partial v}{\partial\lambda} - \frac{\partial}{\partial\varphi}u\cos\varphi\right]$$

respectively. It is easy to see that $\underline{\boldsymbol{\nabla}}\cdot\boldsymbol{\nabla}\psi = \boldsymbol{\nabla}\cdot\underline{\boldsymbol{\nabla}}\psi \equiv 0$ and that $\underline{\boldsymbol{\nabla}}\cdot\boldsymbol{u} = -\boldsymbol{\nabla}\cdot\underline{\boldsymbol{u}}$. Note also that the Jacobian differential operator $\mathcal{J}(a,b)$ is given by $\mathcal{J}(a,b) = \underline{\boldsymbol{\nabla}}a\cdot\boldsymbol{\nabla}b = (\partial a/\partial x)(\partial b/\partial y) - (\partial b/\partial x)(\partial a/\partial y)$.

renaming $p$ and $\boldsymbol{\tau}$, we obtain (using the notation as explained in the box on p. 444)

$$f\underline{\boldsymbol{u}} = -g\nabla\zeta + \frac{\partial\boldsymbol{\tau}}{\partial z} \tag{14.1}$$

$$\nabla\cdot\boldsymbol{u} + \frac{\partial w}{\partial z} = 0 \tag{14.2}$$

Friction is given here by the divergence of a turbulent vertical transport of horizontal momentum represented by the turbulent stress vector $\boldsymbol{\tau} = (\tau^{(x)}, \tau^{(y)})$ with the dynamical boundary condition $\boldsymbol{\tau} = \boldsymbol{\tau}_0$ at the surface, where the windstress $\boldsymbol{\tau}_0$ is prescribed (compare Chapter 13). At the bottom, a corresponding condition $\boldsymbol{\tau} = \boldsymbol{\tau}_\mathrm{b}$ is appropriate. However, whereas the windstress is prescribed as forcing, the bottom stress $\boldsymbol{\tau}_\mathrm{b}$ remains unknown. Therefore, we will use the no-slip condition $\boldsymbol{u} = 0$ at $z = -h$; this condition will then determine $\boldsymbol{\tau}_\mathrm{b}$ from the solution.

In some applications we will augment the momentum balance (14.1) by lateral friction of viscous form, i. e. we assume a simple down-gradient parameterization for the turbulent stress vector,

$$\boldsymbol{\tau} = A_\mathrm{v}\frac{\partial\boldsymbol{u}}{\partial z} \tag{14.3}$$

with a constant turbulent vertical viscosity $A_\mathrm{v}$ (compare Chapter 11.3). A typical value of this viscosity is $A_\mathrm{v} \sim 0.1\,\mathrm{m^2\,s^{-1}}$ near the surface mixed layer or close to the bottom. A much smaller value, i. e. $A_\mathrm{v} \sim 10^{-4}\,\mathrm{m^2\,s^{-1}}$ is assumed in the interior. Note that the upper boundary value for $\boldsymbol{\tau}$, i. e. the surface windstress $\boldsymbol{\tau}_0$ over the world's ocean, is shown in Figure 13.1.

### *14.1.1 The Elementary Current System*

Consider a layer in which friction dominates the pressure term. Using the scaling $f\mathcal{U} \sim A_\mathrm{v}\mathcal{U}/d^2$, where $\mathcal{U}$ is the scale of the velocity, this implies a vertical scale

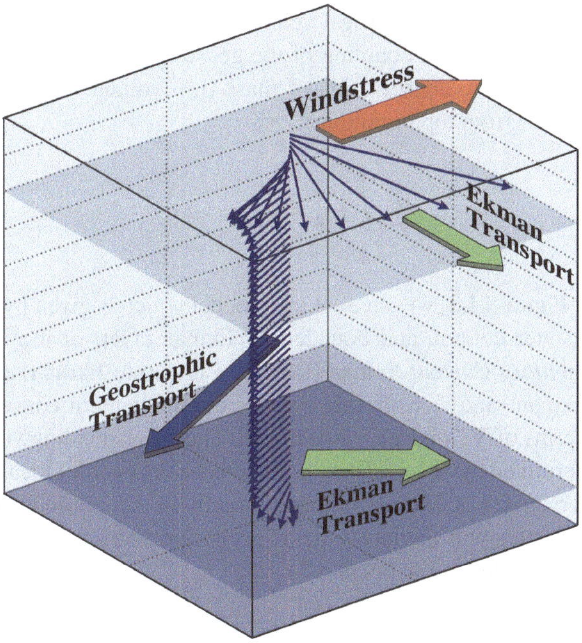


Fig. 14.2 Sketch of the Elementary Current System, which might also be called an Ekman sandwich. The geostrophic part of the current in the interior is stacked in-between boundary layers at the surface and bottom where friction becomes important. Direction and magnitude of the flow are indicated by *small blue arrows* at the respective depth, while the *large red, blue*, and *green arrows* indicate the direction of windstress and volume transport in the surface layer, respectively. The *transparent blue layers* denote the depth of the Ekman surface and bottom layer, respectively

of the layer $d \sim \sqrt{A_v/f}$. Using $f \sim 10^{-4}\,\text{s}^{-1}$ and a vertical turbulent viscosity of $A_v \sim 0.1\,\text{m}^2\,\text{s}^{-1}$, the vertical extent of such a layer is roughly $d \sim 30\,\text{m}$. Typically, there are frictionally dominated boundary layers at the top and the bottom of the ocean where the above scaling applies, while the vertical turbulent viscosity is much smaller in-between so that in the interior a geostrophic balance applies. The frictionally dominated boundary layers are generally called Ekman layers.

It is convenient to split the flow into geostrophic and frictional (Ekman) components, $\boldsymbol{u} = \boldsymbol{u}_g + \boldsymbol{u}_e$ and $w = w_g + w_e$, governed by

$$f \underline{\boldsymbol{u}}_g = -g\nabla\zeta \quad \text{and} \quad f \underline{\boldsymbol{u}}_e = A_v \frac{\partial^2 \boldsymbol{u}_e}{\partial z^2} \tag{14.4}$$

$$\nabla \cdot \boldsymbol{u}_g + \frac{\partial w_g}{\partial z} = 0 \quad \text{and} \quad \nabla \cdot \boldsymbol{u}_e + \frac{\partial w_e}{\partial z} = 0 \tag{14.5}$$

Notice that the decoupling in this form becomes possible because the equations are linear and the (perturbation) pressure field is vertically constant in a homogeneous ocean with $\rho = \text{const}$. It is also obvious that we must assume $f \neq 0$ in the following analysis which, therefore, is not valid near the equator.

We require that $w_e = 0$ at $z = 0$ and consequently have $w_g = 0$ at $z = 0$ as well. Vertical integration of (14.5) yields

$$w_{g,e}(z) = \nabla \cdot \int_z^0 \boldsymbol{u}_{g,e} \mathrm{d}z \tag{14.6}$$

The requirement that the total vertical velocity $w = w_g + w_e$ vanishes at the flat bottom at $z = -h$ leads to a coupling of the geostrophic and Ekman components. This is manifested in the conservation of mass (i. e. volume) obtained by vertical integration of (14.5) from top to bottom,

$$\nabla \cdot \int_{-h}^{0} \left(\boldsymbol{u}_g + \boldsymbol{u}_e\right) \mathrm{d}z = 0 \tag{14.7}$$

As sketched in Figure 14.2, we have set up a model which is given by a geostrophic current in the water column and bounded by Ekman layers at top and bottom. It is termed *Elementary Current System* (the name is due to Ekman) and might also be called an Ekman sandwich. In this system, the momentum conservation (14.1) yields $\boldsymbol{u}_{g,\,e}$ in terms of $\nabla\zeta$ and $\boldsymbol{\tau}_0$, respectively. The corresponding vertical velocity profiles are determined by (14.6). These equations fix the profiles locally, and finally, the mass conservation (14.7) couples them globally in a condition relating $\zeta$ to the forcing $\boldsymbol{\tau}_0$. The latter will obviously be a second order differential equation for $\zeta$ in $x$ and $y$. It will be derived in Section 14.1.5.

### 14.1.2 Ekman Spiral

**69. Derivation of Ekman Spiral and Ekman Depth**

To obtain the Ekman spiral as solutions with simple mathematical manipulations, we write the Ekman part of (14.4) in complex form $f(u_e + \mathrm{i}v_e) = -\mathrm{i}A_v(u_e + \mathrm{i}v_e)_{zz}$ where the index $z$ indicates vertical differentiation. The general solution is

$$u_e + \mathrm{i}v_e = \alpha_+ \exp\left[(\mathrm{i}f/A_v)^{1/2}z\right] + \alpha_- \exp\left[-(\mathrm{i}f/A_v)^{1/2}z\right] \tag{B69.1}$$

with complex constants $\alpha_+$ and $\alpha_-$, which will be determined from the boundary conditions. The behavior of the exponentials depends on the sign of the Coriolis parameter $f$ which differs between the hemispheres. For the moment, we restrict our model to the northern hemisphere where $f$ is positive (the southern hemisphere is discussed in Chapter 16). Then, with the definition $d = \sqrt{2A_v/|f|}$, which is called the *Ekman depth*, we find

$$u_e + \mathrm{i}v_e = \alpha_+ \exp[(\mathrm{i}+1)z/d] + \alpha_- \exp[-(\mathrm{i}+1)z/d] \tag{B69.2}$$

with boundary conditions

$$A_v(u_e + \mathrm{i}v_e)_z = \tau_0^x + \mathrm{i}\tau_0^y \quad \text{at} \quad z = 0 \quad \text{and} \quad u_e + \mathrm{i}v_e = -(u_g + \mathrm{i}v_g) \quad \text{at} \quad z = -h$$

It is apparent that the part of the solution related to $\alpha_+$ decays away from the surface exponentially with a scale $d$ while the part related to $\alpha_-$ decays away from the bottom with the same scale (this equality is due to the assumption that $A_v$ is constant in the water column; it is easy to relax this assumption to two differing values of $A_v$). To make the mathematics simple, we step back from the correct boundary conditions shown above and split the complete solution (describing simultaneously top and bottom Ekman layers by a complicated hyperbolic function) into the 'classical' separate Ekman spirals in the surface and the bottom layers. The first satisfies – with suitable choice of $\alpha_+$ – the stress condition at the surface (and freely decays downward), the second satisfies the no-slip condition at the bottom with correspondingly fixing of $\alpha_-$ (and decays upward). As $d \ll h$ the sum of these independent spirals is, of course, a good approximation to the complete hyperbolic solution.

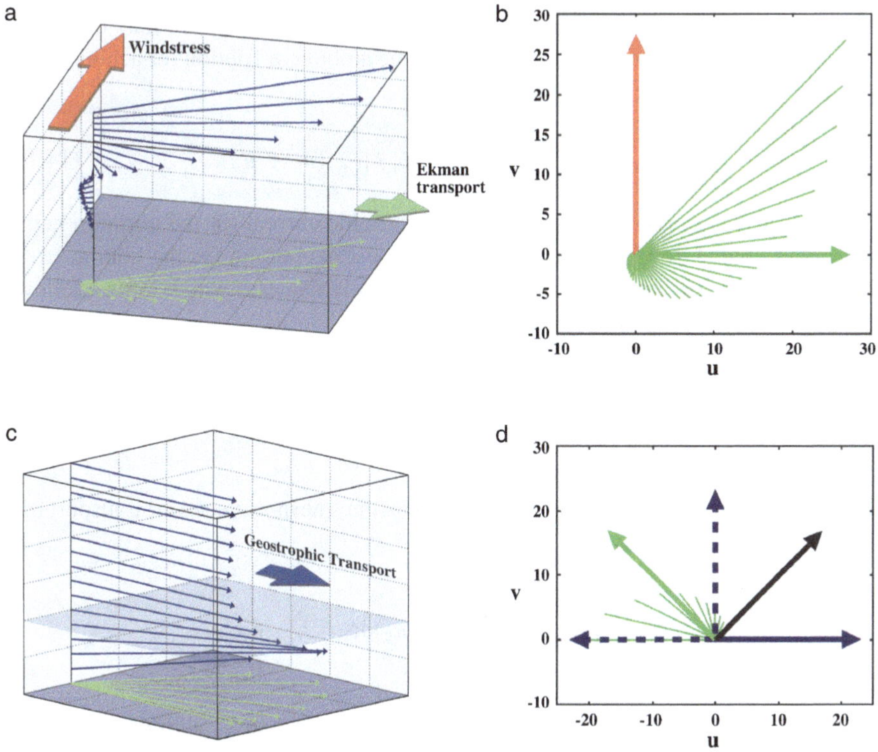


Fig. 14.3 A schematic view of the Ekman spiral in the frictional boundary layers at the surface and the bottom. Shown is the Ekman velocity $\boldsymbol{u}_e$ in the surface Ekman layer (**a**) and the total velocity $\boldsymbol{u} = \boldsymbol{u}_g + \boldsymbol{u}_e$ in the bottom Ekman layer (**c**) as *thin blue arrows* at the respective depths. Also shown are the direction of the surface windstress (*large red arrow*), the Ekman transport in the surface layer (*large green arrow*) and the geostrophic interior transport (*large blue arrow*). *Thin green arrows* denote the velocity $\boldsymbol{u}_e$ and $\boldsymbol{u}$, respectively, projected into the horizontal plane. In **b** and **d** the projected spirals are displayed by the *green lines* (without an *arrow head*). In panel **b** the *fat red arrow* gives the direction of the windstress and the *fat green arrow* the direction of the surface Ekman transport. In panel **d** the *fat arrows* indicate directions as follows: *blue* $\boldsymbol{u}_g$, *blue dashed* $-\boldsymbol{u}_g$ and $\underline{\boldsymbol{u}}_g$, *black* $\boldsymbol{\tau}_b$, *green* $\boldsymbol{U}_e^{bot}$

The geostrophically balanced horizontal current $\boldsymbol{u}_g$ in the Elementary Current System is easily expressed in terms of $\zeta$,

$$\boldsymbol{u}_g = (g/f)\underline{\nabla}\zeta \tag{14.8}$$

Apparently, $\boldsymbol{u}_g$ is vertically constant and pointing at a right angle to the left (in the northern hemisphere; the southern hemisphere is studied in Chapter 16) of the gradient of the surface displacement $\zeta$. The frictionally induced flow in the Elementary Current System yields the *Ekman spiral* as vertical profile. The solution for the top Ekman spiral (see also the box on p. 446 for details how to solve the equations) is given by

$$\boldsymbol{u}_e = d/(2A_v)e^{z/d}\left[(\boldsymbol{\tau}_0 - \underline{\boldsymbol{\tau}_0})\cos z/d + (\boldsymbol{\tau}_0 + \underline{\boldsymbol{\tau}_0})\sin z/d\right] \tag{14.9}$$

with $d = \sqrt{2A_v/|f|}$. For the bottom Ekman spiral we obtain

$$\boldsymbol{u}_e = e^{-(z+h)/d}\left[-\boldsymbol{u}_g\cos(z+h)/d + \underline{\boldsymbol{u}}_g\sin(z+h)/d\right]$$

Both spirals are shown in Figure 14.3. From the latter equation, we may derive the bottom stress which is necessary to establish the amount of friction for achieving the no-slip condition $\boldsymbol{u}(-h) = 0$. We find

$$\boldsymbol{\tau}_{\mathrm{b}} = A_{\mathrm{v}} \frac{\partial \boldsymbol{u}_{\mathrm{e}}}{\partial z}\bigg|_{z=-h} = (d|f|/2)(\boldsymbol{u}_{\mathrm{g}} + \underline{\boldsymbol{u}}_{\mathrm{g}}) \tag{14.10}$$

It is thus oriented $\pi/4$ to the left of the geostrophic current (in the northern hemisphere, see panel d) of Figure 14.3). Note that the expression for the bottom stress is valid because the geostrophic current is vertically constant which in turn follows from the homogeneity of the density.

### 14.1.3 Ekman Transport

We refer to the vertically integrated velocity as the transport vector or transport (dimension $\mathrm{m^2\,s^{-1}}$). It is given by

$$\boldsymbol{U} = \int_{-h}^{0} \boldsymbol{u}\mathrm{d}z = \int_{-h}^{0} (\boldsymbol{u}_{\mathrm{g}} + \boldsymbol{u}_{\mathrm{e}})\mathrm{d}z = \boldsymbol{U}_{\mathrm{g}} + \boldsymbol{U}_{\mathrm{e}}$$

The line integral $\int_A^B \boldsymbol{U} \cdot \mathrm{d}\underline{\boldsymbol{s}}$ between two points $A, B$ on a curve with line increment $\mathrm{d}\boldsymbol{s}$ gives the transport of volume (dimension $\mathrm{m^3\,s^{-1}}$) through the corresponding section. Since the geostrophic current is vertically constant in the present configuration, its transport is $\boldsymbol{U}_{\mathrm{g}} = h\boldsymbol{u}_{\mathrm{g}}$. This is the geostrophic transport in the Ekman sandwich.

The frictionally induced transport is derived directly by integration of $f\underline{\boldsymbol{u}}_{\mathrm{e}} = \partial\boldsymbol{\tau}/\partial z$ and use of the boundary conditions for the stresses at the top and the bottom,

$$\boldsymbol{U}_{\mathrm{e}} = -(\underline{\boldsymbol{\tau}}_0 - \underline{\boldsymbol{\tau}}_{\mathrm{b}})/f$$

It can be split into its two parts. The top Ekman layer contributes $\boldsymbol{U}_{\mathrm{e}}^{\mathrm{top}} = -\underline{\boldsymbol{\tau}}_0/f$. This part is usually called the *Ekman transport* while the total Ekman transport also includes the frictionally induced transport in the bottom layer. The Ekman transport is orthogonal to the wind (to the right in the northern hemisphere, see Figure 14.3). As the windstress is prescribed, the top Ekman transport is independent of the particular parameterization for the turbulent friction. Note that this is a powerful result since the parameterization of turbulent momentum fluxes, which was chosen here as (14.3), is uncertain.

In contrast to the surface layer, the transport in the bottom Ekman layer is not independent of the turbulence closure, as seen in the bottom stress (14.10) which was derived from the parameterization (14.3). The reason are the different boundary conditions, i. e. the bottom stress depends on the turbulence closure, while the surface stress is given. The transport of the bottom Ekman layer is (for the northern hemisphere)

$$\boldsymbol{U}_{\mathrm{e}}^{\mathrm{bot}} = \underline{\boldsymbol{\tau}}_{\mathrm{b}}/f = (d/2)(\underline{\boldsymbol{u}}_{\mathrm{g}} - \boldsymbol{u}_{\mathrm{g}})$$

where the Ekman depth $d$ depends on the turbulent vertical viscosity $A_{\mathrm{v}}$. The orientation of the bottom Ekman transport relative to the geostrophic velocity is depicted in panel d) of Figure 14.3. The total transport of the sandwich thus becomes

$$\boldsymbol{U} = h\boldsymbol{u}_{\mathrm{g}} - (\underline{\boldsymbol{\tau}}_0 - \underline{\boldsymbol{\tau}}_{\mathrm{b}})/f = [h - (d/2)]\boldsymbol{u}_{\mathrm{g}} - \underline{\boldsymbol{\tau}}_0/f + (d/2)\underline{\boldsymbol{u}}_{\mathrm{g}} \tag{14.11}$$

Note that in the first term of the last relation, $d/2$ may be neglected in comparison to $h$. The last term is generally small as well so the total transport approximately becomes

$$\boldsymbol{U} \approx h\boldsymbol{u}_{\mathrm{g}} - \underline{\boldsymbol{\tau}}_0/f$$

### 14.1.4 Ekman Pumping

The Ekman pumping is associated with the frictionally induced vertical velocity $w_{\mathrm{e}}$ which is given by (14.6). This vertical Ekman velocity starts with zero due to the boundary condition at the surface, followed by an exponential pattern within the top Ekman layer, and approaches a constant below. Taking for simplicity a zonal windstress $\boldsymbol{\tau}_0 = (\tau_0^{(x)}(y), 0)$, we find

$$\begin{aligned} w_{\mathrm{e}}^{\mathrm{top}}(z) &= -\frac{\partial}{\partial y}\left(\frac{\tau_0^{(x)}}{f}\right)\int_z^0 \mathrm{e}^{z/d}\left(\cos\frac{z}{d} - \sin\frac{z}{d}\right)\mathrm{d}\left(\frac{z}{d}\right) \\ &= -\frac{\partial}{\partial y}\left(\frac{\tau_0^{(x)}}{f}\right)\left(1 - \cos\frac{z}{d}\mathrm{e}^{z/d}\right) \end{aligned}$$

At depths below the Ekman depth $d$, the profile becomes approximately constant with the value

$$w_{\mathrm{e}}^{\mathrm{top}}(-d) \approx \nabla\cdot\boldsymbol{U}_{\mathrm{e}}^{\mathrm{top}} = \underline{\nabla}\cdot(\boldsymbol{\tau}_0/f) \tag{14.12}$$

representing the amount of volume pumped from below into (or from above out of) the Ekman layer in case that the top Ekman transport is divergent (or convergent). The value of $w_{\mathrm{e}}^{\mathrm{top}}(-d)$ below the surface layer, given by (14.12), is often called the *Ekman pumping*. Below $z = -d$ and above the bottom Ekman layer, the vertical Ekman velocity remains constant.

Figure 14.4 shows the Ekman pumping, calculated from the long-term mean windstress of Figure 13.1. It is negative in the subtropical regions on the order of 20–50 m per year and mostly positive over the subpolar regions. Towards the equator, $f$ goes to zero, and Ekman pumping and Ekman transport become ill-defined in our discussion (which could be resolved including the effect of lateral friction). Therefore, $w_{\mathrm{e}}$ is not shown in an equatorial strip in Figure 14.4.

The bottom Ekman layer behaves similarly to the surface, starting at $z = -h + d$ with $w_{\mathrm{e}}^{\mathrm{bot}} \approx 0$ and increasing to a nonzero value at the bottom $z = -h$. We find

$$\begin{aligned} w_{\mathrm{e}}^{\mathrm{bot}}(z) &= \nabla\cdot d\int_z^0 e^{-(z+h)/d}\left[-\boldsymbol{u}_{\mathrm{g}}\cos(z+h)/d + \underline{\boldsymbol{u}}_{\mathrm{g}}\sin(z+h)/d\right]\mathrm{d}(z/d) \\ &\approx \nabla\cdot(d/2)\left[-(\boldsymbol{u}_{\mathrm{g}} - \underline{\boldsymbol{u}}_{\mathrm{g}})\cos(z+h)/d + (\boldsymbol{u}_{\mathrm{g}} + \underline{\boldsymbol{u}}_{\mathrm{g}})\sin(z+h)/d\right] \\ &\quad \times \mathrm{e}^{-(z+h)/d} \end{aligned}$$

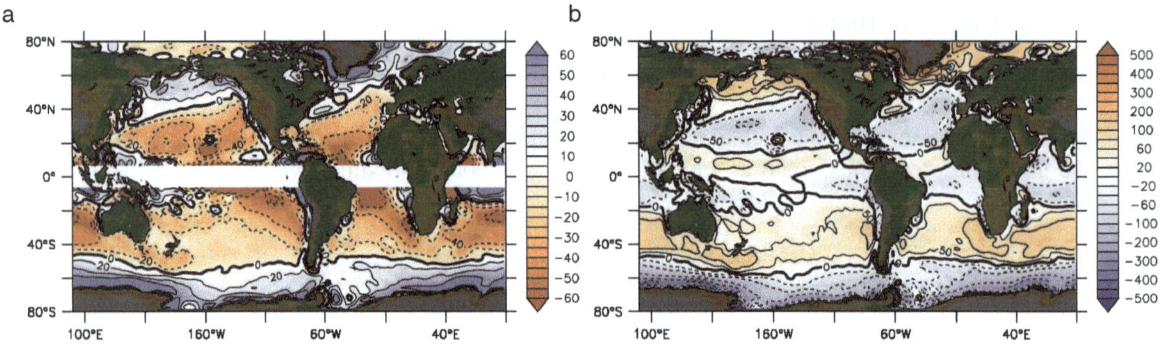

Fig. 14.4 Ekman pumping $w_e^{top}$ in m per year (**a**) and windstress curl in $10^{-12}\,\mathrm{m\,s^{-2}}$ (**b**), calculated from the long-term mean windstress shown in Figure 13.1

neglecting terms $\sim \exp(-h/d)$. At the bottom, we thus obtain

$$w_e^{bot}(-h) = \nabla \cdot (d/2)(\underline{u}_g - u_g) = \nabla \cdot U_e^{bot} = -\underline{\nabla} \cdot (\tau_b/f)$$

Take for simplicity $d = \text{const}$ in this relation. Then

$$w_e^{bot}(-h) \approx (d/2)(\nabla \cdot \underline{u}_g - \nabla \cdot u_g) = (d/2)(\nabla \cdot \underline{u}_g + (\beta/f)v_g) \tag{14.13}$$

Note that the total frictional vertical velocity at the bottom is $w_e^{top}(-d) + w_e^{bot}(-h)$, and both terms are generally nonzero. Finally, the geostrophic vertical velocity is evaluated as

$$w_g(z) = \nabla \cdot \int_z^0 u_g \mathrm{d}z = -z\nabla \cdot u_g = zg(\beta/f^2)\frac{\partial \zeta}{\partial x} = z(\beta/f)\,v_g \tag{14.14}$$

with $\beta = (2\Omega/a)\cos\varphi$. It is thus linear with depth and generally has also a nonzero value $-h(\beta/f)v_g$ at the bottom. The profiles of $w_e$ and $w_g$ are depicted in Figure 14.5 for different signs of the meridional geostrophic velocity $v_g$.

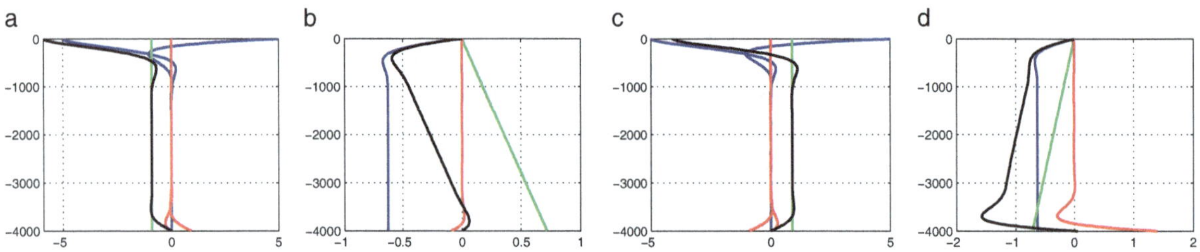

Fig. 14.5 Sketch of the profiles of the horizontal (**a**,**c** in $10^{-3}\,\mathrm{m\,s^{-1}}$) and vertical (**b**,**d** in $10^{-6}\,\mathrm{m\,s^{-1}}$) velocity components. **a** and **c** geostrophic (*green lines*), frictional surface layer (*blue lines*, $u$ and $v$ each), frictional bottom layer (*red lines*, $u$ and $v$ each), total $v$ (*black line*). **b** and **d** same for the vertical velocity. Panels (**a**) and (**b**) are for a negative $v_g$, appropriate for the Sverdrup regime in a northern hemisphere subtropical gyre. Panels (**c**) and (**d**) are for a positive $v_g$, appropriate for the western boundary layer regime in the northern hemisphere. Note that for $v_g < 0$, the bottom frictional layer is irrelevant for balancing the vertical velocities at the bottom (panel (**b**)). This is not the case for $v_g > 0$ (panel (**d**))

### 14.1.5 Equilibrium Wind-Driven Model Regimes

Having established the solutions for the Ekman velocities in the top and bottom Ekman layers and the geostrophic velocity in-between, we proceed now to consider further model ingredients, i. e. a constant or varying $f = f(y)$ and zero or nonzero bottom friction, to distinguish between different regimes with differing dynamical balances. We will explain by this discussion important aspects of the circulation in the interior of the large-scale gyre of the oceans and in particular the intensified western recirculation in boundary currents. In the following, we again restrict the discussion to the northern hemisphere; the main results, however, are valid for $f < 0$ as well.

It might be worth mentioning here that in 1923 V.W. Ekman (and some later partly unpublished work, see Welander, 1985) had most of the equations of the 'Ekman sandwich' presented above. He even had implemented the $\beta$-term which leads to the Stommel model (14.23) below, but his equations were obscured by considering topography and stratification as well. So, oceanographic wisdom had to wait 25 years for the celebrated explanation of the western intensification of ocean currents in the great ocean basins, which arose from (14.23) as discussed below.

#### $f$-Plane Regime

We start our discussion of the model regimes by taking $\beta = 0$. This defines the $f$-plane regime where the Coriolis parameter is considered constant. From (14.14), it follows that the vertical geostrophic velocity vanishes identically, and without a bottom Ekman layer, the surface Ekman flow would pump water through the bottom with $w_e(-d) \neq 0$. To avoid this unacceptable situation, we must consider bottom friction, i. e. $\boldsymbol{\tau}_b \neq 0$. In this case, the frictionally induced vertical velocity $w_e$ starts to deviate from the value of the Ekman pumping $w_e^{top}(-d)$ when entering the bottom Ekman layer (roughly below $z = -h + d$) by picking up contributions from the divergence from the bottom Ekman velocity $\boldsymbol{u}_e$. At the bottom

$$w_e(-h) = \nabla \cdot (\boldsymbol{U}_e^{top} + \boldsymbol{U}_e^{bot}) = w_e^{top}(-d) + w_e^{bot}(-h) = \underline{\nabla} \cdot (\boldsymbol{\tau}_0 - \boldsymbol{\tau}_b)/f$$

so that in order to obtain $w(-h) = w_e(-h) = 0$, the bottom Ekman transport has to converge or diverge at the appropriate rate to balance the Ekman pumping in the top Ekman layer. In other words, there is a pumping $\nabla \cdot \boldsymbol{U}_e^{bot}$ at the bottom, induced by the Ekman bottom velocities, which compensates the pumping induced by the wind-stress. Note that although the divergences of the top and bottom Ekman transports cancel, this does not imply a compensation of these transports themselves.

Note also that the condition $w_e(-h) = 0$ requires a specific horizontal geostrophic current for the $f$-plane case. Using $\underline{\nabla} \cdot \boldsymbol{\tau}_0/f = \underline{\nabla} \cdot \boldsymbol{\tau}_b/f = (d/2)\underline{\nabla} \cdot \boldsymbol{u}_g$ and (14.8) to replace the geostrophic velocity in terms of the surface displacement $\zeta$, we find

$$g(d/2)\nabla^2\zeta = \underline{\nabla} \cdot \boldsymbol{\tau}_0 \tag{14.15}$$

It turns out that this balance is in fact inadequate to determine anything similar to the observed surface displacement: there is no asymmetry in the balance that could account for the observed east-west asymmetry in the circulation. The reason is the neglection of the variations of the Coriolis parameter with latitude, which we will take into account in the next section.

### Sverdrup Regime

Now we consider the case $\beta \neq 0$. If there is no bottom friction and thus no bottom Ekman layer, $w_e(z)$ stays at the value of the Ekman pumping, reaching undiminished to the bottom. The geostrophic $w_g$ increases (for $v_g < 0$ in the northern hemisphere; see Figure 14.5) with depth (or decreases for $v_g > 0$) linearly from its surface value $w_g(0) = 0$ to the bottom where $w_g(-h) = -h(\beta/f)v_g$. It must balance the Ekman pumping there, $w_g(-h) \approx -w_e(-d)$, or

$$gh(\beta/f^2)\frac{\partial\zeta}{\partial x} = h(\beta/f)v_g = (\beta/f)V_g = \underline{\nabla}\cdot(\boldsymbol{\tau}_0/f) \qquad (14.16)$$

to avoid pumping through the bottom. These conditions describe the Sverdrup regime. Note that downward Ekman pumping implies a southward geostrophic velocity in the northern hemisphere. Adding the meridional geostrophic transport $V_g$, arising in this case, to the meridional Ekman transport $V_e^{top} = -\tau_0^{(x)}/f$ of the top layer, we find the famous Sverdrup relation

$$\beta V = \beta(V_g + V_e^{top}) = f\underline{\nabla}\cdot(\boldsymbol{\tau}_0/f) - (\beta/f)\tau_0^{(x)} = \underline{\nabla}\cdot\boldsymbol{\tau}_0 \qquad (14.17)$$

which will be discussed in more detail further below. The total meridional transport $V$ given by (14.17) is called the *Sverdrup transport*. It can be calculated for the World Ocean from the windstress curl, which is shown in Figure 14.4. Remember that the validity of (14.17) only requires that $w(-h) = 0$ and $\boldsymbol{\tau}(-h) = \boldsymbol{\tau}_b = 0$. In fact, $z = -h$ need not be the bottom but could be any level in the water column where $w$ and $\boldsymbol{\tau}$ vanish simultaneously.

As $\nabla\cdot\boldsymbol{U} = 0$ according to (14.7), a stream function $\psi$ may be introduced for the total transport, so that

$$\boldsymbol{U} = \underline{\nabla}\psi \qquad (14.18)$$

The stream function determines the total transport between two points in a direct way, i.e. $\int_A^B \boldsymbol{U}\cdot \mathrm{d}\underline{\mathbf{s}} = \int_A^B \nabla\psi\cdot \mathrm{d}\boldsymbol{s} = \psi(B) - \psi(A)$, where $\mathrm{d}\boldsymbol{s}$ is a line element of the section from $A$ to $B$ and $\mathrm{d}\underline{\mathbf{s}}$ is directed perpendicular to it (see also Appendix A.1.4). Written in terms of the barotropic stream function, the Sverdrup equation (14.17) becomes

$$\beta V = \beta\frac{\partial\psi}{\partial x} = \underline{\nabla}\cdot\boldsymbol{\tau}_0 \qquad (14.19)$$

The stream function $\psi$ corresponds to the mass transport, and (14.19) gives the meridional transport due to wind forcing of the layer between the surface and the depth $z = -h$.

Using (14.19), the stream function can be calculated by integration from the eastern coast $x = x_E(y)$ where $\psi$ can be put to a constant (equal to zero without any

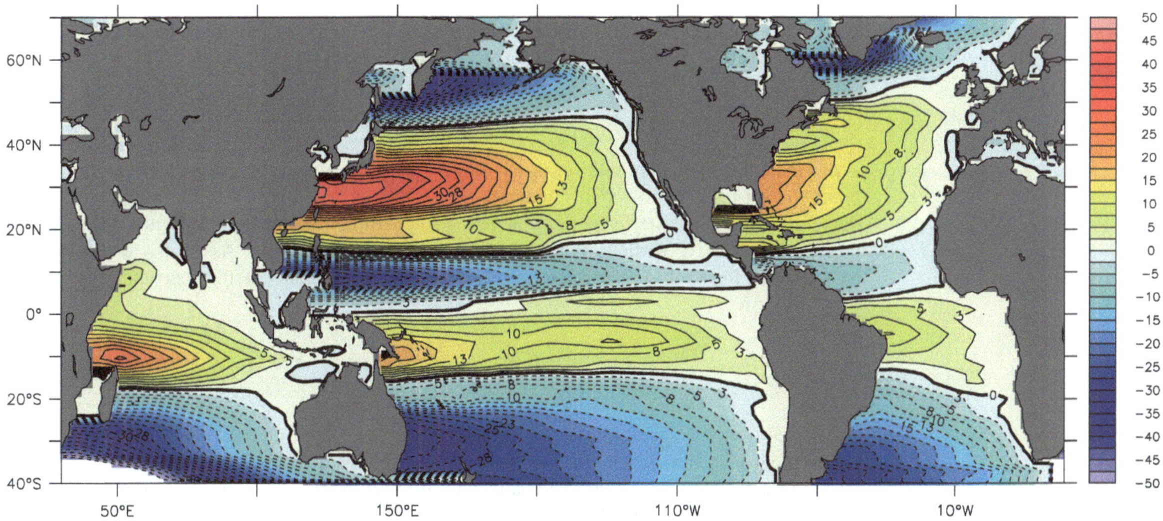


**Fig. 14.6** Stream function $\psi$ in Sv $= 10^6\,\mathrm{m^3\,s^{-1}}$ calculated from the Sverdrup relation (14.19) and the windstress curl shown in Figure 14.4. Westward integration of (14.19) starts at the eastern boundary of each basin or at $30°$ E with $\psi = 0$ as boundary condition. The figure excludes the Southern Ocean because here the Sverdrup regime breaks down as discussed in Chapter 16

restriction) because the flow normal to the coast must vanish. Hence

$$\psi(x, y) = -\frac{1}{\beta} \int_x^{x_E} \underset{\neg}{\nabla} \cdot \boldsymbol{\tau}_0 \mathrm{d}x$$

Figure 14.6 shows the transport stream function $\psi$ calculated with realistic windstress curl as shown in Figure 14.4. One finds indeed the large-scale gyre circulation in the tropical, subtropical, and subpolar ocean basins as in the schematic circulation shown in Figure 14.1. However, note that in the Southern Ocean, the zonal integration of the Sverdrup relation (14.19) does not apply. This issue will be further discussed in Chapter 16. In the example in the box on p. 454, the zonal flow follows the direction of the windstress. This is because of the sinusoidal nature of the windstress profile where $-\partial^2 \tau_0^{(x)}/\partial y^2 \sim \tau_0^{(x)}$. Sverdrup's original work studied the circulation due to the observed equatorial wind field in the Pacific, and Sverdrup was the first to explain why the North Equatorial Counter Current (NECC) runs against the wind. This feature can also be seen in Figure 14.6.

## Stommel Regime

In the subtropical gyre of the northern hemisphere, the Ekman pumping is downward everywhere and thus, from (14.16), $v_g$ is negative in the Sverdrup regime. Likewise, $\underset{\neg}{\nabla} \cdot \boldsymbol{\tau}_0$ is negative, and the total transport $V$ is southward for the Sverdrup solution. Clearly, these conditions cannot hold for the entire basin. Somewhere the bottom Ekman layer must come into play to allow for a northward $v_g$ and a total northward transport $V$. This compensation will take place in a western boundary current.

For a positive $v_g$, the vertical geostrophic pumping, given by (14.14), is downward, and the vertical Ekman velocity in the bottom layer has to compensate it.

**70. Sverdrup Solution for a Box Ocean**

A prototype of application of the above theory is the Sverdrup circulation in a rectangular ocean ($0 < x < B, 0 < y < L$) forced by a zonal windstress

$$\tau_0^{(x)}(y) = -\tau_0 \cos\frac{\pi y}{L}, \quad \tau_0^y = 0 \tag{B70.1}$$

The windstress of this simple set-up is shown for positive $\tau_0$ in the Figure below (a). This prototype set-up aims to roughly represent the midlatitude westerlies in the northern half of the domain and the easterly trade winds in the southern half (compare also Figure 13.1). With positive $\tau_0$, negative vorticity is introduced everywhere in the basin. Assuming for simplicity a constant $\beta$, the Sverdrup transport velocities and the stream function then becomes

$$U = -(B-x)\frac{\tau_0\pi^2}{\beta L^2}\cos\frac{\pi y}{L}, \quad V = -\frac{\tau_0\pi}{\beta L}\sin\frac{\pi y}{L},$$
$$\psi(x,y) = (B-x)\frac{\tau_0\pi}{\beta L}\sin\frac{\pi y}{L}$$

representing a clockwise circulation. Note that $U$ follows the wind direction, and that $V$ is directed southward over the entire domain. Note also that the return flow in the western boundary layer is excluded in the Sverdrup regime.
In the figure below, parameters were chosen as $B = 5{,}000\,\text{km}$, $L = 4{,}000\,\text{km}$, $\tau_0 = 10^{-4}\,\text{m}^2\,\text{s}^{-2}$ and $f = f_0 + \beta y$ with $f_0 = 7\times10^{-5}\,\text{s}^{-1}$ and $\beta = 2\times10^{-11}\,\text{m}^{-1}\,\text{s}^{-1}$. The maximum transport across the basin width is $(\pi B/L)\tau_0/\beta$ and amounts to about 20 Sv (the unit 1 Sv $= 10^6\,\text{m}^3\,\text{s}^{-1}$ is named after H.U. Sverdrup). The maximum volume transport is located at the center latitude because the Ekman transport $V_e = -\tau_0^{(x)}/f$ vanishes there and the flow happens to be entirely geostrophic.

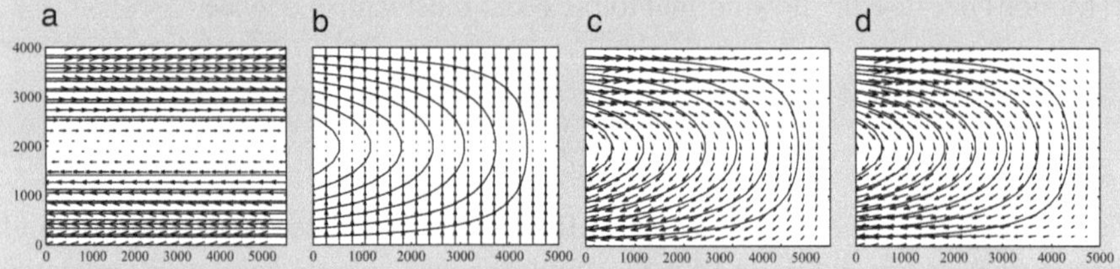


*The figure shows the windstress pattern (a), the transports due to the Ekman layer (b), the geostrophic part (c) with $U_g = U - U_e, V_g = V - V_e$ (note that $U_e = 0$) and the Sverdrup transport (d). The Sverdrup transport stream function $\psi$ is also shown in b), c), and d) as solid lines. By comparison with the circulation scheme in Figure 14.1, it becomes clear that the example models the Sverdrup part of the subtropical gyres occurring in the Atlantic and Pacific oceans in the northern and southern hemisphere. Further gyres can be added, e. g. subpolar gyres at the poleward flanks, with reversed circulation. They are separated by the line where $\underline{\nabla}\cdot\boldsymbol{\tau}_0 = 0$.*

Therefore, we consider the vertical Ekman velocity of the bottom layer, given by (14.13), and compare it with the vertical geostrophic velocity given by (14.14). The first term, $(\beta/f)v_g$, of the right-hand side of (14.13) will always be small compared to $w_g(-h)$ since $h \gg d$. The second term on the right-hand side of (14.13) remains also small as long as $(d/2)\partial v_g/\partial x \ll h\beta v_g/f$, or $R/\beta \ll L$ (originating from the scaling $\partial v_g/\partial x \sim v_g/L$), where $R = (d/2)f/h$ measures the bottom friction by an inverse time scale and $L$ denotes the length scale of the zonal variation of the geostrophic current.

When the length scale $L$ becomes smaller, representing now the variation of the current in a western boundary current where $v_g > 0$, a balance between vertical Ekman and geostrophic velocities might hold. This means that the flow must achieve a scale $L = O(\delta)$ with $\delta = R/\beta$. The scale $\delta$ will be found as width of a western boundary layer where $v_g > 0$ and $V > 0$, which closes the Sverdrup regime to a globally valid solution. Note that the time-scale, associated with bottom friction, satisfies $1/R \gg 1/|f|$ since $d \ll h$. This time-scale can be identified in the spin-down of the circulation by bottom friction in a time-dependent mode.

The general case with nonvanishing $\boldsymbol{\tau}_\mathrm{b}$ and $\beta$ has thus contributions from both Ekman layers and the geostrophically induced $w_\mathrm{g}$. It has more freedom in the balance of vertical velocity at the bottom, i. e. $w_\mathrm{e}(-h) = -w_\mathrm{g}(-h)$, as the previously discussed regimes. Here

$$\nabla \cdot \boldsymbol{U} = \nabla \cdot (\boldsymbol{U}_\mathrm{e} + \boldsymbol{U}_\mathrm{g}) = \underset{\neg}{\nabla} \cdot (\boldsymbol{\tau}_0 - \boldsymbol{\tau}_\mathrm{b})/f - (\beta/f)V_\mathrm{g} = 0 \tag{14.20}$$

This balance may be rewritten in the form

$$\beta(V_\mathrm{g} + V_\mathrm{e}^\mathrm{top} + V_\mathrm{e}^\mathrm{bot}) = \beta V = \underset{\neg}{\nabla} \cdot \boldsymbol{\tau}_0 - \underset{\neg}{\nabla} \cdot \boldsymbol{\tau}_\mathrm{b} \tag{14.21}$$

since $V_\mathrm{e}^\mathrm{top} = -\tau_0^{(x)}/f$ and $V_\mathrm{e}^\mathrm{bot} = \tau_\mathrm{b}^{(x)}/f$. We note in (14.21) the modification of Sverdrup's balance (14.17) by the bottom Ekman layer. Equation (14.21) can also be derived by vertical integration and taking the curl of the momentum balance (14.1).

There are many ways to write (14.20) or (14.21) in other forms and extract further physical insight. Expressing $V_\mathrm{g}$ and $\boldsymbol{\tau}_\mathrm{b}$ in (14.20) by the surface displacement, we find

$$\underset{\neg}{\nabla} \cdot (d/2)(g/f)(\underset{\neg}{\nabla}\zeta - \nabla\zeta) + gh(\beta/f^2)\frac{\partial \zeta}{\partial x} = \underset{\neg}{\nabla} \cdot \boldsymbol{\tau}_0/f$$

and thus, ignoring the $y$-dependence in $d/f$ in the first term, the relation

$$R\nabla^2\zeta + \beta\frac{\partial \zeta}{\partial x} = \frac{f^2}{gh}w_\mathrm{e}(-d) \tag{14.22}$$

results, which combines (14.15) and (14.16). In principle, equation (14.22) may be solved with suitable (von Neumann or Dirichlet or mixed) boundary conditions for $\zeta$. Dynamically consistent boundary conditions are derived from the condition that $\boldsymbol{n} \cdot \boldsymbol{U} = 0$ on solid boundaries, and thus from (14.11),

$$\boldsymbol{n} \cdot \boldsymbol{U} = -gh\frac{\partial \zeta}{\partial s} - \frac{1}{2}Dg\frac{\partial \zeta}{\partial n} - \tau_0^{(s)} = 0$$

where $s, n$ are coordinates associated with the boundary ($s$ along the boundary, $n$ normal to it). The above balance completes the theory of the Ekman sandwich. It shows how the surface displacement reacts with Sverdrup dynamics and bottom friction on the wind forcing, established in this balance via the Ekman pumping.

Expressing $V$ in (14.21) by the barotropic stream function $\psi$, given by (14.18), we obtain

$$\beta\frac{\partial \psi}{\partial x} = \underset{\neg}{\nabla} \cdot \boldsymbol{\tau}_0 - \underset{\neg}{\nabla} \cdot |f|(d/2)(\boldsymbol{u}_\mathrm{g} + \underset{\neg}{\boldsymbol{u}}_\mathrm{g})$$

A closed equation for $\psi$ is only achieved if $\boldsymbol{u}_\mathrm{g}$ is somehow related to $\psi$, or to $\boldsymbol{U} = \underset{\neg}{\nabla}\psi$. In principle, this may be done exactly from (14.11). A simplifying assumption is, however, to neglect the frictional terms in $\boldsymbol{U}$ and set $\boldsymbol{u}_\mathrm{g} \approx \boldsymbol{U}/h$ (only in the friction term of the above balance) so that we arrive at

$$R\nabla^2\psi + \beta\frac{\partial \psi}{\partial x} = \underset{\neg}{\nabla} \cdot \boldsymbol{\tau}_0 \tag{14.23}$$

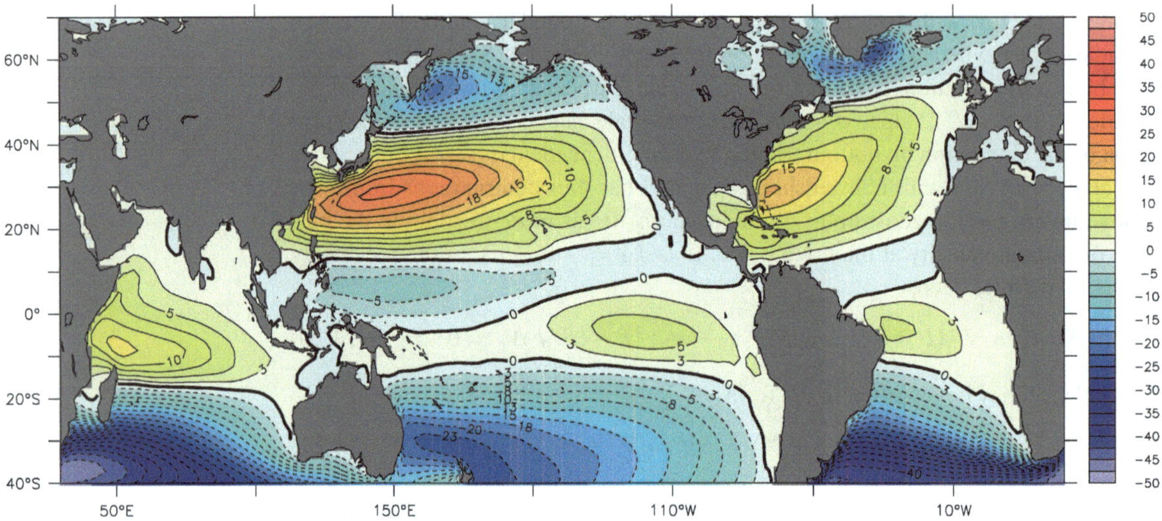


**Fig. 14.7** Stream function $\psi$ in $\mathrm{Sv} = 10^6\,\mathrm{m^3\,s^{-1}}$ calculated from the Stommel equation (14.23) and the realistic windstress curl shown in Figure 14.4. $\psi = 0\,\mathrm{Sv}$ was used as boundary condition and a boundary layer width of $\delta = 100\,\mathrm{km}$

where again $R$ was assumed constant. This is STOMMEL[2]'s equation. It has a much simpler boundary condition, $\psi = \mathrm{const}$, than the equivalent equation (14.22) for the surface displacement. Another property of the Stommel equation concerns the exception of the equator: whereas the Ekman and geostrophic theories, used above to determine the current profile and sea surface height, require nonzero $f$, we may abandon this restriction when considering balance equations for the total transport (this also applies to the Sverdrup theory, see the previous section).

Figure 14.7 shows a numerical solution of the Stommel equation (14.23) for the realistic windstress curl which was also used to display the global Sverdrup solution in Figure 14.6. In contrast to the Sverdrup solution, the Stommel solution now satisfies the boundary condition $\psi = 0$. It does not deviate much from the Sverdrup relation outside boundary layers. However, the tropical gyres in all ocean basins are much weaker due to the impact of the bottom friction. Again, the numerical solution of the Stommel equation (14.23) does not yield realistic results in the Southern Ocean. This issue will be further discussed in Chapter 16. The incorporation of lateral friction leads to the Stommel–Munk model. It is discussed in the box on p. 457.

### 14.1.6 The Western Boundary Current

The total transport in the Sverdrup regime occurs between the eastern edge of the western boundary layer, $x = \delta$, and the eastern coast, $x = 0$. The total transport is there $\psi(x = \delta, y)$ at the latitude $y$. If it is nonzero, the corresponding transport must be returned within the boundary layer. This transport is thus prescribed by the wind system outside the boundary layer, i. e. in the Sverdrup regime. Clearly, because the boundary layer width is much smaller than the basin width, the currents in the boundary layer have to be much stronger than in the Sverdrup regime.

[2] HENRY MELSON STOMMEL, *1920 in Wilmington †1992 in Boston, oceanographer.

**71. The Stommel–Munk Equation**

We derive the Stommel–Munk equation by starting again with the planetary geostrophic equations (14.1) and (14.2) for a homogenous ocean ($\rho = \text{const}$) with a flat bottom. Following Munk (1950), we add lateral friction with the lateral viscosity $A_\text{h}$,

$$f\underset{\neg}{\boldsymbol{u}} = -\nabla p + \frac{\partial \boldsymbol{\tau}}{\partial z} + A_\text{h}\nabla^2 \boldsymbol{u}$$

This lateral friction is meant to be a parameterization for the divergence of the lateral turbulent momentum fluxes related to mesoscale eddies.
For simplicity, we make the rigid lid approximation, i. e. $w = 0$ at the mean sea surface $z = 0$. Vertical integration of the horizontal momentum balance from the flat bottom to the surface and elimination of the pressure gradient term by cross-differentiation yields the *Stommel–Munk model* of the circulation in a homogeneous ocean,

$$R\nabla^2\psi - A_\text{h}\nabla^4\psi + \beta\frac{\partial\psi}{\partial x} = \underset{\neg}{\nabla}\cdot\boldsymbol{\tau}_0 \tag{B71.1}$$

The bottom friction coefficient $R = d|f|/2h$ measures the friction in the bottom Ekman layer of depth $d$, while the second term on the left-hand side derives from lateral diffusion of momentum. Only one of these processes is necessary to extract the momentum imparted by the windstress. If lateral friction is dominant, a simple scaling shows that the boundary layer width is given by $\delta = (A_\text{h}/\beta)^{1/3}$ while for the case in which bottom friction is dominant, the boundary layer scale is given by $\delta = R/\beta$.
Compared to the original Stommel equation (14.23), a higher order derivative is involved in (B71.1), and an additional appropriate dynamical boundary condition is needed. This could be e. g. "no-slip", i. e. zero flow along the boundary which means for the stream function $\partial\psi/\partial n = 0$ at the boundary, where $\partial/\partial n$ denotes a derivative normal to the boundary. An alternative is the so-called "free-slip" boundary condition, i. e. zero normal derivative of the flow along the boundary, which means in fact zero vorticity or $\partial^2\psi/\partial n^2 = 0$ at the boundary.

In an equilibrium view, the boundary layer has a rather passive role: whatever dynamics are present in addition to the terms of the Sverdrup balance, they have to accomplish the return flow. A crucial point is that the new physics added in the boundary layer must allow to satisfy the boundary condition of zero normal flow at the western rim of the basin. In case of the Stommel equation (14.23), the additional term stems from bottom friction.

A central feature of the Stommel solution is the western boundary layer in which an intensification of the current occurs and a northward, strong flow develops. If the prototype ocean forcing, as given by (B70.1), represents the North Atlantic, this flow is the model's Gulf Stream (see also the box on p. 458). Figure 14.7 shows that other oceans have similar strong poleward flows attached to their western coast boundaries, i. e. western boundary currents. The questions arises why these intensified currents always appear on the western side of the oceans and not on the eastern side. We could point at the exact solution (B72.2) which reveals this feature. The Sverdrup theory, however, is built on an *a priori* assumption on the boundary layer's position. The answer has to be found in the governing differential equation (14.23). A little mathematical treatment clearly shows this behavior. Consider (14.23) in the form

$$\delta\left(\frac{\partial^2\psi}{\partial x^2} + \frac{\partial^2\psi}{\partial y^2}\right) + \frac{\partial\psi}{\partial x} = \frac{1}{\beta}\underset{\neg}{\nabla}\cdot\boldsymbol{\tau}_0 = F \tag{14.24}$$

and assume for the beginning a western boundary layer. We introduce a scaled coordinate $\xi = x/\delta$, which runs from 0 to 1 in the boundary layer, and a scaled coordinate $\eta = y/L$ for the meridional direction such that $\psi(x, y) = \tilde{\psi}(\xi, \eta)$ in the boundary

**72. Analytical Solution of the Stommel Problem**

The solution of (14.23) subject to the boundary condition $\psi = 0$ for the prototype windstress (B70.1) can be found as follows. We write the solution by separating the $x$ and $y$ dependence,

$$\psi(x, y) = \varphi(x) \sin \pi y/L$$

which vanishes all along the basin rim if $\varphi(x) = 0$ at $x = 0, B$. Note that this separation works only because all terms in (14.23) become proportional to the sine term. Inserting it into (14.23), we find that the scaled $\tilde{\varphi}(\tilde{x}) = \varphi(x)L\beta/(\pi B\tau_0)$ must satisfy

$$\delta/B\left[\tilde{\varphi}'' - (\pi B/L)^2\tilde{\varphi}\right] + \tilde{\varphi}' = -1 \tag{B72.1}$$

in the scaled range $0 < \tilde{x} = x/B < 1$. The dash in (B72.1) denotes a derivative with respect to $\tilde{x}$. The width of the boundary layer, $\delta = R/\beta$, appears in front of the highest derivative. The term originates from the friction term in (14.23). For a typical value for $R \sim 10^{-7}\,\mathrm{s}^{-1}$ (about 100 d), the width $\delta$ becomes about 100 km and thus $\delta/B \ll 1$. If we neglect the terms $\sim \delta$ in (B72.1), the equation represents the Sverdrup regime where $\varphi \sim B - x$.
Equation (B72.1) is a second order differential equation with constant coefficients, and the exact solution is found by straightforward analysis,

$$\tilde{\varphi}(\tilde{x}) = \frac{L^2}{\delta\pi^2 B}\left[1 - \frac{(1-\mathrm{e}^{k_2})\mathrm{e}^{k_1\tilde{x}} - (1-\mathrm{e}^{k_1})\mathrm{e}^{k_2\tilde{x}}}{\mathrm{e}^{k_1} - \mathrm{e}^{k_2}}\right] \tag{B72.2}$$

where $k_{1,2} = -B/(2\delta)[1 \pm \sqrt{1 + (2\pi\delta/L)^2}]$. The complete solution for $\psi(x, y)$ is shown below for different choices of the boundary layer width $\delta$. All other parameters are chosen as in the prototype example for the Sverdrup solution the box on p. 454. For small $\delta$, the solution is approximated by the Sverdrup solution, except close to the western boundary. But note that a significant deviation between the true solution and the Sverdrup solution is visible everywhere, also outside the boundary layer. In all cases, a strong northward western boundary current compensates the weak interior southward Sverdrup transport.

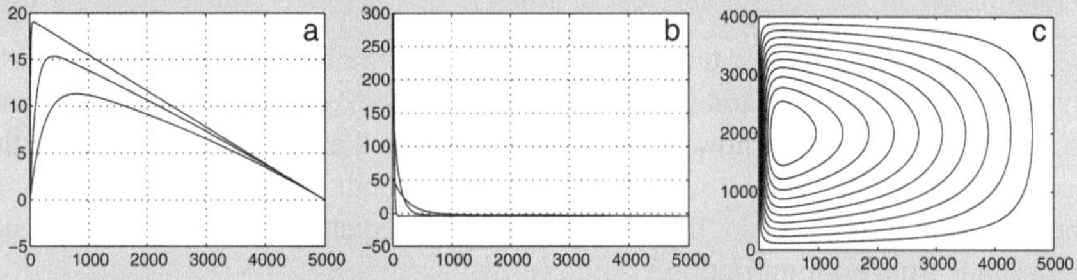


*Stream functions of the Stommel problem for the prototype windstress (B70.1). Shown is $\varphi(x)$ in Sv in (a) and the meridional transport $V = \partial\psi/\partial x$ in $\mathrm{m^2\,s^{-1}}$ in (b) for $\delta = 250$ km, $\delta = 100$ km and $\delta = 10$ km. Also shown is the stream function $\psi$ in (c) for $\delta = 100$ km.*

layer. The $x$-derivatives of $\psi$ are large, but the $\xi$-derivatives of $\tilde{\psi}$ are of order unity, as obvious in the transformation $\partial\tilde{\psi}/\partial\xi = \delta\partial\psi/\partial x$ and $\partial^2\tilde{\psi}/\partial\xi^2 = \delta^2\partial\psi/\partial x^2$. The boundary layer function $\tilde{\psi}$ is governed by

$$\frac{\partial^2\tilde{\psi}}{\partial\xi^2} + \frac{\delta^2}{L^2}\frac{\partial^2\tilde{\psi}}{\partial\eta^2} + \frac{\partial\tilde{\psi}}{\partial\xi} = \delta F$$

derived from (14.24). In the limit $\delta \to 0$,

$$\frac{\partial^2\tilde{\psi}}{\partial\xi^2} + \frac{\partial\tilde{\psi}}{\partial\xi} = 0$$

with the general solution $\tilde{\psi} = C_1 + C_2 e^{-\xi}$ which is exponentially decaying. It must satisfy the boundary condition $\tilde{\psi}(\xi = 0) = 0$ on the coast and match the Sverdrup solution $\psi_{Sv}(x = 0)$ as $\xi \to \infty$. This can be achieved by $C_1 = -C_2 = \psi_{Sv}(0)$, and we recover our results from the previous section. A more detailed mathematical treatment of boundary layer problems in given in Appendix A.2.2.

Can we put a boundary layer at the eastern coast? Redefine the boundary layer coordinate $\lambda = (L - x)/\delta$ appropriate to an eastern boundary layer of width $\delta$ and repeat the corresponding analysis. We find

$$\frac{\partial \hat{\psi}}{\partial \lambda^2} + \frac{\delta^2}{L^2}\frac{\partial \hat{\psi}}{\partial \eta^2} - \frac{\partial \hat{\psi}}{\partial \lambda} = \delta F$$

for $\hat{\psi}(\lambda, \eta) = \psi(x, y)$. Note the minus sign at the $\beta$-term which renders the homogeneous solution exponentially increasing away from the coast, $\hat{\psi} = C_1 + C_2 e^{\lambda}$. For $\lambda \to \infty$, matching is impossible because the boundary layer solution becomes infinite. We have to set $C_2 = 0$. Then the boundary condition yields $C_1 = 0$. Hence there is no eastern boundary layer in this simple model.

The above considerations about the westward intensification have a firm mathematical basis, but there is a simple physical reason as well. It is clear by now that the terms involving derivatives must balance there in the boundary layer, i. e. $-R\partial^2\psi/\partial x^2 = -R\partial V/\partial x \approx \beta\partial\psi/\partial x = \beta V = V\mathrm{d}f/\mathrm{d}y$, which is a balance between bottom friction of (relative) vorticity and advection of planetary vorticity. As the Sverdrup transport is negative in the gyre (with the curl specified by (B70.1)), we find a positive advection term $V\mathrm{d}f/\mathrm{d}y$ in the western boundary current, and hence the friction term $-R\partial V/\partial x$ must be positive as well: it must eliminate positive vorticity. But with a negative $V$ in the Sverdrup regime and a positive $V$ in the boundary layer, we obtain $\partial V/\partial x > 0$ for a potential eastern boundary layer setting, and thus a balance is not possible there. On the other hand, we find $\partial V/\partial x < 0$ for a potential western boundary layer setting, and thus here is the only possibility of an overall balance of vorticity.

## 14.2 The Role of Stratification and Topography

In the previous sections, we have assumed a flat ocean bottom in order to derive the equations of the Sverdrup model, (14.19), the Stommel model (14.23), or the Stommel/Munk model (B71.1) of the wind-driven circulation. This assumption was necessary to eliminate the depth-integrated pressure gradient force when taking the curl of the depth-integrated momentum budget. Although the Stommel or Stommel/Munk model leads to a realistic global flow pattern, as shown in Figure 14.7, it is clear that the assumption of a flat bottom is unrealistic. In this section, we will discuss the influence of topographic variations on the wind-driven circulation. When considering the vorticity balance of the depth-averaged flow, the so-called JEBAR-term (see below) describes the influence of the topography while in the vorticity balance of the depth-integrated flow the so-called bottom pressure torque appears. It will be shown that both effects are equivalent, and they will turn out to be important features in both vorticity balances.

### 14.2.1 The JEBAR Term

Consider an ocean with an arbitrary topography, given by the depth $h(x, y)$, and a density $\rho$ which is no longer assumed constant, i. e. we abandon form now on in this chapter the assumption of a homogeneous ocean. Vertical integration of the horizontal momentum equations in the planetary geostrophic approximation, including the vertical friction term as in the previous section, yields

$$f \underline{U} = -\int_{-h}^{0} \nabla p \mathrm{d}z + \boldsymbol{\tau}_0 - R\boldsymbol{U} \tag{14.25}$$

with the horizontal transport vector $\boldsymbol{U} = \int_{-h}^{0} \boldsymbol{u} \mathrm{d}z$, the windstress $\boldsymbol{\tau}_0$, and the friction coefficient $R$ related to the bottom Ekman layer. Note that the notation as defined in Section 14.1 is used here, and that we will assume a rigid lid (i. e. $w = 0$ at the mean sea surface $z = 0$). Because of the dependency of $h$ on the horizontal position, the integral of the pressure term needs now a bit of care. By partial integration and the use of the hydrostatic relation one obtains

$$\begin{aligned}\int_{-h}^{0} \nabla p \mathrm{d}z &= \nabla \int_{-h}^{0} p \mathrm{d}z - p|_{z=-h} \nabla h = \nabla \left( -\int_{-h}^{0} z \frac{\partial p}{\partial z} \mathrm{d}z + z p|_{z=-h}^{0} \right) - p|_{z=-h} \nabla h \\ &= h \nabla P + \nabla E\end{aligned} \tag{14.26}$$

with the vertically integrated potential energy

$$E = g \int_{-h}^{0} z \rho \mathrm{d}z \tag{14.27}$$

defined with respect to the surface, and the bottom pressure $P = p|_{z=-h}$. Either of the pressure terms in the resulting form

$$f \underline{U} = -h \nabla P - \nabla E + \boldsymbol{\tau}_0 - R\boldsymbol{U} \tag{14.28}$$

of the momentum balance can be eliminated by building the appropriate vorticity balance. That way, either (14.29) or (14.32) will be derived.

If this is first by divided by $h$ and then subjected to the curl operation, equation (14.26) is cast into

$$\nabla \cdot (R/h) \nabla \psi + \underline{\nabla} \psi \cdot \nabla (f/h) + \underline{\nabla}(1/h) \cdot \nabla E = \underline{\nabla} \cdot (\boldsymbol{\tau}_0 / h) \tag{14.29}$$

using again the stream function $\underline{\nabla} \psi = \boldsymbol{U}$. Equation (14.29) still shows some resemblance to the Stommel equation (14.23) for which a flat bottom and constant density was assumed: a torque on the right-hand side of both equations related to the forcing by the curl of the windstress balances a torque introduced by bottom friction (first term on the left-hand side of (14.23) and (14.29)) and a torque related to the change in planetary vorticity when a fluid particle changes position ($\beta \partial \psi / \partial x$ in (14.23) and $\underline{\nabla} \psi \cdot \nabla f / h$ in (14.29)). However, in (14.29) a third term occurs, which is related to the combined effect of topographic variations and stratification, i. e. $\underset{\neg}{\nabla}(1/h) \cdot \nabla E$. It has been called the "JEBAR" (Joint Effect of Baroclinicity and Relief) term, as introduced by Sarkisyan and Ivanov (1971). The JEBAR term often dominates the windstress curl, but its physical meaning can be more difficult to interpret, as demonstrated below. Note that for constant $h$, both terms related to the planetary vorticity in (14.23) and (14.29) become identical and that the JEBAR term is zero, so that the equation (14.29) turns into the Stommel equation (14.23). We might call (14.29) the generalized Stommel equation. Likewise, without the friction term, the equation is the generalized Sverdrup balance.

### 14.2.2 The $f/h$ Contours

When considering (14.29) for the frictionless case (i. e. $R = 0$), for constant density ($\underset{\neg}{\nabla}(1/h) \cdot \nabla E = 0$), and without wind forcing (i. e. $\boldsymbol{\tau}_0 = 0$), we see that it reduces to

$$\underset{\neg}{\nabla}\psi \cdot \nabla(f/h) = 0 \tag{14.30}$$

This is a statement about the balance between the vortex stretching by changes in topography and change of planetary vorticity of a fluid particle representing the depth-averaged flow. Equation (14.30) implies that $\psi$ becomes a function of $f/h$, i. e. streamlines $\psi$ = const and contours of $f/h$ coincide. Thus, for the frictionless case without wind forcing and stratification, the large-scale depth-integrated oceanic flow must follow $f/h$ contours. In mathematical terms, the $f/h$ contours form the characteristics of the differential equation (14.30). The state $\psi = \psi(f/h)$ is often quite loosely referred to as the 'free mode' (even in the presence of friction and forcing, which, of course, must be weak) in the sense that it could somehow be added on top of a 'forced' part of $\psi$. On the other hand, one should note that $\psi(f/h)$ is not 'free', however small friction and forcing become since it is possible to show that this "free" mode is, in fact, entirely determined by friction and forcing (see Section 16.2.4).

Figure 14.8 shows $f/h$ contours in the World Ocean. Note that for constant $h$, the $f/h$ contours would follow latitude circles. However, it is obvious that the topographic variations change this symmetry drastically and put an important constraint on the depth-averaged flow. In general, the influence of topography on the characteristics of the depth-averaged flow is small in the tropics but becomes large in the higher latitudes. In the Atlantic Ocean, the imprint of the Mid-Atlantic ridge can clearly be seen in the region of the subtropical gyres of the North and South Atlantic.

In the presence of windstress, the term related to the windstress curl in (14.29) will drive flow across $f/h$ contours, and we find as balance

$$\underset{\neg}{\nabla}\psi \cdot \nabla(f/h) = \underset{\neg}{\nabla} \cdot (\boldsymbol{\tau}_0/h) \tag{14.31}$$

Here the fluid responds to the injection of vorticity by the wind such that the barotropic flow must cross the $f/h$ contours and give up the balance between vortex stretching and change of planetary vorticity which prevails for unforced steady motions. For constant $h$, we recover the Sverdrup relation (14.19) which relates the meridional depth-integrated transport to the windstress curl forcing. Since Sverdrup

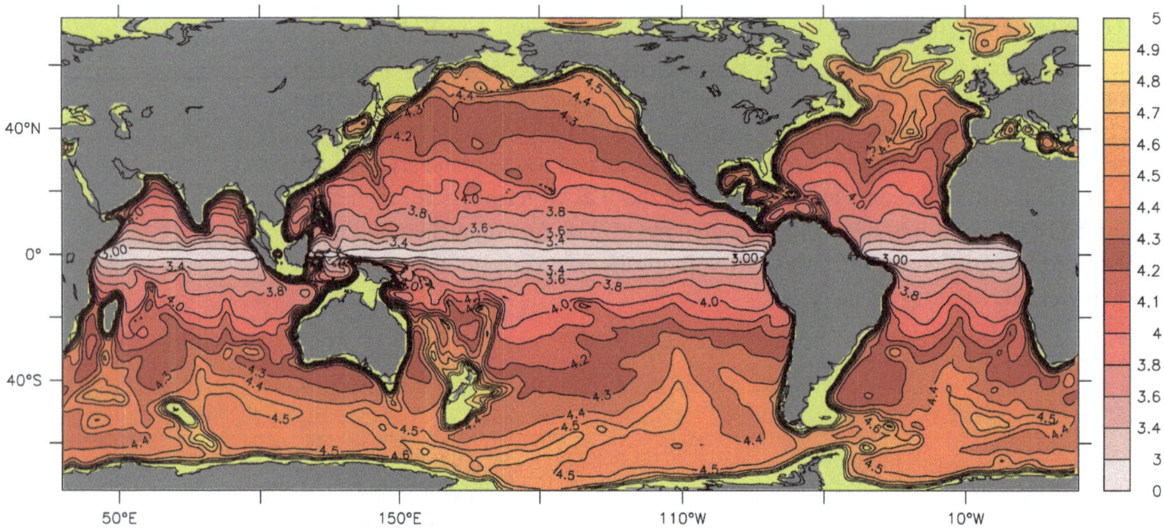

Fig. 14.8 Contours of $f/h$ in the World Ocean. Shown is $\log_{10}(|f|/h[10^{-12}\,\mathrm{m}^{-1}\,\mathrm{s}^{-1}])$. Note the nonequidistant color scale

(1947) succeeded in explaining the existence of the equatorial countercurrent on the basis of (14.19) (for a flat bottom), the Sverdrup relation became classic in physical oceanography to represent the large-scale flow in basin circulations outside frictional boundary layers. However, the remarkable success of the flat-bottom Sverdrup relation (14.19) is surprising noticing the strong effect of topography in (14.31), an issue which will be resolved in the following discussion.

The previous reflections had no particular consideration of closed $f/h$ contours. Evidently, the left-hand side of (14.31) (divided by $|\nabla f/h|$) integrates to zero around a closed $f/h$ contour but the forcing on the right-hand side does in general not share this property. The oceanographic research was aware quite early (e. g. Hasselmann 1982) that the vorticity balance is severely constrained in regions where closed $f/h$ contours exist, as e. g. in the circumpolar region of the Southern Ocean where closed contours occur over the midocean ridges and around Antarctica (see Figure 14.8). Either friction and/or the JEBAR term must gain importance in the vorticity balance (14.29). We will pick up the thread of the discussion about closed $f/h$ contours in Section 16.2.4.

### 14.2.3 Sverdrup's Catastrophe

At first sight, the generalized Sverdrup relation (14.31) for the case with varying topography is apparently more complicated than the flat-bottom Sverdrup relation (14.19). Note that the $f/h$ contours form again the characteristics of the differential equation (14.29) while the characteristics of the flat-bottom Sverdrup relation (14.19) are simply given by latitude circles. We still might integrate the forcing, given by $\underline{\nabla}\cdot(\tau_0/h)$, along $f/h$ contours to obtain the depth-integrated flow normal to these contours. In analogy to the flat-bottom Sverdrup relation, we should start the integration of (14.31) where the $f/h$ contours intersect with the eastern boundary, by setting $\psi = 0$ there.

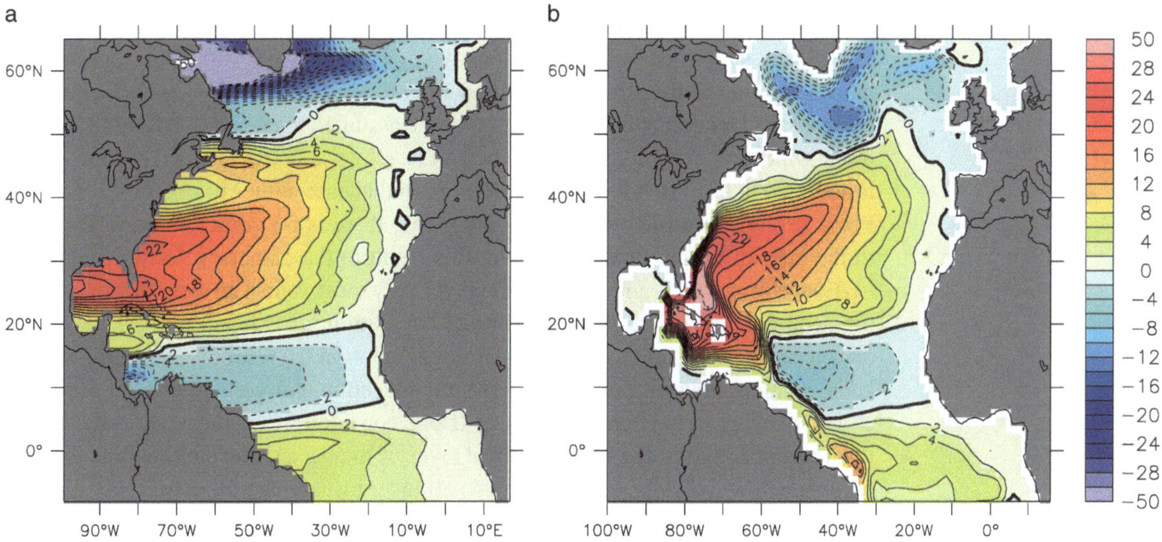


**Fig. 14.9** **a** Stream function $\psi$ in Sv $= 10^6\,\mathrm{m^3\,s^{-1}}$ for the flat-bottom Sverdrup solution (14.19) calculated from realistic windstress. **b** Stream function $\psi$ for a realistic coarse resolution model of the Atlantic Ocean. The windstress driving the model was used to calculate $\psi$ from (14.19) in **a**

When we integrate the generalized Sverdrup relation (14.31) instead of the flat-bottom Sverdrup relation (14.19) along the appropriate characteristics to obtain the stream function $\psi$, we have accounted for the impact of topography, and we might, therefore, expect to get a more realistic depth-integrated circulation. However, the circulation obtained in this way differs very much from the one obtained by the flat-bottom Sverdrup solution (14.19), as shown for instance in Figure 14.6, which we found plausible because it is consistent with what is derived from observations. Indeed, attempts to verify the flat-bottom Sverdrup relation (14.19) from observations are often fairly successful, e. g. in the subtropical North Atlantic, Leetmaa et al. (1977) have obtained a fair agreement between the Sverdrup transport and the sum of geostrophic and Ekman transport, surprisingly without the topographic dependence in (14.31).

To illustrate the success of the flat-bottom Sverdrup solution, Figure 14.9 shows the stream function $\psi$ calculated from the flat-bottom Sverdrup relation (14.19) in the North Atlantic Ocean and compares it with the stream function from a numerical model based on primitive equations. Both $\psi$ pattern were calculated using the same windstress forcing. Although the numerical model contains realistic topography, surface buoyancy forcing and stratification, the flat-bottom Sverdrup solution yields a stream function in the subtropical gyre (except for the western boundary layer) which is very similar to the complex numerical model. Only in the subpolar North Atlantic, the flat-bottom Sverdrup relation overestimates the model's circulation.

A first explanation of this phenomenon has been given by Anderson and Killworth (1977) who considered the spin-up of a two-layer basin with simple topography. They showed that the slowly propagating baroclinic planetary waves which emanate from the eastern coast leave in their wake an essentially steady flat-bottom Sverdrup solution for both the barotropic and baroclinic modes. For a flat-bottom ocean this property is discussed in detail in Section 8.2.6. Thus there is no motion in this model at great depth, and the topography in the steady barotropic response appears to be shielded by the stratification. A simple heuristic way to illustrate this argument is to look at a simplified balance for density $\partial\rho/\partial t + w\partial\rho/\partial z = 0$. This approximate balance applies to the strongly stratified situation $|\partial\rho/\partial z| \gg |\nabla\rho|$, where effects of horizontal advection and turbulent mixing are assumed to be small. After the spin-up of the wind-driven circulation, i. e. when all Rossby waves have passed by, the system is assumed to be in steady state, $\partial\rho/\partial t = 0$. In that case, either $w = 0$ or $\partial\rho/\partial z = 0$. For a strongly stratified ocean, it follows that there is no vertical motion anymore in the interior ocean. Proceeding with this result to the planetary vorticity equation (5.49) we find $v = 0$, and then the zonal velocity must vanish as well, demonstrating the need for a nearly motionless abyssal circulation after the spin-up of the wind-driven circulation and the shielding of the circulation from topography. One should keep in mind, however, that the horizontal advection and vertical mixing may also come into play in the density balance. If one considers vertical mixing only, which is largest near and in the surface mixed layer of the ocean, one can imagine that the complete flow would be confined in a small layer just below the sea surface. Without any mixing in the surface layer, however, the situation describes a singularity, which explains why the above simple argument is sometimes referred to as "Sverdrup's catastrophe" (see Section 8.2.6 for the flat-bottom case).

### 14.2.4 The Bottom Pressure Torque

An alternative way to account for effects of stratification and topographic variations in the Stommel equation is given by eliminating the vertically integrated potential energy $E$ instead of the bottom pressure $P$ from the vertically integrated momentum balance (14.28). Taking the curl of this equation yields

$$R\nabla^2\psi + \beta\frac{\partial\psi}{\partial x} + \nabla P \cdot \underset{\neg}{\nabla} h = \underset{\neg}{\nabla} \cdot \boldsymbol{\tau}_0 \tag{14.32}$$

Note that (14.32) is a vorticity budget for the depth-*integrated* flow instead of the depth-*averaged* flow in (14.29). In (14.32), the torque $\underset{\neg}{\nabla} \cdot \boldsymbol{\tau}_0$ by the windstress can drive flow across lines of constant latitudes (which are the characteristics in this case), but there is also the torque given by the bottom pressure variation on sloping topography, i. e. the bottom pressure torque $\nabla P \cdot \underset{\neg}{\nabla} h$. Its relation to JEBAR is analyzed in the box on p. 465.

Away from boundaries, the component of the transport related to the bottom pressure is generally fairly small. Figure 14.10 shows the components of the depth-integrated vorticity balance in a realistic noneddy-resolving model of the North Atlantic, i. e. the planetary vorticity $\beta\partial\psi/\partial x$ in Figure 14.10a, and the windstress curl $\underset{\neg}{\nabla}\cdot\boldsymbol{\tau}_0$ in (b), together with the bottom pressure torque (d). Clearly, there is a close balance between $\beta\partial\psi/\partial x$ and windstress curl in the interior of the subtropical and tropical Atlantic, as shown in (c). In fact, it is the part of the flow that is discussed in the Anderson–Killworth scenario of Section 14.2.3, and hence the stratified topographic model (B73.1) approximately collapses to the homogeneous flat-bottom model (14.19). A vivid demonstration of this transition has been given in Olbers and Eden (2003) with a numerical primitive equation model and the BARBI model as discussed below in Section 14.2.5.

**73. JEBAR and the Bottom Pressure Torque**

Another way to look at the compensation is the JEBAR term in the depth-averaged vorticity budget (14.29), which comes into play when there is stratification. The complete Sverdrup relation is in fact a three-term balance

$$\underline{\nabla}\psi \cdot \nabla(f/h) + \underline{\nabla}(1/h) \cdot \nabla E = \underline{\nabla} \cdot (\boldsymbol{\tau}_0/h) \qquad \text{(B73.1)}$$

The JEBAR term $\underline{\nabla}(1/h) \cdot \nabla E$ compensates to a large extent the effect of topography in $\underline{\nabla}\psi \cdot \nabla(f/h)$. This can be seen by rearranging the budget (B73.1) to the form

$$\beta \frac{\partial \psi}{\partial x} + \frac{1}{h}\underline{\nabla} h \cdot f \nabla \psi - \frac{1}{h}\underline{\nabla} h \cdot \nabla E = \underline{\nabla} \cdot \boldsymbol{\tau}_0 - \frac{1}{h}\underline{\nabla} h \cdot \boldsymbol{\tau}_0 \qquad \text{(B73.2)}$$

By comparing it with the (frictionless) depth-integrated momentum equation (14.25), rewritten here as $-f\nabla\psi = -h\nabla P - \nabla E + \boldsymbol{\tau}_0$, it becomes clear that the last two terms on the left-hand side and the last on the right-hand side of (B73.2) sum up to the contribution by the bottom pressure $P$ such that we obtain

$$\beta \frac{\partial \psi}{\partial x} + \nabla P \cdot \underline{\nabla} h = \underline{\nabla} \cdot \boldsymbol{\tau}_0 \qquad \text{(B73.3)}$$

It is because of this near cancelation of the two large terms on the left-hand side of (B73.1) that JEBAR is often regarded as being difficult to interpret. The effect of the JEBAR term and the cancelation will be discussed in more detail below in Section 14.2.6.
The bottom pressure torque plays the role of a vortex stretching term which can be seen as follows: the geostrophic bottom velocity $\boldsymbol{u}_g|_{-h}$ is given by $f\underline{\boldsymbol{u}}_g|_{-h} = -\nabla p|_{-h} = -\nabla P + g\rho'|_{-h}\nabla h$, such that the bottom pressure torque is given by the geostrophic bottom flow directed across isobaths, i. e. $\nabla P \cdot \underline{\nabla} h = -f\boldsymbol{u}_g|_{-h} \cdot \nabla h$. Using the kinematic bottom boundary condition, i. e. $\boldsymbol{u} \cdot \nabla h + w = 0$ at $z = -h$ for the geostrophic components $\boldsymbol{u}_g$ and $w_g$, one obtains furthermore

$$\nabla P \cdot \underline{\nabla} h = f w_g|_{-h}$$

The bottom pressure torque is thus given by the vortex stretching by the geostrophically balanced vertical velocity at the ocean bottom. For constant $h$, the bottom pressure torque vanishes, but it will also vanish when the geostrophic flow at the bottom vanishes, i. e. for the case of the Sverdrup catastrophe in the Anderson–Killworth scenario mentioned above. Note that assuming constant density but allowing for topographic variations, the bottom pressure torque will still contribute in (B73.3) due to the variations in the surface pressure. On the other hand, the vorticity equation (14.29) for the depth-average flow does not contain any pressure contribution for the case $h = \text{const}$ and $\rho = \text{const}$ and is, therefore, sufficient to determine the flow.

However, it is also clear from Figure 14.10 a) and b) that towards the western boundary and in the subpolar gyre, the flat-bottom Sverdrup balance breaks down. Here the bottom pressure torque $\nabla P \cdot \underline{\nabla} h$, shown in Figure 14.10 d), becomes important. In fact, in the tropical and subtropical North Atlantic, the dominant vorticity balance appears to be between vortex stretching by the bottom flow and the planetary vorticity $\beta\partial\psi/\partial x$ while the frictional torques play only a minor role. Note that the torque related to the momentum advection, which we have neglected from the beginning by the use of the planetary geostrophic approximation, plays only a minor role on scales larger than the Rossby radius even in high-resolution mesoscale eddy-permitting model simulations, as shown by Hughes and de Cuevas (2001) and Eden and Olbers (2010). The fact that the bottom pressure torque can play a more dominant role than the frictional torque for the vorticity balance in the western boundary current questions the physical relevance of Stommel's original model, which is fur-

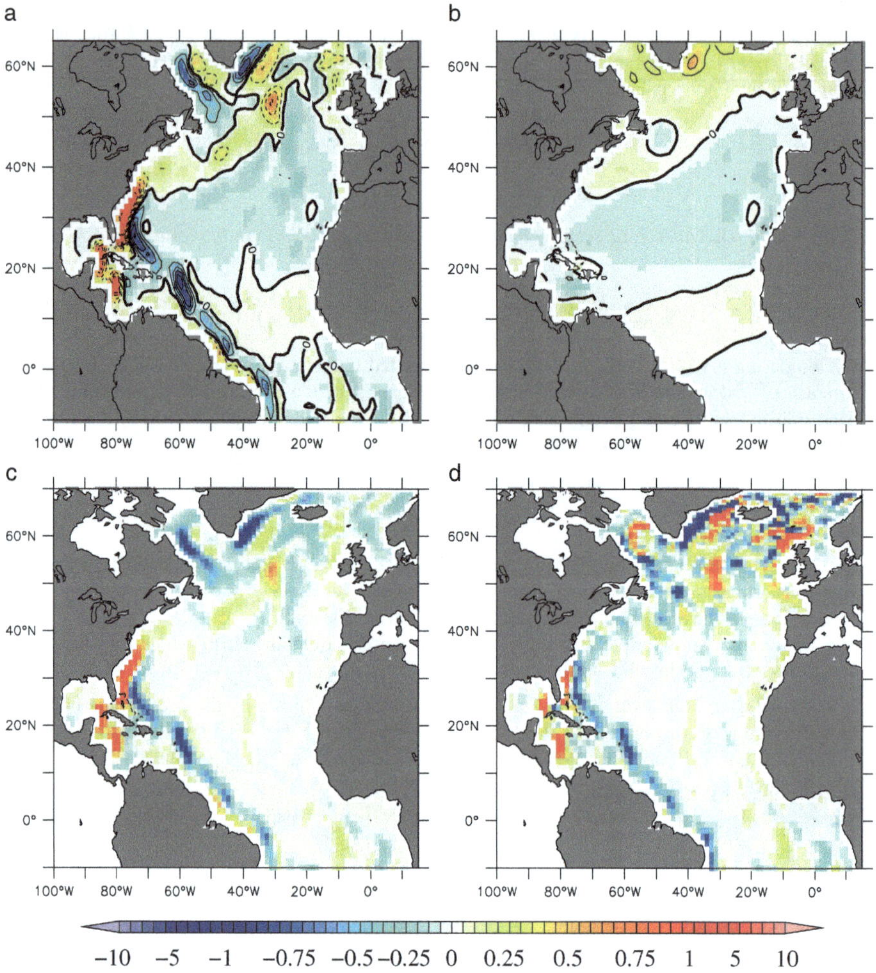


Fig. 14.10 **a** Planetary vorticity term $\beta\partial\psi/\partial x$ in $10^{-9}$ ms$^{-2}$ in a realistic numerical noneddy-resolving model of the Atlantic Ocean, which was also shown in Figure 14.9. **b** Wind stress curl $\underline{\nabla}\cdot\boldsymbol{\tau}_0$ in $10^{-9}$ ms$^{-2}$, driving the numerical model. **c** Difference $\beta\partial\psi/\partial x - \underline{\nabla}\cdot\boldsymbol{\tau}_0$ in $10^{-9}$ ms$^{-2}$. **d** Bottom pressure torque $\boldsymbol{\nabla} P \cdot \underline{\nabla} h$ evaluated in the model in $10^{-9}$ ms$^{-2}$

ther discussed in the next section in the context of the BARBI model. The impact of the bottom pressure torque on the Antarctic Circumpolar Current will be discussed in Section 16.5.

### *14.2.5 A Realistic Application of the BARBI Model*

The BARBI model demonstrates a way to incorporate the combined impact of topographic variations and stratification on the depth-integrated flow via the JEBAR term in a simple model as extension of the Stommel/Munk model (B71.1). The model is derived in Appendix B.2 in general form. Applications in this book are for a simplified two-mode version, given by (B.42)–(B.44). The governing equations are re-

**74. Barotropic and Baroclinic Velocity**

Since BARBI relies on the separation between the barotropic and the baroclinic flow, the concept of a barotropic flow needs some clarification. In Section 2.11, we have defined a barotropic state of a flow as one in which the (three-dimensional) baroclinic vector $(1/\rho^2)\nabla\rho \times \nabla p$ is zero, or equivalently, in which pressure and density surfaces coincide. Furthermore, the Taylor-Proudman theorem, (B16.1), showed that a barotropic, frictionless flow has a horizontal velocity which is constant in the vertical direction (if the rotation vector is vertical as in the primitive equations). This property motivates another less strict definition of a barotropic flow: a state of flow with constant velocity profile $\boldsymbol{u} = \boldsymbol{u}(\boldsymbol{x})$ in the vertical direction. An even looser definition is that of a fluid with constant density where the baroclinic vector is clearly zero and where, because of the hydrostatic pressure state, $\nabla p$ and the geostrophic current must be independent of $z$.
On the other hand, a common procedure in oceanography (and also used in BARBI) is the separation of the velocity profile into the depth-averaged part – which is then called the "barotropic" velocity component – and the shear-containing part – the "baroclinic" velocity component for the definition of barotropic and baroclinic velocity in BARBI. The representation of the three-dimensional circulation of the ocean in terms of barotropic and baroclinic variables has mainly been introduced for time-dependent linear problems such as the spin-up of an ocean basin where wave properties are relevant. The separation has also some advantages in numerical primitive equation models since the boundary conditions for the barotropic stream function are particularly simple (at least for single connected domains) as compared to those of the surface pressure field.

peated here,

$$\frac{\partial}{\partial t}\nabla\cdot\left(\frac{1}{h}\nabla\psi\right) + \underset{\neg}{\nabla}\psi\cdot\nabla\frac{f}{h} = \underset{\neg}{\nabla}E\cdot\nabla\frac{1}{h} + \underset{\neg}{\nabla}\cdot\frac{\boldsymbol{\tau}_0}{h} + A_\mathrm{h}\nabla\cdot\frac{1}{h}\nabla^2\nabla\psi \qquad (14.33)$$

$$\frac{\partial E}{\partial t} + h\underset{\neg}{\nabla}\psi\cdot\nabla\frac{E}{h^2} = \frac{N_0^2}{6}\underset{\neg}{\nabla}\psi\cdot\nabla h^2 + \frac{N_0^2}{2}\nabla\cdot\boldsymbol{u}^* + K_\ell\nabla^2 E \qquad (14.34)$$

$$\frac{\partial \boldsymbol{u}^*}{\partial t} + f\underset{\neg}{\boldsymbol{u}^*} = \frac{h^2}{3}(\nabla E - \boldsymbol{\tau}_0 - A_\mathrm{h}\nabla^2\boldsymbol{U}) + A_\mathrm{h}\nabla^2\boldsymbol{u}^* \qquad (14.35)$$

The variables are the barotropic stream function $\psi$, the vertically integrated potential energy $E$, and a baroclinic velocity $\boldsymbol{u}^*$. Furthermore, $N_0$ characterizes the given background stratification (remember that $N$ is the effective stability frequency), $A_\mathrm{h}$ is an eddy-induced lateral viscosity, and $K_\ell$ the lateral (isopycnal thickness) diffusivity (compare Section 12.1.3). In linearized form, the above equations describe long barotropic Rossby waves, the first baroclinic Rossby waves (long and short), and the the first baroclinic gravity waves, all with the appropriate topographic modifications. A review of the termini "barotropic" and "baroclinic" is given in the box on p. 467.

Figure 14.11 shows a numerical application of the two-mode BARBI model with realistic topography and windstress. The domain is the entire globe, except for the Arctic Ocean, with a horizontal resolution of 2° × 2°. The horizontal viscosity for $\boldsymbol{U}$ and $\boldsymbol{u}^*$ is $A_\mathrm{h} = 4\times10^4\,\mathrm{m^2\,s^{-1}}$. The model is forced with annual mean windstress data from an ECMWF analysis (Barnier et al., 1995). The figure presents results of the BARBI model from three different experiments: an experiment without background stratification (Figure 14.11a), i. e. $N_0 \equiv 0$ but including realistic topographic variations, an experiment with a flat bottom ($h$ = const, Figure 14.11b), and an experiment with topography and background stratification present ($N_0 = 2\times10^{-3}\,\mathrm{s^{-1}}$, Figure 14.11c,d). This suite of experiments follows the by now classical simulations with the early GFDL (MOM) model, described by Bryan and Cox (1972) and Cox (1975). Figure 14.11 shows the barotropic stream functions for the three BARBI experiments in steady state; for the baroclinic case, the baroclinic variable $E$ is shown

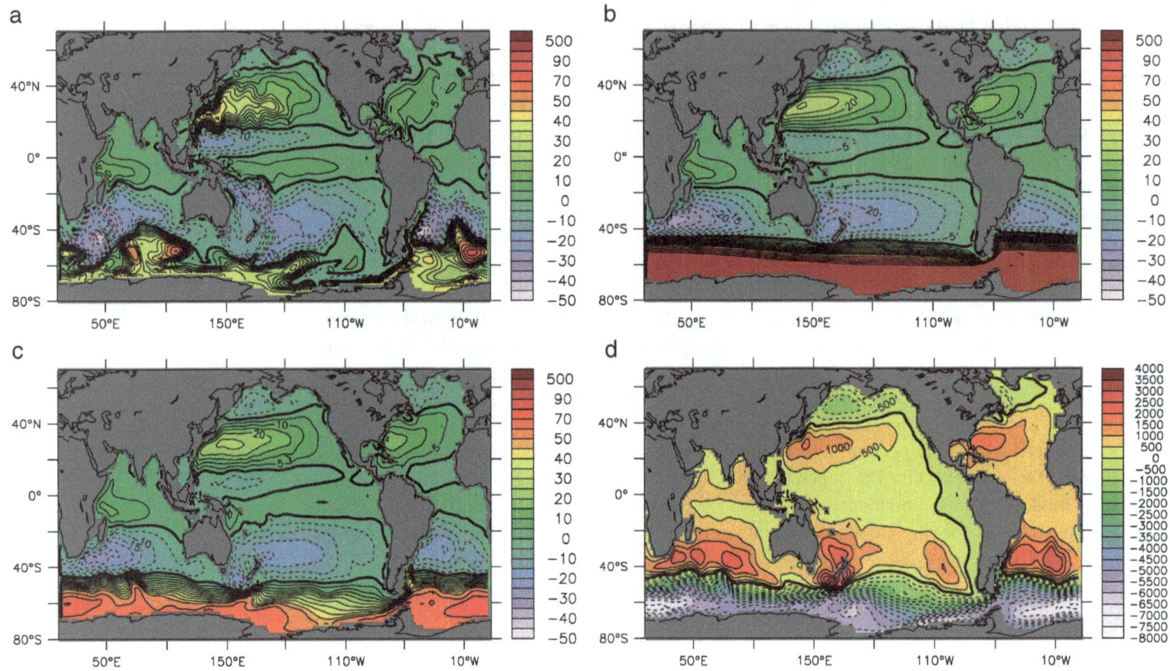

Fig. 14.11 **a** Stream function $\psi$ in Sv $= 10^6\,\mathrm{m}^3\,\mathrm{s}^{-1}$ in the BARBI model for the case without stratification but realistic topography. **b** Same as **a** but for a constant depth $h = 2{,}000$ m. **c** Same as **a** but for a stratification $N_0 = 2 \times 10^{-3}\,\mathrm{s}^{-1}$. **d** Potential energy $E$ in $\mathrm{m}^2\,\mathrm{s}^{-2}$ for the model experiment with stratification

as well. Most notably, the transport of the Antarctic Circumpolar Current (ACC) varies enormously in the three experiments – they will be further discussed in Chapter 16 – but there are also large differences in the subtropical gyres in the Pacific and Atlantic Oceans.

The model behavior in these three configurations reveals in the North Atlantic and the North Pacific the regimes discussed above, i. e. we find the shift from the topographic Sverdrup regime to the flat-bottom one, if baroclinicity is included. However, while the subtropical gyre with maximum transports of about 30 Sv in the North Atlantic for the baroclinic case with topography is within rough bounds of observational estimates, the subpolar North Atlantic shows only a weak cyclonic circulation present in all three experiments, at maximum of 5 Sv in the flat-bottom case. Observational estimates give higher values of about 30–50 Sv (compare also Figure 14.9). This contrast points towards the importance of thermohaline forcing for the strength of the North Atlantic subpolar gyre which we have excluded from our experiments. Overall, however, the experiments in this realistic set-up of topography and stratification confirm that the BARBI model produces the important aspects of the wind-driven large-scale baroclinic circulation over topography.

### 14.2.6 The Baroclinic Stommel Equation

We proceed with some analytical considerations and develop a theoretical basis of the above discussed Sverdrup catastrophe. The discussion is based on the BARBI physics, using now, however, a reduced version given by the barotropic vorticity equation (14.33) and the "baroclinic" vorticity balance (14.36) below: the model is further simplified by filtering waves from the equations by neglecting the time derivative of the baroclinic velocity variable. As a consequence, the equations then contain only long barotropic Rossby waves and the first baroclinic Rossby waves (long and short), while gravity waves are filtered. We also neglect for simplicity the viscous terms (but not the windstress) in (B.44) and write for the divergence of the baroclinic velocity moment

$$\nabla \cdot \boldsymbol{u}^* = \underline{\nabla}[h^2/(3f)] \cdot \nabla E - \underline{\nabla} \cdot [h^2 \boldsymbol{\tau}_0/(3f)]$$

Note that this approximation becomes invalid near the equator. Combining now $\nabla \cdot \boldsymbol{u}^*$ and the budget for $E$ and rewriting the barotropic vorticity budget, this reduced BARBI model (B.43) is governed by the potential energy balance in the form

$$\frac{\partial E}{\partial t} + \frac{N_0^2}{6}\left(\underline{\nabla} E \cdot \nabla \frac{h^2}{f} - \underline{\nabla}\psi \cdot \nabla h^2\right) = -\frac{N_0^2}{6}\underline{\nabla} \cdot \frac{h^2 \boldsymbol{\tau}_0}{f} + K_\ell \nabla^2 E \qquad (14.36)$$

and the vorticity balance (14.33). A wave analysis of this set of equations is presented in Section B.2.6, and in Section B.2.5 it is compared to a two-layer quasi-geostrophic model. The first two terms of (14.36) reveal that $R = N_0 h/(\sqrt{6}|f|)$ is the internal Rossby radius of the BARBI model. In this reduced version of BARBI, two variables remain: the barotropic stream function $\psi$ and the vertically integrated potential energy $E$. The aim of this section is to exploit the similarity of these variables in the full topographic and stratified simulation, shown Figure 14.11c,d.

It is convenient to rephrase the barotropic and baroclinic equations (14.33) and (14.36) in terms of the bottom pressure torque (instead of JEBAR), as done already the box on p. 465 for the barotropic equation, repeated here as

$$\beta \frac{\partial \psi}{\partial x} = \underline{\nabla} \cdot \boldsymbol{\tau}_0 - \nabla P \cdot \underline{\nabla} h + \text{fric} \qquad (14.37)$$

with the (symbolic) friction term, which can stand for bottom friction or lateral viscous friction. Similarly the baroclinic equation (14.36) may be converted to the form

$$\beta \frac{\partial \phi}{\partial x} = \underline{\nabla} \cdot \boldsymbol{\tau}_0 + \frac{\beta}{f}\tau_0^{(x)} + 2\nabla P \cdot \underline{\nabla} h - \frac{K_\ell}{R^2}\nabla^2 \phi \qquad (14.38)$$

with the lateral (isopycnal thickness) diffusivity $K_\ell$. The steady state is considered and $\phi = E/f$ is introduced which acts like a baroclinic stream function, i. e. (14.38) may be viewed as a baroclinic vorticity balance. This nomenclature is justified by realizing that $\phi$ is related to the geostrophic transport relative to the bottom: from the depth-integrated momentum balance (14.28) the total geostrophic transport $\boldsymbol{U}_g$ is given by $f\underline{\boldsymbol{U}}_g = h\nabla P + \nabla E$. Since the geostrophic bottom velocity $\boldsymbol{u}_g|_{-h}$ is evaluated as $f\underline{\boldsymbol{u}}_g|_{-h} = -\nabla p|_{-h} = -\nabla P + g\rho|_{-h}\nabla h$, the geostrophic transport relative to the bottom becomes

$$f(\underline{\boldsymbol{U}}_g - h\underline{\boldsymbol{u}}_g|_{-h}) = -\nabla E - g\rho|_{-h}\nabla h$$

It is often called the baroclinic transport. Hence, besides a small contribution by the perturbation density $\rho|_{-h}$ at the bottom and a small $\beta$ term, $\phi = E/f$ is found to be the stream function of the baroclinic transport.

**75. Cancelation of the Bottom Pressure Torque**

Subtracting the baroclinic vorticity budget (14.38) for the baroclinic stream function $\phi = E/f$ of the BARBI model from the barotropic vorticity budget (14.37) for $\psi$ in order to eliminate the windstress curl and setting the diffusive and frictional terms to zero, one obtains

$$\beta \frac{\partial}{\partial x}(\psi - \phi) = -\tau_0^{(x)}\beta/f - 3\nabla P \cdot \underline{\nabla} h$$

Using now the vertically integrated (steady) zonal momentum budget $f\partial\psi/\partial x = f\partial\phi/\partial x + h\partial P/\partial x - \tau_0^{(x)}$ to eliminate the stream functions, one obtains after some manipulations

$$\nabla P \cdot \underline{\nabla}(f/h^3) = 0$$

This shows that the bottom pressure $P$ becomes constant along contours of $f/h^3$. Together with a boundary condition $P = 0$ somewhere on each such contour, one obtains $P = 0$ everywhere and so the bottom pressure torque is zero everywhere (except for closed contours of $f/h^3$ which are not considered here). Note that by implication, we can also conclude that any deviation of wind-driven flow from the classical Sverdrup balance in steady state is introduced by diffusion and friction. The bottom pressure torque is thus a consequence of diffusion and friction.

It becomes obvious from (14.37) and (14.38) that the bottom pressure torque $\nabla P \cdot \underline{\nabla} h$ is the coupling between the barotropic and baroclinic vorticities. This coupling becomes only active for sloping topography. Note that the bottom pressure torque in the baroclinic vorticity equation can be traced back to the effect of the lifting of the background density by the barotropic and baroclinic flow in the budget of the perturbation density. In the barotropic vorticity budget, we have also related the bottom pressure term to the vortex stretching due to the geostrophically balanced up- or downhill flow at the bottom. The bottom pressure torque plays thus the role of a vorticity change due to stretching in both the barotropic and baroclinic vorticity budgets.

The vorticity budgets in the form (14.37) and (14.38) have far reaching consequences. First notice that within the limits of the approximations used in deriving these equations, the bottom pressure torque vanishes without friction and diffusion. This is proven in the box on p. 470. A vanishing bottom pressure torque, on the other hand, implies that the barotropic and baroclinic stream functions, $\psi$ and $\phi = E/f$, must become similar if diffusion and friction is small. This is proven as follows. With vanishing of the pressure torque and diffusion/friction we find $\beta\partial\psi/\partial x = \underline{\nabla}\cdot\boldsymbol{\tau}_0$ and $\beta\partial\phi/\partial x = \underline{\nabla}\cdot\boldsymbol{\tau}_0 + \beta\tau_0^{(x)}/f$ from the above vorticity budgets, i. e. $\partial\psi/\partial x$ equals the Sverdrup transport and $\partial\phi/\partial x$ equals the difference of the Sverdrup and the Ekman transports (which is the geostrophic transport relative to the bottom; see also (14.17) for the homogeneous case). The difference between $\partial\psi/\partial x$ and $\partial\phi/\partial x$ is thus of order $L/a \ll 1$, where $L$ denotes (the maximum of) the scale of the lateral variation of the windstress, stream functions, and topography, and $a$ is the Earth radius. This regime is appropriate for the formerly identified Sverdrup regime (away from the western boundary region), and we find

$$O(\psi - \phi) \sim \frac{L}{a}\psi \ll \psi$$

because both stream functions must vanish on the eastern coast. The barotropic transport has to return the Sverdrup transport in the western boundary layer and the baro-

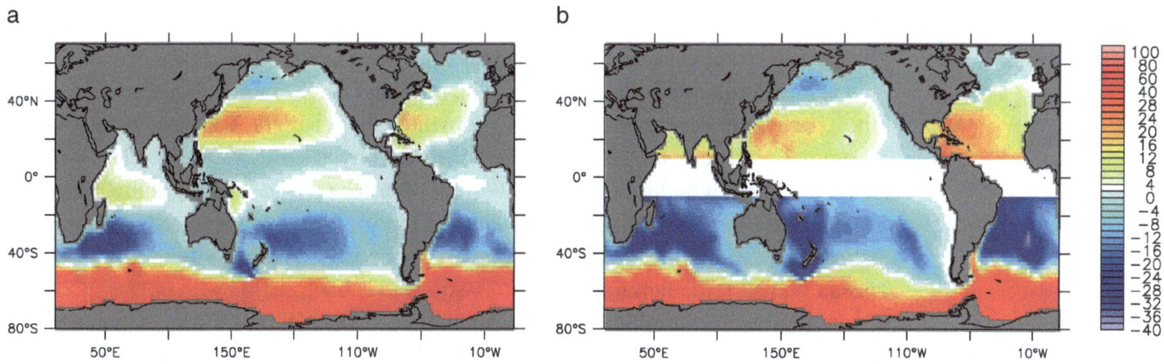


Fig. 14.12 **a** Barotropic stream function $\psi$ in Sv $= 10^6\,\mathrm{m^3\,s^{-1}}$ in the BARBI model with realistic topography and $N_0 = 2 \times 10^{-3}\,\mathrm{s^{-1}}$. **b** Baroclinic stream function $\phi$ also in Sv

clinic one has to return geostrophic transport. We conclude that the above similarity to order $L/a$ must hold in the boundary layer as well. Figure 14.12 shows $\psi$ and $\phi$ in the realistic global BARBI model, which are indeed very similar.

On the other hand, we have seen in the simulations that the bottom pressure torque is large in the western boundary currents, pointing towards the importance of diffusion and possibly friction in the western boundary layer. We assume that friction is small[3], implying that the bottom pressure torque balances the $\beta$-term in (14.37). The analysis of the box on p. 470 can be extended to this regime. We find from the difference equation and zonal momentum balance the relation

$$\frac{\beta h}{f}\frac{\partial P}{\partial x} - 3\nabla P \cdot \underline{\nabla} h + \frac{K_\ell}{R^2}\nabla^2\phi = 0 \tag{14.39}$$

The first term on the left-hand side is $O(L/a)$ so that to this order the bottom pressure torque is given by (one third of) the diffusive term. Inserting this result into the vorticity budgets,

$$\beta\frac{\partial\psi}{\partial x} = \underline{\nabla}\cdot\boldsymbol{\tau}_0 - \frac{1}{3}\frac{K_\ell}{R^2}\nabla^2\phi + O(L/a) = \underline{\nabla}\cdot\boldsymbol{\tau}_0 - \frac{1}{3}\frac{K_\ell}{R^2}\nabla^2\psi + O(L/a) \tag{14.40}$$

$$\begin{aligned}\beta\frac{\partial\phi}{\partial x} &= \underline{\nabla}\cdot\boldsymbol{\tau}_0 + \frac{\beta}{f}\tau_0^{(x)} - \frac{1}{3}\frac{K_\ell}{R^2}\nabla^2\phi + O(L/a)\\ &= \underline{\nabla}\cdot\boldsymbol{\tau}_0 - \frac{1}{3}\frac{K_\ell}{R^2}\nabla^2\phi + O(L/a)\end{aligned} \tag{14.41}$$

we notice that the equations become identical to order $O(L/a)$.

Consider the implication for the barotropic balance, written now as

$$\beta\frac{\partial\psi}{\partial x} = \underline{\nabla}\cdot\boldsymbol{\tau}_0 - \frac{1}{3}\frac{K_\ell}{R^2}\nabla^2\psi \tag{14.42}$$

The equation is formally similar to Stommel's model (14.23) for a flat-bottom ocean, where bottom friction was introduced to balance the return flow in the western

[3] This can be justified a posteriori: comparing the diffusive term with a viscous friction term we obtain $(K_\ell/3\mathcal{R}^2)/(A_h/L^2) = O(10)$, taking the parameters of the BARBI simulation.

boundary layer. Here the parameter combination $K_\ell/(3R^2)$ enters from density diffusion and stratification, representing the inverse of a time-scale on the order of days, but acting similar to bottom friction. The relation (14.42) might, therefore, be called the "baroclinic Stommel equation". Somehow surprisingly, we have rediscovered Stommel's vorticity budget with rather different physics. Equation (14.42) is valid in a stratified-topographic regime and correct to order $O(L/a)$. Topography, however, is shielded by the stratification, the latitude circles have been restored as characteristics, and the forcing restored to the curl of the windstress, all as in a flat-bottom ocean – in fact, the familiar flat-bottom solution is back.

In contrast to Stommel's original model, however, the return flow is balanced in the barotropic vorticity budget (14.37) by the effect of the bottom pressure torque. By referring to (14.39), the bottom pressure torque was replaced to first order approximation in $L/a$ by the effect of the bolus velocity appropriate to the Gent and McWilliams (1990) parameterization in the baroclinic vorticity budget (originally in the density balance, see Appendix B.2).

## 14.3 Main Thermocline Dynamics

While we have focussed on the depth-averaged flow in the preceding sections, we discuss in this section some of the analytical theories of the three-dimensional structure of the wind-driven circulation. A particular feature of the subtropical gyres is the main thermocline, which represents an intrinsic aspect of the large-scale ocean circulation. It is characterized by a strong vertical gradient in temperature (see e. g. Figure 15.2) and density – it is, therefore, also called the pycnocline – at a relatively shallow depth range of a couple of hundred meters below the seasonal mixed layer, which can be found in all ocean basins from equatorial to subtropical regions.

A main factor determining the thermocline is the surface forcing by the wind. As discussed specifically in Section 14.1.4, a main effect of this forcing lies in the vertical transport of mass (Ekman pumping) between a frictional near-surface layer and the ocean interior below, as consequence of convergent or divergent wind-induced mass transports in that layer. The pumping is quantified by the Ekman pumping velocity $w_e$ defined in (14.12). In the subtropical gyres, the downward Ekman pumping is responsible for the bowl-shaped structure of the main thermocline, with deepest isopycnal depth in the center of the gyres, as seen in Figure 14.13. A second crucial factor derives from the fluxes of heat and freshwater at the air-sea boundary. These fluxes are important in determining temperature and salinity in the well-mixed layer. At least in regions where the Ekman-pumping is directed downwards, such as e. g. in the subtropical gyres in ocean basins, it is obvious that the surface mixed-layer properties are transported into the ocean interior and hence determine the density field.

The strong vertical gradients in temperature and salinity associated with the permanent thermocline could not exist in a motionless equilibrium state, where they would be eroded by molecular diffusion on a time scale of $H^2/\kappa \approx 10^6$ years. Although this is a very long time, the ocean has already been existing in its present form for a much longer time, and the very existence of a thermocline constitutes a proof for the presence of a circulation maintaining this structure. We will discuss this circulation in the following.

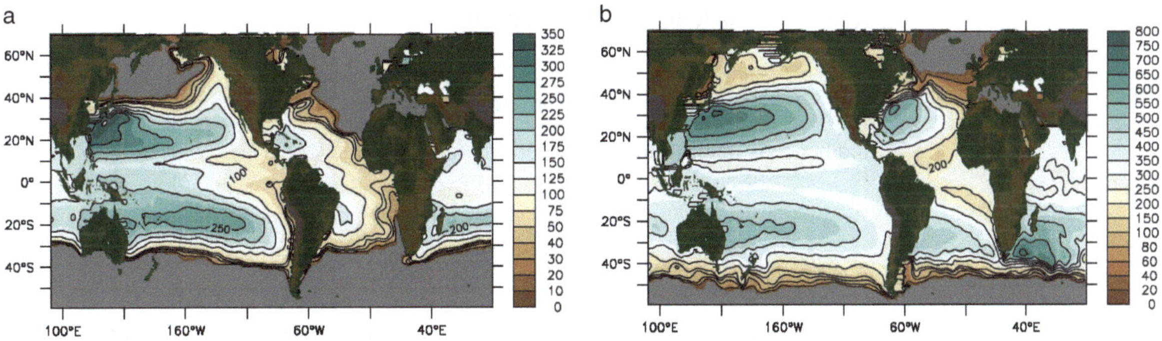


Fig. 14.13 Depth of the isopycnal $\sigma_\theta = 25.5\,\mathrm{kg\,m^{-3}}$ (**a**) and $\sigma_\theta = 26.7\,\mathrm{kg\,m^{-3}}$ (**b**) in m, representative for the main thermocline. Regions where the isopycnals outcrop are *shaded gray*. Data are taken from the World Ocean Atlas 2005 (Antonov et al., 2006; Locarnini et al., 2006)

### 14.3.1 Scaling Considerations

In Section 5.1, the scaling for large-scale motions of small Rossby-number was discussed. This scaling remains relevant for the circulation in the thermocline, the main difference resulting from the scale for the vertical velocity. The effect of friction causes an Ekman boundary layer of several 10 m thickness. At the bottom of the Ekman layer, there is a pumping velocity $w_e$ with magnitude $W_e \approx 10^{-6}\,\mathrm{m\,s^{-1}}$ (compare Figure 14.4a). In the interior ocean below the boundary layer – and outside of the western boundary layer – friction, however, can be neglected. It follows that the momentum balance is linear and in geostrophic balance, and (5.45) and (5.46) are the appropriate equations of motion. For convenience, we will use (5.50) instead of (4.54)–(4.56), so that temperature and salinity are combined into a single density variable, and mixing is represented as vertical diffusion.

With the scaling $u, v \sim U, w \sim W, p \sim P, \rho \sim R, x, y \sim a$ and $z \sim H$, one finds from (5.45), (5.46) and (5.50)

$$P = 2\Omega a\, U\,, \quad U = Wa/H \quad \text{and} \quad R = P/gH$$

or equivalently $gHR = 2\Omega a^2 W/H$. The only way to allow for $W = W_e$ and $R = R_s$, where $W_e$ denotes the magnitude of the Ekman pumping and $R_s$ the (given) density variation in the surface mixed layer, is to adjust the depth scale $H$ such that

$$H = H_a = \left(\frac{2\Omega a^2 W_e}{g R_s}\right)^{\frac{1}{2}} \tag{14.43}$$

With $2\Omega = 1.4 \times 10^{-4}\,\mathrm{s^{-1}}$, $g = 10\,\mathrm{m\,s^{-2}}$, $R_s = 2 \times 10^{-3}$, and $W_e = 10^{-6}\,\mathrm{m\,s^{-1}}$ one obtains $H_a \approx 500\,\mathrm{m}$, and hence a correct order of magnitude for the main thermocline which is deeper than the surface layer, but still considerably smaller than the water depth.

The depth scale $H_a$ reflects an advective balance in the density equation. Diapycnal diffusion of salt and temperature is small below the mixed layer and unlikely to play a major role for the main thermocline. Observations suggest a value of typically $K_v \approx 10^{-5}\,\mathrm{m^2\,s^{-1}}$ in the thermocline. In the density balance (5.50), the magnitude of the diffusion term relative to vertical advection thus is

$$\frac{K_v \partial^2 \rho / \partial z^2}{w \partial \rho / \partial z} = \frac{K_v}{W_e H_a} \approx 0.02$$

and hence is negligible in the interior.

### 14.3.2 Similarity Solutions

The equations governing the main thermocline circulation are nonlinear because of the advection term in the density equation (5.50). In general, therefore, only numerical solutions are possible. However, some particular stationary analytical solutions have been found.

As shown in Section 5.3.1, the system consisting of (5.45), (5.46) and (5.50) can be condensed in one equation for a single variable $M$. This $M$-equation is thought to describe the evolution of the oceanic thermocline in response to pumping of water, provided that appropriate boundary conditions are chosen. In steady state and with $\mathcal{G}_\rho = 0$, it takes the form

$$-\frac{\partial^2 M}{\partial \varphi \partial z}\frac{\partial^3 M}{\partial \lambda \partial z^2} + \frac{\partial^2 M}{\partial \lambda \partial z}\frac{\partial^3 M}{\partial \varphi \partial z^2} + \cot\varphi \frac{\partial M}{\partial \lambda}\frac{\partial^3 M}{\partial z^3} = 0 \qquad (14.44)$$

All other fields ($u$, $v$, $p$, $\rho$ and $w$) can be determined from $M(\lambda, \varphi, z)$ as outlined in Section 5.3.1. Equation (14.44) is valid below the turbulent layer which is immediately influenced by wind and surface fluxes (roughly the upper 50–100 m). The bottom of that layer is assumed to be at a constant depth $z = -d_s$. Here, the conditions $\rho(\lambda, \varphi, -d_s) = \rho_s(\lambda, \varphi)$ and $w(\lambda, \varphi, -d_s) = w_e(\lambda, \varphi)$ have to be imposed. In terms of $M$, the boundary conditions at $z = -d_s$ take the form

$$w = w_e \Rightarrow \frac{\partial M}{\partial \lambda} = 2\Omega a^2 \sin^2 \varphi w_e(\lambda, \varphi) \qquad (14.45)$$

$$\rho = \rho_s \Rightarrow \frac{\partial^2 M}{\partial z^2} = -g\rho_s(\lambda, \varphi) \qquad (14.46)$$

If the thermocline depth is much smaller than the ocean depth, the boundary conditions for $M$ at the bottom can be approximated by requiring that the solution tends to zero or constant values at great depth.

A particular solution of (14.44) is given by

$$M(\lambda, \varphi, z) = A(\lambda, \varphi)\mathrm{e}^{c(z+d_s)/\sin\varphi} \qquad (14.47)$$

with constant $c$ and arbitrary $A(\lambda, \varphi)$. The solution (14.47) was given by Needler (1967) with the boundary condition (14.46) which requires

$$A(\lambda, \varphi) = -\frac{\sin^2 \varphi}{c^2}\rho_s(\lambda, \varphi)$$

With the alternative choice

$$A(\lambda, \varphi) = 2\Omega a^2 \sin^2 \varphi \int_0^\lambda w_e(\lambda', \varphi)\mathrm{d}\lambda'$$

condition (14.45) is satisfied. However, it is not possible to satisfy both conditions simultaneously. As both conditions are independent of each other and physically of equal importance, a solution of (14.47) satisfying only one of them is rather unsatisfactory even though it still may have some realistic aspects. A further shortcoming of (14.47) is that it permits only a latitudinal variation of the vertical scale, i. e. $z_* \sim \sin\varphi$. Therefore, the westward deepening of the thermocline, which is an important aspect of the observed zonal structure, is not reproduced at all.

Note that from (14.47) it follows that

$$Q = (2\Omega \sin\varphi)\frac{\partial\rho}{\partial z} \equiv 2\Omega c \cdot \rho = \text{const} \cdot \rho$$

so that the large-scale potential vorticity is a function of density only and consequently constant on isopycnals.

### 14.3.3 Ideal Fluid Solutions

A situation where all diabatic and frictional terms can be ignored has been referred to as *ideal fluid* (cf. Section 4.1.6). For motions of planetary scale, the ideal fluid equations are given by (5.45), (5.46) and the adiabatic form of (5.50). In steady state, it then follows from (5.50)–(5.52) that the density $\rho$, the potential vorticity $Q = f\partial\rho/\partial z$, and Bernoulli-function $B = p + gz\rho$ all satisfy a conservation equation of the form

$$\boldsymbol{u} \cdot \nabla\chi + w\frac{\partial\chi}{\partial z} = 0$$

for $\chi = \rho, Q$ and $B$. This remarkable property of three conserved quantities has the consequence that the three-dimensional velocity vector $(\boldsymbol{u}, w)$ must lie in all three surfaces $\rho = $ const, $Q = $ const, and $B = $ const. In fact, these iso-surfaces of density, potential vorticity, and Bernoulli function must intersect in lines which are streamlines. Hence a functional relation $\Phi(\rho, Q, B) = 0$ must exist for an ideal fluid system. The function $\Phi$ is in principle determined by conditions on boundaries. In practice, it is often assumed that this function is well behaved and that solutions for one of the variables, e. g. $Q = F(\rho, B)$, are single-valued and continuous, although this is by no means guaranteed. With a given function $F(\rho, B)$, the system is governed by a single equation which follows from vertical differentiation of the hydrostatic equation $\partial p/\partial z = -g\rho$ as

$$f\frac{\partial^2 p}{\partial z^2} + gF\left(\rho, p - z\frac{\partial p}{\partial z}\right) = 0 \tag{14.48}$$

Specifically, Welander (1971) has assumed a linear relation $Q = F(\rho, B) = a\rho + bB$ with $a, b = $ const. Further differentiation of (14.48) with respect to $z$ then yields a second-order differential equation for density

$$f\frac{\partial^2\rho}{\partial z^2} = (a + bgz)\frac{\partial\rho}{\partial z}$$

For $b = 0$, the similarity solution (14.47) is recovered. For $b \neq 0$, the solution is given as

$$\rho = \rho_s(x, y) + c(x, y) \int_{-d_s}^{z} \exp\left[-\frac{(z' + z_\star)^2}{d^2(y)}\right] dz' \tag{14.49}$$

Here, $\rho_s(x, y)$ and $c(x, y)$ are integration constants, $d^2 = -2f/(bg)$ and $z_\star = a/(bg)$. Obviously, $\rho_s(x, y)$ can be chosen to reflect the density below the mixed layer at $z = -d_s$, and $c(x, y)$ such that $\rho(x, y, z) \to 0$ as $z \to -\infty$ so that the solution resembles a pycnocline. However, again there is no freedom to satisfy the additional condition $w = w_e$.

The limitations of the so-called ideal fluid solutions[4] (14.49) are quite analogous to those of similarity solutions, although the derivation is rather different. The reason is here that an appropriate choice for both $\rho_s$ and $w_e$ would result in a particular functional relationship $Q = F(\rho, B)$ which cannot be determined a priori.

### 14.3.4 Thermocline Ventilation in an Isopycnal Layer Model

The main problem with the particular solutions considered in the previous sections is that it was impossible to simultaneously satisfy both forcing conditions $w = w_e$ and $\rho = \rho_s$ at $z = 0$. An alternative which is still approachable with analytical solution techniques has been given by Luyten et al. (1983) and is based on discrete isopycnal layers rather than a continuous density variable (see Appendix B.1.4 for a detailed description of the model).

Each layer (index $n$) is bounded by the interfaces $z = -d_n$ below and $z = -d_{n-1}$ above, and is characterized by a constant density $\rho_n$ (scaled by $\rho_0$), thickness $h_n = d_n - d_{n-1}$, pressure $p_n = M_n - g\rho_n z$, and velocity $\boldsymbol{u}_n, w_n$. The equations of motion are still based on the planetary-geostrophic approximation discussed in Section 5.3. Specifically, the relevant meridional thermal wind, hydrostatic and Sverdrup transport relations, and the conservation of potential vorticity $Q_n = f/h_n$ and the 'Bernoulli function' $M_n$ (which in this approximation coincides with the Montgomery potential, see Appendix B.1.1) take the form

$$f\,(\boldsymbol{u}_n - \boldsymbol{u}_{n+1}) = g' \underset{\neg}{\nabla} d_n \tag{14.50}$$

$$M_n - M_{n+1} = g' d_n \tag{14.51}$$

$$\beta \sum_n v_n h_n = f w_e \tag{14.52}$$

$$\boldsymbol{u}_n \cdot \nabla Q_n = 0 \tag{14.53}$$

$$\boldsymbol{u}_n \cdot \nabla M_n = 0 \tag{14.54}$$

The notation is as in the box on p. 444, with the coordinates $x = a\lambda \cos\varphi$ and $y = a\varphi$, which are referred to as "longitude" and "latitude". To simplify matters, it is assumed here that the density differences between neighboring layers are all identical, i. e. $\rho_{n+1} - \rho_n = \Delta\rho$ independent of $n$. The reduced gravity is denoted

[4] Note that the similarity solutions discussed in Section 14.3.2 are also based on ideal fluid equations.

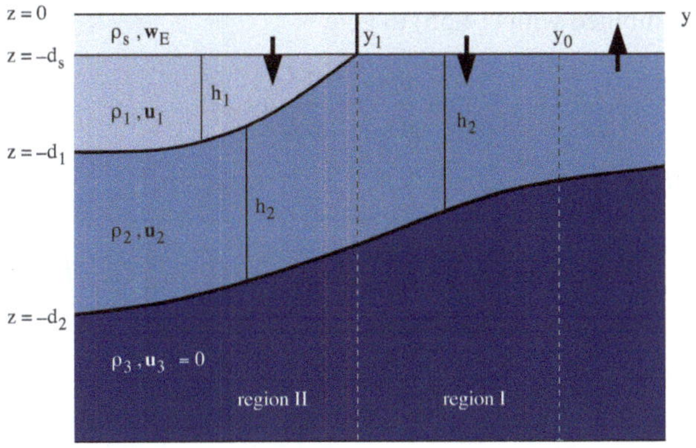


Fig. 14.14 Sketch of the ventilated thermocline model with two moving layers

as $g' = g\Delta\rho$. Note also that (14.53) is valid only in regions where the layer $n$ is unventilated, i. e. not in direct contact with the surface layer.

In the general ideal fluid case considered in the previous section, the existence of three conserved quantities ($\rho$, $B = p + g\rho z$ and $Q$) resulted in a relation $Q = F(\rho, B)$. In the layer situation the corresponding variables are $\rho_n$, $M_n$ and $Q_n$, and therefore

$$Q_n = F(\rho_n, M_n) \equiv F_n(M_n)$$

In each unventilated layer, $Q_n$ is, therefore, determined by $M_n$, and as shown below, it is possible to explicitly construct the functional dependence $F_n(M_n)$.

Consider specifically a situation with at most three layers below a surface layer of constant depth $d_s$, as sketched in Figure 14.14. We assume that the lowest layer (index 3) is at rest, $u_3 = v_3 = 0$. It follows that $\nabla M_3 = 0$, so that $M_3 = \text{const} = 0$.

The boundary conditions are formulated as follows: It is assumed that the surface layer density depends only on latitude, i. e. $\rho_s = \rho_s(y)$, approximately representing zonal mean conditions. Accordingly, the latitude $y_n$ where the interface $d_n$ reaches the surface layer is specified. Forcing by the wind via the Ekman pumping velocity $w_e(x, y)$ is already contained in the Sverdrup transport relation (14.52). The analysis is restricted to regions where $w_e(x, y) < 0$ which is assumed to hold south of a latitude $y_0$. As the number of layers in motion differs in the different regions displayed in Figure 14.14, these regions need to be considered separately.

## Region I

In the region between the latitudes $y_1$ and $y_0$, only layer 2 is in motion. It has the thickness $h_2 \equiv d_2 - d_s$ and is in contact with the surface layer. The Sverdrup transport is

$$\beta v_2 h_2 = f w_e(x, y) \tag{14.55}$$

As $w_e < 0$, the transport in the layer is directed southward. From (14.50) and $\boldsymbol{u}_3 = 0$, one further has

$$\boldsymbol{u}_2 = (g'/f)\underline{\nabla} h_2 \tag{14.56}$$

which can be combined with (14.55) to give

$$h_2 \frac{\partial h_2}{\partial x} = \frac{f^2}{g'\beta} w_e \tag{14.57}$$

Integration from the eastern boundary at $x = x_E$ results in

$$h_2(x, y) = \left[D^2(x, y) + H_I^2(y)\right]^{\frac{1}{2}} \quad \text{with} \quad D^2 = -2\frac{f^2}{g'\beta} \int_x^{x_E} w_e(x', y)\mathrm{d}x' \tag{14.58}$$

Here, $H_I(y) = h_2(x_E, y)$ is an integration constant. The condition $u_2 = 0$ at $x = x_E$ requires that $H_I(y) = H_0 = \text{const.}$

Equation (14.57) respectively (14.58) constitutes a central result of the ventilation theory. The zonal change of thermocline depth is directly related to the wind forcing and corresponds to westward deepening in case of pumping (note that $D^2 > 0$ if $w_e < 0$). If the Ekman pumping velocity depends only on latitude, then $D^2(x, y) \sim (x_E - x)$, so that the increase is linear with westerly longitude and reaches a maximum of $D_{\max}^2 = 2(f^2/g'\beta)|w_e|(x_E - x)$. With $H_0 = 100\,\mathrm{m}$, $g' = 2 \times 10^{-2}\,\mathrm{m\,s^{-2}}$, $w_e = 10^{-6}\,\mathrm{m\,s^{-1}}$, and $x_E - x = 5{,}000\,\mathrm{km}$, one obtains $D_{\max} \approx 250\,\mathrm{m}$ which has the right order of magnitude. Note that the east-west change of thermocline depth according to (14.58) is somewhat smaller than $D_{\max}$, depending on the initial depth $H_0$.

### Region II

In the region south of $y_1$, layers 1 and 2 are in motion. The fluid in layer 2 is subducted at the latitude $y_1$. From (14.51), with $M_3 = 0$ and $d_2 = h_1 + h_2 + d_s$, the pressure in layer 2 is given as $M_2 = g'(h_1 + h_2 + d_s)$. As both $Q_2 = f/h_2$ and $M_2$ are conserved south of $y_1$, there must be a relation

$$f/h_2 = F(h_1 + h_2) \tag{14.59}$$

with an arbitrary $F(\xi)$. Now at the outcrop latitude $y = y_1$, one has $f(y_1) = f_1$ and $h_1 = 0$, and, therefore, $f_1/h_2 = F(h_2)$. As this holds for *all* $x$ along $y = y_1$, the unknown functional form must be $F(\xi) = f_1/\xi$, and it follows from (14.59) that

$$\frac{f}{h_2} = \frac{f_1}{h_1 + h_2} \quad \text{or} \quad \frac{h_1}{h_2} = \frac{f_1}{f(y)} - 1 \tag{14.60}$$

The ratio of both layer thicknesses hence depends only on latitude. From (14.50), and with $d_1 = h_1 + d_s$, one further obtains

$$\boldsymbol{u}_2 = (g'/f)\underline{\nabla}(h_1 + h_2) \quad \text{and} \quad \boldsymbol{u}_1 - \boldsymbol{u}_2 = (g'/f)\underline{\nabla} h_1 \tag{14.61}$$

which can be used to express the layer velocities $v_1, v_2$ in the Sverdrup transport relation $\beta(v_1 h_1 + v_2 h_2) = f w_e$, resulting in

$$h_1 \frac{\partial h_1}{\partial x} + (h_1 + h_2)\frac{\partial}{\partial x}(h_1 + h_2) = \frac{f^2}{\beta g'} w_e$$

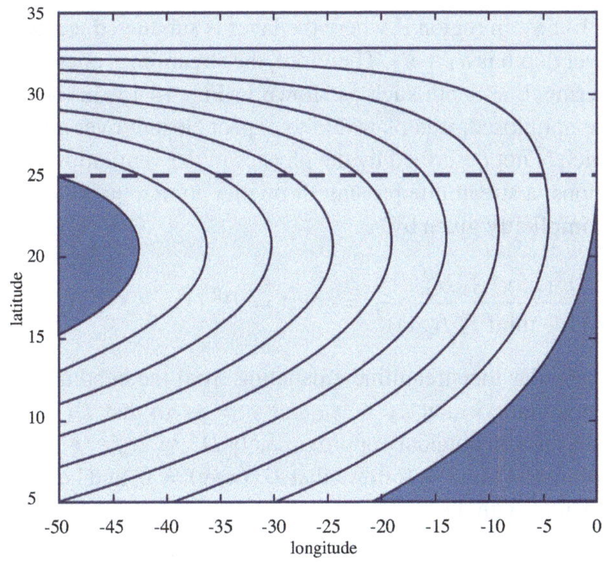


**Fig. 14.15** Streamlines in layer 2 for a basin of $50°$ longitudinal width, ranging from $5–35°$ N. The *dashed line* indicates the latitude where layer 2 is subducted. The Ekman velocity is derived from the windstress as defined in the box on p. 454 and is directed downward south of $32.8°$ N. The thermocline depth at the eastern boundary is $H_0 = 200$ m

which can be integrated to yield $h_1^2 + (h_1 + h_2)^2 = D^2(x, y) + H_{\mathrm{II}}^2(y)$. Here, $D(x, y)$ is defined as in (14.58), and $H_{\mathrm{II}}^2(y) = h_1^2(x_{\mathrm{E}}, y) + [h_1(x_{\mathrm{E}}, y) + h_2(x_{\mathrm{E}}, y)]^2$ is an integration constant. To satisfy the no-flow boundary condition at the eastern boundary, both $h_1$ and $h_1 + h_2$ must be constant at $x = x_{\mathrm{E}}$ so that $H_{\mathrm{II}} = \text{const}$. As $h_1(x_{\mathrm{E}}, y_1) = 0$, it follows that $h_1(x_{\mathrm{E}}, y) \equiv 0$ for all latitudes. From continuity at $y = y_1$, it then follows that $H_{\mathrm{II}} = H_0$. Finally, with (14.60), one obtains

$$h_1(x, y) + h_2(x, y) = \left[\frac{D^2(x, y) + H_0^2}{1 + (1 - f/f_1)^2}\right]^{\frac{1}{2}} \tag{14.62}$$

With (14.62), (14.60), and (14.61), the variables $\boldsymbol{u}_1, \boldsymbol{u}_1, h_1$ and $h_2$ can be determined to complete the solution. Since according to (14.60) the ratio $h_1/h_2$ depends on latitude, it even appears that the only possibility to avoid a conflict would be to require $H_0 = 0$ so that both layers have zero thickness at the eastern boundary. However, as we shall see below, south of subduction latitude the solution (14.62) cannot be valid at the eastern boundary.

The thermocline depth can be defined as $h_2$ in region I, and $h_1 + h_2$ in region II. An exemplary depth distribution is shown in Fig. 14.15. Overall, the thermocline depth increases westward. The latitudinal dependence is mainly governed by the pumping velocity $w_{\mathrm{e}}$, modified by the denominator in (14.62) in region II.

## Streamlines and Unventilated Zones

It is illuminating to consider the fluid trajectories in layer 2 which can readily be determined. In region I, according to (14.56) the streamlines are given by $h_2 = \text{const}$, with $h_2$ from (14.58). In region II where the layer is subducted, according to (14.61) the relevant layer depth is $h_1 + h_2$. Therefore, the streamlines coincide with the lines of constant thermocline depth such as shown in Fig. 14.15. In the west where the streamlines are not closed, it is plausible to expect closure over a boundary current in the west which is not described by the physics in the ventilation model.

In both regions, a streamline passing through a point $x = x_*$ on the subduction line $y = y_1$ is implicitly given by

$$\frac{D^2(x,y) + H_0^2}{1 + (1 - \min(f/f_1, 1))^2} = \text{const} = D^2(x_*, y_1) + H_0^2 \tag{14.63}$$

Consider in particular the streamline emanating from the subduction latitude right at the eastern boundary, i. e. at $x_* = x_E$ and $y = y_1$ so that $D(x_E, y_1) \equiv 0$. Note also that near the eastern boundary approximately $D^2 \approx 2(f^2/g'\beta)|w_e|(x_E - x)$. In region I, according to (14.63) it follows that $D^2(x, y) = 0$, and hence the streamline in region I coincides with the eastern boundary. In region II one has

$$D^2(x,y) = H_0^2 (1 - f/f_1)^2$$

and the streamline $x(y)$ is hence given as

$$x(y) = x_E - \frac{g'\beta H_0^2}{2|w_e|}\left(\frac{1}{f} - \frac{1}{f_1}\right)^2 \approx x_E - \frac{g'\beta^3 H_0^2}{2f_1^2|w_e|}(y - y_1)^2$$

The last expression holds since near the subduction latitude we have $f \approx f_1 + \beta(y - y_1)$. Hence this streamline has a parabolic shape near the eastern boundary. The region east of this streamline has been termed *shadow zone* because it cannot be reached by any streamline from the ventilated region north of $y_1$. Therefore, the relation (14.60) and the solution based on it cannot be valid here. To achieve a more detailed view of the shadow zone, one has to include processes such as diffusion which are absent from the ventilation model. In the context of the ventilation model it is, however, plausible to assume that the shadow zone is at rest, since there is no flow across the boundary and across the streamline.

Another streamline of particular interest passes through the westernmost point on the subduction line. From (14.63), we obtain with $x_* = x_W$

$$D^2(x,y) = H_0^2 (1 - f/f_1)^2 + D^2(x_W, y_1)\left[1 + (1 - f/f_1)^2\right]$$

Near $y = y_1$ where $f \approx f_1$ it follows that $D^2(x(y), y) \approx D^2(x_W, y_1) \approx$ const. Now $D(x, y)$ always increases westward. If it also increases southward (north of the maximum convergence zone), the trajectory $D =$ const must initially go toward the south-east. The origin of this streamline then is not in the ventilated region but rather in the western boundary current, a region where the ventilation model does not hold. Hence a solution for the region to the west of this streamline cannot be obtained with the ventilation model. See Section 14.3.5 for further discussion.

The validity of the ventilation model – in the simple form discussed here – is restricted to regions below the surface layer where $w_e < 0$, outside of western boundary currents, and the density structure is simplified to a small number of layers which have to be connected to ventilated regions. In spite of these limitations, the model succeeds in describing some principal elements of the main thermocline circulation in regions where the surface layer circulation is convergent. It allows to relate the zonal and meridional distribution of thermocline depth to the forcing in a physically insightful way. In particular, the existence of nonventilated zones clearly demonstrates the failure of similarity solutions.

It is in principle possible to apply the ventilation model – with appropriate modifications – to regions where $w_e > 0$; however, less insight is gained because here the role of lateral diffusion due to ocean eddies is more central (see Section 16.3 for the overturning in the Southern Ocean).

### 14.3.5 Circulation in Unventilated Regions

In this section, the circulation in regions that are not connected to ventilated regions is briefly discussed. Specifically, we will consider the situation where streamlines are closed, so that the corresponding trajectory is nowhere subject to surface forcing and remains subducted. The discussion is conveniently based on the quasi-geostrophic approximation discussed in Section 5.2 and Appendix B.1.3. In the absence of forcing, potential vorticity is conserved, and the lines of constant potential vorticity (*geostrophic contours*) coincide with streamlines described by a stream function $\Psi$. In steady state, the conservation is as before

$$\boldsymbol{u} \cdot \nabla Q = J(\Psi, Q) = 0 \tag{14.64}$$

Note that (14.64) implies that $Q = F(\Psi)$. Now it is plausible that even weak forcing can influence the motion since fluid elements are staying along closed streamlines for a long time. Let us, therefore, add weak forcing to (14.64) due to eddy-mixing in the form of

$$J(\Psi, Q) = -\nabla \cdot \overline{\boldsymbol{u}'Q'} = \nabla \cdot K \nabla Q \tag{14.65}$$

with $K > 0$. The dimensionless measure for the relative size of the diffusion term is the Peclet number $\text{Pe} = UL/K$. With $U = 10\,\text{cm}\,\text{s}^{-1}$, $L = 1{,}000\,\text{km}$, and $K = 10^3\,\text{m}^2\,\text{s}^{-1}$, one finds $\text{Pe} = 10^2 \gg 1$. The dominant balance in (14.65) is, therefore, still advective, $J(\Psi, Q) = O(\epsilon)$ with $\epsilon = \text{Pe}^{-1}$, and it follows that

$$Q = F(\Psi) + O(\epsilon) \quad \text{and hence also} \quad \nabla Q = \frac{\partial F}{\partial \Psi} \nabla \Psi + O(\epsilon) \tag{14.66}$$

The functional form $F(\Psi)$ can now be determined by integrating (14.65) over the area $A(\Psi)$ which is limited by a streamline $\Psi = \text{const}$, i. e.

$$\int_A J(\Psi, Q)\mathrm{d}A = \int_A \nabla \cdot K \nabla Q \mathrm{d}A$$

The left-hand side vanishes because

$$\int_A J(\Psi, Q)\mathrm{d}A = \int \mathrm{d}n \oint \mathrm{d}s \left( \frac{\partial \Psi}{\partial n}\frac{\partial Q}{\partial s} - \frac{\partial \Psi}{\partial s}\frac{\partial Q}{\partial n} \right) = \int \mathrm{d}n \frac{\partial \Psi}{\partial n} \oint \mathrm{d}s \frac{\partial Q}{\partial s} = 0$$

(note that $\partial\Psi/\partial s = 0$ along a streamline). Applying the Gaussian theorem (in two dimensions) to the right-hand side yields

$$\int_A \nabla \cdot K\nabla Q \mathrm{d}A = \oint \mathrm{d}s \boldsymbol{e}_\Psi \cdot (K\nabla Q) = 0 \tag{14.67}$$

where $\boldsymbol{e}_\Psi = \nabla\Psi/|\nabla\Psi|$ is the unit vector normal to the streamline. With (14.66) one further has

$$0 = \oint \mathrm{d}s K \frac{\partial F}{\partial \Psi} \nabla\Psi \cdot \frac{\nabla\Psi}{|\nabla\Psi|} + O(\epsilon)$$

Using again the constancy of $\Psi$ along the streamline, this is equivalent to

$$0 = \frac{\partial F}{\partial \Psi} \oint \mathrm{d}s K |\nabla\Psi| + O(\epsilon) \tag{14.68}$$

As the integrand is always positive, (14.68) can only be satisfied if $\partial F/\partial\Psi \equiv 0 + O(\epsilon)$, i. e.

$$F(\Psi) = \text{const} + O(\epsilon) \tag{14.69}$$

It follows that the potential vorticity is constant (*homogenized*) everywhere inside the closed streamline (Rhines and Young, 1982).

Note that (14.69) holds independently of the specific form of $Q$; the only condition is that (14.65) holds everywhere, including the western boundary region when the streamline passes through. The result is even valid for an arbitrary tracer that satisfies (14.65) as long as lateral diffusion is small. It is an interesting paradox that weak diffusion over long time can lead to complete mixing whereas strong diffusion does not necessarily have the same effect.

The result (14.69) has been applied to the ventilation model discussed in the previous section, to close the circulation in the western unventilated zone which is reached only by streamlines from the boundary current region.

# The Meridional Overturning of the Oceans

# 15

This chapter presents a brief overview of available observations and a more detailed discussion of the theory of the meridional overturning circulation (MOC). It should be pointed out that the theory of the MOC is not yet as solid as the theory of the wind-driven circulation. There is a variety of model types – box models of various configurations and zonally averaged models – many have imbedded severe approximations and unjustifiable assumptions which should be taken with care. We made an attempt to discuss these shortcomings and to present better alternatives if possible.

Let us imagine that in 985 when ERIK THE RED[1] was on his way to Greenland he whirled with his boat a small water package at the surface of the North Polar Sea in such a way that this water was chilled by the cold winds and thus sank with its slightly increased density into large depths. Our water package is now a part of the oceans' global meridional overturning circulation, an overturning motion spanning all oceans. The package is driven deep into the Atlantic, travels along the American continent at depths of 2,000–4,000 m, crosses the equator and then flows into the south polar area. There, it forms a part of the Antarctic Circumpolar Current, and then, after leaving the Southern Ocean and after upwelling in the North Pacific, it returns as part of the surface currents through the Indonesian Archipelago into the Indian Ocean and round the Cape of Good Hope again into the Atlantic Ocean and arrives at its origin via the Gulf Stream and the North Atlantic Current. The hypothetical path of the water package is sketched in the simplified schematics of the overturning circulation in Figure 15.1.

By the formation of deep, cold water masses at high latitudes and their subsequent upwelling and warming elsewhere, the implied circulation leads to a continuous ventilation of the deep ocean. It is often called the *thermohaline circulation* (THC), although the term *meridional overturning circulation* (MOC) is also often used (compare the box on p. 484). Another name of the same feature is *global conveyor belt*. We prefer MOC, since the name THC implies that it is the increase of density connected with fluxes of heat and salt that actually *drives* the overturning,

[1] ERIK "THE RED" THORVALDSSON, * about 950 in Jæren, Norway, †about 1003 in Brattahlíd, Greenland, sailor and discoverer.

D. Olbers, J. Willebrand, C. Eden, *Ocean Dynamics*,
DOI 10.1007/978-3-642-23450-7_15, © Springer-Verlag Berlin Heidelberg 2012

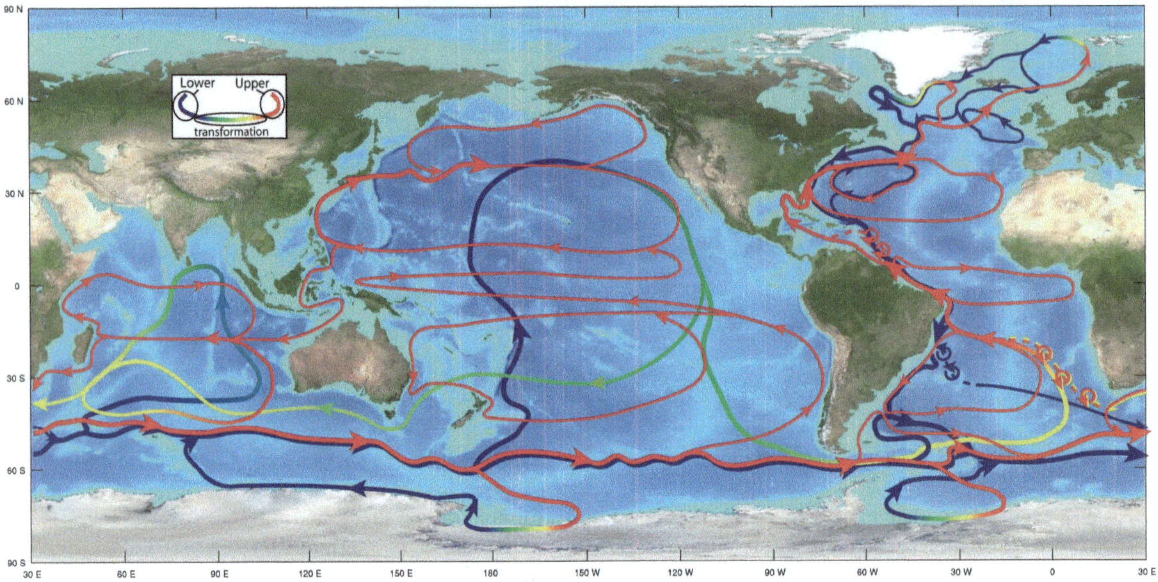


Fig. 15.1 A schematic of the global meridional overturning circulation and some of its recirculation loops according to Rick Lumpkin (pers. communication; see also Richardson (2008)). The surface currents are colored in red, while yellow, green and blue colors depict the deeper circulation (see the inlet positioned on Asia). Directions of the currents are given by arrows. The surface currents are basically those of Figure 14.1. Note the large eddies in the South Atlantic, indicating that the associated transports are not achieved by continuous large-scale currents

which is not entirely true (the "driving" is the process which warms and thus heaves the cold water to the surface again either by turbulent mixing or mechanical energy input, as discussed in Section 15.3.3). The deep, heavy water masses are formed in small areas in the North Atlantic, in the Greenland Sea but also in the minor seas around the Antarctic continent, mainly in the Weddell Sea, and there they sink into deeper layers, partly to the bottom (see also Section 15.1.1). The deep circulation

**76. MOC versus THC**

The meridional overturning circulation (MOC) generally moves surface water poleward, where it sinks and continues in a deep reverse flow with an equatorward direction. It leads to a renewal of deep waters and is controlled by the density differences due to temperature (thermal) and salinity (haline) differences, imprinted by the surface fluxes of heat and freshwater. It is thus often also called the *thermohaline* circulation (THC).
Note, however, that the name THC is often incorrectly used for the entire MOC. In fact, we will highlight in this chapter the thermohaline component of the MOC, but MOC may also include a wind-driven component. Major current systems, which have a strong wind-driven component, are embedded in the overturning circulation, such as the Gulf Stream and the Antarctic Circumpolar Current. For a clearer distinction, we should address the total meridional overturning as the MOC, i. e. the total northward or southward flow in a basin, integrated over longitude. The THC is then only the part of the MOC that would be driven by the heat and freshwater exchange with the atmosphere (and the water-mass transformation in the interior) only. Defined as such, the THC is not directly observable while the MOC as a true ocean current system can be measured (at least in principle). On the other hand, the THC can be modeled by excluding the wind forcing. Alternatively, excluding the thermohaline surface fluxes the wind-driven MOC can be investigated. It is the aim of this chapter, to analyze several of such models for the thermohaline part of the MOC.

consists of intensive boundary currents at the western continental slopes and a shallow poleward recirculation in the single ocean basins, a structure which is caused by the Earth's rotation as in the wind-driven basin gyres (see also Section 15.2).

The MOC is closed by slow upwelling in all ocean basins and returns to the Atlantic in the surface currents in narrow straits through the Indonesian Archipelago and the Indian Ocean round the Cape of Good Hope, as indicated by Figure 15.1. The Antarctic Circumpolar Current connects the flow of the water masses into the other oceans. Note that a similar, lower part of the MOC is started by the outflow of bottom water around Antarctica into the South Polar Seas. It is a complicated system of loops, with downward branches round the Antarctic continent and deep transport ways into all three oceans, where the water packages lose their identity by mixing and upwelling. Hence there is actually a number of single MOCs in each ocean basin, which is highly simplified in Figure 15.1. The picture of the global MOC which we draw above for Erik's water package is, therefore, handy, but extremely simplified. A water package could follow many other less prominent paths, e. g. return from the Pacific to the South Atlantic around Cape Horn. This path is called the *cold-water route*, in contrast to the *warm-water route* round the Cape of Good Hope.

It is important to note that the description of the MOC in terms of a global conveyor is extremely oversimplified, and contradicts many known characteristics of the oceanic circulation (see the discussion by Wunsch, 2002). The history of diagrams of the overturning circulation is consolidated by Richardson (2008). Many of the diagrams like those in Broecker (1991); Gordon (1986); Schmitz Jr. (1995) should be regarded as oversimplification and sometime have only been meant as logos. The more recent Figure 15.1 has been created by Rick Lumpkin on the basis of a global inverse model. The circulation includes the Tasmanian leakage (flow around Tasmania into the Indian Ocean) in addition to the flow through the Indonesian Seas. It also shows that the transports in the South Atlantic are mainly carried by large-scale eddies, shed off the Agulhas retroflection and from deeper western boundary currents north and south of the equator off Brazil. Nevertheless, that concept can be useful to describe some aspects of the large-scale meridional transports of heat and freshwater, and will be used in this sense in the following.

An important aspect of the MOC is that it transports the substances contained in the sea water. On the downward branch in the North Polar Sea, salt, oxygen, carbon dioxide, and other trace substances are transported into deeper layers, thus ventilating the deep ocean. Other examples for transported properties in the ocean are the man-made chloro-fluoro-carbons (CFC) or nuclides, produced by nuclear bomb tests, that have been brought into the atmosphere since the fifties of the last century, or radiocarbon that is produced within the atmosphere by cosmic rays, enters the ocean and decays while being advected. On its way through the oceans, the deep currents also collect everything trickling from above: the excreted and dead biological material from marine organisms populating the sun-lit surface layers of the ocean forms a continuous shower of shells, diatoms, and other remainders. This shower can be noticed in the decreasing oxygen and increasing nutrient concentrations like nitrate and phosphate of the deep water on its way from the surface formation region, since the trickling biological material is remineralized by bacteria. Of course, the global MOC also has a prominent role to play in the meridional transport of heat because warm surface waters and cold abyssal waters generally flow in opposite directions. Some elementary estimates of the THC are reviewed in the box on p. 486.

**77. The Strength of the Meridional Overturning**

A simple calculation yields a scaling for the strength of the MOC. Consider a box ocean as sketched in the figure of this box. The ocean is of size $B \times L$ with two layers, the upper of height $h_1$ surfaces at high latitude, and the deep layer is of height $h_2$, with corresponding meridional velocities $v_i$ and temperatures $T_i$ (salinity is here ignored). South of the upper layer outcrop, the ocean is warmed at the surface at a rate $Q$ and it is cooled by $-Q$ north of the outcrop. The heat balance of the upper layer is

$$\rho c_p B(-v_1 h_1 T_1 + wLT_2) = QBL$$

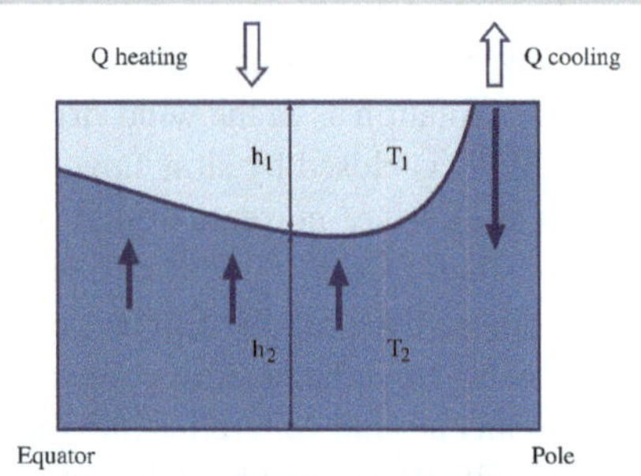


*An ocean box used for a simple scaling of the MOC. A north-south section is shown with a warm pool (light blue) overlying a deep cold ocean.*

The vertical velocity is $w = v_1 h_1 / L$ from continuity. Using numbers typical for the North Atlantic, $B = L = 4{,}000$ km, $h_1 = 1{,}000$ m, $Q = 50\,\mathrm{W\,m^{-2}}$, and a temperature difference $T_1 - T_2 = 10°$C, the size of the required meridional velocity is $v_1 = 0.006\,\mathrm{m\,s^{-1}}$ or $w = 1.2\times10^{-6}\,\mathrm{m\,s^{-1}}$ and implies a production of deep water of a magnitude $wBL = 20$ Sv. This is the strength of the overturning circulation driven by the heat flux. The meridional heat transport in this ocean is

$$H = \rho c_p B(v_1 h_1 T_1 + v_2 h_2 T_2)$$

with $v_1 h_1 + v_2 h_2 = 0$ to conserve mass. We find $H = 10^{15}\,\mathrm{W} \equiv 1$ PW, which is similar to what has been observed (Ganachaud and Wunsch, 2000). Finally, the vertical heat flux by upwelling of cold deep water is balanced by downward diffusion of heat from the surface water (Munk's abyssal recipe, Munk (1966); see also the box on p. 364),

$$w\frac{\partial T}{\partial z} = K_v \frac{\partial^2 T}{\partial z^2}$$

Hence $K_v = wh \sim 10^{-4}\,\mathrm{m^2\,s^{-1}}$, which must be seen as global mean vertical (diapycnal) diffusivity, necessary to maintain the stratification against upwelling.

The dilution and decay of trace substances can be used to estimate how long a certain water mass has been away from the surface of the oceans and, therefore, give information how long the overturning circulation takes for a complete renewal of the deep water. The result is that the overturning circulation takes roundabout a thousand years to complete a cycle. From this number, it becomes immediately evident from a simple scaling, using the sluggish observed velocities in the deep ocean, that the global MOC cannot be entirely advective (taking a path length of 60,000 km which is large, and a velocity of $1\,\mathrm{cm\,s^{-1}}$ which is fast, yields an advective time scale of only 200 years). Slower processes such as mixing by mesoscale eddies and small-scale turbulence must thus also contribute. On the other hand, perturbations of the thermohaline state may propagate much faster than the time scale of thousand years. Advection by swift surface and deep boundary currents is faster, and the ocean wave guide is controlled by very fast Kelvin waves along the ocean's margins and along the equator (see Section 8.4). This way, disturbances from remote areas, e. g. from Labrador Sea convection, are felt in other ocean basins only after some decades. In any case, the fate of Erik the Red's package of water is loss of identity by mixing:

if a thousand years later a research vessel had this package in her CTD bottles[2] on a recent expedition into the North Polar Sea, only a tiny fraction of the molecules from the time of Erik the Red would have remained from the original water and its content of trace substances.

## 15.1 Basic Ingredients of the Meridional Overturning

Before we discuss the fundamental physics and various models of the overturning, we present some general features of water masses and thermohaline forcing.

### 15.1.1 Water Masses of the Ocean

Water masses represent larger bodies of water characterized by nearly the same salinity and temperature, imprinted on them during their formation process. This mostly happens at the ocean's surface where water is injected into deeper areas due to the convergence of currents or when the surface heat and/or moisture loss makes it heavy enough to sink and displace the lighter water below it. By mixing of two or more water masses in the interior ocean, a new one can develop as well.

As an example, Figure 15.2 shows the main water masses in the Atlantic Ocean, in terms of potential temperature and salinity. The deeper region is roughly divided into

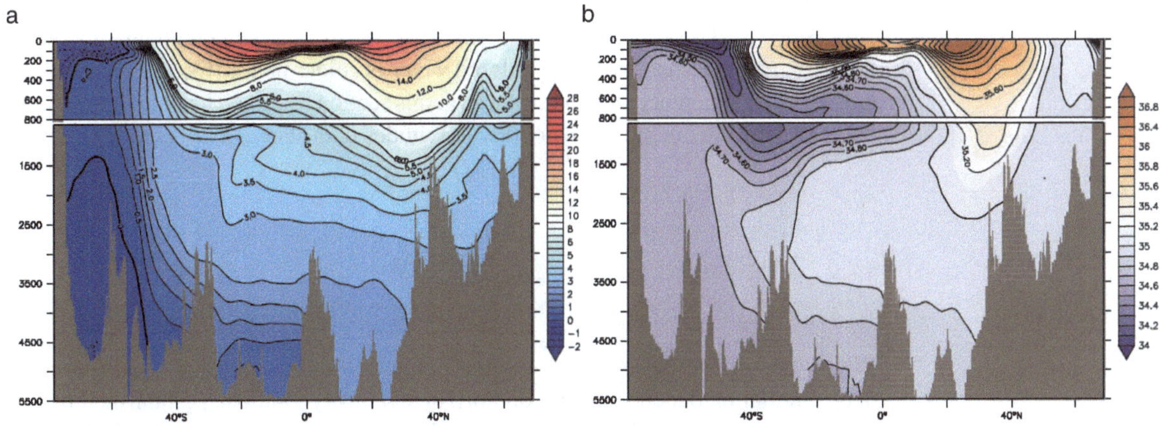

Fig. 15.2 Potential temperature in °C (**a**) and salinity in g kg$^{-1}$ (**b**) as a function of depth in m and latitude approximately along **30°** W in the Atlantic Ocean. Data are taken from the World Ocean Atlas 2005 (Antonov et al., 2006; Locarnini et al., 2006). The upper part (**800** m) of the section is shown enhanced at the top, showing the warm-water sphere with the shallow thermocline near the equator and the deeper one in the subtropics. From the salinity we note the intrusion of Antarctic Intermediate Water (AAIW) from the south in the Atlantic Ocean, while the cold temperature at depth spreading northward represents Antarctic Bottom Water (AABW). In-between these water masses, we find the southward spreading North Atlantic Deep Water (NADW)

[2] CTD is an acronym for Conductivity Temperature Depth, meaning a device lowered from a vessel by a cable into the ocean, continuously measuring salinity (via electric conductivity), temperature, and pressure. During its way down, water samples are also taken at certain depth levels by so-called Nansen bottles.

the North Atlantic Deep Water (NADW), the Antarctic Intermediate Water (AAIW), and the Antarctic Bottom Water (AABW). NADW is formed by cooling and evaporation of the warm surface water flowing to the Greenland Sea through the North Atlantic current into the Irminger, Labrador, and Norwegian Sea. AAIW sinks at the Antarctic Convergence zone, and AABW is formed by cooling and deposition of salt under the sea ice and the shelf ice areas round the Antarctic continent – mainly in the area of the Weddell and Ross Seas. AABW belongs to the heaviest water in the ocean and spreads from its relatively small area of origin to the bottom layers of all oceans. It is astonishing that small, apparently unimportant characteristics of sea-water density decisively regulate this vertical layering of water masses in the ocean. The originating NADW is e. g. even heavier than the AABW if we consider the formation process at the surface, and only the compression caused by the pressure during sinking reverses this relation due to the dependence of the compressibility on temperature and salinity (compare Section 1.2.6).

The spreading of newly formed water masses occurs nearly adiabatically, i. e. by advection, thereby conserving salinity, heat, and the concentrations of passive tracers. The water masses sink into their respective density layers at depth. Diapycnal mixing is very low, but, as we shall discuss later in this chapter, even at this very low level it is important for the MOC. The water masses can thus be identified over large distances as demonstrated in Figure 15.2. There is a large discrepancy in size between the parts of the world's ocean in which the major water masses are formed and the water volume with their specific characteristics – besides temperature and salinity also the content of many trace substances. That is, atmospheric and oceanic conditions in relatively small, preferably polar areas control the characteristics of the deeper cold water masses. About 75% of the oceans' volume with temperatures below 4°C is in contact with about 4% of the global sea surface, water colder than 2°C only with 1%. This can be seen in Figure 15.2 a) showing the potential temperature in the Atlantic Ocean. Dark blue colored water is colder than 2°C , and only in high latitudes it is in contact with the surface (in this section this appears only in the south, but it also appears in the north at other longitudes than shown in this section).

### 15.1.2 The Thermohaline Surface Forcing

While the surface oceanic currents are influenced directly by the windstress, the deep circulation depends mostly on horizontal density gradients, established by the combined effect of surface thermal and haline forcing and by mixing and advection (and by the indirect effects of wind forcing). The pole-to-equator difference of radiative heating in the atmosphere and the ocean produces horizontal density differences between the colder polar and the warmer equatorial sea surface temperature. This forcing alone would generate denser water at higher latitudes and lighter water in tropical regions. However, the excess of evaporation over precipitation towards the equator causes the mean salinity to decrease with latitude so that the thermal effect on density opposes that of salinity, i. e. temperature favors downwelling of dense water at high latitudes and upwelling at the equator, while for salinity the opposite is true.

The surface flux of density is shown in Figure 15.3. It is computed from the net heat flux $\mathcal{J}_{\text{heat}}$ and the freshwater flux $\mathcal{P}-\mathcal{E}$ at the surface (see also Chapter 13) by

$$\mathcal{J}_{\text{dens}} = -\frac{\alpha}{c_p}\mathcal{J}_{\text{heat}} - \rho_0 \gamma S_s(\mathcal{P}-\mathcal{E})$$

where $S_s$ is the surface salinity. Effects of freezing and melting of sea ice are ignored. Generally, the surface density is increased by the density flux at high latitudes and decreased in equatorial regions. The Atlantic, however, shows an exceptional behavior because most of the South Atlantic features a decrease of surface density. Note that it is not the northern-most area of the Atlantic (where the sinking takes place) which has the largest density increase but rather the area of the western boundary currents (this feature can be seen as well in the North Pacific). The density flux is here dominated by the massive loss of heat to the atmosphere due to the northward movement of warm water into a colder environment. Evidently, it is not the pattern of the density flux which determines the areas of sinking in the ocean, but rather the established pattern of temperature and salinity, or density at the surface, shown in Figure 15.4: only in the northern North Atlantic in the Labrador and Greenland Sea and in the southern Southern Ocean near the Antarctic Continent, the surface density is increased to its largest values, corresponding to the densities in the deep ocean. Note, however, that the surface density is also large in the Mediterranean Sea, indicating another sinking area.

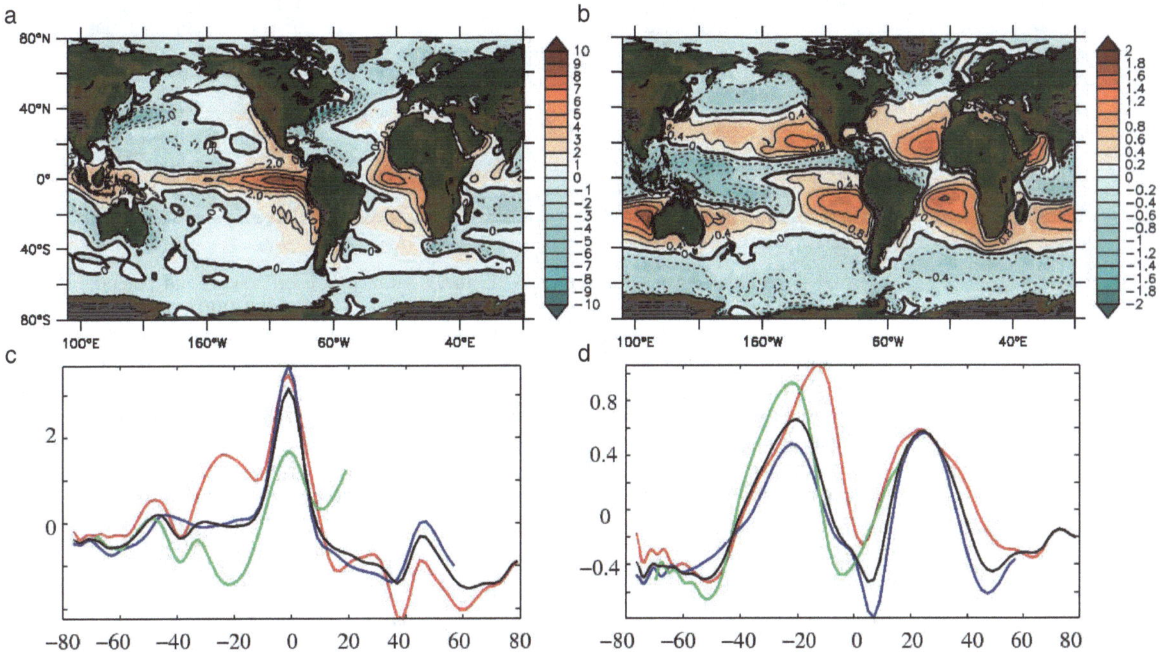


Fig. 15.3 Total surface density flux in $\text{m g m}^{-2}\,\text{s}^{-1}$ out of the ocean (**a**) and its haline contribution $\rho_0\gamma S_{\text{sq}}(\mathcal{E}-\mathcal{P})$ (**b**). Positive values (*brown*) denote regions where the surface density is decreased by the flux, i. e. density is "leaving" the ocean. The density flux is computed from the fluxes of heat and freshwater shown in Figure 13.4. Note that the color scales in **a** and **b** are different, i. e. heat flux is the dominating contribution. **c** and **d** show the long-term zonal mean of the surface density flux (c) and its haline component (**d**). Note also here the different vertical axes. The global mean is denoted by a black line, the zonally averaged components over the Atlantic, Pacific and, Indian Ocean are denoted by *red*, *blue*, and *green* lines, respectively

Fig. 15.4 Annual mean temperature in °C (**a**,**b**), salinity in $\mathrm{g\,kg^{-1}}$ (**c**,**d**) and potential density (referenced to the surface) in $\mathrm{kg\,m^{-3}}$ (**e**,**f**) at the surface (**a**,**c**,**e**) and at 2,000 m depth (**b**,**d**,**f**). Data are taken from the World Ocean Atlas 2005 (Antonov et al., 2006; Locarnini et al., 2006). While the surface temperatures are very similar in the Atlantic and Pacific, there is is marked difference in the surface salinities. The Atlantic is much saltier which results from the atmospheric transport of water vapor from the Atlantic surface waters to the Pacific by the low latitude easterlies. The deep Atlantic is warmer than the Pacific but much saltier, partly due to outflow of salty water from the Mediterranean but also due to salty water in the northern latitudes which at some places is heavy enough to sink. Water in the Atlantic is thus overall much denser than in the Pacific

### 15.1.3 The Asymmetry of the Meridional Overturning

Cooling at the northern latitudes does not necessarily lead to a sinking of the surface water. At low temperatures, roughly below 4°C, the density hardly depends on temperature but is predominantly controlled by salinity, and the surface salinity regulates whether winter cooling can increase the density of the water enough to sink, or whether cooling even to the freezing point leaves surface water still too buoyant to sink. The most conspicuous feature of the global MOC is its geographic asymmetry: water sinks in the North Atlantic (with a rate of 15–20 Sv as discussed in the box on

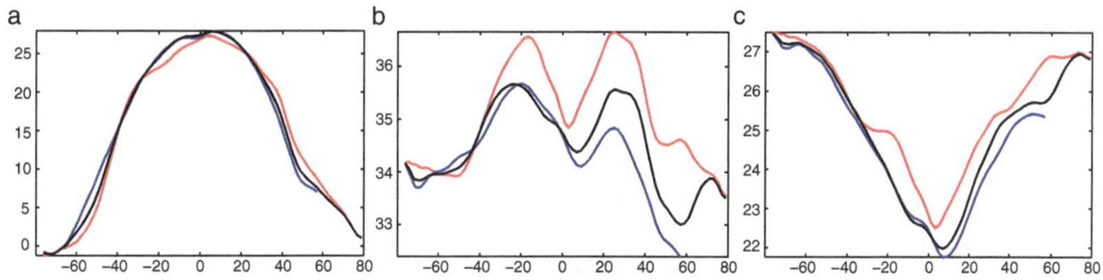


Fig. 15.5 Zonally averaged sea surface temperature (**a** in °C ), salinity (**b** in $g\,kg^{-1}$), and density (**c** in $kg/m^3$) as a function of latitude. *Red* lines denotes averages in the Atlantic Ocean, *blue*q Pacific Ocean and *black* global averages. Data are taken from Levitus and Boyer (1994); Levitus et al. (1994)

p. 486) to depths between 1 and 4 km, but no such deep sinking occurs in the North Pacific. The reason for this difference lies in the different surface salt concentrations (see Figure 15.4); the North Pacific is so low in surface salinity that not even water cooled to the freezing point becomes dense enough for deep sinking.

Meridional profiles of the surface conditions of the temperature, salinity, and density fields are shown in Figure 15.5 for the Atlantic and the Pacific. We note that the difference in density is mainly due to the differences in salt content in the two oceans. According to these observations, the Atlantic Ocean has a higher salinity and higher density than the Pacific Ocean. This arises because water vapor is transported in the atmosphere from the Atlantic to the Pacific by the westward trade winds across the American continent and thus does not return to the Atlantic by precipitation but rather by an interocean transport via the connection in the Southern Ocean. The surface salinity is largely controlled by the difference between evaporation and precipitation. Towards the north in the Atlantic, precipitation exceeds evaporation, and thus the surface salinity decreases from equator to pole. Hence it very much depends on the salt export from equatorial areas by advection towards the pole whether or not a thermohaline circulation with sinking in the North Atlantic exists. We will discuss the interplay between thermohaline forcing, strength of the MOC, and basin geometry below in Section 15.5.

### 15.1.4 The Formation of Water Masses

The deep oceans are cold. This is true even for the deep equatorial waters. This was not always the case: paleo-oceanographic observations suggest that the top-to-bottom temperature difference in the equatorial ocean, which is presently about 25°C, was reduced by almost 20°C before the establishment of the Circumpolar Current in the Southern Ocean, pointing towards a completely different MOC at those times. However, the cold waters in the present ocean must have their origin at high latitudes, where water masses come in contact with the cold atmosphere. In some polar areas, they might also become saltier through the formation of ice and brine rejection during ice formation. These processes lead to comparatively high densities, which then leads eventually to static instability and deep convection, which allows the dense water to sink to depth. These vertical sinking features, so-called *plumes*, partly reach the bottom. During their descent, they entrain lighter water from the surroundings, and so the final depth of the sinking water body depends on mixing processes but also on the density of the neighboring water and details of the equation of state. An excellent overview on theory and observations of oceanic convection has been given by Marshall and Schott (1999). An important effect is the dependence of the thermal expansion coefficient on pressure (thermobaric effect; compare also Section 1.2.6).

In the deep convecting plumes, heat (or better "coldness"), salt and other substances are transferred to great depth, as demonstrated by the plume simulation of Figure 15.6. They occur either in the open ocean (open ocean convection) or along continental slopes (slope convection), mainly around Antarctica and in the Arctic Ocean where the heavy water first fills up some reservoirs on the shelves, which then overflow and form plumes running down the continental slopes into the abyss. The main location of Antarctic slope convection is situated in the Weddell Sea. Open ocean convection has mainly been found in the Labrador Sea and also, but less frequently, in the Greenland Sea. There is also open ocean convection in the Mediterranean Sea and in other smaller basins close to the major oceans.

Ultimately, the deep cold water has to leave the region of formation to become a source of the global MOC. There is, however, a vast separation in the scales of the formation process (the convective plumes) and the outflow from the polar regions. The convection cells in the open ocean have a very small lateral size, less than a few kilometers (see e. g. Figure 15.6). Furthermore, there is no significant downwelling of mass because the vertical mass flux is almost balanced within the plume regime, i. e. although there is rigorous vertical movement within the plumes, there is almost no net vertical transport when averaged over several neighboring plumes and the area which embeds the plumes. Contrary to what is frequently reported, the global-scale overturning is thus *not* directly driven by these convective sinking events of dense surface water to the bottom. The convection does, however, set the water-mass characteristics: it makes the deep water cold, more or less salty, and e. g. rich in oxygen. Slope convection around Antarctica is not related to a significant net mass flux either, but after turbulent entrainment of surrounding water masses it becomes what finally is known as AABW. Its outflow off the Weddell Sea (see right panel of Figure 15.7) and other areas around Antarctica has a significant contribution to the

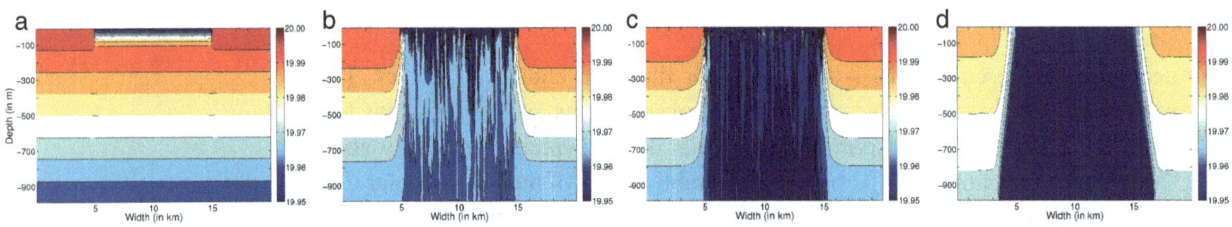


Fig. 15.6 Deep convection with plumes in an idealized three-dimensional model forced by cooling at the surface. The model uses nonhydrostatic, primitive equations and rotation. A box of $20\,\text{km} \times 20\,\text{km} \times 1{,}000\,\text{m}$ extent is considered, which is initialized with a vertical temperature gradient of $0.04$ K from top to bottom. The surface is cooled with a rate of $1{,}000\,\text{W m}^{-2}$ within a disk with a radius of 5 km. The horizontal resolution is 100 m and the vertical resolution is 20 m. The figure shows temperature in °C after 2 h (**a**), 12 h (**b**), 24 h (**c**) and 72 h (**d**) integration, at a section through the center of the box. After Sander et al. (1995)

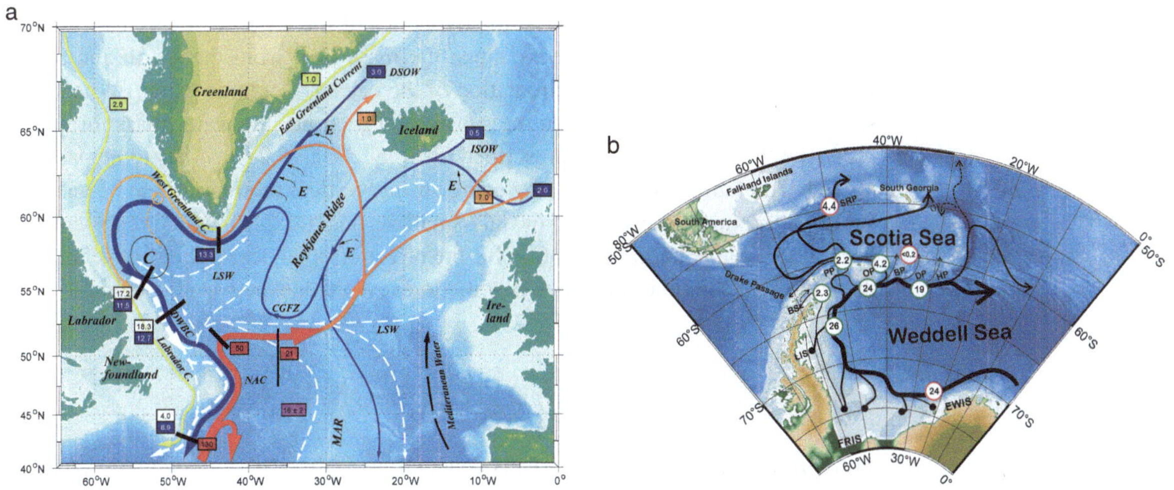


Fig. 15.7 **a** Schematic of the flow of important water masses in the subpolar North Atlantic Ocean, as compiled from many observational studies. It shows the surface flow of the North Atlantic Current (NAC, thick red) crossing the Middle Atlantic Ridge (MAR) above the Charlie Gibbs Fracture Zone (CGFZ), after which it turns yellow and flows either into the Norwegian Sea or recirculates into the western subpolar North Atlantic. The deep flow including the Deep Western Boundary Current (DWBC) is indicated in blue. The light yellow currents are cold and shallow and flow along the shelf break. Open ocean convection (C) takes place in the Labrador Sea, where Labrador Sea Water (LSW) is formed, while entrainment (E) is important for the overflow of Iceland Scotland Overflow Water (ISOW) and the Denmark Strait Overflow Water (DSOW). Note that LSW, ISOW, and DSOW form the NADW further downstream of the lower part of the MOC in the Atlantic Ocean. After Schott and Brandt (2007). **b** Spreading of WSDW (Weddell Sea Deep Water) in a numerical model of the circulation of the Weddell and Scotia Seas. The origin of WSDW on the continental shelf is marked by black dots, and the black curves represent the branches of the flow. Transport rates (in Sv) are given for some branches. Abbreviations: BS – Bransfield Strait, DP – Discovery Passage, EWIS – East Weddell Sea Shelf Ice areas, FRIS – Filchner-Ronne-Shelf Ice, GP – Georgia Passage, HP – Hoyer Passage, LIS – Larsen Shelf Ice, OP – Orkney Passage, PP – Philip Passage, SRP – Shag Rocks Passage. From Schodlok (2002)

MOC. The rate of AABW forming is, however, yet quite uncertain; estimates range between $5 \times 10^6\,\mathrm{m^3\,s^{-1}} \equiv 5{-}15\,\mathrm{Sv}$.

Water that has been cooled and convected in the Greenland and Norwegian Sea partly overflows the sills in the Denmark Strait and the Faroer Bank Channel as small-scale gravity-driven flow and then sinks, forming an initial branch of the North Atlantic MOC (see left panel of Figure 15.7). Although these overflows themselves have only a strength of 1–2 Sv, turbulent entrainment during the downslope flow increases the volume transport of the overflows considerably. The overflow water from the Denmark Strait and the Faroer Bank Channel then forms the deep part of the western boundary current of the Labrador Sea and joins with a deep water outflow out of the central Labrador Sea where deep open ocean convection takes place. Overflow and Labrador convection contribute to about equal to the sum of around 10–15 Sv poleward transport of NADW, but there is strong interannual variability in the open ocean convection (see e. g. Schott et al., 2004).

However, in the present chapter we are neither concerned with the complicated details of the flow of the newly formed water masses after formation and its spreading into the interior ocean, nor with details of the small-scale, patchy, and sporadic open ocean deep convection and slope convection and subsequent local mixing. Instead, we are concerned here with the dynamics of the large-scale circulation *after* the water-mass formation process, since the mass flow from the North and South Polar Seas drives the abyssal circulation of the oceans to a large extent. When constructing models of the MOC in the subsequent sections, the water-mass formation process either will be simply prescribed (as in Section 15.2) or it has to be parameterized. The latter is often done by using strong vertical mixing in case of unstable stratification (and otherwise no or only small vertical mixing), which represents indeed a reasonable parameterization for the effect of open ocean deep convection on the large-scale circulation, which is sometimes called convective adjustment.

## 15.2 The Stommel–Arons Overturning Model

The surface currents in ocean basins concentrate at the western boundaries, as discussed in Chapter 14. It might thus not be surprising that in the lower branch of the MOC, an important part of the flow is also attached to the western boundary. In fact, the Deep Western Boundary Current (DWBC) along the North American coast in the Atlantic was predicted by the Stommel–Arons model (Stommel and Arons, 1960a,b; Stommel et al., 1958) as being part of the MOC. This a singular case where a theoretical study predicted a major ocean current before it was actually observed. An important aspect of the Stommel–Arons model is that it gives a view of the deep circulation, but it is, furthermore, a full three-dimensional overturning model as discussed in this section. The strength of the MOC, however, is prescribed and not part of the model outcome.

The concept of the Stommel–Arons model is the existence of a DWBC, which is forced by the formation of water masses, feeding the abyss and driving an interior recirculation. The sources of the abyssal waters are located in the northern North Atlantic and around Antarctica. The resulting model is rather simple: an abyssal layer of given height $h$ is filled by pumping in water with fixed sources (in $\mathrm{m}^3\,\mathrm{s}^{-1}$) at positions in the northern/southern rim (or the equator, as a thought experiment), whereas the return flow is equally distributed over the whole layer interface with a uniform vertical velocity $w_0$ upward through the thermocline into the upper ocean (see Figure 15.8). The densities of the layers are constant (with a difference of $\Delta\rho$ between the layers), and the bottom of the layer is assumed to be flat.

We first concentrate on the interior solution for the abyssal layer away from the western boundary layer. The boundary layer is treated in detail in the box on p. 497. The ocean basin is assumed to be of a box-shape, with a constant zonal $B = x_\mathrm{E} - x_\mathrm{W}$ and meridional extent $L = y_\mathrm{N} - y_\mathrm{S}$, where $x_\mathrm{E}$, $x_\mathrm{W}$, $y_\mathrm{N}$ and $y_\mathrm{S}$ denote the limiting zonal and meridional coordinates of the box, respectively. As in Chapter 14, we apply the planetary approximation given by (5.45) and (5.46) which is relevant for the oceanic circulation on lateral scales much larger than the Rossby radius, as discussed in Chapter 5.1. We first consider the steady momentum equations and the continuity equation averaged over the layer depth $h$, written here for convenience in Cartesian coordinates,

$$-fv = -\frac{\partial p}{\partial x}\,, \quad fu = -\frac{\partial p}{\partial y}\,, \quad \frac{\partial u}{\partial x} + \frac{\partial v}{\partial y} + \frac{w_0}{h} = 0 \tag{15.1}$$

where $w_0$ denotes the uniform upwelling velocity at the upper interface of the layer. The horizontal velocity is assumed to be constant over the layer (in agreement with

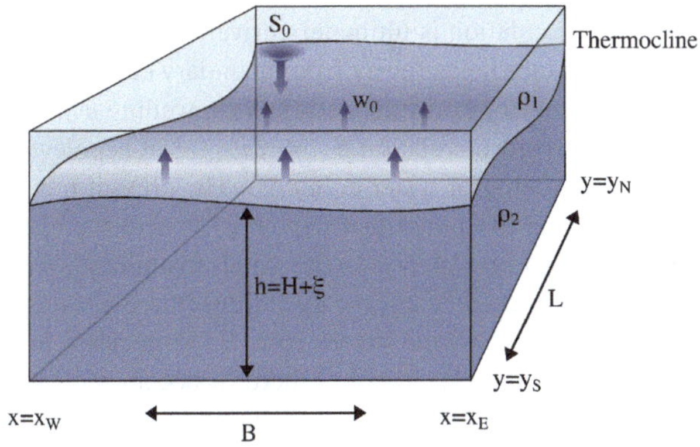


Fig. 15.8 Illustration of the Stommel–Arons layer model with a source $S_0$ of deep water, located at the northern rim of a box-shaped ocean in the western boundary domain, and uniform upwelling $w_0$

a constant density). Note that we use in fact the linearized version of a two-layer model discussed in detail in Appendix B.1.2. The model is linearized by approximating the layer depth $h = H + \xi$ by the constant mean depth $H$, but the pressure[3] $p$ is varying due to the small interface displacement $\xi$ and the surface height displacement $\zeta$, with $|\xi|, |\zeta| \ll h$ (see Figure 15.8).

Aiming at simple models we frequently consider a particular parameterization for the turbulent stress terms in the momentum balances. The flux divergence of the lateral turbulent eddy fluxes of horizontal momentum is replaced by a Rayleigh friction according to **78. Rayleigh Friction**

$$\mathcal{F}_u = -\nabla \cdot \overline{\boldsymbol{u}'u'} = -ru \quad \text{and} \quad \mathcal{F}_v = -\nabla \cdot \overline{\boldsymbol{u}'v'} = -rv$$

with a constant coefficient $r$. Though such a friction is often used, it should be borne in mind that its mathematical simple form is quite unphysical because momentum is not transported but destructed locally. Note also that viscous boundary conditions can no longer be satisfied in the Rayleigh model.

We further assume for simplicity an equatorial $\beta$-plane, i. e. $f(y) = \beta y$. Taking the curl of the momentum equation yields the planetary vorticity (or Sverdrup) relation $\beta v = f w_0/H$, which we met before in Section 5.3 on planetary-scale motion. Note that the Sverdrup relation fully determines the meridional velocity. The zonal velocity then follows from a zonal integration of the continuity equation with the kinematic boundary condition $u(x_\mathrm{E}) = 0$, hence

$$v = \frac{w_0}{H} y \quad \text{and} \quad u = \frac{2w_0}{H}(x_\mathrm{E} - x)$$

The pressure or the interface displacement can be calculated from either of the geostrophic balances; the other one is then trivially satisfied. The pressure is found to be

$$p = \frac{\beta w_0}{H} y^2 (x - x_\mathrm{E}) \tag{15.2}$$

[3] Pressure is scaled by a constant reference density.

Note that although the circulation is ultimately driven by the localized sources, the interior velocity and pressure fields (away from boundary currents) are independent of the location of the source; they only depend on the upwelling $w_0$. Since the interior flow is governed by the Sverdrup relation $\beta v = f w_0/H$, it is poleward everywhere because of $w_0 > 0$ (stretching of water columns). The present model is similar to the wind-driven circulation discussed in Chapter 14, where we also have derived the Sverdrup relation given by (14.17). In the wind-driven case, $w_0$ is the Ekman pumping. Here, it is the imposed uniform upwelling (the constancy of $w_0$ makes the model even simpler than the wind-driven case). In both cases, however, there is no flow across the equator since $f$ vanishes there. It becomes clear that a western boundary current must be added to the interior solution to establish flow across the equator, which we observe for the MOC. The western boundary currents are constructed by closing the mass balance, as in the original Stommel–Arons model and as detailed below, or by introducing viscous or Rayleigh-type boundary layers, as in the wind-driven Stommel model of the box on p. 458. For the Stommel–Arons problem, this is outlined in the box on p. 497.

The water-mass source and the interior upwelling through the interface will not only set the abyssal layer into motion but also the surface layer. The upper layer has a reversed circulation such that the vertically integrated (barotropic) circulation vanishes. Suppose that the layers have a density difference $\Delta\rho$. The upper layer pressure is $p_u = g\zeta$ and the lower layer pressure is $p_\ell = g\zeta + g\Delta\rho\xi$, where $\zeta$ is the sea surface topography and $\xi$ the interface topography. From (15.2), we obtain

$$\zeta = -\frac{\beta w_0}{gH} y^2 (x - x_E) \quad \text{and} \quad \xi = 2\frac{\beta w_0}{g\Delta\rho H} y^2 (x - x_E)$$

where we have assumed for simplicity an equal mean depth $H$ for the two layers. The solution for the circulation is displayed in Figure 15.9. In the lower layer, the flow enters the interior domain on the western side and leaves at the northern boundary.

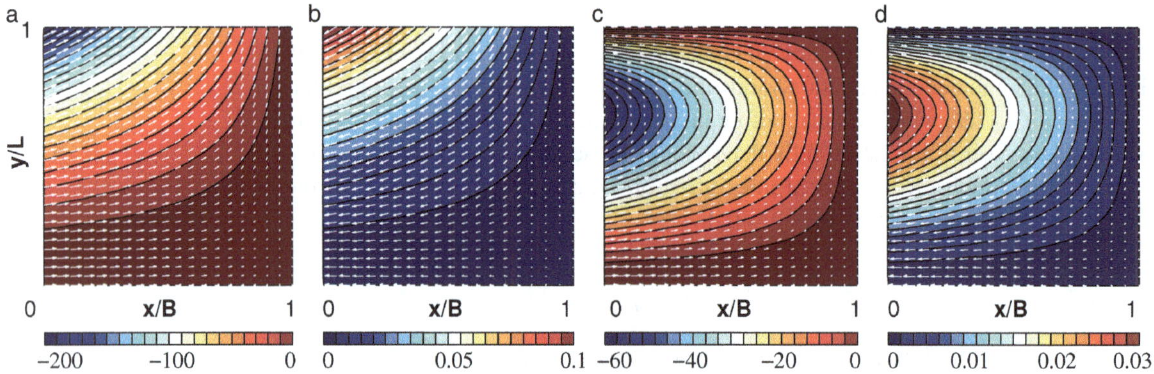


**Fig. 15.9** The Stommel–Arons circulation in a rectangular basin; **a** and **c** are for the lower layer and **b** and **d** for the upper one. The left set of panels **a** and **b** uses a constant upwelling $w_0$, while the right set **c** and **d** has a linear $w_0(y)$ decreasing to zero at $y = y_N = L$. The colors represent the displacement fields for the interface $\xi$ (**a**,**c**) and the surface $\zeta$ (**b**, **d**) in m. The arrows represent the flow field. Both layers have equal thickness of $H = 2{,}000$ m, the zonal and meridional extent of the domain is $B = L = 5{,}000$ km, the upper to lower layer density difference is $\Delta\rho = 10^{-3}$, and the source strength is $S_0 = 20$ Sv

**79. The Boundary Layer of the Stommel–Arons Model**

To include a frictional western boundary current in the model, we add Rayleigh-friction, $-ru$ and $-rv$, to the geostrophic equations (15.1). Rayleigh friction is explained in the box on p. 495. We then solve the momentum equations for $u$ and $v$ and implement the result into the continuity equation in (15.1). An equation for the pressure is found that way. We consider the equation for the case of small friction only, i. e. in the limit $r/f \ll 1$,

$$r\frac{\partial^2 p}{\partial x^2} + f^2\frac{\partial}{\partial y}\left(\frac{r}{f^2}\frac{\partial p}{\partial y}\right) + \beta\frac{\partial p}{\partial x} = \frac{f^2 w_0}{H}$$

which is in fact (8.2.6) from the wind-driven western boundary case discussed in Chapter 14, with less approximations in the frictional terms. In the boundary layer, the first and third terms on the left-hand side dominate, hence

$$\frac{\partial^2 p}{\partial x^2} + \frac{\beta}{r}\frac{\partial p}{\partial x} = 0$$

As in the wind-driven case, the width of the boundary layer is found to be $\delta = r/\beta$. The solution for the pressure is, therefore,

$$p_b(x, y) = p_i(x = 0, y) + P(y)e^{-x/\delta}$$

where $p_i$ is the interior solution, given by (15.2), and $P(y)$ is the amplitude of the boundary layer correction (for details on the mathematical treatment of boundary layers see Appendix A.2.2). Note that we have taken $x_W = 0$ without restriction. The function $P(y)$ follows from mass conservation: the meridional velocity in the boundary is given by $f v_b(y) = \partial p_b/\partial x$ with the transport $T_b(y) = H\int v_b \mathrm{d}x$, integrating from $x = 0$ to the outer edge of the layer, i. e. infinity in the boundary layer coordinate system. The results is

$$P(y) = -\frac{f}{H}T_b(y) = -\frac{\beta y}{H}T_b(y)$$

Note that this solution actually yields the kinematic boundary condition $u = 0$ at the western coast: Using $fu(x = 0) = -\partial P/\partial y - \partial p_i(x = 0)/\partial y - rv(x = 0)$ and implementing $P(y) = 2(\beta w_0/H)By^2$ (for the case of the polar source location), $p_i(x = 0) = -(\beta w_0/H)By^2$ and $v(x = 0) = v_b(x = 0) = \partial p_b(x = 0)/\partial x/f = P(y)/(f\delta)$, we find indeed $u(x = 0) = 0$. The pressure field is shown in the following figure for the polar and the equatorial source locations.

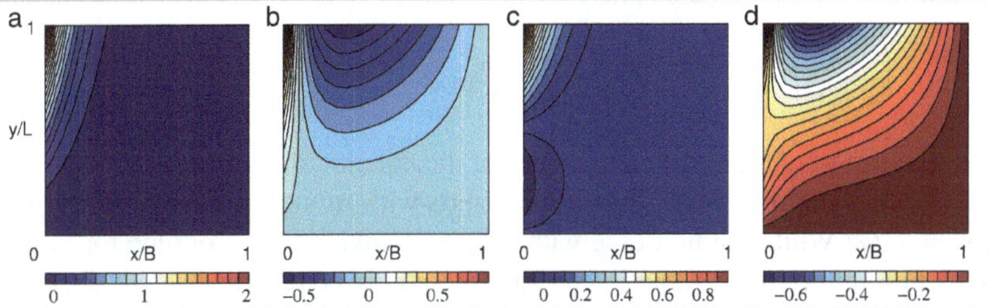


*Pressure distribution of the Stommel–Arons solution with a western boundary layer. For the left two panels (**a,b**) the source is located at the pole, for the right two panels (**c,d**) it is at the equator. The respective first panel of each set shows the boundary layer correction, and the second panel is the total solution.*

While the inflow from the west is completed by inclusion of a frictional western boundary layer, as detailed in the box on p. 497, the northern outflow from the interior can be resolved by abandoning the assumption of uniform upwelling, i. e. making $w_0$ dependent on $y$ there. If $w_0(y_N) = 0$, the meridional velocity goes to zero as well, according to the Sverdrup relation, and thus satisfies the boundary condition. The above derived solution for $v$ and $p$ remains the same but now with a varying $w_0(y)$. The solution is shown in the right two panels of Figure 15.9, using for simplicity a linear $w_0(y) \sim (1 - y/y_\mathrm{N})$. The zonal flow is changed to

$$u = \left(2w_0 + \frac{\mathrm{d}w_0}{\mathrm{d}y}y\right)\frac{x_\mathrm{E} - x}{H}$$

In both cases, however, the northward interior flow must be balanced by a western boundary current of small width and transport $T_\mathrm{b}(y)$ (taken positive northward). Let us place a single source $S_0$ at the northern rim as indicated in Figure 15.8. Requiring mass conservation for the entire basin, we find $S_0 = w_0 BL$. The mass balance for the area north of a latitude $y$ is $T_\mathrm{b}(y) + T_\mathrm{i}(y) + S_0 = T_\mathrm{u}(y)$ where $T_\mathrm{i}(y)$ is the transport across the latitude $y$ in the interior and $T_\mathrm{u}(y)$ is the upwelling over the respective interface. One finds

$$T_\mathrm{i}(y) = H\int_{x_\mathrm{W}}^{x_\mathrm{E}} v\mathrm{d}x = yw_0B \quad \text{and} \quad T_\mathrm{u}(y) = \int_{y}^{y_\mathrm{N}}\int_{x_\mathrm{W}}^{x_\mathrm{E}} w_0\mathrm{d}x\mathrm{d}y = (y_\mathrm{N} - y)w_0B$$

and hence

$$T_\mathrm{b}(y) = w_0B\,(y_\mathrm{N} - y - y - L) = -2yw_0B$$

In the last relation, we have used $y_\mathrm{S} = 0$ for the latitude of the equator so that $y_\mathrm{N} = L$. The transport in the boundary layer is negative, i. e. equatorward. At the northern boundary $y = y_\mathrm{N} = L$, it is twice the strength of the source, $T_\mathrm{b}(y_\mathrm{N}) = -2S_0$, and at the equator it vanishes. The interior transport $T_\mathrm{i}$ vanishes at the equator and increases to a total $S_0$ at $y = y_\mathrm{N}$. Thus the amount $S_0$ just recirculates in the basin (assuming that the outflow at $y = y_\mathrm{N} = L$ somehow joins the western boundary current).

The source $S_0$ can also be placed at other locations. Pumping the water at the equator into a northern hemisphere basin yields

$$T_\mathrm{b}(y) = \frac{S_0}{L}(y_\mathrm{N} - 2y)$$

with a transport of $T_\mathrm{b} = S_0$ to the north at the equator, diminishing to zero at $y = L/2$. From the north a southward current starts with strength $S_0$, which goes to zero at $y = L/2$. While the first case with northern sinking is appropriate for the deep circulation in the North Atlantic, the latter case may be regarded as the situation in the North Pacific.

Note that the whole problem has been solved by geostrophic equations. This is possible because the interior flow vanishes at the equator. Boundary layers must appear at the western rim and the northern rim, which are, of course, ageostrophic (see the box on p. 497). The model is easily written in spherical coordinates and has been extended to a two-hemisphere basin and even to global ocean with several sources, as shown in Figure 15.10.

The Stommel–Arons model captures some characteristics of the abyssal circulation, as e. g. the existence and direction of the DWBC in some ocean basins as well as the interior recirculation, but was never meant to give a realistic description. It fails in many other aspects, in fact all of its ingredients can be questioned. At first, the deep circulation is strongly influenced by submarine topography. It can easily be

Fig. 15.10 The abyssal circulation of the World Ocean according to the Stommel–Arons model, generated by sources of equal strength in the North Atlantic and in the Weddell Sea (indicated by the black dots) and by uniform upwelling elsewhere. Redrawn after Stommel (1958). The strength of the deep circumpolar branch is prescribed

incorporated, but the results would not become more realistic because of a second, more severe, drawback of the model. The upwelling could be very localized rather than given by the uniform distribution in the model, or, as discussed recently, it may be very weak over most of the basin while most of it occurs along the upward sloping isopycnals in the Southern Ocean (see Chapter 16). However, extending the model by these aspects opens a door to all kinds of circulations in models with little value. Many other features are entirely absent which could potentially influence the abyssal circulation: stratification, mesoscale eddy transports and mixing.

## 15.3 Sandström's Inference

The work of SANDSTRÖM[4] (Sandström, 1908) has recently been revisited to argue about the driving of the MOC. Sandström tried to force a circulation in a small water tank by placing heating and cooling sources horizontally separated and at the same or at different depths. Turbulent mixing in the tank was avoided (actually Sandström was unaware of its role in the ocean and even ignored molecular diffusion). He found, in agreement with the circulation theorem derived below, that a fluid heated and cooled from the surface, or cooled (compressed) at a higher pressure (or depth) than that at which it is heated (expanded) does not develop any significant circulation. Defant (1961) supported this 'inference' with thermodynamical arguments in

[4] JOHAN WILHELM SANDSTRÖM, *1874 in Norra Degerfors, Sweden, †1947 in Sweden, oceanographer and meteorologist.

terms of a CARNOT[5] process (see Section 15.3.2), and he was the first[6] to refer to Sandström's inference as 'Sandström's theorem' (Defant, 1961).

### 15.3.1 Consequences from Bjerknes' Theorem

We start with Bjerknes' circulation theorem (see (2.205) in Section 2.11). Consider a loop $\Gamma$ along a streamline of the whole global MOC. The three-dimensional velocity is denoted by $\boldsymbol{u}$, frictional force (per mass) by $\mathcal{F}$, specific volume by $\upsilon = 1/\rho$, and $\mathrm{d}\boldsymbol{s}$ denotes the increment along the loop $\Gamma$. The theorem is then expressed by

$$\frac{\mathrm{D}}{\mathrm{D}t}\oint \boldsymbol{u}\cdot\mathrm{d}\boldsymbol{s} = -\oint \upsilon\mathrm{d}p + \oint \mathcal{F}\cdot\mathrm{d}\boldsymbol{s} \tag{15.3}$$

and states a balance between the rate of change of the circulation (the integral $\oint \boldsymbol{u}\cdot\mathrm{d}\boldsymbol{s}$ is called the 'circulation'), the pressure work term, and the friction term (last term in (15.3)). Unless there is wind forcing, the friction integral acts against an increase of the circulation and must thus be negative. In order to sustain a steady circulation, the pressure work must drive it, i. e. we obtain the condition

$$-\oint \upsilon\mathrm{d}p = \oint p\mathrm{d}\upsilon > 0 \tag{15.4}$$

and end up with the integral which also controls the Carnot process (see also Section 15.3.2). Following (15.4), expansion ($\mathrm{d}\upsilon > 0$) must take place at higher pressure (deeper in the water column) than contraction ($\mathrm{d}\upsilon < 0$) as sketched in the right panel of Figure 15.11. It becomes obvious that there is no driving of the circulation if expansion (e. g. by heating) and contraction (e. g. by cooling) takes place at the same pressure level. In the ocean, heating and cooling takes place at the ocean surface, i. e. indeed at the same depth, at least if molecular and turbulent transports in the interior are ignored (see below). We also ignore the insignificant effect that the tropical sea surface is about one meter higher – and thus at lower pressure – than the sea

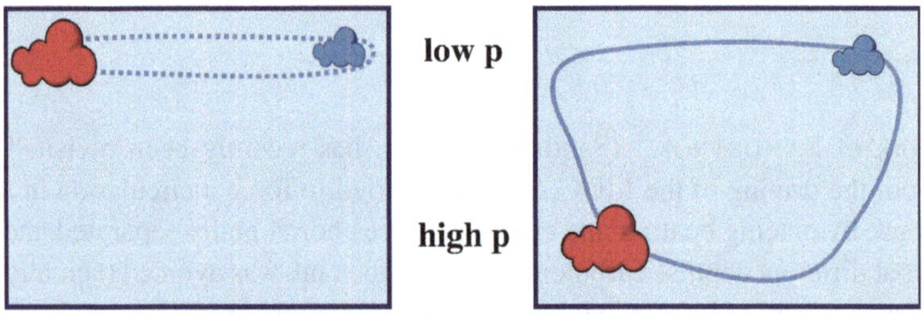


Fig. 15.11 Illustration of Sandström's theorem. The red parcel is expanded by heating, the blue compressed by cooling. The picture can be read in the physical space or in the thermodynamic space $(\upsilon, p)$ with the specific volume $\upsilon$ increasing from right to left

[5] SADI CARNOT, *1796 and †1832 in Paris, physicist and engineer.

[6] See Kuhlbrodt (2008) for a historical consolidation of the Sandström controversy. The controversy is that Sandström did not wait long enough in his experiments to see that, on the long run molecular diffusion will take the heating signal into the interior, eventually below the cooling level: hence some kind of overturning must always develop, even if the actual forcing is applied at the same pressure level.

surface at high latitudes. Hence, if there is no other forcing, the MOC should come to rest due to friction. In order to drive a circulation, heating must occur at higher pressure than cooling, i. e. at greater depth. This occurs e. g. in the atmosphere, which, therefore, is akin to a heat engine. The ocean, on the other hand, must be regarded as a mechanical engine, a fact which will become clearer in the course of the following discussion. Note that the above theorem also implies that neither deep convection in high latitudes nor any other process of deep water formation can drive the MOC by whatever mechanism or strength, since deep convection places the cooling even at higher pressures than the pressure at the surface.

### 15.3.2 Thermodynamics of the Meridional Overturning

Bjerknes' theorem (15.3) can be interpreted as a balance of the work done by the different forces acting on a material fluid parcel along a closed loop in the fluid. Remember that the origin of the integral $\oint \upsilon \mathrm{d}p$ derives from the pressure force (per mass) $\upsilon \nabla p$ causing the work $\upsilon \nabla p \cdot \mathrm{d}\boldsymbol{s}$ on the parcel, when it is displaced by $\mathrm{d}\boldsymbol{s}$ along the circulation loop. The converted form $\oint p \mathrm{d}\upsilon$, on the other hand, hints towards the work involved in contraction or expansion which is a purely thermodynamic process. Their integrals are the same but their physics is totally different.

The work $\delta W = -p\mathrm{d}\upsilon$ appears in the first law of thermodynamics (1.23), formulated for the internal energy $E$,

$$\mathrm{d}E = \delta Q - p\mathrm{d}\upsilon + \frac{\partial H}{\partial S}\mathrm{d}S \tag{15.5}$$

which is a statement of conservation of internal energy of a material fluid parcel. Note that $\delta W = -p\mathrm{d}\upsilon$ is the work done by the environment on the parcel (if $\mathrm{d}\upsilon < 0$ – i. e. contraction – this work increases the parcel's internal energy). Furthermore, if $\delta Q > 0$, there is a supply of heat to the parcel and $\mathrm{d}S > 0$ a supply of salt. As shown in Section 1.2, $\partial H/\partial S = H_\mathrm{s} - H_\mathrm{w}$ is the difference between the specific enthalpies of salt and pure water. However, the contribution of salinity changes to $\mathrm{d}E$ is entirely negligible for realistic conditions (see e. g. the discussion in Section 2.6.2). Likewise, the contraction/expansion work is generally as well a small term in (15.5) in the ocean. We have shown how to combine $\mathrm{d}E$ and $-p\mathrm{d}\upsilon$ to the change of potential temperature which then is balanced by the heating rate in form

**80. Bjerknes Theorem in Boussinesq Form**

We like to add a further note to the discussion of the Sandström inference: most dynamical concepts in oceanography are represented in the Boussinesq framework where the integrals $\oint p\mathrm{d}\upsilon$ or $\oint \upsilon \mathrm{d}p$ do not appear because the contribution of the pressure gradient is multiplied by the inverse reference density and thus integrates to zero. Instead, the buoyancy force remains and yields the theorem

$$\frac{\mathrm{D}}{\mathrm{D}t}\oint \boldsymbol{u}\cdot \mathrm{d}\boldsymbol{s} = -\oint (\tilde{\rho}/\rho_0)\nabla \Phi \cdot \mathrm{d}\boldsymbol{s} + \oint \boldsymbol{\mathcal{F}}\cdot \mathrm{d}\boldsymbol{s}$$

referring to the notation of Section 4.1.2. The integral involving the geopotential $\Phi$ can be converted to $\oint \Phi \mathrm{d}\tilde{\rho}/\rho_0$. The theorem then states that in order to allow for a circulation, expansion (contraction) must occur at high (low) values of the geopotential, i. e. heating must occur at greater depth than cooling.

of the divergence of the heat flux and the mechanical dissipation, referring to (2.74) in the appropriate form a conservation equation, and here again the heating term is locally by far the dominating contribution (see the discussion in Section 2.6). Thus, while $\upsilon dp$ is a major player in the mechanical balance (15.4) all along the circulation loop, the work $-pd\upsilon$ is small in the thermodynamical balance (15.5), applied to the same loop. Here, heating and cooling by heat flux divergences overwhelm the contraction/expansion work and almost balance in the integral.

There is also a lesson from the second law of thermodynamics (see Section 1.2.4). Assume that the ocean (O) is attached to two heat reservoirs, the equatorial/tropical (E) and polar (P) atmosphere, which exchange heat quantities $Q_E$ and $Q_P$ at temperatures $T_E$ and $T_P$, respectively, with the ocean. According to the second law, the change of entropy $d\eta_E + d\eta_O + d\eta_P$ of the total ocean-atmosphere system must be positive, and $d\eta_E \geq -Q_E/T_E$, $d\eta_P \geq Q_P/T_P$. Note that the amounts of heat are treated as positive, $Q_E > 0$, $Q_P > 0$. Furthermore, the processes within the ocean are assumed reversible, and in a complete cycle its entropy change is zero, i. e. $d\eta_O = 0$. We find

$$\frac{Q_P}{T_P} - \frac{Q_E}{T_E} \geq 0 \quad \text{or} \quad \frac{Q_P}{Q_E} \geq \frac{T_P}{T_E}$$

The ocean does not change its internal energy in the cycle, so that the first law yields $W_O = Q_E - Q_P$ for the work done by the ocean to sustain a circulation. This requires $W_O > 0$, and obviously, according to the above inequality, heating must occur at a higher temperature than cooling, i. e. $T_E > T_P$. The system is analogous to the classical Carnot engine[7]. The Carnot factor $W_O/Q_E \leq 1 - T_P/T_E$ measures the efficiency to convert heating to mechanical energy. As a heat engine, the ocean is very ineffective: with $T_E - T_P \approx 15\,\text{K}$, we find $W/Q_E \leq 2\%$. If the processes in the ocean are irreversible, the efficiency must be even smaller, and as a consequence of $W_O/Q_E \to 0$ we obtain the approximate conservation of heat, $Q_E \approx Q_P$. What is not seen in these relations but outlined in Section 15.3.1 is that the heating must occur at the expansion and cooling at the contraction of fluid parcels, i. e. at different pressures ($p_E < p_P$), which is not satisfied if all exchange processes occur at the ocean surface.

### 15.3.3 Energetics of the Meridional Overturning

A closer view on the consequences from Sandström's inference is achieved by considering the conservation of energy. In the deep water sinking areas, potential energy is lost and continuously converted to small-scale turbulent kinetic energy. The latter is subsequently dissipated due to molecular friction. Hence in a steady state, there must be a source of potential energy. Energy must be fed into the system where the deep water returns to the surface, and this source of potential energy must be regarded as the driver of the MOC.

[7] In the Carnot engine we have isothermal expansion at the temperature $T_E$ where the heat $Q_E$ is absorbed, followed by adiabatic expansion where the system cools to $T_P$, then isothermal contraction at $T_P$ where the heat $Q_P$ is released, followed by adiabatic compression where the system becomes warmer again. All parts of the Carnot cycle are reversible and make a closed cycle.

To be specific, consider the energetics in the Boussinesq approximation, as derived in Section 4.1.4. Integration over the whole ocean volume (denoted by an overbar) yields

$$\frac{\mathrm{d}\overline{E_\mathrm{k}}}{\mathrm{d}t} = -g\overline{\tilde{\rho}w} + \overline{\boldsymbol{u}\cdot(\mathcal{F}_\mathrm{e}+\mathcal{F}_\mathrm{b})} - \overline{\epsilon}\,, \qquad \frac{\mathrm{d}\overline{E_\mathrm{p}}}{\mathrm{d}t} = g\overline{\tilde{\rho}w} + g\overline{z\mathcal{G}_\rho} \tag{15.6}$$

with kinetic and potential energies $E_\mathrm{k} = \rho_0\boldsymbol{u}^2/2$ and $E_\mathrm{p} = g\tilde{\rho}z/\rho_0$, respectively, and an external (wind) stress $\mathcal{F}_\mathrm{e}$, a frictional stress $\mathcal{F}_\mathrm{b}$ at solid boundaries (bottom, side walls), and an interior density source term $\mathcal{G}_\rho$ due to diapycnal mixing in the interior and convection in water-mass formation regions. Note that $\tilde{\rho}$ is a perturbation density in the notation of Section 4.1.4. The term $\overline{\boldsymbol{u}\cdot\mathcal{F}_\mathrm{b}}$ is negative, and the dissipation term $\epsilon$ is positive. For sinking of heavy fluid, $\tilde{\rho}w$ is negative, and $\mathcal{G}_\rho$ is positive for cooling. The reference level of the potential energy is at the surface $z = 0$, but what follows is independent of this particular choice: one must assume $\overline{\mathcal{G}_\rho} = 0$ since otherwise there would not be a steady density balance.

In steady state where $\mathrm{d}\bar{E}_\mathrm{k}/\mathrm{d}t = \mathrm{d}\bar{E}_\mathrm{p}/\mathrm{d}t = 0$, the work by external and boundary stresses, the work by mixing and convection, and the viscous dissipation must balance,

$$\overline{\boldsymbol{u}\cdot(\mathcal{F}_\mathrm{e}+\mathcal{F}_\mathrm{b})} - \overline{\epsilon} = -g\overline{z\mathcal{G}_\rho} \tag{15.7}$$

We conclude from (15.7) and the steady version of (15.6):

- In the absence of wind forcing, i. e. if $\mathcal{F}_\mathrm{e} = 0$, the left-hand side of (15.7) is negative, and hence $\overline{z\mathcal{G}_\rho}$ must be positive. This implies that heating must be at lower levels than cooling, as we have found before.
- From (15.6), we conclude from $\overline{z\mathcal{G}_\rho} > 0$, implying $\overline{\rho w} < 0$, that denser water must sink. This generates kinetic energy which overcomes friction and dissipation.

Writing the density source explicitly as sum of a convective and a diffusive term, $\mathcal{G}_\rho = C_\rho + \partial(K_\mathrm{v}\partial\tilde{\rho}/\partial z)/\partial z$, we obtain

$$\overline{z\mathcal{G}_\rho} = \overline{zC_\rho} - \overline{K_\mathrm{v}\frac{\partial\tilde{\rho}}{\partial z}}$$

The density flux at the bottom is assumed to be zero. The surface density flux, equal to $K_\mathrm{v}\partial\tilde{\rho}/\partial z$ at $z = 0$, does not contribute explicitly as an external source to this balance. The convective source term is negative at the top of convective plumes (density is taken away) and positive below (the water is made denser here). Hence $\overline{zC_\rho} < 0$, and we conclude that convection cannot drive an overturning circulation. However, the diffusion term $-K_\mathrm{v}\partial\tilde{\rho}/\partial z$ is positive in a stably stratified ocean: it can drive the overturning against friction.

Of course, there is another mechanism that can break the Sandström constraint: this is direct driving of a large-scale circulation by windstress. In order that an external stress $\mathcal{F}_\mathrm{e}$ drives the MOC, it must play a significant role in Bjerknes' circulation constraint (15.3) and in the energy budgets (15.6) and (15.7). In principle, the upwelling can also be wind-induced (Ekman pumping), but then isopycnals must outcrop at the surface as in the Southern Ocean. But still, without mixing heavy water with lighter water in rising areas or in the surface layers of MOC, the circulation system cannot be closed. Since mixing consumes turbulent mechanical energy which is

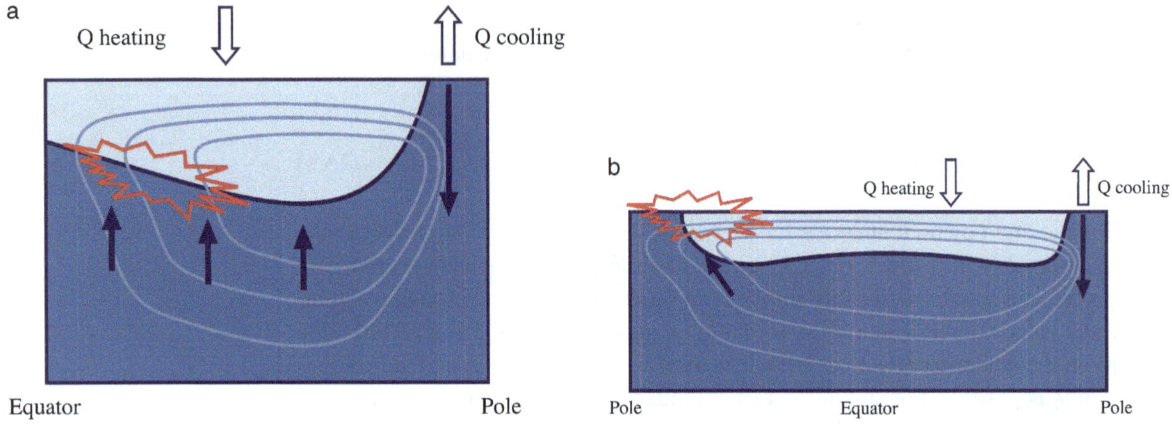


**Fig. 15.12** Two extreme scenarios for the driving of the MOC. The *black* line is the pycnocline, the blue are streamlines of the MOC. Mixing is indicated by the red star, which occurs either in the interior (**a**) or at the surface (**b**). In the former case, the interior mixing drives the MOC, while in the latter case, the windstress and breaking surface waves drive the MOC

generated by wind in the surface layer and/or by tides and breaking of internal waves in the interior, the overturning can only work on the basis of mechanical driving: the ocean thus is a mechanical engine.

One can envision a whole range of scenarios concerning where and how strongly the heaving of the cold water and the mixing occurs. One extreme case that all transfer is within the ocean thermo- or pycnocline is sketched in the left panel of Figure 15.12. Another extreme with minimum mixing in the interior is displayed in the right panel: here the cold water is advected in the deep layers into the Southern Ocean where the isopycnals rise to the surface. The water is pumped upward on isopycnals by wind and eddy forcing (see Chapter 16). Mixing then must occur in the surface layer to allow for a northward flow out of the Southern Ocean.

## 15.4 Scaling Laws for the Meridional Overturning

From a simple scaling of the relevant equations, we now try to infer the dependencies of some important properties of the MOC on externally prescribed parameters. The meridional density gradient and mixing have been highlighted as major constituents controlling the overturning. How these quantities connect is not easily understood. The following scaling analysis is based on geostrophy, i. e. on the thermal wind equation $f\,\partial v/\partial z = -g\,\partial\rho/\partial x$, and assumes $\partial v/\partial z \sim v/h$ and $\partial\rho/\partial x \sim \delta\rho/B$ where $h$ is a vertical scale (normally interpreted as the depth of the thermocline), $B$ is the zonal extent of the basin, and $\delta\rho$ is the east-west difference of the density[8]. One finds

$$v \sim \frac{g}{f}\frac{\delta\rho}{B}h \qquad (15.8)$$

[8] In the present chapter, a zonal difference of a quantity $\phi$ will generally be written as $\delta\phi$ whereas meridional scales or differences are written as $\Delta\phi$.

This relation includes the east-west density difference $\delta\rho$ as a parameter, which is, however, difficult to infer from external parameters or the forcing of the ocean. This would be easier for the north-south difference $\Delta\rho$, which could be thought as imprinted by the surface boundary conditions. Therefore, the density difference $\Delta\rho$ is often used as an independent parameter to assess the strength of the MOC, as proposed for instance by Bryan (1987). Many scaling attempts and also closure schemes for reduced models (see Section 15.5 and Section 15.6) reside on relations as $\delta\rho \sim \Delta\rho$ or $\delta\rho/B \sim \Delta\rho/L$, or $v \sim u$ in one or the other way. Note, however, that these are rather brute force assumptions. We will discuss an alternative, more consistent scaling below in Section 15.4.2 based on the Stommel–Arons model.

### 15.4.1 Conventional Scaling Attempts

A qualitative argument of how the relation $\delta\rho \sim \Delta\rho$ could be established was proposed by Colin de Verdiere (1993). Realizing that the cold waters at high latitudes are denser than those at low latitudes, we expect a decrease of sea level from the equator to the poles (which is actually observed). Note, however, that the wind-driven circulation overlays the THC, and it is questionable to what degree the observed sea surface topography is due to the THC. Associated with this pressure pattern is a zonal geostrophic surface current towards the east, which will produce upwelling at the western coast and downwelling at the eastern coast. The upwelling will bring cold (dense) water from the interior to the surface layers. A secondary high (in the east) and low (in the west) pressure system will then arise in the sea level, and then finally a secondary northward surface circulation forms – the upper branch of the MOC. Hence a *meridional* gradient $\Delta\rho$ generates a *meridional* flow in an indirect way due to the basin geometry, as sketched in Figure 15.13.

The relation $\delta\rho \sim \Delta\rho$ is also inherent to the box models of the MOC, which will be introduced below in Section 15.5, since they only resolve the meridional gradients

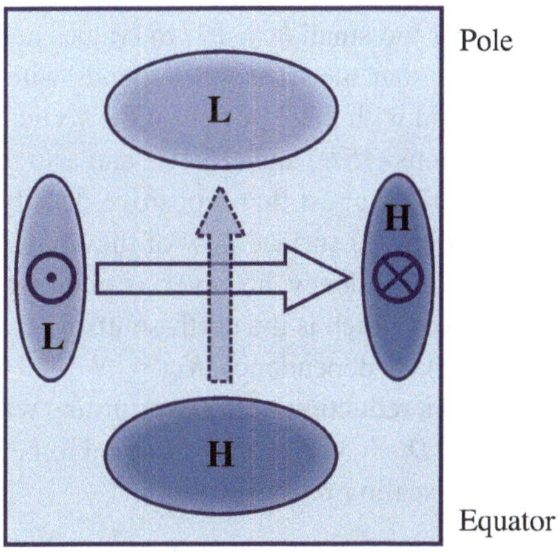


Fig. 15.13 Schematic of the surface flow driven by a north-south density gradient in an ocean basin. The primary north-south gradient – as a result of the surface forcing – is in balance with an eastward geostrophic current which generates a secondary high and low pressure system. This, in turn, drives a northward geostrophic current, the upper branch of the MOC circulation

of density and pressure. A linear relation $vhL = C\Delta\rho$ between the transport and the meridional density gradient is assumed in those models, with a constant $C$ related to friction. A justification may be taken from the linear meridional momentum balance, ignoring the Coriolis force, in which the north-south pressure gradient balances some kind of friction (see also the box on p. 507). Further, some closures for the zonal pressure gradient in zonally averaged models, as discussed in Section 15.6, are also based on the assumption $\delta\rho \sim \Delta\rho$ or $\delta p \sim \Delta p$ in one or the other way.

A further problem for the geostrophic scaling (15.8) is that the thermocline depth $h$ is to be determined together with the scales of $v$, whereas the scales for the north-south variation $L$ and the basin width $B$ can be considered as prescribed. We continue with the 'brute force' assumption and assume in addition to geostrophy an advective-diffusive density budget and also the Sverdrup relation for the scaling, i. e.

$$f\frac{\partial u}{\partial z} = g\frac{\partial \rho}{\partial y}, \quad w\frac{\partial \rho}{\partial z} = K_{\rm v}\frac{\partial^2 \rho}{\partial z^2} \quad \text{and} \quad \frac{\partial w}{\partial z} = \beta v/f$$

This yields after some manipulations (with the Earth radius $a \sim f/\beta$)

$$w \sim K_{\rm v}^{2/3}\left(\frac{g\Delta\rho}{faL}\right)^{1/3}, \quad v \sim (K_{\rm v}a)^{1/3}\left(\frac{g\Delta\rho}{fL}\right)^{2/3}, \quad h \sim \left(\frac{faLK_{\rm v}}{g\Delta\rho}\right)^{1/3} \tag{15.9}$$

and for the overturning transport

$$vhL \sim (K_{\rm v}aL)^{2/3}\left(\frac{g\Delta\rho}{f}\right)^{1/3} \tag{15.10}$$

To obtain this result, the crucial assumption $v \sim u$ (the above mentioned flaw) was made, because only three relations are given for four unknowns. Nevertheless, reasonable scales are obtained; with $\Delta\rho = 4 \times 10^{-3}\,\text{kg}\,\text{m}^{-3}$, $L = 5 \times 10^6$ m, and $K_{\rm v} = 10^{-4}\,\text{m}^2\,\text{s}^{-1}$ for the vertical diffusivity, one finds a reasonable transport $vhL = 15$ Sv, but a relatively high velocity $v = 0.015\,\text{m}\,\text{s}^{-1}$, a too small upwelling $w = 5 \times 10^{-7}\,\text{m}\,\text{s}^{-1}$, and a far too small $h = 197$ m (values in the range 500–800 m are appropriate). Note further that the transport depends with a 1/3 power on the meridional density gradient and with a 2/3 power on the vertical diffusivity.

The resulting scaling relations (15.9) and (15.10), and also the alternative scaling presented in the box on p. 507, suggest that a positive density difference between pole and equator implies a poleward surface flow of the MOC, for which $\Delta\rho$ could be considered as driver. The dependency, however, is quite weak, and the diffusivity $K_{\rm v}$ could also vary with $\Delta\rho$ (which is the north-south difference but roughly also the surface to depth variation). A dependency $K_{\rm v} \sim N^{-2} \sim 1/\Delta\rho$, as frequently discussed, in fact would yield a reduction of the overturning with increasing density difference. If the density flux $D \sim K_{\rm v}\Delta\rho/h$ is assumed to be fixed rather than the density difference, a different scaling results,

$$w \sim K_{\rm v}^{1/2}\left(\frac{gD}{faL}\right)^{1/4}, \quad v \sim \left(\frac{agD}{fL}\right)^{1/2}, \quad h \sim K_{\rm v}^{1/2}\left(\frac{faL}{gD}\right)^{1/4}$$

81. An Alternative Scaling

Ignoring the Coriolis term in the linear and steady meridional momentum balance, with vertical friction implemented, leads, after differentiation and with the hydrostatic equation,

$$-g\frac{\partial\rho}{\partial y} = A_v\frac{\partial^3 v}{\partial z^3}$$

i. e. a balance between the vertically differentiated pressure force and vertical friction which can be regarded as a vorticity balance in the $(y,z)$-plane (Marotzke et al., 1988). Using as before the advective/diffusive density balance $w\,\partial\rho/\partial z = K_v\partial^2\rho/\partial z^2$ and the Sverdrup relation $\partial w/\partial z = \beta v/f$ for scaling, one finds

$$w \sim K_v^{4/5}\left(\frac{g\Delta\rho}{LaA_v}\right)^{1/5}, \quad v \sim (K_v a)^{3/5}\left(\frac{g\Delta\rho}{A_v L}\right)^{2/5},$$

$$h \sim \left(\frac{LaK_vA_v}{g\Delta\rho}\right)^{1/5} \quad \text{and} \quad vhL \sim (K_v aL)^{4/5}\left(\frac{g\Delta\rho}{A_v}\right)^{1/5}$$

Note that in terms of the *Rayleigh number* $\mathrm{Ra} = h^5 g\Delta\rho/(aLK_vA_v)$ which physically can be viewed as the ratio of the buoyancy time scale to frictional and diffusive time scales, the third relation is equivalent to $\mathrm{Ra} \sim 1$. To obtain a reasonable size for the transport, the viscosity must be increased to a large value, e. g. $A_v = 10\,\mathrm{m^2\,s^{-1}}$ yields $vhL = 13\,\mathrm{Sv}$, $v = 0.01\,\mathrm{m\,s^{-1}}$ and $h = 240\,\mathrm{m}$. Again, the zonal momentum balance is ignored. In this physically questionable approach, the transport goes with 1/5 power of the meridional density gradient which is a very weak dependence.

and

$$vhL \sim K_v^{1/2}(aL)^{3/4}\left(\frac{gD}{f}\right)^{1/4} \tag{15.11}$$

which point towards a slightly stronger dependence of the MOC on the diffusivity, i. e. $\sim {K_v}^{1/2}$ (interestingly the meridional velocity is independent of $K_v$ in this model).

The above derived 1/3- and 1/2-power-laws between the overturning strength and the diapycnal diffusivity are confirmed by many numerical experiments with simple ocean geometry (e. g. Bryan, 1987). In more realistic models, in particular those which include a Southern Ocean, the situation is drastically different because then the overturning can be closed by transport along isopycnals and upwelling in the circumpolar area; hence little or no mixing at all is necessary in the ocean interior (see the discussion in Section 15.3.3 and Chapter 16). The upwelling is then driven by windstress instead of the interior mixing and is largely counteracted by an opposing eddy-driven overturning, with eddies arising from the baroclinic instability of the Antarctic Circumpolar Current. Mixing is only required in the surface layer where the northward flow crosses the outcropping isopycnals. In this scenario, processes in the Southern Ocean control the overturning strength, as described in Chapter 16 and in Section 15.5.5.

### 15.4.2 A Frictional Model of the Meridional Overturning

The scaling presented in Section 15.4 was based on the rather ad-hoc assumption $\delta\rho \sim \Delta\rho$ or equivalent relations and thus appears unsatisfactory. We need a better understanding of how the zonal density difference $\delta\rho$ (or pressure difference) relates to the meridional density scale $\Delta\rho$. The model discussed in this section allows to compute this relation in an almost exact way. As in the Stommel–Arons model, we consider a single ocean basin of constant zonal width $B = x_E - x_W$ and meridional extent $2L$, an idealized Atlantic Ocean which is zonally closed at all latitudes. We write the eastern boundary as $x = x_E$ and take the western one at $x_W = 0$. The linear and steady balance of momentum is written for simplicity with Rayleigh friction[9], as explained in the box on p. 495, to represent lateral eddy stresses,

$$-fv = -\frac{\partial p}{\partial x} - ru \quad \text{and} \quad fu = -\frac{\partial p}{\partial y} - rv \tag{15.12}$$

Windstress effects are ignored here but can in principle be added. We define zonal averages (across the whole basin, denoted by an overbar) of all fields, e. g. $\bar{p}(y,z) = (1/B)\int_{x_W}^{x_E} p(x,y,z)\mathrm{d}x$ for the pressure. The zonal averaged form of (15.12) is

$$-f\bar{v} = -\frac{\delta p}{B} - r\bar{u} \quad \text{and} \quad f\bar{u} = -\frac{\partial \bar{p}}{\partial y} - r\bar{v} \tag{15.13}$$

The zonal pressure difference $\delta p(y,z) = p(x_E) - p(x_W)$ appears in the zonally averaged zonal balance in (15.13). It can be related to the friction acting along the boundary of the ocean. In fact, taking the difference of the meridional balance in (15.12) between the eastern and western boundaries yields

$$\frac{\partial \delta p}{\partial y} = r\,(v_W - v_E) \tag{15.14}$$

where $v_W = v(x = x_W)$ and $v_E = v(x = x_E)$ and the vanishing of $u$ on the boundaries was used.

We proceed now as follows. As done for the Stommel–Arons model, we divide the basin into an interior frictionless regime, with variables denoted by the index i, and a western boundary layer, with variables marked by the index b and having width $B_b$. It is given by $B_b = r/\beta$ for the Stommel–Arons model, see the box on p. 497. Clearly, we expect that the contribution from the western boundary current in (15.14) is dominating over the contribution from the eastern side, i. e. $v_W \gg v_E$ in accordance with the Stommel–Arons model. The idea is now to relate the velocity $v_W$ to the basin mean $\bar{v}$ and the velocity $v_i$ in the interior domain. We denote by $\bar{v}_i$ the average of $v_i$ over the range $x = B_b$ to $x = x_E = B$ of the interior domain, and by $\bar{v}_b$ the average of $v_b$ over the range $x = 0$ to $x = B_b$ of the boundary layer domain. Then

$$B\bar{v} = B_b\bar{v}_b + B_i\bar{v}_i \tag{15.15}$$

where $B_i = B - B_b$. The approximation $B_i \simeq B$ will be used in the following.

Assuming now $v_W \simeq \bar{v}_b = B(\bar{v} - \bar{v}_i)/B_b$ according to (15.15), we obtain from (15.14)

$$\frac{\partial \delta p}{\partial y} = \beta B(\bar{v} - \bar{v}_i) \tag{15.16}$$

[9] The analysis of zonal mean equations can be carried out for scale-dependent diffusion as well and yields a similar result with $r \sim A_h/\delta_W^2$, where $\delta_W$ is the boundary layer width. It is worth mentioning that (15.13) is obtained in identical form for the viscous boundary layer width $\delta_W = (A_h/\beta)^{1/3}$.

where $B_b = r/\beta$ was used explicitly. In a second step, we eliminate the interior velocity $\bar{v}_i$, using the frictionless vorticity balance in the interior and its zonal mean,

$$\frac{\partial f v_i}{\partial y} + \frac{\partial f u_i}{\partial x} = 0 \quad \text{and} \quad \frac{\partial (f \bar{v}_i)}{\partial y} = \frac{f u^\star}{B}$$

where $u^\star(y) = u(x = B_b, y)$ is the zonal inflow into the boundary layer at its outer rim. Note that the kinematic condition $u(x = x_E) = 0$ was used at the eastern boundary. We assume that $u^\star$ is geostrophic with respect to the local pressure $p^\star(y) = p(x = B_b, y)$ on the rim of the boundary layer, i. e. $f u^\star = -\partial p^\star / \partial y$. Integrating the zonally averaged interior vorticity budget from the latitude $y = 0$ (the equator; this choice avoids an integration constant) to $y$ yields

$$\bar{v}_i = -\frac{p^\star - p^\star(y = 0)}{Bf}$$

which by implementation into (15.16) leads to the relation

$$\bar{v} = \frac{1}{\beta B} \frac{\partial \delta p}{\partial y} - \frac{p^\star - p^\star(y = 0)}{Bf} \tag{15.17}$$

Note that we have not yet made use of the zonal mean form of the momentum balance (15.13). We write the zonal balance as $\bar{v} = \bar{v}_g + \bar{v}_{ag}$, i. e. we separate between the geostrophic part $\bar{v}_g = \delta p / fB$ and the ageostrophic part $\bar{v}_{ag}$. The relation (15.17) takes then the form

$$\bar{v}_{ag} = \frac{f}{\beta} \frac{\partial}{\partial y} \frac{\delta p}{fB} - \frac{p^\star - p^\star(y = 0)}{Bf} = \frac{f}{\beta} \frac{\partial \bar{v}_g}{\partial y} - \frac{p^\star - p^\star(y = 0)}{Bf} \tag{15.18}$$

The two $\delta p$-terms have been combined into one. We may interpret these fundamental relations in two ways:

1. The equations (15.17) and (15.18) determine the total and the ageostrophic part of the meridional mean flow by the zonal pressure difference $\delta p$ and the meridional difference of the pressure $p^\star$ in the transition region between the frictionless interior and the frictional boundary layer.
2. If $\bar{v}_{ag}$ can be regarded as small (see further discussion), then (15.18) is the sought-after relation between $\delta p$ or $\bar{v}_g$ and the meridional pressure or density structure. We proceed with this option below.

The ageostrophic part is given by $\bar{v}_{ag} = r\bar{u}/f$. The zonal mean velocity has contributions from the interior and the boundary layer, like (15.15). Exploiting that the interior flow is frictionless we find

$$f \bar{v}_{ag} = r \left( B_b \bar{u}_b + B_i \bar{u}_i \right) / B = \frac{r}{B} \left( B_b \bar{u}_b + B_i \frac{r}{f} \frac{\partial \bar{p}_i}{\partial y} \right)$$

Since $\bar{p}_i \sim p^\star$ we note that the interior contribution is small compared to the pressure difference on the right-hand side of (15.18). The contribution from the boundary layer is also fairly small: compared to the first term on the right-hand side of (15.18) it is of order $r^2/f^2 \ll 1$.

In summary, a good approximation of (15.18) is

$$\frac{f}{\beta} \frac{\partial}{\partial y} \frac{\delta p}{fB} = \frac{p^\star - p^\star(y = 0)}{Bf} = \frac{f}{\beta} \frac{\partial \bar{v}_g}{\partial y} \simeq \frac{f}{\beta} \frac{\partial \bar{v}}{\partial y} \tag{15.19}$$

The validity of this relation is verified for the Stommel–Arons model in the box on p. 510. With the continuity equation $\partial\bar{v}/\partial y = -\partial\bar{w}/\partial z$, the relation (15.19) states that the *upwelling* is proportional to the meridional pressure or density difference rather than the *meridional velocity*, as inherent in the previous inconsistent scaling laws presented in Section 15.4 and in the box models discussed in Section 15.5. Note that the difference of pressure or density is not related to a gradient; in fact there is a change of sign across the equator implied by the form of (15.19).

82. Equation (15.19) for the Stommel–Arons Model

The solution of the Stommel–Arons model has been given in Section 15.2. We find for the contribution **from the interior**

$$\frac{\delta p_{\mathrm{i}}}{B} = \frac{\beta w_0}{H} y^2 \quad \text{hence} \quad \frac{f}{\beta}\frac{\partial}{\partial y}\left(\frac{\delta p_{\mathrm{i}}}{fB}\right) = \frac{w_0}{H} y$$

The pressure terms have a part from the boundary layer. Here we take the source at the pole where $T_{\mathrm{B}}(y) = -2w_0 B y$, hence $P(y) = -f T_{\mathrm{B}}/H = (\beta w_0 B/H) 2y^2$. Then the contribution **from the boundary layer** becomes

$$\frac{\delta p_{\mathrm{b}}}{B} = -\frac{P}{B} = -\frac{\beta w_0}{H} 2y^2 \quad \text{hence} \quad \frac{f}{\beta}\frac{\partial}{\partial y}\left(\frac{\delta p_{\mathrm{b}}}{fB}\right) = -\frac{w_0}{H} 2y$$

With $\delta p = \delta p_{\mathrm{i}} + \delta p_{\mathrm{b}}$ and $p^\star/fB = -w_0 y/H$, it is found that the Stommel–Arons model satisfies (15.19). This is also the case for the equatorial source.

The relation (15.19) now allows for a consistent scaling of the overturning problem (without the assumption $v \sim u$ or $v \sim \partial\rho/\partial y$, used in the previous scaling attempts). We use for the scaling the advective-diffusive density balance $w\partial\rho/\partial z = K_{\mathrm{v}}\partial^2\rho/\partial z^2$ as before, the continuity equation $\partial\bar{v}/\partial y = -\partial\bar{w}/\partial z$ and the relation (15.19) together with the hydrostatic relation $\partial p/\partial z = -g\rho$ to replace the pressure with density. We now find, with $f/\beta \sim a$ as before,

$$w \sim K_{\mathrm{v}}^{2/3}\left(\frac{g\Delta\rho}{Bfa}\right)^{1/3}, \quad v \sim K_{\mathrm{v}}^{1/3} L \left(\frac{g\Delta\rho}{Bfa}\right)^{2/3}, \quad h \sim K_{\mathrm{v}}^{1/3}\left(\frac{Bfa}{g\Delta\rho}\right)^{1/3}$$

and for the overturning transport

$$vhL \sim K_{\mathrm{v}}^{2/3} L^2 \left(\frac{g\Delta\rho}{Bfa}\right)^{1/3} \tag{15.20}$$

if the scale $\Delta\rho$ of the meridional density variation is prescribed. The dependency on $K_{\mathrm{v}}$ and $\Delta\rho$ remains as before. A major distinction to the previous scaling is, however, that $\Delta\rho$ is now the difference with respect to the equator rather than a gradient-related scale. Otherwise, the above relations differ only marginally from (15.9) and (15.10) in the placement of the length scales $L$, $B$, and $a$, i. e. we would find now deviations for long ($y$-direction) and narrow ($x$-direction) basins and vice versa. Nevertheless, within the framework of the frictional model presented in this section, the scaling is now physically consistent.

## 15.5 Box Models of the Meridional Overturning

In the present deep circulation of the Atlantic, two prominent meridional overturning cells appear. They are generally classified as thermally direct cells, meaning that the surface heat flux (and interior mixing of heat) is the dominant driving mechanism and that the haline surface fluxes are not as important. We know that there is one meridional overturning cell with sinking at the high latitudes of the northern hemisphere associated with the southward spreading of NADW and one with sinking at the high latitudes of the Southern Hemisphere associated with the northward spreading of AABW. However, paleo-climatic data indicate that as recently as 11,000 years ago, the deep circulation and the downwelling at high latitudes has been much weaker, and that there might have been even times with a collapsed or reversed overturning circulation. We will discuss in this section that there are processes which can in fact give rise to the existence of multiple steady states of the system with greatly different overturning cells and with the possibility of rapid transitions between the equilibria.

### 15.5.1 Motivation and Construction of Box Models

Stommel, in his seminal paper of 1961 (Stommel, 1961), was the first to point out that the MOC may have more than one equilibrium state, seemingly with the same surface forcing of heat and freshwater. This feature, based on simple box models of the MOC, has found ample interest in recent years in context with the ocean's role in climate change. In this section, we will present the most popular box models, namely Stommel's model (Stommel, 1961) and Welander's model (Welander, 1986). Our focus will be on the mechanisms that allow for multiple steady states in these models and on the question how symmetric (with respect to the equator) conditions of the driving can create asymmetric overturning cells.

The principle for the construction of box models for a tracer $\chi$ is outlined in Appendix B.4. Here, we need the balances of temperature and salinity for a particular set-up of boxes. For instance, for a two-box system, like Stommel's model (see below), we may take directly the set of equations (B.65), replacing $\chi_j$ by $S_j$ and $T_j$ for $j = 1, 2$, respectively. The external fluxes $F_{je}$, appearing in (B.65), have to be specified for the heat and freshwater exchange with the atmosphere, either by a prescription of the fluxes or by a restoring type of fluxes (given by (13.12) and (13.13)) or by a combination of such choices. Both choices will be discussed below. Note that the choice of fluxes for a box system should be oriented at the appropriate meridional integrals of the zonally integrated fluxes presented in Chapter 13.

The most important aspect for constructing a box model is to find an expression for the transport variable $q$ (which has the dimension $\mathrm{m}^3\,\mathrm{s}^{-1}$) in terms of the salinity values $S_j$ and the temperatures $T_j$, since these are the only variables resolved by the box system. Usually, a relation $q = C(\rho_2 - \rho_1)$ with a constant $C$ between $q$ and the density differences between the boxes is assumed. This relation is motivated in the original Stommel model as follows: the deep transport is assumed to occur in a pipe connecting the two boxes. The density difference between the boxes causes a pressure drop which accelerates the fluid against friction at the sidewalls in the pipe. The flow compensating the deep transport occurs in the upper layer and follows in its strength from mass conservation. Note that this concept, put forward by Stommel, has guided the physical interpretation of the MOC since then, including the identification of scaling laws of the MOC (see Section 15.4) and the construction of zonally averaged models. It is, however, highly problematic because of the disregard of the Coriolis force. We have discussed in Section 15.4.2 a more consistent way to motivate the scaling laws. Further, with respect to the energetics of the MOC discussed in Section 15.3.3, it is important to note that it is not the density difference between the boxes which drives the overturning in the box models, but the interior mixing within the individual boxes. In the box framework, the mixing efficiency is infinite in the sense that each box is mixed instantaneously to a homogeneous state if advection or fluxes transfer freshwater or heat through the interfaces.

Expressing the density in terms of temperature and salinity, the transport $q$ in the Stommel box model becomes

$$q = C\,[\alpha(T_1 - T_2) - \gamma(S_1 - S_2)] \tag{15.21}$$

with the constant $C$ which takes values of about $5 \times 10^9\,\mathrm{m}^3\,\mathrm{s}^{-1}$ for a reasonably realistic MOC, and with the typical values for the thermal expansion and haline contraction coefficients $\alpha$ and $\gamma$, respectively.

It will be convenient to scale the temperature and salinity budgets of the boxes, and we will work below with the scaled equations only. Consider for example the salt balance of box 1,

$$V_1 \frac{\mathrm{d}S_1}{\mathrm{d}t} = |q|(S_2 - S_1) + (K_\mathrm{h} A_{12}/D)(S_2 - S_1) + A_1 F_1 \tag{15.22}$$

as given by (B.65) from Appendix B.4 for the box configuration in Figure 15.14. Here, $V_1$ is the box volume, $A_{12}$ the area of the interface between box 1 and box 2, $A_1$ the surface area of box 1, $F_1$ the flux of salt[10] into that box through the surface, and $D$ an appropriate meridional scale for the finite difference form of the diffusion between the boxes. We now introduce scales for the temperature and salinity differences, $T_\star$ and $S_\star$, respectively. We further redefine the salinity variable to include the buoyancy ratio $b' = \alpha T_\star/(\gamma S_\star)$, thus $T' = T/T_\star$ and $S' = S/(b'S_\star)$. This scaling yields $\rho' = \rho/(\alpha T_\star)$ with a scaled density $\rho' = S' - T'$. We then take a typical horizontal diffusivity $\tilde{K}_\mathrm{h}$ and consider the scaling $q' = qD/(A_{12}\tilde{K}_\mathrm{h})$ for the transport and obtain

$$q' = C'(\rho_2' - \rho_1')$$

with the scaled constant $C' = CD\alpha T_\star/(A_{12}\tilde{K}_\mathrm{h})$. It has a typical value of 10. With $V_1 = D_1 A_{12}$ and the scaling $t' = t\tilde{K}_\mathrm{h}/D^2$ for the time, (15.22) becomes in scaled form

$$D_1' \frac{\mathrm{d}S_1'}{\mathrm{d}t'} = |q'|(S_2' - S_1') + K(S_2' - S_1') + F_1' \tag{15.23}$$

with $D_1' = D_1/D$, $K = K_\mathrm{h}/\tilde{K}_\mathrm{h}$ and $F' = F_1/(b'S_\star)DA_1/(A_{12}\tilde{K}_\mathrm{h})$. The temperature budget for box 1 and the budgets for box 2 are scaled accordingly. We omit the primes in the following.

### 15.5.2 Stommel's Box Model

We discuss in this section Stommel's box model, for which the configuration of the boxes is depicted in Figure 15.14. There are two boxes lying side by side, and

[10] Here and in the following, with 'salt flux' the equivalent salt flux is meant, as discussed on Section 2.2.2.

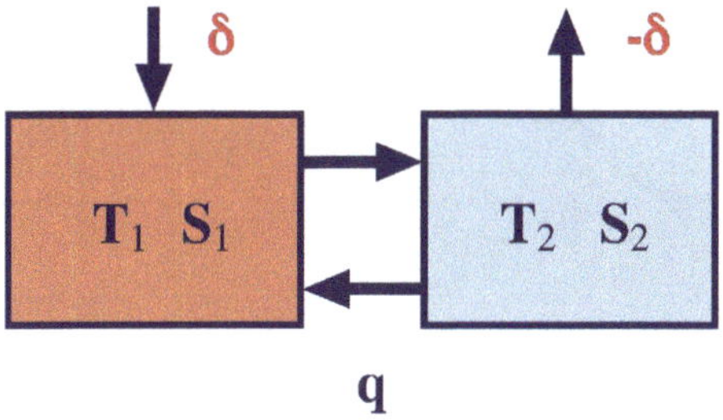


Fig. 15.14 The box model of Stommel (1961). Box 1 represents the equatorial Atlantic and box 2 the subpolar/polar Atlantic. Arrows between boxes indicate the positive direction of the transport $q$, and $\delta$ is a salt flux (a downward arrow/positive $\delta$ is a source of salt for the corresponding box)

we regard them as an equatorial box and a polar box. For simplicity, we assume $D_1 = D_2 = 1$, i. e. identical volumes of the boxes. For a complete description the salt balance (15.23) for box 1 has to be augmented by the one for box 2 and by the two corresponding heat balances of the boxes. Note that to allow for steady state solutions, the salt fluxes have to add to zero, $F_1 + F_2 = 0$, and likewise for the heat fluxes, $H_1 + H_2 = 0$, so we use $\delta = F_1 = -F_2$ and $\eta = H_1 = -H_2$ as surface flux variables. With this choice, a positive $\delta$ implies an increase of salt and likewise a positive $\eta$ an increase of temperature in box 1. It follows that the total salt content is constant, $S_1 + S_2 = S_0 = \text{const}$, and likewise the heat content $T_1 + T_2 = T_0 = \text{const}$. Hence to completely describe the dynamics of the model, it is sufficient to consider the equations for $\Delta S = S_1 - S_2$ and $\Delta T = T_1 - T_2$, given by

$$\frac{1}{2}\frac{\mathrm{d}\Delta S}{\mathrm{d}t} = \delta - |q|\Delta S - K\Delta S \quad \text{and} \quad \frac{1}{2}\frac{\mathrm{d}\Delta T}{\mathrm{d}t} = \eta - |q|\Delta T - K\Delta T \tag{15.24}$$

with the flux relation $q = -C\Delta\rho = C(\Delta T - \Delta S)$. The flux relation couples the two equations and introduces a nonlinearity to the system. The model is driven by a salt source ($\delta > 0$) or sink ($\delta < 0$) in box 1 and a compensating flux in box 2 and corresponding heat fluxes represented by $\eta$ with the same properties.

### Fixed Surface Fluxes

If the salt and heat fluxes $\delta$ and $\eta$ are prescribed, the model (15.24) can further be reduced to the balance of the density difference $\Delta\rho = \rho_1 - \rho_2$,

$$\frac{1}{2}\frac{\mathrm{d}\Delta\rho}{\mathrm{d}t} = \varphi - C|\Delta\rho|\Delta\rho - K\Delta\rho \tag{15.25}$$

which is driven by the fixed density flux $\varphi = \delta - \eta$. A positive salt flux $\delta > 0$ makes box 1 heavier and a positive heat flux $\eta > 0$ makes it lighter; box 2 behaves reversed. It is easy to show that under these 'fixed flux conditions' only a single stationary solution exists, i. e. though the model is nonlinear we do not find any multiple steady states. Note that this property carries over to the models with more than two boxes, considered later in this section. Only the introduction of mixed surface boundary conditions yields multiple steady states. In terms of the transport variable

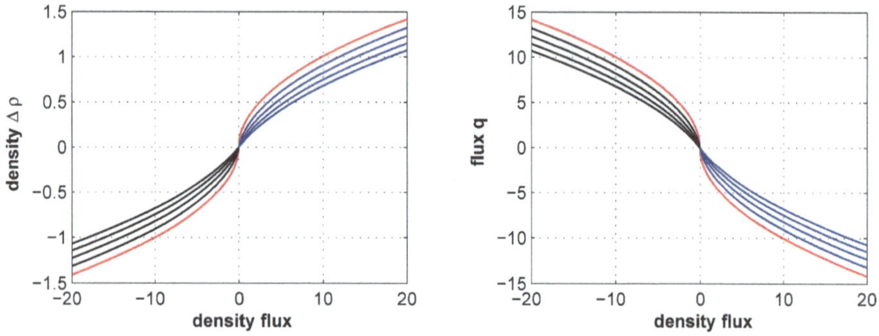


Fig. 15.15 Steady states of the Stommel box model (15.25) as a function of $\varphi$ for fixed fluxes for $C = 10$ and various values of the lateral diffusivity $K = 0, 2, 4, 6, 8$. *Blue* lines denote the branch with $q < 0$, *black* lines the branch with $q > 0$. The *red* curve is for $K = 0$

$q = -C\Delta\rho$, the steady solutions of (15.25) are

$$q = -\frac{1}{2}K + \sqrt{\frac{1}{4}K^2 - C\varphi}\,, \quad \text{for} \quad q > 0 \quad \text{and}$$
$$q = \frac{1}{2}K - \sqrt{\frac{1}{4}K^2 + C\varphi}\,, \quad \text{for} \quad q < 0 \tag{15.26}$$

and are depicted for several values of $K$ in Figure 15.15. As expected, for a negative density flux $\varphi < 0$, the circulation is poleward with sinking in box 2 and rising in box 1, since box 2 is made heavier than box 1. For $\varphi > 0$, the circulation and density difference is reversed. Increasing the lateral diffusivity $K$ leads to weaker overturning strength.

### Mixed Surface Boundary Conditions

Mixed boundary conditions are surface boundary conditions in which the salt flux $\delta$ is prescribed, but the surface heat flux is given by a restoring boundary condition as in (13.12), i. e. by a restoring to an atmospheric temperature $T_1^{\mathrm{a}}$ over box 1 for which the surface heat flux is given by $H_1 = \theta(T_1^{\mathrm{a}} - T_1)$, and likewise $H_2 = \theta(T_2^{\mathrm{a}} - T_2)$ for box 2. The total heat balance $H_1 + H_2 = 0$ required for a steady state, implies that $T_1 + T_2 = T_0 = T_1^{\mathrm{a}} + T_2^{\mathrm{a}}$. From $H_1 = -H_2 = \eta = \theta(T_1^{\mathrm{a}} - T_1) = -\theta(T_2^{\mathrm{a}} - T_2)$ we infer $2\eta = \theta(T_1^{a} - T_2^{\mathrm{a}} - \Delta T)$. The thermal balance becomes

$$\frac{1}{2}\frac{\mathrm{d}\Delta T}{\mathrm{d}t} = \eta^{\mathrm{a}} - |q|\Delta T - \left(K + \frac{1}{2}\theta\right)\Delta T \tag{15.27}$$

with $\eta^{\mathrm{a}} = \theta(T_1^{\mathrm{a}} - T_2^{\mathrm{a}})/2$. The thermal driving is thus achieved by the atmospheric temperature gradient. Note that in the present two-box system, the restoring leads to an increase of the effective diffusivity in the thermal balance. With the fixed flux condition for salt as before there is no such term in the salt balance[11], and the system given by the salt balance from (15.24) and the heat balance (15.27) can no longer be

[11] The original Stommel model has a restoring also in the salt balance (see the box on p. 519) and zero lateral diffusion. This model leads to multiple steady states only if the restoring coefficients for salt and temperature are different. We regard salt restoring (or freshwater restoring) as unphysical for the exchange at the ocean surface and, therefore, use the fixed flux condition.

expressed only by the density. Both variables – temperature and salinity – are needed. The steady states of this coupled system can, of course, be found analytically but we restrict the discussion to a numerical demonstration showing that multiple equilibria exist. Figure 15.16 displays the trajectories in the phase space $(\Delta T, \Delta S)$ for three conditions. With $\theta = 0$, the equations for salinity and temperature can be collapsed to a closed density balance, and as discussed before there is only one equilibrium solution (left panel). Taking $\theta \neq 0$, a proper choice of the forcing $\eta_a$ and $\delta$ leads to either three or one equilibrium (middle and right panels).

On the other hand, a much simpler system, showing the same principal features, is derived if the thermal balance is replaced by prescribing the temperatures in both boxes. Such an assumption can be justified by the relatively fast thermal response of the ocean to the restoring surface boundary condition compared to the much slower, entirely advective-diffusive haline response. Formally, we derive the system by letting $\theta$ approach a very large value such that $\theta/|q|$ and $\theta/K$ become large. Then we obtain $\Delta T = T_1^{\mathrm{a}} - T_2^{\mathrm{a}}$ to lowest order in the steady state. The first order balance yields the heat flux

$$\eta^{(1)} = (|q| + K)\Delta T \tag{15.28}$$

which is necessary to balance this state. The model is then governed by

$$\frac{1}{2}\frac{\mathrm{d}\Delta S}{\mathrm{d}t} = \delta - |q|\Delta S - K\Delta S \quad \text{and} \quad q = C(\Delta T - \Delta S) \tag{15.29}$$

and a fixed $\Delta T$. It is characterized by the possibility of multiple equilibrium solutions, which are outlined in the box on p. 520. Consider first the simpler case without the lateral diffusion. The steady states then follow from $\delta - |q|(\Delta T - q/C) = 0$, obtained by eliminating $\Delta S$ between the two equations in (15.29), and depending on

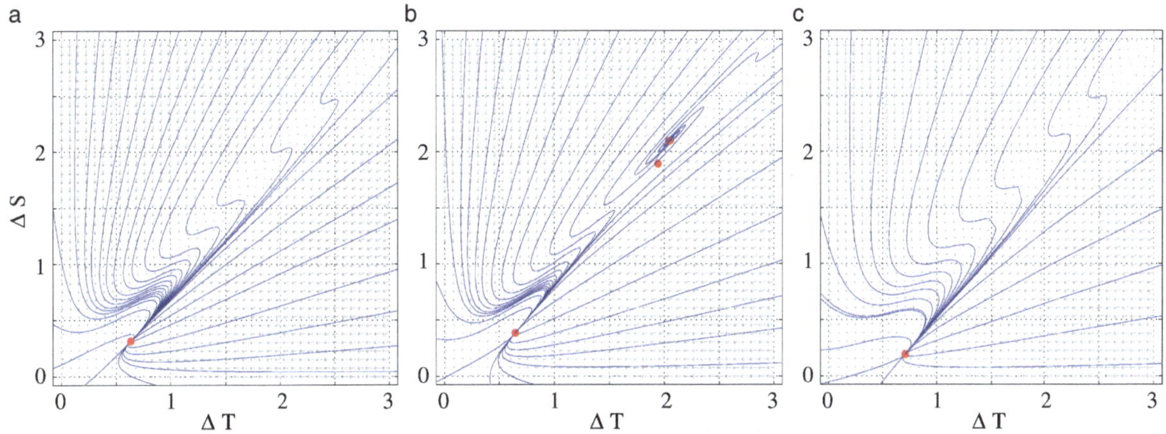


**Fig. 15.16** Phase portrays of the Stommel box model for the mixed surface conditions in the $(\Delta T, \Delta S)$-space. The model equations are (15.24) for salinity and (15.27) for temperature. Trajectories of $\Delta T(t)$ and $\Delta S(t)$ are shown as *blue* lines; arrows show the vector $(\mathrm{d}\Delta T/\mathrm{d}t, \mathrm{d}\Delta S/\mathrm{d}t)$ and steady states by *red* dots. Parameters are in panel **a** $C = 10$, $\delta = 1$, $\eta_a = 2$, $K = 0$ and $\theta = 0$, in panel **b** the same but $\theta = 1$, and in panel **c** as**b** but $\eta_a = 4$. All steady states are attractive (spiral sinks), except the middle one in panel **b** which is an unstable saddle node

the sign of $q$, two parabolic equations arise,

$$\begin{aligned} q^2 - qC\Delta T + C\delta &= 0 \quad \text{for} \quad q > 0 \\ q^2 - qC\Delta T - C\delta &= 0 \quad \text{for} \quad q < 0 \end{aligned} \tag{15.30}$$

The solutions $q(\delta)$ are easily computed or simply plotted as $\delta = \delta(q)$, as done in Figure 15.17. The magenta and green curves display the full parabolas, and the black dashed curve reflects the restrictions to $q > 0$ and $q < 0$, respectively. The intersections of the solution at $\delta = 0$ occur at $q = 0$ and at $q = C\Delta T$, the turning point on the branch of $q > 0$ is at $q = C\Delta T/2$, $\delta = C\Delta T^2/4$, and the interval $0 < \delta < C\Delta T^2/4$ is obviously the one where multiple states appear. There are three steady states in this forcing interval; they will be discussed in detail below. Note that the window of multiple states is empty if $\Delta T = 0$; the problem then collapses on a single state as previously shown for (15.25). Note further that $q > 0$ requires $\Delta T > \Delta S$: the flow is dominated by the temperature contrast. Likewise $q < 0$ requires $\Delta S > \Delta T$: the flow is dominated by the salinity contrast. We may thus speak of the *thermally driven* branch and the *haline driven* branch, respectively.

The three possible steady states can be characterized as follows:

- Solutions with $\Delta T > \Delta S$, i. e. positive $q$, are mainly driven by the thermal contrast between the boxes. There is sinking in the northern box 2; this is called the *thermally driven* solution and is equivalent to the solution $a_1$ in the box on p. 520 and the blue branch in Figure 15.18. For negative $\delta$, where haline forcing supports the thermal forcing, the thermally driven solution has the strongest MOC ($q$), and it is the only possible steady state (left panels in the figure of the box on p. 520). The thermally driven branch continues to exist for positive[12] $\delta \le C(\Delta T + K/C)^2/4$. Note that the thermally driven state with strong positive $q$ (but positive $\delta$) resembles the presently observed state of the MOC in the Atlantic Ocean.
- The solution with the highest salinity difference has a reversed circulation, i. e. a negative $q$. It is the *haline driven* solution with equatorial sinking, and

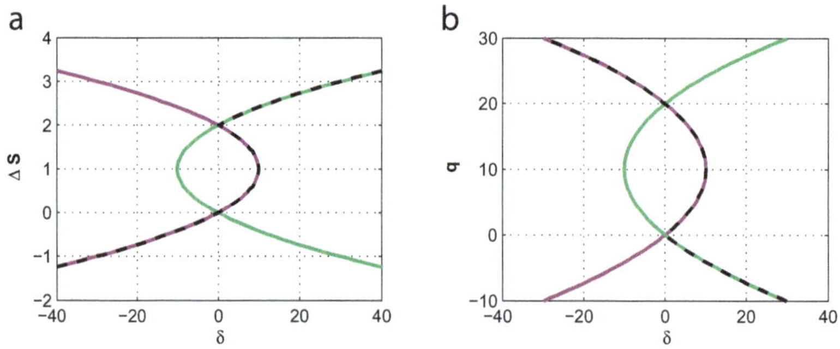


Fig. 15.17 Steady states of the Stommel model (15.29) without diffusion and $C = 10$, $\Delta T = 2$. **b** The functions $\delta(q)$ are plotted according to (15.30). The branch with $q > 0$ is *magenta*, the branch with $q < 0$ is *green*. The *black dashed line* is the solution pieced together as in (15.30). **a** Here, the corresponding functions $\delta(\Delta S)$ are shown which are obtained by eliminating $q = C(\Delta T - \Delta S)$ in the steady version of (15.29)

[12] Here, $\Delta T > K/C$ is assumed; if $\Delta T$ is below this threshold, the limiting $\delta$ becomes $K\Delta T$.

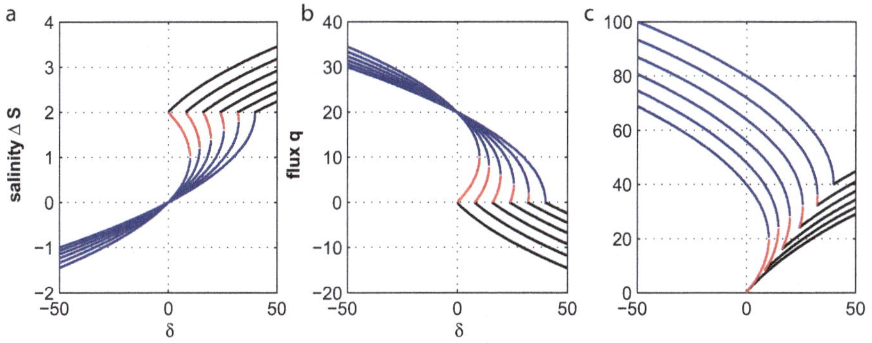


Fig. 15.18 Steady states of the Stommel box model (15.29) as a function of $\delta$ for $C = 10, \Delta T = 2$ and several values of the lateral diffusivity variable $K = 0, 2, 4, 6, 8, 10$. Panel **a** shows the salinity difference $\Delta S$ between the boxes, **b** the overturning transport $q$, and **c** the surface heat flux into box 1. *Blue* lines denote the thermally driven branch, *black* lines the haline driven branch, and *red* lines unstable steady states. The $K = 0$ curve is the one the highest amplitude in $\Delta S$ and $q$ and the smallest heat flux

it is derived from the solution $a_3$ of the box on p. 520 and the black branch in Figure 15.18. For negative $\delta$, it does not occur, but for large positive $\delta$ it is the only steady state and is characterized by a relatively weak and reversed MOC (right panels in the figure of the box on p. 520). It exists for positive $\delta > K\Delta T$ and features equatorial sinking and polar upwelling. The overturning of this solution is driven predominantly by the haline contrast (we have $\Delta S > \Delta T$ and $q < 0$), since the haline (freshwater) forcing is strong enough to overcome the thermal forcing. Note that although the present conditions in the ocean correspond to a positive $\delta$, the steady state of the MOC with equatorial sinking is not observed, i. e. the ocean appears to be in the thermally driven steady state. There is, however, speculation that the MOC might also take its possible steady state with equatorial sinking after a perturbation of the present state, e. g. in a climate change scenario.

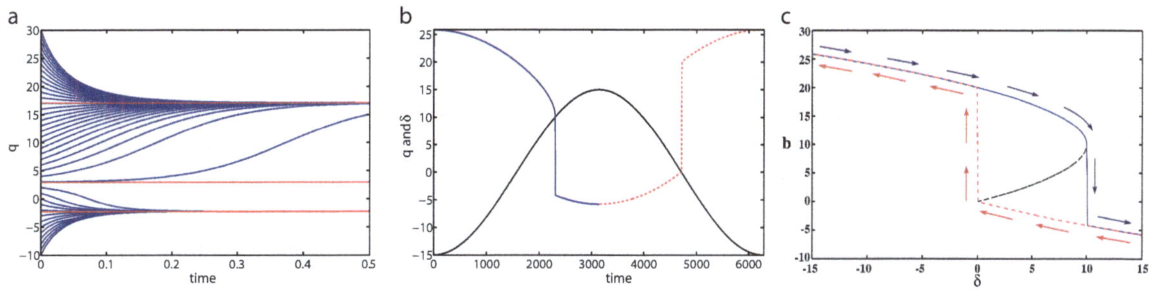


Fig. 15.19 Examples for the time-dependent behavior of the Stommel model. In **a** various initial conditions are used to demonstrate the approach towards the respective equilibrium (*red straight lines*). Here $\delta = 5$ and $C = 10$, $\Delta T = 2$. In **b** and **c** the forcing is made time-dependent as shown in the *black* curve in **b**. The increasing part of $\delta(t)$ yields the *blue* part of the quasi-steady solution $q(t)$, the decreasing part yields the *red-dashed* part. For further explanation see text. The *black* piece of the curve in **c** is the unstable branch, plotted for reference only. The *arrows* indicate the direction which the time-dependent integration follows

- At intermediate positive values of $\delta$, when $K\Delta T < \delta < C(\Delta T + K/C)^2/4$ (see footnote), we find three possible steady states (middle panels in the figure of the box on p. 520). The larger one with positive $q$ is thermally driven (it is the continuation of the thermal branch from the negative $\delta$ range to positive $\delta$), while the one with negative $q$ is haline driven. The intermediate state is also thermally driven but unstable (see below (15.31) and compare with Appendix A.2.1) and related to the solution $a_2$ of the box on p. 520. It is the red branch in Figure 15.18. Since the state is unstable, it is not likely that we will meet the system for a longer period in this state, assuming that there are always small perturbations which will drive the system away from the unstable state.
- The window of multiple states shrinks to zero for $K/C = \Delta T$, where only a single stable steady state survives.

The mathematics of the general case of (15.29) with inclusion of the diffusion term is analyzed in the box on p. 520 in terms of a generic function $f(x)$ which covers the steady states of most box models. For the Stommel model, we find correspondence with the function $f(x)$ taking $d = \delta/C$, $k = K/C$ and $\vartheta = \Delta T$. The variable $x$ and the roots $a_i$ of $f(x)$ are identified with $\Delta S$, and the strength of the advection is given by $q_i = C(\vartheta - a_i)$. The solutions of (15.29) are plotted in Figure 15.18. Hence, as before, there are at most three values of $q$ and $\Delta S$ for a fixed freshwater flux $\delta$. This is an important result since it implies that for a given freshwater flux and $\Delta T$, there could be different steady states of the MOC, which might have been realized during different climates, as suggested by the paleo-oceanographic observations.

The heat flux which has to enter box 1 in order to balance the prescribed temperature difference is shown in the right panel of Figure 15.18. It is evaluated according to (15.28). If diffusion is active, the amount $K\Delta T$ is transferred from the warm box 1 to the cold box 2 by diffusion in addition to the advective part, i. e. with diffusion more heat has to circulate in the system.

The stability condition for a steady state can be derived by considering the equation for a small perturbation $\hat{S}$ about a steady state $\Delta S$, governed by

$$\frac{\mathrm{d}\hat{S}}{\mathrm{d}t} = 2f'(\Delta S)\hat{S} \tag{15.31}$$

with $f' = \mathrm{d}f/\mathrm{d}x$. The steady state $\Delta S$ is stable for $f'$ being negative, since then the perturbation is getting smaller, and unstable for $f' > 0$ because perturbations will then grow (see also the figure of the box on p. 520).

Following (15.31), the approach (or expelling) of a perturbation towards (from) a steady state has an exponential time dependence close to the steady state. This is demonstrated in Figure 15.19a where a range of initial conditions is integrated to equilibrium for a fixed value of the forcing $\delta$. Making $\delta$ time-dependent, the most interesting behavior is that of a very slow cyclic change, leading to the hysteresis shown in the panels b and c. The trajectory starts with $\delta = -15$ (blue curve in Figure 15.19b,c) with some arbitrary initial condition, is immediately (on this long time scale) attracted to the thermally dominated equilibrium and proceeds now with increasing $\delta$ in a quasi-steady way on this branch towards the bifurcation point at $\delta = C\Delta T/2 = 10$. With still increasing $\delta$ it cannot remain on the thermal branch but is attracted to the haline equilibrium, on a fast time scale according to (15.31), where it continues to stay. At $\delta = 15$ the forcing starts decreasing, the trajectorynow follows the haline equilibrium (red dashed curve) to the bifurcation at $\delta = 0$ and jumps there back to the thermal branch. After completing the cycle in the forcing, the trajectory $(\delta(t), q(t))$ forms the hysteresis shown in Figure 15.19c.

**83. Stommel's Original Box Model**

The box model considered originally by Stommel (1961) has restoring conditions for temperature and salinity. Let us thus study

$$\frac{1}{2}\frac{\mathrm{d}\Delta T}{\mathrm{d}t} = \eta^{\mathrm{a}} - |q|\Delta T - \left(K + \frac{1}{2}\theta_{\mathrm{T}}\right)\Delta T \quad \text{and}$$

$$\frac{1}{2}\frac{\mathrm{d}\Delta S}{\mathrm{d}t} = \delta^{\mathrm{a}} - |q|\Delta S - \left(K + \frac{1}{2}\theta_{\mathrm{S}}\right)\Delta S$$

If the salinity restoring is meaningful at all, the corresponding coefficient must be small, i.e. $\theta_{\mathrm{S}} \ll \theta_{\mathrm{T}}$. With $\Delta T = x\lambda_{\mathrm{T}}$, $\Delta S = y\lambda_{\mathrm{T}}$ where $\lambda_{\mathrm{T}} = K + \theta_{\mathrm{T}}/2$, $\lambda_{\mathrm{S}} = K + \theta_{\mathrm{S}}/2$ we obtain a scaled version of the equations, given by

$$\dot{x} = \eta_a - |x - y|x - x \quad \text{and} \quad \dot{y} = \delta_a - |x - y|y - \lambda y$$

where the forcing parameters are written in the form $\eta_a = \eta^{\mathrm{a}}/C\lambda_{\mathrm{T}}^2$, $\delta_a = \delta^{\mathrm{a}}/C\lambda_T^2$, $\lambda = \lambda_{\mathrm{S}}/\lambda_{\mathrm{T}}$ and $2C\lambda_{\mathrm{T}}$ is used as dimensionless time unit. The system leads to a cubic problem for the steady state solution. Temperature and salinity are expressed by $x = \eta_a/(1 + |p|)$ and $y = \delta_a/(\lambda + |p|)$, respectively, in terms of the transport (density) variable $p = x - y$ which follows from

$$p = \frac{\eta_a}{1 + |p|} - \frac{\delta_a}{\lambda + |p|} \quad \text{or} \quad p(1 + |p|)(\lambda + |p|) = \eta_a(\lambda + |p|) - \delta_a(1 + |p|)$$

For the case $\lambda = 1$ (same restoring coefficients for temperature and salinity), only one solution is obtained for any combination of forcing values: as for the case of fixed fluxes, discussed above, the problem collapses on a single equation for density with a unique solution. This case and other values for $\lambda$ are considered in the figure below where $x, y, p$ are plotted as function of $\delta_a$. A window with three steady states emerges and, as before, the thermally dominated and the haline driven state can be identified as in the case of mixed conditions. Only the branches with negative slope in $p(\delta_a)$ are stable.

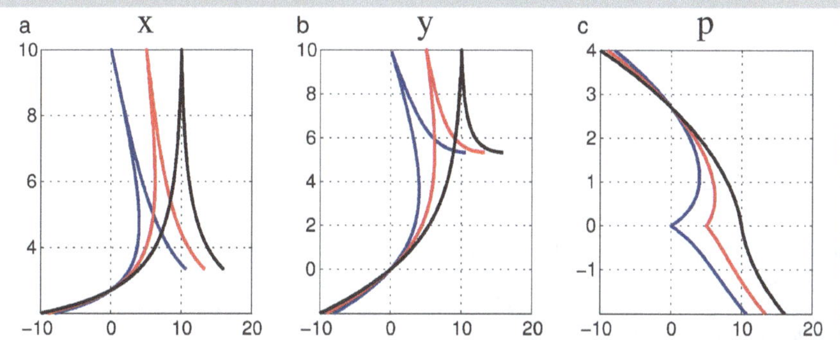


*Steady state solutions of Stommel's original box model with restoring boundary conditions. The steady states of **a** x (temperature), **b** y (salinity), and **c** p (transport) are shown as a function of $\delta_a$ for $\eta_a = 10$ and several values of the coefficient $\lambda = 0, 0.5, 1.0$ (the latter in black; it refers to the right most of the curves where only a single steady state is present).*

**84. A Generic Non-linearity of Box Models**

Stommel's model (and also Welander's model; see next section) is governed by a unique function of salinity contrast. It is easily extracted from the governing equations (15.29) or (15.32) and given by

$$f(x) = d - |\vartheta - x|x - kx$$

with constant coefficients $d$, $\vartheta$ and $k$. The function $f(x)$ is important for both Stommel's and Welander's box model, since it determines the time development of the respective box variables, the roots of the function determine the steady states, and the derivative of $f$ determines their stability. Depending on $\vartheta$ and $d$ the function $f(x)$ has three roots, given by

$$a_{1,2} = \frac{1}{2}\vartheta_+ \left(1 \pm \sqrt{1 - \frac{4d}{\vartheta_+^2}}\right), \quad a_3 = \frac{1}{2}\vartheta_- \left(1 + \sqrt{1 + \frac{4d}{\vartheta_-^2}}\right)$$

with $\vartheta_\pm = \vartheta \pm k$. It is clear that only real values $a_{1,2} \leq \vartheta$ and $a_3 \geq \vartheta$ are admissible. The function $f(x)$ and its roots are displayed in the figure shown below. There is always one solution, and at most there are three. Bifurcations deriving from $f(x)$ are of the saddle-node type (compare Appendix A.2.1). Furthermore, the potential $V(x) = -\int^x f(x')\mathrm{d}x'$ with $\mathrm{d}V/\mathrm{d}x = -f$ is shown in the figure. The roots of $f$ correspond then to extrema of $V$. If the dynamical system is governed by $\dot{x} = f(x) = -\mathrm{d}V/\mathrm{d}x$, then $\dot{V} = \dot{x}\mathrm{d}V/\mathrm{d}x = -(\dot{x})^2 = -(\mathrm{d}V/\mathrm{d}x)^2 \leq 0$, showing that $V(x(t))$ always decreases until $x$ tends to a steady state where $f = -\mathrm{d}V/\mathrm{d}x = 0$.

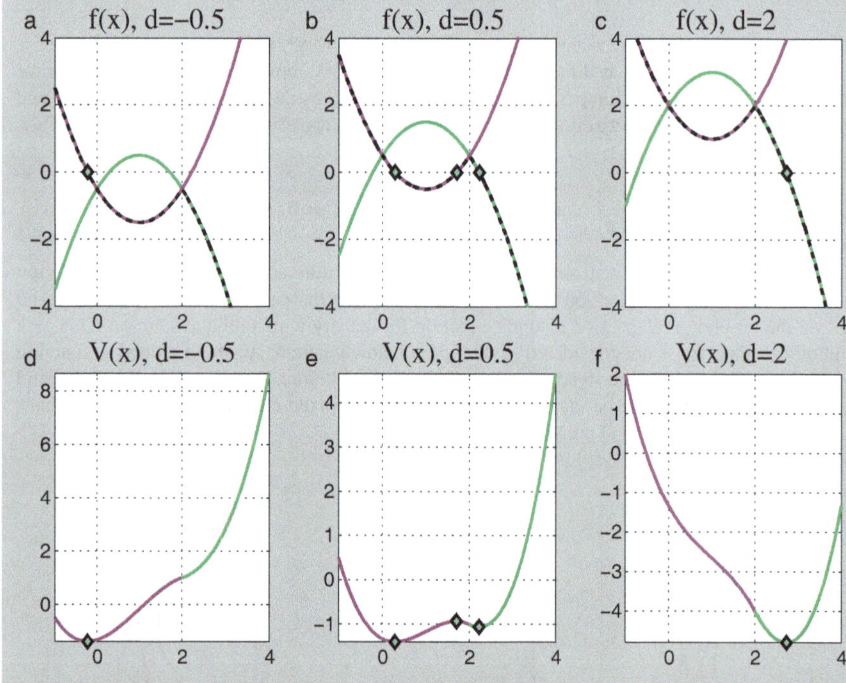


***a–c** The function $f(x)$ (*dashed black curve*; the* green *and* magenta *curves are quadratic extensions, shown for $k = 0$ and $\vartheta = 2$. **a**, **d**: $d = -0.5$, **b**, **e**: $d = 0.5$, **c**, **f**: $d = 2$.) **d–f** The potentials $V(x)$ corresponding to each case of $f(x)$. The roots of $f(x)$ and minima of $V(x)$ are indicated by* diamonds.

### 15.5.3 Welander's Box Model

We present now a double-hemispheric box model which was designed and discussed by WELANDER[13], aiming at the overturning in a single ocean as the Atlantic (Welander, 1986). For simplification, the model has, as before, an inactive temperature, i. e. the temperature differences between the boxes are prescribed as thermal forcing, with the same reasoning as before. Furthermore, the salinity flux is assumed to be prescribed; we are thus using mixed boundary conditions.

The set-up of the boxes is depicted in Figure 15.20. At the interfaces of the boxes, salinity and heat are exchanged, as discussed in Appendix B.4 and the above Section 15.5.1. The equations for the salt budgets of box 1 and box 3 are given by

$$
\begin{aligned}
D_1 \frac{\mathrm{d}S_1}{\mathrm{d}t} &= -F + \delta + |q_1|(S_2 - S_1) + K(S_2 - S_1) \\
D_3 \frac{\mathrm{d}S_3}{\mathrm{d}t} &= -F - \delta + |q_3|(S_2 - S_3) + K(S_2 - S_3)
\end{aligned}
\qquad (15.32)
$$

The advective flow between the boxes is parameterized as in the relation (15.21) and is given by

$$
q_i = C\,[T_2 - T_i - (S_2 - S_i)] \quad \text{for} \quad i = 1, 3
$$

with fixed values $T_2 - T_i$. Note the different directions of the flow rates $q_1$ and $q_3$, as indicated in Figure 15.20. The box dimensions are chosen for convenience as $D_1 = D_3 = 1 - 1/\sqrt{3} = D$ and $D_2 = 2/\sqrt{3} = 2(1 - D)$. The total amount of salt, given by $2S_0$, is again conserved, which allows to use

$$
DS_1 + 2(1 - D)S_2 + DS_3 = 2S_0 = \text{const} \qquad (15.33)
$$

instead of the prognostic salt balance of box 2. The model is driven by the salt fluxes $F$ and $\delta$, representing symmetric and asymmetric (with respect to the equator) contributions to the net salt flux as indicated in Figure 15.20. Furthermore, the equatorial-to-pole temperature difference $\Delta T_{\mathrm{ep}} = T_2 - T_1 = T_2 - T_3$ is assumed equal for both hemispheres (implying $T_1 = T_3$). The only asymmetry in the set-up is then given by $\delta$. In a realistic forcing scenario, $\delta$ should be small compared to the symmetric flux pattern given by $F$.

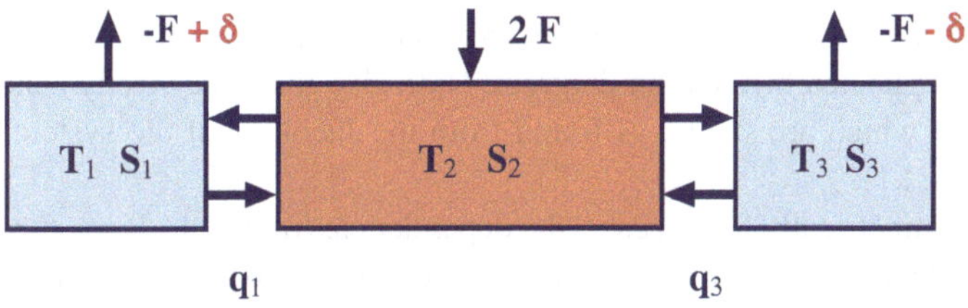


Fig. 15.20 The box model of Welander (1986). Arrows between boxes indicate the positive direction of the fluxes $q_i$ for $i = 1, 3$, and $F$ and $\delta$ denote the symmetric and asymmetric components of the salt flux

[13] PIERRE WELANDER, *1925 in Nice, †1996 in Seattle, oceanographer.

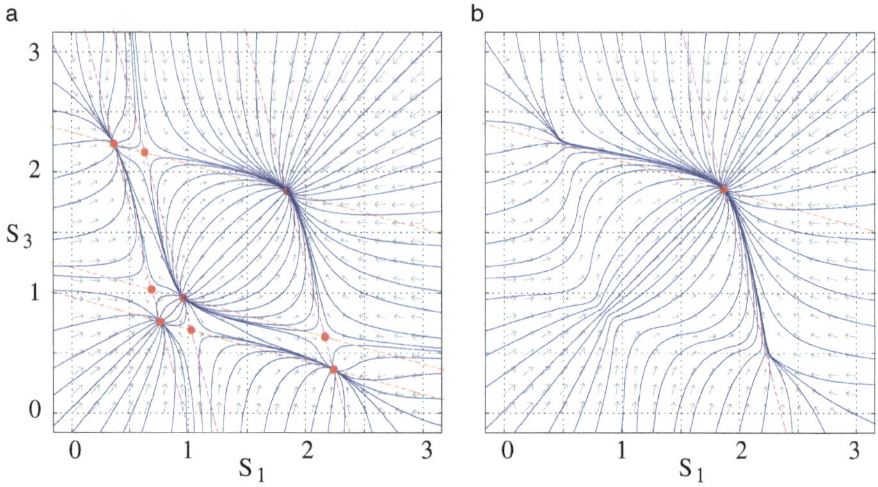


Fig. 15.21 Phase portrays for Welander's three-box model (15.32) for symmetric forcing ($\delta = 0$). Steady states are denoted by *red dots*, trajectories of $S_1(t)$ and $S_3(t)$ by *blue lines*, and *arrows* show the vector $(\mathrm{d}S_1/\mathrm{d}t, \mathrm{d}S_3/\mathrm{d}t)$. **a** is for $K = 0.85$, **b** for $K = 3$, while $C = 10$, $S_0 = 2$, $F = 5$ and $\Delta T_{\mathrm{ep}} = 2$. The four outermost stable solutions in the left picture are spiral sinks, the middle point is a spiral source, and the remaining four points are unstable saddle nodes (compare Appendix A.2.1). For the $K = 3$ case, only one solution survives; it is a spiral sink

Typical phase portrays of the system (15.32) are shown in Figure 15.21, indicating the possibility of nine steady states. To obtain the steady states for the Welander model, we may use the roots of $f(x)$, introduced in the box on p. 520, and solve the linear system of equations

$$S_2 - S_1 = a_i \ , \quad i = 1,2,3 \ , \qquad S_2 - S_3 = a_j \ , \quad j = 1,2,3$$

together with (15.33) for $\epsilon = \Delta T_{\mathrm{ep}}$, $k = K/C$. In $S_2 - S_1$, we have to put $d = d_- = (F - \delta)/C$, and in $S_2 - S_3$ we have to put $d = d_+ = (F + \delta)/C$. These equations define the nine (or less) solutions. The thin magenta straight lines in Figure 15.21 are those given by the above linear relations.

A bifurcation diagram of the system (15.32) (see also Appendix A.2.1) is shown in Figure 15.22, i.e. the steady states as a function of a control parameter, here taken as the strength of the symmetric forcing $F$, while the asymmetric part is set for simplicity to zero, $\delta = 0$ (the impact of the asymmetric forcing $\delta \neq 0$ is discussed in the box on p. 524). It becomes obvious that some of the solutions cannot be differentiated from each other in terms of the fluxes $q_i$ and in terms of the salinity $S_2$ of the middle box, but they can be distinguished in terms of $S_1$ and $S_3$, showing a rather complicated bifurcation diagram. Note also that nine steady states are visible, four of which are stable, as found from the Jacobian of the system (compare Appendix A.2.1).

In the symmetrically forced case, the dependencies of $S_1$ and $S_3$ on the forcing $F$ are identical and likewise the dependency of $q_1$ and $q_3$, but this does not mean that the individual steady states have this correspondence (see also the discussion below): in the range where multiple steady states exist, solutions may have e.g. a positive $q_1$ and a negative $q_3$. It is instructive to look at variables which show the underlying symmetry occurring in a symmetrically forced experiment: $q_1 - q_3$ (also shown in Figure 15.22) is zero for symmetric overturning, i.e. there is polar up- or down-

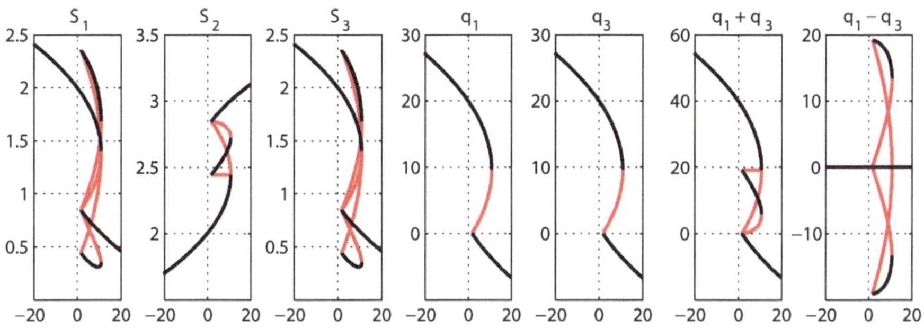


Fig. 15.22 A bifurcation diagram for Welander's three-box model for symmetric forcing, i. e. steady states as a function of the symmetric freshwater forcing $F$, with $\delta = 0$ and $C = 10$, $S_0 = 2$, $K = 0.85$ and $\Delta T_{\text{ep}} = 2$. *Black* lines denote stable steady states while *red* lines indicate unstable ones

welling in both hemispheres and equatorial down- or upwelling, i. e. a two-cell circulation; $q_1 + q_3$ (also shown in Figure 15.22), on the other hand, would be zero for a single-cell overturning including all three boxes, with equal strength of the circulation in both hemispheres. Such a single-cell circulation would correspond to the presently observed MOC in the Atlantic Ocean. Note also that such a state cannot be realized (as a stable state) in the present box model with symmetric temperature forcing. However, there are single-cell circulations possible with e. g. northward flow in both hemispheres, $q_1 > 0$, $q_3 < 0$, but $q_1 > -q_3$. Hence there is still some corresponding equatorial upwelling.

As before in Stommel's box model, the steady solutions of Welander's box model may be classified in an obvious way as predominantly thermally or haline driven states, but here also more complicated states appear. The four stable solutions shown in Figure 15.22 can be classified as follows:

- A two-cell circulation with strong poleward flow in both hemispheres and equatorial upwelling which is thermally driven. Both $q_i > 0$, lying on the upper branch of $q_1 + q_3$ and on the zero branch of $q_1 - q_3$.
- A much weaker two-cell circulation with equatorward flow in both hemispheres and equatorial downwelling which is haline driven. Both $q_i < 0$, lying on the lower branch of $q_1 + q_3$ and on the zero branch of $q_1 - q_3$.
- A single-cell circulation with strong northward flow in the northern hemisphere and weak northward flow in the southern hemisphere ($q_1 > -q_3 > 0$, lying on the upper branch of $q_1 - q_3$), and sinking in the north.
- A single-cell circulation with strong southward flow in the southern hemisphere and weak southward flow in the northern hemisphere ($q_3 > -q_1 > 0$, lying on the lower branch of $q_1 - q_3$), and sinking in the south.

The last two solutions are of particular interest, since they resemble the present asymmetric configuration of the MOC in the Atlantic Ocean. In terms of $(q_1 + q_3)$, shown in Figure 15.22, both single-cell solutions lie on the middle stable curve. They clearly exist only for $F > 0$. Figure 15.23 shows the single-cell solution with northern sinking (in blue) and southern sinking (in green) separately, to demonstrate that both have asymmetric salinities and transports in a seemingly symmetrically forced state (with respect to temperatures and salt fluxes), but the surface heat fluxes, computed from $H_i = (|q_i| + K)\Delta T_{\text{ep}}$ for $i = 1, 3$ (also shown in Figure 15.22), differ. The

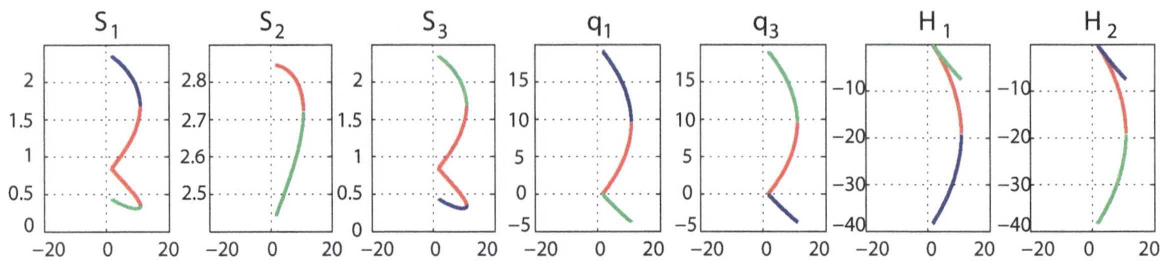


Fig. 15.23 Single-cell solutions in Welander's three-box model for symmetric forcing ($\delta = 0$) as a function of the symmetric freshwater forcing $F$ for $C = 10$, $S_0 = 2$, $K = 0.85$ and $\Delta T_{ep} = 2$. *Blue* lines denote stable single-cell solutions with northern sinking, *green* lines denote stable single-cell solutions with southern sinking, and *red* lines denote unstable solutions. For $S_2$, *green* and *blue* segments overlie each other

forcing of the respective states is thus asymmetric. In the single-cell state with northern sinking, heat is transported vigorously from box 2 to box 1 (this circulation is in the thermally driven mode) but only little heat is transported from box 2 to box 3 (this circulation is in the haline driven mode), and it needs a large heat release to the atmosphere over box 1, much less than over box 3 (vice versa for the single-cell state with southern sinking).

The single-cell solutions, which we consider as relevant for the present MOC in the Atlantic Ocean, can only exist in the limited $F$ range where the function $f(x)$ of the box on p. 520 has three roots. The overlapping range depends on the diffusivity $K$ and the asymmetric forcing $\delta$. The window can be diminished up to the point where overlapping ceases, and the single-cell solutions have to break down to give way

**85. Breaking the Symmetry in Welander's Box Model**

The steady states of Welander's three-box model can be deformed considerably by the implementation of lateral diffusion described by the $K$-terms. As for Stommel's box model, an increase of $K$ shifts the window of multiple steady states towards larger $F$ and shrinks it to a smaller range up to the point where the window closes at a critical $K$. Multiple steady states then do not exist any longer. On this route, the symmetry of the solutions remains valid. More important is the influence of the asymmetric forcing $\delta$, because it breaks the symmetry immediately. Due to the 'modulus advection' in the box models, the fundamental bifurcations are not generic pitchfork bifurcations (see Appendix A.2.1), but they behave similarly when asymmetric forcing is introduced. As demonstrated in the figure below, the asymmetric forcing leads to breaking up the bifurcations on the zero line $q_1 = q_3$, creating an isolated branch which, moreover, shrinks to nothing, if $\delta$ exceeds a critical value.

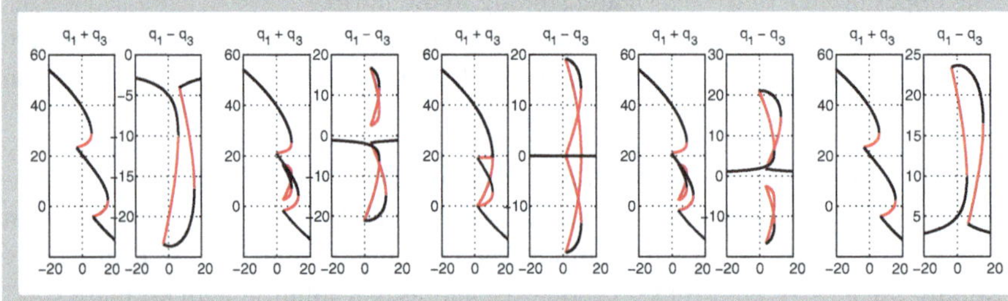


*Steady states of Welander's three-box model as a function of the symmetric freshwater forcing strength $F$ for $C = 10$, $S_0 = 2$, $K = 0.85$, $\Delta T_{ep} = 2$ and $\delta = -5, -2, 0, 2, 5$ in the five double panels from left to right, showing only $q_1 + q_3$ and $q_1 - q_3$.* Black *lines denote stable steady states,* red *lines unstable ones.*

to a two-cell state. Increasing diffusion leads to a smaller range of the existence for single-cell solutions. Likewise, negative $\delta$ narrows the overlapping range while a positive $\delta$ widens the range.

There are obvious deficits in the box model approach presented so far. One problem is that of resolution: the box models discussed above are just barely resolving the most important aspects of the global MOC, i. e. polar up/downwelling forced by an approximately symmetric (with respect to the equator) surface freshwater flux. Clearly, we can add more and more boxes to gain a better resolution. The Welander model is readily expanded to a multibox model of the Atlantic- Pacific system, which has a truly global MOC. The second deficit of the box models in general is the hydraulic pipe law (15.21), since it does not account for the Coriolis force, which we know is important for the large-scale circulation of the ocean. In fact, it turns out that the second deficit yields an inconsistency between the box models and more dynamically complete models like the Stommel–Arons model of Section 15.2. This inconsistency will be discussed in the next section. A third problem of Stommel's and Welander's box model is absence of wind forcing. This deficit is remedied in Section 15.5.5.

### 15.5.4 An Inconsistency of the Box Models

The law (15.21), which was used as the closure for the meridional transport in the box models, is inconsistent with the large-scale ocean dynamics. This was first noted by Straub (1996). To demonstrate this inconsistency, Figure 15.24 shows the result of an integration of the linearized version of the two-layer model discussed in the box on p. 527,

$$
\begin{aligned}
\frac{\partial u}{\partial t} - f v &= -g^* \frac{\partial \xi}{\partial x} - r u \\
\frac{\partial v}{\partial t} + f u &= -g^* \frac{\partial \xi}{\partial y} - r v \\
\frac{\partial \xi}{\partial t} + H \left( \frac{\partial v}{\partial y} + \frac{\partial u}{\partial x} \right) &= Q - \lambda \xi
\end{aligned}
$$

where $\xi$ denotes the thickness variations of the lower layer of a two-layer ocean with mean lower thickness $H$, and with a density difference between the two layers $\Delta\rho$ represented by the reduced gravity $g^* = g\Delta\rho/\rho_0$. The velocities $u$ and $v$ denote the differences between the upper and lower layer velocities.

The model is almost identical to the Stommel–Arons model discussed in Section 15.2. The prescribed deepwater source in the lower layer is denoted by $Q$. The interior upwelling is parameterized by the term $-\lambda\xi$ in the thickness balance of the lower layer. It represents a parameterization for the effect of diapycnal diffusion on the lower layer thickness. In the linearized momentum balance, friction by subgrid-scale processes is represented by Rayleigh friction with coefficient $r$. For the experiments shown in the following, we have used $r = 2 \times 10^{-6}\,\mathrm{s}^{-1}$, $\lambda = 10^{-9}\,\mathrm{s}^{-1}$, $g^* = 0.02\,\mathrm{m\,s}^{-2}$, and $H = 400\,\mathrm{m}$. The system is integrated on an equatorial $\beta$ plane with $\beta = 2.3 \times 10^{-11}\,\mathrm{m}^{-1}\,\mathrm{s}^{-1}$, and the horizontal resolution is 20 km in the zonal and meridional direction. This layer model will be referred to as LM.

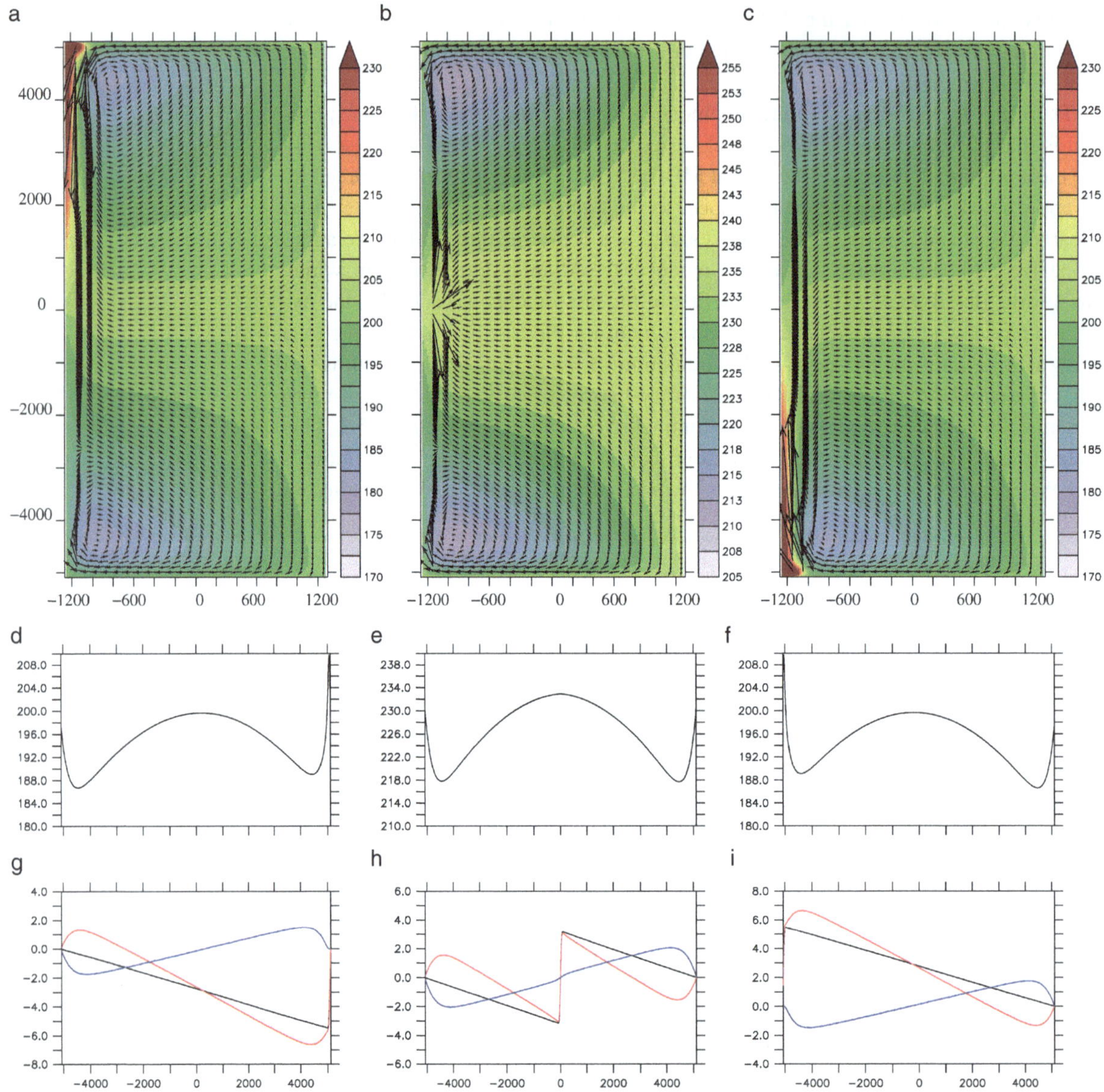

**Fig. 15.24** Numerical simulation of the layer model after 160 years of integration. **a** Layer thickness $\xi$ in m (*contours*) and velocity (*arrows*) in m s$^{-1}$ with the deepwater source $Q$ localized in the north-western corner. **b** Same as **a** but with the deepwater source $Q$ located at the equator $y = 0$ on the western side. **c** Same as **a** but with $Q$ at the south-western corner. **d** Zonally averaged layer thickness $\bar{\xi}$ in m for the experiment shown in **a**. **e** Same as **d** but for the equatorial source. **f** Same as **e** but for the southern source. **g** Total meridional transport in Sv (*black*), transport in the western boundary layer (*red*) and transport in the interior (*blue*) for the experiment shown in **a**. **h** Same as **g**, but for the equatorial source. **i** Same as **g**, but for the southern source

To demonstrate the influence of the transport and the pressure field on the location of the deep water source, $Q$ was placed at three different locations in three different experiments. The results are shown in Figure 15.24a–c where the location of the deepwater source was placed at the north-western corner of the model

**86. Linearized Reduced Gravity Model**

The layer model of Appendix B.1.2 is given in a linearized version by

$$\frac{\partial u_i}{\partial t} - f v_i = -\frac{\partial p_i}{\partial x} + F_i^{(u)}, \quad \frac{\partial v_i}{\partial t} + f u_i = -\frac{\partial p_i}{\partial y} + F_i^{(v)}$$

$$\frac{\partial d_i'}{\partial t} + H_i \left( \frac{\partial v_i}{\partial y} + \frac{\partial u_i}{\partial x} \right) = 0$$

We consider a system with two layers where $i = 1, 2$, and where the layer thickness $d_i$ was decomposed into a large mean value $H_i$ and a perturbation $d_i'$ which was assumed to be small. The upper layer pressure is given by $p_1 = g\zeta$ and the lower layer pressure by $p_2 = g\zeta + g\xi\Delta\rho/\rho_0$, where $\zeta$ is the surface elevation, $\xi$ the interface elevation, and $\Delta\rho/\rho_0$ the (scaled) density difference between the layers. Written in terms of the velocity difference $u' = u_2 - u_1$ and $v' = v_2 - v_1$, we obtain

$$\frac{\partial u'}{\partial t} - f v' = -g^* \frac{\partial \xi}{\partial x} + F^{(u')}, \quad \frac{\partial v'}{\partial t} + f u' = -g^* \frac{\partial \xi}{\partial y} + F^{(v')}$$

with the reduced gravity $g^* = g\Delta\rho/\rho_0$. The upper layer and lower layer thickness are given by $d_1' = \zeta - \xi$ and $d_2' = \xi$, such that the two layer thickness balances are combined as

$$\frac{\partial \xi}{\partial t} - \frac{1}{2}\frac{H_2}{H_1 + H_2}\frac{\partial \zeta}{\partial t} + \frac{1}{2}\frac{H_1 H_2}{H_1 + H_2}\left( \frac{\partial v'}{\partial y} + \frac{\partial u'}{\partial x} \right) = 0$$

Since the surface elevation $\zeta$ is much smaller than the interface displacement, the term related to $\partial\zeta/\partial t$ is usually neglected.

domain, at the equator at the western boundary and at the south-western corner of the model domain. As in the Stommel–Arons model, two dynamically different regimes develop, i. e. a narrow western boundary layer with strong meridional flow and an weak interior flow, which is characterized by uniform upwelling. In the interior, the velocity field and the thickness contours are almost identical in all three cases, and the meridional interior transport is always poleward. This is because in the interior the Sverdrup balance

$$\beta v = -f\left( \frac{\partial u}{\partial x} + \frac{\partial v}{\partial y} \right) = \frac{f}{H}\left( \frac{\partial}{\partial t} + \lambda \right)\xi$$

holds to a very good approximation. We also need to know that $\xi$ is related to $Q$ only in an integral sense, i. e.

$$\lambda \int \xi \mathrm{d}x\mathrm{d}y = \int Q \mathrm{d}x\mathrm{d}y \tag{15.34}$$

in the integral over the whole model domain; however, the right-hand side of (15.34) has only contributions from the western boundary region. The meridional interior transport is driven by the interior upwelling, which is almost identical in each case, i. e. uniform and of similar magnitude. The differences between the experiments can only be seen in the western boundary current which has to balance, on the one hand, the interior flow and the upwelling (in all three cases similar) but, on the other hand, the different inflows of the deep water source $Q$, as discussed in Section 15.2.

Figure 15.24 also shows the zonal averages of the thickness $\xi$ and the meridional transport in the lower layer, i. e. the MOC in the two-layer model. It is clear that the location and strength of the deep water source $Q$ in the thickness budget determines the transport in the lower layer, i. e. when $Q$ is located at the northern-western corner, the total meridional transport is southward in both hemispheres of the domain, see Figure 15.24g (solid line), and it is everywhere northward for a deepwater source $Q$ located at the south-western corner of the domain, see Figure 15.24i, while the total transport is polewards in both hemispheres for an equatorial source, see Figure 15.24h. The magnitude of the source $Q$ drives a transport of about 5 Sv in the vicinity of the source in each case, which linearly reduces due to interior constant upwelling into the upper layer with increasing distance to the source.

The transport in the western boundary layer, also shown in Figure 15.24, is of similar magnitude as the total transport, although the western boundary layer is much smaller than the total width of the basin. It is also of the same direction as the total transport, except for the region $y < -2{,}500$ km, $|y| > 2{,}500$ km and $y > 2{,}500$ km for the experiment with northern, equatorial, and southern source, respectively, where it opposes the total transport. The interior meridional transport, on the other hand, is poleward and identical in each experiment, as already discussed.

Since the western boundary layer is small compared to the width of the interior region, the zonal integral of $\xi$ is dominated by the interior profile to a large extent, and thus the zonally integrated $\xi$ becomes independent of the location of the deep water source $Q$. The result is that $\xi$, and in particular the meridional gradient of $\xi$, becomes independent of the location of $Q$ and thus of the sign and strength of the meridional transport. This is in contrast to Stommel's law (15.21), where the meridional transport between the two boxes is parameterized by the meridional density difference between the boxes.

As a consequence, the parameterization in Stommel's box model is not consistent with the Stommel–Arons model. This inconsistency is transferred to the other box models, which we have discussed in Section 15.5. Moreover, the scaling laws, introduced earlier in Section 15.4, also suffer in this respect (but not the one in Section 15.4.2).

The independence of the meridional gradient of the zonally averaged thickness from the meridional transport is not specific to layered models but is also found in a primitive equation model (PEM) with a configuration similar to that of LM. In the present PEM, we have neglected momentum advection (as before), and, for simplicity, the only tracer is temperature. The model domain is identical to the layered model, but there are 20 vertical levels of 50 m thickness, such that the domain is 1,000 m deep. PEM is forced by relaxation of temperature in the uppermost grid box towards a target temperature, which is zonally and meridionally uniform except for a small region of meridional width $r/\beta$ (equivalent to the western boundary layer width) at the northern or equatorial region with a 3 K smaller target temperature. This way, a northern or equatorial deepwater formation region is introduced as in the layered model. A case with southern source is just a mirror of the one with northern source and, therefore, not further discussed. The time scale of relaxation at the surface is 20 days. Convection in case of unstable stratification is parameterized by setting the vertical diffusivity to very large values. As in LM, there is no wind forcing, i. e. we focus here on the thermohaline circulation. Friction is identical to the LM, except that we introduce in addition lateral and vertical friction with viscosities of $3.2 \times 10^4\,\mathrm{m^2\,s^{-1}}$ and $10^{-3}\,\mathrm{m^2\,s^{-1}}$, respectively. Otherwise unphysical oscillations develop on a short time scale. We use the Quicker advection scheme (Leonard, 1979) for tracers and vertical diffusivity of $10^{-4}\,\mathrm{m^2\,s^{-1}}$ in addition.

The steady solution of PEM, shown in Figure 15.25, indeed has much resemblance to LM. In the experiment with northern source, there is a deep temperature

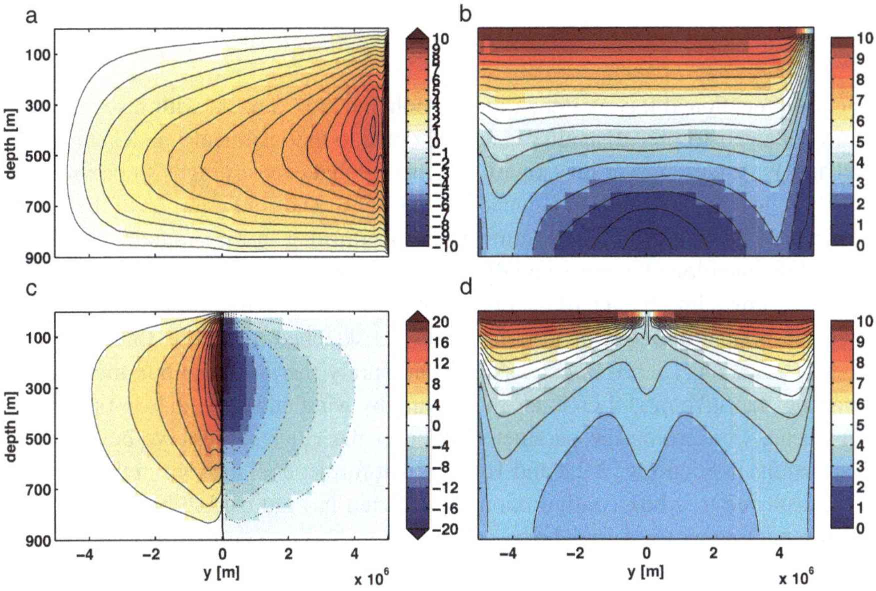


Fig. 15.25 The upper row displays in **a** the meridional overturning stream function $\psi$ (contour interval is 0.5 Sv) and in **b** the zonally averaged temperature (contour interval is $0.5°$C) of PEM with a deepwater formation region at the northern boundary. The meridional overturning stream function (contour interval is 2 Sv) and zonally averaged temperature (contour interval is $0.5°$C) for an equatorial deepwater formation region is shown in **c** and **d**, respectively. The results are time averages over the last 100 years of a 200 years integration

minimum at the equator, and isopycnals below about 500 m depth are symmetric with respect to the equator, bending towards the bottom and towards the poles. A similar 'hill', symmetric around the equator, can be seen in the experiment with the equatorial source, although it is located more to the surface than at depth. Figure 15.25 also shows the meridional transport in both experiments with PEM by the meridional stream function $\Lambda$ with $\bar{v} = -\partial\Lambda/\partial z$ (see (15.47) below). The surface forcing drives a volume transport of a couple of Sv in both cases. In case of the northern source, there is southward flow at depth, almost uniform upwelling in the interior, and northward return flow at the surface. Sign, magnitude and structure of the meridional transport is also very similar to LM in the experiment with equatorial source (see Figure 15.25c,d).

As for LM, the meridional gradient of the zonally averaged density (i. e. temperature) or pressure (not shown) is similar in both experiments with PEM and is of opposite sign in both hemispheres. Their depth dependence differs. The meridional transport, on the other hand, does not show any direct dependency on $\partial\bar{p}/\partial y$, with respect to the individual hemispheres or experiments, proving the downgradient closures based on $\bar{v} \sim \partial\bar{p}/\partial y$ to be wrong in primitive equation models as well. The reason is, of course, the same as in LM, notably the zonally averaged pressure $\bar{p}$ is dominated by the interior zonal mean of $p$, which is in turn governed by the frictionless Sverdrup relation in the interior.

### 15.5.5 A Box Model with Forcing Induced by Wind and Eddies

The aforementioned inconsistency of thermohaline box models with respect to the large-scale geostrophic dynamics is not restricted to the 'classical' models of Stommel and Welander. Newer investigations of the global overturning with box models mostly follow the approach of Gnanadesikan (1999). The configuration of the model is sketched in Figure 15.26a. It aims at the overturning in the Atlantic, and – despite of the mentioned inconsistent dynamics – we will use it to pick up the question where the upwelling branch of the circulation is situated: in the tropical thermocline or in the Southern Ocean or in both places, as sketched in Figure 15.12 and discussed in Section 15.3.3. In extension of the purely thermohaline box models, the Gnanadesikan-like model includes the driving by wind and by eddies in the Southern Ocean. Concerning the parameterization of the latter process we build on the descriptions in Sections 12.2.4 and 16.6 without further explanations in this section.

Transferred to a box configuration, as depicted in Figure 15.26b, the Gnanadesikan model consists of four boxes, the Southern Ocean ($s$), the tropical Atlantic domain ($u$) above the thermocline, the region ($n$) of deep water formation in the North Atlantic, and the deep ocean ($d$) beneath box $u$. The crucial parameterization concerns the transport $q_{\mathrm{n}}$ from tropical box $u$ to the northern box $n$. Gnanadesikan (1999) explicitly assumes a frictional closure as in the previous box models (see Section 15.5.1; see also the comprehensive discussion in the next section). The transport $q_{\mathrm{n}}$ is placed entirely in the western boundary layer (the recirculation is ignored) and a balance between friction and the *overall* meridional pressure gradient (not the local one) is utilized. The argument used in Johnson et al. (2007), on the other hand, avoids friction: the transport $q_{\mathrm{n}}$ across the thermocline in the northern hemisphere is computed from the geostrophic balance $fv = g\Delta\rho\partial\xi/\partial x$, where $\xi$ is the actual thermocline depth and a motionless abyss is assumed (the configuration is presented in the box on p. 527). Note that $\Delta\rho$ is the *vertical* difference of the density at northern latitudes of the tropical region of the model. We find

$$q_{\mathrm{n}} = \int_W^E \xi v \mathrm{d}x = \frac{g\Delta\rho}{2f_{\mathrm{n}}}\left(\xi_E^2 - \xi_W^2\right) \tag{15.35}$$

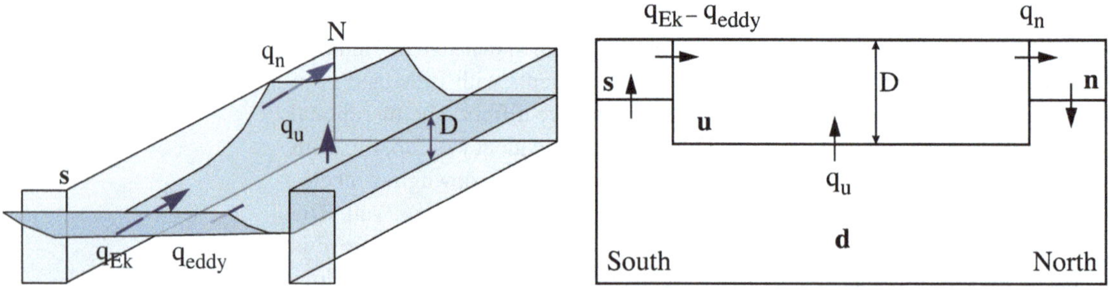


Fig. 15.26 A schematic view of the conceptual model by Gnanadesikan (1999). **a** The thermocline depth $D$ depends upon a balance between the northward Ekman transport $q_{\mathrm{Ek}}$ arising due to southern ocean winds, the eddy-induced southward transport $q_{\mathrm{eddy}}$ in the circumpolar current, diapycnal upwelling $q_{\mathrm{u}}$ in the interior, and northern hemisphere sinking $q_{\mathrm{n}}$. **b** A version with four compartments named $s$, $u$, $n$ and $d$. Redrawn after Johnson et al. (2007)

where the indices refer to the eastern ($E$) and western ($W$) boundary points of the respective section. The crucial step is now to relate the $E - W$ difference of $\xi^2$ to the mean (squared) thermocline depth $D^2$. In the model a unique value of $D$ is used for the entire Atlantic and $\xi_{\rm E}^2 - \xi_{\rm W}^2$ is equated to $D^2$. We are thus facing again the inconsistent relation between an east-west difference and a section mean of the pressure field, as discussed in the previous section. The frictional approach of Gnanadesikan (1999) yields the same dependence $q_{\rm n} \sim \Delta\rho D^2$ as the geostrophic approach of Johnson et al. (2007), only the constant of proportionality differs. Both approaches are thus affected by Straub's dilemma. Note that fixing $D$ to a constant thermocline depth leads to the linear relation between $q_{\rm n}$ and $\Delta\rho$ of the Stommel model.

The parameterizations of the remaining transports are less questionable. The upwelling transport $q_{\rm u}$, situated in the tropical regions, is set by the advective-diffusive Munk balance (see the box on p. 364),

$$q_{\rm u} = \frac{K_{\rm v} A_{\rm u}}{D} \tag{15.36}$$

where $K_{\rm v}$ is the vertical diffusivity and $A_{\rm u}$ the area of upwelling. The transports in the Southern Ocean are discussed later in Section 16.6. We argue there for the expressions

$$q_{\rm Ek} = \frac{\tau_0 X}{|f_{\rm s}|} \quad \text{and} \quad q_{\rm eddy} = -K_\ell X \frac{D}{Y} \tag{15.37}$$

for the wind-induced (Ekman) part and the eddy-induced part, respectively. Here, $X$ is the circumpolar length at Southern Ocean latitudes, and $Y$ is the meridional width of the frontal region. Furthermore, $K_\ell$ is the thickness diffusivity, following Gent and McWilliams (1990), and the isopycnal slope is written as $s = -D/Y$. The mass balance of the volume above the thermocline is given by $q_{\rm Ek} + q_{\rm eddy} + q_{\rm u} - q_{\rm n} = 0$. We proceed with the $q_{\rm n} = (g\Delta\rho)/(2f_{\rm n})D^2$ of Johnson et al. (2007). It yields the cubic equation

$$\frac{K_{\rm v} A_{\rm u}}{D} + \frac{\tau_0 X}{|f_{\rm s}|} - \frac{K_\ell X}{Y} D - \frac{g\Delta\rho}{2f_{\rm n}} D^2 = 0 \tag{15.38}$$

for the thermocline depth $D$.

It is not surprising that the conventional scaling relations (15.9) are recovered from (15.38) if the contributions from the Southern Ocean are ignored – the physics leading to these relations are identical (compare to Section 15.4.1). More specific, the relations

$$D = \left( \frac{2K_{\rm v} f_{\rm n} A_{\rm u}}{\Delta\rho} \right)^{1/3} \quad \text{and} \quad q_{\rm n} = \frac{K_{\rm v} A_{\rm u}}{D} = (K_{\rm v} A_{\rm u})^{2/3} \left( \frac{\Delta\rho}{2f_{\rm n}} \right)^{1/3} \tag{15.39}$$

are obtained and apply to the scenario depicted in Figure 15.12a. To achieve the value $D = 570\,\text{m}$, estimated by Gnanadesikan (1999) from the observed density section in the Atlantic, the vertical diffusivity must be chosen as $K_{\rm v} = 0.4 \times 10^{-4}\,\text{m}^2\,\text{s}^{-1}$ which is a reasonable value[14] for a large-scale mean (see e. g. Section 11.3.3). The northward transport $q_{\rm n}$ becomes 17 Sv which is also reasonable.

[14] Note that the value of $K_{\rm v}$ is smaller than the one used for the estimates reported in Section 15.4.1. On the other hand the lateral scales in that section also differ from the ones used here. We adopt the standard values $X = 30{,}000\,\text{km}$, $Y = 1{,}500\,\text{km}$, $A_{\rm u} = 2.4 \times 10^{14}\,\text{m}^2$, $\tau_0 = 10^{-4}\,\text{m}^2\,\text{s}^{-2}$, $g\Delta\rho = 0.01\,\text{m}\,\text{s}^{-2}$, $f_{\rm n} = |f_{\rm s}| = 10^{-4}\,\text{s}^{-1}$ from Gnanadesikan (1999).

The other interesting limit can easily be evaluated as well, the one depicted in Figure 15.12b where all upwelling occurs in the Southern Ocean. Abandoning the tropical upwelling ($K_v \to 0$), we find a quadratic equation for $D$,

$$\frac{\tau_0 X}{|f_s|} - \frac{K_\ell X}{Y} D - \frac{g \Delta\rho}{2 f_n} D^2 = 0 \tag{15.40}$$

If, for simplicity, the slope $s = -D/Y$ is assumed approximately constant, a square root behavior of the thermocline depth on the windstress in the Southern Ocean results, and, moreover, the northward transport $q_n$ becomes independent on the density gradient. In this scenario, the water sinking in the North Atlantic is replenished by the northward Ekman transport $q_{Ek}$ in the Southern Ocean, and the counteracting eddy-driven transport $q_{eddy}$ must be smaller than the Ekman transport (which is generally the case; see the discussion in Section 16.6). This is also the reason why only one of the two exact solutions of (15.40)

$$D = \frac{K_\ell X f_n}{g \Delta\rho Y} \left( -1 \pm \sqrt{1 + \frac{2\tau_0 Y^2 g \Delta\rho}{K_\ell^2 X f_n |f_s|}} \right) \tag{15.41}$$

is relevant (the negative $D$ corresponds an unphysical inversion of the slope $s$). The above mentioned value of $D$ is obtained by the choice of $K_\ell = 1{,}200\,\mathrm{m^2\,s^{-1}}$ which leads to $q_n = 16$ Sv. The Ekman transport is $q_{Ek} = 30$ Sv, the eddy transport $q_{eddy} =$ 14 Sv (southward). All these values are reasonable and we refer to Chapter 16 on the Southern Ocean circulation for a thorough discussion.

We conclude that the model does not distinguish between either of the two scenarios, i. e. upwelling in the tropics or in the Southern Ocean. Concerning the thermocline depth and water-mass transports, the parameter values yield about the same magnitudes and both scenarios are acceptable. The solution of the full (cubic) problem (15.38) is straightforward (see Gnanadesikan, 1999) but does not bring much additional understanding (there is only one positive solution for $D$). Knowing then the dependence of the transports on the density difference, a coupling to the thermohaline balances of salt and heat can readily be done as for above box models. As no surprise, multiple steady states are found (see e. g. Johnson et al., 2007).

## 15.6 Zonally Averaged Models of the Meridional Overturning

The vertical integral of the equations of motion emphasizes the wind-driven part of the ocean circulation. The effects of stratification appear as an integrated forcing (in form of the bottom torque or the JEBAR term, see Section 14.2) in the equation of the horizontal mass transport. A complementary view is gained from zonal integration, and one might hope to learn from the zonally averaged equations about the meridional flow, the MOC, and thus about the thermohaline circulation (see also the box on p. 484). Much work has been invested during the recent decades to study the MOC in zonally averaged models of the ocean circulation (e. g. Marotzke et al., 1988; Wright and Stocker, 1991; Wright et al., 1995). There are now several coupled Earth system models of intermediate complexity including zonally averaged ocean model components. Because of their low computational costs, such models are often used for paleo-climate simulations and long-term climate projections – several of them are included in the fourth (climate) Assessment Report (Solomon et al., 2007) – while ocean-only versions are used e. g. for studies discussing the stability of the thermohaline circulation. Zonally averaged models serve as an intermediate step between the box models discussed in the previous section and comprehensive three-dimensional ocean models.

### 15.6.1 The Zonally Averaged System of Equations

In this section, we will introduce and discuss the zonally averaged system of equations. We have used such equations before in Section 15.4.2. As there and as in Section 15.2 on the Stommel–Arons model, we work within the planetary approximation, which is relevant for the oceanic circulation on lateral scales much larger than the Rossby radius, as discussed in Chapter 5.1. As before, we consider an ocean basin of constant[15] zonal width $B = x_E - x_W$ and define zonal averages (denoted by an overbar) of all fields, e. g. $\bar{T}(y,z) = (1/B)\int_{x_E}^{x_W} T(x,y,z)\mathrm{d}x$ for the temperature. The zonally averaged equations, written here for convenience in Cartesian coordinates, become

$$\frac{\partial \bar{u}}{\partial t} - f\bar{v} = -\frac{\delta p}{B} + \frac{\partial \bar{\tau}^{(x)}}{\partial z} + \bar{F}^{(x)} \tag{15.42}$$

$$\frac{\partial \bar{v}}{\partial t} + f\bar{u} = -\frac{\partial \bar{p}}{\partial y} + \frac{\partial \bar{\tau}^{(y)}}{\partial z} + \bar{F}^{(y)} \tag{15.43}$$

$$\frac{\partial \bar{p}}{\partial z} = -g\bar{\rho} \tag{15.44}$$

$$\frac{\partial \bar{v}}{\partial y} + \frac{\partial \bar{w}}{\partial z} = 0 \tag{15.45}$$

$$\frac{\partial \bar{\chi}}{\partial t} + \frac{\partial \bar{v}\bar{\chi}}{\partial y} + \frac{\partial \bar{w}\bar{\chi}}{\partial z} = \bar{Q}_\chi - \frac{\partial \overline{v'\chi'}}{\partial y} - \frac{\partial \overline{w'\chi'}}{\partial z} \tag{15.46}$$

where $\chi$ stands for any of the two active tracers salinity $S$ or potential temperature $T$. The zonally averaged density is assumed to satisfy the equation of state in the form $\bar{\rho} = F(\bar{S}, \bar{T}, \bar{p})$, i. e. the impact of nonlinearity of the equation of state on the zonal averages is ignored.

The zonal pressure difference $\delta p(y,z) = p(x_E) - p(x_W)$ between the eastern and the western coasts appears in the zonal balance (15.42). It is unknown but cannot be ignored because of the dominating geostrophic balance in large-scale flows. How to find the zonal pressure difference for a Rayleigh model was already discussed in Section 15.4.2 and is assessed further below when we derive specific model types. The zonally averaged thermohaline balances (15.46) also contain as unknowns the divergence of eddy fluxes $(\overline{v'\chi'}, \overline{w'\chi'})$, arising from the zonal covariances of velocity and $\chi$, which are sometimes called standing eddies (see the box on p. 563). These eddy fluxes are unknown as well. They have to be parameterized by zonally averaged quantities, but are usually simply ignored, as we will do in the following.

In addition, turbulent eddy fluxes, resulting from the previous Reynolds averaging over motions smaller than the Rossby radius, appear in the equations. As usual,

[15] Most of our analysis can easily be generalized to a zonal basin width depending on $(y,z)$. The equations, however, become cumbersome without much more to learn from them.

we have separated the vertical component $\bar{\boldsymbol{\tau}} = (\bar{\tau}^{(x)}, \bar{\tau}^{(y)})$ of the eddy-induced momentum fluxes from the horizontal ones. The terms $\bar{F}^{(x)}$ and $\bar{F}^{(y)}$ denote the zonal means of the horizontal divergences of the eddy-induced fluxes of momentum. As before in Section 15.4.2, we will eventually parameterize them as Rayleigh friction, i. e. $\bar{F}^{(x)} = -r\bar{u}$ and $\bar{F}^{(y)} = -r\bar{v}$, and we will use the usual downgradient closure for the vertical frictional terms, i. e. $\bar{\tau}^{(x)} = A_v \partial\bar{u}/\partial z$ and $\bar{\tau}^{(y)} = A_v \partial\bar{v}/\partial z$ with a vertical viscosity $A_v$. Diabatic sources and sinks in the thermohaline balance (15.46) are summarized by $\bar{Q}_\chi$. They will be expressed in the conventional diffusive form as well, both for the lateral and the vertical direction.

The two-dimensional mass balance (15.45) allows to describe the circulation by the Eulerian stream function $\Lambda$ for the velocity in the meridional-vertical plane. This stream function will be called the MOC stream function and is given by

$$\bar{v} = -\frac{\partial \Lambda}{\partial z} \quad \text{and} \quad \bar{w} = \frac{\partial \Lambda}{\partial y} \tag{15.47}$$

Mass conservation requires that the stream function $\Lambda$ is constant on solid boundaries.

Except for above mentioned unknown pressure difference and the standing-eddy flux, the zonally averaged equations are complete and may be solved if appropriate boundary conditions are specified. Boundary conditions for the vertical eddy fluxes of momentum at the sea surface at $z = 0$ and the bottom at $z = -h$ translate to $\bar{\boldsymbol{\tau}} = \boldsymbol{\tau}_0$ at $z = 0$ and $\bar{\boldsymbol{\tau}} = \boldsymbol{\tau}_b$ at $z = -h$, where $\boldsymbol{\tau}_0$ is the windstress vector and $\boldsymbol{\tau}_b$ the frictional stress at the bottom. The bottom stress condition, given above, is generally replaced by a no-slip condition for $(\bar{u}, \bar{v})$ which then determines the frictional bottom stress $\boldsymbol{\tau}_b$. Free-slip models, on the other hand, use $\boldsymbol{\tau}_b = 0$. At the bottom $z = -h$ we assume zero fluxes of heat and salt. Lateral eddy fluxes of heat and salt across lateral boundaries will be taken zero at solid meridional boundaries.

The wind forcing may, of course, be abandoned altogether to obtain a strictly thermohaline driven model, and we will focus on such models in this section. We would like to point out, however, that most models can be easily be supplemented by wind-driving.

## 15.6.2 The Downgradient Closures

Next we discuss some prominent closures for the zonal pressure difference $\delta p$. A very simple case is a nonrotating ocean. Then the meridional momentum balance (15.43) is sufficient to determine the zonal mean meridional velocity $\bar{v}$, and the zonal balance and knowledge of the zonal flow $\bar{u}$ becomes irrelevant because it does not enter the other equations. In steady state, the meridional pressure gradient

**87. Hagen–Poiseuille Flow**

Flow in a tube under the action of a pressure drop between the ends and friction on the tube walls was studied by Hagen in 1839 and Poiseuille in 1840. In a two-dimensional setting (with the 'tube' along the $y$-direction), the flow is governed by $A_v \partial^2 v/\partial z^2 = \partial p/\partial y$ with no-slip boundary conditions $v = 0$ on the bounding planes at $z = 0$ and $z = -H$. For a constant pressure gradient, the solution is parabolic: $v(z) = (\partial p/\partial y/(2A_v))(z + H)z$ and clearly $v < 0$ for $\partial p/\partial y > 0$. The volume transport is $\int v(z)\mathrm{d}z = -(\partial p/\partial y)H^3/(12A_v)$.

is balanced by friction,

$$\frac{\partial \bar{\tau}^{(y)}}{\partial z} + \bar{F}^{(y)} = \frac{\partial \bar{p}}{\partial y} \tag{15.48}$$

and parameterizing friction in one or the other conventional way (where the friction terms depend solely on $\bar{v}$), we end up with a closed set of equations. Note that in such models $\bar{v}$ is downgradient with respect to the pressure field. A prototype is the HAGEN–POISEUILLE[16] flow, briefly introduced in the box on p. 534. We may refer to the closure models, presented in this section, as Hagen–Poiseuille models.

## Ignoring the Coriolis-term

Marotzke et al. (1988) proposed a closure for the zonally averaged system based on the momentum budget (15.43), in which the Coriolis force was simply ignored (see also the box on p. 507). In its simplest form one has the balance

$$0 = -\frac{\partial \bar{p}}{\partial y} - r\bar{v} \tag{15.49}$$

Note that we have assumed steady state, that we have neglected the vertical stress term $\tau^{(y)}$, and that we have also used Rayleigh friction for the frictional term $F^{(y)}$. Then (15.49) relates the meridional flow with the meridional pressure gradient very similar to the hydraulic law in Stommel's box model (15.21). Motivation for the closure comes from a two-dimensional circulation without rotation which is frictionally controlled as discussed in the box on p. 534.

Originally, Marotzke et al. (1988) proposed to use vertical friction instead of Rayleigh friction. However, the specific choice of the friction does not change the fundamental relation between the meridional pressure gradient and the meridional transport. Using vertical friction with the usual downgradient form $\tau^{(y)} = A_v \partial \bar{v}/\partial z$ instead of Rayleigh friction yields accordingly

$$A_v \frac{\partial^2 \bar{v}}{\partial z^2} = \frac{\partial \bar{p}}{\partial y} \tag{15.50}$$

for simplicity written here with a constant vertical viscosity $A_v$. The horizontal momentum transport by eddies is ignored, i.e. $F^{(y)} = 0$. To obtain the meridional transport, (15.50) has to be vertically integrated twice, using appropriate boundary conditions at top or bottom. To achieve, however, some limited similarity with a rotating three-dimensional ocean, with respect to the strength of the circulation, the vertical viscosity has to be increased, replacing $A_v$ by a much larger effective diffusivity $A'_v$. A formal argument for this closure is given by Wright et al. (1998) showing that $A'_v \sim A_v B/\delta$ where $\delta$ is the width of the western boundary current and $B$ the basin width. Hence the closure by Marotzke et al. (1988) hence effectively describes a super-viscous Hagen–Poiseuille flow.

[16] GOTTHILF HEINRICH LUDWIG HAGEN, *1797 in Königsberg, †1884 in Berlin, engineer, and JEAN LOUIS MARIE POISEUILLE, *1797, †1869 in Paris, physiologist and physicist.

### Parameterizing the Zonal Pressure Difference

A closure with very similar results but with different reasoning was proposed by Wright and Stocker (1991). They choose

$$\frac{\delta p}{B} = -\gamma \sin 2\varphi \frac{\partial \bar{p}}{\partial y} \tag{15.51}$$

where $\varphi$ denotes latitude and $\gamma$ a constant of order 1. The *zonal* pressure difference is expressed in terms of the local *meridional* pressure gradient. The model is of the downgradient nature: using the geostrophic approximation, $\bar{v}$ is found to be proportional to $\partial \bar{p}/\partial y$. Thus the closure is also similar to the hydraulic law of Stommel's box model (15.21) (with a hydraulic constant $C \sim \cos\varphi$ dependent on latitude). Wright and Stocker found some limited empirical support for the closure (15.51) in numerical experiments with a three-dimensional circulation model (however, in a highly diffusive state). Note that the closure is not based on a dynamical concept.

### Parameterizing the Ageostrophic Velocity

Wright et al. (1998) avoid a direct closure for the pressure difference $\delta p$. The zonal momentum balance is entirely abandoned, and the zonally averaged meridional momentum balance is written as

$$f\bar{u} + \frac{\partial \bar{p}}{\partial y} = f(\bar{u} - \bar{u}^{(g)}) = -r\bar{v} \tag{15.52}$$

with the geostrophically balanced zonal mean velocity $\bar{u}^{(g)} = -(1/f)\partial \bar{p}/\partial y$. Note that, for simplicity, the vertical stress term is again neglected and the horizontal stress is parameterized as Rayleigh friction. The original model uses horizontal and vertical viscous friction terms with no principle consequences on the closure.

To determine the meridional velocity $\bar{v}$ related to the frictional terms, the ageostrophic zonal velocity $\bar{u} - \bar{u}^{(g)}$ must be known. For this reason, Wright et al. (1998) divide the zonal extent $B$ of the ocean into a western frictional boundary layer of width $\delta$ and an interior part of width $B_\mathrm{i} = B - \delta \gg \delta$. They further write

$$B(\bar{u} - \bar{u}^{(g)}) = B_\mathrm{i}(\bar{u}_\mathrm{i} - \bar{u}_\mathrm{i}^{(g)}) + \delta(\bar{u}_\mathrm{b} - \bar{u}_\mathrm{b}^{(g)}) \tag{15.53}$$

with the total zonal velocity averaged over the interior and the western boundary layer, $\bar{u}_\mathrm{i}$ and $\bar{u}_\mathrm{b}$, and their geostrophic counterparts $\bar{u}_\mathrm{i}^{(g)}$ and $\bar{u}_\mathrm{b}^{(g)}$, respectively. The interior flow is largely geostrophic, $|\bar{u}_\mathrm{i} - \bar{u}_\mathrm{i}^{(g)})| \ll |\bar{u}_\mathrm{i}^{(g)}|$, and thus the first term on the right-hand side of (15.53) is generally small. In the western boundary layer, however, the flow has both a geostrophic and an ageostrophic component, but the local velocity $u$ vanishes on the continental side of the layer and is largely governed by the geostrophic balance on the offshore edge of the western boundary layer. Wright et al. (1998) further assume that the interior geostrophic component continues only moderately changed into the boundary layer and to the actual boundary. Hence the magnitudes of $\bar{u}_\mathrm{b} - \bar{u}_\mathrm{b}^{(g)}$ and $\bar{u}_\mathrm{b}^{(g)}$ should be similar but of opposing signs in the boundary layer, and roughly

$$B(\bar{u} - \bar{u}^{(g)}) \approx \delta(\bar{u}_\mathrm{b} - \bar{u}_\mathrm{b}^{(g)}) \approx -\delta\bar{u}_\mathrm{i}^{(g)} \approx -\delta\bar{u}^{(g)} \tag{15.54}$$

should hold. Inserting the parameterized ageostrophic velocity given by (15.54) into the meridional momentum balance (15.52) yields

$$\bar{v} = -\frac{\delta}{rB}\frac{\partial \bar{p}}{\partial y} \tag{15.55}$$

which is again similar to Stommel's hydraulic law (15.21) and a closure of Hagen–Poiseuille type. Note that the closure (15.55) is entirely based on geometric arguments: it uses the observed structure of a basin-wide circulation with a narrow western boundary current but no further dynamics.

#### The Inconsistency of the Downgradient Closures

All above mentioned closures express the meridional flow by the meridional pressure gradient. It was shown in Section 15.5.4 that this downgradient relation does not hold in models which contain the zonally resolved dynamics, as for instance the Stommel–Arons model, discussed in Section 15.2, and the LM and PEM simulations, discussed in Section 15.5.4. These considerations cast doubts on the validity of the closures presented so far in this section. We, therefore, continue with two dynamically consistent closures in the following sections.

### 15.6.3 A Vorticity-Based Closure

Wright et al. (1995) propose a dynamically consistent closure by splitting the ocean basin into a western boundary layer and an interior and consider the vorticity budgets averaged separately over both regions. Assuming a frictionless interior and using specific parameterizations for friction in the western boundary layer, they derive a relation between the meridional transport $\bar{v}$ and zonally averaged pressure $\bar{p}$, which is nonlocal and not downgradient. The derivation of the closure in Wright et al. (1995) is unnecessarily complicated. Here we give an alternative simplified derivation with less assumptions to arrive at a similar equation. Instead of the vorticity balance and its integration, we will use the momentum equations directly. The model, in fact, bears on the same dynamics and arguments put forward in Section 15.4.2.

Our analysis starts with the zonal momentum balance (15.12) in the zonally averaged form

$$-f\bar{v}_{\mathrm{b}} = -(p_\delta - p_{\mathrm{W}})/B_{\mathrm{b}} - r\bar{u}_{\mathrm{b}} \tag{15.56}$$

$$f\bar{v}_{\mathrm{i}} = -(p_{\mathrm{E}} - p_\delta)/B_{\mathrm{i}} - r\bar{u}_{\mathrm{i}} \tag{15.57}$$

derived separately for the boundary layer and the interior regime. The pressure values $p_{\mathrm{E}}$, $p_{\mathrm{W}}$, and $p_\delta$ refer to the eastern and western boundary and the outer eastern rim of the boundary layer, respectively. Furthermore, $B$, $B_{\mathrm{b}}$, and $B_{\mathrm{i}}$ denote the total basin width from the western to the eastern boundary, the width of the western boundary layer, and the width of the interior, respectively. The total meridional transport and the zonally averaged pressure can be obtained via

$$B\bar{v} = B_{\mathrm{b}}\bar{v}_{\mathrm{b}} + B_{\mathrm{i}}\bar{v}_{\mathrm{i}} \quad \text{and} \quad B\bar{p} = B_{\mathrm{b}}\bar{p}_{\mathrm{b}} + B_{\mathrm{i}}\bar{p}_{\mathrm{i}} \tag{15.58}$$

Zonal averages of variables over the whole basin are indicated with overbars without an index; zonal averages over the boundary layer or the interior carry an additional index b or i, respectively. The width of the interior domain is $B_\mathrm{i}$, the boundary layer width, given by $\delta = r/\beta$, is denoted by $B_\mathrm{b}$ for consistency, and the total zonal extent of our model domain is $B = B_\mathrm{b} + B_\mathrm{i}$. Obviously, $B_\mathrm{b} \ll B_\mathrm{i}$.

Only a few approximations lead to the closure of Wright et al. (1995). First, the friction term in (15.57) will be neglected. Because of the kinematic boundary condition $u_\mathrm{E} = 0$ at the eastern boundary it follows from (15.12) that $\partial p_\mathrm{E}/\partial y = 0$ or $p_\mathrm{E} = \text{const}$. Second, the pressure perturbation $p_\mathrm{W}$ along the western boundary is eliminated by the meridional velocity $v_\mathrm{W}$ using the steady meridional momentum balance at $x_\mathrm{W}$ in the form

$$0 = -\frac{\partial p_\mathrm{W}}{\partial y} - r v_\mathrm{W} \tag{15.59}$$

and $v_\mathrm{W}$ is parameterized by $\bar{v}_\mathrm{b}$. Note that $u_\mathrm{W} = 0$ was assumed. Next, the friction coefficient $r$ in (15.59) is replaced by $\beta B_\mathrm{b}$ using $B_\mathrm{b} = r/\beta$ as the boundary layer width (note that this 'trick' cannot be done for viscous friction, leading to some complications in Wright et al. (1995)). Integrating (15.59) meridionally starting at $y_0$, (15.56) takes the form

$$f\bar{v}_\mathrm{b} - \beta \int_{y_0}^{y} \bar{v}_\mathrm{b} \mathrm{d}y' = \int_{y_0}^{y} f \frac{\partial \bar{v}_\mathrm{b}}{\partial y} + (f\bar{v}_\mathrm{b})\Big|_{y_0} = (p_\delta - p_\mathrm{W}(y_0))/B_\mathrm{b} + r\bar{u}_\mathrm{b} \tag{15.60}$$

where integration limit $y_0$ is arbitrary. The meridional velocities $\bar{v}_\mathrm{i}$ and $\bar{v}_\mathrm{b}$ then follow from (15.60) and

$$f\bar{v}_\mathrm{i} = (p_\mathrm{E} - p_\delta)/B_\mathrm{i} \tag{15.61}$$

and are seen to be both determined by $p_\delta$, the pressure on the interface of the two regime domains. Note that the frictionless interior balance leads to $p_\mathrm{E} = p_\delta(y = 0)$. Wright et al. (1995) propose the closure $p_\delta = \gamma \bar{p}$ in terms of $\bar{p}$ and neglect the last term in (15.60) related to friction. Both are quite good assumptions for the models LM and PEM in Section 15.5.4 outside the northern and southern boundary layers (not shown). Note, however, that (15.60) only determines the derivative of $\bar{v}_\mathrm{b}$, and thus it is necessary to set an integration constant for $\bar{v}_\mathrm{b}$. One may take $\bar{v}_\mathrm{b}(y_0)$ which by (15.60) is obviously related to the unknown $p_\mathrm{W}(y_0)$. To arrive at the central equation of the model, (15.60) is derived with respect to $y$, divided by $f$ and integrated from $y_0$ to $y$ (in the same hemisphere to avoid the singularity at $y = 0$) to give $\bar{v}_\mathrm{b}$. Using (15.58), the total meridional flow is then governed by

$$B\bar{v} = B_\mathrm{b}\bar{v}_\mathrm{b}(y_0) + \gamma \int_{y_0}^{y} f^{-1}\frac{\partial \bar{p}}{\partial y}\mathrm{d}y' - (\gamma/f)(\bar{p} - \bar{p}(y = 0)) \tag{15.62}$$

involving now the unknown $\bar{v}_\mathrm{b}(y_0)$. As integration constant, Wright et al. (1995) use the boundary transport at the northern and southern boundary, which they relate to the interior flow at the respective boundary (the relation (15.62) is used twice to circumvent the singularity at the equator).

In contrast to the dynamically inconsistent closures of the previous section, the present closure represents a nonlocal relation between the zonally averaged pressure and the meridional flow, and it is based on dynamically consistent arguments. However, there is a problem associated with it: the integration constant $\bar{v}_b(y_0)$ remains to be specified. In the Stommel–Arons model, the interior flow $\bar{v}_i$ is poleward and does not depend on the location of the deepwater source. However, the total meridional flow, $\bar{v}$ does depend on it and can even take a sign different from that of $\bar{v}_i$. It follows that the integration constant $\bar{v}_b(y_0)$ in (15.62) has to contain the information about the sign of the total transport. Nevertheless, it remains unknown as long as we do not know the location of the deep water source. This major drawback of the closure proposed by Wright et al. (1995) is resolved by the closure which we describe in the following section.

### 15.6.4 A Zonally Averaged Model with Consistent Dynamics

In this section, we discuss a closure for zonally averaged models which avoids the specification of integration constants. We simply use separate temperature and salinity and momentum budgets averaged over the interior and the western boundary layer. We also keep all time derivatives. The system of equations averaged separately over the western boundary layer and over the interior is given by

$$\begin{aligned}
\frac{\partial \bar{u}_\alpha}{\partial t} - f\bar{v}_\alpha &= -\delta p_\alpha / B_\alpha + \bar{F}^{\mathrm{u}}_\alpha \\
\frac{\partial \bar{v}_\alpha}{\partial t} + f\bar{u}_\alpha &= -\frac{\partial \bar{p}_\alpha}{\partial y} + \bar{F}^{\mathrm{v}}_\alpha \\
\frac{\partial \bar{\chi}_\alpha}{\partial t} + \frac{\partial \bar{v}_\alpha \bar{\chi}_\alpha}{\partial y} + \frac{\partial \bar{w}_\alpha \bar{\chi}_\alpha}{\partial z} &= \frac{\partial}{\partial z} K_\alpha \frac{\partial \bar{\chi}_\alpha}{\partial z} - \epsilon_\alpha u_\delta \bar{\chi}_\alpha / B_\alpha \\
\frac{\partial \bar{v}_\alpha}{\partial y} + \frac{\partial \bar{w}_\alpha}{\partial z} &= -\epsilon_\alpha u_\delta / B_\alpha
\end{aligned}$$

with $\alpha = \mathrm{b}, \mathrm{i}$ indicating the boundary or interior part, respectively, and $\epsilon_\mathrm{b} = 1$ and $\epsilon_\mathrm{i} = -1$. Momentum advection has been neglected as before. The standing eddy fluxes in the temperature and salinity budget (both denoted here by $\chi$) also have been neglected. Density is given by the equation of state, applied to the averaged temperature and salinity. The (scaled) pressure $\bar{p}_\alpha$ is related to the density $\bar{\rho}_\alpha$ by the hydrostatic relation $\partial \bar{p}_\alpha / \partial z = -g\bar{\rho}_\alpha$. The pressure differences over the western boundary layer and the interior, $\delta p_\mathrm{b} = p(B_\mathrm{b}) - p(x_\mathrm{W})$ and $\delta p_\mathrm{i} = p(x_\mathrm{E}) - p(B_\mathrm{b})$, respectively, and the flow from western boundary into the interior, $u_\delta = u(B_\mathrm{b})$, are unknowns in the system and have to be parameterized.

For these unknowns, the parameterizations

$$\delta p_\mathrm{b} = \gamma_1(\bar{p}_\mathrm{i} - \bar{p}_\mathrm{b}) \,, \quad u_\delta = \gamma_2 \bar{u}_\mathrm{b} \,, \quad \delta p_\mathrm{i} = \delta p_\mathrm{i}(y=0) + \int_0^y f u_\delta \mathrm{d}y' \tag{15.63}$$

will be used. The interior pressure difference at the equator, $\delta p_\mathrm{i}(y = 0)$, is set by the steady zonal momentum balance at the equator. All other aspects of the model are similar to primitive equation models, e. g. the model includes a rigid lid surface boundary condition, a diagnostic relation from the divergence of the vertically integrated momentum budget to find the surface pressure, and convective adjustment to parameterize convection by large vertical diffusivities in case of static instability.

How can the above relations (15.63) be motivated? The pressure difference $\delta p_b$ over the western boundary layer is simply parameterized using the zonally averaged pressure difference between interior and the western boundary layer, the only pressure difference known in the model. For the zonal flow in and out of the western boundary layer, an equally simple parameterization is chosen, i. e. it is set proportional to the zonally averaged zonal velocity in the western boundary layer. The last parameterization, the one for the interior pressure difference $\delta p_i = p_E - p_\delta$, needs a bit more explanation. It may be motivated by the local vorticity balance

$$\frac{\partial f u}{\partial x} + \frac{\partial f v}{\partial y} = -r\left(\frac{\partial v}{\partial x} - \frac{\partial u}{\partial y}\right)$$

which after averaging over the interior domain yields

$$-f\frac{u_\delta}{B_i} + \frac{\partial f \bar{v}_i}{\partial y} = -r\left(\frac{\partial \bar{v}_i}{\partial x} - \frac{\partial \bar{u}_i}{\partial y}\right)$$

Neglection of all friction terms, i. e. abandoning the right-hand side, and using $f\bar{v}_i \simeq \delta p_i/B_i$, and integration then yields the last relation in (15.63).

Figure 15.27 shows $\delta p_b$, $\delta p_i$, and $u_\delta$ diagnosed in a three-dimensionally resolved primitive equation model (PEM) with the same configuration. The figure compares these fields with their parameterizations given by (15.63). There is a good agreement concerning sign and structure of the variables and their parameterizations. Only in the southernmost part, the parameterization for $\delta p_b$ does not have the correct sign. It turns out that an important parameter is the boundary layer width $B_b$, which we have chosen here as $B_b = 3.7\, r/\beta$ since this value seems to match best the boundary layer width in PEM. In fact, the boundary layer is broader than expected from the Rayleigh friction term alone because we also have included harmonic friction in PEM, which leads to a wider boundary layer.

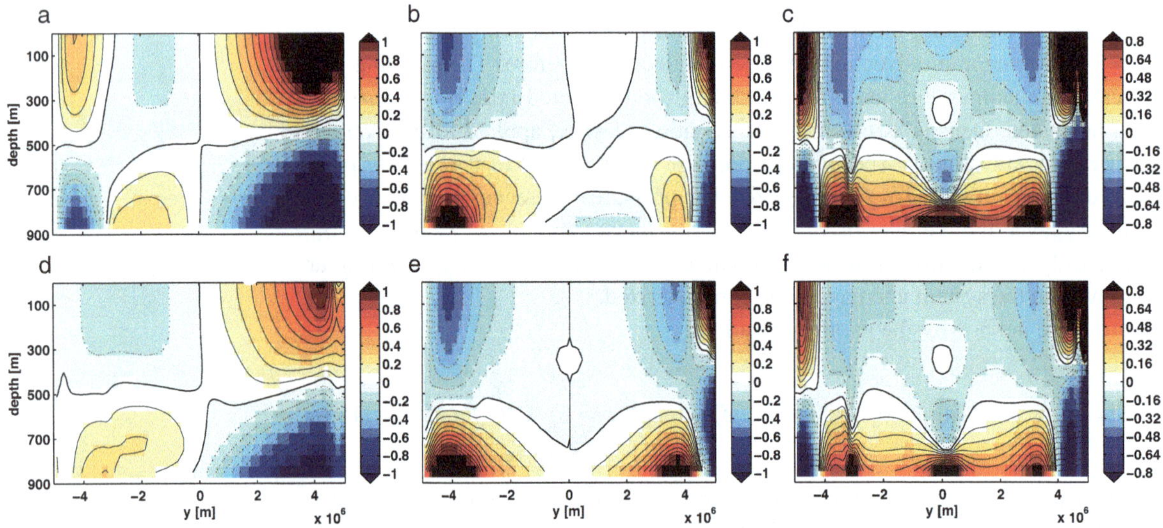


Fig. 15.27 Comparison of parameterized variables with the model result for the primitive equation model. The upper row shows $\delta p_b$ (**a**), $\delta p_i$ (**b**), and $u_\delta$ (**c**) in the zonally resolved model. The lower row shows the respective parameterizations: $\bar{p}_i - \bar{p}_b$ (**d**), $\int_0^y f u_\delta dy'$ (**e**), and $\bar{u}_b$ (**f**). The contour intervals are $0.08\,\text{m}^2\,\text{s}^{-2}$ in (**a**); $0.18\,\text{m}^2\,\text{s}^{-2}$ in (**d**); $0.1\,\text{m}^2\,\text{s}^{-2}$ in (**b**); $0.05\,\text{m}^2\,\text{s}^{-2}$ in (**e**), and $5\times10^{-4}\,\text{m}\,\text{s}^{-1}$ in (**c**) and (**f**)

### 15.6.5 Zonally Averaged Models Versus Box Models

We now compare results from numerical integrations of the dynamically consistent zonally averaged model of the previous Section 15.6.4 with the results obtained with the downgradient closure of Marotzke et al. (1988), presented in Section 15.6.2. We also return to the discussion of Section 15.5 on the box models of the MOC. Zonally resolved model simulations will also be shown. Many predictions of the box models concerning possible steady states, characterized by the dominant thermal or haline surface forcing, are reproduced by the zonally averaged models. However, there are also some new aspects of the numerical solutions, which are not present in the simple box models.

Figure 15.28 shows the results of a simulation of the zonally averaged model of Section 15.6.4 in a single-hemisphere configuration. The corresponding box model is thus the one by Stommel (1961), see Section 15.5.2. The extent of the domain is 5,000 km in meridional and 1,500 m in vertical direction, with a resolution of 320 km and 100 m, respectively. It includes salinity and temperature and a linear equation of state with expansion coefficients $\alpha = -0.2 \times 10^{-3}\,\mathrm{K}^{-1}$ and $\gamma = 0.8 \times 10^{-3}\,(\mathrm{g\,kg}^{-1})^{-1}$. Isopycnal mixing with a diffusivity of $1{,}000\,\mathrm{m}^2\,\mathrm{s}^{-1}$ and vertical diffusivity of $10^{-4}\,\mathrm{m}^2\,\mathrm{s}^{-1}$, lateral friction with viscosity of $8 \times 10^5\,\mathrm{m}^2\,\mathrm{s}^{-1}$, and Rayleigh friction with parameter of $2 \times 10^{-6}\,\mathrm{s}^{-1}$ is used. The model uses an equatorial $\beta$-plane with $f = \beta y$ with $\beta = 2.3 \times 10^{-11}\,(\mathrm{ms})^{-1}$, such that the southern boundary of the domain coincides with the equator. The model is driven by mixed boundary conditions, i. e. by surface relaxation of temperature and a surface freshwater flux. The latter is slowly varied during the integration of about 20,000 years. The maximum of the zonally averaged surface freshwater forcing strength is shown in Figure 15.28e and can be compared with the haline contribution to the surface density flux shown in Figure 15.3. While the subtropical values are roughly comparable to today's conditions, the subpolar freshwater forcing is clearly much larger and resembles an extreme situation.

The configuration of the zonally averaged model is chosen similar to Stommel's box model. In fact, much of the behavior of the numerical solution can now be interpreted using the results of the discussion of the box model. At the start of the integration with vanishing freshwater forcing, the model is in the thermally driven state with northern sinking and a strength of the MOC of a couple of Sv. Increasing the surface freshwater forcing to about 30–40% of its maximum value, the thermally driven MOC eventually collapses and reaches a state of a weak reversed MOC which switches occasionally to a state with very strong MOC, i. e. even stronger than in the thermally driven state, on a time scale of centuries. We might interpret the weak reversed state of the MOC as the haline dominated steady state of Stommel's box model. However, this state is apparently not really stable as in the box model, but it has windows in the strength of the freshwater flux where the solution is of oscillatory nature. This is clearly a feature of the zonally averaged model which is not present in the box model. There might be a series of Hopf bifurcations in the zonally averaged model (compare with Section A.2.1), which is actually found in a low-order model version discussed in the next section. In fact, such oscillations of the MOC on long time scales are often found in numerical ocean models of the MOC. Note that because the freshwater forcing is slowly sweeping across the Hopf points, the oscillations do not develop in a proper form.

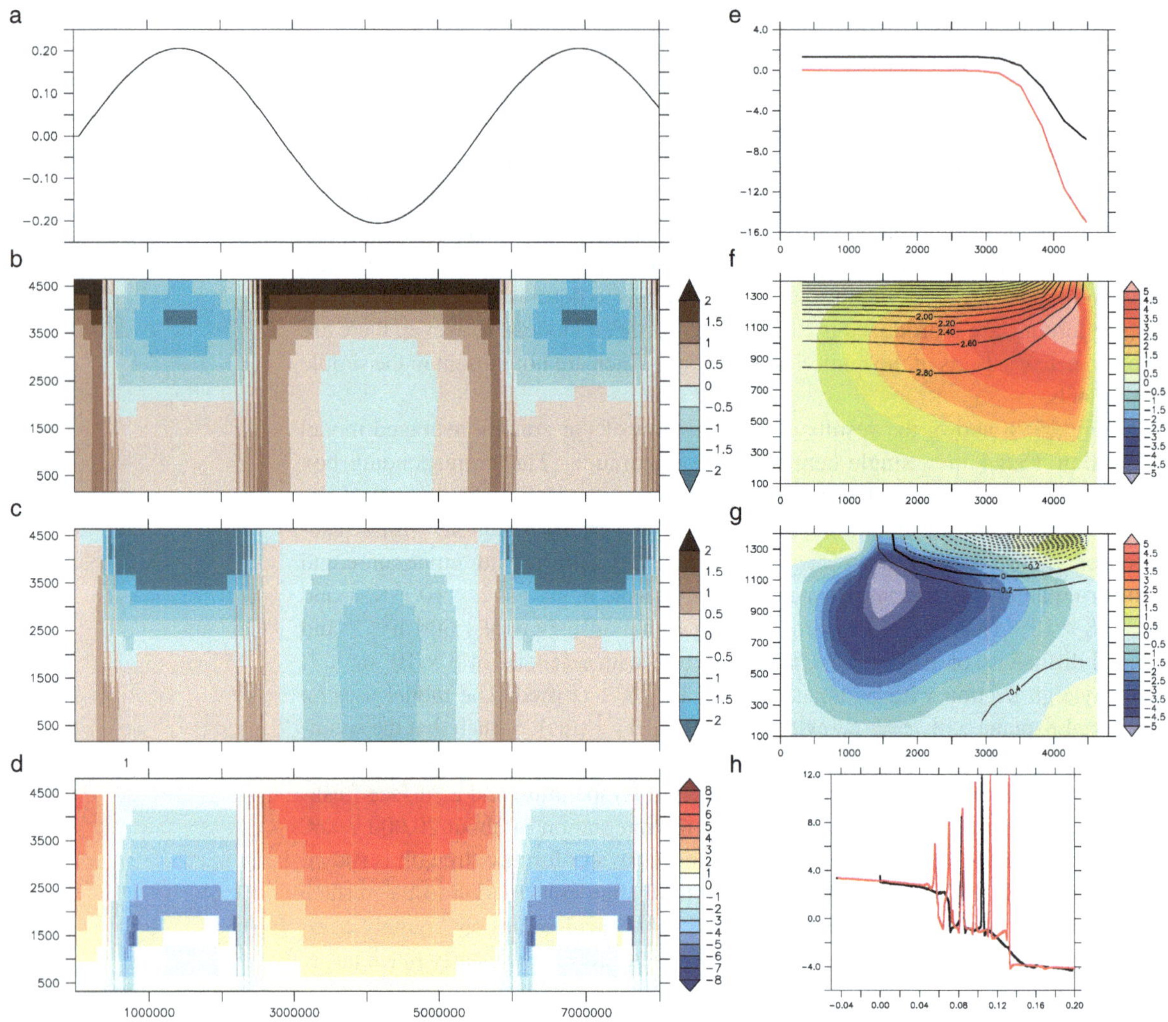

Fig. 15.28 A simulation over about 20,000 years with a single-hemispheric zonally averaged model of the MOC. **a** Amplitude of surface freshwater forcing (positive into the subpolar ocean) as a function of time in days. **b** Near-surface density in kg m$^{-3}$ also as a function of time in days and meridional coordinate in km. **c** Haline contribution to surface density in kg m$^{-3}$. **d** Upper layer northward transport in Sv. **e** Haline surface forcing in mg m$^{-2}$ s$^{-1}$ (*black* line) at its maximum and relaxation temperature in °C (*red line*). **f** MOC stream function defined by (15.47) in Sv at $t = 10^5$ d, i. e. in the thermally driven state. Also shown are contour lines of density with contour interval of $0.2$ kg m$^{-3}$. **g** MOC stream function at $t = 7.5 \times 10^5$ d in the haline driven state. **h** MOC strength in Sv at $y = 4{,}000$ km as a function of freshwater forcing amplitude shown in **a**. The *black* line shows the initial increase of freshwater forcing, the *red* line the decrease

Increasing the freshwater forcing further to more than 50% of its maximum value, the MOC completely reverses and now turns completely into the stable haline driven state, with sinking in the subtropical part of the domain. Note that the direction of the MOC indeed follows the surface density gradient as anticipated by Stommel's hydraulic law (15.21) (which we do not use in the present zonally averaged model). The reverse transition, between the oscillatory state to the thermally driven state, occurs for the decreasing freshwater forcing at later stages of the integration. The dependency of the MOC on the freshwater strength shows resemblance to Stommel's

Box model (Figure 15.28h), although we cannot assume that all possible steady states have been reached during the course of the integration.

Figure 15.29 shows a simulation of another zonally averaged model, using now the closure of Marotzke et al. (1988). There is also a switch from the thermally driven state to the haline driven one around the same integration time as for the model in Figure 15.28. However, the oscillatory state with the periodic strong events in the MOC is only moderately present in this closure version. Only one of such events is hit during the particular integration. The results are becoming, therefore, even more similar to Stommel's Box model with respect to the dependency of the MOC on the freshwater strength shown in Figure 15.29c. As in the box model we find a hysteresis of the MOC, i. e. there are two possible states of the MOC for a certain range of values of the freshwater forcing.

A prominent difference between the consistent zonally averaged model of Section 15.6.4 and the model with the downgradient closure of Marotzke et al. (1988) is given by the MOC in the haline driven state (compare Figure 15.28g with Figure 15.29b). Not surprisingly, it turns out that the former is more realistic than the latter, which is disclosed when comparing with a zonally resolved primitive equation model using identical configuration and forcing. Figure 15.30 shows the MOC

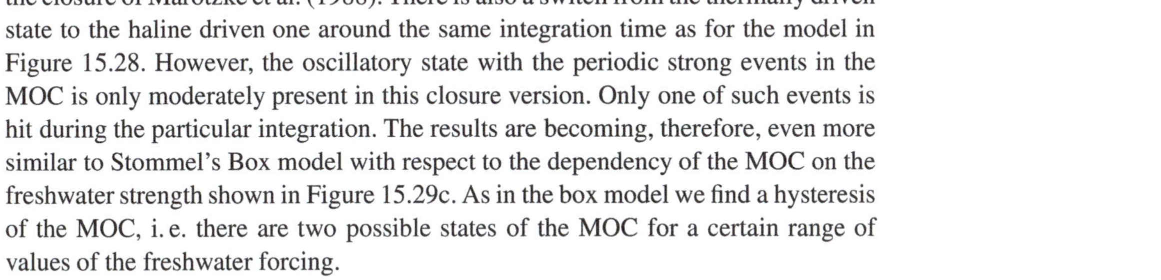

**Fig. 15.29** A simulation with a single-hemispheric zonally averaged model using the closure of Marotzke et al. (1988) using the same forcing as in the simulation shown in Figure 15.28. **a** MOC stream function as defined in (15.47) in Sv at $t = 10^5$ d, i. e. in the thermally driven state. Also shown are contour lines of density with contour interval of $0.2\,\mathrm{kg\,m^{-3}}$. **b** MOC stream function at $t = 7.5 \times 10^5$ d in the haline driven state. **c** MOC strength in Sv at $y = 4{,}000$ km as a function of freshwater forcing amplitude shown in **a**. The *black* line shows the initial increase of freshwater forcing, the *red* line the decrease

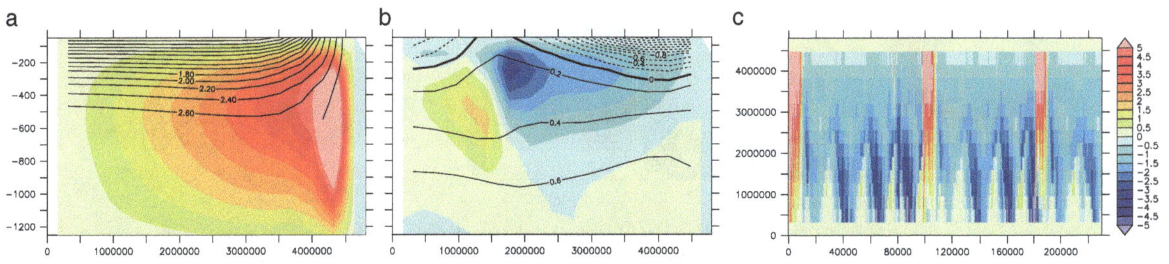

**Fig. 15.30** A simulation with a single-hemispheric zonally resolved model, using the same forcing as in the simulation shown in Figure 15.28. **a** MOC stream function as defined in (15.47) in Sv for vanishing freshwater flux, i. e. in the thermally driven state of the MOC. Also shown are contour lines of density with contour interval of $0.2\,\mathrm{kg\,m^{-3}}$. **b** MOC stream function in the haline driven state, with freshwater forcing corresponding to $t = 7.5 \times 10^5$ d. **c** Time series over about 600 years of the upper layer transport in Sv in the haline driven state

and density structure in the thermally driven and haline driven states of this resolved primitive equation model, simulating only certain time slices of the long-term simulations of the zonally averaged models. The simulations use vanishing freshwater forcing in Figure 15.30a and freshwater forcing corresponding to the haline driven states with oscillations in the dynamically consistent zonally averaged model in Figure 15.30b. The density structure in the thermally driven state is also much closer to the one of the dynamically consistent model than to the one with the downgradient closure. It fact, the zonally resolved model also reveals centennial scale oscillation similar to the dynamically consistent zonally averaged model, as shown in in Figure 15.30c. The time scale of the oscillation is similar to the dynamically consistent zonally averaged model.

Finally, a simulation of a double-hemisphere configuration of the consistent zonally averaged model is analyzed. Results are presented in Figure 15.31. The corresponding box model is the one by Welander (1986), discussed in Section 15.5.3. The temperature and freshwater forcing in the northern hemisphere are identical to Figure 15.28e, while the forcing in the southern hemisphere is chosen symmetric with respect to the equator. The freshwater forcing is varied as in the simulation of Fig-

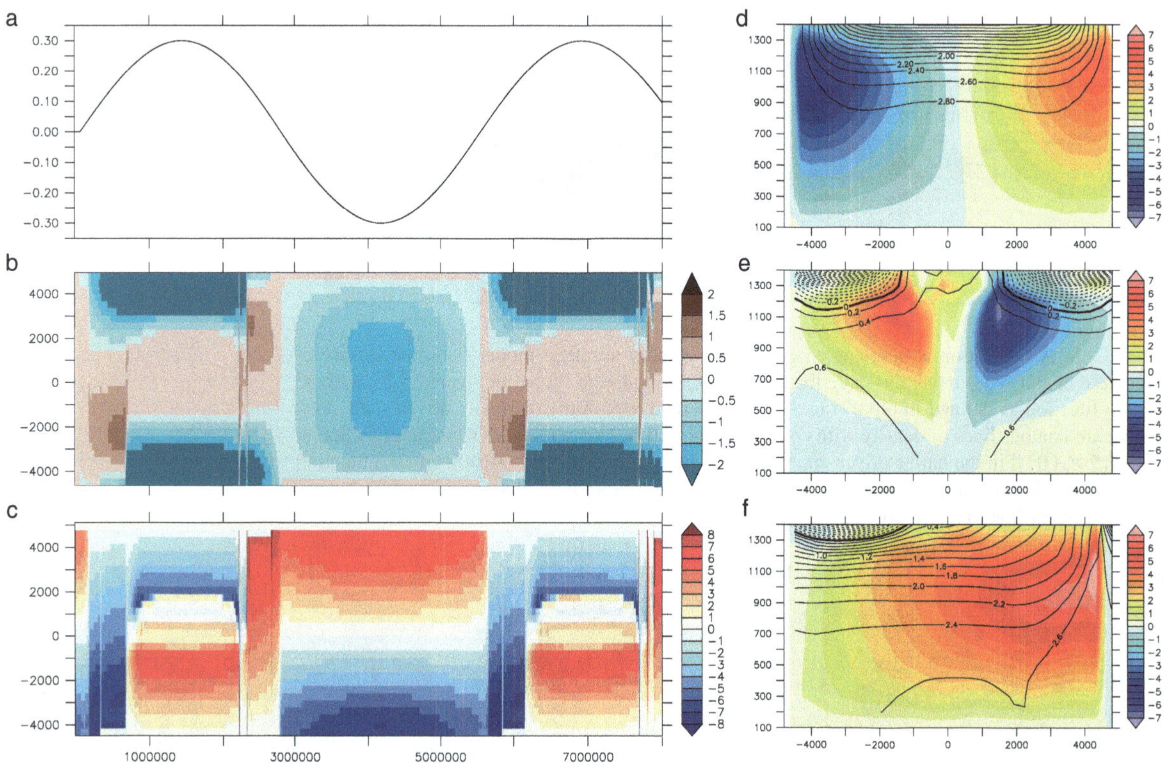

Fig. 15.31 A simulation with a double-hemispheric zonally averaged model with symmetric forcing, which is identical to Figure 15.28e in the northern hemisphere. **a** shows the time series of the freshwater forcing, **b** the haline contribution to the upper layer density in kg m$^{-3}$, **c** the upper layer transport in Sv. **d** shows the MOC stream function at $t = 10^5$ d, i. e. at the beginning of the simulation, where the MOC is in the thermally driven, symmetric double-cell state. **e** shows the MOC stream function at $t = 8 \times 10^5$ d, i. e. in the haline driven, symmetric, double-cell state and **f** show the MOC stream function at $t = 2.5 \times 10^6$ d, i. e. in the asymmetric, single-cell, thermally driven state

ure 15.28 on a long time scale. We find a thermally driven, symmetric double-cell state of the MOC with sinking at the northern and southern boundary, which is realized in the model for vanishing freshwater flux (Figure 15.31d) at the beginning of the simulation or for strongly negative freshwater fluxes. Increasing the freshwater forcing from zero at the beginning of the simulation to about 10–20% of its maximum value, the symmetric double-cell structure collapses to an asymmetric single-cell, which connects both hemispheres (Figure 15.31f). This can be either a state with northern or southern sinking, which both occur during the simulation. However, the single-cell solution is still thermally driven in the northern hemisphere. Increasing the freshwater further to a similar threshold as for the single-hemisphere configuration, the MOC collapses and turns into the haline driven state, which is symmetric with subtropical sinking (Figure 15.31e). Note that all these states are also present in Welander's box model. Reducing the freshwater forcing, the model frequently turns into the symmetric, thermally driven double-cell state, but seems to prefer the asymmetric thermally driven single-cell state. However, reducing the freshwater forcing further to negative values, the model turns into the symmetric, thermally driven double-cell state, in which it stays as long as the freshwater forcing is negative.

Similar to the single-hemisphere configuration, we meet centennial oscillations of the MOC during the transition from the haline to the thermally driven states, which are not present in Welander's box model. These oscillations are not present in a model simulation using the downgradient closure in the zonally averaged model either (not shown here). On the other hand, the model with the downgradient closure shows also a thermally driven symmetric, double-cell MOC, a thermally driven, asymmetric single-cell state, and a symmetric, haline driven state, and is thus fully consistent with Welander's box model. Of course, this not surprising as the dynamics of models are virtually identical.

### 15.6.6 * A Low-Order Model of the Meridional Overturning

We finalize the chapter on the ocean's overturning with an analytical model which bears considerable mathematical work. It describes an analytical solution of the MOC and is based on the zonally averaged system of equations using the downgradient closure of Marotzke et al. (1988) from Section 15.6.2. We have marked this closure as being dynamically not consistent, but found in the integrations of the previous section that it behaves in many respects very similar to the dynamically consistent zonally averaged model of Section 15.6.4. We may consider this similarity as support for the downgradient closures to simulate principle aspects of the MOC correctly. The ultimate reason, however, for returning to the down-gradient closure model is its simplicity: there is only one equation governing the momentum development of the MOC, which is an enormous advantage over the consistent model which has four momentum balances and two continuity equations (there is no stream function for the two regime domains), particularly when the aim is an analytical treatment. The model equations are solved by a spectral low-order approach. The principle procedure to construct such low-order models is outlined in Appendix B.3.

As in discussed Section 15.6.2 for the closure of Marotzke et al. (1988), the dynamics are based on a balance between friction and the meridional pressure gradients given by

$$\frac{\partial \bar{v}}{\partial t} = -\frac{\partial \bar{p}}{\partial y} + A_{\text{v}} \frac{\partial^2 \bar{v}}{\partial z^2}$$

We have used vertical friction, written here for simplicity with a constant vertical viscosity $A_{\text{v}}$. The model is now time-dependent. When we use the MOC stream function $\Lambda$ with $-\partial \Lambda / \partial z = \bar{v}$ and $\partial \Lambda / \partial y = \bar{w}$, the balance becomes

$$\frac{\partial^2 \Lambda}{\partial t \partial z} - A_{\text{v}} \frac{\partial^3 \Lambda}{\partial z^3} = \frac{\partial \bar{p}}{\partial y} = g \frac{\partial \bar{\zeta}}{\partial y} + g \int_z^0 \frac{\partial \bar{\rho}}{\partial y} \mathrm{d}z \tag{15.64}$$

which actually is a vorticity equation. The pressure gradient has been broken into the surface and the baroclinic parts. The vertical integral of this balance is

$$\left( \frac{\partial \Lambda}{\partial t} - A_{\text{v}} \frac{\partial^2 \Lambda}{\partial z^2} \right) \bigg|_{-H}^{0} = \int_{-H}^{0} \frac{\partial \bar{p}}{\partial y} \mathrm{d}z \tag{15.65}$$

Because the net transport must vanish, i. e. $\Lambda = 0$ at top and bottom, the time derivative in (15.65) vanishes, and the balance is between pressure driving and friction at top and bottom of the ocean. Next, we separate the surface part from (15.64), applying a further vertical derivative, and arrive at the equations

$$\frac{\partial^3 \Lambda}{\partial t \partial z^2} - A_{\text{v}} \frac{\partial^4 \Lambda}{\partial z^4} = -g \frac{\partial \bar{\rho}}{\partial y} \quad \text{and} \quad g \frac{\partial \bar{\zeta}}{\partial y} = \left( \frac{\partial^2 \Lambda}{\partial t \partial z} - A_{\text{v}} \frac{\partial^3 \Lambda}{\partial z^3} \right) \bigg|_{z=0} \tag{15.66}$$

The problem is now split into the baroclinic (density-driven) part and the barotropic part. Note that the surface displacement has become a diagnostic variable. Only the baroclinic pressure, effectively the density gradient $\partial \bar{\rho} / \partial y$, acts as prescribed forcing whereas the surface pressure $g\bar{\zeta}$ is part of the response. Hence we continue with the determination of the stream function from the first equation in (15.66).

It will be useful to scale the variables as we did for box models (see Section 15.5). We take a typical density increment $\rho_\star$ to scale the density according to $\bar{\rho}' = \bar{\rho}/\rho_\star$. For the stream function, we use the diffusive scaling $\Lambda' = \Lambda L/(H \tilde{K}_{\text{h}})$ with a typical lateral diffusivity $\tilde{K}_{\text{h}}$. Then, using the scaled coordinates $\xi = z/H$, $\eta = y/L$ and time $\tau = t\tilde{K}_{\text{h}}/L^2$ and dropping the primes, we obtain the scaled version of (15.66),

$$\text{Pr}^{-1} \frac{\partial^3 \Lambda}{\partial \tau \partial \xi^2} - \frac{\partial^4 \Lambda}{\partial \xi^4} = -\text{Ra} \frac{\partial \bar{\rho}}{\partial \eta} \tag{15.67}$$

with the Rayleigh number $\text{Ra} = gH^3 \rho_\star / (A_{\text{v}} \tilde{K}_{\text{h}})$ and the Prandtl number $\text{Pr} = (A_{\text{v}}/\tilde{K}_{\text{h}})(L/H)^2$ as dimensionless coefficients.

The above described momentum dynamics is now augmented by the thermohaline balance (15.46) with the conventional downgradient parameterizations of the eddy heat and salt fluxes which are contained in $\bar{Q}_\chi$. However, to simplify further, we first proceed with the density balance, i. e. $\bar{\chi} = \bar{\rho}$, with the above described diffusive fluxes

$$\bar{Q}_\rho = K_{\text{h}} \frac{\partial^2 \bar{\rho}}{\partial y^2} + K_{\text{v}} \frac{\partial^2 \bar{\rho}}{\partial z^2}$$

We have assumed that the diffusivities $K_\mathrm{h}$ and $K_\mathrm{v}$ are constant. Using the above described scaling, the density $\bar{\rho}$ is governed by

$$\frac{\partial \bar{\rho}}{\partial \tau} + \frac{\partial \Lambda}{\partial \eta}\frac{\partial \bar{\rho}}{\partial \xi} - \frac{\partial \Lambda}{\partial \xi}\frac{\partial \bar{\rho}}{\partial \eta} = \kappa_\mathrm{h}\frac{\partial^2 \bar{\rho}}{\partial \eta^2} + \kappa_\mathrm{v}\frac{\partial^2 \bar{\rho}}{\partial \xi^2} \tag{15.68}$$

where $\kappa_\mathrm{h} = K_\mathrm{h}/\tilde{K}_\mathrm{h}$, $\kappa_\mathrm{v} = L^2 K_\mathrm{v}/(H^2 \tilde{K}_\mathrm{h})$ are the scaled lateral and vertical diffusivities. Later we will distinguish between temperature and salinity. A list of appropriate parameter values can be found in Table 15.1. Remember that $A_\mathrm{v}$ is tuned up to yield a reasonable size for the overturning rate.

Table 15.1 List of parameters and standard values of the low-order model. The density scale $\rho_\star$ is dimensionless (equal to density divided by a reference density). The basin width $L$ and the scale $\tilde{K}_\mathrm{h} H/L$ of the stream function imply a scale of 4 Sv for the basin-wide transport

| | | | |
|---|---|---|---|
| $H, L$ | 4,000 m, 5,000 km | $\Lambda$ scale $\tilde{K}_\mathrm{h} H/L$ | $0.8\,\mathrm{m^2\,s^{-1}}$ |
| $\rho_\star$ | $2.4 \times 10^{-3}$ | time scale $L^2/\tilde{K}_\mathrm{h}$ | 800 years |
| $K_\mathrm{v}$ | $10^{-4}\,\mathrm{m^2\,s^{-1}}$ | $\kappa_\mathrm{v} = (L/H)^2 K_\mathrm{v}/\tilde{K}_\mathrm{h}$ | 0.156 |
| $\tilde{K}_\mathrm{h}$ | $10^3\,\mathrm{m^2\,s^{-1}}$ | $\kappa_\mathrm{h} = K_\mathrm{h}/\tilde{K}_\mathrm{h}$ | 1 |
| $A_\mathrm{v}$ | $6 \times 10^2\,\mathrm{m^2\,s^{-1}}$ | | |
| $\mathrm{Ra} = g\rho_\star H^3/(A_\mathrm{v}\tilde{K}_\mathrm{h})$ | $2.5 \times 10^3$ | $\mathrm{Pr} = (L/H)^2 A_\mathrm{v}/\tilde{K}_\mathrm{h}$ | $9 \times 10^5$ |

In the presented form, the equations describe the overturning in a nonrotating ocean (or a tank experiment). With the large super-viscosity implemented, as proposed by the closure of Marotzke et al. (1988), the model should apply to the large-scale ocean circulation. We have argued in this respect in Section 15.6.2.

## A Diffusive Ocean

In order to correctly implement the surface boundary conditions for heat and salt in the low-order model, it turns out to be useful to consider first a steady state in which the advection of the thermohaline fields is neglected. The ocean is then in an entirely diffusive balance given by

$$\kappa_\mathrm{h}\frac{\partial^2 \bar{\rho}}{\partial \eta^2} + \kappa_\mathrm{v}\frac{\partial \bar{\rho}}{\partial \xi^2} = 0 \tag{15.69}$$

The model domain extends from $\xi = -1$ to $\xi = 0$ and from $\eta = -1$ to $\eta = 1$, which should represent an ocean with two hemispheres as in Welander's box model. Solutions of (15.69) are conveniently constructed using the eigenfunctions $P_n(\eta)$ of the meridional differential operator, defined by

$$P_n'' + \upsilon_\mathrm{n}^2 P_n = 0$$

with suitable boundary conditions[17]. We use $P_n' = 0$ as boundary conditions at the meridional limits of our ocean, because this choice will establish there a vanishing

[17] Derivatives of functions of a single independent variable, either $\eta$ or $\xi$, are denoted by a prime, e. g. $\mathrm{d}P_n(\eta)/\mathrm{d}\eta = P_\mathrm{n}'$ and $\mathrm{d}\rho_\mathrm{n}(\xi)/\mathrm{d}\xi = \rho_\mathrm{n}'$.

flux of heat, salt, and density. There are two sets of functions which together comprise the complete, orthogonal, and normalized set of eigenfunctions. For $n > 1$ these are

$$P_n(\eta) = \sin \upsilon_n \eta \quad \text{for} \quad n = 1, 3, 5, \ldots$$
$$P_n(\eta) = \cos \upsilon_n \eta \quad \text{for} \quad n = 2, 4, 6, \ldots$$

with the eigenvalue $\upsilon_n = n\pi/2$. The set is augmented by $P_0 = 1/\sqrt{2}$ with eigenvalue $\upsilon_0 = 0$. We proceed with the expansion

$$\bar{\rho}(\eta, \xi) = \sum_n \rho_n(\xi) P_n(\eta) \quad \text{for} \quad D^{\text{surf}}(\eta) = \kappa_{\text{v}} \sum_n \mathcal{R}_n P_n(\eta) \tag{15.70}$$

for the density and the surface density flux $D^{\text{surf}}(\eta)$ and consider the boundary condition $\kappa_{\text{v}} \partial\bar{\rho}/\partial\xi = D^{\text{surf}}$ at the top $\xi = 0$ and $\partial\bar{\rho}/\partial\xi = 0$ at the bottom $\xi = -1$. The forcing variable $\mathcal{R}_n$ for the mode $n$ is related to the associated heat and freshwater flux. Inserting this expansion into (15.69), we arrive at the differential equation

$$\rho_n'' - \frac{\upsilon_n^2}{\kappa} \rho_n = 0 \quad \text{with} \quad \rho_n'(0) = \mathcal{R}_n \quad \text{and} \quad \rho_n'(-1) = 0 \tag{15.71}$$

Here, $\kappa = \kappa_{\text{v}}/\kappa_{\text{h}} = (K_{\text{v}}/K_{\text{h}})(L/H)^2$ is the ratio of the vertical and horizontal diffusivities. Finally, the solution of the density state is given by

$$\rho_n(\xi) = \mathcal{R}_n h_n(\xi) \tag{15.72}$$

in terms of the forcing amplitude $\mathcal{R}_n$ and a vertical structure function $h_n(\xi)$ of hyperbolic form, depending on the inverse depth scale $\xi_n = \upsilon_n/\sqrt{\kappa}$ (see the box on p. 549). The MOC stream function $\Lambda(\eta, \xi)$, driven by the meridional density gradient, follows from the stationary form of (15.67). Setting

$$\Lambda(\eta, \xi) = \text{Ra} \sum_n \mathcal{R}_n \ell_n(\xi) P_n'(\eta) \tag{15.73}$$

the solution is completed by solving the fourth order differential equation $\ell_n'''' = h_n$ for the vertical structure function $\ell_n(\xi)$ of the stream function with appropriate boundary conditions. The solution is also described in the box on p. 549 for free-slip at the top and no-slip at the bottom. An example of the structure of the density and stream function as function of $(\eta, \xi)$ is found in Figure 15.32c,f and Figure 15.35a,d, shown below. The boundary conditions imposed on $\ell_n$ cause $\Lambda$ to vanish at the top and the bottom. The vanishing at meridional boundaries, however, results from $P_n' = 0$ at these limits, and this is a consequence of density conservation, requiring $\partial\rho/\partial\eta = 0$ at meridional boundaries.

**88. Structure Functions of the Diffusive Solution**

The vertical structure of the diffusive solutions ($n = 1, 2, \ldots$) is given by the functions

$$h_n(\xi) = \cosh \xi_n(\xi + 1)/(\xi_n \sinh \xi_n)$$
$$\ell_n(\xi) = \left(\cosh \xi_n(\xi + 1) + \alpha_0 + \alpha_1\xi + \alpha_2\xi^2 + \alpha_3\xi^3\right)/(\xi_n^5 \sinh \xi_n)$$

for the expansion coefficients of the density field with $\xi_n = v_n/\sqrt{\kappa}$. The functions $h_n(\xi)$ follow from (15.71), and $\ell_n(\xi)$ is governed by $\ell_n'''' = h_n$. For free-slip at the top and no-slip at the bottom, $\ell_n''(0) = 0, \ell_n'(-1) = 0$, and mass conservation, $\ell_n(0) = \ell_n(-1) = 0$, the coefficients of $\ell_n$ are

$$ll\alpha_0 = -\cosh \xi_n \qquad \alpha_1 = -\frac{1}{4}\left[(6 + \xi_n^2)\cosh \xi_n - 6\right]$$
$$\alpha_2 = -\frac{1}{2}\xi_n^2 \cosh \xi_n \qquad \alpha_3 = -\frac{1}{4}\left[(\xi_n^2 - 2)\cosh \xi_n + 2\right]$$

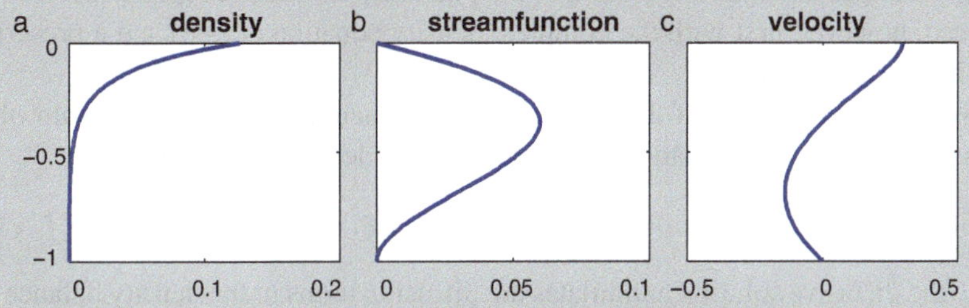


*The vertical structure functions of the diffusive solution for the meridional structure* $n = 2$ *and* $\kappa = 0.15625$. ***a*** $h_2(\xi)$ *for the density.* ***b*** $10^3\ell_2(\xi)$ *for the stream function.* ***c*** $-10^3\ell_2'(\xi)$ *for the meridional velocity.*

## An Advective-Diffusive Ocean

In contrast to the diffusive balance (15.69), the real ocean has a dominance of advection compared to diffusion. According to the discussion in Section 15.3, however, we cannot abandon mixing entirely. Furthermore, the diffusion must be present in our low-order model to couple the fields to the forcing, which we ensure by using the above described diffusive solution also for the complete low-order model with advection.

For realistic conditions, the density flux $D^{\text{surf}}(\eta)$ should be chosen almost symmetric with respect to the equator (at $\eta = 0$) with a dominating $\mathcal{R}_2$-amplitude, which must be a negative number to account for the dominating effect of heating in the density flux (making the equatorial density lighter, see e. g. Figure 15.3). Our low-order model will thus be forced only by the $\mathcal{R}_2$ component of the surface density flux. Let us denote the above derived diffusive solution by $\rho^{\text{d}}(\eta, \xi)$ and $\Lambda^{\text{d}}(\eta, \xi)$. Hence we have

$$\rho^{\text{d}}(\eta, \xi) = \mathcal{R}h_2(\xi)P_2(\eta) \quad \text{and} \quad \Lambda^{\text{d}}(\eta, \xi) = \text{Ra}\mathcal{P}\ell_2(\xi)P_2'(\eta) \tag{15.74}$$

with $\mathcal{R} = \mathcal{R}_2$, $\mathcal{P} = \mathcal{R}_2$ and $\mathcal{P} = \mathcal{R}$ in the presently considered density-forced model (later we will distinguish between density, temperature, and salinity where these quantities will be different). This diffusive solution is not a solution of the complete thermohaline balance (15.68), because the above density field and stream function yield a nonzero advective term

$$\frac{\partial \Lambda^{\mathrm{d}}}{\partial \eta}\frac{\partial \rho^{\mathrm{d}}}{\partial \xi} - \frac{\partial \Lambda^{\mathrm{d}}}{\partial \xi}\frac{\partial \rho^{\mathrm{d}}}{\partial \eta}$$
$$= -\mathrm{Ra}\mathcal{RP}\left[\upsilon_2^2 \ell_2(\xi) h_2'(\xi) P_2(\eta) P_2(\eta) + \ell_2'(\xi) h_2(\xi) P_2'(\eta) P_2'(\eta)\right] \qquad (15.75)$$

which must lead to a change of the density field away from the diffusive solution. Note that the advective term (15.75) is symmetric in the meridional coordinate $\eta$, and thus we expect that it generates only symmetric density perturbations, leading to a double-cell circulation according to the momentum budget (15.67). The route towards an asymmetric density, related to a single-cell overturning as in Welander's box model, is possible, but it is a bit intricate as it runs via the nonlinearity of the density balance and the formulation of the boundary conditions for heat and salt. We proceed, however, first with the complete density equation (15.68) and a prescribed density flux at the surface.

We express the fields of the low-order model including advection as sum of the diffusive solution and deviations due to advection, denoted by a tilde,

$$\bar{\rho}(\eta,\xi,\tau) = \rho^{\mathrm{d}}(\eta,\xi) + \tilde{\rho}(\eta,\xi,\tau) \quad \text{and} \quad \Lambda(\eta,\xi,\tau) = \Lambda^{\mathrm{d}}(\eta,\xi) + \tilde{\Lambda}(\eta,\xi,\tau)$$

Since the diffusive solution annihilates the diffusion terms in the density balance and the viscous term the density torque in the momentum balance, the deviation density $\tilde{\rho}$ and the stream function $\tilde{\Lambda}$ are governed by

$$\frac{\partial \tilde{\rho}}{\partial \tau} - \frac{\partial}{\partial \xi}(\Lambda^{\mathrm{d}} + \tilde{\Lambda})\frac{\partial}{\partial \eta}(\rho^{\mathrm{d}} + \tilde{\rho}) + \frac{\partial}{\partial \eta}(\Lambda^{\mathrm{d}} + \tilde{\Lambda})\frac{\partial}{\partial \xi}(\rho^{\mathrm{d}} + \tilde{\rho}) = \kappa_{\mathrm{h}}\frac{\partial^2 \tilde{\rho}}{\partial \eta^2} + \kappa_{\mathrm{v}}\frac{\partial \tilde{\rho}}{\partial \xi^2}$$
$$\mathrm{Pr}^{-1}\frac{\partial^3 \tilde{\Lambda}}{\partial \xi^2 \partial \tau} - \frac{\partial^4 \tilde{\Lambda}}{\partial \xi^4} = -\mathrm{Ra}\frac{\partial \tilde{\rho}}{\partial \eta} \qquad (15.76)$$

The forcing by the surface density flux enters via the advective terms which includes the diffusive solution. These are terms proportional to either $\mathcal{RP}$ (the one given by (15.75)), to $\mathcal{P}$, or to $\mathcal{R}$.

A spectral low-order model is now obtained by expanding the $\eta$-dependence of the perturbation fields into the meridional eigenfunctions $P_n(\eta)$ and the $\xi$-dependence in a complete orthonormal eigenfunction base $H_q(\xi)$ for the density. We choose the base defined by $H_q'' + \lambda_q^2 H_q = 0$ with $H_q'(0) = H_q'(-1) = 0$ (zero flux of the perturbation density field). For the stream function, we use $F_q(\xi)$ defined by $F_q'''' = H_q$ with free-slip at the top and no-slip at the bottom, $F_q''(0) = 0$, $F_q'(-1) = 0$, and mass conservation requiring $F_q(0) = F_q(-1) = 0$. Note that the $F_q$ are forced functions, not eigenfunctions.

If we insert the complete expansion

$$\tilde{\rho}(\eta,\xi,\tau) = \sum_{nq} \tilde{\rho}_{nq}(\tau) H_q(\xi) P_n(\eta) \quad \text{and}$$
$$\tilde{\Lambda}(\eta,\xi,\tau) = \mathrm{Ra}\sum_{nq} \tilde{\Lambda}_{nq}(\tau) F_q(\xi) P_n'(\eta)$$

into the equations (15.76) and project onto the eigenfunctions $P_n(\eta)$ and $H_q(\xi)$, the resulting model is still exact within the framework of the approximations leading to the zonally averaged equations. It is certainly far too complex to be solved analytically.

We proceed, therefore, with a minimal model resulting from a very heavy truncation. We retain only the modes $\rho_{01}$, $\rho_{10}$, $\rho_{20}$, and $\Lambda_{10}$, $\Lambda_{20}$. This choice reflects a fundamental set of nonlinear interactions among the modes. The truncated model[18] is given by

$$\begin{aligned}\tilde{\rho}(\eta,\xi,\tau)/f_1 &= L(\tau)H_1(\xi)P_0 + 4M(\tau)H_0P_1(\eta) + N(\tau)H_0P_2(\eta)\\ \tilde{\Lambda}(\eta,\xi,\tau)/(f_1\mathrm{Ra}) &= 4U(\tau)F_0(\xi)P_1'(\eta) + V(\tau)F_0(\xi)P_2'(\eta)\end{aligned} \tag{15.77}$$

with the redefined modal amplitudes $U \sim \Lambda_{10}$, $V \sim \Lambda_{20}$, $L \sim \rho_{01}$, $M \sim \rho_{10}$, and $N \sim \rho_{20}$. This yields a low-order system of five ordinary differential equations for the time dependence of these amplitudes – the low-order model

$$\dot{U} = -\nu(U - M) \tag{15.78}$$

$$\dot{V} = -\nu(V - N) \tag{15.79}$$

$$\dot{L} = -4UM - VN + bN\mathcal{P} + aV\mathcal{R} + c\mathcal{R}\mathcal{P} - r_{\mathrm{v}}L \tag{15.80}$$

$$4\dot{M} = UL + dU\mathcal{R} - r_{\mathrm{h}}M \tag{15.81}$$

$$\dot{N} = VL - bL\mathcal{P} - r_{\mathrm{h}}N \tag{15.82}$$

where $\nu = 21f_3\mathrm{Pr}/\mathrm{Ra}$, $r_{\mathrm{v}} = \pi^2 f_3\kappa_{\mathrm{v}}/\mathrm{Ra}$, $r_{\mathrm{h}} = \pi^2 f_3\kappa_h/\mathrm{Ra}$ and coupling coefficients $a$, $b$, $c$, and $d$ appear[19]. The latter are integrals of the vertical structure functions $h_2$, $\ell_2$, $H_0$, $H_1$, and $F_0$, entering as triple products and depending on $\kappa = \kappa_{\mathrm{v}}/\kappa_{\mathrm{h}}$ via the inverse vertical scale $\xi_2 = \pi/\sqrt{\kappa}$ of mode 2 (see the box on p. 549). The low-order model, of course, reflects the linear momentum balance (note that (15.78) and (15.79) are balances between vertical friction and the meridional pressure gradient) and the nonlinear density balance (the balances (15.80)–(15.82) advection and diffusion terms are evident). It is immediately obvious, as mentioned above, that the above model with the symmetric density forcing and linear terms has $M = 0$ in steady state and thus only generates a symmetric density field, here modeled by the amplitudes $L$ and $N$. As we demonstrate now, the nonlinear terms do not break this fundamental symmetry in the density-forced model.

For the steady state, $M = U$ and $N = V$ must hold, and two sets of solutions arise from the density equations. One set has $U = 0$ (inferred from (15.81)) and $V$ determined by a cubic equation (inferred from (15.80) and (15.82)); the other one has a nonzero $U$ determined by a quadratic equation (compute $L$ from (15.81) and combine it with (15.80) and (15.82)). But the latter is actually never real ($U^2$ is always negative; see (15.85) below for $\mathcal{T} = 0$), so the second set does not exist. We are thus stuck to the previous case $U = 0$ and note that only double-cell solutions are possible. After some mathematical manipulations, the cubic equation for $V$ is

[18] Some renormalizations have been made to put the resulting low-order model equations into a simple form: the factors $f_1 = 120$, $f_2 = (\pi^2/1{,}800)/(16 - \pi^2)$, $f_3 = f_1 f_2$ arise from the normalization of the spectral functions and the triple function integrals of the coupling coefficients of the nonlinear terms. We have absorbed some factors in a new time scaling: the dot-time derivative implies a time scaling $\tau' = \tau\mathrm{Ra}/f_3$. The structure functions in (15.74) are found in the box on p. 549, and those in (15.77) are $H_0 = 1$, $H_1 = \sqrt{2}\cos\pi\xi$, $F_0 = (1/24)(\xi^4 + (2/3)\xi^3 - (1/2)\xi)$. To perform all the integrations to arrive at (15.78)–(15.82), a tool like MAPLE is extremely helpful.

[19] For $\kappa_{\mathrm{v}} = 0.15625$, $\kappa_{\mathrm{h}} = 1$, we find $a = -6.1\times10^{-5}$, $b = -8.8\times10^{-5}$, $c = -6.2\times10^{-9}$, $d = 1.4\times10^{-4}$. We use $\nu = 811$, $r_{\mathrm{v}} = 6.6\times10^{-5}$, $r_{\mathrm{h}} = 4.2\times10^{-4}$.

written as[20]

$$(V/\mathcal{R})^3 - (a+2b)(V/\mathcal{R})^2 + \left[b(a-b) - c + \frac{r_\mathrm{v} r_\mathrm{h}}{\mathcal{R}^2}\right] V/\mathcal{R} + bc = 0 \quad (15.83)$$

Note that it is easy to solve in a graphical way, plotting $\mathcal{R}^2$ versus $V/\mathcal{R}$. Because the discriminant is positive for the negative $a, b, c$ obtained in the model, we arrive at the conclusion that there is only a single double-cell solution $U = 0$ and $V \neq 0$ for symmetric density forcing in the density-driven case. This results corresponds to what we have learned from the box models in Section 15.5 for the case of a prescribed density forcing. The equilibria are shown[21] in Figure 15.32 for four values of the Rayleigh number Ra. Note that all solutions are stable. The results show that the diffusive solutions (Figure 15.32c,f), are strengthened by the nonlinear contribution in their equatorial upwelling and polar downwelling (for density forcing with negative $\mathcal{R}$).

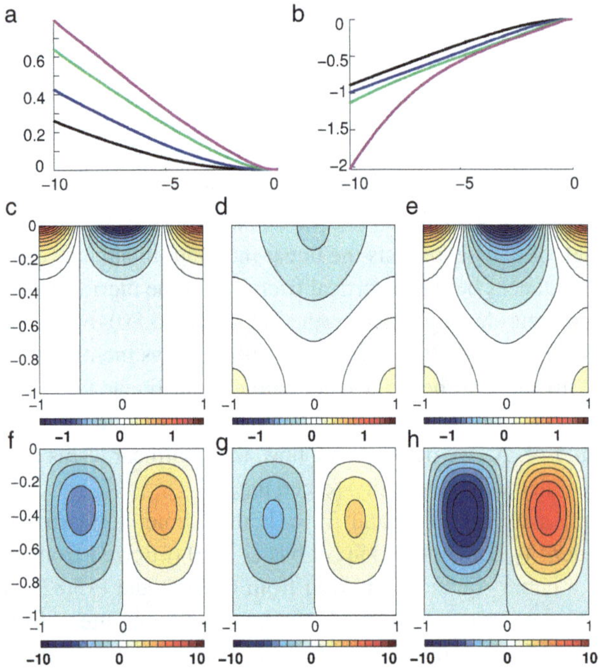

**Fig. 15.32** Steady states of the density-driven low-order model (15.78)–(15.82) in terms of **a** $V$ and **b** $L$ (in $10^{-3}$) as a function of the forcing amplitude $\mathcal{R}$ and for the Rayleigh numbers Ra = 500 (*black*), Ra = 1,000 (*blue*), Ra = 2,500 (*green*) Ra = 5,000 (*magenta*). The variables $U$ and $M$ are always zero, i.e. there is only the double-cell MOC state, while $N$ is identical to $V$. The solutions are stable everywhere. The starting point of the continuation integration is at the known analytical solution $U = V = M = L = N = 0$ for $\mathcal{R} = 0$. Panels **c**, **d**, and **e** show the density structure related to the diffusive solution $\rho^\mathrm{d}$ (**c**), the perturbation $\tilde{\rho}$ (**d**) and the total density (**e**) for Ra = 2,500 and $\mathcal{R} = -10$. Panels **f**, **g**, and **h** show the corresponding stream functions $\Lambda^\mathrm{d}$ (**f**), $\tilde{\Lambda}$ (**g**), and total $\Lambda$ (**h**)

[20] Remember that $r_\mathrm{v}$ and $r_\mathrm{h}$ are inversely proportional to the Rayleigh number Ra.

[21] Figure 15.32 and also the following, are not produced from (15.83), but by use of the numerical continuation tool `bifurk.m` written by Christoph Völker, AWI Bremerhaven, which evaluates also the stability. More details on stability are given in Appendix A.2.1.

### Mixed Boundary Conditions

In this section, we discuss the implementation of mixed boundary conditions for heat and salt in the low-order model. We have seen that mixed boundary conditions lead in the box models to the possibility of multiple steady states, a feature which we also expect here. This implies to consider the advective-diffusive balances of heat and salt instead of the single density balance (15.68). The analysis is a bit lengthy, but it is – in extension of the already discussed density-driven model – in most aspects straightforward, as briefly outlined below.

- We replace the density balance (15.68) by independent equations for heat (temperature $\bar{T}$) and salt (salinity $\bar{S}$). Using the scaling introduced above and in Section 15.5, the density is $\bar{\rho} = \bar{S} - \bar{T}$ which enters the momentum balance (15.67) via the meridional gradient of $\bar{\rho}$.
- We consider the surface flux condition $\kappa_v \partial \bar{S}/\partial \xi = F^{\text{surf}}$ for salinity and, for simplicity, a prescribed temperature, $\bar{T} = T^{\text{surf}}$ at the surface $\xi = 0$, as we did for the box models. At the bottom, we take zero flux conditions for both fields. The forcing functions, i. e. the surface temperature $T^{\text{surf}}$ and the salt flux $F^{\text{surf}}$, will be represented the eigenfunctions $P_n(\eta)$,

$$T^{\text{surf}}(\mu) = \sum_n \mathcal{T}_n P_n(\mu)\,, \quad F^{\text{surf}}(\mu) = \kappa_v \sum_n \mathcal{S}_n P_n(\mu) \tag{15.84}$$

  The salt flux is written in terms of a salinity forcing variable $\mathcal{S}_n$ (which, therefore, represents a flux and not a concentration). In the following, only the the basic symmetric component $n = 2$ will be used, as before for the density flux-driven model. Because the meridional structure function $P_2(\eta)$ has a central maximum (at the equator), we shall take $\mathcal{T}_2 > 0$ to establish a warm equator, and $\mathcal{S}_2 > 0$ to have salt input in this region.
- Diffusive solutions $S^{\text{d}}(\eta, \xi) = \mathcal{S}_2 h_2(\xi) P_2(\eta)$ and $T^{\text{d}}(\eta, \xi) = \mathcal{T}_2 g_2(\xi) P_2(\eta)$ are constructed as we did for the density case. The salinity case is identical to the above given density solution, because the boundary conditions are the same; the temperature has a different structure function $g_n(\xi) = \cosh \xi_n(\xi + 1)/\cosh \xi_n$. The diffusive solution for the stream function thus becomes

$$\Lambda^{\text{d}}(\eta, \xi) = \text{Ra}\,[\mathcal{S}_2 - \mathcal{T}_2 \xi_2 \tanh \xi_2]\, \ell_2(\xi) P_2'(\eta)$$

- We proceed to write the nonlinear solution as $\bar{S} = S^{\text{d}} + \tilde{S}$, $\bar{T} = T^{\text{d}} + \tilde{T}$ and construct the complete low-order model, as done above for density, but now with three salinity variables resulting from $\tilde{S}$ and three temperature variables resulting from $\tilde{T}$. However, as for the box models, we confine the analysis to the simplified case of a prescribed temperature (the complete model with active temperature is derived analyzed in Olbers and Zhang, 2008). Then the deviation field $\tilde{T}$ is set to zero, and only $T^{\text{d}}$ enters via the implied advection by $\Lambda^{\text{d}}$. The model is thus identical to (15.78)–(15.82) where $L, M, N$ are now salinity variables (using the same notation as before) and $\mathcal{R} = \mathcal{S}_2$, $\mathcal{P} = \mathcal{S}_2 - \mathcal{T}_2 \xi_2 \tanh \xi_2$. We will use the abbreviation $\mathcal{T} = \mathcal{T}_2 \xi_2 \tanh \xi_2$. Note that $\mathcal{P}$ and $\mathcal{R}$ are now different.

The discussion of the steady states of the model with mixed boundary conditions can thus be smoothly continued from what has been found for the case of a prescribed density flux. The set with nonzero $U$ follows from[22]

$$4U^2 = (\mathcal{R}-\mathcal{T})(d\mathcal{R}-r_h)\frac{b}{d}\left[a + \frac{b}{\mathcal{R}}(\mathcal{R}-\mathcal{T}) - \frac{b}{d\mathcal{R}^2}(d\mathcal{R}-r_h)(\mathcal{R}-\mathcal{T})\right] + c\mathcal{R}(\mathcal{R}-\mathcal{T}) - r_v(r_h - d\mathcal{R}) \quad (15.85)$$

and is no longer empty if the temperature $\mathcal{T} \sim \mathcal{T}_2$ exceeds a certain critical value which depends on the Rayleigh number Ra. The functional dependence $U = U(\mathcal{T}, \mathcal{R}, \mathrm{Ra})$ is shown in Figure 15.33 where the salt flux $\mathcal{R} = \mathcal{S}_2$ is the abscissa and various values for Ra and $\mathcal{T}$ are used. The solution represents a single-cell overturning, and we note that it exists in a limited interval of the salt flux $\mathcal{R}$. Both limits of this interval are pitchfork bifurcations of the model, as can be seen in the further discussion. It becomes evident in Figure 15.33 that the principal pitchfork bifurcation (see Appendix A.2.1), which opens the two branches of a single-cell overturning, occurs at ever-decreasing values of the salt flux $\mathcal{S}_2$ if Ra is increased. The $\mathcal{S}_2$-domain, where these two solutions exist, actually shrinks, i. e. for each Ra an upper bound of $\mathcal{T}_2$ exists where the single-cell overturning collapses. These properties can be easily revealed in analytical form by solving (15.85) for $U = 0$, and like we did for the Welander box model, we can derive the range of existence of the single-cell solutions for the low-order model.

The complete bifurcations of the thermohaline driven model, shown in Figure 15.34 with the continuation parameter $\mathcal{S}_2$, depict the well-known sequence of a double-cell (i. e. $U = 0$ with nonzero $V$) thermally driven branch at low freshwater forcing, which turns at a pitchfork bifurcation (marked by the green circle) into two single-cell branches (i. e. $U \neq 0$) with northern or southern sinking. For very large freshwater forcing, the branches join again to a haline driven branch with a double-cell circulation (i. e. $U = 0$ again). For sufficiently large Ra, the two single-cell branches run with increasing $\mathcal{S}_2$ into Hopf bifurcations (marked by a blue square; for Hopf bifurcations see Appendix A.2.1), which then open a window where all branches are unstable. The window with three unstable branches shrinks

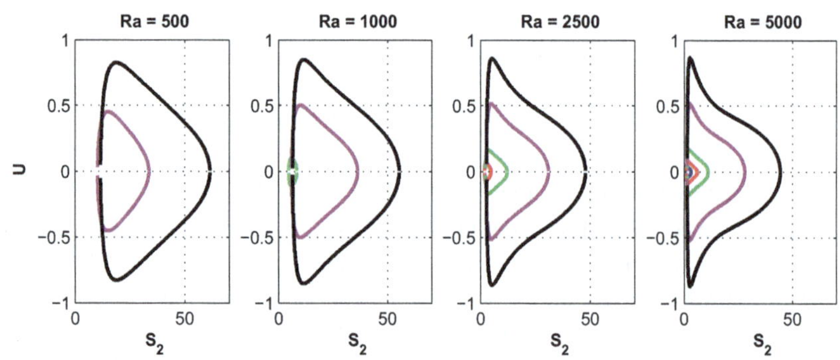


Fig. 15.33 Steady states of the low-order model with mixed boundary conditions in terms of $U$ (in $10^{-3}$) for various values of the Rayleigh number Ra (from 500 to 5,000) and $\mathcal{T}_2 = 0.3$, 0.5, 1.0, 3.0, 5.0 in the colors *blue*, *red*, *green*, *magenta*, *black* in each panel. This result is obtained by analytical solution of the steady state equations, and stability is not considered. This will be shown in Figure 15.34

[22] Remember that $r_v$ and $r_h$ are inversely proportional to the Rayleigh number Ra.

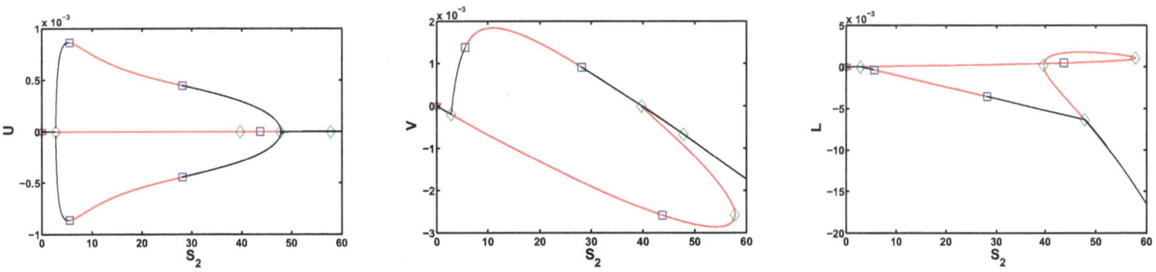

Fig. 15.34 Bifurcation diagrams $U$, $V$ and $L$ as function of $\mathcal{S}_2$ for the low-order model with mixed boundary conditions for Ra $= 2{,}500$ and $\mathcal{T}_2 = 5.0$. The diagrams have the complete information of the bifurcation types: the initial point is indicated by a *magenta star*, a bifurcation point by a *green square*, a Hopf bifurcation by a *blue square*. *Black lines* denote stable branches, *red lines* unstable ones

with smaller Ra and eventually vanishes (not shown). Figure 15.35 shows the pattern of the salinity and stream function fields on the single-cell branch close to the first Hopf point in Figure 15.34 (right panels). The double-cell, thermally and haline driven branches are always stable, and there are always stable single-cell solutions in some limited $\mathcal{S}_2$-window. Similar to the box models (see the box on p. 519), the symmetry of the low-order model is lost, and the pitchfork bifurcations are broken up if asymmetric forcing components are included (not shown).

A general similarity of the present low-order model and Welander's box model of Section 15.5 is obvious. Both share the property of a single solution for a density-driven overturning and the existence of multiple steady states in form of thermally and haline driven double-cell and single-cell states, with the basic pitchfork bifurcations. What is unique to the low-order model are the Hopf bifurcations and the

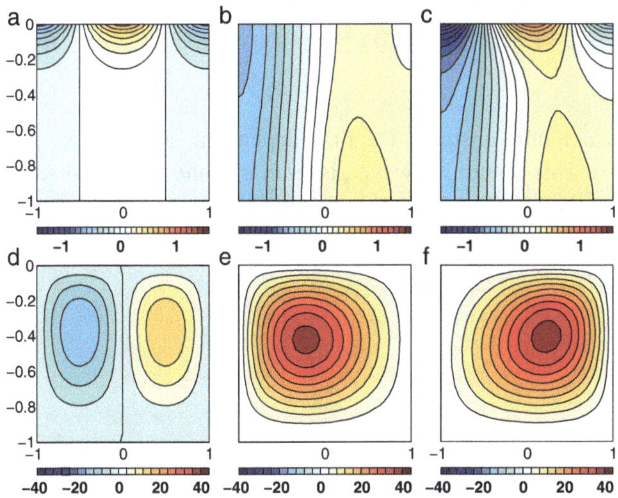

Fig. 15.35 Salinity (**a**,**b**,**c**) and stream function (**d**,**e**,**f**) of the low-order model with mixed boundary conditions for Ra $= 2{,}500$, $\mathcal{T}_2 = 5.0$ and $\mathcal{S}_2 = 5.916$ corresponding to the first Hopf bifurcation on the single-cell branch with northern sinking. Note that the state with southern sinking is a mirror image of the northern sinking one. **a** and **d** show the diffusive part of the solutions, **b** and **e** the perturbation and **c** and **f** the total solutions

**89. The Hopf Bifurcation of the Low-order Model**

The model, described by (15.78)–(15.82), contains in the tracer balances nonlinearities which are typical for advective terms in spectral low-order models. They give rise to a quadratic invariant. Because $\nu \gg 1$, we may assume for simplicity that the momentum balances (15.78) and (15.79) are in steady state, thus $U = M, V = N$. We then obtain

$$\frac{\mathrm{d}}{\mathrm{d}t}\left[L^2 + 16M^2 + N^2\right] = (aN + c\mathcal{P})L\mathcal{R} - \left[r_\mathrm{v}L^2 + 4(r_\mathrm{h} - d\mathcal{R})M^2 + r_\mathrm{h}N^2\right]$$

Steady states thus lie on a quadratic surface in the $(L, M, N)$-space, which may be ellipsoidal or hyperbolical, depending on the forcing amplitudes $\mathcal{R}$ and $\mathcal{P}$.
The Hopf bifurcations open the unstable window initially with a simple limit cycle (see panel below panel for $\mathcal{S}_2 = 6$). With increasing forcing amplitudes $\mathcal{S}_2$, a number of period doubling occurs (for $\mathcal{S}_2 = 7$ and $\mathcal{S}_2 = 8$). Eventually, the system then shows chaotic behavior ($\mathcal{S}_2 = 9$ and $\mathcal{S}_2 = 18$), however, with several interruptions for certain values of $\mathcal{S}_2$ (shown for $\mathcal{S}_2 = 17$) where simple limit cycles again occur. The time scale of these unsteady oscillations is some thousand years.

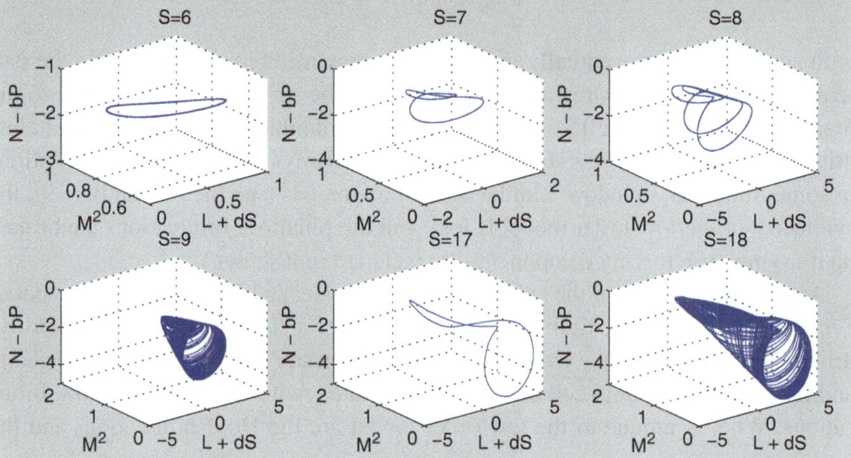


*The attractor in $(L, M, N)$-space for various values of the forcing $\mathcal{S}_2$ (the values are given in the title of the respective panel) and $\mathcal{T}_2 = 5.0$, $Ra = 2{,}500$. Note that $\mathcal{S}_2$ is named $S$ in the panels; likewise $P$ stands for $\mathcal{P}$.*

windows in which only unstable, i. e. time-dependent, solutions are present (see the box on p. 556). This feature, however, is similar to the more realistic zonally averaged models of the MOC, discussed in Section 15.6.

# 16 The Circulation of the Southern Ocean

In this chapter, we discuss the circulation of the Southern Ocean, which is partly unbounded in the zonal direction around Antarctica and which is, therefore, governed by unique dynamics. We start with a description of the special features of the Southern Ocean and a discussion of the dynamics of a homogeneous, wind-driven case with and without topography. We then turn to the role of mesoscale eddies for the meridional overturning circulation of a stratified Southern Ocean and the complete dynamics of the Southern Ocean in a zonally and also in a vertically averaged framework. Based on this knowledge, we construct several simplified models of the Southern Ocean circulation, driven by wind and surface density flux.

Roughly 75 % of the World Ocean volume has temperatures below 4 °C, connected with only 2 % of the ocean surface (at polar latitudes). Paleoceanographic data have revealed that this was not always the case. Before Drake Passage opened the gateway between South America and the Antarctic continent due to continental drift about 30 Myr ago, the climate of the ocean was considerably warmer. Paleo records suggest that in the course of the establishment of the Southern Ocean[1] in its present shape, the difference between surface and bottom temperatures in equatorial regions changed from about 7 °C to its present value of about 26 °C. The polar climate of the southern hemisphere got increasingly colder by the growth of glacial ice on the Antarctic continent and the gradual development of the sea ice cover around it, leading to the formation of deep cold water masses spreading as Antarctic Bottom Water (AABW) to the adjacent northern ocean basins. The opening of Drake Passage also established the strongest and longest current system in the World Ocean, the Antarctic Circumpolar Current (ACC), extending around the globe with a length of roughly 24,000 km. The Antarctic water ring ranges from the Antarctic continent to about 50° S. As the most important link between the ocean basins of the Atlantic, Pacific, and Indian Oceans, the ACC serves as a conduit of all active and passive oceanic tracers which affect Earth's climate, notably heat and salt which strongly in-

[1] Many oceanographers refer to the region around the continent of Antarctica as the Southern Ocean. The International Hydrographic Organisation (IHO), which is the authority responsible for the naming of oceanic features, does not recognize a subregion of the World Ocean of that name but includes its various parts into the other three oceans. From an oceanographic point of view, subdivisions of the World Ocean should reflect regional differences in its dynamics. The Southern Ocean certainly deserves its own name on that ground (Tomczak and Godfrey, 2003).

D. Olbers, J. Willebrand, C. Eden, *Ocean Dynamics*,
DOI 10.1007/978-3-642-23450-7_16, © Springer-Verlag Berlin Heidelberg 2012

fluence the oceanic mass stratification, circulation, and consequently the ocean heat transport, and the greenhouse gas carbon dioxide and other chemical and biological components. But in contrast to this strong zonal exchange brought about by the deep-reaching and strong zonal current, these same characteristics of the ACC act to limit meridional exchange and tend to isolate the ocean to the south from heat and substance sources in the rest of the World Ocean.

## 16.1 Basic Ingredients of Southern Ocean Dynamics

Before the detailed discussion of the physical processes and models, which are relevant for the dynamics of the Southern Ocean, we start with a general overview. Most of the dynamical features will be taken up in the later sections.

### *16.1.1 The Antarctic Circumpolar Current*

The ACC is the World Ocean's most intensive current system, flowing round the Antarctic continent with a volume transport of $130\pm10$ Sv (that is about 1,000 times the Amazon discharge; it would empty the Baltic Sea in three days). This is the largest transport rate in the ocean and can largely be explained by the mostly westerly, very intense winds (see Figure 13.1) in the Southern Ocean. The wind-driven dynamics of the ACC and the importance of the thermohaline surface fluxes over the Southern Ocean will be discussed subsequently in this chapter. The schematic of the global conveyor belt (Figure 15.1) shows that the deep structure of the ACC has a specialty: while in all other large-scale current systems the deep current flows against the surface current, the ACC flows completely in one direction: the surface as well as the bottom currents are almost everywhere eastward.

In the north, the Antarctic water ring is limited by the southern extent of the Subtropical Convergence which winds itself around the globe as a nearly continuous but diffuse water belt, a broad zone of transition between the tropical/temperate and the polar/cold oceans where the permanent thermocline reaches the surface. There are other (and more pronounced) such fronts sketched in the upper left panel of Figure 16.1 and in more detail in the lower panel. The most important are the Subantarctic Front and the Polar Front at about 60° S. These fronts escort the ACC which is not a single current but is composed of several filaments. The fronts are traced by the regionally and temporally highly variable surface temperature gradient displayed in Figure 16.1 (lower panel), which shows that the ACC is, in fact, a fragmented system of more or less intense jet streams. The thermal fronts have a close correspondence in density and extend to great depths, in most places to the bottom (see Figure 16.2 below), but can also be correlated with surface elevations as detected in satellite altimetry data. At these fronts, the ocean water masses are subject to an abrupt change in their physical, chemical, and biological properties. We must not imagine them as static lines; they are always shifting, extending, withdrawing, sometimes forming convexities and shedding intensive mesoscale eddies. This mesoscale turbulence, the ocean weather, accompanies the current and has a strong influence on it.

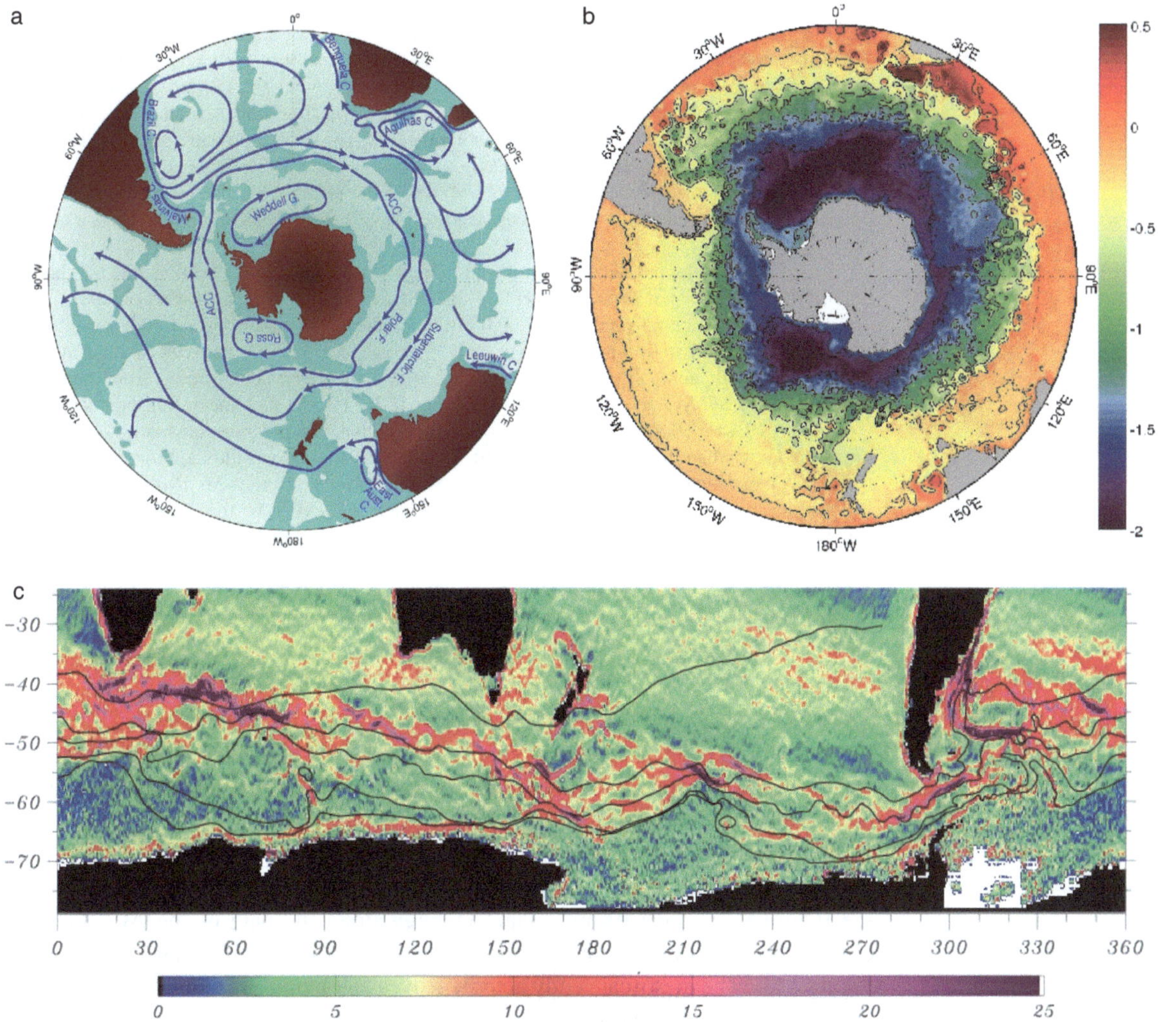


Fig. 16.1 **a** Schematic map of major currents in the southern hemisphere oceans south of 20° S. Depths shallower than 3,500 m are shaded. The two major cores of the Antarctic Circumpolar Current (ACC) encircling Antarctica are shown: the Subantarctic Front and Polar Front. F stands for front, C for Current, and G for gyre. From (Rintoul et al., 2001). **b** Instantaneous sea surface height [in meter] from the regional model of the Southern Ocean described in Figure 12.7. Notice the drop of the mean sea surface by about 1.5 m across the ACC. **c** Magnitude of the mean sea surface temperature gradient in mK/km from satellite observations. Superimposed are positions of (from north to south) the Subtropical Front, Subantarctic Front, Polar Front, South ACC Front, and southern boundary of the ACC. From Hughes and Ash (2001)

### 16.1.2 Mesoscale Eddies

The zonal periodicity of the Southern Ocean creates a circumpolar pathway of water masses circling the globe and allowing the ACC to play a major part in the meridional overturning circulation. But the zonality also acts as a brake. In the ocean basins which are zonally blocked by continents, we find a meridional exchange of heat accomplished by the time mean gyre current systems. There is no significant transport of heat by mean currents across the latitudes of the ACC (DeSzoeke and

**90. Observational Estimates of the ACC Transport**

The meridional momentum balance of the ACC is basically geostrophic, i. e. the zonal current velocity (at each geopotential level) is related to the meridional pressure gradient. This gradient results from a dip of about 1.5 m (from north to south, see Figure 16.1) of the sea surface $z = \zeta$ across the current system, and the gradient of density in the fronts, which can be inferred e. g. from the hydrographic section shown in Figure 16.2. The surface pressure gradient yields an overall eastward geostrophic surface velocity $u_g|_{z=0} = -(g/f)\partial\zeta/\partial y$, and the stratification yields a positive shear $\partial u_g/\partial z = (g/f)\partial\rho/\partial y$ of the geostrophic part of the current. The zonal velocity thus reduces with depth but generally not as strongly as to imply a reversal of the flow.

The above 'thermal wind relation' is utilized to infer from hydrographic section data the 'baroclinic' transport normal to the section and referred to a common depth. Based on six hydrographic sections, Cunningham et al. (2003) estimated the average transport referred to common deepest level of all station pairs through Drake Passage, the gateway between South America and the Antarctic continent, as $136.7 \pm 7.8$ Sv, with about equal contributions from the Polar Front ($57.5 \pm 5.7$ Sv) and the Subantarctic Front ($53 \pm 10$ Sv). The mean transport south of Australia is $147 \pm 10$ Sv (Rintoul and Sokolov, 2001). Note that the transport south of Australia must be larger than that at Drake Passage to balance the Indonesian throughflow and the Tasman leakage, which are believed to be of order 10 Sv. However, given the remaining uncertainty in the barotropic flow at both locations, the agreement is likely to be fortuitous. Variability of transport south of Australia has been estimated in a 6 years record of a repeated hydrographic section (Rintoul et al., 2002) as fairly small (1–3 Sv). Note that the geostrophic transport is not the total transport. The total volume transport through a section also contains the Ekman transports (due to the windstress and the frictional bottom stress) and other contributions induced by nonlinearities and lateral friction.

Levine, 1981). Instead, heat and other tracers must be carried across the current by smaller-scale and time-varying features in the current field, usually summarized as the mesoscale eddy field.

Transient eddies with scales of tens to a hundred km – i. e. several times larger than the baroclinic Rossby radius which is of order 10 km in the Southern Ocean, see Figure 8.2 – are very prominent features along the path of the ACC. This is e. g. illustrated in the numerical experiment of the Southern Ocean circulation of Figure 12.7. The mesoscale eddy activity leaves an obvious imprint. The upper right panel of Figure 16.1 shows an instantaneous snapshot of the sea surface height in a model of the Southern Ocean circulation, and the lower panel displays the gradient of the sea surface temperature from satellite data. These properties of the flow explicitly show the single branches of the Circumpolar Current; however, there are considerable torsions with scales of several tens of kilometers, caused by and being carried along in the mesoscale eddies. This eddy activity reaches from the surface to great depths and is characteristic for all large current systems. In the Circumpolar Current, however, mesoscale eddies are particularly pronounced. They are responsible for the poleward heat transport across the current and the vertical transport of momentum into the deeper layers of the ocean. Estimates of the meridional eddy heat flux from a number of moored instruments and further analysis confirmed the southward transfer with sufficient magnitude to close the overall heat budget.

### *16.1.3 The Meridional Overturning Circulation*

Figure 16.2 shows water-mass properties on a section between Australia and Antarctica (roughly along 140° E). The figure makes visible that water-mass properties do

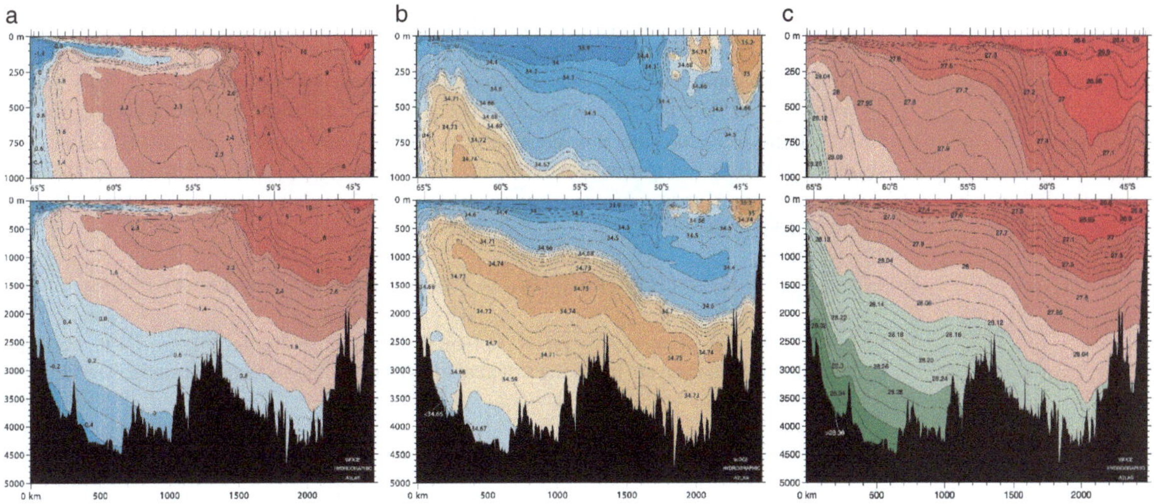

Fig. 16.2 **a** Potential temperature in °C, **b** salinity in $\mathrm{g\,kg^{-1}}$, and **c** neutral density in $\mathrm{kg\,m^{-3}}$ versus pressure in dbar along a hydrographic section between Australia and Antarctica (roughly at **140°** E). The section (WOCE SR3) was occupied in September 1996, and the figures are taken from the Southern Ocean Hydrographic Atlas (Orsi and Whitworth III, 2005)

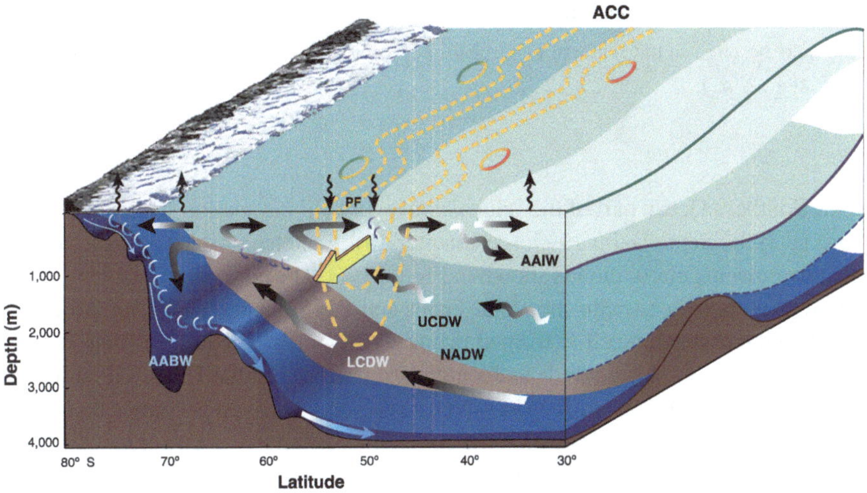


Fig. 16.3 A sketch of the ACC system showing the zonal flow (*yellow lines*) and the meridional overturning circulation (*dark arrows*) and water masses as indicated in the figure. Antarctica is at the *left side*. The east-west section displays the isopycnal and sea surface tilts in relation to submarine ridges. The *curly vertical arrows* at the surface indicate the buoyancy flux; the *small light arrows* attached to the isopycnals represent turbulent mixing. Redrawn from Olbers and Visbeck (2005)

penetrate across the ACC and, in fact, there is a prominent meridional overturning circulation associated with the predominantly zonal ACC. The figure implies this overturning circulation by the distribution of salinity and temperature which reveals the prominent water masses: Antarctic Intermediate Water (AAIW) is seen in the blue core of low salinity, Circumpolar Deep Water (CDW) is seen in the deeper brown core of high salinity and AABW in the deep blue core of low potential temperature. The meridional overturning circulation was described as early as 1933 by Sverdrup (see Sverdrup, 1933; Sverdrup et al., 1942) and has later been interpreted

**91. Ekman Spiral and Transport in the Southern Hemisphere**

The Ekman solutions for the southern hemisphere follow directly from the discussion in Chapter 14 and in the box on p. 446 and are given here for reference. The horizontal flow was decomposed into a geostrophically balanced part and a frictional component defined as $\boldsymbol{u} = \boldsymbol{u}_g + \boldsymbol{u}_e$, with $f\underline{\boldsymbol{u}}_e = A_v \partial^2 \boldsymbol{u}_e / \partial z^2$, where $A_v$ denotes a vertical turbulent viscosity related to small-scale turbulence. The latter flow component becomes important in the surface and bottom Ekman layers. Note that $\underline{\boldsymbol{u}}$ denotes a by 90° (anticlockwise) rotated vector as introduced in the box on p. 444. While (14.9) gives the solution for $\boldsymbol{u}_e$ in the surface Ekman layer with $f > 0$, the solution becomes

$$\boldsymbol{u}_e = d/(2A_v)\mathrm{e}^{z/d}\left[(\boldsymbol{\tau}_0 + \underline{\boldsymbol{\tau}}_0)\ \cos z/d + (\boldsymbol{\tau}_0 - \underline{\boldsymbol{\tau}}_0)\ \sin z/d\right]$$

for $f < 0$, where $d = \sqrt{2A_v/|f|}$ is the Ekman depth (which may be different for the surface and bottom Ekman layers by taking different $A_v$'s) and $\boldsymbol{\tau}_0$ the surface windstress. The bottom Ekman spiral becomes for $f < 0$

$$\boldsymbol{u}_e = \mathrm{e}^{-(z+h)/d}\left[-\boldsymbol{u}_g \cos(z+h)/d - \underline{\boldsymbol{u}}_g \sin(z+h)/d\right]$$

while for the bottom stress, (14.10) changes for $f < 0$ to

$$\boldsymbol{\tau}_b = A_v \frac{\partial \boldsymbol{u}_e}{\partial z}\Big|_{z=-h} = d|f|/2(\boldsymbol{u}_g - \underline{\boldsymbol{u}}_g)$$

The volume transport $\boldsymbol{U}_e = \int \boldsymbol{u}_e \mathrm{d}z$ integrated over the surface Ekman layer, i. e. the Ekman transport is given in the southern hemisphere by $\boldsymbol{U}_e^{\mathrm{top}} = -\underline{\boldsymbol{\tau}}_0/f$, while the bottom Ekman transport becomes in the southern hemisphere $\boldsymbol{U}_e^{\mathrm{bot}} = \underline{\boldsymbol{\tau}}_b/f = d|f|/(2f)(\underline{\boldsymbol{u}}_g + \boldsymbol{u}_g) = -d/2(\underline{\boldsymbol{u}}_g + \boldsymbol{u}_g)$.

as the Southern Ocean part of the 'global conveyor belt' circulation (Broecker, 1991; Gordon, 1986; Schmitz Jr., 1995, see Chapter 15). A sketch of the zonal and meridional overturning circulation is shown in Figure 16.3. The core of the eastward flowing ACC is associated with the Polar Front and the Subantarctic Front. An upper meridional overturning cell is formed primarily by northward Ekman transport beneath the strong westerly winds and southward transport in the Upper Circumpolar Deep Water (UCDW) layer. A lower overturning cell is driven primarily by formation of dense AABW near the Antarctic continent and inflowing Lower Circumpolar Deep Water (LCDW or NADW). This deep, relatively saline water spreads poleward and wells up towards the sea surface. Shallowing of the isopycnals is evident as the deep water rises up towards the sea surface. There it is partly transported southward, it cools and sinks, flooding the bottom layers with waters colder than 0 °C. This cold bottom water spreads well into the global oceans as AABW. The inflow of CDW is balanced by a northward flow of lower salinity waters near 1,000 m (AAIW) and by sinking of slightly lower salinity water along the continental slope of Antarctica. This process (salty water in, fresher water out) removes the slight excess of regional precipitation from the Southern Ocean.

## 16.2 Homogeneous Wind-Driven Models of the Southern Ocean

In this section, we explore the momentum balance and the flow characteristics of a wind-driven current in simple configurations of the Southern Ocean without stratification for different conditions, as e. g. a flat-bottom ocean compared to an ocean with topographic variations, or different scenarios of vertical compared to lateral viscous momentum transport. In most of what follows, we will employ the Elementary Current System which was introduced in Chapter 14. Now, however, it

**92. Transient and Standing Eddies**

In the description of the global atmospheric circulation, it is custom to reduce the information contained in observations by considering zonally averaged time-mean fields and deviations from it (see e. g. Peixoto and Oort (1992)). Such averaging will also be used in the present text, but it has to be used with care. Let us denote the time average of a field $v$ by $\langle v\rangle$ and split $v$ into its time average and the deviation $v^* = v - \langle v\rangle$. The deviation will be referred to the *transient eddy component*. Likewise, the zonal average of the time-mean variable may be defined by $\overline{\langle v\rangle}$ with the zonal deviations $v^\dagger = \langle v\rangle - \overline{\langle v\rangle}$, called the *standing eddy component*. Hence we obtain the representations

$$v = \langle v\rangle + v^* = \overline{\langle v\rangle} + v^\dagger + v^* = \overline{\langle v\rangle} + v'$$

Only the transient component $v^*$ has the character of a turbulent (time dependent) eddy field while the steady zonal deviation $v^\dagger$ must be considered as part of the time-mean circulation. Its separation from the time and zonal mean $\overline{\langle v\rangle}$ is somehow artificial, since the separation is usually not accompanied by a difference in the dynamics. Summarizing the transient and standing eddies according to $v' = v^\dagger + v^*$, on the other hand, attributes to the eddying motion a nonlocal character because it also includes in that case the deviation from the zonal mean.

The time-mean meridional flux of temperature $\theta$ is $\langle v\theta\rangle = \langle v\rangle\langle\theta\rangle + \langle v^*\theta^*\rangle$ since $\langle\langle\alpha\rangle\beta\rangle = \langle\alpha\rangle\langle\beta\rangle$ and $\langle\beta^*\rangle = 0$ (compare Section 2.8.2). This identifies a flux by the time-mean fields and a flux carried by the covariance of the deviation fields, the "eddy" flux $\langle v^*\theta^*\rangle$ induced by the transient eddies. Note that dividing each term by $\rho_0 c_p$ yields a mean heat flux and an 'eddy" heat flux. Likewise, we have $\overline{\langle v\theta\rangle} = \overline{\langle v\rangle}\;\overline{\langle\theta\rangle} + \overline{v^\dagger\theta^\dagger} + \overline{\langle v^*\theta^*\rangle}$ with the total flux and eddy-induced contributions from standing and transient eddies.

The application of this separation to the belt of latitudes passing Drake Passage reveals that indeed standing eddy and transient eddy components arise, and that the standing eddy component is often in fact larger than the transient one. To avoid this complication, an average oriented along mean streamlines or the contours of the time-mean sea surface height (SSH) is sometimes used instead. This kind of averaging often drastically reduces the standing eddy part (not completely, however, since the current may veer with depth) and shows the true nature of the transient eddy field to transport properties across mean streamlines.

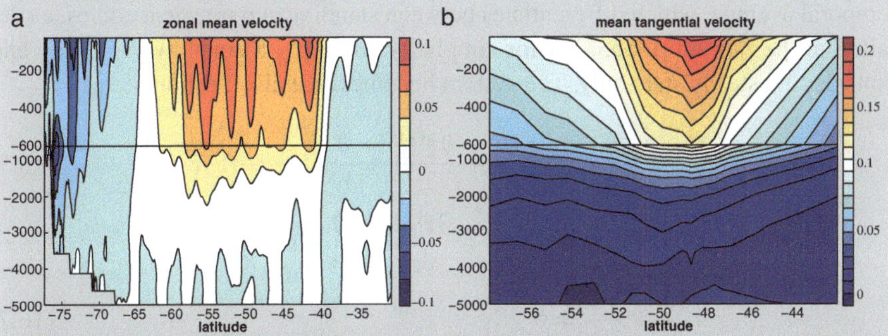


*The zonal average zonal velocity (in* m s$^{-1}$, ***a*** *and the path averaged tangential velocity (in* m s$^{-1}$, ***b****, following the time-mean SSH contours in a high resolution model of the Southern Ocean (the POP model; see Olbers and Ivchenko (2001)). In* ***b*** *the latitudinal coordinate is given by the mean latitude of the SSH contours. Only contours passing Drake Passage are used, and the path-averaged structure of the ACC becomes visible which is complete disguised in the zonal average.*

is used for southern hemisphere conditions where $f < 0$. The Ekman spiral and transport for the southern hemisphere are detailed for reference in the box on p. 562.

For simplicity, we consider an ocean with a southern solid boundary at $y = y_S$. It might be open or closed (then we have a channel) at a northern latitude $y = y_N$, we will denote the meridional extent of the domain by $Y = y_N - y_S$. The zonal extent is denoted by $X$. We assume here that the surface windstress $\boldsymbol{\tau}_0 = (\tau_0^{(x)}, \tau_0^{(y)})$ is zonal (i. e. $\tau_0^{(y)} \equiv 0$). In some examples, we will use a sinusoidal form where

$$\tau_0^{(x)}(y) = \tau_s \sin\left(\pi \frac{y - y_S}{Y}\right) \tag{16.1}$$

and zero for $y > y_N$. The configuration of this idealized homogeneous Southern Ocean and the resulting circulation is sketched in Figure 16.4 below.

We start with the equations of motion and continuity equation in the Boussinesq and hydrostatic approximation (the primitive equations as discussed in Chapter 4.2.5), written for convenience in Cartesian notation as

$$\frac{\partial u}{\partial t} + \nabla \cdot (\boldsymbol{u} u) - f v = -\frac{\partial p}{\partial x} + \frac{\partial \tau^{(x)}}{\partial z} \tag{16.2}$$

$$\frac{\partial v}{\partial t} + \nabla \cdot (\boldsymbol{u} v) + f u = -\frac{\partial p}{\partial y} + \frac{\partial \tau^{(y)}}{\partial z} \tag{16.3}$$

$$\frac{\partial u}{\partial x} + \frac{\partial v}{\partial y} + \frac{\partial w}{\partial z} = 0 \tag{16.4}$$

together with the hydrostatic relation $\partial p/\partial z = -g\tilde{\rho}/\rho_0$. Here $\boldsymbol{u} = (u, v, w)$ and $\nabla$ are three-dimensional. Note that $p$ denotes a scaled pressure, i. e. pressure $\tilde{p}$ divided by the reference density $\rho_0$. We have assumed that the turbulent (and molecular) stress transports horizontal momentum predominantly vertically. Thus, friction due to three-dimensional small-scale turbulence (and molecular processes) is contained in the flux divergences $\partial \tau^{(x)}/\partial z$ and $\partial \tau^{(y)}/\partial z$. Note that $\boldsymbol{\tau} = (\tau^{(x)}, \tau^{(y)})$ connects at the surface to the windstress $\boldsymbol{\tau}_0$. Rigid-lid conditions will be used for the sea surface.

We now apply a zonal average to the equations. Zonally averaged variables are indicated by an overbar. Note that the average could also be a combined zonal and temporal average, which differentiates between standing and transient eddies, as detailed in the box on p. 563, but for simplicity we apply a zonal average only and consider the steady state only. The system becomes (see also Section 15.6.1)

$$-f\bar{v} = -\frac{\delta p}{X} + \frac{\partial \bar{\tau}^{(x)}}{\partial z} - \frac{\partial}{\partial y}\overline{vu} - \frac{\partial}{\partial z}\overline{wu} \tag{16.5}$$

$$f\bar{u} = -\frac{\partial \bar{p}}{\partial y} + \frac{\partial \bar{\tau}^{(y)}}{\partial z} - \frac{\partial}{\partial y}\overline{vv} - \frac{\partial}{\partial z}\overline{wv} \tag{16.6}$$

$$\frac{\partial \bar{v}}{\partial y} + \frac{\partial \bar{w}}{\partial z} = 0 \tag{16.7}$$

The pressure force in the zonal momentum budget, $-\delta p/X$, with $X$ being the zonal extent of the Southern Ocean, only occurs in the presence of topography. It arises from the deep pressure difference in the valleys between topographic peaks along the zonal path and will be explained in the box on p. 568. Integrating the hydrostatic relation, the zonal mean pressure $\bar{p}$ is written as sum of a part given by the surface elevation and a part related to the stratification, the baroclinic part,

$$\bar{p} = g\bar{\zeta} + g\int_z^0 (\bar{\tilde{\rho}}/\rho_0)\,\mathrm{d}z = g\bar{\zeta} + p^{\mathrm{clin}} \tag{16.8}$$

Since we have assumed a homogeneous ocean where $\tilde{\rho}$ = const, there will be no contribution by $p^{\mathrm{clin}}$, and the pressure gradients are only related to the surface height $\zeta$ in this section.

The zonal average of the momentum advection on the right-hand side of (16.5) and (16.6) is often decomposed into a momentum advection by the zonally averaged flow ($\bar{v}$ and $\bar{w}$) and an eddy flux of momentum, e. g. as $\overline{vu} = \bar{v}\bar{u} + \overline{v'u'}$ and similar for the other terms, as discussed in the box on p. 563. We will neglect the mean advection terms of momentum. The vertical advection of momentum is small in a geostrophic scaling (see Section 5.1), and the mean meridional advection velocity, $\bar{v}$, vanishes to zero order if zonal mean zonal pressure gradient is vanishing. Likewise, we may argue with a small Rossby number appropriate for a large-scale flow. Fluctuations of $v$ and $u$, on the other hand, can play a role for the meridional (eddy) transports of momentum on scales approaching the Rossby radius. Hence, in some of the educational examples below, we will specify the eddy momentum fluxes $\overline{v'u'}$ and $\overline{v'v'}$ as a downgradient diffusion of horizontal momentum, although it is generally agreed that diffusion is not an appropriate representation of the lateral eddy momentum transport.

The continuity equation implies by vertical integration that in steady state $\partial U/\partial x + \partial V/\partial y = 0$ with $(U, V) = \int_{-h}^0 (u, v)\mathrm{d}z$. Zonal averaging yields $\partial\bar{V}/\partial y = 0$, hence $\bar{V}$ = const. In an ocean with a southern boundary where $\bar{v} = 0$ we thus have

$$\bar{V} = \int_{-h}^0 \bar{v}\,\mathrm{d}z = 0 \tag{16.9}$$

We now continue to construct several simple homogeneous models of the Southern Ocean and discuss the predicted values of the ACC transport and other quantities. None of these models is entirely realistic, since variations of density $\tilde{\rho}$ are missing, but they are still useful to learn about some principal aspects of the dynamics of a circumpolar ocean.

### 16.2.1 A Homogeneous Model with a Flat Bottom and Bottom Friction

First we consider a flat-bottom ocean, i. e. $h$ = const. The pressure difference term in (16.5) vanishes for a flat bottom. We also neglect the mesoscale eddy momentum fluxes $\overline{v'u'}$ so that no effect of the momentum advection remains in (16.5). The zonal current $\bar{u}$ is then geostrophic outside Ekman layers at the top and the bottom. Taking the vertical integral of (16.5) from the bottom $z = -h$ to the surface at (approxi-

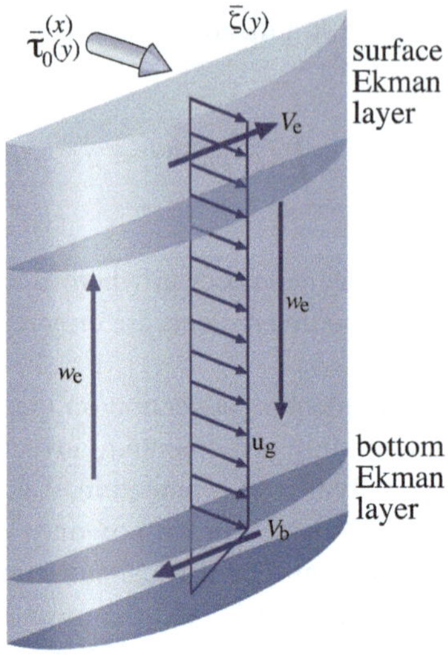


Fig. 16.4 Sketch of the solution of the homogeneous flat-bottom case. Above the frictional bottom Ekman layer, the zonal current has no vertical shear and is in geostrophic balance with a sea surface tilt. The meridional and vertical transports are due to the surface Ekman transport, Ekman pumping in the interior, and a frictional bottom current in the bottom Ekman layer. Note that all fields are independent of longitude

mately) $z = 0$ yields

$$f \int_{-h}^{0} \bar{v} \mathrm{d}z = 0 = \int_{-h}^{0} \frac{\partial \bar{\tau}^{(x)}}{\partial z} \mathrm{d}z = \bar{\tau}_0^{(x)} - \bar{\tau}_\mathrm{b}^{(x)} \tag{16.10}$$

We notice the importance of a frictional bottom stress, $\boldsymbol{\tau}_\mathrm{b} = (\tau_\mathrm{b}^{(x)}, \tau_\mathrm{b}^{(y)}) = \boldsymbol{\tau}(z = -h)$, to balance the zonal momentum input by the wind. Although we will work with this balance in the following analysis and also occasionally later in this chapter for didactic purposes, it should be made clear that a frictional bottom stress of the magnitude of the windstress is unrealistically large. Its role will later be replaced by the bottom formstress as detailed in the box on p. 568 and further below.

To sustain that stress, a current must be present rubbing on the floor, which can only be a geostrophic current $\bar{u}_\mathrm{g} = -(\partial \bar{p}/\partial y)/f = -(g/f)\partial\bar{\zeta}/\partial y$ associated with the meridional tilt of the sea surface (see Section 14.1.1 on the Elementary Current System), while $\bar{v}_\mathrm{g} = 0$. For $\tilde{\rho} = \text{const}$, $\bar{u}_\mathrm{g}$ penetrates undiminished through the whole water column and implies a bottom stress given by (see the box on p. 562)

$$\bar{\boldsymbol{\tau}}_\mathrm{b} = \begin{pmatrix} 1 \\ -1 \end{pmatrix} \frac{1}{2} g d_\mathrm{b} \frac{\partial \bar{\zeta}}{\partial y}$$

Hence we find from (16.10)

$$\frac{\partial \bar{\zeta}}{\partial y} = 2\bar{\tau}_0^{(x)}/(g d_\mathrm{b}) \quad \text{or} \quad \bar{\boldsymbol{u}}_\mathrm{g} = \begin{pmatrix} -2\bar{\tau}_0^{(x)}/(d_\mathrm{b} f) \\ 0 \end{pmatrix} \tag{16.11}$$

The sea surface is sloping upward to the north for westerly winds ($\bar{\tau}_0^{(x)} > 0$). The geostrophic flow occurs at all depths and flows into the direction of the windstress.

**Table 16.1** List of parameters and standard values used for scale estimates in this section: $X$, $Y = y_N - y_S$ are the zonal and meridional extent of the Southern Ocean, respectively, $N$ is the stability or Brunt–Väisälä frequency, $R_i = Nh/|f|$ is the internal Rossby radius, $A_h$ is a lateral eddy-induced viscosity, $K_\ell$ the thickness diffusivity, $d_e$ is the depth of the surface Ekman layer and $d_b$ likewise for the ocean bottom

| | | | | | |
|---|---|---|---|---|---|
| $f$ | $-1.25 \times 10^{-4}\,\mathrm{s}^{-1}$ | $\beta = \mathrm{d}f/\mathrm{d}y$ | $1.15 \times 10^{-11}\,\mathrm{m}^{-1}\,\mathrm{s}^{-1}$ | $A_h$ | $10^4\,\mathrm{m}^2\,\mathrm{s}^{-1}$ |
| $Y$ | 1,000 km | $X$ | 20,000 km | $h$ | 4,000 m |
| $d_e$ | 100 m | $d_b$ | 100 m | $\tau_0^{(x)}$ | $10^{-4}\,\mathrm{m}^2\,\mathrm{s}^{-2}$ |
| $N$ | $1.2 \times 10^{-3}\,\mathrm{s}^{-1}$ | $R_i$ | 12.6 km | $K_\ell$ | $1{,}000\,\mathrm{m}^2\,\mathrm{s}^{-1}$ |

In addition, we have a meridional transport $\bar{V}_e$ in the surface Ekman layer and zonal and meridional transports $(\bar{U}_b, \bar{V}_b)$ in the bottom Ekman layer, given by (from the box on p. 562)

$$\overline{\boldsymbol{U}}_e = \begin{pmatrix} \bar{U}_e \\ \bar{V}_e \end{pmatrix} = \begin{pmatrix} 0 \\ -\bar{\tau}_0^{(x)}/f \end{pmatrix} \quad \text{and} \quad \overline{\boldsymbol{U}}_b = \begin{pmatrix} \bar{U}_b \\ \bar{V}_b \end{pmatrix} = \begin{pmatrix} 1 \\ 1 \end{pmatrix} \bar{\tau}_0^{(x)}/f \tag{16.12}$$

Hence we find $\bar{V} = \bar{V}_e + \bar{V}_b = 0$, as required by the mass balance. The implied meridional circulation is closed by Ekman pumping arising from the windstress below the surface layer and from frictional bottom stress, in the same manner as explained in Section 14.1.1. Note that the zonal geostrophic transport $d_b\bar{u}_g = -2\bar{\tau}_0^{(x)}/f$ in the bottom layer is opposed to and exceeds the frictional transport $\bar{U}_b$. Assuming that $h \gg d_b$, the total transport $\boldsymbol{U}$ is thus almost identical to the geostrophic transport $h\boldsymbol{u}_g$. The zonal transports and the rest of the circulation in this simple homogeneous model follows by specifying the zonal windstress $\bar{\tau}_0^{(x)}$, and is sketched for the westerly winds characteristic for the Southern Ocean in Figure 16.4.

Table 16.1 lists scales for the relevant parameters representative for the Southern Ocean, from which we can deduce the consequences of our simple flat-bottom and homogeneous model. The standard parameters yield indeed fairly reasonable values for some of the quantities, i. e. for ACC strength $\bar{u}_g \sim 1.6 \times 10^{-2}\,\mathrm{m\,s}^{-1}$, ACC transport $Yh\bar{u}_g \sim 64$ Sv, frictional ACC transport $Y\bar{U}_b \sim -0.8$ Sv and meridional overturning transport $X\bar{V}_e = -X\bar{V}_b = 16$ Sv. An exception is the surface tilt: we find only $\partial\bar{\zeta}/\partial y \sim 2 \times 10^{-7}$ which is an order of magnitude too small (we know from satellite observations that the value of the surface slope in the ACC is roughly $1.5\,\mathrm{m}/1{,}000\,\mathrm{km} = 1.5 \times 10^{-6}$). Clearly, we can increase $\partial\bar{\zeta}/\partial y$ of the model by choosing a smaller Ekman depth for the bottom layer, but this would increase the transport as well. We are facing here a similar dilemma as described in the next paragraph, giving us a hint that something is missing in our model.

### 16.2.2 Hidaka's Dilemma with Lateral Friction

Next let us consider the presence of mesoscale eddies by adding the common diffusive representation of eddy momentum fluxes to the zonal and meridional momentum balances (16.5) and (16.6), i. e. by assuming $\overline{v'u'} = -A_h\partial\bar{u}/\partial y$ and $\overline{v'v'} = -A_h\partial\bar{v}/\partial y$, which yields

$$-f\bar{v} = -\frac{\delta p}{X} + \frac{\partial\bar{\tau}^{(x)}}{\partial z} + A_h\frac{\partial^2\bar{u}}{\partial y^2}, \quad f\bar{u} = -\frac{\partial\bar{p}}{\partial y} + \frac{\partial\bar{\tau}^{(y)}}{\partial z} + A_h\frac{\partial^2\bar{v}}{\partial y^2}$$

**93. Bottom Formstress**

The pressure difference $\delta p$ in the zonally averaged zonal momentum balance (16.5) is a shorthand notation of the sum over all ridges blocking the zonal path at the depth $z$, of the pressure difference $\delta_i p = p_i^W(y,z) - p_i^E(y,z)$ associated with the individual ridge numbered by $i$ (note that $E/W$ refer to the eastern/western side of the ridge, not of the intermediate valleys). Hence the cumulative pressure difference

$$\delta p(y,z) = \sum_i \delta_i p(y,z)$$

vanishes above the highest topography, assumed to block the path at a depth $z = -D(y)$ and latitude $y$. The vertical integral of the cumulative pressure difference is given by

$$\mathcal{F}(z) = -\int_z^0 \frac{\delta p}{X} \mathrm{d}z = -\int_z^0 \frac{1}{X} \sum_i \delta_i p(y,z) \mathrm{d}z = -\frac{1}{X} \sum_i \int_{x_i^W(z)}^{x_i^E(z)} P \frac{\partial h}{\partial x} \mathrm{d}x$$

with the total zonal extent $X$ of the circumpolar path and the bottom pressure $P = p(z = -h)$. The quantity $\mathcal{F}(z)$ is zero for $z > -D$ and accumulates at the greatest depth (occurring in the deepest valley at the latitude $y$) to the total bottom formstress

$$\mathcal{F}_\mathrm{b} = \mathcal{F}(-h) = \frac{1}{X} \oint P \frac{\partial h}{\partial x} \mathrm{d}x = \overline{P \frac{\partial h}{\partial x}} = -\overline{h \frac{\partial P}{\partial x}}$$

Note that the formstress becomes negative (taking eastward momentum out of the water and transferring it to the solid earth) if on average $p_\mathrm{W} > p_\mathrm{E}$. In this case, the geostrophic current within the valleys is southward.

where we have assumed a constant lateral turbulent viscosity $A_\mathrm{h}$ for simplicity. We emphasize again that this diffusive assumption about the nature of the eddy momentum fluxes is rather unphysical for large-scale ocean dynamics and that we introduce lateral momentum diffusion for educational purpose only in this section. For a flat bottom and $\tilde{\rho} = $ const, the vertically integrated balances of zonal and meridional momentum become

$$0 = \bar{\tau}_0^{(x)} - \bar{\tau}_\mathrm{b}^{(x)} + A_\mathrm{h} \frac{\partial^2 \bar{U}}{\partial y^2}, \quad f\bar{U} = -gh\frac{\partial \bar{\zeta}}{\partial y} - \bar{\tau}_\mathrm{b}^{(y)} \qquad (16.13)$$

Momentum can now be exported in the meridional direction from the wind patch towards the southern boundary where friction on the wall can work, or it can also be exported towards the north. The bottom friction is no longer needed to obtain a balance, but can we find a reasonable solution for free-slip at the bottom, i. e. for $\boldsymbol{\tau}_\mathrm{b} \equiv 0$?

The system (16.13) with zero bottom friction seems to be straightforward, but it turns out to be quite unphysical. From the zonal balance, we infer the scaling $\bar{U} \sim Y^2 \tau_0^{(x)}/A_\mathrm{h}$ so that the zonal transport scales as $Y\bar{U} \sim Y^3 \tau_0^{(x)}/A_\mathrm{h}$. With our standard parameters of Table 16.1, this yields values exceeding $10^4$ Sv, which is totally unrealistic. Using the sinusoidal windstress (16.1) yields a total transport $\int \bar{U} \mathrm{d}y = \tau_0^{(x)} Y^3 (4 + \pi^2)/(2\pi^3 A_\mathrm{h})$, modifying the above large value by the small factor $(4 + \pi^2)/(2\pi^3)$ to a somewhat smaller size which still amounts to 2240 Sv for the parameter values of Table 16.1. The relation is obtained for the boundary conditions $\bar{U} = 0$ at $y = y_\mathrm{S}$ and $\partial \bar{U}/\partial y = 0$ at $y = y_\mathrm{N}$.

To arrive at a realistic size of 140 Sv for the transport, we need to increase $A_\mathrm{h}$ by more than a factor of 15. This is obviously an unrealistic magnitude for the momentum diffusivity as we can see also from the following scaling: a diffusivity $A_h$ relates to the mesoscale eddy size $\ell'$ and eddy velocity $u'$ by $A_h \sim \ell' u'$. For $A_h = 1.5 \times 10^5\,\mathrm{m^2\,s^{-1}}$ and $v' \sim 0.1\,\mathrm{m\,s^{-1}}$, we find a far too large mesoscale eddy size of $\ell' \sim 1.5 \times 10^6\,\mathrm{m} \sim 1{,}000\,\mathrm{km}$. From the geostrophic balance in the meridional momentum equation of (16.13), it follows that the sea surface would increase by 6 m across the ACC which is a factor four too large. Hidaka's work (Hidaka and Tsuchiya, 1953) made the dilemma obvious of either choosing a reasonable value for the diffusivity $A_h$ in a numerical model of the Southern Ocean, leading to an unrealistically large simulated ACC transport, or choosing an unrealistically high value for $A_h$ to arrive at the correct size of about 140 Sv.

The dilemma is mitigated if we return to no-slip at the bottom. The geostrophic current $\bar{u}_g = -(g/f)\partial\bar{\zeta}/\partial y$ is, however, no longer the only driving of the frictional bottom layer, but it is rather the interior velocity $(\bar{u}_i, \bar{v}_i)$ at the top of that layer. Assuming $\partial\boldsymbol{\tau}/\partial z = 0$ for the interior, it follows

$$-f\bar{v}_i = A_h \frac{\partial^2 \bar{u}_i}{\partial y^2}, \qquad f\bar{u}_i = -g\frac{\partial\bar{\zeta}}{\partial y} + A_h \frac{\partial^2 \bar{v}_i}{\partial y^2}$$

We may solve the entire problem for the sinusoidal wind, but note here that the lateral Ekman number $A_h/(fY^2) \sim 10^{-4}$ is very small, and thus the interior current is well approximated by the geostrophic part, i. e. $\boldsymbol{u}_i \approx \boldsymbol{u}_g$. This leads to $\bar{\tau}_b^{(x)} = (gd_b/2)\partial\bar{\zeta}/\partial y$ as before in the case considered in Section 16.2.1 and when we insert now $\partial\bar{\zeta}/\partial y = -f\bar{U}/(gh)$, derived from (16.13) with $\bar{\tau}_b^{(y)} = -(gd_b/2)\partial\bar{\zeta}/\partial y$ and $d_b \ll h$, the balance (16.13) of the zonal flow becomes

$$\bar{\tau}_0^{(x)} + \frac{1}{2}\frac{d_b f}{h}\bar{U} + A_h \frac{\partial \bar{U}}{\partial y} = 0$$

The second term, resulting from the frictional bottom stress, generally overcomes the lateral momentum diffusion by one to two orders of magnitude so that the latter only plays a minor role, except in lateral boundary layers (compare the box on p. 458). We find $\bar{U} \sim -2\bar{\tau}_0^{(x)}h/(d_b f)$ and $Y\bar{U} \sim 64\,\mathrm{Sv}$ and $\partial\bar{\zeta}/\partial y \sim 10^{-7}$, as in the case of Section 16.2.1, i. e. a realistic ACC transport but too small sea surface slopes across the ACC. Note that the bottom stress implies a frictional transport $\bar{\boldsymbol{U}}_b = -(d_b/2h)\bar{U}(1,1)$. The same scaling argument used for the interior flow is also valid for the upper layer, and we conclude that the flow in that layer is almost contained in the Ekman transport $\bar{V}_e = -\bar{\tau}_0^x/f$. It balances the frictional transport $\bar{V}_b$ in the bottom layer. In fact, the lateral momentum diffusion is of no significance in this solution.

### 16.2.3 Homogenous Southern Ocean with Topography

Though formulated as a zonal average, the flat-bottom and homogeneous models of the ACC are valid in three dimensions as well, since in the absence of zonal perturbations the flow must be strictly independent of longitude. The introduction of a zonally varying topography $h = h(x, y)$ breaks the zonal symmetry, and it leads to a zonally undulating current. Further, the presence of zonal deep pressure perturbations enter the balance of the flow as bottom formstress as detailed in the

**94. A Homogeneous Southern Ocean Model**

The model is based on the equations (16.2)–(16.3) where the momentum advection $\nabla \cdot (\boldsymbol{uu})$ is neglected in agreement with the planetary geostrophic approximation (see Chapter 5.1). Lateral friction is included by viscous diffusion. Note that in this homogeneous model the pressure gradient force is related to the sea surface elevation $\zeta$ only, since the density $\tilde{\rho}$ is held constant. The model is integrated numerically to a steady state of the circulation around Antarctica (south of 20° S) with a horizontal resolution $2° \times 1°$, the horizontal viscosity is $A_h = 4 \times 10^4\,\mathrm{m^2\,s^{-1}}$, and there is no bottom friction. The model is forced with realistic annual mean windstress data shown in the figure below. The topography is either flat with a constant depth of 5,000 m (case FLAT) or realistic but slightly smoothed with a two-dimensional symmetric filter (case TOPO). The Drake Passage of the model is open between 62.5 and 55.5° S. The figure (a) below shows the $f/h$ contours for case TOPO (compare Section 14.2) and the zonal (b) and meridional (c) components of the windstress driving both configurations.

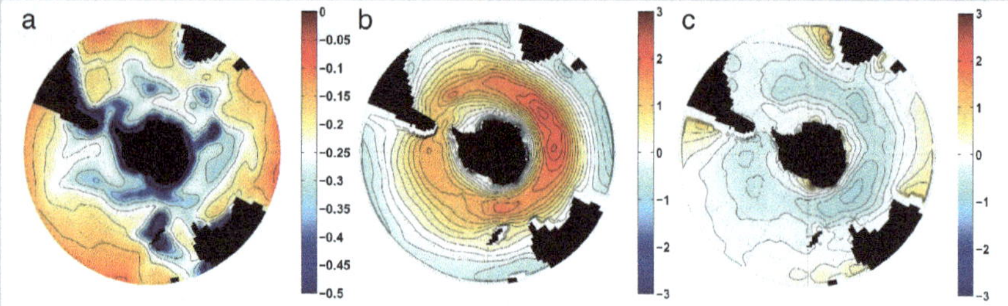


*Topography and forcing of the homogeneous model of the Southern Ocean.* ***a*** *geostrophic contours* $f/h$*, contour interval is* $5 \times 10^{-9}\,\mathrm{m^{-1}\,s^{-1}}$*.* ***b*** $\tau_0^{(x)}$ *and* ***c*** $\tau_0^{(y)}$ *in* $10^{-4}\,\mathrm{m^2\,s^{-2}}$*. Contour interval for* ***b*** *and* ***c*** *is* $10^{-4}\,\mathrm{m^2\,s^{-2}}$*.*

For a rigid-lid condition, the depth-integrated flow $\boldsymbol{U}$ is divergence-free and the introduction of a stream function $\psi$ for $\boldsymbol{U}$ becomes possible. A barotropic version of the model can be formulated which is governed by the vorticity equation

$$\frac{\partial}{\partial t}\boldsymbol{\nabla} \cdot \frac{1}{h}\boldsymbol{\nabla}\psi + \underset{\neg}{\boldsymbol{\nabla}}\psi \cdot \boldsymbol{\nabla}\frac{f}{h} = \underset{\neg}{\boldsymbol{\nabla}} \cdot \frac{\boldsymbol{\tau}_0}{h} + \underset{\neg}{\boldsymbol{\nabla}} \cdot \frac{\boldsymbol{\mathcal{R}}}{h}$$

with the usual boundary kinematic and dynamic conditions, augmented by an integral constraint which is needed to define the values of the stream function on the different continents in the multiconnected domain of the Southern Ocean. The path integral

$$\oint_C \mathrm{d}\boldsymbol{s} \cdot \boldsymbol{\nabla} P = \oint_C \mathrm{d}\boldsymbol{s} \cdot \frac{1}{h}\left[-\underset{\neg}{\boldsymbol{\nabla}}\frac{\partial \psi}{\partial t} + f\boldsymbol{\nabla}\psi + \boldsymbol{\tau}_0 - \boldsymbol{\tau}_b + \boldsymbol{\mathcal{R}}\right]$$

of the barotropic momentum balance (see (14.28)) has to vanish for any path $C$ around Antarctica. Note that if it vanishes for one particular path, then it vanishes for all paths due to the validity of the vorticity balance. Here we use a general friction term ($\boldsymbol{\mathcal{R}}$ is the negative vertically integrated Reynolds stress divergence; see (16.14)) and augment the equation by the time tendency term.

box on p. 568. Keeping the eddy momentum fluxes, the vertically integrated zonal momentum balance (16.5) becomes

$$\bar{\tau}_0^{(x)} - \bar{\tau}_b^{(x)} + \mathcal{F}_b + \mathcal{R}_b^{(x)} = 0 \qquad (16.14)$$

where $\mathcal{R}^{(x)}(z) = -\int_z^0 (\partial \overline{v'u'}/\partial y)\mathrm{d}z$ denotes the vertically integrated Reynolds stress divergence for which $\mathcal{R}_b^{(x)} = \mathcal{R}^{(x)}(-h)$ is the bottom value. Note that we still neglect vertical momentum advection and advection by $\bar{v}$, which are both small in the planetary-geostrophic approximation. The bottom formstress $\mathcal{F}_b$ is defined in the box on p. 568. It depends on the zonal pressure difference $\delta p(y,z)$, appearing

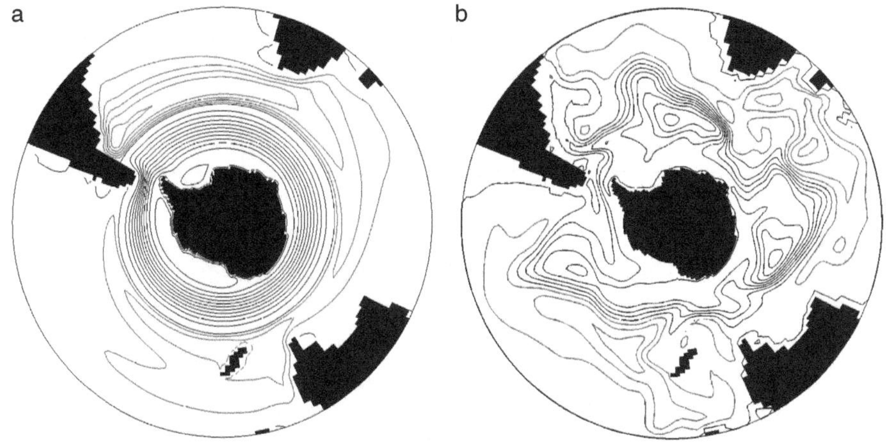


Fig. 16.5 Stream function $\psi$ in Sv for the numerical homogeneous model of the Southern Ocean (see the box on p. 570) for the case FLAT (**a**) and TOPO (**b**). Positive contours are *black*, negative *red* (the value $\psi = 0$ is on the South American continent). Contour intervals for FLAT are 100 Sv for positive and 20 Sv for negative contours. The contour interval for TOPO is 10 Sv

in (16.5). Neither the bottom formstress $\mathcal{F}_b$ nor $\mathcal{R}_b^{(x)}$ can be determined from zonal mean equations. In order to obtain a nonzero $\delta p$ and $\mathcal{F}_b$, there must be a nonvanishing geostrophic current $\boldsymbol{u}_g$ at depth, and thus variations in the bottom pressure $P$, and hence some functional dependence $\delta p = \delta p[\bar{\boldsymbol{u}}, y, z]$ must exist. With knowledge of this relation the model would be closed and could be solved. We will work out examples below in Section 16.7. In general, however, the relation will be too complicated or not even accessible.

In the remainder of this section, we discuss two simulations with a numerical model, a flat-bottom case FLAT and a topographic case TOPO. The configuration of both cases is summarized in the box on p. 570. The model has the realistic geometry, topography, and wind forcing of the Southern Ocean: the figures in the box show the geostrophic contours $f/h$ (compare Section 14.2) and the windstress forcing over the Southern Ocean. Since the model employs the rigid-lid assumption, it is convenient to introduce a stream function $\psi$ for the depth-integrated flow $\boldsymbol{U} = (U, V) = \int_{-h}^{0} \boldsymbol{u} \mathrm{d}z$ with $U = -\partial\psi/\partial y$ and $V = \partial\psi/\partial x$. Note that $\psi$ determines the volume transport between two points, i. e. $\int_A^B \nabla\psi \cdot \mathrm{d}\boldsymbol{s} = \psi(B) - \psi(A)$, where $\mathrm{d}\boldsymbol{s}$ is a line element of the section from $A$ to $B$.

The stream functions for the experiments FLAT and TOPO are shown in Figure 16.5. The flat-bottom, homogeneous case (experiment FLAT) has an unrealistically high ACC transport of about 1100 Sv, while the homogeneous ocean with topography (experiment TOPO) has a transport of only 33 Sv that is far too low to represent ACC conditions.

The flat-bottom case FLAT has an almost zonal ACC. There is a slight squeezing of streamlines in Drake Passage and a marked northward shift in the path of the current behind that obstacle. Since the bottom formstress cannot operate and the bottom friction is set to zero, lateral friction is the only momentum sink: with the unrealistically high ACC transport and the moderate lateral viscosity $A_h$, we are indeed facing Hidaka's dilemma, as discussed in the previous section.

The topographic case TOPO, on the other hand, establishes a bottom formstress arising from the pressure being out-of-phase with the submarine barriers of the flow,

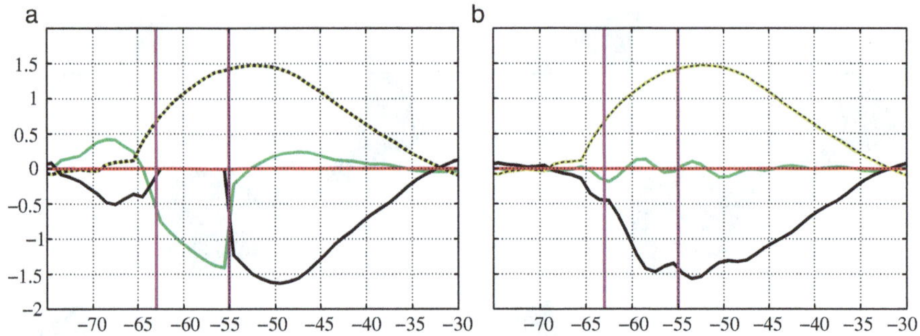


**Fig. 16.6** Zonally and vertically integrated zonal momentum balance as a function of latitude (N) for the cases FLAT (**a**) and TOPO (**b**). Shown are windstress $\bar{\tau}_0^{(x)}$ (yellow black dashed line), bottom formstress $\overline{h\partial P/\partial x}$ (*black line*) and lateral friction $A_{\rm h}\partial^2\bar{U}/\partial y^2$ (*green line*) in $10^{-4}$ m$^2$ s$^{-2}$. Drake Passage is indicated by the *vertical purple lines*

as discussed below. The depth-integrated flow is mostly parallel to the geostrophic contours $f/h$ = const (compare with the figure (a) in the box on p. 570). Referring the Section 14.2.2, it appears that neither the windstress forcing in the vorticity balance (14.31) nor the lateral friction (not contained in (14.31)) is very efficient in driving significant flow across the $f/h$ contours. Furthermore, since there are not many $f/h$ contours running through Drake Passage, the ACC transport remains weak. Embedded in the circumpolar flow, however, are huge closed circulation cells in areas with closed $f/h$ contours, namely above the Mid-Atlantic Ridge and around the Kerguelen plateau, with transports exceeding 100 Sv. These features will be analyzed in the next section.

The zonally averaged zonal momentum balances of FLAT and TOPO are shown in Figure 16.6. The balance (16.14) applies, however, there is no bottom friction in the present experiments. In the case FLAT, the dominant balance in the belt of latitudes passing through Drake Passage is given by lateral friction and windstress, since there is no other balance possible. In TOPO, friction is only minor in the zonally unbounded region, instead we find as dominant terms the windstress $\bar{\tau}_0^{(x)}$ and the bottom formstress $\mathcal{F}_{\rm b}$, i. e. the momentum balance of TOPO reduces to

$$\bar{\tau}_0^{(x)} - \overline{h\frac{\partial P}{\partial x}} = 0 \tag{16.15}$$

which is a balance between the applied windstress and transfer of zonal momentum into the bottom, occurring independently at each latitude. The bottom formstress acts as a 'form drag', avoiding Hidaka's dilemma even for a case without bottom friction. In a homogeneous ocean, as discussed in this section, the bottom pressure is entirely due to a tilt of the sea surface. Note that to balance a windstress of $10^{-4}$ m$^2$ s$^{-2}$, only a few centimeters are required across a ridge of a width of 1,000 km, with higher sea surface on the upstream side. Munk and Palmén (1951) were the first to discuss the balance (16.15) for the ACC, but surprisingly much of the research on the ACC after Munk and Palmen's article forgot the importance of the bottom formstress for some time and tried frictional balances, e. g. Hidaka and Tsuchiya (1953) and Gill (1968).

According to the box on p. 568, the bottom formstress $\mathcal{F}_{\rm b}$ arises from a systematic phase shift of the bottom pressure $P$ with respect to the topography $h$. The bottom

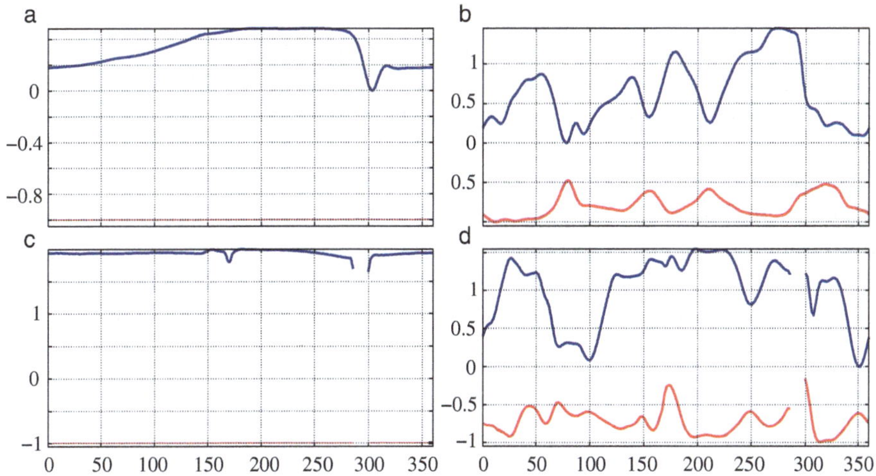


**Fig. 16.7** Bottom pressure $P$ and depth $h$ along two latitudes in the numerical homogeneous model of the Southern Ocean (see the box on p. 570) for the cases FLAT (left) and TOPO (right). The *panels* show the bottom pressure $P$ (*blue lines*, in the scaled form $P/\max(|P|)+1$) and the topography $h$ (*red lines*, in the scaled form $-h/\max(h)$ to separate the two curves) at the latitude $59.5°$ S (**a,b**), running through Drake Passage, and at the latitude $46.5°$ S (**c,d**), which is located north of Drake Passage. The passage is located roughly at the longitude $300°$ E (see the gap in the *curves* of the **c** and **d**)

pressure equals $P = g\zeta + g(\tilde{\rho}/\rho_0)h$ where the baroclinic term has no formstress effect since there are no density variations, and $h$ and $\partial h/\partial x$ are exactly out of phase. To obtain a net westward acceleration of the eastward current, i. e. a brake effect on the flow, highs of $P$ must appear to the west and lows in $P$ to the east of topographic barriers in the path of the flow. In fact, the flow organizes itself such that this sink of eastward momentum can become effective. Figure 16.7 displays the bottom pressure at a central latitude through Drake Passage together with the corresponding ocean depth, and in a similar way for a latitude north of Drake Passage. The shaping of the formstress in case TOPO is apparent: in the Drake Passage belt, we notice a westward shift of roughly 10° of the maxima in $P$ with respect to the maxima in $h$ and no such (or a much less correlated) pattern at latitudes to the north. In case FLAT, high pressure appears as well west of Drake Passage, but in contrast to TOPO the zonal gradients are much smaller. There is a gradual increase all around Antarctica which is balanced by the drop in Drake Passage. The bottom formstress in FLAT is, of course, zero as $h$ = const. Outside the zonally unbounded belt, the balance (16.15) is also valid in TOPO, and friction still does not play an important role. For zonally bounded regions, the bottom formstress is mainly given by the pressure difference $h\Delta P$ on the continents. This, in fact, is the principle balance of the zonal mean zonal momentum balance[2] in the basin-wide gyre circulations embedded between continents (see Chapter 14). Considering FLAT in the zonally bounded parts of the domain, we note that lateral friction plays a more important role than in TOPO.

Nevertheless, both simulations of the homogeneous model of the Southern Ocean are clearly quite unrealistic for the ACC. What is missing in the homogeneous

[2] Note that bottom or lateral friction is still necessary to close the energy budget because formstress is not an energy sink. This can be seen because pressure does not show up in the basin integrated kinetic energy budget (12.17).

model? In the following discussion, we demonstrate that stratification puts the system into a very different flow regime. At first, varying $\tilde{\rho}$ contributes to the meridional pressure gradient $\partial \bar{p}/\partial y$ such that an increasing compensation of the surface pressure gradient occurs with increasing depth. The geostrophic current thus diminishes with depth (see also the box on p. 560). The bottom formstress gets a baroclinic component as well, acting generally against the barotropic one, and thus accelerates the eastward flow. Most important, however, is the influence of baroclinicity on the overturning circulation. In a stratified medium, the current cannot easily cross isopycnals, and hence the zonal average, highlighted in the previous discussion, cannot reflect the local overturning anymore. We discuss these processes in Section 16.3.

### 16.2.4 The Barotropic Circulation over Closed $f/h$ Contours

The occurrence of regions in the Southern Ocean with closed $f/h$ contours was briefly pointed out in Section 14.2.2 (see Figure 14.8 and the box on p. 570). Prominent examples are the Mid-Atlantic Ridge, the Mid-Pacific Ridge, the region around Kerguelen Island, and Antarctica as a whole, however, the latter to a lesser degree, because only few contours completely cycle around. Problems in the vorticity balance become obvious in the particular form (14.31) where the left-hand side (divided by $|\nabla f/h|$) integrates to zero around a closed $f/h$ contour, but the windstress as a prescribed forcing cannot adjust to such a constraint. In this section we develop a solution for the barotropic circulation in regions of closed $f/h$ contours, attempting to understand the huge circulation cells appearing in the numerical solution TOPO in the previous section (see Figures 16.5 and 16.8).

We would like to point out that the barotropic case is on the whole an academic problem because the strong circulation cells over closed $f/h$ contours mostly disappear, when stratification is included and the circulation is simulated to higher degree of realism. This important property of stratification was presented and explained in Section 14.2 and will reopened specifically for the Southern Ocean in Section 16.5.

As before we will use for simplicity bottom friction in the form implemented in the vorticity balance (14.29). We consider a barotropic condition and thus cancel the

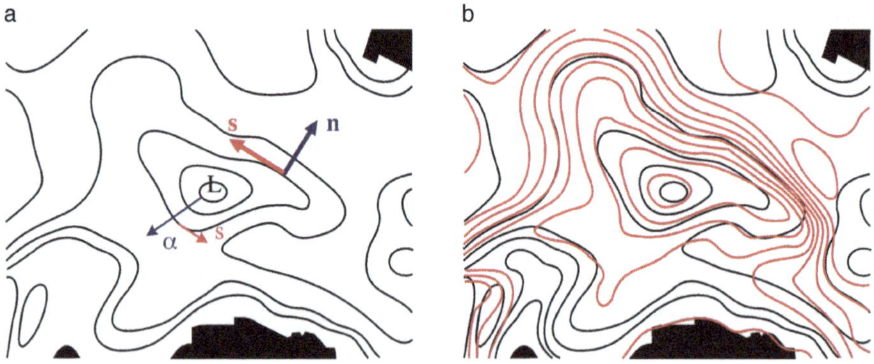


Fig. 16.8 The figures show the region east of Drake Passage. **a** $(\alpha, s)$ coordinates with normal vectors $\boldsymbol{n} = \nabla\alpha/|\nabla\alpha|$ and $\boldsymbol{s}$, here depicted for $\alpha$ increasing outward. **b** stream function $\psi$ (*red curves*) and $f/h$ (*black curves*) in the experiment TOPO

JEBAR term in this equation, which is then expressed by

$$\nabla \cdot (R/h)\nabla\psi + \underset{\neg}{\nabla}\psi \cdot \nabla(f/h) = \underset{\neg}{\nabla} \cdot (\boldsymbol{\tau}_0/h) \tag{16.16}$$

We write this equation in a coordinate system ($\alpha = f/h, s$), attached to the region with closed contours, where $s$ is an angle-type coordinate running cyclic along $\alpha = f/h = \text{const}$ (see Fig. 16.8 for the Mid-Atlantic Ridge situation and Appendix A.4 for the mathematical details of curvilinear coordinates). We may assume that $\oint \mathrm{d}s = 1$ without restriction. The metric coefficients are $\nu = |\nabla\alpha|$ and $\sigma = |\nabla s|$ which are generally functions of $\alpha$ and $s$. All terms in (16.16) are proportional to the product $\nu\sigma$, e. g. for the curl of the windstress we find

$$\underset{\neg}{\nabla} \cdot (\boldsymbol{\tau}_0/h) = \nu\sigma \left[ \frac{\partial}{\partial s}\frac{\tau_\perp}{\nu h} - \frac{\partial}{\partial\alpha}\frac{\tau_\parallel}{\sigma h} \right]$$

where $\tau_\perp = \boldsymbol{n}\cdot\boldsymbol{\tau}$ and $\tau_\parallel = \boldsymbol{s}\cdot\boldsymbol{\tau}$ are the components of the windstress that are normal and parallel to the $\alpha$ contours, respectively. Abbreviating for the moment the terms in (16.16) by $\nu\sigma W = \underset{\neg}{\nabla}\cdot(\boldsymbol{\tau}_0/h)$ (the vorticity source by windstress), $\nu\sigma F[\psi] = \nabla \cdot (R/h)\nabla\psi$ (the friction), and $\nu\sigma B[\psi] = \underset{\neg}{\nabla}\psi \cdot \nabla(f/h)$ (the topographic-planetary term), their integrated values must balance for each $\alpha$ closed contour. Hence,

$$\oint_{f/h=\text{const}} (W - F[\psi])\,\mathrm{d}s = 0 \tag{16.17}$$

because the contribution from the topographic-planetary term $B = \partial\psi/\partial s$ cancels exactly. We conclude that a frictionless flow over closed $f/h$ contours cannot exist in a homogeneous ocean, if a forcing $W$ with $\oint W \mathrm{d}s \neq 0$ is present. The integral of $W$ in (16.17) is evaluated as

$$\langle W \rangle = \oint W \mathrm{d}s = -\frac{\partial}{\partial\alpha}\oint \frac{\tau_\parallel}{\sigma h}\mathrm{d}s = -\frac{\partial}{\partial\alpha}\left\langle \frac{\tau_\parallel}{\sigma h} \right\rangle$$

and found to depend only the along-contour component of the windstress. The average over a closed $\alpha$ contour will henceforth be denoted by the cornered brackets. If the wind field is large-scale compared to the undulations of the $\alpha$ contours, then $\tau_\parallel$ is more or less in one direction on one side of the contours and reverses on the other side. This large-scale property of the windstress curl enters the closed-contour problem (16.17).

The aim is now to define a contour mean stream function $\bar{\psi}(\alpha)$ – in some sense the lowest spatial mode in the $(\alpha, s)$ coordinates – which is driven by the contour-averaged forcing $\langle W \rangle$, while the deviation $\psi' = \psi - \bar{\psi}$, representing higher modes, is driven by the deviation $W - \langle W \rangle$. With $\bar{\psi} = \langle \psi \rangle$ this is generally not achieved because then $\bar{\psi}$ and $\psi'$ are still coupled. A specific definition of the contour mean, however, allows for a partial decoupling. We arrive at such a decomposition with the following considerations. Writing the stream function $\psi = \psi(\alpha, s)$ in the local coordinates $(\alpha, s)$, the vorticity balance becomes

$$\frac{\partial\psi}{\partial s} + R\left( \frac{\partial}{\partial\alpha}\frac{\nu}{\sigma h}\frac{\partial}{\partial\alpha} + \frac{\partial}{\partial s}\frac{\sigma}{\nu h}\frac{\partial}{\partial s} \right)\psi = W$$

and integration around an $\alpha$ contour leads to

$$R\frac{\partial}{\partial\alpha}\oint \frac{\nu}{\sigma h}\frac{\partial\psi}{\partial\alpha}\mathrm{d}s = \langle W \rangle$$

which is (16.17), written now in the specific coordinates. We insert $\psi = \bar{\psi} + \psi'$ with a yet undefined separation of the components and obtain

$$R\frac{\partial}{\partial\alpha}\left(\oint \frac{\nu}{\sigma h}\mathrm{d}s\right)\frac{\partial\bar{\psi}}{\partial\alpha} + R\frac{\partial}{\partial\alpha}\oint \frac{\nu}{\sigma h}\frac{\partial\psi'}{\partial\alpha}\,\mathrm{d}s = \langle W\rangle$$

In order to reach our goal of decoupled equations, the $\psi'$-term on the left-hand side must vanish. The mean must thus defined by

$$\frac{\partial\bar{\psi}}{\partial\alpha} = \oint \frac{\nu}{\sigma h}\frac{\partial\psi}{\partial\alpha}\mathrm{d}s \Big/ \oint \frac{\nu}{\sigma h}\mathrm{d}s \tag{16.18}$$

Note that $\nu\partial\psi/\partial\alpha = \boldsymbol{s}\cdot\boldsymbol{U} = U_{\|}$, hence the above definition refers to the contour mean of $U_{\|}/h$. Note further that we have to distinguish between the contour integral, denoted by the cornered brackets, and the weighted contour mean, given above and denoted by an overbar. The vorticity balance is now separated into the contour mean and the deviation, governed by

$$R\frac{\partial}{\partial\alpha}\left\langle\frac{\nu}{\sigma h}\right\rangle\frac{\partial\bar{\psi}}{\partial\alpha} = \langle W\rangle$$

$$\frac{\partial\psi'}{\partial s} + R\left(\frac{\partial}{\partial\alpha}\frac{\nu}{\sigma h}\frac{\partial}{\partial\alpha} + \frac{\partial}{\partial s}\frac{\sigma}{\nu h}\frac{\partial}{\partial s}\right)\psi' + R\frac{\partial}{\partial\alpha}\left(\frac{\nu}{\sigma h} - \left\langle\frac{\nu}{\sigma h}\right\rangle\right)\frac{\partial\bar{\psi}}{\partial\alpha} = W - \langle W\rangle \tag{16.19}$$

The definition of the weighted mean (16.18) allows to treat $\bar{\psi}$ independently from $\psi'$ but not vice versa. Note, however, that the coupling of the mean to the deviation stream function, given by the third term on the left-hand side of (16.19), may be small because the specific coupling coefficient is a deviation term.

The equation for the mean $\bar{\psi}$ is readily solved. First notice that by integration the momentum balance results,

$$R\left\langle\frac{\nu}{\sigma h}\right\rangle\frac{\partial\bar{\psi}}{\partial\alpha} = -\left\langle\frac{\tau_{\|}}{\sigma h}\right\rangle \tag{16.20}$$

which is the balance for the component parallel to and averaged around $\alpha$ contours. An interesting feature of (16.20) is that for constant $f$, where $h = f/\alpha = h(\alpha)$ and thus $h$ and $\alpha$ contours coincide, the actual depth profile drops out. It is a consequence of the simple form of the bottom friction applied in the present model. Equation (16.20) is ready for integration to give $\bar{\psi}(\alpha)$, but we postpone the discussion to the specific example outlined below.

The 'closed contour dilemma' may now be rephrased as follows. If the contour averaged forcing $\langle W\rangle$ is nonzero, then a resonantly forced component $\bar{\psi}(\alpha)$ exists which necessarily requires friction to adjust: there is no free-mode solution in such a region, referring to the discussion in Section 14.2.2. On the other hand, the deviation $\psi' = \psi - \bar{\psi}$ does not necessarily require friction. In fact, $\psi'$ could exist in a Sverdrup-type balance over the region with closed contours. As explained above, the forcing $W - \langle W\rangle$ for the $f/h$ islands over the midocean ridges in the Southern Ocean may become small, so that $\psi'$ is only a small perturbation compared to $\bar{\psi}$: the circulation is then mostly along the $f/h$ contours.

The solution for the deviation field $\psi'$ for a general situation is not possible. If friction, acting on the $\psi'$ field, is negligible the integration becomes trivial. Nevertheless, this approximation does not survive if boundary conditions have to be satisfied on the outmost $\alpha$ contour, connecting the circulation over the closed-contour region with the outside world. In this case, the friction term in the $\psi'$-equation must be revived. Below we give an example for circulation over circular $\alpha$ contours.

### Example: Circular Geostrophic Contours

Assume polar coordinates $(r, \theta)$ and take $\alpha = \alpha(r)$ circular. Hence with $f = f(r, \theta) = f_0 + \beta r \sin\theta$ we find for the ocean depth $h = h(r, \theta) = f(r, \theta)/\alpha(r)$. Note that the $\theta$-dependence of the depth occurs for this specific case only via the $\beta$-effect. The equations are conveniently rewritten in $r = r(\alpha)$, instead of $\alpha(r)$, and $\theta = 2\pi s$ (the latter is for $\partial\alpha/\partial r > 0$, e. g. a hill in the southern hemisphere; for a valley $\theta = -2\pi s$). We find $\nu = |\nabla\alpha| = |\partial\alpha/\partial r|, \sigma = |\nabla s| = 1/r$. In these coordinates, the vorticity balance is of the form (the proper form of the equation in polar coordinates),

$$
\begin{aligned}
&R\frac{1}{r}\left(\frac{\partial}{\partial r}\frac{r}{h}\frac{\partial\psi}{\partial r} + \frac{1}{r}\frac{\partial}{\partial\theta}\frac{1}{h}\frac{\partial\psi}{\partial\theta}\right) - \frac{1}{r}\frac{\partial\alpha}{\partial r}\frac{\partial\psi}{\partial\theta} = \underset{\neg}{\nabla}\cdot(\boldsymbol{\tau}_0/h) = \frac{1}{r}W \\
&W = \frac{\partial(r\tau^{(\theta)}/h)}{\partial r} - \frac{\partial(\tau^{(r)}/h)}{\partial\theta}
\end{aligned}
\tag{16.21}
$$

with $(\tau^{(r)}, \tau^{(\theta)})$ as components of $\boldsymbol{\tau}_0$ in the polar coordinates (these are the $\tau_\perp$ and $\tau_\parallel$ from the general case discussed above). For simplicity, we take now a uni-directional windstress $\boldsymbol{\tau}_0 = (\tau^{(x)}(y), 0)$ along the $x$-axis with a constant curl, i.e. $\tau^{(x)}(y) = a + cy$ with constants $a, c$. Then, with $y = r\sin\theta$ we have $\tau^{(x)} = a + cr\sin\theta, \tau^{(r)} = \tau^{(x)}\cos\theta, \tau^{(\theta)} = -\tau^{(x)}\sin\theta$, so that

$$
W = r\underset{\neg}{\nabla}\cdot(\boldsymbol{\tau}_0/h) = -\frac{cr}{h(r)} + \frac{a + cr\sin\theta}{h^2(r)}\left(r\sin\theta\frac{\partial h}{\partial r} + \cos\theta\frac{\partial h}{\partial\theta}\right) \tag{16.22}
$$

Remember that $\partial h/\partial\theta \sim \beta$ arises in this configuration only through the $\beta$-effect.

The basic features can be learned from a $f$-plane situation where $\beta = 0$ and, with $\alpha$ being circular, we also have $h = h(r) = f_0/\alpha(r)$ with circular contours (the general case is treated in the box on p. 579). Then

$$
\langle W\rangle = -\frac{1}{2}c\frac{\partial}{\partial r}\left(\frac{r^2}{h(r)}\right)
$$

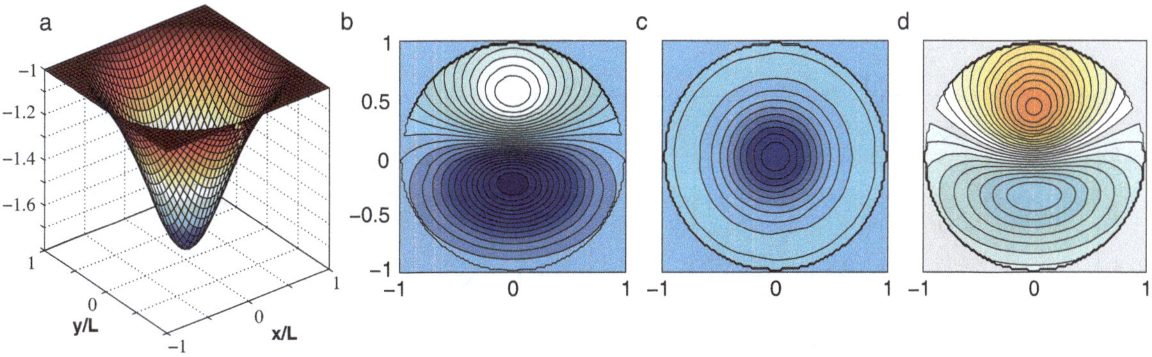


Fig. 16.9 **a** The valley of $\alpha(r) = f_0/h(r)$ for a Gaussian hill, shown in the $(x/L, y/L)$-plane. Note that $\alpha$ is scaled by the constant $|f_0|/H_0$. **b** The curl $W/r$ of the windstress, **c** the contour mean $\langle W\rangle/r$, and **d** the deviation $(W - \langle W\rangle)/r$

We illustrate the pattern of $W/r$ and its mean $\langle W\rangle/r$ in Figure 16.9, using a Gaussian hill, leading to the ocean depth $h(r) = H_0 - H_1 \exp(-r^2/r_0^2)$.

For the present case, the averaging procedures for the overbar and the cornered brackets collapse because the metric coefficients and the depth are function of only $r$. Hence we average over the angle $\theta$, defining $\bar{\psi} = \int \psi d\theta/(2\pi)$ and arrive at the equation

$$R\frac{\partial}{\partial r}\frac{r}{h}\frac{\partial \bar{\psi}}{\partial r} = -\frac{1}{2}c\frac{\partial}{\partial r}\left(\frac{r^2}{h(r)}\right)$$

which determines the contour-averaged stream function $\bar{\psi}$ as

$$\bar{\psi}(r) = \bar{\psi}(r_0) - \frac{1}{4}\frac{c}{R}\left(r^2 - r_0^2\right) \tag{16.23}$$

Here $r = r_0$ is conveniently the location of the outer rim of the topography. As mentioned before, the solution is independent of the specific depth profile $h(r)$. The solution is displayed in Figure 16.10.

The equation for the deviation stream function $\psi' = \psi - \bar{\psi}$ becomes

$$\begin{aligned} R\left(\frac{\partial}{\partial r}\frac{r}{h}\frac{\partial \psi'}{\partial r} + \frac{1}{hr}\frac{\partial^2 \psi'}{\partial \theta^2}\right) - \frac{\partial \alpha}{\partial r}\frac{\partial \psi'}{\partial \theta} &= W - \langle W\rangle \\ &= \frac{r}{h^2}\frac{\partial h}{\partial r}\left[a \sin\theta + cr\left(\sin^2\theta - \frac{1}{2}\right)\right] \end{aligned} \tag{16.24}$$

Note that $\psi'$ is here strictly decoupled from the mean $\bar{\psi}$. Assume now for the moment that the friction terms in this balance are small. Then $\psi'$ is determined as

$$\psi'(r,\theta) = -\frac{r\cos\theta}{f}\left(a + \frac{1}{2}cr\sin\theta\right) \tag{16.25}$$

Again the actual profile of $h(r)$ drops out. If it happens that $\bar{\psi}(r_0) + \psi'(r_0,\theta)$ matches continuously to the flow outside the area of closed contours, the deviation field can remain frictionless and given by (16.25). We depict such a case in Figure 16.10. The general case, however, requires friction in the $\psi'$-balance (16.24). Then a boundary condition at $r = r_0$ can be posed.

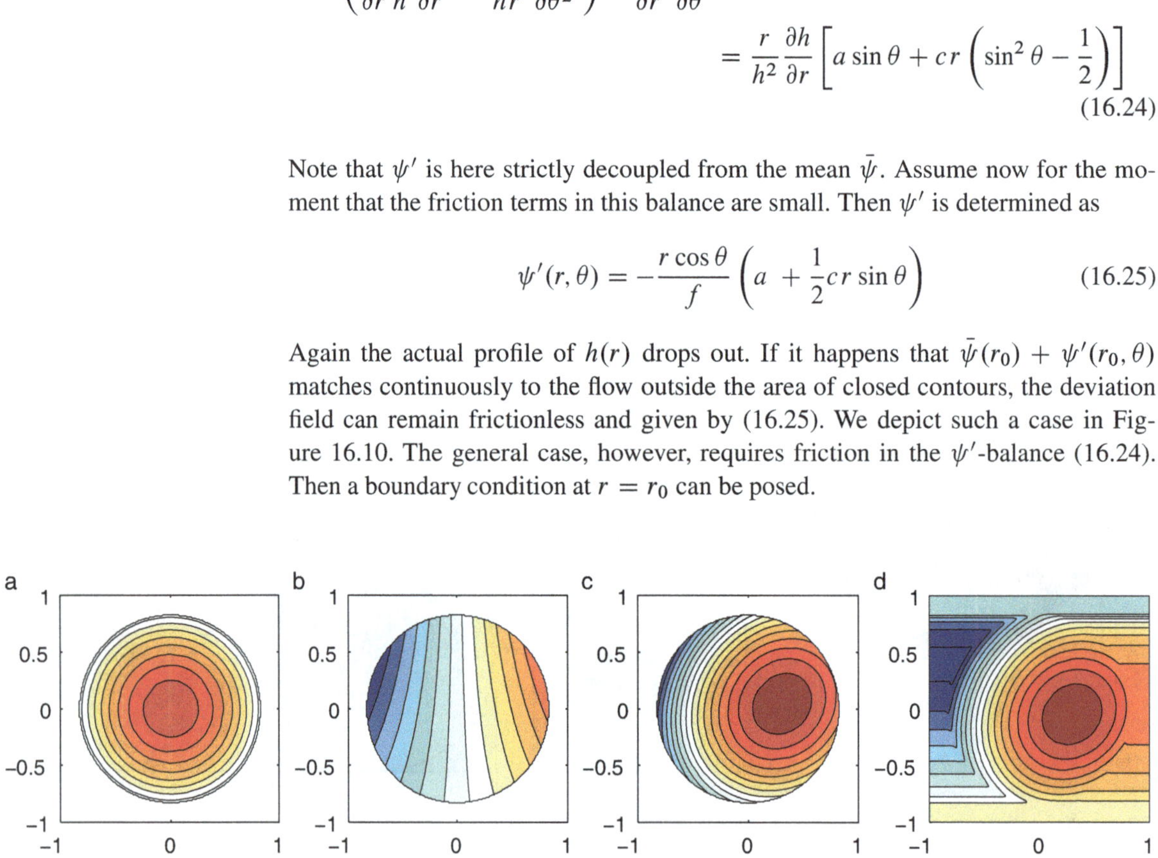


Fig. 16.10 The solution parts **a** $\bar{\psi}$ from (16.23), **b** $\psi'$ for the frictionless approximation (16.25), **c** the sum of both, and d) the sum in background flow which matches to the flow on the rim of the closed contour region

95. Circular $f/h$ contour with $\beta \neq 0$

We average (16.21) with the specific windstress curl (16.22) over the angle $\theta$, denoted as before by cornered brackets, and arrive at

$$R\frac{\partial}{\partial r}\left\langle\frac{r}{h}\frac{\partial\psi}{\partial r}\right\rangle = R\frac{\partial}{\partial r}\left\langle\frac{r}{h}\right\rangle\frac{\partial\bar{\psi}}{\partial r} = \langle W\rangle =$$
$$= \frac{\partial}{\partial r}r\alpha\left(cr\left\langle\frac{\sin^2\theta}{f}\right\rangle - a\left\langle\frac{\sin\theta}{f}\right\rangle\right)$$

Integrating once with respect to $r$ yields

$$R\frac{\partial\bar{\psi}}{\partial r} = \left(cr\left\langle\frac{\sin^2\theta}{f}\right\rangle - a\left\langle\frac{\sin\theta}{f}\right\rangle\right)\Big/\left\langle\frac{1}{f}\right\rangle \tag{B95.1}$$

which determines the contour-averaged stream function $\bar{\psi}$. We have to evaluate the contour-averaged coefficients appearing on the right-hand side. They are found in term of elementary square-root functions for $f_0 < 0$, $f_0^2 > (\beta r)^2$ (not given here). The equation (B95.1) then becomes

$$R\frac{\partial\bar{\psi}}{\partial r} = (cf_0/\beta + a)\frac{1-\sqrt{1-(\beta r/f_0)^2}}{\beta r/f_0}$$

which is solved by

$$\bar{\psi}(r) = \frac{f_0}{\beta}\frac{cf_0/\beta + a}{R}$$
$$\times\left(-\sqrt{1-(\beta r/f_0)^2} - \frac{1}{2}\ln\frac{1-\sqrt{1-(\beta r/f_0)^2}}{1+\sqrt{1-(\beta r/f_0)^2}} + \ln\beta r/f_0 + 1 - \frac{1}{2}\ln 4\right) \tag{B95.2}$$

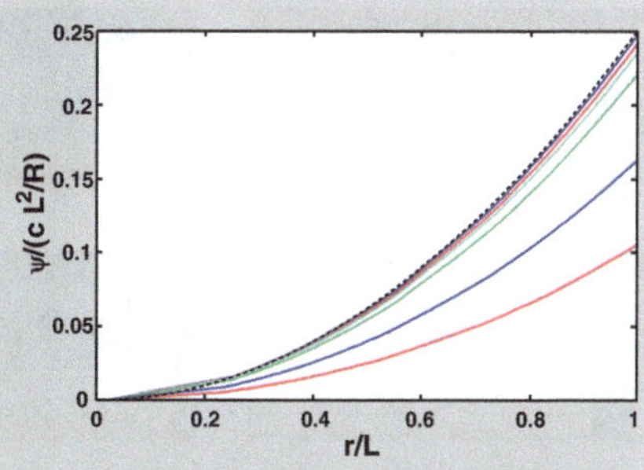


*The solution (16.23) for the case $\beta = 0$ [dashed] is here compared to (B95.2) for various values of $\beta = (0.2, 0.6, 1.0, 2.0, 6.0, 10.0) \times 10^{-12}\,\mathrm{m^{-1}\,s^{-1}}$. Shown are the scaled stream functions $\bar{\psi}/(cL^2/R)$ as function of $r/L$.*

The integration constant is adjusted such that $\bar{\psi}(0) = 0$. It is interesting that the limit $\beta \to 0$, implying $f = $ const and $h = h(r)$, leads to the simple solution $\bar{\psi}(r) = -cr^2/(4R)$ discussed in the text. The above figure compares these solutions and demonstrates the agreement in the limit $\beta \to 0$. The balance of the deviation stream function $\psi'$ for case $\beta \neq 0$ can be solved in similar way as discussed in the text.

### Some Numerical Solutions

Finally we corroborate our findings with some numerical solutions (produced with a finite-element code of the BARBI model, implemented here for homogeneous density conditions; courtesy of Sergey Danilov). The mark, that a closed contour region leaves in the overall circulation, very much depends on details of the forcing and the position of the mountain in the forcing field, supporting the contour-averaged windstress curl. Also the position of the mountain relative to the upstream circulation matters, finding its expression in the boundary condition of the above discussed analytical solution on the outer rim. Figure 16.11 displays four cases of wind-driven circulations in southern hemisphere channel (extent from 45 to 30° S and length 60°; the depth is 4,000 m, perturbed by a ridge and a mountain as can be seen in the $f/h$

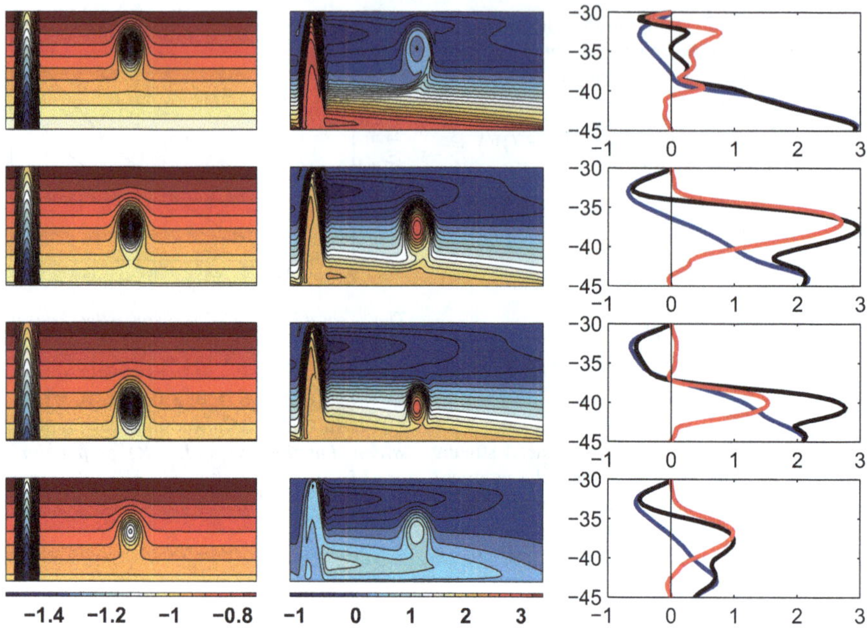


Fig. 16.11 Wind-driven barotropic flow in a zonal channel with a ridge and a mountain, the latter has different positions for the four cases, which can be inferred from the $f/h$ contours shown in the *first column* of panels. The *second column* displays the stream function, and the *third column* displays the meridional profiles of the stream function west of the mountain (*blue curves*), along the center of it (*black curves*), and their difference (*red curves*). Further details of the experiments are given in the text. Units for the stream function are 100 Sv, CI = 20 Sv

contours in the panels of the left column). The windstress is sinusoidal and eastward with amplitude $2 \times 10^{-4}\,\mathrm{m^2\,s^{-2}}$ and zero on the northern and southern channel rims. The ridge has a height of 1,000 m and mountain is 1,500 m high with a circular Gaussian shape. Its position in the channel changes from a northern place in the first experiment to a central one in the second and a more southern place in the third experiment. The $f/h$ contours in these three cases are not blocked by the ridge – they close in the southern domain of the channel, unlike the ones in the fourth case where the ridge height is increased to 1,400 m and all $f/h$ become blocked. Consequently, the circulation of this case is weakest (shown in the fourth row of Figure 16.11 and the profiles in the right column). Note that in absence of the hill, we have an eastward flow with lobes which reverse at the northern and southern boundaries.

The case with the northern most position of the hill leads to a negative contour averaged windstress curl $\langle W \rangle$ because the wind on the northern slope of the hill is weaker than the one the southern slope. This forcing, therefore, drives a reversed (counterclockwise) circulation around the closed contours of the hill (uppermost row of Figure 16.11). The next two cases (central and southern positions, shown in the second and third rows) yield a massive clockwise contour-averaged circulation over the hill, which sticks out in a prominent way in the overall central eastward flow. In the last case (blocked $f/h$ contours, hill central; fourth row) the contour-averaged flow is still clockwise but much weaker than in the previous cases. All these features are clearly visible in the circulation patterns, shown in the respective second column panels.

The left columns of Figure 16.11 illustrate the stream function of the central longitude of the hill relative to the upstream meridional profile just outside the western slope of the respective hills, i. e. $\psi_{\rm cent}(y) - \psi_{\rm ups}(y)$ is plotted (red curves). The upstream profile $\psi_{\rm ups}(y)$ (blue) and the central profile $\psi_{\rm cent}(y)$ (black) are displayed displayed as well.

## 16.3 The Meridional Overturning of the Southern Ocean

Consider again the zonally averaged zonal momentum budget (16.5) and, for simplicity, focus for the moment on the interior flow with $\delta p = 0$ and neglect momentum advection. Evidently, the meridional flow is only driven by the frictional vertical stress $\bar{\tau}^{(x)}$ which is small in the interior, and thus the meridional velocity $\bar{v}$ in the interior must be small compared to the Ekman current in the surface layer. In the idealized cases of Section 16.2, we have assumed absence of interior small-scale turbulence, $\tau^{(x)} = 0$, and a vanishing meridional current in the interior was the consequence. The interior circulation in the meridional/vertical plane, i. e. the *meridional overturning circulation*, is then strictly vertical, supported by surface (and bottom) Ekman transports and the vertical pumping $\bar{w}_{\rm e} = -\partial(\bar{\tau}_0^{(x)}/f)/\partial y$ out of the surface (and bottom) Ekman layer as summarized in Figure 16.4. While this picture is certainly valid for a homogeneous ocean, the situation should change for a stratified ocean, as sketched in Figure 16.3. Here, the interior meridional overturning circulation is not vertical, but it is believed to closely follow the isopycnals. We discuss the meridional overturning circulation in a stratified ocean in this section in more detail.

### 16.3.1 The Eulerian Meridional Overturning Circulation

The stream function $\Lambda(y, z)$ for the meridional overturning circulation may be defined as $\partial\Lambda/\partial z = -\bar{v}$, $\partial\Lambda/\partial y = \bar{w}$ and $\Lambda(y, z = 0) = 0$, hence

$$\Lambda(y, z) = \int_z^0 \bar{v}{\rm d}z \tag{16.26}$$

is vanishing also on the bottom $z = -h$ by (16.9). The stream function $\Lambda$ will be called the *Eulerian stream function* (averaged on constant geopotentials, $z =$ const), in contrast to a Lagrangian one (averaged on isopycnals, $\tilde{\rho} =$ const), which yields a better description of the meridional overturning circulation in the Southern Ocean, as discussed below in this section.

We find the balance of the meridional overturning in terms of the stream function by vertical integration of (16.5) from the level $z$ to the surface $z = 0$,

$$-f\Lambda = \bar{\tau}_0^{(x)} - \bar{\tau}^{(x)} + \mathcal{F} - \mathcal{R}^{(x)} \tag{16.27}$$

now written for a nonzero $\delta p$. The Reynolds stress term $\mathcal{R}^{(x)}(z) = -\int_z^0 \partial\overline{v'u'}/\partial y{\rm d}z$ is induced by standing and transient eddies (compare the box on p. 563) and usually very small. The bottom formstress $\mathcal{F}(z)$ arise through $\delta p$ and is defined in

### 96. An Idealized Southern Ocean/Atlantic Basin Model

We use an eddy-permitting primitive equation model consisting of a zonally reentrant channel (the Southern Ocean part) which is connected to a northern ocean basin enclosed by land at the eastern, western and northern sides (the 'Atlantic basin' part, see below). The equations are formulated in Cartesian coordinates. The domain of the idealized model extends over 2,500 km in zonal and meridional direction with 20 km horizontal resolution and 20 vertical levels with 50 m thickness (the basin depth is 900 m). The model includes only temperature, which is proportional to density or buoyancy (with $\partial\rho/\partial T = -0.2\,\mathrm{kg\,m^{-3}\,K^{-1}}$). The model uses a southern hemisphere $\beta$-plane. Relevant model parameters are: vertical viscosity $A_\mathrm{v} = 10^{-3}\,\mathrm{m^2\,s^{-1}}$, vertical diffusivity $K_\mathrm{v} = 10^{-4}\,\mathrm{m^2\,s^{-1}}$, biharmonic viscosity $A_\mathrm{hbi} = 10^{12}\,\mathrm{m^4\,s^{-1}}$, bottom friction coefficient $r = 10^{-5}\,\mathrm{s^{-1}}$. The circulation in the model is driven by a sinusoidal eastward windstress over the Southern Ocean part only and a relaxation boundary condition at the surface for temperature $T$ similar to (13.12) with a relaxation time scale of 30 days and target buoyancy $T^*$, as given in the following figure.

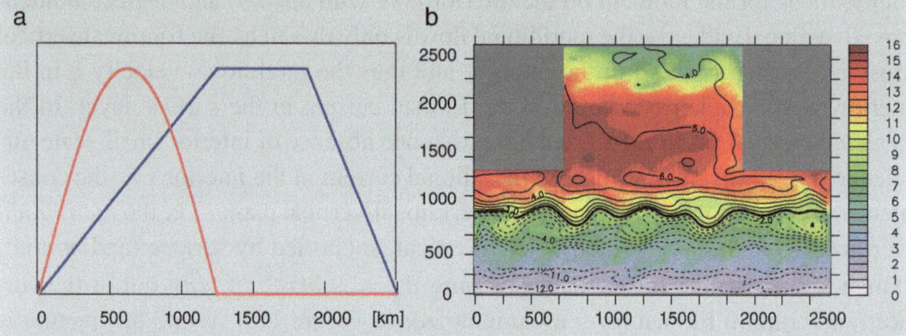


*a Meridional structure of the target temperature $T^*$ (blue) and zonal windstress $\tau_0^{(x)}$ (red) as a function of $y$ in km. Amplitudes are $15\,°\mathrm{C}$ for $T^*$ and $10^{-4}\,\mathrm{m^2\,s^{-2}}$ for the windstress. b Snapshot of temperature $T$ in °C (color) and pressure $p$ (contour lines) at the uppermost level of the model. Grey areas indicate land mass.*

The closed region north of the reentrant channel plays the role of boundary condition for the channel model; it is not meant as a realistic model by its own. Water-mass formation in the closed region (the 'North Atlantic' part of the idealized model) takes places at low latitudes instead of high latitudes as in the real North Atlantic (the Coriolis parameter varies from about $-0.6 \times 10^{-4}\,\mathrm{s^{-1}}$ at the southern boundary to about $0.1 \times 10^{-4}\,\mathrm{s^{-1}}$ at the northern boundary). However, the role of water-mass formation in the closed part of the domain for the dynamics of the Southern Ocean is well captured by the configuration.
The model produces a strong zonal current in the Southern Ocean part. This zonal flow is predominantly eastward with a strong jet slightly northward of the windstress maximum (at about $y = 800$ km) and then monotonically decreasing to the north, south, and also with depth. The zonal flow extends through the whole water depth. The model produces deep water formation and sinking close the northern edge of the Atlantic basin part of the domain, which then wells up in the Southern Ocean, which mimics the global conveyor belt of the World Ocean. We will discuss simulations with a flat bottom and with an idealized ridge system. Note that the zonal symmetry of the flat-bottom version of the model leads in general to an absence of standing eddies, although there is a small amount due to the continents at the transition from the channel to the basin geometry.

the box on p. 568. Vertical and meridional advection of zonal momentum by the (zero order, geostrophic) mean flow is neglected, as explained before in Section 16.2.

With a small vertical stress $\bar{\tau}^{(x)}$ below the mixed layer and small Reynolds stresses $\mathcal{R}^{(x)}$, the Eulerian stream function $\Lambda$ is generally dominated by the northward Ekman transport $\bar{V}_\mathrm{e} = -\bar{\tau}_0^{(x)}/f$ in the surface layer and a deep geostrophic return flow (associated with $\mathcal{F}$) in the valleys between the topography peaks along the particular latitude, or a frictional return flow $\bar{V}_\mathrm{b}$ in a bottom boundary layer if the

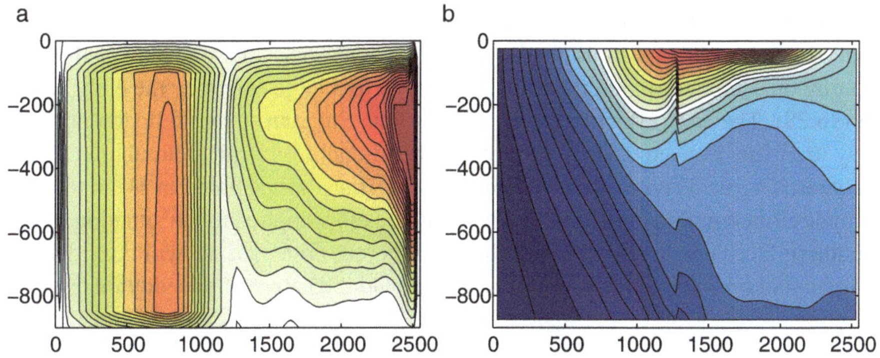


Fig. 16.12 **a** The Eulerian stream function $\Lambda$ in the idealized model of the box on p. 582 as a function of $z$ in m and $y$ in km. Contour interval is 0.5 Sv, positive values are red (note the negative cell tightly attached to the southern boundary). **b** The zonal (and time) mean buoyancy $\bar{b}$ with a contour interval of $0.001\ \mathrm{m\,s^{-2}}$. The peculiar jump in buoyancy near the connection of the channel and the closed part of the domain is related to standing eddy features. This can also be observed in the isopycnal and residual stream function shown in Figures 16.13 and 16.15

model ocean is flat and $\mathcal{F} \equiv 0$. Note that the Eulerian stream function $\Lambda$ above topographic features (or the bottom Ekman layer) depends on the zonal windstress $\bar{\tau}_0^{(x)}$ only, regardless of whether the ocean is homogeneous as in the previous section or not. This is why the Eulerian stream function does not well represent the real nature of the meridional overturning circulation in a stratified ocean, which is believed to be directed along isopycnals, as sketched in Figure 16.3, and not vertical.

We exemplify this behavior with an idealized flat-bottom eddy-permitting numerical model of the Southern Ocean (Viebahn and Eden, 2010), described in the box on p. 582. Figure 16.12a) shows the Eulerian stream function $\Lambda$ of the model. In the interior of the Southern Ocean part of the model, $\Lambda$ is indeed vertically constant and thus has a vanishing meridional velocity, but with a meridionally varying windstress (and $f$) there is vertical (Ekman) pumping. The zonal mean buoyancy $\bar{b}$ is shown in Figure 16.12b), with relatively small changes in the slopes of isolines of $\bar{b}$ (isopycnals) in the North Atlantic. In the Southern Ocean part of the domain we find strongly inclined isopycnals towards the southern boundary of the domain with outcrop at the surface, similar to the situation in the Southern Ocean sketched in Figure 16.3. The zonally averaged buoyancy budget is given by

$$\frac{\partial \bar{b}}{\partial t} + \bar{v}\frac{\partial \bar{b}}{\partial y} + \bar{w}\frac{\partial \bar{b}}{\partial z} = \bar{Q} - \frac{\partial}{\partial y}\overline{v'b'} - \frac{\partial}{\partial z}\overline{w'b'} \tag{16.28}$$

where $Q$ denotes diabatic processes such as small-scale turbulent mixing. Ignoring the eddy fluxes $\overline{v'b'}$ and $\overline{w'b'}$ for the moment, a mean flow across isolines of $\bar{b}$ can only be balanced in steady state by $\bar{Q}$, i. e. by small-scale (diapycnal) mixing. The Eulerian overturning stream function $\Lambda$, which implies strong vertical transports across lines of constant $\bar{b}$, i. e. across isopycnals, would therefore indicate a very strong diapycnal mixing. This is in contrast to our belief that the circulation in the ocean interior is more or less adiabatic with transports of active and passive tracers predominately along isopycnals. It becomes clear that the big southern cell in Figure 16.12 (sometimes called DEACON[3] cell) is a construct of the zonally averaging

[3] SIR GEORGE DEACON, *1906 in Leicester, †1984 in Southampton, oceanographer and chemist.

on $z$-levels, i. e. an artifact of the Eulerian stream function. Since $Q$ in the model is indeed much smaller than implied by $\Lambda$, it becomes clear that the diapycnal transport by the Eulerian stream function is balanced to a large extent by the eddy fluxes in (16.28). The eddy effect has to be added to the Eulerian stream function in some way to obtain a meaningful picture of the meridional overturning circulation in the Southern Ocean. This consideration leads to the Transformed Eulerian Mean formalism, described in Section 12.2.3, which will be below applied to the situation in the Southern Ocean in Section 16.3.4. However, before doing so, we will consider the Lagrangian or isopycnal framework to understand the nature of the eddy fluxes.

### *16.3.2 The Isopycnal Overturning Circulation*

One may view the same system from a different perspective. Because the zonally averaged mean flow does not reflect the meridional overturning in a stratified ocean, it turns out to be meaningful to perform the zonal average going along a longitude $x$ on isopycnals $\rho = \text{const}$. To capture the eddy component, we have to consider time-dependent quantities. The meridional velocity, expressed in isopycnal coordinates $(x, y, \rho)$ (see Appendix B.1.1), is written as $v(x, y, \rho, t)$. The transport between two isopycnals $\rho = \rho_1$ and $\rho = \rho_2$ becomes $\int_{\rho_2}^{\rho_1} v(x, y, \rho, t) h_\rho \mathrm{d}\rho$ where $h_\rho = -\partial z/\partial \rho$ measures the infinitesimal thickness between isopycnals. The real thickness between two infinitesimal close isopycnals $\rho_1$ and $\rho_2$ is $\eta = h_\rho \Delta\rho$ with $\Delta\rho = \rho_2 - \rho_1$. A stream function only exist for the zonal and time mean (see the mass conservation equation (B.6) for isopycnal coordinates; a detailed derivation and discussion of different stream functions and their relations is given in Section 12.3.5). The zonal mean transport below a certain density surface is given by

$$\oint \int_{\rho}^{\rho_0} v(x, y, \rho', t) h_\rho(x, y, \rho', t) \mathrm{d}\rho' \mathrm{d}x = \int_{\rho}^{\rho_0} \oint v(x, y, \rho', t) h_\rho(x, y, \rho', t) \mathrm{d}x \mathrm{d}\rho'$$

It measures the transport between an arbitrary isopycnal $\rho$ and a reference isopycnal $\rho_0$. The *isopycnal stream function* $\psi(y, \rho)$ for this configuration is defined by the time-mean of the above transport. Note that the isopycnal stream function resembles a zonal and temporal correlation between $v$ and $h_\rho$. The time average of the above transport may thus be split into a part resulting from the time-mean quantities and an eddy part (see the box on p. 563). This property distinguishes the isopycnal stream function from the Eulerian one, but we will not discuss this issue further and will ignore the differentiation between standing and transient eddies for simplicity.

The isopycnal stream function $\psi(y, \rho)$ from the idealized model of the box on p. 582 is shown in Figure 16.13a, while Figure 16.13b shows $\psi$ as a function of a representative depth variable instead of $\rho$, i. e. the depth of the "isopycnally averaged density" $\tilde{\rho}$ (see the box on p. 417) for better comparison with the Eulerian stream function $\Lambda(y, z)$. It becomes obvious that $\psi$ captures very different physics than the Eulerian projection $\Lambda$: the meridional overturning is now seen to be predominantly along isopycnals in the interior of both the Southern Ocean and the Atlantic part of the domain (compare Figures 16.12 and 16.13). Note that in particular the large, diapycnal overturning cell in the Southern Ocean (the Deacon cell) of the model is absent in the isopycnal stream function, whereas the difference between the

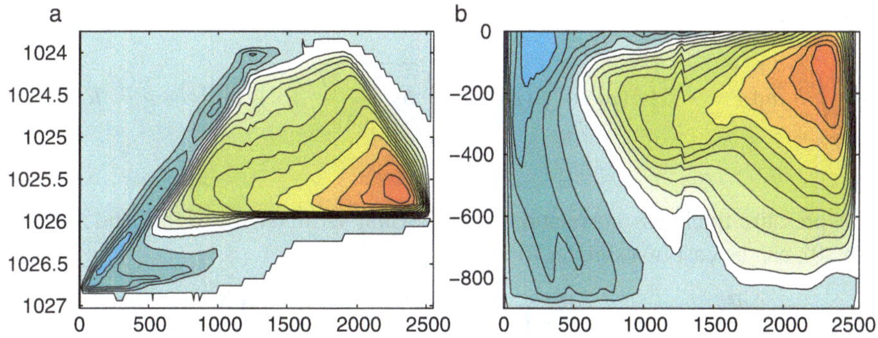


**Fig. 16.13** Isopycnal stream function $\psi$ in Sv in the experiment with the idealized model of the box on p. 582. **a** $\psi$ as a function of $y$ in km and $\rho$ in kg m$^{-3}$. Note that the $\rho$ axis increases inversely. **b** Interpolated $\psi$ as a function of $y$ and depth $z(y, \tilde{\rho})$ in m where $\tilde{\rho}$ is the "isopycnally averaged density", i. e. $\psi$ was transformed back to a representative $z$ coordinate (see also the box on p. 417). Contour interval is 0.5 Sv, *red color* denote positive values of $\psi$, *blue* negative

stream functions is not so large in the North Atlantic part. At the surface, water-mass transformation is clearly visible. The Ekman transport is now contained in a shallow clockwise cell in the upper mixed layer which embeds besides the Ekman flow a counterclockwise recirculation in the surface boundary layer. Note that the isopycnal stream function is very similar to the residual stream function which will be introduced in Section 16.3.4 below. However, before discussing the residual stream function, we focus on the zonally averaged equations of motion in isopycnal coordinates and their interpretation.

### 16.3.3 Interfacial Formstress and Vertical Transfer of Momentum

In Appendix B.1.1, we introduce the equation of motions in isopycnal coordinates. The zonally averaged zonal momentum budget (B.5) becomes

$$-f\overline{h_\rho v} = -\overline{h_\rho \frac{\partial M}{\partial x_\rho}} + \frac{\partial}{\partial \rho}\left(\bar{\tau}^{(x)} + \hat{\mathcal{F}} - \hat{\mathcal{R}}^{(x)}\right) \tag{16.29}$$

Note that the overbar denotes in this section a zonal mean taken at constant $\rho$ (instead of constant $z$, as in Section 16.3.1). The bottom formstress $\hat{\mathcal{F}}$ and the Reynolds stress divergence $\hat{\mathcal{R}}^{(x)}$ take the same meaning as before in Section 16.3.1, but here they appear in isopycnal coordinates; there is, however, no need to further specify them here. It should be kept in mind that all quantities in this section are functions of $\rho$ instead of $z$.

The Montgomery potential $M(x, y, \rho, t) = p + g\rho z$ is derived in Appendix B.1.1. It stands for the pressure in isopycnal coordinates, and we focus our discussion on this quantity. Imagine a stack of well-mixed isopycnal layers (an isopycnal model) and consider one particular layer bounded by the interfaces $\rho = \rho_1$ at $z = -\eta_1$ and

$\rho = \rho_2$ at $z = -\eta_2$ below. We integrate (16.29) over this layer, which yields

$$-f \int_{\rho_2}^{\rho_1} \overline{h_\rho v} \mathrm{d}\rho = -f \left[\psi(\rho_1) - \psi(\rho_2)\right] = \int_{\rho_2}^{\rho_1} \overline{h_\rho \frac{\partial - M}{\partial x_\rho}\, d\rho} + \left(\bar{\tau}^{(x)} + \hat{\mathcal{F}} - \hat{\mathcal{R}}^{(x)}\right)\Big|_{\rho_2}^{\rho_1} \tag{16.30}$$

The pressure-thickness correlation, i. e. the first term on the right-hand side of (16.30), can be cast in a simpler form:

$$h_\rho \frac{\partial M}{\partial x_\rho} = -\frac{\partial}{\partial \rho}\left(z \frac{\partial M}{\partial x_\rho}\right) + z \frac{\partial}{\partial \rho}\frac{\partial M}{\partial x_\rho} = -\frac{\partial}{\partial \rho}\left(z \frac{\partial M}{\partial x_\rho}\right) + gz \frac{\partial z}{\partial x_\rho}$$

using the isopycnal thickness $h_\rho = -\partial z/\partial \rho$ and the hydrostatic relation $\partial M/\partial \rho = gz$ from Appendix B.1.1. The zonal average eliminates the last contribution since it can be written as $gz\partial z/\partial x_\rho = g\partial(z^2/2)/\partial x_\rho$. The other contribution yields

$$\int_{\rho_2}^{\rho_1} \overline{h_\rho \frac{\partial M}{\partial x_\rho} \mathrm{d}\rho} = \overline{\eta_2 \frac{\partial M_2}{\partial x_\rho}} - \overline{\eta_1 \frac{\partial M_1}{\partial x_\rho}} = \overline{\eta_2 \frac{\partial p_2}{\partial x}} - \overline{\eta_1 \frac{\partial p_1}{\partial x}} \tag{16.31}$$

where $M_i = M|_{\rho=\rho_i}$, $\eta_i = z(\rho_i)$ and $p_i = p|_{z=\eta_i}$ with $i = 1, 2$. The right-hand side of (16.31) may be interpreted as an integrated vertical divergence of a stress $\overline{\eta \partial p/\partial x}$, fluxing zonal momentum through the layer interfaces (remember that the terms appear in the balance of zonal momentum). The layer of ocean gains zonal momentum by the amount $\overline{\eta_1 \partial p_1/\partial x}$ from the fluid above $z = -\eta_1(x)$ and loses $\overline{\eta_2 \partial p_2/\partial x}$ to the fluid below $z = -\eta_2(x)$.

It becomes clear that for infinitesimally distant isopycnals, the vertical divergence of the interfacial formstress $\overline{\eta \partial p/\partial x}$ enters the zonally averaged zonal momentum balance. The similarity of the construction of $\overline{\eta \partial p/\partial x}$ to the bottom formstress is evident (see the box on p. 568), but note that here the isopycnal layer depth $\eta$ and the pressure $p$ are also time-dependent. Evidently, only deviations of $\eta$ and $p$ from their zonal mean $\overline{\eta}$ and $\overline{p}$ contribute to the correlation $\overline{\eta \partial p/\partial x} = -\overline{p \partial \eta/\partial x}$. This latter form reveals that the isopycnal must be inclined for the stress to be nonzero. Though frequently viewed as a vertical transport of momentum – because isopycnal layers are generally stacked vertically – the process is actually associated with a lateral transport of momentum across the inclined isopycnals. To become a vertical transport, the injected momentum must be redistributed within the layer by small-scale turbulence: the layer must be effectively mixed which is implicitly assumed by the constant density within the layer.

The above described stress is called *interfacial formstress*[4]. Using the geostrophically balanced velocity perturbations $f v'_g = \partial p'/\partial x$ and approximating the layer depth fluctuation with the density anomaly, $\eta' = \rho'/(\partial\bar{\rho}/\partial z)$, we obtain to leading order

$$\overline{\eta' \frac{\partial p'}{\partial x}} \approx f \frac{\overline{\rho' v'}}{\partial \bar{\rho}/\partial z} \tag{16.32}$$

When we ignore salinity contributions in this relation, the density can be replaced by temperature, and thus a poleward eddy flux of heat is equivalent to a downward

[4] More precisely, the name interfacial formstress refers to the zonally and temporally averaged equations, from which also a standing and a transient component may result (see the box on p. 563).

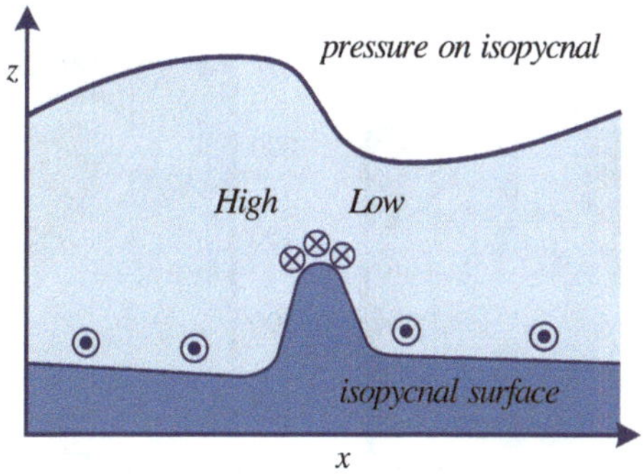


Fig. 16.14 Schematic sketch demonstrating the interfacial formstress for an isopycnal interface (shown is the $(x, z)$-section). There is higher pressure (*top curve*) at the depth of the density surface where it is rising to the east compared with where it is deepening to the east. This results in an eastward pressure force (interfacial formstress) on the water below. This is related to the fact that the northward flow occurs where the vertical thickness of water above the density surface is small, and southward flow occurs where the thickness is large; so there is a net southward mass flux at lighter densities due to the geostrophic flow. The same kind of pressure force acting on the sloping bottom topography leads to the bottom formstress. Redrawn from Rintoul et al. (2001)

eddy transport of zonal momentum by the interfacial formstress in the water column. Both processes – eddy heat transport and eddy momentum transport – are, therefore, strictly coupled. We will find a relation similar to (16.32), which is obtained here in the isopycnal coordinate framework, also for $z$-level averages using the residual mean framework below in Section 16.3.4.

In summary, we note that the horizontal pressure gradients can establish a transfer of horizontal momentum since they transport horizontal momentum across tilted surfaces from one isopycnal layer of the ocean to another. A layer bounded by tilted isopycnals is thus forced by interfacial formstresses at the bounding top and bottom surfaces, in the same way as the Ekman layer is driven by frictional stresses at top and bottom. To obtain a nonzero interfacial formstress, the pressure must vary at the isopycnal depth in a way that a nonzero correlation with the isopycnal depth variations occurs, as elucidated in the sketch of Figure 16.14. Note that in deriving the relation (16.31), it was assumed that the isopycnal layer does not run into the bottom nor that it touches the sea surface. If this situation occurs, additional pressure terms arise from the bounding isopycnal outcrops. These terms represent a flux of horizontal momentum through these boundaries into the isopycnal layer. For a layer intersecting the bottom, the corresponding flux is part of the bottom formstress discussed in the box on p. 568.

### 16.3.4 The Residual Overturning Circulation

The Eulerian view of the meridional overturning discussed in Section 16.3.1 can be extended to a close correspondence with the isopycnal framework, using the TEM approach (Transformed Eulerian Mean, Andrews and McIntyre, 1976, introduced in Section 12.2.3). It acknowledges that the transport (advection) of zonally averaged tracers is not only performed by the zonal mean Eulerian stream function $\Lambda$, but that the eddy fluxes also act to advect density and tracers. This property of the eddy fluxes

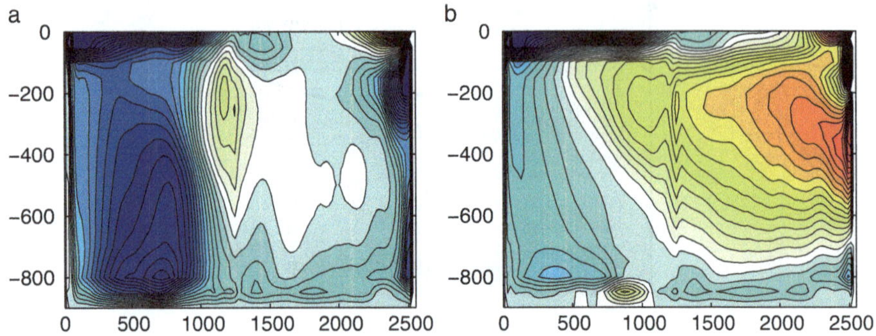


Fig. 16.15 Residual mean eddy stream function $-\boldsymbol{B}$ (**a**) and residual mean stream function $\phi = \Lambda - \boldsymbol{B}$ (**b**) in Sv in the model of the box on p. 582. Contour interval is 0.5 Sv in each case, positive values are *red*, negative ones *blue*

can be inferred from the projection of the eddy density flux $\overline{\boldsymbol{u}'\rho'}$ into the component normal and along the mean isopycnal,

$$\overline{\boldsymbol{u}'\rho'} = B\underline{\nabla}\bar{\rho} - K_\mathrm{d}\nabla\bar{\rho} \tag{16.33}$$

with $\boldsymbol{u} = (v, w)$, $\nabla = (\partial/\partial y, \partial/\partial z)$ and $\underline{\nabla} = (-\partial/\partial z, \partial/\partial y)$ as defined in the box on p. 385. Note that the overbar is again a zonal mean on $z$-levels in this section. Taking the divergence of the decomposition (16.33) of the eddy fluxes reveals that $K_\mathrm{d}$ is an eddy-induced *diapycnal diffusivity* (if positive), and $-B$ is an *eddy stream function* advecting the density $\bar{\rho}(y, z)$ in addition to the Eulerian flow. The advection of $\bar{\rho}$ is thus given by the *residual stream function* $\phi = \Lambda - B$, and the diapycnal transport is represented by the diffusivity $K_\mathrm{d}$. This decomposition is illustrated in Figure 16.15 while the Eulerian stream function $\Lambda$ is shown in Figure 16.12. It becomes obvious that the eddy-driven overturning is partly compensating the mean Eulerian overturning in the Southern Ocean, with the effect that the sum, i. e. the residual stream function $\phi$ is very similar to the isopycnal stream function $\psi$. Note that the differences between $\phi$ and $\psi$, most prominent at the surface, the bottom, and the lateral boundaries, are related either to rotational eddy fluxes, for which the original TEM framework (16.33) does not account, or to 'real' diapycnal eddy fluxes related in turn to irreversible mixing of (zonally averaged) density. How to implement rotational eddy fluxes is shown in Section 12.3.1.

Adding $fB$ on both sides of (16.27) yields the balance of the zonal residual mean momentum, expressed in terms of the residual stream function $\phi$,

$$-f\phi = fB + \bar{\tau}_0^{(x)} - \bar{\tau}^{(x)} + \mathcal{F} - \mathcal{R}^{(x)} = -f\frac{\overline{\rho'v'}}{\partial\bar{\rho}/\partial z} + \bar{\tau}_0^{(x)} - \bar{\tau}^{(x)} + \mathcal{F} - \mathcal{R}^{(x)} \tag{16.34}$$

The eddy term $fB$, computed from (16.33), is $f\overline{\boldsymbol{u}'\rho'} \cdot \underline{\nabla}\bar{\rho}/|\nabla\bar{\rho}|^2$. In the second form of the balance (16.34), the term follows from the approximation (12.27), assuming that $\partial\bar{\rho}/\partial z \gg \partial\bar{\rho}/\partial y$ and discussed in Section 12.2.3 on TEM. We find here a resemblance to the interfacial formstress (16.31) and (16.32). It shows again that the eddies accomplish a vertical transport of zonal momentum by the meridional eddy density flux $\overline{v'\rho'}$. We shall refer to the Eulerian form $f\overline{\rho'v'}/(\partial\bar{\rho}/\partial z)$ of the flux also as interfacial formstress. Comparing (16.34) with its isopycnal counterpart (16.30), we note that the residual stream function is constructed to resemble the isopycnal one.

Using a downgradient parameterization for the meridional eddy density flux, $\overline{v'\rho'} \sim -K_\ell \partial\bar{\rho}/\partial y$ with the isopycnal thickness diffusivity $K_\ell$ as discussed in Section 16.6, and the thermal wind relation to replace $\partial\bar{\rho}/\partial y$, we arrive at

$$-f \frac{\partial}{\partial z} \frac{\overline{\rho' v'}}{\partial\bar{\rho}/\partial z} = \frac{\partial}{\partial z}\left(K_\ell \frac{f^2}{N^2}\frac{\partial\bar{u}}{\partial z}\right)$$

This term appears in the residual zonal momentum budget ($\partial/\partial z$ of (16.34)). The interfacial formstress, transporting momentum vertically as discussed above, is thus also formally associated with vertical friction with an effective vertical diffusivity $K_\ell(f/N)^2$ with a typical (very large) size of $10\,\mathrm{m^2\,s^{-1}}$ (see e. g. Olbers et al., 1985; Rhines and Young, 1982). The value is obtained from the standard parameter values in Table 16.1.

## 16.4 The Zonal Mean Dynamics

The zonal mean of the zonal momentum balance is central in the dynamics of the ACC because the main external forcing – the windstress – is predominantly zonal and the flow is more or less zonal as well. The balance has been written in various forms in the preceding sections, namely in the Eulerian form (16.27), in the isopycnal form (16.29) and in the residual form (16.34). All aspects of the zonally averaged flow are now combined and discussed further in this section by a simple conceptual model.

Consider a strip of ocean from Antarctica to the northern rim of the ACC and split the water column into three layers (which may still be stratified), separated by isopycnals (see the sketch in Figures 16.3 and 16.16). The upper layer reaches from the sea surface $z = \zeta$ to some isopycnal at depth $z = -\eta_1$ and includes the surface Ekman layer. The intermediate layer lies below with its base at $z = -\eta_2$ which is above the highest topography in the Drake Passage belt (this is the range of latitudes which run through Drake Passage). These two layers are 'unblocked' by topography and feel no bottom formstress. The lower layer reaches from $z = -\eta_2$ to the ocean bottom at $z = -h$ and is thus 'blocked' beneath the depth of the highest topography. As before, we apply a zonal average to the balance equation (16.29) of zonal momentum in isopycnal coordinates and integrate over the depth and the zonal extent for the three layers to obtain

$$\begin{aligned}
-f\bar{V}_1 &= -\overline{\eta_1' \frac{\partial p_1'}{\partial x}} + \bar{\tau}_0^{(x)} - \bar{\tau}_1^{(x)} - \mathcal{R}_1^{(x)} \\
-f\bar{V}_2 &= \overline{\eta_1' \frac{\partial p_1'}{\partial x}} - \overline{\eta_2' \frac{\partial p_2'}{\partial x}} + \bar{\tau}_1^{(x)} - \bar{\tau}_2^{(x)} - \mathcal{R}_2^{(x)} \\
-f\bar{V}_3 &= \overline{\eta_2' \frac{\partial p_2'}{\partial x}} - \overline{h\frac{\partial P}{\partial x}} + \bar{\tau}_2^{(x)} - \bar{\tau}_b^{(x)} - \mathcal{R}_3^{(x)}
\end{aligned} \tag{16.35}$$

The northward volume flux in each layer is denoted by $\bar{V}_i, i = 1, 2, 3$ (layer $3 = b$ is the bottom layer). Furthermore, $p_i$ is the pressure at the respective layer depths, $P$ the bottom pressure, and the overbar denotes a zonal mean. In this case, the coordinate $x$ is along the specific path. As before, $\bar{\tau}_0^{(x)}$ denotes the windstress, $\bar{\tau}_i^{(x)}$ the frictional stresses related to small-scale turbulence between the isopycnal layers,

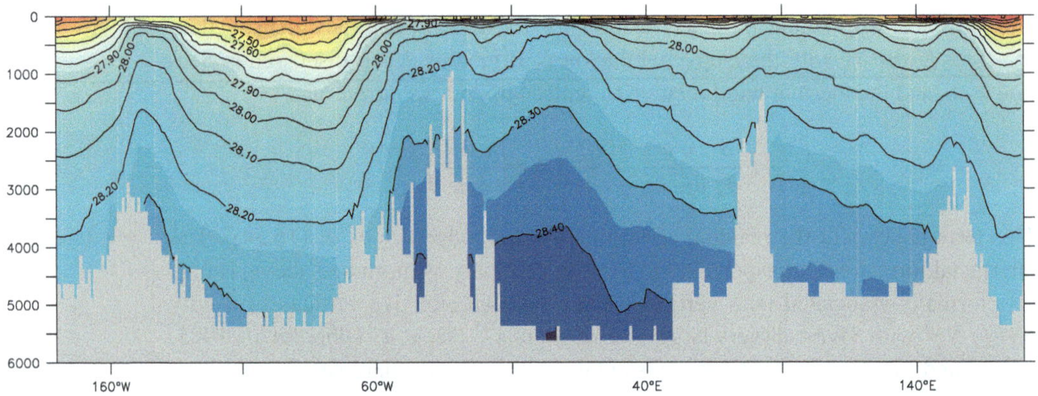

Fig. 16.16 A zonal section of the observed potential density $\rho$ (in the form $\rho - 1{,}000\,\mathrm{kg\,m^{-3}}$) in the Southern Ocean at $60°$ S as a function of depth in m and longitude, taken from the Hydrographic Atlas of the Southern Ocean (Olbers et al., 1992). The section is viewed from the south

$\bar{\tau}_{\mathrm{b}}^{(x)}$ the frictional bottom stress, and $\mathcal{R}_i^{(x)}$ the integrated divergence of the lateral eddy momentum fluxes in the respective layers. Note that the surface term $\overline{\zeta \partial p_0/\partial x}$ drops out in the first equation because the surface pressure is given by the surface displacement, $p_0 = g\zeta$.

Dividing the zonal mean momentum balances in the three layers by $f$, we recognize the components of the meridional overturning circulation: eddy-driven and windstress-driven parts as well as the geostrophic contribution associated with the bottom formstress. The contributions from interfacial friction and Reynolds stresses are usually small and will thus be mostly neglected in the following. The wind-driven component $-\bar{\tau}_0/f$ (the Ekman transport) in the top layer and the geostrophic component in the bottom layer, $\overline{h(\partial P/\partial x)}/f$, also appear if the flow is averaged between geopotential levels (constant depth) instead of isopycnals as done here. As discussed above, these Eulerian quantities form the Deacon cell of the Southern Ocean meridional overturning (see Section 16.3).

### 16.4.1 Complete Balance of Zonal Mean Zonal Momentum

Since $\sum_i \bar{V}_i = 0$ is valid at any latitude by mass balance (neglecting the very small effect of precipitation and evaporation on the mass balance), the overall balance of zonal momentum is between the applied windstress, the bottom formstress, the frictional stress on the bottom, and the vertical integral of the Reynolds stress contributions,

$$\bar{\tau}_0^{(x)} - \bar{\tau}_{\mathrm{b}}^{(x)} - \overline{h\frac{\partial P}{\partial x}} - \sum_i \mathcal{R}_i^{(x)} = 0 \qquad (16.36)$$

The frictional bottom stress is generally small in the ocean. Lateral eddy momentum fluxes in the ACC turned out to be rather small (compared to the windstress) as well and indifferent in sign in various investigations of observed data (e. g. Hughes and Ash, 2001; Morrow et al., 1992; Phillips and Rintoul, 2000). Hence the momentum put into the ACC by the windstress is transferred to the solid Earth predominantly

97. BARBI Model of the Southern Ocean

The BARBI model is derived in Appendix B.2 and used in a two-mode version in Section 14.2.5. The application of this BARBI version, governed by (14.33)–(14.35), to the Southern Ocean has realistic topography and surface windstress forcing identical to the homogeneous model of the box on p. 570, but it includes a laterally constant prescribed background stratification representative of the Southern Ocean (see also Olbers and Lettmann, 2007). Resolution and horizontal friction of the BARBI model are identical homogeneous model of the box on p. 570, further parameters are $N_0 = 1.5 \times 10^{-3}\,\mathrm{s}^{-1}$, $K_\ell = 2 \times 10^3\,\mathrm{m}^2\,\mathrm{s}^{-1}$.

The figure in this box shows the stream function $\psi$ for the depth-averaged flow $\boldsymbol{U} = \int_{-h}^{0} \boldsymbol{u}\,\mathrm{d}z$, with $U = -\partial\psi/\partial y$ and $V = \partial\psi/\partial x$, the baroclinic potential energy $E = g\int_{-h}^{0} \rho z\,\mathrm{d}z$, and the bottom pressure $P$. In contrast to the homogeneous simulations TOPO and FLAT of Figure 16.5, the BARBI model now shows a considerable amount of realism. The zonal transport through Drake Passage has a realistic size of 140 Sv, and a structured sized circumpolar flow appears with little apparent influence of the underlying topography. Note that the potential energy $E$ follows quite closely the streamlines while the bottom pressure $P$ has a tendency to follow the $f/h$ contours, as was also found for the homogeneous case TOPO (see Figure 16.5). Nevertheless, the baroclinicity introduced by the baroclinic potential energy $E$ – in the momentum balance as a pressure gradient and in the vorticity as the JEBAR term (see Section 14.2) – breaks the constraint for the flow to follow $f/h$ contours.

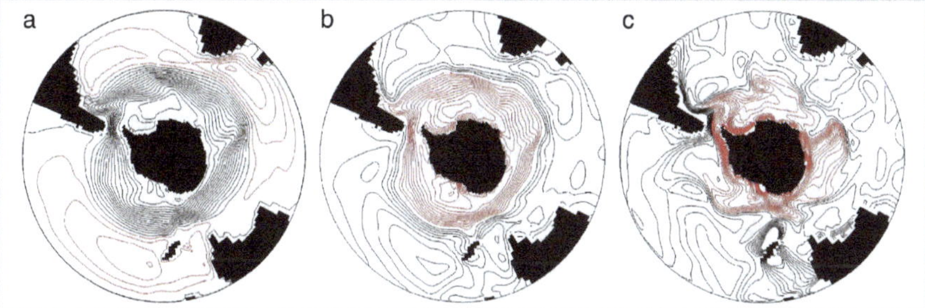


*a Stream function $\psi$ for the depth-averaged flow in* Sv*, contour interval is* $10$ Sv. *b Baroclinic potential energy $E$ with contour interval of* $1{,}000\,\mathrm{m}^3\,\mathrm{s}^{-2}$. *c Bottom pressure $P$ with contour interval of* $0.1\,\mathrm{m}^2\,\mathrm{s}^{-2}$. *Positive values are black, negative ones red.*

by the bottom formstress at the same latitude. We have argued for this balance in (16.15); it has been confirmed in most numerical models which include submarine topographic barriers in the zonal flow and have a realistic (small) magnitude of the Reynolds stress divergence. Eddy effects appear thus rather unimportant in the vertically integrated balance. It is worth mentioning that some coarse ocean circulation models do not confirm (16.36). The reason is that such models use very large lateral viscosities so that the parameterized Reynolds stresses become large, even though the simulated current is broad and smooth. In models with a flat ocean bottom, either bottom friction may become important and/or the Reynolds terms could come into play (compare Section 16.2.2).

We illustrate the elements of the total momentum balance (16.36) of the Southern Ocean with two rather different models. The BARBI model, described in detail in the Appendix B.2 and in Section 14.2.5, is applied to the Southern Ocean, as discussed in the box on p. 591. In addition we will use below the idealized model of the Southern Ocean, introduced in the box on p. 582, but in configurations including idealized topographic variations instead of a flat bottom.

The zonally averaged balance of zonal momentum for the BARBI model of the Southern Ocean of the box on p. 591 is shown in Figure 16.17a and indeed reveals the dominant balance between surface windstress and bottom formstress in the latitude

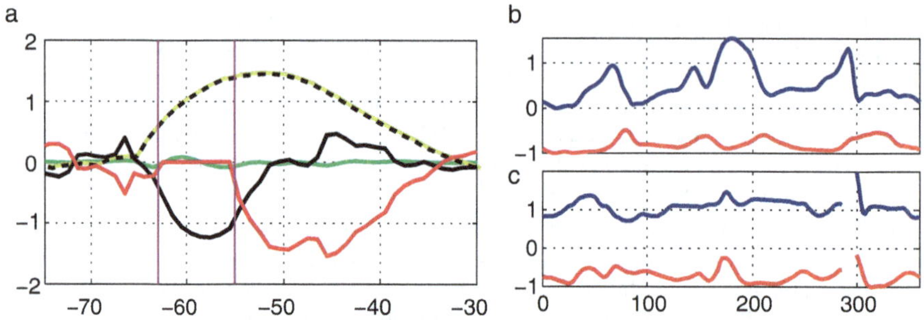


**Fig. 16.17** **a**: Zonally and vertically integrated momentum balance as function of latitude for the BARBI model of the Southern Ocean the box on p. 591. The latitude band of Drake Passage is indicated by *vertical purple lines*. The *yellow–black dashed line* denotes windstress $\bar{\tau}_0^{(x)}$, *black line* the bottom formstress $\overline{h\partial P/\partial x}$, *green line* the lateral friction, and the *red line* the pressure difference on the continental boundaries $\Delta E|_{\text{continents}}$. Units are $10^{-4}\,\text{m}^2\,\text{s}^{-2}$. **b** and **c**: Bottom pressure $P$ (*blue lines*, in $P/\max(|P|)+1$) and depth $h$ (*red lines* in $-h/\max(h)$) as a function of longitude for $59.5°$ S in the Drake Passage (**b**), and a more northern latitude at $46.5°$ S (**c**)

band of Drake Passage with very little contribution from friction (notice that the value $A_\text{h} = 4 \times 10^4\,\text{m}^2\,\text{s}^{-1}$ of the lateral eddy viscosity is moderate but not small). The figure presents the total meridional extent of the model, and the pressure force acting on the continents north and south of Drake Passage latitudes comes into play. This term is disregarded in the conceptual model (16.35) and, therefore, not present in (16.36) which must be replaced by

$$\bar{\tau}_0^{(x)} - \bar{\tau}_\text{b}^{(x)} - \overline{h\frac{\partial P}{\partial x}} - \Delta E\big|_{\text{continents}} - \sum_i \mathcal{R}_i^{(x)} = 0 \tag{16.37}$$

if continents are blocking the averaging path in addition to submarine topography. Windstress and the two pressure terms are the principle contributors to the balance of momentum in the basin-wide gyre circulations embedded between continents. Figure 16.17b shows the bottom pressure-depth correlation for the BARBI model of the Southern Ocean in the same manner as Figure 16.7 for the homogeneous simulations FLAT and TOPO. As in TOPO, we find high pressure upstream of ridges and low pressure downstream in the Drake Passage band, leading to a sink of eastward momentum.

Figure 16.18 shows the zonally and vertically integrated momentum balance from the eddy-permitting idealized model of the Southern Ocean of the box on p. 582. In this configuration, however, the model is integrated with topographic variations instead of a flat bottom, i. e. with a meridional ridge in the 'Drake Passage' and another one stretching through the middle of the attached basin and the southern channel. The depth of the domain is now 2,000 m instead of 1,000 m and the topographic variations are below the upper 1,000 m. The figure shows that the topography drives northward excursions of the ACC over the ridges and southward excursions downstream of the ridges. The zonal mean momentum balance (Figure 16.18c) is again dominated by the input of momentum by the surface windstress and the bottom formstress in the Southern Ocean part of the model. Here, however, the Reynolds stress term also becomes significant at the northward flank of the zonal jet. This is because of the zonal excursion of the zonal flow, which yields a nonvanishing correlation $\overline{v'u'}$. Note that this feature can be related to the standing eddy signal and not to tran-

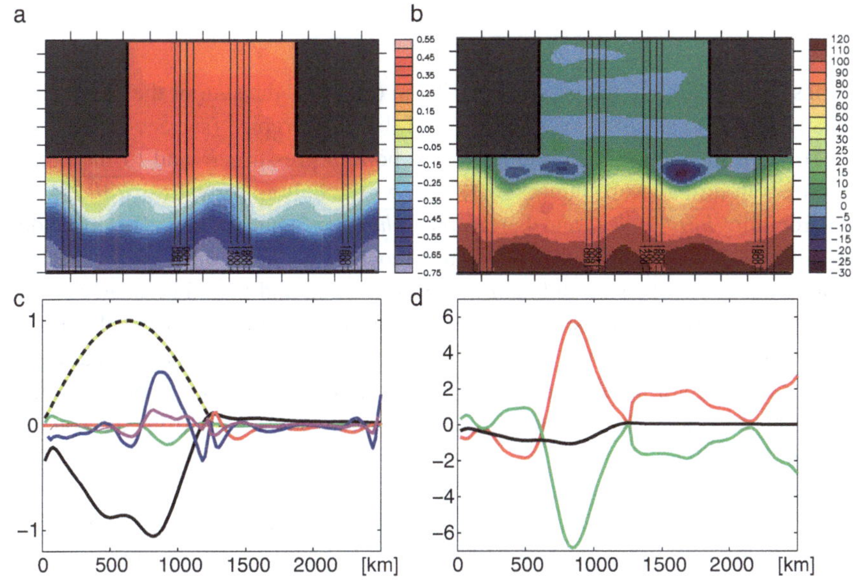


Fig. 16.18 Time average of the sea surface height (**a**, in m) and the stream function $\psi$ of the vertically integrated transport (**b**, in Sv) of the idealized model of the Southern Ocean (the box on p. 582), but here with topographic variations as indicated by the *black contour lines* ($h$ in m, contour interval is 200 m) with meridional ridges in "Drake passage" and in the middle of the domain. **c**: Zonal mean and vertically integrated zonal momentum balance, i. e. windstress $\bar{\tau}_0^{(x)}$ (*yellow–black dashed*), bottom formstress $\overline{h\partial P/\partial x}$ (*black*), bottom friction (*green*), pressure difference on continents (*red*), Reynolds stress $\mathcal{R}_{\mathrm{b}}^{(x)}$ (*blue*), and mean advection $\bar{v}\partial\bar{u}/\partial x$ (magenta) in $10^{-4}\,\mathrm{m}^2\,\mathrm{s}^{-2}$ as a function of $y$ in km. **d**: Baroclinic formstress $\overline{h\partial P^{\mathrm{clin}}/\partial x}$ (*red*), barotropic formstress $g\overline{h\partial\zeta/\partial x}$ (*green*), and total bottom formstress $\overline{h\partial P/\partial x}$ (*black*) in $10^{-4}\,\mathrm{m}^2\,\mathrm{s}^{-2}$

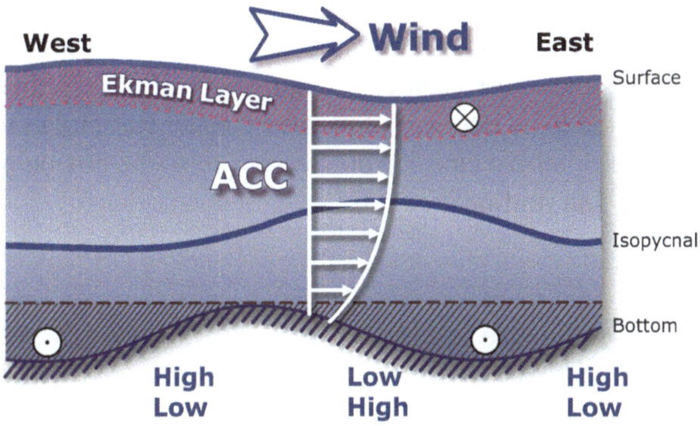


Fig. 16.19 Sketch of the zonal balance of momentum for the ACC. The system is viewed from Antarctica. The flow establishes a high of barotropic (surface) pressure and a low of baroclinic (density related) pressure upstream of a zonal ridge and a corresponding low/high downstream. The associated barotropic formstress and the baroclinic bottom formstress almost balance; their residual counteracts the windstress. The wind drives a northward Ekman transport (⊗) in the surface layer. Corresponding to the bottom formstress there is southward geostrophic return flow (⊙) in the valleys between the ridges which partly block the zonal path

sient eddies as discussed in the box on p. 563, i. e. can be regarded as an artifact of the zonal averaging since the Reynolds stress due to the transient eddies remains small.

### 16.4.2 Barotropic and Baroclinic Bottom Formstress

It is instructive to write the pressure $p$, like in (16.8), as the sum of the "baroclinic" (density-related) part $p^{\text{clin}} = g \int_z^0 \rho dz$ and the "barotropic" (surface-related) part $g\zeta$. This decomposition also applies to the bottom pressure $P$ and to the bottom formstress in turn as well. Figure 16.18d shows the bottom formstress $\overline{h\partial P/\partial x}$ and its baroclinic, $\overline{h\partial P^{\text{clin}}/\partial x}$, and barotropic, $\overline{gh\partial\zeta/\partial x}$, components. The individual components of the bottom formstress are much larger than the zonal windstress by about an order of magnitude but of opposite sign, and thus they nearly cancel. While the total bottom formstress $\overline{h\partial P/\partial x}$ and the barotropic component clearly take out the momentum put in the ocean by windstress, we note that the baroclinic part $\overline{h\partial P^{\text{clin}}/\partial x}$ does not have the corresponding sign: the baroclinic bottom formstress accelerates the eastward current (the center is roughly 800 km from the southern boundary).

Such a depth-pressure correlation can in fact be seen in circumpolar hydrographic sections passing through Drake Passage around Antarctica, e. g. as shown in Figure 16.16. From the density $\rho$, we can infer the baroclinic pressure $p^{\text{clin}}$ contained in the mass stratification. It is obvious in the above section that there is more lighter water to the west of the submarine ridges than to the east. The bottom formstress derived from such a pattern accelerates the eastward current, acting thus in cooperation with the eastward windstress. A summary of the balance of zonal momentum in the ACC is shown in the sketch of Figure 16.19.

## 16.5 The Vertically Averaged Dynamics of the Southern Ocean

The previous discussion in this chapter mainly dealt with zonally averaged properties and balances of the Southern Ocean circulation. We continue in this section with a particular view on the *local* balances of momentum and vorticity in the light of the BARBI model of the Southern Ocean from Section 16.4 (the box on p. 591), which includes topography and stratification.

### 16.5.1 The Vertically Averaged Momentum Budget

The vertically integrated balance of momentum can be written as

$$\begin{aligned} -fV &= -h\frac{\partial P}{\partial x} - \frac{\partial E}{\partial x} + \tau_0^{(x)} - \tau_b^{(x)} - \mathcal{R}^{(x)} \quad \text{and} \\ fU &= -h\frac{\partial P}{\partial y} - \frac{\partial E}{\partial y} + -\tau_0^{(y)}\tau_b^{(y)} - \mathcal{R}^{(y)} \end{aligned} \tag{16.38}$$

Both pressure terms, the bottom pressure $P$ and the vertically integrated potential energy $E = g \int_{-h}^0 \rho z dz$, originate from the vertically integrated pressure gradient (compare with (14.26)). The other terms have the same meaning as before in this section, but are not zonally averaged. Note that the Reynolds stress terms $\mathcal{R}^{(x)}$ and $\mathcal{R}^{(y)}$ are parameterized in the BARBI model by lateral diffusion as in Section 16.2.2.

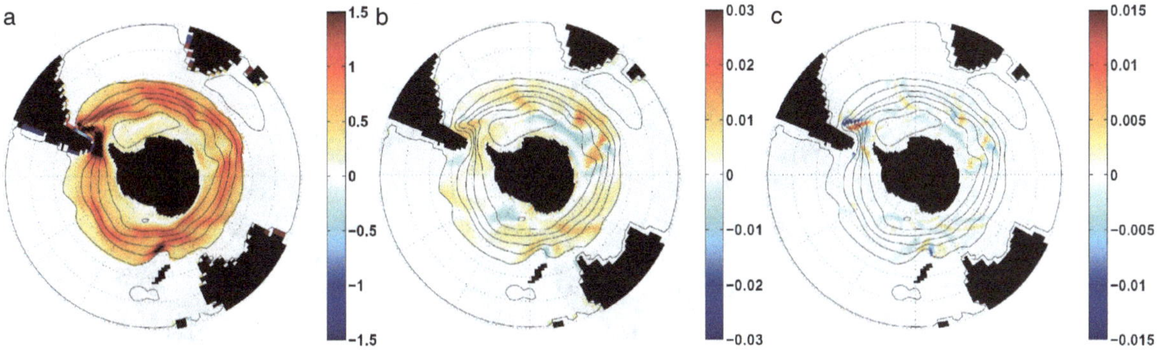


Fig. 16.20 The vertically integrated momentum balance, (16.39), normal to streamlines for the BARBI model of the Southern Ocean (see the box on p. 591). **a** Normal gradient of bottom pressure and baroclinic potential energy $-h\partial P/\partial n - \partial E/\partial n$. The corresponding Coriolis force $fU^{(\|)}$ is indistinguishable from this figure. **b** Normal component of windstress $\tau_0^{(\perp)}$. **c** Normal component of friction term $\mathcal{R}^{(\perp)}$. Note the different color scales. The contour lines show the stream function $\psi$ with contour interval of 25 Sv. Units: $10^{-2}\ \mathrm{m}^2\,\mathrm{s}^{-2}$

Although we know that this downgradient parameterization is unphysical, it is commonly used for numerical reasons. It plays, however, only a minor role in the analysis discussed below.

The flow is, of course, predominantly in a geostrophic balance. Thus, deviations from this state are of interest. These are revealed by writing the equations in natural coordinates $(s, n)$ oriented at the flow direction with $s$ parallel to $\boldsymbol{U}$ and $n$ normal to $\boldsymbol{U}$. Since $U^{(\perp)} \equiv 0$, we obtain from (16.38)

$$fU^{(\|)} = -h\frac{\partial P}{\partial n} - \frac{\partial E}{\partial n} + \tau_0^{(\perp)} - \mathcal{R}^{(\perp)} \quad \text{and} \quad 0 = -h\frac{\partial P}{\partial s} - \frac{\partial E}{\partial s} + \tau_0^{(\|)} - \mathcal{R}^{(\|)} \tag{16.39}$$

where bottom friction has been neglected (there is none in BARBI). Figure 16.20 shows the terms in the momentum balance normal to $\boldsymbol{U}$. The geostrophic terms $fU^{(\|)}$ and $-h\partial/P\partial n - \partial E/\partial n$ are dominating that balance by more than two orders of magnitude. Windstress and the lateral friction are much smaller; the latter occurs only very localized. The complementary balance of the parallel component of momentum is dominated by the pressure gradients parallel to $\boldsymbol{U}$ and the windstress (not shown), while the Coriolis force vanishes. Overall, there is surprisingly little effect from the friction $\mathcal{R}^{(\perp)}$ and $\mathcal{R}^{(\|)}$, although the lateral viscosity is not small (see the box on p. 570 and the box on p. 591).

The partition between the two pressure gradients related to the bottom pressure and the potential energy $E$ is of interest. The normal gradients $h\partial P/\partial n$ and $\partial E/\partial n$ are of similar size. However, the bottom pressure gradient is very localized, occurring in worm-like features oriented parallel to $\boldsymbol{U}$, whereas $\partial E/\partial n$ shows a broad scale structure. It has an imbedded filamented pattern which compensates the bottom pressure worms, so that the total pressure force becomes smoother, as shown in Figure 16.20a. The along-streamline pressure gradients $h\partial P/\partial s$ and $\partial E/\partial s$ are much smaller than the normal components and shown in Figure 16.21. They have a similar pattern and size but of mostly an opposing sign such that the net pressure term in the parallel balance is an order of magnitude smaller. Note that $-h\partial P/\partial s$ generates

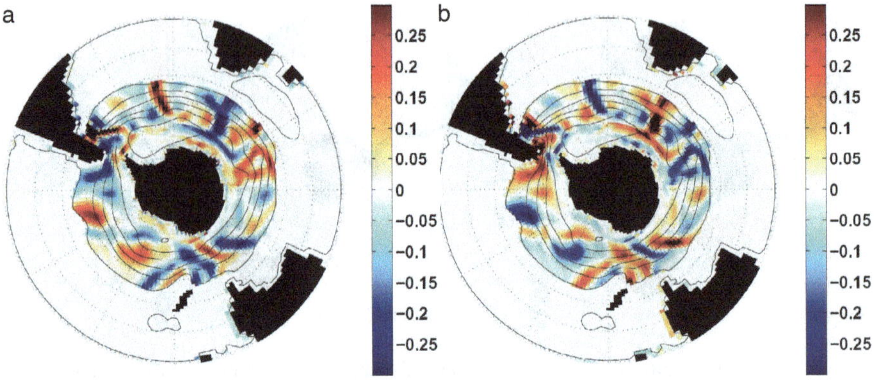


**Fig. 16.21** The vertically integrated momentum balance parallel to streamlines, (16.39), for the BARBI model of the Southern Ocean (see the box on p. 591) in $10^{-2}\,\mathrm{m^2\,s^{-2}}$. **a** $-h\partial P/\partial s$. **b** $-\partial E/\partial s$. The contour lines show the stream function $\psi$ with contour interval of 25 Sv

the bottom formstress in a "zonal" mean following streamlines, balancing the along-stream mean of the parallel windstress in the overall balance of momentum, whereas $\partial E/\partial s$ integrates to zero.

### *16.5.2 The Vertically Averaged Vorticity Budget*

The pressure forces enter the balance of vorticity as torques. As discussed in the box on p. 465, there are two forms of vorticity, i. e. the vorticity for the depth-averaged and the one of the depth-integrated flow. The difference is the appearance of either the bottom pressure $P$ or the potential energy $E$ in the vorticity budgets. However, both forms are equivalent and discussed in detail in Section 14.2. The (steady) vorticity balance of the depth integrated momentum $\boldsymbol{U}$ is given by (14.32) which we rewrite here as

$$\beta\frac{\partial\psi}{\partial x} = \nabla P \cdot \underset{\neg}{\nabla} h + \underset{\neg}{\nabla}\cdot\boldsymbol{\tau}_0 - \underset{\neg}{\nabla}\cdot\mathcal{R} \tag{16.40}$$

with a general friction term. Note that we follow here the vector notation introduced in the box on p. 444, i. e. $\underset{\neg}{\nabla} = (-\partial/\partial y, \partial/\partial x)$. The *bottom pressure torque* $\nabla P \cdot \underset{\neg}{\nabla} h$ appears here as source or sink of vorticity. In the BARBI experiment all terms in this balance contribute with a similar overall size of magnitude but different patterns (not shown). The planetary vorticity term $\beta\partial\psi/\partial x$ is most important after the current leaves Drake Passage, heading northward, and at some other areas where the current must deviate from being almost zonal due to islands or other massive changes in the topography. The bottom pressure torque, arising from depth gradients, has a spotty structure but has locally the highest amplitudes. Friction is again of minor importance. Note that the ACC flow is definitely not in a Sverdrup balance (14.19) anywhere.

The other form of vorticity equation is built from the momentum balance by elimination of the bottom pressure $P$, resulting in (14.29), rewritten here as

$$\underset{\neg}{\nabla}\psi \cdot \nabla\frac{f}{h} = \underset{\neg}{\nabla} E \cdot \nabla\frac{1}{h} + \underset{\neg}{\nabla}\cdot\frac{\boldsymbol{\tau}_0}{h} - \underset{\neg}{\nabla}\cdot\frac{\mathcal{R}}{h} \tag{16.41}$$

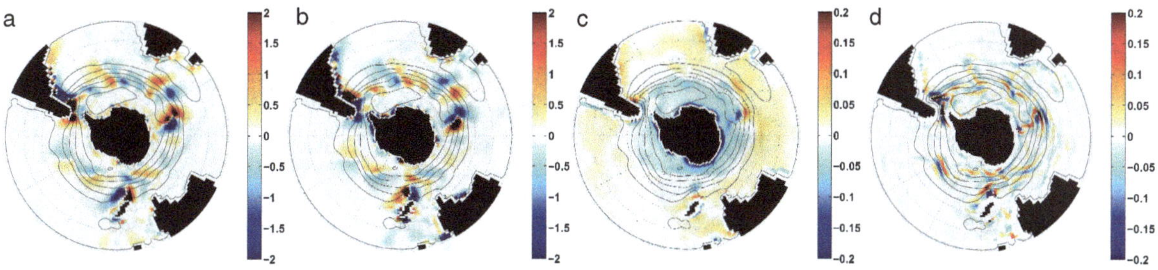


Fig. 16.22 Vorticity balance of the depth-averaged flow for the BARBI model of the Southern Ocean (see the box on p. 591). **a** Topographic-planetary vorticity $\underline{\nabla}\psi \cdot \nabla(f/h)$. **b** JEBAR term $-\underline{\nabla}E \cdot \nabla(1/h)$. **c** Curl of windstress $-\underline{\nabla} \cdot (\boldsymbol{\tau}_0/h)$. **d** Frictional torque $\underline{\nabla} \cdot (\boldsymbol{\mathcal{R}}/h)$ in $10^{-12}\,\mathrm{s}^{-2}$. The contour lines show the stream function $\psi$ with contour interval of 25 Sv. Note the different color scales

It describes the balance of vorticity of the depth-averaged velocity $\boldsymbol{U}/h$ and is forced by another torque-like term, the JEBAR term $\underline{\nabla}E \cdot \nabla(1/h)$. The dominant terms in the balance (16.41) are – by at least an order of magnitude – the topographic-planetary vorticity $\underline{\nabla}\psi \cdot \nabla(f/h)$ and the JEBAR term $\underline{\nabla}E \cdot \nabla(1/h)$. They largely oppose and cancel each other, as obvious from Figure 16.22. Wind stress and friction are an order of magnitude smaller. The approximate compensation of the two terms in (16.41) may be traced back to the predominance of the geostrophic terms in the balance of momentum, or likewise to an approximate compensation of the barotropic component (associated with the planetary vorticity) and the baroclinic component (associated with JEBAR) of the vertical velocity.

Remember that (16.41), with the time tendency term retained, is one of the prognostic equations of the BARBI model (see Appendix B.2 and Section 14.2.5). The other dynamical equation is the balance of baroclinic potential energy, written here in the reduced form

$$\frac{\partial E}{\partial t} + \frac{N^2}{6}\left[\underline{\nabla}E \cdot \nabla\frac{h^2}{f} - \underline{\nabla}\psi \cdot \nabla h^2\right] = -\frac{N^2}{6}\underline{\nabla} \cdot \frac{h^2\boldsymbol{\tau}_0}{f} + K_\ell \nabla^2 E \qquad (16.42)$$

To derive this form from the complete BARBI model, the balance of baroclinic momentum is taken for simplicity as geostrophic, and advection of the perturbation density, buoyancy forcing, and dissipation are ignored (see also Section 14.2.6). The terms of the balance (16.42) are shown in Figure 16.23, revealing again the dominance of two terms, namely the two terms in the brackets on the left-hand side of (16.42). Like the vorticity balance, the potential energy balance suffers from the dominance of the geostrophic terms. While the geostrophic terms entered there as almost compensating torques, they come into the balance of potential energy as integrals of vertical velocities which appear in (16.42) as pumping up/down the background stratification. We may rephrase the statement about the dominance of the geostrophic terms as one about a partial compensation of the bottom-induced vertical barotropic velocity and the geostrophic vertical baroclinic velocity. Both act on the background stratification and thus generate potential energy.

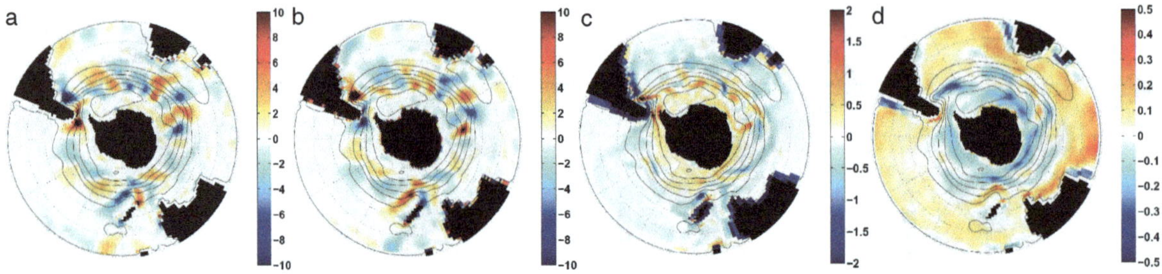


Fig. 16.23 Potential energy balance (16.42) for the BARBI model of the Southern Ocean (see the box on p. 591). **a** Barotropic divergence term $-(N^2/6)\underline{\nabla} E \cdot \nabla(h^2/f)$. **b** Baroclinic divergence term $(N^2/6)\underline{\nabla}\psi \cdot \nabla(h^2)$. **c** Eddy diffusion term $K\nabla^2 E$. **d** Ekman pumping term $(N^2/6)\underline{\nabla} \cdot (\boldsymbol{\tau}_0 h^2/f)$. Units are $10^{-4}$ m$^3$ s$^{-3}$. The contour lines show the stream function $\psi$ with contour interval of 25 Sv. Note the different color scales

## *16.5.3 The Baroclinic Stommel Model of the ACC*

In Section 14.2.6, we have derived the *baroclinic Stommel equation*, using the similarity of the barotropic stream function $\psi$ and the baroclinic stream function $E/f$, and the occurrence of the bottom torque term in both BARBI balances (vorticity and potential energy). Here we gain the same result: $\psi \approx E/f$ derives from the dominance of the Jacobians in (16.41) and (16.42). Expressing these balances in the equivalent form

$$\begin{aligned} \frac{\beta}{h}\frac{\partial \psi}{\partial x} + f\underline{\nabla}\psi \cdot \nabla\frac{1}{h} &= \underline{\nabla} E \cdot \nabla\frac{1}{h} + \underline{\nabla} \cdot \frac{\boldsymbol{\tau}_0}{h} + \underline{\nabla} \cdot \frac{\boldsymbol{\mathcal{R}}}{h} \\ \frac{1}{2hf}\left(\beta\frac{\partial E}{\partial x} + \kappa\nabla^2 E\right) + \underline{\nabla} E \cdot \nabla\frac{1}{h} &= f\underline{\nabla}\psi \cdot \nabla\frac{1}{h} + \frac{f}{2h^3}\underline{\nabla} \cdot \frac{h^2\boldsymbol{\tau}_0}{f} \end{aligned} \tag{16.43}$$

to have identical Jacobians (see the notations in the box on p. 444), eliminating them and using $\psi \approx E/f$, we find

$$\frac{3}{2}\beta\frac{\partial \psi}{\partial x} + \frac{1}{2}\kappa\nabla^2\psi = \frac{f}{2h^2}\underline{\nabla} \cdot \frac{h^2\boldsymbol{\tau}_0}{f} + h\underline{\nabla} \cdot \frac{\boldsymbol{\tau}_0}{h} + h\underline{\nabla} \cdot \frac{\boldsymbol{\mathcal{R}}}{h} \tag{16.44}$$

In these relations we have introduced $\kappa = K_\ell/R^2$ where $R = Nh/(\sqrt{6}|f|)$ is the local internal Rossby radius in the BARBI framework. The viscous term is small and may be abandoned, likewise the $\beta$-term arising in the wind forcing (the topographic terms cancel exactly). Then we arrive at the baroclinic Stommel equation[5]

$$\beta\frac{\partial \psi}{\partial x} + \frac{1}{3}\kappa\nabla^2\psi = \underline{\nabla} \cdot \boldsymbol{\tau}_0 \tag{16.45}$$

The relation immediately allows to derive an equation for the ACC transport. Taking the zonal mean of it, we obtain

$$\frac{1}{3}\kappa\frac{\partial \bar{U}}{\partial y} \approx \frac{\partial \bar{\tau}_0^{(x)}}{\partial y} \tag{16.46}$$

[5] It worth mentioning that 2/3 of the windstress forcing in this equation result from the vorticity balance and $1/3$ from the potential energy balance. The latter arises from Ekman pumping acting on the background stratification.

which may be integrated twice to yield the total transport, roughly $Y\bar{U} \sim (1/2)Y(hN/f)^2\tau_0^{(x)}/K_\ell$. Surprisingly, this is identical to the estimate from the adiabatic model in Section 16.6 below. However, the agreement is accidental because the last equation yields the total transport in a model which regards the presence of standing waves and surface displacement, both of which are absent in the baroclinic part of (16.54), introduced in the next section.

## 16.6 Simple Models of the Zonally Averaged Southern Ocean

Consider again the layered model of the Southern Ocean given by (16.35) and the sketch in Figure 16.3. If the flow conserves potential density, i. e. if there is no diapycnal mixing or thermohaline surface forcing anywhere, then there cannot be transport across isopycnals and, by mass conservation, the meridional transport in each layer must vanish, i. e. $\bar{V}_i = 0$ in the isopycnal setting of (16.35). Neglecting all frictional terms, we find that the interfacial formstress $\overline{\eta_i' \partial p_i'/\partial x}$ is vertically constant and equal to the windstress $\bar{\tau}_0^{(x)}$,

$$\overline{\eta_i' \frac{\partial p_i'}{\partial x}} \simeq \bar{\tau}_0^{(x)} \tag{16.47}$$

and to the bottom formstress (in models with a flat bottom we must replace the bottom formstress by the frictional bottom stress). Then, in each layer, the meridional mass fluxes induced by windstress and pressure gradients are compensated such that $\bar{V}_i = 0$. We will proceed with this scenario of "constant vertical momentum flux" in the next Section 16.6.1 and build a simple model of the Southern Ocean on this base. In passing, we note that many layer models with vanishing diapycnal mixing and thermohaline forcing, even in a mesoscale eddy-permitting configuration, approximately satisfy (16.47) if the Reynolds stress divergence is small. This applies particularly to the quasi-geostrophic models where, at eddy-permitting resolution, Reynolds stresses are found to be small and often even upgradient.

The real ocean is diabatic, i. e. there is mixing across isopycnals by small-scale turbulence and air-sea heat and freshwater fluxes, leading to water-mass conversions. Note that it is still under debate if this water-mass conversion occurs predominantly between the outcropping isopycnals at the surface or in the interior as well (see Section 15.3). In any case, the meridional overturning transports $V_i$ at a certain latitude circle can be nonzero only if there is exchange of mass between the layers south of the respective latitude — implying conversion of water masses south of the ACC. In fact, by mass conservation, the $V_i$ must be equal to the net exchange with the neighboring layers over the area south of the respective latitude. At the same time, referring to (16.35) again, the overturning transports imply a Coriolis force in the individual isopycnal layers, which is in balance with the vertical divergence of the interfacial formstress. Mesoscale eddy effects at the respective latitude and diabatic interior effects of small-scale turbulence, occurring to the south, must thus adjust according to mass and momentum requirements of the zonal current and the meridional overturning. The isopycnal analysis of the zonal momentum balance in eddy-resolving models, as e. g. the isopycnal analysis of the idealized model of the Southern Ocean (see the box on p. 582), exemplify the importance of diabatic processes and the inapplicability of the adiabatic regime: there is a net meridional circulation at all depths in balance with a divergent interfacial formstress and the windstress. This more realistic regime is further analyzed in Section 16.6.2, and we also build simple models of the Southern Ocean on this scenario.

### 16.6.1 A Model of Vanishing Residual Overturning Circulation

The ACC is the outstanding example in the ocean circulation for diapycnal transport of momentum by eddies in the form of interfacial formstresses. At intermediate depths where the bottom formstress does not yet work and small-scale turbulence is marginal, the residual circulation – the meridional overturning circulation – is driven by windstress and the divergence of interfacial formstress, which is immediately apparent from (16.34) or (16.35). In the above described adiabatic regime, these processes are in exact balance, and the residual circulation vanishes, i. e. $\bar{v} = 0, \phi \equiv 0$. Of course, this is a highly unrealistic situation, and we may weaken the condition to the assumption that $-f\bar{v}$ and $-f\phi$ are small in the respective balances (16.34) or (16.35). We proceed with the level-averaged form here and neglect the Reynolds stress term so that

$$-f\frac{\overline{v'\rho'}}{\partial\bar{\rho}/\partial z} + \bar{\tau}_0^{(x)} - \bar{\tau}^{(x)} + \mathcal{F} = 0 \tag{16.48}$$

The zonal current is geostrophic, i. e. $f\bar{u} = f\bar{u}_g = -\partial\bar{p}/\partial y$. A simple model of the ACC and the meridional overturning follows from these equations. Note that the vanishing of the interior meridional overturning circulation in this model may be questioned. A more complete model is discussed in the next section, but it is instructive to consider the model with vanishing overturning first.

For the intermediate layer, defined in Section 16.4, neglecting the small-scale turbulence $\bar{\tau}^{(x)}$ and omitting the bottom formstress term $\mathcal{F}$, we obtain the Johnson–Bryden relation (Johnson and Bryden, 1989),

$$\frac{\overline{v'\rho'}}{\partial\bar{\rho}/\partial z} = \bar{\tau}_0^{(x)} \tag{16.49}$$

Note that the standing eddy component will be neglected in this section[6]. According to (16.49), the northward eddy density flux $\overline{v'\rho'}$ in the circumpolar belt of the ACC is proportional to the zonal windstress $\bar{\tau}_0^{(x)}$. In a first step, Johnson and Bryden (1989) parameterize the transient lateral eddy flux by a down-gradient form with the isopycnal thickness diffusivity[7] $K$,

$$\overline{v'\rho'} = -K\frac{\partial\bar{\rho}}{\partial y} \tag{16.50}$$

similar to Section 16.3.4, and find that the windstress and $K$ constrain the slope of the isopycnals, $s = -(\partial\bar{\rho}/\partial y)/(\partial\bar{\rho}/\partial z) = \bar{\tau}_0^{(x)}/(fK)$. Such a relation is roughly

[6] This is a severe assumption because the standing eddy contribution may exceed the transient component in realistic conditions, see also the box on p. 563. We could instead interpret the mean as the average along the current path, $\bar{\tau}_0^{(x)}$ is then not the zonal windstress but rather the path-following component.

[7] In this section the thickness diffusivity $K_\ell$ will be denoted by $K$ to avoid multiple indices.

consistent with the observed slopes in the ACC belt if the eddy diffusivity is of order $K \sim 10^3\,\mathrm{m^2\,s^{-1}}$ (using $s = 10^{-3}$, $\bar{\tau}_0^{(x)} = 10^{-4}\,\mathrm{m^2\,s^{-2}}$). Replacing the lateral density gradient using the thermal wind relation, $f\partial\bar{u}_g/\partial z = g\partial\bar{\rho}/\partial y$, we obtain

$$K\frac{f^2}{N^2}\frac{\partial \bar{u}_g}{\partial z} = \bar{\tau}_0^{(x)} \tag{16.51}$$

where the vertical density gradient is replaced by the squared Brunt–Väisälä frequency frequency $N^2 = -g\partial\bar{\rho}/\partial z$. Apparently, $K(f/N)^2$ defines an equivalent diffusivity for the vertical momentum transfer which is achieved by lateral density diffusion, as mentioned earlier. We notice here the same equivalence between vertical momentum transfer and horizontal heat transfer by eddies as in Sections 12.2.6 and 16.3.

The Johnson–Bryden relation (16.51) is dynamically incomplete as it does not satisfy a closed momentum balance. Integrating the momentum balance from top to bottom (with vanishing bottom formstress $\mathcal{F} = 0$), we note that the balance $\bar{\tau}_0^{(x)} = \bar{\tau}_b^{(x)}$ between windstress and frictional bottom stress must hold, as for the homogeneous model in Section 16.2. The present model can be straightforwardly extended to include a frictional bottom boundary layer, along the way outlined in Section 16.2. Integrating (16.51), the current profile becomes

$$\bar{u}_g(z) = -\frac{g}{f}\frac{\partial\bar{\zeta}}{\partial y} - \frac{\bar{\tau}_0^{(x)}}{f^2}\int_z^0 \frac{N^2}{K}\mathrm{d}z' = -\frac{g}{f}\frac{\partial\bar{\zeta}}{\partial y} + z\frac{N^2\bar{\tau}_0^{(x)}}{f^2 K} \tag{16.52}$$

The second relation is found, taking for simplicity all coefficients constant, for which we obtain a linear profile of $u_g$. Note that the model is easily extended with a realistic $N(z)$ profile and a given $K(z)$. The velocity at the top of the bottom boundary layer at $z = -h + d_b$ is used to evaluate the frictional bottom stress as

$$\tau_b^{(x)} = \frac{1}{2}d_b|f|\bar{u}_g(-h + db) = -\frac{1}{2}d_b f\left[-\frac{g}{f}\frac{\partial\bar{\zeta}}{\partial y} - (h - d_b)\frac{N^2\tau_0^{(x)}}{f^2 K}\right]$$

and equating this result with $\tau_0^{(x)}$, the associated gradient of the surface displacement follows as

$$\frac{\partial\bar{\zeta}}{\partial y} = \frac{2\tau_0^{(x)}}{g d_b}\left[1 - \frac{1}{2}(h - d_b)d_b\frac{N^2}{fK}\right] \tag{16.53}$$

The first term is equal to the slope resulting from the homogeneous model in Section 16.2. A reasonable size of $\partial\bar{\zeta}/\partial y \sim 0.7 \times 10^{-6}$ is found for the complete expression from our standard parameters (see Table 16.1). Note that the surface slope relates to the isopycnal slope $s = \tau_0^{(x)}/(fK)$ of the adiabatic regime roughly by $(\partial\bar{\zeta}/\partial y)/s \sim -hN^2/g \sim -6\times 10^{-4}$. The transport in this balanced model becomes

$$Y\int_{-h}^{0}\bar{u}_g\mathrm{d}z = Y\left[-\frac{2h}{fd_b} + \frac{1}{2}\frac{(hN/f)^2}{K}\right]\bar{\tau}_0^{(x)} \tag{16.54}$$

which can be evaluated as 145 Sv using the parameters of Table 16.1. The first term is identical to transport in the homogeneous model (see (16.11)) and yields 64 Sv;

the second contribution results from eddy processes and yields 81 Sv. In addition to the geostrophic transport, there is also a frictional transport arising from the bottom stress which is identical to the one in (16.12) of the homogeneous model and thus negligible. Note that the diffusivity $K$ appears as $K/R_i^2$ in the above expression where $R_i = hN/|f|$ is the internal Rossby radius, which is identical to the transport implied by the baroclinic Stommel equation in Section 16.5.3.

A further extension of the Johnson–Bryden model is to include bottom topography and the associated bottom formstress $\mathcal{F}$. We have previously argued that in this case, the frictional bottom stress may be abandoned, and the momentum balance takes the form (16.15). In lack of a physical parameterization of the stress $\mathcal{F}$, we write $\mathcal{F}(z) = \mathcal{F}[\bar{u}_g, z]$ for the stress at the depth $z$, accounting for a functional dependence of the formstress on the current field. Simple examples will be discussed in Section 16.7. Using the geostrophic relation (16.52), the momentum balance (16.15) results in a condition $\tau_0^{(x)} + \mathcal{F}[-(g/f)\partial\bar{\zeta}/\partial y + \bar{u}_g^{\text{clin}}, -h] = 0$, determining the gradient of the surface displacement, as before in the frictional model. In general, however, the functional dependence of the bottom formstress on the current velocity is unknown.

### 16.6.2 Models Driven by Surface Windstress and Thermohaline Forcing

Much of the recent perception of the circulation in the Southern Ocean is focussed on the meridional overturning and ventilation of water masses. The classical view (Sverdrup et al., 1942) of water-mass storage and spreading is updated in Figure 16.3, where the role of eddies in the unblocked part of the water column is highlighted. We have pointed out in the previous sections that eddies and turbulent mixing might accomplish a major task in shaping and balancing the overturning circulation, and it remains to put up some simple models to demonstrate how it might work. For this task, we present in this section two zonally averaged models which include forcing by windstress and by water mass conversion. They are the most realistic analytical models presented in this chapter, although they still neglect bottom formstress and standing eddy contributions.

We assume that all mixing and water-mass formation processes take place in an upper layer of the ocean – basically a turbulent layer where the Ekman transport and pumping is established by the wind. Buoyancy is imprinted on the surface waters by heat and freshwater exchange with the overlying atmosphere. The ocean interior is void of small-scale turbulence, but mesoscale eddies are present that transport and mix substances along isopycnals. We regard this scenario as an extreme. The real ocean might have substantial diapycnal mixing by small-scale turbulence in the interior and furthermore, mesoscale eddies might contribute by a diapycnal flux to water-mass formation as well (see also the box on p. 390). However, the simplified view allows for an analytical treatment, which we now explore. A sketch of the system is shown in Figure 16.24.

The dynamical part of the model is given by the zonal average of the zonal momentum balance, written as (16.34) for the residual stream function, and is augmented by the zonally averaged balance of density. The zonal current is geostrophic.

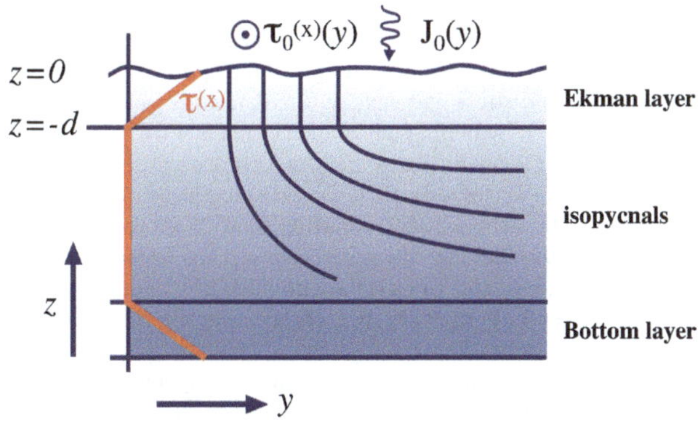


Fig. 16.24 Set-up of the zonal mean overturning model. There is an Ekman layer at the top where the stress is modeled by a body force, spreading the windstress into a layer of constant depth $d$. The stress in the interior is zero. A bottom layer is included in which either a frictional stress or a bottom formstress becomes active. The Ekman layer coincides with the surface mixed layer in which the density is vertically constant. The interior density field shows sloping isopycnals. The ocean system interacts with the overlying atmosphere via the zonal windstress $\bar{\tau}_0^{(x)}$ and the density flux $J_0$

The system is thus governed by

$$-f\phi = fB + \bar{\tau}_0^{(x)} - \bar{\tau}^{(x)} + \mathcal{F} \tag{16.55}$$

$$\frac{\partial \bar{\rho}}{\partial t} + \underset{\neg}{\nabla}\phi \cdot \nabla\bar{\rho} = \nabla \cdot K_{\mathrm{d}}\nabla\bar{\rho} + \frac{\partial J}{\partial z} \tag{16.56}$$

$$f\bar{u} = -\frac{\partial \bar{p}}{\partial y} \tag{16.57}$$

Note that the notation in this section follows the box on p. 385, i.e. $\nabla = (\partial/\partial y, \partial/\partial z)$ and $\underset{\neg}{\nabla} = (-\partial/\partial z, \partial/\partial y)$. All terms take the same meaning as before in this chapter. In particular the eddy stream function $B$ and the residual overturning stream function $\phi$ are introduced in Section 16.3. The Reynolds stress is neglected since we have learned that it is small. Finally, $J$ is a vertical flux of density carried by small-scale turbulence. It is only relevant in the upper near-surface layer where, at the air-sea interface, it equals the surface density flux, $J(y, z = 0) = J_0(y)$.

The density equation, derived in Section 12.2.3 on the Transformed Eulerian Mean (TEM) theory, will be used in its steady form. The eddy stream function $B$ and the diapycnal diffusivity $K_{\mathrm{d}}$ are related to the eddy density fluxes $\overline{v'\rho'}$ and $\overline{w'\rho'}$ by

$$B = -\frac{\overline{v'\rho'}\,(\partial\bar{\rho}/\partial z) - \overline{w'\rho'}(\partial\bar{\rho}/\partial y)}{|\nabla\bar{\rho}|^2} \quad \text{and} \quad K_{\mathrm{d}} = -\frac{\overline{v'\rho'}\,(\partial\bar{\rho}/\partial y) + \overline{w'\rho'}(\partial\bar{\rho}/\partial z)}{|\nabla\bar{\rho}|^2} \tag{16.58}$$

Note that the approximation used in (16.34), $|\partial\bar{\rho}/\partial y| \ll |\partial\bar{\rho}/\partial z|$, becomes invalid in the mixed layer. The meridional density flux is parameterized as in (16.50), with the isopycnal thickness diffusivity $K$, i.e. $\overline{v'\rho'} = -K\partial\bar{\rho}/\partial y$. Remember that $K$ is the Gent and McWilliams diffusivity (see Section 12.2.4). Using this parameterization, one finds from (16.58)

$$B = (K_{\mathrm{d}} - K)s \quad \text{and} \quad \overline{w'\rho'}\Big/\frac{\partial\bar{\rho}_y}{\partial y} = \frac{K}{s} + \left(1 + \frac{1}{s^2}\right)B = -Ks + \frac{K_{\mathrm{d}}}{s}(1 + s^2) \tag{16.59}$$

**98. The zonally average forcing of the Southern Ocean**

To develop our model of the mean density structure, we need the size and pattern of the surface momentum and density fluxes. The estimates, shown in the figure below, are based on the data discussed in Chapter 13. Wind stress and density flux data were obtained by averaging along mean ACC streamlines, defined by finding the latitude of the 3 °C isotherm at a depth of 200 m. All meridional sections were then shifted by their departure from the zonally averaged ACC position, sorted along their new "latitude circles", and subsequently averaged. The maximum Ekman transport in these data is about 30 Sv, occurring at 51° S with upwelling to the south and downwelling to the north of this latitude. The surface density loss depends on the combination of the air-sea heat and freshwater fluxes. Consistent with other estimates, we find an oceanic heat gain of about $10\,\mathrm{W\,m^{-2}}$ over the core of the ACC and cooling of a similar magnitude north and south of the stream. The freshwater flux at the surface shows a net freshening south of 50° S of about $10^{-5}\,\mathrm{kg\,m^{-2}\,s^{-1}}$ and about a similar amount of net evaporation around 30° S, causing a significant modification of the surface density. Only in the most northern part is the freshwater contribution to the density flux smaller than the air-sea heat flux. Our final estimate of air-sea density flux yields a reduction of the surface density south of 48° S and an increase to the north of 48° S. However, the expected error on the surface density flux as a difference of two not well-known quantities could be rather large.

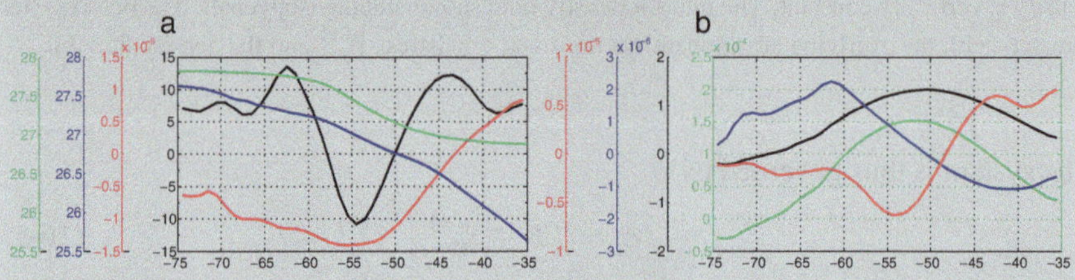


*a Net heat flux (black in* $\mathrm{W\,m^{-2}}$*) and evaporation minus precipitation (red in* $\mathrm{kg\,m^{-2}\,s^{-1}}$*). The densities (surface in blue, 500 m depth in green, both in* $\mathrm{kg\,m^{-3}}$*) are taken from the WOCE climatology. b Wind stress (green in* $\mathrm{m^2\,s^{-2}}$*), Ekman transport (black in* $\mathrm{m\,s^{-1}}$*), Ekman pumping (blue in* $\mathrm{m\,s^{-1}}$*), and surface density flux (red in* $\mathrm{kg\,m^{-2}\,s^{-1}}$*). The data are obtained by an ACC-path following average. From Olbers and Visbeck (2005).*

written in terms of the slope $s = -(\partial\bar{\rho}/\partial y)/(\partial\bar{\rho}/\partial z)$ of the isopycnals. Note that $K$ generally differs much from the diapycnal diffusivity $K_\mathrm{d}$. Both these diffusivities must in turn be parameterized in terms of resolved fields, but in the following we simply assume their spatial structure as given functions of $(y, z)$ (for simplicity either taken constant or taken from a numerical model).

We follow the most simple concepts which have partly been used before. The interior ocean is assumed adiabatic with respect to small-scale mixing, i. e. $J \equiv 0$. We also assume here a vanishing diapycnal eddy flux, hence $K_\mathrm{d} \equiv 0$, so that the density balance becomes adiabatic, $\underset{\neg}{\nabla}\phi \cdot \nabla\bar{\rho} = 0$. The consequence is that isolines of the residual stream function and the density coincide. Furthermore, from $K_\mathrm{d} = 0$, it follows that the eddy density flux is strictly along isopycnals, i. e. $\overline{v'\rho'}(\partial\bar{\rho}/\partial y) + \overline{w'\rho'}(\partial\bar{\rho}/\partial z) = 0$. Then $B = -\overline{v'\rho'}/(\partial\bar{\rho}/\partial z) = -Ks$. The stress $\bar{\tau}^{(x)}$ is assumed nonzero only in the surface Ekman layer, which is embedded in the upper ocean layer. If the bottom formstress $\mathcal{F}$ is absent, there is also a frictional bottom layer. Hence in the interior part of the ocean, the stream function is given by

$$\phi = Ks + M \quad \text{(interior for } \mathcal{F} = 0) \tag{16.60}$$

with $M = -\bar{\tau}_0^{(x)}/f$ denoting henceforth the Ekman transport. Note that in this model, the Ekman transport $M$ equals the Eulerian overturning stream function $\Lambda$ of (16.26) in the interior.

In the upper layer, the diapycnal diffusivity $K_d$ is in general nonzero. In fact, if the layer is vertically mixed ($\partial\bar{\rho}/\partial z = 0, s = \infty$), the diapycnal direction is horizontal and then $K_d \equiv K$. Furthermore, $B = \overline{w'\rho'}/(\partial\bar{\rho}/\partial y)$. We will treat the small-scale stress in the surface Ekman layer as a body force $\bar{\tau}^{(x)} = \bar{\tau}_0^{(x)} E(z)$ where $E(z = 0) = 1$ and zero at the mixed layer base (taken identical to the Ekman layer, see Figure 16.24). Then

$$\phi = -\overline{w'\rho'} \Big/ \frac{\partial\bar{\rho}}{\partial y} + MT(z) \quad \text{(mixed layer)} \tag{16.61}$$

where $T(z) = 1 - E(z)$, which equals 1 in the interior. Note that $\phi$ is basically unknown unless the vertical eddy density flux is parameterized. As discussed in the previous sections, the balance of zonal momentum (16.55) requires a stress layer above the bottom. The stress may appear either in the form of a frictional bottom stress or a bottom formstress (or both). We treat this problem at the end of this section.

## A Diagnostic Model

Marshall and Radko (2003) put forward a partly diagnostic model of the overturning of the Southern Ocean. The upper layer between $z = 0$ and $z = -d$ is assumed to be vertically mixed, and its meridional density profile $\bar{\rho}_m(y)$ is assumed to be given[8]. The fields in the interior, however, are predicted.

In this diagnostic model, by vertical integration of the density balance (16.56) in the mixed layer, the stream function $\phi_m(y) = \phi(y, z = -d)$ at each latitude along the mixed layer base is found from

$$\phi_m \frac{\partial\bar{\rho}_m}{\partial y} = J_0 + \frac{\partial}{\partial y}\left( \int_{-d}^{0} K \mathrm{d}z \frac{\partial\bar{\rho}_m}{\partial y} \right) = J_{\mathrm{eff}} \tag{16.62}$$

and may be inferred from the local values of the air-sea flux $J_0$, the prescribed mixed layer density gradient $\partial\bar{\rho}_m/\partial y$ and the diffusivity $K$. By matching to the interior representation (16.60) at $z = -d$, the slope $s_m(y) = (\phi_m(y) - M(y))/K_m(y)$ of isopycnals can be computed along the mixed base.

The direction of the residual circulation is thus found to be governed by the signs of the (modified) surface flux $J_{\mathrm{eff}}$ and the gradient $\partial\bar{\rho}_m/\partial y$ of the surface density. Following the observations, shown in the box on p. 604, the gradient is negative. We expect a negative $J_{\mathrm{eff}}$ because $J_0 < 0$ in the main area of interest, and because the diffusive contribution is likely small if $\partial\bar{\rho}_m/\partial y$ and $K$ are close to being constant. Then we find $\phi_m > 0$, a northward residual circulation (remember that $\phi(z = 0) = 0$ and the residual meridional velocity is $\bar{v}_{\mathrm{res}} = -\partial\phi/\partial z$). Furthermore, with a stable stratification the slope $s_m$ must be negative, and hence the eddy-driven circulation is southward, $Ks < 0$, compensating the Ekman circulation $M > 0$ in parts. We return to this compensation effect later in the prognostic model.

The depth-latitude dependence $z_\rho = z_\rho(y^\star, y)$ of an isopycnal starting at a point $y = y^\star, z = -d$ on the mixed layer base, with the stream function value $\phi_m(y^\star)$

[8] The index $m$ stands for the value of the respective quantity at the mixed layer base $z = -d$. Since the density is mixed, we have $\bar{\rho}(y, z) = \bar{\rho}_m(y)$ in the mixed layer.

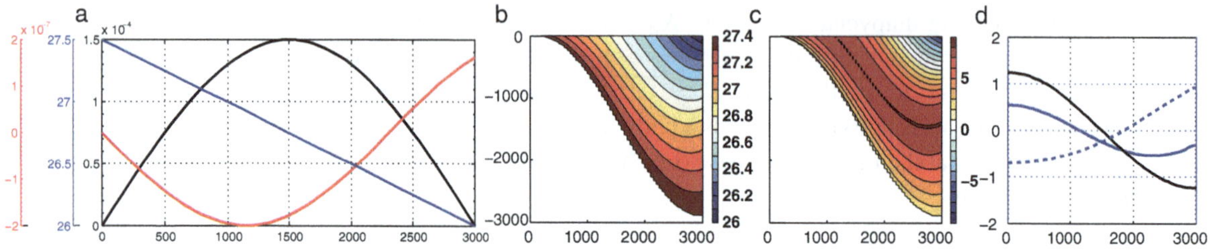

Fig. 16.25 **a** Idealized surface forcing for the diagnostic model as a function of $y$ in km: wind-stress $\bar{\tau}_0^{(x)}$ (*black*) in $\mathrm{m^2\,s^{-2}}$, surface density flux $J_0$ (*red*) in $\mathrm{kg\,m^{-2}\,s^{-1}}$, and surface density $\bar{\rho}_m$ (*blue*) in $\mathrm{kg\,m^{-3}}$. **b** Density $\rho$ in $\mathrm{kg\,m^{-3}}$ of the model using a constant thickness diffusivity $K = 800\,\mathrm{m^2\,s^{-1}}$ and idealized forcing field as in **a**. **c** Residual stream function in Sv using using the zonal path length at $60°$. **d** Residual vertical velocity $\partial\phi/\partial y$ (*blue*), eddy induced vertical velocity $\partial(Ks)/\partial y$ (*blue dashed*) at the base of the mixed layer, and Ekman pumping $\partial M/\partial y$ (*black*) in $10^{-6}\,\mathrm{m\,s^{-1}}$

**99. Characteristics of the Interior Problem**

The density field in the interior is governed by

$$\underset{\neg}{\nabla}\phi \cdot \nabla\bar{\rho} = \frac{\partial\phi}{\partial y}\frac{\partial\bar{\rho}}{\partial z} - \frac{\partial\phi}{\partial z}\frac{\partial\bar{\rho}}{\partial y} = 0$$

The *characteristic curves* $(y(t), z(t))$ of this differential equation are determined by

$$\frac{\mathrm{d}y}{\mathrm{d}t} = -\frac{\partial\phi}{\partial z} \quad \frac{\mathrm{d}z}{\mathrm{d}t} = \frac{\partial\phi}{\partial y}$$

representing a system of ordinary differential equations of Hamiltonian form (see the box on p. 98). The "Hamiltonian" $\phi$ is conserved along the characteristics, $\mathrm{d}\phi/\mathrm{d}t = \dot{y}(\partial\phi/\partial y) + \dot{z}(\partial\phi/\partial z) = 0$. Likewise, the density is constant on these curves: from the first equation in this box we find $\mathrm{d}\rho/\mathrm{d}t = \dot{y}(\partial\rho/\partial y) + \dot{z}(\partial\rho/\partial z) = 0$. Hence knowledge of the characteristics is equivalent to solving the interior problem.
To solve the second set of equations, $\phi = Ks + M$ must be inserted, leading to a complicated problem. Finding the characteristics, however, is much simpler if the previous density equation is rewritten as

$$\frac{\partial\phi}{\partial y} + s\frac{\partial\phi}{\partial z} = 0$$

and the characteristic curves $(y'(t), z'(t))$ of this differential equation are determined by

$$\frac{\mathrm{d}y'}{\mathrm{d}t} = 1 \quad \frac{\mathrm{d}z'}{\mathrm{d}t} = s$$

As before, from $\mathrm{d}y' - s\mathrm{d}z' = 0$, we find that density is constant along these curves, and the derivative $\mathrm{d}\phi/\mathrm{d}t$ of $\phi(y'(t), z'(t))$ along such a characteristic is zero, so $\phi = Ks + M =$ const on each characteristic. Integration of the last equation requires initial conditions $y'(t = 0) = y^\star, s(t = 0) = s_m$. Then $Ks + M(y) = K(y^\star, -d)s_m(y^\star) + M(y^\star)$. Initial data for $s$ are required on a nonisopycnal curve, in our case the depth level $z = -d$. Elimination of $t$ from this problem yields (16.63).

attached to it, can be found by integrating the ordinary differential equation

$$s = -\frac{\partial\bar{\rho}}{\partial y}\Big/\frac{\partial\bar{\rho}}{\partial z} = \frac{\partial}{\partial\rho}\left(\frac{\mathrm{d}z}{\mathrm{d}y}\right) = \frac{\phi_m(y^\star) - M(y)}{K} \qquad (16.63)$$

which can be performed analytically if $K$ is constant (the problem is also easily solved for a factorized diffusivity $K(y, z) = K_1(y)K_2(z)$). With the isopycnal structure determined, the density field and the residual circulation are determined as well.

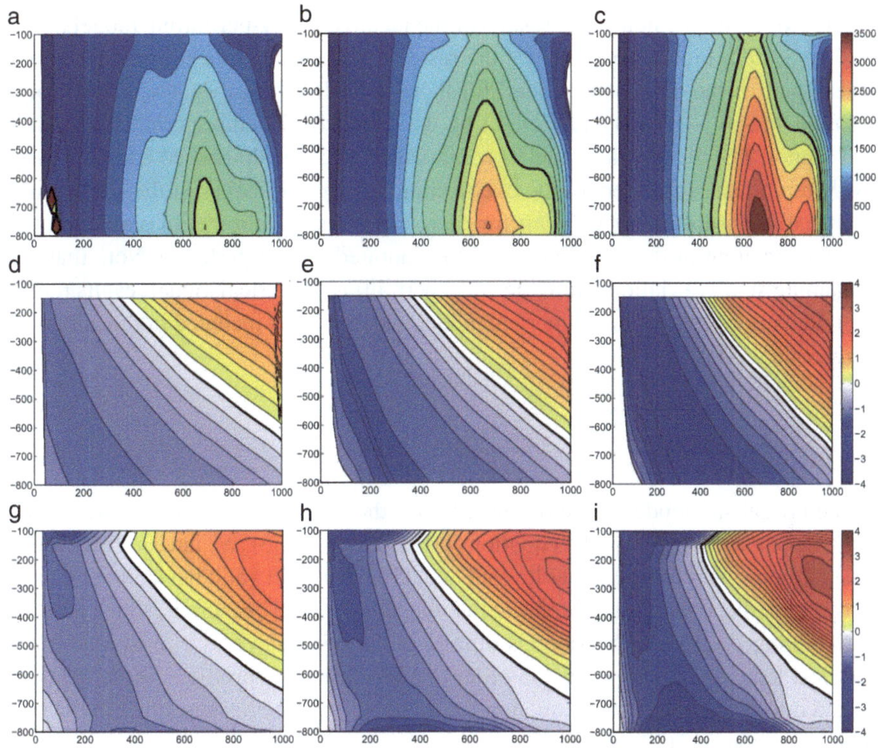

Fig. 16.26 **a–c** Thickness diffusivity $K$ (in $\mathrm{m^2\,s^{-1}}$) diagnosed from the idealized eddy-permitting model of the Southern Ocean described in the box on p. 582 driven by three different amplitudes of the sinusoidal windstress forcing, i. e. 0.5 (**a**), 0.75 (**b**) and $\mathbf{1.0 \times 10^{-4}}\,\mathrm{m^2\,s^{-2}}$ (**c**). The contour intervals for the diffusivities are $\mathbf{200}\,\mathrm{m^2\,s^{-1}}$, and the $\mathbf{2{,}000}\,\mathrm{m^2\,s^{-1}}$ *line is thick*. **d–f** Residual mean stream function $\phi$ in Sv for the diagnostic model with $K$ and surface fluxes identical to the eddy-permitting model. The contour intervals are 0.2 Sv, the *thick line* is the zero contour. **g–i** Residual mean stream function in Sv for the eddy-permitting model

We exemplify this diagnostic model in Figure 16.25, where the driving fluxes $J_0$ and $\bar{\tau}_0^{(x)}$ are taken as sinusoidal and the surface density gradient as constant, with amplitudes and scales resembling the observed fields shown in the box on p. 604 in the latitude range from 70 to 40° S. The thickness diffusivity is assumed constant with a value of $K = 800\,\mathrm{m^2\,s^{-1}}$. The figure shows a realistic density field (compare Figure 16.25b) with Figure 16.2) and upwelling of deep water (NADW) of about 10 Sv, coming from below 1,500 m at the northern rim of the modeled domain. In the mixed layer, the water is made lighter and descends, leaving the domain above 1,500 m depth in the north as AAIW. The vertical velocities at the mixed layer base (right panel) reflect the approximate compensation of the Eulerian and the eddy-induced circulation. It should be mentioned that this solution is sensitive to changes in any of the prescribed fields, particularly to the thickness diffusivity $K$. Lowering (increasing) $K$ makes the isopycnals flatter (steeper). It may occur that isopycnals cross each other, in which case the model with zero diapycnal diffusion, of course, becomes unrealistic. Note also that the negative cell (related to the formation of AABW) is not reproduced in the model.

Finally, we demonstrate that the conceptual framework of the interior residual overturning, given by the diagnostic model, is appropriate as long as the correct eddy diffusivity structure is taken into account. We compare the application of equation (16.63) with the numerical results of the idealized eddy-permitting model of the box on p. 582. Figure 16.26 shows the residual stream function $\phi$ in the Southern Ocean part of the diagnostic model, for three different windstress amplitudes, prescribing $\phi(z = -150\,\mathrm{m})$ and $K(y,z)$ by the results of the eddy-permitting numerical model. The diagnostic model in combination with the diagnosed eddy diffusivity $K$ reproduces the extent of the overturning cell and the slopes of the streamlines in the eddy-permitting model well for each windstress amplitude. In particular, the slight deepening of the zero line (thick) is captured almost perfectly. Note that the remaining small deviations from the eddy-permitting numerical model results represent diabatic effects in this numerical model which are not included in the adiabatic diagnostic model.

## A Prognostic Model

In the prognostic model, to be discussed now, the surface density is not prescribed but will be calculated from the equations (16.55) and (16.56). We have to solve these balances now also for the upper layer. Then, (16.62) is still valid but cannot be used to compute the stream function $\phi_m$ at the mixed layer base because it is the equation determining $\bar{\rho}_m(y)$. As the slope is infinite in the mixed layer, we must load more physical features into the model in order to provide a slope profile at some depth to be fed into the characteristic equation for the interior, either as described in the box on p. 606 or via $\phi_m(y^\star)$ in (16.63).

We derive an equation determining the slope $s_m$ at the mixed base. The mixed layer is vertically completely mixed (hence $K_\mathrm{d} = K$) and absorbs the entire small-scale stress $\bar{\tau}^{(x)}$ as before assumed. Immediately below the mixed layer, before entering the adiabatic interior, the eddy flux can still be diapycnal in part so $K_\mathrm{d}$ is nonzero and generally not equal to $K$. Expressing $\phi_m$ in the vertically integrated density balance (16.62) of the mixed layer by the general form (16.61), this balance is now written as

$$M\frac{\partial \bar{\rho}_m}{\partial y} - \overline{w'\rho'}\big|_{-d} = J_0 + \frac{\partial}{\partial y}\left(\mathcal{K}\frac{\partial \bar{\rho}_m}{\partial y}\right) \tag{16.64}$$

Here, the integral $\mathcal{K}$ of $K$ over the mixed layer is introduced, appearing also in (16.62). We divide by the density gradient $\partial\bar{\rho}/\partial z$, appropriate just below the mixed layer base, and evaluate the vertical density flux $\overline{w'\rho'}$ just below the base using (16.59). Then we obtain

$$(M + K_m s_m)s_m - K_\mathrm{d}(1 + s_m^2) = -J_0\Big/\left(\frac{\partial\bar{\rho}}{\partial z}\right) - \frac{\partial}{\partial y}\left(\mathcal{K}\frac{\partial \bar{\rho}_m}{\partial y}\right)\Big/\left(\frac{\partial\bar{\rho}}{\partial z}\right) \tag{16.65}$$

where $s_m = -(\partial\bar{\rho}_m/\partial y)/(\partial\bar{\rho}/\partial z)$ and $K_m = K(y, z = -d)$. To convert (16.65) to an equation for the slope, a fairly "mild" assumption must be made: in the diffusion term the $y$-dependence of $\partial\bar{\rho}/\partial z$ is ignored,

$$\frac{\partial}{\partial y}\left(\mathcal{K}\frac{\partial \bar{\rho}_m}{\partial y}\right)\Big/\left(\frac{\partial\bar{\rho}}{\partial z}\right) = -\frac{\partial}{\partial y}\left(\mathcal{K}\frac{\partial\bar{\rho}}{\partial z}s_m\right)\Big/\left(\frac{\partial\bar{\rho}}{\partial z}\right) \approx -\frac{\partial}{\partial y}\left(\mathcal{K}s_m\right)$$

Further, in the density forcing term, $\partial\bar{\rho}/\partial z$ is replaced by a the squared Brunt–Väisälä frequency $N^2 = -g(\partial\bar{\rho}/\partial z)/\rho_0$ below the mixed layer which later will

**100. Approximate Solution of the Equation (16.66)**

To gain some insight into the behavior of (16.66) we assume all coefficients $M, \mathcal{K}, K_m, K_d$, and $Q = gJ_0/(\rho_0 N^2)$ to have an identical $y$-dependence, however, with different amplitudes $M_0, \mathcal{K}_0, K_0, K_{d,0}$, and $Q_0$. Allowing for the slight inconsistency to take the diffusion coefficient outside the derivative, an equation with constant coefficients is obtained

$$\mathcal{K}_0 s' = M_0 s + K_0 s^2 - K_{d,0}(1+s^2) - Q_0$$

which can be solved analytically. Depending on $\Delta = M_0^2 + 4(K_0 - K_{d,0})(Q_0 + K_{d,0})$ the solution is given by

$$s = -\frac{1}{2}\frac{M_0}{K_0 - K_{d,0}} \begin{cases} +\frac{\sqrt{-\Delta}}{K_0 - K_{d,0}} \tan\left(\frac{\sqrt{-\Delta}}{2\mathcal{K}_0}(y-y_0)\right) & \text{for} \quad \Delta < 0 \\ -\frac{\sqrt{\Delta}}{K_0 - K_{d,0}} \tanh\left(\frac{\sqrt{\Delta}}{2\mathcal{K}_0}(y-y_0)\right) & \text{for} \quad \Delta > 0 \end{cases}$$

For a realistic situation, $Q_0$ must be negative. If $\Delta < 0$, a 'blow up' develops in the tan-solution, the slope tends to infinity at finite $y$, and obviously, a realistic model demands that the diapycnal mixing increases, i. e. $K_{d,0}$ must depend on $s$.

For $\Delta > 0$, the solution is exponential. We notice that diffusivity $\mathcal{K}_0$ is attached only to the scale of the solution process but not the overall magnitude of the slope. This increases (becomes more negative) with the increase of the Ekman transport and the diapycnal diffusion and decreases with a larger (more negative) surface density flux.

be regarded as prescribed. We arrive at the equation

$$\frac{\partial}{\partial y}(\mathcal{K} s_d) = (M + K_m s_m)\, s_m - K_d(1+s_m^2) - \frac{gJ_0/\rho_0}{N^2} \tag{16.66}$$

determining the slope at (i. e. just below) the mixed base. The eddy-induced advection (second term on the right-hand side) and the surface flux (last term) tend to increase the slope towards positive numbers (if $J_0$ is negative), and only the Ekman advection (if $M$ is positive; first term) and diapycnal mixing (third term) can counteract this tendency. In a realistic situation where $s_m = -O(10^{-3})$ and $|M s_m| \sim K_m s_m^2 = O(10^{-3})$, the diapycnal diffusivity must be fairly small, i. e. $K_d \sim O(10^{-3})$ or smaller. This requirement hints at a dependence of $K_d$ on the slope: for the infinite slope in the mixed layer we have $K_d = K$, reducing to much smaller values just below the mixed layer base where the slope becomes finite and then approaching zero in the adiabatic interior.

In the following, we regard $M$ and $Q = gJ_0/(\rho_0 N^2)$ as given functions of latitude $y$ and, for simplicity, the mixing coefficients $\mathcal{K} = dK_m, K_m, K_d$ as constant. The differential equation (16.66) for $s_m$ is of the Riccati type (see e. g. Zwillinger, 1998). Though some analytical techniques exist to reduce the nonlinear Riccati differential equation to simpler linear equations of second order or to find the general solution if a particular solution is known, exact analytical solutions for a reasonable meridional dependence of the above coefficient functions could not be found. An approximate analytical solution is discussed in the box on p. 609.

We now discuss some numerical solutions of (16.66), shown in Figure 16.27. First consider the case $J_0 = K_d = 0$ shown in Figure 16.27a–d so that the right-hand side of (16.66) is given by $\phi_m s_m$, written in terms of the residual stream function $\phi_m = M + K_m s_m$. The zero line of $\phi_m$ is shown as the black dashed line in Figure 16.27a, e, i, and m. If $s_m$ is above this line (but negative) and $\phi_m$ thus northward, $s_m$ becomes more negative until it hits $\phi_m = 0$ and then turns to increase. The residual circulation reverses here to a southward direction. Depending on $K_m$ and $\mathcal{K}$, this reversion may occur at different latitudes, as demonstrated in the Figure 16.27a–d.

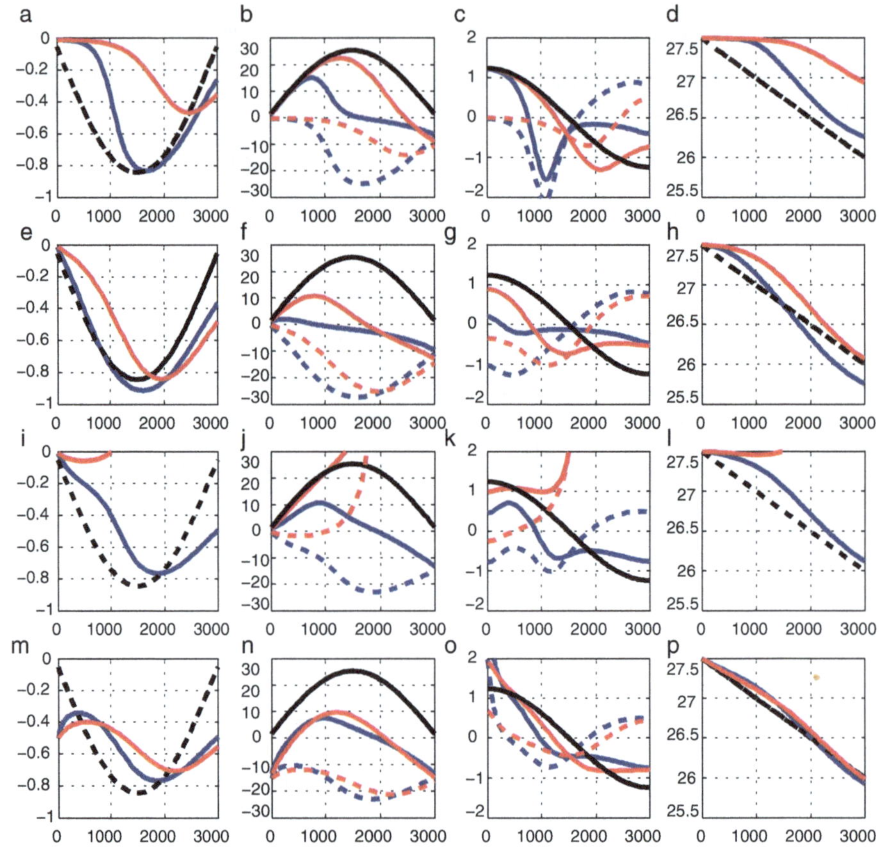

Fig. 16.27 Solutions of the prognostic model (16.66) for $\mathcal{K} = 100 \times 1{,}500\,\mathrm{m^3\,s^{-1}}$ (*blue curves*) and $\mathcal{K} = 300 \times 1{,}500\,\mathrm{m^3\,s^{-1}}$ (*red curves*). All quantities are displayed as a function of $y$ in km. Columns: **a**, **e**, **i**, and **m** show the isopycnal slopes $s_m$ just below the mixed layer (*red* and *blue*; in $10^3$), and $\phi_m = M(y) + K_m s_m = 0$ (*dashed*). **b**, **f**, **j**, and **n** show the residual (*blue* and *red solid lines*) and eddy (*blue* and *red dashed lines*) stream function in Sv, the *black solid line* shows the Ekman transport in Sv. **c**, **g**, **k**, and **o** show vertical eddy (*blue* and *red dashed*) and residual (*blue* and *red solid*) velocities at the base of the mixed layer in $10^{-6}\,\mathrm{m\,s^{-1}}$. The *black solid line* shows the Ekman pumping. **d**, **h**, **l**, and **p** show the simulated surface density $\rho_m$ (*blue* and *red solid*) inferred from the slope and the observed surface density (*black dashed*) in $\mathrm{kg\,m^{-3}}$. *Rows*: **a** to **d** are for $J_O = 0$ and $K_d = 0$, **e** to **h** are for $J_O = 0$ and $K_d = 10^{-4}\,\mathrm{m^2\,s^{-1}}$, and **i** to **l** are for $J_O$ as shown in Figure 16.25 and $K_d = 10^{-4}\,\mathrm{m^2\,s^{-1}}$. **m** to **p** is the same as **i** to **l** but with a different initial conditions as discussed in the text. The Brunt–Väisälä frequency below the mixed layer, entering the density forcing, is chosen as $N = 3 \times 10^{-3}\,\mathrm{s^{-1}}$

Introduction of a small diapycnal diffusivity yields a minor modification of this behavior, as shown in Figure 16.27e–h. The residual stream function may thus start from an initial value being northward or southward and then must reverse as described above. In a southward case, the Eulerian circulation is overcompensated by the eddy-induced component.

The model becomes meaningless if the slope proceeds towards positive numbers. For the first two cases in Figure 16.27, this would occur north of the integration interval; for the third case we find this behavior inside because the surface forcing is turned on and overcomes the negative tendency of the $\phi_m s_m$-term for small slope values (see the red lines applying to the larger $\mathcal{K}$). Here the blow-up singularity,

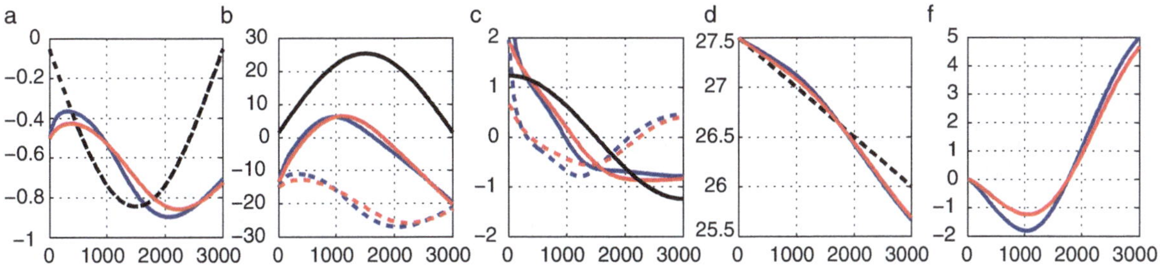


Fig. 16.28 Solutions of the prognostic model (16.66), using the restoring flux (16.67) with $\tilde{J}_0 = 0$, $\alpha_s = 1.5 \times 10^{-6}\,\mathrm{m\,s^{-1}}$ and for $\bar{\rho}_s$ the 'observed' density depicted in **d**) by the *black dashed line*. Two choices for the diffusivities are displayed: $\mathcal{K} = 100 \times 1{,}500\,\mathrm{m^3\,s^{-1}}$ (*blue curves*) and $\mathcal{K} = 300 \times 1{,}500\,\mathrm{m^3\,s^{-1}}$ (*red curves*). **a** to **d** are as in Figure 16.27. **f** shows the diagnosed density flux of the corresponding solutions. All other parameters are identical to the case shown Figure 16.27m to p

explained in the box on p. 609, comes into play. With a smaller $\mathcal{K}$ or a larger (negative) initial value (see Figure 16.27m–p, this behavior can be avoided in the present integration interval, but what is really required is a physically motivated parameterization in the model to counteract such a behavior. With a positive slope, the surface density would increase towards the north, and the air-sea exchange of density would have to react (see below).

Knowing the slope $s_m = -(\partial \rho_m/\partial y)/(\partial \rho/\partial z)$, we can determine density in the mixed layer by approximating the vertical density gradient $\partial\bar{\rho}/\partial z$ by the given $N^2$, so far only done for the forcing term in (16.66). The result is depicted in Figure 16.27d, h, l, and p, showing a reasonable success: for comparison, the approximate profile of the observed surface density (see the box on p. 604 or Figure 16.25) is included by the black dashed line.

The above model is still unsatisfactory for two reasons. First, the model depends on a specified Brunt–Väisälä frequency below the mixed layer, and second, as already demonstrated above, prescription of the density flux is not acceptable. It should not be the modeler but rather the "atmosphere" to react on the ocean's surface condition and choose the flux. To implement a feed-back in a simple way, a restoring-type flux

$$J_0(y) = \tilde{J}_0(y) + \alpha_s[\bar{\rho}_s(y) - \bar{\rho}_m(y)] \tag{16.67}$$

similar to (13.12) may be used. A tunable parameter $\alpha_s$ (with dimension $\mathrm{m\,s^{-1}}$) and an "atmospheric" restoring density $\bar{\rho}_s(y)$ then enter the model. A prescribed climatological flux $\tilde{J}_0$ may be taken into account (e. g. as zero or the one used in the previous examples). The performance of this model is demonstrated in Figure 16.28. For the "blue" solution of these cases, the interior density, residual stream function, zonal velocity (see next paragraph), and pumping velocities are shown in Figure 16.29.

Giving up to prescribe $N^2$ and proceeding to a fully prognostic model, i. e. resolving the transition layer between the mixed layer and the adiabatic interior with appropriate parameterizations, is a fairly complicated task. The reader is referred to Olbers and Visbeck (2005) where prognostic equations are discussed for the slope in this transition layer.

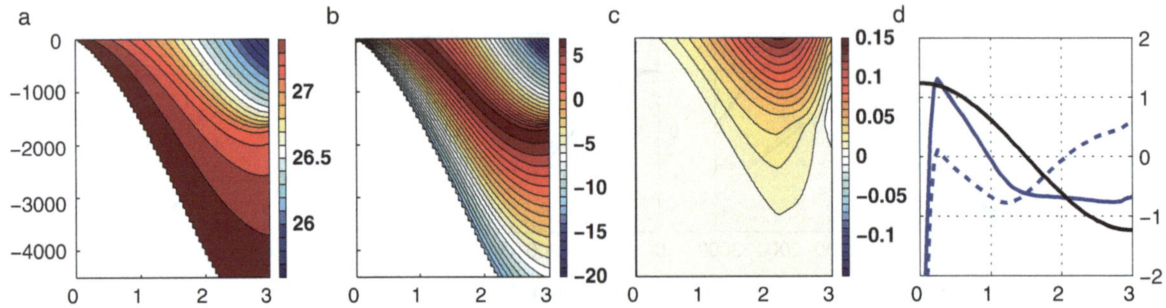

Fig. 16.29 Solution of the prognostic model (16.66), using the restoring flux (16.67) as in Figure 16.28 and $\mathcal{K} = 100 \times 1{,}500\,\mathrm{m^3\,s^{-1}}$. **a** shows the density in $\mathrm{kg\,m^{-3}}$ as a function of depth in m and $y$ in $10^3$ km. **b** shows the residual stream function in Sv. **c** shows the zonal velocity in $\mathrm{m\,s^{-1}}$ and **d** shows the the vertical eddy (*blue dashed*) and residual (*blue solid*) velocities at the base of the mixed layer in $10^{-6}\,\mathrm{m\,s^{-1}}$. The *black solid line* shows the Ekman pumping in $10^{-6}\,\mathrm{m\,s^{-1}}$

### Balance of Zonal Momentum and the Zonal Transport

The model is completed by the determination of the zonal current which has a geostrophic balance, given by (16.57). The baroclinic pressure $\bar{p}^{\mathrm{clin}}$ is determined by the interior density field (see Figure 16.29). The gradient $g\partial\bar{\zeta}/\partial y$ of the surface pressure follows from the vertically integrated balance of zonal momentum, as in Section 16.2. In case of a flat bottom, we have to work with the balance between the applied windstress and the frictional bottom stress, $\bar{\tau}_0^{(x)} = \bar{\tau}_{\mathrm{b}}^{(x)}$, and use the equations of the bottom Ekman layer. In the presence of topography, we have the balance between the windstress and the bottom formstress, $\bar{\tau}_0^{(x)} = -\mathcal{F}_{\mathrm{b}}$, but we can proceed only if a parameterization of $\mathcal{F}_{\mathrm{b}}$ is given. Both routes result in a condition on $\partial\bar{\zeta}/\partial y$. Given this gradient, the velocity profile $\bar{u}(z)$ is determined, and the total transport can be calculated. A "cheap" way to work around the problem of determining the surface pressure is to assume a level of no motion at the bottom, i. e. to calculate the transport as described in the box on p. 560. This solution is depicted in Figure 16.29c.

## 16.7 * Simple Models with Bottom Formstress

The above considerations have so far missed an analytical treatment of the influence of topography (loosely speaking, the formula relating the bottom formstress $\mathcal{F}_{\mathrm{b}}$ to the velocity of the zonal current). The formstress in the equations (16.14) and (16.15) contains the part of the bottom pressure which is out of phase with variations of the topography along the zonal path of integration. Some insight into the mechanism how the flow generates the bottom formstress can be obtained from a truncated part of the full dynamics, called low-order models, where the flow fields are represented by only a few spectral components (see Appendix B.3), i. e. by only a few "waves", which are, of course, Rossby waves in case of the large-scale circulation of the Southern Ocean. A prominent example of such low-order models is the one by Charney and DeVore (1979) (CDV in the following) which resolves only one Rossby wave. We present the CDV model in the next section and a baroclinic extension in Section 16.7.2. This latter section can be viewed as summary of most processes discussed in the chapter on Southern Ocean dynamics, of course represented by only very rudimentary physics.

**101. Derivation of the Homogeneous Charney–DeVore Model**

The homogeneous Charney–DeVore (CDV) low-order model follows from the quasi-geostrophic potential vorticity equation (5.32) for homogeneous density, supplemented by the global zonal momentum balance,

$$\frac{\partial}{\partial t}\nabla^2\psi + \boldsymbol{u}\cdot\boldsymbol{\nabla}\left[\nabla^2\psi + \beta y + \frac{f_0 b}{H}\right] = \epsilon\nabla^2(\psi^* - \psi) \tag{B101.1}$$

$$\frac{\partial T}{\partial t} = \epsilon(T^* - T) + \frac{f_0}{H}\overline{b\frac{\partial\psi}{\partial x}} \tag{B101.2}$$

governing a two-dimensional, zonally unbounded flow over a sinusoidal topography given by $b$ in a zonal channel. Here, $\psi$ is the stream function for the velocity $\boldsymbol{u} = (u, v)$ with $u = -\partial\psi/\partial y$ and $v = \partial\psi/\partial x$, $T$ is the zonally and meridionally averaged zonal velocity (transport). Furthermore, $\epsilon\nabla^2\psi^*$ is the vorticity and $\epsilon T^*$ the zonal momentum imparted (per unit of time) into the system. It represents the external forcing of the system by e. g. thermal forcing or by windstress. $\epsilon$ is a coefficient for linear bottom friction and $f_0$ a constant Coriolis parameter. The last term in the global momentum budget (B101.2) is the bottom formstress. Note that the overbar denotes the zonal and meridional average over the total domain (therefore we refer to (B101.2) as a 'global' balance). The momentum input $\epsilon T^*$ is thus balanced by bottom friction and bottom formstress. $T^*$ is usually taken as constant, so that there is no vorticity forcing.
The total depth of the fluid is $H - b$, and the topography height $b$ is taken sinusoidal, $b = b_0 \cos kx \sin ky$ with $k = 2\pi/X$ where $X$ is the length and $Y = X/2$ the width of the channel. To derive the CDV system (16.36), a truncated spectral representation of $\psi$ given by

$$\psi = -Ty + \frac{1}{k}[T_c \cos kx + T_s \sin kx]\sin ky$$

is used, representing the flow in terms of the zonal mean $U$ and a wave component with cosine and sine amplitudes $T_c$ and $T_s$, respectively. To derive the equation for $U$, the representation is implemented into the zonal momentum balance and integrated over the whole channel domain. To derive the equation for $T_c$ (or $T_s$), insert the representation into the vorticity equation, multiply by $\cos kx$ (or $\sin kx$) and integrate over the whole channel domain (see also the general discussion of spectral models in Appendix B.3).

## 16.7.1 A Homogeneous Charney–DeVore Model

The homogeneous (or barotropic) Charney–DeVore (CDV) low-order model describes some important aspects of the dynamics of a zonal flow with homogeneous density over topography. As discussed in the box on p. 613, the model is based on the quasi-geostrophic potential vorticity equation on a $\beta$-plane, a sinusoidal topography and a truncated spectral representation of the flow in a channel representing the Southern Ocean. A Rossby wave is generated with amplitude $T_c$, which is in phase with the topography, i. e. a cosine component in view of the cosine-topography chosen in the box on p. 613, and amplitude $T_s$, which is out of phase with the topography, i. e. a sine component. The homogeneous CDV model consists of three coupled ordinary differential equations for the amplitudes $T_c$ and $T_s$ and the horizontally averaged zonal flow $T$,

$$\begin{aligned}\dot{T} &= \epsilon(T^* - T) + \frac{1}{4}\delta T_s \\ \dot{T}_c &= -kT_s(T - c_R) - \epsilon T_c \\ \dot{T}_s &= kT_c(T - c_R) - \delta T - \epsilon T_s\end{aligned} \tag{16.68}$$

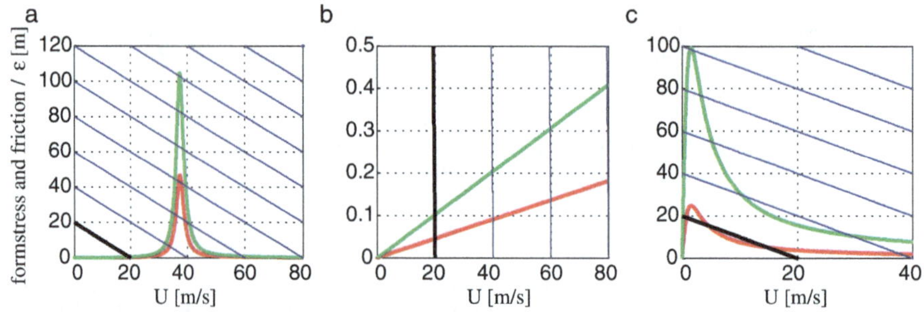


Fig. 16.30 Steady solutions of the homogeneous CDV model showing the negative bottom formstress $-\mathcal{F}_b/\epsilon$ (*red* and *green lines*) in m and $T^* - T$ (*full black* and *dashed blue lines*) versus $T$ in $\mathrm{m\,s^{-1}}$. Various values for $T^*$ are used: the *black line* represents the momentum balance for typical ACC conditions ($T^* = 20\,\mathrm{m\,s^{-1}}$), for the *blue lines* $T^*$ was increased by a factor of $2, 4, 6 \ldots$, respectively. **a** applies to the complete CDV model, **b** applies to a version neglecting the relative vorticity in (B101.1), leading to a cancelation of the $T_s T$ and $T_c T$ terms in (16.68), and **c** applies to the $f$-plane ($\beta = 0, c_R = 0$). The bottom formstress $\mathcal{F}_b/\epsilon$ is shown for $b_0 = 100\,\mathrm{m}$ (*red lines*) and $b_0 = 150\,\mathrm{m}$ (*green lines*) in **a** and **b** and in **c** for $b_0 = 500\,\mathrm{m}$ (*red*) and $b_0 = 1{,}000\,\mathrm{m}$ (*green*). Other relevant parameter values are $f_0 = -1.1 \times 10^{-4}\,\mathrm{s^{-1}}$, $\beta = 1.5 \times 10^{-11}\,\mathrm{m^{-1}\,s^{-1}}$, $\epsilon = 10^{-6}\,\mathrm{s^{-1}}$, $H = 5\,\mathrm{km}$ and $L = 10^4\,\mathrm{km}$. Notice the difference in the *vertical* and *horizontal* axes

where $c_R = \beta/k^2$ is the speed of the planetary barotropic Rossby wave[9] and $\delta = f_0 b_0/H$ measures the topography height. Each balance has a frictional term $\sim \epsilon$, deriving from bottom friction (a lateral viscosity term yields a similar form). The zonal current is forced by a zonal stress, written for simplicity as $\epsilon T^*$. Note that the zonal momentum balance in (16.68) is the time-dependent version of (16.36). In the following we will consider steady solutions of the model (16.68).

The wave induces a pressure field which acts against the zonal acceleration. At the upstream side of the hills, the fluid must be lifted, thus generating high pressure. At the downstream side, a pressure low follows. The westward propagating wave gets stationary by eastward advection in the zonal current and by friction: it is locked in resonance with the mean flow and produces a bottom formstress which becomes a nonlinear functional of the zonal velocity,

$$\mathcal{F}_b[T] = \frac{1}{4}\delta T_s = -\frac{1}{4}\frac{\delta^2 \epsilon T}{\epsilon^2 + k^2(T - c_R)^2} \tag{16.69}$$

derived from the two 'wave' equations in (16.68) for steady state conditions. The total momentum balance in (16.68), written in steady state now as

$$\epsilon(T^* - T) + \mathcal{F}_b[T] = 0 \tag{16.70}$$

then determines the zonal transport $T$. Figure 16.30 displays the solution by plotting $T - T^*$ and $-\mathcal{F}_b[T]/\epsilon$ as function of $T$ in the left panel. Three equilibria are found if $T^*$ is well above $c_R$ but not too large; two of them are stable circulation regimes. For solutions in the resonant range ($T$ close to $c_R$), the friction in the momentum balance is negligible, these solutions are balanced by bottom formstress. The off-resonant solutions are controlled by friction. It is remarkable that the bottom

[9] The solution of the linearized form of the homogeneous CDV model (16.68) is an oscillation with the frequency $\pm\sqrt{(kc_R)^2 + \delta^2/4}$, representing a mixed planetary-topographic wave.

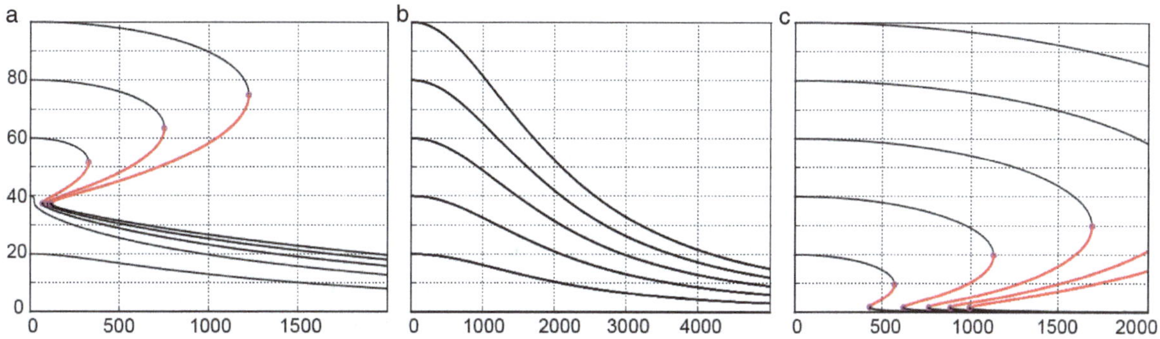


Fig. 16.31 The bifurcation of $T(b_0)$ for the cases shown in Figure 16.30, using $T^* = 20, 40, 60, 80, 100\,\mathrm{m\,s^{-1}}$. **a** complete CDV model. **b** model with cancelation of the relative vorticity. **c** model for $\beta = 0$. The *black branches* are stable, the *red branches* are unstable. Notice the difference in the *horizontal* axes

formstress amplitude $T_s$ is proportional to $\epsilon$ (as shown in (16.69)): friction is essential to shift the pressure field out-of-phase with respect to the topography. Of course, for zero friction the flow would not become steady. But it is important to note that friction plays this twofold role in the balance of the mean zonal flow: there is a direct frictional effect on $T$, manifested here by the bottom friction, and an indirect effect through the feedback by the topographically induced waves where friction allows to build up the phase shift and to generate bottom formstress.

Significant sizes of the bottom formstress can only arise if the topography is sufficiently high, if Rossby waves propagation is present (i. e. $\beta \neq 0$), and if the forcing is sufficiently strong (see left panel of Figure 16.30). The three possible steady states which then exist can be classified according to the size of the mean flow $T$ compared to the wave amplitudes. The *high zonal index regime* is frictionally controlled, the flow is intense and the wave amplitudes are low. The *low zonal index regime* is controlled by bottom formstress, the mean flow is weak, and the wave is intense. The intermediate state is transitional, it is unstable to perturbations. A small perturbation $T'$ is governed by

$$\frac{\partial T'}{\partial t} = -\epsilon T' + \frac{\partial \mathcal{F}_b}{\partial T} T'$$

and becomes unstable if $\partial \mathcal{F}_b / \partial T > \epsilon$. This "formdrag instability" obviously works when the slope of the resonance curve $-\mathcal{F}_b[T]$ is negative and steeper than the one associated with friction (see Figure 16.30) so that a perturbation must run away from the steady state. Apparently, the criterion of instability is always satisfied for the intermediate state whereas the other two states are always stable.

Two ingredients are important to generate a significant resonance. It weakens when the planetary Rossby waves are filtered (which yields the $f$-plane approximation). In this case, only topographic waves are present, as in Figure 16.30c. If the relative vorticity is neglected, the resonance disappears completely, and the momentum balance of this homogeneous model is linear in $T$, see Figure 16.30b. Then multiple steady states do not exist. In these cases, the bottom formstress is very small compared to the forcing, and the flow is directly controlled by the bottom friction: we find $T \cong T^*$, hinting at the Hidaka dilemma (see Section 16.2.2).

Bifurcation diagrams of $T$ as function of the topography height $b_0$ are shown in Figure 16.31 for the three cases. These plots were produced by use of a numerical continuation tool which evaluates the equilibrium solutions and also the stability[10]. Black lines $T[b_0]$ are stable steady states; red ones are unstable. The above discussion on the type and number of solution, there based on fixed heights $b_0$, can here be confirmed in a more general view of a larger range of $b_0$-values. It is evident that the existence of planetary Rossby waves (in the left panel) is essential for the resonance to occur at large values of the zonal current, in contrast to mere topographic waves (in the right panel).

It is worth mentioning that much of the interesting dynamics of the CDV model – such as the occurrence of multiple steady states – is lost by incorporation of more and more spectral components arising for more complex topographies, e. g. a Gaussian hill. The decrease of transport with increase of the topography height, however, remains as a general feature.

Charney and DeVore (1979) have developed this model for atmospheric flow regimes. In the ocean, the resonant solutions cannot be realized because zonal flow (the ACC) speeds resulting for realistic forcing are much less than speeds of barotropic Rossby waves. The forcing $T^*$ relates to the windstress by $\epsilon T^* = \bar{\tau}_0^{(x)}/H$, and reasonable values for the windstress and the bottom friction allow only for a frictionally controlled solution in which nonlinearities are irrelevant. The solution then becomes

$$\mathcal{F}_b[T] = -\frac{1}{4}\epsilon T(b_0/H)^2(ak)^2 \quad HT = \frac{\tau_0^x/\epsilon}{1+\frac{1}{4}(b_0/H)^2(ak)^2} \tag{16.71}$$

where $a = |f_0/\beta|$ is the Earth radius times tangent of latitude, and $\epsilon^2 \ll (\beta/k)^2$ was assumed for simplicity. The transport in this homogeneous model decays away from the frictional solution $HT = \bar{\tau}_0^{(x)}/\epsilon$ (with hundreds of Sv transport) with increasing height $\delta$ of the topography (see Figure 16.31c). The drag of the bottom formstress increases quadratically with the height of the topography and, because $(ak)^2 \gg 1$, attains higher values than friction for moderately sized submarine ridges.

## 16.7.2 A Baroclinic Charney–DeVore Model

In a baroclinic extension of the model, the Charney–DeVore resonance can operate in conditions more representative for the Southern Ocean than the homogeneous CDV model. This will be discussed in the following. A two-layer zonal channel with quasigeostrophic dynamics (see Section B.1.3) is considered in this model, however, with a sinusoidal topography only in the zonal direction. The stream functions of the flow in the two layers are expressed by suitably chosen sinusoidal structure functions as detailed in the box on p. 618. The resulting system is written in terms of six amplitudes representing the barotropic transport $T$ (upper layer plus lower layer transport), the baroclinic transport $S$ (upper layer minus lower layer transport), and respective barotropic and baroclinic sine and cosine components, $T_s$, $S_s$, and $T_c$, $S_c$ (more accurately, the $T$ and $S$ quantities are based on barotropic and baroclinic velocities). The cosine variables are out of phase with respect to the topography and thus generate the barotropic and baroclinic bottom formstress parts. The effect of transient eddies is a downward transfer of momentum by the interfacial formstress

[10] The continuation is performed with a MATLAB code `bifurk.m` written by Christoph Völker.

as discussed in Section 16.3 and is parameterized by friction acting on the interface of the two layers with a scaled coefficient $\kappa$. In addition, we have lateral diffusion of momentum with a scaled viscosity $\epsilon$.

Scaling these variables, as detailed in the box on p. 618, the dimensionless form of the governing differential equations becomes

$$\dot{T} = W_0 + \delta\,(T_c - hS_c) - \epsilon T \tag{16.72}$$

$$\dot{S} = \frac{\lambda^2 W_0}{h} + \frac{Q_0}{1-h} - \frac{\lambda^2}{1-h}\delta\,(T_c - hS_c) + \ell\,[T_c S_s - T_s S_c] - \left(\kappa + \lambda^2\epsilon\right) S \tag{16.73}$$

$$\dot{T}_c = \beta T_s - 2\delta\,(T - hS) - m\,[T\,T_s + h\,(1-h)\,S\,S_s] - \epsilon T_c \tag{16.74}$$

$$\dot{S}_c = \beta\lambda^2 S_s + \frac{2\lambda^2}{1-h}\delta\,(T - hS) - n\,[TS_s - ST_s] - \left(\kappa + \lambda^2\epsilon\right) S_c \tag{16.75}$$

$$\dot{T}_s = -\beta T_c + m\,[T\,T_c + h\,(1-h)\,S\,S_c] - \epsilon T_s \tag{16.76}$$

$$\dot{S}_s = -\beta\lambda^2 S_c + n\,[TS_c - ST_c] - \left(\kappa + \lambda^2\epsilon\right) S_s \tag{16.77}$$

The forcing of the system is given by $W_0$ (windstress; note that we have switched from the $\epsilon T^*$ notation to $W_0$) and $Q_0$ (external heating). Terms derived from nonlinear advection are found in the cornered brackets (the factors $\ell, m, n$ are numerical coupling coefficients depending only on the channel dimensions, see the box on p. 618). The respective term in the zonal baroclinic momentum balance (16.73), the $\ell$-term, is the interfacial formstress induced by the standing eddies. The term related to $\kappa$ in (16.73) is the corresponding transient eddy interfacial formstress. The baroclinicity enters via the scaled internal Rossby radius $\lambda$, and the scaled topography height is $\delta/\pi$. Lateral friction (the $\epsilon$-terms) operates in the barotropic equations (for $T$, $T_c$, and $T_s$) and interfacial friction (the $\kappa$-terms) in addition in the baroclinic equations (for $S$, $S_c$, and $S_s$). Note that generally $\kappa \gg \epsilon\lambda^2$. The sine and cosine equations (16.74)–(16.77) describe planetary-topographic Rossby waves with the same zonal wave number as the topography; $\beta$ is the scaled planetary coefficient $\mathrm{d}f/\mathrm{d}y$. There are terms in these equations arising from nonlinearities (advection; $n$- and $m$-terms) and diffusion. There is no direct forcing term in the wave equations because the external heating function is assumed independent on the zonal coordinate. Note that (16.72) is the balance of vertically integrated momentum. We can identify the barotropic and baroclinic contributions to the bottom formstress, $\delta T_c$ and $-\delta h S_c$, respectively.

In the following discussion, steady state solutions of the model are investigated for various regimes of the flow. The equations for the total transport $T$ and the shear transport $S$ are found in (16.72) and (16.73), respectively. Adding these equations in a suitable way to eliminate the bottom formstress parts, we find the balance of upper layer transport $h[T + (1-h)S]$, given for steady state by

$$W_0 + \frac{hQ_0}{\lambda^2} - \epsilon h\,(T + (1-h)S) = \frac{h(1-h)}{\lambda^2}\,(\kappa S - \ell\,[T_c S_s - T_s S_c]) \tag{16.78}$$

and eliminating the wind forcing between (16.72) and (16.73), we get the balance

$$-\frac{hQ_0}{\lambda^2} + \delta(T_c - hS_c) - \epsilon(1-h)\,(T - hS) = -\frac{h(1-h)}{\lambda^2}\,(\kappa S - \ell\,[T_c S_s - T_s S_c]) \tag{16.79}$$

**102. Derivation of the Baroclinic Charney–DeVore Model**

A channel of zonal length $X$ and meridional width $Y$ is considered with two layers of fluid with densities $\rho_1, \rho_2$, mean layer thicknesses $H_1, H_2, H = H_1 + H_2$, and reduced gravity $g^\star = g(1 - \rho_1/\rho_2)$. The flow is governed by the quasi-geostrophic dynamics and given by the equations described in Appendix B.1.3. We use the formulation (B.15) and (B.16) in terms of the barotropic and baroclinic stream functions $\psi_e$ and $\psi_i$, respectively. Boundary conditions of zero mass flux and no-slip at the vertical walls of the channel are taken into consideration. Auxiliary integral constraints are incorporated for both layers (see e. g. McWilliams, 1977, and the box on p. 570).

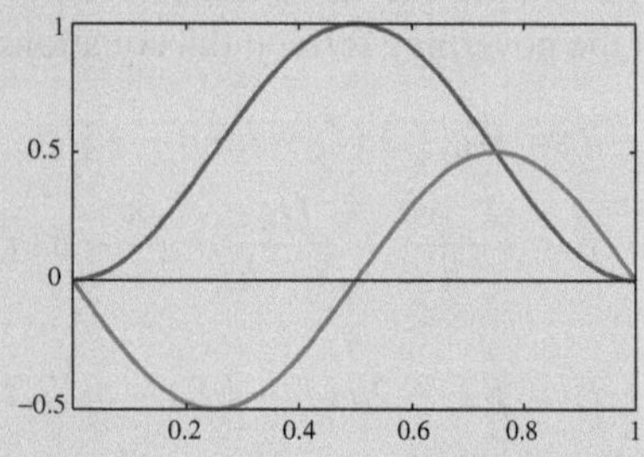


*Windstress [blue] and heating rate [green] of the low-order model.*

The topography is taken as $b(x,y) = b_0 \sin(2\pi x/X)$. The flow is forced by the zonal windstress $\tau = \tau_0 \sin^2(\pi y/Y)$ and a local heating rate $Q = -\hat{Q}_0 \cos(\pi y/Y)\sin(\pi y/Y)$. Here, $\tau$ and $Q$ are dimensioned $\mathrm{m^2\,s^{-2}}$ and $\mathrm{W\,m^{-3}}$, respectively. The heating amplitude $\hat{Q}_0$ enters the model equations in the form $B_0 = \alpha g \hat{Q}_0 H_2/(\rho_1 c_p)$ (in $\mathrm{m^2\,s^{-3}}$) where $\alpha$ is the thermal expansion coefficient and $c_p$ the specific heat. The profiles of the forcing function are displayed in the following figure: $\tau/\tau_0$ as blue curve, $Q/\hat{Q}_0$ as green curve, both as function of $y/Y$. Note that the $Q(y)$ profile is chosen to coincide with the profile of Ekman pumping $-\partial(\tau/f_0)/\partial y$.

The truncated representation for both stream functions is given by

$$\begin{aligned}
\psi_e(x,y,t) &= T(t)\cos(\pi y/Y) + T_c(t)\cos(2\pi x/X)\sin(\pi y/Y)\\
&\quad + T_s(t)\sin(2\pi x/X)\sin(\pi y/Y)\\
\psi_i(x,y,t) &= S(t)\cos(\pi y/Y) + S_c(t)\cos(2\pi x/X)\sin(\pi y/Y)\\
&\quad + S_s(t)\sin(2\pi x/X)\sin(\pi y/Y)
\end{aligned}$$

It is implemented into the equations (B.15) and (B.16) and then projected onto the modal functions, as described in Appendix B.3. The resulting equations are then made dimensionless. The scaling and coefficients are as follows: all transport variables are scaled by $(|f_0|Y^2)/\pi^2$ to yield dimensionless $T, S, T_c, T_s, S_c, S_s$. Parameters are $a = Y/(\pi X)$, $\epsilon = \pi^2 A_h/(a|f_0|Y^2), \kappa = \pi^2 K/(a|f_0|Y^2), \lambda = \pi R/Y, \beta = 2Y\mathrm{d}f/\mathrm{d}y/|f_0|$ where $R = \sqrt{(g^*/f_0^2)H_1H_2/H}$ is the internal Rossby radius. Furthermore, $h = H_1/H$ is a thickness ratio. Coupling coefficients are $\ell = 3\pi^2/8, m = 64\pi^2 a^2/3, n = 16/3$. The scaled depth is $1 + (\delta/\pi)\sin(2\pi x/X)$; thus $\delta/\pi$ is the relative height of the topography. The scaled forcing amplitudes are $W_0 = (3/8)\pi^2\tau_0/(aHYf_0^2), Q_0 = (3/32)\pi^3 B_0/(a|f_0|^3Y^2)$. Time is scaled by $1/(|f_0|a)$.

| | | | |
|---|---|---|---|
| $\tau_0 = 10^{-4}\,\mathrm{m^2\,s^{-2}}$ | $B_0 = 7.19 \times 10^{-7}\,\mathrm{m^2\,s^{-3}}$ | $Y = 1{,}000\,\mathrm{km}$ | $X = 4{,}000\,\mathrm{km}$ |
| $A_h = 10^4\,\mathrm{m^2\,s^{-1}}$ | $K_h = 800\,\mathrm{m^2\,s^{-1}}$ | $g^\star = 5.3 \times 10^{-3}\,\mathrm{m\,s^{-2}}$ | $R = 20.0\,\mathrm{km}$ |
| $f_0 = -1 \times 10^{-4}\,\mathrm{s^{-1}}$ | $\mathrm{d}f/\mathrm{d}y = 2 \times 10^{-11}\,\mathrm{m^{-1}\,s^{-1}}$ | $H_1 = 1{,}000\,\mathrm{m}$ | $H_2 = 3{,}000\,\mathrm{m}$ |
| $W_0 = 1.17 \times 10^{-4}$ | $Q_0 = 1.31 \times 10^{-6}$ | $\epsilon = 1.24 \times 10^{-2}$ | $\kappa = 9.92 \times 10^{-4}$ |
| $a = 7.96 \times 10^{-2}$ | $\lambda = 6.50 \times 10^{-2}$ | $\beta = 0.40$ | $h = 0.25$ |

*List of parameters and standard values for the baroclinic CDV model. The value for $B_0$ corresponds to a heat flux of $0.05\,\mathrm{W\,m^{-3}}$. The coupling coefficients have the values $\ell = 3.70, m = 1.33, n = 5.33$.*

for the lower layer transport $(1-h)(T-hS)$. On the right-hand side we have put the interfacial formstress due to transient eddies (the $\kappa$-terms) and standing eddies (the $\ell$-terms). Quite obviously, these are the only processes which couple the two layers. The balances (16.78) and (16.79) for the individual layers are used below.

Considering the steady state of the above model, we recover the most important physical mechanisms which we have outlined in the previous discussion of this chapter, namely:

### The Barotropic Formstress Drag

A barotropic state is obtained for $\lambda = Q_0 = 0$. Then all $S$-fields are identically zero, and the barotropic solution emerges as described by the barotropic CDV model in Section 16.7.1, in which the transport decays quadratically with increasing topography height due to the drag of the barotropic formstress. For a flat bottom, we are facing the Hidaka dilemma addressed in Section 16.2.2.

### Johnson–Bryden Dynamics

If lateral friction is small and the influence of topography and interfacial formstress by standing eddies are ignored, we regain from (16.73) the Johnson–Bryden state (here we refer to Section 16.6.1),

$$\frac{\kappa}{\lambda^2} S = \frac{W_0}{h} + \frac{Q_0}{\lambda^2(1-h)}$$

in which the transport contained in the shear current is governed by windstress and – in extension of the original Johnson–Bryden model – also by the external heating. Quite obvious in the present setting is the fact that the Johnson–Bryden approach does not lead to the total transport $T$. This can only be done with suitable assumption about the lower layer transport, e. g. a vanishing of the lower layer transport.

### Hidaka's Dilemma in the Surface Layer

Suppose for the moment that the interfacial formstress components are small or the Rossby radius is large (strong stratification). The balance (16.78) for the upper layer turns into

$$W_0 + \frac{hQ_0}{\lambda^2} - \epsilon h\,[T + (1-h)S] \simeq 0 \tag{16.80}$$

and we arrive at the obvious statement that the transport $h(T + (1-h)S)$ in the surface layer is decoupled from the topography: it is given by the windstress and buoyancy flux acting in the above combination against lateral friction. The surface layer transport is then in a "Hidaka"-type state (inversely proportional to the eddy viscosity $\epsilon$ and thus large for a reasonably sized $\epsilon$). This problem was addressed in Section 16.2.2 in a barotropic setting. This Hidaka dilemma in the surface layer, however, must be resolved by the interfacial formstress induced by transient or/and standing eddies. These processes couple the upper layer to the lower one and feed momentum to the bottom formstress acting in the lower layer. Only then the vertically integrated momentum balance appears in the full form (16.72). For a reasonable size of the topography height $\delta$, the transport may take reasonable values. We return to this issue later in a more general frame.

### The Bottom Formstress

The formstress components $\delta T_c$ and $\delta h S_c$ arise by the response of the barotropic and baroclinic wave system, described by the four wave equations (16.74)–(16.77), to the zonal flow crossing the topography. A reasonable solution of the model should yield transports in the range $T \ll \beta = 0.4$ and $S \gtrsim \beta\lambda^2 = 0.0016$ (the numbers are for the standard parameter values), meaning that the flow velocities are much less than the speed of barotropic Rossby waves but supercritical with respect to the baroclinic waves. Such conditions are representative for the Southern Ocean. Then the nonlinearities in the barotropic wave equations (16.74) and (16.76) are small, and these equations represent a linear barotropic planetary-topographic wave. The barotropic formstress $\delta T_c$ can thus be explained by long linear Rossby waves generated by the deep current, $T - hS$ in the present model, crossing the large ridges blocking the circumpolar path of the ACC. In contrast, the baroclinic formstress $\delta h S_c$ might be governed by a nonlinear response, according to $S \gtrsim \beta\lambda^2$ in the baroclinic wave equations (16.75) and (16.77). We first discuss, however, a completely linear model. Nonlinear consequences follow further below.

### Bottom and Interfacial Formstress as Response to Linear Waves

Assuming a linear wave state, i. e. neglecting the advection terms in the wave equations (16.74) to (16.77), the wave amplitudes become

$$T_c = -\frac{2\delta\epsilon}{\beta^2 + \epsilon^2}(T - hS) \quad \text{and} \quad S_c = \frac{2\delta(\kappa/\lambda^2 + \epsilon)}{\beta^2 + (\kappa/\lambda^2 + \epsilon)^2}\,\frac{T - hS}{1 - h} \tag{16.81}$$

and $T_s = -\beta T_c/\epsilon$, $S_s = -\beta S_s/(\kappa/\lambda^2 + \epsilon)$. The barotropic formstress $\delta T_c$ is thus supported by lateral friction ($\epsilon$), the baroclinic one $-\delta h S_c$ by interfacial ($\kappa$) and lateral ($\epsilon$) friction and by stratification ($\lambda$). Both formstress components extract eastward momentum from the flow if the deep layer is moving eastward (remember that the barotropic formstress is $\delta T_c$ and the baroclinic one is $-\delta S_c$). The waves make a total bottom formstress

$$\delta(T_c - hS_c) = -2\delta^2\eta(T - hS) \tag{16.82}$$

where $\eta = \epsilon/(\beta^2 + \epsilon^2) + h'(\kappa/\lambda^2 + \epsilon)/(\beta^2 + (\kappa/\lambda^2 + \epsilon)^2)$ and $h' = h/(1 - h) = H_1/H_2$. The main contribution to $\eta$ comes by far from second term, in fact $\eta \simeq h'(\kappa/\lambda^2)/(\beta^2 + (\kappa/\lambda^2)^2)$ for the standard values given in the box on p. 618. The stress is quadratic in $\delta$ and obviously overcomes the friction term in the barotropic balance (16.72) if $\delta^2\eta > \epsilon$ or $\delta^2 > \delta_0^2 = \epsilon/\eta$. To overcome interfacial friction, it is necessary that $h'\delta^2\eta > \kappa/\lambda^2 + \epsilon$ or $\delta^2 > \delta_1^2 = (\kappa/\lambda^2 + \epsilon)/(\eta h')$. For the standard values of the model parameters (see the box on p. 618), we find $\eta = 0.46, \delta_0/\pi = 0.05, \delta_1/\pi = 0.42$ (remember that the scaled height of the topography is given by $\delta/\pi$).

Note that there is still baroclinic formstress if the interfacial friction is zero, i. e. $\kappa = 0$. In such a state, it is found from (16.80) that the lower layer transport vanishes unless directly forced by baroclinic processes, i. e. unless $Q_0 \neq 0$. It is a simple task to compute $T - hS$ from (16.80) and (16.81) in terms of the baroclinic forcing $Q_0$ and topography height $\delta$: a quadratic decay law with $\delta$ is found as found above for the barotropic CDV model.

Of more concern is, of course, the state with nonzero $\kappa$. Interfacial friction carries momentum downward if the shear $S$ is positive. What is the condition of the term $[T_c S_s - T_s S_c]$ of the standing eddies? We evaluate it from the linear wave solution which yields after some manipulations

$$T_c S_s - T_s S_c = -\frac{1}{\beta^2+\epsilon^2}\frac{\kappa/\lambda^2}{\beta^2+(\kappa/\lambda^2+\epsilon)^2}\frac{4\delta^2}{1-h}(T-hS)^2 \tag{16.83}$$

The standing wave interfacial formstress is – in this quasi-linear approximation – always negative: it must transfer eastward momentum downward. It requires that topography is present and the bottom flow nonvanishing. It is worth noticing that it is supported entirely by interfacial friction: the diffusivity $\kappa$ must be nonzero, the viscosity $\epsilon$ can be zero.

## The Linear Model and the Baroclinic Stommel Regime

We proceed with the linearized version of the model. We thus neglect the interfacial formstress by the standing eddies in (16.73) and the advection terms in the wave equations. Using then the linear solution (16.82) for the bottom formstress, the transport variables are readily evaluated as expressions of $\delta, \epsilon, \kappa, \lambda$ and the forcing amplitudes $W_0$ and $Q_0$. One finds

$$\begin{aligned}
T &= \frac{\left(1+2(\delta/\delta_1)^2\right)W_0/\epsilon + 2(\delta/\delta_0)^2\left(W_0+h'Q_0/\lambda^2\right)/\theta}{1+2(\delta/\delta_0)^2+2(\delta/\delta_1)^2}\\
hS &= \frac{2(\delta/\delta_1)^2 W_0/\epsilon + \left(1+2(\delta/\delta_0)^2\right)\left(W_0+h'Q_0/\lambda^2\right)/\theta}{1+2(\delta/\delta_0)^2+2(\delta/\delta_1)^2}\\
T-hS &= \frac{W_0/\epsilon - \left(W_0+h'Q_0/\lambda^2\right)/\theta}{1+2(\delta/\delta_0)^2+2(\delta/\delta_1)^2}
\end{aligned}$$

where $\theta = \kappa/\lambda^2+\epsilon$ is used and the above mentioned barotropic critical heights $\delta_0 = \sqrt{\epsilon/\eta}$ and the baroclinic one $\delta_1 = \sqrt{\theta/(h'\eta)}$ are implemented. Remembering that $\delta_0 \ll \delta_1$, we notice that the denominator of the above expressions is governed by the $\delta_0$-term.

The transport variables $T$ (total transport, blue), $h[T+(1-h)S]$ (upper layer, black), and $(1-h)(T-hS)$ (lower layer, red) are displayed in the left panels of Figure 16.32 as function of the height $\delta/\pi$ and otherwise typical parameters. For a purely wind-driven state (upper row), the transports are all eastward (in the wind direction). For small topography heights, $\delta/\pi < \delta_0/\pi$, the transport is frictionally controlled, and the Hidaka dilemma is present with $T \sim W_0/\epsilon$.

A purely heat-driven state (lower row), using the heating profile specified in the box on p. 618 with heating in the north and cooling in the south, yields an eastward moving upper layer and a reversed flow in the lower layer; the total transport is, however, also eastward. There is a fairly small effect from friction and bottom formstress, both counteracting the westward flow in the lower layer. The pure heat-driven case resides entirely on the balance between the heat input and the interfacial formstress (which, as explained before, is equivalent to lateral diffusion of heat).

All transports, however, flatten out into a plateau at large topography heights well above the critical height $\delta/\pi = \delta_0/\pi$. The transports in the wind-driven case attain

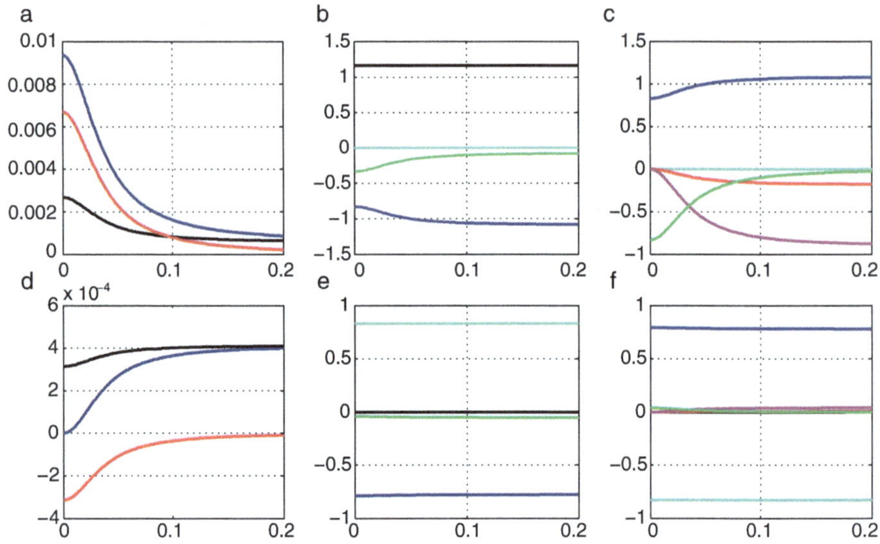


Fig. 16.32 Transports and momentum balance of the linear transport model as function of the topography height $\delta/\pi$. The upper panels **a**, **b**, and **c** are for the case of wind forcing only; the lower panels **d**, **e**, and **f** are for the case of forcing by heating. The *left panel* (**a**, **d**) in each row shows the transport variables: total transport $T$ (*blue*), upper layer transport $h[T + (1-h)S]$ (*black*), and lower layer transport $(1-h)(T-hS)$ (*red*). The *middle panel* (**b**, **e**) shows the balance of momentum in the upper layer; the *right panel* (**c**, **f**) is the balance for the lower layer: wind forcing (*black*), heating (*cyan*), viscous friction (*green*), interfacial friction (*blue*), bottom formstress (*red* for the barotropic component, *magenta* for the baroclinic component). All variables are scaled and parameters are as specified in the box on p. 618. These yield almost identical forcing amplitudes $W_0 = 1.17 \times 10^{-4}$ and $h'Q_0/\lambda^2 = 1.11 \times 10^{-4}$. The critical heights of the topography have values $\delta_0/\pi = 0.05, \delta_1/\pi = 0.42$

fairly small values, and the ones in the heat-driven case approach their respective maxima. In this large height regime, we find for the wind-driven case

$$T, hS \simeq (\delta_0/\delta_1)^2 W_0/\epsilon + W_0/\theta \simeq \frac{W_0/\theta}{1-h} \simeq \frac{\lambda^2 W_0/\kappa}{1-h}$$

The transports $T$ and $hS$ are thus governed by interfacial friction rather lateral viscous friction. The bottom layer becomes increasingly quiescent, i.e. $T - hS \sim (\delta_0/\delta)^2$. However, the bottom formstress approaches a finite value,

$$-2\delta^2\eta(T - hS) \simeq \epsilon(W_0/\epsilon - W_0/\theta) \simeq W_0$$

which balances the interfacial stress $-h(1-h)\kappa S/\lambda^2 \simeq -W_0$. Apparently, this regime is the one discussed in connection with the 'baroclinic Stommel equation' (see Section 16.5.3).

The momentum balance of the model, given by (16.78) and (16.79) and shown in Figure 16.32 (middle and right panels) for the two layers, reveals as well that the wind forcing is transferred almost undiminished by interfacial friction to the lower layer where it is taken up by bottom formstress. Viscous friction is irrelevant. For the present parameters, the baroclinic formstress is dominating. For the Hidaka regime, at small height values $\delta/\pi < \delta_0/\pi$, lateral friction is necessary, at least in the lower layer.

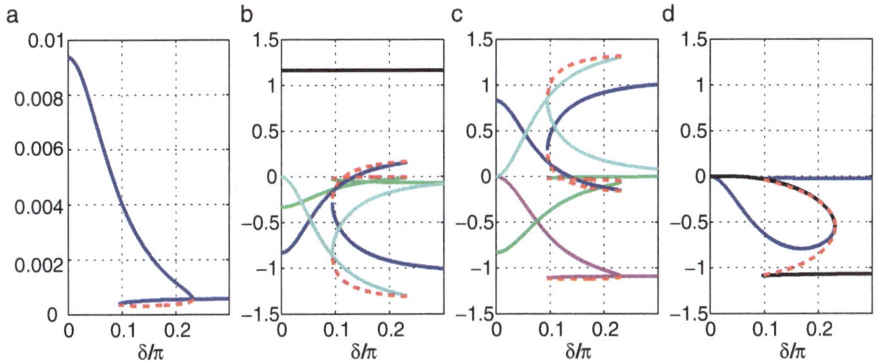


Fig. 16.33 Steady solutions of the nonlinear baroclinic CDV model as function of the topography height $\delta/\pi$, obtained with the continuation tool. Standard values of the parameters are used (the same as for the linear solution shown in Figure 16.32). **a** the total transport $T$. **b** and **c** terms of the momentum balance for the upper and lower layers. **d** barotropic (*blue*) and baroclinic (*black*) parts of the bottom formstress. *Colors* are as follows: wind forcing (*black*), viscous friction (*green*), transient eddy stress (interfacial friction, *blue*), standing eddy stress (*cyan*), bottom formstress (*magenta*). In all variables the unstable part of the solution is indicated by the *red dashed segments*

## Nonlinear Extensions

There are two severe deficits in the linear view of the model. Firstly, evaluating the interfacial formstress induced by standing eddies, even in the quasi-linear approximation (16.83), it is found to be not small: it may attain the same size as the transient one and may generally even be larger than the bottom formstress. Secondly, the neglected advection terms in the wave equations are not small.

The advective terms lead to the baroclinic resonance mentioned above, the existence of which can easily be disclosed: evaluating the baroclinic wave amplitudes from (16.75) and (16.77), including now the advection terms, we find the nonlinear version of the baroclinic bottom formstress

$$\delta h S_c = 2(\delta/\lambda)^2 \frac{\theta X + (\beta - nT/\lambda^2) Y}{(\beta - nT/\lambda^2)^2 + \theta^2}(T - hS)$$

$$\text{with} \quad X = \lambda^2 h' + \frac{n\beta}{\beta^2 + \epsilon^2} hS \quad \text{and} \quad Y = \frac{n\epsilon}{\beta^2 + \epsilon^2} hS$$

revealing immediately the 'baroclinic formstress resonance' at $nT = \beta\lambda^2$, i.e. at a flow speed which equals the speed of the baroclinic Rossby wave. This resonance opens the road to multiple steady states.

We proceed and finalize this section with a numerical solution, obtained with the continuation tool used above for the barotropic model. The complete six-dimensional low-order model (16.72) to (16.77) is implemented, and the equilibrium solutions, including their bifurcations and stability, are determined. A variety of regimes may be realized by varying the basic parameters. However, we restrict the presentation to the nonlinear counterpart to the linear wind-driven model, shown in Figure 16.32 (upper row). Figure 16.33 displays the transport $T$ in the left panel and the terms of the momentum balance for the upper and lower layer in the other panels. Unstable parts of the solution branches are plotted in red; the rest of the model variables are the colored curves as explained in the figure caption. The general dependence of the transport $T(\delta)$ on the topography height is similar in the linear and nonlinear models.

The latter, however, runs with increasing $\delta$ into a multiple state regime (there is a fold bifurcation at the right limit and a Hopf bifurcation at the left limit of the unstable window for the standard values of the parameters). It is worth mentioning that – varying the parameters – no regime was found in which an entirely unstable window existed, i. e. the model has always stable solutions.

The momentum balance of the nonlinear model is more complicated that the linear one. The interfacial stress by standing eddies, the $\ell$-term in (16.78) and (16.79), is now alive and seen to be the main process transferring eastward momentum from the surface to the deep layer at intermediate topography heights. The transient stress, the $\kappa$-term, has importance at low and very large heights and even changes sign at intermediate topography. The bottom formstress is carried by the barotropic part at low heights and by the baroclinic part at high topography. Note that 'low' and 'high' is always meant with respect to the critical height $\delta_0/\pi$.

## Further Reading

The book *Ocean Circulation and Climate* edited by Siedler et al. (2001) contains a collection of review articles on observations and modeling which describe results of the World Ocean Circulation Experiment carried out in the 1990s.

An introduction into regional aspects of the ocean circulation is given in *Regional Oceanography: an Introduction* by Tomczak and Godfrey (2003).

Theories of the large-scale circulation, and in particular a thorough discussion of thermocline ventilation issues, can be found in *Ocean Circulation Theory* by Pedlosky (1998).

*Nonlinear Physical Oceanography* by Dijkstra (2005) focusses on nonlinear aspects and simple models of certain aspects of the large-scale circulation.

Part IV of *Atmospheric and Oceanic Fluid Dynamics* by Vallis (2006) contains a discussion of the large-scale ocean circulation in terms of wind and buoyancy forcing.

The book *The Oceanic Thermohaline Circulation: an Introduction* by Aken (2007) is a good review of many aspects of the thermohaline circulation and the dominating branches of the global meridional overturning. It covers the basic physics and a wide range of observational data and develops the classical suite of THC models.

The physics of the circulation in the Southern Ocean is reviewed by Rintoul et al. (2001) and Olbers et al. (2004).

# Part VI
# Appendix

# Mathematical Basics A

The description of oceanic motion needs some basic mathematical tools, such as vector and tensor analysis, differential equations, statistical concepts, and also consideration of certain coordinate systems. In the following, these issues are briefly touched. More extensive material may be found in classical textbooks such as e. g. *Methods of Mathematical Physics* by Courant and Hilbert (1953) and *Methods of Theoretical Physics* by Morse and Feshbach (1953). Vector and tensor formalisms are also nicely introduced e. g. in *Fundamentals of Atmospheric Dynamics and Thermodynamics* by Riegel (1992) and *Vectors, tensors, and the basic equations of fluid mechanics* by Aris (1989).

## A.1 Representation of Hydrodynamic Fields

In this section, the mathematical representation of scalar, vector and tensor fields for the description of oceanic flows is discussed.

### A.1.1 Scalar and Vector Fields

In the fluid dynamical context, *scalar* fields are single-valued variables, such as pressure, density, salinity, temperature, or internal energy, depending on time and position,

$$\psi = \psi(x, y, z, t) = \psi(x_1, x_2, x_3, t) = \psi(\boldsymbol{x}, t)$$

The manifold of positions where $\psi(x, y, z, t) = \text{const}$ defines a surface in the three-dimensional space. Depending on the field type, these surfaces may have special names, e. g. an isobaric surface (for pressure), an isopycnic surface (for density), an isothermal surface (for temperature), or an isentropic surface (for entropy). Surfaces with different values cannot intersect.

D. Olbers, J. Willebrand, C. Eden, *Ocean Dynamics*,
DOI 10.1007/978-3-642-23450-7_17, © Springer-Verlag Berlin Heidelberg 2012

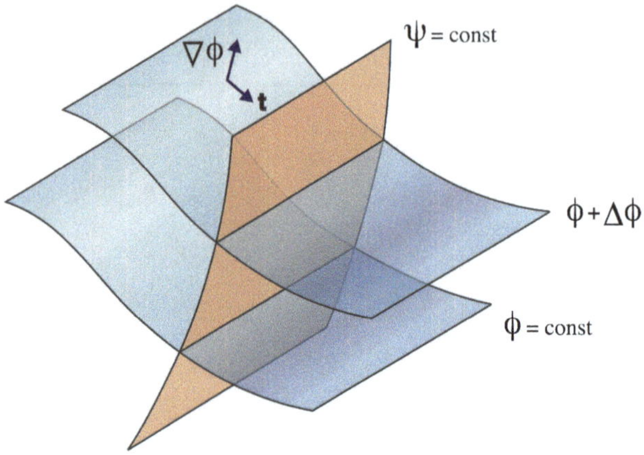


**Fig. A.1** Illustration of a stack of equi-surfaces with corresponding gradient (normal to the surface) and tangential vector

Changing from a position $\boldsymbol{x}$ to $\boldsymbol{x} + \Delta\boldsymbol{x}$ with a small $\Delta\boldsymbol{x} = (\Delta x, \Delta y, \Delta z)$, the value of $\psi$ changes by

$$\begin{aligned}\Delta\psi &= \psi(x + \Delta x, y + \Delta y, z + \Delta z) - \psi(x, y, z) \\ &= \frac{\partial\psi}{\partial x}\Delta x + \frac{\partial\psi}{\partial y}\Delta y + \frac{\partial\psi}{\partial z}\Delta z + O(\Delta x^2)\end{aligned}$$

In vector notation, this can be expressed with the *gradient*- or *nabla*-operator $\boldsymbol{\nabla}$ as

$$\Delta\psi = \sum_{i=1}^{3} \frac{\partial\psi}{\partial x_i}\Delta x_i \equiv \Delta\boldsymbol{x} \cdot \boldsymbol{\nabla}\psi \tag{A.1}$$

$$\text{with} \quad \boldsymbol{\nabla} = \boldsymbol{i}\frac{\partial}{\partial x} + \boldsymbol{j}\frac{\partial}{\partial y} + \boldsymbol{k}\frac{\partial}{\partial z} = \sum_{i=1}^{3} \boldsymbol{e}_i \frac{\partial}{\partial x_i} \equiv \boldsymbol{e}_i \frac{\partial}{\partial x_i} \tag{A.2}$$

Here $\boldsymbol{i}, \boldsymbol{j}, \boldsymbol{k} = \boldsymbol{e}_1, \boldsymbol{e}_2, \boldsymbol{e}_3$ are mutually orthogonal unit vectors in the corresponding directions. Note that the definition (A.2) is valid only for Cartesian coordinates (for other coordinates see Appendix A.4). We will use the sum convention according to which summation is implied if an index occurs twice within one term. In correct forms of this rule, an index cannot occur more than twice.

If $\Delta\boldsymbol{x}$ remains in the surface $\psi = \text{const}$, it follows that $\Delta\psi = \boldsymbol{\nabla}\psi \cdot \Delta\boldsymbol{x} = 0$. Therefore, the gradient stands normal to the equi-scalar surface. The gradient $\boldsymbol{\nabla}\psi$ always points into the direction of increasing values of $\psi$. With two scalars $\psi$ and $\phi$, we can consider the direction of both gradients $\boldsymbol{\nabla}\psi$ and $\boldsymbol{\nabla}\phi$. They are parallel if $\boldsymbol{\nabla}\psi \times \boldsymbol{\nabla}\phi = 0$. If they are parallel *everywhere*, their equi-scalar surfaces are also parallel, i. e. on a surface $\psi = \text{const}$ the scalar $\phi$ is also constant. It then follows that there must be a functional relation $\phi = F(\psi)$ or more generally $F(\psi, \phi) = 0$, and the fields $\psi$ and $\phi$ are then called *homotropic* (barotropic, if one of them is referring to pressure). In general, however, the gradients will not be parallel, i. e. $\boldsymbol{\nabla}\psi \times \boldsymbol{\nabla}\phi \neq 0$. Such fields are called *homoclinic* (resp. baroclinic; see Figure A.1). Thus the equi-scalar surfaces of a baroclinic field are not parallel to pressure surfaces.

Vectors are marked by magnitude and direction, such as e. g. position vector, velocity, force, etc. With the unit vectors $\boldsymbol{i}, \boldsymbol{j}, \boldsymbol{k}$ or $\boldsymbol{e}_1, \boldsymbol{e}_2, \boldsymbol{e}_3$ in the three coordinate directions, every vector field $\boldsymbol{a}(\boldsymbol{x},t)$ can be represented by its components referred to the particular coordinate system,

$$\boldsymbol{a}(\boldsymbol{x},t) = a(x,y,z,t)\boldsymbol{i} + b(x,y,z,t)\boldsymbol{j} + c(x,y,z,t)\boldsymbol{k} = a_i(\boldsymbol{x},t)\boldsymbol{e}_i \quad \text{(A.3)}$$

As $\boldsymbol{i}, \boldsymbol{j}, \boldsymbol{k}$ are orthogonal we obviously have $a = \boldsymbol{i} \cdot \boldsymbol{a}$ etc. Here, the $\boldsymbol{e}_i, i = 1, \ldots, 3$ constitute an orthogonal system as well.

## A.1.2 Divergence and Gauss' Integral Theorem

Consider a volume $V$ which is bounded by a closed surface $A$. A small (infinitesimal) element of the surface is represented by the area $\mathrm{d}A$ of the surface element and the unit vector $\boldsymbol{n}$ which is outward normal to the surface. It is convenient to combine area and direction of the surface element into a vector $\mathrm{d}\boldsymbol{A} = \boldsymbol{n}\mathrm{d}A$. The flux of a vector $\boldsymbol{u}$ across the surface element is $\boldsymbol{u} \cdot \mathrm{d}\boldsymbol{A} = \boldsymbol{u} \cdot \boldsymbol{n}\mathrm{d}A = u_n \mathrm{d}A$. If $\boldsymbol{u}$ is the current velocity, then this flux is the volume transport (in $\mathrm{m}^3\,\mathrm{s}^{-1}$) that crosses the surface element per unit time. The total flux out of the closed volume is given by the integral over the surface $\oint_A \boldsymbol{u} \cdot \mathrm{d}\boldsymbol{A}$ and describes the excess of the fluid outpouring from the volume per time. In the limiting case of a very small volume $V$, the *divergence* (yield, volumetric rate) of $\boldsymbol{u}$ is defined as the scalar

$$\operatorname{div} \boldsymbol{u} = \lim_{V \to 0} \frac{1}{V} \oint_A \boldsymbol{u} \cdot \mathrm{d}\boldsymbol{A} \quad \text{(A.4)}$$

The dimension is $\mathrm{s}^{-1}$ if $\boldsymbol{u}$ is the velocity. Choosing the volume $V$ as a small box parallel to the coordinate axes, the Cartesian form

$$\operatorname{div} \boldsymbol{u} = \frac{\partial u}{\partial x} + \frac{\partial v}{\partial y} + \frac{\partial w}{\partial z} = \nabla \cdot \boldsymbol{u} \quad \text{(A.5)}$$

results from (A.4). It is formally the scalar product of nabla-operator and velocity vector. For a *finite* volume $V$, the Gaussian integral theorem[1] results from (A.4),

$$\int_V \nabla \cdot \boldsymbol{u} \mathrm{d}V = \oint_A \boldsymbol{u} \cdot \mathrm{d}\boldsymbol{A} \quad \text{(A.6)}$$

A proof is obtained by cutting $V$ into infinitesimal (e. g. rectangular) boxes: inner surface integrals cancel because of different signs of the surface normal. Of course, the relations (A.4)–(A.6) are valid for any vector field.

## A.1.3 Rotation and Stokes' Integral Theorem

Apart from the surface integral, also the *line integral* of a vector field along a curve is important. If $\mathrm{d}\boldsymbol{s}$ is the line element (tangential to the curve), we obtain $\boldsymbol{u} \cdot \mathrm{d}\boldsymbol{s} = u_t \mathrm{d}s$

[1] The theorem was first discovered by Lagrange in 1762, then later independently rediscovered by Gauss in 1813, by Green in 1825, and by Ostrogradsky in 1831 (Darrigol, 2009).

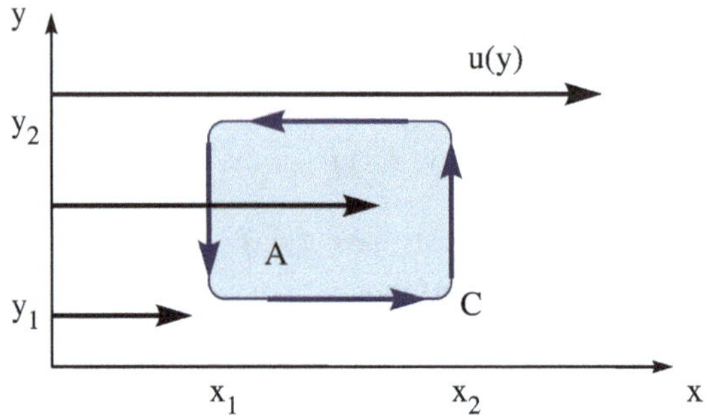


Fig. A.2 Illustration of the integration path in (A.7) for a shear current $\boldsymbol{u} = (u(y), 0, 0)$

($u_t$ is the tangential component of the vector). Look at a closed curve $C$, surrounding a surface $A$ (not necessarily an even curve and surface). The line integral

$$\Gamma = \oint_C \boldsymbol{u} \cdot \mathrm{d}\boldsymbol{s} = \oint_C u_t \mathrm{d}s$$

is called *circulation* around the surface enclosed by $C$. The circulation depends on the vector field $\boldsymbol{u}$ and the surface's bounding curve.

For a (infinitesimally) small surface element $\Delta \boldsymbol{A}$ with the rim given by the curve $C$, the vector $\boldsymbol{n} = \Delta \boldsymbol{A}/|\Delta \boldsymbol{A}|$ defines the normal unit vector. The component of the *rotation* (curl) of $\boldsymbol{u}$ in the direction of $\boldsymbol{n}$ is defined by

$$\boldsymbol{n} \cdot \operatorname{curl} \boldsymbol{u} = \lim_{\Delta A \to 0} \frac{1}{|\Delta \boldsymbol{A}|} \oint_C \boldsymbol{u} \cdot \mathrm{d}\boldsymbol{s} \tag{A.7}$$

For a shear current $\boldsymbol{u} = (u(y), 0, 0)$ (see Figure A.2), we obtain for the circulation

$$\Gamma = (u(y_1) - u(y_2))(x_2 - x_1) = -\frac{u(y_2) - u(y_1)}{y_2 - y_1}(x_2 - x_1)(y_2 - y_1)$$

and the component of $\operatorname{curl} \boldsymbol{u}$ vertical to considered plane is given by $(\operatorname{curl} \boldsymbol{u})_3 = -\partial u/\partial y$. In general, choosing a rectangle parallel to the coordinate surfaces as a surface element, we obtain the Cartesian form

$$\operatorname{curl} \boldsymbol{u} = \boldsymbol{i}\left(\frac{\partial w}{\partial y} - \frac{\partial v}{\partial z}\right) + \boldsymbol{j}\left(\frac{\partial u}{\partial z} - \frac{\partial w}{\partial x}\right) + \boldsymbol{k}\left(\frac{\partial v}{\partial x} - \frac{\partial u}{\partial y}\right)$$

which is formally the vector product of nabla-operator and vector $\boldsymbol{u}$,

$$\operatorname{curl} \boldsymbol{u} = \nabla \times \boldsymbol{u} \tag{A.8}$$

When (A.7) is integrated over a *finite* surface $A$, Stokes' integral theorem[2]

$$\int_A \nabla \times \boldsymbol{u} \cdot \mathrm{d}\boldsymbol{A} = \oint_C \boldsymbol{u} \cdot \mathrm{d}\boldsymbol{s} \tag{A.9}$$

results. This is valid for any vector field, not only for velocity fields. A proof is obtained by decomposing $A$ into infinitesimal surface elements and noting that every inside curve is passed through two times with a different sign.

[2] The theorem was first discovered by William Thomson (Lord Kelvin) who communicated it to Stokes in July 1850. Stokes set the theorem as a question on the 1854 Smith's Prize exam, which led to the theorem bearing his name (Darrigol, 2009).

### A.1.4 Velocity Potential and Stream Function

The description of certain flows can be simplified by the introduction of a velocity potential and a stream function. If the flow is *irrotational*, i. e. $\nabla \times \boldsymbol{u} = 0$, the vector $\boldsymbol{u}$ can be expressed by a *velocity potential* $\phi$ according to

$$\boldsymbol{u} = \nabla\phi$$

Instead of three quantities $(u, v, w)$ we are dealing then only with one quantity, the velocity potential $\phi$. The proof of existence of a potential is via with Stokes' theorem: because

$$\oint_C \boldsymbol{u} \cdot \mathrm{d}\boldsymbol{x} = \int (\nabla \times \boldsymbol{u}) \cdot \boldsymbol{n} \mathrm{d}A = 0$$

the path integral $\int_0^P \boldsymbol{u} \cdot \mathrm{d}\boldsymbol{x}$ is thus independent of the path and the potential $\phi$ is to be calculated from

$$\phi(\boldsymbol{x}_P) = \phi(\boldsymbol{x}_0) + \int_0^P \boldsymbol{u} \cdot \mathrm{d}\boldsymbol{x}$$

If the current is in addition *divergence-free*, i. e. $\nabla \cdot \boldsymbol{u} = 0$, the potential $\phi$ satisfies the LAPLACE[3] equation,

$$\nabla^2\phi = 0$$

The representation of flows which are two-dimensional and divergence-free can be simplified by a *stream function*. If

$$\boldsymbol{u} = (u, v, 0) \quad \text{and} \quad \frac{\partial u}{\partial x} + \frac{\partial v}{\partial y} = 0$$

the vector $\boldsymbol{u}, \boldsymbol{v}$ can be expressed by a stream function $\psi(x, y, t)$ according to

$$\boldsymbol{u} = -\frac{\partial\psi}{\partial y}\,, \quad \boldsymbol{v} = \frac{\partial\psi}{\partial x} \quad \text{or} \quad \boldsymbol{u} = \boldsymbol{k} \times \nabla\psi\,, \quad \boldsymbol{k} = (0, 0, 1)$$

$\boldsymbol{u}$ is thus parallel to the isolines of $\psi$, which represent streamlines Note that this is only valid for divergence-free currents. $\boldsymbol{u}$ is the larger, the closer the isolines lie. Integration results in

$$\psi(\boldsymbol{x}_P) - \psi(\boldsymbol{x}_0) = \int_0^P \nabla\psi \cdot \mathrm{d}\boldsymbol{x} = -\int_0^P (\boldsymbol{k} \times \nabla\psi) \cdot \mathrm{d}\boldsymbol{x} = \int \boldsymbol{u} \cdot \mathrm{d}\boldsymbol{n}$$

i. e. $\psi(\boldsymbol{x}_P) - \psi(\boldsymbol{x}_0)$ describes the volume transport (in $\mathrm{m^2\,s^{-1}}$) between $\boldsymbol{x}_0$ and $\boldsymbol{x}_P$.

[3] PIERRE-SIMON, MARQUIS DE LAPLACE, * 1749 in Beaumont-en-Auge, Normandy, † 1827 in Paris, mathematician and physicist.

### A.1.5 Integral Theorems in Two Dimensions

For a two-dimensional vector field $\boldsymbol{u} = (u, v, 0)$ the Gaussian theorem remains valid if the volume $V$ is replaced by a surface $A$ and the surface $A$ is replaced by a closed curve $C$ with the line element $\mathrm{d}\boldsymbol{n} = \boldsymbol{n}\mathrm{d}s$ where $\boldsymbol{n}$ is normal to the curve $C$. Then Gauss' theorem becomes

$$\int_A \nabla \cdot \boldsymbol{u} \mathrm{d}A = \oint_C \boldsymbol{u} \cdot \mathrm{d}\boldsymbol{n} \tag{A.10}$$

Defining the rotated vectors ($\pi/2$ counterclockwise, see the box on p. 444)

$$\underset{\neg}{\boldsymbol{u}} = (-v, u, 0)\,, \quad \underset{\neg}{\nabla} = \left(-\frac{\partial}{\partial y}, \frac{\partial}{\partial x}, 0\right) \tag{A.11}$$

we obtain

$$\nabla \cdot \boldsymbol{u} = \frac{\partial u}{\partial x} + \frac{\partial v}{\partial y} \quad \text{and} \quad \nabla \times \boldsymbol{u} = \left(0, 0, \frac{\partial v}{\partial x} - \frac{\partial u}{\partial y}\right) = (0, 0, -\nabla \cdot \underset{\neg}{\boldsymbol{u}}) \tag{A.12}$$

and the above Gaussian theorem applied for the vector $\underset{\neg}{\boldsymbol{u}}$ becomes

$$\int_A \nabla \cdot \underset{\neg}{\boldsymbol{u}} \mathrm{d}A = \oint_C \underset{\neg}{\boldsymbol{u}} \cdot \mathrm{d}\boldsymbol{n} \tag{A.13}$$

or

$$-\int_A \nabla \times \boldsymbol{u} \cdot \mathrm{d}\boldsymbol{A} = -\oint_C \boldsymbol{u} \cdot \mathrm{d}\underset{\neg}{\boldsymbol{n}} = -\oint_C \boldsymbol{u} \cdot \mathrm{d}\boldsymbol{s} \tag{A.14}$$

i. e. in two dimensions the integral theorems by Gauss and Stokes are identical.

### A.1.6 Tensor Fields

Apart from the scalar and the vector products, a further possibility to combine two vectors $\boldsymbol{a}$ and $\boldsymbol{b}$ by multiplication is the *dyadic* (or *tensorial*) product which is given by the $3 \times 3$ ordered pairs $a_i b_j$ in the form

$$\boldsymbol{ab} = \begin{pmatrix} a_1 b_1 & a_1 b_2 & a_1 b_3 \\ a_2 b_1 & a_2 b_2 & a_2 b_3 \\ a_3 b_1 & a_3 b_2 & a_3 b_3 \end{pmatrix} \tag{A.15}$$

written without period (nonuniform in the literature). With (A.3), the statements $\boldsymbol{a} = a_i \boldsymbol{e}_i$ and $\boldsymbol{b} = b_i \boldsymbol{e}_i$ can be used to obtain the representation

$$\boldsymbol{ab} = \boldsymbol{e}_i a_i b_j \boldsymbol{e}_j \tag{A.16}$$

for the dyadic product. More generally, quantities consisting of $3 \times 3$ elements $A_{ij}$, $i, j = 1, 2, 3$ are called tensors of second order[4]. We write them in capitals, e.g.

[4] Strictly speaking, not any collection of 9 elements is a tensor. This property is linked to the behavior when a transformation to other coordinates is considered. A simple rule is the 'quotient rule': if $\boldsymbol{a} = (a_i)$ is a vector and the result of $A_{ij} \boldsymbol{a}_j$ is also a vector, then $\mathsf{A} = (A_{ij})$ is a tensor. For more details we refer to text books as e. g. Aris (1989).

$\mathsf{A} = (A_{ij})$. Analogously to (A.16), every tensor can be represented with the aid of the three unit vectors $\boldsymbol{e}_i$ as

$$\mathsf{A} = \boldsymbol{e}_i A_{ij} \boldsymbol{e}_j$$

It follows that scalar multiplication of two tensors is given as

$$\mathsf{C} = \mathsf{A} \cdot \mathsf{B} = \boldsymbol{e}_i A_{ij} \boldsymbol{e}_j \cdot \boldsymbol{e}_m B_{mn} \boldsymbol{e}_n = \boldsymbol{e}_i A_{ij} B_{jn} \boldsymbol{e}_n \quad \Rightarrow \quad C_{ij} = A_{ik} B_{kj}$$

as in case of matrices. The operation is usually not commutative. Multiplication with a vector is accordingly given as $\boldsymbol{a} = \mathsf{A} \cdot \boldsymbol{b}$ or $a_i = A_{ij} b_j$. Finally, another product combination, the so-called double-scalar product, occurs. It is the total contraction of two tensors,

$$\mathsf{A} : \mathsf{B} = \mathrm{tr}(\mathsf{A} \cdot \mathsf{B}) = A_{ij} B_{ji}$$

where $\mathrm{tr}(\mathsf{C}) = C_{ii}$ terms the trace of a square tensor, i.e. the sum of its diagonal elements.

Every tensor $\mathsf{A}$ has a *transposed* tensor $\mathsf{A}^\dagger$ defined by $A^\dagger_{ij} = A_{ji}$. A tensor is called *symmetric* if $A_{ij} = A_{ji}$ and *antisymmetric* if $A_{ij} = -A_{ji}$. For symmetric tensors, $\mathsf{A} \cdot \boldsymbol{b} = \boldsymbol{b} \cdot \mathsf{A}$ is valid, for antisymmetric tensors $\mathsf{A} \cdot \boldsymbol{b} = -\boldsymbol{b} \cdot \mathsf{A}$. Every tensor can be depicted as the sum of a symmetric tensor $\mathsf{A}^s$ and an antisymmetric tensor $\mathsf{A}^a$:

$$\mathsf{A} = \mathsf{A}^s + \mathsf{A}^a \quad \text{with} \quad \mathsf{A}^s = \frac{1}{2}(\mathsf{A} + \mathsf{A}^\dagger)\,, \quad \mathsf{A}^a = \frac{1}{2}(\mathsf{A} - \mathsf{A}^\dagger) \tag{A.17}$$

### Symmetric tensors

are determined by 6 values (3 diagonal and 3 nondiagonal elements). The equation

$$\boldsymbol{x} \cdot \mathsf{A} \cdot \boldsymbol{x} = A_{ij} x_i x_j \equiv \frac{1}{2}(A_{ij} + A_{ji}) x_i x_j = \text{const}$$

constitutes a quadratic form and defines a surface in a three-dimensional space, the so-called tensor ellipsoid (if, as a rule, the form is positive). Here, only the symmetric part of $\mathsf{A}$ is relevant. The tensor ellipsoid can be expressed in the form $\lambda_1 x_1'^2 + \lambda_2 x_2'^2 + \lambda_3 x_3'^2 = \text{const}$ by an orthogonal transformation (principal-axis transformation), where $x_1', \ldots, x_3'$ are the coordinates into the three principal-axis directions.

### Antisymmetric tensors

are determined by 3 nondiagonal elements (the diagonal elements vanish). The product of an antisymmetric tensor and a vector corresponds to the product of two vectors. Defining the vector

$$\boldsymbol{\xi} = \begin{pmatrix} -A_{23} \\ A_{13} \\ -A_{12} \end{pmatrix} \quad \text{corresponding to the matrix} \quad \mathsf{A} = \begin{pmatrix} 0 & A_{12} & A_{13} \\ -A_{12} & 0 & A_{23} \\ -A_{13} & -A_{23} & 0 \end{pmatrix}$$

we obtain $\mathsf{A} \cdot \boldsymbol{b} \equiv \boldsymbol{\xi} \times \boldsymbol{b}$. Note that $\boldsymbol{\xi}$ is strictly a pseudovector since it has different transformation properties than a vector (e.g., under reflection it changes its sign).

### A.1.7 Gradient Formation in a Vector Field

The question how much the scalar $\psi$ changes when slightly altering the position by $\Delta \boldsymbol{x}$ led to the gradient formulation. The analogous question for the spatial change in a vector field $\boldsymbol{u}$, i. e.

$$\begin{aligned}\Delta \boldsymbol{u} &= \boldsymbol{u}(\boldsymbol{x} + \Delta \boldsymbol{x}, t) - \boldsymbol{u}(\boldsymbol{x}, t) \quad \text{or in tensor notation} \\ \Delta u_i &= u_i(\boldsymbol{x} + \Delta \boldsymbol{x}, t) - u_i(\boldsymbol{x}, t)\end{aligned}$$

leads to the concept of a vector gradient. Every component of the vector changes as in the scalar in equation (A.1), i. e. except for terms $O(\Delta x^2)$ we have

$$\Delta u_i = \Delta x_j \frac{\partial u_i}{\partial x_j} \tag{A.18}$$

The quantity $A_{ij} = \partial u_i / \partial x_j$ is an important example for a tensor of second order and can be formally written as the dyadic product of the vectors $\nabla$ and $\boldsymbol{u}$:

$$\mathsf{A} = \boldsymbol{e}_i \frac{\partial u_i}{\partial x_j} \boldsymbol{e}_j = \left( \boldsymbol{e}_j \frac{\partial}{\partial x_j} \right) (\boldsymbol{e}_i u_i) = (\nabla \boldsymbol{u})^\dagger$$

which is the *vector gradient* of $\boldsymbol{u}$. This results in

$$\Delta \boldsymbol{u} = \mathsf{A} \cdot \Delta \boldsymbol{x} = \Delta \boldsymbol{x} \cdot \nabla \boldsymbol{u} \tag{A.19}$$

The concept of a vector gradient is fundamental in the balance of momentum.

### A.1.8 Some Useful Differential and Integral Formulas

The following identities of the gradient and divergence operators are often used in the book. We present them without proof:

$$\begin{aligned}
\nabla \times (\nabla \phi) &= 0 \\
\nabla \cdot (\nabla \times \boldsymbol{a}) &= 0 \\
\nabla \psi \phi &= \psi \nabla \phi + \phi \nabla \psi \\
\nabla \cdot \psi \boldsymbol{a} &= \psi \nabla \cdot \boldsymbol{a} + \boldsymbol{a} \cdot \nabla \psi \\
\nabla \times \psi \boldsymbol{a} &= \psi \nabla \times \boldsymbol{a} + \nabla \psi \times \boldsymbol{a} \\
\nabla \cdot (\boldsymbol{a} \times \boldsymbol{b}) &= \boldsymbol{b} \cdot \nabla \times \boldsymbol{a} - \boldsymbol{a} \cdot \nabla \times \boldsymbol{b} \\
\nabla \times (\boldsymbol{a} \times \boldsymbol{b}) &= (\boldsymbol{b} \cdot \nabla)\boldsymbol{a} - (\boldsymbol{a} \cdot \nabla)\boldsymbol{b} + \boldsymbol{a} \nabla \cdot \boldsymbol{b} - \boldsymbol{b} \nabla \cdot \boldsymbol{a} \\
\nabla (\boldsymbol{a} \cdot \boldsymbol{b}) &= \boldsymbol{a} \times (\nabla \times \boldsymbol{b}) + \boldsymbol{b} \times (\nabla \times \boldsymbol{a}) + (\boldsymbol{a} \cdot \nabla)\boldsymbol{b} + (\boldsymbol{b} \cdot \nabla)\boldsymbol{a} \\
\frac{1}{2} \nabla \boldsymbol{a} \cdot \boldsymbol{a} &= \boldsymbol{a} \times (\nabla \times \boldsymbol{a}) + (\boldsymbol{a} \cdot \nabla)\boldsymbol{a} \\
\nabla \times (\nabla \times \boldsymbol{a}) &= \nabla (\nabla \cdot \boldsymbol{a}) - \nabla^2 \boldsymbol{a}
\end{aligned}$$

## A.2 Differential Equations

The physical laws describing oceanic motions are generally formulated in terms of differential equations (henceforth DEQs). For general systems with temporal and spatial dependence, these equations are *partial* differential equations because they contain derivatives with respect to time and the spatial coordinates. In many cases, however, important aspects of the treatment can be reduced to one-dimensional differential equations, with either time or a spatial coordinate as independent variable.

Throughout this book, we have implicitly assumed that the reader is familiar with the problems arising when one is concerned with differential equations. However, two specific aspects which may be less familiar are briefly discussed in the following, namely i) issues related to the stability of stationary solutions, and ii) problems leading to boundary layers. In both cases, it suffices to consider ordinary DEQs to highlight the principal issues. For more extensive information on the mathematics on differential equations, the reader is directed to books such as *Nonlinear differential equations and dynamical systems* by Verhulst (1996) (a real must!), and the *Handbook of differential equations* by Zwillinger (1998). Chapter 9 of *Asymptotic methods and perturbation theory* by Bender and Orszag (1978) provides an extended discussion of boundary-layer theory. More along physical problems is *From calculus to chaos* by Acheson (1997). For studying the theory of dynamical systems we recommend Drazin (1992) and Perko (2001).

### *A.2.1 Dynamical Systems: Fix Soints, Stability, and Bifurcations*

Ordinary DEQs describing the evolution of a system in time can be written as as a set of first-order equations,

$$\frac{\mathrm{d}x_i}{\mathrm{d}t} = \dot{x}_i = f_i(x_1, \ldots, x_N)\,, \quad n = 1, \ldots, N \quad \text{or in vector notation} \quad \dot{\boldsymbol{x}} = \boldsymbol{f}(\boldsymbol{x}) \tag{A.20}$$

with (real) vector functions $\boldsymbol{x}(t) = (x_1(t), x_2(t), \ldots, x_N(t))$ of time $t$. The DEQ problem (A.20) is often referred to as a *dynamical system*. We will assume that the problem is *autonomous*, i. e. that $\boldsymbol{f} = \boldsymbol{f}(\boldsymbol{x})$ is not explicitly dependent on time which is the case for many relevant systems.

At a fixed time $t$ the vector $\boldsymbol{x}(t)$ may be viewed as a point in the $N$-dimensional space (the *phase space*), and in time $\boldsymbol{x}(t)$ traces out a *trajectory* in phase space. To obtain a unique solution to (A.20), initial conditions $\boldsymbol{x}(t = 0) = \boldsymbol{x}_0$ must be specified. The vector $\boldsymbol{x}_0$ is a constant, a given point in phase space where the trajectory should start. Of course, the trajectory starting at a later time $t = t_0 > 0$ at the same point $x_0$ is an identical curve in phase space: it is only followed at a lagged time which means that autonomous DEQs are invariant under the transformation $t \to t + t_0$. Instead of time we may use one of the components of $\boldsymbol{x}$ as new independent variable. Assuming that $f_1(x) \neq 0$, then the $N - 1$ DEQs

$$\frac{\mathrm{d}x_k}{\mathrm{d}x_1} = \frac{f_k(x)}{f_1(x)} \qquad k = 2, \ldots, N$$

determine the *orbit* in phase space.

It is important to know that a DEQ problem of the kind (A.20) has a unique solution for quite 'mild' conditions imposed on the function $\boldsymbol{f}(\boldsymbol{x})$: only one trajectory starts at the point $\boldsymbol{x}_0$. This implies that trajectories with different initial points cannot cross. They can, however, run into the same point $\boldsymbol{x}'$ and stop there (or come out of it): this point is an *equilibrium point* or *fix point*, $\boldsymbol{f}(\boldsymbol{x}') = 0$.

In our applications, $\boldsymbol{f}$ will be differentiable so that the Jacobian (an $N \times N$-matrix)

$$\boldsymbol{J} = (J_{ij}) = \left(\frac{\partial f_i}{\partial x_j}\right) \tag{A.21}$$

exists at each point $\boldsymbol{x}$ in the domain of interest of solution of (A.20). In the most interesting cases, $\boldsymbol{f}$ will depend on a set of parameters $\boldsymbol{p}$ (which can be a vector of some dimension $M$), so $\boldsymbol{f} = \boldsymbol{f}(\boldsymbol{x}, \boldsymbol{p})$ and the solution will also depend on $\boldsymbol{p}$, i. e. $\boldsymbol{x} = \boldsymbol{x}(t, \boldsymbol{p})$ and also the Jacobian $\boldsymbol{J} = \boldsymbol{J}(\boldsymbol{x}, \boldsymbol{p})$.

Plotting the vector field $\dot{\boldsymbol{x}} = \boldsymbol{f}(\boldsymbol{x})$ as function of the phase space coordinates (which is tangential to the orbits), we obtain the *flow in phase space*: since trajectories do not intersect, a 2d or 3d flow looks like a stationary flow of a fluid. Important to note is the property that the phase space flow is generally compressible: a volume consisting of points which move with the velocity $\boldsymbol{f}(\boldsymbol{x})$ is generally not preserved, it might shrink or expand depending on the convergence or divergence of $\boldsymbol{f}(\boldsymbol{x})$. This property is measured (as in case of physical flows) by the divergence

$$\mathrm{div}(\boldsymbol{f}) = \sum_k \frac{\partial f_k}{\partial x_k} = \mathrm{tr}(\boldsymbol{J})$$

As an example, consider the harmonic oscillator, governed by $\ddot{x} + \omega^2 x + k\dot{x} = 0$, where $\mathrm{tr}(\boldsymbol{J}) = -k$ is easily deduced (rewrite the problem as a set of first order DEQs: $\dot{x} = y$, $\dot{y} = -\omega^2 x - ky$). Since $k > 0$ describes the friction of the system, the phase flow of the oscillator is converging everywhere.

### Stability of Fix Points

A point $\boldsymbol{x}_0$ satisfying $\boldsymbol{f}(\boldsymbol{x}_0) = 0$ is an equilibrium solution, or fix point, of (A.20) which is independent of time, i. e. it will not move in phase space. Note that there are DEQs with no stationary solutions (e. g. $\boldsymbol{f}(\boldsymbol{x}) = \boldsymbol{a} \neq 0$). In the vicinity of a fix point, the behavior of the nonlinear system (A.20) can be derived from the linearized equation

$$\dot{\boldsymbol{\xi}} = \boldsymbol{f}(\boldsymbol{x}_0 + \boldsymbol{\xi}) - \boldsymbol{f}(\boldsymbol{x}_0) \approx \boldsymbol{J}(\boldsymbol{x}_0) \cdot \boldsymbol{\xi} \tag{A.22}$$

where $\boldsymbol{\xi}$ is the deviation from $\boldsymbol{x}_0$, and the Jacobian matrix $\boldsymbol{J}$ is defined as in (A.21). Obviously, linear (or infinitesimal) stability of the system (A.22) at $\boldsymbol{x}_0$ is governed by the set of eigenvalues (see box on p. 9) of $\boldsymbol{J}(\boldsymbol{x}_0)$. Note that $\boldsymbol{J}$ is not necessarily symmetric and may thus have complex eigenvalues. Clearly, if there is any eigenvalue with a positive real part, the system is unstable (an initial perturbation with a component in the direction of the corresponding eigenvector will grow exponentially). On the other hand, if $\mathrm{Re}(\lambda_i) < 0 \; \forall i$, any trajectory $\boldsymbol{x}(t) = \boldsymbol{x}_0 + \boldsymbol{\xi}(t)$ with a small deviation from the fix point will asymptotically approach $\boldsymbol{x}_0$, and in this case $\boldsymbol{x}_0$ is a stable fix point.

More generally, we may ask whether a trajectory $\boldsymbol{x}_r(t)$, starting at $\boldsymbol{x}_r(t_0)$ at some time $t_0$, is stable, meaning whether a neighboring trajectory $\boldsymbol{x}(t)$ starting at $t_0$ in a small distance from $\boldsymbol{x}_r(t_0)$ does not deviate too much from $x_r(t)$. In mathematical terms, the solution $\boldsymbol{x}_r(t)$ is *uniformly stable* if there exists a $\delta(\epsilon) > 0$ for any $\epsilon > 0$ such that any other solution $\boldsymbol{x}(t)$ with $|\boldsymbol{x}(t_0) - \boldsymbol{x}_r(t_0)| < \delta(\epsilon)$ satisfies $|\boldsymbol{x}(t) - \boldsymbol{x}_r(t)| < \epsilon$ for all $t \geq t_0$. The solutions must thus stay close to each other for all times. We have *asymptotic stability* of $\boldsymbol{x}_r(t)$ if in addition to uniform stability

$$\lim_{t\to\infty} |\boldsymbol{x}(t) - \boldsymbol{x}_r(t)| = 0$$

for any $\boldsymbol{x}(t)$. A further concept is that of *orbital stability*. Here we choose a small pipe of radius $\epsilon$ around the reference solution $\boldsymbol{x}_r(t)$ and must find an entire neighborhood of radius $\delta$ at the initial time $t_0$ such that all trajectories starting from there remain in the pipe for all times but must not be close to each other at the same time.

## Bifurcations in One-Dimensional Systems

One-dimensional autonomous DEQs $\dot{x} = f(x)$ are quite restrictive because the trajectory $x(t)$ can only run along the real axis and pass a point only once. One-dimensional systems can only have fix points as attractors, there are no periodic solutions. Between fix points the solution runs monotonic. The one-dimensional autonomous problem is solved by simple quadrature, $\int \mathrm{d}x/f(x) = \int \mathrm{d}t$, if the $x$-integral can be expressed by elementary functions. The only equilibrium points $x_0$ are attractive or repellant points where either $\mathrm{d}f/\mathrm{d}x$ is negative or positive. Hence linear stability is not a complicated issue.

However, if $f = f(x, p)$ has a control parameter $p$, the equilibrium points are functions of the parameter, $x_0 = x_0(p)$, and it is of interest to follow the number of fix points and possible changes of stability properties as function of $p$. The fix point is determined from $f(x_0, p) = 0$. Note that $x_0 = x_0(p)$ is a single-valued analytical function of $p$ as long as the Jacobian $\boldsymbol{J}(x, p) = \partial f/\partial x \neq 0$ at $x_0(p)$. Bifurcations occur where $\boldsymbol{J}(x_0, p)$ becomes zero. Hence the bifurcation points $x_0(p_0)$, $p_0$ follow from

$$f(x_0, p) = 0\,, \qquad \left.\frac{\partial f(x, p)}{\partial x}\right|_{x=x_0} = 0$$

The linear stability is evaluated by $\partial f/\partial x$ and thus changes sign at the bifurcation point. We illustrate this for the saddle node bifurcation (where $f(x, p) = p + \delta x^2$; see below): the first condition yields $x_0(p) = \pm\sqrt{p}$ for $\delta = -1$, the second yields $-2x_0(p) = \mp 2\sqrt{p} = 0$, hence $p = p_0 = 0$.

The multitude of possible bifurcations of a one-dimensional system is discussed by generic (= 'usually occurring') forms for $f(x, p)$. In the following examples, $\delta = \pm 1$ is fixed and not treated as a parameter. The bifurcation diagrams, showing the steady states $x_0(p)$ as function of the control parameter $p$, are displayed in Figure A.3.

1. **Pitchfork**: $f(x, p) = px + \delta x^3$.
   For $\delta = -1$ and $p < 0$ there is only one solution, $x_0 = 0$, which branches at $p = 0$ into three solutions $x_0 = 0$ and $x_0 = \pm\sqrt{p}$ for $p > 0$. Correspondingly, for $\delta = 1$ three solutions exist if $p < 0$, and they collapse to one which exists for

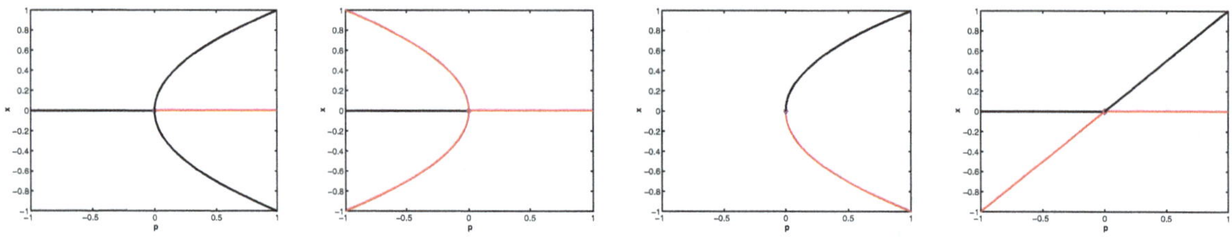

Fig. A.3 Diagrams for the pitchfork (both forms), saddle, and transcritical bifurcations (from left to right). Black indicates stable, red unstable fix point

$p > 0$. The case $\delta = -1$ is called *supercritical pitchfork*, the case $\delta = 1$ is called *subcritical pitchfork*.

2. **Saddle node or fold**: $f(x, p) = p + \delta x^2$.
   For $\delta = -1$ solutions only exist if $p > 0$. These are $x_0(p) = \pm\sqrt{p}$. Likewise, $x_0(p) = \pm\sqrt{-p}$ exists for $\delta = 1$ and $p < 0$. In these cases, a pair of two solutions 'branches' out of nothing.
3. **Transcritical**: $f(x, p) = px + \delta x^2$.
   $x_0 = 0$ is a solution for $p$ and $\delta = \pm 1$, likewise $x_0 = -p/\delta$.

Of the above described three bifurcation types, the saddle node bifurcation is the most basic and most common. The pitchfork and transcritical bifurcations will only occur in a special setting. The reason is quite clear: they need a certain symmetry because the zero solution $\dot{x} = 0$, $x = 0$ must be present for all parameter values. Small perturbations of the model equation would destroy this property. It is easy to see that addition of a constant – even of infinitesimal size – to $f(x, p)$ would lead to a cancelation of the pitchfork and transcritical bifurcations, but not so for the saddle node. The latter is not affected by addition of higher order terms $x^n$, $n > 2$ either, as occurring in a Taylor expansion of a more complex model equation. Such terms would generate a deformation away from the bifurcation point and eventually result in more bifurcations.

Remarkably, the pitchfork bifurcation occurs in two structurally different forms. Depending on the sign of the cubic term, a unique stable solution gives way to a triplet of solutions with two stable ones branching off and an intermediate repelling equilibrium, or there is a stable solution which collides with two unstable ones and leaves only an unstable one to continue for larger parameter values (see Figure A.3).

### Two-dimensional systems

The motion of the phase point $(x_1, x_2)$ of a two-dimensional system in a two-dimensional plane allows for more complicated attractors than one-dimensional systems. Apart from fix points, we can have periodic orbits which are closed solutions with $x_1(t) = x_1(t+T)$, $x_2(t) = x_2(t+T)$ with a period $T > 0$ and all times $t$. Since two-dimensional systems are easily visualized (the phase space is a two-dimensional plane), we give them a more extensive treatment. We write the DEQ with two dependent variables $x_1 = x$, $x_2 = y$ as

$$\dot{x} = P(x, y)\,, \quad \dot{y} = Q(x, y) \tag{A.23}$$

The equilibrium (or fix) points $(x_0, y_0)$ of (A.23) follow from $P(x, y) = Q(x, y) = 0$, and without loss of generality we consider one at $x_0 = 0, y_0 = 0$. Linearization about this fix point yields

$$P(x, y) = ax + by + u(x, y)\ , \quad Q(x, y) = cx + dy + v(x, y)$$

where $u$ and $v$ are at least quadratic in $x, y$. The behavior in the vicinity of the fix point is described by the four parameters $a, b, c, d$ which form the Jacobian matrix

$$\boldsymbol{J} = \begin{pmatrix} a & b \\ c & d \end{pmatrix} \tag{A.24}$$

Stability depends on the two eigenvalues of $J$ which are given by

$$\lambda = \lambda^{\pm} = \alpha \pm \sqrt{\alpha^2 - \beta}$$

with $\alpha = (a + d)/2 = \text{tr}(\boldsymbol{J})/2$, $\beta = ad - bc = \det(J)$. We assume that $\boldsymbol{J}$ is not singular, i. e. an inverse must exist which is the case if $\beta \neq 0$. The eigenvalues are real for $\alpha^2 \geq \beta$ and complex conjugate if $\alpha^2 < \beta$.

The eigenvectors of $\boldsymbol{J}$ can be used to solve the linear initial value problem with $x(0) = x_0$, $y(0) = y_0$. If $\lambda^+ \neq \lambda^-$ (which is the case for $\alpha^2 \neq \beta$), the solution is

$$x(t) = \frac{1}{\lambda^+ - \lambda^-}\{[(a - \lambda^-)x_0 + by_0]e^{\lambda^+ t} - [(a - \lambda^+)x_0 + by_0]e^{\lambda^- t}\}$$
$$y(t) = \frac{1}{\lambda^+ - \lambda^-}\{[cx_0 + (d - \lambda^-)y_0]e^{\lambda^+ t} - [cx_0 + (d - \lambda^+)y_0]e^{\lambda^- t}\}$$

where the real part is to be taken. Both eigenvalues are equal if $\alpha^2 = \beta$. The solution may then be obtained considering the corresponding limit in the above expressions: write $\lambda^{\pm} = \alpha \pm \epsilon$ and use $(1 - e^{-2\epsilon t})/(2\epsilon) \to te^{-2\epsilon t} \to t$ for $\epsilon \to 0$. Then

$$x(t) = x_0 e^{\alpha t} + \left[\frac{1}{2}(a - d)x_0 + by_0\right] te^{\alpha t}$$
$$y(t) = y_0 e^{\alpha t} + \left[cx_0 - \frac{1}{2}(a - d)y_0\right] te^{\alpha t}$$

In any case, if we start initially at $x_0 = 0$, $y_0 = 0$, the solution stays at this fix point. It is asymptotically stable if the eigenvalues have a negative real part. The imaginary part is nonzero for $\beta > \alpha^2$. Then $\lambda^{\pm} = \alpha \pm i\sqrt{\beta - \alpha^2}$, and oscillatory behavior occurs.

The fix points $(x_0, y_0)$ can be classified according to the eigenvalues of the corresponding Jacobian matrix (A.24) evaluated at $(x_0, y_0)$. Excluding zero eigenvalues, one distinguishes the following cases:

1. **Improper Node**: $\alpha^2 > \beta > 0$, two different real eigenvalues. The prototype of a node is given by

   $$\dot{x} = \text{sign}(A)x\ , \quad \dot{y} = Ay$$

   Eigenvalues are $\lambda = \text{sign}(A)$, $A$. Stable if $A < 0$ (or $\alpha < 0$), and unstable if $A > 0$ (or $\alpha > 0$). The stable case is visualized in Figure A.4 (upper left).
2. **Proper Node**: $\alpha^2 = \beta$, two equal real eigenvalues. See previous case with $A = \pm 1$. Unstable if $A = 1$ (or $\alpha > 0$), stable if $A = -1$ (or $\alpha < 0$, upper middle in Figure A.4).

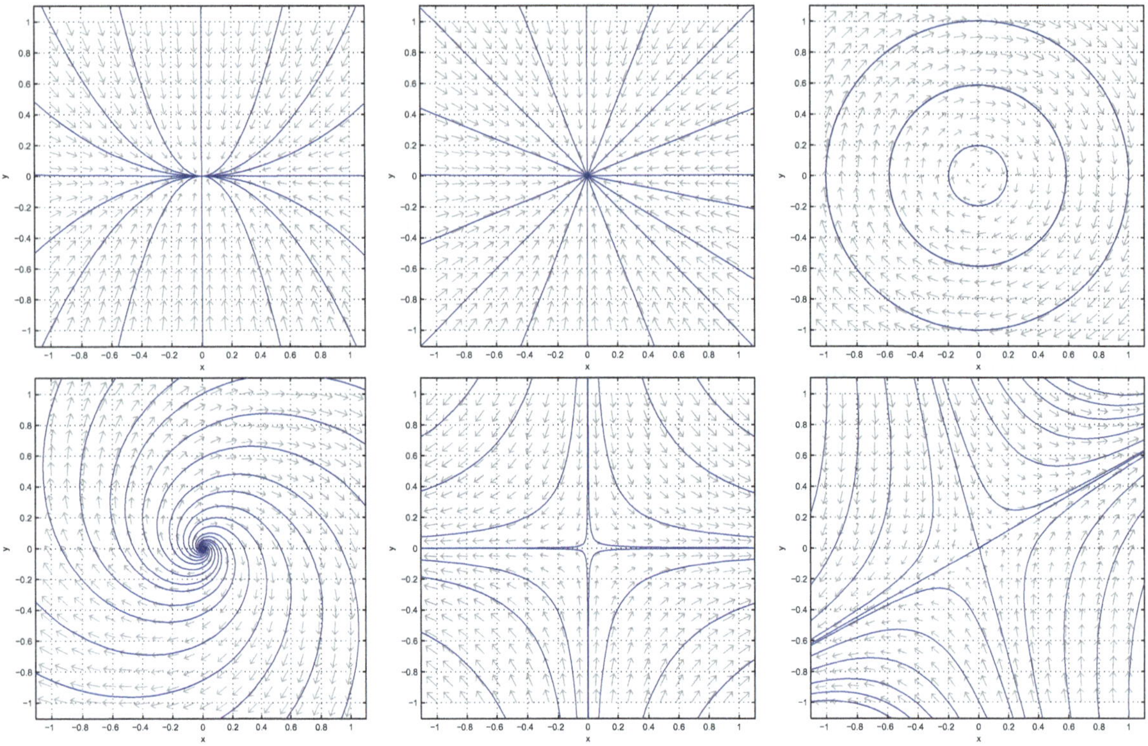

Fig. A.4 Stability of two-dimensional dynamical systems: Improper node, proper node, center (*upper row*, from left). Spiral and two versions of saddle (*lower row*, from left). See text for explanation

3. **Center**: $\alpha = 0$, $\beta > 0$, two conjugate imaginary eigenvalues $\lambda^{\pm} = \pm \mathrm{i}\sqrt{\beta}$. The prototype of a center is given by

$$\dot{x} = y\,, \quad \dot{y} = -x$$

with eigenvalues $\lambda^{\pm} = \pm\mathrm{i}$ (upper right in Figure A.4).
4. **Spiral (focus)**: $\alpha^2 < \beta$, two conjugate eigenvalues. The prototype of a spiral is given by $(A < 0)$

$$\dot{x} = Ax + y\,, \quad \dot{y} = -x + Ay$$

Eigenvalues are $\lambda^{\pm} = A \pm \mathrm{i}$. In polar coordinates this becomes $\dot{r} = Ar, \dot{\phi} = -1$. Stable if $A < 0$ (or $\alpha < 0$), unstable if $A < 0$ (or $\alpha < 0$). The unstable case is displayed in Figure A.4 (lower left).
5. **Saddle**: $\beta < 0$, two real eigenvalues of different sign, therefore, always unstable. The prototype of a saddle is given by

$$\dot{x} = x\,, \quad \dot{y} = -y$$

with eigenvalues $\lambda^{\pm} = \pm 1$ (lower middle in Figure A.4).

Note that each of the prototype cases has been chosen such that both eigenvectors are orthogonal. Orthogonality requires that the Jacobian matrix (A.24) satisfies $\boldsymbol{J}^{\dagger} \cdot \boldsymbol{J} = \boldsymbol{J} \cdot \boldsymbol{J}^{\dagger}$, or equivalently that either $b = c$, or $b = -c$ and $a = d$. As an example, Figure A.4 (lower right) shows a saddle fix point with nonorthogonal eigenvectors.

In a two-dimensional system, a specific bifurcation may occur which adds to the suite of generic bifurcations introduced above. This is the *Hopf bifurcation* where the character of the solution changes rather than the number of solutions. Suppose that we have dependence of $P(x, y)$ and $Q(x, y)$ on a parameter $p$ which carries over to $p$-dependent eigenvalues, $\lambda = \lambda(p) = \alpha(p) + \mathrm{i}\omega(p)$. Changing $p$, one may arrive at a value at which the classification of the above described fix points changes. The most important case is shift from a spiral (stable or unstable) via a center to a new spiral with interchanged stability. Clearly, for this to occur, a pair of conjugate eigenvalues will have to cross the imaginary axis. For instance, we have a stable spiral fix point $x_0(p)$ for $p < p^*$, having thus $\alpha(p) < 0$. If the value of $\alpha$ goes through zero at $p = p^*$ but $\omega$ remains nonzero, we encounter a switch from a stable fix point to an unstable one. For a linear system (like the examples in the above classification), nothing more would happen. In the presence of nonlinearities a new feature may arise. Consider thus

$$\dot{x} = px - \omega y + u(x, y)\ , \quad \dot{y} = \omega x + py + v(x, y)$$

with eigenvalues $\lambda = p \pm i\omega$ so that the bifurcation appears at $p = 0$. Converting to polar coordinates, we obtain

$$\dot{r} = pr + \frac{1}{r}[xu + yv]\ , \quad \dot{\theta} = \omega + \frac{1}{r^2}[xv - yu]$$

If $u$ and $v$ have third-order powers in a polynomial expansion, the generic form for a Hopf bifurcation is $u \sim xr^2$, $v \sim yr^2$. Then

$$\dot{r} = pr + \delta r^3\ , \quad \dot{\theta} = \omega$$

so that a pitchfork bifurcation occurs in the radial component at $p = 0$: for $\delta < 0$ and $p < 0$, only one stable fix point is present (it is the stable fix point of the original problem), but at $p > 0$ there are three fix points in the radial equation (with $r = \sqrt{-p/\delta}$). Since $\theta$ is spinning with frequency $\omega \neq 0$, the solution in $r, \theta$ and, of course, in the original $x, y$ is not stationary: it is a periodic solution of the original equation. The transition from a steady solution to a periodic one occurs when $p$ passes from negative values through zero to positive. The solution in the $x, y$ plane can be described as follows: for $p \leq 0$, all solutions spiral into the fix point $x = 0, y = 0$ as $t \to \infty$; for $p > 0$, the origin becomes an unstable focus, solutions starting there spiral outward to a periodic solution with radius $r = \sqrt{-p/\delta}$ about the origin. As this orbit is also approached when starting from outside, we have a *limit cycle*.

The above scenario (for $\delta < 0$) corresponds to a supercritical Hopf bifurcation with a stable limit cycle. For $\delta > 0$, we have a subcritical Hopf bifurcation branching off at $p = 0$ but occurring then at $p < 0$. This results in an unstable limit cycle.

## Higher-Dimensional Systems

So far we have discussed the bifurcation of one-dimensional and two-dimensional systems which arise when a single parameter is changed. They are called bifurcations of codimension 1. Of course, dynamical systems may possess more than one control parameter, and the theory may thus be extended to study bifurcation of codimension 2 or higher. Another extension concerns the dimension of the system. A forced oscillator has already three independent variables, and many important systems in Earth Science cannot be described adequately by one or two degrees of freedom. New features, as e. g. chaotic behavior, are known to occur only with three or more system variables. It is easy to understand that the one-dimensional and two-dimensional bifurcations occur as well in higher order systems, even more important: they are the generic cases of bifurcations in systems of all dimensions, so what has been learned from one-dimensional and two-dimensional systems can readily be transferred to higher order systems.

The basic concept is the *center manifold* of a dynamical system. Suppose we are studying an $n$-dimensional system $\dot{\boldsymbol{x}} = \boldsymbol{f}(\boldsymbol{x}, p)$ with a control parameter $p$. The fix points $\boldsymbol{x}_0 = \boldsymbol{x}_0(p)$ and the Jacobian $\boldsymbol{J}(\boldsymbol{x}_0) = \boldsymbol{J}(p)$ are then functions of $p$, and consequently the eigenvalues and eigenvectors are $p$-dependent as well, $\lambda_j = \lambda_j(p), a_j = a_j(p)$. Remember that complex eigenvalues come in pairs, $\lambda_j = \lambda_j(p) = \alpha_j(p) + \mathrm{i}\omega_j(p)$ and $\lambda_{j+1} = \lambda_{j+1}(p) = \alpha_j(p) - \mathrm{i}\omega_j(p)$. Suppose that for a particular value $p$ all eigenvalues are in the negative half-plane so that the system is stable. As $p$ is changed, the simplest way of losing stability occurs when one of the real eigenvalues, say $\lambda_k(p)$, moves or a pair of complex ones proceed into the positive half-plane. The value of $p = p_0$ where $\lambda_k(p) = 0$ marks then either a fold or a Hopf bifurcation. As $p$ is changed, the fix point $\boldsymbol{x}_0(p)$ moves in the $n$-dimensional phase space. Clearly, the bifurcation at $p = p_0$ is found in the components of $\boldsymbol{x}_0$ that lie along the direction of the eigenvectors of the critical eigenvalues.

Assume that at a certain parameter value there are $k$ eigenvalues that have a negative real part, $\ell$ have zero real part and $m$ have a positive real part. In the $k$ and $m$ dimensional subspaces, spanned by the corresponding eigenvectors, the phase flow is stable and unstable, respectively, and bifurcations occur for the system variables in the $\ell$-dimensional subspace. The reduction of the system equations to the $\ell$-dimensional subsystem is the aim of the *center manifold theory*. For most practical purposes where the bifurcation of a system is analyzed by numerical techniques, it should be clear that the behavior must be monitored using variables which have a nonzero projection in the $\ell$-dimensional subspace.

### A.2.2 Boundary Value Problems

Boundary value problems are usually occurring in oceanographic problems for functions of a spatial coordinate. We denote this coordinate by $z$ and are then concerned with a (vector) function $\varphi(z)$. Frequently, the differential problems are advective-diffusive DEQs, possibly with a source-sink term arising from chemical reactions or phase transitions. A typical boundary value problem is then

$$\frac{\mathrm{d}}{\mathrm{d}z}(u\varphi) = \frac{\mathrm{d}}{\mathrm{d}z}\left(\kappa\frac{\mathrm{d}\varphi}{\mathrm{d}z}\right) + S \tag{A.25}$$

where $\kappa(z)$ is a diffusivity (usually independent of $\varphi$), $u(z)$ is a velocity which advects the property $\varphi$, and $S(z)$ is a source term forcing the system in the interior of the domain ($S$ may depend on $\varphi$). If the boundary is solid and $u$ vanishes there, the system is driven by the source term $S$ and the diffusive flux $F = -\kappa \mathrm{d}\varphi/\mathrm{d}z$ through the boundaries. The natural boundary condition for (A.25) thus prescribes the flux at the bounding points $z_1, z_2$, i. e. $F(z_i) = Q_i$, $i = 1, 2$. Obviously, the fluxes $Q_i$ and the source must be consistent,

$$Q_2 - Q_1 + \int_{z_1}^{z_2} S\mathrm{d}z = 0$$

to allow for a solution of (A.25). Otherwise, the boundary conditions are wrongly formulated – or the assumption of stationarity is invalid: a rate of change term $\partial\varphi/\partial t$ must be included in (A.25). It becomes clear that boundary conditions cannot be put arbitrarily.

There are problems of nonadvective character, e. g. the equilibration of wave propagation against friction or diffusion, in which case a form

$$A\varphi + B\frac{\partial\varphi}{\partial z} = \frac{\mathrm{d}}{\mathrm{d}z}\left(\kappa\frac{\mathrm{d}\varphi}{\mathrm{d}z}\right) + S \tag{A.26}$$

might arise where $A$ and $B$ are linear (matrix) operators.

If the velocity $u$ in problem (A.25) is one of the unknown variables, i. e. part of the vector $\varphi$, the problem is inherently nonlinear. General properties and existence theorems for nonlinear problems are difficult to obtain. Solutions of linear boundary value problems generally exist if the conditions in the interior of the domain and at the boundaries are consistent, as e. g. arising from considerations of physical systems. A nonlinear differential problem describing a physical system can be expected to have a unique solution.

Boundary layers appear in solutions of DEQs if the highest derivative is multiplied with a small parameter (denoted by $\epsilon$ in this section). Usually, problems with spatial dependence are considered so that $y(x)$ is governed by

$$\epsilon y^{(n)} = F(x, y, y', \ldots, y^{(n-1)}, \epsilon) \tag{A.27}$$

on an interval $0 \le x \le 1$ with $n$ boundary conditions on $y$ and its derivatives. Note that $F$ may depend on $\epsilon$ as well. In the solution $y(x|\epsilon = 0)$ of (A.27) with $\epsilon = 0$, only $n-1$ of these conditions can be satisfied, and thus $y(x|\epsilon = 0)$ will deviate from $y(x|\epsilon)$ approaching one of the boundaries where the remaining condition should apply. A narrow region – the boundary layer – appears where the highest derivative becomes very large in order to help $\epsilon y^{(n)}$ to participate in the balance of the DEQ and mediate between the behavior of $y(x|\epsilon = 0)$ and the true solution with correct boundary conditions. Because the highest derivatives in fluid mechanical equations arise usually from viscous or diffusive processes, the boundary layers in fluid problems are usually due to such physics.

As a simple example to elucidate this behavior, consider the first order problem for $y(x)$

$$\epsilon\frac{\mathrm{d}y}{\mathrm{d}x} + y = f(x) \tag{A.28}$$

with constant $\epsilon \ll 1$, arbitrary $f(x)$ and the boundary condition $y(0) = y_0$. For small $\epsilon$, it is tempting to abandon the first term in (A.28) and take

$$y(x) \approx f(x) \tag{A.29}$$

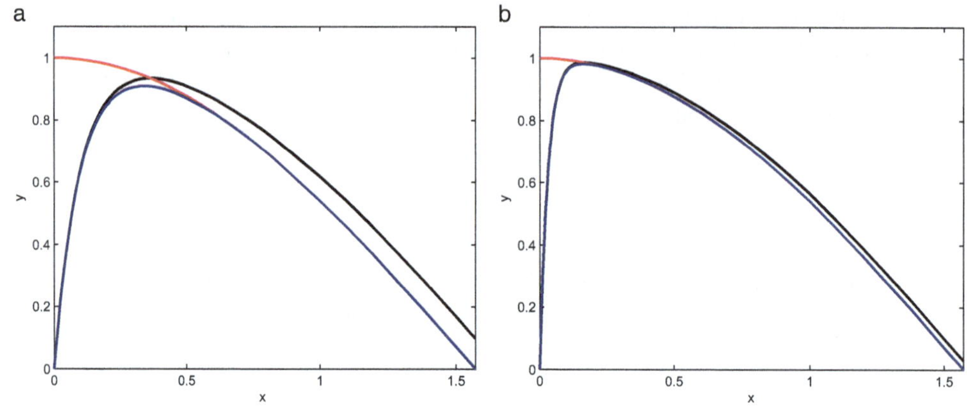


Fig. A.5 Solutions to boundary layer problem (A.28), for $f(x) = \cos x$: First approximation (A.29) (*red*); boundary layer approximation $y^a(x)$ from (A.31) (*blue*), and exact solution (A.30) (*black*). **a** $\epsilon = 0.1$, **b** $\epsilon = 0.03$

as a first approximation. However, with this approximation the boundary condition $y(0) = y_0$ cannot be satisfied, and only if $f(0) - y_0$ is accidentally zero or small, we obtain a meaningful solution. Indeed, the form of the exact solution

$$y(x) = \left[ y_0 + \frac{1}{\epsilon} \int_0^x e^{\xi/\epsilon} f(\xi) d\xi \right] e^{-x/\epsilon} \tag{A.30}$$

makes it evident that an expansion in terms of $\epsilon$ is impossible. The difference between (A.29) and the exact solution (A.30) can be estimated by partial integration of (A.30),

$$y(x) = y_0 e^{-x/\epsilon} + f(x) - f(0) e^{-x/\epsilon} - \int_0^x e^{(\xi - x)/\epsilon} f'(\xi) d\xi$$

and further

$$\int_0^x e^{(\xi - x)/\epsilon} f'(\xi) d\xi = \epsilon e^{(\xi - x)/\epsilon} f'(\xi) \Big|_0^x - \epsilon \int_0^x e^{(\xi - x)/\epsilon} f''(\xi) d\xi = O(\epsilon)$$

Hence

$$y(x) = f(x) + [y_0 - f(0)] e^{-x/\epsilon} + O(\epsilon) = y^a(x) + O(\epsilon) \tag{A.31}$$

Evidently, $y(x) \approx f(x) + O(\epsilon)$ for $x \gg \epsilon$, but in the boundary layer $0 \le x \le \epsilon$ the approximation (A.29) is inappropriate. The boundary layer approximation $y^a(x)$ exactly satisfies $y^a(0) = y_0$ and is good to order $\epsilon$ everywhere. Notice, however, that with increasing distance from the boundary layer, $y^a(x)$ does not necessarily converge toward the exact solution (see e. g. Figure A.5).

With the exponentially decaying solution of the homogeneous part $\epsilon y' + y = 0$ of (A.28) the boundary layer behavior only appears on the left-hand side of the interval. Putting the boundary condition on the right boundary, a mismatch between $f(1)$ and $y(1) = y_1$ grows exponentially $e^{(1-x)/\epsilon}$ to very large values in the interior of the interval, overwhelming $f(x)$ for sufficiently small $\epsilon$. This demonstrates that a DEQ with a small coefficient at the highest derivative does not naturally imply a boundary layer behavior.

There is a collection of methods to solve boundary layer problems in case that exact solutions are not available. A perturbation analysis determines the slowly varying solution in the outer region – outside the boundary layer – by an expansion of the DEQ in terms of $\epsilon$, and yields a leading-order approximation for small $\epsilon$ within the layer – the inner region. Here the derivatives of $y(x)$ are large, but the coefficients of the DEQ vary little and can be approximated by constants. Various complications, however, may occur (but will not be treated here): the outer problem may be nonlinear and have a number of different solutions; boundary layers may occur at both boundaries, but also layers with strong variations may be embedded within the outer region, i. e. not attached to the physical boundaries of the system (so-called internal boundary layers); and finally, the width of the boundary layer may be different from $\epsilon$, and layers of different decreasing widths may be embedded within each other.

We discuss two similar methods to find an approximate solution to boundary layer problems.

### The correction method

extends the outer solution $y^{\mathrm{o}}$ into the boundary layer and searches for an additive correction $y^{\mathrm{b}}$ so that the approximate solution is

$$y(x) = y^{\mathrm{o}}(x) + y^{\mathrm{b}}(x) \tag{A.32}$$

everywhere. The correction must vanish in the outer region. The outer solution is obtained from an expansion in $\epsilon$ and usually only the lowest order is necessary. Thus, for the above example,

$$y^{\mathrm{o}}(x) = f(x) \tag{A.33}$$

$$\epsilon \frac{\mathrm{d}y^{\mathrm{b}}}{\mathrm{d}x} + y^{\mathrm{b}} = 0 \tag{A.34}$$

with the boundary condition $y^{\mathrm{b}}(0) = y_0 - y^{\mathrm{o}}(0) = y_0 - f(0)$. It follows that $y^{\mathrm{b}}(x) = [y_0 - f(0)]\mathrm{e}^{-x/\epsilon}$, and hence the solution is

$$y(x) = y^{\mathrm{o}}(x) + y^{\mathrm{b}}(x) \approx f(x) + [y_0 - f(0)]\mathrm{e}^{-x/\epsilon}$$

which is identical to the approximation $y^{\mathrm{a}}(x)$ derived heuristically in (A.31). The approximation is acceptable if $\epsilon \mathrm{d}y^{\mathrm{o}}/\mathrm{d}x = \epsilon f' \ll y^{\mathrm{o}} = f$ is small. The correction decays exponentially on the scale of the boundary layer, and thus no additional action must be taken.

### The matching method

considers the outer solution $y^{\mathrm{o}} = f(x)$ only in the outer region. In the boundary layer, the complete solution $y^{\mathrm{i}}(x)$ – the inner solution – is to be determined and matched to the outer solution by the requirement (matching condition)

$$y^{\mathrm{i}}(x \to \infty) = y^{\mathrm{o}}(x \to 0)$$

To determine $y^{\mathrm{i}}$, the complete DEQ must be used in the interior region – the boundary layer – but because of the thinness of the layer and the smoothness of all coefficients of the DEQ these may be replaced by constants. For our problem, we would thus consider

$$\epsilon \frac{\mathrm{d}y^{\mathrm{i}}}{\mathrm{d}x} + y^{\mathrm{i}} = f(0) \tag{A.35}$$

with boundary condition $y^{\mathrm{i}}(0) = y_0$. The solution is

$$y^{\mathrm{i}}(x) = f(0) + [y_0 - f(0)]\mathrm{e}^{-x/\epsilon}$$

and the matching in this case is trivial because $y^{\mathrm{i}}(x \to \infty) = f(0)$. More complicated matching is considered below in another example.

The above case has its merits in the simplicity of a linear first-order DEQ with constant coefficients. Let us consider a more complicated case, a second-order DEQ with arbitrary coefficients,

$$\epsilon y'' + a(x)y' + b(x)y = f(x) \tag{A.36}$$

in $0 \le x \le 1$ with boundary conditions $y(0) = A$, $y(1) = B$ and small $\epsilon$. For the beginning, let $f \equiv 0$ and $a(x) \neq 0$ and positive in the interval. The outer solution is determined by

$$a(x)y^{\mathrm{o}\prime} + b(x)y^{\mathrm{o}} = 0$$

Because of the exponential behavior of the first two terms in (A.36), we expect a boundary layer at $x = 0$ for positive $a(x)$, and hence we satisfy the right boundary condition $y(1) = B$ with the outer solution,

$$y^{\mathrm{o}}(x) = B \exp \int_x^1 \frac{b(\xi)}{a(\xi)} \mathrm{d}\xi$$

It becomes evident why we have assumed $a(x) \neq 0$: otherwise the outer solution would become singular and would not be bounded.

Unless the outer solution incidentally meets the left boundary condition, $y^{\mathrm{o}}(0) = A$, a boundary layer develops at the left side $x = 0$ of the interval. Its thickness is $O(\epsilon)$, and we determine the inner solution from (A.36) with constants $\alpha = a(0)$, $\beta = b(0)$ in place of the varying coefficients. As $y'$ is getting large in the boundary layer, we expect $\alpha y' \gg \beta y$ and hence consider

$$\epsilon y^{\mathrm{i}\prime\prime} + \alpha y^{\mathrm{i}\prime} = 0$$

for the inner problem. The general solution is

$$y^{\mathrm{i}}(x) = C_1 + C_2 \mathrm{e}^{-\alpha x/\epsilon}$$

Table A.1 Types of solution for the DEQ (A.36)

| Conditions on $a(x)$ | Type of Solution |
|---|---|
| $a(x) \neq 0$ | |
| $a(x) > 0$ | boundary layer at $x = 0$ |
| $a(x) < 0$ | boundary layer at $x = 1$ |
| $a(x) = 0$ | |
| $b(x) < 0$ | boundary layers at $x = 0$ and $x = 1$ |
| $b(x) > 0$ | rapid oscillating solution |
| $b(x)$ changes sign | turning point behavior |
| $a' \neq b, a(x_0) = 0$ only at one interior point $x_0$ | |
| $a'(x_0) > 0$ | no boundary layers<br>interior layer at $x_0$ |
| $a'(x_0) < 0$ | boundary layers at $x = 0$ and $x = 1$<br>no interior layer at $x_0$ |

and the boundary condition at $x = 0$ requires $C_1 = A - C_2$. We thus have one coefficient free to establish the matching $y^{\mathrm{i}}(x \to \infty) = y^{\mathrm{o}}(x \to 0)$ at the outer rim of the boundary layer which yields

$$A - C_2 = B \exp \int_0^1 \frac{b(\xi)}{a(\xi)} d\xi$$

which determines $C_2$.

Note that the boundary layer is at $x = 0$ because $a(x)$ was assumed positive, and this condition also abandons any possibility for such a layer at $x = 1$. Without expanding further details, we summarize the types of boundary layers and solutions of (A.36) in Table A.1.

## A.3 Description of Random Fields

As we have seen in Part IV, turbulent flows are described in terms of *random*, or *stochastic*, fields which are briefly introduced in the following. More information can be found in *Statistical analysis in climate research* by Storch and Zwiers (2002) and also in Chapter 3 of *Turbulent flows* by Pope (2000).

### A.3.1 Random Variables

A random variable $\xi$, such as e. g. the throw of a dice, cannot be assigned a numerical value. Only after an experiment is made, a particular realization of $\xi$ is known. However, it is well possible to make statements about the likelihood, or *probability*, for certain realizations of $\xi$. The probability is normalized as the relative frequency of the outcome of the experiment, and hence ranges from 0 (impossible event) to 1 (certain event).

Any random variable $\xi$ is completely characterized by its *cumulative probability function* $F(x)$ which is defined as

$$F(x) = P(\xi < x)$$

and specifies the probability $P$ that $\xi$ is smaller as $x$. For the throw of a dice, one has e. g. $P(\xi < 1.5) = F(1.5) = 1/6$. It is clear that $F$ is a monotonically increasing function, with $F(-\infty) = 0$ and $F(\infty) = 1$. For most practical applications, the *probability density function* $f(x)$ defined as $f(x) = \partial F(x)/\partial x$ is more useful. The probability density function (PDF) is proportional to the probability that $\xi$ will take a value in a small interval around $x$, i. e. that $x < \xi < x + \Delta$ since

$$P(\xi < x + \Delta) - P(\xi < x) = F(x + \Delta) - F(x) \approx F(x) + \left.\frac{\partial F}{\partial x}\right|_x \Delta - F(x)$$
$$= f(x)\Delta$$

Note that $\int_{-\infty}^{\infty} f(x)\mathrm{d}x = 1$.

### Moments of Random Variables

The average $\overline{\xi}$ over many realizations $\xi^{(j)}$, $j = 1, \ldots, N$ of a random variable $\xi$ is referred to as its *mean value*, or *expectation*, and given as

$$\overline{\xi} = \lim_{N\to\infty} \frac{1}{N} \sum_{j=1}^{N} \xi^{(j)} = \int_{-\infty}^{\infty} x f(x)\mathrm{d}x$$

Likewise, the expectation for an arbitrary function $g(\xi)$ is given as

$$\overline{g(\xi)} = \int_{-\infty}^{\infty} g(x) f(x)\mathrm{d}x$$

Specifically, the expectation $\overline{\xi^n}$ of the $n$-th power of $\xi$ is called the *moment* of order $n$, the first moment being the mean $\overline{\xi}$. For moments of higher order, it is convenient to consider central moments defined as $\overline{(\xi - \overline{\xi})^n}$. The second central moment is the variance $c^2$, with $c$ being the standard deviation. Occasionally of importance are further the third central moment (skewness when normalized by $c^3$) which is a measure for the asymmetry of the PDF around the mean, and the fourth central moment (flatness when normalized by $c^4$). A random vector variable $\boldsymbol{\xi} = (\xi_1, \xi_2, \ldots, \xi_N)$ is associated with the multivariate probability density function $f(\boldsymbol{x}) = f(x_1, x_2, \ldots, x_N)$. The expectation is

$$\begin{aligned}
&\overline{g(\xi_1, \xi_2, \ldots, \xi_N)} \\
&= \int_{-\infty}^{\infty}\int_{-\infty}^{\infty} \cdots \int_{-\infty}^{\infty} g(x_1, x_2, \ldots, x_N) f(x_1, x_2, \ldots, x_N)\mathrm{d}x_1\mathrm{d}x_2 \ldots \mathrm{d}x_N \\
&= \int g(\boldsymbol{x}) f(\boldsymbol{x})\mathrm{d}\boldsymbol{x}
\end{aligned}$$

The second-order central moments constitute the covariance matrix $\boldsymbol{C} = (C_{mn})$ with elements

$$C_{mn} = \overline{(\xi_m - \overline{\xi_m})(\xi_n - \overline{\xi_n})}\,, \quad m, n = 1, \ldots, N$$

The $\xi_n$ are independent if the PDF falls into products of the individual PDF's, i. e. if

$$f(x_1, x_2, \ldots, x_N) = f_1(x_1) f_2(x_2) \times \cdots \times f_N(x_N)$$

Independence of the $\xi_n$ implies that the $\xi_1, \xi_2, \ldots, \xi_N$ are uncorrelated, i. e. that $\overline{(\xi_i - \bar{\xi}_i)(\xi_j - \bar{\xi}_j)} = \delta_{i,j}$.

### Normal Distribution

The most important PDF is the *normal* or Gaussian distribution which for a scalar random variable is defined as

$$f(x) = \frac{1}{c\sqrt{2\pi}} e^{-\frac{1}{2}\left(\frac{x-a}{c}\right)^2}$$

with mean $a$ and variance $c^2$. For a vector variable $\xi_1, \xi_2, \ldots, \xi_N$, the normal PDF is given as

$$f(x) = \left[\frac{\det(\boldsymbol{W})}{(2\pi)^N}\right]^{\frac{1}{2}} e^{-\frac{1}{2}\sum_{i,j} W_{i,j}(x_i - a_i)(x_j - a_j)}$$

with mean $\bar{\xi}_i = a_i$ and covariance matrix $\boldsymbol{C} = \boldsymbol{W}^{-1}$. The normal distribution is important because of the *central limit theorem* which states:

> If the random variable $\eta = \sum_{n=1}^{N} \xi_n$ is given as the sum of many independent random variables $\xi_n$ with arbitrary PDF and equal variance, then $\eta$ is normally distributed in the limit $N \to \infty$.

In practice, the distribution of $\eta$ is already well approximated by the normal distribution for relatively small values of $N$ (i. e. for $N \gtrsim 10$).

## A.3.2 Random Functions

A random function is a scalar or vector variable which depends on time and/or space, such as e. g. the turbulent velocity $\boldsymbol{u}(\boldsymbol{x}, t)$. Similar to random variables, it is not possible to assign a certain value for a fixed point in space and/or time to the random function: there are only certain realizations and instead a probability density function (PDF) which depends now on space and/or time. For simplicity of notation, in the following only a random function of time $\xi(t)$ is explicitly considered.

Obviously, it is necessary to know the PDF $f(x; t)$ of $\xi(t)$which must depend on time. Moreover, for different times $t_n, n = 1, \ldots, N$ the $\xi(t_n) = \xi_n$ can be considered as a vector random variable; hence the combined PDF $f(x_1, x_2, \ldots, x_N; t_1, t_2, \ldots, t_N)$ is necessary to describe the $\xi_n$, for any value of $N$.

The random function $\xi(t)$ is *stationary* if the PDF is invariant against translation of time, i. e. if

$$f(x_1, x_2, \ldots, x_N; t_1, t_2, \ldots, t_N) = f(x_1, x_2, \ldots, x_N; t_1 + \tau, t_2 + \tau, \ldots, t_N + \tau)$$

It is homogeneous if the PDF is invariant against translations in space. Therefore, stationarity or homogeneity considerably reduces the necessary statistical information to describe a random function $\xi$.

### Moments of Random Functions

The moments of random functions are defined as above for random variables. The mean is given by

$$\overline{\xi(t)} = a(t)$$

In the following, it is assumed that $\xi$ is stationary so that $\partial a/\partial t = 0$, and it is often convenient to redefine $\xi$ so that $a = 0$. The second moment is then given by

$$\overline{\xi(t)\xi(t')} = R(t, t') = R(t' - t) = R(\tau)$$

where $R$ denotes the covariance function, and $\tau = t - t'$ is the time lag. For a stationary vector random function $\boldsymbol{\xi} = (\xi_1, \xi_2, \ldots, \xi_N)$ the covariance tensor is likewise given by

$$\overline{\xi_i(t)\xi_j(t')} = R_{ij}(\tau)$$

For a physical variable, one can normally assume that $R(\tau) \to 0$ for $\tau \to \infty$. The covariance function has the properties

$$R(\tau) = R(-\tau)\,, \quad R(0) > 0 \quad \text{and} \quad |R(\tau)| \leq R(0)$$

Similarly, for vector random functions one has

$$R_{ij}(\tau) = R_{ji}(-\tau), \; \sum_{ij} a_i a_j R_{ij}(0) = 0 \; \text{ for any } a_i \; \text{ and } \; |R_{ij}(\tau)| < \sqrt{R_{ii}(0)R_{jj}(0)}$$

The derivative $\eta = \partial\xi/\partial t$ of a stationary random function $\xi(t)$ with covariance function $R(\tau)$ is also stationary with the covariance function $R_\eta = -\partial^2 R(\tau)/\partial\tau^2$.

It is useful to define the *integral time scale* $T_{\text{int}}$ as

$$T_{\text{int}} = R(0)^{-1} \int_0^\infty R(\tau)\mathrm{d}\tau \tag{A.37}$$

which is positive and denotes the time over which the respective random function remains correlated with itself, i. e. after which $R(\tau)$ approaches zero (see e. g. Section 11.2.2). Equivalently, the integral length scale $L_{\text{int}}$ is defined if a spatial coordinate instead of time is considered (see Section 11.1.3).

### Ergodic Theorem

The analysis of stationary random functions is very much facilitated by use of the *ergodic theorem* which states that a temporal average over a long time interval $T$ will eventually (for $T \to \infty$) coincide with the statistical mean which is defined as the average over many different realizations. Because of the importance of the ergodic theorem, the proof is sketched in the following.

Let $\tilde{\xi}(T) = T^{-1}\int_0^T \xi(t)\mathrm{d}t$ denote the temporal average which obviously satisfies $\overline{\tilde{\xi}(T)} = \overline{\xi} = a$. Consider now the statistical mean of the variance of $\xi$, i. e. $(\tilde{\xi}(T) - a)^2$. One finds

$$\overline{(\tilde{\xi}(T) - a)^2} = \overline{\left(T^{-1}\int_0^T \xi(t)\mathrm{d}t - a\right)^2} = T^{-2}\int_0^T\int_0^T \overline{(\xi(t') - a)(\xi(t) - a)}\mathrm{d}t'\mathrm{d}t$$

$$= 2T^{-2}\int_0^T\int_0^t R(t - t')\mathrm{d}t'\mathrm{d}t = -2T^{-2}\int_0^T\int_t^0 R(\tau)\mathrm{d}\tau\mathrm{d}t$$

$$= 2T^{-2}\int_0^T\int_0^t R(\tau)\mathrm{d}\tau\mathrm{d}t$$

with $\tau = t - t'$ so that $\mathrm{d}\tau = -\mathrm{d}t'$. Now for large $T$, $\int_0^t R(\tau)\mathrm{d}\tau \approx \int_0^\infty R(\tau)\mathrm{d}\tau$. With (A.37), one obtains for large $T$ the approximation

$$\overline{(\tilde{\xi}(T) - a)^2} \approx 2T^{-2}\int_0^T R(0)T_{\mathrm{int}}\mathrm{d}t = 2R(0)\frac{T_{\mathrm{int}}}{T} \tag{A.38}$$

It follows that $\overline{(\tilde{\xi}(T) - a)^2} \to 0$, and hence $\tilde{\xi}(T) \to a$, for $T \to \infty$. Furthermore, (A.38) shows that in practice an average over a finite time interval is a good approximation provided that $T \gg T_{\mathrm{int}}$.

### Spectral Description

A periodic, real function $f(t)$ can be expressed by a Fourier series

$$f(t) = \sum_{n=0}^{N}(c_n \mathrm{e}^{\mathrm{i}\omega_n t} + c_n^* \mathrm{e}^{-\mathrm{i}\omega_n t})$$

where $\omega_n = 2\pi n/T$, $T$ denotes the interval of periodicity and $c_n$ complex Fourier components. $c_0 = 0$ if the mean of $f$ over the period $T$ vanishes. It is clear that for arbitrary $f$ it follows that $N \to \infty$, and for nonperiodic $f$ it follows that $T \to \infty$ and that the description converges to a Fourier integral. The following expression is equivalent but more convenient

$$f(t) = \sum_{n=-N}^{N} c_n \mathrm{e}^{\mathrm{i}\omega_n t}$$

with $c_{-n} = c_n^*$ and $\omega_{-n} = -\omega_n$. Random functions $\xi(t)$ can be expressed in a similar way as a Fourier series

$$\xi(t) = \sum_{n=-N}^{N} \eta_n \mathrm{e}^{\mathrm{i}\omega_n t}$$

where the amplitudes $\eta_n$ are now random variables, but the frequencies $\omega_n$ remain deterministic as before. The mean is given by

$$\overline{\xi(t)} = \sum_{n=-N}^{N} \overline{\eta_n} e^{i\omega_n t}$$

For a stationary random function $\xi(t)$, the mean has to be constant in time which is the case only if $\overline{\eta_n} = 0$. For the covariance function

$$R(\tau) = \overline{\xi(t)\xi(t+\tau)} = \sum_{n,m} \overline{\eta_n \eta_m} e^{i(\omega_n+\omega_m)t} e^{i\omega_m \tau}$$

holds; thus $R(\tau)$ does not depend on time if $\overline{\eta_n \eta_m} = \overline{\eta_n \eta_m}\, \delta_{n,-m}$, thus

$$R(\tau) = \sum_{n} \overline{\eta_n \eta_n^*} e^{-i\omega_n \tau}$$

Introducing the definition $\overline{\eta_n \eta_n^*} = E_n$ and taking the limit towards a Fourier integral (not explicitly discussed here), one obtains finally

$$R(\tau) = \int_{-\infty}^{\infty} E(\omega) e^{-i\omega\tau} d\omega$$

$E(\omega)$ denotes the spectral density of $\xi(t)$, or energy spectrum[5] and is obviously always positive. On the other hand

$$E(\omega) = \frac{1}{2\pi} \int_{-\infty}^{\infty} R(\tau) e^{i\omega\tau} d\tau$$

which means that the energy spectrum can be computed from the covariance function which is actually often the way how $E(\omega)$ is obtained numerically. If $\xi$ depends on space and if $\xi(\boldsymbol{x})$ is homogeneous, one obtains

$$\overline{\xi(\boldsymbol{x})\xi(\boldsymbol{x}+\boldsymbol{r})} = R(\boldsymbol{r}) = \int E(\boldsymbol{k}) e^{i\boldsymbol{k}\cdot\boldsymbol{r}} d\boldsymbol{k}$$

where the energy spectrum $E(\boldsymbol{k})$ with respect to the wave number $\boldsymbol{k}$ appears. Likewise, for homogeneous vector random functions

$$\overline{\xi_i(\boldsymbol{x})\xi_j(\boldsymbol{x}+\boldsymbol{r})} = \mathsf{R}_{ij}(\boldsymbol{r}) = \int \mathsf{E}_{ij}(\boldsymbol{k}) e^{i\boldsymbol{k}\cdot\boldsymbol{r}} d\boldsymbol{k}$$

holds, where $\mathsf{E}_{ij}(\boldsymbol{k})$ denotes the spectral tensor.

## A.4 Coordinate Systems

We briefly outline the mathematics of curvilinear coordinates and exemplify them by introducing oblate spherical and spherical coordinates.

[5] In GFD applications, $\int E(\omega) d\omega$ is usually closely related to a form of physical energy.

### A.4.1 General Curvilinear Orthogonal Coordinates

A transformation from Cartesian coordinates $(x, y, z)$ to curvilinear coordinates $(\lambda, \psi, \mu)$ is defined by

$$\begin{aligned} x &= x(\lambda, \psi, \mu) \\ y &= y(\lambda, \psi, \mu) \\ z &= z(\lambda, \psi, \mu) \end{aligned} \tag{A.39}$$

Only *orthogonal* coordinates are considered where the surfaces $\lambda = \text{const}$, $\psi = \text{const}$ and $\mu = \text{const}$ are *locally* mutually orthogonal. To relate small changes $\mathrm{d}\boldsymbol{x} = (\mathrm{d}x, \mathrm{d}y, \mathrm{d}z)$ to changes $\mathrm{d}\boldsymbol{\lambda} = (\mathrm{d}\lambda, \mathrm{d}\psi, \mathrm{d}\mu)$, differentiation of (A.39) results in

$$\mathrm{d}\boldsymbol{x} = \boldsymbol{D} \cdot \mathrm{d}\boldsymbol{\lambda} = \begin{pmatrix} \frac{\partial x}{\partial \lambda} & \frac{\partial x}{\partial \psi} & \frac{\partial x}{\partial \mu} \\ \frac{\partial y}{\partial \lambda} & \frac{\partial y}{\partial \psi} & \frac{\partial y}{\partial \mu} \\ \frac{\partial z}{\partial \lambda} & \frac{\partial z}{\partial \psi} & \frac{\partial z}{\partial \mu} \end{pmatrix} \cdot \mathrm{d}\boldsymbol{\lambda} \tag{A.40}$$

From (A.40) it follows that the length of a line element is given by

$$(\mathrm{d}\boldsymbol{x})^2 = (\mathrm{d}x)^2 + (\mathrm{d}y)^2 + (\mathrm{d}z)^2 = \mathrm{d}\boldsymbol{\lambda} \cdot \boldsymbol{D}^\dagger \cdot \boldsymbol{D} \cdot \mathrm{d}\boldsymbol{\lambda}$$

where $\boldsymbol{D}^\dagger$ is the transposed of $\boldsymbol{D}$. For the transformation to be orthogonal, this expression must be of the form $(\mathrm{d}x)^2 + (\mathrm{d}y)^2 + (\mathrm{d}z)^2 = h_\lambda^2(\mathrm{d}\lambda)^2 + h_\psi^2(\mathrm{d}\psi)^2 + h_\mu^2(\mathrm{d}\mu)^2$. Hence $\boldsymbol{D}^\dagger \cdot \boldsymbol{D}$ must be a diagonal matrix, with the diagonal elements

$$\begin{aligned} h_\lambda^2 &= \left(\frac{\partial x}{\partial \lambda}\right)^2 + \left(\frac{\partial y}{\partial \lambda}\right)^2 + \left(\frac{\partial z}{\partial \lambda}\right)^2 \\ h_\psi^2 &= \left(\frac{\partial x}{\partial \psi}\right)^2 + \left(\frac{\partial y}{\partial \psi}\right)^2 + \left(\frac{\partial z}{\partial \psi}\right)^2 \\ h_\mu^2 &= \left(\frac{\partial x}{\partial \mu}\right)^2 + \left(\frac{\partial y}{\partial \mu}\right)^2 + \left(\frac{\partial z}{\partial \mu}\right)^2 \end{aligned} \tag{A.41}$$

The normalized matrix $\boldsymbol{T}$ is defined as

$$\boldsymbol{T} = \begin{pmatrix} \frac{1}{h_\lambda}\frac{\partial x}{\partial \lambda} & \frac{1}{h_\psi}\frac{\partial x}{\partial \psi} & \frac{1}{h_\mu}\frac{\partial x}{\partial \mu} \\ \frac{1}{h_\lambda}\frac{\partial y}{\partial \lambda} & \frac{1}{h_\psi}\frac{\partial y}{\partial \psi} & \frac{1}{h_\mu}\frac{\partial y}{\partial \mu} \\ \frac{1}{h_\lambda}\frac{\partial z}{\partial \lambda} & \frac{1}{h_\psi}\frac{\partial z}{\partial \psi} & \frac{1}{h_\mu}\frac{\partial z}{\partial \mu} \end{pmatrix} \tag{A.42}$$

Obviously, $\boldsymbol{T}^\dagger \cdot \boldsymbol{T}$ is also diagonal, and its diagonal elements are unity. Hence $\boldsymbol{T}$ is *orthogonal*, i. e. its inverse is identical with the transposed matrix, i. e. $\boldsymbol{T}^{-1} = \boldsymbol{T}^\dagger$. The columns of $\boldsymbol{T}$ are the unit vectors $\boldsymbol{e}_\lambda, \boldsymbol{e}_\psi, \boldsymbol{e}_\mu$ in the new coordinate system. Note that the direction of the unit vectors is not constant. Any vector can be represented as

$$\begin{aligned} &\boldsymbol{u} = u_\lambda \boldsymbol{e}_\lambda + u_\psi \boldsymbol{e}_\psi + u_\mu \boldsymbol{e}_\mu \\ &\text{with} \quad u_\lambda = \boldsymbol{u} \cdot \boldsymbol{e}_\lambda\,, \quad u_\psi = \boldsymbol{u} \cdot \boldsymbol{e}_\psi\,, \quad u_\mu = \boldsymbol{u} \cdot \boldsymbol{e}_\mu \end{aligned} \tag{A.43}$$

The scaling factors $h_\lambda, h_\psi, h_\mu$ define the length of a line element result from a small change $\mathrm{d}\boldsymbol{\lambda}$ in the new coordinates. With the unit vectors, (A.40) can be written as

$\mathrm{d}\boldsymbol{x} = h_\lambda \boldsymbol{e}_\lambda \mathrm{d}\lambda + h_\psi \boldsymbol{e}_\psi \mathrm{d}\psi + h_\mu \boldsymbol{e}_\mu \mathrm{d}\mu$. A volume element is given as $\mathrm{d}x\mathrm{d}y\mathrm{d}z = h_\lambda h_\psi h_\mu \mathrm{d}\lambda \mathrm{d}\psi \mathrm{d}\mu$. The gradient of a scalar $\chi$ takes the form

$$\nabla\chi = \frac{\boldsymbol{e}_\lambda}{h_\lambda}\frac{\partial\chi}{\partial\lambda} + \frac{\boldsymbol{e}_\psi}{h_\psi}\frac{\partial\chi}{\partial\psi} + \frac{\boldsymbol{e}_\mu}{h_\mu}\frac{\partial\chi}{\partial\mu} \tag{A.44}$$

The expressions for divergence, curl, and Laplace operator are given as

$$\nabla\cdot\boldsymbol{u} = \frac{1}{h_\lambda h_\psi h_\mu}\left[\frac{\partial u_\lambda h_\psi h_\mu}{\partial\lambda} + \frac{\partial u_\psi h_\lambda h_\mu}{\partial\psi} + \frac{\partial u_\mu h_\lambda h_\psi}{\partial\mu}\right] \tag{A.45}$$

$$\begin{aligned}\nabla\times\boldsymbol{u} = {} & \frac{\boldsymbol{e}_\lambda}{h_\psi h_\mu}\left[\frac{\partial h_\mu u_\mu}{\partial\psi} - \frac{\partial h_\psi u_\psi}{\partial\mu}\right] + \frac{\boldsymbol{e}_\psi}{h_\mu h_\lambda}\left[\frac{\partial h_\lambda u_\lambda}{\partial\mu} - \frac{\partial h_\mu u_\mu}{\partial\lambda}\right] \\ & + \frac{\boldsymbol{e}_\mu}{h_\lambda h_\psi}\left[\frac{\partial h_\psi u_\psi}{\partial\lambda} - \frac{\partial h_\lambda u_\lambda}{\partial\psi}\right]\end{aligned} \tag{A.46}$$

$$\nabla^2\chi = \frac{1}{h_\lambda h_\psi h_\mu}\left[\frac{\partial}{\partial\lambda}\left(\frac{h_\psi h_\mu}{h_\lambda}\frac{\partial\chi}{\partial\lambda}\right) + \frac{\partial}{\partial\psi}\left(\frac{h_\mu h_\lambda}{h_\psi}\frac{\partial\chi}{\partial\psi}\right) + \frac{\partial}{\partial\mu}\left(\frac{h_\lambda h_\psi}{h_\mu}\frac{\partial\chi}{\partial\mu}\right)\right] \tag{A.47}$$

The substantial derivative $\mathrm{D}/\mathrm{D}t = \partial/\partial t + \boldsymbol{u}\cdot\nabla$ as appearing in scalar equations is easily expressed in the new coordinates. With (A.44), one obtains

$$\frac{\mathrm{D}}{\mathrm{D}t} = \frac{\partial}{\partial t} + \frac{1}{h_\lambda}\frac{\partial}{\partial\lambda} + \frac{1}{h_\psi}\frac{\partial}{\partial\psi} + \frac{1}{h_\mu}\frac{\partial}{\partial\mu} \tag{A.48}$$

In vector equations such as e. g. the momentum equations, the substantial derivative involves a vector gradient which also needs to be expressed in the new coordinates. By projection of the vector expression $\mathrm{D}\boldsymbol{u}/\mathrm{D}t$ onto the new coordinate axes, one obtains e. g. for the $\mu$–direction

$$\boldsymbol{e}_\mu\cdot\frac{\mathrm{D}}{\mathrm{D}t}\boldsymbol{u} = \boldsymbol{e}_\mu\cdot\frac{\partial\boldsymbol{u}}{\partial t} + \frac{\boldsymbol{e}_\mu}{h_\lambda}\cdot\frac{\partial\boldsymbol{u}}{\partial\lambda} + \frac{\boldsymbol{e}_\mu}{h_\psi}\cdot\frac{\partial\boldsymbol{u}}{\partial\psi} + \frac{\boldsymbol{e}_\mu}{h_\mu}\cdot\frac{\partial\boldsymbol{u}}{\partial\mu} \tag{A.49}$$

The scalar products in (A.49) are evaluated by partial differentiation, e. g. the last term in (A.49) according to

$$\boldsymbol{e}_\mu\cdot\frac{\partial\boldsymbol{u}}{\partial\lambda} = \frac{\partial}{\partial\lambda}(\boldsymbol{e}_\mu\cdot\boldsymbol{u}) - \boldsymbol{u}\cdot\frac{\partial\boldsymbol{e}_\mu}{\partial\lambda}$$

With $\boldsymbol{e}_\mu\cdot\partial\boldsymbol{u}/\partial t = \partial(\boldsymbol{e}_\mu\cdot\boldsymbol{u})/\partial t = \partial u_\mu/\partial t$, one obtains

$$\boldsymbol{e}_\mu\cdot\frac{\mathrm{D}}{\mathrm{D}t}\boldsymbol{u} = \frac{\mathrm{D}u_\mu}{\mathrm{D}t} - \left(\frac{\boldsymbol{u}}{h_\lambda}\cdot\frac{\partial\boldsymbol{e}_\mu}{\partial\lambda} + \frac{\boldsymbol{u}}{h_\psi}\cdot\frac{\partial\boldsymbol{e}_\mu}{\partial\psi} + \frac{\boldsymbol{u}}{h_\mu}\cdot\frac{\partial\boldsymbol{e}_\mu}{\partial\mu}\right) \tag{A.50}$$

and likewise for the other two components. The terms in brackets in (A.50) – so-called metric terms – originate from the change in unit vectors with position. Since the unit vectors are constructed from (A.42), the dependency $\boldsymbol{e}_\mu = \boldsymbol{e}_\mu(\lambda,\psi,\mu)$ is known. Hence the partial derivatives of $\boldsymbol{e}_\mu$ can easily be evaluated and expressed again in terms of all unit vectors with (A.43). The scalar products with $\boldsymbol{u}$ can then be expressed in terms of the components $u_\lambda, u_\psi, u_\mu$.

## A.4.2 Oblate Spheroidal Coordinates

Oblate spheroidal coordinates can be defined as

$$\begin{aligned} x &= (c^2+\mu^2)^{\frac{1}{2}} \cos\psi \cos\lambda \\ y &= (c^2+\mu^2)^{\frac{1}{2}} \cos\psi \sin\lambda \\ z &= \mu \sin\psi \end{aligned} \tag{A.51}$$

Here, $\lambda$ is the eastward longitude, the surfaces $\psi = \text{const}$ are hyperboloids, and the surfaces $\mu = \text{const}$ are oblate spheroids (rotation ellipsoids; see Figure 4.2). It is useful to introduce the abbreviations

$$\begin{aligned} A(\psi,\mu) &= (c^2 \sin^2\psi + \mu^2)^{\frac{1}{2}} \\ B(\mu) &= (c^2+\mu^2)^{\frac{1}{2}} \\ C(\psi,\mu) &= \frac{c^2 \sin\psi \cos\psi}{BA^2} \end{aligned} \tag{A.52}$$

The matrix $\boldsymbol{D}$ of derivatives according to (A.40) is computed from (A.51) as

$$\boldsymbol{D} = \begin{pmatrix} -B\cos\psi\ \sin\lambda & -B\sin\psi\ \cos\lambda & (\mu/B)\cos\psi\ \cos\lambda \\ B\cos\psi\ \cos\lambda & -B\sin\psi\ \sin\lambda & (\rho/B)\cos\psi\ \sin\lambda \\ 0 & \mu\cos\psi & \sin\psi \end{pmatrix} \tag{A.53}$$

The scale factors then follow from (A.53) and the definition (A.41) as $h_\lambda = B\cos\psi$, $h_\psi = A$, $h_\mu = A/B$ so that a volume element is given as $A^2\cos\psi \mathrm{d}\lambda \mathrm{d}\psi \mathrm{d}\mu$. With the scale factors and (A.53), the matrix $\boldsymbol{T}$ can be determined according to (A.42), and from the columns of $\boldsymbol{T}$ one finds the unit vectors in the three coordinate directions as

$$\boldsymbol{e}_\lambda = \begin{pmatrix} -\sin\lambda \\ \cos\lambda \\ 0 \end{pmatrix}, \quad \boldsymbol{e}_\psi = \begin{pmatrix} -(B/A)\sin\psi\ \cos\lambda \\ -(B/A)\sin\psi\ \sin\lambda \\ (\mu/A)\cos\psi \end{pmatrix}, \quad \boldsymbol{e}_\mu = \begin{pmatrix} (\mu/A)\cos\psi\ \cos\lambda \\ (\mu/A)\cos\psi\ \sin\lambda \\ (B/A)\sin\psi \end{pmatrix} \tag{A.54}$$

The unit vectors are mutually orthogonal but change their direction in space, according to

$$\begin{aligned} &\frac{\partial}{\partial\lambda}\boldsymbol{e}_\lambda = \frac{B}{A}\sin\psi\boldsymbol{e}_\psi - \frac{\mu}{A}\cos\psi\boldsymbol{e}_\mu\,, && \frac{\partial}{\partial\psi}\boldsymbol{e}_\lambda = 0\,, && \frac{\partial}{\partial\mu}\boldsymbol{e}_\lambda = 0 \\ &\frac{\partial}{\partial\lambda}\boldsymbol{e}_\psi = -\frac{B}{A}\sin\psi\boldsymbol{e}_\lambda\,, && \frac{\partial}{\partial\psi}\boldsymbol{e}_\psi = -\frac{\mu B}{A^2}\boldsymbol{e}_\mu\,, && \frac{\partial}{\partial\mu}\boldsymbol{e}_\psi = C\boldsymbol{e}_\mu \\ &\frac{\partial}{\partial\lambda}\boldsymbol{e}_\mu = \frac{\rho}{A}\cos\psi\boldsymbol{e}_\lambda\,, && \frac{\partial}{\partial\psi}\boldsymbol{e}_\mu = \frac{\mu B}{A^2}\boldsymbol{e}_\psi\,, && \frac{\partial}{\partial\mu}\boldsymbol{e}_\mu = -C\boldsymbol{e}_\psi \end{aligned} \tag{A.55}$$

where $A, B, C$ are given by (A.52). From (A.44), the gradient differential operator is obtained as

$$\nabla = \frac{\boldsymbol{e}_\lambda}{B\cos\psi}\frac{\partial}{\partial\lambda} + \frac{\boldsymbol{e}_\psi}{A}\frac{\partial}{\partial\psi} + \frac{\boldsymbol{e}_\mu}{A/B}\frac{\partial}{\partial\mu} \tag{A.56}$$

With (A.43), the vector $\boldsymbol{u}$ is expressed in terms of its components $u_\lambda, u_\psi, u_\mu$. Then, divergence, curl, and Laplacian are obtained from (A.45)–(A.47) as

$$\nabla \cdot \boldsymbol{u} = \frac{1}{B\cos\psi}\frac{\partial u_\lambda}{\partial \lambda} + \frac{1}{A^2\cos\psi}\frac{\partial}{\partial\psi}(u_\psi A\cos\psi) + \frac{1}{A^2}\frac{\partial}{\partial\mu}(u_\mu AB) \tag{A.57}$$

$$\begin{aligned}\nabla \times \boldsymbol{u} = {} & \frac{\boldsymbol{e}_\lambda}{A^2/B}\left[\frac{\partial}{\partial\psi}(u_\mu A/B) - \frac{\partial}{\partial\mu}(u_\psi A)\right] \\ & + \frac{\boldsymbol{e}_\psi}{A\cos\psi}\left[\cos\psi\frac{\partial}{\partial\mu}(u_\lambda B) - \frac{A}{B}\frac{\partial u_\mu}{\partial\lambda}\right] \\ & + \frac{\boldsymbol{e}_\mu}{AB\cos\psi}\left[A\frac{\partial u_\psi}{\partial\lambda} - \frac{\partial}{\partial\psi}(u_\lambda B\cos\psi)\right]\end{aligned} \tag{A.58}$$

$$\nabla^2\chi = \frac{1}{B^2\cos^2\psi}\frac{\partial^2\chi}{\partial\lambda^2} + \frac{1}{A^2\cos\psi}\frac{\partial}{\partial\psi}\left(\cos\psi\frac{\partial\chi}{\partial\psi}\right) + \frac{1}{A^2}\frac{\partial}{\partial\mu}\left(B^2\frac{\partial\chi}{\partial\mu}\right) \tag{A.59}$$

The substantial derivative $\mathrm{D}/\mathrm{D}t = \partial/\partial t + \boldsymbol{u}\cdot\nabla$ is given as

$$\frac{\mathrm{D}}{\mathrm{D}t} = \frac{\partial}{\partial t} + \frac{u_\lambda}{B\cos\psi}\frac{\partial}{\partial\lambda} + \frac{u_\psi}{A}\frac{\partial}{\partial\psi} + \frac{u_\mu}{A/B}\frac{\partial}{\partial\mu} \tag{A.60}$$

Projection of the substantial derivative in the equations of motion onto the new coordinates according to (A.50) with the use of (A.55) results in

$$\boldsymbol{e}_\lambda \cdot \frac{\mathrm{D}}{\mathrm{D}t}\boldsymbol{u} = \frac{\mathrm{D}}{\mathrm{D}t}u_\lambda - \frac{u_\lambda u_\psi}{A}\tan\psi + \frac{\mu u_\lambda u_\mu}{AB} \tag{A.61}$$

$$\boldsymbol{e}_\psi \cdot \frac{\mathrm{D}}{\mathrm{D}t}\boldsymbol{u} = \frac{\mathrm{D}}{\mathrm{D}t}u_\psi + \frac{u_\lambda^2}{A}\tan\psi + \frac{\mu u_\psi u_\mu}{A^2} - \frac{CBu_\mu^2}{A} \tag{A.62}$$

$$\boldsymbol{e}_\mu \cdot \frac{\mathrm{D}}{\mathrm{D}t}\boldsymbol{u} = \frac{\mathrm{D}}{\mathrm{D}t}u_\mu - \frac{\mu}{AB}u_\lambda^2 - \frac{\mu}{A^2}u_\psi^2 + \frac{CB}{A}u_\psi u_\mu \tag{A.63}$$

with $\mathrm{D}/\mathrm{D}t$ from (A.60). The projection of the Coriolis acceleration $2\boldsymbol{\Omega}\times\boldsymbol{u}$ is found as

$$2\boldsymbol{\Omega}\times\boldsymbol{u} = 2\begin{pmatrix}\boldsymbol{e}_\lambda\cdot\Omega\times\boldsymbol{u}\\ \boldsymbol{e}_\psi\cdot\Omega\times\boldsymbol{u}\\ \boldsymbol{e}_\mu\cdot\Omega\times\boldsymbol{u}\end{pmatrix} = 2\Omega\begin{pmatrix}-(B/A)u_\psi\sin\psi + (\mu/A)u_\mu\cos\psi\\ (B/A)u_\lambda\sin\psi\\ -(\mu/A)u_\lambda\cos\psi\end{pmatrix} \tag{A.64}$$

### A.4.3 Spherical Coordinates

Spherical coordinates are defined by

$$\begin{aligned} x &= r\cos\varphi\cos\lambda \\ y &= r\cos\varphi\sin\lambda \\ z &= r\sin\varphi \end{aligned} \tag{A.65}$$

Here, $\lambda$ is the eastward positive longitude, $\varphi = \pi - \vartheta$ the northward positive latitude ($\vartheta$ is the pole distance, counted from the North Pole), and $r$ is the radial distance from the Earth's center (see Figure 4.3). All properties of the spherical coordinate system follow from the general case shown in Section A.4.1. By differentiation of (A.65), the matrix $\boldsymbol{D}$ of derivatives according to (A.40) is obtained as

$$\boldsymbol{D} = \begin{pmatrix} -r\cos\varphi\sin\lambda & -r\sin\varphi\cos\lambda & \cos\varphi\cos\lambda \\ r\cos\varphi\cos\lambda & -r\sin\varphi\sin\lambda & \cos\varphi\sin\lambda \\ 0 & r\cos\varphi & \sin\varphi \end{pmatrix}$$

The scale factors then follow as $h_\lambda = r\cos\varphi$, $h_\varphi = r$, $h_r = 1$ so that a volume element is $r^2\cos\varphi \mathrm{d}\lambda\mathrm{d}\varphi\mathrm{d}r$. The unit vectors in the corresponding coordinate directions are

$$\boldsymbol{e}_\lambda = \begin{pmatrix} -\sin\lambda \\ \cos\lambda \\ 0 \end{pmatrix}, \quad \boldsymbol{e}_\varphi = \begin{pmatrix} -\sin\varphi\cos\lambda \\ -\sin\varphi\sin\lambda \\ \cos\varphi \end{pmatrix}, \quad \boldsymbol{e}_r = \begin{pmatrix} \cos\varphi\cos\lambda \\ \cos\varphi\sin\lambda \\ \sin\varphi \end{pmatrix} \tag{A.66}$$

Their change with coordinate direction is

$$\begin{aligned}
&\frac{\partial}{\partial r}\boldsymbol{e}_r = 0\,, && \frac{\partial}{\partial r}\boldsymbol{e}_\varphi = 0\,, && \frac{\partial}{\partial r}\boldsymbol{e}_\lambda = 0 \\
&\frac{\partial}{\partial \varphi}\boldsymbol{e}_r = \boldsymbol{e}_\varphi\,, && \frac{\partial}{\partial \varphi}\boldsymbol{e}_\varphi = -\boldsymbol{e}_r\,, && \frac{\partial}{\partial \varphi}\boldsymbol{e}_\lambda = 0 \\
&\frac{\partial}{\partial \lambda}\boldsymbol{e}_r = \cos\varphi\boldsymbol{e}_\lambda\,, && \frac{\partial}{\partial \lambda}\boldsymbol{e}_\varphi = -\sin\varphi\boldsymbol{e}_\lambda\,, && \frac{\partial}{\partial \lambda}\boldsymbol{e}_\lambda = -\cos\varphi\boldsymbol{e}_r + \sin\varphi\boldsymbol{e}_\varphi
\end{aligned} \tag{A.67}$$

The gradient differential operator takes the form

$$\nabla\chi = \boldsymbol{e}_r\frac{\partial\chi}{\partial r} + \frac{1}{r}\boldsymbol{e}_\varphi\frac{\partial\chi}{\partial\varphi} + \frac{1}{r\cos\varphi}\boldsymbol{e}_\lambda\frac{\partial\chi}{\partial\lambda}$$

When we express the vector $\boldsymbol{u}$ by its components $(u_\lambda, u_\varphi, u_r) \equiv (u, v, w)$, the vector operators divergence and rotation and the Laplace operator are obtained as

$$\begin{aligned}
\nabla\cdot\boldsymbol{u} &= \frac{1}{r\cos\varphi}\left(\frac{\partial u}{\partial\lambda} + \frac{\partial\cos\varphi\, v}{\partial\varphi}\right) + \frac{1}{r^2}\frac{\partial r^2 w}{\partial r} \\
\nabla\times\boldsymbol{u} &= \frac{1}{r^2\cos\varphi}\left[r\cos\varphi\left(\frac{\partial w}{\partial\varphi} - \frac{\partial vr}{\partial r}\right)\boldsymbol{e}_\lambda + r\left(\frac{\partial ur\cos\varphi}{\partial r} - \frac{\partial w}{\partial\lambda}\right)\boldsymbol{e}_\varphi\right. \\
&\quad \left. + \left(\frac{\partial vr}{\partial\lambda} - \frac{\partial ur\cos\varphi}{\partial\varphi}\right)\boldsymbol{e}_r\right] \\
\nabla^2\chi &= \frac{1}{r^2\cos^2\varphi}\left(\frac{\partial^2\chi}{\partial\lambda^2} + \cos\varphi\frac{\partial}{\partial\varphi}\cos\varphi\frac{\partial\chi}{\partial\varphi}\right) + \frac{1}{r^2}\frac{\partial}{\partial r}r^2\frac{\partial\chi}{\partial r}
\end{aligned}$$

The substantial derivative $\mathrm{D}/\mathrm{D}t = \partial/\partial t + \boldsymbol{u}\cdot\nabla$ is given as

$$\frac{\mathrm{D}}{\mathrm{D}t} = \frac{\partial}{\partial t} + \frac{u}{\cos\varphi}\frac{\partial}{\partial\lambda} + \frac{v}{r}\frac{\partial}{\partial\varphi} + w\frac{\partial}{\partial r} \tag{A.68}$$

Projection of the substantial derivative in the equations of motion for $u, v, w$ onto spherical coordinates follows the derivation in Section A.4.1 and leads to

$$\boldsymbol{e}_\lambda\cdot\frac{\mathrm{D}}{\mathrm{D}t}\boldsymbol{u} = \frac{\mathrm{D}u}{\mathrm{D}t} + \frac{uw}{r} - \frac{uv}{r}\tan\varphi \tag{A.69}$$

$$\boldsymbol{e}_\varphi\cdot\frac{\mathrm{D}}{\mathrm{D}t}\boldsymbol{u} = \frac{\mathrm{D}v}{\mathrm{D}t} + \frac{wv}{r} + \frac{u^2}{r}\tan\varphi \tag{A.70}$$

$$\boldsymbol{e}_r\cdot\frac{\mathrm{D}}{\mathrm{D}t}\boldsymbol{u} = \frac{\mathrm{D}w}{\mathrm{D}t} - \frac{u^2+v^2}{r} \tag{A.71}$$

with $\mathrm{D}/\mathrm{D}t$ from (A.68). The projection of the Coriolis acceleration $2\boldsymbol{\Omega} \times \boldsymbol{u}$ is found as

$$2\boldsymbol{\Omega} \times \boldsymbol{u} = 2\begin{pmatrix} \boldsymbol{e}_\lambda \cdot \Omega \times \boldsymbol{u} \\ \boldsymbol{e}_\varphi \cdot \Omega \times \boldsymbol{u} \\ \boldsymbol{e}_r \cdot \Omega \times \boldsymbol{u} \end{pmatrix} = 2\Omega \begin{pmatrix} -v \sin\varphi + w \cos\varphi \\ u \sin\varphi \\ -u \cos\varphi \end{pmatrix} \tag{A.72}$$

# B Models of the Ocean Circulation

In this book, the field of numerical ocean modeling has not been touched, and we recommend the excellent book *Fundamentals of Ocean Climate Models* by Griffies (2004) for an overview on this issue. We have, however, explicitly considered several "models", i. e. systems of equations based on certain approximations suitable to compute solutions. Models based on isopycnal coordinates are standard in oceanography and are discussed briefly. The BARBI model is based on a different representation of the vertical coordinate and is introduced in more detail. Furthermore, a brief overview about model types with a small number of degrees of freedom is given.

## B.1 Models Based on Isopycnal Coordinates

Advection in the ocean occurs mainly along density surfaces, and it is, therefore, often useful to work with a coordinate system where isopycnal surfaces are coordinate surfaces. This coordinate can be continuous, or be represented by a series of layers of constant density.

### B.1.1 Equations of Motion in Isopycnal Coordinates

The introduction of locally orthogonal coordinates aligned to isopycnal surfaces is straightforward, if somewhat cumbersome (see e. g. Chapter 6 of Griffies, 2004). However, isopycnal coordinates are traditionally obtained in a different way, by simply replacing the vertical coordinate $z$ by the density $\rho$, and result in a coordinate system which is not orthogonal.

Ignoring any complications due of the nonlinear equation of state, the appropriate density variable $\rho$ is the potential density which satisfies the budget

$$\frac{\partial \rho}{\partial t} + \boldsymbol{u} \cdot \nabla \rho + w \frac{\partial \rho}{\partial z} = \mathcal{G}_\rho \tag{B.1}$$


DOI 10.1007/978-3-642-23450-7_18, 

Here $\boldsymbol{u}$ and $\nabla = (\partial/\partial x, \partial/\partial y)$ are horizontal velocity and horizontal gradient, and the diabatic term $\mathcal{G}_\rho$ collects small-scale turbulent mixing and other irreversible changes of density. When transforming from Cartesian coordinates $(\boldsymbol{x}, z, t)$ to isopycnal coordinates $(\boldsymbol{x}, \rho, t)$, the isopycnal $\rho(\boldsymbol{x}, z, t) = \text{const}$ is characterized by its depth $z = z(\boldsymbol{x}, t, \rho)$. The gradients (and other derivatives) of any field $\chi(\boldsymbol{x}, z, t) = \chi(\boldsymbol{x}, z(\boldsymbol{x}, \rho, t), t) = \chi(\boldsymbol{x}, \rho, t)$, taken either on a geopotential ($z = \text{const}$) or on an isopycnal ($\rho = \text{const}$), are related by

$$\frac{\partial \chi}{\partial t_\rho} = \frac{\partial \chi}{\partial t} + \frac{\partial \chi}{\partial z}\frac{\partial}{\partial t_\rho} z\,, \quad \nabla_\rho \chi = \nabla \chi + \frac{\partial \chi}{\partial z}\nabla_\rho z \quad \text{and} \quad \frac{\partial \chi}{\partial z} = \frac{\partial \chi}{\partial \rho}\frac{\partial \rho}{\partial z} \tag{B.2}$$

Here the time derivative $\partial/\partial t_\rho$ and lateral gradient $\nabla_\rho = (\partial/\partial x_\rho, \partial/\partial y_\rho)$ are taken holding $\rho = \text{const}$. For the pressure $p$ one finds

$$\nabla p = \nabla_\rho p - \frac{\partial p}{\partial z}\nabla_\rho z = \nabla_\rho (p + g\rho z) \equiv \nabla_\rho M$$

using the hydrostatic relation $\partial p/\partial z = -g\rho$. The *Montgomery potential* $M = p + g\rho z$ serves as a stream function for a geostrophically balanced flow on isopycnals. Note that $\partial M/\partial \rho = gz$ replaces the hydrostatic relation in isopycnal coordinates. Applying (B.2) to the velocity components $\boldsymbol{u} = (u, v)$ yields

$$\left(\frac{\partial}{\partial t} + \boldsymbol{u}\cdot\nabla\right) u = \left(\frac{\partial}{\partial t_\rho} + \boldsymbol{u}\cdot\nabla_\rho\right) u - \frac{\partial u}{\partial z}\left(\frac{\partial}{\partial t_\rho} + \boldsymbol{u}\cdot\nabla_\rho\right) z \tag{B.3}$$

and likewise for $v$. Using (B.1) one further obtains

$$\left(\frac{\partial}{\partial t_\rho} + \boldsymbol{u}\cdot\nabla_\rho\right) z = w - h_\rho \mathcal{G}_\rho \tag{B.4}$$

The quantity $h_\rho = -\partial z/\partial \rho$ (note that $h_\rho > 0$ for stable stratification) is called the (specific) *isopycnal thickness* since the layer between two infinitesimally close isopycnals with densities $\rho$ and $\rho + \Delta\rho$ has the thickness $h_\rho \Delta\rho$. With (B.3), (B.4), (16.2) and (16.3), the momentum balance in isopycnal coordinates follows as

$$\left(\frac{\partial}{\partial t_\rho} + \boldsymbol{u}\cdot\nabla_\rho + \mathcal{G}_\rho \frac{\partial}{\partial \rho}\right)\boldsymbol{u} + f\,\underline{\boldsymbol{u}}_{\neg} = -\nabla_\rho M - \frac{1}{h_\rho}\frac{\partial \boldsymbol{\tau}}{\partial \rho} + \boldsymbol{\mathcal{F}} \tag{B.5}$$

The density budget (B.1) and the continuity equation (16.4) are combined to the isopycnal thickness budget

$$\frac{\partial h_\rho}{\partial t_\rho} + \nabla_\rho \cdot (h_\rho \boldsymbol{u}) + \frac{\partial (h_\rho \mathcal{G}_\rho)}{\partial \rho} = 0 \tag{B.6}$$

As remarked above, the coordinate systems $(\boldsymbol{x}, \rho)$ obviously is not orthogonal, a fact which is relevant for the interpretation of the transformed equations. In particular, the interpretation of the density source $\mathcal{G}_\rho$ as diapycnal 'velocity' (in physical space, that velocity is $h_\rho \mathcal{G}_\rho$), which is suggested by (B.5) and (B.6), is not unique since the lateral velocity $\boldsymbol{u}$ is still along geopotentials, and, therefore, in general it has a component across isopycnal surfaces. Only in the limiting (and mainly interesting) case of small isopycnal slopes, $|\nabla_\rho z| \ll 1$, the coordinates become approximately orthogonal.

### B.1.2 Layer Models

The principal idea in isopycnal layer models is to represent the density stratification by a finite number of layers of constant density. While not necessarily a good approximation to the oceanic situation, such a system can be physically realized, at least in principle. To avoid complications, we will assume that the layers do not intersect each other and with the ocean floor.

Each layer (index $n$) is bounded by the interfaces $z = -d_n$ below and $z = -d_{n-1}$ above, and is characterized by a constant density $\rho_n$, thickness $h_n = d_n - d_{n-1}$, pressure $p_n$, horizontal velocity $\boldsymbol{u}_n$, assumed to be vertically constant within the layer, and vertical velocity $w_n$ (see the illustration in Figure B.1). Integration of the hydrostatic equation $\partial p_n / z = -g\rho_n$ yields

$$p_n(\boldsymbol{x}, z) = M_n(\boldsymbol{x}) - g\rho_n z \tag{B.7}$$

where $M_n$ is the discrete form of the Montgomery potential introduced in the previous section. Continuity of pressure at the interface $z = -d_n$ requires

$$M_{n+1} - M_n = -g(\rho_{n+1} - \rho_n) d_n \tag{B.8}$$

Note that $\nabla p_n \equiv \nabla M_n$ so that $M_n$ instead of $p_n$ can be used in the momentum balance.

In the absence of diapycnal mixing, the isopycnals are material surfaces. More generally, they satisfy

$$\left(\frac{\partial}{\partial t} + \boldsymbol{u}_n \cdot \nabla\right) d_n + w_n|_{z=-d_n} = w_n^\star \tag{B.9}$$

where $w_n^\star$ denotes a diapycnal velocity reflecting the term $\mathcal{G}_\rho$ in the budget (B.1). With (B.9), integration of the continuity equation $\nabla \cdot \boldsymbol{u}_n + \partial w_n / \partial z = 0$ over layer $n$ yields

$$\frac{\partial h_n}{\partial t} + \nabla \cdot h_n \boldsymbol{u}_n = w_n^\star - w_{n-1}^\star \tag{B.10}$$

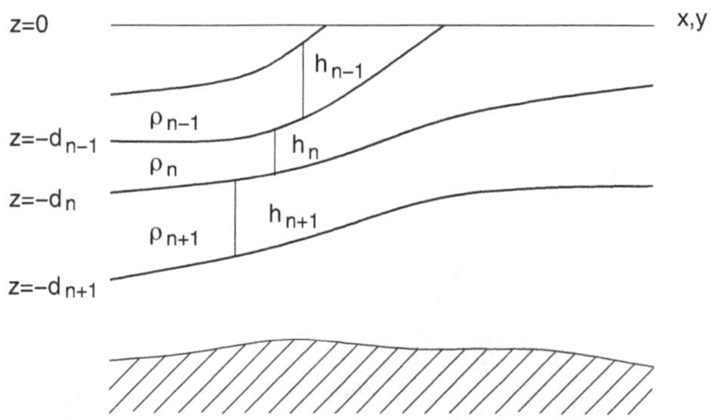


Fig. B.1 Sketch of a generic isopycnal layer model. Each layer (index $n$) is bounded by the interfaces $z = -d_n$ below and $z = -d_{n-1}$ above, and is characterized by a constant density $\rho_n$, thickness $h_n = d_n - d_{n-1}$, pressure $p_n$ or Montgomery function $M_n$, and velocity $u_n$, $w_n$. Note that the number of layers in motion can depend on the horizontal position since density layers can have contact to the surface which is referred to as layer mass conservation. Note that (B.10) physically also reflects density (i. e. , heat and salt) conservation. Integration of (B.5) over the layer $n$ leads to the momentum equation

$$\frac{\partial \boldsymbol{u}_\mathrm{n}}{\partial t} + \boldsymbol{u}_\mathrm{n} \cdot \nabla \boldsymbol{u}_\mathrm{n} + f \, \underline{\mathbf{u}}_\mathrm{n} = -\nabla M_\mathrm{n} + (\boldsymbol{\tau}_{\mathrm{n},0} - \boldsymbol{\tau}_{\mathrm{n,b}})/h_\mathrm{n} + \mathcal{F}_\mathrm{n} \tag{B.11}$$

Here $\boldsymbol{\tau}_{\mathrm{n},0}, \boldsymbol{\tau}_{\mathrm{n,b}}$ denote the stress and top and bottom of layer $n$. Equations (B.10) and (B.11) directly correspond to the continuous form (B.6) and (B.5), for $G_\rho = 0$ (the effect of the diapycnal velocity on momentum advection has been neglected).

### B.1.3 A Two-Layer Quasi-Geostrophic Model

We present some details of the two-layer quasi-geostrophic (QG) model used in this book. It includes topography and diabatic forcing. Layer 1 extends from the surface at $z = \zeta$ to a variable interface at $z = -d_1(x, y, t) = -H_1 + \xi(x, y, t)$; layer 2 extends from the interface to the bottom at $z = -d_2(x, y) = -H + b(x, y)$ where $H = H_1 + H_2$ is the total mean depth, and $H_1$ and $H_2$ are the mean thicknesses of the layers (see Figure B.2). In this notation the interface displacements $\zeta, \xi$ and the bottom displacement $b$ are taken positive if upward.

The pressure fields follow from hydrostatics as $M_1 = \rho_1 g \zeta$ and $M_2 = \rho_1 g \zeta + (\rho_2 - \rho_1) g \xi$. Scaling the pressure by $\rho_1$, we obtain $M_1 = g\zeta, M_2 = g\zeta + g^* \xi$ with the reduced gravity $g^* = g(\rho_2 - \rho_1)/\rho_1$. The horizontal velocities $\boldsymbol{u}_i$ satisfy (B.11). The friction terms refer to the stresses at the top and bottom of the respective layer, and $\mathcal{F}_\mathrm{n}$ denotes the divergence of lateral stresses. For frictionless conditions, the conservation of the potential vorticity $(\eta_\mathrm{n} + f)/h_\mathrm{n}$ with relative vorticity $\eta_\mathrm{n} = \partial v_\mathrm{n}/\partial x - \partial u_\mathrm{n}/\partial y$ is easily proven from (B.10) to (B.11). Straightforward perturbation theory yields the layered version of the quasi-geostrophic potential vorticity theory.

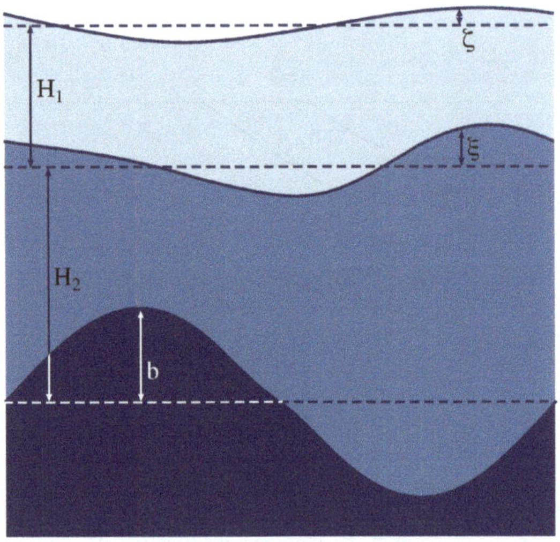


Fig. B.2 Geometry of the quasi-geostrophic two-layer model. The *dark blue* area indicates the bottom. Above are the two layers with a density difference $\Delta\rho = \rho_2 - \rho_1$

Under similar conditions as in Section 5.2 (i. e. assuming $\beta$-plane, rigid-lid approximation, small Rossby number and small vertical excursions of the interface and bottom topography so that the actual layer thicknesses $h_1 = H_1 + \eta - \xi$ and $h_2 = H_2 + \xi - b$ are close to the mean thicknesses), analogously to (5.32) one can derive a quasi-geostrophic vorticity equation for each layer as

$$\frac{\partial Q_1}{\partial t} + \mathcal{J}(\psi_1, Q_1) = \frac{f_0}{H_1}\left(w_{\text{ek}} - w^\star\right) + \operatorname{curl}\boldsymbol{\mathcal{F}}_1 \tag{B.12}$$

$$\frac{\partial Q_2}{\partial t} + \mathcal{J}(\psi_2, Q_2) = \frac{f_0}{H_2} w^\star + \operatorname{curl}\boldsymbol{\mathcal{F}}_2 \tag{B.13}$$

Here $\psi_1 = M_1/f_0$ and $\psi_2 = M_2/f_0$ are the quasi-geostrophic stream functions of the two layers, and the quasi-geostrophic potential vorticities are

$$Q_1 = \nabla^2\psi_1 + \beta y - \frac{f_0}{H_1}(\zeta - \xi) \quad \text{and} \quad Q_2 = \nabla^2\psi_2 + \beta y - \frac{f_0}{H_2}(\xi - b) \tag{B.14}$$

The horizontal velocities are $\boldsymbol{u}_j = (u_j, v_j)$ with $u_j = -\partial\psi_j/\partial y, v_j = \partial\psi_j/\partial x$, $j = 1, 2$, and $\beta$ is the meridional gradient of the Coriolis frequency. The surface displacement is $\zeta = (f_0/g)\psi_1$, and the interface height $\xi$ can be written as $\xi = (f_0/g^*)(\psi_2 - \psi_1)$. The internal Rossby radius of this configuration is $R = (g^* H_1 H_2 / f_0^2 H)^{1/2}$. The Jacobian differential operator $\mathcal{J}$ is used to represent the advective terms, $\mathcal{J}(a, b) = (\partial a/\partial x)(\partial b/\partial y) - (\partial b/\partial x)(\partial a/\partial y)$.

The QG model does not have an Ekman layer. Instead, the wind forcing is represented by an Ekman pumping velocity $w_{\text{ek}}$ at the top of layer 1,

$$w_{\text{ek}} = \operatorname{curl}\frac{\boldsymbol{\tau}_0}{f_0}$$

where $\boldsymbol{\tau}_0$ is the surface windstress. The frictional terms are usually written as down-gradient mixing of zonal momentum with the horizontal viscosity $A_{\text{h}}$, i. e. $\operatorname{curl}\boldsymbol{\mathcal{F}}_j = A_{\text{h}}\nabla^4\psi_j$. In the present model, the diabatic processes, represented by $w_1^\star = w^\star$ as introduced in (B.9), are expressed as the sum of an external 'heating' $Q$ (in $\text{W m}^{-3}$) and a diffusion of the interface height $\xi$ with a diffusivity $K$,

$$w^\star = \frac{\alpha Q}{c_p(-\partial\rho/\partial z)} - K\nabla^2\xi$$

where $\alpha$ is the coefficient of thermal expansion and $c_p$ the specific heat of constant pressure. The diffusion of the interface height may be considered as being equivalent to the diffusion of layer thickness of the GM parameterization (see Section 12.2.4): thus mesoscale eddy processes are to some extent represented in $w^\star$. Writing the vertical density gradient as $\partial\rho/\partial z = -2(\rho_2 - \rho_1)/H$ and inserting the expression for the interface height $\xi$ results in

$$w^\star = \frac{1}{2}\frac{\alpha H Q}{c_p(\rho_2 - \rho_1)} + \frac{K f_0}{g^*}\nabla^2(\psi_1 - \psi_2)$$

It is sometimes useful to consider the barotropic (external) and baroclinic (internal) stream functions defined by

$$\psi_{\text{e}} = (H_1/H)\psi_1 + (H_2/H)\psi_2 \quad \text{and} \quad \psi_{\text{i}} = \psi_1 - \psi_2$$
$$\psi_1 = \psi_{\text{e}} + (H_2/H)\psi_{\text{i}} \quad \text{and} \quad \psi_2 = \psi_{\text{e}} - (H_1/H)\psi_{\text{i}}$$

They are governed by

$$\frac{\partial \nabla^2 \psi_e}{\partial t} + \mathcal{J}\left(\psi_e, \nabla^2\psi_e + \beta y + \frac{f_0}{H}b\right) + \mathcal{J}\left(\psi_i, \frac{H_1 H_2}{H^2}\nabla^2\psi_i + \frac{f_0 H_1}{H^2}b\right)$$
$$= \frac{f_0 w_{ek}}{H} + A_h \nabla^4 \psi_e \tag{B.15}$$

$$\frac{\partial}{\partial t}\left(\nabla^2 - \frac{1}{R^2}\right)\psi_i + \mathcal{J}\left[\psi_e, \left(\nabla^2 - \frac{1}{R^2}\right)\psi_i + \frac{f_0}{H_2}b\right]$$
$$+ \mathcal{J}\left(\psi_i, \nabla^2\psi_e + \frac{H_2 - H_1}{H}\nabla^2\psi_i + \beta y + \frac{f_0 H_1}{H_2 H}b\right)$$
$$= \frac{f_0 w_{ek}}{H_1} - \frac{1}{2}\frac{\alpha g H_2 Q}{f_0 c_p \rho_1 R^2}\frac{H}{H_2} - \frac{K}{R^2}\nabla^2\psi_i + A_h\nabla^4\psi_i \tag{B.16}$$

The 'heating' appears are now as $H_2 Q$ and may be interpreted as the vertically integrated rate (in W m$^{-2}$) for the bottom layer. The QG model is usually applied to a zonal strip (a channel) of $\beta$-plane or a closed basin away from the equator.

### B.1.4 A Planetary-Geostrophic Isopycnal Layer Model

A different application of layer models is the steady-state planetary-geostrophic model used in Section 14.3.4. A main difference to the QG model is that the isopycnals are not restricted to small excursions from a mean depth but can undergo substantial changes, and in particular can intersect with the surface. Under the conditions discussed in Section 5.3, (B.11) reduces to

$$-f \boldsymbol{u}_n = -\underset{\neg}{\nabla} M_n \tag{B.17}$$

Note that the layer Montgomery stream function $M_n = p_n + g\rho_n z$ coincides with the Bernoulli function in the planetary-geostrophic approximation (see Section 5.3). Several flow properties follow from the geostrophic balance (B.17). Obviously, $\boldsymbol{u}_n \cdot \nabla M_n = 0$, so that $M_n$ is constant along streamlines, and hence a conserved variable. From geostrophy and (B.8) one further obtains

$$f(\boldsymbol{u}_n - \boldsymbol{u}_{n+1}) = g(\rho_{n+1} - \rho_n)\underset{\neg}{\nabla} d_n \tag{B.18}$$

which is identified as discrete form of the thermal wind relation. Finally, the Sverdrup relation $\beta v_n = f\partial w_n/\partial z$ for each layer follows from geostrophy and continuity. Integration over the layer, and subsequent summation over all layers results in

$$\beta \sum_n v_n h_n = f w_{ek} \tag{B.19}$$

as the discrete form of the Sverdrup transport relation (cf. Section 14.1.5).

Layers which are in contact with the surface (ventilated layers) are subject to wind forcing through an Ekman pumping velocity $w_{ek}$. Assuming that no diapycnal velocities other than at the surface occur, it follows from the steady state mass budget (B.10) that

$$\nabla \cdot (h_n \boldsymbol{u}_n) = \begin{cases} 0 & \text{(unventilated layers)} \\ -w_{ek} & \text{(ventilated layers)} \end{cases} \tag{B.20}$$

The potential vorticity of layer $n$ is defined as $Q_{\rm n} = f/h_{\rm n}$, and its conservation equation becomes

$$\boldsymbol{u}_{\rm n} \cdot \nabla \frac{f}{h_{\rm n}} = \frac{\beta v_{\rm n}}{h_{\rm n}} - \frac{f}{h_{\rm n}^2} \boldsymbol{u}_{\rm n} \cdot \nabla h_{\rm n} = \frac{1}{h_{\rm n}} \left( \beta v_{\rm n} + f \nabla \cdot \boldsymbol{u}_{\rm n} - \frac{f}{h_{\rm n}} \nabla \cdot h_{\rm n} \boldsymbol{u}_{\rm n} \right)$$

With the Sverdrup relation $\beta v_{\rm n} + f \nabla \cdot \boldsymbol{u}_{\rm n} = 0$ and the mass conservation (B.20) one obtains the conservation of potential vorticity in unventilated layers as

$$\boldsymbol{u}_{\rm n} \cdot \nabla Q_{\rm n} = 0 \qquad \text{(B.21)}$$

## B.2 BARBI: A Model of the Wind-Driven Circulation

In Section 14.1 we derived simple equations describing the barotropic flow. We were lead to the Stommel–Munk equation (B71.1) which is able to describe many aspects of the observed large-scale depth-integrated circulation. The Stommel–Munk model is closed in the way that it contains no other variables besides the barotropic stream function, but it is valid for a flat-bottom ocean only. On the other hand, because there are topographic variations in the ocean, the generalized Stommel equation (14.29) appears more appropriate. However, (14.29) contains a further forcing term – the JEBAR torque, introducing a further variable, the vertically integrated potential density $E$ (referred to the surface). Alternatively, the vorticity of the depth-integrated flow (B73.3) contains the unknown bottom pressure $P$. In this respect, the generalized Stommel equation does not represent a closed model, i. e. one needs an additional equation for $E$ (or $P$). The BARBI (BARrotopic Baroclinic Interaction) model contains such an equation for $E$ and generalizes Stommel's concept for stratified flow over varying topography. BARBI describes the depth-averaged circulation, i. e. the barotropic flow, and its interaction with topography and the stratification. In principle, BARBI could also contain thermohaline forcing, but its main purpose is to describe the wind-driven circulation with a prescribed mean stratification. The BARBI model was developed in Olbers and Eden (2003).

### *B.2.1 Derivation of BARBI*

The BARBI model is intended for the oceanic circulation on scales much larger than the Rossby radius, as discussed in Chapter 5.1. For the momentum balance the nonlinear advection in the primitive equations (4.50) and (4.51) can be neglected. It is, however, convenient to retain the local time derivative, so that both gravity waves and short Rossby waves are allowed in the system. With the notation as introduced in Section 14.1, we thus start from the system

$$\frac{\partial \boldsymbol{u}}{\partial t} + f \underset{\neg}{\boldsymbol{u}} = -\nabla p + \frac{\partial \boldsymbol{\tau}}{\partial z} + \boldsymbol{\mathcal{F}} \tag{B.22}$$

$$\frac{\partial p}{\partial z} = -g\rho \tag{B.23}$$

$$\nabla \cdot \boldsymbol{u} + \frac{\partial w}{\partial z} = 0 \tag{B.24}$$

$$\frac{\partial \rho}{\partial t} + \boldsymbol{u} \cdot \nabla \rho + w\frac{\partial \rho}{\partial z} = w\frac{N^2}{g} + \mathcal{G}_\rho \tag{B.25}$$

The frictional terms are written as a vertical divergence of a stress $\boldsymbol{\tau}$ of horizontal momentum and a lateral divergence $\boldsymbol{\mathcal{F}}$ of the turbulent fluxes of horizontal momentum (a factor $\rho_0$ is absorbed here).

For simplification, the balances of heat and salt have been combined into a thermohaline balance for the density $\rho$ which here is the deviation from a background density $\rho_b$. The background density was already defined in (5.7) and represents the mean vertical stratification. The detailed derivation of the separation from the background state is not given here; it is completely analogous to the discussion in Section 5.2, and (B.25) is identical to (5.15), with the stability frequency $N$ defined as in (5.9). The density source $\mathcal{G}_\rho$ is related to turbulent mesoscale mixing and stirring, diapycnal turbulent microscale processes and the nonlinear equation of state, and will be specified below. Note also that effects of compressibility of the perturbation density are neglected in (B.25) which puts the thermohaline balance in BARBI on a similar footing as that usually implemented in simplified dynamics such as quasi-geostrophic or layered reduced gravity models.

We separate the total horizontal flow $\boldsymbol{u}$ into the depth-averaged "barotropic" velocity $\boldsymbol{U}/h$, and deviations ($\boldsymbol{u}'$) which will be called the "baroclinic" velocity ("barotropic/baroclinic" in the sense discussed in the box on p. 467) according to

$$\boldsymbol{u} = \boldsymbol{U}/h + \boldsymbol{u}' \quad \text{with} \quad \boldsymbol{U} = \int_{-h}^{0} \boldsymbol{u}\,\mathrm{d}z \quad \text{and} \quad \boldsymbol{u}' = \boldsymbol{u} - \boldsymbol{U}/h \tag{B.26}$$

From (B.24) and the kinematic boundary conditions for a rigid lid at the top and the bottom we have

$$w = 0 \quad \text{at} \quad z = 0 \quad \text{and} \quad w + \boldsymbol{u} \cdot \nabla h = 0 \quad \text{at} \quad z = -h \tag{B.27}$$

and the corresponding barotropic ($W$) and baroclinic ($w'$) vertical velocities are defined by

$$w = W + w' \quad \text{with} \quad W = \frac{z}{h^2}\boldsymbol{U} \cdot \nabla h \quad \text{and} \quad w' = -\nabla \cdot \int_{-h}^{z} \boldsymbol{u}'\mathrm{d}z' \tag{B.28}$$

Note that barotropic and baroclinic vertical velocities satisfy the rigid-lid condition independently. As detailed in Section 14.2, vertical integration of the horizontal momentum equation (B.22), partial integration and use of the hydrostatic relation (B.23) leads to

$$\frac{\partial \boldsymbol{U}}{\partial t} + f \underset{\neg}{\boldsymbol{U}} = -h\nabla P - \nabla E + \boldsymbol{\tau}_0 - \boldsymbol{\tau}_b + \int_{-h}^{0} \boldsymbol{\mathcal{F}}\mathrm{d}z \tag{B.29}$$

with the windstress $\boldsymbol{\tau}_0$, bottom stress $\boldsymbol{\tau}_b$, vertically integrated potential energy $E = g\int_{-h}^{0} z\rho\mathrm{d}z$ and bottom pressure $P$. The latter can be eliminated from the momentum equation by dividing by $h$ and taking the curl

$$\frac{\partial}{\partial t}\nabla\cdot\left(\frac{1}{h}\nabla\psi\right)+\underset{\neg}{\nabla}\psi\cdot\nabla\frac{f}{h}+\underset{\neg}{\nabla}\left(\frac{1}{h}\right)\cdot\nabla E=\underset{\neg}{\nabla}\cdot\frac{\boldsymbol{\tau}_0-\boldsymbol{\tau}_b}{h}+\underset{\neg}{\nabla}\cdot\frac{1}{h}\int_{-h}^{0}\boldsymbol{\mathcal{F}}\mathrm{d}z \tag{B.30}$$

where the (barotropic) stream function $\psi$ is introduced by $\underset{\neg}{\nabla}\psi = \boldsymbol{U}$. Knowledge of $E$ (and appropriate viscous closures for $\boldsymbol{\mathcal{F}}$ and $\boldsymbol{\tau}_b$) would allow the computation of the stream function $\psi$. Using the density budget (B.25), it is in fact possible to derive a tendency equation for $E$, as discussed now.

The baroclinic potential energy $E$ is defined in terms of the perturbation density because the part deriving from the background density cancels between the two pressure terms in (B.29) and generates no contribution to the JEBAR term as well. In analogy to statistical moments (compare Section A.3), we will now call $E$ the *first vertical moment* of density. In fact, it will turn out below that higher order vertical moments of density are needed as well. Thus we define

$$E_n = g\int_{-h}^{0} z^n\rho\mathrm{d}z \quad\text{and}\quad \mathcal{E}_n = g\int_{-h}^{0} z^n\rho_b(z)\mathrm{d}z\ , \quad n=0,1,2,\ldots \tag{B.31}$$

hence $E_1 \equiv E$. For convenience we have also define vertical density moments for the background density $\mathcal{E}_n$. Note that $\mathcal{E}_n$ depends only on the horizontal position via the dependency on $h(x, y)$.

To obtain a tendency equation for the vertical density moment $E_n$, the density budget (B.25) is multiplied with $gz^n$ and integrated over depth from bottom to top. To distinguish between barotropic and baroclinic density advection, the total advective flow $\boldsymbol{u}$ ($w$) in (B.25) is split into a barotropic part $\boldsymbol{U}$ ($W$) and a baroclinic part $\boldsymbol{u}'$ ($w'$) as defined in (B.26) and (B.28). Consider all terms in (B.25) separately. The first on the left-hand side of (B.25) yields indeed the change in the vertical density moment

$$g\int_{-h}^{0} z^n\frac{\partial\rho}{\partial t}\mathrm{d}z = \frac{\partial E_n}{\partial t}$$

The second term on the left-hand side of (B.25) splits into the horizontal advection by the barotropic flow and the horizontal advection by the baroclinic flow. Consider the former

$$g\int_{-h}^{0} z^n\boldsymbol{U}/h\cdot\nabla\rho\mathrm{d}z$$

$$=\boldsymbol{U}/h\cdot g\int_{-h}^{0}\nabla z^n\rho\mathrm{d}z=\boldsymbol{U}/h\cdot\nabla E_n-(-1)^n gh^{n-1}\rho(-h)\boldsymbol{U}\cdot\nabla h$$

since $\nabla \int_{-h}^{0} z^n \rho \mathrm{d}z = \int_{-h}^{0} \nabla(z^n \rho)\mathrm{d}z - z^n \rho|_{-h} \nabla(-h)$. The third term on the left-hand side of (B.25) splits into the vertical advection by the barotropic flow and the vertical advection by the baroclinic flow. For the former we find by partial integration

$$g \int_{-h}^{0} z^n W \frac{\partial \rho}{\partial z} \mathrm{d}z = (-1)^n g h^{n-1} \rho(-h) \boldsymbol{U} \cdot \nabla h - h^{-2}(n+1) E_n \boldsymbol{U} \cdot \nabla h$$

Adding both contributions yields

$$g \int_{-h}^{0} z^n \boldsymbol{U}/h \cdot \nabla \rho \mathrm{d}z + g \int_{-h}^{0} z^n W \frac{\partial \rho}{\partial z} \mathrm{d}z = h^n \boldsymbol{U} \cdot \nabla \frac{E_n}{h^{n+1}}$$

It is obvious that the same result must be obtained by replacing the perturbation density by the background density, i. e. the barotropic advection of the background density is given by

$$-\int_{-h}^{0} z^n W N^2 \mathrm{d}z = h^n \boldsymbol{U} \cdot \nabla \frac{\mathcal{E}_n}{h^{n+1}}$$

Collecting these results now yields for the tendency budget of the vertical density moments

$$\frac{\partial E_n}{\partial t} + h^n \boldsymbol{U} \cdot \nabla \frac{E_n}{h^{n+1}} = \bar{S}_n + S_n' + S_n'' + g \int_{-h}^{0} z^n \mathcal{G}_\rho \mathrm{d}z \tag{B.32}$$

with the advective source terms

$$\bar{S}_n = -h^n \boldsymbol{U} \cdot \nabla \frac{\mathcal{E}_n}{h^{n+1}} ,$$

$$S_n' = \int_{-h}^{0} z^n w' N^2(z) \mathrm{d}z , \quad S_n'' = -g \int_{-h}^{0} z^n \left( \nabla \cdot \boldsymbol{u}' \rho + \frac{\partial w' \rho}{\partial z} \right) \mathrm{d}z$$

According to (B.32), changes in the vertical density moments $E_n$ are caused by advection of the background density $\rho_\mathrm{b}$ by the barotropic flow (which is given by $\bar{S}_n$), advection of $\rho_\mathrm{b}$ by the baroclinic flow ($S_n'$) and advection of perturbation density $\rho$ by the baroclinic flow ($S_n''$). The second term on the left-hand side of (B.32) results from horizontal and vertical advection of $\rho$ by the barotropic flow. In addition to the advective sources, there is the diabatic source term and the rate of change.

The close the BARBI model, all advective source terms have to be evaluated or parameterized. First, consider the effects of barotropic advection given by $\bar{S}_n$. Similar to the second term on the left hand side in (B.32), it is evident that the effect of the lifting of the background mass field by the barotropic flow ($\bar{S}_n$) can be expressed entirely in terms of $\boldsymbol{U}$ and the (given) background stratification $\rho_\mathrm{b}(z)$, in form of its corresponding vertical moments $\mathcal{E}_n$. For, e. g. a constant Brunt–Väisälä frequency $N = N_0$, the term takes the simple form

$$\bar{S}_n = -(-1)^n \frac{N_0^2 h^n}{n+2} \boldsymbol{U} \cdot \nabla h$$

Next consider the source term $S'_n$, representing the production of $E_n$ due to lifting of the background mass field by the baroclinic flow $\boldsymbol{u}'$. For simplicity, we assume again a constant Brunt–Väisälä frequency. By partial integration, $S'_n$ is then found to be given by

$$S'_n = \frac{N_0^2}{n+1} \nabla \cdot \int_{-h}^{0} z^{n+1} \boldsymbol{u}' \mathrm{d}z$$

The form of $S'_n$ suggests that the balance of the baroclinic momentum $\boldsymbol{u}'$ should be projected in the same way as (B.25) (on $z^{n+1}$) to obtain prognostic equations for the *baroclinic velocity moments*. Consequently, we define the baroclinic velocity moments as

$$\boldsymbol{u}'_{n+1} = \int_{-h}^{0} z^{n+1} \boldsymbol{u}' \mathrm{d}z \quad \text{for} \quad n = 0, 1, 2, \ldots \quad \text{and} \quad \boldsymbol{u}'_0 = 0 \tag{B.33}$$

and add them to the set of prognostic variables in BARBI. The baroclinic momentum balance is the difference between the complete momentum balance (B.22) and the barotropic one (B.29), and is given by

$$\frac{\partial \boldsymbol{u}'}{\partial t} + f \underset{\neg}{\boldsymbol{u}'} = g \nabla \int_{-h}^{z} \rho \mathrm{d}z + \frac{1}{h} \nabla E_1 + \mathcal{F}' \tag{B.34}$$

with the frictional force $\mathcal{F}' = \partial \tau / \partial z + \mathcal{F} - (\tau_0 - \tau_b + \int_h^0 \mathcal{F} \mathrm{d}z)/h$. Note that the pressure gradient simplifies, since from vertical integration of the hydrostatic relation (B.23) we find $p - P = -g \int_{-h}^{z} \rho \mathrm{d}z$. Multiplication of (B.34) with $z^{n+1}$, vertical integration and integration by parts yields

$$\frac{\partial}{\partial t} \boldsymbol{u}'_{n+1} + f \underset{\neg}{\boldsymbol{u}'}_{n+1} = -\frac{1}{n+2} \left[ (-1)^n h^{n+1} \nabla E_1 + \nabla E_{n+2} \right] + \int_{-h}^{0} z^{n+1} \mathcal{F}' \mathrm{d}z \tag{B.35}$$

An equation for the baroclinic velocity moments $\boldsymbol{u}'_{n+1}$ is thus found, with a quite simple expression of the pressure gradient in terms of gradients of $E_n$. Note that no further assumptions have been made to derive (B.35). The divergence of $\boldsymbol{u}'_{n+1}$ allows us in turn to calculate $S'_n$ in the balance of $E_n$. However, at this point it turns out that indeed all (odd) vertical density moments are needed and not just $E_1$, since the interaction of potential energy $E_1$ with the baroclinic fields via $S'_1$ and thus $\boldsymbol{u}'_2$ couples the third moment $E_3$ to the balance of $E_1$. For $E_3$ we need $S'_3$, thus $\boldsymbol{u}'_4$, therefore, $E_5$, etc. Apparently, one needs a truncation of this coupled hierarchy at some order, which is discussed below. Note also that starting with the balance of potential energy, only the odd moments $E_n$ ($n = 1, 3, 5, \ldots$) are involved.

We also have to evaluate the source term $S''_n$, stemming from advection (both vertical and horizontal) of perturbation density by the baroclinic flow. It is easy to show by scaling analysis that $S''_n$ is small compared to $S'_n$ and $\bar{S}_n$, as long as the density perturbation $\rho$ is small compared to the background density $\rho_b(z)$. Therefore, the source term $S''_n$ is simply neglected in BARBI. However, the term $S''_n$ may become large if the density perturbation becomes large, as e. g. in configurations with strong thermohaline forcing.

### B.2.2 Eddy Parameterization for the BARBI Model

The closures for the subgrid effects represented by $\mathcal{G}_\rho$, $\mathcal{F}$, $\boldsymbol{\tau}$ and $\boldsymbol{\tau}_\mathrm{b}$ in the BARBI model equations (B.30), (B.32) and (B.35) need to be specified. The turbulent density mixing $\mathcal{G}_\rho$ in (B.32) can be split into diapycnal mixing effect due to small-scale isotropic turbulence and an advective effect by mesoscale eddies, neglecting any effects from the nonlinear equation of state. For the advective effect of mesoscale eddy mixing on density, we employ the Gent and McWilliams (1990) parameterization (compare Section 12.2.4) in which the eddy advection (bolus) velocity $(\boldsymbol{v}_\mathrm{e}, w_\mathrm{e})$ is given by a vector stream function $\boldsymbol{B}$, parameterized as

$$\boldsymbol{B} = K_\ell \underset{\neg}{\boldsymbol{s}} \quad \text{with} \quad \boldsymbol{v}_\mathrm{e} = \frac{\partial}{\partial z} \underset{\neg}{\boldsymbol{B}} \quad \text{and} \quad w_\mathrm{e} = \underset{\neg}{\nabla} \cdot \boldsymbol{B}$$

where $K_\ell$ is equivalent to the thickness diffusivity of the Gent and McWilliams parameterization. Here $\boldsymbol{s} = g\nabla\rho/N^2$ denotes the (approximate) vector of the isopycnal slopes, and $\boldsymbol{v}_\mathrm{e}$ and $w_\mathrm{e}$ are the horizontal and vertical components of the bolus velocity respectively. The barotropic part of the bolus velocity vanishes because of the boundary conditions $\boldsymbol{B} = 0$ at top and bottom. In agreement to the treatment in (B.32), we consider the lifting of the background stratification by only the baroclinic vertical bolus velocity, i. e. we assume that

$$\int_{-h}^{0} z\mathcal{G}_\rho \mathrm{d}z \approx -\int_{-h}^{0} zN^2 \underset{\neg}{\nabla} \cdot \boldsymbol{B} \mathrm{d}z \approx g \int_{-h}^{0} \nabla \cdot K_\ell \nabla z\rho \mathrm{d}z \approx K_\ell \nabla^2 E \qquad \text{(B.36)}$$

For simplicity, a flat bottom[1] and constant $K_\ell$ was assumed in (B.36). It turns out that the dominant effect of advective mesoscale eddy density mixing on vertically integrated potential energy $E$ is given by diffusion of $E$ with a diffusivity identical to the thickness diffusivity of the Gent and McWilliams parameterization.

In an integral of the potential energy balance (B.32) over a closed domain there is no sink term due to the mesoscale eddy density mixing if vanishing gradients of $E$ normal to the boundaries are specified. On the other hand, we know that the net effect of the parameterization is given by a conversion of mean (available) potential energy to eddy energy. Indeed, it is possible to show that

$$\int_V z \left\{ \nabla \cdot [\boldsymbol{v}_\mathrm{e}(\rho + \rho_\mathrm{b})] + \frac{\partial}{\partial z} [w_\mathrm{e}(\rho + \rho_\mathrm{b})] \right\} \mathrm{d}V$$
$$\approx -\int_A \left( \nabla \cdot K_\ell \nabla E - \int_{-h}^{0} K_\ell s^2 N^2 \right) \mathrm{d}A = \int_V K_\ell s^2 N^2 \mathrm{d}V \qquad \text{(B.37)}$$

where $V(A)$ denotes a closed volume (area) and where again a flat bottom was assumed and where $s = |\boldsymbol{s}|$. The term on the right-hand side of (B.37) is sign definite and a sink for $E$, resembling the conversion of mean potential energy to eddy energy, mimicking the effect of baroclinic instability. It is possible to parameterize this exchange with eddy energy as linear damping of $E$. However, it turns out that this exchange term is locally much less important than the diffusive term related to

[1] Or, equivalently, vanishing diffusivity $K_\ell$ at top and bottom.

the lifting of the background stratification by the vertical bolus velocity, allowing to simply neglect the exchange term here. Note that this neglect is consistent with the neglect of the effect of advection of perturbation density by the baroclinic flow in BARBI, resembling a second order term when density perturbations remain small compared to the background.

For the frictional closures it is less clear how to proceed. We assume that vertical friction can be neglected in the interior, it is only important near the surface where the vertical turbulent stress vector $\boldsymbol{\tau}$ connects to the surface windstress $\boldsymbol{\tau}_0$. We also neglect the bottom stress $\boldsymbol{\tau}_{\mathrm{b}}$, because we have already seen that in many cases the deep flow over the bottom topography will vanish; in that case the bottom stress would also vanish. On the other hand, our model needs some kind of dissipation of kinetic energy which becomes clear considering the energy cycle in BARBI . The barotropic kinetic energy is given by $K_{\mathrm{b}} = \boldsymbol{U}^2/(2h)$ and by using (B.32) and (B.29), the global domain integrals of $K_{\mathrm{b}}$ and $E$ can be derived,

$$\frac{\partial}{\partial t}\int \mathrm{d}x^2 K_{\mathrm{b}} = -\int \mathrm{d}x^2 E h^{-2}\boldsymbol{U}\cdot\nabla h + \int \mathrm{d}x^2 h^{-1}\boldsymbol{U}\cdot\boldsymbol{\tau}_0 + \int \mathrm{d}x^2 h^{-1}\boldsymbol{U}\cdot\int_{-h}^{0}\boldsymbol{\mathcal{F}}\mathrm{d}z$$

$$\frac{\partial}{\partial t}\int \mathrm{d}x^2 E = \int \mathrm{d}x^2 E h^{-2}\boldsymbol{U}\cdot\nabla h$$

Exchange between $E$ and $K_{\mathrm{b}}$ is given by barotropic density advection over topography, dissipation is due to friction and energy is supplied by the wind work. Note that baroclinic kinetic energy is not considered here, since it can be shown that it does not exchange energy with $E$ as long as the effect of advection of perturbation density by the baroclinic flow is set to zero in the budget for $E$. This effect was assumed and shown to be locally negligible in BARBI as long as the perturbation density remains small compared to the background density. Note also that there is no energy exchange with eddy energy, since we have neglected the dissipation effect by the Gent and McWilliams parameterization of $E$ in (B.37).

It becomes clear that in the balance of total mechanical energy $K_{\mathrm{b}} + E$, a steady state can only be established when the wind work is balanced by some kind of mechanical dissipation. Therefore, we implement in BARBI harmonic lateral friction in combination with a no-slip boundary condition, i. e. $\boldsymbol{\mathcal{F}} = A_{\mathrm{h}}\nabla^2\boldsymbol{u}$. The integrals of $\boldsymbol{\mathcal{F}}$ and $\boldsymbol{\mathcal{F}}'$, appearing in the balance (B.30) of the barotropic vorticity and in the balance (B.35) of the baroclinic velocity moment, are approximated as

$$\int_{-h}^{0}\boldsymbol{\mathcal{F}}\mathrm{d}z \approx A_{\mathrm{h}}\nabla^2\boldsymbol{U} \quad \text{and}$$

$$\int_{-h}^{0} z^{n+1}\boldsymbol{\mathcal{F}}'\mathrm{d}z \approx A_{\mathrm{h}}\nabla^2\boldsymbol{u}_{n+1} + (-1)^n\frac{h^{n+1}}{n+2}(\boldsymbol{\tau}_0 + A_{\mathrm{h}}\nabla^2\boldsymbol{U})$$

We have again neglected the spatial dependency of $h$ for simplicity. However, it should be mentioned again that harmonic lateral friction is problematic, since it is well known that mean momentum is not mixing down mean momentum gradients in turbulent geophysical flows. On the other hand, the formulation using harmonic friction is convenient for numerical reasons since it provides a sink for grid noise such that we follow usual practice and use it here as well.

Finally, the diapycnal density mixing in $\mathcal{G}_\rho$ needs consideration, which we parameterize as vertical diffusion with turbulent diffusivity $K_\text{v}$. We find by integration

$$S_n^{\text{dia}} = g \int_{-h}^{0} z^n \frac{\partial}{\partial z}\left[K_\text{v}\frac{\partial}{\partial z}(\rho + \rho_\text{b})\right]\text{d}z = -gn \int_{-h}^{0} z^{n-1} K_\text{v}\frac{\partial}{\partial z}(\rho + \rho_\text{b})\text{d}z$$
$$\approx -(-1)^n K_\text{v} N^2 h^n$$

assuming zero vertical turbulent density flux through the bottom and $|\partial\rho/\partial z| \ll N^2/g$. Note that $K_\text{v}$ is very small in the interior of the ocean. It becomes only large for convective situations, which we exclude from the model. We might, therefore, neglect $S_n^{\text{dia}}$.

To summarize, BARBI is formulated in terms of the dependent variables $\psi$, $E_n$ and $\boldsymbol{u}'_{n+1}$ which are defined by (B.26), (B.31) and (B.33), respectively, and are determined by the equations

$$\frac{\partial}{\partial t}\nabla\frac{1}{h}\cdot\nabla\psi + \underset{\neg}{\nabla}\psi\cdot\nabla\frac{f}{h} = -\frac{1}{h^2}\underset{\neg}{\nabla}E_1\cdot\nabla h + \underset{\neg}{\nabla}\cdot\frac{\boldsymbol{\tau}_0}{h} + A_\text{h}\nabla\cdot\frac{1}{h}\nabla^2\nabla\psi \tag{B.38}$$

$$\frac{\partial E_n}{\partial t} + h^n\boldsymbol{U}\cdot\nabla\frac{E_n}{h^{n+1}} = -(-1)^n\frac{N_0^2 h^n}{n+2}\boldsymbol{U}\cdot\nabla h + \frac{N_0^2}{n+1}\nabla\cdot\boldsymbol{u}'_{n+1} + K_\ell\nabla^2 E_n \tag{B.39}$$

$$\frac{\partial}{\partial t}\boldsymbol{u}'_{n+1} + f\underset{\neg}{\boldsymbol{u}'}_{n+1} = -\frac{1}{n+2}\left[(-1)^n h^{n+1}\nabla E_1 + \nabla E_{n+2}\right]$$
$$+ (-1)^n\frac{h^{n+1}}{n+2}(\boldsymbol{\tau}_0 + A_\text{h}\nabla^2\boldsymbol{U}) + A_\text{h}\nabla^2\boldsymbol{u}_{n+1} \tag{B.40}$$

for $n = 0, 1, 2, \ldots$ Note that $E_0$ is the bottom pressure which not needed in the BARBI code. All variables in BARBI depend only on the horizontal coordinates (and time). With the projection onto the vertical moments of the baroclinic variables density and baroclinic velocity according to (B.31) and (B.33), BARBI can be viewed as a spectral model in the vertical coordinate. This approach is a natural consequence of the occurrence of the potential energy $E_1$ in the barotropic vorticity equation (JEBAR-term) and of $\boldsymbol{u}'_{n+1}$ in the tendency equation for $E_n$. It was assumed that $\rho$ remains small compared to $\rho_\text{b}$ (such that $S''_n \ll S'_n, \bar{S}_n$). For convenience, it has been assumed that the background density is given by a linear profile, i. e. $N^2 = \text{const.}$ BARBI with a nonconstant Brunt–Väisälä frequency is described in Olbers and Eden (2003).

### B.2.3 The Closure of the Infinite Hierarchy

In the above form, BARBI appears as a spectral model with an infinite number of equations and moments involved. For constant $N$, only odd density moments are present, and a parameterization of $E_{L+2}$ in terms of $E_n, n \leq L$is used, since, we need $E_{L+2}$ in the pressure gradient forcing of $\boldsymbol{u}'_{L+1}$, showing up in turn in the equation for $E_L$. Consideration of wave properties in BARBI is the guide for the closure,

i. e. we aim to construct the parameterization such that a truncated model has the correct gravity and geostrophic wave speeds. BARBI has barotropic and baroclinic geostrophic waves and baroclinic gravity waves. It is in fact possible to gain the correct dispersion relations for all baroclinic (flat-bottom and topographic) waves (and baroclinic modes) in the limit of infinite number of vertical modes (see Olbers and Eden, 2003). For practical purposes, a truncation of the coupled hierarchy of the vertical moments is thus needed.

We describe here how to close BARBI to represent two vertical wave modes, the (long) barotropic Rossby wave and the baroclinic Rossby and gravity waves. For this purpose we start with the simplest BARBI model including the barotropic vorticity balance and only the $E_1$ and $\boldsymbol{u}_2'$ equations for the flat-bottom case ($h$ = const). We neglect horizontal advection in the budget for $E_1$, i. e. we linearize the model for a state of rest, and we neglect friction and diabatic sources in order to obtain the linear wave solution. Note that for $h$ = const baroclinic and barotropic flow become decoupled and we first discuss the baroclinic equations which become

$$\frac{\partial \boldsymbol{u}_2'}{\partial t} + f \underline{\boldsymbol{u}}'_2 = -\frac{1}{3}\nabla E_1 h^2(\gamma - 1) \quad \text{and} \quad \frac{\partial E_1}{\partial t} - \frac{N_0^2}{2}\nabla \cdot \boldsymbol{u}_2' = 0 \qquad \text{(B.41)}$$

The parameterization $E_3 = \gamma h^2 E_1$ of the third order density moment in terms of the first order one was used. We have to find a proper value for the closure parameter $\gamma$. If we restrict (B.41) to an $f$-plane, the dispersion relation of gravity waves is obtained,

$$\omega^2 = f_0^2 + c^2 k_h^2 \quad \text{with} \quad c^2 = \frac{1-\gamma}{6}(N_0 h)^2$$

where the fields are assumed proportional to $\exp i(\boldsymbol{k}_\mathrm{h} \cdot \boldsymbol{x} - \omega t)$ and $\boldsymbol{k}_\mathrm{h}$ denotes the wave vector and $c$ the gravity wave speed. In Section 8.1 we found for constant $N = N_0$, constant $h$ and a rigid lid the baroclinic eigenmodes are $\Phi^\nu(z) = \cos \nu\pi z/h$ with eigenvalues $(\nu\pi)^2$, Rossby radii $N_0 h/(|f|\nu\pi)$ and gravity wave speeds $N_0 h/(\nu\pi)$ with $\nu = 1, 2, \ldots$ To gain the correct phase speed of the first mode gravity wave speed in our BARBI model we must hence choose $(1-\gamma)/6 = 1/\pi^2$ or $\gamma \approx 0.3921$.

The wave analysis of barotropic vorticity balance for $h$ = const simply yields long barotropic Rossby waves. We refer to Section B.2.6 for a wave analysis of BARBI for the two-mode version. In this book we use the BARBI model only with the two modes described here. It is described in more detail in Section B.2.4. For versions with a higher number of density moments we refer to Olbers and Eden (2003).

### B.2.4 The Two-Mode BARBI Model

The simplest BARBI model is based on $\psi$, $E = E_1$ and $\boldsymbol{u}^* = \boldsymbol{u}_2'$ with two modes present, a barotropic and a baroclinic one. It can be derived in a slightly simpler route of arguments than the complete hierarchy, presented in the preceding section: derive (B.32) and (B.35) for $n = 1$, neglect $E_3$ in the latter equation and tune the stability frequency $N_0$ to obtain the correct phase speeds. For that case, $N_0$ does not represent the true stability frequency anymore, but an appropriate effective one. For this closure, the simplest model becomes

$$\frac{\partial}{\partial t}\nabla \cdot \left(\frac{1}{h}\nabla\psi\right) + \underset{\neg}{\nabla}\psi \cdot \nabla\frac{f}{h} = \underset{\neg}{\nabla}E \cdot \nabla\frac{1}{h} + \underset{\neg}{\nabla} \cdot \frac{\boldsymbol{\tau}_0}{h} + A_h \nabla \cdot \frac{1}{h}\nabla^2\nabla\psi \tag{B.42}$$

$$\frac{\partial E}{\partial t} + h\underset{\neg}{\nabla}\psi \cdot \nabla\frac{E}{h^2} = \frac{N_0^2}{6}\underset{\neg}{\nabla}\psi \cdot \nabla h^2 + \frac{N_0^2}{2}\nabla \cdot \boldsymbol{u}^* + K_\ell \nabla^2 E \tag{B.43}$$

$$\frac{\partial \boldsymbol{u}^*}{\partial t} + f\underset{\neg}{\mathbf{u}}^* = \frac{h^2}{3}(\nabla E - \boldsymbol{\tau}_0 - A_{\mathrm{h}}\nabla^2 \boldsymbol{U}) + A_{\mathrm{h}}\nabla^2\boldsymbol{u}^* \tag{B.44}$$

All numerical BARBI applications in this book are obtained with this version. In linearized form, it describes the barotropic Rossby wave, the first baroclinic Rossby wave, and the the first baroclinic gravity wave, all with the appropriate topographic modifications.

### B.2.5 Comparison of BARBI and QG

With some straightforward approximations, the two-layer QG equations, considered in Section B.1.3, and the BARBI model equations become similar. The low-frequency version (B.42) and (14.36) of the two-mode BARBI's governing equations is the coupled set of the barotropic vorticity balance and the balance of baroclinic potential energy which may be cast into the form

$$\frac{\partial}{\partial t}\nabla \cdot \frac{1}{h}\nabla\psi + \frac{\beta}{h}\frac{\partial\psi}{\partial x} - \frac{f}{h^2}\mathcal{J}(\psi, h) + \frac{1}{h^2}\mathcal{J}(E, h) = \operatorname{curl}\frac{\boldsymbol{\tau}}{h} + A_{\mathrm{h}}\nabla \cdot \frac{1}{h}\nabla^2\nabla\psi$$

$$\frac{1}{R^2}\frac{\partial E}{\partial t} + \frac{h}{R^2}\mathcal{J}\left(\psi, \frac{E}{h^2}\right) - \beta\frac{\partial E}{\partial x} + 2\frac{f}{h}\mathcal{J}(E, h) - 2\frac{f^2}{h}\mathcal{J}(\psi, h)$$
$$= -\frac{f^2}{h^2}\operatorname{curl}\frac{h^2\boldsymbol{\tau}}{f} + \frac{\mathcal{Q} - \mathcal{D}}{R^2} + \frac{K}{R^2}\nabla^2 E$$

where $R = Nh/(\sqrt{6}|f|)$ is the local internal Rossby radius in the BARBI frame. We have introduced the Jacobian and curl notation as in the QG section Section B.1.3. There is a term-wise correspondence with the QG equations (B.15) and (B.16), realizing, however, that in BARBI advection of vorticity is neglected. Hence we trim the barotropic QG balance accordingly,

$$\frac{\partial\nabla^2\psi_{\mathrm{e}}}{\partial t} + \beta\frac{\partial\psi_{\mathrm{e}}}{\partial x} - \frac{f_0}{H}\mathcal{J}(\psi_{\mathrm{e}}, h) + \frac{f_0 H_1}{H^2}\mathcal{J}(\psi_i, h) = \frac{f_0 w_{\mathrm{E}}}{H} + A\nabla^4\psi_{\mathrm{e}}$$

In the baroclinic BARBI balance there is no advection by the baroclinic field and no effect from viscosity, hence the equivalent QG form is

$$\frac{1}{R^2}\frac{\partial\psi_{\mathrm{i}}}{\partial t} + \frac{1}{R^2}\mathcal{J}(\psi_{\mathrm{e}}, \psi_{\mathrm{i}}) - \beta\frac{\partial\psi_{\mathrm{i}}}{\partial x} - \frac{f_0}{H_2}\mathcal{J}(\psi_{\mathrm{e}}, h) + \frac{f_0 H_1}{H_2 H}\mathcal{J}(\psi_{\mathrm{i}}, h)$$

$$= -\frac{f_0 w_{\mathrm{E}}}{H_1} + \frac{1}{2}\frac{\alpha f_0 H^2 Q}{c_p(\rho_2 - \rho_1)H_1 H_2} + \frac{K}{R^2}\nabla^2\psi_{\mathrm{i}}$$

A principal difference between both models is that BARBI is valid for arbitrary topography while QG requires infinitesimal slopes. The general form of BARBI,

i. e. without the assumption of geostrophy in the baroclinic momentum balance used to obtain (14.36), is valid on the entire sphere and not just a $\beta$-plane away from the equator, as inherent in the QG approximation. Furthermore, it is worth mentioning that the deficits of BARBI concerning advection can be mediated.

### B.2.6 Waves in BARBI

We look for a WKBJ solution of the equations (B.42) and (14.36) for the BARBI model in the reduced form where gravity waves are filtered. The equations then contain only long barotropic Rossby waves and the first baroclinic Rossby waves (long and short) (see Section 14.2.6 and the previous section). For a wave solution, the ansatz $(f\psi, E) \sim \boldsymbol{R} \exp \mathrm{i}(\boldsymbol{K} \cdot \boldsymbol{x} - \omega t)$ is made with wave vector $\boldsymbol{K} = (k, \ell)$ and eigenvector $\boldsymbol{R}$ (polarization vector). A wave with $\boldsymbol{R} = (1, 0)^{\mathrm{T}}$ is barotropic and with $\boldsymbol{R} = (0, 1)^{\mathrm{T}}$ baroclinic. The gradients $\boldsymbol{\alpha} = \nabla h$, $\boldsymbol{\beta} = (0, \beta)$, and $\boldsymbol{\gamma} = \nabla(f/h)$ are assumed constant in a WKB sense. Note that $\boldsymbol{\gamma} = -(f/h^2)\boldsymbol{\alpha} + \boldsymbol{\beta}/h$. The wave problem is then given by the vorticity balance and the linearized balance of potential energy in the form

$$(1 - R_0^2 \nabla^2) \frac{\partial f \psi}{\partial t} + h R_0^2 \underline{\boldsymbol{\gamma}} \cdot \nabla f \psi + \frac{f}{h} R_0^2 \underline{\boldsymbol{\alpha}} \cdot \nabla E = 0 \tag{B.45}$$

$$(1 - R^2 \nabla^2) \frac{\partial E}{\partial t} + R^2 \left[ h \underline{\boldsymbol{\gamma}} - \frac{f}{h} \underline{\boldsymbol{\alpha}} \right] \cdot \nabla E + 2R^2 \frac{f}{h} \underline{\boldsymbol{\alpha}} \cdot \nabla f \psi = 0 \tag{B.46}$$

where a rotated vector notation is used as before: $\underline{\boldsymbol{\gamma}} = (-\gamma^y, \gamma^x)$ is the rotated vector of $\boldsymbol{\gamma} = (\gamma^x, \gamma^y)$ (anticlockwise by $\pi/2$). We have introduced the term with the external Rossby radius $R_0 = \sqrt{gh/f^2}$ for completeness whereas $R = N_0 h/(\sqrt{6}|f|)$ is the internal Rossby radius. Inserting the wave ansatz yields

$$\begin{pmatrix} \omega - \omega_{\mathrm{R}} & \omega_{\mathrm{T}} \\ 2\omega_{\mathrm{A}} & \omega - 2\omega_{\mathrm{A}} - \omega_{\mathrm{C}} \end{pmatrix} \begin{pmatrix} f\psi \\ E \end{pmatrix} = 0 \tag{B.47}$$

The frequencies $\omega_{\mathrm{T}}$ and $\omega_{\mathrm{A}}$ describe pure topographic waves, $\omega_{\mathrm{R}}$ is a mixed barotropic topographic-planetary wave and $\omega_{\mathrm{C}}$ the flat-bottom baroclinic Rossby wave. They are given by

$$\omega_{\mathrm{T}} = -\frac{f R_0^2}{h} \frac{\boldsymbol{K} \cdot \underline{\boldsymbol{\alpha}}}{1 + K^2 R_0^2}, \quad \omega_{\mathrm{A}} = -\frac{f R^2}{h} \frac{\boldsymbol{K} \cdot \underline{\boldsymbol{\alpha}}}{1 + K^2 R^2}$$

$$\omega_{\mathrm{R}} = h R_0^2 \frac{\boldsymbol{K} \cdot \underline{\boldsymbol{\gamma}}}{1 + K^2 R_0^2}, \quad \omega_{\mathrm{C}} = -\frac{\beta R^2 k}{1 + K^2 R^2}$$

Note that $\omega_{\mathrm{R}} = \omega_{\mathrm{T}} + \omega_{\mathrm{P}}$ and $\omega_{\mathrm{L}} = \omega_{\mathrm{A}} + \omega_{\mathrm{C}}$ where

$$\omega_{\mathrm{P}} = -\frac{\beta R_0^2 k}{1 + K^2 R_0^2}, \quad \omega_{\mathrm{L}} = h R^2 \frac{\boldsymbol{K} \cdot \underline{\boldsymbol{\gamma}}}{1 + K^2 R^2} \tag{B.48}$$

are associated with the flat-bottom barotropic planetary wave and the mixed baroclinic mixed topographic-planetary wave, respectively.

### Barotropic case

For $R = 0$ we find $E \equiv 0$ and $\omega = \omega_R$ as the only solution. This is the barotropic mixed topographic-planetary Rossby wave. Note that the group velocity of long waves, $(KR_0)^2 \ll 1$, is along $f/h$ contours.

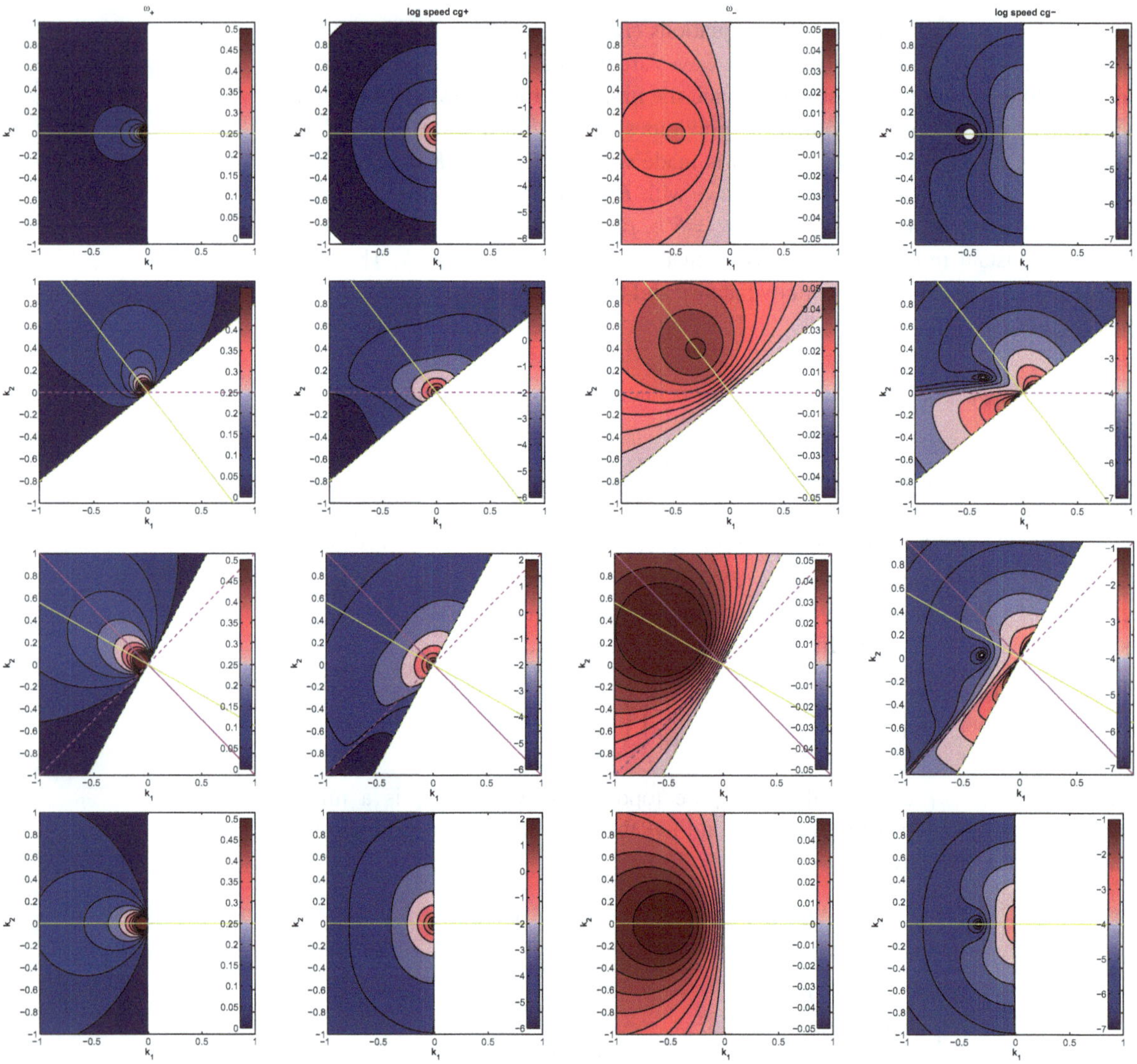


**Fig. B.3** Dispersion relation and group velocity (speed). Left two panels in each row show the +solution, right two panels show the −solution. Upper row: for flat bottom. Lower rows: for $\nabla h = (n, m) \times 10^{-3}$ for $nm = 10, 11$ and $nm = 01$. The graphs show only the functions in the half-plane with positive wave number along $f/h = $ const. The axes are scaled by $K_{\max} = 2/R$. The frequency is in cpd, the maximum for +solution is 0.5 cpd, for −solution 0.05 cpd. The logarithmical of group velocity in $\mathrm{m\,s^{-1}}$ is displayed. The range for the +solution is $10^{-6}$–$10^{2}\ \mathrm{m\,s^{-1}}$, for the −solution $10^{-7}$–$10^{-1}\ \mathrm{m\,s^{-1}}$. The *yellow* cross is oriented at $\nabla(f/h)$, the *purple* cross at $\nabla h$ (the *dashed* lines are along the gradients, the *full* lines are along $f/h = $ const and $h = $ const, respectively)

### Flat-bottom case

Here the vorticity and potential energy balance are decoupled, so that we find $\omega_1 = \omega_P$ and $\omega_2 = \omega_C$ with eigenvectors $\boldsymbol{R}_1 = (1,0)^T$, $\boldsymbol{R}_2 = (0,1)^T$. These are the familiar flat-bottom barotropic and first baroclinic modes (see first row of Figure B.3 and left two panels of Figure B.4).

### Topographic case

The solutions of (B.47) for the eigenfrequencies $\omega_i(\boldsymbol{K})$ with corresponding eigenvectors $\boldsymbol{R}_i(\boldsymbol{K})$ are real. One easily finds

$$\begin{aligned}\omega_\pm(\boldsymbol{K}) &= \frac{1}{2}\left[\omega_A + \omega_L + \omega_R \pm \sqrt{(\omega_R - \omega_C)^2 + 4\omega_A(\omega_L + \omega_T - \omega_P)}\right] \\ \boldsymbol{R}_\pm(\boldsymbol{K}) &= \begin{pmatrix} f\psi \\ E \end{pmatrix}_\pm = \frac{\mathrm{i}}{\sqrt{\omega_T^2 + (\omega_\pm - \omega_R)^2}}\begin{pmatrix} \omega_T \\ \omega_R - \omega^\pm \end{pmatrix}\end{aligned} \qquad \text{(B.49)}$$

Approximately, $\omega_+ \approx \omega_R$ and $\omega_- \approx \omega_L + \omega_A(\omega_P - \omega_T)/\omega_R$. All specific frequencies $\omega_I$, $I$ = ACLPRT, are asymmetric with respect to the wave vector, i. e. $\omega_I(\boldsymbol{K}) = -\omega_I(-\boldsymbol{K})$. We thus need the eigenfrequencies and eigenvectors only on a half-plane because $\omega_\pm(\boldsymbol{K}) = -\omega_\mp(-\boldsymbol{K})$ and correspondingly for the eigenvectors, $\boldsymbol{R}_\pm(\boldsymbol{K}) = (\boldsymbol{R}_\mp(-\boldsymbol{K}))^*$. It appears sensible to choose the half-plane $\boldsymbol{K} \cdot \underset{\neg}{\gamma} > 0$ (i. e. positive component of the wave vector along the $f/h$ contours, implying $k < 0$ for flat bottom). This choice is established in Figure B.3 which show dispersion relations and group velocities a few cases of gradients of topography. Figure B.4 clearly demonstrates the mixed character of planetary-topographic wave modes: for nonzero topographic slopes the modes are neither barotropic nor baroclinic but 'mixed'. A particular detailed discussion of this feature is found in Hallberg (1997) who used a two-layer quasigeostrophic model.

A few properties of these solutions are worth mentioning for use in the main body of the paper:

- Even for moderate slopes the topographic $\beta$ can be much larger than the planetary one, and the periods of topographic waves are much smaller than those of flat-bottom waves. The maximum period of the baroclinic topographic mode

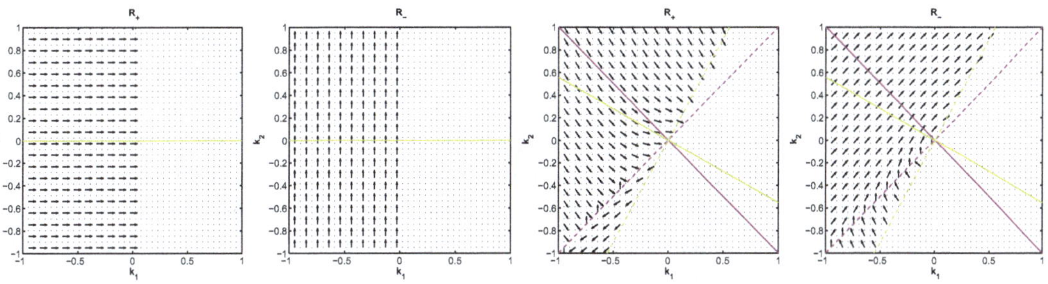


Fig. B.4 Eigenvectors of the + and −modes. The left two panels are for the flat-bottom case (associated with the upper row of Figure B.3), the right two panels are for a topographic case (third row of Figure B.3)

(−wave) is about 100 days (see lower three rows of Figure B.3) compared to about 600 days for flat bottom.

- The +waves propagate most rapidly for wave vectors along $f/h$ = const. For the -waves this occurs for wave vectors across $f/h$ = const. The speeds of topographic waves exceed those of flat-bottom waves considerably (a factor of 10 is easily achieved). Speeds for +wave exceeds typically 10 m s$^{-1}$, for the −wave we find values a large as 0.1 m s$^{-1}$ (note that the scale in Figure B.3 is logarithmic).

## B.3 Spectral Models

Hydrodynamical model are described by partial differential equations, some cases are even nonlinear. Analytical solutions are known only for the most simple, fairly restrictive conditions. In some cases even numerical solutions are difficult to obtain. To gain insight into the behavior of a fluid system on a more qualitative level, low-order models are developed. They resolve the spatial structures in a truncated aspect but allow nonlinearities to be considered in detail. The construction is simple: the spatial structure of the fields is represented by a set of prescribed structure functions with time dependent amplitudes. Projection of the evolution equations then yields a set of coupled ordinary differential equations for the amplitudes. Proper selection of these spatial functions is of course the most delicate and important problem in the construction of a low-order model. Most of such models apply to atmospheric systems. The oceans are embedded in rather irregular basins and even simple rectangular configurations develop dynamically important boundary layers (as the Gulf Stream) which defies representation by simple structure functions. Nevertheless, we have some oceanic low-order models in this book as well.

We start with an evolution equation for a function $\phi(x,t)$ of the form

$$\frac{\partial \phi}{\partial t} + \mathcal{L}[\phi] + \mathcal{N}[\phi,\phi] = F \tag{B.50}$$

The spatial coordinate may be multidimensional, the field a vector function, and for simplicity we have restricted the balance to have only quadratic nonlinearities. The balance must equipped with boundary conditions we we loosely write as $BC[\phi] = 0$.

The function $\phi(x,t)$ represents the spatial (and temporal) dependence of the field $\phi$. It may be expanded in any complete function set $\varphi_n(x)$ with a orthonormality condition

$$\langle \varphi_n \varphi_m \rangle = \int w(x)\varphi_n(x)\varphi_m(x)\mathrm{d}x = \delta_{nm} \tag{B.51}$$

We will assume, however, that the $\varphi_n$ are eigenfunctions of the linear operator $\mathcal{L}$, i. e.

$$\mathcal{L}\varphi_n = \lambda_n \varphi_n \tag{B.52}$$

which simplifies the following equations. The weighting function $w(x)$ in the scalar product (B.51) is then specific for the operator $\mathcal{L}$. For simplicity, we have labeled the set by the single index $n$, which may, however, be multidimensional as well and may include continuous parameters. The expansion of the dynamical variable $\phi$ in terms of the $\varphi_n$ is given by

$$\phi(x,t) = \sum_n \phi_n(t)\varphi_n(x) \tag{B.53}$$

and the inversion is found as

$$\phi_n(t) = \langle \phi \varphi_n \rangle \tag{B.54}$$

We refer to $\varphi_n$ as the $n$-th mode of the system and regard the $\phi_n(t)$ as the modal (or spectral) representation of the field $\phi$. Inserting now the representation (B.53) into the governing equation (B.50) and projecting on the mode $n$ by use of (B.51) we find

$$\frac{\partial \phi_n}{\partial t} + \lambda_n \phi_n + \sum_{k\ell} C^n_{k\ell} \phi_k \phi_\ell = 0 \tag{B.55}$$

with coefficients

$$C^n_{k\ell} = \langle \varphi_n \, \mathcal{N}[\varphi_k, \varphi_\ell] \rangle \tag{B.56}$$

which are called 'coupling coefficients' because they couple the modes according to (B.55). Notice that the modal or spectral model equation (B.55) is completely equivalent to the equation (B.50) governing the field in the physical space.

The advantage of (B.55) over (B.50) is that it is an algebraic equation whereas $\mathcal{L}$ and $\mathcal{N}$ in general contain spatial differential operators. The disadvantage of the spectral model is that we are facing an infinite set of equations so that in practice truncation to a finite system is necessary. Furthermore, before arriving at (B.55) we must find the modal set $\varphi_n(x)$ which maybe very complicated in irregular physical domains with complicated boundary conditions. In practice therefore, spectral models are usual constructed only for simple domains, e. g. periodic channels or unbounded double periodic two-dimensional domains or spherical domains.

## B.4 Box Models

Box models are constructed by considering the budget of the respective property $\chi$ (in units of $\chi$ per volume[2]) in terms of input from external sources and transport across the box boundaries. The aim is to find – with reasonable assumptions – a closed set of equations for the resolved dependent variable which usually is some measure of the total content of $\chi$ in the box, e. g. the mean

$$\bar{\chi} = \frac{1}{V} \int_V \chi \mathrm{d}V \tag{B.57}$$

We start with a conservation equation of the general form (see Section 2.1)

$$\frac{\partial \chi}{\partial t} + \nabla \cdot \boldsymbol{u}\chi = -\nabla \cdot \boldsymbol{F}_\mathrm{d} + \mathcal{S} \tag{B.58}$$

where $\boldsymbol{F}_\mathrm{d}$ denotes a diffusive flux and $\mathcal{S}$ a source or sink term. For a single isolated box the resulting equation is of course trivial because $\bar{\chi}$ would just increase in time in reaction to the integrated source. Hence, we consider a 'box' (actually of arbitrary

[2] For example, for salinity we have $\chi = \rho S$. If the density may be approximated by a constant $\rho_0$, the equations can be divided by $\rho_0$ in which case $\chi$ is a dimensionless concentration.

shape) which is open over some part $A_0$ of its surface area to let through the property $\chi$ by advection and diffusion, i. e. $\boldsymbol{n}\cdot\boldsymbol{u}\neq 0$, $\boldsymbol{n}\cdot\boldsymbol{F}_{\mathrm{d}}\neq 0$ where $\boldsymbol{n}$ is the outward normal vector on $A_0$. Integration then yields

$$V\frac{\mathrm{d}\bar{\chi}}{\mathrm{d}t}+\int_{A_0}\boldsymbol{n}\cdot(\boldsymbol{u}\chi+\boldsymbol{F}_{\mathrm{d}})\mathrm{d}A=\int_{V}\mathcal{S}\mathrm{d}V=V\bar{\mathcal{S}} \tag{B.59}$$

by Gauss' theorem. To become more specific let $A_0$ be made up from

- a part $A_{\mathrm{e}}$ where the box opens to the external environment (for an ocean box this could be the contact area to the overlying atmosphere). Here we assume $\boldsymbol{n}\cdot\boldsymbol{u}=0$ and $\boldsymbol{n}\cdot\boldsymbol{F}_{\mathrm{d}}=-F_{\mathrm{e}}$ where $F_e$ is a prescribed flux of the property $\chi$ (positive if inward).
- and a number of interfaces $A_j$ where the box is in contact to other boxes of the same kind and labeled with the index $j$. Let $v_j=\boldsymbol{n}\cdot\boldsymbol{u}$ be the outward normal velocity on these areas and $F_{j\mathrm{d}}=\boldsymbol{n}\cdot\boldsymbol{F}_{\mathrm{d}}$ the normal diffusive fluxes.

We then obtain

$$V\frac{\mathrm{d}\bar{\chi}}{\mathrm{d}t}+\sum_j\int_{A_j}(v_j\chi+F_{j\mathrm{d}})\mathrm{d}A=V\bar{\mathcal{S}}+\int_{A_{\mathrm{e}}}F_{\mathrm{e}}\mathrm{d}S \tag{B.60}$$

Since the mean $\bar{\chi}$ in the box is the only resolved variable in a box system, the integrated advection and diffusion terms in (B.60) must somehow be expressed in terms of $\bar{\chi}$ of the specific box and the neighboring boxes.

We now denote the reference box with label $i$ and evaluate the exchange with the neighboring box denoted by $j$ with the mean property $\bar{\chi}_{\mathrm{j}}$ and the common contact area $A_{\mathrm{ij}}$ to the box $j$. A reasonable assumption about the diffusive flux across $A_{\mathrm{ij}}$ is a down-gradient parametrization with a diffusivity $\kappa_{\mathrm{ij}}$. Since only the mean value of $\chi$ of the respective boxes is known we write

$$\int_{A_{\mathrm{ij}}}F_{\mathrm{jd}}\mathrm{d}A=-A_{\mathrm{ij}}\kappa_{\mathrm{ij}}\frac{\bar{\chi}_{\mathrm{j}}-\bar{\chi}_{\mathrm{i}}}{\Delta_{\mathrm{ij}}} \tag{B.61}$$

where $\Delta_{\mathrm{ij}}$ is some appropriate distance along the outward normal $\boldsymbol{n}$ over which the gradient is to be evaluated as a difference.

For the advection term the procedure is relatively straightforward as well. We separate the flow velocity $v_{\mathrm{ij}}$ into its positive (outward from box $i$) and negative components, defining $v_{\mathrm{ij}}^{\pm}=(v_{\mathrm{ij}}\pm|v_{\mathrm{ij}}|)/2$ so that $v_{\mathrm{ij}}=v_{\mathrm{ij}}^{+}+v_{\mathrm{ij}}^{-}$. The volume transports

$$q_{\mathrm{ij}}^{\pm}=\int_{A_{\mathrm{ij}}}v_{\mathrm{ij}}^{\pm}\mathrm{d}A \tag{B.62}$$

are now used to model the advective transport of $\chi$ across $A_{\mathrm{ij}}$ in the following upstream form

$$\int_{A_{\mathrm{ij}}}v_{\mathrm{ij}}\chi\mathrm{d}A=q_{\mathrm{ij}}^{+}\bar{\chi}_{\mathrm{i}}+q_{\mathrm{ij}}^{-}\bar{\chi}_{\mathrm{j}} \tag{B.63}$$

because $q_{ij}^{+}$ transports the property $\bar{\chi}_i$ out of the box $i$ whereas $q_{ij}^{-}$ transports $\bar{\chi}_j$ from the neighboring box with label $j$ into the box $i$. Hence, we arrive at

$$V_i \frac{d\bar{\chi}_i}{dt} = -\sum_j \left(q_{ij}^{+}\bar{\chi}_i + q_{ij}^{-}\bar{\chi}_j\right) + \sum_j K_{ij}\left(\bar{\chi}_j - \bar{\chi}_i\right) + V_i\bar{S}_i + A_{ie}\bar{F}_{ie} \qquad \text{(B.64)}$$

where $K_{ij} = \kappa_{ij}A_{ij}/\Delta_{ij}$ is the coefficient governing the diffusion between box $i$ and box $j$, $\bar{F}_{ie}$ is the integrated external flux normalized by the respective area $A_{ie}$ (negative outward). Note that diffusion always tends to relax $\bar{\chi}_i$ towards the $\chi$-content in the neighboring boxes. To close the model, the transports $q_{ij}^{\pm}$, the diffusivities $K_{ij}$, the forcing terms $\bar{S}_i$ and the fluxes $\bar{F}_{ie}$ must be specified. Note that the overall budget for the property $\chi$ is obtained by summing over $i$. Here, the advection and diffusion terms cancel. Note, furthermore, that the matrix $(K_{ij})$ of diffusion coefficients is symmetric. A sketch of the box system with the advective and external fluxes is shown in Figure B.5.

For a box having only one interface to other boxes, as e. g. in a two-box system of boxes 1 and 2, we find $q_{12}^{+} = -q_{12}^{-} = |q|$. A two-box system is thus governed by

$$\begin{aligned} V_1\frac{d\bar{\chi}_1}{dt} &= |q|(\bar{\chi}_2 - \bar{\chi}_1) + K(\bar{\chi}_2 - \bar{\chi}_1) + V_1\bar{S}_1 + A_{1e}\bar{F}_{1e} \\ V_2\frac{d\bar{\chi}_2}{dt} &= |q|(\bar{\chi}_1 - \bar{\chi}_2) + K(\bar{\chi}_1 - \bar{\chi}_2) + V_2\bar{S}_2 + A_{2e}\bar{F}_{2e} \end{aligned} \qquad \text{(B.65)}$$

Box systems are thus governed by relatively simple equations like (B.64) and (B.65). We would like to emphasize, however, that this simplicity is gained by quite severe and partly contradictory assumptions. Because only the mean values of the property $\chi$ in the boxes is resolved, we must express the values on the box boundaries by its mean value $\bar{\chi}$, relevant for advection and diffusion, which in essence assumes a fast (compared to the other resolved processes) mixing process within the box, horizontally as well as vertically. But this effective mixing is artificially stopped at the box boundaries as otherwise the entire box system would be homogeneous (box models often even use zero diffusion between boxes). The problem can be mitigated by implementation of more resolution, e. g. using more boxes or more complex dynamics within a box. The latter concept could relax the complete mixing within the box and imply then the need for a relation between the boundary values and the box mean, i. e. $\chi(A_{ij})$ and $\chi(A_{ie})$ must be expressed by $\bar{\chi}_i$.

The coefficients $q_{ij}^{+}$ and $q_{ij}^{-}$ model the advection of the properties in the system, however, they are also demanded to reflect the conservation of mass respectively volume. The corresponding matrices are sparsely populated: only entries for box pairs

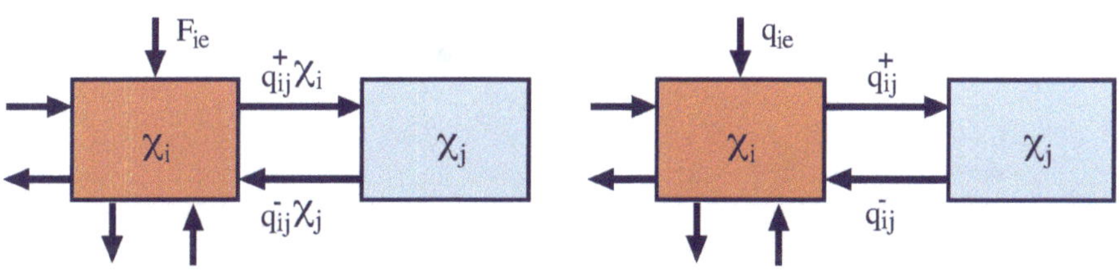


Fig. B.5 Sketch of box system with fluxes. The reference box is labeled $i$. **a**: Fluxes associated with the conservation of the property $\chi$. **b**: Fluxes associated with the conservation of volume. Note that we take the fluxes $F_{ie}$ and $q_{ie}$ at the open surface as positive in the inward direction. These fluxes may be due to input at the ocean surface or by river discharge with a common interface are nonzero. Furthermore, the diagonal is empty, $q_{\mathrm{ii}}^{\pm} = 0$, and quite obviously, the volume fluxes are related by $q_{\mathrm{ij}}^{+} = -q_{\mathrm{ji}}^{-}$. For each box $i$, mass (volume) conservation demands that the total volume flux across its boundary vanishes, i. e.

$$\sum_j \left( q_{\mathrm{ij}}^{+} + q_{\mathrm{ij}}^{-} \right) = q_{\mathrm{ie}} \tag{B.66}$$

where $q_{\mathrm{ie}}$ is the external flux.

# References


Abramowitz M, Stegun IA (1984) Handbook of mathematical functions with formulas, graphs, and mathematical tables. Dover Publ. Inc., New York

Acheson D (1997) From calculus to chaos: an introduction to dynamics. Oxford Univ. Press

Adcroft A, Scott JR, Marotzke J (2001) Impact of geothermal heating on the global ocean circulation. Geophys Res Letters 28:1735–1738

Aken Hv (2007) The oceanic thermohaline circulation: an introduction. Atmospheric and oceanographic sciences library, Springer Verlag

Alford M (2001) Internal swell generation: The spatial distribution of energy flux from the wind to mixed layer near-inertial motions. J Phys Oceanogr 31(8):2359–2368

Anderson D, Gill A (1979) Beta-dispersion of inertial waves. J Geophys Res 84:1836–1842

Anderson DLT, Gill AE (1975) Spin-up of a stratified ocean, with applications to upwelling. Deep-Sea Res 22:583–586

Anderson DLT, Killworth PD (1977) Spin-up of a stratified ocean, with topography. J Phys Oceanogr 24:709–732

Anderson PS (2009) Measurement of Prandtl number as a function of Richardson number avoiding self-correlation. Boundary-layer meteorology 131(3):345–362

Andrews DG, McIntyre ME (1976) Planetary waves in horizontal and vertical shear: The generalized Eliassen–Palm relation and the zonal mean acceleration. J Atmos Sci 33:2031–2048

Andrews DG, McIntyre ME (1978) An exact theory of nonlinear waves on a Lagrangian-mean flow. J Fluid Mech 89:609–646

Andrews DG, Holton JR, Leovy CB (1987) Middle atmosphere dynamics. Academic Press, Orlando

Antonov JI, Locarnini RA, Boyer TP, Mishonov AV, Garcia HE, Levitus S (2006) World Ocean Atlas 2005 Volume 2: Salinity. NOAA Atlas NESDIS 62

Apel JR (1987) Principles of Ocean Physics. Academic Press

Aris R (1989) Vectors, tensors, and the basic equations of fluid mechanics. Dover Publications

Bacon S, Fofonoff N (1996) Oceanic heat flux calculation. J Phys Oceanogr 13:1327–1329



D. Olbers, J. Willebrand, C. Eden, *Ocean Dynamics*,
DOI 10.1007/978-3-642-23450-7, © Springer-Verlag Berlin Heidelberg 2012

Barnier B, Siefridt L, Marchesiello P (1995) Thermal forcing for a global ocean circulation model using a three year climatology of ECMWF analysis. J Mar Sys 6:363–380

Batchelor GK (1977) An introduction to fluid dynamics. Cambridge Univ. Press, Cambridge

Batchelor GK (1990) The theory of homogeneous turbulence. Cambridge Univ. Press, Cambridge

Baumert H, Simpson J, Sündermann J (2005) Marine turbulence: Theories, Observations, and Models. Bd. 1, Cambridge Univ. Press, Cambridge

Bell Jr TH (1975) Topographically generated internal waves in the open ocean. J Geophys Res 80:320–327

Bender CM, Orszag SA (1978) Advanced mathematical methods for scientists and engineers: Asymptotic methods and perturbation theory. McGraw-Hill Book Company, New York

Bjerknes VW (1898) Über einen hydrodynamischen Fundamentalsatz und seine Anwendung, besonders auf die Mechanik der Atmosphäre und des Weltmeeres. Kgl Svenska Vet Akad Handlingar, Stockholm 31:1–35

Brekhovskikh L, Lysanov I (2003) Fundamentals of Ocean Acoustics. AIP series in modern acoustics and signal processing, AIP Press/Springer Verlag

Bretherton F (1966) Critical layer instability in baroclinic flows. Quart J Royal Met Soc 92:325–334

Bretherton FP (1970) A note on Hamilton's principle for perfect fluids. J Fluid Mech 44:19–31

Brethouwer G, Billant P, Lindborg E, Chomaz JM (2007) Scaling analysis and simulation of strongly stratified turbulent flows. J Fluid Mech 585:343–368

Broecker WJ (1991) The great ocean conveyor. Oceanography 4:79–89

Bryan F (1987) Parameter sensitivity of primitive equation ocean general circulation models. J Phys Oceanogr 17:970–985

Bryan K, Cox MD (1972) The circulation of the World Ocean : a numerical study. Part I, A homogeneous model. J Phys Oceanogr 2:319–335

Bühler O (2009) Waves and mean flows. Cambridge monographs on mechanics, Cambridge Univ. Press, Cambridge

Burger AP (1958) Scale consideration of planetary motions of the atmosphere. Tellus 10:195–205

Charney JG, DeVore JG (1979) Multiple flow equilibra in the atmosphere and blocking. J Atmos Sci 36:1205 – 1216

Chelton DB, deSzoeke RA, Schlax MG, Naggar KE, Siwertz N (1998) Geographical variability of the first baroclinic Rossby radius of deformation. J Phys Oceanogr 28:433–460

Courant R, Hilbert D (1953) Methods of mathematical physics. Interscience Publishers

Cox MD (1975) A baroclinic numerical model of the World Ocean: preliminary results. In: Numerical models of ocean circulation, Durham, New Hampshire, Oct. 17 – 20, 1972, pp 107–120

Craik ADD (1988) Wave interactions and fluid flows. Cambridge Univ. Press, Cambridge

Cunningham SA, Alderson SG, King BA, Brandon MA (2003) Transport and variability of the Antarctic Circumpolar Current in Drake Passage. J Geophys Res 108:1–17

Curry J, Webster P (1999) Thermodynamics of atmospheres and oceans. International geophysics series, Academic Press
Cushman-Roisin B (1994) Introduction to geophysical fluid dynamics. Prentice Hall
Danilov S, Gurarie D (2002) Rhines scale and spectra of the $\beta$-plane turbulence with bottom drag. Physical Review E 65
Darrigol O (2009) Worlds of flow: a history of hydrodynamics from the Bernoullis to Prandtl. Cambridge Univ. Press, Cambridge
De Groot SR, Mazur P (1984) Non-equilibrium thermodynamics. Dover Publ. Inc., New York
Deardorff JW (1966) The counter-gradient heat flux in the lower atmosphere and in the laboratory. J Atmos Sci 23:503–506
Defant A (1961) Physical oceanography, vol. 1. Pergamon Press, New York
DeSzoeke RA, Levine MD (1981) The advective flux of heat by mean geostrophic motions in the Southern Ocean. Deep-Sea Res 28:1057–1085
Dijkstra H (2005) Nonlinear physical oceanography: a dynamical systems approach to the large scale ocean circulation and El Niño. Atmospheric and oceanographic sciences library, Springer Verlag
Drazin PG (1992) Nonlinear systems. 10, Cambridge Univ. Prress, Cambridge
Dutton JA (1976) The ceaseless wind: an introduction to the theory of atmospheric motion. Dover Publ. Inc., New York
Eady ET (1949) Long waves and cyclone waves. Tellus 1:33–52
Eden C (2010) Parameterising meso-scale eddy momentum fluxes based on potential vorticity mixing and a gauge term. Ocean Modelling 32(1-2):58–71
Eden C, Greatbatch RJ (2008) Diapycnal mixing by mesoscale eddies. Ocean Modelling 23:113–120
Eden C, Greatbatch RJ (2009) A diagnosis of isopycnal mixing by meso-scale eddies. Ocean Modelling 27:98–106
Eden C, Olbers D (2010) Why western boundary currents are diffusive: A link between bottom pressure torque and bolus velocity. Ocean Modelling 32:14–24
Eden C, Willebrand J (1999) Neutral density revisited. Deep-Sea Res II 46:33–54
Eden C, Greatbatch RJ, Olbers D (2007) Interpreting eddy fluxes. J Phys Oceanogr 37:1282–1296
Ekman VW (1905) On the influence of the earth's rotation on ocean currents. Arkiv för Mat Astr och Fysik 2:1–53
Ekman VW (1923) Über Horizontalzirkulation bei winderzeugten Meeresströmungen. Arkiv för Mat Astr och Fysik 17
Emile-Geay J, Madec G (2009) Geothermal heating, diapycnal mixing and the abyssal circulation. Ocean Sci 5:203–217
Eriksen CC (1982) Observations of internal wave reflection off sloping bottoms. J Geophys Res 87:525–538
Ertel H (1942) Ein neuer hydrodynamischer Erhaltungssatz. Meteor Z 59:277–281
Feistel R (2003) A new extended Gibbs thermodynamic potential of seawater. Prog Oceanogr 58:43–114
Feistel R (2008) A Gibbs function for seawater thermodynamics for -6 to 80 °C and salinity up to 120 $\mathrm{gkg}^{-1}$. Deep-Sea Res 55:1639–1671
Fennel W, Lass HU (1989) Analytical theory of forced oceanic waves. Akademie-Verlag, Berlin
Fofonoff NP, Millard Jr RC (1983) Algorithms for computation of fundamental properties of water. Technical papers in marine science, UNESCO

Frisch U (1995) Turbulence. Cambridge Univ. Press, Cambridge

Ganachaud A, Wunsch C (2000) Improved estimates of global ocean circulation, heat transport and mixing from hydrographic data. Nature 408:453–457

Garrett C, Munk W (1975) Space-time scales of internal waves: A progress report. J Geophys Res 80:291–297

Gaspar P, Gregoris Y, Lefevre JM (1990) A simple eddy kinetic energy model for simulations of the oceanic vertical mixing: tests at station PAPA and Long-Term Upper Ocean Study site. J Geophys Res 95:16,179–16,193

Gates WL (2004) Derivation of the equations of atmospheric motion in oblate spheroidal coordinates. J Atmos Sci pp 2478–2487

Gent P, McWilliams J (1996) Eliassen–palm fluxes and the momentum equation in non-eddy-resolving ocean circulation models. J Phys Oceanogr 26:2539–2546

Gent PR, McWilliams JC (1990) Isopycnal mixing in ocean circulation models. J Phys Oceanogr 20:150–155

Gent PR, Willebrand J, McDougall TJ, McWilliams JC (1995) Parameterizing eddy-induced tracer transports in ocean circulation models. J Phys Oceanogr 25:463–474

Gerkema T, Shrira VI (2005) Near-inertial waves on the "nontraditional" $\beta$ plane. J Geophys Res 110:C01,003

Gerkema T, Zimmerman JTF, Maas LRM, van Haren H (2008) Geophysical and astrophysical fluid dynamics beyond the traditional approximation. Rev Geophys 46:RG2004

Gill AE (1968) A linear model of the Antarctic Circumpolar Current. J Fluid Mech 32:465 – 488

Gill AE (1974) The stability of planetary waves on an infinite beta-plane. Geophys Fluid Dynamics 6:29–47

Gill AE (1982) Atmosphere-Ocean dynamics. Academic Press, New York

Gnanadesikan A (1999) A simple predictive model for the structure of the oceanic pycnocline. Science 283(5410):2077

Goldstein S (1931) On the stability of superposed streams of fluids of different densities. Proc Roy Soc A 132:524–554

Gordon AL (1986) Interocean exchange of thermocline water. J Geophys Res 91:5037–5046

Gradshteyn IS, Ryzhik IM (2000) Table of integrals, series, and products. Translated from the Russian. Translation edited and with a preface by Alan Jeffrey and Daniel Zwillinger. 6th ed. San Diego, CA: Academic Press. xlvii, 1163 p.

Greatbatch RJ, Lamb K (1990) On parameterizing vertical mixing of momentum in non-eddy-resolving ocean models. J Phys Oceanogr 20:1634–1637

Griffies SM (2004) Fundamentals of ocean climate models. Princeton Univ. Press

Haine T, Marshall J (1998) Gravitational, symmetric and baroclinic instability of the ocean mixed layer. J Phys Oceanogr 28:634–658

Hallberg R (1997) Localized coupling between surface- and bottom-intensified flow over topography. J Phys Oceanogr 27:977–998

Hasselmann K (1966) Feynman diagrams and interaction rules of wave-wave scattering processes. Rev Geophys 4:1–32

Hasselmann K (1967a) A criterion for nonlinear wave stability. J Fluid Mech 30:737–739

Hasselmann K (1967b) Nonlinear interactions treated by the methods of theoretical physics (with application to the generation of waves by wind). Proceedings of the

Royal Society of London Series A Mathematical and Physical Sciences 299:77–100

Hasselmann K (1982) An ocean model for climate variability studies. Prog Oceanogr 11:69–92

Haynes PH, McIntyre ME (1990) On the conservation and impermeability theorem for potential vorticity. J Atmos Sci 47:2021–2031

Hesselberg T (1926) Die Gesetze der ausgeglichenen Bewegungen. Beitr z Physik d freien Atmosph 12:141–160

Hidaka K, Tsuchiya M (1953) On the Antarctic Circumpolar Current. J Mar Res 12:214–222

Howard L (1961) Note on a paper of John W. Miles. J Fluid Mech 10:509–512

Huang K (1987) Statistic mechanics. Wiley, New York

Hughes CW, Ash ER (2001) Eddy forcing of the mean flow in the Southern Ocean. J Geophys Res 106:2713–2722

Hughes CW, de Cuevas BA (2001) Why western boundary currents in realistic oceans are inviscid: A link between form stress and bottom pressure torques. J Phys Oceanogr 31:2871–2885

Hunt JCR, Phillips OM, Williams D (1991) Turbulence and stochastic processes: Kolmogorov's ideas 50 years on. Scholium Intl

IOC, SCOR and IAPSO (2010) The international thermodynamic equation of seawater–2010: Calculation and use of thermodynamic properties. Intergovernmental Oceanographic Commission, Manuals and Guides No. 56, UNESCO

Jackett D, McDougall T, Feistel R, Wright D, Griffies S (2006) Algorithms for density, potential temperature, conservative temperature and freezing temperature of seawater. J Atm Ocean Techn 23:163–171

Jackett DR, McDougall TJ (1997) A neutral density variable for the World's Oceans. J Phys Oceanogr 27:237–263

Johnson GC, Bryden HL (1989) On the size of the Antarctic Circumpolar Current. Deep-Sea Res 36:39–53

Johnson H, Marshall D, Sproson D (2007) Reconciling theories of a mechanically driven meridional overturning circulation with thermohaline forcing and multiple equilibria. Climate Dynamics 29(7):821–836

Kalnay E, Kanamitsu M, Kistler R, Collins W, Deaven D, Gandin L, Iredell M, Saha S, White G, Woollen J, Zhu Y, Chelliah M, Ebiszusaki W, Higgins W, Janowiak J, Mo K, Ropelewski C, Wang J, Leetmaa A, Reynolds R, Jenne R, Joseph D (1996) The NCEP/NCAR 40-years reanalysis project. Bull Amer Meteor Soc 77:437–471

Kamenkovich VM (1977) Fundamentals of ocean dynamics. Elsevier, Amsterdam

Kantha L, Clayson C (2000) Small scale processes in geophysical fluid flows. International geophysics series, Academic Press

Killworth P, Chelton D, de Szoeke R (1997) The speed of observed and theoretical long extra-tropical planetary waves. J Phys Oceanogr 27:1946–1966

Killworth PD (1998) Eddy parameterisation in large scale flow. In: Chassignet E, Verron J (eds) Ocean modeling and parameterization, Kluver Academic Publishers, Dordrecht, pp 253–268

Klein P, Hua BL, Lapeyre G, Capet X, Gentil SL, Sasaki H (2008) Upper ocean turbulence from high-resolution 3D simulations. J Phys Oceanogr 38:1748–1763

Kolmogorov AN (1941) Dissipation of energy in the locally isotropic turbulence (English translation). Dokl Akad Nauk SSSR 32:19–21

Komen GJ, Cavaleri L, Donelan M, Hasselmann K, Hasselmann S, Janssen P (1994) Dynamics and modelling of ocean waves. Cambridge Univ. Press, New York

Kraus EB, Turner JS (1966) A one-dimensional model of the seasonal thermocline. Tellus 19:88–97

Kraus EB, Turner JS (1967) A one-dimensional model of the seasonal thermocline II, the general theory and its consequences. Tellus 19:98–105

Kuhlbrodt T (2008) On Sandström's inferences from his tank experiments: a hundred years later. Tellus 60:819–836

Kundu P, Cohen I, Hu H (2004) Fluid mechanics. Elsevier Academic Press

Landau LD, Lifshitz EM (1982) Mechanics, course of theoretical physics Vol. 1. Pergamon Press Ltd.

Landau LD, Lifshitz EM (1987) Fluid mechanics, course of theoretical physics Vol. 6. Pergamon Press Ltd.

Langer R (1949) The asymptotic solutions of ordinary linear differential equations of the second order, with special reference to a turning point. Transactions of the American Mathematical Society 47:461–490

Large WG (1998) Modeling and parameterizing the ocean planetary boundary layer. In: Chassignet E, Verron J (eds) Ocean Modeling and Parameterization, Kluver Academic Publishers, Dordrecht, pp 81–120

Large WG, McWilliams JC, Doney SC (1994) Oceanic vertical mixing: a review and a model with nonlocal boundary layer parameterization. Rev Geophys 32:363–403

LeBlond P, Mysak L (1980) Waves in the ocean. Oceanography Series, Vol 20, Elsevier

Leetmaa A, Niiler P, Stommel H (1977) Does the Sverdrup relation account for the mid-Atlantic circulation? J Mar Res 35:1–10

Leonard BP (1979) A stable and accurate convective modelling procedure based on quadratic upstream interpolation. Computer Methods in Applied Mechanics and Engineering 19:59–98

Levitus S, Boyer TP (1994) World Ocean Atlas 1994, Volume 4: Temperature. U.S. Dept. of Commerce, Washington, DC

Levitus S, Burgett R, Boyer TP (1994) World Ocean Atlas 1994, Volume 3: Salinity. U.S. Dept. of Commerce, Washington, DC

Li L, Ingersoll AP, Jiang X, Feldman D, Yung YL (2007) Lorenz energy cycle of the global atmosphere based on reanalysis datasets. Geophys Res Letters 34:16,813

Lighthill M (1958) Introduction to Fourier analysis and generalised functions. Cambridge Univ. Press, Cambridge

Lighthill MJ (1978) Waves in fluids. Cambridge Univ. Press, Cambridge

Lin CC (1963) Liquid helium. Proceedings of the Enrico Fermi International School of Physics, Course XXI, G Careri, ed, Academic Press, New York pp 93–146

Locarnini RA, Mishonov AV, Antonov JI, Boyer TP, Garcia HE (2006) World Ocean Atlas 2005, Volume 1: Temperature. NOAA Atlas NESDIS 61:1–182

Lorenz EN (1955) Available potential energy and the maintenance of the general circulation. Tellus 7:157–167

Luyten JR, Pedlosky J, Stommel H (1983) The ventilated thermocline. J Phys Oceanogr 23:292–309

Manley JM, Rowe HE (1956) Some general properties of nonlinear elements. Part I. General energy relations. Proceedings of the IRE 44:904–913

Marotzke J, Welander P, Willebrand J (1988) Instability and multiple steady states in a meridional-plane model of the thermohaline circulation. Tellus 40A:162–172

Marshall J, Radko T (2003) Residual-mean solutions for the Antarctic Circumpolar Current and its associated overturning circulation. J Phys Oceanogr 33:2341–2354
Marshall J, Schott F (1999) Open-ocean convection: Observations, theory, and models. Rev Geophys 37(1):1–64
Marshall J, Shutts G (1981) A note on rotational and divergent eddy fluxes. J Phys Oceanogr 11:1677–1680
Marshall JC (1981) On the parameterization of geostrophic eddies in the ocean. J Phys Oceanogr 11:1257–1271
McComas CH (1977) Equilibrium mechanisms within the oceanic internal wave field. J Phys Oceanogr 7:836–845
McComas CH, Bretherton FP (1977) Resonant interaction of oceanic internal waves. J Geophys Res 82:1397–1412
McDougall T (2003) Potential enthalpy: A conservative oceanic variable for evaluating heat content and heat fluxes. J Phys Oceanogr 33:945–963
McDougall TJ (1987) Neutral surfaces. J Phys Oceanogr 17:1950–1964
McDougall TJ (1995) The influence of ocean mixing on the absolute velocity vector. J Phys Oceanogr 25:705–725
McDougall TJ (1998) Three-dimensional residual-mean theory. In: Chassignet E, Verron J (eds) Ocean Modeling and Parameterization, Kluver Academic Publishers, Dordrecht, pp 269–302
McDougall TJ (2003) Potential enthalpy: A conservative oceanic variable for evaluating heat content and heat fluxes. J Phys Oceanogr 33:945–963
McDougall TJ, McIntosh PC (1996) The temporal-residual-mean velocity. Part I: Derivation and the scalar conservation equations. J Phys Oceanogr 26:2653–2665
McDougall TJ, McIntosh PC (2001) The temporal-residual-mean velocity. Part II: Isopycnal interpretation and the tracer and momentum equations. J Phys Oceanogr 31:1222–1246
McIntyre M (1968) On stationary topography-induced Rossby-wave patterns in a barotropic zonal current. DHZ – Ocean Dynamics 21:203–214
McWilliams J (2006) Fundamentals of geophysical fluid dynamics. Cambridge Univ. Press, Cambridge
McWilliams JC (1977) A note on a consistent quasigeostrophic model in a multiply connected domain. Dyn Atmos Oceans 1:427–441
Mellor GL, Yamada L (1982) Development of a turbulence closure model for geophysical fluid problems. Rev Geophys 20:851–875
Merian J (1828) Ueber die Bewegung tropfbarer Flüssigkeiten in Gefässen. Schweighauser, Basel. Reproduced by K. Vonder Mühll, 1886. Math. Annalen 27, 575-600.
Mironov DV, Gryanik VM, Moeng CH, Olbers DJ, Warncke TH (2000) Vertical turbulence structure and second-moment budgets in convection with rotation: A large-eddy simulation study. Quart J Royal Met Soc 126:477–515, part B
Miropolsky YZ (2001) Dynamics of internal gravity waves in the ocean. Kluwer Academic Publishers, Dordrecht
Morrow R, Church J, Coleman R, Chelton D, White N (1992) Eddy momentum flux and its contribution to the Southern Ocean momentum balance. Nature 357:482–484
Morse PM, Feshbach H (1953) Methods of theoretical physics, Part I. McGraw-Hill, New York

Mowbray D, Rarity B (1967) A theoretical and experimental investigation of the phase configuration of internal waves of small amplitude in a density stratified liquid. J Fluid Mech 28:1–16

Müller P, Olbers DJ (1975) On the dynamics of internal waves in the deep ocean. J Geophys Res 80:3848–3860

Müller P, Olbers DJ, Willebrand J (1978) The IWEX spectrum. J Geophys Res 83:479–500

Müller P, Holloway G, Henyey F, Pomphrey N (1986) Nonlinear interactions among internal gravity waves. Rev Geophys 24:493–536

Munk W (1981) Internal waves and small-scale processes. In: Warren BA, Wunsch C (eds) Evolution of physical oceanography, MIT Press, Cambridge, MA, pp 264–291

Munk W, Worcester P, Wunsch C (1995) Ocean acoustic tomography. Cambridge monographs on mechanics, Cambridge Univ. Press, Cambridge

Munk WH (1950) On the wind-driven ocean circulation. J Atmos Sci 7:80–93

Munk WH (1966) Abyssal recipes. Deep-Sea Res 13:707–730

Munk WH, Palmén E (1951) Note on the dynamics of the Antarctic Circumpolar Current. Tellus 3:53–55

Natarov A, Müller P (2005) A dissipation function for the internal wave radiative balance equation. J Atm Ocean Techn 22:1782–1796

Needler GT (1967) A model for thermohaline circulation in an ocean of finite depth. J Mar Res 25:329–342

Needler GT (1985) The absolute velocity as a function of conserved measurable quantities. Prog Oceanogr 14:421–429

Niiler PP, Kraus EB (1977) One-dimensional models of the upper ocean. In: EB Kraus (ed) Modeling and prediction of the upper layers of the ocean, Pergamon, pp 143–172

Nurser AJG, Lee MM (2004) Isopycnal averaging at constant height: Part I: the formulation and a case study. J Phys Oceanogr 34:2721–2739

Obukhov A (1962) On the dynamics of a stratified liquid. Dokl Acad Nauk SSSR 145:1239–1242, English transl. in *Soviet Physics, Dokl.*, 7, 682–684

Ogura Y, Phillips N (1962) Scale analysis of deep and shallow convection in the atmosphere. J Atmos Sci 19:173–179

Olbers D, Eden C (2003) A model with simplified circulation dynamics for a baroclinic ocean with topography. Part I: Waves and wind-driven circulations. J Phys Oceanogr 33:2719–2737

Olbers D, Ivchenko VO (2001) On the meridional circulation and balance of momentum in the Southern Ocean of POP. Ocean Dyn 52:79–93

Olbers D, Lettmann K (2007) Barotropic and baroclinic processes in the transport variability of the Antarctic Circumpolar Current. Ocean Dyn 57:559–578

Olbers D, Visbeck M (2005) A zonally averaged model of the meridional overturning in the Southern Ocean. J Phys Oceanogr 35:1190–1205

Olbers D, Zhang J (2008) The global thermohaline circulation in box and spectral low-order models. Part 1: single basin models. Ocean Dyn 58:311–334, DOI {10.1007/s10236-008-0156-3}

Olbers D, Gouretski V, Seiß G, Schröter J (1992) Hydrographic Atlas of the Southern Ocean. Alfred Wegener Institute for Polar and Marine Research, Bremerhaven

Olbers D, Borowski D, Völker C, Wolff JO (2004) The dynamical balance, transport and circulation of the Antarctic Circumpolar Current. Antarctic Science 16:439–470

Olbers DJ (1974) On the energy balance of small-scale internal waves in the deep sea. Hamburger Geophysikalische Einzelschriften 24

Olbers DJ (1976) Nonlinear energy transfer and the energy balance of the internal wave field in the deep ocean. J Fluid Mech 74:375–399

Olbers DJ (1981) The propagation of internal waves in a geostrophic current. J Phys Oceanogr 11:1224–1233

Olbers DJ (1983) Models of the oceanic internal wave field. Rev Geophys Space Physics 21:1567–1606

Olbers DJ (1986) Internal waves. In: Sündermann, J (ed) Landolt-Börnstein - Numerical data and functional relationships in science and technology - New Series, Group V, Volume 3a, Springer Verlag, Berlin, pp 37–82

Olbers DJ, Herterich K (1979) The spectral energy transfer from surface waves to internal waves. J Fluid Mech 92:349–379

Olbers DJ, M Wenzel M, Willebrand J (1985) The inference of North Atlantic circulation patterns from climatological hydrographic data. Rev Geophys 23:313–356

Onsager L (1931a) Reciprocal relations in irreversible processes. I. Physical Review 37:405–426

Onsager L (1931b) Reciprocal relations in irreversible processes. II. Physical Review 38:2265–2279

Open University OCT (1989) Ocean circulation, vol 3. Pergamon

Orsi AH, Whitworth III T (2005) Southern Ocean. Vol. 1, Hydrographic Atlas of the World Ocean Circulation Experiment (WOCE) (eds. M. Sparrow, P. Chapman and J. Gould). WOCE International Project Office, Univ. of Southampton

Osborn TR (1980) Estimates of the local rate of vertical diffusion from dissipation measurements. J Phys Oceanogr 10:83–89

Osborn TR, Cox CS (1972) Oceanic fine structure. Geoph Astroph Fluid Dyn 3:321–345

Pacanowski RC, Philander SGH (1981) Parametrization of vertical mixing in numerical models of tropical oceans. J Phys Oceanogr 11:1443–1451

Pedlosky J (1987) Geophysical fluid dynamics. Springer Verlag, Berlin

Pedlosky J (1998) Ocean circulation theory. New York and Berlin, Springer Verlag

Pedlosky J (2003) Waves in the ocean and atmosphere: Introduction to wave dynamics. Springer Verlag

Peixoto JP, Oort AH (1992) Physics of climate. American Institute of Physics, New York

Perko L (2001) Differential equations and dynamical systems. Springer Verlag, Berlin and Heidelberg

Philander S (1990) El Niño, La Niña, and the Southern Oscillation. Academic Press, San Diego

Philipps OM (1977) The dynamics of the upper ocean. Cambridge Univ. Press, Cambridge

Phillips HE, Rintoul SR (2000) Eddy variability and energetics from direct current measurements in the Antarctic Circumpolar Current south of Australia. J Phys Oceanogr 30:3050–3076

Phillips NA (1954) Energy transformations an meridional circulations associated with simple baroclinic waves in a two-level, quasigeostrophic model. Tellus 6:273–286

Phillips NA (1966) The equations of motion for a shallow rotating atmosphere and the 'traditional approximation'. J Atmos Sci 23:626–627

Plumb R (1990) A nonacceleration theorem for transient quasi-geostrophic eddies on a three-dimensional time-mean flow. J Atmos Sci 47(15):1825–1836

Polzin KL (2009) An abyssal recipe. Ocean Modelling 30(4):298–309

Polzin KL, Toole JM, Ledwell JR, Schmitt RW (1997) Spatial variability of turbulent mixing in the abyssal ocean. Science 276:93–96

Pope S (2000) Turbulent flows. Cambridge Univ. Press, Cambridge

Prandtl L (1925) Bericht über Untersuchungen zur ausgebildeten Turbulenz. Zeitschrift für angewandte Mathematik und Mechanik 5:136–139

Prigogine I (1962) Non-equilibrium statistical mechanics. Interscience Publishers

Pugh DT (1987) Tides, surge and mean sea-level. John Wiley and Sons, Chichester

Reynolds O (1883) An experimental investigation of the circumstances which determine whether the motion of water shall be direct or sinuous, and of the law of resistance in parallel channels. Proceedings of the Royal Society of London 35(224-226):84

Rhines P (1975) Waves and turbulence on a beta-plane. J Fluid Mech 69:417–443

Rhines P (1977) The dynamics of unsteady currents. In: Goldberg E (ed) The Sea, Vol. VI, Wiley, New York, pp 189–318

Rhines P (1993) Oceanic general circulation: wave- and advection dynamics. In: Willebrand J, Anderson DLT (eds) Modelling ocean climate interactions, Springer Verlag, Berlin, pp 67–149

Rhines PB (1982) Basic dynamics of the large-scale geostrophic circulation. Summer Study Program in Geophysical Fluid Dynamics, Woods Hole Oceanographic Institution pp 1–47

Rhines PB, Young WR (1982) A theory of the wind-driven circulation. I. Mid-Ocean gyres. J Mar Res 40 (Suppl.):559–596

Richardson P (2008) On the history of meridional overturning circulation schematic diagrams. Prog Oceanogr 76(4):466–486

Riegel C (1992) Fundamentals of atmospheric dynamics and thermodynamics. Edited by A.F.C. Bridger. World Scientific

Rintoul S, Sokolov S, Church J (2002) A 6 year record of baroclinic transport variability of the Antarctic Circumpolar Current at 140 E derived from expendable bathythermograph and altimeter measurements. J Geophys Res 107(C10):3155

Rintoul SR, Sokolov S (2001) Baroclinic transport variability of the Antarctic Circumpolar Current south of Australia (WOCE repeat section SR3). J Geophys Res 106:2815–2832

Rintoul SR, Hughes C, Olbers D (2001) The Antarctic Circumpolar Current system. In: G Siedler, J Church and J Gould (ed) Ocean circulation and climate, Academic Press, New York, pp 271–302

Rummel R, Rapp RH (1976) The influence of the atmosphere on geoid and potential coefficient determinations from gravity data. J Geophys Res 81:5639–5642

Salmon R (1998) Lectures on geophysical fluid dynamics. Oxford Univ. Press, Oxford

Sander J, Wolf-Gladrow D, Olbers D (1995) Numerical studies of open deep convection. J Phys Oceanogr 100:20,579–20,600

Sandström JW (1908) Dynamische Versuche mit Meerwasser. Ann Hydrogr Mar Meteorol 36:6–23

Sarkisyan AS, Ivanov VF (1971) Joint effect of baroclinicity and bottom relief as an important factor in the dynamics of sea currents. Akad Nauk Atmopsh Oceanic Phys 7:173–188

Schmitz Jr WJ (1995) On the interbasin-scale thermohaline circulation. Rev Geophys 33:151–173

Schmitz Jr WJ (1996) On the World Ocean circulation: Volume I. Some global features/North Atlantic circulation. Tech. Rep. WHOI-96-03, Woods Hole Oceanographic Institution, Woods Hole, MA, 140 pp.

Schodlok M (2002) Über die Tiefenwasserausbreitung im Weddellmeer und in der Scotia-See: Numerische Untersuchungen der Transport-und Austauschprozesse in der Weddell-Scotia-Konfluenz-Zone. PhD thesis, Univ. Bremen

Schott F, Brandt P (2007) Circulation and deep water export of the subpolar North Atlantic during the 1990's. In: Schmittner A, Chiang JCH, Hemming SR (eds) Ocean Circulation: Mechanisms and Impacts, Geophysical Monograph Series 173, AGU, Washington D. C., pp 91–118

Schott F, Zantopp R, Stramma L, Dengler M, Fischer J, Wibaux M (2004) Circulation and deep-water export at the western exit of the subpolar North Atlantic. J Phys Oceanogr 34:817–843

Seliger RL, Whitham GB (1968) Variational principles in continuum mechanics. Proceedings of the Royal Society of London Series A Mathematical and Physical Sciences 305:1–25

Serrin J (1959) Mathematical principles of classical fluid mechanics. Handbuch der Physik, Part 1 8:125–263

Sharqawy MH, Lienhard V, Zubair SM (2010) Thermophysical properties of seawater: a review of existing correlations and data. Desalination and Water Treatment 16:354–380

Siedler G, Church J, Gould J (2001) Ocean circulation and climate: observing and modelling the global ocean. International geophysics series, Academic

Solomon S, Qin D, Manning M, Chen Z, Marquis M, Averyt KB, Tignor M, Miller HL (2007) Climate change 2007: the physical science basis. Cambridge Univ. Press, Cambridge

St Laurent L, Garrett C (2002) The role of internal tides in mixing the deep ocean. J Phys Oceanogr 32:2882–2899

Stacey FD (1992) Physics of the Earth. Brookfield Press, Brisbane

Starr VP (1968) Physics of negative vicosity phenomena. McGraw-Hill, New York, USA

Stommel H (1948) The westward intensification of wind-driven currents. Trans Am Geophys Union 29:202–206

Stommel H (1958) The abyssal circulation. Deep-Sea Res 5:80–82

Stommel H (1961) Thermohaline convection with two stable regimes of flow. Tellus 13:224–230

Stommel H, Arons AB (1960a) On the abyssal circulation of the world ocean. Part I: Stationary planetary flow patterns on a sphere. Deep-Sea Res 6:140–154

Stommel H, Arons AB (1960b) On the abyssal circulation of the world ocean. Part II: An idealized model of circulation pattern and amplitude in oceanic basins. Deep-Sea Res 6:217–233

Stommel H, Schott F (1977) The beta spiral and the determination of the absolute velocity field from hydrographic station data. Deep-Sea Res 24:325–329

Stommel H, Arons AB, Faller AJ (1958) Some examples of stationary planetary flow patterns in bounded basins. Tellus 10:179–187

Storch H, Zwiers F (2002) Statistical analysis in climate research. Cambridge Univ. Press, Cambridge

Straub DN (1996) An inconsistency between two classical models of the ocean buoyancy driven circulation. Tellus 48:477–481

Sverdrup HU (1933) On vertical circulation in the ocean due to the action of the wind with application to conditions within the Antarctic Circumpolar Current. Discovery Reports VII:139–170

Sverdrup HU (1947) Wind-driven currents in a baroclinic ocean; with application to the equatorial currents of the eastern pacific. Proceedings of the National Academy of Sciences 33:318–326

Sverdrup HU, Johnson MW, Fleming RH (1942) The oceans: their physics, chemistry and general biology. Prentice-Hall, New York

Tandon A, Garrett C (1996) On a recent parameterization of mesoscale eddies. J Phys Oceanogr 26:406–416

Taylor G (1931) Effect of variation in density on the stability of superposed streams of fluid. Proc Roy Soc A 132:499–523

Taylor GI (1915) Eddy motion in the atmosphere. Philos Trans Roy Soc London A215:1–23

Taylor GI (1921) Diffusion by continuous movements. Proceedings of the London Mathematical Society 20:196–211

Thomson, W (Lord Kelvin) (1869) On vortex motion. Trans R Soc Edin 25:217–260

Thorpe AJ, Volkert H, Ziemianski MJ (2003) The Bjerknes' circulation theorem. Bull Amer Meteor Soc 84:471–480

Thorpe SA (2005) The turbulent ocean. Cambridge Univ. Press, Cambridge

Tomczak M, Godfrey JS (2003) Regional oceanography: an introduction. Daya Books, New Delhi

Toole JM (1998) Turbulent mixing in the ocean. In: Chassignet E, Verron J (eds) Ocean modeling and parameterization, Kluver Academic Publishers, Dordrecht, pp 171–190

Treguier AM, Held IM, Larichev VD (1997) Parameterization of quasigeostrophic eddies in primitive equation ocean models. J Phys Oceanogr 27:567–580

Umlauf L, Burchard H (2005) Second-order turbulence models for geophysical boundary layers. A review of recent work. Cont Shelf Res 25:795–827

Vallis GK (2006) Atmospheric and oceanic fluid dynamics. Fundamentals and large-scale circulation. Cambridge Univ. Press, Cambridge

Colin de Verdiere A (1993) On the thermohaline circulation. In: Willebrand J, Anderson DLT (eds) Modelling ocean climate interactions, Springer Verlag, Berlin, pp 151–183

Verhulst F (1996) Nonlinear differential equations and dynamical systems. Springer Verlag, Berlin and Heidelberg

Veronis G (1973) Large scale ocean circulation. In: Adv. Appl. Mech., Academic Press, New York, vol 13, pp 1–92

Viebahn J, Eden C (2010) Towards the impact of eddies on the response of the Southern Ocean to climate change. Ocean Modelling 34(3-4):150–165

Warren BA (2006) The first law of thermodynamics in a salty ocean. Prog Oceanogr 70:149–167

Weber E, Weber W (1825) Über die Wellen tropfbarer Flüssigkeiten mit Anwendung auf die Schall-und Lichtwellen. Bey Gerhard Fleischer

Webster C (1964) An experimental study of turbulence in a density-stratified shear flow. Journal of Fluid Mechanics 19(02):221–245

Welander P (1971) Some exact solutions to the equations describing the ideal fluid thermocline. J Mar Res 29:60–68

Welander P (1973) Lateral friction in the oceans as an effect of potential vorticity mixing. Geophysical and Astrophysical Fluid Dynamics 5(1):173–189

Welander P (1985) Introduction to translation of V.W. Ekman's "Outline of a unified ocean current theory". Tellus 37A:378–379

Welander P (1986) Thermohaline effects in the ocean circulation and related simple models. In: Willebrand J, Anderson D (eds) Large-scale transport processes in oceans and atmosphere, D. Reidel, pp 163–200

White AA (2002) A view of the equations of meteorological dynamics and various approximations. In: Norbury J, Roulstone I (eds) Large-Scale Atmosphere-Ocean Dynamics I: Analytical methods and numerical models, Cambridge Univ. Press, Cambridge, pp 1–100

White AA, Hoskins BJ, Roulstone I, Staniforth A (2005) Consistent approximate models of the global atmosphere: shallow, deep, hydrostatic, quasi-hydrostatic and non-hydrostatic. Quart J Royal Met Soc 131:2081–2107

White AA, Staniforth A, Wood N (2008) Spheroidal coordinate systems for modelling global atmospheres. Quart J Royal Met Soc 134:261–270

Whitehead J (1998) Topographic control of ocean flows in deep passages and straits. Rev Geophys 36:423–440

Whitham GB (1974) Linear and nonlinear waves. Wiley, New York

Wright DG, Stocker TF (1991) A zonally averaged ocean model for the thermohaline circulation. Part I: Model development and flow dynamics. J Phys Oceanogr 12:1713–1724

Wright DG, Vreugdenhil CB, Hughes TMC (1995) Vorticity dynamics and zonally averaged ocean circulation models. J Phys Oceanogr 25:2141–2154

Wright DG, Stocker TF, Mercer D (1998) Closures used in zonally averaged ocean models. J Phys Oceanogr 28:791–804

Wunsch C (2002) Ocean observations and the climate forecast problem. In: Pearce RP (ed) Meteorology at the Millennium, Academic, London, pp 217–224

Wunsch C, Ferrari R (2004) Vertical mixing, energy and the general circulation of the oceans. Annual Review of Fluid Mechanics 36:281–314

Zwillinger D (1998) Handbook of differential equations. Academic Press, London

# Index



D. Olbers, J. Willebrand, C. Eden, *Ocean Dynamics*,
DOI 10.1007/978-3-642-23450-7, © Springer-Verlag Berlin Heidelberg 2012

S

**Y**



**Z**